$\int$

STUDENT SOLUTIONS MANUAL

for

Stewart's

ESSENTIAL CALCULUS

SECOND EDITION

BROOKS/COLE
CENGAGE Learning™

Australia · Brazil · Japan · Korea · Mexico · Singapore · Spain · United Kingdom · United States

ISBN-13: 978-1-133-49094-4
ISBN-10: 1-133-49094-8

Brooks/Cole
20 Davis Drive
Belmont, CA 94002-3098
USA

Cengage Learning is a leading provider of customized learning solutions with office locations around the globe, including Singapore, the United Kingdom, Australia, Mexico, Brazil, and Japan. Locate your local office at **www.cengage.com/global**

Cengage Learning products are represented in Canada by Nelson Education, Ltd.

To learn more about Brooks/Cole, visit **www.cengage.com/brookscole**

Purchase any of our products at your local college store or at our preferred online store **www.cengagebrain.com**

Printed in the United States of America
2 3 4 5 6 7 17 16 15 14 13

ABBREVIATIONS AND SYMBOLS

CD	concave downward
CU	concave upward
D	the domain of f
FDT	First Derivative Test
HA	horizontal asymptote(s)
I	interval of convergence
IP	inflection point(s)
R	radius of convergence
VA	vertical asymptote(s)
$\overset{CAS}{=}$	indicates the use of a computer algebra system.
$\overset{H}{=}$	indicates the use of l'Hospital's Rule.
$\overset{j}{=}$	indicates the use of Formula j in the Table of Integrals in the back endpapers.
$\overset{s}{=}$	indicates the use of the substitution $\{u = \sin x, \, du = \cos x \, dx\}$.
$\overset{c}{=}$	indicates the use of the substitution $\{u = \cos x, \, du = -\sin x \, dx\}$.

∫ CONTENTS

☐ DIAGNOSTIC TESTS

Test A   Algebra

1. (a) $(-3)^4 = (-3)(-3)(-3)(-3) = 81$

(b) $-3^4 = -(3)(3)(3)(3) = -81$

(c) $3^{-4} = \dfrac{1}{3^4} = \dfrac{1}{81}$

(d) $\dfrac{5^{23}}{5^{21}} = 5^{23-21} = 5^2 = 25$

(e) $\left(\frac{2}{3}\right)^{-2} = \left(\frac{3}{2}\right)^2 = \frac{9}{4}$

(f) $16^{-3/4} = \dfrac{1}{16^{3/4}} = \dfrac{1}{\left(\sqrt[4]{16}\,\right)^3} = \dfrac{1}{2^3} = \dfrac{1}{8}$

2. (a) Note that $\sqrt{200} = \sqrt{100 \cdot 2} = 10\sqrt{2}$ and $\sqrt{32} = \sqrt{16 \cdot 2} = 4\sqrt{2}$. Thus $\sqrt{200} - \sqrt{32} = 10\sqrt{2} - 4\sqrt{2} = 6\sqrt{2}$.

(b) $(3a^3b^3)(4ab^2)^2 = 3a^3b^3 16a^2b^4 = 48a^5b^7$

(c) $\left(\dfrac{3x^{3/2}y^3}{x^2y^{-1/2}}\right)^{-2} = \left(\dfrac{x^2y^{-1/2}}{3x^{3/2}y^3}\right)^2 = \dfrac{(x^2y^{-1/2})^2}{(3x^{3/2}y^3)^2} = \dfrac{x^4y^{-1}}{9x^3y^6} = \dfrac{x^4}{9x^3y^6y} = \dfrac{x}{9y^7}$

3. (a) $3(x+6) + 4(2x-5) = 3x + 18 + 8x - 20 = 11x - 2$

(b) $(x+3)(4x-5) = 4x^2 - 5x + 12x - 15 = 4x^2 + 7x - 15$

(c) $\left(\sqrt{a} + \sqrt{b}\,\right)\left(\sqrt{a} - \sqrt{b}\,\right) = \left(\sqrt{a}\,\right)^2 - \sqrt{a}\sqrt{b} + \sqrt{a}\sqrt{b} - \left(\sqrt{b}\,\right)^2 = a - b$

Or: Use the formula for the difference of two squares to see that $\left(\sqrt{a} + \sqrt{b}\,\right)\left(\sqrt{a} - \sqrt{b}\,\right) = \left(\sqrt{a}\,\right)^2 - \left(\sqrt{b}\,\right)^2 = a - b$.

(d) $(2x+3)^2 = (2x+3)(2x+3) = 4x^2 + 6x + 6x + 9 = 4x^2 + 12x + 9$.

Note: A quicker way to expand this binomial is to use the formula $(a+b)^2 = a^2 + 2ab + b^2$ with $a = 2x$ and $b = 3$:

$(2x+3)^2 = (2x)^2 + 2(2x)(3) + 3^2 = 4x^2 + 12x + 9$

(e) See Reference Page 1 for the binomial formula $(a+b)^3 = a^3 + 3a^2b + 3ab^2 + b^3$. Using it, we get

$(x+2)^3 = x^3 + 3x^2(2) + 3x(2^2) + 2^3 = x^3 + 6x^2 + 12x + 8$.

4. (a) Using the difference of two squares formula, $a^2 - b^2 = (a+b)(a-b)$, we have

$4x^2 - 25 = (2x)^2 - 5^2 = (2x+5)(2x-5)$.

(b) Factoring by trial and error, we get $2x^2 + 5x - 12 = (2x-3)(x+4)$.

(c) Using factoring by grouping and the difference of two squares formula, we have

$x^3 - 3x^2 - 4x + 12 = x^2(x-3) - 4(x-3) = (x^2 - 4)(x-3) = (x-2)(x+2)(x-3)$.

(d) $x^4 + 27x = x(x^3 + 27) = x(x+3)(x^2 - 3x + 9)$

This last expression was obtained using the sum of two cubes formula, $a^3 + b^3 = (a+b)(a^2 - ab + b^2)$ with $a = x$ and $b = 3$. [See Reference Page 1 in the textbook.]

(e) The smallest exponent on x is $-\frac{1}{2}$, so we will factor out $x^{-1/2}$.

$3x^{3/2} - 9x^{1/2} + 6x^{-1/2} = 3x^{-1/2}(x^2 - 3x + 2) = 3x^{-1/2}(x-1)(x-2)$

(f) $x^3y - 4xy = xy(x^2 - 4) = xy(x-2)(x+2)$

5. (a) $\dfrac{x^2 + 3x + 2}{x^2 - x - 2} = \dfrac{(x+1)(x+2)}{(x+1)(x-2)} = \dfrac{x+2}{x-2}$

(b) $\dfrac{2x^2 - x - 1}{x^2 - 9} \cdot \dfrac{x+3}{2x+1} = \dfrac{(2x+1)(x-1)}{(x-3)(x+3)} \cdot \dfrac{x+3}{2x+1} = \dfrac{x-1}{x-3}$

(c) $\dfrac{x^2}{x^2 - 4} - \dfrac{x+1}{x+2} = \dfrac{x^2}{(x-2)(x+2)} - \dfrac{x+1}{x+2} = \dfrac{x^2}{(x-2)(x+2)} - \dfrac{x+1}{x+2} \cdot \dfrac{x-2}{x-2} = \dfrac{x^2 - (x+1)(x-2)}{(x-2)(x+2)}$

$= \dfrac{x^2 - (x^2 - x - 2)}{(x+2)(x-2)} = \dfrac{x+2}{(x+2)(x-2)} = \dfrac{1}{x-2}$

(d) $\dfrac{\dfrac{y}{x} - \dfrac{x}{y}}{\dfrac{1}{y} - \dfrac{1}{x}} = \dfrac{\dfrac{y}{x} - \dfrac{x}{y}}{\dfrac{1}{y} - \dfrac{1}{x}} \cdot \dfrac{xy}{xy} = \dfrac{y^2 - x^2}{x - y} = \dfrac{(y-x)(y+x)}{-(y-x)} = \dfrac{y+x}{-1} = -(x+y)$

6. (a) $\dfrac{\sqrt{10}}{\sqrt{5} - 2} = \dfrac{\sqrt{10}}{\sqrt{5} - 2} \cdot \dfrac{\sqrt{5} + 2}{\sqrt{5} + 2} = \dfrac{\sqrt{50} + 2\sqrt{10}}{\left(\sqrt{5}\right)^2 - 2^2} = \dfrac{5\sqrt{2} + 2\sqrt{10}}{5 - 4} = 5\sqrt{2} + 2\sqrt{10}$

(b) $\dfrac{\sqrt{4+h} - 2}{h} = \dfrac{\sqrt{4+h} - 2}{h} \cdot \dfrac{\sqrt{4+h} + 2}{\sqrt{4+h} + 2} = \dfrac{4 + h - 4}{h(\sqrt{4+h} + 2)} = \dfrac{h}{h(\sqrt{4+h} + 2)} = \dfrac{1}{\sqrt{4+h} + 2}$

7. (a) $x^2 + x + 1 = \left(x^2 + x + \frac{1}{4}\right) + 1 - \frac{1}{4} = \left(x + \frac{1}{2}\right)^2 + \frac{3}{4}$

(b) $2x^2 - 12x + 11 = 2(x^2 - 6x) + 11 = 2(x^2 - 6x + 9 - 9) + 11 = 2(x^2 - 6x + 9) - 18 + 11 = 2(x-3)^2 - 7$

8. (a) $x + 5 = 14 - \frac{1}{2}x \iff x + \frac{1}{2}x = 14 - 5 \iff \frac{3}{2}x = 9 \iff x = \frac{2}{3} \cdot 9 \iff x = 6$

(b) $\dfrac{2x}{x+1} = \dfrac{2x-1}{x} \implies 2x^2 = (2x-1)(x+1) \iff 2x^2 = 2x^2 + x - 1 \iff x = 1$

(c) $x^2 - x - 12 = 0 \iff (x+3)(x-4) = 0 \iff x + 3 = 0 \text{ or } x - 4 = 0 \iff x = -3 \text{ or } x = 4$

(d) By the quadratic formula, $2x^2 + 4x + 1 = 0 \iff$

$x = \dfrac{-4 \pm \sqrt{4^2 - 4(2)(1)}}{2(2)} = \dfrac{-4 \pm \sqrt{8}}{4} = \dfrac{-4 \pm 2\sqrt{2}}{4} = \dfrac{2(-2 \pm \sqrt{2})}{4} = \dfrac{-2 \pm \sqrt{2}}{2} = -1 \pm \frac{1}{2}\sqrt{2}.$

(e) $x^4 - 3x^2 + 2 = 0 \iff (x^2 - 1)(x^2 - 2) = 0 \iff x^2 - 1 = 0 \text{ or } x^2 - 2 = 0 \iff x^2 = 1 \text{ or } x^2 = 2 \iff$

$x = \pm 1 \text{ or } x = \pm\sqrt{2}$

(f) $3|x - 4| = 10 \iff |x - 4| = \frac{10}{3} \iff x - 4 = -\frac{10}{3} \text{ or } x - 4 = \frac{10}{3} \iff x = \frac{2}{3} \text{ or } x = \frac{22}{3}$

(g) Multiplying through $2x(4 - x)^{-1/2} - 3\sqrt{4 - x} = 0$ by $(4 - x)^{1/2}$ gives $2x - 3(4 - x) = 0 \iff$

$2x - 12 + 3x = 0 \iff 5x - 12 = 0 \iff 5x = 12 \iff x = \frac{12}{5}.$

9. (a) $-4 < 5 - 3x \le 17 \iff -9 < -3x \le 12 \iff 3 > x \ge -4 \text{ or } -4 \le x < 3.$

In interval notation, the answer is $[-4, 3)$.

(b) $x^2 < 2x + 8 \iff x^2 - 2x - 8 < 0 \iff (x+2)(x-4) < 0.$ Now, $(x+2)(x-4)$ will change sign at the critical

values $x = -2$ and $x = 4$. Thus the possible intervals of solution are $(-\infty, -2)$, $(-2, 4)$, and $(4, \infty)$. By choosing a

single test value from each interval, we see that $(-2, 4)$ is the only interval that satisfies the inequality.

(c) The inequality $x(x-1)(x+2) > 0$ has critical values of $-2, 0$, and 1. The corresponding possible intervals of solution are $(-\infty, -2)$, $(-2, 0)$, $(0, 1)$ and $(1, \infty)$. By choosing a single test value from each interval, we see that both intervals $(-2, 0)$ and $(1, \infty)$ satisfy the inequality. Thus, the solution is the union of these two intervals: $(-2, 0) \cup (1, \infty)$.

(d) $|x - 4| < 3 \iff -3 < x - 4 < 3 \iff 1 < x < 7$. In interval notation, the answer is $(1, 7)$.

(e) $\dfrac{2x - 3}{x + 1} \leq 1 \iff \dfrac{2x - 3}{x + 1} - 1 \leq 0 \iff \dfrac{2x - 3}{x + 1} - \dfrac{x + 1}{x + 1} \leq 0 \iff \dfrac{2x - 3 - x - 1}{x + 1} \leq 0 \iff \dfrac{x - 4}{x + 1} \leq 0.$

Now, the expression $\dfrac{x - 4}{x + 1}$ may change signs at the critical values $x = -1$ and $x = 4$, so the possible intervals of solution are $(-\infty, -1)$, $(-1, 4]$, and $[4, \infty)$. By choosing a single test value from each interval, we see that $(-1, 4]$ is the only interval that satisfies the inequality.

10. (a) False. In order for the statement to be true, it must hold for all real numbers, so, to show that the statement is false, pick $p = 1$ and $q = 2$ and observe that $(1 + 2)^2 \neq 1^2 + 2^2$. In general, $(p + q)^2 = p^2 + 2pq + q^2$.

(b) True as long as a and b are nonnegative real numbers. To see this, think in terms of the laws of exponents:
$$\sqrt{ab} = (ab)^{1/2} = a^{1/2} b^{1/2} = \sqrt{a}\,\sqrt{b}.$$

(c) False. To see this, let $p = 1$ and $q = 2$, then $\sqrt{1^2 + 2^2} \neq 1 + 2$.

(d) False. To see this, let $T = 1$ and $C = 2$, then $\dfrac{1 + 1(2)}{2} \neq 1 + 1$.

(e) False. To see this, let $x = 2$ and $y = 3$, then $\dfrac{1}{2 - 3} \neq \dfrac{1}{2} - \dfrac{1}{3}$.

(f) True since $\dfrac{1/x}{a/x - b/x} \cdot \dfrac{x}{x} = \dfrac{1}{a - b}$, as long as $x \neq 0$ and $a - b \neq 0$.

Test B   Analytic Geometry

1. (a) Using the point $(2, -5)$ and $m = -3$ in the point-slope equation of a line, $y - y_1 = m(x - x_1)$, we get
$$y - (-5) = -3(x - 2) \implies y + 5 = -3x + 6 \implies y = -3x + 1.$$

(b) A line parallel to the x-axis must be horizontal and thus have a slope of 0. Since the line passes through the point $(2, -5)$, the y-coordinate of every point on the line is -5, so the equation is $y = -5$.

(c) A line parallel to the y-axis is vertical with undefined slope. So the x-coordinate of every point on the line is 2 and so the equation is $x = 2$.

(d) Note that $2x - 4y = 3 \implies -4y = -2x + 3 \implies y = \frac{1}{2}x - \frac{3}{4}$. Thus the slope of the given line is $m = \frac{1}{2}$. Hence, the slope of the line we're looking for is also $\frac{1}{2}$ (since the line we're looking for is required to be parallel to the given line). So the equation of the line is $y - (-5) = \frac{1}{2}(x - 2) \implies y + 5 = \frac{1}{2}x - 1 \implies y = \frac{1}{2}x - 6$.

2. First we'll find the distance between the two given points in order to obtain the radius, r, of the circle:
$$r = \sqrt{[3 - (-1)]^2 + (-2 - 4)^2} = \sqrt{4^2 + (-6)^2} = \sqrt{52}.$$ Next use the standard equation of a circle, $(x - h)^2 + (y - k)^2 = r^2$, where (h, k) is the center, to get $(x + 1)^2 + (y - 4)^2 = 52$.

3. We must rewrite the equation in standard form in order to identify the center and radius. Note that

$x^2 + y^2 - 6x + 10y + 9 = 0 \implies x^2 - 6x + 9 + y^2 + 10y = 0$. For the left-hand side of the latter equation, we

factor the first three terms and complete the square on the last two terms as follows: $x^2 - 6x + 9 + y^2 + 10y = 0 \implies$

$(x - 3)^2 + y^2 + 10y + 25 = 25 \implies (x - 3)^2 + (y + 5)^2 = 25$. Thus, the center of the circle is $(3, -5)$ and the radius is 5.

4. (a) $A(-7, 4)$ and $B(5, -12) \implies m_{AB} = \dfrac{-12 - 4}{5 - (-7)} = \dfrac{-16}{12} = -\dfrac{4}{3}$

(b) $y - 4 = -\frac{4}{3}[x - (-7)] \implies y - 4 = -\frac{4}{3}x - \frac{28}{3} \implies 3y - 12 = -4x - 28 \implies 4x + 3y + 16 = 0$. Putting $y = 0$,

we get $4x + 16 = 0$, so the x-intercept is -4, and substituting 0 for x results in a y-intercept of $-\frac{16}{3}$.

(c) The midpoint is obtained by averaging the corresponding coordinates of both points: $\left(\dfrac{-7+5}{2}, \dfrac{4+(-12)}{2} \right) = (-1, -4)$.

(d) $d = \sqrt{[5 - (-7)]^2 + (-12 - 4)^2} = \sqrt{12^2 + (-16)^2} = \sqrt{144 + 256} = \sqrt{400} = 20$

(e) The perpendicular bisector is the line that intersects the line segment $\overline{AB}$ at a right angle through its midpoint. Thus the

perpendicular bisector passes through $(-1, -4)$ and has slope $\frac{3}{4}$ [the slope is obtained by taking the negative reciprocal of

the answer from part (a)]. So the perpendicular bisector is given by $y + 4 = \frac{3}{4}[x - (-1)]$ or $3x - 4y = 13$.

(f) The center of the required circle is the midpoint of $\overline{AB}$, and the radius is half the length of $\overline{AB}$, which is 10. Thus, the

equation is $(x + 1)^2 + (y + 4)^2 = 100$.

5. (a) Graph the corresponding horizontal lines (given by the equations $y = -1$ and

$y = 3$) as solid lines. The inequality $y \geq -1$ describes the points (x, y) that lie

on or *above* the line $y = -1$. The inequality $y \leq 3$ describes the points (x, y)

that lie on or *below* the line $y = 3$. So the pair of inequalities $-1 \leq y \leq 3$

describes the points that lie on or *between* the lines $y = -1$ and $y = 3$.

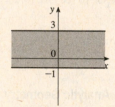

(b) Note that the given inequalities can be written as $-4 < x < 4$ and $-2 < y < 2$,

respectively. So the region lies between the vertical lines $x = -4$ and $x = 4$ and

between the horizontal lines $y = -2$ and $y = 2$. As shown in the graph, the

region common to both graphs is a rectangle (minus its edges) centered at the

origin.

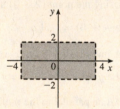

(c) We first graph $y = 1 - \frac{1}{2}x$ as a dotted line. Since $y < 1 - \frac{1}{2}x$, the points in the

region lie *below* this line.

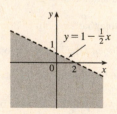

(d) We first graph the parabola $y = x^2 - 1$ using a solid curve. Since $y \geq x^2 - 1$, the points in the region lie on or *above* the parabola.

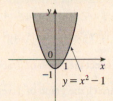

$y = x^2 - 1$

(e) We graph the circle $x^2 + y^2 = 4$ using a dotted curve. Since $\sqrt{x^2 + y^2} < 2$, the region consists of points whose distance from the origin is less than 2, that is, the points that lie *inside* the circle.

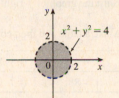

$x^2 + y^2 = 4$

(f) The equation $9x^2 + 16y^2 = 144$ is an ellipse centered at $(0,0)$. We put it in standard form by dividing by 144 and get $\dfrac{x^2}{16} + \dfrac{y^2}{9} = 1$. The x-intercepts are located at a distance of $\sqrt{16} = 4$ from the center while the y-intercepts are a distance of $\sqrt{9} = 3$ from the center (see the graph).

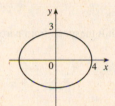

Test C   Functions

1. (a) Locate -1 on the x-axis and then go down to the point on the graph with an x-coordinate of -1. The corresponding y-coordinate is the value of the function at $x = -1$, which is -2. So, $f(-1) = -2$.

   (b) Using the same technique as in part (a), we get $f(2) \approx 2.8$.

   (c) Locate 2 on the y-axis and then go left and right to find all points on the graph with a y-coordinate of 2. The corresponding x-coordinates are the x-values we are searching for. So $x = -3$ and $x = 1$.

   (d) Using the same technique as in part (c), we get $x \approx -2.5$ and $x \approx 0.3$.

   (e) The domain is all the x-values for which the graph exists, and the range is all the y-values for which the graph exists. Thus, the domain is $[-3, 3]$, and the range is $[-2, 3]$.

2. Note that $f(2 + h) = (2 + h)^3$ and $f(2) = 2^3 = 8$. So the difference quotient becomes

   $$\frac{f(2 + h) - f(2)}{h} = \frac{(2 + h)^3 - 8}{h} = \frac{8 + 12h + 6h^2 + h^3 - 8}{h} = \frac{12h + 6h^2 + h^3}{h} = \frac{h(12 + 6h + h^2)}{h} = 12 + 6h + h^2.$$

3. (a) Set the denominator equal to 0 and solve to find restrictions on the domain: $x^2 + x - 2 = 0 \Rightarrow$ $(x - 1)(x + 2) = 0 \Rightarrow x = 1$ or $x = -2$. Thus, the domain is all real numbers except 1 or -2 or, in interval notation, $(-\infty, -2) \cup (-2, 1) \cup (1, \infty)$.

   (b) Note that the denominator is always greater than or equal to 1, and the numerator is defined for all real numbers. Thus, the domain is $(-\infty, \infty)$.

   (c) Note that the function h is the sum of two root functions. So h is defined on the intersection of the domains of these two root functions. The domain of a square root function is found by setting its radicand greater than or equal to 0. Now,

$4 - x \geq 0 \ \Rightarrow \ x \leq 4$ and $x^2 - 1 \geq 0 \ \Rightarrow \ (x-1)(x+1) \geq 0 \ \Rightarrow \ x \leq -1$ or $x \geq 1$. Thus, the domain of h is $(-\infty, -1] \cup [1, 4]$.

4. (a) Reflect the graph of f about the x-axis.

(b) Stretch the graph of f vertically by a factor of 2, then shift 1 unit downward.

(c) Shift the graph of f right 3 units, then up 2 units.

5. (a) Make a table and then connect the points with a smooth curve:

x	-2	-1	0	1	2
y	-8	-1	0	1	8

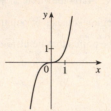

(b) Shift the graph from part (a) left 1 unit.

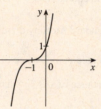

(c) Shift the graph from part (a) right 2 units and up 3 units.

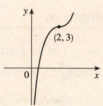

(d) First plot $y = x^2$. Next, to get the graph of $f(x) = 4 - x^2$,

reflect f about the x-axis and then shift it upward 4 units.

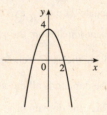

(e) Make a table and then connect the points with a smooth curve:

x	0	1	4	9
y	0	1	2	3

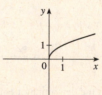

(f) Stretch the graph from part (e) vertically by a factor of two.

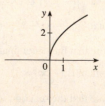

(g) First plot $y = 2^x$. Next, get the graph of $y = -2^x$ by reflecting the graph of

$y = 2^x$ about the x-axis.

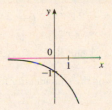

(h) Note that $y = 1 + x^{-1} = 1 + 1/x$. So first plot $y = 1/x$ and then shift it

upward 1 unit.

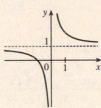

6. (a) $f(-2) = 1 - (-2)^2 = -3$ and $f(1) = 2(1) + 1 = 3$

(b) For $x \le 0$ plot $f(x) = 1 - x^2$ and, on the same plane, for $x > 0$ plot the graph

of $f(x) = 2x + 1$.

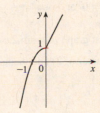

7. (a) $(f \circ g)(x) = f(g(x)) = f(2x - 3) = (2x - 3)^2 + 2(2x - 3) - 1 = 4x^2 - 12x + 9 + 4x - 6 - 1 = 4x^2 - 8x + 2$

(b) $(g \circ f)(x) = g(f(x)) = g(x^2 + 2x - 1) = 2(x^2 + 2x - 1) - 3 = 2x^2 + 4x - 2 - 3 = 2x^2 + 4x - 5$

(c) $(g \circ g \circ g)(x) = g(g(g(x))) = g(g(2x - 3)) = g(2(2x - 3) - 3) = g(4x - 9) = 2(4x - 9) - 3$

$= 8x - 18 - 3 = 8x - 21$

Test D   Trigonometry

1. (a) $300° = 300° \left(\dfrac{\pi}{180°} \right) = \dfrac{300\pi}{180} = \dfrac{5\pi}{3}$ (b) $-18° = -18° \left(\dfrac{\pi}{180°} \right) = -\dfrac{18\pi}{180} = -\dfrac{\pi}{10}$

2. (a) $\dfrac{5\pi}{6} = \dfrac{5\pi}{6} \left(\dfrac{180}{\pi} \right)^\circ = 150°$ (b) $2 = 2 \left(\dfrac{180}{\pi} \right)^\circ = \left(\dfrac{360}{\pi} \right)^\circ \approx 114.6°$

3. We will use the arc length formula, $s = r\theta$, where s is arc length, r is the radius of the circle, and θ is the measure of the

central angle in radians. First, note that $30° = 30° \left(\dfrac{\pi}{180°} \right) = \dfrac{\pi}{6}$. So $s = (12) \left(\dfrac{\pi}{6} \right) = 2\pi$ cm.

4. (a) $\tan(\pi/3) = \sqrt{3}$ [You can read the value from a right triangle with sides 1, 2, and $\sqrt{3}$.]

(b) Note that $7\pi/6$ can be thought of as an angle in the third quadrant with reference angle $\pi/6$. Thus, $\sin(7\pi/6) = -\frac{1}{2}$,

since the sine function is negative in the third quadrant.

(c) Note that $5\pi/3$ can be thought of as an angle in the fourth quadrant with reference angle $\pi/3$. Thus,

$\sec(5\pi/3) = \dfrac{1}{\cos(5\pi/3)} = \dfrac{1}{1/2} = 2$, since the cosine function is positive in the fourth quadrant.

5. $\sin\theta = a/24 \;\Rightarrow\; a = 24\sin\theta$ and $\cos\theta = b/24 \;\Rightarrow\; b = 24\cos\theta$

6. $\sin x = \frac{1}{3}$ and $\sin^2 x + \cos^2 x = 1 \;\Rightarrow\; \cos x = \sqrt{1 - \frac{1}{9}} = \frac{2\sqrt{2}}{3}$. Also, $\cos y = \frac{4}{5} \;\Rightarrow\; \sin y = \sqrt{1 - \frac{16}{25}} = \frac{3}{5}$.

So, using the sum identity for the sine, we have

$$\sin(x + y) = \sin x \cos y + \cos x \sin y = \frac{1}{3} \cdot \frac{4}{5} + \frac{2\sqrt{2}}{3} \cdot \frac{3}{5} = \frac{4 + 6\sqrt{2}}{15} = \frac{1}{15}\left(4 + 6\sqrt{2}\right)$$

7. (a) $\tan\theta\,\sin\theta + \cos\theta = \dfrac{\sin\theta}{\cos\theta}\sin\theta + \cos\theta = \dfrac{\sin^2\theta}{\cos\theta} + \dfrac{\cos^2\theta}{\cos\theta} = \dfrac{1}{\cos\theta} = \sec\theta$

(b) $\dfrac{2\tan x}{1 + \tan^2 x} = \dfrac{2\sin x/(\cos x)}{\sec^2 x} = 2\,\dfrac{\sin x}{\cos x}\cos^2 x = 2\sin x\,\cos x = \sin 2x$

8. $\sin 2x = \sin x \;\Leftrightarrow\; 2\sin x\,\cos x = \sin x \;\Leftrightarrow\; 2\sin x\,\cos x - \sin x = 0 \;\Leftrightarrow\; \sin x\,(2\cos x - 1) = 0 \;\Leftrightarrow\;$

$\sin x = 0$ or $\cos x = \frac{1}{2} \;\Rightarrow\; x = 0, \frac{\pi}{3}, \pi, \frac{5\pi}{3}, 2\pi$.

9. We first graph $y = \sin 2x$ (by compressing the graph of $\sin x$ by a factor of 2) and then shift it upward 1 unit.

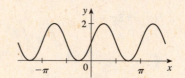

1 ☐ FUNCTIONS AND LIMITS

1.1 Functions and Their Representations

In exercises requiring estimations or approximations, your answers may vary slightly from the answers given here.

1. The functions $f(x) = x + \sqrt{2-x}$ and $g(u) = u + \sqrt{2-u}$ give exactly the same output values for every input value, so f and g are equal.

3. (a) The point $(1, 3)$ is on the graph of f, so $f(1) = 3$.

   (b) When $x = -1$, y is about -0.2, so $f(-1) \approx -0.2$.

   (c) $f(x) = 1$ is equivalent to $y = 1$. When $y = 1$, we have $x = 0$ and $x = 3$.

   (d) A reasonable estimate for x when $y = 0$ is $x = -0.8$.

   (e) The domain of f consists of all x-values on the graph of f. For this function, the domain is $-2 \le x \le 4$, or $[-2, 4]$.

   The range of f consists of all y-values on the graph of f. For this function, the range is $-1 \le y \le 3$, or $[-1, 3]$.

   (f) As x increases from -2 to 1, y increases from -1 to 3. Thus, f is increasing on the interval $[-2, 1]$.

5. No, the curve is not the graph of a function because a vertical line intersects the curve more than once. Hence, the curve fails the Vertical Line Test.

7. Yes, the curve is the graph of a function because it passes the Vertical Line Test. The domain is $[-3, 2]$ and the range is $[-3, -2) \cup [-1, 3]$.

9. The person's weight increased to about 160 pounds at age 20 and stayed fairly steady for 10 years. The person's weight dropped to about 120 pounds for the next 5 years, then increased rapidly to about 170 pounds. The next 30 years saw a gradual increase to 190 pounds. Possible reasons for the drop in weight at 30 years of age: diet, exercise, health problems.

11. The water will cool down almost to freezing as the ice melts. Then, when the ice has melted, the water will slowly warm up to room temperature.

13. Of course, this graph depends strongly on the geographical location!

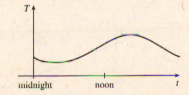

15. As the price increases, the amount sold decreases.

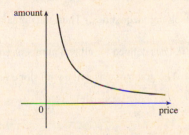

17.

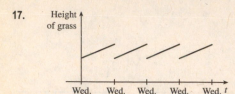

19. $f(x) = 3x^2 - x + 2$.

$f(2) = 3(2)^2 - 2 + 2 = 12 - 2 + 2 = 12$.

$f(-2) = 3(-2)^2 - (-2) + 2 = 12 + 2 + 2 = 16$.

$f(a) = 3a^2 - a + 2$.

$f(-a) = 3(-a)^2 - (-a) + 2 = 3a^2 + a + 2$.

$f(a+1) = 3(a+1)^2 - (a+1) + 2 = 3(a^2 + 2a + 1) - a - 1 + 2 = 3a^2 + 6a + 3 - a + 1 = 3a^2 + 5a + 4$.

$2f(a) = 2 \cdot f(a) = 2(3a^2 - a + 2) = 6a^2 - 2a + 4$.

$f(2a) = 3(2a)^2 - (2a) + 2 = 3(4a^2) - 2a + 2 = 12a^2 - 2a + 2$.

$f(a^2) = 3(a^2)^2 - (a^2) + 2 = 3(a^4) - a^2 + 2 = 3a^4 - a^2 + 2$.

$$[f(a)]^2 = [3a^2 - a + 2]^2 = (3a^2 - a + 2)(3a^2 - a + 2)$$
$$= 9a^4 - 3a^3 + 6a^2 - 3a^3 + a^2 - 2a + 6a^2 - 2a + 4 = 9a^4 - 6a^3 + 13a^2 - 4a + 4.$$

$f(a+h) = 3(a+h)^2 - (a+h) + 2 = 3(a^2 + 2ah + h^2) - a - h + 2 = 3a^2 + 6ah + 3h^2 - a - h + 2$.

21. $f(x) = 4 + 3x - x^2$, so $f(3+h) = 4 + 3(3+h) - (3+h)^2 = 4 + 9 + 3h - (9 + 6h + h^2) = 4 - 3h - h^2$,

and $\dfrac{f(3+h) - f(3)}{h} = \dfrac{(4 - 3h - h^2) - 4}{h} = \dfrac{h(-3-h)}{h} = -3 - h$.

23. $\dfrac{f(x) - f(a)}{x - a} = \dfrac{\dfrac{1}{x} - \dfrac{1}{a}}{x - a} = \dfrac{\dfrac{a - x}{xa}}{x - a} = \dfrac{a - x}{xa(x - a)} = \dfrac{-1(x - a)}{xa(x - a)} = -\dfrac{1}{ax}$

25. $f(x) = (x+4)/(x^2 - 9)$ is defined for all x except when $0 = x^2 - 9 \iff 0 = (x+3)(x-3) \iff x = -3$ or 3, so the domain is $\{x \in \mathbb{R} \mid x \neq -3, 3\} = (-\infty, -3) \cup (-3, 3) \cup (3, \infty)$.

27. $F(p) = \sqrt{2 - \sqrt{p}}$ is defined when $p \geq 0$ and $2 - \sqrt{p} \geq 0$. Since $2 - \sqrt{p} \geq 0 \implies 2 \geq \sqrt{p} \implies \sqrt{p} \leq 2 \implies 0 \leq p \leq 4$, the domain is $[0, 4]$.

29. $h(x) = 1 / \sqrt[4]{x^2 - 5x}$ is defined when $x^2 - 5x > 0 \iff x(x - 5) > 0$. Note that $x^2 - 5x \neq 0$ since that would result in division by zero. The expression $x(x - 5)$ is positive if $x < 0$ or $x > 5$. (See *Review of Algebra* at www.stewartcalculus.com for methods for solving inequalities.) Thus, the domain is $(-\infty, 0) \cup (5, \infty)$.

31. $f(x) = 2 - 0.4x$ is defined for all real numbers, so the domain is $\mathbb{R}$, or $(-\infty, \infty)$. The graph of f is a line with slope -0.4 and y-intercept 2.

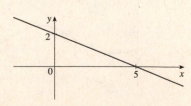

33. $f(t) = 2t + t^2$ is defined for all real numbers, so the domain is $\mathbb{R}$, or

$(-\infty, \infty)$. The graph of f is a parabola opening upward since the

coefficient of t^2 is positive. To find the t-intercepts, let $y = 0$ and solve

for t. $0 = 2t + t^2 = t(2 + t) \implies t = 0$ or $t = -2$. The t-coordinate of

the vertex is halfway between the t-intercepts, that is, at $t = -1$. Since

$f(-1) = 2(-1) + (-1)^2 = -2 + 1 = -1$, the vertex is $(-1, -1)$.

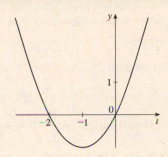

35. $g(x) = \sqrt{x - 5}$ is defined when $x - 5 \geq 0$ or $x \geq 5$, so the domain is $[5, \infty)$.

Since $y = \sqrt{x - 5} \implies y^2 = x - 5 \implies x = y^2 + 5$, we see that g is the

top half of a parabola.

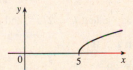

37. $G(x) = \dfrac{3x + |x|}{x}$. Since $|x| = \begin{cases} x & \text{if } x \geq 0 \\ -x & \text{if } x < 0 \end{cases}$, we have

$$G(x) = \begin{cases} \dfrac{3x + x}{x} & \text{if } x > 0 \\ \dfrac{3x - x}{x} & \text{if } x < 0 \end{cases} = \begin{cases} \dfrac{4x}{x} & \text{if } x > 0 \\ \dfrac{2x}{x} & \text{if } x < 0 \end{cases} = \begin{cases} 4 & \text{if } x > 0 \\ 2 & \text{if } x < 0 \end{cases}$$

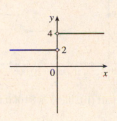

Note that G is not defined for $x = 0$. The domain is $(-\infty, 0) \cup (0, \infty)$.

39. $f(x) = \begin{cases} x + 2 & \text{if } x < 0 \\ 1 - x & \text{if } x \geq 0 \end{cases}$

The domain is $\mathbb{R}$.

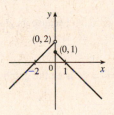

41. $f(x) = \begin{cases} x + 2 & \text{if } x \leq -1 \\ x^2 & \text{if } x > -1 \end{cases}$

Note that for $x = -1$, both $x + 2$ and x^2 are equal to 1.

The domain is $\mathbb{R}$.

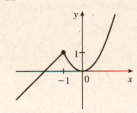

43. Recall that the slope m of a line between the two points (x_1, y_1) and (x_2, y_2) is $m = \dfrac{y_2 - y_1}{x_2 - x_1}$ and an equation of the line

connecting those two points is $y - y_1 = m(x - x_1)$. The slope of the line segment joining the points $(1, -3)$ and $(5, 7)$ is

$\dfrac{7 - (-3)}{5 - 1} = \dfrac{5}{2}$, so an equation is $y - (-3) = \frac{5}{2}(x - 1)$. The function is $f(x) = \frac{5}{2}x - \frac{11}{2}$, $1 \leq x \leq 5$.

45. We need to solve the given equation for y. $x + (y - 1)^2 = 0 \iff (y - 1)^2 = -x \iff y - 1 = \pm\sqrt{-x} \iff$

$y = 1 \pm \sqrt{-x}$. The expression with the positive radical represents the top half of the parabola, and the one with the negative

radical represents the bottom half. Hence, we want $f(x) = 1 - \sqrt{-x}$. Note that the domain is $x \leq 0$.

47. Let the length and width of the rectangle be L and W. Then the perimeter is $2L + 2W = 20$ and the area is $A = LW$.

Solving the first equation for W in terms of L gives $W = \dfrac{20 - 2L}{2} = 10 - L$. Thus, $A(L) = L(10 - L) = 10L - L^2$. Since lengths are positive, the domain of A is $0 < L < 10$. If we further restrict L to be larger than W, then $5 < L < 10$ would be the domain.

49. Let the length of a side of the equilateral triangle be x. Then by the Pythagorean Theorem, the height y of the triangle satisfies $y^2 + \left(\frac{1}{2}x\right)^2 = x^2$, so that $y^2 = x^2 - \frac{1}{4}x^2 = \frac{3}{4}x^2$ and $y = \frac{\sqrt{3}}{2}x$. Using the formula for the area A of a triangle, $A = \frac{1}{2}(\text{base})(\text{height})$, we obtain $A(x) = \frac{1}{2}(x)\left(\frac{\sqrt{3}}{2}x\right) = \frac{\sqrt{3}}{4}x^2$, with domain $x > 0$.

51. Let each side of the base of the box have length x, and let the height of the box be h. Since the volume is 2, we know that $2 = hx^2$, so that $h = 2/x^2$, and the surface area is $S = x^2 + 4xh$. Thus, $S(x) = x^2 + 4x(2/x^2) = x^2 + (8/x)$, with domain $x > 0$.

53. (a)

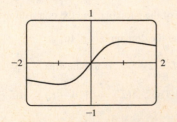

(b) On \$14,000, tax is assessed on \$4000, and 10%(\$4000) = \$400.

On \$26,000, tax is assessed on \$16,000, and

10%(\$10,000) + 15%(\$6000) = \$1000 + \$900 = \$1900.

(c) As in part (b), there is \$1000 tax assessed on \$20,000 of income, so the graph of T is a line segment from $(10{,}000, 0)$ to $(20{,}000, 1000)$. The tax on \$30,000 is \$2500, so the graph of T for $x > 20{,}000$ is the ray with initial point $(20{,}000, 1000)$ that passes through $(30{,}000, 2500)$.

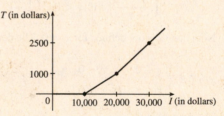

55. f is an odd function because its graph is symmetric about the origin. g is an even function because its graph is symmetric with respect to the y-axis.

57. (a) Because an even function is symmetric with respect to the y-axis, and the point $(5, 3)$ is on the graph of this even function, the point $(-5, 3)$ must also be on its graph.

(b) Because an odd function is symmetric with respect to the origin, and the point $(5, 3)$ is on the graph of this odd function, the point $(-5, -3)$ must also be on its graph.

59. $f(x) = \dfrac{x}{x^2 + 1}$.

$f(-x) = \dfrac{-x}{(-x)^2 + 1} = \dfrac{-x}{x^2 + 1} = -\dfrac{x}{x^2 + 1} = -f(x)$.

So f is an odd function.

61. $f(x) = \dfrac{x}{x + 1}$, so $f(-x) = \dfrac{-x}{-x + 1} = \dfrac{x}{x - 1}$.

Since this is neither $f(x)$ nor $-f(x)$, the function f is neither even nor odd.

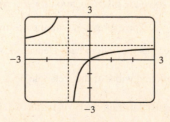

63. $f(x) = 1 + 3x^2 - x^4$.

$f(-x) = 1 + 3(-x)^2 - (-x)^4 = 1 + 3x^2 - x^4 = f(x)$.

So f is an even function.

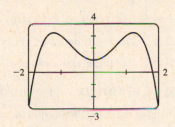

65. (i) If f and g are both even functions, then $f(-x) = f(x)$ and $g(-x) = g(x)$. Now

$(f+g)(-x) = f(-x) + g(-x) = f(x) + g(x) = (f+g)(x)$, so $f+g$ is an *even* function.

(ii) If f and g are both odd functions, then $f(-x) = -f(x)$ and $g(-x) = -g(x)$. Now

$(f+g)(-x) = f(-x) + g(-x) = -f(x) + [-g(x)] = -[f(x) + g(x)] = -(f+g)(x)$, so $f+g$ is an *odd* function.

(iii) If f is an even function and g is an odd function, then $(f+g)(-x) = f(-x) + g(-x) = f(x) + [-g(x)] = f(x) - g(x)$,

which is not $(f+g)(x)$ nor $-(f+g)(x)$, so $f+g$ is *neither* even nor odd. (Exception: if f is the zero function, then

$f+g$ will be *odd*. If g is the zero function, then $f+g$ will be *even*.)

1.2 A Catalog of Essential Functions

1. (a) An equation for the family of linear functions with slope 2

is $y = f(x) = 2x + b$, where b is the y-intercept.

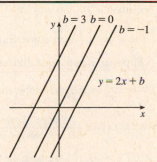

(b) $f(2) = 1$ means that the point $(2, 1)$ is on the graph of f. We can use the

point-slope form of a line to obtain an equation for the family of linear

functions through the point $(2, 1)$. $y - 1 = m(x - 2)$, which is equivalent

to $y = mx + (1 - 2m)$ in slope-intercept form.

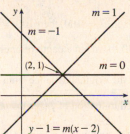

(c) To belong to both families, an equation must have slope $m = 2$, so the

equation in part (b), $y = mx + (1 - 2m)$, becomes $y = 2x - 3$. It is the

only function that belongs to both families.

3. All members of the family of linear functions $f(x) = c - x$ have graphs

that are lines with slope -1. The y-intercept is c.

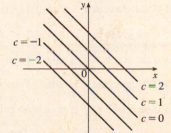

5. Since $f(-1) = f(0) = f(2) = 0$, f has zeros of $-1, 0$, and 2, so an equation for f is $f(x) = a[x - (-1)](x - 0)(x - 2)$,

or $f(x) = ax(x + 1)(x - 2)$. Because $f(1) = 6$, we'll substitute 1 for x and 6 for $f(x)$.

$6 = a(1)(2)(-1) \Rightarrow -2a = 6 \Rightarrow a = -3$, so an equation for f is $f(x) = -3x(x + 1)(x - 2)$.

7. (a) $D = 200$, so $c = 0.0417D(a + 1) = 0.0417(200)(a + 1) = 8.34a + 8.34$. The slope is 8.34, which represents the

change in mg of the dosage for a child for each change of 1 year in age.

(b) For a newborn, $a = 0$, so $c = 8.34$ mg.

9. (a)

(b) The slope of $\frac{9}{5}$ means that F increases $\frac{9}{5}$ degrees for each increase

of $1°C$. (Equivalently, F increases by 9 when C increases by 5

and F decreases by 9 when C decreases by 5.) The F-intercept of

32 is the Fahrenheit temperature corresponding to a Celsius

temperature of 0.

11. (a) Using N in place of x and T in place of y, we find the slope to be $\dfrac{T_2 - T_1}{N_2 - N_1} = \dfrac{80 - 70}{173 - 113} = \dfrac{10}{60} = \dfrac{1}{6}$. So a linear

equation is $T - 80 = \frac{1}{6}(N - 173) \Leftrightarrow T - 80 = \frac{1}{6}N - \frac{173}{6} \Leftrightarrow T = \frac{1}{6}N + \frac{307}{6}$ $\left[\frac{307}{6} = 51.1\overline{6}\right]$.

(b) The slope of $\frac{1}{6}$ means that the temperature in Fahrenheit degrees increases one-sixth as rapidly as the number of cricket

chirps per minute. Said differently, each increase of 6 cricket chirps per minute corresponds to an increase of $1°F$.

(c) When $N = 150$, the temperature is given approximately by $T = \frac{1}{6}(150) + \frac{307}{6} = 76.1\overline{6}°F \approx 76°F$.

13. (a) We are given $\dfrac{\text{change in pressure}}{10 \text{ feet change in depth}} = \dfrac{4.34}{10} = 0.434$. Using P for pressure and d for depth with the point

$(d, P) = (0, 15)$, we have the slope-intercept form of the line, $P = 0.434d + 15$.

(b) When $P = 100$, then $100 = 0.434d + 15 \Leftrightarrow 0.434d = 85 \Leftrightarrow d = \frac{85}{0.434} \approx 195.85$ feet. Thus, the pressure is

100 lb/in^2 at a depth of approximately 196 feet.

15. If x is the original distance from the source, then the illumination is $f(x) = kx^{-2} = k/x^2$. Moving halfway to the lamp gives

us an illumination of $f\left(\frac{1}{2}x\right) = k\left(\frac{1}{2}x\right)^{-2} = k(2/x)^2 = 4(k/x^2)$, so the light is 4 times as bright.

17. (a) If the graph of f is shifted 3 units upward, its equation becomes $y = f(x) + 3$.

(b) If the graph of f is shifted 3 units downward, its equation becomes $y = f(x) - 3$.

(c) If the graph of f is shifted 3 units to the right, its equation becomes $y = f(x - 3)$.

(d) If the graph of f is shifted 3 units to the left, its equation becomes $y = f(x + 3)$.

(e) If the graph of f is reflected about the x-axis, its equation becomes $y = -f(x)$.

(f) If the graph of f is reflected about the y-axis, its equation becomes $y = f(-x)$.

(g) If the graph of f is stretched vertically by a factor of 3, its equation becomes $y = 3f(x)$.

(h) If the graph of f is shrunk vertically by a factor of 3, its equation becomes $y = \frac{1}{3}f(x)$.

19. (a) (graph 3) The graph of f is shifted 4 units to the right and has equation $y = f(x - 4)$.

(b) (graph 1) The graph of f is shifted 3 units upward and has equation $y = f(x) + 3$.

(c) (graph 4) The graph of f is shrunk vertically by a factor of 3 and has equation $y = \frac{1}{3}f(x)$.

(d) (graph 5) The graph of f is shifted 4 units to the left and reflected about the x-axis. Its equation is $y = -f(x + 4)$.

(e) (graph 2) The graph of f is shifted 6 units to the left and stretched vertically by a factor of 2. Its equation is
$y = 2f(x + 6)$.

21. (a) To graph $y = f(2x)$ we shrink the graph of f horizontally by a factor of 2.

The point $(4, -1)$ on the graph of f corresponds to the point $\left(\frac{1}{2} \cdot 4, -1\right) = (2, -1)$.

(b) To graph $y = f\left(\frac{1}{2}x\right)$ we stretch the graph of f horizontally by a factor of 2.

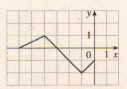

The point $(4, -1)$ on the graph of f corresponds to the point $(2 \cdot 4, -1) = (8, -1)$.

(c) To graph $y = f(-x)$ we reflect the graph of f about the y-axis.

The point $(4, -1)$ on the graph of f corresponds to the point $(-1 \cdot 4, -1) = (-4, -1)$.

(d) To graph $y = -f(-x)$ we reflect the graph of f about the y-axis, then about the x-axis.

The point $(4, -1)$ on the graph of f corresponds to the point $(-1 \cdot 4, -1 \cdot -1) = (-4, 1)$.

23. $y = \dfrac{1}{x + 2}$: Start with the graph of the reciprocal function $y = 1/x$ and shift 2 units to the left.

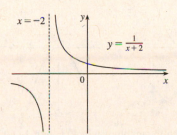

25. $y = -\sqrt[3]{x}$: Start with the graph of $y = \sqrt[3]{x}$ and reflect about the x-axis.

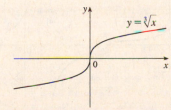

27. $y = \sqrt{x-2} - 1$: Start with the graph of $y = \sqrt{x}$, shift 2 units to the right, and then shift 1 unit downward.

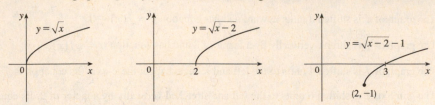

29. $y = \sin(x/2)$: Start with the graph of $y = \sin x$ and stretch horizontally by a factor of 2.

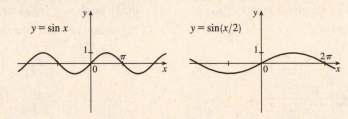

31. $y = \frac{1}{2}(1 - \cos x)$: Start with the graph of $y = \cos x$, reflect about the x-axis, shift 1 unit upward, and then shrink vertically by a factor of 2.

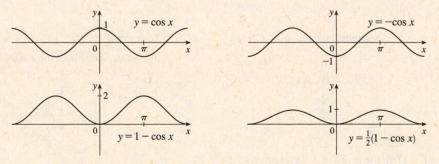

33. $y = 1 - 2x - x^2 = -(x^2 + 2x) + 1 = -(x^2 + 2x + 1) + 2 = -(x+1)^2 + 2$: Start with the graph of $y = x^2$, reflect about the x-axis, shift 1 unit to the left, and then shift 2 units upward.

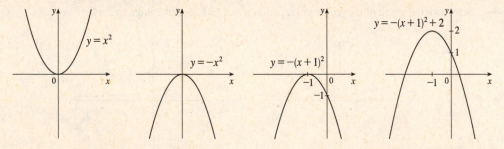

35. $y = 2/(x+1)$: Start with the graph of $y = 1/x$, shift 1 unit to the left, and then stretch vertically by a factor of 2.

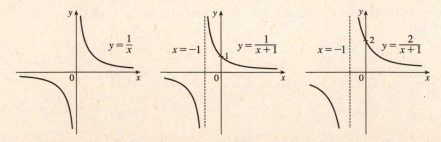

37. $f(x) = x^3 + 2x^2$; $g(x) = 3x^2 - 1$. $D = \mathbb{R}$ for both f and g.

(a) $(f + g)(x) = (x^3 + 2x^2) + (3x^2 - 1) = x^3 + 5x^2 - 1$, $D = \mathbb{R}$.

(b) $(f - g)(x) = (x^3 + 2x^2) - (3x^2 - 1) = x^3 - x^2 + 1$, $D = \mathbb{R}$.

(c) $(fg)(x) = (x^3 + 2x^2)(3x^2 - 1) = 3x^5 + 6x^4 - x^3 - 2x^2$, $D = \mathbb{R}$.

(d) $\left(\dfrac{f}{g}\right)(x) = \dfrac{x^3 + 2x^2}{3x^2 - 1}$, $D = \left\{x \mid x \neq \pm\dfrac{1}{\sqrt{3}}\right\}$ since $3x^2 - 1 \neq 0$.

39. $f(x) = x^2 - 1$, $D = \mathbb{R}$; $g(x) = 2x + 1$, $D = \mathbb{R}$.

(a) $(f \circ g)(x) = f(g(x)) = f(2x + 1) = (2x + 1)^2 - 1 = (4x^2 + 4x + 1) - 1 = 4x^2 + 4x$, $D = \mathbb{R}$.

(b) $(g \circ f)(x) = g(f(x)) = g(x^2 - 1) = 2(x^2 - 1) + 1 = (2x^2 - 2) + 1 = 2x^2 - 1$, $D = \mathbb{R}$.

(c) $(f \circ f)(x) = f(f(x)) = f(x^2 - 1) = (x^2 - 1)^2 - 1 = (x^4 - 2x^2 + 1) - 1 = x^4 - 2x^2$, $D = \mathbb{R}$.

(d) $(g \circ g)(x) = g(g(x)) = g(2x + 1) = 2(2x + 1) + 1 = (4x + 2) + 1 = 4x + 3$, $D = \mathbb{R}$.

41. $f(x) = 1 - 3x$; $g(x) = \cos x$. $D = \mathbb{R}$ for both f and g, and hence for their composites.

(a) $(f \circ g)(x) = f(g(x)) = f(\cos x) = 1 - 3\cos x$.

(b) $(g \circ f)(x) = g(f(x)) = g(1 - 3x) = \cos(1 - 3x)$.

(c) $(f \circ f)(x) = f(f(x)) = f(1 - 3x) = 1 - 3(1 - 3x) = 1 - 3 + 9x = 9x - 2$.

(d) $(g \circ g)(x) = g(g(x)) = g(\cos x) = \cos(\cos x)$ [Note that this is *not* $\cos x \cdot \cos x$.]

43. $f(x) = x + \dfrac{1}{x}$, $D = \{x \mid x \neq 0\}$; $g(x) = \dfrac{x + 1}{x + 2}$, $D = \{x \mid x \neq -2\}$

(a) $(f \circ g)(x) = f(g(x)) = f\left(\dfrac{x + 1}{x + 2}\right) = \dfrac{x + 1}{x + 2} + \dfrac{1}{\dfrac{x + 1}{x + 2}} = \dfrac{x + 1}{x + 2} + \dfrac{x + 2}{x + 1}$

$$= \dfrac{(x + 1)(x + 1) + (x + 2)(x + 2)}{(x + 2)(x + 1)} = \dfrac{(x^2 + 2x + 1) + (x^2 + 4x + 4)}{(x + 2)(x + 1)} = \dfrac{2x^2 + 6x + 5}{(x + 2)(x + 1)}$$

Since $g(x)$ is not defined for $x = -2$ and $f(g(x))$ is not defined for $x = -2$ and $x = -1$,

the domain of $(f \circ g)(x)$ is $D = \{x \mid x \neq -2, -1\}$.

(b) $(g \circ f)(x) = g(f(x)) = g\left(x + \dfrac{1}{x}\right) = \dfrac{\left(x + \dfrac{1}{x}\right) + 1}{\left(x + \dfrac{1}{x}\right) + 2} = \dfrac{\dfrac{x^2 + 1 + x}{x}}{\dfrac{x^2 + 1 + 2x}{x}} = \dfrac{x^2 + x + 1}{x^2 + 2x + 1} = \dfrac{x^2 + x + 1}{(x + 1)^2}$

Since $f(x)$ is not defined for $x = 0$ and $g(f(x))$ is not defined for $x = -1$,

the domain of $(g \circ f)(x)$ is $D = \{x \mid x \neq -1, 0\}$.

(c) $(f \circ f)(x) = f(f(x)) = f\left(x + \dfrac{1}{x}\right) = \left(x + \dfrac{1}{x}\right) + \dfrac{1}{x + \dfrac{1}{x}} = x + \dfrac{1}{x} + \dfrac{1}{\dfrac{x^2 + 1}{x}} = x + \dfrac{1}{x} + \dfrac{x}{x^2 + 1}$

$$= \dfrac{x(x)(x^2 + 1) + 1(x^2 + 1) + x(x)}{x(x^2 + 1)} = \dfrac{x^4 + x^2 + x^2 + 1 + x^2}{x(x^2 + 1)}$$

$$= \dfrac{x^4 + 3x^2 + 1}{x(x^2 + 1)}, \quad D = \{x \mid x \neq 0\}$$

(d) $(g \circ g)(x) = g(g(x)) = g\left(\dfrac{x+1}{x+2}\right) = \dfrac{\dfrac{x+1}{x+2}+1}{\dfrac{x+1}{x+2}+2} = \dfrac{\dfrac{x+1+1(x+2)}{x+2}}{\dfrac{x+1+2(x+2)}{x+2}} = \dfrac{x+1+x+2}{x+1+2x+4} = \dfrac{2x+3}{3x+5}$

Since $g(x)$ is not defined for $x = -2$ and $g(g(x))$ is not defined for $x = -\frac{5}{3}$,

the domain of $(g \circ g)(x)$ is $D = \left\{x \mid x \neq -2, -\frac{5}{3}\right\}$.

45. $(f \circ g \circ h)(x) = f(g(h(x))) = f(g(x^3+2)) = f[(x^3+2)^2]$

$\quad = f(x^6 + 4x^3 + 4) = \sqrt{(x^6 + 4x^3 + 4) - 3} = \sqrt{x^6 + 4x^3 + 1}$

47. Let $g(x) = 2x + x^2$ and $f(x) = x^4$. Then $(f \circ g)(x) = f(g(x)) = f(2x + x^2) = (2x + x^2)^4 = F(x)$.

49. Let $g(t) = t^2$ and $f(t) = \sec t \tan t$. Then $(f \circ g)(t) = f(g(t)) = f(t^2) = \sec(t^2) \tan(t^2) = v(t)$.

51. Let $h(x) = \sqrt{x}$, $g(x) = x - 1$, and $f(x) = \sqrt{x}$. Then

$\quad (f \circ g \circ h)(x) = f(g(h(x))) = f(g(\sqrt{x})) = f(\sqrt{x} - 1) = \sqrt{\sqrt{x} - 1} = R(x)$.

53. Let $h(x) = \sqrt{x}$, $g(x) = \sec x$, and $f(x) = x^4$. Then

$\quad (f \circ g \circ h)(x) = f(g(h(x))) = f(g(\sqrt{x})) = f(\sec \sqrt{x}) = (\sec \sqrt{x})^4 = \sec^4(\sqrt{x}) = H(x)$.

55. (a) $g(2) = 5$, because the point $(2, 5)$ is on the graph of g. Thus, $f(g(2)) = f(5) = 4$, because the point $(5, 4)$ is on the

graph of f.

(b) $g(f(0)) = g(0) = 3$

(c) $(f \circ g)(0) = f(g(0)) = f(3) = 0$

(d) $(g \circ f)(6) = g(f(6)) = g(6)$. This value is not defined, because there is no point on the graph of g that has

x-coordinate 6.

(e) $(g \circ g)(-2) = g(g(-2)) = g(1) = 4$

(f) $(f \circ f)(4) = f(f(4)) = f(2) = -2$

57. (a) Using the relationship *distance = rate · time* with the radius r as the distance, we have $r(t) = 60t$.

(b) $A = \pi r^2 \quad \Rightarrow \quad (A \circ r)(t) = A(r(t)) = \pi(60t)^2 = 3600\pi t^2$. This formula gives us the extent of the rippled area

(in cm^2) at any time t.

59. (a)

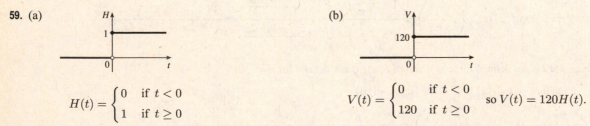

$H(t) = \begin{cases} 0 & \text{if } t < 0 \\ 1 & \text{if } t \geq 0 \end{cases}$

(b)

$V(t) = \begin{cases} 0 & \text{if } t < 0 \\ 120 & \text{if } t \geq 0 \end{cases}$ so $V(t) = 120H(t)$.

(c)

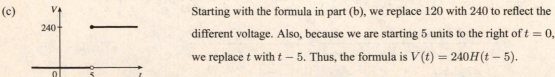

Starting with the formula in part (b), we replace 120 with 240 to reflect the

different voltage. Also, because we are starting 5 units to the right of $t = 0$,

we replace t with $t - 5$. Thus, the formula is $V(t) = 240H(t - 5)$.

61. If $f(x) = m_1 x + b_1$ and $g(x) = m_2 x + b_2$, then

$$(f \circ g)(x) = f(g(x)) = f(m_2 x + b_2) = m_1(m_2 x + b_2) + b_1 = m_1 m_2 x + m_1 b_2 + b_1.$$

So $f \circ g$ is a linear function with slope $m_1 m_2$.

63. (a) By examining the variable terms in g and h, we deduce that we must square g to get the terms $4x^2$ and $4x$ in h. If we let

$$f(x) = x^2 + c, \text{ then } (f \circ g)(x) = f(g(x)) = f(2x + 1) = (2x + 1)^2 + c = 4x^2 + 4x + (1 + c). \text{ Since}$$

$h(x) = 4x^2 + 4x + 7$, we must have $1 + c = 7$. So $c = 6$ and $f(x) = x^2 + 6$.

(b) We need a function g so that $f(g(x)) = 3(g(x)) + 5 = h(x)$. But

$$h(x) = 3x^2 + 3x + 2 = 3(x^2 + x) + 2 = 3(x^2 + x - 1) + 5, \text{ so we see that } g(x) = x^2 + x - 1.$$

65. We need to examine $h(-x)$.

$$h(-x) = (f \circ g)(-x) = f(g(-x)) = f(g(x)) \quad [\text{because } g \text{ is even}] \quad = h(x)$$

Because $h(-x) = h(x)$, h is an even function.

1.3 The Limit of a Function

1. (a) $y = y(t) = 40t - 16t^2$. At $t = 2$, $y = 40(2) - 16(2)^2 = 16$. The average velocity between times 2 and $2 + h$ is

$$v_{\text{ave}} = \frac{y(2 + h) - y(2)}{(2 + h) - 2} = \frac{\left[40(2 + h) - 16(2 + h)^2\right] - 16}{h} = \frac{-24h - 16h^2}{h} = -24 - 16h, \text{ if } h \neq 0.$$

  (i) $[2, 2.5]$: $h = 0.5$, $v_{\text{ave}} = -32$ ft/s   (ii) $[2, 2.1]$: $h = 0.1$, $v_{\text{ave}} = -25.6$ ft/s

  (iii) $[2, 2.05]$: $h = 0.05$, $v_{\text{ave}} = -24.8$ ft/s   (iv) $[2, 2.01]$: $h = 0.01$, $v_{\text{ave}} = -24.16$ ft/s

(b) The instantaneous velocity when $t = 2$ (h approaches 0) is -24 ft/s.

3. (a) As x approaches 1, the values of $f(x)$ approach 2, so $\lim\limits_{x \to 1} f(x) = 2$.

(b) As x approaches 3 from the left, the values of $f(x)$ approach 1, so $\lim\limits_{x \to 3^-} f(x) = 1$.

(c) As x approaches 3 from the right, the values of $f(x)$ approach 4, so $\lim\limits_{x \to 3^+} f(x) = 4$.

(d) $\lim\limits_{x \to 3} f(x)$ does not exist since the left-hand limit does not equal the right-hand limit.

(e) When $x = 3$, $y = 3$, so $f(3) = 3$.

5. (a) $\lim\limits_{t \to 0^-} g(t) = -1$ (b) $\lim\limits_{t \to 0^+} g(t) = -2$

(c) $\lim\limits_{t \to 0} g(t)$ does not exist because the limits in part (a) and part (b) are not equal.

(d) $\lim\limits_{t \to 2^-} g(t) = 2$ (e) $\lim\limits_{t \to 2^+} g(t) = 0$

(f) $\lim\limits_{t \to 2} g(t)$ does not exist because the limits in part (d) and part (e) are not equal.

(g) $g(2) = 1$ (h) $\lim\limits_{t \to 4} g(t) = 3$

7. $\lim\limits_{x\to 0^-} f(x) = -1,\quad \lim\limits_{x\to 0^+} f(x) = 2,\quad f(0) = 1$

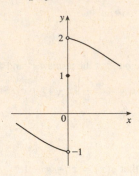

9. $\lim\limits_{x\to 3^+} f(x) = 4,\quad \lim\limits_{x\to 3^-} f(x) = 2,\ \lim\limits_{x\to -2} f(x) = 2,$

$f(3) = 3,\quad f(-2) = 1$

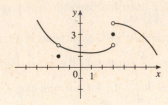

11. For $f(x) = \dfrac{x^2 - 2x}{x^2 - x - 2}$:

x	$f(x)$
2.5	0.714286
2.1	0.677419
2.05	0.672131
2.01	0.667774
2.005	0.667221
2.001	0.666778

x	$f(x)$
1.9	0.655172
1.95	0.661017
1.99	0.665552
1.995	0.666110
1.999	0.666556

It appears that $\lim\limits_{x\to 2} \dfrac{x^2 - 2x}{x^2 - x - 2} = 0.\overline{6} = \frac{2}{3}$.

13. For $f(x) = \dfrac{\sin x}{x + \tan x}$:

x	$f(x)$
± 1	0.329033
± 0.5	0.458209
± 0.2	0.493331
± 0.1	0.498333
± 0.05	0.499583
± 0.01	0.499983

It appears that $\lim\limits_{x\to 0} \dfrac{\sin x}{x + \tan x} = 0.5 = \dfrac{1}{2}$.

15. For $f(x) = \dfrac{\sqrt{x + 4} - 2}{x}$:

x	$f(x)$
1	0.236068
0.5	0.242641
0.1	0.248457
0.05	0.249224
0.01	0.249844

x	$f(x)$
-1	0.267949
-0.5	0.258343
-0.1	0.251582
-0.05	0.250786
-0.01	0.250156

It appears that $\lim\limits_{x\to 0} \dfrac{\sqrt{x + 4} - 2}{x} = 0.25 = \frac{1}{4}$.

17. For $f(x) = \dfrac{x^6 - 1}{x^{10} - 1}$:

x	$f(x)$
0.5	0.985337
0.9	0.719397
0.95	0.660186
0.99	0.612018
0.999	0.601200

x	$f(x)$
1.5	0.183369
1.1	0.484119
1.05	0.540783
1.01	0.588022
1.001	0.598800

It appears that $\lim\limits_{x\to 1} \dfrac{x^6 - 1}{x^{10} - 1} = 0.6 = \frac{3}{5}$.

19. (a) From the graphs, it seems that $\lim\limits_{x\to 0} \dfrac{\cos 2x - \cos x}{x^2} = -1.5$.

(b)

x	$f(x)$
± 0.1	-1.493759
± 0.01	-1.499938
± 0.001	-1.499999
± 0.0001	-1.500000

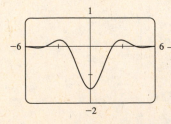

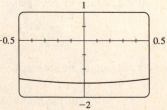

21. For $f(x) = x^2 - (2^x/1000)$:

(a)

x	$f(x)$
1	0.998000
0.8	0.638259
0.6	0.358484
0.4	0.158680
0.2	0.038851
0.1	0.008928
0.05	0.001465

It appears that $\lim\limits_{x \to 0} f(x) = 0$.

(b)

x	$f(x)$
0.04	0.000572
0.02	−0.000614
0.01	−0.000907
0.005	−0.000978
0.003	−0.000993
0.001	−0.001000

It appears that $\lim\limits_{x \to 0} f(x) = -0.001$.

23. The leftmost question mark is the solution of $\sqrt{x} = 1.6$ and the rightmost, $\sqrt{x} = 2.4$. So the values are $1.6^2 = 2.56$ and $2.4^2 = 5.76$. On the left side, we need $|x - 4| < |2.56 - 4| = 1.44$. On the right side, we need $|x - 4| < |5.76 - 4| = 1.76$. To satisfy both conditions, we need the more restrictive condition to hold—namely, $|x - 4| < 1.44$. Thus, we can choose $\delta = 1.44$, or any smaller positive number.

25.

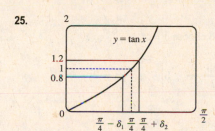

From the graph, we find that $y = \tan x = 0.8$ when $x \approx 0.675$, so

$\frac{\pi}{4} - \delta_1 \approx 0.675 \quad \Rightarrow \quad \delta_1 \approx \frac{\pi}{4} - 0.675 \approx 0.1106$. Also, $y = \tan x = 1.2$

when $x \approx 0.876$, so $\frac{\pi}{4} + \delta_2 \approx 0.876 \quad \Rightarrow \quad \delta_2 = 0.876 - \frac{\pi}{4} \approx 0.0906$.

Thus, we choose $\delta = 0.0906$ (or any smaller positive number) since this is the smaller of δ_1 and δ_2.

27. (a) $A = \pi r^2$ and $A = 1000 \text{ cm}^2 \quad \Rightarrow \quad \pi r^2 = 1000 \quad \Rightarrow \quad r^2 = \frac{1000}{\pi} \quad \Rightarrow \quad r = \sqrt{\frac{1000}{\pi}} \quad (r > 0) \quad \approx 17.8412 \text{ cm}$.

(b) $|A - 1000| \le 5 \quad \Rightarrow \quad -5 \le \pi r^2 - 1000 \le 5 \quad \Rightarrow \quad 1000 - 5 \le \pi r^2 \le 1000 + 5 \quad \Rightarrow$

$\sqrt{\frac{995}{\pi}} \le r \le \sqrt{\frac{1005}{\pi}} \quad \Rightarrow \quad 17.7966 \le r \le 17.8858$. $\sqrt{\frac{1000}{\pi}} - \sqrt{\frac{995}{\pi}} \approx 0.04466$ and $\sqrt{\frac{1005}{\pi}} - \sqrt{\frac{1000}{\pi}} \approx 0.04455$. So

if the machinist gets the radius within 0.0445 cm of 17.8412, the area will be within 5 cm^2 of 1000.

(c) x is the radius, $f(x)$ is the area, a is the target radius given in part (a), L is the target area (1000), ε is the tolerance in the area (5), and δ is the tolerance in the radius given in part (b).

29. Given $\varepsilon > 0$, we need $\delta > 0$ such that if $0 < |x - 3| < \delta$, then

$\left|(1 + \frac{1}{3}x) - 2\right| < \varepsilon$. But $\left|(1 + \frac{1}{3}x) - 2\right| < \varepsilon \quad \Leftrightarrow \quad \left|\frac{1}{3}x - 1\right| < \varepsilon \quad \Leftrightarrow$

$\left|\frac{1}{3}\right||x - 3| < \varepsilon \quad \Leftrightarrow \quad |x - 3| < 3\varepsilon$. So if we choose $\delta = 3\varepsilon$, then

$0 < |x - 3| < \delta \quad \Rightarrow \quad \left|(1 + \frac{1}{3}x) - 2\right| < \varepsilon$. Thus, $\lim\limits_{x \to 3}(1 + \frac{1}{3}x) = 2$ by

the definition of a limit.

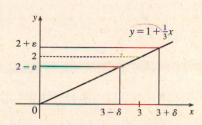

31. Given $\varepsilon > 0$, we need $\delta > 0$ such that if $0 < |x - (-3)| < \delta$, then

$|(1 - 4x) - 13| < \varepsilon$. But $|(1 - 4x) - 13| < \varepsilon$ $\Leftrightarrow$

$|-4x - 12| < \varepsilon$ $\Leftrightarrow$ $|-4|\,|x + 3| < \varepsilon$ $\Leftrightarrow$ $|x - (-3)| < \varepsilon/4$. So if

we choose $\delta = \varepsilon/4$, then $0 < |x - (-3)| < \delta$ $\Rightarrow$ $|(1 - 4x) - 13| < \varepsilon$.

Thus, $\lim_{x \to -3}(1 - 4x) = 13$ by the definition of a limit.

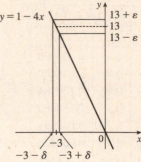

33. Given $\varepsilon > 0$, we need $\delta > 0$ such that if $0 < |x - 1| < \delta$, then $\left|\dfrac{2 + 4x}{3} - 2\right| < \varepsilon$. But $\left|\dfrac{2 + 4x}{3} - 2\right| < \varepsilon$ $\Leftrightarrow$

$\left|\dfrac{4x - 4}{3}\right| < \varepsilon$ $\Leftrightarrow$ $\left|\dfrac{4}{3}\right|\,|x - 1| < \varepsilon$ $\Leftrightarrow$ $|x - 1| < \frac{3}{4}\varepsilon$. So if we choose $\delta = \frac{3}{4}\varepsilon$, then $0 < |x - 1| < \delta$ $\Rightarrow$

$\left|\dfrac{2 + 4x}{3} - 2\right| < \varepsilon$. Thus, $\lim_{x \to 1}\dfrac{2 + 4x}{3} = 2$ by the definition of a limit.

35. Given $\varepsilon > 0$, we need $\delta > 0$ such that if $0 < |x - 2| < \delta$, then $\left|\dfrac{x^2 + x - 6}{x - 2} - 5\right| < \varepsilon$ $\Leftrightarrow$

$\left|\dfrac{(x + 3)(x - 2)}{x - 2} - 5\right| < \varepsilon$ $\Leftrightarrow$ $|x + 3 - 5| < \varepsilon$ $[x \neq 2]$ $\Leftrightarrow$ $|x - 2| < \varepsilon$. So choose $\delta = \varepsilon$.

Then $0 < |x - 2| < \delta$ $\Rightarrow$ $|x - 2| < \varepsilon$ $\Rightarrow$ $|x + 3 - 5| < \varepsilon$ $\Rightarrow$ $\left|\dfrac{(x + 3)(x - 2)}{x - 2} - 5\right| < \varepsilon$ $[x \neq 2]$ $\Rightarrow$

$\left|\dfrac{x^2 + x - 6}{x - 2} - 5\right| < \varepsilon$. By the definition of a limit, $\lim_{x \to 2}\dfrac{x^2 + x - 6}{x - 2} = 5$.

37. Given $\varepsilon > 0$, we need $\delta > 0$ such that if $0 < |x - a| < \delta$, then $|x - a| < \varepsilon$. So $\delta = \varepsilon$ will work.

39. Given $\varepsilon > 0$, we need $\delta > 0$ such that if $0 < |x - 0| < \delta$, then $\left|x^2 - 0\right| < \varepsilon$ $\Leftrightarrow$ $x^2 < \varepsilon$ $\Leftrightarrow$ $|x| < \sqrt{\varepsilon}$. Take $\delta = \sqrt{\varepsilon}$.

Then $0 < |x - 0| < \delta$ $\Rightarrow$ $\left|x^2 - 0\right| < \varepsilon$. Thus, $\lim_{x \to 0}x^2 = 0$ by the definition of a limit.

41. Given $\varepsilon > 0$, we need $\delta > 0$ such that if $0 < |x - 0| < \delta$, then $\big||x| - 0\big| < \varepsilon$. But $\big||x|\big| = |x|$. So this is true if we pick $\delta = \varepsilon$.

Thus, $\lim_{x \to 0}|x| = 0$ by the definition of a limit.

43. Given $\varepsilon > 0$, we need $\delta > 0$ such that if $0 < |x - 3| < \delta$, then $\left|x^2 - 9\right| < \varepsilon$ $\Leftrightarrow$ $|(x - 3)(x + 3)| < \varepsilon$. Notice that if

$|x - 3| < 1$, then $-1 < x - 3 < 1$ $\Rightarrow$ $5 < x + 3 < 7$ $\Rightarrow$ $|x + 3| < 7$. So take $\delta = \min\{1, \varepsilon/7\}$. Then

$0 < |x - 3| < \delta$ $\Leftrightarrow$ $|(x - 3)(x + 3)| < |7(x - 3)| = 7 \cdot |x - 3| < 7\delta \leq \varepsilon$. Thus, $\lim_{x \to 3}x^2 = 9$ by the definition

of a limit.

45. (a) The points of intersection in the graph are $(x_1, 2.6)$ and $(x_2, 3.4)$

with $x_1 \approx 0.891$ and $x_2 \approx 1.093$. Thus, we can take δ to be the

smaller of $1 - x_1$ and $x_2 - 1$. So $\delta = x_2 - 1 \approx 0.093$.

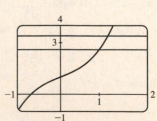

(b) Solving $x^3 + x + 1 = 3 + \varepsilon$ gives us two nonreal complex roots and

one real root, which is $x(\varepsilon) = \dfrac{\left(216 + 108\varepsilon + 12\sqrt{336 + 324\varepsilon + 81\varepsilon^2}\,\right)^{2/3} - 12}{6\left(216 + 108\varepsilon + 12\sqrt{336 + 324\varepsilon + 81\varepsilon^2}\,\right)^{1/3}}$. Thus, $\delta = x(\varepsilon) - 1$.

(c) If $\varepsilon = 0.4$, then $x(\varepsilon) \approx 1.093\,272\,342$ and $\delta = x(\varepsilon) - 1 \approx 0.093$, which agrees with our answer in part (a).

1.4   Calculating Limits

1. (a) $\displaystyle\lim_{x\to 2}\,[f(x) + 5g(x)] = \lim_{x\to 2} f(x) + \lim_{x\to 2}\,[5g(x)]$ [Limit Law 1]

$\qquad = \displaystyle\lim_{x\to 2} f(x) + 5\lim_{x\to 2} g(x)$ [Limit Law 3]

$\qquad = 4 + 5(-2) = -6$

(b) $\displaystyle\lim_{x\to 2}\,[g(x)]^3 = \left[\lim_{x\to 2} g(x)\right]^3$ [Limit Law 6]

$\qquad = (-2)^3 = -8$

(c) $\displaystyle\lim_{x\to 2}\sqrt{f(x)} = \sqrt{\lim_{x\to 2} f(x)}$ [Limit Law 11]

$\qquad = \sqrt{4} = 2$

(d) $\displaystyle\lim_{x\to 2}\frac{3f(x)}{g(x)} = \frac{\displaystyle\lim_{x\to 2}\,[3f(x)]}{\displaystyle\lim_{x\to 2} g(x)}$ [Limit Law 5]

$\qquad = \dfrac{3\displaystyle\lim_{x\to 2} f(x)}{\displaystyle\lim_{x\to 2} g(x)}$ [Limit Law 3]

$\qquad = \dfrac{3(4)}{-2} = -6$

(e) Because the limit of the denominator is 0, we can't use Limit Law 5. The given limit, $\displaystyle\lim_{x\to 2}\frac{g(x)}{h(x)}$, does not exist because the denominator approaches 0 while the numerator approaches a nonzero number.

(f) $\displaystyle\lim_{x\to 2}\frac{g(x)\,h(x)}{f(x)} = \frac{\displaystyle\lim_{x\to 2}\,[g(x)\,h(x)]}{\displaystyle\lim_{x\to 2} f(x)}$ [Limit Law 5]

$\qquad = \dfrac{\displaystyle\lim_{x\to 2} g(x)\cdot\lim_{x\to 2} h(x)}{\displaystyle\lim_{x\to 2} f(x)}$ [Limit Law 4]

$\qquad = \dfrac{-2\cdot 0}{4} = 0$

3. $\displaystyle\lim_{x\to 3}\,(5x^3 - 3x^2 + x - 6) = \lim_{x\to 3}\,(5x^3) - \lim_{x\to 3}\,(3x^2) + \lim_{x\to 3} x - \lim_{x\to 3} 6$ [Limit Laws 2 and 1]

$\qquad = 5\displaystyle\lim_{x\to 3} x^3 - 3\lim_{x\to 3} x^2 + \lim_{x\to 3} x - \lim_{x\to 3} 6$ [3]

$\qquad = 5(3^3) - 3(3^2) + 3 - 6$ [9, 8, and 7]

$\qquad = 105$

5. $\displaystyle\lim_{t\to -2}\frac{t^4 - 2}{2t^2 - 3t + 2} = \frac{\displaystyle\lim_{t\to -2}\,(t^4 - 2)}{\displaystyle\lim_{t\to -2}\,(2t^2 - 3t + 2)}$ [Limit Law 5]

$\qquad = \dfrac{\displaystyle\lim_{t\to -2} t^4 - \lim_{t\to -2} 2}{2\displaystyle\lim_{t\to -2} t^2 - 3\lim_{t\to -2} t + \lim_{t\to -2} 2}$ [1, 2, and 3]

$\qquad = \dfrac{16 - 2}{2(4) - 3(-2) + 2}$ [9, 7, and 8]

$\qquad = \dfrac{14}{16} = \dfrac{7}{8}$

7. $\displaystyle\lim_{x\to 8}\left(1+\sqrt[3]{x}\,\right)\left(2-6x^2+x^3\right) = \lim_{x\to 8}\left(1+\sqrt[3]{x}\,\right)\cdot\lim_{x\to 8}\left(2-6x^2+x^3\right)$ [Limit Law 4]

$$= \left(\lim_{x\to 8}1+\lim_{x\to 8}\sqrt[3]{x}\,\right)\cdot\left(\lim_{x\to 8}2-6\lim_{x\to 8}x^2+\lim_{x\to 8}x^3\right) \quad \text{[1, 2, and 3]}$$

$$= \left(1+\sqrt[3]{8}\,\right)\cdot\left(2-6\cdot 8^2+8^3\right) \quad \text{[7, 10, 9]}$$

$$= (3)(130) = 390$$

9. $\displaystyle\lim_{\theta\to\pi/2}\theta\sin\theta = \left(\lim_{\theta\to\pi/2}\theta\right)\left(\lim_{\theta\to\pi/2}\sin\theta\right)$ [4]

$$= \frac{\pi}{2}\cdot\sin\frac{\pi}{2} \quad \text{[8 and Direct Substitution Property]}$$

$$= \frac{\pi}{2}$$

11. $\displaystyle\lim_{x\to 5}\frac{x^2-6x+5}{x-5} = \lim_{x\to 5}\frac{(x-5)(x-1)}{x-5} = \lim_{x\to 5}(x-1) = 5-1 = 4$

13. $\displaystyle\lim_{x\to 5}\frac{x^2-5x+6}{x-5}$ does not exist since $x-5\to 0$, but $x^2-5x+6\to 6$ as $x\to 5$.

15. $\displaystyle\lim_{t\to -3}\frac{t^2-9}{2t^2+7t+3} = \lim_{t\to -3}\frac{(t+3)(t-3)}{(2t+1)(t+3)} = \lim_{t\to -3}\frac{t-3}{2t+1} = \frac{-3-3}{2(-3)+1} = \frac{-6}{-5} = \frac{6}{5}$

17. $\displaystyle\lim_{h\to 0}\frac{(-5+h)^2-25}{h} = \lim_{h\to 0}\frac{(25-10h+h^2)-25}{h} = \lim_{h\to 0}\frac{-10h+h^2}{h} = \lim_{h\to 0}\frac{h(-10+h)}{h} = \lim_{h\to 0}(-10+h) = -10$

19. By the formula for the sum of cubes, we have

$$\lim_{x\to -2}\frac{x+2}{x^3+8} = \lim_{x\to -2}\frac{x+2}{(x+2)(x^2-2x+4)} = \lim_{x\to -2}\frac{1}{x^2-2x+4} = \frac{1}{4+4+4} = \frac{1}{12}.$$

21. $\displaystyle\lim_{h\to 0}\frac{\sqrt{9+h}-3}{h} = \lim_{h\to 0}\frac{\sqrt{9+h}-3}{h}\cdot\frac{\sqrt{9+h}+3}{\sqrt{9+h}+3} = \lim_{h\to 0}\frac{\left(\sqrt{9+h}\,\right)^2-3^2}{h\left(\sqrt{9+h}+3\right)} = \lim_{h\to 0}\frac{(9+h)-9}{h\left(\sqrt{9+h}+3\right)}$

$$= \lim_{h\to 0}\frac{h}{h\left(\sqrt{9+h}+3\right)} = \lim_{h\to 0}\frac{1}{\sqrt{9+h}+3} = \frac{1}{3+3} = \frac{1}{6}$$

23. $\displaystyle\lim_{x\to 16}\frac{4-\sqrt{x}}{16x-x^2} = \lim_{x\to 16}\frac{(4-\sqrt{x})(4+\sqrt{x})}{(16x-x^2)(4+\sqrt{x})} = \lim_{x\to 16}\frac{16-x}{x(16-x)(4+\sqrt{x})}$

$$= \lim_{x\to 16}\frac{1}{x(4+\sqrt{x})} = \frac{1}{16\left(4+\sqrt{16}\right)} = \frac{1}{16(8)} = \frac{1}{128}$$

25. $\displaystyle\lim_{x\to -4}\frac{\dfrac{1}{4}+\dfrac{1}{x}}{4+x} = \lim_{x\to -4}\frac{\dfrac{x+4}{4x}}{4+x} = \lim_{x\to -4}\frac{x+4}{4x(4+x)} = \lim_{x\to -4}\frac{1}{4x} = \frac{1}{4(-4)} = -\frac{1}{16}$

27. $\displaystyle\lim_{h\to 0}\frac{(x+h)^3-x^3}{h} = \lim_{h\to 0}\frac{(x^3+3x^2h+3xh^2+h^3)-x^3}{h} = \lim_{h\to 0}\frac{3x^2h+3xh^2+h^3}{h}$

$$= \lim_{h\to 0}\frac{h(3x^2+3xh+h^2)}{h} = \lim_{h\to 0}(3x^2+3xh+h^2) = 3x^2$$

29. (a)

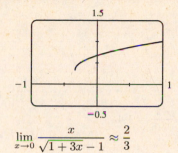

$$\lim_{x \to 0} \frac{x}{\sqrt{1+3x}-1} \approx \frac{2}{3}$$

(b)

x	$f(x)$
-0.001	0.6661663
-0.0001	0.6666167
-0.00001	0.6666617
-0.000001	0.6666662
0.000001	0.6666672
0.00001	0.6666717
0.0001	0.6667167
0.001	0.6671663

The limit appears to be $\dfrac{2}{3}$.

(c) $\displaystyle\lim_{x \to 0} \left(\frac{x}{\sqrt{1+3x}-1} \cdot \frac{\sqrt{1+3x}+1}{\sqrt{1+3x}+1} \right) = \lim_{x \to 0} \frac{x(\sqrt{1+3x}+1)}{(1+3x)-1} = \lim_{x \to 0} \frac{x(\sqrt{1+3x}+1)}{3x}$

$\displaystyle \qquad\qquad = \frac{1}{3} \lim_{x \to 0} \left(\sqrt{1+3x}+1 \right)$ \qquad\qquad [Limit Law 3]

$\displaystyle \qquad\qquad = \frac{1}{3} \left[\sqrt{\lim_{x \to 0}(1+3x)} + \lim_{x \to 0} 1 \right]$ \qquad\qquad [1 and 11]

$\displaystyle \qquad\qquad = \frac{1}{3} \left(\sqrt{\lim_{x \to 0} 1 + 3 \lim_{x \to 0} x} + 1 \right)$ \qquad\qquad [1, 3, and 7]

$\displaystyle \qquad\qquad = \frac{1}{3} \left(\sqrt{1 + 3 \cdot 0} + 1 \right)$ \qquad\qquad [7 and 8]

$\displaystyle \qquad\qquad = \frac{1}{3}(1+1) = \frac{2}{3}$

31. Let $f(x) = -x^2$, $g(x) = x^2 \cos 20\pi x$ and $h(x) = x^2$. Then

$-1 \le \cos 20\pi x \le 1 \;\Rightarrow\; -x^2 \le x^2 \cos 20\pi x \le x^2 \;\Rightarrow\; f(x) \le g(x) \le h(x)$.

So since $\displaystyle\lim_{x \to 0} f(x) = \lim_{x \to 0} h(x) = 0$, by the Squeeze Theorem we have

$\displaystyle\lim_{x \to 0} g(x) = 0$.

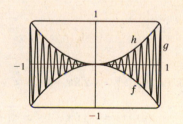

33. We have $\displaystyle\lim_{x \to 4} (4x - 9) = 4(4) - 9 = 7$ and $\displaystyle\lim_{x \to 4} (x^2 - 4x + 7) = 4^2 - 4(4) + 7 = 7$. Since $4x - 9 \le f(x) \le x^2 - 4x + 7$

for $x \ge 0$, $\displaystyle\lim_{x \to 4} f(x) = 7$ by the Squeeze Theorem.

35. $-1 \le \cos(2/x) \le 1 \;\Rightarrow\; -x^4 \le x^4 \cos(2/x) \le x^4$. Since $\displaystyle\lim_{x \to 0} (-x^4) = 0$ and $\displaystyle\lim_{x \to 0} x^4 = 0$, we have

$\displaystyle\lim_{x \to 0} \left[x^4 \cos(2/x) \right] = 0$ by the Squeeze Theorem.

37. $|x - 3| = \begin{cases} x - 3 & \text{if } x - 3 \ge 0 \\ -(x-3) & \text{if } x - 3 < 0 \end{cases} = \begin{cases} x - 3 & \text{if } x \ge 3 \\ 3 - x & \text{if } x < 3 \end{cases}$

Thus, $\displaystyle\lim_{x \to 3^+} (2x + |x - 3|) = \lim_{x \to 3^+} (2x + x - 3) = \lim_{x \to 3^+} (3x - 3) = 3(3) - 3 = 6$ and

$\displaystyle\lim_{x \to 3^-} (2x + |x - 3|) = \lim_{x \to 3^-} (2x + 3 - x) = \lim_{x \to 3^-} (x + 3) = 3 + 3 = 6$. Since the left and right limits are equal,

$\displaystyle\lim_{x \to 3} (2x + |x - 3|) = 6$.

39. $\left|2x^3 - x^2\right| = \left|x^2(2x-1)\right| = \left|x^2\right| \cdot \left|2x-1\right| = x^2 \left|2x-1\right|$

$$\left|2x-1\right| = \begin{cases} 2x-1 & \text{if } 2x-1 \geq 0 \\ -(2x-1) & \text{if } 2x-1 < 0 \end{cases} = \begin{cases} 2x-1 & \text{if } x \geq 0.5 \\ -(2x-1) & \text{if } x < 0.5 \end{cases}$$

So $\left|2x^3 - x^2\right| = x^2[-(2x-1)]$ for $x < 0.5$.

Thus, $\displaystyle\lim_{x \to 0.5^-} \frac{2x-1}{\left|2x^3 - x^2\right|} = \lim_{x \to 0.5^-} \frac{2x-1}{x^2[-(2x-1)]} = \lim_{x \to 0.5^-} \frac{-1}{x^2} = \frac{-1}{(0.5)^2} = \frac{-1}{0.25} = -4$.

41. Since $\left|x\right| = -x$ for $x < 0$, we have $\displaystyle\lim_{x \to 0^-} \left(\frac{1}{x} - \frac{1}{\left|x\right|}\right) = \lim_{x \to 0^-} \left(\frac{1}{x} - \frac{1}{-x}\right) = \lim_{x \to 0^-} \frac{2}{x}$, which does not exist since the

denominator approaches 0 and the numerator does not.

43. (a) (i) $\displaystyle\lim_{x \to 2^+} g(x) = \lim_{x \to 2^+} \frac{x^2 + x - 6}{\left|x - 2\right|} = \lim_{x \to 2^+} \frac{(x+3)(x-2)}{\left|x - 2\right|}$

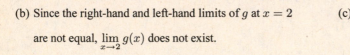

$$= \lim_{x \to 2^+} \frac{(x+3)(x-2)}{x - 2} \quad [\text{since } x - 2 > 0 \text{ if } x \to 2^+]$$

$$= \lim_{x \to 2^+} (x+3) = 5$$

(ii) The solution is similar to the solution in part (i), but now $\left|x - 2\right| = 2 - x$ since $x - 2 < 0$ if $x \to 2^-$.

Thus, $\displaystyle\lim_{x \to 2^-} g(x) = \lim_{x \to 2^-} -(x+3) = -5$.

(b) Since the right-hand and left-hand limits of g at $x = 2$

are not equal, $\displaystyle\lim_{x \to 2} g(x)$ does not exist.

(c)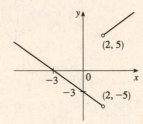

45. (a) (i) $[\![x]\!] = -2$ for $-2 \leq x < -1$, so $\displaystyle\lim_{x \to -2^+} [\![x]\!] = \lim_{x \to -2^+} (-2) = -2$

(ii) $[\![x]\!] = -3$ for $-3 \leq x < -2$, so $\displaystyle\lim_{x \to -2^-} [\![x]\!] = \lim_{x \to -2^-} (-3) = -3$.

The right and left limits are different, so $\displaystyle\lim_{x \to -2} [\![x]\!]$ does not exist.

(iii) $[\![x]\!] = -3$ for $-3 \leq x < -2$, so $\displaystyle\lim_{x \to -2.4} [\![x]\!] = \lim_{x \to -2.4} (-3) = -3$.

(b) (i) $[\![x]\!] = n - 1$ for $n - 1 \leq x < n$, so $\displaystyle\lim_{x \to n^-} [\![x]\!] = \lim_{x \to n^-} (n-1) = n - 1$.

(ii) $[\![x]\!] = n$ for $n \leq x < n + 1$, so $\displaystyle\lim_{x \to n^+} [\![x]\!] = \lim_{x \to n^+} n = n$.

(c) $\displaystyle\lim_{x \to a} [\![x]\!]$ exists $\Leftrightarrow$ a is not an integer.

47. The graph of $f(x) = [\![x]\!] + [\![-x]\!]$ is the same as the graph of $g(x) = -1$ with holes at each integer, since $f(a) = 0$ for any

integer a. Thus, $\displaystyle\lim_{x \to 2^-} f(x) = -1$ and $\displaystyle\lim_{x \to 2^+} f(x) = -1$, so $\displaystyle\lim_{x \to 2} f(x) = -1$. However,

$f(2) = [\![2]\!] + [\![-2]\!] = 2 + (-2) = 0$, so $\displaystyle\lim_{x \to 2} f(x) \neq f(2)$.

49. $\displaystyle\lim_{x\to 0} \frac{\sin 3x}{x} = \lim_{x\to 0} \frac{3\sin 3x}{3x}$ [multiply numerator and denominator by 3]

$\displaystyle\qquad = 3\lim_{3x\to 0} \frac{\sin 3x}{3x}$ [as $x \to 0$, $3x \to 0$]

$\displaystyle\qquad = 3\lim_{\theta\to 0} \frac{\sin \theta}{\theta}$ [let $\theta = 3x$]

$\qquad = 3(1)$ [Equation 2]

$\qquad = 3$

51. $\displaystyle\lim_{t\to 0} \frac{\tan 6t}{\sin 2t} = \lim_{t\to 0}\left(\frac{\sin 6t}{t} \cdot \frac{1}{\cos 6t} \cdot \frac{t}{\sin 2t}\right) = \lim_{t\to 0}\frac{6\sin 6t}{6t} \cdot \lim_{t\to 0}\frac{1}{\cos 6t} \cdot \lim_{t\to 0}\frac{2t}{2\sin 2t}$

$\displaystyle\qquad = 6\lim_{t\to 0}\frac{\sin 6t}{6t} \cdot \lim_{t\to 0}\frac{1}{\cos 6t} \cdot \frac{1}{2}\lim_{t\to 0}\frac{2t}{\sin 2t} = 6(1) \cdot \frac{1}{1} \cdot \frac{1}{2}(1) = 3$

53. $\displaystyle\lim_{x\to 0} \frac{\sin 3x}{5x^3 - 4x} = \lim_{x\to 0}\left(\frac{\sin 3x}{3x} \cdot \frac{3}{5x^2 - 4}\right) = \lim_{x\to 0}\frac{\sin 3x}{3x} \cdot \lim_{x\to 0}\frac{3}{5x^2 - 4} = 1 \cdot \left(\frac{3}{-4}\right) = -\frac{3}{4}$

55. Divide numerator and denominator by θ.   ($\sin \theta$ also works.)

$$\lim_{\theta\to 0} \frac{\sin \theta}{\theta + \tan \theta} = \lim_{\theta\to 0} \frac{\dfrac{\sin \theta}{\theta}}{1 + \dfrac{\sin \theta}{\theta} \cdot \dfrac{1}{\cos \theta}} = \frac{\displaystyle\lim_{\theta\to 0}\frac{\sin \theta}{\theta}}{1 + \displaystyle\lim_{\theta\to 0}\frac{\sin \theta}{\theta}\lim_{\theta\to 0}\frac{1}{\cos \theta}} = \frac{1}{1 + 1\cdot 1} = \frac{1}{2}$$

57. Since $p(x)$ is a polynomial, $p(x) = a_0 + a_1 x + a_2 x^2 + \cdots + a_n x^n$. Thus, by the Limit Laws,

$$\lim_{x\to a} p(x) = \lim_{x\to a}\left(a_0 + a_1 x + a_2 x^2 + \cdots + a_n x^n\right) = a_0 + a_1\lim_{x\to a} x + a_2\lim_{x\to a} x^2 + \cdots + a_n\lim_{x\to a} x^n$$

$$= a_0 + a_1 a + a_2 a^2 + \cdots + a_n a^n = p(a)$$

Thus, for any polynomial p, $\displaystyle\lim_{x\to a} p(x) = p(a)$.

59. $\displaystyle\lim_{h\to 0}\sin(a + h) = \lim_{h\to 0}(\sin a \cos h + \cos a \sin h) = \lim_{h\to 0}(\sin a \cos h) + \lim_{h\to 0}(\cos a \sin h)$

$\displaystyle\qquad = \left(\lim_{h\to 0}\sin a\right)\left(\lim_{h\to 0}\cos h\right) + \left(\lim_{h\to 0}\cos a\right)\left(\lim_{h\to 0}\sin h\right) = (\sin a)(1) + (\cos a)(0) = \sin a$

61. $\displaystyle\lim_{x\to 1}[f(x) - 8] = \lim_{x\to 1}\left[\frac{f(x) - 8}{x - 1} \cdot (x - 1)\right] = \lim_{x\to 1}\frac{f(x) - 8}{x - 1} \cdot \lim_{x\to 1}(x - 1) = 10 \cdot 0 = 0.$

Thus, $\displaystyle\lim_{x\to 1} f(x) = \lim_{x\to 1}\{[f(x) - 8] + 8\} = \lim_{x\to 1}[f(x) - 8] + \lim_{x\to 1} 8 = 0 + 8 = 8.$

Note: The value of $\displaystyle\lim_{x\to 1}\frac{f(x) - 8}{x - 1}$ does not affect the answer since it's multiplied by 0. What's important is that

$\displaystyle\lim_{x\to 1}\frac{f(x) - 8}{x - 1}$ exists.

63. Let $f(x) = [\![x]\!]$ and $g(x) = -[\![x]\!]$. Then $\displaystyle\lim_{x\to 3} f(x)$ and $\displaystyle\lim_{x\to 3} g(x)$ do not exist   [Example 8]

but $\displaystyle\lim_{x\to 3}[f(x) + g(x)] = \lim_{x\to 3}([\![x]\!] - [\![x]\!]) = \lim_{x\to 3} 0 = 0.$

65. Since the denominator approaches 0 as $x \to -2$, the limit will exist only if the numerator also approaches

0 as $x \to -2$. In order for this to happen, we need $\lim\limits_{x \to -2} \left(3x^2 + ax + a + 3\right) = 0 \iff$

$3(-2)^2 + a(-2) + a + 3 = 0 \iff 12 - 2a + a + 3 = 0 \iff a = 15$. With $a = 15$, the limit becomes

$$\lim_{x \to -2} \frac{3x^2 + 15x + 18}{x^2 + x - 2} = \lim_{x \to -2} \frac{3(x+2)(x+3)}{(x-1)(x+2)} = \lim_{x \to -2} \frac{3(x+3)}{x-1} = \frac{3(-2+3)}{-2-1} = \frac{3}{-3} = -1.$$

1.5 Continuity

1. From Definition 1, $\lim\limits_{x \to 4} f(x) = f(4)$.

3. (a) f is discontinuous at -4 since $f(-4)$ is not defined and at -2, 2, and 4 since the limit does not exist (the left and right

limits are not the same).

(b) f is continuous from the left at -2 since $\lim\limits_{x \to -2^-} f(x) = f(-2)$. f is continuous from the right at 2 and 4 since

$\lim\limits_{x \to 2^+} f(x) = f(2)$ and $\lim\limits_{x \to 4^+} f(x) = f(4)$. It is continuous from neither side at -4 since $f(-4)$ is undefined.

5. The graph of $y = f(x)$ must have a discontinuity at

$x = 2$ and must show that $\lim\limits_{x \to 2^+} f(x) = f(2)$.

7. The graph of $y = f(x)$ must have a removable

discontinuity (a hole) at $x = 3$ and a jump discontinuity

at $x = 5$.

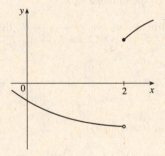

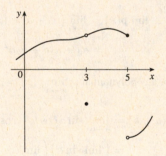

9. (a) The toll is \$7 between 7:00 AM and 10:00 AM and between 4:00 PM and 7:00 PM.

(b) The function T has jump discontinuities at $t = 7$, 10, 16, and 19. Their

significance to someone who uses the road is that, because of the sudden jumps in

the toll, they may want to avoid the higher rates between $t = 7$ and $t = 10$ and

between $t = 16$ and $t = 19$ if feasible.

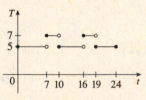

11. If f and g are continuous and $g(2) = 6$, then $\lim\limits_{x \to 2} [3f(x) + f(x) g(x)] = 36 \implies$

$3 \lim\limits_{x \to 2} f(x) + \lim\limits_{x \to 2} f(x) \cdot \lim\limits_{x \to 2} g(x) = 36 \implies 3f(2) + f(2) \cdot 6 = 36 \implies 9f(2) = 36 \implies f(2) = 4$.

13. $\lim\limits_{x \to -1} f(x) = \lim\limits_{x \to -1} \left(x + 2x^3\right)^4 = \left(\lim\limits_{x \to -1} x + 2 \lim\limits_{x \to -1} x^3\right)^4 = \left[-1 + 2(-1)^3\right]^4 = (-3)^4 = 81 = f(-1)$.

By the definition of continuity, f is continuous at $a = -1$.

15. $f(x) = \dfrac{1}{x+2}$ is discontinuous at $a = -2$ because $f(-2)$ is undefined.

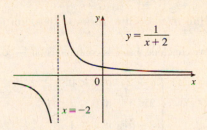

17. $f(x) = \begin{cases} 1 - x^2 & \text{if } x < 1 \\ 1/x & \text{if } x \geq 1 \end{cases}$

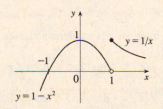

The left-hand limit of f at $a = 1$ is

$\displaystyle\lim_{x \to 1^-} f(x) = \lim_{x \to 1^-} (1 - x^2) = 0$. The right-hand limit of f at $a = 1$ is

$\displaystyle\lim_{x \to 1^+} f(x) = \lim_{x \to 1^+} (1/x) = 1$. Since these limits are not equal, $\displaystyle\lim_{x \to 1} f(x)$

does not exist and f is discontinuous at 1.

19. $F(x) = \dfrac{2x^2 - x - 1}{x^2 + 1}$ is a rational function, so it is continuous on its domain, $(-\infty, \infty)$, by Theorem 5(b).

21. $x^3 - 2 = 0 \ \Rightarrow \ x^3 = 2 \ \Rightarrow \ x = \sqrt[3]{2}$, so $Q(x) = \dfrac{\sqrt[3]{x-2}}{x^3 - 2}$ has domain $\left(-\infty, \sqrt[3]{2}\right) \cup \left(\sqrt[3]{2}, \infty\right)$. Now $x^3 - 2$ is

continuous everywhere by Theorem 5(a) and $\sqrt[3]{x-2}$ is continuous everywhere by Theorems 5(a), 7, and 9. Thus, Q is

continuous on its domain by part 5 of Theorem 4.

23. $M(x) = \sqrt{1 + \dfrac{1}{x}} = \sqrt{\dfrac{x+1}{x}}$ is defined when $\dfrac{x+1}{x} \geq 0 \ \Rightarrow \ x + 1 \geq 0$ and $x > 0$ or $x + 1 \leq 0$ and $x < 0 \ \Rightarrow \ x > 0$

or $x \leq -1$, so M has domain $(-\infty, -1] \cup (0, \infty)$. M is the composite of a root function and a rational function, so it is

continuous at every number in its domain by Theorems 7 and 9.

25.

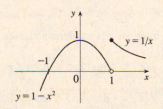

$y = \dfrac{1}{1 + \sin x}$ is undefined and hence discontinuous when

$1 + \sin x = 0 \ \Leftrightarrow \ \sin x = -1 \ \Leftrightarrow \ x = -\dfrac{\pi}{2} + 2\pi n$, n an

integer. The figure shows discontinuities for $n = -1, 0$, and 1; that

is, $-\dfrac{5\pi}{2} \approx -7.85$, $-\dfrac{\pi}{2} \approx -1.57$, and $\dfrac{3\pi}{2} \approx 4.71$.

27. Because we are dealing with root functions, $5 + \sqrt{x}$ is continuous on $[0, \infty)$, $\sqrt{x + 5}$ is continuous on $[-5, \infty)$, so the

quotient $f(x) = \dfrac{5 + \sqrt{x}}{\sqrt{5 + x}}$ is continuous on $[0, \infty)$. Since f is continuous at $x = 4$, $\displaystyle\lim_{x \to 4} f(x) = f(4) = \dfrac{7}{3}$.

29. $f(x) = \begin{cases} x^2 & \text{if } x < 1 \\ \sqrt{x} & \text{if } x \geq 1 \end{cases}$

By Theorem 5, since $f(x)$ equals the polynomial x^2 on $(-\infty, 1)$, f is continuous on $(-\infty, 1)$. By Theorem 7, since $f(x)$

equals the root function $\sqrt{x}$ on $(1, \infty)$, f is continuous on $(1, \infty)$. At $x = 1$, $\displaystyle\lim_{x \to 1^-} f(x) = \lim_{x \to 1^-} x^2 = 1$ and

$\lim\limits_{x \to 1^+} f(x) = \lim\limits_{x \to 1^+} \sqrt{x} = 1$. Thus, $\lim\limits_{x \to 1} f(x)$ exists and equals 1. Also, $f(1) = \sqrt{1} = 1$. Thus, f is continuous at $x = 1$.

We conclude that f is continuous on $(-\infty, \infty)$.

31. $f(x) = \begin{cases} x + 2 & \text{if } x < 0 \\ 2x^2 & \text{if } 0 \le x \le 1 \\ 2 - x & \text{if } x > 1 \end{cases}$

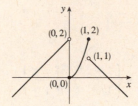

f is continuous on $(-\infty, 0)$, $(0, 1)$, and $(1, \infty)$ since on each of

these intervals it is a polynomial. Now $\lim\limits_{x \to 0^-} f(x) = \lim\limits_{x \to 0^-} (x + 2) = 2$ and

$\lim\limits_{x \to 0^+} f(x) = \lim\limits_{x \to 0^+} 2x^2 = 0$, so f is discontinuous at 0. Since $f(0) = 0$, f is continuous from the right at 0. Also

$\lim\limits_{x \to 1^-} f(x) = \lim\limits_{x \to 1^-} 2x^2 = 2$ and $\lim\limits_{x \to 1^+} f(x) = \lim\limits_{x \to 1^+} (2 - x) = 1$, so f is discontinuous at 1. Since $f(1) = 2$,

f is continuous from the left at 1.

33. $f(x) = \begin{cases} cx^2 + 2x & \text{if } x < 2 \\ x^3 - cx & \text{if } x \ge 2 \end{cases}$

f is continuous on $(-\infty, 2)$ and $(2, \infty)$. Now $\lim\limits_{x \to 2^-} f(x) = \lim\limits_{x \to 2^-} (cx^2 + 2x) = 4c + 4$ and

$\lim\limits_{x \to 2^+} f(x) = \lim\limits_{x \to 2^+} (x^3 - cx) = 8 - 2c$. So f is continuous $\Leftrightarrow$ $4c + 4 = 8 - 2c$ $\Leftrightarrow$ $6c = 4$ $\Leftrightarrow$ $c = \frac{2}{3}$. Thus, for f

to be continuous on $(-\infty, \infty)$, $c = \frac{2}{3}$.

35. (a) $f(x) = \dfrac{x^4 - 1}{x - 1} = \dfrac{(x^2 + 1)(x^2 - 1)}{x - 1} = \dfrac{(x^2 + 1)(x + 1)(x - 1)}{x - 1} = (x^2 + 1)(x + 1)$ $\quad$ [or $x^3 + x^2 + x + 1$]

for $x \ne 1$. The discontinuity is removable and $g(x) = x^3 + x^2 + x + 1$ agrees with f for $x \ne 1$ and is continuous on $\mathbb{R}$.

(b) $f(x) = \dfrac{x^3 - x^2 - 2x}{x - 2} = \dfrac{x(x^2 - x - 2)}{x - 2} = \dfrac{x(x - 2)(x + 1)}{x - 2} = x(x + 1)$ $\quad$ [or $x^2 + x$] $\quad$ for $x \ne 2$. The discontinuity

is removable and $g(x) = x^2 + x$ agrees with f for $x \ne 2$ and is continuous on $\mathbb{R}$.

(c) $\lim\limits_{x \to \pi^-} f(x) = \lim\limits_{x \to \pi^-} [\![\sin x]\!] = \lim\limits_{x \to \pi^-} 0 = 0$ and $\lim\limits_{x \to \pi^+} f(x) = \lim\limits_{x \to \pi^+} [\![\sin x]\!] = \lim\limits_{x \to \pi^+} (-1) = -1$, so $\lim\limits_{x \to \pi} f(x)$ does not

exist. The discontinuity at $x = \pi$ is a jump discontinuity.

37. $f(x) = x^2 + 10 \sin x$ is continuous on the interval $[31, 32]$, $f(31) \approx 957$, and $f(32) \approx 1030$. Since $957 < 1000 < 1030$,

there is a number c in $(31, 32)$ such that $f(c) = 1000$ by the Intermediate Value Theorem. *Note:* There is also a number c in

$(-32, -31)$ such that $f(c) = 1000$.

39. $f(x) = x^4 + x - 3$ is continuous on the interval $[1, 2]$, $f(1) = -1$, and $f(2) = 15$. Since $-1 < 0 < 15$, there is a number c

in $(1, 2)$ such that $f(c) = 0$ by the Intermediate Value Theorem. Thus, there is a root of the equation $x^4 + x - 3 = 0$ in the

interval $(1, 2)$.

41. $f(x) = \cos x - x$ is continuous on the interval $[0, 1]$, $f(0) = 1$, and $f(1) = \cos 1 - 1 \approx -0.46$. Since $-0.46 < 0 < 1$, there

is a number c in $(0, 1)$ such that $f(c) = 0$ by the Intermediate Value Theorem. Thus, there is a root of the equation

$\cos x - x = 0$, or $\cos x = x$, in the interval $(0, 1)$.

43. (a) $f(x) = \cos x - x^3$ is continuous on the interval $[0, 1]$, $f(0) = 1 > 0$, and $f(1) = \cos 1 - 1 \approx -0.46 < 0$. Since

$1 > 0 > -0.46$, there is a number c in $(0, 1)$ such that $f(c) = 0$ by the Intermediate Value Theorem. Thus, there is a root

of the equation $\cos x - x^3 = 0$, or $\cos x = x^3$, in the interval $(0, 1)$.

(b) $f(0.86) \approx 0.016 > 0$ and $f(0.87) \approx -0.014 < 0$, so there is a root between 0.86 and 0.87, that is, in the interval

$(0.86, 0.87)$.

45. (a) Let $f(x) = x^5 - x^2 - 4$. Then $f(1) = 1^5 - 1^2 - 4 = -4 < 0$ and $f(2) = 2^5 - 2^2 - 4 = 24 > 0$. So by the

Intermediate Value Theorem, there is a number c in $(1, 2)$ such that $f(c) = c^5 - c^2 - 4 = 0$.

(b) We can see from the graphs that, correct to three decimal places, the root is $x \approx 1.434$.

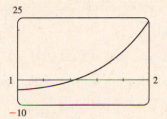   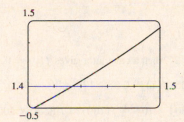

47. If there is such a number, it satisfies the equation $x^3 + 1 = x \Leftrightarrow x^3 - x + 1 = 0$. Let the left-hand side of this equation be

called $f(x)$. Now $f(-2) = -5 < 0$, and $f(-1) = 1 > 0$. Note also that $f(x)$ is a polynomial, and thus continuous. So by the

Intermediate Value Theorem, there is a number c between -2 and -1 such that $f(c) = 0$, so that $c = c^3 + 1$.

49. $f(x) = x^4 \sin(1/x)$ is continuous on $(-\infty, 0) \cup (0, \infty)$ since it is the product of a polynomial and a composite of a

trigonometric function and a rational function. Now since $-1 \le \sin(1/x) \le 1$, we have $-x^4 \le x^4 \sin(1/x) \le x^4$. Because

$\lim\limits_{x \to 0}(-x^4) = 0$ and $\lim\limits_{x \to 0} x^4 = 0$, the Squeeze Theorem gives us $\lim\limits_{x \to 0}(x^4 \sin(1/x)) = 0$, which equals $f(0)$. Thus, f is

continuous at 0 and, hence, on $(-\infty, \infty)$.

51. Define $u(t)$ to be the monk's distance from the monastery, as a function of time t (in hours), on the first day, and define $d(t)$

to be his distance from the monastery, as a function of time, on the second day. Let D be the distance from the monastery to

the top of the mountain. From the given information we know that $u(0) = 0$, $u(12) = D$, $d(0) = D$ and $d(12) = 0$. Now

consider the function $u - d$, which is clearly continuous. We calculate that $(u - d)(0) = -D$ and $(u - d)(12) = D$.

So by the Intermediate Value Theorem, there must be some time t_0 between 0 and 12 such that $(u - d)(t_0) = 0 \Leftrightarrow$

$u(t_0) = d(t_0)$. So at time t_0 after 7:00 AM, the monk will be at the same place on both days.

1.6   Limits Involving Infinity

1. (a) $\lim\limits_{x \to \infty} f(x) = -2$

(b) $\lim\limits_{x \to -\infty} f(x) = 2$

(c) $\lim\limits_{x \to 1} f(x) = \infty$

(d) $\lim\limits_{x \to 3} f(x) = -\infty$

(e) Vertical: $x = 1$, $x = 3$; horizontal: $y = -2$, $y = 2$

3. $\lim\limits_{x\to 0} f(x) = -\infty$,

$\lim\limits_{x\to -\infty} f(x) = 5$,

$\lim\limits_{x\to \infty} f(x) = -5$

5. $\lim\limits_{x\to 2} f(x) = -\infty$, $\lim\limits_{x\to \infty} f(x) = \infty$,

$\lim\limits_{x\to -\infty} f(x) = 0$, $\lim\limits_{x\to 0^+} f(x) = \infty$,

$\lim\limits_{x\to 0^-} f(x) = -\infty$

7. $f(0) = 3$, $\lim\limits_{x\to 0^-} f(x) = 4$,

$\lim\limits_{x\to 0^+} f(x) = 2$,

$\lim\limits_{x\to -\infty} f(x) = -\infty$, $\lim\limits_{x\to 4^-} f(x) = -\infty$,

$\lim\limits_{x\to 4^+} f(x) = \infty$, $\lim\limits_{x\to \infty} f(x) = 3$

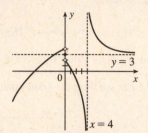

9. If $f(x) = x^2/2^x$, then a calculator gives $f(0) = 0$, $f(1) = 0.5$, $f(2) = 1$, $f(3) = 1.125$, $f(4) = 1$, $f(5) = 0.78125$,

$f(6) = 0.5625$, $f(7) = 0.3828125$, $f(8) = 0.25$, $f(9) = 0.158203125$, $f(10) = 0.09765625$, $f(20) \approx 0.00038147$,

$f(50) \approx 2.2204 \times 10^{-12}$, $f(100) \approx 7.8886 \times 10^{-27}$.

It appears that $\lim\limits_{x\to \infty} \left(x^2/2^x\right) = 0$.

11. Vertical: $x \approx -1.62$, $x \approx 0.62$, $x = 1$;

Horizontal: $y = 1$

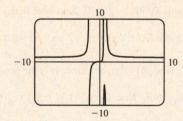

13. $\lim\limits_{x\to -3^+} \dfrac{x+2}{x+3} = -\infty$ since the numerator is negative and the denominator approaches 0 from the positive side as $x \to -3^+$.

15. $\lim\limits_{x\to 1} \dfrac{2-x}{(x-1)^2} = \infty$ since the numerator is positive and the denominator approaches 0 through positive values as $x \to 1$.

17. $\lim\limits_{x\to 2\pi^-} x \csc x = \lim\limits_{x\to 2\pi^-} \dfrac{x}{\sin x} = -\infty$ since the numerator is positive and the denominator approaches 0 through negative

values as $x \to 2\pi^-$.

19. $\lim\limits_{x\to \infty} \dfrac{3x-2}{2x+1} = \lim\limits_{x\to \infty} \dfrac{(3x-2)/x}{(2x+1)/x} = \lim\limits_{x\to \infty} \dfrac{3-2/x}{2+1/x} = \dfrac{\lim\limits_{x\to \infty} 3 - 2\lim\limits_{x\to \infty} 1/x}{\lim\limits_{x\to \infty} 2 + \lim\limits_{x\to \infty} 1/x} = \dfrac{3-2(0)}{2+0} = \dfrac{3}{2}$

21. $\lim\limits_{t\to \infty} \dfrac{\sqrt{t}+t^2}{2t-t^2} = \lim\limits_{t\to \infty} \dfrac{(\sqrt{t}+t^2)/t^2}{(2t-t^2)/t^2} = \lim\limits_{t\to \infty} \dfrac{1/t^{3/2}+1}{2/t-1} = \dfrac{0+1}{0-1} = -1$

23. $\lim\limits_{x\to \infty} \dfrac{(2x^2+1)^2}{(x-1)^2(x^2+x)} = \lim\limits_{x\to \infty} \dfrac{(2x^2+1)^2/x^4}{[(x-1)^2(x^2+x)]/x^4} = \lim\limits_{x\to \infty} \dfrac{[(2x^2+1)/x^2]^2}{[(x^2-2x+1)/x^2][(x^2+x)/x^2]}$

$= \lim\limits_{x\to \infty} \dfrac{(2+1/x^2)^2}{(1-2/x+1/x^2)(1+1/x)} = \dfrac{(2+0)^2}{(1-0+0)(1+0)} = 4$

25. $\lim\limits_{x \to \infty} \left(\sqrt{9x^2 + x} - 3x\right) = \lim\limits_{x \to \infty} \dfrac{\left(\sqrt{9x^2 + x} - 3x\right)\left(\sqrt{9x^2 + x} + 3x\right)}{\sqrt{9x^2 + x} + 3x} = \lim\limits_{x \to \infty} \dfrac{\left(\sqrt{9x^2 + x}\right)^2 - (3x)^2}{\sqrt{9x^2 + x} + 3x}$

$$= \lim\limits_{x \to \infty} \dfrac{(9x^2 + x) - 9x^2}{\sqrt{9x^2 + x} + 3x} = \lim\limits_{x \to \infty} \dfrac{x}{\sqrt{9x^2 + x} + 3x} \cdot \dfrac{1/x}{1/x}$$

$$= \lim\limits_{x \to \infty} \dfrac{x/x}{\sqrt{9x^2/x^2 + x/x^2} + 3x/x} = \lim\limits_{x \to \infty} \dfrac{1}{\sqrt{9 + 1/x} + 3} = \dfrac{1}{\sqrt{9} + 3} = \dfrac{1}{3 + 3} = \dfrac{1}{6}$$

27. $\lim\limits_{x \to \infty} \dfrac{x^4 - 3x^2 + x}{x^3 - x + 2} = \lim\limits_{x \to \infty} \dfrac{(x^4 - 3x^2 + x)/x^3}{(x^3 - x + 2)/x^3} \quad \begin{bmatrix} \text{divide by the highest power} \\ \text{of } x \text{ in the denominator} \end{bmatrix} = \lim\limits_{x \to \infty} \dfrac{x - 3/x + 1/x^2}{1 - 1/x^2 + 2/x^3} = \infty$

since the numerator increases without bound and the denominator approaches 1 as $x \to \infty$.

29. $\lim\limits_{x \to \infty} \cos x$ does not exist because as x increases $\cos x$ does not approach any one value, but oscillates between 1 and -1.

31. $\lim\limits_{x \to \infty} \left(x - \sqrt{x}\right) = \lim\limits_{x \to \infty} \sqrt{x}\left(\sqrt{x} - 1\right) = \infty$ since $\sqrt{x} \to \infty$ and $\sqrt{x} - 1 \to \infty$ as $x \to \infty$.

33. $\lim\limits_{x \to -\infty} \left(x^4 + x^5\right) = \lim\limits_{x \to -\infty} x^5\left(\frac{1}{x} + 1\right)$ [factor out the largest power of x] $= -\infty$ because $x^5 \to -\infty$ and $1/x + 1 \to 1$

as $x \to -\infty$.

Or: $\lim\limits_{x \to -\infty} \left(x^4 + x^5\right) = \lim\limits_{x \to -\infty} x^4\left(1 + x\right) = -\infty$.

35. $\lim\limits_{x \to \infty} \dfrac{2x^2 + x - 1}{x^2 + x - 2} = \lim\limits_{x \to \infty} \dfrac{\dfrac{2x^2 + x - 1}{x^2}}{\dfrac{x^2 + x - 2}{x^2}} = \lim\limits_{x \to \infty} \dfrac{2 + \dfrac{1}{x} - \dfrac{1}{x^2}}{1 + \dfrac{1}{x} - \dfrac{2}{x^2}} = \dfrac{\lim\limits_{x \to \infty}\left(2 + \dfrac{1}{x} - \dfrac{1}{x^2}\right)}{\lim\limits_{x \to \infty}\left(1 + \dfrac{1}{x} - \dfrac{2}{x^2}\right)}$

$$= \dfrac{\lim\limits_{x \to \infty} 2 + \lim\limits_{x \to \infty} \dfrac{1}{x} - \lim\limits_{x \to \infty} \dfrac{1}{x^2}}{\lim\limits_{x \to \infty} 1 + \lim\limits_{x \to \infty} \dfrac{1}{x} - 2\lim\limits_{x \to \infty} \dfrac{1}{x^2}} = \dfrac{2 + 0 - 0}{1 + 0 - 2(0)} = 2, \quad \text{so } y = 2 \text{ is a horizontal asymptote.}$$

$y = f(x) = \dfrac{2x^2 + x - 1}{x^2 + x - 2} = \dfrac{(2x - 1)(x + 1)}{(x + 2)(x - 1)}$, so $\lim\limits_{x \to -2^-} f(x) = \infty$,

$\lim\limits_{x \to -2^+} f(x) = -\infty, \ \lim\limits_{x \to 1^-} f(x) = -\infty, \ \text{and} \ \lim\limits_{x \to 1^+} f(x) = \infty.$ Thus, $x = -2$

and $x = 1$ are vertical asymptotes. The graph confirms our work.

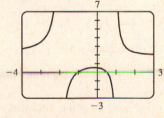

37. (a)

From the graph of $f(x) = \sqrt{x^2 + x + 1} + x$, we

estimate the value of $\lim\limits_{x \to -\infty} f(x)$ to be -0.5.

(b)

x	$f(x)$
$-10{,}000$	-0.4999625
$-100{,}000$	-0.4999962
$-1{,}000{,}000$	-0.4999996

From the table, we estimate the limit to be -0.5.

(c) $\displaystyle\lim_{x \to -\infty} \left(\sqrt{x^2 + x + 1} + x \right) = \lim_{x \to -\infty} \left(\sqrt{x^2 + x + 1} + x \right) \left[\dfrac{\sqrt{x^2 + x + 1} - x}{\sqrt{x^2 + x + 1} - x} \right] = \lim_{x \to -\infty} \dfrac{(x^2 + x + 1) - x^2}{\sqrt{x^2 + x + 1} - x}$

$\displaystyle = \lim_{x \to -\infty} \dfrac{(x+1)(1/x)}{\left(\sqrt{x^2 + x + 1} - x \right)(1/x)} = \lim_{x \to -\infty} \dfrac{1 + (1/x)}{-\sqrt{1 + (1/x) + (1/x^2)} - 1}$

$\displaystyle = \dfrac{1 + 0}{-\sqrt{1 + 0 + 0} - 1} = -\dfrac{1}{2}$

Note that for $x < 0$, we have $\sqrt{x^2} = |x| = -x$, so when we divide the radical by x, with $x < 0$, we get

$\dfrac{1}{x}\sqrt{x^2 + x + 1} = -\dfrac{1}{\sqrt{x^2}}\sqrt{x^2 + x + 1} = -\sqrt{1 + (1/x) + (1/x^2)}$.

39. From the graph, it appears $y = 1$ is a horizontal asymptote.

$\displaystyle \lim_{x \to \infty} \dfrac{3x^3 + 500x^2}{x^3 + 500x^2 + 100x + 2000} = \lim_{x \to \infty} \dfrac{\dfrac{3x^3 + 500x^2}{x^3}}{\dfrac{x^3 + 500x^2 + 100x + 2000}{x^3}} = \lim_{x \to \infty} \dfrac{3 + (500/x)}{1 + (500/x) + (100/x^2) + (2000/x^3)}$

$\displaystyle = \dfrac{3 + 0}{1 + 0 + 0 + 0} = 3$, so $y = 3$ is a horizontal asymptote.

The discrepancy can be explained by the choice of the viewing window. Try $[-100{,}000, 100{,}000]$ by $[-1, 4]$ to get a graph that lends credibility to our calculation that $y = 3$ is a horizontal asymptote.

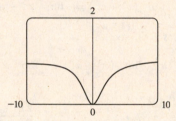

41. Let's look for a rational function.

(1) $\displaystyle\lim_{x \to \pm\infty} f(x) = 0 \;\; \Rightarrow \;\;$ degree of numerator $<$ degree of denominator

(2) $\displaystyle\lim_{x \to 0} f(x) = -\infty \;\; \Rightarrow \;\;$ there is a factor of x^2 in the denominator (not just x, since that would produce a sign change at $x = 0$), and the function is negative near $x = 0$.

(3) $\displaystyle\lim_{x \to 3^-} f(x) = \infty$ and $\displaystyle\lim_{x \to 3^+} f(x) = -\infty \;\; \Rightarrow \;\;$ vertical asymptote at $x = 3$; there is a factor of $(x - 3)$ in the denominator.

(4) $f(2) = 0 \;\; \Rightarrow \;\;$ 2 is an x-intercept; there is at least one factor of $(x - 2)$ in the numerator.

Combining all of this information and putting in a negative sign to give us the desired left- and right-hand limits gives us

$f(x) = \dfrac{2 - x}{x^2(x - 3)}$ as one possibility.

43. (a) We must first find the function f. Since f has a vertical asymptote $x = 4$ and x-intercept $x = 1$, $x - 4$ is a factor of the denominator and $x - 1$ is a factor of the numerator. There is a removable discontinuity at $x = -1$, so $x - (-1) = x + 1$ is a factor of both the numerator and denominator. Thus, f now looks like this: $f(x) = \dfrac{a(x - 1)(x + 1)}{(x - 4)(x + 1)}$, where a is still to

be determined. Then $\displaystyle\lim_{x \to -1} f(x) = \lim_{x \to -1} \dfrac{a(x - 1)(x + 1)}{(x - 4)(x + 1)} = \lim_{x \to -1} \dfrac{a(x - 1)}{x - 4} = \dfrac{a(-1 - 1)}{(-1 - 4)} = \dfrac{2}{5}a$, so $\dfrac{2}{5}a = 2$, and

$a = 5$. Thus $f(x) = \dfrac{5(x-1)(x+1)}{(x-4)(x+1)}$ is a ratio of quadratic functions satisfying all the given conditions and

$$f(0) = \frac{5(-1)(1)}{(-4)(1)} = \frac{5}{4}.$$

(b) $\displaystyle \lim_{x \to \infty} f(x) = 5 \lim_{x \to \infty} \frac{x^2 - 1}{x^2 - 3x - 4} = 5 \lim_{x \to \infty} \frac{(x^2/x^2) - (1/x^2)}{(x^2/x^2) - (3x/x^2) - (4/x^2)} = 5 \frac{1 - 0}{1 - 0 - 0} = 5(1) = 5$

45. Divide the numerator and the denominator by the highest power of x in $Q(x)$.

    (a) If $\deg P < \deg Q$, then the numerator $\to 0$ but the denominator doesn't. So $\displaystyle \lim_{x \to \infty} [P(x)/Q(x)] = 0$.

    (b) If $\deg P > \deg Q$, then the numerator $\to \pm\infty$ but the denominator doesn't, so $\displaystyle \lim_{x \to \infty} [P(x)/Q(x)] = \pm\infty$

    (depending on the ratio of the leading coefficients of P and Q).

47. $\displaystyle \lim_{x \to \infty} \frac{4x - 1}{x} = \lim_{x \to \infty} \left(4 - \frac{1}{x}\right) = 4$ and $\displaystyle \lim_{x \to \infty} \frac{4x^2 + 3x}{x^2} = \lim_{x \to \infty} \left(4 + \frac{3}{x}\right) = 4$. Therefore, by the Squeeze Theorem,

$\displaystyle \lim_{x \to \infty} f(x) = 4$.

49. (a) After t minutes, $25t$ liters of brine with 30 g of salt per liter has been pumped into the tank, so it contains

    $(5000 + 25t)$ liters of water and $25t \cdot 30 = 750t$ grams of salt. Therefore, the salt concentration at time t will be

$$C(t) = \frac{750t}{5000 + 25t} = \frac{30t}{200 + t} \frac{\text{g}}{\text{L}}.$$

    (b) $\displaystyle \lim_{t \to \infty} C(t) = \lim_{t \to \infty} \frac{30t}{200 + t} = \lim_{t \to \infty} \frac{30t/t}{200/t + t/t} = \frac{30}{0 + 1} = 30$. So the salt concentration approaches that of the brine

    being pumped into the tank.

51. $\dfrac{1}{(x+3)^4} > 10{,}000 \iff (x+3)^4 < \dfrac{1}{10{,}000} \iff |x+3| < \dfrac{1}{\sqrt[4]{10{,}000}} \iff |x - (-3)| < \dfrac{1}{10}$

53. Let $N < 0$ be given. Then, for $x < -1$, we have $\dfrac{5}{(x+1)^3} < N \iff \dfrac{5}{N} < (x+1)^3 \iff \sqrt[3]{\dfrac{5}{N}} < x + 1$. Let

$\delta = -\sqrt[3]{\dfrac{5}{N}}$. Then $-1 - \delta < x < -1 \implies \sqrt[3]{\dfrac{5}{N}} < x + 1 < 0 \implies \dfrac{5}{(x+1)^3} < N$, so $\displaystyle \lim_{x \to -1^-} \frac{5}{(x+1)^3} = -\infty$.

55. Let $g(x) = \dfrac{3x^2 + 1}{2x^2 + x + 1}$ and $f(x) = |g(x) - 1.5|$. Note that

$\displaystyle \lim_{x \to \infty} g(x) = \frac{3}{2}$ and $\displaystyle \lim_{x \to \infty} f(x) = 0$. We are interested in finding the

x-value at which $f(x) < 0.05$. From the graph, we find that $x \approx 14.804$,

so we choose $N = 15$ (or any larger number).

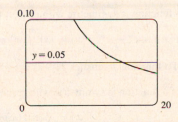

57. (a) $1/x^2 < 0.0001 \iff x^2 > 1/0.0001 = 10\,000 \iff x > 100 \quad (x > 0)$

    (b) If $\varepsilon > 0$ is given, then $1/x^2 < \varepsilon \iff x^2 > 1/\varepsilon \iff x > 1/\sqrt{\varepsilon}$. Let $N = 1/\sqrt{\varepsilon}$.

    Then $x > N \implies x > \dfrac{1}{\sqrt{\varepsilon}} \implies \left|\dfrac{1}{x^2} - 0\right| = \dfrac{1}{x^2} < \varepsilon$, so $\displaystyle \lim_{x \to \infty} \frac{1}{x^2} = 0$.

59. Suppose that $\lim\limits_{x \to \infty} f(x) = L$. Then for every $\varepsilon > 0$ there is a corresponding positive number N such that $|f(x) - L| < \varepsilon$

whenever $x > N$. If $t = 1/x$, then $x > N \iff 0 < 1/x < 1/N \iff 0 < t < 1/N$. Thus, for every $\varepsilon > 0$ there is a

corresponding $\delta > 0$ (namely $1/N$) such that $|f(1/t) - L| < \varepsilon$ whenever $0 < t < \delta$. This proves that

$$\lim_{t \to 0^+} f(1/t) = L = \lim_{x \to \infty} f(x).$$

Now suppose that $\lim\limits_{x \to -\infty} f(x) = L$. Then for every $\varepsilon > 0$ there is a corresponding negative number N such that

$|f(x) - L| < \varepsilon$ whenever $x < N$. If $t = 1/x$, then $x < N \iff 1/N < 1/x < 0 \iff 1/N < t < 0$. Thus, for every

$\varepsilon > 0$ there is a corresponding $\delta > 0$ (namely $-1/N$) such that $|f(1/t) - L| < \varepsilon$ whenever $-\delta < t < 0$. This proves that

$$\lim_{t \to 0^-} f(1/t) = L = \lim_{x \to -\infty} f(x).$$

1   Review

CONCEPT CHECK

1. (a) A **function** f is a rule that assigns to each element x in a set D exactly one element, called $f(x)$, in a set E. The set D is
called the **domain** of the function. The **range** of f is the set of all possible values of $f(x)$ as x varies throughout the
domain.

   (b) If f is a function with domain D, then its **graph** is the set of ordered pairs $\{(x, f(x)) \mid x \in D\}$.

   (c) Use the Vertical Line Test on page 4.

2. The four ways to represent a function are: verbally, numerically, visually, and algebraically. An example of each is given
below.

   Verbally: An assignment of students to chairs in a classroom (a description in words)

   Numerically: A tax table that assigns an amount of tax to an income (a table of values)

   Visually: A graphical history of the Dow Jones average (a graph)

   Algebraically: A relationship between distance, rate, and time: $d = rt$ (an explicit formula)

3. (a) If a function f satisfies $f(-x) = f(x)$ for every number x in its domain, then f is called an **even function**. If the graph of
a function is symmetric with respect to the y-axis, then f is even. Examples of an even function: $f(x) = x^2$,
$f(x) = x^4 + x^2$, $f(x) = |x|$, $f(x) = \cos x$.

   (b) If a function f satisfies $f(-x) = -f(x)$ for every number x in its domain, then f is called an **odd function**. If the graph
of a function is symmetric with respect to the origin, then f is odd. Examples of an odd function: $f(x) = x^3$,
$f(x) = x^3 + x^5$, $f(x) = \sqrt[3]{x}$, $f(x) = \sin x$.

4. A function f is called **increasing** on an interval I if $f(x_1) < f(x_2)$ whenever $x_1 < x_2$ in I.

5. A **mathematical model** is a mathematical description (often by means of a function or an equation) of a real-world
phenomenon.

6. (a) Linear function: $f(x) = 2x + 1$, $f(x) = ax + b$

(b) Power function: $f(x) = x^2$, $f(x) = x^a$

(c) Exponential function: $f(x) = 2^x$, $f(x) = a^x$

(d) Quadratic function: $f(x) = x^2 + x + 1$, $f(x) = ax^2 + bx + c$

(e) Polynomial of degree 5: $f(x) = x^5 + 2$

(f) Rational function: $f(x) = \dfrac{x}{x+2}$, $f(x) = \dfrac{P(x)}{Q(x)}$ where $P(x)$ and $Q(x)$ are polynomials

7.

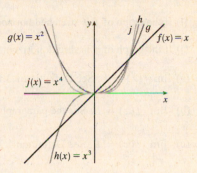
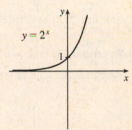

8. (a)

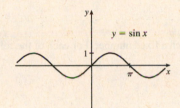

(b)

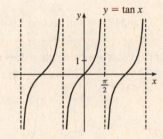

(c)

(d)

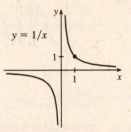

(e)

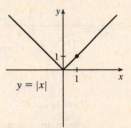

(f)

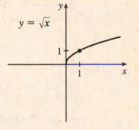

9. (a) The domain of $f + g$ is the intersection of the domain of f and the domain of g; that is, $A \cap B$.

(b) The domain of fg is also $A \cap B$.

(c) The domain of f/g must exclude values of x that make g equal to 0; that is, $\{x \in A \cap B \mid g(x) \neq 0\}$.

10. Given two functions f and g, the **composite** function $f \circ g$ is defined by $(f \circ g)(x) = f(g(x))$. The domain of $f \circ g$ is the set of all x in the domain of g such that $g(x)$ is in the domain of f.

11. (a) If the graph of f is shifted 2 units upward, its equation becomes $y = f(x) + 2$.

(b) If the graph of f is shifted 2 units downward, its equation becomes $y = f(x) - 2$.

(c) If the graph of f is shifted 2 units to the right, its equation becomes $y = f(x - 2)$.

(d) If the graph of f is shifted 2 units to the left, its equation becomes $y = f(x + 2)$.

(e) If the graph of f is reflected about the x-axis, its equation becomes $y = -f(x)$.

(f) If the graph of f is reflected about the y-axis, its equation becomes $y = f(-x)$.

(g) If the graph of f is stretched vertically by a factor of 2, its equation becomes $y = 2f(x)$.

(h) If the graph of f is shrunk vertically by a factor of 2, its equation becomes $y = \frac{1}{2}f(x)$.

(i) If the graph of f is stretched horizontally by a factor of 2, its equation becomes $y = f(\frac{1}{2}x)$.

(j) If the graph of f is shrunk horizontally by a factor of 2, its equation becomes $y = f(2x)$.

12. (a) $\lim_{x \to a} f(x) = L$: See Definition 1.3.1 and Figures 1 and 2 in Section 1.3.

(b) $\lim_{x \to a^+} f(x) = L$: See the paragraph after Definition 1.3.2 and Figure 9(b) in Section 1.3.

(c) $\lim_{x \to a^-} f(x) = L$: See Definition 1.3.2 and Figure 9(a) in Section 1.3.

(d) $\lim_{x \to a} f(x) = \infty$: See Definition 1.6.1 and Figure 2 in Section 1.6.

(e) $\lim_{x \to \infty} f(x) = L$: See Definition 1.6.3 and Figure 8 in Section 1.6.

13. In general, the limit of a function fails to exist when the function does not approach a fixed number. For each of the following functions, the limit fails to exist at $x = 2$.

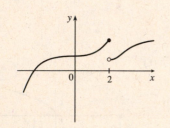

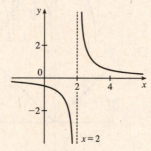

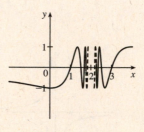

The left- and right-hand
limits are not equal.

There is an
infinite discontinuity.

There are an infinite
number of oscillations.

14. (a)–(g) See the statements of Limit Laws 1–6 and 11 in Section 1.4.

15. See Theorem 4 in Section 1.4.

16. (a) A function f is continuous at a number a if $f(x)$ approaches $f(a)$ as x approaches a; that is, $\lim_{x \to a} f(x) = f(a)$.

(b) A function f is continuous on the interval $(-\infty, \infty)$ if f is continuous at every real number a. The graph of such a function has no breaks and every vertical line crosses it.

17. See Theorem 1.5.9.

18. (a) See Definition 1.6.2 and Figures 2–4 in Section 1.6.

(b) See Definition 1.6.4 and Figures 8 and 9 in Section 1.6.

TRUE-FALSE QUIZ

1. False. Let $f(x) = x^2$, $s = -1$, and $t = 1$. Then $f(s + t) = (-1 + 1)^2 = 0^2 = 0$, but
$f(s) + f(t) = (-1)^2 + 1^2 = 2 \neq 0 = f(s + t)$.

3. False. Let $f(x) = x^2$. Then $f(3x) = (3x)^2 = 9x^2$ and $3f(x) = 3x^2$. So $f(3x) \neq 3f(x)$.

5. True. See the Vertical Line Test.

7. False. Limit Law 2 applies only if the individual limits exist (these don't).

9. True. Limit Law 5 applies.

11. False. Consider $\lim\limits_{x \to 5} \dfrac{x(x-5)}{x-5}$ or $\lim\limits_{x \to 5} \dfrac{\sin(x-5)}{x-5}$. The first limit exists and is equal to 5. By Equation 1.4.6, we know that the latter limit exists (and it is equal to 1).

13. True. A polynomial is continuous everywhere, so $\lim\limits_{x \to b} p(x)$ exists and is equal to $p(b)$.

15. True. See Figure 10 in Section 1.6.

17. False. Consider $f(x) = \begin{cases} 1/(x-1) & \text{if } x \neq 1 \\ 2 & \text{if } x = 1 \end{cases}$

19. False. For example, if $x = -3$, then $\sqrt{(-3)^2} = \sqrt{9} = 3$, not -3.

21. True. Use Theorem 1.5.7 with $a = 2$, $b = 5$, and $g(x) = 4x^2 - 11$. Note that $f(4) = 3$ is not needed.

23. True, by the definition of a limit with $\varepsilon = 1$.

25. True. See Exercise 50(b) in Section 1.5.

EXERCISES

1. (a) When $x = 2$, $y \approx 2.7$. Thus, $f(2) \approx 2.7$.   (b) $f(x) = 3 \Rightarrow x \approx 2.3, 5.6$

(c) The domain of f is $-6 \leq x \leq 6$, or $[-6, 6]$.   (d) The range of f is $-4 \leq y \leq 4$, or $[-4, 4]$.

(e) f is increasing on $[-4, 4]$, that is, on $-4 \leq x \leq 4$.

(f) f is odd since its graph is symmetric about the origin.

3. $f(x) = 2/(3x - 1)$.   Domain: $3x - 1 \neq 0 \Rightarrow 3x \neq 1 \Rightarrow x \neq \frac{1}{3}$. $D = \left(-\infty, \frac{1}{3}\right) \cup \left(\frac{1}{3}, \infty\right)$

                  Range: all reals except 0 ($y = 0$ is the horizontal asymptote for f.) $R = (-\infty, 0) \cup (0, \infty)$

5. $y = 1 + \sin x$.   Domain: $\mathbb{R}$

                  Range: $-1 \leq \sin x \leq 1 \Rightarrow 0 \leq 1 + \sin x \leq 2 \Rightarrow 0 \leq y \leq 2$.

7. (a) To obtain the graph of $y = f(x) + 8$, we shift the graph of $y = f(x)$ up 8 units.

(b) To obtain the graph of $y = f(x + 8)$, we shift the graph of $y = f(x)$ left 8 units.

(c) To obtain the graph of $y = 1 + 2f(x)$, we stretch the graph of $y = f(x)$ vertically by a factor of 2, and then shift the resulting graph 1 unit upward.

(d) To obtain the graph of $y = f(x - 2) - 2$, we shift the graph of $y = f(x)$ right 2 units (for the "-2" inside the parentheses), and then shift the resulting graph 2 units downward.

(e) To obtain the graph of $y = -f(x)$, we reflect the graph of $y = f(x)$ about the x-axis.

(f) To obtain the graph of $y = 3 - f(x)$, we reflect the graph of $y = f(x)$ about the x-axis, and then shift the resulting graph 3 units upward.

9. $y = -\sin 2x$: Start with the graph of $y = \sin x$, compress horizontally by a factor of 2, and reflect about the x-axis.

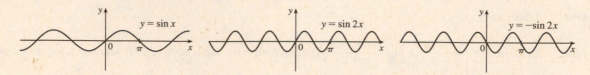

11. $y = 1 + \frac{1}{2}x^3$: Start with the graph of $y = x^3$, compress vertically by a factor of 2, and shift 1 unit upward.

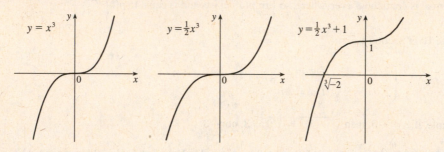

13. $f(x) = \dfrac{1}{x+2}$:

Start with the graph of $f(x) = 1/x$

and shift 2 units to the left.

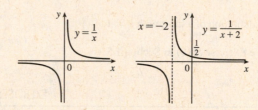

15. (a) The terms of f are a mixture of odd and even powers of x, so f is neither even nor odd.

(b) The terms of f are all odd powers of x, so f is odd.

(c) $f(-x) = \cos\left((-x)^2\right) = \cos\left(x^2\right) = f(x)$, so f is even.

(d) $f(-x) = 1 + \sin(-x) = 1 - \sin x$. Now $f(-x) \neq f(x)$ and $f(-x) \neq -f(x)$, so f is neither even nor odd.

17. $f(x) = \sqrt{x}, D = [0, \infty)$; $g(x) = \sin x, D = \mathbb{R}$.

(a) $(f \circ g)(x) = f(g(x)) = f(\sin x) = \sqrt{\sin x}$. For $\sqrt{\sin x}$ to be defined, we must have $\sin x \geq 0 \iff x \in [0, \pi]$,
$[2\pi, 3\pi], [-2\pi, -\pi], [4\pi, 5\pi], [-4\pi, -3\pi], \ldots$, so $D = \{x \mid x \in [2n\pi, \pi + 2n\pi]$, where n is an integer$\}$.

(b) $(g \circ f)(x) = g(f(x)) = g(\sqrt{x}) = \sin \sqrt{x}$. x must be greater than or equal to 0 for $\sqrt{x}$ to be defined, so $D = [0, \infty)$.

(c) $(f \circ f)(x) = f(f(x)) = f(\sqrt{x}) = \sqrt{\sqrt{x}} = \sqrt[4]{x}$. $D = [0, \infty)$.

(d) $(g \circ g)(x) = g(g(x)) = g(\sin x) = \sin(\sin x)$. $D = \mathbb{R}$.

19. (a) (i) $\lim\limits_{x \to 2^+} f(x) = 3$ (ii) $\lim\limits_{x \to -3^+} f(x) = 0$

(iii) $\lim\limits_{x \to -3} f(x)$ does not exist since the left and right limits are not equal. (The left limit is -2.)

(iv) $\lim\limits_{x \to 4} f(x) = 2$

(v) $\lim\limits_{x \to 0} f(x) = \infty$ (vi) $\lim\limits_{x \to 2^-} f(x) = -\infty$

(vii) $\lim\limits_{x \to \infty} f(x) = 4$ (viii) $\lim\limits_{x \to -\infty} f(x) = -1$

(b) The equations of the horizontal asymptotes are $y = -1$ and $y = 4$.

(c) The equations of the vertical asymptotes are $x = 0$ and $x = 2$.

(d) f is discontinuous at $x = -3, 0, 2$, and 4. The discontinuities are jump, infinite, infinite, and removable, respectively.

21. $\lim\limits_{x \to 0} \cos(x + \sin x) = \cos\left[\lim\limits_{x \to 0}(x + \sin x)\right]$ [by Theorem 1.5.7] $= \cos 0 = 1$

23. $\lim\limits_{x \to -3} \dfrac{x^2 - 9}{x^2 + 2x - 3} = \lim\limits_{x \to -3} \dfrac{(x+3)(x-3)}{(x+3)(x-1)} = \lim\limits_{x \to -3} \dfrac{x-3}{x-1} = \dfrac{-3-3}{-3-1} = \dfrac{-6}{-4} = \dfrac{3}{2}$

25. $\lim\limits_{h \to 0} \dfrac{(h-1)^3 + 1}{h} = \lim\limits_{h \to 0} \dfrac{(h^3 - 3h^2 + 3h - 1) + 1}{h} = \lim\limits_{h \to 0} \dfrac{h^3 - 3h^2 + 3h}{h} = \lim\limits_{h \to 0} (h^2 - 3h + 3) = 3$

Another solution: Factor the numerator as a sum of two cubes and then simplify.

$\lim\limits_{h \to 0} \dfrac{(h-1)^3 + 1}{h} = \lim\limits_{h \to 0} \dfrac{(h-1)^3 + 1^3}{h} = \lim\limits_{h \to 0} \dfrac{[(h-1) + 1]\left[(h-1)^2 - 1(h-1) + 1^2\right]}{h}$

$\qquad = \lim\limits_{h \to 0} \left[(h-1)^2 - h + 2\right] = 1 - 0 + 2 = 3$

27. $\lim\limits_{r \to 9} \dfrac{\sqrt{r}}{(r-9)^4} = \infty$ since $(r-9)^4 \to 0^+$ as $r \to 9$ and $\dfrac{\sqrt{r}}{(r-9)^4} > 0$ for $r \neq 9$.

29. $\lim\limits_{s \to 16} \dfrac{4 - \sqrt{s}}{s - 16} = \lim\limits_{s \to 16} \dfrac{4 - \sqrt{s}}{(\sqrt{s} + 4)(\sqrt{s} - 4)} = \lim\limits_{s \to 16} \dfrac{-1}{\sqrt{s} + 4} = \dfrac{-1}{\sqrt{16} + 4} = -\dfrac{1}{8}$

31. $\lim\limits_{x \to \infty} \dfrac{1 + 2x - x^2}{1 - x + 2x^2} = \lim\limits_{x \to \infty} \dfrac{\left(1 + 2x - x^2\right)/x^2}{\left(1 - x + 2x^2\right)/x^2} = \lim\limits_{x \to \infty} \dfrac{1/x^2 + 2/x - 1}{1/x^2 - 1/x + 2} = \dfrac{0 + 0 - 1}{0 - 0 + 2} = -\dfrac{1}{2}$

33. $\lim\limits_{x \to \infty} \left(\sqrt{x^2 + 4x + 1} - x\right) = \lim\limits_{x \to \infty} \left[\dfrac{\sqrt{x^2 + 4x + 1} - x}{1} \cdot \dfrac{\sqrt{x^2 + 4x + 1} + x}{\sqrt{x^2 + 4x + 1} + x}\right] = \lim\limits_{x \to \infty} \dfrac{(x^2 + 4x + 1) - x^2}{\sqrt{x^2 + 4x + 1} + x}$

$\qquad = \lim\limits_{x \to \infty} \dfrac{(4x + 1)/x}{\left(\sqrt{x^2 + 4x + 1} + x\right)/x} \qquad \left[\text{divide by } x = \sqrt{x^2} \text{ for } x > 0\right]$

$\qquad = \lim\limits_{x \to \infty} \dfrac{4 + 1/x}{\sqrt{1 + 4/x + 1/x^2} + 1} = \dfrac{4 + 0}{\sqrt{1 + 0 + 0} + 1} = \dfrac{4}{2} = 2$

35. $\lim\limits_{x \to 0} \dfrac{\cot 2x}{\csc x} = \lim\limits_{x \to 0} \dfrac{\cos 2x \sin x}{\sin 2x} = \lim\limits_{x \to 0} \cos 2x \left[\dfrac{(\sin x)/x}{(\sin 2x)/x}\right] = \lim\limits_{x \to 0} \cos 2x \left[\dfrac{\lim\limits_{x \to 0}[(\sin x)/x]}{2\lim\limits_{x \to 0}[(\sin 2x)/2x]}\right] = 1 \cdot \dfrac{1}{2 \cdot 1} = \dfrac{1}{2}$

37. From the graph of $y = \left(\cos^2 x\right)/x^2$, it appears that $y = 0$ is the horizontal

asymptote and $x = 0$ is the vertical asymptote. Now $0 \leq (\cos x)^2 \leq 1 \ \Rightarrow$

$\dfrac{0}{x^2} \leq \dfrac{\cos^2 x}{x^2} \leq \dfrac{1}{x^2} \ \Rightarrow \ 0 \leq \dfrac{\cos^2 x}{x^2} \leq \dfrac{1}{x^2}$. But $\lim\limits_{x \to \pm\infty} 0 = 0$ and

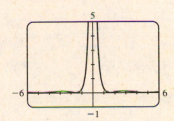

$\lim\limits_{x \to \pm\infty} \dfrac{1}{x^2} = 0$, so by the Squeeze Theorem, $\lim\limits_{x \to \pm\infty} \dfrac{\cos^2 x}{x^2} = 0$.

Thus, $y = 0$ is the horizontal asymptote. $\lim\limits_{x \to 0} \dfrac{\cos^2 x}{x^2} = \infty$ because $\cos^2 x \to 1$ and $x^2 \to 0^+$ as $x \to 0$, so $x = 0$ is the

vertical asymptote.

39. Since $2x - 1 \leq f(x) \leq x^2$ for $0 < x < 3$ and $\lim_{x \to 1} (2x - 1) = 1 = \lim_{x \to 1} x^2$, we have $\lim_{x \to 1} f(x) = 1$ by the Squeeze Theorem.

41. Given $\varepsilon > 0$, we need $\delta > 0$ such that if $0 < |x - 2| < \delta$, then $|(14 - 5x) - 4| < \varepsilon$. But $|(14 - 5x) - 4| < \varepsilon \iff$

$|-5x + 10| < \varepsilon \iff |-5| \, |x - 2| < \varepsilon \iff |x - 2| < \varepsilon/5$. So if we choose $\delta = \varepsilon/5$, then $0 < |x - 2| < \delta \implies$

$|(14 - 5x) - 4| < \varepsilon$. Thus, $\lim_{x \to 2} (14 - 5x) = 4$ by the definition of a limit.

43. If $\varepsilon > 0$ is given, then $1/x^4 < \varepsilon \iff x^4 > 1/\varepsilon \iff x > 1/\sqrt[4]{\varepsilon}$. Let $N = 1/\sqrt[4]{\varepsilon}$.

Then $x > N \implies x > \dfrac{1}{\sqrt[4]{\varepsilon}} \implies \left| \dfrac{1}{x^4} - 0 \right| = \dfrac{1}{x^4} < \varepsilon$, so $\lim_{x \to \infty} \dfrac{1}{x^4} = 0$.

45. (a) $f(x) = \sqrt{-x}$ if $x < 0$, $f(x) = 3 - x$ if $0 \leq x < 3$, $f(x) = (x - 3)^2$ if $x > 3$.

    (i) $\lim_{x \to 0^+} f(x) = \lim_{x \to 0^+} (3 - x) = 3$ (ii) $\lim_{x \to 0^-} f(x) = \lim_{x \to 0^-} \sqrt{-x} = 0$

    (iii) Because of (i) and (ii), $\lim_{x \to 0} f(x)$ does not exist.   (iv) $\lim_{x \to 3^-} f(x) = \lim_{x \to 3^-} (3 - x) = 0$

    (v) $\lim_{x \to 3^+} f(x) = \lim_{x \to 3^+} (x - 3)^2 = 0$ (vi) Because of (iv) and (v), $\lim_{x \to 3} f(x) = 0$.

(b) f is discontinuous at 0 since $\lim_{x \to 0} f(x)$ does not exist.   (c)

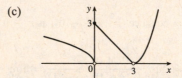

    f is discontinuous at 3 since $f(3)$ does not exist.

47. $f(x) = x^5 - x^3 + 3x - 5$ is continuous on the interval $[1, 2]$, $f(1) = -2$, and $f(2) = 25$. Since $-2 < 0 < 25$, there is a

number c in $(1, 2)$ such that $f(c) = 0$ by the Intermediate Value Theorem. Thus, there is a root of the equation

$x^5 - x^3 + 3x - 5 = 0$ in the interval $(1, 2)$.

2 □ DERIVATIVES

2.1 Derivatives and Rates of Change

1. (a) (i) Using Definition 1 with $f(x) = 4x - x^2$ and $P(1, 3)$,

$$m = \lim_{x \to a} \frac{f(x) - f(a)}{x - a} = \lim_{x \to 1} \frac{(4x - x^2) - 3}{x - 1}$$

$$= \lim_{x \to 1} \frac{-(x^2 - 4x + 3)}{x - 1} = \lim_{x \to 1} \frac{-(x - 1)(x - 3)}{x - 1}$$

$$= \lim_{x \to 1} (3 - x) = 3 - 1 = 2$$

(ii) Using Equation 2 with $f(x) = 4x - x^2$ and $P(1, 3)$,

$$m = \lim_{h \to 0} \frac{f(a + h) - f(a)}{h} = \lim_{h \to 0} \frac{f(1 + h) - f(1)}{h}$$

$$= \lim_{h \to 0} \frac{[4(1 + h) - (1 + h)^2] - 3}{h}$$

$$= \lim_{h \to 0} \frac{4 + 4h - 1 - 2h - h^2 - 3}{h} = \lim_{h \to 0} \frac{-h^2 + 2h}{h}$$

$$= \lim_{h \to 0} \frac{h(-h + 2)}{h} = \lim_{h \to 0} (-h + 2) = 2$$

(b) An equation of the tangent line is $y - f(a) = f'(a)(x - a) \Rightarrow y - f(1) = f'(1)(x - 1) \Rightarrow y - 3 = 2(x - 1)$,

or $y = 2x + 1$.

(c)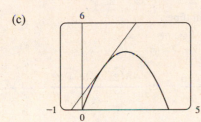

The graph of $y = 2x + 1$ is tangent to the graph of $y = 4x - x^2$ at the point $(1, 3)$. Now zoom in toward the point $(1, 3)$ until the parabola and the tangent line are indistiguishable.

3. Using (1) with $f(x) = 4x - 3x^2$ and $P(2, -4)$ [we could also use (2)],

$$m = \lim_{x \to a} \frac{f(x) - f(a)}{x - a} = \lim_{x \to 2} \frac{(4x - 3x^2) - (-4)}{x - 2} = \lim_{x \to 2} \frac{-3x^2 + 4x + 4}{x - 2}$$

$$= \lim_{x \to 2} \frac{(-3x - 2)(x - 2)}{x - 2} = \lim_{x \to 2} (-3x - 2) = -3(2) - 2 = -8$$

Tangent line: $y - (-4) = -8(x - 2) \Leftrightarrow y + 4 = -8x + 16 \Leftrightarrow y = -8x + 12$.

5. Using (1), $m = \lim_{x \to 1} \frac{\sqrt{x} - \sqrt{1}}{x - 1} = \lim_{x \to 1} \frac{(\sqrt{x} - 1)(\sqrt{x} + 1)}{(x - 1)(\sqrt{x} + 1)} = \lim_{x \to 1} \frac{x - 1}{(x - 1)(\sqrt{x} + 1)} = \lim_{x \to 1} \frac{1}{\sqrt{x} + 1} = \frac{1}{2}$.

Tangent line: $y - 1 = \frac{1}{2}(x - 1) \Leftrightarrow y = \frac{1}{2}x + \frac{1}{2}$

7. (a) Using (2) with $y = f(x) = 3 + 4x^2 - 2x^3$,

$$m = \lim_{h \to 0} \frac{f(a+h) - f(a)}{h} = \lim_{h \to 0} \frac{3 + 4(a+h)^2 - 2(a+h)^3 - (3 + 4a^2 - 2a^3)}{h}$$

$$= \lim_{h \to 0} \frac{3 + 4(a^2 + 2ah + h^2) - 2(a^3 + 3a^2h + 3ah^2 + h^3) - 3 - 4a^2 + 2a^3}{h}$$

$$= \lim_{h \to 0} \frac{3 + 4a^2 + 8ah + 4h^2 - 2a^3 - 6a^2h - 6ah^2 - 2h^3 - 3 - 4a^2 + 2a^3}{h}$$

$$= \lim_{h \to 0} \frac{8ah + 4h^2 - 6a^2h - 6ah^2 - 2h^3}{h} = \lim_{h \to 0} \frac{h(8a + 4h - 6a^2 - 6ah - 2h^2)}{h}$$

$$= \lim_{h \to 0} (8a + 4h - 6a^2 - 6ah - 2h^2) = 8a - 6a^2$$

(b) At $(1, 5)$: $m = 8(1) - 6(1)^2 = 2$, so an equation of the tangent line

is $y - 5 = 2(x - 1) \iff y = 2x + 3$.

At $(2, 3)$: $m = 8(2) - 6(2)^2 = -8$, so an equation of the tangent

line is $y - 3 = -8(x - 2) \iff y = -8x + 19$.

(c)

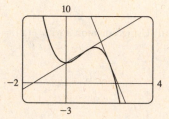

9. (a) Since the slope of the tangent at $t = 0$ is 0, the car's initial velocity was 0.

(b) The slope of the tangent is greater at C than at B, so the car was going faster at C.

(c) Near A, the tangent lines are becoming steeper as x increases, so the velocity was increasing, so the car was speeding up.

Near B, the tangent lines are becoming less steep, so the car was slowing down. The steepest tangent near C is the one at

C, so at C the car had just finished speeding up, and was about to start slowing down.

(d) Between D and E, the slope of the tangent is 0, so the car did not move during that time.

11. Let $s(t) = 40t - 16t^2$.

$$v(2) = \lim_{t \to 2} \frac{s(t) - s(2)}{t - 2} = \lim_{t \to 2} \frac{(40t - 16t^2) - 16}{t - 2} = \lim_{t \to 2} \frac{-16t^2 + 40t - 16}{t - 2} = \lim_{t \to 2} \frac{-8(2t^2 - 5t + 2)}{t - 2}$$

$$= \lim_{t \to 2} \frac{-8(t - 2)(2t - 1)}{t - 2} = -8 \lim_{t \to 2} (2t - 1) = -8(3) = -24$$

Thus, the instantaneous velocity when $t = 2$ is -24 ft/s.

13. $v(a) = \lim_{h \to 0} \dfrac{s(a+h) - s(a)}{h} = \lim_{h \to 0} \dfrac{\frac{1}{(a+h)^2} - \frac{1}{a^2}}{h} = \lim_{h \to 0} \dfrac{\frac{a^2 - (a+h)^2}{a^2(a+h)^2}}{h} = \lim_{h \to 0} \dfrac{a^2 - (a^2 + 2ah + h^2)}{ha^2(a+h)^2}$

$$= \lim_{h \to 0} \frac{-(2ah + h^2)}{ha^2(a+h)^2} = \lim_{h \to 0} \frac{-h(2a + h)}{ha^2(a+h)^2} = \lim_{h \to 0} \frac{-(2a + h)}{a^2(a+h)^2} = \frac{-2a}{a^2 \cdot a^2} = \frac{-2}{a^3} \text{ m/s}$$

So $v(1) = \dfrac{-2}{1^3} = -2$ m/s, $v(2) = \dfrac{-2}{2^3} = -\dfrac{1}{4}$ m/s, and $v(3) = \dfrac{-2}{3^3} = -\dfrac{2}{27}$ m/s.

15. $g'(0)$ is the only negative value. The slope at $x = 4$ is smaller than the slope at $x = 2$ and both are smaller than the slope at $x = -2$. Thus, $g'(0) < 0 < g'(4) < g'(2) < g'(-2)$.

17. For the tangent line $y = 4x - 5$: when $x = 2$, $y = 4(2) - 5 = 3$ and its slope is 4 (the coefficient of x). At the point of tangency, these values are shared with the curve $y = f(x)$; that is, $f(2) = 3$ and $f'(2) = 4$.

19. We begin by drawing a curve through the origin with a slope of 3 to satisfy $f(0) = 0$ and $f'(0) = 3$. Since $f'(1) = 0$, we will round off our figure so that there is a horizontal tangent directly over $x = 1$. Last, we make sure that the curve has a slope of -1 as we pass over $x = 2$. Two of the many possibilities are shown.

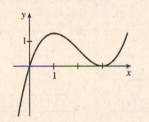 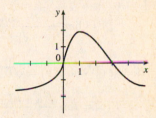

21. Using (4) with $f(x) = 3x^2 - x^3$ and $a = 1$,

$$f'(1) = \lim_{h \to 0} \frac{f(1+h) - f(1)}{h} = \lim_{h \to 0} \frac{[3(1+h)^2 - (1+h)^3] - 2}{h}$$

$$= \lim_{h \to 0} \frac{(3 + 6h + 3h^2) - (1 + 3h + 3h^2 + h^3) - 2}{h} = \lim_{h \to 0} \frac{3h - h^3}{h} = \lim_{h \to 0} \frac{h(3 - h^2)}{h}$$

$$= \lim_{h \to 0} (3 - h^2) = 3 - 0 = 3$$

Tangent line: $y - 2 = 3(x - 1)$ $\Leftrightarrow$ $y - 2 = 3x - 3$ $\Leftrightarrow$ $y = 3x - 1$

23. (a) Using (4) with $F(x) = 5x/(1 + x^2)$ and the point $(2, 2)$, we have

(b)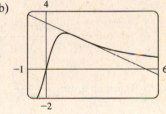

$$F'(2) = \lim_{h \to 0} \frac{F(2+h) - F(2)}{h} = \lim_{h \to 0} \frac{\dfrac{5(2+h)}{1 + (2+h)^2} - 2}{h}$$

$$= \lim_{h \to 0} \frac{\dfrac{5h + 10}{h^2 + 4h + 5} - 2}{h} = \lim_{h \to 0} \frac{\dfrac{5h + 10 - 2(h^2 + 4h + 5)}{h^2 + 4h + 5}}{h}$$

$$= \lim_{h \to 0} \frac{-2h^2 - 3h}{h(h^2 + 4h + 5)} = \lim_{h \to 0} \frac{h(-2h - 3)}{h(h^2 + 4h + 5)} = \lim_{h \to 0} \frac{-2h - 3}{h^2 + 4h + 5} = \frac{-3}{5}$$

So an equation of the tangent line at $(2, 2)$ is $y - 2 = -\frac{3}{5}(x - 2)$ or $y = -\frac{3}{5}x + \frac{16}{5}$.

25. Use (4) with $f(x) = 3x^2 - 4x + 1$.

$$f'(a) = \lim_{h \to 0} \frac{f(a+h) - f(a)}{h} = \lim_{h \to 0} \frac{[3(a+h)^2 - 4(a+h) + 1] - (3a^2 - 4a + 1)]}{h}$$

$$= \lim_{h \to 0} \frac{3a^2 + 6ah + 3h^2 - 4a - 4h + 1 - 3a^2 + 4a - 1}{h} = \lim_{h \to 0} \frac{6ah + 3h^2 - 4h}{h}$$

$$= \lim_{h \to 0} \frac{h(6a + 3h - 4)}{h} = \lim_{h \to 0} (6a + 3h - 4) = 6a - 4$$

27. Use (4) with $f(t) = (2t+1)/(t+3)$.

$$f'(a) = \lim_{h \to 0} \frac{f(a+h) - f(a)}{h} = \lim_{h \to 0} \frac{\dfrac{2(a+h)+1}{(a+h)+3} - \dfrac{2a+1}{a+3}}{h} = \lim_{h \to 0} \frac{(2a+2h+1)(a+3) - (2a+1)(a+h+3)}{h(a+h+3)(a+3)}$$

$$= \lim_{h \to 0} \frac{(2a^2 + 6a + 2ah + 6h + a + 3) - (2a^2 + 2ah + 6a + a + h + 3)}{h(a+h+3)(a+3)}$$

$$= \lim_{h \to 0} \frac{5h}{h(a+h+3)(a+3)} = \lim_{h \to 0} \frac{5}{(a+h+3)(a+3)} = \frac{5}{(a+3)^2}$$

29. Use (4) with $f(x) = \sqrt{1-2x}$.

$$f'(a) = \lim_{h \to 0} \frac{f(a+h) - f(a)}{h} = \lim_{h \to 0} \frac{\sqrt{1-2(a+h)} - \sqrt{1-2a}}{h}$$

$$= \lim_{h \to 0} \frac{\sqrt{1-2(a+h)} - \sqrt{1-2a}}{h} \cdot \frac{\sqrt{1-2(a+h)} + \sqrt{1-2a}}{\sqrt{1-2(a+h)} + \sqrt{1-2a}} = \lim_{h \to 0} \frac{\left(\sqrt{1-2(a+h)}\right)^2 - \left(\sqrt{1-2a}\right)^2}{h\left(\sqrt{1-2(a+h)} + \sqrt{1-2a}\right)}$$

$$= \lim_{h \to 0} \frac{(1-2a-2h) - (1-2a)}{h\left(\sqrt{1-2(a+h)} + \sqrt{1-2a}\right)} = \lim_{h \to 0} \frac{-2h}{h\left(\sqrt{1-2(a+h)} + \sqrt{1-2a}\right)}$$

$$= \lim_{h \to 0} \frac{-2}{\sqrt{1-2(a+h)} + \sqrt{1-2a}} = \frac{-2}{\sqrt{1-2a} + \sqrt{1-2a}} = \frac{-2}{2\sqrt{1-2a}} = \frac{-1}{\sqrt{1-2a}}$$

Note that the answers to Exercises 31 – 36 are not unique.

31. By (4), $\displaystyle\lim_{h \to 0} \frac{(1+h)^{10} - 1}{h} = f'(1)$, where $f(x) = x^{10}$ and $a = 1$.

Or: By (4), $\displaystyle\lim_{h \to 0} \frac{(1+h)^{10} - 1}{h} = f'(0)$, where $f(x) = (1+x)^{10}$ and $a = 0$.

33. By Equation 5, $\displaystyle\lim_{x \to 5} \frac{2^x - 32}{x - 5} = f'(5)$, where $f(x) = 2^x$ and $a = 5$.

35. By (4), $\displaystyle\lim_{h \to 0} \frac{\cos(\pi + h) + 1}{h} = f'(\pi)$, where $f(x) = \cos x$ and $a = \pi$.

Or: By (4), $\displaystyle\lim_{h \to 0} \frac{\cos(\pi + h) + 1}{h} = f'(0)$, where $f(x) = \cos(\pi + x)$ and $a = 0$.

37. The sketch shows the graph for a room temperature of $72°$ and a refrigerator temperature of $38°$. The initial rate of change is greater in magnitude than the rate of change after an hour.

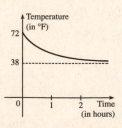

39. (a) (i) $[2002, 2006]$: $\dfrac{N(2006) - N(2002)}{2006 - 2002} = \dfrac{233 - 141}{4} = \dfrac{92}{4} = 23$ millions of cell phone subscribers per year

(ii) [2002, 2004]: $\dfrac{N(2004) - N(2002)}{2004 - 2002} = \dfrac{182 - 141}{2} = \dfrac{41}{2} = 20.5$ millions of cell phone subscribers per year

(iii) [2000, 2002]: $\dfrac{N(2002) - N(2000)}{2002 - 2000} = \dfrac{141 - 109}{2} = \dfrac{32}{2} = 16$ millions of cell phone subscribers per year

(b) Using the values from (ii) and (iii), we have $\dfrac{20.5 + 16}{2} = 18.25$ millions of cell phone subscribers per year.

(c) Estimating A as $(2000, 107)$ and B as $(2004, 175)$, the slope at 2002

is $\dfrac{175 - 107}{2004 - 2000} = \dfrac{68}{4} = 17$ millions of cell phone subscribers per

year.

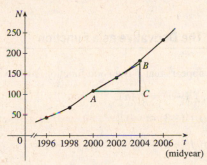

41. (a) (i) $\dfrac{\Delta C}{\Delta x} = \dfrac{C(105) - C(100)}{105 - 100} = \dfrac{6601.25 - 6500}{5} = \$20.25/\text{unit.}$

(ii) $\dfrac{\Delta C}{\Delta x} = \dfrac{C(101) - C(100)}{101 - 100} = \dfrac{6520.05 - 6500}{1} = \$20.05/\text{unit.}$

(b) $\dfrac{C(100 + h) - C(100)}{h} = \dfrac{\left[5000 + 10(100 + h) + 0.05(100 + h)^2\right] - 6500}{h} = \dfrac{20h + 0.05h^2}{h}$

$= 20 + 0.05h, \ h \neq 0$

So the instantaneous rate of change is $\displaystyle\lim_{h \to 0} \dfrac{C(100 + h) - C(100)}{h} = \lim_{h \to 0}(20 + 0.05h) = \$20/\text{unit.}$

43. (a) $f'(x)$ is the rate of change of the production cost with respect to the number of ounces of gold produced. Its units are

dollars per ounce.

(b) After 800 ounces of gold have been produced, the rate at which the production cost is increasing is $\$17/\text{ounce}$. So the cost

of producing the 800th (or 801st) ounce is about $\$17$.

(c) In the short term, the values of $f'(x)$ will decrease because more efficient use is made of start-up costs as x increases. But

eventually $f'(x)$ might increase due to large-scale operations.

45. $T'(8)$ is the rate at which the temperature is changing at 8:00 AM. To estimate the value of $T'(8)$, we will average the

difference quotients obtained using the times $t = 6$ and $t = 10$.

Let $A = \dfrac{T(6) - T(8)}{6 - 8} = \dfrac{75 - 84}{-2} = 4.5$ and $B = \dfrac{T(10) - T(8)}{10 - 8} = \dfrac{90 - 84}{2} = 3$. Then

$T'(8) = \displaystyle\lim_{t \to 8} \dfrac{T(t) - T(8)}{t - 8} \approx \dfrac{A + B}{2} = \dfrac{4.5 + 3}{2} = 3.75°\text{F/h.}$

47. (a) $S'(T)$ is the rate at which the oxygen solubility changes with respect to the water temperature. Its units are $(\text{mg/L})/°\text{C}$.

(b) For $T = 16°\text{C}$, it appears that the tangent line to the curve goes through the points $(0, 14)$ and $(32, 6)$. So

$S'(16) \approx \dfrac{6 - 14}{32 - 0} = -\dfrac{8}{32} = -0.25 \ (\text{mg/L})/°\text{C}$. This means that as the temperature increases past $16°\text{C}$, the oxygen

solubility is decreasing at a rate of $0.25 \ (\text{mg/L})/°\text{C}$.

49. Since $f(x) = x\sin(1/x)$ when $x \neq 0$ and $f(0) = 0$, we have

$$f'(0) = \lim_{h \to 0} \frac{f(0 + h) - f(0)}{h} = \lim_{h \to 0} \frac{h\sin(1/h) - 0}{h} = \lim_{h \to 0} \sin(1/h). \text{ This limit does not exist since } \sin(1/h) \text{ takes the}$$

values -1 and 1 on any interval containing 0. (Compare with Example 5 in Section 1.3.)

2.2   The Derivative as a Function

1. It appears that f is an odd function, so f' will be an even function—that

is, $f'(-a) = f'(a)$.

(a) $f'(-3) \approx -0.2$

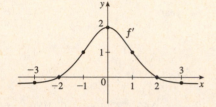

(b) $f'(-2) \approx 0$ (c) $f'(-1) \approx 1$ (d) $f'(0) \approx 2$

(e) $f'(1) \approx 1$ (f) $f'(2) \approx 0$ (g) $f'(3) \approx -0.2$

3. (a)$' =$ II, since from left to right, the slopes of the tangents to graph (a) start out negative, become 0, then positive, then 0, then

negative again. The actual function values in graph II follow the same pattern.

(b)$' =$ IV, since from left to right, the slopes of the tangents to graph (b) start out at a fixed positive quantity, then suddenly

become negative, then positive again. The discontinuities in graph IV indicate sudden changes in the slopes of the tangents.

(c)$' =$ I, since the slopes of the tangents to graph (c) are negative for $x < 0$ and positive for $x > 0$, as are the function values of

graph I.

(d)$' =$ III, since from left to right, the slopes of the tangents to graph (d) are positive, then 0, then negative, then 0, then

positive, then 0, then negative again, and the function values in graph III follow the same pattern.

Hints for Exercises 4 –11: First plot x-intercepts on the graph of f' for any horizontal tangents on the graph of f. Look for any corners on the graph
of f—there will be a discontinuity on the graph of f'. On any interval where f has a tangent with positive (or negative) slope, the graph of f' will be
positive (or negative). If the graph of the function is linear, the graph of f' will be a horizontal line.

5.

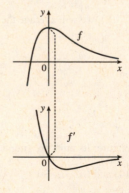

7.

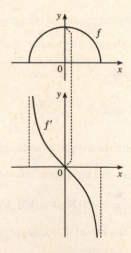

9.

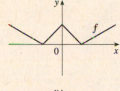

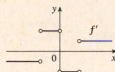

11.

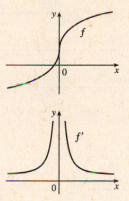

13. (a) $C'(t)$ is the instantaneous rate of change of percentage of full capacity with respect to elapsed time in hours.

   (b) The graph of $C'(t)$ tells us that the rate of change of percentage of full capacity is decreasing and approaching 0.

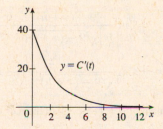

15. It appears that there are horizontal tangents on the graph of M for $t = 1963$ and $t = 1971$. Thus, there are zeros for those values of t on the graph of M'. The derivative is negative for the years 1963 to 1971.

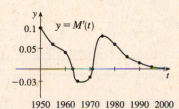

17. (a) By zooming in, we estimate that $f'(0) = 0$, $f'\left(\frac{1}{2}\right) = 1$, $f'(1) = 2$, and $f'(2) = 4$.

   (b) By symmetry, $f'(-x) = -f'(x)$. So $f'\left(-\frac{1}{2}\right) = -1$, $f'(-1) = -2$, and $f'(-2) = -4$.

   (c) It appears that $f'(x)$ is twice the value of x, so we guess that $f'(x) = 2x$.

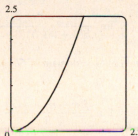

   (d) $f'(x) = \lim_{h \to 0} \dfrac{f(x+h) - f(x)}{h} = \lim_{h \to 0} \dfrac{(x+h)^2 - x^2}{h}$

   $= \lim_{h \to 0} \dfrac{(x^2 + 2hx + h^2) - x^2}{h} = \lim_{h \to 0} \dfrac{2hx + h^2}{h} = \lim_{h \to 0} \dfrac{h(2x + h)}{h} = \lim_{h \to 0} (2x + h) = 2x$

19. $f'(x) = \lim_{h \to 0} \dfrac{f(x+h) - f(x)}{h} = \lim_{h \to 0} \dfrac{\left[\frac{1}{2}(x+h) - \frac{1}{3}\right] - \left(\frac{1}{2}x - \frac{1}{3}\right)}{h} = \lim_{h \to 0} \dfrac{\frac{1}{2}x + \frac{1}{2}h - \frac{1}{3} - \frac{1}{2}x + \frac{1}{3}}{h}$

   $= \lim_{h \to 0} \dfrac{\frac{1}{2}h}{h} = \lim_{h \to 0} \frac{1}{2} = \frac{1}{2}$

Domain of f = domain of $f' = \mathbb{R}$.

21. $f'(x) = \lim\limits_{h \to 0} \dfrac{f(x+h) - f(x)}{h} = \lim\limits_{h \to 0} \dfrac{[(x+h)^2 - 2(x+h)^3] - (x^2 - 2x^3)}{h}$

$= \lim\limits_{h \to 0} \dfrac{x^2 + 2xh + h^2 - 2x^3 - 6x^2h - 6xh^2 - 2h^3 - x^2 + 2x^3}{h}$

$= \lim\limits_{h \to 0} \dfrac{2xh + h^2 - 6x^2h - 6xh^2 - 2h^3}{h} = \lim\limits_{h \to 0} \dfrac{h(2x + h - 6x^2 - 6xh - 2h^2)}{h}$

$= \lim\limits_{h \to 0} (2x + h - 6x^2 - 6xh - 2h^2) = 2x - 6x^2$

Domain of f = domain of f' = $\mathbb{R}$.

23. $g'(x) = \lim\limits_{h \to 0} \dfrac{g(x+h) - g(x)}{h} = \lim\limits_{h \to 0} \dfrac{\sqrt{9 - (x+h)} - \sqrt{9-x}}{h} \left[\dfrac{\sqrt{9 - (x+h)} + \sqrt{9-x}}{\sqrt{9 - (x+h)} + \sqrt{9-x}} \right]$

$= \lim\limits_{h \to 0} \dfrac{[9 - (x+h)] - (9-x)}{h\left[\sqrt{9 - (x+h)} + \sqrt{9-x}\right]} = \lim\limits_{h \to 0} \dfrac{-h}{h\left[\sqrt{9 - (x+h)} + \sqrt{9-x}\right]}$

$= \lim\limits_{h \to 0} \dfrac{-1}{\sqrt{9 - (x+h)} + \sqrt{9-x}} = \dfrac{-1}{2\sqrt{9-x}}$

Domain of $g = (-\infty, 9]$, domain of $g' = (-\infty, 9)$.

25. $G'(t) = \lim\limits_{h \to 0} \dfrac{G(t+h) - G(t)}{h} = \lim\limits_{h \to 0} \dfrac{\dfrac{1 - 2(t+h)}{3 + (t+h)} - \dfrac{1 - 2t}{3 + t}}{h}$

$= \lim\limits_{h \to 0} \dfrac{\dfrac{[1 - 2(t+h)](3+t) - [3 + (t+h)](1 - 2t)}{[3 + (t+h)](3+t)}}{h}$

$= \lim\limits_{h \to 0} \dfrac{3 + t - 6t - 2t^2 - 6h - 2ht - (3 - 6t + t - 2t^2 + h - 2ht)}{h[3 + (t+h)](3+t)} = \lim\limits_{h \to 0} \dfrac{-6h - h}{h(3 + t + h)(3 + t)}$

$= \lim\limits_{h \to 0} \dfrac{-7h}{h(3 + t + h)(3 + t)} = \lim\limits_{h \to 0} \dfrac{-7}{(3 + t + h)(3 + t)} = \dfrac{-7}{(3+t)^2}$

Domain of G = domain of $G' = (-\infty, -3) \cup (-3, \infty)$.

27. $f'(x) = \lim\limits_{h \to 0} \dfrac{f(x+h) - f(x)}{h} = \lim\limits_{h \to 0} \dfrac{(x+h)^4 - x^4}{h} = \lim\limits_{h \to 0} \dfrac{(x^4 + 4x^3h + 6x^2h^2 + 4xh^3 + h^4) - x^4}{h}$

$= \lim\limits_{h \to 0} \dfrac{4x^3h + 6x^2h^2 + 4xh^3 + h^4}{h} = \lim\limits_{h \to 0} \left(4x^3 + 6x^2h + 4xh^2 + h^3\right) = 4x^3$

Domain of f = domain of f' = $\mathbb{R}$.

29. (a) $f'(x) = \lim\limits_{h \to 0} \dfrac{f(x+h) - f(x)}{h} = \lim\limits_{h \to 0} \dfrac{[(x+h)^4 + 2(x+h)] - (x^4 + 2x)}{h}$

$= \lim\limits_{h \to 0} \dfrac{x^4 + 4x^3h + 6x^2h^2 + 4xh^3 + h^4 + 2x + 2h - x^4 - 2x}{h}$

$= \lim\limits_{h \to 0} \dfrac{4x^3h + 6x^2h^2 + 4xh^3 + h^4 + 2h}{h} = \lim\limits_{h \to 0} \dfrac{h(4x^3 + 6x^2h + 4xh^2 + h^3 + 2)}{h}$

$= \lim\limits_{h \to 0} (4x^3 + 6x^2h + 4xh^2 + h^3 + 2) = 4x^3 + 2$

(b) Notice that $f'(x) = 0$ when f has a horizontal tangent, $f'(x)$ is positive when the tangents have positive slope, and $f'(x)$ is negative when the tangents have negative slope.

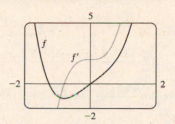

31. (a) $U'(t)$ is the rate at which the unemployment rate is changing with respect to time. Its units are percent per year.

(b) To find $U'(t)$, we use $\lim\limits_{h \to 0} \dfrac{U(t+h) - U(t)}{h} \approx \dfrac{U(t+h) - U(t)}{h}$ for small values of h.

For 1999: $U'(1999) \approx \dfrac{U(2000) - U(1999)}{2000 - 1999} = \dfrac{4.0 - 4.2}{1} = -0.2$

For 2000: We estimate $U'(2000)$ by using $h = -1$ and $h = 1$, and then average the two results to obtain a final estimate.

$h = -1 \quad \Rightarrow \quad U'(2000) \approx \dfrac{U(1999) - U(2000)}{1999 - 2000} = \dfrac{4.2 - 4.0}{-1} = -0.2;$

$h = 1 \quad \Rightarrow \quad U'(2000) \approx \dfrac{U(2001) - U(2000)}{2001 - 2000} = \dfrac{4.7 - 4.0}{1} = 0.7.$

So we estimate that $U'(2000) \approx \frac{1}{2}[(-0.2) + 0.7] = 0.25$.

t	1999	2000	2001	2002	2003	2004	2005	2006	2007	2008
$U'(t)$	−0.2	0.25	0.9	0.65	−0.15	−0.45	−0.45	−0.25	0.6	1.2

33. f is not differentiable at $x = -4$, because the graph has a corner there, and at $x = 0$, because there is a discontinuity there.

35. f is not differentiable at $x = -1$, because the graph has a vertical tangent there, and at $x = 4$, because the graph has a corner there.

37. As we zoom in toward $(-1, 0)$, the curve appears more and more like a straight line, so $f(x) = x + \sqrt{|x|}$ is differentiable at $x = -1$. But no matter how much we zoom in toward the origin, the curve doesn't straighten out—we can't eliminate the sharp point (a cusp). So f is not differentiable at $x = 0$.

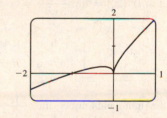

39. $a = f$, $b = f'$, $c = f''$. We can see this because where a has a horizontal tangent, $b = 0$, and where b has a horizontal tangent, $c = 0$. We can immediately see that c can be neither f nor f', since at the points where c has a horizontal tangent, neither a nor b is equal to 0.

41. We can immediately see that a is the graph of the acceleration function, since at the points where a has a horizontal tangent, neither c nor b is equal to 0. Next, we note that $a = 0$ at the point where b has a horizontal tangent, so b must be the graph of the velocity function, and hence, $b' = a$. We conclude that c is the graph of the position function.

43. $f'(x) = \lim\limits_{h \to 0} \dfrac{f(x+h) - f(x)}{h} = \lim\limits_{h \to 0} \dfrac{[3(x+h)^2 + 2(x+h) + 1] - (3x^2 + 2x + 1)}{h}$

$\quad = \lim\limits_{h \to 0} \dfrac{(3x^2 + 6xh + 3h^2 + 2x + 2h + 1) - (3x^2 + 2x + 1)}{h} = \lim\limits_{h \to 0} \dfrac{6xh + 3h^2 + 2h}{h}$

$\quad = \lim\limits_{h \to 0} \dfrac{h(6x + 3h + 2)}{h} = \lim\limits_{h \to 0} (6x + 3h + 2) = 6x + 2$

$f''(x) = \lim\limits_{h \to 0} \dfrac{f'(x+h) - f'(x)}{h} = \lim\limits_{h \to 0} \dfrac{[6(x+h) + 2] - (6x + 2)}{h} = \lim\limits_{h \to 0} \dfrac{(6x + 6h + 2) - (6x + 2)}{h}$

$\quad = \lim\limits_{h \to 0} \dfrac{6h}{h} = \lim\limits_{h \to 0} 6 = 6$

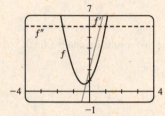

We see from the graph that our answers are reasonable because the graph of f' is that of a linear function and the graph of f'' is that of a constant function.

45. (a) Note that we have factored $x - a$ as the difference of two cubes in the third step.

$f'(a) = \lim\limits_{x \to a} \dfrac{f(x) - f(a)}{x - a} = \lim\limits_{x \to a} \dfrac{x^{1/3} - a^{1/3}}{x - a} = \lim\limits_{x \to a} \dfrac{x^{1/3} - a^{1/3}}{(x^{1/3} - a^{1/3})(x^{2/3} + x^{1/3}a^{1/3} + a^{2/3})}$

$\quad = \lim\limits_{x \to a} \dfrac{1}{x^{2/3} + x^{1/3}a^{1/3} + a^{2/3}} = \dfrac{1}{3a^{2/3}}$ or $\frac{1}{3}a^{-2/3}$

(b) $f'(0) = \lim\limits_{h \to 0} \dfrac{f(0+h) - f(0)}{h} = \lim\limits_{h \to 0} \dfrac{\sqrt[3]{h} - 0}{h} = \lim\limits_{h \to 0} \dfrac{1}{h^{2/3}}$. This function increases without bound, so the limit does not exist, and therefore $f'(0)$ does not exist.

(c) $\lim\limits_{x \to 0} |f'(x)| = \lim\limits_{x \to 0} \dfrac{1}{3x^{2/3}} = \infty$ and f is continuous at $x = 0$ (root function), so f has a vertical tangent at $x = 0$.

47. $f(x) = |x - 6| = \begin{cases} x - 6 & \text{if } x - 6 \geq 6 \\ -(x - 6) & \text{if } x - 6 < 0 \end{cases} = \begin{cases} x - 6 & \text{if } x \geq 6 \\ 6 - x & \text{if } x < 6 \end{cases}$

So the right-hand limit is $\lim\limits_{x \to 6^+} \dfrac{f(x) - f(6)}{x - 6} = \lim\limits_{x \to 6^+} \dfrac{|x - 6| - 0}{x - 6} = \lim\limits_{x \to 6^+} \dfrac{x - 6}{x - 6} = \lim\limits_{x \to 6^+} 1 = 1$, and the left-hand limit

is $\lim\limits_{x \to 6^-} \dfrac{f(x) - f(6)}{x - 6} = \lim\limits_{x \to 6^-} \dfrac{|x - 6| - 0}{x - 6} = \lim\limits_{x \to 6^-} \dfrac{6 - x}{x - 6} = \lim\limits_{x \to 6^-} (-1) = -1$. Since these limits are not equal,

$f'(6) = \lim\limits_{x \to 6} \dfrac{f(x) - f(6)}{x - 6}$ does not exist and f is not differentiable at 6.

However, a formula for f' is $f'(x) = \begin{cases} 1 & \text{if } x > 6 \\ -1 & \text{if } x < 6 \end{cases}$

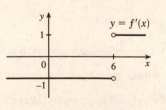

Another way of writing the formula is $f'(x) = \dfrac{x - 6}{|x - 6|}$.

49. (a) If f is even, then

$$f'(-x) = \lim_{h \to 0} \frac{f(-x+h) - f(-x)}{h} = \lim_{h \to 0} \frac{f[-(x-h)] - f(-x)}{h}$$

$$= \lim_{h \to 0} \frac{f(x-h) - f(x)}{h} = -\lim_{h \to 0} \frac{f(x-h) - f(x)}{-h} \qquad [\text{let } \Delta x = -h]$$

$$= -\lim_{\Delta x \to 0} \frac{f(x + \Delta x) - f(x)}{\Delta x} = -f'(x)$$

Therefore, f' is odd.

(b) If f is odd, then

$$f'(-x) = \lim_{h \to 0} \frac{f(-x+h) - f(-x)}{h} = \lim_{h \to 0} \frac{f[-(x-h)] - f(-x)}{h}$$

$$= \lim_{h \to 0} \frac{-f(x-h) + f(x)}{h} = \lim_{h \to 0} \frac{f(x-h) - f(x)}{-h} \qquad [\text{let } \Delta x = -h]$$

$$= \lim_{\Delta x \to 0} \frac{f(x + \Delta x) - f(x)}{\Delta x} = f'(x)$$

Therefore, f' is even.

51.

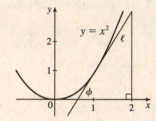

In the right triangle in the diagram, let Δy be the side opposite angle ϕ and Δx the side adjacent to angle ϕ. Then the slope of the tangent line ℓ is $m = \Delta y / \Delta x = \tan \phi$. Note that $0 < \phi < \frac{\pi}{2}$. We know (see Exercise 17) that the derivative of $f(x) = x^2$ is $f'(x) = 2x$. So the slope of the tangent to the curve at the point $(1, 1)$ is 2. Thus, ϕ is the angle between 0 and $\frac{\pi}{2}$ whose tangent is 2; that is, $\phi = \tan^{-1} 2 \approx 63°$.

2.3 Basic Differentiation Formulas

1. $f(x) = 2^{40}$ is a constant function, so its derivative is 0, that is, $f'(x) = 0$.

3. $f(t) = 2 - \frac{2}{3}t \;\Rightarrow\; f'(t) = 0 - \frac{2}{3} = -\frac{2}{3}$

5. $f(x) = x^3 - 4x + 6 \;\Rightarrow\; f'(x) = 3x^2 - 4(1) + 0 = 3x^2 - 4$

7. $f(x) = 3x^2 - 2\cos x \;\Rightarrow\; f'(x) = 6x - 2(-\sin x) = 6x + 2\sin x$

9. $g(x) = x^2(1 - 2x) = x^2 - 2x^3 \;\Rightarrow\; g'(x) = 2x - 2(3x^2) = 2x - 6x^2$

11. $g(t) = 2t^{-3/4} \;\Rightarrow\; g'(t) = 2(-\frac{3}{4}t^{-7/4}) = -\frac{3}{2}t^{-7/4}$

13. $A(s) = -\dfrac{12}{s^5} = -12s^{-5} \;\Rightarrow\; A'(s) = -12(-5s^{-6}) = 60s^{-6}$ or $60/s^6$

15. $R(a) = (3a + 1)^2 = 9a^2 + 6a + 1 \;\Rightarrow\; R'(a) = 9(2a) + 6(1) + 0 = 18a + 6$

17. $S(p) = \sqrt{p} - p = p^{1/2} - p \;\Rightarrow\; S'(p) = \frac{1}{2}p^{-1/2} - 1$ or $\dfrac{1}{2\sqrt{p}} - 1$

19. $y = \dfrac{x^2 + 4x + 3}{\sqrt{x}} = x^{3/2} + 4x^{1/2} + 3x^{-1/2} \;\; \Rightarrow$

$y' = \frac{3}{2}x^{1/2} + 4(\frac{1}{2})x^{-1/2} + 3(-\frac{1}{2})x^{-3/2} = \frac{3}{2}\sqrt{x} + \dfrac{2}{\sqrt{x}} - \dfrac{3}{2x\sqrt{x}}$ $\left[\text{note that } x^{3/2} = x^{2/2} \cdot x^{1/2} = x\sqrt{x}\right]$

The last expression can be written as $\dfrac{3x^2}{2x\sqrt{x}} + \dfrac{4x}{2x\sqrt{x}} - \dfrac{3}{2x\sqrt{x}} = \dfrac{3x^2 + 4x - 3}{2x\sqrt{x}}$.

21. $v = t^2 - \dfrac{1}{\sqrt[4]{t^3}} = t^2 - t^{-3/4} \;\; \Rightarrow \;\; v' = 2t - (-\frac{3}{4})t^{-7/4} = 2t + \dfrac{3}{4t^{7/4}} = 2t + \dfrac{3}{4t\sqrt[4]{t^3}}$

23. $z = \dfrac{A}{y^{10}} + B\cos y = Ay^{-10} + B\cos y \;\; \Rightarrow \;\; \dfrac{dz}{dy} = A(-10)y^{-11} + B(-\sin y) = -\dfrac{10A}{y^{11}} - B\sin y$

25. We first expand using the Binomial Theorem (see Reference Page 1).

$H(x) = (x + x^{-1})^3 = x^3 + 3x^2 x^{-1} + 3x(x^{-1})^2 + (x^{-1})^3 = x^3 + 3x + 3x^{-1} + x^{-3} \;\; \Rightarrow$

$H'(x) = 3x^2 + 3 + 3(-1x^{-2}) + (-3x^{-4}) = 3x^2 + 3 - 3x^{-2} - 3x^{-4}$

27. $y = 6\cos x \;\; \Rightarrow \;\; y' = -6\sin x$. At $(\pi/3, 3)$, $y' = -6\sin(\pi/3) = -6(\sqrt{3}/2) = -3\sqrt{3}$ and an equation of the tangent

line is $y - 3 = -3\sqrt{3}\,(x - \pi/3)$ or $y = -3\sqrt{3}x + 3 + \pi\sqrt{3}$. The slope of the normal line is $1/(3\sqrt{3})$ (the negative

reciprocal of $-3\sqrt{3}$) and an equation of the normal line is $y - 3 = \dfrac{1}{3\sqrt{3}}\left(x - \dfrac{\pi}{3}\right)$ or $y = \dfrac{1}{3\sqrt{3}}x + 3 - \dfrac{\pi}{9\sqrt{3}}$.

29. $y = 3x^2 - x^3 \;\; \Rightarrow \;\; y' = 6x - 3x^2$.

At $(1, 2)$, $y' = 6 - 3 = 3$, so an equation of the tangent line is

$y - 2 = 3(x - 1)$ or $y = 3x - 1$.

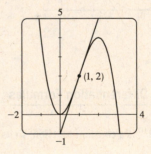

31. $f(x) = x^4 - 3x^3 + 16x \;\; \Rightarrow \;\; f'(x) = 4x^3 - 9x^2 + 16 \;\; \Rightarrow \;\; f''(x) = 12x^2 - 18x$

33. $g(t) = 2\cos t - 3\sin t \;\; \Rightarrow \;\; g'(t) = -2\sin t - 3\cos t \;\; \Rightarrow \;\; g''(t) = -2\cos t + 3\sin t$

35. $\dfrac{d}{dx}(\sin x) = \cos x \;\; \Rightarrow \;\; \dfrac{d^2}{dx^2}(\sin x) = -\sin x \;\; \Rightarrow \;\; \dfrac{d^3}{dx^3}(\sin x) = -\cos x \;\; \Rightarrow \;\; \dfrac{d^4}{dx^4}(\sin x) = \sin x$.

The derivatives of $\sin x$ occur in a cycle of four. Since $99 = 4(24) + 3$, we have $\dfrac{d^{99}}{dx^{99}}(\sin x) = \dfrac{d^3}{dx^3}(\sin x) = -\cos x$.

37. $f(x) = x + 2\sin x$ has a horizontal tangent when $f'(x) = 0 \;\; \Leftrightarrow \;\; 1 + 2\cos x = 0 \;\; \Leftrightarrow \;\; \cos x = -\frac{1}{2} \;\; \Leftrightarrow$

$x = \frac{2\pi}{3} + 2\pi n$ or $\frac{4\pi}{3} + 2\pi n$, where n is an integer. Note that $\frac{4\pi}{3}$ and $\frac{2\pi}{3}$ are $\pm\frac{\pi}{3}$ units from π. This allows us to write the

solutions in the more compact equivalent form $(2n + 1)\pi \pm \frac{\pi}{3}$, n an integer.

39. $y = 6x^3 + 5x - 3 \;\; \Rightarrow \;\; m = y' = 18x^2 + 5$, but $x^2 \geq 0$ for all x, so $m \geq 5$ for all x.

41. The slope of $y = x^2 - 5x + 4$ is given by $m = y' = 2x - 5$. The slope of $x - 3y = 5$ $\Leftrightarrow$ $y = \frac{1}{3}x - \frac{5}{3}$ is $\frac{1}{3}$,

so the desired normal line must have slope $\frac{1}{3}$, and hence, the tangent line to the parabola must have slope -3. This occurs if

$2x - 5 = -3$ $\Rightarrow$ $2x = 2$ $\Rightarrow$ $x = 1$. When $x = 1$, $y = 1^2 - 5(1) + 4 = 0$, and an equation of the normal line is

$y - 0 = \frac{1}{3}(x - 1)$ or $y = \frac{1}{3}x - \frac{1}{3}$.

43. (a) $s = t^3 - 3t$ $\Rightarrow$ $v(t) = s'(t) = 3t^2 - 3$ $\Rightarrow$ $a(t) = v'(t) = 6t$

(b) $a(2) = 6(2) = 12 \text{ m/s}^2$

(c) $v(t) = 3t^2 - 3 = 0$ when $t^2 = 1$, that is, $t = 1$ $[t \geq 0]$ and $a(1) = 6 \text{ m/s}^2$.

45. (a) $s = f(t) = t^3 - 12t^2 + 36t$ (in feet) $\Rightarrow$ $v(t) = f'(t) = 3t^2 - 24t + 36$ (in ft/s)

(b) $v(3) = 27 - 72 + 36 = -9 \text{ ft/s}$

(c) The particle is at rest when $v(t) = 0$. $3t^2 - 24t + 36 = 0$ $\Leftrightarrow$ $3(t - 2)(t - 6) = 0$ $\Leftrightarrow$ $t = 2$ s or 6 s.

(d) The particle is moving in the positive direction when $v(t) > 0$. $3(t - 2)(t - 6) > 0$ $\Leftrightarrow$ $0 \leq t < 2$ or $t > 6$.

(e) Since the particle is moving in the positive direction and in the

negative direction, we need to calculate the distance traveled in the

intervals $[0, 2]$, $[2, 6]$, and $[6, 8]$ separately.

$|f(2) - f(0)| = |32 - 0| = 32$.

$|f(6) - f(2)| = |0 - 32| = 32$.

$|f(8) - f(6)| = |32 - 0| = 32$.

The total distance is $32 + 32 + 32 = 96$ ft.

(f)

$$t = 8, \quad s = 32$$
$$t = 6, \quad s = 0$$
$$t = 0, \quad s = 0$$
$$t = 2, \quad s = 32$$

(g) $v(t) = 3t^2 - 24t + 36$ $\Rightarrow$

$a(t) = v'(t) = 6t - 24$.

$a(3) = 6(3) - 24 = -6 \text{ (ft/s)/s}$ or ft/s^2.

(h)

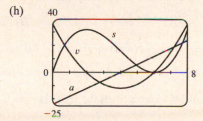

47. (a) $s(t) = t^3 - 4.5t^2 - 7t$ $\Rightarrow$ $v(t) = s'(t) = 3t^2 - 9t - 7 = 5$ $\Leftrightarrow$ $3t^2 - 9t - 12 = 0$ $\Leftrightarrow$

$3(t - 4)(t + 1) = 0$ $\Leftrightarrow$ $t = 4$ or -1. Since $t \geq 0$, the particle reaches a velocity of 5 m/s at $t = 4$ s.

(b) $a(t) = v'(t) = 6t - 9 = 0$ $\Leftrightarrow$ $t = 1.5$. The acceleration changes from negative to positive, so the velocity changes

from decreasing to increasing. Thus, at $t = 1.5$ s, the velocity has its minimum value.

49. (a) $h(t) = 15t - 1.86t^2$ $\Rightarrow$ $v(t) = h'(t) = 15 - 3.72t$. The velocity after 2 s is $v(2) = 15 - 3.72(2) = 7.56 \text{ m/s}$.

(b) $25 = h$ $\Leftrightarrow$ $1.86t^2 - 15t + 25 = 0$ $\Leftrightarrow$ $t = \dfrac{15 \pm \sqrt{15^2 - 4(1.86)(25)}}{2(1.86)}$ $\Leftrightarrow$ $t = t_1 \approx 2.35$ or $t = t_2 \approx 5.71$.

The velocities are $v(t_1) = 15 - 3.72t_1 \approx 6.24 \text{ m/s}$ [upward] and $v(t_2) = 15 - 3.72t_2 \approx -6.24 \text{ m/s}$ [downward].

51. (a) $C(x) = 1200 + 12x - 0.1x^2 + 0.0005x^3 \Rightarrow C'(x) = 12 - 0.2x + 0.0015x^2$ \$/yard, which is the marginal cost

function.

(b) $C'(200) = 12 - 0.2(200) + 0.0015(200)^2 = \$32$/yard, and this is the rate at which costs are increasing with respect to

the production level when $x = 200$. $C'(200)$ predicts the cost of producing the 201st yard.

(c) The cost of manufacturing the 201st yard of fabric is $C(201) - C(200) = 3632.2005 - 3600 \approx \$32.20$, which is

approximately $C'(200)$.

53. $S(r) = 4\pi r^2 \Rightarrow S'(r) = 8\pi r \Rightarrow$

(a) $S'(1) = 8\pi$ ft^2/ft   (b) $S'(2) = 16\pi$ ft^2/ft   (c) $S'(3) = 24\pi$ ft^2/ft

As the radius increases, the surface area grows at an increasing rate. In fact, the rate of change is linear with respect to the

radius.

55. (a) $P = \dfrac{k}{V}$ and $P = 50$ when $V = 0.106$, so $k = PV = 50(0.106) = 5.3$. Thus, $P = \dfrac{5.3}{V}$ and $V = \dfrac{5.3}{P}$.

(b) $V = 5.3P^{-1} \Rightarrow \dfrac{dV}{dP} = 5.3(-1P^{-2}) = -\dfrac{5.3}{P^2}$. When $P = 50$, $\dfrac{dV}{dP} = -\dfrac{5.3}{50^2} = -0.00212$. The derivative is the

instantaneous rate of change of the volume with respect to the pressure at $25\,^\circ$C. Its units are m^3/kPa.

57. $f'(x) = \lim\limits_{h \to 0} \dfrac{f(x+h) - f(x)}{h} = \lim\limits_{h \to 0} \dfrac{\dfrac{1}{x+h} - \dfrac{1}{x}}{h} = \lim\limits_{h \to 0} \dfrac{x - (x+h)}{hx(x+h)} = \lim\limits_{h \to 0} \dfrac{-h}{hx(x+h)} = \lim\limits_{h \to 0} \dfrac{-1}{x(x+h)} = -\dfrac{1}{x^2}$

59. $y = A\sin x + B\cos x \Rightarrow y' = A\cos x - B\sin x \Rightarrow y'' = -A\sin x - B\cos x$. Substituting these

expressions for y, y', and y'' into the given differential equation $y'' + y' - 2y = \sin x$ gives us

$(-A\sin x - B\cos x) + (A\cos x - B\sin x) - 2(A\sin x + B\cos x) = \sin x \Leftrightarrow$

$-3A\sin x - B\sin x + A\cos x - 3B\cos x = \sin x \Leftrightarrow (-3A - B)\sin x + (A - 3B)\cos x = 1\sin x$, so we must have

$-3A - B = 1$ and $A - 3B = 0$ (since 0 is the coefficient of $\cos x$ on the right side). Solving for A and B, we add the first

equation to three times the second to get $B = -\frac{1}{10}$ and $A = -\frac{3}{10}$.

61.

Let (a, a^2) be a point on the parabola at which the tangent line passes

through the point $(0, -4)$. The tangent line has slope $2a$ and equation

$y - (-4) = 2a(x - 0) \Leftrightarrow y = 2ax - 4$. Since (a, a^2) also lies on the

line, $a^2 = 2a(a) - 4$, or $a^2 = 4$. So $a = \pm 2$ and the points are $(2, 4)$

and $(-2, 4)$.

63. $y = f(x) = ax^2 \Rightarrow f'(x) = 2ax$. So the slope of the tangent to the parabola at $x = 2$ is $m = 2a(2) = 4a$. The slope

of the given line, $2x + y = b \Leftrightarrow y = -2x + b$, is seen to be -2, so we must have $4a = -2 \Leftrightarrow a = -\frac{1}{2}$. So when

$x = 2$, the point in question has y-coordinate $-\frac{1}{2} \cdot 2^2 = -2$. Now we simply require that the given line, whose equation is

$2x + y = b$, pass through the point $(2, -2)$: $2(2) + (-2) = b$ ⇔ $b = 2$. So we must have $a = -\frac{1}{2}$ and $b = 2$.

65. $y = f(x) = ax^3 + bx^2 + cx + d$ ⇒ $f'(x) = 3ax^2 + 2bx + c$. The point $(-2, 6)$ is on f, so $f(-2) = 6$ ⇒

$-8a + 4b - 2c + d = 6$ **(1)**. The point $(2, 0)$ is on f, so $f(2) = 0$ ⇒ $8a + 4b + 2c + d = 0$ **(2)**. Since there are

horizontal tangents at $(-2, 6)$ and $(2, 0)$, $f'(\pm 2) = 0$. $f'(-2) = 0$ ⇒ $12a - 4b + c = 0$ **(3)** and $f'(2) = 0$ ⇒

$12a + 4b + c = 0$ **(4)**. Subtracting equation **(3)** from **(4)** gives $8b = 0$ ⇒ $b = 0$. Adding **(1)** and **(2)** gives $8b + 2d = 6$,

so $d = 3$ since $b = 0$. From **(3)** we have $c = -12a$, so **(2)** becomes $8a + 4(0) + 2(-12a) + 3 = 0$ ⇒ $3 = 16a$ ⇒

$a = \frac{3}{16}$. Now $c = -12a = -12\left(\frac{3}{16}\right) = -\frac{9}{4}$ and the desired cubic function is $y = \frac{3}{16}x^3 - \frac{9}{4}x + 3$.

67. Substituting $x = 1$ and $y = 1$ into $y = ax^2 + bx$ gives us $a + b = 1$ **(1)**. The slope of the tangent line $y = 3x - 2$ is 3 and the

slope of the tangent to the parabola at (x, y) is $y' = 2ax + b$. At $x = 1$, $y' = 3$ ⇒ $3 = 2a + b$ **(2)**. Subtracting **(1)** from

(2) gives us $2 = a$ and it follows that $b = -1$. The parabola has equation $y = 2x^2 - x$.

69. *Solution 1:*   Let $f(x) = x^{1000}$. Then, by the definition of a derivative, $f'(1) = \lim\limits_{x \to 1} \dfrac{f(x) - f(1)}{x - 1} = \lim\limits_{x \to 1} \dfrac{x^{1000} - 1}{x - 1}$.

But this is just the limit we want to find, and we know (from the Power Rule) that $f'(x) = 1000x^{999}$, so

$f'(1) = 1000(1)^{999} = 1000$. So $\lim\limits_{x \to 1} \dfrac{x^{1000} - 1}{x - 1} = 1000$.

Solution 2:   Note that $(x^{1000} - 1) = (x - 1)(x^{999} + x^{998} + x^{997} + \cdots + x^2 + x + 1)$. So

$$\lim_{x \to 1} \frac{x^{1000} - 1}{x - 1} = \lim_{x \to 1} \frac{(x - 1)(x^{999} + x^{998} + x^{997} + \cdots + x^2 + x + 1)}{x - 1} = \lim_{x \to 1}(x^{999} + x^{998} + x^{997} + \cdots + x^2 + x + 1)$$

$$= \underbrace{1 + 1 + 1 + \cdots + 1 + 1 + 1}_{\text{1000 ones}} = 1000, \text{ as above.}$$

71. $y = x^2$ ⇒ $y' = 2x$, so the slope of a tangent line at the point (a, a^2) is $y' = 2a$ and the slope of a normal line is $-1/(2a)$,

for $a \neq 0$. The slope of the normal line through the points (a, a^2) and $(0, c)$ is $\dfrac{a^2 - c}{a - 0}$, so $\dfrac{a^2 - c}{a} = -\dfrac{1}{2a}$ ⇒

$a^2 - c = -\frac{1}{2}$ ⇒ $a^2 = c - \frac{1}{2}$. The last equation has two solutions if $c > \frac{1}{2}$, one solution if $c = \frac{1}{2}$, and no solution if

$c < \frac{1}{2}$. Since the y-axis is normal to $y = x^2$ regardless of the value of c (this is the case for $a = 0$), we have three normal lines

if $c > \frac{1}{2}$ and one normal line if $c \leq \frac{1}{2}$.

2.4   The Product and Quotient Rules

1. Product Rule: $f(x) = (1 + 2x^2)(x - x^2)$ ⇒

$$f'(x) = (1 + 2x^2)(1 - 2x) + (x - x^2)(4x) = 1 - 2x + 2x^2 - 4x^3 + 4x^2 - 4x^3 = 1 - 2x + 6x^2 - 8x^3.$$

Multiplying first: $f(x) = (1 + 2x^2)(x - x^2) = x - x^2 + 2x^3 - 2x^4$ ⇒ $f'(x) = 1 - 2x + 6x^2 - 8x^3$ (equivalent).

3. $g(t) = t^3 \cos t$ ⇒ $g'(t) = t^3(-\sin t) + (\cos t) \cdot 3t^2 = 3t^2 \cos t - t^3 \sin t$ or $t^2(3 \cos t - t \sin t)$

5. $F(y) = \left(\dfrac{1}{y^2} - \dfrac{3}{y^4}\right)(y + 5y^3) = (y^{-2} - 3y^{-4})(y + 5y^3) \overset{PR}{\Rightarrow}$

$\quad F'(y) = (y^{-2} - 3y^{-4})(1 + 15y^2) + (y + 5y^3)(-2y^{-3} + 12y^{-5})$

$\qquad = (y^{-2} + 15 - 3y^{-4} - 45y^{-2}) + (-2y^{-2} + 12y^{-4} - 10 + 60y^{-2})$

$\qquad = 5 + 14y^{-2} + 9y^{-4} \text{ or } 5 + 14/y^2 + 9/y^4$

7. $f(x) = \sin x + \frac{1}{2}\cot x \Rightarrow f'(x) = \cos x - \frac{1}{2}\csc^2 x$

9. $h(\theta) = \theta\csc\theta - \cot\theta \Rightarrow h'(\theta) = \theta(-\csc\theta\cot\theta) + (\csc\theta)\cdot 1 - (-\csc^2\theta) = \csc\theta - \theta\csc\theta\cot\theta + \csc^2\theta$

The notations $\overset{PR}{\Rightarrow}$ and $\overset{QR}{\Rightarrow}$ indicate the use of the Product and Quotient Rules, respectively.

11. $g(x) = \dfrac{1 + 2x}{3 - 4x} \overset{QR}{\Rightarrow} g'(x) = \dfrac{(3 - 4x)(2) - (1 + 2x)(-4)}{(3 - 4x)^2} = \dfrac{6 - 8x + 4 + 8x}{(3 - 4x)^2} = \dfrac{10}{(3 - 4x)^2}$

13. $y = \dfrac{x^3}{1 - x^2} \overset{QR}{\Rightarrow} y' = \dfrac{(1 - x^2)(3x^2) - x^3(-2x)}{(1 - x^2)^2} = \dfrac{x^2(3 - 3x^2 + 2x^2)}{(1 - x^2)^2} = \dfrac{x^2(3 - x^2)}{(1 - x^2)^2}$

15. $y = \dfrac{v^3 - 2v\sqrt{v}}{v} = v^2 - 2\sqrt{v} = v^2 - 2v^{1/2} \Rightarrow y' = 2v - 2\left(\frac{1}{2}\right)v^{-1/2} = 2v - v^{-1/2}.$

We can change the form of the answer as follows: $2v - v^{-1/2} = 2v - \dfrac{1}{\sqrt{v}} = \dfrac{2v\sqrt{v} - 1}{\sqrt{v}} = \dfrac{2v^{3/2} - 1}{\sqrt{v}}$

17. $f(t) = \dfrac{2t}{2 + \sqrt{t}} \overset{QR}{\Rightarrow} f'(t) = \dfrac{(2 + t^{1/2})(2) - 2t\left(\frac{1}{2}t^{-1/2}\right)}{(2 + \sqrt{t})^2} = \dfrac{4 + 2t^{1/2} - t^{1/2}}{(2 + \sqrt{t})^2} = \dfrac{4 + t^{1/2}}{(2 + \sqrt{t})^2} \text{ or } \dfrac{4 + \sqrt{t}}{(2 + \sqrt{t})^2}$

19. $y = \dfrac{x}{2 - \tan x} \Rightarrow y' = \dfrac{(2 - \tan x)(1) - x(-\sec^2 x)}{(2 - \tan x)^2} = \dfrac{2 - \tan x + x\sec^2 x}{(2 - \tan x)^2}$

21. $f(\theta) = \dfrac{\sec\theta}{1 + \sec\theta} \Rightarrow$

$\quad f'(\theta) = \dfrac{(1 + \sec\theta)(\sec\theta\tan\theta) - (\sec\theta)(\sec\theta\tan\theta)}{(1 + \sec\theta)^2} = \dfrac{(\sec\theta\tan\theta)\left[(1 + \sec\theta) - \sec\theta\right]}{(1 + \sec\theta)^2} = \dfrac{\sec\theta\tan\theta}{(1 + \sec\theta)^2}$

23. $y = \dfrac{t\sin t}{1 + t} \Rightarrow$

$\quad y' = \dfrac{(1 + t)(t\cos t + \sin t) - t\sin t(1)}{(1 + t)^2} = \dfrac{t\cos t + \sin t + t^2\cos t + t\sin t - t\sin t}{(1 + t)^2} = \dfrac{(t^2 + t)\cos t + \sin t}{(1 + t)^2}$

25. $f(x) = \dfrac{x}{x + c/x} \Rightarrow f'(x) = \dfrac{(x + c/x)(1) - x(1 - c/x^2)}{\left(x + \dfrac{c}{x}\right)^2} = \dfrac{x + c/x - x + c/x}{\left(\dfrac{x^2 + c}{x}\right)^2} = \dfrac{2c/x}{\dfrac{(x^2 + c)^2}{x^2}} \cdot \dfrac{x^2}{x^2} = \dfrac{2cx}{(x^2 + c)^2}$

27. $y = \dfrac{x^2 - 1}{x^2 + x + 1} \Rightarrow$

$\quad y' = \dfrac{(x^2 + x + 1)(2x) - (x^2 - 1)(2x + 1)}{(x^2 + x + 1)^2} = \dfrac{2x^3 + 2x^2 + 2x - 2x^3 - x^2 + 2x + 1}{(x^2 + x + 1)^2} = \dfrac{x^2 + 4x + 1}{(x^2 + x + 1)^2}.$

At $(1, 0)$, $y' = \dfrac{6}{3^2} = \dfrac{2}{3}$, and an equation of the tangent line is $y - 0 = \frac{2}{3}(x - 1)$, or $y = \frac{2}{3}x - \frac{2}{3}$.

29. $y = \cos x - \sin x \;\Rightarrow\; y' = -\sin x - \cos x$, so $y'(\pi) = -\sin \pi - \cos \pi = 0 - (-1) = 1.$ An equation of the tangent line to the curve $y = \cos x - \sin x$ at the point $(\pi, -1)$ is $y - (-1) = 1(x - \pi)$ or $y = x - \pi - 1.$

31. (a) $y = f(x) = \dfrac{1}{1 + x^2} \;\Rightarrow$

$f'(x) = \dfrac{(1 + x^2)(0) - 1(2x)}{(1 + x^2)^2} = \dfrac{-2x}{(1 + x^2)^2}.$ So the slope of the

tangent line at the point $\left(-1, \tfrac{1}{2}\right)$ is $f'(-1) = \dfrac{2}{2^2} = \tfrac{1}{2}$ and its

equation is $y - \tfrac{1}{2} = \tfrac{1}{2}(x + 1)$ or $y = \tfrac{1}{2}x + 1.$

(b)

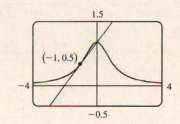

33. $f(x) = \dfrac{x^2}{1 + x} \;\Rightarrow\; f'(x) = \dfrac{(1 + x)(2x) - x^2(1)}{(1 + x)^2} = \dfrac{2x + 2x^2 - x^2}{(1 + x)^2} = \dfrac{x^2 + 2x}{x^2 + 2x + 1} \;\Rightarrow$

$f''(x) = \dfrac{(x^2 + 2x + 1)(2x + 2) - (x^2 + 2x)(2x + 2)}{(x^2 + 2x + 1)^2} = \dfrac{(2x + 2)(x^2 + 2x + 1 - x^2 - 2x)}{[(x+1)^2]^2}$

$= \dfrac{2(x + 1)(1)}{(x + 1)^4} = \dfrac{2}{(x + 1)^3},$

so $f''(1) = \dfrac{2}{(1 + 1)^3} = \dfrac{2}{8} = \dfrac{1}{4}.$

35. $H(\theta) = \theta \sin \theta \;\Rightarrow\; H'(\theta) = \theta(\cos \theta) + (\sin \theta) \cdot 1 = \theta \cos \theta + \sin \theta \;\Rightarrow$

$H''(\theta) = \theta(-\sin \theta) + (\cos \theta) \cdot 1 + \cos \theta = -\theta \sin \theta + 2 \cos \theta$

37. $\dfrac{d}{dx}(\csc x) = \dfrac{d}{dx}\left(\dfrac{1}{\sin x}\right) = \dfrac{(\sin x)(0) - 1(\cos x)}{\sin^2 x} = \dfrac{-\cos x}{\sin^2 x} = -\dfrac{1}{\sin x} \cdot \dfrac{\cos x}{\sin x} = -\csc x \cot x$

39. $\dfrac{d}{dx}(\cot x) = \dfrac{d}{dx}\left(\dfrac{\cos x}{\sin x}\right) = \dfrac{(\sin x)(-\sin x) - (\cos x)(\cos x)}{\sin^2 x} = -\dfrac{\sin^2 x + \cos^2 x}{\sin^2 x} = -\dfrac{1}{\sin^2 x} = -\csc^2 x$

41. We are given that $f(5) = 1$, $f'(5) = 6$, $g(5) = -3$, and $g'(5) = 2.$

(a) $(fg)'(5) = f(5)g'(5) + g(5)f'(5) = (1)(2) + (-3)(6) = 2 - 18 = -16$

(b) $\left(\dfrac{f}{g}\right)'(5) = \dfrac{g(5)f'(5) - f(5)g'(5)}{[g(5)]^2} = \dfrac{(-3)(6) - (1)(2)}{(-3)^2} = -\dfrac{20}{9}$

(c) $\left(\dfrac{g}{f}\right)'(5) = \dfrac{f(5)g'(5) - g(5)f'(5)}{[f(5)]^2} = \dfrac{(1)(2) - (-3)(6)}{(1)^2} = 20$

43. (a) From the graphs of f and g, we obtain the following values: $f(1) = 2$ since the point $(1, 2)$ is on the graph of f;

$g(1) = 1$ since the point $(1, 1)$ is on the graph of g; $f'(1) = 2$ since the slope of the line segment between $(0, 0)$ and

$(2, 4)$ is $\dfrac{4 - 0}{2 - 0} = 2$; $g'(1) = -1$ since the slope of the line segment between $(-2, 4)$ and $(2, 0)$ is $\dfrac{0 - 4}{2 - (-2)} = -1.$

Now $u(x) = f(x)g(x)$, so $u'(1) = f(1)g'(1) + g(1)f'(1) = 2 \cdot (-1) + 1 \cdot 2 = 0.$

(b) $v(x) = f(x)/g(x)$, so $v'(5) = \dfrac{g(5)f'(5) - f(5)g'(5)}{[g(5)]^2} = \dfrac{2\left(-\tfrac{1}{3}\right) - 3 \cdot \tfrac{2}{3}}{2^2} = \dfrac{-\tfrac{8}{3}}{4} = -\dfrac{2}{3}$

45. (a) $y = xg(x) \quad \Rightarrow \quad y' = xg'(x) + g(x) \cdot 1 = xg'(x) + g(x)$

(b) $y = \dfrac{x}{g(x)} \quad \Rightarrow \quad y' = \dfrac{g(x) \cdot 1 - xg'(x)}{[g(x)]^2} = \dfrac{g(x) - xg'(x)}{[g(x)]^2}$

(c) $y = \dfrac{g(x)}{x} \quad \Rightarrow \quad y' = \dfrac{xg'(x) - g(x) \cdot 1}{(x)^2} = \dfrac{xg'(x) - g(x)}{x^2}$

47. If $y = f(x) = \dfrac{x}{x+1}$, then $f'(x) = \dfrac{(x+1)(1) - x(1)}{(x+1)^2} = \dfrac{1}{(x+1)^2}$. When $x = a$, the equation of the tangent line is

$y - \dfrac{a}{a+1} = \dfrac{1}{(a+1)^2}(x - a)$. This line passes through $(1, 2)$ when $2 - \dfrac{a}{a+1} = \dfrac{1}{(a+1)^2}(1 - a) \quad \Leftrightarrow$

$2(a+1)^2 - a(a+1) = 1 - a \quad \Leftrightarrow \quad 2a^2 + 4a + 2 - a^2 - a - 1 + a = 0 \quad \Leftrightarrow \quad a^2 + 4a + 1 = 0$.

The quadratic formula gives the roots of this equation as $a = \dfrac{-4 \pm \sqrt{4^2 - 4(1)(1)}}{2(1)} = \dfrac{-4 \pm \sqrt{12}}{2} = -2 \pm \sqrt{3}$,

so there are two such tangent lines. Since

$$f\left(-2 \pm \sqrt{3}\right) = \dfrac{-2 \pm \sqrt{3}}{-2 \pm \sqrt{3} + 1} = \dfrac{-2 \pm \sqrt{3}}{-1 \pm \sqrt{3}} \cdot \dfrac{-1 \mp \sqrt{3}}{-1 \mp \sqrt{3}}$$

$$= \dfrac{2 \pm 2\sqrt{3} \mp \sqrt{3} - 3}{1 - 3} = \dfrac{-1 \pm \sqrt{3}}{-2} = \dfrac{1 \mp \sqrt{3}}{2},$$

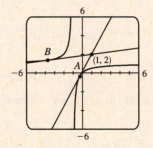

the lines touch the curve at $A\left(-2 + \sqrt{3}, \frac{1 - \sqrt{3}}{2}\right) \approx (-0.27, -0.37)$

and $B\left(-2 - \sqrt{3}, \frac{1 + \sqrt{3}}{2}\right) \approx (-3.73, 1.37)$.

49. $R = \dfrac{f}{g} \quad \Rightarrow \quad R' = \dfrac{gf' - fg'}{g^2}$. For $f(x) = x - 3x^3 + 5x^5$, $f'(x) = 1 - 9x^2 + 25x^4$,

and for $g(x) = 1 + 3x^3 + 6x^6 + 9x^9$, $g'(x) = 9x^2 + 36x^5 + 81x^8$.

Thus, $R'(0) = \dfrac{g(0)f'(0) - f(0)g'(0)}{[g(0)]^2} = \dfrac{1 \cdot 1 - 0 \cdot 0}{1^2} = \dfrac{1}{1} = 1$.

51. (a) $x(t) = 8 \sin t \quad \Rightarrow \quad v(t) = x'(t) = 8 \cos t \quad \Rightarrow \quad a(t) = x''(t) = -8 \sin t$

(b) The mass at time $t = \frac{2\pi}{3}$ has position $x\left(\frac{2\pi}{3}\right) = 8 \sin \frac{2\pi}{3} = 8\left(\frac{\sqrt{3}}{2}\right) = 4\sqrt{3}$, velocity $v\left(\frac{2\pi}{3}\right) = 8 \cos \frac{2\pi}{3} = 8\left(-\frac{1}{2}\right) = -4$,

and acceleration $a\left(\frac{2\pi}{3}\right) = -8 \sin \frac{2\pi}{3} = -8\left(\frac{\sqrt{3}}{2}\right) = -4\sqrt{3}$. Since $v\left(\frac{2\pi}{3}\right) < 0$, the particle is moving to the left. Because

v and a have the same sign, the particle is speeding up.

53. $PV = nRT \quad \Rightarrow \quad T = \dfrac{PV}{nR} = \dfrac{PV}{(10)(0.0821)} = \dfrac{1}{0.821}(PV)$. Using the Product Rule, we have

$\dfrac{dT}{dt} = \dfrac{1}{0.821}\left[P(t)V'(t) + V(t)P'(t)\right] = \dfrac{1}{0.821}\left[(8)(-0.15) + (10)(0.10)\right] \approx -0.2436 \text{ K/min}$.

55. (a) $(fgh)' = [(fg)h]' = (fg)'h + (fg)h' = (f'g + fg')h + (fg)h' = f'gh + fg'h + fgh'$

(b) $y = x \sin x \cos x$ $\Rightarrow$ $\dfrac{dy}{dx} = \sin x \cos x + x \cos x \cos x + x \sin x \, (-\sin x) = \sin x \cos x + x \cos^2 x - x \sin^2 x$

57. (a) $\dfrac{d}{dx}\left(\dfrac{1}{g(x)}\right) = \dfrac{g(x) \cdot \dfrac{d}{dx}(1) - 1 \cdot \dfrac{d}{dx}[g(x)]}{[g(x)]^2}$ [Quotient Rule] $= \dfrac{g(x) \cdot 0 - 1 \cdot g'(x)}{[g(x)]^2} = \dfrac{0 - g'(x)}{[g(x)]^2} = -\dfrac{g'(x)}{[g(x)]^2}$

(b) $y = \dfrac{1}{x^4 + x^2 + 1}$ $\Rightarrow$ $y' = -\dfrac{2x(2x^2 + 1)}{(x^4 + x^2 + 1)^2}$

(c) $\dfrac{d}{dx}(x^{-n}) = \dfrac{d}{dx}\left(\dfrac{1}{x^n}\right) = -\dfrac{(x^n)'}{(x^n)^2}$ [by the Reciprocal Rule] $= -\dfrac{nx^{n-1}}{x^{2n}} = -nx^{n-1-2n} = -nx^{-n-1}$

2.5   The Chain Rule

1. Let $u = g(x) = 1 + 4x$ and $y = f(u) = \sqrt[3]{u}$. Then $\dfrac{dy}{dx} = \dfrac{dy}{du}\dfrac{du}{dx} = \left(\tfrac{1}{3}u^{-2/3}\right)(4) = \dfrac{4}{3\sqrt[3]{(1+4x)^2}}$.

3. Let $u = g(x) = \pi x$ and $y = f(u) = \tan u$. Then $\dfrac{dy}{dx} = \dfrac{dy}{du}\dfrac{du}{dx} = (\sec^2 u)(\pi) = \pi \sec^2 \pi x$.

5. Let $u = g(x) = \sin x$ and $y = f(u) = \sqrt{u}$. Then $\dfrac{dy}{dx} = \dfrac{dy}{du}\dfrac{du}{dx} = \tfrac{1}{2}u^{-1/2}\cos x = \dfrac{\cos x}{2\sqrt{u}} = \dfrac{\cos x}{2\sqrt{\sin x}}$.

7. $F(x) = (x^4 + 3x^2 - 2)^5$ $\Rightarrow$ $F'(x) = 5(x^4 + 3x^2 - 2)^4 \cdot \dfrac{d}{dx}(x^4 + 3x^2 - 2) = 5(x^4 + 3x^2 - 2)^4(4x^3 + 6x)$

$\left[\text{or } 10x(x^4 + 3x^2 - 2)^4(2x^2 + 3)\right]$

9. $F(x) = \sqrt{1 - 2x} = (1 - 2x)^{1/2}$ $\Rightarrow$ $F'(x) = \tfrac{1}{2}(1 - 2x)^{-1/2}(-2) = -\dfrac{1}{\sqrt{1 - 2x}}$

11. $f(z) = \dfrac{1}{z^2 + 1} = (z^2 + 1)^{-1}$ $\Rightarrow$ $f'(z) = -1(z^2 + 1)^{-2}(2z) = -\dfrac{2z}{(z^2 + 1)^2}$

13. $y = \cos(a^3 + x^3)$ $\Rightarrow$ $y' = -\sin(a^3 + x^3) \cdot 3x^2$ [a^3 is just a constant] $= -3x^2 \sin(a^3 + x^3)$

15. Use the Product Rule. $y = x \sec kx$ $\Rightarrow$ $y' = x\,(\sec kx \, \tan kx \cdot k) + \sec kx \cdot 1 = \sec kx \,(kx \tan kx + 1)$

17. $f(x) = (2x - 3)^4(x^2 + x + 1)^5$ $\Rightarrow$

$f'(x) = (2x - 3)^4 \cdot 5(x^2 + x + 1)^4(2x + 1) + (x^2 + x + 1)^5 \cdot 4(2x - 3)^3 \cdot 2$

$\qquad = (2x - 3)^3(x^2 + x + 1)^4[(2x - 3) \cdot 5(2x + 1) + (x^2 + x + 1) \cdot 8]$

$\qquad = (2x - 3)^3(x^2 + x + 1)^4(20x^2 - 20x - 15 + 8x^2 + 8x + 8) = (2x - 3)^3(x^2 + x + 1)^4(28x^2 - 12x - 7)$

19. $h(t) = (t + 1)^{2/3}(2t^2 - 1)^3$ $\Rightarrow$

$h'(t) = (t + 1)^{2/3} \cdot 3(2t^2 - 1)^2 \cdot 4t + (2t^2 - 1)^3 \cdot \tfrac{2}{3}(t + 1)^{-1/3} = \tfrac{2}{3}(t + 1)^{-1/3}(2t^2 - 1)^2[18t(t + 1) + (2t^2 - 1)]$

$\qquad = \tfrac{2}{3}(t + 1)^{-1/3}(2t^2 - 1)^2(20t^2 + 18t - 1)$

21. $y = \left(\dfrac{x^2+1}{x^2-1}\right)^3 \;\Rightarrow$

$$y' = 3\left(\frac{x^2+1}{x^2-1}\right)^2 \cdot \frac{d}{dx}\left(\frac{x^2+1}{x^2-1}\right) = 3\left(\frac{x^2+1}{x^2-1}\right)^2 \cdot \frac{(x^2-1)(2x)-(x^2+1)(2x)}{(x^2-1)^2}$$

$$= 3\left(\frac{x^2+1}{x^2-1}\right)^2 \cdot \frac{2x[x^2-1-(x^2+1)]}{(x^2-1)^2} = 3\left(\frac{x^2+1}{x^2-1}\right)^2 \cdot \frac{2x(-2)}{(x^2-1)^2} = \frac{-12x(x^2+1)^2}{(x^2-1)^4}$$

23. $y = \sin(x\cos x) \;\Rightarrow\; y' = \cos(x\cos x)\cdot[x(-\sin x)+\cos x\cdot 1] = (\cos x - x\sin x)\cos(x\cos x)$

25. $y = \dfrac{r}{\sqrt{r^2+1}} \;\Rightarrow$

$$y' = \frac{\sqrt{r^2+1}\,(1) - r\cdot\frac12(r^2+1)^{-1/2}(2r)}{\left(\sqrt{r^2+1}\right)^2} = \frac{\sqrt{r^2+1}-\dfrac{r^2}{\sqrt{r^2+1}}}{\left(\sqrt{r^2+1}\right)^2} = \frac{\dfrac{\sqrt{r^2+1}\sqrt{r^2+1}-r^2}{\sqrt{r^2+1}}}{\left(\sqrt{r^2+1}\right)^2}$$

$$= \frac{(r^2+1)-r^2}{\left(\sqrt{r^2+1}\right)^3} = \frac{1}{(r^2+1)^{3/2}} \text{ or } (r^2+1)^{-3/2}$$

Another solution: Write y as a product and make use of the Product Rule. $y = r(r^2+1)^{-1/2} \;\Rightarrow$

$$y' = r\cdot -\tfrac12(r^2+1)^{-3/2}(2r) + (r^2+1)^{-1/2}\cdot 1 = (r^2+1)^{-3/2}[-r^2+(r^2+1)^1] = (r^2+1)^{-3/2}(1) = (r^2+1)^{-3/2}.$$

The step that students usually have trouble with is factoring out $(r^2+1)^{-3/2}$. But this is no different than factoring out x^2 from x^2+x^5; that is, we are just factoring out a factor with the *smallest* exponent that appears on it. In this case, $-\frac32$ is smaller than $-\frac12$.

27. $y = \sin\sqrt{1+x^2} \;\Rightarrow\; y' = \cos\sqrt{1+x^2}\cdot\frac12(1+x^2)^{-1/2}\cdot 2x = \left(x\cos\sqrt{1+x^2}\right)/\sqrt{1+x^2}$

29. $y = \sin(\tan 2x) \;\Rightarrow\; y' = \cos(\tan 2x)\cdot\frac{d}{dx}(\tan 2x) = \cos(\tan 2x)\cdot\sec^2(2x)\cdot\frac{d}{dx}(2x) = 2\cos(\tan 2x)\sec^2(2x)$

31. $y = \sec^2 x + \tan^2 x = (\sec x)^2 + (\tan x)^2 \;\Rightarrow$

$$y' = 2(\sec x)(\sec x\tan x) + 2(\tan x)(\sec^2 x) = 2\sec^2 x\tan x + 2\sec^2 x\tan x = 4\sec^2 x\tan x$$

33. $y = \left(\dfrac{1-\cos 2x}{1+\cos 2x}\right)^4 \;\Rightarrow$

$$y' = 4\left(\frac{1-\cos 2x}{1+\cos 2x}\right)^3 \cdot \frac{(1+\cos 2x)(2\sin 2x)+(1-\cos 2x)(-2\sin 2x)}{(1+\cos 2x)^2}$$

$$= 4\left(\frac{1-\cos 2x}{1+\cos 2x}\right)^3 \cdot \frac{2\sin 2x\,(1+\cos 2x+1-\cos 2x)}{(1+\cos 2x)^2} = \frac{4(1-\cos 2x)^3}{(1+\cos 2x)^3}\,\frac{2\sin 2x\,(2)}{(1+\cos 2x)^2} = \frac{16\sin 2x\,(1-\cos 2x)^3}{(1+\cos 2x)^5}$$

35. $y = \cot^2(\sin\theta) = [\cot(\sin\theta)]^2 \;\Rightarrow$

$$y' = 2[\cot(\sin\theta)]\cdot\frac{d}{d\theta}[\cot(\sin\theta)] = 2\cot(\sin\theta)\cdot[-\csc^2(\sin\theta)\cdot\cos\theta] = -2\cos\theta\cot(\sin\theta)\csc^2(\sin\theta)$$

37. $y = [x^2 + (1 - 3x)^5]^3 \implies$

$y' = 3[x^2 + (1 - 3x)^5]^2(2x + 5(1 - 3x)^4(-3)) = 3[x^2 + (1 - 3x)^5]^2[2x - 15(1 - 3x)^4]$

39. $g(x) = (2r \sin rx + n)^p \implies g'(x) = p(2r \sin rx + n)^{p-1}(2r \cos rx \cdot r) = p(2r \sin rx + n)^{p-1}(2r^2 \cos rx)$

41. $y = \cos \sqrt{\sin(\tan \pi x)} = \cos(\sin(\tan \pi x))^{1/2} \implies$

$y' = -\sin(\sin(\tan \pi x))^{1/2} \cdot \dfrac{d}{dx}(\sin(\tan \pi x))^{1/2} = -\sin(\sin(\tan \pi x))^{1/2} \cdot \tfrac{1}{2}(\sin(\tan \pi x))^{-1/2} \cdot \dfrac{d}{dx}(\sin(\tan \pi x))$

$= \dfrac{-\sin \sqrt{\sin(\tan \pi x)}}{2\sqrt{\sin(\tan \pi x)}} \cdot \cos(\tan \pi x) \cdot \dfrac{d}{dx}\tan \pi x = \dfrac{-\sin \sqrt{\sin(\tan \pi x)}}{2\sqrt{\sin(\tan \pi x)}} \cdot \cos(\tan \pi x) \cdot \sec^2(\pi x) \cdot \pi$

$= \dfrac{-\pi \cos(\tan \pi x) \sec^2(\pi x) \sin \sqrt{\sin(\tan \pi x)}}{2\sqrt{\sin(\tan \pi x)}}$

43. $y = \cos(x^2) \implies y' = -\sin(x^2) \cdot 2x = -2x \sin(x^2) \implies$

$y'' = -2x \cos(x^2) \cdot 2x + \sin(x^2) \cdot (-2) = -4x^2 \cos(x^2) - 2\sin(x^2)$

45. $H(t) = \tan 3t \implies H'(t) = 3\sec^2 3t \implies$

$H''(t) = 2 \cdot 3\sec 3t \dfrac{d}{dt}(\sec 3t) = 6\sec 3t\,(3\sec 3t \tan 3t) = 18\sec^2 3t \tan 3t$

47. $y = \sin(\sin x) \implies y' = \cos(\sin x) \cdot \cos x$. At $(\pi, 0)$, $y' = \cos(\sin \pi) \cdot \cos \pi = \cos(0) \cdot (-1) = 1(-1) = -1$, and an

equation of the tangent line is $y - 0 = -1(x - \pi)$, or $y = -x + \pi$.

49. (a) $y = f(x) = \tan\left(\tfrac{\pi}{4}x^2\right) \implies f'(x) = \sec^2\left(\tfrac{\pi}{4}x^2\right)\left(2 \cdot \tfrac{\pi}{4}x\right)$.

(b)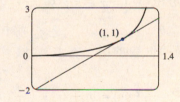

The slope of the tangent at $(1, 1)$ is thus

$f'(1) = \sec^2 \tfrac{\pi}{4}\left(\tfrac{\pi}{2}\right) = 2 \cdot \tfrac{\pi}{2} = \pi$, and its equation

is $y - 1 = \pi(x - 1)$ or $y = \pi x - \pi + 1$.

51. For the tangent line to be horizontal, $f'(x) = 0$. $f(x) = 2\sin x + \sin^2 x \implies f'(x) = 2\cos x + 2\sin x \cos x = 0 \iff$

$2\cos x(1 + \sin x) = 0 \iff \cos x = 0$ or $\sin x = -1$, so $x = \tfrac{\pi}{2} + 2n\pi$ or $\tfrac{3\pi}{2} + 2n\pi$, where n is any integer. Now

$f\left(\tfrac{\pi}{2}\right) = 3$ and $f\left(\tfrac{3\pi}{2}\right) = -1$, so the points on the curve with a horizontal tangent are $\left(\tfrac{\pi}{2} + 2n\pi, 3\right)$ and $\left(\tfrac{3\pi}{2} + 2n\pi, -1\right)$,

where n is any integer.

53. $F(x) = f(g(x)) \implies F'(x) = f'(g(x)) \cdot g'(x)$, so $F'(5) = f'(g(5)) \cdot g'(5) = f'(-2) \cdot 6 = 4 \cdot 6 = 24$

55. (a) $h(x) = f(g(x)) \implies h'(x) = f'(g(x)) \cdot g'(x)$, so $h'(1) = f'(g(1)) \cdot g'(1) = f'(2) \cdot 6 = 5 \cdot 6 = 30$.

(b) $H(x) = g(f(x)) \implies H'(x) = g'(f(x)) \cdot f'(x)$, so $H'(1) = g'(f(1)) \cdot f'(1) = g'(3) \cdot 4 = 9 \cdot 4 = 36$.

57. (a) $u(x) = f(g(x)) \implies u'(x) = f'(g(x))g'(x)$. So $u'(1) = f'(g(1))g'(1) = f'(3)g'(1)$. To find $f'(3)$, note that f is

linear from $(2, 4)$ to $(6, 3)$, so its slope is $\dfrac{3 - 4}{6 - 2} = -\dfrac{1}{4}$. To find $g'(1)$, note that g is linear from $(0, 6)$ to $(2, 0)$, so its slope

is $\dfrac{0 - 6}{2 - 0} = -3$. Thus, $f'(3)g'(1) = \left(-\tfrac{1}{4}\right)(-3) = \tfrac{3}{4}$.

(b) $v(x) = g(f(x))$ $\Rightarrow$ $v'(x) = g'(f(x))f'(x)$. So $v'(1) = g'(f(1))f'(1) = g'(2)f'(1)$, which does not exist since

$g'(2)$ does not exist.

(c) $w(x) = g(g(x))$ $\Rightarrow$ $w'(x) = g'(g(x))g'(x)$. So $w'(1) = g'(g(1))g'(1) = g'(3)g'(1)$. To find $g'(3)$, note that g is

linear from $(2, 0)$ to $(5, 2)$, so its slope is $\dfrac{2-0}{5-2} = \dfrac{2}{3}$. Thus, $g'(3)g'(1) = \left(\frac{2}{3}\right)(-3) = -2$.

59. The point $(3, 2)$ is on the graph of f, so $f(3) = 2$. The tangent line at $(3, 2)$ has slope $\dfrac{\Delta y}{\Delta x} = \dfrac{-4}{6} = -\dfrac{2}{3}$.

$g(x) = \sqrt{f(x)}$ $\Rightarrow$ $g'(x) = \frac{1}{2}[f(x)]^{-1/2} \cdot f'(x)$ $\Rightarrow$

$g'(3) = \frac{1}{2}[f(3)]^{-1/2} \cdot f'(3) = \frac{1}{2}(2)^{-1/2}(-\frac{2}{3}) = -\dfrac{1}{3\sqrt{2}}$ or $-\frac{1}{6}\sqrt{2}$.

61. $r(x) = f(g(h(x)))$ $\Rightarrow$ $r'(x) = f'(g(h(x))) \cdot g'(h(x)) \cdot h'(x)$, so

$r'(1) = f'(g(h(1))) \cdot g'(h(1)) \cdot h'(1) = f'(g(2)) \cdot g'(2) \cdot 4 = f'(3) \cdot 5 \cdot 4 = 6 \cdot 5 \cdot 4 = 120$

63. In general, if $y = f(2x)$, then the Chain Rule gives $y' = 2f'(2x)$, $y' = 4f''(2x)$, $y''' = 8f'''(2x)$, $\ldots$, $y^{(n)} = 2^n f^{(n)}(2x)$.

Here $f(x) = \cos x$ and $50 = 4(12) + 2$, so $f^{(50)}(x) = f^{(2)}(x) = -\cos 2x$ and $y^{(50)} = -2^{50}\cos 2x$.

65. (a) $B(t) = 4.0 + 0.35\sin\dfrac{2\pi t}{5.4}$ $\Rightarrow$ $\dfrac{dB}{dt} = \left(0.35\cos\dfrac{2\pi t}{5.4}\right)\left(\dfrac{2\pi}{5.4}\right) = \dfrac{0.7\pi}{5.4}\cos\dfrac{2\pi t}{5.4} = \dfrac{7\pi}{54}\cos\dfrac{2\pi t}{5.4}$

(b) At $t = 1$, $\dfrac{dB}{dt} = \dfrac{7\pi}{54}\cos\dfrac{2\pi}{5.4} \approx 0.16$.

67. With $m = m_0\left(1 - \dfrac{v^2}{c^2}\right)^{-1/2}$,

$F = \dfrac{d}{dt}(mv) = m\dfrac{d}{dt}(v) + v\dfrac{d}{dt}(m) = m_0\left(1 - \dfrac{v^2}{c^2}\right)^{-1/2} \cdot a + v \cdot m_0\left[-\dfrac{1}{2}\left(1 - \dfrac{v^2}{c^2}\right)^{-3/2}\right]\left(-\dfrac{2v}{c^2}\right)\dfrac{d}{dt}(v)$

$= m_0\left(1 - \dfrac{v^2}{c^2}\right)^{-3/2} \cdot a\left[\left(1 - \dfrac{v^2}{c^2}\right) + \dfrac{v^2}{c^2}\right] = \dfrac{m_0 a}{(1 - v^2/c^2)^{3/2}}$

Note that we factored out $(1 - v^2/c^2)^{-3/2}$ since $-3/2$ was the lesser exponent. Also note that $\dfrac{d}{dt}(v) = a$.

69. By the Chain Rule, $a(t) = \dfrac{dv}{dt} = \dfrac{dv}{ds}\dfrac{ds}{dt} = \dfrac{dv}{ds}v(t) = v(t)\dfrac{dv}{ds}$. The derivative dv/dt is the rate of change of the velocity

with respect to time (in other words, the acceleration) whereas the derivative dv/ds is the rate of change of the velocity with

respect to the displacement.

71. (a) $\dfrac{d}{dx}(\sin^n x \cos nx) = n\sin^{n-1} x \cos x \cos nx + \sin^n x\,(-n\sin nx)$ [Product Rule]

$= n\sin^{n-1} x\,(\cos nx \cos x - \sin nx \sin x)$ [factor out $n\sin^{n-1} x$]

$= n\sin^{n-1} x \cos(nx + x)$ [Addition Formula for cosine]

$= n\sin^{n-1} x \cos[(n+1)x]$ [factor out x]

(b) $\dfrac{d}{dx}\left(\cos^n x \cos nx\right) = n\cos^{n-1}x\left(-\sin x\right)\cos nx + \cos^n x\left(-n\sin nx\right)$ [Product Rule]

$\qquad = -n\cos^{n-1}x\left(\cos nx\,\sin x + \sin nx\,\cos x\right)$ [factor out $-n\cos^{n-1}x$]

$\qquad = -n\cos^{n-1}x\sin(nx+x)$ [Addition Formula for sine]

$\qquad = -n\cos^{n-1}x\sin[(n+1)x]$ [factor out x]

73. Since $\theta^\circ = \left(\frac{\pi}{180}\right)\theta$ rad, we have $\dfrac{d}{d\theta}\left(\sin\theta^\circ\right) = \dfrac{d}{d\theta}\left(\sin\frac{\pi}{180}\theta\right) = \frac{\pi}{180}\cos\frac{\pi}{180}\theta = \frac{\pi}{180}\cos\theta^\circ$.

75. $F(x) = f(3f(4f(x))) \quad\Rightarrow$

$\qquad F'(x) = f'(3f(4f(x)))\cdot\dfrac{d}{dx}\big(3f(4f(x))\big) = f'(3f(4f(x)))\cdot 3f'(4f(x))\cdot\dfrac{d}{dx}\big(4f(x)\big)$

$\qquad\quad = f'(3f(4f(x)))\cdot 3f'(4f(x))\cdot 4f'(x),\quad$ so

$\qquad F'(0) = f'(3f(4f(0)))\cdot 3f'(4f(0))\cdot 4f'(0) = f'(3f(4\cdot 0))\cdot 3f'(4\cdot 0)\cdot 4\cdot 2 = f'(3\cdot 0)\cdot 3\cdot 2\cdot 4\cdot 2 = 2\cdot 3\cdot 2\cdot 4\cdot 2 = 96.$

77. $\dfrac{d^2y}{dx^2} = \dfrac{d}{dx}\left(\dfrac{dy}{dx}\right)$ [Leibniz notation for the second derivative]

$\qquad = \dfrac{d}{dx}\left(\dfrac{dy}{du}\dfrac{du}{dx}\right)$ [Chain Rule]

$\qquad = \dfrac{dy}{du}\cdot\dfrac{d}{dx}\left(\dfrac{du}{dx}\right) + \dfrac{du}{dx}\cdot\dfrac{d}{dx}\left(\dfrac{dy}{du}\right)$ [Product Rule]

$\qquad = \dfrac{dy}{du}\cdot\dfrac{d^2u}{dx^2} + \dfrac{du}{dx}\cdot\dfrac{d}{du}\left(\dfrac{dy}{du}\right)\cdot\dfrac{du}{dx}$ [dy/du is a function of u]

$\qquad = \dfrac{dy}{du}\dfrac{d^2u}{dx^2} + \dfrac{d^2y}{du^2}\left(\dfrac{du}{dx}\right)^2$

Or: Using function notation for $y = f(u)$ and $u = g(x)$, we have $y = f(g(x))$, so

$y' = f'(g(x))\cdot g'(x)$ [by the Chain Rule] $\quad\Rightarrow$

$(y')' = [f'(g(x))\cdot g'(x)]' = f'(g(x))\cdot g''(x) + g'(x)\cdot f''(g(x))\cdot g'(x) = f'(g(x))\cdot g''(x) + f''(g(x))\cdot [g'(x)]^2.$

2.6 Implicit Differentiation

1. (a) $\dfrac{d}{dx}\left(9x^2 - y^2\right) = \dfrac{d}{dx}(1) \quad\Rightarrow\quad 18x - 2y\,y' = 0 \quad\Rightarrow\quad 2y\,y' = 18x \quad\Rightarrow\quad y' = \dfrac{9x}{y}$

(b) $9x^2 - y^2 = 1 \quad\Rightarrow\quad y^2 = 9x^2 - 1 \quad\Rightarrow\quad y = \pm\sqrt{9x^2 - 1}$, so $y' = \pm\frac{1}{2}(9x^2-1)^{-1/2}(18x) = \pm\dfrac{9x}{\sqrt{9x^2-1}}$.

(c) From part (a), $y' = \dfrac{9x}{y} = \dfrac{9x}{\pm\sqrt{9x^2-1}}$, which agrees with part (b).

3. $\dfrac{d}{dx}\left(x^3 + y^3\right) = \dfrac{d}{dx}(1) \quad\Rightarrow\quad 3x^2 + 3y^2\cdot y' = 0 \quad\Rightarrow\quad 3y^2\,y' = -3x^2 \quad\Rightarrow\quad y' = -\dfrac{x^2}{y^2}$

5. $\dfrac{d}{dx}\left(x^2 + xy - y^2\right) = \dfrac{d}{dx}\left(4\right) \;\Rightarrow\; 2x + x \cdot y' + y \cdot 1 - 2y\,y' = 0 \;\Rightarrow$

$xy' - 2y\,y' = -2x - y \;\Rightarrow\; (x - 2y)\,y' = -2x - y \;\Rightarrow\; y' = \dfrac{-2x - y}{x - 2y} = \dfrac{2x + y}{2y - x}$

7. $\dfrac{d}{dx}\left(y\cos x\right) = \dfrac{d}{dx}\left(x^2 + y^2\right) \;\Rightarrow\; y(-\sin x) + \cos x \cdot y' = 2x + 2y\,y' \;\Rightarrow\; \cos x \cdot y' - 2y\,y' = 2x + y\sin x \;\Rightarrow$

$y'(\cos x - 2y) = 2x + y\sin x \;\Rightarrow\; y' = \dfrac{2x + y\sin x}{\cos x - 2y}$

9. $\dfrac{d}{dx}\left(4\cos x\sin y\right) = \dfrac{d}{dx}(1) \;\Rightarrow\; 4\left[\cos x \cdot \cos y \cdot y' + \sin y \cdot (-\sin x)\right] = 0 \;\Rightarrow$

$y'(4\cos x\cos y) = 4\sin x\sin y \;\Rightarrow\; y' = \dfrac{4\sin x\sin y}{4\cos x\cos y} = \tan x\tan y$

11. $\dfrac{d}{dx}\left[\tan(x/y)\right] = \dfrac{d}{dx}(x + y) \;\Rightarrow\; \sec^2(x/y) \cdot \dfrac{y \cdot 1 - x \cdot y'}{y^2} = 1 + y' \;\Rightarrow$

$y\sec^2(x/y) - x\sec^2(x/y) \cdot y' = y^2 + y^2 y' \;\Rightarrow\; y\sec^2(x/y) - y^2 = y^2 y' + x\sec^2(x/y) \;\Rightarrow$

$y\sec^2(x/y) - y^2 = \left[y^2 + x\sec^2(x/y)\right] \cdot y' \;\Rightarrow\; y' = \dfrac{y\sec^2(x/y) - y^2}{y^2 + x\sec^2(x/y)}$

13. $\sqrt{xy} = 1 + x^2 y \;\Rightarrow\; \tfrac{1}{2}(xy)^{-1/2}(xy' + y \cdot 1) = 0 + x^2 y' + y \cdot 2x \;\Rightarrow\; \dfrac{x}{2\sqrt{xy}}\,y' + \dfrac{y}{2\sqrt{xy}} = x^2 y' + 2xy \;\Rightarrow$

$y'\left(\dfrac{x}{2\sqrt{xy}} - x^2\right) = 2xy - \dfrac{y}{2\sqrt{xy}} \;\Rightarrow\; y'\left(\dfrac{x - 2x^2\sqrt{xy}}{2\sqrt{xy}}\right) = \dfrac{4xy\sqrt{xy} - y}{2\sqrt{xy}} \;\Rightarrow\; y' = \dfrac{4xy\sqrt{xy} - y}{x - 2x^2\sqrt{xy}}$

15. $\dfrac{d}{dx}\left(y\cos x\right) = \dfrac{d}{dx}\left(1 + \sin(xy)\right) \;\Rightarrow\; y(-\sin x) + \cos x \cdot y' = \cos(xy) \cdot (xy' + y \cdot 1) \;\Rightarrow$

$\cos x \cdot y' - x\cos(xy) \cdot y' = y\sin x + y\cos(xy) \;\Rightarrow\; \left[\cos x - x\cos(xy)\right]y' = y\sin x + y\cos(xy) \;\Rightarrow$

$y' = \dfrac{y\sin x + y\cos(xy)}{\cos x - x\cos(xy)}$

17. $\dfrac{d}{dx}\left\{f(x) + x^2[f(x)]^3\right\} = \dfrac{d}{dx}(10) \;\Rightarrow\; f'(x) + x^2 \cdot 3[f(x)]^2 \cdot f'(x) + [f(x)]^3 \cdot 2x = 0.$ If $x = 1$, we have

$f'(1) + 1^2 \cdot 3[f(1)]^2 \cdot f'(1) + [f(1)]^3 \cdot 2(1) = 0 \;\Rightarrow\; f'(1) + 1 \cdot 3 \cdot 2^2 \cdot f'(1) + 2^3 \cdot 2 = 0 \;\Rightarrow$

$f'(1) + 12f'(1) = -16 \;\Rightarrow\; 13f'(1) = -16 \;\Rightarrow\; f'(1) = -\tfrac{16}{13}.$

19. $x^2 + xy + y^2 = 3 \;\Rightarrow\; 2x + x\,y' + y \cdot 1 + 2yy' = 0 \;\Rightarrow\; x\,y' + 2y\,y' = -2x - y \;\Rightarrow\; y'(x + 2y) = -2x - y \;\Rightarrow$

$y' = \dfrac{-2x - y}{x + 2y}.$ When $x = 1$ and $y = 1$, we have $y' = \dfrac{-2 - 1}{1 + 2 \cdot 1} = \dfrac{-3}{3} = -1$, so an equation of the tangent line is

$y - 1 = -1(x - 1)$ or $y = -x + 2.$

21. $x^2 + y^2 = (2x^2 + 2y^2 - x)^2$ $\Rightarrow$ $2x + 2y\,y' = 2(2x^2 + 2y^2 - x)(4x + 4y\,y' - 1)$.   When $x = 0$ and $y = \frac{1}{2}$, we have

$0 + y' = 2(\frac{1}{2})(2y' - 1)$ $\Rightarrow$ $y' = 2y' - 1$ $\Rightarrow$ $y' = 1$, so an equation of the tangent line is $y - \frac{1}{2} = 1(x - 0)$

or $y = x + \frac{1}{2}$.

23. $2(x^2 + y^2)^2 = 25(x^2 - y^2)$ $\Rightarrow$ $4(x^2 + y^2)(2x + 2y\,y') = 25(2x - 2y\,y')$ $\Rightarrow$

$4(x + y\,y')(x^2 + y^2) = 25(x - y\,y')$ $\Rightarrow$ $4y\,y'(x^2 + y^2) + 25y y' = 25x - 4x(x^2 + y^2)$ $\Rightarrow$

$y' = \dfrac{25x - 4x(x^2 + y^2)}{25y + 4y(x^2 + y^2)}$.   When $x = 3$ and $y = 1$, we have $y' = \dfrac{75 - 120}{25 + 40} = -\dfrac{45}{65} = -\dfrac{9}{13}$,

so an equation of the tangent line is $y - 1 = -\frac{9}{13}(x - 3)$ or $y = -\frac{9}{13}x + \frac{40}{13}$.

25. $9x^2 + y^2 = 9$ $\Rightarrow$ $18x + 2y\,y' = 0$ $\Rightarrow$ $2y\,y' = -18x$ $\Rightarrow$ $y' = -9x/y$ $\Rightarrow$

$y'' = -9\left(\dfrac{y \cdot 1 - x \cdot y'}{y^2}\right) = -9\left(\dfrac{y - x(-9x/y)}{y^2}\right) = -9 \cdot \dfrac{y^2 + 9x^2}{y^3} = -9 \cdot \dfrac{9}{y^3}$ [since x and y must satisfy the

original equation, $9x^2 + y^2 = 9$]. Thus, $y'' = -81/y^3$.

27. $x^3 + y^3 = 1$ $\Rightarrow$ $3x^2 + 3y^2 y' = 0$ $\Rightarrow$ $y' = -\dfrac{x^2}{y^2}$ $\Rightarrow$

$y'' = -\dfrac{y^2(2x) - x^2 \cdot 2y\,y'}{(y^2)^2} = -\dfrac{2xy^2 - 2x^2 y(-x^2/y^2)}{y^4} = -\dfrac{2xy^4 + 2x^4 y}{y^6} = -\dfrac{2xy(y^3 + x^3)}{y^6} = -\dfrac{2x}{y^5}$,

since x and y must satisfy the original equation, $x^3 + y^3 = 1$.

29. (a) $y^2 = 5x^4 - x^2$ $\Rightarrow$ $2y\,y' = 5(4x^3) - 2x$ $\Rightarrow$ $y' = \dfrac{10x^3 - x}{y}$.

(b)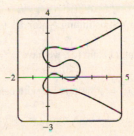

So at the point $(1, 2)$ we have $y' = \dfrac{10(1)^3 - 1}{2} = \dfrac{9}{2}$, and an equation

of the tangent line is $y - 2 = \frac{9}{2}(x - 1)$ or $y = \frac{9}{2}x - \frac{5}{2}$.

31. (a) There are eight points with horizontal tangents: four at $x \approx 1.57735$ and

four at $x \approx 0.42265$.

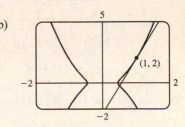

(b) $y' = \dfrac{3x^2 - 6x + 2}{2(2y^3 - 3y^2 - y + 1)}$ $\Rightarrow$ $y' = -1$ at $(0, 1)$ and $y' = \frac{1}{3}$ at $(0, 2)$.

Equations of the tangent lines are $y = -x + 1$ and $y = \frac{1}{3}x + 2$.

(c) $y' = 0$ $\Rightarrow$ $3x^2 - 6x + 2 = 0$ $\Rightarrow$ $x = 1 \pm \frac{1}{3}\sqrt{3}$

(d) By multiplying the right side of the equation by $x - 3$, we obtain the first graph. By modifying the equation in other ways, we can generate the other graphs.

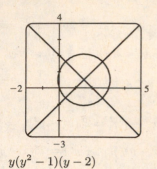

$$y(y^2 - 1)(y - 2)$$
$$= x(x - 1)(x - 2)(x - 3)$$

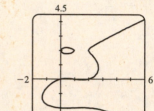

$$y(y^2 - 4)(y - 2)$$
$$= x(x - 1)(x - 2)$$

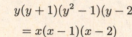
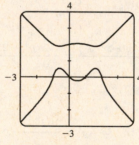

$$y(y + 1)(y^2 - 1)(y - 2)$$
$$= x(x - 1)(x - 2)$$

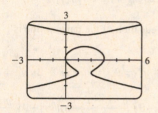
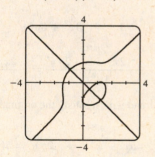

$$(y + 1)(y^2 - 1)(y - 2)$$
$$= (x - 1)(x - 2)$$

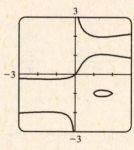

$$x(y + 1)(y^2 - 1)(y - 2)$$
$$= y(x - 1)(x - 2)$$

$$y(y^2 + 1)(y - 2)$$
$$= x(x^2 - 1)(x - 2)$$

$$y(y + 1)(y^2 - 2)$$
$$= x(x - 1)(x^2 - 2)$$

33. From Exercise 23, a tangent to the lemniscate will be horizontal if $y' = 0 \Rightarrow 25x - 4x(x^2 + y^2) = 0 \Rightarrow$

$x[25 - 4(x^2 + y^2)] = 0 \Rightarrow x^2 + y^2 = \frac{25}{4}$ **(1)**. (Note that when x is 0, y is also 0, and there is no horizontal tangent at the

origin.) Substituting $\frac{25}{4}$ for $x^2 + y^2$ in the equation of the lemniscate, $2(x^2 + y^2)^2 = 25(x^2 - y^2)$, we get $x^2 - y^2 = \frac{25}{8}$ **(2)**.

Solving **(1)** and **(2)**, we have $x^2 = \frac{75}{16}$ and $y^2 = \frac{25}{16}$, so the four points are $\left(\pm \frac{5\sqrt{3}}{4}, \pm \frac{5}{4} \right)$.

35. $x^2 + y^2 = r^2$ is a circle with center O and $ax + by = 0$ is a line through O [assume a

and b are not both zero]. $x^2 + y^2 = r^2 \Rightarrow 2x + 2yy' = 0 \Rightarrow y' = -x/y$, so the

slope of the tangent line at P_0 (x_0, y_0) is $-x_0/y_0$. The slope of the line OP_0 is y_0/x_0,

which is the negative reciprocal of $-x_0/y_0$. Hence, the curves are orthogonal, and the

families of curves are orthogonal trajectories of each other.

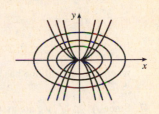

37. $y = cx^2 \Rightarrow y' = 2cx$ and $x^2 + 2y^2 = k$ [assume $k > 0$] $\Rightarrow 2x + 4yy' = 0 \Rightarrow$

$2yy' = -x \Rightarrow y' = -\dfrac{x}{2(y)} = -\dfrac{x}{2(cx^2)} = -\dfrac{1}{2cx}$, so the curves are orthogonal if

$c \neq 0$. If $c = 0$, then the horizontal line $y = cx^2 = 0$ intersects $x^2 + 2y^2 = k$ orthogonally

at $\left(\pm\sqrt{k}, 0\right)$, since the ellipse $x^2 + 2y^2 = k$ has vertical tangents at those two points.

39. (a) $\left(P + \dfrac{n^2 a}{V^2}\right)(V - nb) = nRT \Rightarrow PV - Pnb + \dfrac{n^2 a}{V} - \dfrac{n^3 ab}{V^2} = nRT \Rightarrow$

$\dfrac{d}{dP}(PV - Pnb + n^2 aV^{-1} - n^3 abV^{-2}) = \dfrac{d}{dP}(nRT) \Rightarrow$

$PV' + V \cdot 1 - nb - n^2 aV^{-2} \cdot V' + 2n^3 abV^{-3} \cdot V' = 0 \Rightarrow V'(P - n^2 aV^{-2} + 2n^3 abV^{-3}) = nb - V \Rightarrow$

$V' = \dfrac{nb - V}{P - n^2 aV^{-2} + 2n^3 abV^{-3}}$ or $\dfrac{dV}{dP} = \dfrac{V^3(nb - V)}{PV^3 - n^2 aV + 2n^3 ab}$

(b) Using the last expression for dV/dP from part (a), we get

$$\frac{dV}{dP} = \frac{(10\text{ L})^3[(1\text{ mole})(0.04267\text{ L/mole}) - 10\text{ L}]}{\begin{bmatrix}(2.5\text{ atm})(10\text{ L})^3 - (1\text{ mole})^2(3.592\text{ L}^2\text{-atm/ mole}^2)(10\text{ L}) \\ + 2(1\text{ mole})^3(3.592\text{ L}^2\text{-atm/ mole}^2)(0.04267\text{ L/ mole})\end{bmatrix}}$$

$$= \frac{-9957.33\text{ L}^4}{2464.386541\text{ L}^3\text{-atm}} \approx -4.04\text{ L/ atm}.$$

41. If the circle has radius r, its equation is $x^2 + y^2 = r^2 \Rightarrow 2x + 2yy' = 0 \Rightarrow y' = -\dfrac{x}{y}$, so the slope of the tangent line

at $P(x_0, y_0)$ is $-\dfrac{x_0}{y_0}$. The negative reciprocal of that slope is $\dfrac{-1}{-x_0/y_0} = \dfrac{y_0}{x_0}$, which is the slope of OP, so the tangent line at

P is perpendicular to the radius OP.

43. To find the points at which the ellipse $x^2 - xy + y^2 = 3$ crosses the x-axis, let $y = 0$ and solve for x.

$y = 0 \Rightarrow x^2 - x(0) + 0^2 = 3 \Leftrightarrow x = \pm\sqrt{3}$. So the graph of the ellipse crosses the x-axis at the points $\left(\pm\sqrt{3}, 0\right)$.

Using implicit differentiation to find y', we get $2x - xy' - y + 2yy' = 0 \Rightarrow y'(2y - x) = y - 2x \Leftrightarrow y' = \dfrac{y - 2x}{2y - x}$.

So y' at $\left(\sqrt{3}, 0\right)$ is $\dfrac{0 - 2\sqrt{3}}{2(0) - \sqrt{3}} = 2$ and y' at $\left(-\sqrt{3}, 0\right)$ is $\dfrac{0 + 2\sqrt{3}}{2(0) + \sqrt{3}} = 2$. Thus, the tangent lines at these points are parallel.

45. $x^2 y^2 + xy = 2 \Rightarrow x^2 \cdot 2yy' + y^2 \cdot 2x + x \cdot y' + y \cdot 1 = 0 \Leftrightarrow y'(2x^2 y + x) = -2xy^2 - y \Leftrightarrow$

$y' = -\dfrac{2xy^2 + y}{2x^2 y + x}$. So $-\dfrac{2xy^2 + y}{2x^2 y + x} = -1 \Leftrightarrow 2xy^2 + y = 2x^2 y + x \Leftrightarrow y(2xy + 1) = x(2xy + 1) \Leftrightarrow$

$y(2xy + 1) - x(2xy + 1) = 0 \Leftrightarrow (2xy + 1)(y - x) = 0 \Leftrightarrow xy = -\frac{1}{2}$ or $y = x$. But $xy = -\frac{1}{2} \Rightarrow$

$x^2y^2 + xy = \frac{1}{4} - \frac{1}{2} \neq 2$, so we must have $x = y$. Then $x^2y^2 + xy = 2 \Rightarrow x^4 + x^2 = 2 \Leftrightarrow x^4 + x^2 - 2 = 0 \Leftrightarrow$

$(x^2 + 2)(x^2 - 1) = 0$. So $x^2 = -2$, which is impossible, or $x^2 = 1 \Leftrightarrow x = \pm 1$. Since $x = y$, the points on the curve

where the tangent line has a slope of -1 are $(-1, -1)$ and $(1, 1)$.

47. Since $A^2 < a^2$, we are assured that there are four points of intersection.

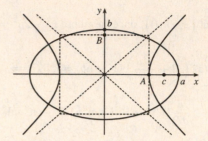

(1) $\dfrac{x^2}{a^2} + \dfrac{y^2}{b^2} = 1 \Rightarrow \dfrac{2x}{a^2} + \dfrac{2yy'}{b^2} = 0 \Rightarrow$

$$\dfrac{yy'}{b^2} = -\dfrac{x}{a^2} \Rightarrow y' = m_1 = -\dfrac{xb^2}{ya^2}.$$

(2) $\dfrac{x^2}{A^2} - \dfrac{y^2}{B^2} = 1 \Rightarrow \dfrac{2x}{A^2} - \dfrac{2yy'}{B^2} = 0 \Rightarrow$

$$\dfrac{yy'}{B^2} = \dfrac{x}{A^2} \Rightarrow y' = m_2 = \dfrac{xB^2}{yA^2}.$$

Now $m_1 m_2 = -\dfrac{xb^2}{ya^2} \cdot \dfrac{xB^2}{yA^2} = -\dfrac{b^2B^2}{a^2A^2} \cdot \dfrac{x^2}{y^2}$ **(3)**. Subtracting equations, **(1)** − **(2)**, gives us $\dfrac{x^2}{a^2} + \dfrac{y^2}{b^2} - \dfrac{x^2}{A^2} + \dfrac{y^2}{B^2} = 0 \Rightarrow$

$\dfrac{y^2}{b^2} + \dfrac{y^2}{B^2} = \dfrac{x^2}{A^2} - \dfrac{x^2}{a^2} \Rightarrow \dfrac{y^2B^2 + y^2b^2}{b^2B^2} = \dfrac{x^2a^2 - x^2A^2}{A^2a^2} \Rightarrow \dfrac{y^2(b^2 + B^2)}{b^2B^2} = \dfrac{x^2(a^2 - A^2)}{a^2A^2}$ **(4)**. Since

$a^2 - b^2 = A^2 + B^2$, we have $a^2 - A^2 = b^2 + B^2$. Thus, equation **(4)** becomes $\dfrac{y^2}{b^2B^2} = \dfrac{x^2}{A^2a^2} \Rightarrow \dfrac{x^2}{y^2} = \dfrac{A^2a^2}{b^2B^2}$, and

substituting for $\dfrac{x^2}{y^2}$ in equation **(3)** gives us $m_1 m_2 = -\dfrac{b^2B^2}{a^2A^2} \cdot \dfrac{a^2A^2}{b^2B^2} = -1$. Hence, the ellipse and hyperbola are orthogonal

trajectories.

49. (a) $y = J(x)$ and $xy'' + y' + xy = 0 \Rightarrow xJ''(x) + J'(x) + xJ(x) = 0$. If $x = 0$, we have $0 + J'(0) + 0 = 0$,

so $J'(0) = 0$.

(b) Differentiating $xy'' + y' + xy = 0$ implicitly, we get $xy''' + y'' \cdot 1 + y'' + xy' + y \cdot 1 = 0 \Rightarrow$

$xy''' + 2y'' + xy' + y = 0$, so $xJ'''(x) + 2J''(x) + xJ'(x) + J(x) = 0$. If $x = 0$, we have

$0 + 2J''(0) + 0 + 1$ [$J(0) = 1$ is given] $= 0 \Rightarrow 2J''(0) = -1 \Rightarrow J''(0) = -\frac{1}{2}$.

2.7   Related Rates

1. $V = x^3 \Rightarrow \dfrac{dV}{dt} = \dfrac{dV}{dx}\dfrac{dx}{dt} = 3x^2\dfrac{dx}{dt}$

3. Let s denote the side of a square. The square's area A is given by $A = s^2$. Differentiating with respect to t gives us

$\dfrac{dA}{dt} = 2s\dfrac{ds}{dt}$. When $A = 16$, $s = 4$. Substitution 4 for s and 6 for $\dfrac{ds}{dt}$ gives us $\dfrac{dA}{dt} = 2(4)(6) = 48$ cm^2/s.

5. $V = \pi r^2 h = \pi(5)^2 h = 25\pi h \Rightarrow \dfrac{dV}{dt} = 25\pi\dfrac{dh}{dt} \Rightarrow 3 = 25\pi\dfrac{dh}{dt} \Rightarrow \dfrac{dh}{dt} = \dfrac{3}{25\pi}$ m/min.

7. (a) $y = \sqrt{2x+1}$ and $\dfrac{dx}{dt} = 3 \;\Rightarrow\; \dfrac{dy}{dt} = \dfrac{dy}{dx}\dfrac{dx}{dt} = \dfrac{1}{2}(2x+1)^{-1/2} \cdot 2 \cdot 3 = \dfrac{3}{\sqrt{2x+1}}$. When $x = 4$, $\dfrac{dy}{dt} = \dfrac{3}{\sqrt{9}} = 1$.

(b) $y = \sqrt{2x+1} \;\Rightarrow\; y^2 = 2x+1 \;\Rightarrow\; 2x = y^2 - 1 \;\Rightarrow\; x = \frac{1}{2}y^2 - \frac{1}{2}$ and $\dfrac{dy}{dt} = 5 \;\Rightarrow$

$\dfrac{dx}{dt} = \dfrac{dx}{dy}\dfrac{dy}{dt} = y \cdot 5 = 5y$. When $x = 12$, $y = \sqrt{25} = 5$, so $\dfrac{dx}{dt} = 5(5) = 25$.

9. $\dfrac{d}{dt}(x^2 + y^2 + z^2) = \dfrac{d}{dt}(9) \;\Rightarrow\; 2x\dfrac{dx}{dt} + 2y\dfrac{dy}{dt} + 2z\dfrac{dz}{dt} = 0 \;\Rightarrow\; x\dfrac{dx}{dt} + y\dfrac{dy}{dt} + z\dfrac{dz}{dt} = 0$. If $\dfrac{dx}{dt} = 5$, $\dfrac{dy}{dt} = 4$ and

$(x, y, z) = (2, 2, 1)$, then $2(5) + 2(4) + 1\dfrac{dz}{dt} = 0 \;\Rightarrow\; \dfrac{dz}{dt} = -18$.

11. (a) Given: the rate of decrease of the surface area is 1 cm²/min. If we let t be

time (in minutes) and S be the surface area (in cm²), then we are given that

$dS/dt = -1$ cm²/s.

(b) Unknown: the rate of decrease of the diameter when the diameter is 10 cm.

If we let x be the diameter, then we want to find dx/dt when $x = 10$ cm.

(c)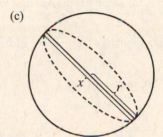

(d) If the radius is r and the diameter $x = 2r$, then $r = \frac{1}{2}x$ and

$S = 4\pi r^2 = 4\pi\left(\frac{1}{2}x\right)^2 = \pi x^2 \;\Rightarrow\; \dfrac{dS}{dt} = \dfrac{dS}{dx}\dfrac{dx}{dt} = 2\pi x\dfrac{dx}{dt}$.

(e) $-1 = \dfrac{dS}{dt} = 2\pi x\dfrac{dx}{dt} \;\Rightarrow\; \dfrac{dx}{dt} = -\dfrac{1}{2\pi x}$. When $x = 10$, $\dfrac{dx}{dt} = -\dfrac{1}{20\pi}$. So the rate of decrease is $\dfrac{1}{20\pi}$ cm/min.

13. (a) Given: a plane flying horizontally at an altitude of 1 mi and a speed of 500 mi/h passes directly over a radar station.

If we let t be time (in hours) and x be the horizontal distance traveled by the plane (in mi), then we are given

that $dx/dt = 500$ mi/h.

(b) Unknown: the rate at which the distance from the plane to the station is increasing

when it is 2 mi from the station. If we let y be the distance from the plane to the station,

then we want to find dy/dt when $y = 2$ mi.

(c)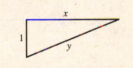

(d) By the Pythagorean Theorem, $y^2 = x^2 + 1 \;\Rightarrow\; 2y\,(dy/dt) = 2x\,(dx/dt)$.

(e) $\dfrac{dy}{dt} = \dfrac{x}{y}\dfrac{dx}{dt} = \dfrac{x}{y}(500)$. Since $y^2 = x^2 + 1$, when $y = 2$, $x = \sqrt{3}$, so $\dfrac{dy}{dt} = \dfrac{\sqrt{3}}{2}(500) = 250\sqrt{3} \approx 433$ mi/h.

15.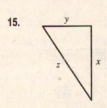

We are given that $\dfrac{dx}{dt} = 60$ mi/h and $\dfrac{dy}{dt} = 25$ mi/h. $z^2 = x^2 + y^2 \;\Rightarrow$

$2z\dfrac{dz}{dt} = 2x\dfrac{dx}{dt} + 2y\dfrac{dy}{dt} \;\Rightarrow\; z\dfrac{dz}{dt} = x\dfrac{dx}{dt} + y\dfrac{dy}{dt} \;\Rightarrow\; \dfrac{dz}{dt} = \dfrac{1}{z}\left(x\dfrac{dx}{dt} + y\dfrac{dy}{dt}\right)$.

After 2 hours, $x = 2(60) = 120$ and $y = 2(25) = 50 \;\Rightarrow\; z = \sqrt{120^2 + 50^2} = 130$,

so $\dfrac{dz}{dt} = \dfrac{1}{z}\left(x\dfrac{dx}{dt} + y\dfrac{dy}{dt}\right) = \dfrac{120(60) + 50(25)}{130} = 65$ mi/h.

17. 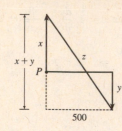 We are given that $\dfrac{dx}{dt} = 4$ ft/s and $\dfrac{dy}{dt} = 5$ ft/s. $z^2 = (x+y)^2 + 500^2 \Rightarrow$

$2z\dfrac{dz}{dt} = 2(x+y)\left(\dfrac{dx}{dt} + \dfrac{dy}{dt}\right)$. 15 minutes after the woman starts, we have

$x = (4 \text{ ft/s})(20 \text{ min})(60 \text{ s/min}) = 4800$ ft and $y = 5 \cdot 15 \cdot 60 = 4500 \Rightarrow$

$z = \sqrt{(4800 + 4500)^2 + 500^2} = \sqrt{86{,}740{,}000}$, so

$\dfrac{dz}{dt} = \dfrac{x+y}{z}\left(\dfrac{dx}{dt} + \dfrac{dy}{dt}\right) = \dfrac{4800 + 4500}{\sqrt{86{,}740{,}000}}(4+5) = \dfrac{837}{\sqrt{8674}} \approx 8.99$ ft/s.

19. $A = \frac{1}{2}bh$, where b is the base and h is the altitude. We are given that $\dfrac{dh}{dt} = 1$ cm/min and $\dfrac{dA}{dt} = 2$ cm^2/min. Using the

Product Rule, we have $\dfrac{dA}{dt} = \dfrac{1}{2}\left(b\dfrac{dh}{dt} + h\dfrac{db}{dt}\right)$. When $h = 10$ and $A = 100$, we have $100 = \frac{1}{2}b(10) \Rightarrow \frac{1}{2}b = 10 \Rightarrow$

$b = 20$, so $2 = \dfrac{1}{2}\left(20 \cdot 1 + 10\dfrac{db}{dt}\right) \Rightarrow 4 = 20 + 10\dfrac{db}{dt} \Rightarrow \dfrac{db}{dt} = \dfrac{4-20}{10} = -1.6$ cm/min.

21. 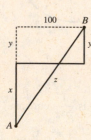 We are given that $\dfrac{dx}{dt} = 35$ km/h and $\dfrac{dy}{dt} = 25$ km/h. $z^2 = (x+y)^2 + 100^2 \Rightarrow$

$2z\dfrac{dz}{dt} = 2(x+y)\left(\dfrac{dx}{dt} + \dfrac{dy}{dt}\right)$. At 4:00 PM, $x = 4(35) = 140$ and $y = 4(25) = 100 \Rightarrow$

$z = \sqrt{(140 + 100)^2 + 100^2} = \sqrt{67{,}600} = 260$, so

$\dfrac{dz}{dt} = \dfrac{x+y}{z}\left(\dfrac{dx}{dt} + \dfrac{dy}{dt}\right) = \dfrac{140 + 100}{260}(35 + 25) = \dfrac{720}{13} \approx 55.4$ km/h.

23. Using Q for the origin, we are given $\dfrac{dx}{dt} = -2$ ft/s and need to find $\dfrac{dy}{dt}$ when $x = -5$.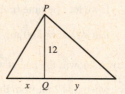

Using the Pythagorean Theorem twice, we have $\sqrt{x^2 + 12^2} + \sqrt{y^2 + 12^2} = 39$,

the total length of the rope. Differentiating with respect to t, we get

$\dfrac{x}{\sqrt{x^2 + 12^2}}\dfrac{dx}{dt} + \dfrac{y}{\sqrt{y^2 + 12^2}}\dfrac{dy}{dt} = 0$, so $\dfrac{dy}{dt} = -\dfrac{x\sqrt{y^2 + 12^2}}{y\sqrt{x^2 + 12^2}}\dfrac{dx}{dt}$.

Now when $x = -5$, $39 = \sqrt{(-5)^2 + 12^2} + \sqrt{y^2 + 12^2} = 13 + \sqrt{y^2 + 12^2} \Leftrightarrow \sqrt{y^2 + 12^2} = 26$, and

$y = \sqrt{26^2 - 12^2} = \sqrt{532}$. So when $x = -5$, $\dfrac{dy}{dt} = -\dfrac{(-5)(26)}{\sqrt{532}\,(13)}(-2) = -\dfrac{10}{\sqrt{133}} \approx -0.87$ ft/s.

So cart B is moving towards Q at about 0.87 ft/s.

25. By similar triangles, $\dfrac{3}{1} = \dfrac{b}{h}$, so $b = 3h$. The trough has volume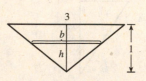

$V = \frac{1}{2}bh(10) = 5(3h)h = 15h^2 \Rightarrow 12 = \dfrac{dV}{dt} = 30h\dfrac{dh}{dt} \Rightarrow \dfrac{dh}{dt} = \dfrac{2}{5h}$.

When $h = \frac{1}{2}$, $\dfrac{dh}{dt} = \dfrac{2}{5 \cdot \frac{1}{2}} = \dfrac{4}{5}$ ft/min.

27. We are given that $\dfrac{dV}{dt} = 30$ ft^3/min. $V = \dfrac{1}{3}\pi r^2 h = \dfrac{1}{3}\pi \left(\dfrac{h}{2}\right)^2 h = \dfrac{\pi h^3}{12} \quad \Rightarrow$

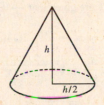

$\dfrac{dV}{dt} = \dfrac{dV}{dh}\dfrac{dh}{dt} \quad \Rightarrow \quad 30 = \dfrac{\pi h^2}{4}\dfrac{dh}{dt} \quad \Rightarrow \quad \dfrac{dh}{dt} = \dfrac{120}{\pi h^2}$.

When $h = 10$ ft, $\dfrac{dh}{dt} = \dfrac{120}{10^2\pi} = \dfrac{6}{5\pi} \approx 0.38$ ft/min.

29. $A = \frac{1}{2}bh$, but $b = 5$ m and $\sin\theta = \dfrac{h}{4} \quad \Rightarrow \quad h = 4\sin\theta$, so $A = \frac{1}{2}(5)(4\sin\theta) = 10\sin\theta$.

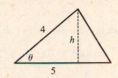

We are given $\dfrac{d\theta}{dt} = 0.06$ rad/s, so $\dfrac{dA}{dt} = \dfrac{dA}{d\theta}\dfrac{d\theta}{dt} = (10\cos\theta)(0.06) = 0.6\cos\theta$.

When $\theta = \dfrac{\pi}{3}$, $\dfrac{dA}{dt} = 0.6\left(\cos\dfrac{\pi}{3}\right) = (0.6)\left(\frac{1}{2}\right) = 0.3$ m^2/s.

31. From the figure and given information, we have $x^2 + y^2 = L^2$, $\dfrac{dy}{dt} = -0.15$ m/s, and

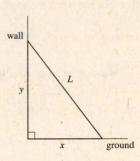

$\dfrac{dx}{dt} = 0.2$ m/s when $x = 3$ m. Differentiating implicitly with respect to t, we get

$x^2 + y^2 = L^2 \quad \Rightarrow \quad 2x\dfrac{dx}{dt} + 2y\dfrac{dy}{dt} = 0 \quad \Rightarrow \quad y\dfrac{dy}{dt} = -x\dfrac{dx}{dt}$. Substituting the given

information gives us $y(-0.15) = -3(0.2) \quad \Rightarrow \quad y = 4$ m. Thus, $3^2 + 4^2 = L^2 \quad \Rightarrow$

$L^2 = 25 \quad \Rightarrow \quad L = 5$ m.

33. Differentiating both sides of $PV = C$ with respect to t and using the Product Rule gives us $P\dfrac{dV}{dt} + V\dfrac{dP}{dt} = 0 \quad \Rightarrow$

$\dfrac{dV}{dt} = -\dfrac{V}{P}\dfrac{dP}{dt}$. When $V = 600$, $P = 150$ and $\dfrac{dP}{dt} = 20$, so we have $\dfrac{dV}{dt} = -\dfrac{600}{150}(20) = -80$. Thus, the volume is

decreasing at a rate of 80 cm^3/min.

35. With $R_1 = 80$ and $R_2 = 100$, $\dfrac{1}{R} = \dfrac{1}{R_1} + \dfrac{1}{R_2} = \dfrac{1}{80} + \dfrac{1}{100} = \dfrac{180}{8000} = \dfrac{9}{400}$, so $R = \dfrac{400}{9}$. Differentiating $\dfrac{1}{R} = \dfrac{1}{R_1} + \dfrac{1}{R_2}$

with respect to t, we have $-\dfrac{1}{R^2}\dfrac{dR}{dt} = -\dfrac{1}{R_1^2}\dfrac{dR_1}{dt} - \dfrac{1}{R_2^2}\dfrac{dR_2}{dt} \quad \Rightarrow \quad \dfrac{dR}{dt} = R^2\left(\dfrac{1}{R_1^2}\dfrac{dR_1}{dt} + \dfrac{1}{R_2^2}\dfrac{dR_2}{dt}\right)$. When $R_1 = 80$ and

$R_2 = 100$, $\dfrac{dR}{dt} = \dfrac{400^2}{9^2}\left[\dfrac{1}{80^2}(0.3) + \dfrac{1}{100^2}(0.2)\right] = \dfrac{107}{810} \approx 0.132$ Ω/s.

37. (a) By the Pythagorean Theorem, $4000^2 + y^2 = \ell^2$. Differentiating with respect to t,

we obtain $2y\dfrac{dy}{dt} = 2\ell\dfrac{d\ell}{dt}$. We know that $\dfrac{dy}{dt} = 600$ ft/s, so when $y = 3000$ ft,

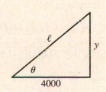

$\ell = \sqrt{4000^2 + 3000^2} = \sqrt{25{,}000{,}000} = 5000$ ft

and $\dfrac{d\ell}{dt} = \dfrac{y}{\ell}\dfrac{dy}{dt} = \dfrac{3000}{5000}(600) = \dfrac{1800}{5} = 360$ ft/s.

(b) Here $\tan\theta = \dfrac{y}{4000} \quad \Rightarrow \quad \dfrac{d}{dt}(\tan\theta) = \dfrac{d}{dt}\left(\dfrac{y}{4000}\right) \quad \Rightarrow \quad \sec^2\theta\dfrac{d\theta}{dt} = \dfrac{1}{4000}\dfrac{dy}{dt} \quad \Rightarrow \quad \dfrac{d\theta}{dt} = \dfrac{\cos^2\theta}{4000}\dfrac{dy}{dt}$. When

$y = 3000$ ft, $\dfrac{dy}{dt} = 600$ ft/s, $\ell = 5000$ and $\cos\theta = \dfrac{4000}{\ell} = \dfrac{4000}{5000} = \dfrac{4}{5}$, so $\dfrac{d\theta}{dt} = \dfrac{(4/5)^2}{4000}(600) = 0.096$ rad/s.

39. We are given that $\dfrac{dx}{dt} = 300$ km/h. By the Law of Cosines,

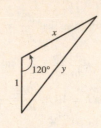

$$y^2 = x^2 + 1^2 - 2(1)(x)\cos 120° = x^2 + 1 - 2x\left(-\tfrac{1}{2}\right) = x^2 + x + 1, \text{ so}$$

$$2y\dfrac{dy}{dt} = 2x\dfrac{dx}{dt} + \dfrac{dx}{dt} \;\Rightarrow\; \dfrac{dy}{dt} = \dfrac{2x+1}{2y}\dfrac{dx}{dt}. \text{ After 1 minute, } x = \tfrac{300}{60} = 5 \text{ km} \;\Rightarrow$$

$$y = \sqrt{5^2 + 5 + 1} = \sqrt{31} \text{ km} \;\Rightarrow\; \dfrac{dy}{dt} = \dfrac{2(5)+1}{2\sqrt{31}}(300) = \dfrac{1650}{\sqrt{31}} \approx 296 \text{ km/h.}$$

41. Let the distance between the runner and the friend be ℓ. Then by the Law of Cosines,

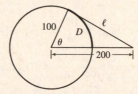

$$\ell^2 = 200^2 + 100^2 - 2 \cdot 200 \cdot 100 \cdot \cos\theta = 50{,}000 - 40{,}000\cos\theta \;\;(\star). \text{ Differentiating}$$

implicitly with respect to t, we obtain $2\ell\dfrac{d\ell}{dt} = -40{,}000(-\sin\theta)\dfrac{d\theta}{dt}$. Now if D is the

distance run when the angle is θ radians, then by the formula for the length of an arc

on a circle, $s = r\theta$, we have $D = 100\theta$, so $\theta = \dfrac{1}{100}D \;\Rightarrow\; \dfrac{d\theta}{dt} = \dfrac{1}{100}\dfrac{dD}{dt} = \dfrac{7}{100}$. To substitute into the expression for

$\dfrac{d\ell}{dt}$, we must know $\sin\theta$ at the time when $\ell = 200$, which we find from $(\star)$: $200^2 = 50{,}000 - 40{,}000\cos\theta \;\Leftrightarrow$

$\cos\theta = \tfrac{1}{4} \;\Rightarrow\; \sin\theta = \sqrt{1 - \left(\tfrac{1}{4}\right)^2} = \dfrac{\sqrt{15}}{4}$. Substituting, we get $2(200)\dfrac{d\ell}{dt} = 40{,}000\dfrac{\sqrt{15}}{4}\left(\tfrac{7}{100}\right) \;\Rightarrow$

$d\ell/dt = \dfrac{7\sqrt{15}}{4} \approx 6.78$ m/s. Whether the distance between them is increasing or decreasing depends on the direction in which

the runner is running.

2.8 Linear Approximations and Differentials

1. $f(x) = x^4 + 3x^2 \;\Rightarrow\; f'(x) = 4x^3 + 6x$, so $f(-1) = 4$ and $f'(-1) = -10$.

Thus, $L(x) = f(-1) + f'(-1)(x - (-1)) = 4 + (-10)(x + 1) = -10x - 6$.

3. $f(x) = \sqrt{x} \;\Rightarrow\; f'(x) = \tfrac{1}{2}x^{-1/2} = 1/(2\sqrt{x})$, so $f(4) = 2$ and $f'(4) = \tfrac{1}{4}$. Thus,

$L(x) = f(4) + f'(4)(x - 4) = 2 + \tfrac{1}{4}(x - 4) = 2 + \tfrac{1}{4}x - 1 = \tfrac{1}{4}x + 1$.

5. $f(x) = \sqrt{1 - x} \;\Rightarrow\; f'(x) = \dfrac{-1}{2\sqrt{1-x}}$, so $f(0) = 1$ and $f'(0) = -\tfrac{1}{2}$.

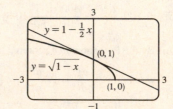

Therefore,

$$\sqrt{1 - x} = f(x) \approx f(0) + f'(0)(x - 0) = 1 + \left(-\tfrac{1}{2}\right)(x - 0) = 1 - \tfrac{1}{2}x.$$

So $\sqrt{0.9} = \sqrt{1 - 0.1} \approx 1 - \tfrac{1}{2}(0.1) = 0.95$

and $\sqrt{0.99} = \sqrt{1 - 0.01} \approx 1 - \tfrac{1}{2}(0.01) = 0.995$.

7. $f(x) = \sqrt[4]{1 + 2x} \;\Rightarrow\; f'(x) = \tfrac{1}{4}(1 + 2x)^{-3/4}(2) = \tfrac{1}{2}(1 + 2x)^{-3/4}$, so

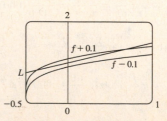

$f(0) = 1$ and $f'(0) = \tfrac{1}{2}$. Thus, $f(x) \approx f(0) + f'(0)(x - 0) = 1 + \tfrac{1}{2}x$.

We need $\sqrt[4]{1 + 2x} - 0.1 < 1 + \tfrac{1}{2}x < \sqrt[4]{1 + 2x} + 0.1$, which is true when

$-0.368 < x < 0.677$.

9. $f(x) = \dfrac{1}{(1+2x)^4} = (1+2x)^{-4}$ $\Rightarrow$

$f'(x) = -4(1+2x)^{-5}(2) = \dfrac{-8}{(1+2x)^5}$, so $f(0) = 1$ and $f'(0) = -8$.

Thus, $f(x) \approx f(0) + f'(0)(x-0) = 1 + (-8)(x-0) = 1 - 8x$.

We need $1/(1+2x)^4 - 0.1 < 1 - 8x < 1/(1+2x)^4 + 0.1$, which is true

when $-0.045 < x < 0.055$.

11. To estimate $(1.999)^4$, we'll find the linearization of $f(x) = x^4$ at $a = 2$. Since $f'(x) = 4x^3$, $f(2) = 16$, and

$f'(2) = 32$, we have $L(x) = 16 + 32(x-2)$. Thus, $x^4 \approx 16 + 32(x-2)$ when x is near 2, so

$(1.999)^4 \approx 16 + 32(1.999 - 2) = 16 - 0.032 = 15.968$.

13. To estimate $(8.06)^{2/3}$, we'll find the linearization of $f(x) = x^{2/3}$ at $a = 8$. Since $f'(x) = \frac{2}{3}x^{-1/3} = 2/\left(3\sqrt[3]{x}\right)$, $f(8) = 4$,

and $f'(8) = \frac{1}{3}$, we have $L(x) = 4 + \frac{1}{3}(x-8) = \frac{1}{3}x + \frac{4}{3}$. Thus, $x^{2/3} \approx \frac{1}{3}x + \frac{4}{3}$ when x is near 8, so

$(8.06)^{2/3} \approx \frac{1}{3}(8.06) + \frac{4}{3} = \frac{12.06}{3} = 4.02$.

15. $y = f(x) = \sec x$ $\Rightarrow$ $f'(x) = \sec x \tan x$, so $f(0) = 1$ and $f'(0) = 1 \cdot 0 = 0$. The linear approximation of f at 0 is

$f(0) + f'(0)(x-0) = 1 + 0(x) = 1$. Since 0.08 is close to 0, approximating $\sec 0.08$ with 1 is reasonable.

17. (a) For $y = f(t) = \tan\sqrt{t}$, $f'(t) = \sec^2\sqrt{t} \cdot \frac{1}{2}t^{-1/2} = \dfrac{\sec^2\sqrt{t}}{2\sqrt{t}}$, so $dy = \dfrac{\sec^2\sqrt{t}}{2\sqrt{t}}\,dt$.

(b) For $y = f(v) = \dfrac{1-v^2}{1+v^2}$,

$f'(v) = \dfrac{(1+v^2)(-2v) - (1-v^2)(2v)}{(1+v^2)^2} = \dfrac{-2v[(1+v^2)+(1-v^2)]}{(1+v^2)^2} = \dfrac{-2v(2)}{(1+v^2)^2} = \dfrac{-4v}{(1+v^2)^2}$,

so $dy = \dfrac{-4v}{(1+v^2)^2}\,dv$.

19. (a) $y = \tan x$ $\Rightarrow$ $dy = \sec^2 x\,dx$

(b) When $x = \pi/4$ and $dx = -0.1$, $dy = [\sec(\pi/4)]^2(-0.1) = \left(\sqrt{2}\right)^2(-0.1) = -0.2$.

$\Delta y = f(x + \Delta x) - f(x) = \tan\left(\dfrac{\pi}{4} - 0.1\right) - \tan\left(\dfrac{\pi}{4}\right) \approx -0.18237$.

21. (a) If x is the edge length, then $V = x^3$ $\Rightarrow$ $dV = 3x^2\,dx$. When $x = 30$ and $dx = 0.1$, $dV = 3(30)^2(0.1) = 270$, so the

maximum possible error in computing the volume of the cube is about 270 cm^3. The relative error is calculated by dividing

the change in V, ΔV, by V. We approximate ΔV with dV.

Relative error $= \dfrac{\Delta V}{V} \approx \dfrac{dV}{V} = \dfrac{3x^2\,dx}{x^3} = 3\,\dfrac{dx}{x} = 3\left(\dfrac{0.1}{30}\right) = 0.01$.

Percentage error $=$ relative error $\times 100\% = 0.01 \times 100\% = 1\%$.

(b) $S = 6x^2$ $\Rightarrow$ $dS = 12x\,dx$. When $x = 30$ and $dx = 0.1$, $dS = 12(30)(0.1) = 36$, so the maximum possible error in

computing the surface area of the cube is about 36 cm^2.

Relative error $= \dfrac{\Delta S}{S} \approx \dfrac{dS}{S} = \dfrac{12x\,dx}{6x^2} = 2\,\dfrac{dx}{x} = 2\left(\dfrac{0.1}{30}\right) = 0.00\overline{6}$.

Percentage error $=$ relative error $\times 100\% = 0.00\overline{6} \times 100\% = 0.\overline{6}\%$.

23. (a) For a sphere of radius r, the circumference is $C = 2\pi r$ and the surface area is $S = 4\pi r^2$, so

$$r = \frac{C}{2\pi} \quad \Rightarrow \quad S = 4\pi\left(\frac{C}{2\pi}\right)^2 = \frac{C^2}{\pi} \quad \Rightarrow \quad dS = \frac{2}{\pi}C\,dC. \text{ When } C = 84 \text{ and } dC = 0.5, \, dS = \frac{2}{\pi}(84)(0.5) = \frac{84}{\pi},$$

so the maximum error is about $\dfrac{84}{\pi} \approx 27 \text{ cm}^2$. Relative error $\approx \dfrac{dS}{S} = \dfrac{84/\pi}{84^2/\pi} = \dfrac{1}{84} \approx 0.012 = 1.2\%$

(b) $V = \dfrac{4}{3}\pi r^3 = \dfrac{4}{3}\pi\left(\dfrac{C}{2\pi}\right)^3 = \dfrac{C^3}{6\pi^2} \quad \Rightarrow \quad dV = \dfrac{1}{2\pi^2}C^2\,dC.$ When $C = 84$ and $dC = 0.5$,

$dV = \dfrac{1}{2\pi^2}(84)^2(0.5) = \dfrac{1764}{\pi^2}$, so the maximum error is about $\dfrac{1764}{\pi^2} \approx 179 \text{ cm}^3$.

The relative error is approximately $\dfrac{dV}{V} = \dfrac{1764/\pi^2}{(84)^3/(6\pi^2)} = \dfrac{1}{56} \approx 0.018 = 1.8\%$.

25. $V = RI \quad \Rightarrow \quad I = \dfrac{V}{R} \quad \Rightarrow \quad dI = -\dfrac{V}{R^2}\,dR.$ The relative error in calculating I is $\dfrac{\Delta I}{I} \approx \dfrac{dI}{I} = \dfrac{-(V/R^2)\,dR}{V/R} = -\dfrac{dR}{R}$.

Hence, the relative error in calculating I is approximately the same (in magnitude) as the relative error in R.

27. $F = kR^4 \quad \Rightarrow \quad dF = 4kR^3\,dR \quad \Rightarrow \quad \dfrac{dF}{F} = \dfrac{4kR^3\,dR}{kR^4} = 4\left(\dfrac{dR}{R}\right).$ Thus, the relative change in F is about 4 times the

relative change in R. So a 5% increase in the radius corresponds to a 20% increase in blood flow.

29. (a) The graph shows that $f'(1) = 2$, so $L(x) = f(1) + f'(1)(x - 1) = 5 + 2(x - 1) = 2x + 3$.

$f(0.9) \approx L(0.9) = 4.8$ and $f(1.1) \approx L(1.1) = 5.2$.

(b) From the graph, we see that $f'(x)$ is positive and decreasing. This means that the slopes of the tangent lines are positive, but the tangents are becoming less steep. So the tangent lines lie *above* the curve. Thus, the estimates in part (a) are too large.

2 Review

CONCEPT CHECK

1. See Definition 2.1.1.

2. See the paragraph containing Formula 3 in Section 2.1.

3. See Definition 2.1.4. The pages following the definition discuss interpretations of $f'(a)$ as the slope of a tangent line to the graph of f at $x = a$ and as an instantaneous rate of change of $f(x)$ with respect to x when $x = a$.

4. (a) The average rate of change of y with respect to x over the interval $[x_1, x_2]$ is $\dfrac{f(x_2) - f(x_1)}{x_2 - x_1}$.

(b) The instantaneous rate of change of y with respect to x at $x = x_1$ is $\displaystyle\lim_{x_2 \to x_1} \dfrac{f(x_2) - f(x_1)}{x_2 - x_1}$.

5. See the paragraphs before and after Example 7 in Section 2.2.

6. (a) A function f is differentiable at a number a if its derivative f' exists at $x = a$; that is, if $f'(a)$ exists.

(b) See Theorem 2.2.4. This theorem also tells us that if f is *not* continuous at a, then f is *not* differentiable at a.

(c)

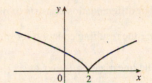

7. See the discussion and Figure 7 on page 89.

8. (a) The Power Rule: If n is any real number, then $\dfrac{d}{dx}(x^n) = nx^{n-1}$. The derivative of a variable base raised to a constant power is the power times the base raised to the power minus one.

(b) The Constant Multiple Rule: If c is a constant and f is a differentiable function, then $\dfrac{d}{dx}[cf(x)] = c\dfrac{d}{dx}f(x)$.

The derivative of a constant times a function is the constant times the derivative of the function.

(c) The Sum Rule: If f and g are both differentiable, then $\dfrac{d}{dx}[f(x) + g(x)] = \dfrac{d}{dx}f(x) + \dfrac{d}{dx}g(x)$. The derivative of a sum of functions is the sum of the derivatives.

(d) The Difference Rule: If f and g are both differentiable, then $\dfrac{d}{dx}[f(x) - g(x)] = \dfrac{d}{dx}f(x) - \dfrac{d}{dx}g(x)$. The derivative of a difference of functions is the difference of the derivatives.

(e) The Product Rule: If f and g are both differentiable, then $\dfrac{d}{dx}[f(x)\,g(x)] = f(x)\dfrac{d}{dx}g(x) + g(x)\dfrac{d}{dx}f(x)$. The derivative of a product of two functions is the first function times the derivative of the second function plus the second function times the derivative of the first function.

(f) The Quotient Rule: If f and g are both differentiable, then $\dfrac{d}{dx}\left[\dfrac{f(x)}{g(x)}\right] = \dfrac{g(x)\dfrac{d}{dx}f(x) - f(x)\dfrac{d}{dx}g(x)}{[g(x)]^2}$.

The derivative of a quotient of functions is the denominator times the derivative of the numerator minus the numerator times the derivative of the denominator, all divided by the square of the denominator.

(g) The Chain Rule: If f and g are both differentiable and $F = f \circ g$ is the composite function defined by $F(x) = f(g(x))$, then F is differentiable and F' is given by the product $F'(x) = f'(g(x))\,g'(x)$. The derivative of a composite function is the derivative of the outer function evaluated at the inner function times the derivative of the inner function.

9. (a) $y = x^n \;\Rightarrow\; y' = nx^{n-1}$

(b) $y = \sin x \;\Rightarrow\; y' = \cos x$

(c) $y = \cos x \;\Rightarrow\; y' = -\sin x$

(d) $y = \tan x \;\Rightarrow\; y' = \sec^2 x$

(e) $y = \csc x \;\Rightarrow\; y' = -\csc x \cot x$

(f) $y = \sec x \;\Rightarrow\; y' = \sec x \tan x$

(g) $y = \cot x \;\Rightarrow\; y' = -\csc^2 x$

10. Implicit differentiation consists of differentiating both sides of an equation involving x and y with respect to x, and then solving the resulting equation for y'.

11. (a) The linearization L of f at $x = a$ is $L(x) = f(a) + f'(a)(x - a)$.

(b) If $y = f(x)$, then the differential dy is given by $dy = f'(x)\,dx$.

(c) See Figure 5 in Section 2.8.

TRUE-FALSE QUIZ

1. False. See the note after Theorem 4 in Section 2.2.

3. False. See the warning before the Product Rule.

5. True. $\dfrac{d}{dx}\sqrt{f(x)} = \dfrac{d}{dx}[f(x)]^{1/2} = \dfrac{1}{2}[f(x)]^{-1/2}\,f'(x) = \dfrac{f'(x)}{2\sqrt{f(x)}}$

7. False. $f(x) = |x^2 + x| = x^2 + x$ for $x \geq 0$ or $x \leq -1$ and $|x^2 + x| = -(x^2 + x)$ for $-1 < x < 0$. So $f'(x) = 2x + 1$ for $x > 0$ or $x < -1$ and $f'(x) = -(2x + 1)$ for $-1 < x < 0$. But $|2x + 1| = 2x + 1$ for $x \geq -\frac{1}{2}$ and $|2x + 1| = -2x - 1$ for $x < -\frac{1}{2}$.

9. True. $g(x) = x^5 \Rightarrow g'(x) = 5x^4 \Rightarrow g'(2) = 5(2)^4 = 80$, and by the definition of the derivative,

$$\lim_{x \to 2} \frac{g(x) - g(2)}{x - 2} = g'(2) = 80.$$

11. False. A tangent line to the parabola $y = x^2$ has slope $dy/dx = 2x$, so at $(-2, 4)$ the slope of the tangent is $2(-2) = -4$ and an equation of the tangent line is $y - 4 = -4(x + 2)$. [The given equation, $y - 4 = 2x(x + 2)$, is not even linear!]

EXERCISES

1. Estimating the slopes of the tangent lines at $x = 2$, 3, and 5, we obtain approximate values 0.4, 2, and 0.1. Since the graph is concave downward at $x = 5$, $f''(5)$ is negative. Arranging the numbers in increasing order, we have:

$f''(5) < 0 < f'(5) < f'(2) < 1 < f'(3)$.

3. (a) $f'(r)$ is the rate at which the total cost changes with respect to the interest rate. Its units are dollars/(percent per year).

(b) The total cost of paying off the loan is increasing by $\$1200$/(percent per year) as the interest rate reaches 10%. So if the interest rate goes up from 10% to 11%, the cost goes up approximately $\$1200$.

(c) As r increases, C increases. So $f'(r)$ will always be positive.

5.

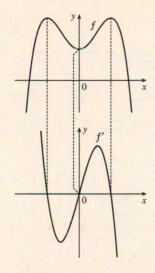

7. The graph of a has tangent lines with positive slope for $x < 0$ and negative slope for $x > 0$, and the values of c fit this pattern, so c must be the graph of the derivative of the function for a. The graph of c has horizontal tangent lines to the left and right of the x-axis and b has zeros at these points. Hence, b is the graph of the derivative of the function for c. Therefore, a is the graph of f, c is the graph of f', and b is the graph of f''.

9. $C'(1990)$ is the rate at which the total value of US currency in circulation is changing in billions of dollars per year. To estimate the value of $C'(1990)$, we will average the difference quotients obtained using the times $t = 1985$ and $t = 1995$.

Let $A = \dfrac{C(1985) - C(1990)}{1985 - 1990} = \dfrac{187.3 - 271.9}{-5} = \dfrac{-84.6}{-5} = 16.92$ and

$B = \dfrac{C(1995) - C(1990)}{1995 - 1990} = \dfrac{409.3 - 271.9}{5} = \dfrac{137.4}{5} = 27.48$. Then

$C'(1990) = \lim\limits_{t \to 1990} \dfrac{C(t) - C(1990)}{t - 1990} \approx \dfrac{A + B}{2} = \dfrac{16.92 + 27.48}{2} = \dfrac{44.4}{2} = 22.2$ billion dollars/year.

11. $f(x) = x^3 + 5x + 4 \quad \Rightarrow$

$f'(x) = \lim\limits_{h \to 0} \dfrac{f(x+h) - f(x)}{h} = \lim\limits_{h \to 0} \dfrac{(x+h)^3 + 5(x+h) + 4 - (x^3 + 5x + 4)}{h}$

$= \lim\limits_{h \to 0} \dfrac{3x^2 h + 3xh^2 + h^3 + 5h}{h} = \lim\limits_{h \to 0} (3x^2 + 3xh + h^2 + 5) = 3x^2 + 5$

13. $y = (x^2 + x^3)^4 \quad \Rightarrow \quad y' = 4(x^2 + x^3)^3(2x + 3x^2) = 4(x^2)^3(1 + x)^3 x(2 + 3x) = 4x^7(x + 1)^3(3x + 2)$

15. $y = \dfrac{x^2 - x + 2}{\sqrt{x}} = x^{3/2} - x^{1/2} + 2x^{-1/2} \quad \Rightarrow \quad y' = \dfrac{3}{2} x^{1/2} - \dfrac{1}{2} x^{-1/2} - x^{-3/2} = \dfrac{3}{2} \sqrt{x} - \dfrac{1}{2\sqrt{x}} - \dfrac{1}{\sqrt{x^3}}$

17. $y = x^2 \sin \pi x \quad \Rightarrow \quad y' = x^2(\cos \pi x)\pi + (\sin \pi x)(2x) = x(\pi x \cos \pi x + 2 \sin \pi x)$

19. $y = \dfrac{t^4 - 1}{t^4 + 1} \quad \Rightarrow \quad y' = \dfrac{(t^4 + 1)4t^3 - (t^4 - 1)4t^3}{(t^4 + 1)^2} = \dfrac{4t^3[(t^4 + 1) - (t^4 - 1)]}{(t^4 + 1)^2} = \dfrac{8t^3}{(t^4 + 1)^2}$

21. $y = \tan\sqrt{1 - x} \quad \Rightarrow \quad y' = (\sec^2\sqrt{1 - x})\left(\dfrac{1}{2\sqrt{1 - x}}\right)(-1) = -\dfrac{\sec^2\sqrt{1 - x}}{2\sqrt{1 - x}}$

23. $\dfrac{d}{dx}(xy^4 + x^2 y) = \dfrac{d}{dx}(x + 3y) \quad \Rightarrow \quad x \cdot 4y^3 y' + y^4 \cdot 1 + x^2 \cdot y' + y \cdot 2x = 1 + 3y' \quad \Rightarrow$

$y'(4xy^3 + x^2 - 3) = 1 - y^4 - 2xy \quad \Rightarrow \quad y' = \dfrac{1 - y^4 - 2xy}{4xy^3 + x^2 - 3}$

25. $y = \dfrac{\sec 2\theta}{1 + \tan 2\theta}$ $\Rightarrow$

$y' = \dfrac{(1 + \tan 2\theta)(\sec 2\theta \, \tan 2\theta \cdot 2) - (\sec 2\theta)(\sec^2 2\theta \cdot 2)}{(1 + \tan 2\theta)^2} = \dfrac{2 \sec 2\theta \left[(1 + \tan 2\theta) \tan 2\theta - \sec^2 2\theta\right]}{(1 + \tan 2\theta)^2}$

$= \dfrac{2 \sec 2\theta \left(\tan 2\theta + \tan^2 2\theta - \sec^2 2\theta\right)}{(1 + \tan 2\theta)^2} = \dfrac{2 \sec 2\theta \left(\tan 2\theta - 1\right)}{(1 + \tan 2\theta)^2}$ $\quad [1 + \tan^2 x = \sec^2 x]$

27. $y = (1 - x^{-1})^{-1}$ $\Rightarrow$

$y' = -1(1 - x^{-1})^{-2}[-(-1x^{-2})] = -(1 - 1/x)^{-2}x^{-2} = -((x-1)/x)^{-2}x^{-2} = -(x-1)^{-2}$

29. $\sin(xy) = x^2 - y$ $\Rightarrow$ $\cos(xy)(xy' + y \cdot 1) = 2x - y'$ $\Rightarrow$ $x\cos(xy)y' + y' = 2x - y\cos(xy)$ $\Rightarrow$

$y'[x\cos(xy) + 1] = 2x - y\cos(xy)$ $\Rightarrow$ $y' = \dfrac{2x - y\cos(xy)}{x\cos(xy) + 1}$

31. $y = \cot(3x^2 + 5)$ $\Rightarrow$ $y' = -\csc^2(3x^2 + 5)(6x) = -6x\csc^2(3x^2 + 5)$

33. $y = \sqrt{x}\cos\sqrt{x}$ $\Rightarrow$

$y' = \sqrt{x}\left(\cos\sqrt{x}\right)' + \cos\sqrt{x}\left(\sqrt{x}\right)' = \sqrt{x}\left[-\sin\sqrt{x}\left(\tfrac{1}{2}x^{-1/2}\right)\right] + \cos\sqrt{x}\left(\tfrac{1}{2}x^{-1/2}\right)$

$= \tfrac{1}{2}x^{-1/2}\left(-\sqrt{x}\sin\sqrt{x} + \cos\sqrt{x}\right) = \dfrac{\cos\sqrt{x} - \sqrt{x}\sin\sqrt{x}}{2\sqrt{x}}$

35. $y = \tan^2(\sin\theta) = [\tan(\sin\theta)]^2$ $\Rightarrow$ $y' = 2[\tan(\sin\theta)] \cdot \sec^2(\sin\theta) \cdot \cos\theta$

37. $y = (x\tan x)^{1/5}$ $\Rightarrow$ $y' = \tfrac{1}{5}(x\tan x)^{-4/5}(\tan x + x\sec^2 x)$

39. $y = \sin(\tan\sqrt{1 + x^3})$ $\Rightarrow$ $y' = \cos(\tan\sqrt{1 + x^3})(\sec^2\sqrt{1 + x^3})\left[3x^2/(2\sqrt{1 + x^3})\right]$

41. $f(t) = \sqrt{4t + 1}$ $\Rightarrow$ $f'(t) = \tfrac{1}{2}(4t + 1)^{-1/2} \cdot 4 = 2(4t + 1)^{-1/2}$ $\Rightarrow$

$f''(t) = 2(-\tfrac{1}{2})(4t + 1)^{-3/2} \cdot 4 = -4/(4t + 1)^{3/2}$, so $f''(2) = -4/9^{3/2} = -\tfrac{4}{27}$.

43. $x^6 + y^6 = 1$ $\Rightarrow$ $6x^5 + 6y^5 y' = 0$ $\Rightarrow$ $y' = -x^5/y^5$ $\Rightarrow$

$y'' = -\dfrac{y^5(5x^4) - x^5(5y^4 y')}{(y^5)^2} = -\dfrac{5x^4 y^4\left[y - x(-x^5/y^5)\right]}{y^{10}} = -\dfrac{5x^4\left[(y^6 + x^6)/y^5\right]}{y^6} = -\dfrac{5x^4}{y^{11}}$

45. $y = 4\sin^2 x$ $\Rightarrow$ $y' = 4 \cdot 2\sin x\cos x$. At $\left(\tfrac{\pi}{6}, 1\right)$, $y' = 8 \cdot \tfrac{1}{2} \cdot \tfrac{\sqrt{3}}{2} = 2\sqrt{3}$, so an equation of the tangent line

is $y - 1 = 2\sqrt{3}\left(x - \tfrac{\pi}{6}\right)$, or $y = 2\sqrt{3}\,x + 1 - \pi\sqrt{3}/3$.

47. $y = \sqrt{1 + 4\sin x}$ $\Rightarrow$ $y' = \tfrac{1}{2}(1 + 4\sin x)^{-1/2} \cdot 4\cos x = \dfrac{2\cos x}{\sqrt{1 + 4\sin x}}$.

At $(0, 1)$, $y' = \dfrac{2}{\sqrt{1}} = 2$, so an equation of the tangent line is $y - 1 = 2(x - 0)$, or $y = 2x + 1$.

The slope of the normal line is $-\tfrac{1}{2}$, so an equation of the normal line is $y - 1 = -\tfrac{1}{2}(x - 0)$, or $y = -\tfrac{1}{2}x + 1$.

49. $y = \sin x + \cos x \;\Rightarrow\; y' = \cos x - \sin x = 0 \;\Leftrightarrow\; \cos x = \sin x$ and $0 \le x \le 2\pi \;\Leftrightarrow\; x = \frac{\pi}{4}$ or $\frac{5\pi}{4}$, so the points

are $\left(\frac{\pi}{4}, \sqrt{2}\,\right)$ and $\left(\frac{5\pi}{4}, -\sqrt{2}\,\right)$.

51. (a) $h(x) = f(x)\,g(x) \;\Rightarrow\; h'(x) = f(x)\,g'(x) + g(x)\,f'(x) \;\Rightarrow$

$\qquad h'(2) = f(2)\,g'(2) + g(2)\,f'(2) = (3)(4) + (5)(-2) = 12 - 10 = 2$

$\quad$ (b) $F(x) = f(g(x)) \;\Rightarrow\; F'(x) = f'(g(x))\,g'(x) \;\Rightarrow\; F'(2) = f'(g(2))\,g'(2) = f'(5)(4) = 11 \cdot 4 = 44$

53. $f(x) = x^2 g(x) \;\Rightarrow\; f'(x) = x^2 g'(x) + g(x)(2x) = x[x g'(x) + 2g(x)]$

55. $f(x) = [g(x)]^2 \;\Rightarrow\; f'(x) = 2[g(x)] \cdot g'(x) = 2g(x)\,g'(x)$

57. $f(x) = g(g(x)) \;\Rightarrow\; f'(x) = g'(g(x))\,g'(x)$

59. $f(x) = g(\sin x) \;\Rightarrow\; f'(x) = g'(\sin x) \cdot \cos x$

61. $h(x) = \dfrac{f(x)\,g(x)}{f(x) + g(x)} \;\Rightarrow$

$\qquad h'(x) = \dfrac{[f(x) + g(x)]\,[f(x)\,g'(x) + g(x)\,f'(x)] - f(x)\,g(x)\,[f'(x) + g'(x)]}{[f(x) + g(x)]^2}$

$\qquad\quad = \dfrac{[f(x)]^2\,g'(x) + f(x)\,g(x)\,f'(x) + f(x)\,g(x)\,g'(x) + [g(x)]^2\,f'(x) - f(x)\,g(x)\,f'(x) - f(x)\,g(x)\,g'(x)}{[f(x) + g(x)]^2}$

$\qquad\quad = \dfrac{f'(x)\,[g(x)]^2 + g'(x)\,[f(x)]^2}{[f(x) + g(x)]^2}$

63. f is not differentiable: at $x = -4$ because f is not continuous, at $x = -1$ because f has a corner, at $x = 2$ because f is not

continuous, and at $x = 5$ because f has a vertical tangent.

65. (a) $y = t^3 - 12t + 3 \;\Rightarrow\; v(t) = y' = 3t^2 - 12 \;\Rightarrow\; a(t) = v'(t) = 6t$

$\quad$ (b) $v(t) = 3(t^2 - 4) > 0$ when $t > 2$, so it moves upward

$\qquad$ when $t > 2$ and downward when $0 \le t < 2$.

$\quad$ (c) Distance upward $= y(3) - y(2) = -6 - (-13) = 7$,

$\qquad$ Distance downward $= y(0) - y(2) = 3 - (-13) = 16$.

$\qquad$ Total distance $= 7 + 16 = 23$.

$\quad$ (d)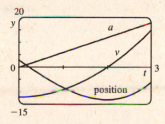

67. If $x =$ edge length, then $V = x^3 \;\Rightarrow\; dV/dt = 3x^2\,dx/dt = 10 \;\Rightarrow\; dx/dt = 10/(3x^2)$ and $S = 6x^2 \;\Rightarrow$

$\qquad dS/dt = (12x)\,dx/dt = 12x[10/(3x^2)] = 40/x$. When $x = 30$, $dS/dt = \frac{40}{30} = \frac{4}{3}$ cm^2/min.

69. Given $dh/dt = 5$ and $dx/dt = 15$, find dz/dt. $z^2 = x^2 + h^2 \;\Rightarrow$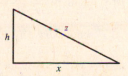

$\qquad 2z\,\dfrac{dz}{dt} = 2x\,\dfrac{dx}{dt} + 2h\,\dfrac{dh}{dt} \;\Rightarrow\; \dfrac{dz}{dt} = \dfrac{1}{z}(15x + 5h)$. When $t = 3$,

$\qquad h = 45 + 3(5) = 60$ and $x = 15(3) = 45 \;\Rightarrow\; z = \sqrt{45^2 + 60^2} = 75$,

$\qquad$ so $\dfrac{dz}{dt} = \dfrac{1}{75}[15(45) + 5(60)] = 13$ ft/s.

71. We are given $d\theta/dt = -0.25$ rad/h. $\tan\theta = 400/x$ $\Rightarrow$

$x = 400\cot\theta$ $\Rightarrow$ $\dfrac{dx}{dt} = -400\csc^2\theta\,\dfrac{d\theta}{dt}$. When $\theta = \dfrac{\pi}{6}$,

$\dfrac{dx}{dt} = -400(2)^2(-0.25) = 400$ ft/h.

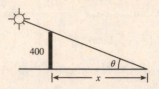

73. (a) $f(x) = \sqrt[3]{1+3x} = (1+3x)^{1/3}$ $\Rightarrow$ $f'(x) = (1+3x)^{-2/3}$, so the linearization of f at $a = 0$ is

$L(x) = f(0) + f'(0)(x-0) = 1^{1/3} + 1^{-2/3}x = 1 + x$. Thus, $\sqrt[3]{1+3x} \approx 1 + x$ $\Rightarrow$

$\sqrt[3]{1.03} = \sqrt[3]{1+3(0.01)} \approx 1 + (0.01) = 1.01$.

(b) The linear approximation is $\sqrt[3]{1+3x} \approx 1 + x$, so for the required accuracy

we want $\sqrt[3]{1+3x} - 0.1 < 1 + x < \sqrt[3]{1+3x} + 0.1$. From the graph,

it appears that this is true when $-0.235 < x < 0.401$.

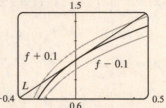

75. $A = x^2 + \frac{1}{2}\pi\left(\frac{1}{2}x\right)^2 = \left(1 + \frac{\pi}{8}\right)x^2$ $\Rightarrow$ $dA = \left(2 + \frac{\pi}{4}\right)x\,dx$. When $x = 60$

and $dx = 0.1$, $dA = \left(2 + \frac{\pi}{4}\right)60(0.1) = 12 + \frac{3\pi}{2}$, so the maximum error is

approximately $12 + \frac{3\pi}{2} \approx 16.7$ cm^2.

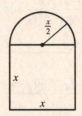

77. $\displaystyle\lim_{h\to 0}\frac{\sqrt[4]{16+h}-2}{h} = \left[\frac{d}{dx}\sqrt[4]{x}\right]_{x=16} = \frac{1}{4}x^{-3/4}\bigg|_{x=16} = \frac{1}{4\left(\sqrt[4]{16}\right)^3} = \frac{1}{32}$

79. $\displaystyle\lim_{x\to 0}\frac{\sqrt{1+\tan x}-\sqrt{1+\sin x}}{x^3} = \lim_{x\to 0}\frac{\left(\sqrt{1+\tan x}-\sqrt{1+\sin x}\right)\left(\sqrt{1+\tan x}+\sqrt{1+\sin x}\right)}{x^3\left(\sqrt{1+\tan x}+\sqrt{1+\sin x}\right)}$

$\displaystyle = \lim_{x\to 0}\frac{(1+\tan x)-(1+\sin x)}{x^3\left(\sqrt{1+\tan x}+\sqrt{1+\sin x}\right)} = \lim_{x\to 0}\frac{\sin x\,(1/\cos x - 1)}{x^3\left(\sqrt{1+\tan x}+\sqrt{1+\sin x}\right)}\cdot\frac{\cos x}{\cos x}$

$\displaystyle = \lim_{x\to 0}\frac{\sin x\,(1-\cos x)}{x^3\left(\sqrt{1+\tan x}+\sqrt{1+\sin x}\right)\cos x}\cdot\frac{1+\cos x}{1+\cos x}$

$\displaystyle = \lim_{x\to 0}\frac{\sin x\cdot\sin^2 x}{x^3\left(\sqrt{1+\tan x}+\sqrt{1+\sin x}\right)\cos x\,(1+\cos x)}$

$\displaystyle = \left(\lim_{x\to 0}\frac{\sin x}{x}\right)^3\lim_{x\to 0}\frac{1}{\left(\sqrt{1+\tan x}+\sqrt{1+\sin x}\right)\cos x\,(1+\cos x)}$

$\displaystyle = 1^3\cdot\frac{1}{\left(\sqrt{1}+\sqrt{1}\right)\cdot 1\cdot(1+1)} = \frac{1}{4}$

81. Using $f'(a) = \displaystyle\lim_{x\to a}\frac{f(x)-f(a)}{x-a}$, we recognize the given expression, $f(x) = \displaystyle\lim_{t\to x}\frac{\sec t - \sec x}{t-x}$, as $g'(x)$

with $g(x) = \sec x$. Now $f'\left(\frac{\pi}{4}\right) = g''\left(\frac{\pi}{4}\right)$, so we will find $g''(x)$. $g'(x) = \sec x\tan x$ $\Rightarrow$

$g''(x) = \sec x\sec^2 x + \tan x\sec x\tan x = \sec x(\sec^2 x + \tan^2 x)$, so $g''\left(\frac{\pi}{4}\right) = \sqrt{2}(\sqrt{2}^2 + 1^2) = \sqrt{2}(2+1) = 3\sqrt{2}$.

3 □ APPLICATIONS OF DIFFERENTIATION

3.1 Maximum and Minimum Values

1. A function f has an **absolute minimum** at $x = c$ if $f(c)$ is the smallest function value on the entire domain of f, whereas f has a **local minimum** at c if $f(c)$ is the smallest function value when x is near c.

3. Absolute maximum at s, absolute minimum at r, local maximum at c, local minima at b and r, neither a maximum nor a minimum at a and d.

5. Absolute maximum value is $f(4) = 5$; there is no absolute minimum value; local maximum values are $f(4) = 5$ and $f(6) = 4$; local minimum values are $f(2) = 2$ and $f(1) = f(5) = 3$.

7. Absolute minimum at 2, absolute maximum at 3, local minimum at 4

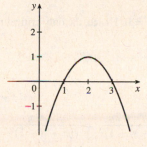

9. Absolute maximum at 5, absolute minimum at 2, local maximum at 3, local minima at 2 and 4

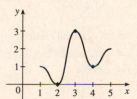

11. (a)

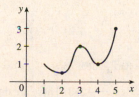

(b)

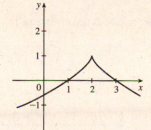

(c)

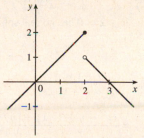

13. (a) *Note:* By the Extreme Value Theorem, f must *not* be continuous; because if it were, it would attain an absolute minimum.

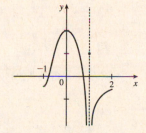

(b)

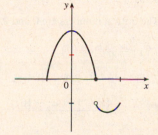

15. $f(x) = \frac{1}{2}(3x - 1)$, $x \le 3$. Absolute maximum

$f(3) = 4$; no local maximum. No absolute or local

minimum.

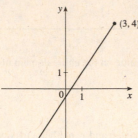

17. $f(x) = \sin x$, $0 \le x < \pi/2$. No absolute or local

maximum. Absolute minimum $f(0) = 0$; no local

minimum.

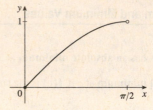

19. $f(x) = 1 + (x + 1)^2$, $-2 \le x < 5$. No absolute or local

maximum. Absolute and local minimum $f(-1) = 1$.

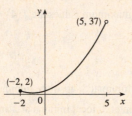

21. $f(x) = 1 - \sqrt{x}$. Absolute maximum $f(0) = 1$;

no local maximum. No absolute or local minimum.

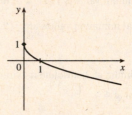

23. $f(x) = 4 + \frac{1}{3}x - \frac{1}{2}x^2 \Rightarrow f'(x) = \frac{1}{3} - x$. $f'(x) = 0 \Rightarrow x = \frac{1}{3}$. This is the only critical number.

25. $f(x) = 2x^3 - 3x^2 - 36x \Rightarrow f'(x) = 6x^2 - 6x - 36 = 6(x^2 - x - 6) = 6(x + 2)(x - 3)$.

$f'(x) = 0 \Leftrightarrow x = -2, 3$. These are the only critical numbers.

27. $g(t) = t^4 + t^3 + t^2 + 1 \Rightarrow g'(t) = 4t^3 + 3t^2 + 2t = t(4t^2 + 3t + 2)$. Using the quadratic formula, we see that

$4t^2 + 3t + 2 = 0$ has no real solution (its discriminant is negative), so $g'(t) = 0$ only if $t = 0$. Hence, the only critical number

is 0.

29. $g(y) = \dfrac{y - 1}{y^2 - y + 1} \Rightarrow$

$g'(y) = \dfrac{(y^2 - y + 1)(1) - (y - 1)(2y - 1)}{(y^2 - y + 1)^2} = \dfrac{y^2 - y + 1 - (2y^2 - 3y + 1)}{(y^2 - y + 1)^2} = \dfrac{-y^2 + 2y}{(y^2 - y + 1)^2} = \dfrac{y(2 - y)}{(y^2 - y + 1)^2}$.

$g'(y) = 0 \Rightarrow y = 0, 2$. The expression $y^2 - y + 1$ is never equal to 0, so $g'(y)$ exists for all real numbers.

The critical numbers are 0 and 2.

31. $F(x) = x^{4/5}(x - 4)^2 \Rightarrow$

$F'(x) = x^{4/5} \cdot 2(x - 4) + (x - 4)^2 \cdot \frac{4}{5}x^{-1/5} = \frac{1}{5}x^{-1/5}(x - 4)[5 \cdot x \cdot 2 + (x - 4) \cdot 4]$

$= \dfrac{(x - 4)(14x - 16)}{5x^{1/5}} = \dfrac{2(x - 4)(7x - 8)}{5x^{1/5}}$

$F'(x) = 0 \Rightarrow x = 4, \frac{8}{7}$. $F'(0)$ does not exist. Thus, the three critical numbers are 0, $\frac{8}{7}$, and 4.

33. $f(\theta) = 2\cos\theta + \sin^2\theta \;\Rightarrow\; f'(\theta) = -2\sin\theta + 2\sin\theta\cos\theta.$ $f'(\theta) = 0 \;\Rightarrow\; 2\sin\theta(\cos\theta - 1) = 0 \;\Rightarrow\; \sin\theta = 0$

or $\cos\theta = 1 \;\Rightarrow\; \theta = n\pi$ [n an integer] or $\theta = 2n\pi$. The solutions $\theta = n\pi$ include the solutions $\theta = 2n\pi$, so the critical

numbers are $\theta = n\pi$.

35. $f(x) = 12 + 4x - x^2,\; [0, 5].$ $f'(x) = 4 - 2x = 0 \;\Leftrightarrow\; x = 2.$ $f(0) = 12,\, f(2) = 16,$ and $f(5) = 7.$

So $f(2) = 16$ is the absolute maximum value and $f(5) = 7$ is the absolute minimum value.

37. $f(x) = 2x^3 - 3x^2 - 12x + 1,\; [-2, 3].$ $f'(x) = 6x^2 - 6x - 12 = 6(x^2 - x - 2) = 6(x - 2)(x + 1) = 0 \;\Leftrightarrow\;$

$x = 2, -1.$ $f(-2) = -3,\, f(-1) = 8,\, f(2) = -19,$ and $f(3) = -8.$ So $f(-1) = 8$ is the absolute maximum value and

$f(2) = -19$ is the absolute minimum value.

39. $f(x) = 3x^4 - 4x^3 - 12x^2 + 1,\, [-2, 3].$ $f'(x) = 12x^3 - 12x^2 - 24x = 12x(x^2 - x - 2) = 12x(x + 1)(x - 2) = 0 \;\Leftrightarrow\;$

$x = -1, 0, 2.$ $f(-2) = 33,\, f(-1) = -4,\, f(0) = 1,\, f(2) = -31,$ and $f(3) = 28.$ So $f(-2) = 33$ is the absolute maximum

value and $f(2) = -31$ is the absolute minimum value.

41. $f(t) = t\sqrt{4 - t^2},\; [-1, 2].$

$\quad f'(t) = t \cdot \frac{1}{2}(4 - t^2)^{-1/2}(-2t) + (4 - t^2)^{1/2} \cdot 1 = \dfrac{-t^2}{\sqrt{4 - t^2}} + \sqrt{4 - t^2} = \dfrac{-t^2 + (4 - t^2)}{\sqrt{4 - t^2}} = \dfrac{4 - 2t^2}{\sqrt{4 - t^2}}.$

$\quad f'(t) = 0 \;\Rightarrow\; 4 - 2t^2 = 0 \;\Rightarrow\; t^2 = 2 \;\Rightarrow\; t = \pm\sqrt{2},$ but $t = -\sqrt{2}$ is not in the given interval, $[-1, 2].$

$\quad f'(t)$ does not exist if $4 - t^2 = 0 \;\Rightarrow\; t = \pm 2,$ but -2 is not in the given interval. $f(-1) = -\sqrt{3},\, f(\sqrt{2}) = 2,$ and

$\quad f(2) = 0.$ So $f(\sqrt{2}) = 2$ is the absolute maximum value and $f(-1) = -\sqrt{3}$ is the absolute minimum value.

43. $f(t) = 2\cos t + \sin 2t,\; [0, \pi/2].$

$\quad f'(t) = -2\sin t + \cos 2t \cdot 2 = -2\sin t + 2(1 - 2\sin^2 t) = -2(2\sin^2 t + \sin t - 1) = -2(2\sin t - 1)(\sin t + 1).$

$\quad f'(t) = 0 \;\Rightarrow\; \sin t = \frac{1}{2}$ or $\sin t = -1 \;\Rightarrow\; t = \frac{\pi}{6}.$ $f(0) = 2,\, f(\frac{\pi}{6}) = \sqrt{3} + \frac{1}{2}\sqrt{3} = \frac{3}{2}\sqrt{3} \approx 2.60,$ and $f(\frac{\pi}{2}) = 0.$

So $f(\frac{\pi}{6}) = \frac{3}{2}\sqrt{3}$ is the absolute maximum value and $f(\frac{\pi}{2}) = 0$ is the absolute minimum value.

45. $f(x) = x^a(1 - x)^b,\; 0 \le x \le 1,\, a > 0,\, b > 0.$

$\quad f'(x) = x^a \cdot b(1 - x)^{b-1}(-1) + (1 - x)^b \cdot ax^{a-1} = x^{a-1}(1 - x)^{b-1}[x \cdot b(-1) + (1 - x) \cdot a]$

$\quad\quad = x^{a-1}(1 - x)^{b-1}(a - ax - bx)$

At the endpoints, we have $f(0) = f(1) = 0$ [the minimum value of f]. In the interval $(0, 1),$ $f'(x) = 0 \;\Leftrightarrow\; x = \dfrac{a}{a + b}.$

$\quad f\!\left(\dfrac{a}{a + b}\right) = \left(\dfrac{a}{a + b}\right)^a\left(1 - \dfrac{a}{a + b}\right)^b = \dfrac{a^a}{(a + b)^a}\left(\dfrac{a + b - a}{a + b}\right)^b = \dfrac{a^a}{(a + b)^a} \cdot \dfrac{b^b}{(a + b)^b} = \dfrac{a^a b^b}{(a + b)^{a+b}}.$

So $f\!\left(\dfrac{a}{a + b}\right) = \dfrac{a^a b^b}{(a + b)^{a+b}}$ is the absolute maximum value.

47. (a)

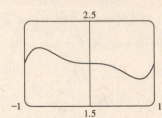

From the graph, it appears that the absolute maximum value is about

$f(-0.77) = 2.19$, and the absolute minimum value is about $f(0.77) = 1.81$.

(b) $f(x) = x^5 - x^3 + 2 \;\Rightarrow\; f'(x) = 5x^4 - 3x^2 = x^2(5x^2 - 3)$. So $f'(x) = 0 \;\Rightarrow\; x = 0, \pm\sqrt{\tfrac{3}{5}}$.

$$f\left(-\sqrt{\tfrac{3}{5}}\right) = \left(-\sqrt{\tfrac{3}{5}}\right)^5 - \left(-\sqrt{\tfrac{3}{5}}\right)^3 + 2 = -\left(\tfrac{3}{5}\right)^2\sqrt{\tfrac{3}{5}} + \tfrac{3}{5}\sqrt{\tfrac{3}{5}} + 2$$

$$= \left(\tfrac{3}{5} - \tfrac{9}{25}\right)\sqrt{\tfrac{3}{5}} + 2 = \tfrac{6}{25}\sqrt{\tfrac{3}{5}} + 2 \quad \text{(maximum)}$$

and similarly, $f\left(\sqrt{\tfrac{3}{5}}\right) = -\tfrac{6}{25}\sqrt{\tfrac{3}{5}} + 2$ (minimum).

49. (a)

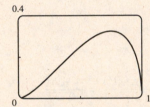

From the graph, it appears that the absolute maximum value is about

$f(0.75) = 0.32$, and the absolute minimum value is $f(0) = f(1) = 0$;

that is, at both endpoints.

(b) $f(x) = x\sqrt{x - x^2} \;\Rightarrow\; f'(x) = x \cdot \dfrac{1 - 2x}{2\sqrt{x - x^2}} + \sqrt{x - x^2} = \dfrac{(x - 2x^2) + (2x - 2x^2)}{2\sqrt{x - x^2}} = \dfrac{3x - 4x^2}{2\sqrt{x - x^2}}$.

So $f'(x) = 0 \;\Rightarrow\; 3x - 4x^2 = 0 \;\Rightarrow\; x(3 - 4x) = 0 \;\Rightarrow\; x = 0$ or $\tfrac{3}{4}$.

$f(0) = f(1) = 0$ (minimum), and $f\left(\tfrac{3}{4}\right) = \tfrac{3}{4}\sqrt{\tfrac{3}{4} - \left(\tfrac{3}{4}\right)^2} = \tfrac{3}{4}\sqrt{\tfrac{3}{16}} = \tfrac{3\sqrt{3}}{16}$ (maximum).

51. The density is defined as $\rho = \dfrac{\text{mass}}{\text{volume}} = \dfrac{1000}{V(T)}$ (in g/cm^3). But a critical point of ρ will also be a critical point of V

[since $\dfrac{d\rho}{dT} = -1000V^{-2}\dfrac{dV}{dT}$ and V is never 0], and V is easier to differentiate than ρ.

$V(T) = 999.87 - 0.06426T + 0.0085043T^2 - 0.0000679T^3 \;\Rightarrow\; V'(T) = -0.06426 + 0.0170086T - 0.0002037T^2$.

Setting this equal to 0 and using the quadratic formula to find T, we get

$$T = \frac{-0.0170086 \pm \sqrt{0.0170086^2 - 4 \cdot 0.0002037 \cdot 0.06426}}{2(-0.0002037)} \approx 3.9665°\text{C} \text{ or } 79.5318°\text{C}. \text{ Since we are only interested}$$

in the region $0°\text{C} \le T \le 30°\text{C}$, we check the density ρ at the endpoints and at $3.9665°\text{C}$: $\rho(0) \approx \dfrac{1000}{999.87} \approx 1.00013$;

$\rho(30) \approx \dfrac{1000}{1003.7628} \approx 0.99625$; $\rho(3.9665) \approx \dfrac{1000}{999.7447} \approx 1.000255$. So water has its maximum density at

about $3.9665°\text{C}$.

53. Let $a = -0.000\,032\,37$, $b = 0.000\,903\,7$, $c = -0.008\,956$, $d = 0.03629$, $e = -0.04458$, and $f = 0.4074$.

Then $S(t) = at^5 + bt^4 + ct^3 + dt^2 + et + f$ and $S'(t) = 5at^4 + 4bt^3 + 3ct^2 + 2dt + e$.

We now apply the Closed Interval Method to the continuous function S on the interval $0 \leq t \leq 10$. Since S' exists for all t,

the only critical numbers of S occur when $S'(t) = 0$. We use a rootfinder on a CAS (or a graphing device) to find that

$S'(t) = 0$ when $t_1 \approx 0.855$, $t_2 \approx 4.618$, $t_3 \approx 7.292$, and $t_4 \approx 9.570$. The values of S at these critical numbers are

$S(t_1) \approx 0.39$, $S(t_2) \approx 0.43645$, $S(t_3) \approx 0.427$, and $S(t_4) \approx 0.43641$. The values of S at the endpoints of the interval are

$S(0) \approx 0.41$ and $S(10) \approx 0.435$. Comparing the six numbers, we see that sugar was most expensive at $t_2 \approx 4.618$

(corresponding roughly to March 1998) and cheapest at $t_1 \approx 0.855$ (June 1994).

55. (a) $v(r) = k(r_0 - r)r^2 = kr_0r^2 - kr^3 \quad \Rightarrow \quad v'(r) = 2kr_0r - 3kr^2$.

$v'(r) = 0 \quad \Rightarrow \quad kr(2r_0 - 3r) = 0 \quad \Rightarrow \quad r = 0$ or $\frac{2}{3}r_0$ (but 0 is not in the

interval). Evaluating v at $\frac{1}{2}r_0$, $\frac{2}{3}r_0$, and r_0, we get $v\left(\frac{1}{2}r_0\right) = \frac{1}{8}kr_0^3$,

$v\left(\frac{2}{3}r_0\right) = \frac{4}{27}kr_0^3$, and $v(r_0) = 0$. Since $\frac{4}{27} > \frac{1}{8}$, v attains its maximum

value at $r = \frac{2}{3}r_0$. This supports the statement in the text.

(b)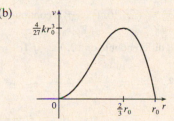

(c) From part (a), the maximum value of v is $\frac{4}{27}kr_0^3$.

57. $f(x) = x^{101} + x^{51} + x + 1 \quad \Rightarrow \quad f'(x) = 101x^{100} + 51x^{50} + 1 \geq 1$ for all x, so $f'(x) = 0$ has no solution. Thus, $f(x)$

has no critical number, so $f(x)$ can have no local maximum or minimum.

59. If f has a local minimum at c, then $g(x) = -f(x)$ has a local maximum at c, so $g'(c) = 0$ by the case of Fermat's Theorem

proved in the text. Thus, $f'(c) = -g'(c) = 0$.

3.2   The Mean Value Theorem

1. $f(x) = 5 - 12x + 3x^2$, $[1, 3]$.   Since f is a polynomial, it is continuous and differentiable on $\mathbb{R}$, so it is continuous on $[1, 3]$

and differentiable on $(1, 3)$. Also $f(1) = -4 = f(3)$. $f'(c) = 0 \quad \Leftrightarrow \quad -12 + 6c = 0 \quad \Leftrightarrow \quad c = 2$, which is in the open

interval $(1, 3)$, so $c = 2$ satisfies the conclusion of Rolle's Theorem.

3. $f(x) = \sqrt{x} - \frac{1}{3}x$, $[0, 9]$. f, being the difference of a root function and a polynomial, is continuous and differentiable

on $[0, \infty)$, so it is continuous on $[0, 9]$ and differentiable on $(0, 9)$. Also, $f(0) = 0 = f(9)$. $f'(c) = 0 \quad \Leftrightarrow$

$\dfrac{1}{2\sqrt{c}} - \dfrac{1}{3} = 0 \quad \Leftrightarrow \quad 2\sqrt{c} = 3 \quad \Leftrightarrow \quad \sqrt{c} = \dfrac{3}{2} \quad \Rightarrow \quad c = \dfrac{9}{4}$, which is in the open interval $(0, 9)$, so $c = \dfrac{9}{4}$ satisfies the

conclusion of Rolle's Theorem.

5. $f(x) = 1 - x^{2/3}$. $f(-1) = 1 - (-1)^{2/3} = 1 - 1 = 0 = f(1)$. $f'(x) = -\frac{2}{3}x^{-1/3}$, so $f'(c) = 0$ has no solution. This

does not contradict Rolle's Theorem, since $f'(0)$ does not exist, and so f is not differentiable on $(-1, 1)$.

7. $f'(c) = \dfrac{f(8) - f(0)}{8 - 0} = \dfrac{3 - 0}{8} = \dfrac{3}{8}$. It appears that

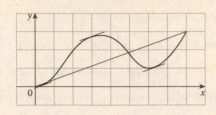

$f'(c) = \frac{3}{8}$ when $c \approx 0.3$, 3, and 6.3.

9. $f(x) = 2x^2 - 3x + 1$, $[0, 2]$. f is continuous on $[0, 2]$ and differentiable on $(0, 2)$ since polynomials are continuous and

differentiable on $\mathbb{R}$. $f'(c) = \dfrac{f(b) - f(a)}{b - a} \iff 4c - 3 = \dfrac{f(2) - f(0)}{2 - 0} = \dfrac{3 - 1}{2} = 1 \iff 4c = 4 \iff c = 1$, which

is in $(0, 2)$.

11. $f(x) = \sqrt[3]{x}$, $[0, 1]$. f is continuous on $\mathbb{R}$ and differentiable on $(-\infty, 0) \cup (0, \infty)$, so f is continuous on $[0, 1]$

and differentiable on $(0, 1)$. $f'(c) = \dfrac{f(b) - f(a)}{b - a} \iff \dfrac{1}{3c^{2/3}} = \dfrac{f(1) - f(0)}{1 - 0} \iff \dfrac{1}{3c^{2/3}} = \dfrac{1 - 0}{1} \iff$

$3c^{2/3} = 1 \iff c^{2/3} = \frac{1}{3} \iff c^2 = \left(\frac{1}{3}\right)^3 = \frac{1}{27} \iff c = \pm\sqrt{\frac{1}{27}} = \pm\frac{\sqrt{3}}{9}$, but only $\frac{\sqrt{3}}{9}$ is in $(0, 1)$.

13. $f(x) = \sqrt{x}$, $[0, 4]$. $f'(c) = \dfrac{f(4) - f(0)}{4 - 0} \iff \dfrac{1}{2\sqrt{c}} = \dfrac{2 - 0}{4} \iff$

$\dfrac{1}{2\sqrt{c}} = \dfrac{1}{2} \iff \sqrt{c} = 1 \iff c = 1$. The secant line and the tangent line

are parallel.

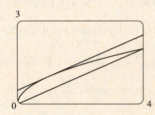

15. $f(x) = (x - 3)^{-2} \Rightarrow f'(x) = -2(x - 3)^{-3}$. $f(4) - f(1) = f'(c)(4 - 1) \Rightarrow \dfrac{1}{1^2} - \dfrac{1}{(-2)^2} = \dfrac{-2}{(c - 3)^3} \cdot 3 \Rightarrow$

$\dfrac{3}{4} = \dfrac{-6}{(c - 3)^3} \Rightarrow (c - 3)^3 = -8 \Rightarrow c - 3 = -2 \Rightarrow c = 1$, which is not in the open interval $(1, 4)$. This does not

contradict the Mean Value Theorem since f is not continuous at $x = 3$.

17. Let $f(x) = 2x + \cos x$. Then $f(-\pi) = -2\pi - 1 < 0$ and $f(0) = 1 > 0$. Since f is the sum of the polynomial $2x$ and the

trignometric function $\cos x$, f is continuous and differentiable for all x. By the Intermediate Value Theorem, there is a number

c in $(-\pi, 0)$ such that $f(c) = 0$. Thus, the given equation has at least one real root. If the equation has distinct real roots a and

b with $a < b$, then $f(a) = f(b) = 0$. Since f is continuous on $[a, b]$ and differentiable on (a, b), Rolle's Theorem implies that

there is a number r in (a, b) such that $f'(r) = 0$. But $f'(r) = 2 - \sin r > 0$ since $\sin r \leq 1$. This contradiction shows that the

given equation can't have two distinct real roots, so it has exactly one root.

19. Let $f(x) = x^3 - 15x + c$ for x in $[-2, 2]$. If f has two real roots a and b in $[-2, 2]$, with $a < b$, then $f(a) = f(b) = 0$. Since

the polynomial f is continuous on $[a, b]$ and differentiable on (a, b), Rolle's Theorem implies that there is a number r in (a, b)

such that $f'(r) = 0$. Now $f'(r) = 3r^2 - 15$. Since r is in (a, b), which is contained in $[-2, 2]$, we have $|r| < 2$, so $r^2 < 4$.

It follows that $3r^2 - 15 < 3 \cdot 4 - 15 = -3 < 0$. This contradicts $f'(r) = 0$, so the given equation can't have two real roots

in $[-2, 2]$. Hence, it has at most one real root in $[-2, 2]$.

21. (a) Suppose that a cubic polynomial $P(x)$ has roots $a_1 < a_2 < a_3 < a_4$, so $P(a_1) = P(a_2) = P(a_3) = P(a_4)$.

By Rolle's Theorem there are numbers c_1, c_2, c_3 with $a_1 < c_1 < a_2$, $a_2 < c_2 < a_3$ and $a_3 < c_3 < a_4$ and

$P'(c_1) = P'(c_2) = P'(c_3) = 0$. Thus, the second-degree polynomial $P'(x)$ has three distinct real roots, which is

impossible.

(b) We prove by induction that a polynomial of degree n has at most n real roots. This is certainly true for $n = 1$. Suppose

that the result is true for all polynomials of degree n and let $P(x)$ be a polynomial of degree $n + 1$. Suppose that $P(x)$ has

more than $n + 1$ real roots, say $a_1 < a_2 < a_3 < \cdots < a_{n+1} < a_{n+2}$. Then $P(a_1) = P(a_2) = \cdots = P(a_{n+2}) = 0$.

By Rolle's Theorem there are real numbers $c_1, \ldots, c_{n+1}$ with $a_1 < c_1 < a_2, \ldots, a_{n+1} < c_{n+1} < a_{n+2}$ and

$P'(c_1) = \cdots = P'(c_{n+1}) = 0$. Thus, the nth degree polynomial $P'(x)$ has at least $n + 1$ roots. This contradiction shows

that $P(x)$ has at most $n + 1$ real roots.

23. By the Mean Value Theorem, $f(4) - f(1) = f'(c)(4 - 1)$ for some $c \in (1, 4)$. But for every $c \in (1, 4)$ we have

$f'(c) \geq 2$. Putting $f'(c) \geq 2$ into the above equation and substituting $f(1) = 10$, we get

$f(4) = f(1) + f'(c)(4 - 1) = 10 + 3f'(c) \geq 10 + 3 \cdot 2 = 16$. So the smallest possible value of $f(4)$ is 16.

25. Suppose that such a function f exists. By the Mean Value Theorem there is a number $0 < c < 2$ with

$f'(c) = \dfrac{f(2) - f(0)}{2 - 0} = \dfrac{5}{2}$. But this is impossible since $f'(x) \leq 2 < \frac{5}{2}$ for all x, so no such function can exist.

27. We use Exercise 26 with $f(x) = \sqrt{1 + x}$, $g(x) = 1 + \frac{1}{2}x$, and $a = 0$. Notice that $f(0) = 1 = g(0)$ and

$f'(x) = \dfrac{1}{2\sqrt{1 + x}} < \dfrac{1}{2} = g'(x)$ for $x > 0$. So by Exercise 26, $f(b) < g(b) \quad \Rightarrow \quad \sqrt{1 + b} < 1 + \frac{1}{2}b$ for $b > 0$.

Another method: Apply the Mean Value Theorem directly to either $f(x) = 1 + \frac{1}{2}x - \sqrt{1 + x}$ or $g(x) = \sqrt{1 + x}$ on $[0, b]$.

29. Let $f(x) = \sin x$ and let $b < a$. Then $f(x)$ is continuous on $[b, a]$ and differentiable on (b, a). By the Mean Value Theorem,

there is a number $c \in (b, a)$ with $\sin a - \sin b = f(a) - f(b) = f'(c)(a - b) = (\cos c)(a - b)$. Thus,

$|\sin a - \sin b| \leq |\cos c| \, |b - a| \leq |a - b|$. If $a < b$, then $|\sin a - \sin b| = |\sin b - \sin a| \leq |b - a| = |a - b|$. If $a = b$, both

sides of the inequality are 0.

31. For $x > 0$, $f(x) = g(x)$, so $f'(x) = g'(x)$. For $x < 0$, $f'(x) = (1/x)' = -1/x^2$ and $g'(x) = (1 + 1/x)' = -1/x^2$, so

again $f'(x) = g'(x)$. However, the domain of $g(x)$ is not an interval [it is $(-\infty, 0) \cup (0, \infty)$] so we cannot conclude that

$f - g$ is constant (in fact it is not).

33. Let $g(t)$ and $h(t)$ be the position functions of the two runners and let $f(t) = g(t) - h(t)$. By hypothesis,

$f(0) = g(0) - h(0) = 0$ and $f(b) = g(b) - h(b) = 0$, where b is the finishing time. Then by the Mean Value Theorem,

there is a time c, with $0 < c < b$, such that $f'(c) = \dfrac{f(b) - f(0)}{b - 0}$. But $f(b) = f(0) = 0$, so $f'(c) = 0$. Since

$f'(c) = g'(c) - h'(c) = 0$, we have $g'(c) = h'(c)$. So at time c, both runners have the same speed $g'(c) = h'(c)$.

3.3 Derivatives and the Shapes of Graphs

Abbreviations: CU = Concave upward, CD = Concave downward, IP = Inflection point, HA = Horizontal asymptote, VA = Vertical asymptote.

1. (a) $f(x) = 2x^3 + 3x^2 - 36x \Rightarrow f'(x) = 6x^2 + 6x - 36 = 6(x^2 + x - 6) = 6(x+3)(x-2)$.

We don't need to include the "6" in the chart to determine the sign of $f'(x)$.

Interval	$x+3$	$x-2$	$f'(x)$	f
$x < -3$	$-$	$-$	$+$	increasing on $(-\infty, -3)$
$-3 < x < 2$	$+$	$-$	$-$	decreasing on $(-3, 2)$
$x > 2$	$+$	$+$	$+$	increasing on $(2, \infty)$

(b) f changes from increasing to decreasing at $x = -3$ and from decreasing to increasing at $x = 2$. Thus, $f(-3) = 81$ is a local maximum value and $f(2) = -44$ is a local minimum value.

(c) $f'(x) = 6x^2 + 6x - 36 \Rightarrow f''(x) = 12x + 6$. $f''(x) = 0$ at $x = -\frac{1}{2}$, $f''(x) > 0 \Leftrightarrow x > -\frac{1}{2}$, and $f''(x) < 0 \Leftrightarrow x < -\frac{1}{2}$. Thus, f is concave upward on $\left(-\frac{1}{2}, \infty\right)$ and concave downward on $\left(-\infty, -\frac{1}{2}\right)$. There is an inflection point at $\left(-\frac{1}{2}, f\left(-\frac{1}{2}\right)\right) = \left(-\frac{1}{2}, \frac{37}{2}\right)$.

3. (a) $f(x) = x^4 - 2x^2 + 3 \Rightarrow f'(x) = 4x^3 - 4x = 4x(x^2 - 1) = 4x(x+1)(x-1)$.

Interval	$x+1$	x	$x-1$	$f'(x)$	f
$x < -1$	$-$	$-$	$-$	$-$	decreasing on $(-\infty, -1)$
$-1 < x < 0$	$+$	$-$	$-$	$+$	increasing on $(-1, 0)$
$0 < x < 1$	$+$	$+$	$-$	$-$	decreasing on $(0, 1)$
$x > 1$	$+$	$+$	$+$	$+$	increasing on $(1, \infty)$

(b) f changes from increasing to decreasing at $x = 0$ and from decreasing to increasing at $x = -1$ and $x = 1$. Thus, $f(0) = 3$ is a local maximum value and $f(\pm 1) = 2$ are local minimum values.

(c) $f''(x) = 12x^2 - 4 = 12\left(x^2 - \frac{1}{3}\right) = 12\left(x + 1/\sqrt{3}\right)\left(x - 1/\sqrt{3}\right)$. $f''(x) > 0 \Leftrightarrow x < -1/\sqrt{3}$ or $x > 1/\sqrt{3}$ and $f''(x) < 0 \Leftrightarrow -1/\sqrt{3} < x < 1/\sqrt{3}$. Thus, f is concave upward on $\left(-\infty, -\sqrt{3}/3\right)$ and $\left(\sqrt{3}/3, \infty\right)$ and concave downward on $\left(-\sqrt{3}/3, \sqrt{3}/3\right)$. There are inflection points at $\left(\pm\sqrt{3}/3, \frac{22}{9}\right)$.

5. (a) $f(x) = \sin x + \cos x$, $0 \leq x \leq 2\pi$. $f'(x) = \cos x - \sin x = 0 \Rightarrow \cos x = \sin x \Rightarrow 1 = \dfrac{\sin x}{\cos x} \Rightarrow \tan x = 1 \Rightarrow x = \frac{\pi}{4}$ or $\frac{5\pi}{4}$. Thus, $f'(x) > 0 \Leftrightarrow \cos x - \sin x > 0 \Leftrightarrow \cos x > \sin x \Leftrightarrow 0 < x < \frac{\pi}{4}$ or $\frac{5\pi}{4} < x < 2\pi$ and $f'(x) < 0 \Leftrightarrow \cos x < \sin x \Leftrightarrow \frac{\pi}{4} < x < \frac{5\pi}{4}$. So f is increasing on $\left(0, \frac{\pi}{4}\right)$ and $\left(\frac{5\pi}{4}, 2\pi\right)$ and f is decreasing on $\left(\frac{\pi}{4}, \frac{5\pi}{4}\right)$.

(b) f changes from increasing to decreasing at $x = \frac{\pi}{4}$ and from decreasing to increasing at $x = \frac{5\pi}{4}$. Thus, $f\left(\frac{\pi}{4}\right) = \sqrt{2}$ is a local maximum value and $f\left(\frac{5\pi}{4}\right) = -\sqrt{2}$ is a local minimum value.

(c) $f''(x) = -\sin x - \cos x = 0$ $\Rightarrow$ $-\sin x = \cos x$ $\Rightarrow$ $\tan x = -1$ $\Rightarrow$ $x = \frac{3\pi}{4}$ or $\frac{7\pi}{4}$. Divide the interval

$(0, 2\pi)$ into subintervals with these numbers as endpoints and complete a second derivative chart.

Interval	$f''(x) = -\sin x - \cos x$	Concavity
$\left(0, \frac{3\pi}{4}\right)$	$f''\left(\frac{\pi}{2}\right) = -1 < 0$	downward
$\left(\frac{3\pi}{4}, \frac{7\pi}{4}\right)$	$f''(\pi) = 1 > 0$	upward
$\left(\frac{7\pi}{4}, 2\pi\right)$	$f''\left(\frac{11\pi}{6}\right) = \frac{1}{2} - \frac{1}{2}\sqrt{3} < 0$	downward

There are inflection points at $\left(\frac{3\pi}{4}, 0\right)$ and $\left(\frac{7\pi}{4}, 0\right)$.

7. $f(x) = 1 + 3x^2 - 2x^3$ $\Rightarrow$ $f'(x) = 6x - 6x^2 = 6x(1 - x)$.

   First Derivative Test: $f'(x) > 0$ $\Rightarrow$ $0 < x < 1$ and $f'(x) < 0$ $\Rightarrow$ $x < 0$ or $x > 1$. Since f' changes from negative

   to positive at $x = 0$, $f(0) = 1$ is a local minimum value; and since f' changes from positive to negative at $x = 1$, $f(1) = 2$ is

   a local maximum value.

   Second Derivative Test: $f''(x) = 6 - 12x$. $f'(x) = 0$ $\Leftrightarrow$ $x = 0, 1$. $f''(0) = 6 > 0$ $\Rightarrow$ $f(0) = 1$ is a local

   minimum value. $f''(1) = -6 < 0$ $\Rightarrow$ $f(1) = 2$ is a local maximum value.

   Preference: For this function, the two tests are equally easy.

9. (a) By the Second Derivative Test, if $f'(2) = 0$ and $f''(2) = -5 < 0$, f has a local maximum at $x = 2$.

   (b) If $f'(6) = 0$, we know that f has a horizontal tangent at $x = 6$. Knowing that $f''(6) = 0$ does not provide any additional

   information since the Second Derivative Test fails. For example, the first and second derivatives of $y = (x - 6)^4$,

   $y = -(x - 6)^4$, and $y = (x - 6)^3$ all equal zero for $x = 6$, but the first has a local minimum at $x = 6$, the second has a

   local maximum at $x = 6$, and the third has an inflection point at $x = 6$.

11. (a) There is an IP at $x = 3$ because the graph of f changes from CD to CU there. There is an IP at $x = 5$ because the graph

   of f changes from CU to CD there.

   (b) There is an IP at $x = 2$ and at $x = 6$ because $f'(x)$ has a maximum value there, and so $f''(x)$ changes from positive to

   negative there. There is an IP at $x = 4$ because $f'(x)$ has a minimum value there and so $f''(x)$ changes from negative to

   positive there.

   (c) There is an inflection point at $x = 1$ because $f''(x)$ changes from negative to positive there, and so the graph of f changes

   from concave downward to concave upward. There is an inflection point at $x = 7$ because $f''(x)$ changes from positive to

   negative there, and so the graph of f changes from concave upward to concave downward.

13. The function must be always decreasing (since the first derivative is always

   negative) and concave downward (since the second derivative is always

   negative).

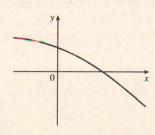

15. $f'(0) = f'(2) = f'(4) = 0 \Rightarrow$ horizontal tangents at $x = 0, 2, 4$.

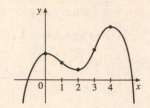

$f'(x) > 0$ if $x < 0$ or $2 < x < 4 \Rightarrow$ f is increasing on $(-\infty, 0)$ and $(2, 4)$.

$f'(x) < 0$ if $0 < x < 2$ or $x > 4 \Rightarrow$ f is decreasing on $(0, 2)$ and $(4, \infty)$.

$f''(x) > 0$ if $1 < x < 3 \Rightarrow$ f is concave upward on $(1, 3)$.

$f''(x) < 0$ if $x < 1$ or $x > 3 \Rightarrow$ f is concave downward on $(-\infty, 1)$

and $(3, \infty)$. There are inflection points when $x = 1$ and 3.

17. $f'(x) > 0$ if $|x| < 2 \Rightarrow$ f is increasing on $(-2, 2)$.

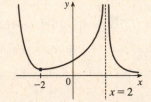
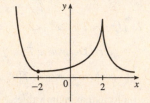

$f'(x) < 0$ if $|x| > 2 \Rightarrow$ f is decreasing on $(-\infty, -2)$

and $(2, \infty)$. $f'(-2) = 0 \Rightarrow$ horizontal tangent at

$x = -2$. $\lim\limits_{x \to 2} |f'(x)| = \infty \Rightarrow$ there is a vertical

asymptote or vertical tangent (cusp) at $x = 2$. $f''(x) > 0$ if $x \neq 2 \Rightarrow$ f is concave upward on $(-\infty, 2)$ and $(2, \infty)$.

19. (a) f is increasing where f' is positive, that is, on $(0, 2)$, $(4, 6)$, and $(8, \infty)$; and decreasing where f' is negative, that is, on

(2, 4) and $(6, 8)$.

(b) f has local maxima where f' changes from positive to negative, at $x = 2$ and at $x = 6$, and local minima where f' changes

from negative to positive, at $x = 4$ and at $x = 8$.

(c) f is concave upward (CU) where f' is increasing, that is, on $(3, 6)$ and $(6, \infty)$, and concave downward (CD) where f' is

decreasing, that is, on $(0, 3)$.

(d) There is a point of inflection where f changes from   (e)

being CD to being CU, that is, at $x = 3$.

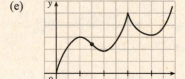

21. (a) $f(x) = x^3 - 12x + 2 \Rightarrow$ $f'(x) = 3x^2 - 12 = 3(x^2 - 4) = 3(x + 2)(x - 2)$. $f'(x) > 0 \Leftrightarrow x < -2$ or $x > 2$

and $f'(x) < 0 \Leftrightarrow -2 < x < 2$. So f is increasing on $(-\infty, -2)$ and $(2, \infty)$ and f is decreasing on $(-2, 2)$.

(b) f changes from increasing to decreasing at $x = -2$, so $f(-2) = 18$ is a   (c)

local maximum value. f changes from decreasing to increasing at $x = 2$,

so $f(2) = -14$ is a local minimum value.

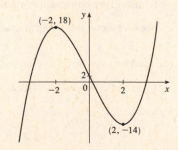

(d) $f''(x) = 6x$. $f''(x) = 0 \Leftrightarrow x = 0$. $f''(x) > 0$ on $(0, \infty)$ and

$f''(x) < 0$ on $(-\infty, 0)$. So f is concave upward on $(0, \infty)$ and f is

concave downward on $(-\infty, 0)$. There is an inflection point at $(0, 2)$.

23. (a) $f(x) = 2 + 2x^2 - x^4 \Rightarrow f'(x) = 4x - 4x^3 = 4x(1 - x^2) = 4x(1 + x)(1 - x)$. $f'(x) > 0 \Leftrightarrow x < -1$ or

$0 < x < 1$ and $f'(x) < 0 \Leftrightarrow -1 < x < 0$ or $x > 1$. So f is increasing on $(-\infty, -1)$ and $(0, 1)$ and f is decreasing

on $(-1, 0)$ and $(1, \infty)$.

(b) f changes from increasing to decreasing at $x = -1$ and $x = 1$, so $f(-1) = 3$ and $f(1) = 3$ are local maximum values.

f changes from decreasing to increasing at $x = 0$, so $f(0) = 2$ is a local minimum value.

(c) $f''(x) = 4 - 12x^2 = 4(1 - 3x^2)$. $f''(x) = 0 \Leftrightarrow 1 - 3x^2 = 0 \Leftrightarrow$ (d)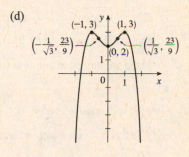

$x^2 = \frac{1}{3} \Leftrightarrow x = \pm 1/\sqrt{3}$. $f''(x) > 0$ on $\left(-1/\sqrt{3}, 1/\sqrt{3}\right)$ and $f''(x) < 0$

on $\left(-\infty, -1/\sqrt{3}\right)$ and $\left(1/\sqrt{3}, \infty\right)$. So f is concave upward on

$\left(-1/\sqrt{3}, 1/\sqrt{3}\right)$ and f is concave downward on $\left(-\infty, -1/\sqrt{3}\right)$ and

$\left(1/\sqrt{3}, \infty\right)$. $f\left(\pm 1/\sqrt{3}\right) = 2 + \frac{2}{3} - \frac{1}{9} = \frac{23}{9}$. There are points of inflection

at $\left(\pm 1/\sqrt{3}, \frac{23}{9}\right)$.

25. (a) $h(x) = (x + 1)^5 - 5x - 2 \Rightarrow h'(x) = 5(x + 1)^4 - 5$. $h'(x) = 0 \Leftrightarrow 5(x + 1)^4 = 5 \Leftrightarrow (x + 1)^4 = 1 \Rightarrow$

$(x + 1)^2 = 1 \Rightarrow x + 1 = 1$ or $x + 1 = -1 \Rightarrow x = 0$ or $x = -2$. $h'(x) > 0 \Leftrightarrow x < -2$ or $x > 0$ and

$h'(x) < 0 \Leftrightarrow -2 < x < 0$. So h is increasing on $(-\infty, -2)$ and $(0, \infty)$ and h is decreasing on $(-2, 0)$.

(b) $h(-2) = 7$ is a local maximum value and $h(0) = -1$ is a local minimum value. (c)

(d) $h''(x) = 20(x + 1)^3 = 0 \Leftrightarrow x = -1$. $h''(x) > 0 \Leftrightarrow x > -1$ and

$h''(x) < 0 \Leftrightarrow x < -1$, so h is CU on $(-1, \infty)$ and h is CD on $(-\infty, -1)$.

There is a point of inflection at $(-1, h(-1)) = (-1, 3)$.

27. (a) $F(x) = x\sqrt{6 - x} \Rightarrow$

$F'(x) = x \cdot \frac{1}{2}(6 - x)^{-1/2}(-1) + (6 - x)^{1/2}(1) = \frac{1}{2}(6 - x)^{-1/2}[-x + 2(6 - x)] = \frac{-3x + 12}{2\sqrt{6 - x}}$.

$F'(x) > 0 \Leftrightarrow -3x + 12 > 0 \Leftrightarrow x < 4$ and $F'(x) < 0 \Leftrightarrow 4 < x < 6$. So F is increasing on $(-\infty, 4)$ and F is

decreasing on $(4, 6)$.

(b) F changes from increasing to decreasing at $x = 4$, so $F(4) = 4\sqrt{2}$ is a local maximum value. There is no local minimum

value.

(c) $F'(x) = -\frac{3}{2}(x - 4)(6 - x)^{-1/2} \Rightarrow$ (d)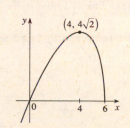

$F''(x) = -\frac{3}{2}\left[(x - 4)\left(-\frac{1}{2}(6 - x)^{-3/2}(-1)\right) + (6 - x)^{-1/2}(1)\right]$

$= -\frac{3}{2} \cdot \frac{1}{2}(6 - x)^{-3/2}[(x - 4) + 2(6 - x)] = \frac{3(x - 8)}{4(6 - x)^{3/2}}$

$F''(x) < 0$ on $(-\infty, 6)$, so F is CD on $(-\infty, 6)$. There is no inflection point.

29. (a) $C(x) = x^{1/3}(x+4) = x^{4/3} + 4x^{1/3}$ ⇒ $C'(x) = \frac{4}{3}x^{1/3} + \frac{4}{3}x^{-2/3} = \frac{4}{3}x^{-2/3}(x+1) = \frac{4(x+1)}{3\sqrt[3]{x^2}}$. $C'(x) > 0$ if

$-1 < x < 0$ or $x > 0$ and $C'(x) < 0$ for $x < -1$, so C is increasing on $(-1, \infty)$ and C is decreasing on $(-\infty, -1)$.

(b) $C(-1) = -3$ is a local minimum value.

(c)

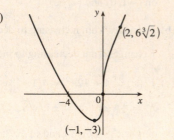

(d) $C''(x) = \frac{4}{9}x^{-2/3} - \frac{8}{9}x^{-5/3} = \frac{4}{9}x^{-5/3}(x-2) = \frac{4(x-2)}{9\sqrt[3]{x^5}}$.

$C''(x) < 0$ for $0 < x < 2$ and $C''(x) > 0$ for $x < 0$ and $x > 2$, so C is

concave downward on $(0, 2)$ and concave upward on $(-\infty, 0)$ and $(2, \infty)$.

There are inflection points at $(0, 0)$ and $\left(2, 6\sqrt[3]{2}\right) \approx (2, 7.56)$.

31. (a) $f(\theta) = 2\cos\theta + \cos^2\theta$, $0 \leq \theta \leq 2\pi$ ⇒ $f'(\theta) = -2\sin\theta + 2\cos\theta(-\sin\theta) = -2\sin\theta(1 + \cos\theta)$.

$f'(\theta) = 0$ ⇔ $\theta = 0, \pi$, and 2π. $f'(\theta) > 0$ ⇔ $\pi < \theta < 2\pi$ and $f'(\theta) < 0$ ⇔ $0 < \theta < \pi$. So f is increasing

on $(\pi, 2\pi)$ and f is decreasing on $(0, \pi)$.

(b) $f(\pi) = -1$ is a local minimum value.

(c) $f'(\theta) = -2\sin\theta(1 + \cos\theta)$ ⇒

$$f''(\theta) = -2\sin\theta(-\sin\theta) + (1 + \cos\theta)(-2\cos\theta) = 2\sin^2\theta - 2\cos\theta - 2\cos^2\theta$$

$$= 2(1 - \cos^2\theta) - 2\cos\theta - 2\cos^2\theta = -4\cos^2\theta - 2\cos\theta + 2$$

$$= -2(2\cos^2\theta + \cos\theta - 1) = -2(2\cos\theta - 1)(\cos\theta + 1)$$

Since $-2(\cos\theta + 1) < 0$ [for $\theta \neq \pi$], $f''(\theta) > 0$ ⇒ $2\cos\theta - 1 < 0$ ⇒ $\cos\theta < \frac{1}{2}$ ⇒ $\frac{\pi}{3} < \theta < \frac{5\pi}{3}$ and

$f''(\theta) < 0$ ⇒ $\cos\theta > \frac{1}{2}$ ⇒ $0 < \theta < \frac{\pi}{3}$ or $\frac{5\pi}{3} < \theta < 2\pi$. So f is CU on $\left(\frac{\pi}{3}, \frac{5\pi}{3}\right)$ and f is CD on $\left(0, \frac{\pi}{3}\right)$ and

$\left(\frac{5\pi}{3}, 2\pi\right)$. There are points of inflection at $\left(\frac{\pi}{3}, f\left(\frac{\pi}{3}\right)\right) = \left(\frac{\pi}{3}, \frac{5}{4}\right)$ and $\left(\frac{5\pi}{3}, f\left(\frac{5\pi}{3}\right)\right) = \left(\frac{5\pi}{3}, \frac{5}{4}\right)$.

(d)

33. $f(x) = 1 + \frac{1}{x} - \frac{1}{x^2}$ has domain $(-\infty, 0) \cup (0, \infty)$.

(a) $\lim\limits_{x \to \pm\infty}\left(1 + \frac{1}{x} - \frac{1}{x^2}\right) = 1$, so $y = 1$ is a HA. $\lim\limits_{x \to 0^+}\left(1 + \frac{1}{x} - \frac{1}{x^2}\right) = \lim\limits_{x \to 0^+}\left(\frac{x^2 + x - 1}{x^2}\right) = -\infty$ since

$(x^2 + x - 1) \to -1$ and $x^2 \to 0$ as $x \to 0^+$ [a similar argument can be made for $x \to 0^-$], so $x = 0$ is a VA.

(b) $f'(x) = -\frac{1}{x^2} + \frac{2}{x^3} = -\frac{1}{x^3}(x-2)$. $f'(x) = 0$ ⇔ $x = 2$. $f'(x) > 0$ ⇔ $0 < x < 2$ and $f'(x) < 0$ ⇔ $x < 0$

or $x > 2$. So f is increasing on $(0, 2)$ and f is decreasing on $(-\infty, 0)$ and $(2, \infty)$.

(c) f changes from increasing to decreasing at $x = 2$, so $f(2) = \frac{5}{4}$ is a local

maximum value. There is no local minimum value.

(e)

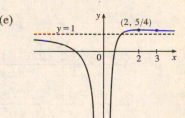

(d) $f''(x) = \frac{2}{x^3} - \frac{6}{x^4} = \frac{2}{x^4}(x - 3)$. $f''(x) = 0 \Leftrightarrow x = 3$. $f''(x) > 0 \Leftrightarrow$

$x > 3$ and $f''(x) < 0 \Leftrightarrow x < 0$ or $0 < x < 3$. So f is CU on $(3, \infty)$ and

f is CD on $(-\infty, 0)$ and $(0, 3)$. There is an inflection point at $\left(3, \frac{11}{9}\right)$.

35. (a) $\lim\limits_{x \to -\infty} \left(\sqrt{x^2 + 1} - x\right) = \infty$ and

$\lim\limits_{x \to \infty} \left(\sqrt{x^2 + 1} - x\right) = \lim\limits_{x \to \infty} \left(\sqrt{x^2 + 1} - x\right) \dfrac{\sqrt{x^2 + 1} + x}{\sqrt{x^2 + 1} + x} = \lim\limits_{x \to \infty} \dfrac{1}{\sqrt{x^2 + 1} + x} = 0$, so $y = 0$ is a HA.

(b) $f(x) = \sqrt{x^2 + 1} - x \Rightarrow f'(x) = \dfrac{x}{\sqrt{x^2 + 1}} - 1$. Since $\dfrac{x}{\sqrt{x^2 + 1}} < 1$ for all x, $f'(x) < 0$, so f is decreasing on $\mathbb{R}$.

(c) No minimum or maximum

(d) $f''(x) = \dfrac{(x^2 + 1)^{1/2}(1) - x \cdot \frac{1}{2}(x^2 + 1)^{-1/2}(2x)}{\left(\sqrt{x^2 + 1}\right)^2}$

(e)

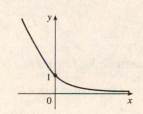

$= \dfrac{(x^2 + 1)^{1/2} - \dfrac{x^2}{(x^2 + 1)^{1/2}}}{x^2 + 1} = \dfrac{(x^2 + 1) - x^2}{(x^2 + 1)^{3/2}} = \dfrac{1}{(x^2 + 1)^{3/2}} > 0,$

so f is CU on $\mathbb{R}$. No IP

37. The nonnegative factors $(x + 1)^2$ and $(x - 6)^4$ do not affect the sign of $f'(x) = (x + 1)^2(x - 3)^5(x - 6)^4$.

So $f'(x) > 0 \Rightarrow (x - 3)^5 > 0 \Rightarrow x - 3 > 0 \Rightarrow x > 3$. Thus, f is increasing on the interval $(3, \infty)$.

39. (a)

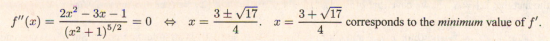

From the graph, we get an estimate of $f(1) \approx 1.41$ as a local maximum

value, and no local minimum value.

$f(x) = \dfrac{x + 1}{\sqrt{x^2 + 1}} \Rightarrow f'(x) = \dfrac{1 - x}{(x^2 + 1)^{3/2}}.$

$f'(x) = 0 \Leftrightarrow x = 1$. $f(1) = \frac{2}{\sqrt{2}} = \sqrt{2}$ is the exact value.

(b) From the graph in part (a), f increases most rapidly somewhere between $x = -\frac{1}{2}$ and $x = -\frac{1}{4}$. To find the exact value,

we need to find the maximum value of f', which we can do by finding the critical numbers of f'.

$f''(x) = \dfrac{2x^2 - 3x - 1}{(x^2 + 1)^{5/2}} = 0 \Leftrightarrow x = \dfrac{3 \pm \sqrt{17}}{4}$. $x = \dfrac{3 + \sqrt{17}}{4}$ corresponds to the *minimum* value of f'.

The maximum value of f' occurs at $x = \frac{3 - \sqrt{17}}{4} \approx -0.28$.

41. $f(x) = ax^3 + bx^2 + cx + d \Rightarrow f'(x) = 3ax^2 + 2bx + c$.

We are given that $f(1) = 0$ and $f(-2) = 3$, so $f(1) = a + b + c + d = 0$ and

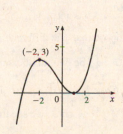

$f(-2) = -8a + 4b - 2c + d = 3$. Also $f'(1) = 3a + 2b + c = 0$ and

$f'(-2) = 12a - 4b + c = 0$ by Fermat's Theorem. Solving these four equations, we get

$a = \frac{2}{9}, b = \frac{1}{3}, c = -\frac{4}{3}, d = \frac{7}{9}$, so the function is $f(x) = \frac{1}{9}\left(2x^3 + 3x^2 - 12x + 7\right)$.

43. $y = \dfrac{1+x}{1+x^2}$ $\Rightarrow$ $y' = \dfrac{(1+x^2)(1) - (1+x)(2x)}{(1+x^2)^2} = \dfrac{1 - 2x - x^2}{(1+x^2)^2}$ $\Rightarrow$

$y'' = \dfrac{(1+x^2)^2(-2-2x) - (1-2x-x^2) \cdot 2(1+x^2)(2x)}{[(1+x^2)^2]^2} = \dfrac{2(1+x^2)[(1+x^2)(-1-x) - (1-2x-x^2)(2x)]}{(1+x^2)^4}$

$= \dfrac{2(-1 - x - x^2 - x^3 - 2x + 4x^2 + 2x^3)}{(1+x^2)^3} = \dfrac{2(x^3 + 3x^2 - 3x - 1)}{(1+x^2)^3} = \dfrac{2(x-1)(x^2 + 4x + 1)}{(1+x^2)^3}$

So $y'' = 0$ $\Rightarrow$ $x = 1, -2 \pm \sqrt{3}$. Let $a = -2 - \sqrt{3}$, $b = -2 + \sqrt{3}$, and $c = 1$. We can show that $f(a) = \frac{1}{4}(1 - \sqrt{3})$,

$f(b) = \frac{1}{4}(1 + \sqrt{3})$, and $f(c) = 1$. To show that these three points of inflection lie on one straight line, we'll show that the

slopes m_{ac} and m_{bc} are equal.

$$m_{ac} = \frac{f(c) - f(a)}{c - a} = \frac{1 - \frac{1}{4}(1 - \sqrt{3})}{1 - (-2 - \sqrt{3})} = \frac{\frac{3}{4} + \frac{1}{4}\sqrt{3}}{3 + \sqrt{3}} = \frac{1}{4}$$

$$m_{bc} = \frac{f(c) - f(b)}{c - b} = \frac{1 - \frac{1}{4}(1 + \sqrt{3})}{1 - (-2 + \sqrt{3})} = \frac{\frac{3}{4} - \frac{1}{4}\sqrt{3}}{3 - \sqrt{3}} = \frac{1}{4}$$

45. Let the cubic function be $f(x) = ax^3 + bx^2 + cx + d$ $\Rightarrow$ $f'(x) = 3ax^2 + 2bx + c$ $\Rightarrow$ $f''(x) = 6ax + 2b$.

So f is CU when $6ax + 2b > 0$ $\Leftrightarrow$ $x > -b/(3a)$, CD when $x < -b/(3a)$, and so the only point of inflection occurs

when $x = -b/(3a)$. If the graph has three x-intercepts x_1, x_2 and x_3, then the expression for $f(x)$ must factor as

$f(x) = a(x - x_1)(x - x_2)(x - x_3)$. Multiplying these factors together gives us

$$f(x) = a[x^3 - (x_1 + x_2 + x_3)x^2 + (x_1 x_2 + x_1 x_3 + x_2 x_3)x - x_1 x_2 x_3]$$

Equating the coefficients of the x^2-terms for the two forms of f gives us $b = -a(x_1 + x_2 + x_3)$. Hence, the x-coordinate of

the point of inflection is $-\dfrac{b}{3a} = -\dfrac{-a(x_1 + x_2 + x_3)}{3a} = \dfrac{x_1 + x_2 + x_3}{3}$.

47. By hypothesis $g = f'$ is differentiable on an open interval containing c. Since $(c, f(c))$ is a point of inflection, the concavity

changes at $x = c$, so $f''(x)$ changes signs at $x = c$. Hence, by the First Derivative Test, f' has a local extremum at $x = c$.

Thus, by Fermat's Theorem $f''(c) = 0$.

49. Using the fact that $|x| = \sqrt{x^2}$, we have that $g(x) = x|x| = x\sqrt{x^2}$ $\Rightarrow$ $g'(x) = \sqrt{x^2} + \sqrt{x^2} = 2\sqrt{x^2} = 2|x|$ $\Rightarrow$

$g''(x) = 2x(x^2)^{-1/2} = \dfrac{2x}{|x|} < 0$ for $x < 0$ and $g''(x) > 0$ for $x > 0$, so $(0, 0)$ is an inflection point. But $g''(0)$ does not

exist.

51. Suppose that f is differentiable on an interval I and $f'(x) > 0$ for all x in I except $x = c$. To show that f is increasing on I,

let x_1, x_2 be two numbers in I with $x_1 < x_2$.

Case 1 $x_1 < x_2 < c$. Let J be the interval $\{x \in I \mid x < c\}$. By applying the Increasing/Decreasing Test to f

on J, we see that f is increasing on J, so $f(x_1) < f(x_2)$.

Case 2 $c < x_1 < x_2$. Apply the Increasing/Decreasing Test to f on $K = \{x \in I \mid x > c\}$.

Case 3 $x_1 < x_2 = c$. Apply the proof of the Increasing/Decreasing Test, using the Mean Value Theorem (MVT) on the interval $[x_1, x_2]$ and noting that the MVT does not require f to be differentiable at the endpoints of $[x_1, x_2]$.

Case 4 $c = x_1 < x_2$. Same proof as in Case 3.

Case 5 $x_1 < c < x_2$. By Cases 3 and 4, f is increasing on $[x_1, c]$ and on $[c, x_2]$, so $f(x_1) < f(c) < f(x_2)$.

In all cases, we have shown that $f(x_1) < f(x_2)$. Since x_1, x_2 were any numbers in I with $x_1 < x_2$, we have shown that f is increasing on I.

3.4   Curve Sketching

1. $y = f(x) = x^3 - 12x^2 + 36x = x(x^2 - 12x + 36) = x(x-6)^2$ **A.** f is a polynomial, so $D = \mathbb{R}$.
B. x-intercepts are 0 and 6, y-intercept $= f(0) = 0$ **C.** No symmetry   **D.** No asymptote

E. $f'(x) = 3x^2 - 24x + 36 = 3(x^2 - 8x + 12) = 3(x-2)(x-6) < 0$ $\Leftrightarrow$
$2 < x < 6$, so f is decreasing on $(2, 6)$ and increasing on $(-\infty, 2)$ and $(6, \infty)$.
F. Local maximum value $f(2) = 32$, local minimum value $f(6) = 0$

G. $f''(x) = 6x - 24 = 6(x-4) > 0$ $\Leftrightarrow$ $x > 4$, so f is CU on $(4, \infty)$ and
CD on $(-\infty, 4)$. IP at $(4, 16)$

H.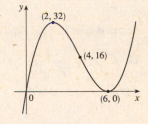

3. $y = f(x) = x^4 - 4x = x(x^3 - 4)$ **A.** $D = \mathbb{R}$ **B.** x-intercepts are 0 and $\sqrt[3]{4}$,
y-intercept $= f(0) = 0$ **C.** No symmetry   **D.** No asymptote

E. $f'(x) = 4x^3 - 4 = 4(x^3 - 1) = 4(x-1)(x^2 + x + 1) > 0$ $\Leftrightarrow$ $x > 1$, so
f is increasing on $(1, \infty)$ and decreasing on $(-\infty, 1)$.   **F.** Local minimum value
$f(1) = -3$, no local maximum   **G.** $f''(x) = 12x^2 > 0$ for all x, so f is CU on
$(-\infty, \infty)$. No IP

H.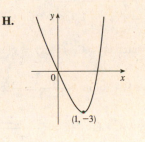

5. $y = f(x) = x(x-4)^3$ **A.** $D = \mathbb{R}$ **B.** x-intercepts are 0 and 4, y-intercept $f(0) = 0$ **C.** No symmetry

D. No asymptote

H.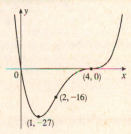

E. $f'(x) = x \cdot 3(x-4)^2 + (x-4)^3 \cdot 1 = (x-4)^2[3x + (x-4)]$
$\qquad = (x-4)^2(4x - 4) = 4(x-1)(x-4)^2 > 0$ $\Leftrightarrow$

$x > 1$, so f is increasing on $(1, \infty)$ and decreasing on $(-\infty, 1)$.

F. Local minimum value $f(1) = -27$, no local maximum value

G. $f''(x) = 4[(x-1) \cdot 2(x-4) + (x-4)^2 \cdot 1] = 4(x-4)[2(x-1) + (x-4)]$
$\qquad = 4(x-4)(3x - 6) = 12(x-4)(x-2) < 0$ $\Leftrightarrow$

$2 < x < 4$, so f is CD on $(2, 4)$ and CU on $(-\infty, 2)$ and $(4, \infty)$. IPs at $(2, -16)$ and $(4, 0)$

7. $y = f(x) = \frac{1}{5}x^5 - \frac{8}{3}x^3 + 16x = x\left(\frac{1}{5}x^4 - \frac{8}{3}x^2 + 16\right)$ **A.** $D = \mathbb{R}$ **B.** x-intercept 0, y-intercept $= f(0) = 0$

C. $f(-x) = -f(x)$, so f is odd; the curve is symmetric about the origin.   **D.** No asymptote

E. $f'(x) = x^4 - 8x^2 + 16 = (x^2 - 4)^2 = (x+2)^2(x-2)^2 > 0$ for all x

except ± 2, so f is increasing on $\mathbb{R}$.   **F.** There is no local maximum or

minimum value.

G. $f''(x) = 4x^3 - 16x = 4x(x^2 - 4) = 4x(x+2)(x-2) > 0$ ⇔

$-2 < x < 0$ or $x > 2$, so f is CU on $(-2, 0)$ and $(2, \infty)$, and f is CD on

$(-\infty, -2)$ and $(0, 2)$. IP at $\left(-2, -\frac{256}{15}\right)$, $(0, 0)$, and $\left(2, \frac{256}{15}\right)$

H.

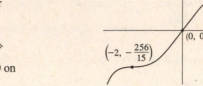

9. $y = f(x) = x/(x-1)$ **A.** $D = \{x \mid x \neq 1\} = (-\infty, 1) \cup (1, \infty)$ **B.** x-intercept $= 0$, y-intercept $= f(0) = 0$

C. No symmetry   **D.** $\lim\limits_{x \to \pm\infty} \dfrac{x}{x-1} = 1$, so $y = 1$ is a HA. $\lim\limits_{x \to 1^-} \dfrac{x}{x-1} = -\infty$, $\lim\limits_{x \to 1^+} \dfrac{x}{x-1} = \infty$, so $x = 1$ is a VA.

E. $f'(x) = \dfrac{(x-1) - x}{(x-1)^2} = \dfrac{-1}{(x-1)^2} < 0$ for $x \neq 1$, so f is

decreasing on $(-\infty, 1)$ and $(1, \infty)$.   **F.** No extreme values

G. $f''(x) = \dfrac{2}{(x-1)^3} > 0$ ⇔ $x > 1$, so f is CU on $(1, \infty)$ and

CD on $(-\infty, 1)$. No IP

H.

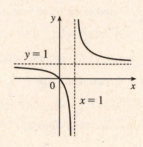

11. $y = f(x) = 1/(x^2 - 9)$ **A.** $D = \{x \mid x \neq \pm 3\} = (-\infty, -3) \cup (-3, 3) \cup (3, \infty)$ **B.** y-intercept $= f(0) = -\frac{1}{9}$, no

x-intercept   **C.** $f(-x) = f(x)$ ⇒ f is even; the curve is symmetric about the y-axis.   **D.** $\lim\limits_{x \to \pm\infty} \dfrac{1}{x^2 - 9} = 0$, so $y = 0$

is a HA. $\lim\limits_{x \to 3^-} \dfrac{1}{x^2 - 9} = -\infty$, $\lim\limits_{x \to 3^+} \dfrac{1}{x^2 - 9} = \infty$, $\lim\limits_{x \to -3^-} \dfrac{1}{x^2 - 9} = \infty$, $\lim\limits_{x \to -3^+} \dfrac{1}{x^2 - 9} = -\infty$, so $x = 3$ and $x = -3$

are VA.   **E.** $f'(x) = -\dfrac{2x}{(x^2 - 9)^2} > 0$ ⇔ $x < 0$ $(x \neq -3)$ so f is increasing on $(-\infty, -3)$ and $(-3, 0)$ and

decreasing on $(0, 3)$ and $(3, \infty)$.   **F.** Local maximum value $f(0) = -\frac{1}{9}$.

H.

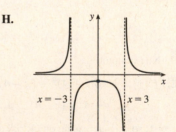

G. $y'' = \dfrac{-2(x^2 - 9)^2 + (2x)2(x^2 - 9)(2x)}{(x^2 - 9)^4} = \dfrac{6(x^2 + 3)}{(x^2 - 9)^3} > 0$ ⇔

$x^2 > 9$ ⇔ $x > 3$ or $x < -3$, so f is CU on $(-\infty, -3)$ and $(3, \infty)$ and

CD on $(-3, 3)$. No IP

13. $y = f(x) = x/(x^2 + 9)$ **A.** $D = \mathbb{R}$ **B.** y-intercept: $f(0) = 0$; x-intercept: $f(x) = 0$ ⇔ $x = 0$

C. $f(-x) = -f(x)$, so f is odd and the curve is symmetric about the origin.   **D.** $\lim\limits_{x \to \pm\infty} [x/(x^2 + 9)] = 0$, so $y = 0$ is a

HA; no VA   **E.** $f'(x) = \dfrac{(x^2 + 9)(1) - x(2x)}{(x^2 + 9)^2} = \dfrac{9 - x^2}{(x^2 + 9)^2} = \dfrac{(3 + x)(3 - x)}{(x^2 + 9)^2} > 0$ ⇔ $-3 < x < 3$, so f is increasing on $(-3, 3)$ and decreasing on $(-\infty, -3)$ and $(3, \infty)$. **F.** Local minimum value $f(-3) = -\frac{1}{6}$, local maximum value $f(3) = \frac{1}{6}$

G. $f''(x) = \dfrac{(x^2+9)^2(-2x) - (9-x^2) \cdot 2(x^2+9)(2x)}{[(x^2+9)^2]^2} = \dfrac{(2x)(x^2+9)\left[-(x^2+9) - 2(9-x^2)\right]}{(x^2+9)^4}$

$= \dfrac{2x(x^2-27)}{(x^2+9)^3} = 0 \iff x = 0, \pm\sqrt{27} = \pm 3\sqrt{3}$

H.

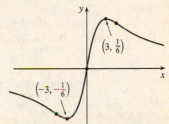

$f''(x) > 0 \iff -3\sqrt{3} < x < 0$ or $x > 3\sqrt{3}$, so f is CU on $\left(-3\sqrt{3}, 0\right)$ and $(3\sqrt{3}, \infty)$, and CD on $\left(-\infty, -3\sqrt{3}\right)$ and $\left(0, 3\sqrt{3}\right)$. There are three inflection points: $(0, 0)$ and $\left(\pm 3\sqrt{3}, \pm\frac{1}{12}\sqrt{3}\right)$.

15. $y = f(x) = \dfrac{x-1}{x^2}$ **A.** $D = \{x \mid x \neq 0\} = (-\infty, 0) \cup (0, \infty)$ **B.** No y-intercept; x-intercept: $f(x) = 0 \iff x = 1$

C. No symmetry **D.** $\displaystyle\lim_{x\to\pm\infty} \dfrac{x-1}{x^2} = 0$, so $y = 0$ is a HA. $\displaystyle\lim_{x\to 0} \dfrac{x-1}{x^2} = -\infty$, so $x = 0$ is a VA.

E. $f'(x) = \dfrac{x^2 \cdot 1 - (x-1) \cdot 2x}{(x^2)^2} = \dfrac{-x^2 + 2x}{x^4} = \dfrac{-(x-2)}{x^3}$, so $f'(x) > 0 \iff 0 < x < 2$ and $f'(x) < 0 \iff$

$x < 0$ or $x > 2$. Thus, f is increasing on $(0, 2)$ and decreasing on $(-\infty, 0)$ and $(2, \infty)$. **F.** No local minimum, local maximum value $f(2) = \frac{1}{4}$.

H.

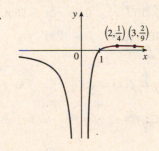

G. $f''(x) = \dfrac{x^3 \cdot (-1) - [-(x-2)] \cdot 3x^2}{(x^3)^2} = \dfrac{2x^3 - 6x^2}{x^6} = \dfrac{2(x-3)}{x^4}$.

$f''(x)$ is negative on $(-\infty, 0)$ and $(0, 3)$ and positive on $(3, \infty)$, so f is CD on $(-\infty, 0)$ and $(0, 3)$ and CU on $(3, \infty)$. IP at $\left(3, \frac{2}{9}\right)$

17. $y = f(x) = x\sqrt{5-x}$ **A.** The domain is $\{x \mid 5 - x \geq 0\} = (-\infty, 5]$ **B.** y-intercept: $f(0) = 0$; x-intercepts: $f(x) = 0 \iff x = 0, 5$ **C.** No symmetry **D.** No asymptote

E. $f'(x) = x \cdot \frac{1}{2}(5-x)^{-1/2}(-1) + (5-x)^{1/2} \cdot 1 = \frac{1}{2}(5-x)^{-1/2}[-x + 2(5-x)] = \dfrac{10-3x}{2\sqrt{5-x}} > 0 \iff x < \frac{10}{3}$,

so f is increasing on $\left(-\infty, \frac{10}{3}\right)$ and decreasing on $\left(\frac{10}{3}, 5\right)$.

F. Local maximum value $f\left(\frac{10}{3}\right) = \frac{10}{9}\sqrt{15} \approx 4.3$; no local minimum

H.

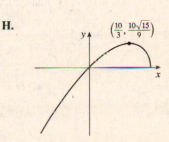

G. $f''(x) = \dfrac{2(5-x)^{1/2}(-3) - (10-3x) \cdot 2\left(\frac{1}{2}\right)(5-x)^{-1/2}(-1)}{\left(2\sqrt{5-x}\right)^2}$

$= \dfrac{(5-x)^{-1/2}[-6(5-x) + (10-3x)]}{4(5-x)} = \dfrac{3x-20}{4(5-x)^{3/2}}$

$f''(x) < 0$ for $x < 5$, so f is CD on $(-\infty, 5)$. No IP

19. $y = f(x) = x/\sqrt{x^2 + 1}$ **A.** $D = \mathbb{R}$ **B.** y-intercept: $f(0) = 0$; x-intercepts: $f(x) = 0 \Rightarrow x = 0$

C. $f(-x) = -f(x)$, so f is odd; the graph is symmetric about the origin.

D. $\displaystyle\lim_{x\to\infty} f(x) = \lim_{x\to\infty} \frac{x}{\sqrt{x^2+1}} = \lim_{x\to\infty} \frac{x/x}{\sqrt{x^2+1}/x} = \lim_{x\to\infty} \frac{x/x}{\sqrt{x^2+1}/\sqrt{x^2}} = \lim_{x\to\infty} \frac{1}{\sqrt{1+1/x^2}} = \frac{1}{\sqrt{1+0}} = 1$

and

$\displaystyle\lim_{x\to-\infty} f(x) = \lim_{x\to-\infty} \frac{x}{\sqrt{x^2+1}} = \lim_{x\to-\infty} \frac{x/x}{\sqrt{x^2+1}/x} = \lim_{x\to-\infty} \frac{x/x}{\sqrt{x^2+1}/\left(-\sqrt{x^2}\right)} = \lim_{x\to-\infty} \frac{1}{-\sqrt{1+1/x^2}}$

$\qquad = \dfrac{1}{-\sqrt{1+0}} = -1$ so $y = \pm 1$ are HA.

No VA.

E. $f'(x) = \dfrac{\sqrt{x^2+1} - x \cdot \dfrac{2x}{2\sqrt{x^2+1}}}{[(x^2+1)^{1/2}]^2} = \dfrac{x^2+1-x^2}{(x^2+1)^{3/2}} = \dfrac{1}{(x^2+1)^{3/2}} > 0$ for all x, so f is increasing on $\mathbb{R}$.

F. No extreme values

G. $f''(x) = -\frac{3}{2}(x^2+1)^{-5/2} \cdot 2x = \dfrac{-3x}{(x^2+1)^{5/2}}$, so $f''(x) > 0$ for $x < 0$

and $f''(x) < 0$ for $x > 0$. Thus, f is CU on $(-\infty, 0)$ and CD on $(0, \infty)$.

IP at $(0, 0)$

H.

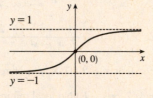

21. $y = f(x) = \sqrt{1-x^2}/x$ **A.** $D = \{x \mid |x| \le 1, x \ne 0\} = [-1, 0) \cup (0, 1]$ **B.** x-intercepts ± 1, no y-intercept

C. $f(-x) = -f(x)$, so the curve is symmetric about $(0, 0)$. **D.** $\displaystyle\lim_{x\to 0^+} \frac{\sqrt{1-x^2}}{x} = \infty$, $\displaystyle\lim_{x\to 0^-} \frac{\sqrt{1-x^2}}{x} = -\infty$,

so $x = 0$ is a VA. **E.** $f'(x) = \dfrac{\left(-x^2/\sqrt{1-x^2}\right) - \sqrt{1-x^2}}{x^2} = -\dfrac{1}{x^2\sqrt{1-x^2}} < 0$, so f is decreasing

on $(-1, 0)$ and $(0, 1)$. **F.** No extreme values

G. $f''(x) = \dfrac{2-3x^2}{x^3(1-x^2)^{3/2}} > 0 \Leftrightarrow -1 < x < -\sqrt{\frac{2}{3}}$ or $0 < x < \sqrt{\frac{2}{3}}$, so

f is CU on $\left(-1, -\sqrt{\frac{2}{3}}\right)$ and $\left(0, \sqrt{\frac{2}{3}}\right)$ and CD on $\left(-\sqrt{\frac{2}{3}}, 0\right)$ and $\left(\sqrt{\frac{2}{3}}, 1\right)$.

IP at $\left(\pm\sqrt{\frac{2}{3}}, \pm\frac{1}{\sqrt{2}}\right)$

H.

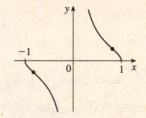

23. $y = f(x) = x - 3x^{1/3}$ **A.** $D = \mathbb{R}$ **B.** y-intercept: $f(0) = 0$; x-intercepts: $f(x) = 0 \Rightarrow x = 3x^{1/3} \Rightarrow$

$x^3 = 27x \Rightarrow x^3 - 27x = 0 \Rightarrow x(x^2 - 27) = 0 \Rightarrow x = 0, \pm 3\sqrt{3}$ **C.** $f(-x) = -f(x)$, so f is odd;

the graph is symmetric about the origin. **D.** No asymptote **E.** $f'(x) = 1 - x^{-2/3} = 1 - \dfrac{1}{x^{2/3}} = \dfrac{x^{2/3} - 1}{x^{2/3}}$.

$f'(x) > 0$ when $|x| > 1$ and $f'(x) < 0$ when $0 < |x| < 1$, so f is increasing on $(-\infty, -1)$ and $(1, \infty)$, and decreasing on $(-1, 0)$ and $(0, 1)$ [hence decreasing on $(-1, 1)$ since f is

continuous on $(-1, 1)$]. **F.** Local maximum value $f(-1) = 2$, local minimum

value $f(1) = -2$ **G.** $f''(x) = \frac{2}{3}x^{-5/3} < 0$ when $x < 0$ and $f''(x) > 0$

when $x > 0$, so f is CD on $(-\infty, 0)$ and CU on $(0, \infty)$. IP at $(0, 0)$

H.

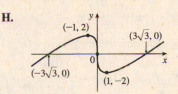

25. $y = f(x) = \sqrt[3]{x^2 - 1}$ **A.** $D = \mathbb{R}$ **B.** y-intercept: $f(0) = -1$; x-intercepts: $f(x) = 0$ $\Leftrightarrow$ $x^2 - 1 = 0$ $\Leftrightarrow$

$x = \pm 1$ **C.** $f(-x) = f(x)$, so the curve is symmetric about the y-axis. **D.** No asymptote

E. $f'(x) = \frac{1}{3}(x^2 - 1)^{-2/3}(2x) = \dfrac{2x}{3\sqrt[3]{(x^2 - 1)^2}}$. $f'(x) > 0$ $\Leftrightarrow$ $x > 0$ and $f'(x) < 0$ $\Leftrightarrow$ $x < 0$, so f is

increasing on $(0, \infty)$ and decreasing on $(-\infty, 0)$. **F.** Local minimum value $f(0) = -1$

G. $f''(x) = \dfrac{2}{3} \cdot \dfrac{(x^2 - 1)^{2/3}(1) - x \cdot \frac{2}{3}(x^2 - 1)^{-1/3}(2x)}{[(x^2 - 1)^{2/3}]^2}$

H.

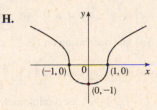

$\qquad = \dfrac{2}{9} \cdot \dfrac{(x^2 - 1)^{-1/3}[3(x^2 - 1) - 4x^2]}{(x^2 - 1)^{4/3}} = -\dfrac{2(x^2 + 3)}{9(x^2 - 1)^{5/3}}$

$f''(x) > 0$ $\Leftrightarrow$ $-1 < x < 1$ and $f''(x) < 0$ $\Leftrightarrow$ $x < -1$ or $x > 1$, so

f is CU on $(-1, 1)$ and f is CD on $(-\infty, -1)$ and $(1, \infty)$. IP at $(\pm 1, 0)$

27. $y = f(x) = \sin^3 x$ **A.** $D = \mathbb{R}$ **B.** x-intercepts: $f(x) = 0$ $\Leftrightarrow$ $x = n\pi$, n an integer; y-intercept $= f(0) = 0$

C. $f(-x) = -f(x)$, so f is odd and the curve is symmetric about the origin. Also, $f(x + 2\pi) = f(x)$, so f is periodic

with period 2π, and we determine **E–G** for $0 \le x \le \pi$. Since f is odd, we can reflect the graph of f on $[0, \pi]$ about the

origin to obtain the graph of f on $[-\pi, \pi]$, and then since f has period 2π, we can extend the graph of f for all real numbers.

D. No asymptote **E.** $f'(x) = 3\sin^2 x \cos x > 0$ $\Leftrightarrow$ $\cos x > 0$ and $\sin x \ne 0$ $\Leftrightarrow$ $0 < x < \frac{\pi}{2}$, so f is increasing on

$\left(0, \frac{\pi}{2}\right)$ and f is decreasing on $\left(\frac{\pi}{2}, \pi\right)$. **F.** Local maximum value $f\left(\frac{\pi}{2}\right) = 1$ [local minimum value $f\left(-\frac{\pi}{2}\right) = -1$]

G. $f''(x) = 3\sin^2 x \, (-\sin x) + 3\cos x \, (2\sin x \cos x) = 3\sin x \, (2\cos^2 x - \sin^2 x)$

$\qquad = 3\sin x \, [2(1 - \sin^2 x) - \sin^2 x] = 3\sin x (2 - 3\sin^2 x) > 0$ $\Leftrightarrow$

$\sin x > 0$ and $\sin^2 x < \frac{2}{3}$ $\Leftrightarrow$ $0 < x < \pi$ and $0 < \sin x < \sqrt{\frac{2}{3}}$ $\Leftrightarrow$ $0 < x < \sin^{-1}\sqrt{\frac{2}{3}}$ $\left[\text{let } \alpha = \sin^{-1}\sqrt{\frac{2}{3}}\right]$ or

$\pi - \alpha < x < \pi$, so f is CU on $(0, \alpha)$ and $(\pi - \alpha, \pi)$, and f is CD on $(\alpha, \pi - \alpha)$. There are inflection points at $x = 0, \pi, \alpha$,

and $x = \pi - \alpha$.

H.

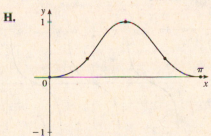

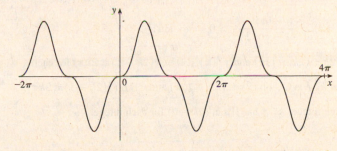

29. $y = f(x) = x \tan x$, $-\frac{\pi}{2} < x < \frac{\pi}{2}$ **A.** $D = \left(-\frac{\pi}{2}, \frac{\pi}{2}\right)$ **B.** Intercepts are 0 **C.** $f(-x) = f(x)$, so the curve is

symmetric about the y-axis. **D.** $\lim\limits_{x \to (\pi/2)^-} x \tan x = \infty$ and $\lim\limits_{x \to -(\pi/2)^+} x \tan x = \infty$, so $x = \frac{\pi}{2}$ and $x = -\frac{\pi}{2}$ are VA.

E. $f'(x) = \tan x + x \sec^2 x > 0 \iff 0 < x < \frac{\pi}{2}$, so f increases on $\left(0, \frac{\pi}{2}\right)$ **H.**

and decreases on $\left(-\frac{\pi}{2}, 0\right)$. **F.** Absolute and local minimum value $f(0) = 0$.

G. $y'' = 2 \sec^2 x + 2x \tan x \sec^2 x > 0$ for $-\frac{\pi}{2} < x < \frac{\pi}{2}$, so f is

CU on $\left(-\frac{\pi}{2}, \frac{\pi}{2}\right)$. No IP

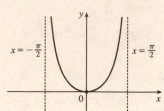

31. $y = f(x) = \frac{1}{2}x - \sin x$, $0 < x < 3\pi$ **A.** $D = (0, 3\pi)$ **B.** No y-intercept. (We don't need to find the x-intercept

because the equation $\sin x = \frac{1}{2}x$ can't be solved exactly. See Section 3.6.) **C.** No symmetry **D.** No asymptote

E. $f'(x) = \frac{1}{2} - \cos x > 0 \iff \cos x < \frac{1}{2} \iff \frac{\pi}{3} < x < \frac{5\pi}{3}$ or $\frac{7\pi}{3} < x < 3\pi$, so f is increasing on $\left(\frac{\pi}{3}, \frac{5\pi}{3}\right)$ and

$\left(\frac{7\pi}{3}, 3\pi\right)$ and decreasing on $\left(0, \frac{\pi}{3}\right)$ and $\left(\frac{5\pi}{3}, \frac{7\pi}{3}\right)$.

F. Local minimum value $f\left(\frac{\pi}{3}\right) = \frac{\pi}{6} - \frac{\sqrt{3}}{2}$, local maximum value **H.**

$f\left(\frac{5\pi}{3}\right) = \frac{5\pi}{6} + \frac{\sqrt{3}}{2}$, local minimum value $f\left(\frac{7\pi}{3}\right) = \frac{7\pi}{6} - \frac{\sqrt{3}}{2}$

G. $f''(x) = \sin x > 0 \iff 0 < x < \pi$ or $2\pi < x < 3\pi$, so f is CU on

$(0, \pi)$ and $(2\pi, 3\pi)$ and CD on $(\pi, 2\pi)$. IPs at $\left(\pi, \frac{\pi}{2}\right)$ and $(2\pi, \pi)$

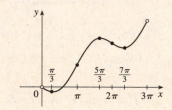

33. $y = f(x) = \dfrac{\sin x}{1 + \cos x} \left[\overset{\text{when}}{\underset{\cos x \neq 1}{=}} \dfrac{\sin x}{1 + \cos x} \cdot \dfrac{1 - \cos x}{1 - \cos x} = \dfrac{\sin x\,(1 - \cos x)}{\sin^2 x} = \dfrac{1 - \cos x}{\sin x} = \csc x - \cot x \right]$

A. The domain of f is the set of all real numbers except odd integer multiples of π; that is, all reals except $(2n + 1)\pi$, where

n is an integer. **B.** y-intercept: $f(0) = 0$; x-intercepts: $x = 2n\pi$, n an integer. **C.** $f(-x) = -f(x)$, so f is an odd

function; the graph is symmetric about the origin and has period 2π. **D.** When n is an odd integer,

$\lim\limits_{x \to (n\pi)^-} f(x) = \infty$ and $\lim\limits_{x \to (n\pi)^+} f(x) = -\infty$, so $x = n\pi$ is a VA for each odd integer n. No HA.

E. $f'(x) = \dfrac{(1 + \cos x) \cdot \cos x - \sin x(-\sin x)}{(1 + \cos x)^2} = \dfrac{1 + \cos x}{(1 + \cos x)^2} = \dfrac{1}{1 + \cos x}$. $f'(x) > 0$ for all x except odd multiples of

π, so f is increasing on $((2k - 1)\pi, (2k + 1)\pi)$ for each integer k. **F.** No extreme values

G. $f''(x) = \dfrac{\sin x}{(1 + \cos x)^2} > 0 \Rightarrow \sin x > 0 \Rightarrow$ **H.**

$x \in (2k\pi, (2k + 1)\pi)$ and $f''(x) < 0$ on $((2k - 1)\pi, 2k\pi)$ for each

integer k. f is CU on $(2k\pi, (2k + 1)\pi)$ and CD on $((2k - 1)\pi, 2k\pi)$

for each integer k. f has IPs at $(2k\pi, 0)$ for each integer k.

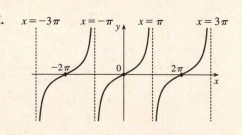

35. $m = f(v) = \dfrac{m_0}{\sqrt{1 - v^2/c^2}}$. The m-intercept is $f(0) = m_0$. There are no v-intercepts. $\lim\limits_{v \to c^-} f(v) = \infty$, so $v = c$ is a VA.

$$f'(v) = -\tfrac{1}{2} m_0 (1 - v^2/c^2)^{-3/2}(-2v/c^2) = \dfrac{m_0 v}{c^2(1 - v^2/c^2)^{3/2}} = \dfrac{m_0 v}{\dfrac{c^2(c^2 - v^2)^{3/2}}{c^3}} = \dfrac{m_0 cv}{(c^2 - v^2)^{3/2}} > 0, \text{ so } f \text{ is}$$

increasing on $(0, c)$. There are no local extreme values.

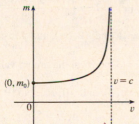

$$f''(v) = \dfrac{(c^2 - v^2)^{3/2}(m_0 c) - m_0 cv \cdot \tfrac{3}{2}(c^2 - v^2)^{1/2}(-2v)}{[(c^2 - v^2)^{3/2}]^2}$$

$$= \dfrac{m_0 c(c^2 - v^2)^{1/2}[(c^2 - v^2) + 3v^2]}{(c^2 - v^2)^3} = \dfrac{m_0 c(c^2 + 2v^2)}{(c^2 - v^2)^{5/2}} > 0,$$

so f is CU on $(0, c)$. There are no inflection points.

37. $y = -\dfrac{W}{24EI}x^4 + \dfrac{WL}{12EI}x^3 - \dfrac{WL^2}{24EI}x^2 = -\dfrac{W}{24EI}x^2(x^2 - 2Lx + L^2)$

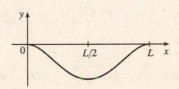

$$= \dfrac{-W}{24EI}x^2(x - L)^2 = cx^2(x - L)^2$$

where $c = -\dfrac{W}{24EI}$ is a negative constant and $0 \le x \le L$. We sketch

$f(x) = cx^2(x - L)^2$ for $c = -1$. $f(0) = f(L) = 0$.

$f'(x) = cx^2[2(x - L)] + (x - L)^2(2cx) = 2cx(x - L)[x + (x - L)] = 2cx(x - L)(2x - L)$. So for $0 < x < L$,

$f'(x) > 0 \iff x(x - L)(2x - L) < 0$ [since $c < 0$] $\iff L/2 < x < L$ and $f'(x) < 0 \iff 0 < x < L/2$.

Thus, f is increasing on $(L/2, L)$ and decreasing on $(0, L/2)$, and there is a local and absolute minimum at the

point $(L/2, f(L/2)) = (L/2, cL^4/16)$. $f'(x) = 2c[x(x - L)(2x - L)] \implies$

$f''(x) = 2c[1(x - L)(2x - L) + x(1)(2x - L) + x(x - L)(2)] = 2c(6x^2 - 6Lx + L^2) = 0 \iff$

$x = \dfrac{6L \pm \sqrt{12L^2}}{12} = \tfrac{1}{2}L \pm \dfrac{\sqrt{3}}{6}L$, and these are the x-coordinates of the two inflection points.

39. $y = f(x) = \dfrac{x^2}{x - 1} = x + 1 + \dfrac{1}{x - 1}$ **A.** $D = (-\infty, 1) \cup (1, \infty)$ **B.** x-intercept: $f(x) = 0 \iff x = 0$;

y-intercept: $f(0) = 0$ **C.** No symmetry   **D.** $\lim\limits_{x \to 1^-} f(x) = -\infty$ and $\lim\limits_{x \to 1^+} f(x) = \infty$, so $x = 1$ is a VA.

$\lim\limits_{x \to \pm\infty}[f(x) - (x + 1)] = \lim\limits_{x \to \pm\infty} \dfrac{1}{x - 1} = 0$, so the line $y = x + 1$ is a SA.

E. $f'(x) = 1 - \dfrac{1}{(x - 1)^2} = \dfrac{(x - 1)^2 - 1}{(x - 1)^2} = \dfrac{x^2 - 2x}{(x - 1)^2} = \dfrac{x(x - 2)}{(x - 1)^2} > 0$ for   **H.**

$x < 0$ or $x > 2$, so f is increasing on $(-\infty, 0)$ and $(2, \infty)$, and f is decreasing

on $(0, 1)$ and $(1, 2)$.   **F.** Local maximum value $f(0) = 0$, local minimum value

$f(2) = 4$ **G.** $f''(x) = \dfrac{2}{(x - 1)^3} > 0$ for $x > 1$, so f is CU on $(1, \infty)$ and f

is CD on $(-\infty, 1)$. No IP

41. $y = f(x) = \dfrac{x^3 + 4}{x^2} = x + \dfrac{4}{x^2}$ **A.** $D = (-\infty, 0) \cup (0, \infty)$ **B.** x-intercept: $f(x) = 0 \iff x = -\sqrt[3]{4}$; no y-intercept

C. No symmetry **D.** $\lim\limits_{x \to 0} f(x) = \infty$, so $x = 0$ is a VA. $\lim\limits_{x \to \pm\infty} [f(x) - x] = \lim\limits_{x \to \pm\infty} \dfrac{4}{x^2} = 0$, so $y = x$ is a SA.

E. $f'(x) = 1 - \dfrac{8}{x^3} = \dfrac{x^3 - 8}{x^3} > 0$ for $x < 0$ or $x > 2$, so f is increasing on **H.**

$(-\infty, 0)$ and $(2, \infty)$, and f is decreasing on $(0, 2)$. **F.** Local minimum value

$f(2) = 3$, no local maximum value **G.** $f''(x) = \dfrac{24}{x^4} > 0$ for $x \neq 0$, so f is CU

on $(-\infty, 0)$ and $(0, \infty)$. No IP

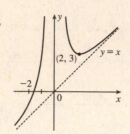

43. $y = f(x) = \sqrt{4x^2 + 9} \Rightarrow f'(x) = \dfrac{4x}{\sqrt{4x^2 + 9}} \Rightarrow$

$f''(x) = \dfrac{\sqrt{4x^2 + 9} \cdot 4 - 4x \cdot 4x/\sqrt{4x^2 + 9}}{4x^2 + 9} = \dfrac{4(4x^2 + 9) - 16x^2}{(4x^2 + 9)^{3/2}} = \dfrac{36}{(4x^2 + 9)^{3/2}}.$ f is defined on $(-\infty, \infty)$.

$f(-x) = f(x)$, so f is even, which means its graph is symmetric about the y-axis. The y-intercept is $f(0) = 3$. There are no x-intercepts since $f(x) > 0$ for all x.

$\lim\limits_{x \to \infty} \left(\sqrt{4x^2 + 9} - 2x\right) = \lim\limits_{x \to \infty} \dfrac{\left(\sqrt{4x^2 + 9} - 2x\right)\left(\sqrt{4x^2 + 9} + 2x\right)}{\sqrt{4x^2 + 9} + 2x}$

$= \lim\limits_{x \to \infty} \dfrac{(4x^2 + 9) - 4x^2}{\sqrt{4x^2 + 9} + 2x} = \lim\limits_{x \to \infty} \dfrac{9}{\sqrt{4x^2 + 9} + 2x} = 0$

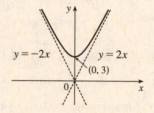

and, similarly, $\lim\limits_{x \to -\infty} \left(\sqrt{4x^2 + 9} + 2x\right) = \lim\limits_{x \to -\infty} \dfrac{9}{\sqrt{4x^2 + 9} - 2x} = 0,$

so $y = \pm 2x$ are slant asymptotes. f is decreasing on $(-\infty, 0)$ and increasing on $(0, \infty)$ with local minimum $f(0) = 3$.

$f''(x) > 0$ for all x, so f is CU on $\mathbb{R}$.

45. $f(x) = 4x^4 - 32x^3 + 89x^2 - 95x + 29 \Rightarrow f'(x) = 16x^3 - 96x^2 + 178x - 95 \Rightarrow f''(x) = 48x^2 - 192x + 178.$

$f(x) = 0 \iff x \approx 0.5, 1.60$; $f'(x) = 0 \iff x \approx 0.92, 2.5, 2.58$ and $f''(x) = 0 \iff x \approx 1.46, 2.54.$

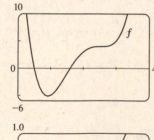

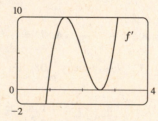

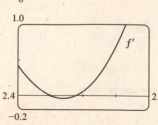

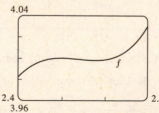

From the graphs of f', we estimate that $f' < 0$ and that f is decreasing on $(-\infty, 0.92)$ and $(2.5, 2.58)$, and that $f' > 0$ and f

is increasing on $(0.92, 2.5)$ and $(2.58, \infty)$ with local minimum values $f(0.92) \approx -5.12$ and $f(2.58) \approx 3.998$ and local

maximum value $f(2.5) = 4$. The graphs of f' make it clear that f has a maximum and a minimum near $x = 2.5$, shown more

clearly in the fourth graph.

From the graph of f'', we estimate that $f'' > 0$ and that f is CU on

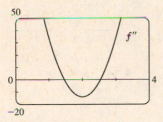

$(-\infty, 1.46)$ and $(2.54, \infty)$, and that $f'' < 0$ and f is CD on $(1.46, 2.54)$.

There are inflection points at about $(1.46, -1.40)$ and $(2.54, 3.999)$.

47. $f(x) = 6 \sin x + \cot x$, $-\pi \leq x \leq \pi$ $\Rightarrow$ $f'(x) = 6 \cos x - \csc^2 x$ $\Rightarrow$ $f''(x) = -6 \sin x + 2 \csc^2 x \cot x$

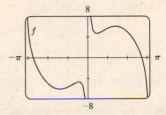 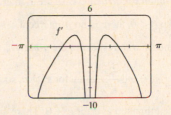 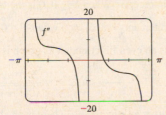

From the graph of f, we see that there are VAs at $x = 0$ and $x = \pm\pi$. f is an odd function, so its graph is symmetric about

the origin. From the graph of f', we estimate that f is decreasing on $(-\pi, -1.40)$, increasing on $(-1.40, -0.44)$, decreasing

on $(-0.44, 0)$, decreasing on $(0, 0.44)$, increasing on $(0.44, 1.40)$, and decreasing on $(1.40, \pi)$, with local minimum values

$f(-1.40) \approx -6.09$ and $f(0.44) \approx 4.68$, and local maximum values $f(-0.44) \approx -4.68$ and $f(1.40) \approx 6.09$.

From the graph of f'', we estimate that f is CU on $(-\pi, -0.77)$, CD on $(-0.77, 0)$, CU on $(0, 0.77)$, and CD on

$(0.77, \pi)$. There are IPs at about $(-0.77, -5.22)$ and $(0.77, 5.22)$.

49. $f(x) = 1 + \dfrac{1}{x} + \dfrac{8}{x^2} + \dfrac{1}{x^3}$ $\Rightarrow$ $f'(x) = -\dfrac{1}{x^2} - \dfrac{16}{x^3} - \dfrac{3}{x^4} = -\dfrac{1}{x^4}(x^2 + 16x + 3)$ $\Rightarrow$

$f''(x) = \dfrac{2}{x^3} + \dfrac{48}{x^4} + \dfrac{12}{x^5} = \dfrac{2}{x^5}(x^2 + 24x + 6)$.

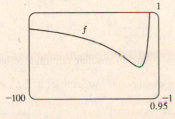 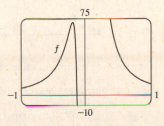

From the graphs, it appears that f increases on $(-15.8, -0.2)$ and decreases on $(-\infty, -15.8)$, $(-0.2, 0)$, and $(0, \infty)$; that f

has a local minimum value of $f(-15.8) \approx 0.97$ and a local maximum value of $f(-0.2) \approx 72$; that f is CD on $(-\infty, -24)$

and $(-0.25, 0)$ and is CU on $(-24, -0.25)$ and $(0, \infty)$; and that f has IPs at $(-24, 0.97)$ and $(-0.25, 60)$.

To find the exact values, note that $f' = 0$ $\Rightarrow$ $x = \dfrac{-16 \pm \sqrt{256 - 12}}{2} = -8 \pm \sqrt{61}$ $[\approx -0.19$ and $-15.81]$.

f' is positive (f is increasing) on $(-8 - \sqrt{61}, -8 + \sqrt{61})$ and f' is negative (f is decreasing) on $(-\infty, -8 - \sqrt{61})$,

$(-8+\sqrt{61}, 0)$, and $(0,\infty)$. $f''=0 \Rightarrow x = \dfrac{-24 \pm \sqrt{576-24}}{2} = -12 \pm \sqrt{138}$ $[\approx -0.25$ and $-23.75]$. f'' is

positive (f is CU) on $(-12-\sqrt{138}, -12+\sqrt{138})$ and $(0,\infty)$ and f'' is negative (f is CD) on $(-\infty, -12-\sqrt{138})$

and $(-12+\sqrt{138}, 0)$.

51. $f(x) = \sqrt{x^4 + cx^2} = |x|\sqrt{x^2+c} \Rightarrow f'(x) = \dfrac{x(2x^2+c)}{\sqrt{x^4+cx^2}} \Rightarrow$

$f''(x) = \dfrac{x^4(2x^2+3c)}{(x^4+cx^2)^{3/2}}$

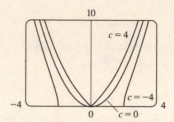

x-intercepts: When $c \geq 0$, 0 is the only *x*-intercept. When $c < 0$, the *x*-intercepts are $\pm\sqrt{-c}$.

y-intercept $= f(0) = 0$ when $c \geq 0$. When $c < 0$, there is no *y*-intercept.

f is an even function, so its graph is symmetric with respect to the *y*-axis.

$f'(x) = \dfrac{x(2x^2+c)}{|x|\sqrt{x^2+c}} = -\dfrac{2x^2+c}{\sqrt{x^2+c}}$ for $x < 0$ and $\dfrac{2x^2+c}{\sqrt{x^2+c}}$ for $x > 0$, so f has a corner or "point" at $x = 0$ that gets

sharper as c increases. There is an absolute minimum at 0 for $c \geq 0$. There are no other maximums nor minimums.

$f''(x) = \dfrac{x^4(2x^2+3c)}{|x|^3(x^2+c)^{3/2}}$, so $f''(x) = 0 \Rightarrow x = \pm\sqrt{-3c/2}$ [only for $c < 0$]. f'' changes sign at $\pm\sqrt{-3c/2}$, so f is

CU on $\left(-\infty, -\sqrt{-3c/2}\right)$ and $\left(\sqrt{-3c/2}, \infty\right)$, and f is CD on $\left(-\sqrt{-3c/2}, -\sqrt{-c}\right)$ and $\left(\sqrt{-c}, \sqrt{-3c/2}\right)$. There are

IPs at $\left(\pm\sqrt{-3c/2}, \sqrt{3c^2/4}\right)$. The more negative c becomes, the farther the IPs move from the origin. The only transitional

value is $c = 0$.

53. Note that $c = 0$ is a transitional value at which the graph consists of the *x*-axis. Also, we can see that if we substitute $-c$ for c,

the function $f(x) = \dfrac{cx}{1+c^2x^2}$ will be reflected in the *x*-axis, so we investigate only positive values of c (except $c = -1$, as a

demonstration of this reflective property). Also, f is an odd function. $\displaystyle\lim_{x\to\pm\infty} f(x) = 0$, so $y = 0$ is a horizontal asymptote

for all c. We calculate $f'(x) = \dfrac{(1+c^2x^2)c - cx(2c^2x)}{(1+c^2x^2)^2} = -\dfrac{c(c^2x^2-1)}{(1+c^2x^2)^2}$. $f'(x) = 0 \Leftrightarrow c^2x^2 - 1 = 0 \Leftrightarrow$

$x = \pm 1/c$. So there is an absolute maximum value of $f(1/c) = \frac{1}{2}$ and an absolute minimum value of $f(-1/c) = -\frac{1}{2}$.

These extrema have the same value regardless of c, but the maximum points move closer to the *y*-axis as c increases.

$f''(x) = \dfrac{(-2c^3x)(1+c^2x^2)^2 - (-c^3x^2+c)[2(1+c^2x^2)(2c^2x)]}{(1+c^2x^2)^4}$

$\quad = \dfrac{(-2c^3x)(1+c^2x^2) + (c^3x^2-c)(4c^2x)}{(1+c^2x^2)^3} = \dfrac{2c^3x(c^2x^2-3)}{(1+c^2x^2)^3}$

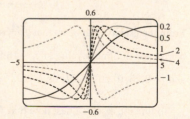

$f''(x) = 0 \Leftrightarrow x = 0$ or $\pm\sqrt{3}/c$, so there are inflection points at $(0,0)$ and

at $\left(\pm\sqrt{3}/c, \pm\sqrt{3}/4\right)$. Again, the *y*-coordinate of the inflection points does not depend on c, but as c increases, both inflection

points approach the *y*-axis.

55. $f(x) = cx + \sin x \quad \Rightarrow \quad f'(x) = c + \cos x \quad \Rightarrow \quad f''(x) = -\sin x$

$f(-x) = -f(x)$, so f is an odd function and its graph is symmetric with respect to the origin.

$f(x) = 0 \quad \Leftrightarrow \quad \sin x = -cx$, so 0 is always an x-intercept.

$f'(x) = 0 \quad \Leftrightarrow \quad \cos x = -c$, so there is no critical number when $|c| > 1$. If $|c| \le 1$, then there are infinitely many critical numbers. If x_1 is the unique solution of $\cos x = -c$ in the interval $[0, \pi]$, then the critical numbers are $2n\pi \pm x_1$, where n ranges over the integers. (Special cases: When $c = -1$, $x_1 = 0$; when $c = 0$, $x = \frac{\pi}{2}$; and when $c = 1$, $x_1 = \pi$.)

$f''(x) < 0 \quad \Leftrightarrow \quad \sin x > 0$, so f is CD on intervals of the form $(2n\pi, (2n+1)\pi)$. f is CU on intervals of the form $((2n-1)\pi, 2n\pi)$. The inflection points of f are the points $(n\pi, n\pi c)$, where n is an integer.

If $c \ge 1$, then $f'(x) \ge 0$ for all x, so f is increasing and has no extremum. If $c \le -1$, then $f'(x) \le 0$ for all x, so f is decreasing and has no extremum. If $|c| < 1$, then $f'(x) > 0 \quad \Leftrightarrow \quad \cos x > -c \quad \Leftrightarrow \quad x$ is in an interval of the form $(2n\pi - x_1, 2n\pi + x_1)$ for some integer n. These are the intervals on which f is increasing. Similarly, we find that f is decreasing on the intervals of the form $(2n\pi + x_1, 2(n+1)\pi - x_1)$. Thus, f has local maxima at the points $2n\pi + x_1$, where f has the values $c(2n\pi + x_1) + \sin x_1 = c(2n\pi + x_1) + \sqrt{1 - c^2}$, and f has local minima at the points $2n\pi - x_1$, where we have $f(2n\pi - x_1) = c(2n\pi - x_1) - \sin x_1 = c(2n\pi - x_1) - \sqrt{1 - c^2}$.

The transitional values of c are -1 and 1. The inflection points move vertically, but not horizontally, when c changes.

When $|c| \ge 1$, there is no extremum. For $|c| < 1$, the maxima are spaced 2π apart horizontally, as are the minima. The horizontal spacing between maxima and adjacent minima is regular (and equals π) when $c = 0$, but the horizontal space between a local maximum and the nearest local minimum shrinks as $|c|$ approaches 1.

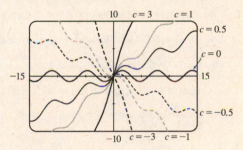

3.5   Optimization Problems

1. (a)

First Number	Second Number	Product
1	22	22
2	21	42
3	20	60
4	19	76
5	18	90
6	17	102
7	16	112
8	15	120
9	14	126
10	13	130
11	12	132

We needn't consider pairs where the first number is larger than the second, since we can just interchange the numbers in such cases. The answer appears to be 11 and 12, but we have considered only integers in the table.

(b) Call the two numbers x and y. Then $x + y = 23$, so $y = 23 - x$. Call the product P. Then

$P = xy = x(23 - x) = 23x - x^2$, so we wish to maximize the function $P(x) = 23x - x^2$. Since $P'(x) = 23 - 2x$,

we see that $P'(x) = 0 \iff x = \frac{23}{2} = 11.5$. Thus, the maximum value of P is $P(11.5) = (11.5)^2 = 132.25$ and it

occurs when $x = y = 11.5$.

Or: Note that $P''(x) = -2 < 0$ for all x, so P is everywhere concave downward and the local maximum at $x = 11.5$

must be an absolute maximum.

3. The two numbers are x and $\dfrac{100}{x}$, where $x > 0$. Minimize $f(x) = x + \dfrac{100}{x}$. $f'(x) = 1 - \dfrac{100}{x^2} = \dfrac{x^2 - 100}{x^2}$. The critical

number is $x = 10$. Since $f'(x) < 0$ for $0 < x < 10$ and $f'(x) > 0$ for $x > 10$, there is an absolute minimum at $x = 10$.

The numbers are 10 and 10.

5. Let the vertical distance be given by $v(x) = (x + 2) - x^2$, $-1 \le x \le 2$.

$v'(x) = 1 - 2x = 0 \iff x = \frac{1}{2}$. $v(-1) = 0$, $v\left(\frac{1}{2}\right) = \frac{9}{4}$, and $v(2) = 0$, so

there is an absolute maximum at $x = \frac{1}{2}$. The maximum distance is

$v\left(\frac{1}{2}\right) = \frac{1}{2} + 2 - \frac{1}{4} = \frac{9}{4}$.

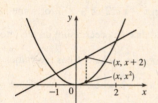

7. If the rectangle has dimensions x and y, then its perimeter is $2x + 2y = 100$ m, so $y = 50 - x$. Thus, the area is

$A = xy = x(50 - x)$. We wish to maximize the function $A(x) = x(50 - x) = 50x - x^2$, where $0 < x < 50$. Since

$A'(x) = 50 - 2x = -2(x - 25)$, $A'(x) > 0$ for $0 < x < 25$ and $A'(x) < 0$ for $25 < x < 50$. Thus, A has an absolute

maximum at $x = 25$, and $A(25) = 25^2 = 625$ m^2. The dimensions of the rectangle that maximize its area are $x = y = 25$ m.

(The rectangle is a square.)

9. (a)

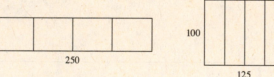

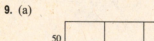

The areas of the three figures are 12,500, 12,500, and 9000 ft^2. There appears to be a maximum area of at least 12,500 ft^2.

(b) Let x denote the length of each of two sides and three dividers.

Let y denote the length of the other two sides.

(c) Area $A = \text{length} \times \text{width} = y \cdot x$

(d) Length of fencing $= 750 \Rightarrow 5x + 2y = 750$

(e) $5x + 2y = 750 \Rightarrow y = 375 - \frac{5}{2}x \Rightarrow A(x) = \left(375 - \frac{5}{2}x\right)x = 375x - \frac{5}{2}x^2$

(f) $A'(x) = 375 - 5x = 0 \Rightarrow x = 75$. Since $A''(x) = -5 < 0$ there is an absolute maximum when $x = 75$. Then

$y = \frac{375}{2} = 187.5$. The largest area is $75\left(\frac{375}{2}\right) = 14{,}062.5$ ft^2. These values of x and y are between the values in the first

and second figures in part (a). Our original estimate was low.

11. Let b be the length of the base of the box and h the height. The surface area is $1200 = b^2 + 4hb$ $\Rightarrow$ $h = (1200 - b^2)/(4b)$.

The volume is $V = b^2 h = b^2(1200 - b^2)/4b = 300b - b^3/4$ $\Rightarrow$ $V'(b) = 300 - \frac{3}{4}b^2$.

$V'(b) = 0$ $\Rightarrow$ $300 = \frac{3}{4}b^2$ $\Rightarrow$ $b^2 = 400$ $\Rightarrow$ $b = \sqrt{400} = 20$. Since $V'(b) > 0$ for $0 < b < 20$ and $V'(b) < 0$ for

$b > 20$, there is an absolute maximum when $b = 20$ by the First Derivative Test for Absolute Extreme Values (see page 176).

If $b = 20$, then $h = (1200 - 20^2)/(4 \cdot 20) = 10$, so the largest possible volume is $b^2 h = (20)^2(10) = 4000$ cm^3.

13. (a) Let the rectangle have sides x and y and area A, so $A = xy$ or $y = A/x$. The problem is to minimize the

perimeter $= 2x + 2y = 2x + 2A/x = P(x)$. Now $P'(x) = 2 - 2A/x^2 = 2(x^2 - A)/x^2$. So the critical number is

$x = \sqrt{A}$. Since $P'(x) < 0$ for $0 < x < \sqrt{A}$ and $P'(x) > 0$ for $x > \sqrt{A}$, there is an absolute minimum at $x = \sqrt{A}$.

The sides of the rectangle are $\sqrt{A}$ and $A/\sqrt{A} = \sqrt{A}$, so the rectangle is a square.

(b) Let p be the perimeter and x and y the lengths of the sides, so $p = 2x + 2y$ $\Rightarrow$ $2y = p - 2x$ $\Rightarrow$ $y = \frac{1}{2}p - x$.

The area is $A(x) = x\left(\frac{1}{2}p - x\right) = \frac{1}{2}px - x^2$. Now $A'(x) = 0$ $\Rightarrow$ $\frac{1}{2}p - 2x = 0$ $\Rightarrow$ $2x = \frac{1}{2}p$ $\Rightarrow$ $x = \frac{1}{4}p$. Since

$A''(x) = -2 < 0$, there is an absolute maximum for A when $x = \frac{1}{4}p$ by the Second Derivative Test. The sides of the

rectangle are $\frac{1}{4}p$ and $\frac{1}{2}p - \frac{1}{4}p = \frac{1}{4}p$, so the rectangle is a square.

15. The distance d from the origin $(0,0)$ to a point $(x, 2x+3)$ on the line is given by $d = \sqrt{(x-0)^2 + (2x+3-0)^2}$ and the

square of the distance is $S = d^2 = x^2 + (2x+3)^2$. $S' = 2x + 2(2x+3)2 = 10x + 12$ and $S' = 0$ $\Leftrightarrow$ $x = -\frac{6}{5}$. Now

$S'' = 10 > 0$, so we know that S has a minimum at $x = -\frac{6}{5}$. Thus, the y-value is $2\left(-\frac{6}{5}\right) + 3 = \frac{3}{5}$ and the point is $\left(-\frac{6}{5}, \frac{3}{5}\right)$.

17.

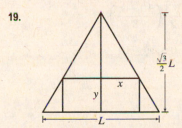

From the figure, we see that there are two points that are farthest away from

$A(1, 0)$. The distance d from A to an arbitrary point $P(x, y)$ on the ellipse is

$d = \sqrt{(x-1)^2 + (y-0)^2}$ and the square of the distance is

$S = d^2 = x^2 - 2x + 1 + y^2 = x^2 - 2x + 1 + (4 - 4x^2) = -3x^2 - 2x + 5$.

$S' = -6x - 2$ and $S' = 0$ $\Rightarrow$ $x = -\frac{1}{3}$. Now $S'' = -6 < 0$, so we know

that S has a maximum at $x = -\frac{1}{3}$. Since $-1 \le x \le 1$, $S(-1) = 4$,

$S\left(-\frac{1}{3}\right) = \frac{16}{3}$, and $S(1) = 0$, we see that the maximum distance is $\sqrt{\frac{16}{3}}$. The corresponding y-values are

$y = \pm\sqrt{4 - 4\left(-\frac{1}{3}\right)^2} = \pm\sqrt{\frac{32}{9}} = \pm\frac{4}{3}\sqrt{2} \approx \pm 1.89$. The points are $\left(-\frac{1}{3}, \pm\frac{4}{3}\sqrt{2}\right)$.

19.

The height h of the equilateral triangle with sides of length L is $\frac{\sqrt{3}}{2}L$,

since $h^2 + (L/2)^2 = L^2$ $\Rightarrow$ $h^2 = L^2 - \frac{1}{4}L^2 = \frac{3}{4}L^2$ $\Rightarrow$

$h = \frac{\sqrt{3}}{2}L$. Using similar triangles, $\dfrac{\frac{\sqrt{3}}{2}L - y}{x} = \dfrac{\frac{\sqrt{3}}{2}L}{L/2} = \sqrt{3}$ $\Rightarrow$

$\sqrt{3}x = \frac{\sqrt{3}}{2}L - y$ $\Rightarrow$ $y = \frac{\sqrt{3}}{2}L - \sqrt{3}x$ $\Rightarrow$ $y = \frac{\sqrt{3}}{2}(L - 2x)$.

The area of the inscribed rectangle is $A(x) = (2x)y = \sqrt{3}\,x(L - 2x) = \sqrt{3}\,Lx - 2\sqrt{3}\,x^2$, where $0 \le x \le L/2$. Now

$0 = A'(x) = \sqrt{3}\,L - 4\sqrt{3}\,x \;\Rightarrow\; x = \sqrt{3}\,L/(4\sqrt{3}) = L/4$. Since $A(0) = A(L/2) = 0$, the maximum occurs when

$x = L/4$, and $y = \frac{\sqrt{3}}{2}L - \frac{\sqrt{3}}{4}L = \frac{\sqrt{3}}{4}L$, so the dimensions are $L/2$ and $\frac{\sqrt{3}}{4}L$.

21.

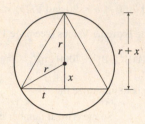

The area of the triangle is

$A(x) = \frac{1}{2}(2t)(r + x) = t(r + x) = \sqrt{r^2 - x^2}\,(r + x)$. Then

$0 = A'(x) = r\dfrac{-2x}{2\sqrt{r^2 - x^2}} + \sqrt{r^2 - x^2} + x\dfrac{-2x}{2\sqrt{r^2 - x^2}}$

$= -\dfrac{x^2 + rx}{\sqrt{r^2 - x^2}} + \sqrt{r^2 - x^2} \;\Rightarrow$

$\dfrac{x^2 + rx}{\sqrt{r^2 - x^2}} = \sqrt{r^2 - x^2} \;\Rightarrow\; x^2 + rx = r^2 - x^2 \;\Rightarrow\; 0 = 2x^2 + rx - r^2 = (2x - r)(x + r) \;\Rightarrow$

$x = \frac{1}{2}r$ or $x = -r$. Now $A(r) = 0 = A(-r) \;\Rightarrow$ the maximum occurs where $x = \frac{1}{2}r$, so the triangle has

height $r + \frac{1}{2}r = \frac{3}{2}r$ and base $2\sqrt{r^2 - \left(\frac{1}{2}r\right)^2} = 2\sqrt{\frac{3}{4}r^2} = \sqrt{3}\,r$.

23.

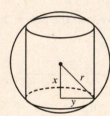

The cylinder has volume $V = \pi y^2(2x)$. Also $x^2 + y^2 = r^2 \;\Rightarrow\; y^2 = r^2 - x^2$, so

$V(x) = \pi(r^2 - x^2)(2x) = 2\pi(r^2x - x^3)$, where $0 \le x \le r$.

$V'(x) = 2\pi(r^2 - 3x^2) = 0 \;\Rightarrow\; x = r/\sqrt{3}$. Now $V(0) = V(r) = 0$, so there is a

maximum when $x = r/\sqrt{3}$ and $V(r/\sqrt{3}) = \pi(r^2 - r^2/3)(2r/\sqrt{3}) = 4\pi r^3/(3\sqrt{3})$.

25.

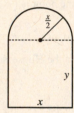

Perimeter $= 30 \;\Rightarrow\; 2y + x + \pi\left(\dfrac{x}{2}\right) = 30 \;\Rightarrow$

$y = \frac{1}{2}\left(30 - x - \dfrac{\pi x}{2}\right) = 15 - \dfrac{x}{2} - \dfrac{\pi x}{4}$. The area is the area of the rectangle plus the area of

the semicircle, or $xy + \frac{1}{2}\pi\left(\dfrac{x}{2}\right)^2$, so $A(x) = x\left(15 - \dfrac{x}{2} - \dfrac{\pi x}{4}\right) + \frac{1}{8}\pi x^2 = 15x - \frac{1}{2}x^2 - \frac{\pi}{8}x^2$.

$A'(x) = 15 - \left(1 + \frac{\pi}{4}\right)x = 0 \;\Rightarrow\; x = \dfrac{15}{1 + \pi/4} = \dfrac{60}{4 + \pi}$. $A''(x) = -\left(1 + \dfrac{\pi}{4}\right) < 0$, so this gives a maximum.

The dimensions are $x = \dfrac{60}{4 + \pi}$ ft and $y = 15 - \dfrac{30}{4 + \pi} - \dfrac{15\pi}{4 + \pi} = \dfrac{60 + 15\pi - 30 - 15\pi}{4 + \pi} = \dfrac{30}{4 + \pi}$ ft, so the height of the

rectangle is half the base.

27.

Let x be the length of the wire used for the square. The total area is

$A(x) = \left(\dfrac{x}{4}\right)^2 + \dfrac{1}{2}\left(\dfrac{10 - x}{3}\right)\dfrac{\sqrt{3}}{2}\left(\dfrac{10 - x}{3}\right)$

$= \frac{1}{16}x^2 + \frac{\sqrt{3}}{36}(10 - x)^2,\; 0 \le x \le 10$

$A'(x) = \frac{1}{8}x - \frac{\sqrt{3}}{18}(10 - x) = 0 \;\Leftrightarrow\; \frac{9}{72}x + \frac{4\sqrt{3}}{72}x - \frac{40\sqrt{3}}{72} = 0 \;\Leftrightarrow\; x = \dfrac{40\sqrt{3}}{9 + 4\sqrt{3}}$.

Now $A(0) = \left(\dfrac{\sqrt{3}}{36}\right)100 \approx 4.81$, $A(10) = \dfrac{100}{16} = 6.25$ and $A\left(\dfrac{40\sqrt{3}}{9 + 4\sqrt{3}}\right) \approx 2.72$, so

(a) The maximum area occurs when $x = 10$ m, and all the wire is used for the square.

(b) The minimum area occurs when $x = \frac{40\sqrt{3}}{9+4\sqrt{3}} \approx 4.35$ m.

29.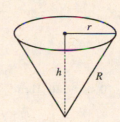

$h^2 + r^2 = R^2 \quad\Rightarrow\quad V = \frac{\pi}{3}r^2h = \frac{\pi}{3}(R^2 - h^2)h = \frac{\pi}{3}(R^2h - h^3)$.

$V'(h) = \frac{\pi}{3}(R^2 - 3h^2) = 0$ when $h = \frac{1}{\sqrt{3}}R$. This gives an absolute maximum, since

$V'(h) > 0$ for $0 < h < \frac{1}{\sqrt{3}}R$ and $V'(h) < 0$ for $h > \frac{1}{\sqrt{3}}R$. The maximum volume is

$V\left(\frac{1}{\sqrt{3}}R\right) = \frac{\pi}{3}\left(\frac{1}{\sqrt{3}}R^3 - \frac{1}{3\sqrt{3}}R^3\right) = \frac{2}{9\sqrt{3}}\pi R^3$.

31.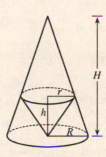

By similar triangles, $\frac{H}{R} = \frac{H-h}{r}$ **(1)**. The volume of the inner cone is $V = \frac{1}{3}\pi r^2 h$,

so we'll solve **(1)** for h. $\frac{Hr}{R} = H - h \quad\Rightarrow$

$h = H - \frac{Hr}{R} = \frac{HR - Hr}{R} = \frac{H}{R}(R - r)$ **(2)**.

Thus, $V(r) = \frac{\pi}{3}r^2 \cdot \frac{H}{R}(R - r) = \frac{\pi H}{3R}(Rr^2 - r^3) \quad\Rightarrow$

$V'(r) = \frac{\pi H}{3R}(2Rr - 3r^2) = \frac{\pi H}{3R}r(2R - 3r)$.

$V'(r) = 0 \quad\Rightarrow\quad r = 0$ or $2R = 3r \quad\Rightarrow\quad r = \frac{2}{3}R$ and from **(2)**, $h = \frac{H}{R}\left(R - \frac{2}{3}R\right) = \frac{H}{R}\left(\frac{1}{3}R\right) = \frac{1}{3}H$.

$V'(r)$ changes from positive to negative at $r = \frac{2}{3}R$, so the inner cone has a maximum volume of

$V = \frac{1}{3}\pi r^2 h = \frac{1}{3}\pi\left(\frac{2}{3}R\right)^2\left(\frac{1}{3}H\right) = \frac{4}{27}\cdot\frac{1}{3}\pi R^2 H$, which is approximately 15% of the volume of the larger cone.

33. $P(R) = \frac{E^2 R}{(R+r)^2} \quad\Rightarrow$

$P'(R) = \frac{(R+r)^2 \cdot E^2 - E^2 R \cdot 2(R+r)}{[(R+r)^2]^2} = \frac{(R^2 + 2Rr + r^2)E^2 - 2E^2R^2 - 2E^2Rr}{(R+r)^4}$

$= \frac{E^2r^2 - E^2R^2}{(R+r)^4} = \frac{E^2(r^2 - R^2)}{(R+r)^4} = \frac{E^2(r+R)(r-R)}{(R+r)^4} = \frac{E^2(r-R)}{(R+r)^3}$

$P'(R) = 0 \quad\Rightarrow\quad R = r \quad\Rightarrow\quad P(r) = \frac{E^2r}{(r+r)^2} = \frac{E^2r}{4r^2} = \frac{E^2}{4r}$.

The expression for $P'(R)$ shows that $P'(R) > 0$ for $R < r$ and $P'(R) < 0$ for $R > r$. Thus, the maximum value of the

power is $E^2/(4r)$, and this occurs when $R = r$.

35. $S = 6sh - \frac{3}{2}s^2\cot\theta + 3s^2\frac{\sqrt{3}}{2}\csc\theta$

(a) $\frac{dS}{d\theta} = \frac{3}{2}s^2\csc^2\theta - 3s^2\frac{\sqrt{3}}{2}\csc\theta\cot\theta$ or $\frac{3}{2}s^2\csc\theta\left(\csc\theta - \sqrt{3}\cot\theta\right)$.

(b) $\frac{dS}{d\theta} = 0$ when $\csc\theta - \sqrt{3}\cot\theta = 0 \quad\Rightarrow\quad \frac{1}{\sin\theta} - \sqrt{3}\frac{\cos\theta}{\sin\theta} = 0 \quad\Rightarrow\quad \cos\theta = \frac{1}{\sqrt{3}}$. The First Derivative Test shows

that the minimum surface area occurs when $\theta = \cos^{-1}\left(\frac{1}{\sqrt{3}}\right) \approx 55°$.

(c)

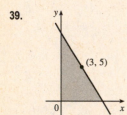

If $\cos\theta = \frac{1}{\sqrt{3}}$, then $\cot\theta = \frac{1}{\sqrt{2}}$ and $\csc\theta = \frac{\sqrt{3}}{\sqrt{2}}$, so the surface area is

$$S = 6sh - \frac{3}{2}s^2\frac{1}{\sqrt{2}} + 3s^2\frac{\sqrt{3}}{2}\frac{\sqrt{3}}{\sqrt{2}} = 6sh - \frac{3}{2\sqrt{2}}s^2 + \frac{9}{2\sqrt{2}}s^2$$

$$= 6sh + \frac{6}{2\sqrt{2}}s^2 = 6s\left(h + \frac{1}{2\sqrt{2}}s\right)$$

37.

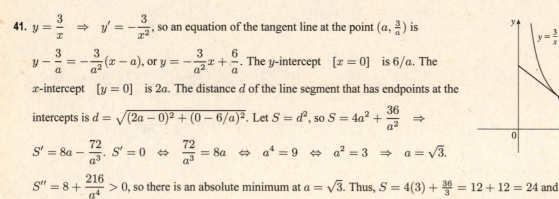

The total illumination is $I(x) = \frac{3k}{x^2} + \frac{k}{(10-x)^2}$, $0 < x < 10$. Then

$$I'(x) = \frac{-6k}{x^3} + \frac{2k}{(10-x)^3} = 0 \;\Rightarrow\; 6k(10-x)^3 = 2kx^3 \;\Rightarrow$$

$3(10-x)^3 = x^3 \;\Rightarrow\; \sqrt[3]{3}\,(10-x) = x \;\Rightarrow\; 10\sqrt[3]{3} - \sqrt[3]{3}\,x = x \;\Rightarrow\; 10\sqrt[3]{3} = x + \sqrt[3]{3}\,x \;\Rightarrow$

$10\sqrt[3]{3} = \left(1 + \sqrt[3]{3}\right)x \;\Rightarrow\; x = \dfrac{10\sqrt[3]{3}}{1 + \sqrt[3]{3}} \approx 5.9$ ft. This gives a minimum since $I''(x) > 0$ for $0 < x < 10$.

39.

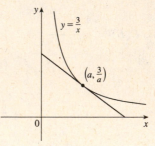

The line with slope m (where $m < 0$) through $(3, 5)$ has equation $y - 5 = m(x - 3)$ or

$y = mx + (5 - 3m)$. The y-intercept is $5 - 3m$ and the x-intercept is $-5/m + 3$. So the

triangle has area $A(m) = \frac{1}{2}(5 - 3m)(-5/m + 3) = 15 - 25/(2m) - \frac{9}{2}m$. Now

$$A'(m) = \frac{25}{2m^2} - \frac{9}{2} = 0 \;\Leftrightarrow\; m^2 = \frac{25}{9} \;\Rightarrow\; m = -\frac{5}{3} \text{ (since } m < 0\text{)}.$$

$A''(m) = -\dfrac{25}{m^3} > 0$, so there is an absolute minimum when $m = -\frac{5}{3}$. Thus, an equation of the line is $y - 5 = -\frac{5}{3}(x - 3)$

or $y = -\frac{5}{3}x + 10$.

41. $y = \dfrac{3}{x} \;\Rightarrow\; y' = -\dfrac{3}{x^2}$, so an equation of the tangent line at the point $\left(a, \frac{3}{a}\right)$ is

$y - \dfrac{3}{a} = -\dfrac{3}{a^2}(x - a)$, or $y = -\dfrac{3}{a^2}x + \dfrac{6}{a}$. The y-intercept $[x = 0]$ is $6/a$. The

x-intercept $[y = 0]$ is $2a$. The distance d of the line segment that has endpoints at the

intercepts is $d = \sqrt{(2a - 0)^2 + (0 - 6/a)^2}$. Let $S = d^2$, so $S = 4a^2 + \dfrac{36}{a^2} \;\Rightarrow$

$S' = 8a - \dfrac{72}{a^3}$. $S' = 0 \;\Leftrightarrow\; \dfrac{72}{a^3} = 8a \;\Leftrightarrow\; a^4 = 9 \;\Leftrightarrow\; a^2 = 3 \;\Rightarrow\; a = \sqrt{3}$.

$S'' = 8 + \dfrac{216}{a^4} > 0$, so there is an absolute minimum at $a = \sqrt{3}$. Thus, $S = 4(3) + \frac{36}{3} = 12 + 12 = 24$ and

hence, $d = \sqrt{24} = 2\sqrt{6}$.

43. (a) If $c(x) = \dfrac{C(x)}{x}$, then, by the Quotient Rule, we have $c'(x) = \dfrac{xC'(x) - C(x)}{x^2}$. Now $c'(x) = 0$ when

$xC'(x) - C(x) = 0$ and this gives $C'(x) = \dfrac{C(x)}{x} = c(x)$. Therefore, the marginal cost equals the average cost.

(b) (i) $C(x) = 16{,}000 + 200x + 4x^{3/2}$, $C(1000) = 16{,}000 + 200{,}000 + 40{,}000\sqrt{10} \approx 216{,}000 + 126{,}491$, so

$C(1000) \approx \$342{,}491$. $c(x) = C(x)/x = \dfrac{16{,}000}{x} + 200 + 4x^{1/2}$, $c(1000) \approx \$342.49$/unit. $C'(x) = 200 + 6x^{1/2}$,

$C'(1000) = 200 + 60\sqrt{10} \approx \$389.74$/unit.

(ii) We must have $C'(x) = c(x)$ $\Leftrightarrow$ $200 + 6x^{1/2} = \dfrac{16{,}000}{x} + 200 + 4x^{1/2}$ $\Leftrightarrow$ $2x^{3/2} = 16{,}000$ $\Leftrightarrow$

$x = (8{,}000)^{2/3} = 400$ units. To check that this is a minimum, we calculate

$c'(x) = \dfrac{-16{,}000}{x^2} + \dfrac{2}{\sqrt{x}} = \dfrac{2}{x^2}\left(x^{3/2} - 8000\right)$. This is negative for $x < (8000)^{2/3} = 400$, zero at $x = 400$,

and positive for $x > 400$, so c is decreasing on $(0, 400)$ and increasing on $(400, \infty)$. Thus, c has an absolute minimum

at $x = 400$. [*Note:* $c''(x)$ is *not* positive for all $x > 0$.]

(iii) The minimum average cost is $c(400) = 40 + 200 + 80 = \$320/\text{unit}$.

45. (a) We are given that the demand function p is linear and $p(27{,}000) = 10$, $p(33{,}000) = 8$, so the slope is

$\dfrac{10 - 8}{27{,}000 - 33{,}000} = -\dfrac{1}{3000}$ and an equation of the line is $y - 10 = \left(-\dfrac{1}{3000}\right)(x - 27{,}000)$ $\Rightarrow$

$y = p(x) = -\dfrac{1}{3000}x + 19 = 19 - (x/3000)$.

(b) The revenue is $R(x) = xp(x) = 19x - (x^2/3000)$ $\Rightarrow$ $R'(x) = 19 - (x/1500) = 0$ when $x = 28{,}500$. Since

$R''(x) = -1/1500 < 0$, the maximum revenue occurs when $x = 28{,}500$ $\Rightarrow$ the price is $p(28{,}500) = \$9.50$.

47. (a) As in Example 6, we see that the demand function p is linear. We are given that $p(1000) = 450$ and deduce that

$p(1100) = 440$, since a \$10 reduction in price increases sales by 100 per week. The slope for p is $\dfrac{440 - 450}{1100 - 1000} = -\dfrac{1}{10}$,

so an equation is $p - 450 = -\dfrac{1}{10}(x - 1000)$ or $p(x) = -\dfrac{1}{10}x + 550$.

(b) $R(x) = xp(x) = -\dfrac{1}{10}x^2 + 550x$. $R'(x) = -\dfrac{1}{5}x + 550 = 0$ when $x = 5(550) = 2750$.

$p(2750) = 275$, so the rebate should be $450 - 275 = \$175$.

(c) $C(x) = 68{,}000 + 150x$ $\Rightarrow$ $P(x) = R(x) - C(x) = -\dfrac{1}{10}x^2 + 550x - 68{,}000 - 150x = -\dfrac{1}{10}x^2 + 400x - 68{,}000$,

$P'(x) = -\dfrac{1}{5}x + 400 = 0$ when $x = 2000$. $p(2000) = 350$. Therefore, the rebate to maximize profits should be

$450 - 350 = \$100$.

49.

Every line segment in the first quadrant passing through (a, b) with endpoints on the x-
and y-axes satisfies an equation of the form $y - b = m(x - a)$, where $m < 0$. By setting
$x = 0$ and then $y = 0$, we find its endpoints, $A\left(0, b - am\right)$ and $B\left(a - \dfrac{b}{m}, 0\right)$. The
distance d from A to B is given by $d = \sqrt{\left[\left(a - \dfrac{b}{m}\right) - 0\right]^2 + \left[0 - (b - am)\right]^2}$.

It follows that the square of the length of the line segment, as a function of m, is given by

$S(m) = \left(a - \dfrac{b}{m}\right)^2 + (am - b)^2 = a^2 - \dfrac{2ab}{m} + \dfrac{b^2}{m^2} + a^2m^2 - 2abm + b^2$. Thus,

$$S'(m) = \dfrac{2ab}{m^2} - \dfrac{2b^2}{m^3} + 2a^2m - 2ab = \dfrac{2}{m^3}\left(abm - b^2 + a^2m^4 - abm^3\right)$$

$$= \dfrac{2}{m^3}\left[b(am - b) + am^3(am - b)\right] = \dfrac{2}{m^3}(am - b)(b + am^3)$$

Thus, $S'(m) = 0$ $\Leftrightarrow$ $m = b/a$ or $m = -\sqrt[3]{\dfrac{b}{a}}$. Since $b/a > 0$ and $m < 0$, m must equal $-\sqrt[3]{\dfrac{b}{a}}$. Since $\dfrac{2}{m^3} < 0$, we see that $S'(m) < 0$ for $m < -\sqrt[3]{\frac{b}{a}}$ and $S'(m) > 0$ for $m > -\sqrt[3]{\frac{b}{a}}$. Thus, S has its absolute minimum value when $m = -\sqrt[3]{\frac{b}{a}}$.

That value is

$$S\left(-\sqrt[3]{\frac{b}{a}}\right) = \left(a + b\sqrt[3]{\frac{a}{b}}\right)^2 + \left(-a\sqrt[3]{\frac{b}{a}} - b\right)^2 = \left(a + \sqrt[3]{ab^2}\right)^2 + \left(\sqrt[3]{a^2 b} + b\right)^2$$

$$= a^2 + 2a^{4/3}b^{2/3} + a^{2/3}b^{4/3} + a^{4/3}b^{2/3} + 2a^{2/3}b^{4/3} + b^2 = a^2 + 3a^{4/3}b^{2/3} + 3a^{2/3}b^{4/3} + b^2$$

The last expression is of the form $x^3 + 3x^2 y + 3xy^2 + y^3 \quad [= (x + y)^3] \quad$ with $x = a^{2/3}$ and $y = b^{2/3}$,

so we can write it as $(a^{2/3} + b^{2/3})^3$ and the shortest such line segment has length $\sqrt{S} = (a^{2/3} + b^{2/3})^{3/2}$.

51.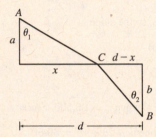

The total time is

$$T(x) = (\text{time from } A \text{ to } C) + (\text{time from } C \text{ to } B)$$

$$= \frac{\sqrt{a^2 + x^2}}{v_1} + \frac{\sqrt{b^2 + (d - x)^2}}{v_2}, \ 0 < x < d$$

$$T'(x) = \frac{x}{v_1 \sqrt{a^2 + x^2}} - \frac{d - x}{v_2 \sqrt{b^2 + (d - x)^2}} = \frac{\sin \theta_1}{v_1} - \frac{\sin \theta_2}{v_2}$$

The minimum occurs when $T'(x) = 0 \quad \Rightarrow \quad \dfrac{\sin \theta_1}{v_1} = \dfrac{\sin \theta_2}{v_2}$.

[*Note:* $T''(x) > 0$]

53.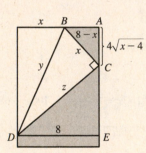

$y^2 = x^2 + z^2$, but triangles CDE and BCA are similar, so

$z/8 = x/\left(4\sqrt{x - 4}\right) \quad \Rightarrow \quad z = 2x/\sqrt{x - 4}$. Thus, we minimize

$f(x) = y^2 = x^2 + 4x^2/(x - 4) = x^3/(x - 4), \ 4 < x \le 8$.

$f'(x) = \dfrac{(x - 4)(3x^2) - x^3}{(x - 4)^2} = \dfrac{x^2[3(x - 4) - x]}{(x - 4)^2} = \dfrac{2x^2(x - 6)}{(x - 4)^2} = 0$

when $x = 6$. $f'(x) < 0$ when $x < 6$, $f'(x) > 0$ when $x > 6$, so the minimum

occurs when $x = 6$ in.

55.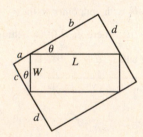

In the small triangle with sides a and c and hypotenuse W, $\sin \theta = \dfrac{a}{W}$ and

$\cos \theta = \dfrac{c}{W}$. In the triangle with sides b and d and hypotenuse L, $\sin \theta = \dfrac{d}{L}$ and

$\cos \theta = \dfrac{b}{L}$. Thus, $a = W \sin \theta$, $c = W \cos \theta$, $d = L \sin \theta$, and $b = L \cos \theta$, so the

area of the circumscribed rectangle is

$$A(\theta) = (a + b)(c + d) = (W \sin \theta + L \cos \theta)(W \cos \theta + L \sin \theta)$$

$$= W^2 \sin \theta \cos \theta + WL \sin^2 \theta + LW \cos^2 \theta + L^2 \sin \theta \cos \theta$$

$$= LW \sin^2 \theta + LW \cos^2 \theta + (L^2 + W^2) \sin \theta \cos \theta$$

$$= LW(\sin^2 \theta + \cos^2 \theta) + (L^2 + W^2) \cdot \tfrac{1}{2} \cdot 2 \sin \theta \cos \theta = LW + \tfrac{1}{2}(L^2 + W^2) \sin 2\theta, \ 0 \le \theta \le \tfrac{\pi}{2}$$

This expression shows, without calculus, that the maximum value of $A(\theta)$ occurs when $\sin 2\theta = 1 \quad \Leftrightarrow \quad 2\theta = \tfrac{\pi}{2} \quad \Rightarrow$

$\theta = \tfrac{\pi}{4}$. So the maximum area is $A\left(\tfrac{\pi}{4}\right) = LW + \tfrac{1}{2}(L^2 + W^2) = \tfrac{1}{2}(L^2 + 2LW + W^2) = \tfrac{1}{2}(L + W)^2$.

57. (a) $I(x) \propto \dfrac{\text{strength of source}}{(\text{distance from source})^2}$. Adding the intensities from the left and right lightbulbs,

$$I(x) = \frac{k}{x^2 + d^2} + \frac{k}{(10 - x)^2 + d^2} = \frac{k}{x^2 + d^2} + \frac{k}{x^2 - 20x + 100 + d^2}.$$

(b) The magnitude of the constant k won't affect the location of the point of maximum intensity, so for convenience we take

$$k = 1. \quad I'(x) = -\frac{2x}{(x^2 + d^2)^2} - \frac{2(x - 10)}{(x^2 - 20x + 100 + d^2)^2}.$$

Substituting $d = 5$ into the equations for $I(x)$ and $I'(x)$, we get

$$I_5(x) = \frac{1}{x^2 + 25} + \frac{1}{x^2 - 20x + 125} \quad \text{and} \quad I_5'(x) = -\frac{2x}{(x^2 + 25)^2} - \frac{2(x - 10)}{(x^2 - 20x + 125)^2}$$

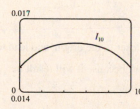
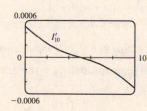

From the graphs, it appears that $I_5(x)$ has a minimum at $x = 5$ m.

(c) Substituting $d = 10$ into the equations for $I(x)$ and $I'(x)$ gives

$$I_{10}(x) = \frac{1}{x^2 + 100} + \frac{1}{x^2 - 20x + 200} \quad \text{and} \quad I_{10}'(x) = -\frac{2x}{(x^2 + 100)^2} - \frac{2(x - 10)}{(x^2 - 20x + 200)^2}$$

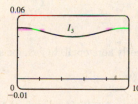
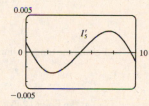

From the graphs, it seems that for $d = 10$, the intensity is minimized at the endpoints, that is, $x = 0$ and $x = 10$. The midpoint is now the most brightly lit point!

(d) From the first figures in parts (b) and (c), we see that the minimal illumination changes from the midpoint ($x = 5$ with $d = 5$) to the endpoints ($x = 0$ and $x = 10$ with $d = 10$).

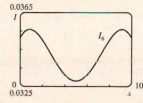
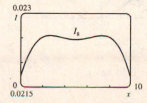
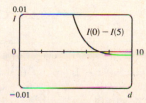

So we try $d = 6$ (see the first figure) and we see that the minimum value still occurs at $x = 5$. Next, we let $d = 8$ (see the second figure) and we see that the minimum value occurs at the endpoints. It appears that for some value of d between 6 and 8, we must have minima at both the midpoint and the endpoints, that is, $I(5)$ must equal $I(0)$. To find this value of d, we solve $I(0) = I(5)$ (with $k = 1$):

$$\frac{1}{d^2} + \frac{1}{100 + d^2} = \frac{1}{25 + d^2} + \frac{1}{25 + d^2} = \frac{2}{25 + d^2} \quad \Rightarrow \quad (25 + d^2)(100 + d^2) + d^2(25 + d^2) = 2d^2(100 + d^2) \quad \Rightarrow$$

$$2500 + 125d^2 + d^4 + 25d^2 + d^4 = 200d^2 + 2d^4 \quad \Rightarrow \quad 2500 = 50d^2 \quad \Rightarrow \quad d^2 = 50 \quad \Rightarrow \quad d = 5\sqrt{2} \approx 7.071$$

[for $0 \le d \le 10$]. The third figure, a graph of $I(0) - I(5)$ with d independent, confirms that $I(0) - I(5) = 0$, that is, $I(0) = I(5)$, when $d = 5\sqrt{2}$. Thus, the point of minimal illumination changes abruptly from the midpoint to the endpoints when $d = 5\sqrt{2}$.

3.6 Newton's Method

1. (a)

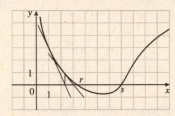

The tangent line at $x = 1$ intersects the x-axis at $x \approx 2.3$, so $x_2 \approx 2.3$. The tangent line at $x = 2.3$ intersects the x-axis at $x \approx 3$, so $x_3 \approx 3.0$.

(b) $x_1 = 5$ would *not* be a better first approximation than $x_1 = 1$ since the tangent line is nearly horizontal. In fact, the second approximation for $x_1 = 5$ appears to be to the left of $x = 1$.

3. Since the tangent line $y = 9 - 2x$ is tangent to the curve $y = f(x)$ at the point $(2, 5)$, we have $x_1 = 2$, $f(x_1) = 5$, and $f'(x_1) = -2$ [the slope of the tangent line]. Thus, by Equation 2,

$$x_2 = x_1 - \frac{f(x_1)}{f'(x_1)} = 2 - \frac{5}{-2} = \frac{9}{2}$$

Note that geometrically $\frac{9}{2}$ represents the x-intercept of the tangent line $y = 9 - 2x$.

5. The initial approximations $x_1 = a, b,$ and c will work, resulting in a second approximation closer to the origin, and lead to the root of the equation $f(x) = 0$, namely, $x = 0$. The initial approximation $x_1 = d$ will not work because it will result in successive approximations farther and farther from the origin.

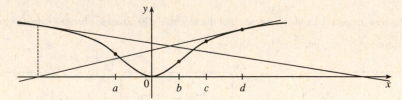

7. $f(x) = x^5 - x - 1 \ \Rightarrow \ f'(x) = 5x^4 - 1$, so $x_{n+1} = x_n - \dfrac{x_n^5 - x_n - 1}{5x_n^4 - 1}$. Now $x_1 = 1 \ \Rightarrow$

$$x_2 = 1 - \frac{1 - 1 - 1}{5 - 1} = 1 - \left(-\frac{1}{4}\right) = 1.25 \ \Rightarrow \ x_3 = 1.25 - \frac{(1.25)^5 - 1.25 - 1}{5(1.25)^4 - 1} \approx 1.1785.$$

9. $f(x) = x^3 + x + 3 \ \Rightarrow \ f'(x) = 3x^2 + 1$, so $x_{n+1} = x_n - \dfrac{x_n^3 + x_n + 3}{3x_n^2 + 1}$.

Now $x_1 = -1 \ \Rightarrow$

$$x_2 = -1 - \frac{(-1)^3 + (-1) + 3}{3(-1)^2 + 1} = -1 - \frac{-1 - 1 + 3}{3 + 1} = -1 - \frac{1}{4} = -1.25.$$

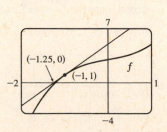

Newton's method follows the tangent line at $(-1, 1)$ down to its intersection with the x-axis at $(-1.25, 0)$, giving the second approximation $x_2 = -1.25$.

11. To approximate $x = \sqrt[5]{20}$ (so that $x^5 = 20$), we can take $f(x) = x^5 - 20$. So $f'(x) = 5x^4$, and thus,

$x_{n+1} = x_n - \dfrac{x_n^5 - 20}{5x_n^4}$. Since $\sqrt[5]{32} = 2$ and 32 is reasonably close to 20, we'll use $x_1 = 2$. We need to find approximations

until they agree to eight decimal places. $x_1 = 2 \;\Rightarrow\; x_2 = 1.85, x_3 \approx 1.82148614, x_4 \approx 1.82056514,$

$x_5 \approx 1.82056420 \approx x_6$. So $\sqrt[5]{20} \approx 1.82056420$, to eight decimal places.

Here is a quick and easy method for finding the iterations for Newton's method on a programmable calculator.

(The screens shown are from the TI-84 Plus, but the method is similar on other calculators.) Assign $f(x) = x^5 - 20$

to Y_1, and $f'(x) = 5x^4$ to Y_2. Now store $x_1 = 2$ in X and then enter $X - Y_1/Y_2 \rightarrow X$ to get $x_2 = 1.85$. By successively

pressing the ENTER key, you get the approximations $x_3, x_4, \ldots$.

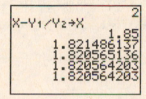

In Derive, load the utility file SOLVE. Enter NEWTON(x^5-20,x,2) and then APPROXIMATE to get

[2, 1.85, 1.82148614, 1.82056514, 1.82056420]. You can request a specific iteration by adding a fourth argument. For

example, NEWTON(x^5-20,x,2,2) gives [2, 1.85, 1.82148614].

In Maple, make the assignments $f := x \to x\hat{\ }5 - 20;, g := x \to x - f(x)/D(f)(x);$, and $x := 2.;$. Repeatedly execute

the command $x := g(x);$ to generate successive approximations.

In Mathematica, make the assignments $f[x\_] := x\hat{\ }5 - 20, g[x\_] := x - f[x]/f'[x]$, and $x = 2$. Repeatedly execute the

command $x = g[x]$ to generate successive approximations.

13. $f(x) = x^4 - 2x^3 + 5x^2 - 6 \;\Rightarrow\; f'(x) = 4x^3 - 6x^2 + 10x \;\Rightarrow\; x_{n+1} = x_n - \dfrac{x_n^4 - 2x_n^3 + 5x_n^2 - 6}{4x_n^3 - 6x_n^2 + 10x_n}$. We need to find

approximations until they agree to six decimal places. We'll let x_1 equal the midpoint of the given interval, $[1, 2]$.

$x_1 = 1.5 \;\Rightarrow\; x_2 = 1.2625, x_3 \approx 1.218808, x_4 \approx 1.217563, x_5 \approx 1.217562 \approx x_6$. So the root is 1.217562 to six decimal

places.

15.

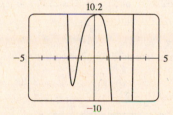

$f(x) = x^6 - x^5 - 6x^4 - x^2 + x + 10 \;\Rightarrow\;$

$f'(x) = 6x^5 - 5x^4 - 24x^3 - 2x + 1 \;\Rightarrow\;$

$x_{n+1} = x_n - \dfrac{x_n^6 - x_n^5 - 6x_n^4 - x_n^2 + x_n + 10}{6x_n^5 - 5x_n^4 - 24x_n^3 - 2x_n + 1}.$

From the graph of f, there appear to be roots near -1.9, -1.2, 1.1, and 3.

$x_1 = -1.9$	$x_1 = -1.2$	$x_1 = 1.1$	$x_1 = 3$
$x_2 \approx -1.94278290$	$x_2 \approx -1.22006245$	$x_2 \approx 1.14111662$	$x_2 \approx 2.99$
$x_3 \approx -1.93828380$	$x_3 \approx -1.21997997 \approx x_4$	$x_3 \approx 1.13929741$	$x_3 \approx 2.98984106$
$x_4 \approx -1.93822884$		$x_4 \approx 1.13929375 \approx x_5$	$x_4 \approx 2.98984102 \approx x_5$
$x_5 \approx -1.93822883 \approx x_6$			

To eight decimal places, the roots of the equation are -1.93822883, -1.21997997, 1.13929375, and 2.98984102.

17.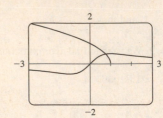

Solving $\dfrac{x}{x^2+1} = \sqrt{1-x}$ is the same as solving

$$f(x) = \frac{x}{x^2+1} - \sqrt{1-x} = 0. \quad f'(x) = \frac{1-x^2}{(x^2+1)^2} + \frac{1}{2\sqrt{1-x}} \quad \Rightarrow$$

$$x_{n+1} = x_n - \frac{\dfrac{x_n}{x_n^2+1} - \sqrt{1-x_n}}{\dfrac{1-x_n^2}{(x_n^2+1)^2} + \dfrac{1}{2\sqrt{1-x_n}}}.$$

From the graph, we see that the curves intersect at about 0.8. $x_1 = 0.8 \quad \Rightarrow \quad x_2 \approx 0.76757581$, $x_3 \approx 0.76682610$, $x_4 \approx 0.76682579 \approx x_5$. To eight decimal places, the root of the equation is 0.76682579.

19. From the graph, $y = x^2\sqrt{2-x-x^2}$ and $y = 1$ intersect twice, at $x \approx -2$ and at $x \approx -1$.

$$f(x) = x^2\sqrt{2-x-x^2} - 1 \quad \Rightarrow \quad f'(x) = x^2 \cdot \tfrac{1}{2}(2-x-x^2)^{-1/2}(-1-2x) + (2-x-x^2)^{1/2} \cdot 2x$$

$$= \tfrac{1}{2}x(2-x-x^2)^{-1/2}\left[x(-1-2x) + 4(2-x-x^2)\right] = \frac{x(8-5x-6x^2)}{2\sqrt{(2+x)(1-x)}},$$

so $x_{n+1} = x_n - \dfrac{x_n^2\sqrt{2-x_n-x_n^2} - 1}{\dfrac{x_n(8-5x_n-6x_n^2)}{2\sqrt{(2+x_n)(1-x_n)}}}$. Trying $x_1 = -2$ won't work because $f'(-2)$ is undefined, so we try $x_1 = -1.95$.

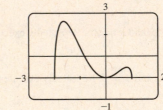

$x_1 = -1.95$	$x_1 = -0.8$
$x_2 \approx -1.98580357$	$x_2 \approx -0.82674444$
$x_3 \approx -1.97899778$	$x_3 \approx -0.82646236$
$x_4 \approx -1.97807848$	$x_4 \approx -0.82646233 \approx x_5$
$x_5 \approx -1.97806682$	
$x_6 \approx -1.97806681 \approx x_7$	

To eight decimal places, the roots of the equation are -1.97806681 and -0.82646233.

21. (a) $f(x) = x^2 - a \quad \Rightarrow \quad f'(x) = 2x$, so Newton's method gives

$$x_{n+1} = x_n - \frac{x_n^2 - a}{2x_n} = x_n - \frac{1}{2}x_n + \frac{a}{2x_n} = \frac{1}{2}x_n + \frac{a}{2x_n} = \frac{1}{2}\left(x_n + \frac{a}{x_n}\right).$$

(b) Using (a) with $a = 1000$ and $x_1 = \sqrt{900} = 30$, we get $x_2 \approx 31.666667$, $x_3 \approx 31.622807$, and $x_4 \approx 31.622777 \approx x_5$.

So $\sqrt{1000} \approx 31.622777$.

23. $f(x) = x^3 - 3x + 6 \quad \Rightarrow \quad f'(x) = 3x^2 - 3$. If $x_1 = 1$, then $f'(x_1) = 0$ and the tangent line used for approximating x_2 is horizontal. Attempting to find x_2 results in trying to divide by zero.

25. For $f(x) = x^{1/3}$, $f'(x) = \frac{1}{3}x^{-2/3}$ and

$$x_{n+1} = x_n - \frac{f(x_n)}{f'(x_n)} = x_n - \frac{x_n^{1/3}}{\frac{1}{3}x_n^{-2/3}} = x_n - 3x_n = -2x_n.$$

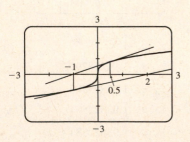

Therefore, each successive approximation becomes twice as large as the previous one in absolute value, so the sequence of approximations fails to converge to the root, which is 0. In the figure, we have $x_1 = 0.5$, $x_2 = -2(0.5) = -1$, and $x_3 = -2(-1) = 2$.

27.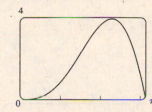

$$y = x^2 \sin x \implies y' = x^2 \cos x + (\sin x)(2x) \implies$$

$$y'' = x^2(-\sin x) + (\cos x)(2x) + (\sin x)(2) + 2x \cos x$$

$$= -x^2 \sin x + 4x \cos x + 2 \sin x \implies$$

$$y''' = -x^2 \cos x + (\sin x)(-2x) + 4x(-\sin x) + (\cos x)(4) + 2 \cos x$$

$$= -x^2 \cos x - 6x \sin x + 6 \cos x.$$

From the graph of $y = x^2 \sin x$, we see that $x = 1.5$ is a reasonable guess for the x-coordinate of the inflection point. Using Newton's method with $g(x) = y''$ and $g'(x) = y'''$, we get $x_1 = 1.5 \implies x_2 \approx 1.520092, x_3 \approx 1.519855 \approx x_4$. The inflection point is about $(1.519855, 2.306964)$.

29. In this case, $A = 18{,}000$, $R = 375$, and $n = 5(12) = 60$. So the formula $A = \dfrac{R}{i}[1 - (1 + i)^{-n}]$ becomes

$$18{,}000 = \frac{375}{x}[1 - (1 + x)^{-60}] \iff 48x = 1 - (1 + x)^{-60} \quad [\text{multiply each term by } (1 + x)^{60}] \iff$$

$48x(1 + x)^{60} - (1 + x)^{60} + 1 = 0$. Let the LHS be called $f(x)$, so that

$$f'(x) = 48x(60)(1 + x)^{59} + 48(1 + x)^{60} - 60(1 + x)^{59}$$

$$= 12(1 + x)^{59}[4x(60) + 4(1 + x) - 5] = 12(1 + x)^{59}(244x - 1)$$

$x_{n+1} = x_n - \dfrac{48x_n(1 + x_n)^{60} - (1 + x_n)^{60} + 1}{12(1 + x_n)^{59}(244x_n - 1)}$. An interest rate of 1% per month seems like a reasonable estimate for

$x = i$. So let $x_1 = 1\% = 0.01$, and we get $x_2 \approx 0.0082202$, $x_3 \approx 0.0076802$, $x_4 \approx 0.0076291$, $x_5 \approx 0.0076286 \approx x_6$. Thus, the dealer is charging a monthly interest rate of 0.76286% (or 9.55% per year, compounded monthly).

3.7 Antiderivatives

1. $f(x) = \frac{1}{2} + \frac{3}{4}x^2 - \frac{4}{5}x^3 \implies F(x) = \frac{1}{2}x + \dfrac{3}{4}\dfrac{x^{2+1}}{2+1} - \dfrac{4}{5}\dfrac{x^{3+1}}{3+1} + C = \frac{1}{2}x + \frac{1}{4}x^3 - \frac{1}{5}x^4 + C$

Check: $F'(x) = \frac{1}{2} + \frac{1}{4}(3x^2) - \frac{1}{5}(4x^3) + 0 = \frac{1}{2} + \frac{3}{4}x^2 - \frac{4}{5}x^3 = f(x)$

3. $f(x) = 7x^{2/5} + 8x^{-4/5} \implies F(x) = 7\left(\frac{5}{7}x^{7/5}\right) + 8(5x^{1/5}) + C = 5x^{7/5} + 40x^{1/5} + C$

5. $f(x) = 3\sqrt{x} - 2\sqrt[3]{x} = 3x^{1/2} - 2x^{1/3} \implies F(x) = 3\left(\frac{2}{3}x^{3/2}\right) - 2\left(\frac{3}{4}x^{4/3}\right) + C = 2x^{3/2} - \frac{3}{2}x^{4/3} + C$

7. $g(t) = \dfrac{1 + t + t^2}{\sqrt{t}} = t^{-1/2} + t^{1/2} + t^{3/2} \implies G(t) = 2t^{1/2} + \frac{2}{3}t^{3/2} + \frac{2}{5}t^{5/2} + C$

9. $h(\theta) = 2\sin\theta - \sec^2\theta \implies H(\theta) = -2\cos\theta - \tan\theta + C_n$ on the interval $\left(n\pi - \frac{\pi}{2}, n\pi + \frac{\pi}{2}\right)$.

11. $f(t) = 2\sec t \tan t + \frac{1}{2}t^{-1/2}$ has domain $\left(0, \frac{\pi}{2}\right)$ and $\left(n\pi - \frac{\pi}{2}, n\pi + \frac{\pi}{2}\right)$ for integers $n \geq 1$. The antiderivative is

$F(t) = 2\sec t + t^{1/2} + C_0$ on the interval $\left(0, \frac{\pi}{2}\right)$ or $F(t) = 2\sec t + t^{1/2} + C_n$ on the interval $\left(n\pi - \frac{\pi}{2}, n\pi + \frac{\pi}{2}\right)$ for integers $n \geq 1$.

13. $f(x) = \dfrac{x^5 - x^4 + 2x}{x^4} = x - 1 + \dfrac{2}{x^3} = x - 1 + 2x^{-3} \;\Rightarrow\; F(x) = \dfrac{x^2}{2} - x + 2\left(\dfrac{x^{-3+1}}{-3+1}\right) + C_1 = \frac{1}{2}x^2 - x - \dfrac{1}{x^2} + C_1$

on $(0, \infty)$ and $F(x) = \frac{1}{2}x^2 - x - \dfrac{1}{x^2} + C_2$ on $(-\infty, 0)$

15. $f(x) = 5x^4 - 2x^5 \;\Rightarrow\; F(x) = 5 \cdot \dfrac{x^5}{5} - 2 \cdot \dfrac{x^6}{6} + C = x^5 - \frac{1}{3}x^6 + C.$

$F(0) = 4 \;\Rightarrow\; 0^5 - \frac{1}{3} \cdot 0^6 + C = 4 \;\Rightarrow\; C = 4$, so $F(x) = x^5 - \frac{1}{3}x^6 + 4.$

The graph confirms our answer since $f(x) = 0$ when F has a local maximum, f is

positive when F is increasing, and f is negative when F is decreasing.

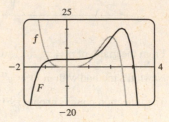

17. $f''(x) = 20x^3 - 12x^2 + 6x \;\Rightarrow\; f'(x) = 20\left(\dfrac{x^4}{4}\right) - 12\left(\dfrac{x^3}{3}\right) + 6\left(\dfrac{x^2}{2}\right) + C = 5x^4 - 4x^3 + 3x^2 + C \;\Rightarrow$

$f(x) = 5\left(\dfrac{x^5}{5}\right) - 4\left(\dfrac{x^4}{4}\right) + 3\left(\dfrac{x^3}{3}\right) + Cx + D = x^5 - x^4 + x^3 + Cx + D$

19. $f''(x) = \frac{2}{3}x^{2/3} \;\Rightarrow\; f'(x) = \frac{2}{3}\left(\dfrac{x^{5/3}}{5/3}\right) + C = \frac{2}{5}x^{5/3} + C \;\Rightarrow\; f(x) = \frac{2}{5}\left(\dfrac{x^{8/3}}{8/3}\right) + Cx + D = \frac{3}{20}x^{8/3} + Cx + D$

21. $f'''(t) = \cos t \;\Rightarrow\; f''(t) = \sin t + C_1 \;\Rightarrow\; f'(t) = -\cos t + C_1 t + D \;\Rightarrow\; f(t) = -\sin t + Ct^2 + Dt + E,$

where $C = \frac{1}{2}C_1.$

23. $f'(x) = 1 + 3\sqrt{x} \;\Rightarrow\; f(x) = x + 3\left(\frac{2}{3}x^{3/2}\right) + C = x + 2x^{3/2} + C. \quad f(4) = 4 + 2(8) + C$ and $f(4) = 25 \;\Rightarrow$

$20 + C = 25 \;\Rightarrow\; C = 5$, so $f(x) = x + 2x^{3/2} + 5.$

25. $f'(x) = \sqrt{x}(6 + 5x) = 6x^{1/2} + 5x^{3/2} \;\Rightarrow\; f(x) = 4x^{3/2} + 2x^{5/2} + C.$

$f(1) = 6 + C$ and $f(1) = 10 \;\Rightarrow\; C = 4$, so $f(x) = 4x^{3/2} + 2x^{5/2} + 4.$

27. $f'(t) = 2\cos t + \sec^2 t \;\Rightarrow\; f(t) = 2\sin t + \tan t + C$ because $-\pi/2 < t < \pi/2.$

$f\left(\frac{\pi}{3}\right) = 2\left(\sqrt{3}/2\right) + \sqrt{3} + C = 2\sqrt{3} + C$ and $f\left(\frac{\pi}{3}\right) = 4 \;\Rightarrow\; C = 4 - 2\sqrt{3}$, so $f(t) = 2\sin t + \tan t + 4 - 2\sqrt{3}.$

29. $f''(x) = -2 + 12x - 12x^2 \;\Rightarrow\; f'(x) = -2x + 6x^2 - 4x^3 + C. \quad f'(0) = C$ and $f'(0) = 12 \;\Rightarrow\; C = 12$, so

$f'(x) = -2x + 6x^2 - 4x^3 + 12$ and hence, $f(x) = -x^2 + 2x^3 - x^4 + 12x + D. \quad f(0) = D$ and $f(0) = 4 \;\Rightarrow\; D = 4,$

so $f(x) = -x^2 + 2x^3 - x^4 + 12x + 4.$

31. $f''(\theta) = \sin\theta + \cos\theta \;\Rightarrow\; f'(\theta) = -\cos\theta + \sin\theta + C. \quad f'(0) = -1 + C$ and $f'(0) = 4 \;\Rightarrow\; C = 5$, so

$f'(\theta) = -\cos\theta + \sin\theta + 5$ and hence, $f(\theta) = -\sin\theta - \cos\theta + 5\theta + D. \quad f(0) = -1 + D$ and $f(0) = 3 \;\Rightarrow\; D = 4,$

so $f(\theta) = -\sin\theta - \cos\theta + 5\theta + 4.$

33. $f''(x) = x^{-2},\, x > 0 \;\Rightarrow\; f'(x) = -1/x + C \;\Rightarrow\; f(x) = -\ln|x| + Cx + D = -\ln x + Cx + D$ [since $x > 0$].

$f(1) = 0 \;\Rightarrow\; C + D = 0$ and $f(2) = 0 \;\Rightarrow\; -\ln 2 + 2C + D = 0 \;\Rightarrow\; -\ln 2 + 2C - C = 0$ [since $D = -C$] $\;\Rightarrow$

$-\ln 2 + C = 0 \;\Rightarrow\; C = \ln 2$ and $D = -\ln 2.$ So $f(x) = -\ln x + (\ln 2)x - \ln 2.$

35. Given $f'(x) = 2x + 1$, we have $f(x) = x^2 + x + C$. Since f passes through $(1, 6)$, $f(1) = 6$ $\Rightarrow$ $1^2 + 1 + C = 6$ $\Rightarrow$ $C = 4$. Therefore, $f(x) = x^2 + x + 4$ and $f(2) = 2^2 + 2 + 4 = 10$.

37. b is the antiderivative of f. For small x, f is negative, so the graph of its antiderivative must be decreasing. But both a and c are increasing for small x, so only b can be f's antiderivative. Also, f is positive where b is increasing, which supports our conclusion.

39. $v(t) = s'(t) = \sin t - \cos t$ $\Rightarrow$ $s(t) = -\cos t - \sin t + C$. $s(0) = -1 + C$ and $s(0) = 0$ $\Rightarrow$ $C = 1$, so $s(t) = -\cos t - \sin t + 1$.

41. $a(t) = v'(t) = 10 \sin t + 3 \cos t$ $\Rightarrow$ $v(t) = -10 \cos t + 3 \sin t + C$ $\Rightarrow$ $s(t) = -10 \sin t - 3 \cos t + Ct + D$. $s(0) = -3 + D = 0$ and $s(2\pi) = -3 + 2\pi C + D = 12$ $\Rightarrow$ $D = 3$ and $C = \frac{6}{\pi}$. Thus, $s(t) = -10 \sin t - 3 \cos t + \frac{6}{\pi} t + 3$.

43. (a) We first observe that since the stone is dropped 450 m above the ground, $v(0) = 0$ and $s(0) = 450$.
$v'(t) = a(t) = -9.8$ $\Rightarrow$ $v(t) = -9.8t + C$. Now $v(0) = 0$ $\Rightarrow$ $C = 0$, so $v(t) = -9.8t$ $\Rightarrow$ $s(t) = -4.9t^2 + D$. Last, $s(0) = 450$ $\Rightarrow$ $D = 450$ $\Rightarrow$ $s(t) = 450 - 4.9t^2$.

(b) The stone reaches the ground when $s(t) = 0$. $450 - 4.9t^2 = 0$ $\Rightarrow$ $t^2 = 450/4.9$ $\Rightarrow$ $t_1 = \sqrt{450/4.9} \approx 9.58$ s.

(c) The velocity with which the stone strikes the ground is $v(t_1) = -9.8\sqrt{450/4.9} \approx -93.9$ m/s.

(d) This is just reworking parts (a) and (b) with $v(0) = -5$. Using $v(t) = -9.8t + C$, $v(0) = -5$ $\Rightarrow$ $0 + C = -5$ $\Rightarrow$ $v(t) = -9.8t - 5$. So $s(t) = -4.9t^2 - 5t + D$ and $s(0) = 450$ $\Rightarrow$ $D = 450$ $\Rightarrow$ $s(t) = -4.9t^2 - 5t + 450$.
Solving $s(t) = 0$ by using the quadratic formula gives us $t = \left(5 \pm \sqrt{8845}\right)/(-9.8)$ $\Rightarrow$ $t_1 \approx 9.09$ s.

45. By Exercise 44 with $a = -9.8$, $s(t) = -4.9t^2 + v_0 t + s_0$ and $v(t) = s'(t) = -9.8t + v_0$. So
$[v(t)]^2 = (-9.8t + v_0)^2 = (9.8)^2 t^2 - 19.6v_0 t + v_0^2 = v_0^2 + 96.04t^2 - 19.6v_0 t = v_0^2 - 19.6\left(-4.9t^2 + v_0 t\right)$.
But $-4.9t^2 + v_0 t$ is just $s(t)$ without the s_0 term; that is, $s(t) - s_0$. Thus, $[v(t)]^2 = v_0^2 - 19.6\left[s(t) - s_0\right]$.

47. Using Exercise 44 with $a = -32$, $v_0 = 0$, and $s_0 = h$ (the height of the cliff), we know that the height at time t is $s(t) = -16t^2 + h$. $v(t) = s'(t) = -32t$ and $v(t) = -120$ $\Rightarrow$ $-32t = -120$ $\Rightarrow$ $t = 3.75$, so $0 = s(3.75) = -16(3.75)^2 + h$ $\Rightarrow$ $h = 16(3.75)^2 = 225$ ft.

49. Taking the upward direction to be positive we have that for $0 \le t \le 10$ (using the subscript 1 to refer to $0 \le t \le 10$),
$a_1(t) = -(9 - 0.9t) = v_1'(t)$ $\Rightarrow$ $v_1(t) = -9t + 0.45t^2 + v_0$, but $v_1(0) = v_0 = -10$ $\Rightarrow$
$v_1(t) = -9t + 0.45t^2 - 10 = s_1'(t)$ $\Rightarrow$ $s_1(t) = -\frac{9}{2}t^2 + 0.15t^3 - 10t + s_0$. But $s_1(0) = 500 = s_0$ $\Rightarrow$
$s_1(t) = -\frac{9}{2}t^2 + 0.15t^3 - 10t + 500$. $s_1(10) = -450 + 150 - 100 + 500 = 100$, so it takes more than 10 seconds for the raindrop to fall. Now for $t > 10$, $a(t) = 0 = v'(t)$ $\Rightarrow$
$v(t) = \text{constant} = v_1(10) = -9(10) + 0.45(10)^2 - 10 = -55$ $\Rightarrow$ $v(t) = -55$.
At 55 m/s, it will take $100/55 \approx 1.8$ s to fall the last 100 m. Hence, the total time is $10 + \frac{100}{55} = \frac{130}{11} \approx 11.8$ s.

51. $a(t) = k$, the initial velocity is $30 \text{ mi/h} = 30 \cdot \frac{5280}{3600} = 44 \text{ ft/s}$, and the final velocity (after 5 seconds) is

$50 \text{ mi/h} = 50 \cdot \frac{5280}{3600} = \frac{220}{3} \text{ ft/s}$. So $v(t) = kt + C$ and $v(0) = 44 \;\Rightarrow\; C = 44$. Thus, $v(t) = kt + 44 \;\Rightarrow\;$

$v(5) = 5k + 44$. But $v(5) = \frac{220}{3}$, so $5k + 44 = \frac{220}{3} \;\Rightarrow\; 5k = \frac{88}{3} \;\Rightarrow\; k = \frac{88}{15} \approx 5.87 \text{ ft/s}^2$.

53. Let the acceleration be $a(t) = k \text{ km/h}^2$. We have $v(0) = 100 \text{ km/h}$ and we can take the initial position $s(0)$ to be 0.

We want the time t_f for which $v(t) = 0$ to satisfy $s(t) < 0.08 \text{ km}$. In general, $v'(t) = a(t) = k$, so $v(t) = kt + C$,

where $C = v(0) = 100$. Now $s'(t) = v(t) = kt + 100$, so $s(t) = \frac{1}{2}kt^2 + 100t + D$, where $D = s(0) = 0$.

Thus, $s(t) = \frac{1}{2}kt^2 + 100t$. Since $v(t_f) = 0$, we have $kt_f + 100 = 0$ or $t_f = -100/k$, so

$$s(t_f) = \frac{1}{2}k\left(-\frac{100}{k}\right)^2 + 100\left(-\frac{100}{k}\right) = 10{,}000\left(\frac{1}{2k} - \frac{1}{k}\right) = -\frac{5{,}000}{k}. \text{ The condition } s(t_f) \text{ must satisfy is}$$

$-\dfrac{5{,}000}{k} < 0.08 \;\Rightarrow\; -\dfrac{5{,}000}{0.08} > k \quad [k \text{ is negative}] \;\Rightarrow\; k < -62{,}500 \text{ km/h}^2$, or equivalently,

$k < -\frac{3125}{648} \approx -4.82 \text{ m/s}^2$.

55. (a) First note that $90 \text{ mi/h} = 90 \times \frac{5280}{3600} \text{ ft/s} = 132 \text{ ft/s}$. Then $a(t) = 4 \text{ ft/s}^2 \;\Rightarrow\; v(t) = 4t + C$, but $v(0) = 0 \;\Rightarrow\;$

$C = 0$. Now $4t = 132$ when $t = \frac{132}{4} = 33 \text{ s}$, so it takes 33 s to reach 132 ft/s. Therefore, taking $s(0) = 0$, we have

$s(t) = 2t^2$, $0 \le t \le 33$. So $s(33) = 2178 \text{ ft}$. 15 minutes $= 15(60) = 900 \text{ s}$, so for $33 < t \le 933$ we have

$v(t) = 132 \text{ ft/s} \;\Rightarrow\; s(933) = 132(900) + 2178 = 120{,}978 \text{ ft} = 22.9125 \text{ mi}$.

(b) As in part (a), the train accelerates for 33 s and travels 2178 ft while doing so. Similarly, it decelerates for 33 s and travels

2178 ft at the end of its trip. During the remaining $900 - 66 = 834 \text{ s}$ it travels at 132 ft/s, so the distance traveled is

$132 \cdot 834 = 110{,}088 \text{ ft}$. Thus, the total distance is $2178 + 110{,}088 + 2178 = 114{,}444 \text{ ft} = 21.675 \text{ mi}$.

(c) 45 mi $= 45(5280) = 237{,}600 \text{ ft}$. Subtract $2(2178)$ to take care of the speeding up and slowing down, and we have

$233{,}244 \text{ ft}$ at 132 ft/s for a trip of $233{,}244/132 = 1767 \text{ s}$ at 90 mi/h. The total time is

$1767 + 2(33) = 1833 \text{ s} = 30 \text{ min } 33 \text{ s} = 30.55 \text{ min}$.

(d) $37.5(60) = 2250 \text{ s}$. $2250 - 2(33) = 2184 \text{ s}$ at maximum speed. $2184(132) + 2(2178) = 292{,}644$ total feet or

$292{,}644/5280 = 55.425 \text{ mi}$.

3 Review

CONCEPT CHECK

1. A function f has an **absolute maximum** at $x = c$ if $f(c)$ is the largest function value on the entire domain of f, whereas f has

a **local maximum** at c if $f(c)$ is the largest function value when x is near c. See Figure 4 in Section 3.1.

2. (a) See the Extreme Value Theorem on page 146.

(b) See the Closed Interval Method on page 149.

3. (a) See Fermat's Theorem on page 147.

(b) See the definition of a critical number on page 149.

4. (a) See Rolle's Theorem on page 152.

(b) See the Mean Value Theorem on page 154. Geometric interpretation—there is some point P on the graph of a function f [on the interval (a, b)] where the tangent line is parallel to the secant line that connects $(a, f(a))$ and $(b, f(b))$.

5. (a) See the Increasing/Decreasing (I/D) Test on page 159.

(b) A function f is concave upward on an interval I if the graph of f lies above all of its tangent lines on I.

(c) See the Concavity Test on page 162.

(d) An inflection point is a point where a curve changes its direction of concavity. They can be found by determining the points at which the second derivative changes sign.

6. (a) See the First Derivative Test on page 160.

(b) See the Second Derivative Test on page 163.

(c) See the note before Example 6 in Section 3.3.

7. Without calculus you could get misleading graphs that fail to show the most interesting features of a function. See the third paragraph in Section 3.4.

8. (a) See Figure 3 in Section 3.6.

(b) $x_2 = x_1 - \dfrac{f(x_1)}{f'(x_1)}$

(c) $x_{n+1} = x_n - \dfrac{f(x_n)}{f'(x_n)}$

(d) Newton's method is likely to fail or to work very slowly when $f'(x_1)$ is close to 0. It also fails when $f'(x_i)$ is undefined, such as with $f(x) = 1/x - 2$ and $x_1 = 1$.

9. (a) See the definition at the beginning of Section 3.7.

(b) If F_1 and F_2 are both antiderivatives of f on an interval I, then they differ by a constant.

TRUE-FALSE QUIZ

1. False.   For example, take $f(x) = x^3$, then $f'(x) = 3x^2$ and $f'(0) = 0$, but $f(0) = 0$ is not a maximum or minimum; $(0, 0)$ is an inflection point.

3. False.   For example, $f(x) = x$ is continuous on $(0, 1)$ but attains neither a maximum nor a minimum value on $(0, 1)$. Don't confuse this with f being continuous on the *closed* interval $[a, b]$, which would make the statement true.

5. True.   This is an example of part (b) of the I/D Test.

7. False. $f'(x) = g'(x) \;\Rightarrow\; f(x) = g(x) + C$. For example, if $f(x) = x + 2$ and $g(x) = x + 1$, then $f'(x) = g'(x) = 1$, but $f(x) \neq g(x)$.

9. True. The graph of one such function is sketched.

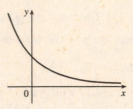

11. True. Let $x_1 < x_2$ where $x_1, x_2 \in I$. Then $f(x_1) < f(x_2)$ and $g(x_1) < g(x_2)$ [since f and g are increasing on I],

so $(f+g)(x_1) = f(x_1) + g(x_1) < f(x_2) + g(x_2) = (f+g)(x_2)$.

13. False. Take $f(x) = x$ and $g(x) = x - 1$. Then both f and g are increasing on $(0, 1)$. But $f(x)\,g(x) = x(x-1)$ is not

increasing on $(0, 1)$.

15. True. Let $x_1, x_2 \in I$ and $x_1 < x_2$. Then $f(x_1) < f(x_2)$ [f is increasing] $\Rightarrow$ $\dfrac{1}{f(x_1)} > \dfrac{1}{f(x_2)}$ [f is positive] $\Rightarrow$

$g(x_1) > g(x_2)$ $\Rightarrow$ $g(x) = 1/f(x)$ is decreasing on I.

17. True. If f is periodic, then there is a number p such that $f(x + p) = f(p)$ for all x. Differentiating gives

$f'(x) = f'(x + p) \cdot (x + p)' = f'(x + p) \cdot 1 = f'(x + p)$, so f' is periodic.

19. True. By the Mean Value Theorem, there exists a number c in $(0, 1)$ such that $f(1) - f(0) = f'(c)(1 - 0) = f'(c)$.

Since $f'(c)$ is nonzero, $f(1) - f(0) \neq 0$, so $f(1) \neq f(0)$.

EXERCISES

1. $f(x) = x^3 - 6x^2 + 9x + 1$, $[2, 4]$. $f'(x) = 3x^2 - 12x + 9 = 3(x^2 - 4x + 3) = 3(x - 1)(x - 3)$. $f'(x) = 0$ $\Rightarrow$

$x = 1$ or $x = 3$, but 1 is not in the interval. $f'(x) > 0$ for $3 < x < 4$ and $f'(x) < 0$ for $2 < x < 3$, so $f(3) = 1$ is a local

minimum value. Checking the endpoints, we find $f(2) = 3$ and $f(4) = 5$. Thus, $f(3) = 1$ is the absolute minimum value and

$f(4) = 5$ is the absolute maximum value.

3. $f(x) = \dfrac{3x - 4}{x^2 + 1}$, $[-2, 2]$. $f'(x) = \dfrac{(x^2 + 1)(3) - (3x - 4)(2x)}{(x^2 + 1)^2} = \dfrac{-(3x^2 - 8x - 3)}{(x^2 + 1)^2} = \dfrac{-(3x + 1)(x - 3)}{(x^2 + 1)^2}$.

$f'(x) = 0$ $\Rightarrow$ $x = -\frac{1}{3}$ or $x = 3$, but 3 is not in the interval. $f'(x) > 0$ for $-\frac{1}{3} < x < 2$ and $f'(x) < 0$ for

$-2 < x < -\frac{1}{3}$, so $f\left(-\frac{1}{3}\right) = \frac{-5}{10/9} = -\frac{9}{2}$ is a local minimum value. Checking the endpoints, we find $f(-2) = -2$ and

$f(2) = \frac{2}{5}$. Thus, $f\left(-\frac{1}{3}\right) = -\frac{9}{2}$ is the absolute minimum value and $f(2) = \frac{2}{5}$ is the absolute maximum value.

5. $f(0) = 0$, $f'(-2) = f'(1) = f'(9) = 0$, $\displaystyle\lim_{x \to \infty} f(x) = 0$, $\displaystyle\lim_{x \to 6} f(x) = -\infty$,

$f'(x) < 0$ on $(-\infty, -2)$, $(1, 6)$, and $(9, \infty)$, $f'(x) > 0$ on $(-2, 1)$ and $(6, 9)$,

$f''(x) > 0$ on $(-\infty, 0)$ and $(12, \infty)$, $f''(x) < 0$ on $(0, 6)$ and $(6, 12)$

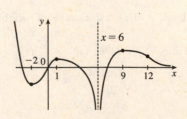

7. f is odd, $f'(x) < 0$ for $0 < x < 2$, $\quad f'(x) > 0$ for $x > 2$,

$f''(x) > 0$ for $0 < x < 3$, $\quad f''(x) < 0$ for $x > 3$, $\quad \lim\limits_{x \to \infty} f(x) = -2$

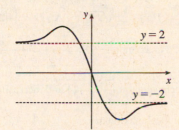

9. $y = f(x) = 2 - 2x - x^3$ **A.** $D = \mathbb{R}$ **B.** y-intercept: $f(0) = 2$.

The x-intercept (approximately 0.770917) can be found using Newton's

Method. **C.** No symmetry **D.** No asymptote

E. $f'(x) = -2 - 3x^2 = -(3x^2 + 2) < 0$, so f is decreasing on $\mathbb{R}$.

F. No extreme value **G.** $f''(x) = -6x < 0$ on $(0, \infty)$ and $f''(x) > 0$ on

$(-\infty, 0)$, so f is CD on $(0, \infty)$ and CU on $(-\infty, 0)$. There is an IP at $(0, 2)$.

H.

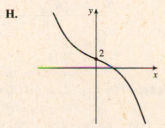

11. $y = f(x) = x^4 - 3x^3 + 3x^2 - x = x(x-1)^3$ **A.** $D = \mathbb{R}$ **B.** y-intercept: $f(0) = 0$; x-intercepts: $f(x) = 0$ $\Leftrightarrow$

$x = 0$ or $x = 1$ **C.** No symmetry **D.** f is a polynomial function and hence, it has no asymptote.

E. $f'(x) = 4x^3 - 9x^2 + 6x - 1$. Since the sum of the coefficients is 0, 1 is a root of f', so

$f'(x) = (x-1)(4x^2 - 5x + 1) = (x-1)^2(4x-1)$. $f'(x) < 0$ $\Rightarrow$ $x < \frac{1}{4}$, so f is decreasing on $\left(-\infty, \frac{1}{4}\right)$

and f is increasing on $\left(\frac{1}{4}, \infty\right)$. **F.** $f'(x)$ does not change sign at $x = 1$, so **H.**

there is not a local extremum there. $f\left(\frac{1}{4}\right) = -\frac{27}{256}$ is a local minimum value.

G. $f''(x) = 12x^2 - 18x + 6 = 6(2x - 1)(x - 1)$. $f''(x) = 0$ $\Leftrightarrow$ $x = \frac{1}{2}$

or 1. $f''(x) < 0$ $\Leftrightarrow$ $\frac{1}{2} < x < 1$ $\Rightarrow$ f is CD on $\left(\frac{1}{2}, 1\right)$ and CU on

$\left(-\infty, \frac{1}{2}\right)$ and $(1, \infty)$. There are inflection points at $\left(\frac{1}{2}, -\frac{1}{16}\right)$ and $(1, 0)$.

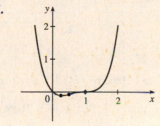

13. $y = f(x) = \dfrac{1}{x(x-3)^2}$ **A.** $D = \{x \mid x \neq 0, 3\} = (-\infty, 0) \cup (0, 3) \cup (3, \infty)$ **B.** No intercepts. **C.** No symmetry.

D. $\lim\limits_{x \to \pm\infty} \dfrac{1}{x(x-3)^2} = 0$, so $y = 0$ is a HA. $\lim\limits_{x \to 0^+} \dfrac{1}{x(x-3)^2} = \infty$, $\lim\limits_{x \to 0^-} \dfrac{1}{x(x-3)^2} = -\infty$, $\lim\limits_{x \to 3} \dfrac{1}{x(x-3)^2} = \infty$,

so $x = 0$ and $x = 3$ are VA. **E.** $f'(x) = -\dfrac{(x-3)^2 + 2x(x-3)}{x^2(x-3)^4} = \dfrac{3(1-x)}{x^2(x-3)^3}$ $\Rightarrow$ $f'(x) > 0$ $\Leftrightarrow$ $1 < x < 3$,

so f is increasing on $(1, 3)$ and decreasing on $(-\infty, 0)$, $(0, 1)$, and $(3, \infty)$. **H.**

F. Local minimum value $f(1) = \frac{1}{4}$ **G.** $f''(x) = \dfrac{6(2x^2 - 4x + 3)}{x^3(x-3)^4}$.

Note that $2x^2 - 4x + 3 > 0$ for all x since it has negative discriminant.

So $f''(x) > 0$ $\Leftrightarrow$ $x > 0$ $\Rightarrow$ f is CU on $(0, 3)$ and $(3, \infty)$ and

CD on $(-\infty, 0)$. No IP

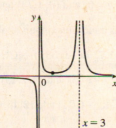

15. $y = f(x) = x\sqrt{2+x}$ **A.** $D = [-2, \infty)$ **B.** y-intercept: $f(0) = 0$; x-intercepts: -2 and 0 **C.** No symmetry

D. No asymptote **E.** $f'(x) = \dfrac{x}{2\sqrt{2+x}} + \sqrt{2+x} = \dfrac{1}{2\sqrt{2+x}}[x + 2(2+x)] = \dfrac{3x+4}{2\sqrt{2+x}} = 0$ when $x = -\frac{4}{3}$, so f is

decreasing on $\left(-2, -\frac{4}{3}\right)$ and increasing on $\left(-\frac{4}{3}, \infty\right)$. **F.** Local minimum value $f\left(-\frac{4}{3}\right) = -\frac{4}{3}\sqrt{\frac{2}{3}} = -\frac{4\sqrt{6}}{9} \approx -1.09$,

no local maximum

H.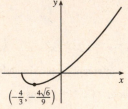

G. $f''(x) = \dfrac{2\sqrt{2+x}\cdot 3 - (3x+4)\dfrac{1}{\sqrt{2+x}}}{4(2+x)} = \dfrac{6(2+x) - (3x+4)}{4(2+x)^{3/2}}$

$= \dfrac{3x+8}{4(2+x)^{3/2}}$

$\left(-\frac{4}{3}, -\frac{4\sqrt{6}}{9}\right)$

$f''(x) > 0$ for $x > -2$, so f is CU on $(-2, \infty)$. No IP

17. $y = f(x) = \sin^2 x - 2\cos x$ **A.** $D = \mathbb{R}$ **B.** y-intercept: $f(0) = -2$ **C.** $f(-x) = f(x)$, so f is symmetric with respect

to the y-axis. f has period 2π. **D.** No asymptote **E.** $y' = 2\sin x \cos x + 2\sin x = 2\sin x(\cos x + 1)$. $y' = 0$ $\Leftrightarrow$

$\sin x = 0$ or $\cos x = -1$ $\Leftrightarrow$ $x = n\pi$ or $x = (2n+1)\pi$. $y' > 0$ when $\sin x > 0$, since $\cos x + 1 \geq 0$ for all x.

Therefore, $y' > 0$ [and so f is increasing] on $(2n\pi, (2n+1)\pi)$; $y' < 0$ [and so f is decreasing] on $((2n-1)\pi, 2n\pi)$.

F. Local maximum values are $f((2n+1)\pi) = 2$; local minimum values are $f(2n\pi) = -2$.

G. $y' = \sin 2x + 2\sin x$ $\Rightarrow$ $y'' = 2\cos 2x + 2\cos x = 2(2\cos^2 x - 1) + 2\cos x = 4\cos^2 x + 2\cos x - 2$

$= 2(2\cos^2 x + \cos x - 1) = 2(2\cos x - 1)(\cos x + 1)$

$y'' = 0$ $\Leftrightarrow$ $\cos x = \frac{1}{2}$ or -1 $\Leftrightarrow$ $x = 2n\pi \pm \frac{\pi}{3}$ or $x = (2n+1)\pi$. **H.**

$y'' > 0$ [and so f is CU] on $\left(2n\pi - \frac{\pi}{3}, 2n\pi + \frac{\pi}{3}\right)$; $y'' \leq 0$ [and so f is CD]

on $\left(2n\pi + \frac{\pi}{3}, 2n\pi + \frac{5\pi}{3}\right)$. There are inflection points at $\left(2n\pi \pm \frac{\pi}{3}, -\frac{1}{4}\right)$.

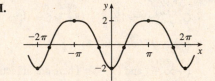

19. $f(x) = \dfrac{x^2 - 1}{x^3}$ $\Rightarrow$ $f'(x) = \dfrac{x^3(2x) - (x^2 - 1)3x^2}{x^6} = \dfrac{3 - x^2}{x^4}$ $\Rightarrow$

$f''(x) = \dfrac{x^4(-2x) - (3 - x^2)4x^3}{x^8} = \dfrac{2x^2 - 12}{x^5}$

Estimates: From the graphs of f' and f'', it appears that f is increasing on

$(-1.73, 0)$ and $(0, 1.73)$ and decreasing on $(-\infty, -1.73)$ and $(1.73, \infty)$;

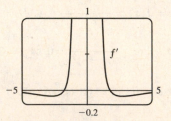

f has a local maximum of about $f(1.73) = 0.38$ and a local minimum of about

$f(-1.7) = -0.38$; f is CU on $(-2.45, 0)$ and $(2.45, \infty)$, and CD on

$(-\infty, -2.45)$ and $(0, 2.45)$; and f has inflection points at about

$(-2.45, -0.34)$ and $(2.45, 0.34)$.

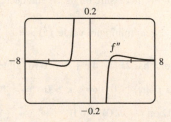

Exact: Now $f'(x) = \dfrac{3 - x^2}{x^4}$ is positive for $0 < x^2 < 3$, that is, f is increasing

on $(-\sqrt{3}, 0)$ and $(0, \sqrt{3})$; and $f'(x)$ is negative (and so f is decreasing) on

$(-\infty, -\sqrt{3})$ and $(\sqrt{3}, \infty)$. $f'(x) = 0$ when $x = \pm\sqrt{3}$.

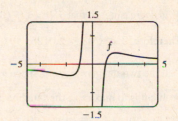

f' goes from positive to negative at $x = \sqrt{3}$, so f has a local maximum of

$f(\sqrt{3}) = \dfrac{(\sqrt{3})^2 - 1}{(\sqrt{3})^3} = \dfrac{2\sqrt{3}}{9}$; and since f is odd, we know that maxima on the

interval $(0, \infty)$ correspond to minima on $(-\infty, 0)$, so f has a local minimum of

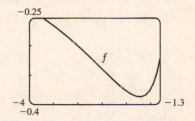

$f(-\sqrt{3}) = -\dfrac{2\sqrt{3}}{9}$. Also, $f''(x) = \dfrac{2x^2 - 12}{x^5}$ is positive (so f is CU) on

$(-\sqrt{6}, 0)$ and $(\sqrt{6}, \infty)$, and negative (so f is CD) on $(-\infty, -\sqrt{6})$ and

$(0, \sqrt{6})$. There are IP at $\left(\sqrt{6}, \dfrac{5\sqrt{6}}{36}\right)$ and $\left(-\sqrt{6}, -\dfrac{5\sqrt{6}}{36}\right)$.

21. $f(x) = 3x^6 - 5x^5 + x^4 - 5x^3 - 2x^2 + 2 \quad\Rightarrow\quad f'(x) = 18x^5 - 25x^4 + 4x^3 - 15x^2 - 4x \quad\Rightarrow$

$f''(x) = 90x^4 - 100x^3 + 12x^2 - 30x - 4$

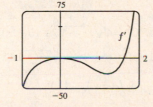
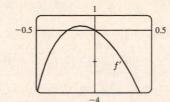
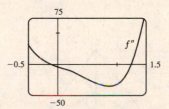

From the graphs of f' and f'', it appears that f is increasing on $(-0.23, 0)$ and $(1.62, \infty)$ and decreasing on $(-\infty, -0.23)$

and $(0, 1.62)$; f has a local maximum of $f(0) = 2$ and local minima of about $f(-0.23) = 1.96$ and $f(1.62) = -19.2$;

f is CU on $(-\infty, -0.12)$ and $(1.24, \infty)$ and CD on $(-0.12, 1.24)$; and f has inflection points at about $(-0.12, 1.98)$ and

$(1.24, -12.1)$.

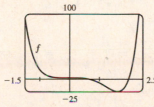
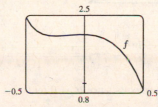

23. Let $f(x) = 3x + 2\cos x + 5$. Then $f(0) = 7 > 0$ and $f(-\pi) = -3\pi - 2 + 5 = -3\pi + 3 = -3(\pi - 1) < 0$, and since f is

continuous on $\mathbb{R}$ (hence on $[-\pi, 0]$), the Intermediate Value Theorem assures us that there is at least one zero of f in $[-\pi, 0]$.

Now $f'(x) = 3 - 2\sin x > 0$ implies that f is increasing on $\mathbb{R}$, so there is exactly one zero of f, and hence, exactly one real

root of the equation $3x + 2\cos x + 5 = 0$.

25. Since f is continuous on $[32, 33]$ and differentiable on $(32, 33)$, then by the Mean Value Theorem there exists a number c in

$(32, 33)$ such that $f'(c) = \frac{1}{5}c^{-4/5} = \dfrac{\sqrt[5]{33} - \sqrt[5]{32}}{33 - 32} = \sqrt[5]{33} - 2$, but $\frac{1}{5}c^{-4/5} > 0 \quad\Rightarrow\quad \sqrt[5]{33} - 2 > 0 \quad\Rightarrow\quad \sqrt[5]{33} > 2$. Also

f' is decreasing, so that $f'(c) < f'(32) = \frac{1}{5}(32)^{-4/5} = 0.0125 \;\Rightarrow\; 0.0125 > f'(c) = \sqrt[5]{33} - 2 \;\Rightarrow\; \sqrt[5]{33} < 2.0125$.

Therefore, $2 < \sqrt[5]{33} < 2.0125$.

27. Call the two integers x and y. Then $x + 4y = 1000$, so $x = 1000 - 4y$. Their product is $P = xy = (1000 - 4y)y$, so our

problem is to maximize the function $P(y) = 1000y - 4y^2$, where $0 < y < 250$ and y is an integer. $P'(y) = 1000 - 8y$, so

$P'(y) = 0 \;\Leftrightarrow\; y = 125$. $P''(y) = -8 < 0$, so $P(125) = 62{,}500$ is an absolute maximum. Since the optimal y turned

out to be an integer, we have found the desired pair of numbers, namely $x = 1000 - 4(125) = 500$ and $y = 125$.

29.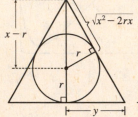

By similar triangles, $\dfrac{y}{x} = \dfrac{r}{\sqrt{x^2 - 2rx}}$, so the area of the triangle is

$$A(x) = \tfrac{1}{2}(2y)x = xy = \frac{rx^2}{\sqrt{x^2 - 2rx}} \;\Rightarrow$$

$$A'(x) = \frac{2rx\sqrt{x^2 - 2rx} - rx^2(x - r)/\sqrt{x^2 - 2rx}}{x^2 - 2rx} = \frac{rx^2(x - 3r)}{(x^2 - 2rx)^{3/2}} = 0$$

when $x = 3r$.

$A'(x) < 0$ when $2r < x < 3r$, $A'(x) > 0$ when $x > 3r$. So $x = 3r$ gives a minimum and $A(3r) = \dfrac{r(9r^2)}{\sqrt{3}\,r} = 3\sqrt{3}\,r^2$.

31.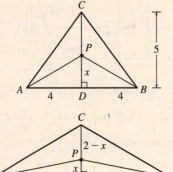

We minimize $L(x) = |PA| + |PB| + |PC| = 2\sqrt{x^2 + 16} + (5 - x)$,

$0 \le x \le 5$. $L'(x) = 2x/\sqrt{x^2 + 16} - 1 = 0 \;\Leftrightarrow\; 2x = \sqrt{x^2 + 16} \;\Leftrightarrow$

$4x^2 = x^2 + 16 \;\Leftrightarrow\; x = \frac{4}{\sqrt{3}}$. $L(0) = 13$, $L\!\left(\frac{4}{\sqrt{3}}\right) \approx 11.9$, $L(5) \approx 12.8$, so the

minimum occurs when $x = \frac{4}{\sqrt{3}} \approx 2.3$.

If $|CD| = 2$, $L(x)$ changes from $(5 - x)$ to $(2 - x)$ with $0 \le x \le 2$. But we still

get $L'(x) = 0 \;\Leftrightarrow\; x = \frac{4}{\sqrt{3}}$, which isn't in the interval $[0, 2]$. Now $L(0) = 10$

and $L(2) = 2\sqrt{20} = 4\sqrt{5} \approx 8.9$. The minimum occurs when $P = C$.

33. $v = K\sqrt{\dfrac{L}{C} + \dfrac{C}{L}} \;\Rightarrow\; \dfrac{dv}{dL} = \dfrac{K}{2\sqrt{(L/C) + (C/L)}}\left(\dfrac{1}{C} - \dfrac{C}{L^2}\right) = 0 \;\Leftrightarrow\; \dfrac{1}{C} = \dfrac{C}{L^2} \;\Leftrightarrow\; L^2 = C^2 \;\Leftrightarrow\; L = C$.

This gives the minimum velocity since $v' < 0$ for $0 < L < C$ and $v' > 0$ for $L > C$.

35. Let x denote the number of \$1 decreases in ticket price. Then the ticket price is \$12 $-$ \$1(x), and the average attendance is

$11{,}000 + 1000(x)$. Now the revenue per game is

$$R(x) = (\text{price per person}) \times (\text{number of people per game})$$

$$= (12 - x)(11{,}000 + 1000x) = -1000x^2 + 1000x + 132{,}000$$

for $0 \le x \le 4$ [since the seating capacity is $15{,}000$] $\;\Rightarrow\; R'(x) = -2000x + 1000 = 0 \;\Leftrightarrow\; x = 0.5$. This is a

maximum since $R''(x) = -2000 < 0$ for all x. Now we must check the value of $R(x) = (12 - x)(11{,}000 + 1000x)$ at

$x = 0.5$ and at the endpoints of the domain to see which value of x gives the maximum value of R.

$R(0) = (12)(11,000) = 132,000$, $R(0.5) = (11.5)(11,500) = 132,250$, and $R(4) = (8)(15,000) = 120,000$. Thus, the maximum revenue of \$132,250 per game occurs when the average attendance is 11,500 and the ticket price is \$11.50.

37. $f(t) = \cos t + t - t^2 \Rightarrow f'(t) = -\sin t + 1 - 2t$. $f'(t)$ exists for all

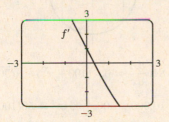

t, so to find the maximum of f, we can examine the zeros of f'.

From the graph of f', we see that a good choice for t_1 is $t_1 = 0.3$.

Use $g(t) = -\sin t + 1 - 2t$ and $g'(t) = -\cos t - 2$ to obtain

$t_2 \approx 0.33535293$, $t_3 \approx 0.33541803 \approx t_4$. Since $f''(t) = -\cos t - 2 < 0$

for all t, $f(0.33541803) \approx 1.16718557$ is the absolute maximum.

39. $f(x) = (x+1)(2x-1) = 2x^2 + x - 1 \Rightarrow F(x) = 2\left(\frac{1}{3}x^3\right) + \frac{1}{2}x^2 - x + C = \frac{2}{3}x^3 + \frac{1}{2}x^2 - x + C$

41. $f'(t) = 2t - 3\sin t \Rightarrow f(t) = t^2 + 3\cos t + C$.

$f(0) = 3 + C$ and $f(0) = 5 \Rightarrow C = 2$, so $f(t) = t^2 + 3\cos t + 2$.

43. $f''(x) = 1 - 6x + 48x^2 \Rightarrow f'(x) = x - 3x^2 + 16x^3 + C$. $f'(0) = C$ and $f'(0) = 2 \Rightarrow C = 2$, so

$f'(x) = x - 3x^2 + 16x^3 + 2$ and hence, $f(x) = \frac{1}{2}x^2 - x^3 + 4x^4 + 2x + D$.

$f(0) = D$ and $f(0) = 1 \Rightarrow D = 1$, so $f(x) = \frac{1}{2}x^2 - x^3 + 4x^4 + 2x + 1$.

45. $v(t) = s'(t) = 2t - \sin t \Rightarrow s(t) = t^2 + \cos t + C$.

$s(0) = 0 + 1 + C = C + 1$ and $s(0) = 3 \Rightarrow C + 1 = 3 \Rightarrow C = 2$, so $s(t) = t^2 + \cos t + 2$.

47. Choosing the positive direction to be upward, we have $a(t) = -9.8 \Rightarrow v(t) = -9.8t + v_0$, but $v(0) = 0 = v_0 \Rightarrow$

$v(t) = -9.8t = s'(t) \Rightarrow s(t) = -4.9t^2 + s_0$, but $s(0) = s_0 = 500 \Rightarrow s(t) = -4.9t^2 + 500$. When $s = 0$,

$-4.9t^2 + 500 = 0 \Rightarrow t_1 = \sqrt{\frac{500}{4.9}} \approx 10.1 \Rightarrow v(t_1) = -9.8\sqrt{\frac{500}{4.9}} \approx -98.995$ m/s. Since the canister has been

designed to withstand an impact velocity of 100 m/s, the canister will *not burst*.

49. (a)

The cross-sectional area of the rectangular beam is

$A = 2x \cdot 2y = 4xy = 4x\sqrt{100 - x^2}$, $0 \le x \le 10$, so

$$\frac{dA}{dx} = 4x\left(\frac{1}{2}\right)(100 - x^2)^{-1/2}(-2x) + (100 - x^2)^{1/2} \cdot 4$$

$$= \frac{-4x^2}{(100 - x^2)^{1/2}} + 4(100 - x^2)^{1/2} = \frac{4[-x^2 + (100 - x^2)]}{(100 - x^2)^{1/2}}.$$

$\frac{dA}{dx} = 0$ when $-x^2 + (100 - x^2) = 0 \Rightarrow x^2 = 50 \Rightarrow x = \sqrt{50} \approx 7.07 \Rightarrow y = \sqrt{100 - \left(\sqrt{50}\right)^2} = \sqrt{50}$.

Since $A(0) = A(10) = 0$, the rectangle of maximum area is a square.

(b)

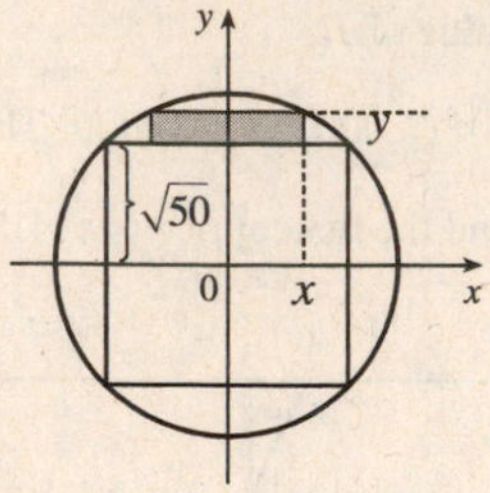

The cross-sectional area of each rectangular plank (shaded in the figure) is

$$A = 2x\left(y - \sqrt{50}\right) = 2x\left[\sqrt{100 - x^2} - \sqrt{50}\right], \quad 0 \le x \le \sqrt{50}, \text{ so}$$

$$\frac{dA}{dx} = 2\left(\sqrt{100 - x^2} - \sqrt{50}\right) + 2x\left(\tfrac{1}{2}\right)(100 - x^2)^{-1/2}(-2x)$$

$$= 2(100 - x^2)^{1/2} - 2\sqrt{50} - \frac{2x^2}{(100 - x^2)^{1/2}}$$

Set $\dfrac{dA}{dx} = 0$: $(100 - x^2) - \sqrt{50}\,(100 - x^2)^{1/2} - x^2 = 0 \;\Rightarrow\; 100 - 2x^2 = \sqrt{50}\,(100 - x^2)^{1/2} \;\Rightarrow$

$10{,}000 - 400x^2 + 4x^4 = 50(100 - x^2) \;\Rightarrow\; 4x^4 - 350x^2 + 5000 = 0 \;\Rightarrow\; 2x^4 - 175x^2 + 2500 = 0 \;\Rightarrow$

$x^2 = \dfrac{175 \pm \sqrt{10{,}625}}{4} \approx 69.52 \text{ or } 17.98 \;\Rightarrow\; x \approx 8.34 \text{ or } 4.24.$ But $8.34 > \sqrt{50}$, so $x_1 \approx 4.24 \;\Rightarrow$

$y - \sqrt{50} = \sqrt{100 - x_1^2} - \sqrt{50} \approx 1.99.$ Each plank should have dimensions about $8\tfrac{1}{2}$ inches by 2 inches.

(c) From the figure in part (a), the width is $2x$ and the depth is $2y$, so the strength is

$S = k(2x)(2y)^2 = 8kxy^2 = 8kx(100 - x^2) = 800kx - 8kx^3, \; 0 \le x \le 10. \; dS/dx = 800k - 24kx^2 = 0$ when

$24kx^2 = 800k \;\Rightarrow\; x^2 = \frac{100}{3} \;\Rightarrow\; x = \frac{10}{\sqrt{3}} \;\Rightarrow\; y = \sqrt{\frac{200}{3}} = \frac{10\sqrt{2}}{\sqrt{3}} = \sqrt{2}\,x.$ Since $S(0) = S(10) = 0$, the

maximum strength occurs when $x = \frac{10}{\sqrt{3}}$. The dimensions should be $\frac{20}{\sqrt{3}} \approx 11.55$ inches by $\frac{20\sqrt{2}}{\sqrt{3}} \approx 16.33$ inches.

4 □ INTEGRALS

4.1 Areas and Distances

1. (a) Since f is *increasing*, we can obtain a *lower* estimate by using *left* endpoints. We are instructed to use five rectangles, so $n = 5$.

$$L_5 = \sum_{i=1}^{5} f(x_{i-1}) \Delta x \quad [\Delta x = \tfrac{b-a}{n} = \tfrac{10-0}{5} = 2]$$

$$= f(x_0) \cdot 2 + f(x_1) \cdot 2 + f(x_2) \cdot 2 + f(x_3) \cdot 2 + f(x_4) \cdot 2$$

$$= 2\left[f(0) + f(2) + f(4) + f(6) + f(8) \right]$$

$$\approx 2(1 + 3 + 4.3 + 5.4 + 6.3) = 2(20) = 40$$

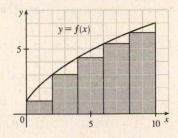

Since f is *increasing*, we can obtain an *upper* estimate by using *right* endpoints.

$$R_5 = \sum_{i=1}^{5} f(x_i) \Delta x$$

$$= 2\left[f(x_1) + f(x_2) + f(x_3) + f(x_4) + f(x_5) \right]$$

$$= 2\left[f(2) + f(4) + f(6) + f(8) + f(10) \right]$$

$$\approx 2(3 + 4.3 + 5.4 + 6.3 + 7) = 2(26) = 52$$

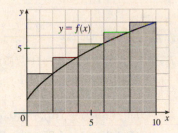

Comparing R_5 to L_5, we see that we have added the area of the rightmost upper rectangle, $f(10) \cdot 2$, to the sum and subtracted the area of the leftmost lower rectangle, $f(0) \cdot 2$, from the sum.

(b) $$L_{10} = \sum_{i=1}^{10} f(x_{i-1}) \Delta x \quad [\Delta x = \tfrac{10-0}{10} = 1]$$

$$= 1\left[f(x_0) + f(x_1) + \cdots + f(x_9) \right]$$

$$= f(0) + f(1) + \cdots + f(9)$$

$$\approx 1 + 2.1 + 3 + 3.7 + 4.3 + 4.9 + 5.4 + 5.8 + 6.3 + 6.7$$

$$= 43.2$$

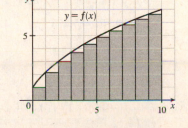

$$R_{10} = \sum_{i=1}^{10} f(x_i) \Delta x = f(1) + f(2) + \cdots + f(10)$$

$$= L_{10} + 1 \cdot f(10) - 1 \cdot f(0) \quad \begin{bmatrix} \text{add rightmost upper rectangle,} \\ \text{subtract leftmost lower rectangle} \end{bmatrix}$$

$$= 43.2 + 7 - 1 = 49.2$$

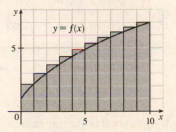

3. (a) $R_4 = \sum_{i=1}^{4} f(x_i)\,\Delta x \quad \left[\Delta x = \dfrac{\pi/2 - 0}{4} = \dfrac{\pi}{8}\right] \quad = \left[\sum_{i=1}^{4} f(x_i)\right]\Delta x$

$\qquad = [f(x_1) + f(x_2) + f(x_3) + f(x_4)]\,\Delta x$

$\qquad = \left[\cos\frac{\pi}{8} + \cos\frac{2\pi}{8} + \cos\frac{3\pi}{8} + \cos\frac{4\pi}{8}\right]\frac{\pi}{8}$

$\qquad \approx (0.9239 + 0.7071 + 0.3827 + 0)\frac{\pi}{8} \approx 0.7908$

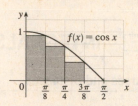

Since f is *decreasing* on $[0, \pi/2]$, an *underestimate* is obtained by using the *right* endpoint approximation, R_4.

(b) $L_4 = \sum_{i=1}^{4} f(x_{i-1})\,\Delta x = \left[\sum_{i=1}^{4} f(x_{i-1})\right]\Delta x$

$\qquad = [f(x_0) + f(x_1) + f(x_2) + f(x_3)]\,\Delta x$

$\qquad = \left[\cos 0 + \cos\frac{\pi}{8} + \cos\frac{2\pi}{8} + \cos\frac{3\pi}{8}\right]\frac{\pi}{8}$

$\qquad \approx (1 + 0.9239 + 0.7071 + 0.3827)\frac{\pi}{8} \approx 1.1835$

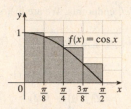

L_4 is an overestimate. Alternatively, we could just add the area of the leftmost upper rectangle and subtract the area of the rightmost lower rectangle; that is, $L_4 = R_4 + f(0) \cdot \frac{\pi}{8} - f\left(\frac{\pi}{2}\right) \cdot \frac{\pi}{8}$.

5. (a) $f(x) = 1 + x^2$ and $\Delta x = \dfrac{2 - (-1)}{3} = 1 \quad \Rightarrow$

$\qquad R_3 = 1 \cdot f(0) + 1 \cdot f(1) + 1 \cdot f(2) = 1 \cdot 1 + 1 \cdot 2 + 1 \cdot 5 = 8.$

$\qquad \Delta x = \dfrac{2 - (-1)}{6} = 0.5 \quad \Rightarrow$

$\qquad R_6 = 0.5[f(-0.5) + f(0) + f(0.5) + f(1) + f(1.5) + f(2)]$

$\qquad\quad = 0.5(1.25 + 1 + 1.25 + 2 + 3.25 + 5)$

$\qquad\quad = 0.5(13.75) = 6.875$

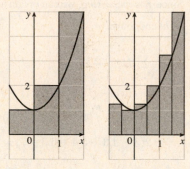

(b) $L_3 = 1 \cdot f(-1) + 1 \cdot f(0) + 1 \cdot f(1) = 1 \cdot 2 + 1 \cdot 1 + 1 \cdot 2 = 5$

$\qquad L_6 = 0.5[f(-1) + f(-0.5) + f(0) + f(0.5) + f(1) + f(1.5)]$

$\qquad\quad = 0.5(2 + 1.25 + 1 + 1.25 + 2 + 3.25)$

$\qquad\quad = 0.5(10.75) = 5.375$

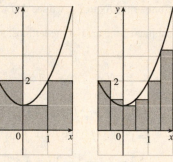

(c) $M_3 = 1 \cdot f(-0.5) + 1 \cdot f(0.5) + 1 \cdot f(1.5)$

$\qquad\quad = 1 \cdot 1.25 + 1 \cdot 1.25 + 1 \cdot 3.25 = 5.75$

$\qquad M_6 = 0.5[f(-0.75) + f(-0.25) + f(0.25)$

$\qquad\qquad\quad + f(0.75) + f(1.25) + f(1.75)]$

$\qquad\quad = 0.5(1.5625 + 1.0625 + 1.0625 + 1.5625 + 2.5625 + 4.0625)$

$\qquad\quad = 0.5(11.875) = 5.9375$

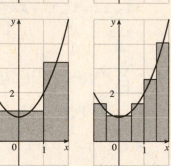

(d) M_6 appears to be the best estimate.

7. $f(x) = 2 + \sin x$, $0 \le x \le \pi$, $\Delta x = \pi/n$.

$n = 2$: The maximum values of f on both subintervals occur at $x = \frac{\pi}{2}$, so

$$upper\;sum = f\left(\tfrac{\pi}{2}\right) \cdot \tfrac{\pi}{2} + f\left(\tfrac{\pi}{2}\right) \cdot \tfrac{\pi}{2}$$
$$= 3 \cdot \tfrac{\pi}{2} + 3 \cdot \tfrac{\pi}{2} = 3\pi \approx 9.42.$$

The minimum values of f on the subintervals occur at $x = 0$ and $x = \pi$, so

$$lower\;sum = f(0) \cdot \tfrac{\pi}{2} + f(\pi) \cdot \tfrac{\pi}{2}$$
$$= 2 \cdot \tfrac{\pi}{2} + 2 \cdot \tfrac{\pi}{2} = 2\pi \approx 6.28.$$

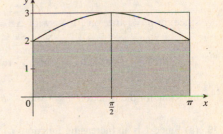

$n = 4$: $upper\;sum = \left[f\left(\tfrac{\pi}{4}\right) + f\left(\tfrac{\pi}{2}\right) + f\left(\tfrac{\pi}{2}\right) + f\left(\tfrac{3\pi}{4}\right)\right]\left(\tfrac{\pi}{4}\right)$

$$= \left[\left(2 + \tfrac{1}{2}\sqrt{2}\right) + (2+1) + (2+1) + \left(2 + \tfrac{1}{2}\sqrt{2}\right)\right]\left(\tfrac{\pi}{4}\right)$$
$$= \left(10 + \sqrt{2}\right)\left(\tfrac{\pi}{4}\right) \approx 8.96$$

$lower\;sum = \left[f(0) + f\left(\tfrac{\pi}{4}\right) + f\left(\tfrac{3\pi}{4}\right) + f(\pi)\right]\left(\tfrac{\pi}{4}\right)$

$$= \left[(2+0) + \left(2 + \tfrac{1}{2}\sqrt{2}\right) + \left(2 + \tfrac{1}{2}\sqrt{2}\right) + (2+0)\right]\left(\tfrac{\pi}{4}\right)$$
$$= \left(8 + \sqrt{2}\right)\left(\tfrac{\pi}{4}\right) \approx 7.39$$

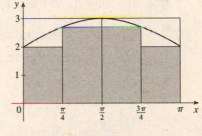

$n = 8$: $upper\;sum = \left[f\left(\tfrac{\pi}{8}\right) + f\left(\tfrac{\pi}{4}\right) + f\left(\tfrac{3\pi}{8}\right) + f\left(\tfrac{\pi}{2}\right) + f\left(\tfrac{\pi}{2}\right)\right.$
$$\left. + f\left(\tfrac{5\pi}{8}\right) + f\left(\tfrac{3\pi}{4}\right) + f\left(\tfrac{7\pi}{8}\right)\right]\left(\tfrac{\pi}{8}\right)$$
$$\approx 8.65$$

$lower\;sum = \left[f(0) + f\left(\tfrac{\pi}{8}\right) + f\left(\tfrac{\pi}{4}\right) + f\left(\tfrac{3\pi}{8}\right) + f\left(\tfrac{5\pi}{8}\right)\right.$
$$\left. + f\left(\tfrac{3\pi}{4}\right) + f\left(\tfrac{7\pi}{8}\right) + f(\pi)\right]\left(\tfrac{\pi}{8}\right)$$
$$\approx 7.86$$

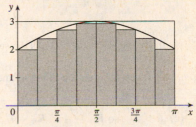

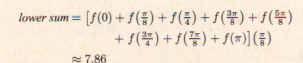

9. Since v is an increasing function, L_6 will give us a lower estimate and R_6 will give us an upper estimate.

$$L_6 = (0 \text{ ft/s})(0.5 \text{ s}) + (6.2)(0.5) + (10.8)(0.5) + (14.9)(0.5) + (18.1)(0.5) + (19.4)(0.5) = 0.5(69.4) = 34.7 \text{ ft}$$

$$R_6 = 0.5(6.2 + 10.8 + 14.9 + 18.1 + 19.4 + 20.2) = 0.5(89.6) = 44.8 \text{ ft}$$

11. Lower estimate for oil leakage: $R_5 = (7.6 + 6.8 + 6.2 + 5.7 + 5.3)(2) = (31.6)(2) = 63.2 \text{ L}$.

Upper estimate for oil leakage: $L_5 = (8.7 + 7.6 + 6.8 + 6.2 + 5.7)(2) = (35)(2) = 70 \text{ L}$.

13. For a decreasing function, using left endpoints gives us an overestimate and using right endpoints results in an underestimate.

We will use M_6 to get an estimate. $\Delta t = 1$, so

$$M_6 = 1[v(0.5) + v(1.5) + v(2.5) + v(3.5) + v(4.5) + v(5.5)] \approx 55 + 40 + 28 + 18 + 10 + 4 = 155 \text{ ft}$$

For a very rough check on the above calculation, we can draw a line from $(0, 70)$ to $(6, 0)$ and calculate the area of the

triangle: $\frac{1}{2}(70)(6) = 210$. This is clearly an overestimate, so our midpoint estimate of 155 is reasonable.

15. $f(x) = \dfrac{2x}{x^2 + 1}$, $1 \le x \le 3$. $\Delta x = (3-1)/n = 2/n$ and $x_i = 1 + i\Delta x = 1 + 2i/n$.

$$A = \lim_{n \to \infty} R_n = \lim_{n \to \infty} \sum_{i=1}^{n} f(x_i)\Delta x = \lim_{n \to \infty} \sum_{i=1}^{n} \frac{2(1 + 2i/n)}{(1 + 2i/n)^2 + 1} \cdot \frac{2}{n}.$$

17. $\lim\limits_{n\to\infty}\sum\limits_{i=1}^{n}\dfrac{\pi}{4n}\tan\dfrac{i\pi}{4n}$ can be interpreted as the area of the region lying under the graph of $y=\tan x$ on the interval $\left[0,\frac{\pi}{4}\right]$,

since for $y=\tan x$ on $\left[0,\frac{\pi}{4}\right]$ with $\Delta x=\dfrac{\pi/4-0}{n}=\dfrac{\pi}{4n}$, $x_i=0+i\,\Delta x=\dfrac{i\pi}{4n}$, and $x_i^*=x_i$, the expression for the area is

$A=\lim\limits_{n\to\infty}\sum\limits_{i=1}^{n}f\left(x_i^*\right)\Delta x=\lim\limits_{n\to\infty}\sum\limits_{i=1}^{n}\tan\left(\dfrac{i\pi}{4n}\right)\dfrac{\pi}{4n}$. Note that this answer is not unique, since the expression for the area is

the same for the function $y=\tan(x-k\pi)$ on the interval $\left[k\pi,k\pi+\frac{\pi}{4}\right]$, where k is any integer.

19. (a) Since f is an increasing function, L_n is an underestimate of A [lower sum] and R_n is an overestimate of A [upper sum].

Thus, A, L_n, and R_n are related by the inequality $L_n<A<R_n$.

(b)
$$R_n=f(x_1)\Delta x+f(x_2)\Delta x+\cdots+f(x_n)\Delta x$$
$$L_n=f(x_0)\Delta x+f(x_1)\Delta x+\cdots+f(x_{n-1})\Delta x$$
$$R_n-L_n=f(x_n)\Delta x-f(x_0)\Delta x$$
$$=\Delta x[f(x_n)-f(x_0)]$$
$$=\dfrac{b-a}{n}[f(b)-f(a)]$$

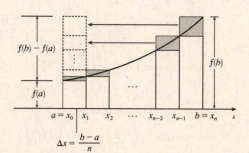

In the diagram, R_n-L_n is the sum of the areas of the shaded rectangles. By sliding the shaded rectangles to the left so

that they stack on top of the leftmost shaded rectangle, we form a rectangle of height $f(b)-f(a)$ and width $\dfrac{b-a}{n}$.

(c) $A>L_n$, so $R_n-A<R_n-L_n$; that is, $R_n-A<\dfrac{b-a}{n}[f(b)-f(a)]$.

21. (a) $y=f(x)=x^5$. $\Delta x=\dfrac{2-0}{n}=\dfrac{2}{n}$ and $x_i=0+i\,\Delta x=\dfrac{2i}{n}$.

$A=\lim\limits_{n\to\infty}R_n=\lim\limits_{n\to\infty}\sum\limits_{i=1}^{n}f(x_i)\,\Delta x=\lim\limits_{n\to\infty}\sum\limits_{i=1}^{n}\left(\dfrac{2i}{n}\right)^5\cdot\dfrac{2}{n}=\lim\limits_{n\to\infty}\sum\limits_{i=1}^{n}\dfrac{32i^5}{n^5}\cdot\dfrac{2}{n}=\lim\limits_{n\to\infty}\dfrac{64}{n^6}\sum\limits_{i=1}^{n}i^5$.

(b) $\sum\limits_{i=1}^{n}i^5\overset{\text{CAS}}{=}\dfrac{n^2(n+1)^2(2n^2+2n-1)}{12}$

(c) $\lim\limits_{n\to\infty}\dfrac{64}{n^6}\cdot\dfrac{n^2(n+1)^2(2n^2+2n-1)}{12}=\dfrac{64}{12}\lim\limits_{n\to\infty}\dfrac{(n^2+2n+1)(2n^2+2n-1)}{n^2\cdot n^2}$

$=\dfrac{16}{3}\lim\limits_{n\to\infty}\left(1+\dfrac{2}{n}+\dfrac{1}{n^2}\right)\left(2+\dfrac{2}{n}-\dfrac{1}{n^2}\right)=\dfrac{16}{3}\cdot1\cdot2=\dfrac{32}{3}$

23. $y=f(x)=\cos x$. $\Delta x=\dfrac{b-0}{n}=\dfrac{b}{n}$ and $x_i=0+i\,\Delta x=\dfrac{bi}{n}$.

$A=\lim\limits_{n\to\infty}R_n=\lim\limits_{n\to\infty}\sum\limits_{i=1}^{n}f(x_i)\,\Delta x=\lim\limits_{n\to\infty}\sum\limits_{i=1}^{n}\cos\left(\dfrac{bi}{n}\right)\cdot\dfrac{b}{n}\overset{\text{CAS}}{=}\lim\limits_{n\to\infty}\left[\dfrac{b\sin\left(b\left(\frac{1}{2n}+1\right)\right)}{2n\sin\left(\dfrac{b}{2n}\right)}-\dfrac{b}{2n}\right]\overset{\text{CAS}}{=}\sin b$

If $b=\frac{\pi}{2}$, then $A=\sin\frac{\pi}{2}=1$. [So the estimate $a\approx1.006$ that we obtained in Example 3(b) is very close to the exact area.]

4.2 The Definite Integral

1. $f(x) = 3 - \frac{1}{2}x$, $2 \le x \le 14$. $\Delta x = \dfrac{b-a}{n} = \dfrac{14-2}{6} = 2$.

Since we are using left endpoints, $x_i^* = x_{i-1}$.

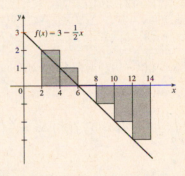

$$L_6 = \sum_{i=1}^{6} f(x_{i-1})\,\Delta x$$

$$= (\Delta x)\,[f(x_0) + f(x_1) + f(x_2) + f(x_3) + f(x_4) + f(x_5)]$$

$$= 2[f(2) + f(4) + f(6) + f(8) + f(10) + f(12)]$$

$$= 2[2 + 1 + 0 + (-1) + (-2) + (-3)] = 2(-3) = -6$$

The Riemann sum represents the sum of the areas of the two rectangles above the x-axis minus the sum of the areas of the three rectangles below the x-axis; that is, the *net area* of the rectangles with respect to the x-axis.

3. $M_5 = \sum_{i=1}^{5} f(\overline{x}_i)\,\Delta x \quad [x_i^* = \overline{x}_i = \frac{1}{2}(x_{i-1} + x_i)$ is a midpoint and $\Delta x = 1]$

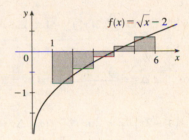

$$= 1\,[f(1.5) + f(2.5) + f(3.5)$$
$$+ f(4.5) + f(5.5)] \quad [f(x) = \sqrt{x} - 2]$$

$$\approx -0.856759$$

The Riemann sum represents the sum of the areas of the two rectangles above the x-axis minus the sum of the areas of the three rectangles below the x-axis.

5. $f(x) = x^3$. $\displaystyle\sum_{i=1}^{n} f(x_i^*)\,\Delta x_i = \frac{1}{2}\sum_{i=1}^{n} f(x_i^*) = \frac{1}{2}[(-1)^3 + (-0.4)^3 + (0.2)^3 + 1^3] = -0.028$

7. (a) $\displaystyle\int_0^{10} f(x)\,dx \approx R_5 = [f(2) + f(4) + f(6) + f(8) + f(10)]\,\Delta x$

$$= [-1 + 0 + (-2) + 2 + 4](2) = 3(2) = 6$$

(b) $\displaystyle\int_0^{10} f(x)\,dx \approx L_5 = [f(0) + f(2) + f(4) + f(6) + f(8)]\,\Delta x$

$$= [3 + (-1) + 0 + (-2) + 2](2) = 2(2) = 4$$

(c) $\displaystyle\int_0^{10} f(x)\,dx \approx M_5 = [f(1) + f(3) + f(5) + f(7) + f(9)]\,\Delta x$

$$= [0 + (-1) + (-1) + 0 + 3](2) = 1(2) = 2$$

9. Since f is increasing, $L_5 \le \int_{10}^{30} f(x)\,dx \le R_5$.

Lower estimate $= L_5 = \displaystyle\sum_{i=1}^{5} f(x_{i-1})\Delta x = 4[f(10) + f(14) + f(18) + f(22) + f(26)]$

$$= 4[-12 + (-6) + (-2) + 1 + 3] = 4(-16) = -64$$

Upper estimate $= R_5 = \displaystyle\sum_{i=1}^{5} f(x)\Delta x = 4[f(14) + f(18) + f(22) + f(26) + f(30)]$

$$= 4[-6 + (-2) + 1 + 3 + 8] = 4(4) = 16$$

11. $\Delta x = (8 - 0)/4 = 2$, so the endpoints are 0, 2, 4, 6, and 8, and the midpoints are 1, 3, 5, and 7. The Midpoint Rule gives

$$\int_0^8 \sin \sqrt{x} \, dx \approx \sum_{i=1}^{4} f(\bar{x}_i) \, \Delta x = 2\left(\sin \sqrt{1} + \sin \sqrt{3} + \sin \sqrt{5} + \sin \sqrt{7}\,\right) \approx 2(3.0910) = 6.1820.$$

13. $\Delta x = (2 - 0)/5 = \frac{2}{5}$, so the endpoints are 0, $\frac{2}{5}$, $\frac{4}{5}$, $\frac{6}{5}$, $\frac{8}{5}$, and 2, and the midpoints are $\frac{1}{5}$, $\frac{3}{5}$, $\frac{5}{5}$, $\frac{7}{5}$ and $\frac{9}{5}$. The Midpoint Rule gives

$$\int_0^2 \frac{x}{x+1} \, dx \approx \sum_{i=1}^{5} f(\bar{x}_i) \, \Delta x = \frac{2}{5}\left(\frac{\frac{1}{5}}{\frac{1}{5}+1} + \frac{\frac{3}{5}}{\frac{3}{5}+1} + \frac{\frac{5}{5}}{\frac{5}{5}+1} + \frac{\frac{7}{5}}{\frac{7}{5}+1} + \frac{\frac{9}{5}}{\frac{9}{5}+1}\right) = \frac{2}{5}\left(\frac{127}{56}\right) = \frac{127}{140} \approx 0.9071.$$

15. On $[2, 6]$, $\displaystyle\lim_{n \to \infty} \sum_{i=1}^{n} \frac{1 - x_i^2}{4 + x_i^2} \, \Delta x = \int_2^6 \frac{1 - x^2}{4 + x^2} \, dx.$

17. On $[2, 7]$, $\displaystyle\lim_{n \to \infty} \sum_{i=1}^{n} [5(x_i^*)^3 - 4x_i^*] \, \Delta x = \int_2^7 (5x^3 - 4x) \, dx.$

19. Note that $\Delta x = \dfrac{5 - 2}{n} = \dfrac{3}{n}$ and $x_i = 2 + i\,\Delta x = 2 + \dfrac{3i}{n}$.

$$\int_2^5 (4 - 2x) \, dx = \lim_{n \to \infty} \sum_{i=1}^{n} f(x_i) \, \Delta x = \lim_{n \to \infty} \sum_{i=1}^{n} f\left(2 + \frac{3i}{n}\right)\frac{3}{n} = \lim_{n \to \infty} \frac{3}{n} \sum_{i=1}^{n} \left[4 - 2\left(2 + \frac{3i}{n}\right)\right]$$

$$= \lim_{n \to \infty} \frac{3}{n} \sum_{i=1}^{n} \left[-\frac{6i}{n}\right] = \lim_{n \to \infty} \frac{3}{n}\left(-\frac{6}{n}\right)\sum_{i=1}^{n} i = \lim_{n \to \infty} \left(-\frac{18}{n^2}\right)\left[\frac{n(n+1)}{2}\right]$$

$$= \lim_{n \to \infty} \left(-\frac{18}{2}\right)\left(\frac{n+1}{n}\right) = -9\lim_{n \to \infty}\left(1 + \frac{1}{n}\right) = -9(1) = -9$$

21. Note that $\Delta x = \dfrac{0 - (-2)}{n} = \dfrac{2}{n}$ and $x_i = -2 + i\,\Delta x = -2 + \dfrac{2i}{n}$.

$$\int_{-2}^0 (x^2 + x) \, dx = \lim_{n \to \infty} \sum_{i=1}^{n} f(x_i) \, \Delta x = \lim_{n \to \infty} \sum_{i=1}^{n} f\left(-2 + \frac{2i}{n}\right)\frac{2}{n} = \lim_{n \to \infty} \frac{2}{n} \sum_{i=1}^{n} \left[\left(-2 + \frac{2i}{n}\right)^2 + \left(-2 + \frac{2i}{n}\right)\right]$$

$$= \lim_{n \to \infty} \frac{2}{n} \sum_{i=1}^{n} \left[4 - \frac{8i}{n} + \frac{4i^2}{n^2} - 2 + \frac{2i}{n}\right] = \lim_{n \to \infty} \frac{2}{n} \sum_{i=1}^{n} \left(\frac{4i^2}{n^2} - \frac{6i}{n} + 2\right)$$

$$= \lim_{n \to \infty} \frac{2}{n}\left[\frac{4}{n^2}\sum_{i=1}^{n} i^2 - \frac{6}{n}\sum_{i=1}^{n} i + \sum_{i=1}^{n} 2\right] = \lim_{n \to \infty} \left[\frac{8}{n^3}\frac{n(n+1)(2n+1)}{6} - \frac{12}{n^2}\frac{n(n+1)}{2} + \frac{2}{n}\cdot n(2)\right]$$

$$= \lim_{n \to \infty} \left[\frac{4}{3}\frac{(n+1)(2n+1)}{n^2} - 6\frac{n+1}{n} + 4\right] = \lim_{n \to \infty} \left[\frac{4}{3}\frac{n+1}{n}\frac{2n+1}{n} - 6\left(1 + \frac{1}{n}\right) + 4\right]$$

$$= \lim_{n \to \infty} \left[\frac{4}{3}\left(1 + \frac{1}{n}\right)\left(2 + \frac{1}{n}\right) - 6\left(1 + \frac{1}{n}\right) + 4\right] = \frac{4}{3}(1)(2) - 6(1) + 4 = \frac{2}{3}$$

23. Note that $\Delta x = \dfrac{1-0}{n} = \dfrac{1}{n}$ and $x_i = 0 + i\,\Delta x = \dfrac{i}{n}$.

$$\int_0^1 (x^3 - 3x^2)\,dx = \lim_{n\to\infty} \sum_{i=1}^n f(x_i)\,\Delta x = \lim_{n\to\infty} \sum_{i=1}^n f\!\left(\frac{i}{n}\right)\Delta x = \lim_{n\to\infty} \sum_{i=1}^n \left[\left(\frac{i}{n}\right)^3 - 3\left(\frac{i}{n}\right)^2\right]\frac{1}{n}$$

$$= \lim_{n\to\infty} \frac{1}{n}\sum_{i=1}^n \left[\frac{i^3}{n^3} - \frac{3i^2}{n^2}\right] = \lim_{n\to\infty} \frac{1}{n}\left[\frac{1}{n^3}\sum_{i=1}^n i^3 - \frac{3}{n^2}\sum_{i=1}^n i^2\right]$$

$$= \lim_{n\to\infty} \left\{\frac{1}{n^4}\left[\frac{n(n+1)}{2}\right]^2 - \frac{3}{n^3}\frac{n(n+1)(2n+1)}{6}\right\} = \lim_{n\to\infty}\left[\frac{1}{4}\frac{n+1}{n}\frac{n+1}{n} - \frac{1}{2}\frac{n+1}{n}\frac{2n+1}{n}\right]$$

$$= \lim_{n\to\infty}\left[\frac{1}{4}\left(1+\frac{1}{n}\right)\left(1+\frac{1}{n}\right) - \frac{1}{2}\left(1+\frac{1}{n}\right)\left(2+\frac{1}{n}\right)\right] = \frac{1}{4}(1)(1) - \frac{1}{2}(1)(2) = -\frac{3}{4}$$

25. $f(x) = \dfrac{x}{1+x^5}$, $a = 2$, $b = 6$, and $\Delta x = \dfrac{6-2}{n} = \dfrac{4}{n}$. Using Theorem 4, we get $x_i^* = x_i = 2 + i\,\Delta x = 2 + \dfrac{4i}{n}$,

so $\displaystyle\int_2^6 \frac{x}{1+x^5}\,dx = \lim_{n\to\infty} R_n = \lim_{n\to\infty}\sum_{i=1}^n \frac{2+\dfrac{4i}{n}}{1+\left(2+\dfrac{4i}{n}\right)^5}\cdot\frac{4}{n}$.

27. $\Delta x = (\pi - 0)/n = \pi/n$ and $x_i^* = x_i = \pi i/n$.

$$\int_0^\pi \sin 5x\,dx = \lim_{n\to\infty}\sum_{i=1}^n (\sin 5x_i)\left(\frac{\pi}{n}\right) = \lim_{n\to\infty}\sum_{i=1}^n \left(\sin\frac{5\pi i}{n}\right)\frac{\pi}{n} \overset{\text{CAS}}{=} \pi \lim_{n\to\infty}\frac{1}{n}\cot\left(\frac{5\pi}{2n}\right) \overset{\text{CAS}}{=} \pi\left(\frac{2}{5\pi}\right) = \frac{2}{5}$$

29. (a) Think of $\int_0^2 f(x)\,dx$ as the area of a trapezoid with bases 1 and 3 and height 2. The area of a trapezoid is $A = \frac{1}{2}(b+B)h$,

so $\int_0^2 f(x)\,dx = \frac{1}{2}(1+3)2 = 4$.

(b) $\int_0^5 f(x)\,dx = \underset{\text{trapezoid}}{\int_0^2 f(x)\,dx} + \underset{\text{rectangle}}{\int_2^3 f(x)\,dx} + \underset{\text{triangle}}{\int_3^5 f(x)\,dx}$

$$= \tfrac{1}{2}(1+3)2 + \quad 3\cdot 1 \quad + \quad \tfrac{1}{2}\cdot 2\cdot 3 \quad = 4+3+3 = 10$$

(c) $\int_5^7 f(x)\,dx$ is the negative of the area of the triangle with base 2 and height 3. $\int_5^7 f(x)\,dx = -\frac{1}{2}\cdot 2\cdot 3 = -3$.

(d) $\int_7^9 f(x)\,dx$ is the negative of the area of a trapezoid with bases 3 and 2 and height 2, so it equals

$-\frac{1}{2}(B+b)h = -\frac{1}{2}(3+2)2 = -5$. Thus,

$$\int_0^9 f(x)\,dx = \int_0^5 f(x)\,dx + \int_5^7 f(x)\,dx + \int_7^9 f(x)\,dx = 10 + (-3) + (-5) = 2.$$

31. $\int_{-1}^2 (1-x)\,dx$ can be interpreted as the difference of the areas of the two

shaded triangles; that is, $\frac{1}{2}(2)(2) - \frac{1}{2}(1)(1) = 2 - \frac{1}{2} = \frac{3}{2}$.

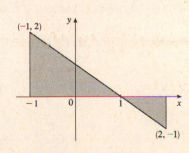

33. $\int_{-3}^{0}\left(1+\sqrt{9-x^2}\,\right)dx$ can be interpreted as the area under the graph of

$f(x)=1+\sqrt{9-x^2}$ between $x=-3$ and $x=0$. This is equal to one-quarter

the area of the circle with radius 3, plus the area of the rectangle, so

$\int_{-3}^{0}\left(1+\sqrt{9-x^2}\,\right)dx=\frac{1}{4}\pi\cdot 3^2+1\cdot 3=3+\frac{9}{4}\pi.$

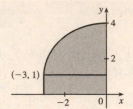

35. $\int_{-1}^{2}|x|\,dx$ can be interpreted as the sum of the areas of the two shaded

triangles; that is, $\frac{1}{2}(1)(1)+\frac{1}{2}(2)(2)=\frac{1}{2}+\frac{4}{2}=\frac{5}{2}.$

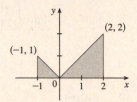

37. $\int_{\pi}^{\pi}\sin^2 x\cos^4 x\,dx=0$ since the limits of integration are equal.

39. $\int_{-2}^{2}f(x)\,dx+\int_{2}^{5}f(x)\,dx-\int_{-2}^{-1}f(x)\,dx=\int_{-2}^{5}f(x)\,dx+\int_{-1}^{-2}f(x)\,dx$ [by Property 5 and reversing limits]

$$=\int_{-1}^{5}f(x)\,dx \qquad \text{[Property 5]}$$

41. $\int_{0}^{9}[2f(x)+3g(x)]\,dx=2\int_{0}^{9}f(x)\,dx+3\int_{0}^{9}g(x)\,dx=2(37)+3(16)=122$

43. $\int_{0}^{1}(5-6x^2)\,dx=\int_{0}^{1}5\,dx-6\int_{0}^{1}x^2\,dx=5(1-0)-6\left(\frac{1}{3}\right)=5-2=3$

45. $I=\int_{-4}^{2}[f(x)+2x+5]\,dx=\int_{-4}^{2}f(x)\,dx+2\int_{-4}^{2}x\,dx+\int_{-4}^{2}5\,dx=I_1+2I_2+I_3$

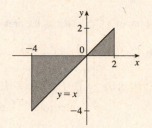

$I_1=-3$ [area below x-axis] $+3-3=-3$

$I_2=-\frac{1}{2}(4)(4)$ [area of triangle, see figure] $+\frac{1}{2}(2)(2)$

$\quad=-8+2=-6$

$I_3=5[2-(-4)]=5(6)=30$

Thus, $I=-3+2(-6)+30=15.$

47. If $-1\le x\le 1$, then $0\le x^2\le 1$ and $1\le 1+x^2\le 2$, so $1\le\sqrt{1+x^2}\le\sqrt{2}$ and

$1[1-(-1)]\le\int_{-1}^{1}\sqrt{1+x^2}\,dx\le\sqrt{2}\,[1-(-1)]$ [Property 8]; that is, $2\le\int_{-1}^{1}\sqrt{1+x^2}\,dx\le 2\sqrt{2}.$

49. If $1\le x\le 2$, then $\frac{1}{2}\le\dfrac{1}{x}\le 1$, so $\frac{1}{2}(2-1)\le\int_{1}^{2}\dfrac{1}{x}\,dx\le 1(2-1)$ or $\frac{1}{2}\le\int_{1}^{2}\dfrac{1}{x}\,dx\le 1.$

51. If $\frac{\pi}{4}\le x\le\frac{\pi}{3}$, then $1\le\tan x\le\sqrt{3}$, so $1\left(\frac{\pi}{3}-\frac{\pi}{4}\right)\le\int_{\pi/4}^{\pi/3}\tan x\,dx\le\sqrt{3}\left(\frac{\pi}{3}-\frac{\pi}{4}\right)$ or $\frac{\pi}{12}\le\int_{\pi/4}^{\pi/3}\tan x\,dx\le\frac{\pi}{12}\sqrt{3}.$

53. $\displaystyle\lim_{n\to\infty}\sum_{i=1}^{n}\frac{i^4}{n^5}=\lim_{n\to\infty}\sum_{i=1}^{n}\frac{i^4}{n^4}\cdot\frac{1}{n}=\lim_{n\to\infty}\sum_{i=1}^{n}\left(\frac{i}{n}\right)^4\frac{1}{n}.$ At this point, we need to recognize the limit as being of the form

$\displaystyle\lim_{n\to\infty}\sum_{i=1}^{n}f(x_i)\,\Delta x$, where $\Delta x=(1-0)/n=1/n$, $x_i=0+i\,\Delta x=i/n$, and $f(x)=x^4$. Thus, the definite integral

is $\int_{0}^{1}x^4\,dx.$

4.3   Evaluating Definite Integrals

1. $\int_{-2}^{3}(x^2-3)\,dx = \left[\frac{1}{3}x^3 - 3x\right]_{-2}^{3} = (9-9) - \left(-\frac{8}{3}+6\right) = \frac{8}{3} - \frac{18}{3} = -\frac{10}{3}$

3. $\int_{-2}^{0}\left(\frac{1}{2}t^4 + \frac{1}{4}t^3 - t\right)dt = \left[\frac{1}{10}t^5 + \frac{1}{16}t^4 - \frac{1}{2}t^2\right]_{-2}^{0} = 0 - \left[\frac{1}{10}(-32) + \frac{1}{16}(16) - \frac{1}{2}(4)\right] = -\left(-\frac{16}{5} + 1 - 2\right) = \frac{21}{5}$

5. $\int_{0}^{2}(2x-3)(4x^2+1)\,dx = \int_{0}^{2}(8x^3 - 12x^2 + 2x - 3)\,dx = \left[2x^4 - 4x^3 + x^2 - 3x\right]_{0}^{2} = (32 - 32 + 4 - 6) - 0 = -2$

7. $\int_{0}^{\pi}(4\sin\theta - 3\cos\theta)\,d\theta = \left[-4\cos\theta - 3\sin\theta\right]_{0}^{\pi} = (4-0) - (-4-0) = 8$

9. $\displaystyle\int_{1}^{4}\left(\frac{4+6u}{\sqrt{u}}\right)du = \int_{1}^{4}\left(\frac{4}{\sqrt{u}} + \frac{6u}{\sqrt{u}}\right)du = \int_{1}^{4}(4u^{-1/2} + 6u^{1/2})\,du = \left[8u^{1/2} + 4u^{3/2}\right]_{1}^{4}$

$\qquad\qquad = (16 + 32) - (8 + 4) = 36$

11. $\int_{0}^{1}x\left(\sqrt[3]{x} + \sqrt[4]{x}\right)dx = \int_{0}^{1}(x^{4/3} + x^{5/4})\,dx = \left[\frac{3}{7}x^{7/3} + \frac{4}{9}x^{9/4}\right]_{0}^{1} = \left(\frac{3}{7} + \frac{4}{9}\right) - 0 = \frac{55}{63}$

13. $\int_{1}^{4}\sqrt{5/x}\,dx = \sqrt{5}\int_{1}^{4}x^{-1/2}\,dx = \sqrt{5}\left[2\sqrt{x}\right]_{1}^{4} = \sqrt{5}\,(2\cdot 2 - 2\cdot 1) = 2\sqrt{5}$

15. $\int_{0}^{\pi/4}\sec^2 t\,dt = \left[\tan t\right]_{0}^{\pi/4} = \tan\frac{\pi}{4} - \tan 0 = 1 - 0 = 1$

17. $\displaystyle\int_{0}^{\pi/4}\frac{1+\cos^2\theta}{\cos^2\theta}\,d\theta = \int_{0}^{\pi/4}\left(\frac{1}{\cos^2\theta} + \frac{\cos^2\theta}{\cos^2\theta}\right)d\theta = \int_{0}^{\pi/4}(\sec^2\theta + 1)\,d\theta$

$\qquad\qquad = \left[\tan\theta + \theta\right]_{0}^{\pi/4} = \left(\tan\frac{\pi}{4} + \frac{\pi}{4}\right) - (0+0) = 1 + \frac{\pi}{4}$

19. $\int_{1}^{2}(1+2y)^2\,dy = \int_{1}^{2}(1+4y+4y^2)\,dy = \left[y + 2y^2 + \frac{4}{3}y^3\right]_{1}^{2} = \left(2 + 8 + \frac{32}{3}\right) - \left(1 + 2 + \frac{4}{3}\right) = \frac{62}{3} - \frac{13}{3} = \frac{49}{3}$

21. $\displaystyle\int_{1}^{64}\frac{1+\sqrt[3]{x}}{\sqrt{x}}\,dx = \int_{1}^{64}\left(\frac{1}{x^{1/2}} + \frac{x^{1/3}}{x^{1/2}}\right)dx = \int_{1}^{64}\left(x^{-1/2} + x^{(1/3)-(1/2)}\right)dx = \int_{1}^{64}(x^{-1/2} + x^{-1/6})\,dx$

$\qquad\qquad = \left[2x^{1/2} + \frac{6}{5}x^{5/6}\right]_{1}^{64} = \left(16 + \frac{192}{5}\right) - \left(2 + \frac{6}{5}\right) = 14 + \frac{186}{5} = \frac{256}{5}$

23. $\int_{0}^{1}\left(\sqrt[4]{x^5} + \sqrt[5]{x^4}\right)dx = \int_{0}^{1}(x^{5/4} + x^{4/5})\,dx = \left[\frac{x^{9/4}}{9/4} + \frac{x^{9/5}}{9/5}\right]_{0}^{1} = \left[\frac{4}{9}x^{9/4} + \frac{5}{9}x^{9/5}\right]_{0}^{1} = \frac{4}{9} + \frac{5}{9} - 0 = 1$

25. $\int_{1}^{4}\sqrt{t}\,(1+t)\,dt = \int_{1}^{4}(t^{1/2} + t^{3/2})\,dt = \left[\frac{2}{3}t^{3/2} + \frac{2}{5}t^{5/2}\right]_{1}^{4} = \left(\frac{16}{3} + \frac{64}{5}\right) - \left(\frac{2}{3} + \frac{2}{5}\right) = \frac{14}{3} + \frac{62}{5} = \frac{256}{15}$

27. $|x-3| = \begin{cases} x-3 & \text{if } x-3\geq 0 \\ -(x-3) & \text{if } x-3 < 0 \end{cases} = \begin{cases} x-3 & \text{if } x\geq 3 \\ 3-x & \text{if } x<3 \end{cases}$

Thus, $\qquad\qquad \int_{2}^{5}|x-3|\,dx = \int_{2}^{3}(3-x)\,dx + \int_{3}^{5}(x-3)\,dx = \left[3x - \frac{1}{2}x^2\right]_{2}^{3} + \left[\frac{1}{2}x^2 - 3x\right]_{3}^{5}$

$\qquad\qquad\qquad = \left(9 - \frac{9}{2}\right) - (6-2) + \left(\frac{25}{2} - 15\right) - \left(\frac{9}{2} - 9\right) = \frac{5}{2}$

29. $\int_{-1}^{2}(x - 2|x|)\,dx = \int_{-1}^{0}[x - 2(-x)]\,dx + \int_{0}^{2}[x - 2(x)]\,dx = \int_{-1}^{0}3x\,dx + \int_{0}^{2}(-x)\,dx = 3\left[\frac{1}{2}x^2\right]_{-1}^{0} - \left[\frac{1}{2}x^2\right]_{0}^{2}$

$\qquad\qquad = 3\left(0 - \frac{1}{2}\right) - (2-0) = -\frac{7}{2} = -3.5$

31. $f(x) = 1/x^2$ is not continuous on the interval $[-1, 3]$, so the Evaluation Theorem does not apply. In fact, f has an infinite

discontinuity at $x = 0$, so $\int_{-1}^{3} (1/x^2)\, dx$ does not exist.

33. $y = 1 - x^2 = 0$ when $x = \pm 1$, so the curve lies above the x-axis when $-1 \le x \le 1$. The area is

$$A = \int_{-1}^{1} (1 - x^2)\, dx = 2\int_{0}^{1} (1 - x^2)\, dx = 2\left[x - \tfrac{1}{3}x^3\right]_0^1 = \tfrac{4}{3}$$

35. It appears that the area under the graph is about $\tfrac{2}{3}$ of the area of the viewing

rectangle, or about $\tfrac{2}{3}\pi \approx 2.1$. The actual area is

$$\int_{0}^{\pi} \sin x\, dx = [-\cos x]_0^\pi = (-\cos\pi) - (-\cos 0) = -(-1) + 1 = 2.$$

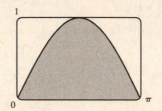

37. $\int_{-1}^{2} x^3\, dx = \left[\tfrac{1}{4}x^4\right]_{-1}^{2} = 4 - \tfrac{1}{4} = \tfrac{15}{4} = 3.75$

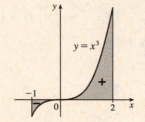

39. $\dfrac{d}{dx}\left[\sqrt{x^2 + 1} + C\right] = \dfrac{d}{dx}\left[(x^2 + 1)^{1/2} + C\right] = \tfrac{1}{2}(x^2 + 1)^{-1/2} \cdot 2x = \dfrac{x}{\sqrt{x^2 + 1}}$

41. $\int x\sqrt{x}\, dx = \int x^{3/2}\, dx = \tfrac{2}{5}x^{5/2} + C.$

The members of the family in the figure correspond to $C = 5, 3, 0, -2,$ and -4.

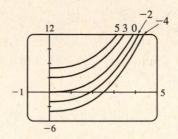

43. $\displaystyle\int (x^2 + x^{-2})\, dx = \dfrac{x^3}{3} + \dfrac{x^{-1}}{-1} + C = \dfrac{1}{3}x^3 - \dfrac{1}{x} + C$

45. $\displaystyle\int (u + 4)(2u + 1)\, du = \int (2u^2 + 9u + 4)\, du = 2\dfrac{u^3}{3} + 9\dfrac{u^2}{2} + 4u + C = \dfrac{2}{3}u^3 + \dfrac{9}{2}u^2 + 4u + C$

47. $\displaystyle\int \dfrac{\sin x}{1 - \sin^2 x}\, dx = \int \dfrac{\sin x}{\cos^2 x}\, dx = \int \dfrac{1}{\cos x} \cdot \dfrac{\sin x}{\cos x}\, dx = \int \sec x \tan x\, dx = \sec x + C$

49. $A = \int_{0}^{2} (2y - y^2)\, dy = \left[y^2 - \tfrac{1}{3}y^3\right]_0^2 = \left(4 - \tfrac{8}{3}\right) - 0 = \tfrac{4}{3}$

51. If $w'(t)$ is the rate of change of weight in pounds per year, then $w(t)$ represents the weight in pounds of the child at age t. We know from the Net Change Theorem that $\int_5^{10} w'(t)\,dt = w(10) - w(5)$, so the integral represents the increase in the child's weight (in pounds) between the ages of 5 and 10.

53. Since $r(t)$ is the rate at which oil leaks, we can write $r(t) = -V'(t)$, where $V(t)$ is the volume of oil at time t. [Note that the minus sign is needed because V is decreasing, so $V'(t)$ is negative, but $r(t)$ is positive.] Thus, by the Net Change Theorem, $\int_0^{120} r(t)\,dt = -\int_0^{120} V'(t)\,dt = -\left[V(120) - V(0)\right] = V(0) - V(120)$, which is the number of gallons of oil that leaked from the tank in the first two hours (120 minutes).

55. By the Net Change Theorem, $\int_{1000}^{5000} R'(x)\,dx = R(5000) - R(1000)$, so it represents the increase in revenue when production is increased from 1000 units to 5000 units.

57. In general, the unit of measurement for $\int_a^b f(x)\,dx$ is the product of the unit for $f(x)$ and the unit for x. Since $f(x)$ is measured in newtons and x is measured in meters, the units for $\int_0^{100} f(x)\,dx$ are newton-meters. (A newton-meter is abbreviated N·m.)

59. (a) Displacement $= \int_0^3 (3t - 5)\,dt = \left[\frac{3}{2}t^2 - 5t\right]_0^3 = \frac{27}{2} - 15 = -\frac{3}{2}$ m

(b) Distance traveled $= \int_0^3 |3t - 5|\,dt = \int_0^{5/3}(5 - 3t)\,dt + \int_{5/3}^3 (3t - 5)\,dt$

$$= \left[5t - \frac{3}{2}t^2\right]_0^{5/3} + \left[\frac{3}{2}t^2 - 5t\right]_{5/3}^3 = \frac{25}{3} - \frac{3}{2} \cdot \frac{25}{9} + \frac{27}{2} - 15 - \left(\frac{3}{2} \cdot \frac{25}{9} - \frac{25}{3}\right) = \frac{41}{6}\ \text{m}$$

61. (a) $v'(t) = a(t) = t + 4 \ \Rightarrow \ v(t) = \frac{1}{2}t^2 + 4t + C \ \Rightarrow \ v(0) = C = 5 \ \Rightarrow \ v(t) = \frac{1}{2}t^2 + 4t + 5$ m/s

(b) Distance traveled $= \int_0^{10} |v(t)|\,dt = \int_0^{10}\left|\frac{1}{2}t^2 + 4t + 5\right|\,dt = \int_0^{10}\left(\frac{1}{2}t^2 + 4t + 5\right)\,dt = \left[\frac{1}{6}t^3 + 2t^2 + 5t\right]_0^{10}$

$$= \frac{500}{3} + 200 + 50 = 416\tfrac{2}{3}\ \text{m}$$

63. Let s be the position of the car. We know from Equation 2 that $s(100) - s(0) = \int_0^{100} v(t)\,dt$. We use the Midpoint Rule for $0 \le t \le 100$ with $n = 5$. Note that the length of each of the five time intervals is 20 seconds $= \frac{20}{3600}$ hour $= \frac{1}{180}$ hour.

So the distance traveled is

$$\int_0^{100} v(t)\,dt \approx \frac{1}{180}\left[v(10) + v(30) + v(50) + v(70) + v(90)\right] = \frac{1}{180}(38 + 58 + 51 + 53 + 47) = \frac{247}{180} \approx 1.4\ \text{miles.}$$

65. By the Net Change Theorem, the amount of water that flows from the tank during the first 10 minutes is

$$\int_0^{10} r(t)\,dt = \int_0^{10}(200 - 4t)\,dt = \left[200t - 2t^2\right]_0^{10} = (2000 - 200) - 0 = 1800\ \text{liters.}$$

4.4   The Fundamental Theorem of Calculus

1. (a) $g(x) = \int_0^x f(t)\,dt$,

$g(0) = \int_0^0 f(t)\,dt = 0$,

$g(1) = \int_0^1 f(t)\,dt = 1 \cdot 2 = 2$ [rectangle],

$g(2) = \int_0^2 f(t)\,dt = \int_0^1 f(t)\,dt + \int_1^2 f(t)\,dt = g(1) + \int_1^2 f(t)\,dt$

$\qquad = 2 + 1\cdot 2 + \frac{1}{2}\cdot 1\cdot 2 = 5$ \qquad [rectangle plus triangle],

$g(3) = \int_0^3 f(t)\,dt = g(2) + \int_2^3 f(t)\,dt = 5 + \frac{1}{2}\cdot 1\cdot 4 = 7,$

$g(6) = g(3) + \int_3^6 f(t)\,dt$ \qquad [the integral is negative since f lies under the x-axis]

$\qquad = 7 + \left[-\left(\frac{1}{2}\cdot 2\cdot 2 + 1\cdot 2\right) \right] = 7 - 4 = 3$

(b) g is increasing on $(0, 3)$ because as x increases from 0 to 3, we keep adding more area.

(c)

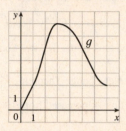

(d) g has a maximum value when we start subtracting area; that is, at $x = 3$.

3.

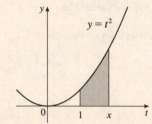

(a) By FTC1 with $f(t) = t^2$ and $a = 1$, $g(x) = \int_1^x t^2\,dt \Rightarrow$

$\qquad g'(x) = f(x) = x^2.$

(b) Using FTC2, $g(x) = \int_1^x t^2\,dt = \left[\frac{1}{3}t^3\right]_1^x = \frac{1}{3}x^3 - \frac{1}{3} \Rightarrow g'(x) = x^2.$

5. $f(t) = \dfrac{1}{t^3 + 1}$ and $g(x) = \displaystyle\int_1^x \dfrac{1}{t^3 + 1}\,dt$, so by FTC1, $g'(x) = f(x) = \dfrac{1}{x^3 + 1}$. Note that the lower limit, 1, could be any

real number greater than -1 and not affect this answer.

7. $f(t) = (t - t^2)^8$ and $g(s) = \displaystyle\int_5^s (t - t^2)^8\,dt$, so by FTC1, $g'(s) = f(s) = (s - s^2)^8.$

9. Let $u = \dfrac{1}{x}$. Then $\dfrac{du}{dx} = -\dfrac{1}{x^2}$. Also, $\dfrac{dh}{dx} = \dfrac{dh}{du}\dfrac{du}{dx}$, so

$h'(x) = \dfrac{d}{dx}\displaystyle\int_2^{1/x} \sin^4 t\,dt = \dfrac{d}{du}\displaystyle\int_2^u \sin^4 t\,dt \cdot \dfrac{du}{dx} = \sin^4 u\,\dfrac{du}{dx} = \dfrac{-\sin^4(1/x)}{x^2}.$

11. Let $u = \tan x$. Then $\dfrac{du}{dx} = \sec^2 x$. Also, $\dfrac{dy}{dx} = \dfrac{dy}{du}\dfrac{du}{dx}$, so

$y' = \dfrac{d}{dx}\displaystyle\int_0^{\tan x} \sqrt{t + \sqrt{t}}\,dt = \dfrac{d}{du}\displaystyle\int_0^u \sqrt{t + \sqrt{t}}\,dt \cdot \dfrac{du}{dx} = \sqrt{u + \sqrt{u}}\,\dfrac{du}{dx} = \sqrt{\tan x + \sqrt{\tan x}}\,\sec^2 x.$

13. $g(x) = \displaystyle\int_{2x}^{3x} \dfrac{u^2 - 1}{u^2 + 1}\,du = \displaystyle\int_{2x}^0 \dfrac{u^2 - 1}{u^2 + 1}\,du + \displaystyle\int_0^{3x} \dfrac{u^2 - 1}{u^2 + 1}\,du = -\displaystyle\int_0^{2x} \dfrac{u^2 - 1}{u^2 + 1}\,du + \displaystyle\int_0^{3x} \dfrac{u^2 - 1}{u^2 + 1}\,du \Rightarrow$

$g'(x) = -\dfrac{(2x)^2 - 1}{(2x)^2 + 1} \cdot \dfrac{d}{dx}(2x) + \dfrac{(3x)^2 - 1}{(3x)^2 + 1} \cdot \dfrac{d}{dx}(3x) = -2\cdot\dfrac{4x^2 - 1}{4x^2 + 1} + 3\cdot\dfrac{9x^2 - 1}{9x^2 + 1}$

15. $g_{\text{ave}} = \dfrac{1}{b - a}\int_a^b g(x)\,dx = \dfrac{1}{8 - 1}\int_1^8 \sqrt[3]{x}\,dx = \dfrac{1}{7}\left[\dfrac{3}{4}x^{4/3}\right]_1^8 = \dfrac{3}{28}(16 - 1) = \dfrac{45}{28}$

17. $g_{\text{ave}} = \dfrac{1}{\frac{\pi}{2} - 0}\int_0^{\pi/2} \cos x\,dx = \dfrac{2}{\pi}\left[\sin x\right]_0^{\pi/2} = \dfrac{2}{\pi}(1 - 0) = \dfrac{2}{\pi}$

19. (a) $f_{\text{ave}} = \frac{1}{5-2}\int_2^5 (x-3)^2\,dx = \frac{1}{3}\int_2^5 (x^2 - 6x + 9)\,dx$

(b)

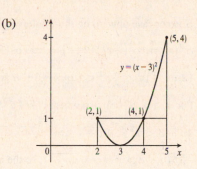

$$= \frac{1}{3}\left[\frac{1}{3}x^3 - 3x^2 + 9x\right]_2^5$$

$$= \frac{1}{3}\left[\frac{1}{3}(125) - 3(25) + 9(5) - \frac{1}{3}(8) + 3(4) - 9(2)\right] = 1$$

(c) $f(c) = f_{\text{ave}} \quad\Leftrightarrow\quad (c-3)^2 = 1 \quad\Leftrightarrow$

$c - 3 = \pm 1 \quad\Leftrightarrow\quad c = 2$ or 4

21. Use geometric interpretations to find the values of the integrals.

$$\int_0^8 f(x)\,dx = \int_0^1 f(x)\,dx + \int_1^2 f(x)\,dx + \int_2^3 f(x)\,dx + \int_3^4 f(x)\,dx + \int_4^6 f(x)\,dx + \int_6^7 f(x)\,dx + \int_7^8 f(x)\,dx$$

$$= -\tfrac{1}{2} + \tfrac{1}{2} + \tfrac{1}{2} + 1 + 4 + \tfrac{3}{2} + 2 = 9$$

Thus, the average value of f on $[0, 8] = f_{\text{ave}} = \frac{1}{8-0}\int_0^8 f(x)\,dx = \frac{1}{8}(9) = \frac{9}{8}$.

23. $y = \int_0^x \dfrac{t^2}{t^2 + t + 2}\,dt \;\Rightarrow\; y' = \dfrac{x^2}{x^2 + x + 2} \;\Rightarrow$

$$y'' = \frac{(x^2 + x + 2)(2x) - x^2(2x + 1)}{(x^2 + x + 2)^2} = \frac{2x^3 + 2x^2 + 4x - 2x^3 - x^2}{(x^2 + x + 2)^2} = \frac{x^2 + 4x}{(x^2 + x + 2)^2} = \frac{x(x + 4)}{(x^2 + x + 2)^2}.$$

The curve y is concave downward when $y'' < 0$; that is, on the interval $(-4, 0)$.

25. (a) By FTC1, $g'(x) = f(x)$. So $g'(x) = f(x) = 0$ at $x = 1, 3, 5, 7$, and 9. g has local maxima at $x = 1$ and 5 (since $f = g'$ changes from positive to negative there) and local minima at $x = 3$ and 7. There is no local maximum or minimum at $x = 9$, since f is not defined for $x > 9$.

(b) We can see from the graph that $\left|\int_0^1 f\,dt\right| < \left|\int_1^3 f\,dt\right| < \left|\int_3^5 f\,dt\right| < \left|\int_5^7 f\,dt\right| < \left|\int_7^9 f\,dt\right|$. So $g(1) = \left|\int_0^1 f\,dt\right|$,

$g(5) = \int_0^5 f\,dt = g(1) - \left|\int_1^3 f\,dt\right| + \left|\int_3^5 f\,dt\right|$, and $g(9) = \int_0^9 f\,dt = g(5) - \left|\int_5^7 f\,dt\right| + \left|\int_7^9 f\,dt\right|$. Thus,

$g(1) < g(5) < g(9)$, and so the absolute maximum of $g(x)$ occurs at $x = 9$.

(c) g is concave downward on those intervals where $g'' < 0$. But $g'(x) = f(x)$, so $g''(x) = f'(x)$, which is negative on (approximately) $\left(\frac{1}{2}, 2\right)$, $(4, 6)$ and $(8, 9)$. So g is concave downward on these intervals.

(d)

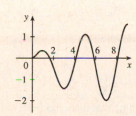

27. By FTC2, $\int_1^4 f'(x)\,dx = f(4) - f(1)$, so $17 = f(4) - 12 \;\Rightarrow\; f(4) = 17 + 12 = 29$.

29. (a) The Fresnel function $S(x) = \int_0^x \sin\left(\frac{\pi}{2}t^2\right)dt$ has local maximum values where $0 = S'(x) = \sin\left(\frac{\pi}{2}t^2\right)$ and

S' changes from positive to negative. For $x > 0$, this happens when $\frac{\pi}{2}x^2 = (2n - 1)\pi$ [odd multiples of π] $\Leftrightarrow$

$x^2 = 2(2n - 1) \quad\Leftrightarrow\quad x = \sqrt{4n - 2}$, n any positive integer. For $x < 0$, S' changes from positive to negative where

$\frac{\pi}{2}x^2 = 2n\pi$ [even multiples of π] $\Leftrightarrow\quad x^2 = 4n \quad\Leftrightarrow\quad x = -2\sqrt{n}$. S' does not change sign at $x = 0$.

(b) S is concave upward on those intervals where $S''(x) > 0$. Differentiating our expression for $S'(x)$, we get

$S''(x) = \cos\left(\frac{\pi}{2}x^2\right)\left(2\frac{\pi}{2}x\right) = \pi x \cos\left(\frac{\pi}{2}x^2\right)$. For $x > 0$, $S''(x) > 0$ where $\cos\left(\frac{\pi}{2}x^2\right) > 0$ $\Leftrightarrow$ $0 < \frac{\pi}{2}x^2 < \frac{\pi}{2}$ or

$\left(2n - \frac{1}{2}\right)\pi < \frac{\pi}{2}x^2 < \left(2n + \frac{1}{2}\right)\pi$, n any integer $\Leftrightarrow$ $0 < x < 1$ or $\sqrt{4n-1} < x < \sqrt{4n+1}$, n any positive integer.

For $x < 0$, $S''(x) > 0$ where $\cos\left(\frac{\pi}{2}x^2\right) < 0$ $\Leftrightarrow$ $\left(2n - \frac{3}{2}\right)\pi < \frac{\pi}{2}x^2 < \left(2n - \frac{1}{2}\right)\pi$, n any integer $\Leftrightarrow$

$4n - 3 < x^2 < 4n - 1$ $\Leftrightarrow$ $\sqrt{4n-3} < |x| < \sqrt{4n-1}$ $\Rightarrow$ $\sqrt{4n-3} < -x < \sqrt{4n-1}$ $\Rightarrow$

$-\sqrt{4n-3} > x > -\sqrt{4n-1}$, so the intervals of upward concavity for $x < 0$ are $\left(-\sqrt{4n-1}, -\sqrt{4n-3}\right)$, n any

positive integer. To summarize: S is concave upward on the intervals $(0, 1)$, $(-\sqrt{3}, -1)$, $(\sqrt{3}, \sqrt{5})$, $(-\sqrt{7}, -\sqrt{5})$,

$(\sqrt{7}, 3), \ldots$.

(c) In Maple, we use `plot({int(sin(Pi*t^2/2),t=0..x),0.2},x=0..2);`. Note that Maple recognizes the

Fresnel function, calling it `FresnelS(x)`. In Mathematica, we use

`Plot[{Integrate[Sin[Pi*t^2/2],{t,0,x}],0.2},{x,0,2}]`. In Derive, we load the utility file

`FRESNEL` and plot `FRESNEL_SIN(x)`. From the graphs, we see that $\int_0^x \sin\left(\frac{\pi}{2}t^2\right)dt = 0.2$ at $x \approx 0.74$.

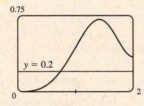
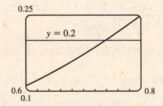

31. Using FTC1, we differentiate both sides of $6 + \int_a^x \dfrac{f(t)}{t^2}\,dt = 2\sqrt{x}$ to get $\dfrac{f(x)}{x^2} = 2\dfrac{1}{2\sqrt{x}}$ $\Rightarrow$ $f(x) = x^{3/2}$.

To find a, we substitute $x = a$ in the original equation to obtain $6 + \int_a^a \dfrac{f(t)}{t^2}\,dt = 2\sqrt{a}$ $\Rightarrow$ $6 + 0 = 2\sqrt{a}$ $\Rightarrow$

$3 = \sqrt{a}$ $\Rightarrow$ $a = 9$.

33. (a) Let $F(t) = \int_0^t f(s)\,ds$. Then, by FTC1, $F'(t) = f(t) =$ rate of depreciation, so $F(t)$ represents the loss in value over the

interval $[0, t]$.

(b) $C(t) = \dfrac{1}{t}\left[A + \displaystyle\int_0^t f(s)\,ds\right] = \dfrac{A + F(t)}{t}$ represents the average expenditure per unit of t during the interval $[0, t]$,

assuming that there has been only one overhaul during that time period. The company wants to minimize average

expenditure.

(c) $C(t) = \dfrac{1}{t}\left[A + \displaystyle\int_0^t f(s)\,ds\right]$. Using FTC1, we have $C'(t) = -\dfrac{1}{t^2}\left[A + \displaystyle\int_0^t f(s)\,ds\right] + \dfrac{1}{t}f(t)$.

$C'(t) = 0$ $\Rightarrow$ $t\,f(t) = A + \displaystyle\int_0^t f(s)\,ds$ $\Rightarrow$ $f(t) = \dfrac{1}{t}\left[A + \displaystyle\int_0^t f(s)\,ds\right] = C(t)$.

4.5   The Substitution Rule

1. Let $u = \pi x$. Then $du = \pi\, dx$ and $\frac{1}{\pi}\, du = dx$, so $\int \sin \pi x\, dx = \int \sin u\, \left(\frac{1}{\pi}\, du\right) = \frac{1}{\pi}(-\cos u) + C = -\frac{1}{\pi} \cos \pi x + C$.

3. Let $u = x^3 + 1$. Then $du = 3x^2\, dx$ and $x^2\, dx = \frac{1}{3}\, du$, so
$$\int x^2 \sqrt{x^3 + 1}\, dx = \int \sqrt{u}\, \left(\tfrac{1}{3}\, du\right) = \frac{1}{3} \frac{u^{3/2}}{3/2} + C = \frac{1}{3} \cdot \frac{2}{3} u^{3/2} + C = \frac{2}{9}(x^3 + 1)^{3/2} + C.$$

5. Let $u = \cos \theta$. Then $du = -\sin \theta\, d\theta$ and $\sin \theta\, d\theta = -du$, so
$$\int \cos^3 \theta \sin \theta\, d\theta = \int u^3 (-du) = -\frac{u^4}{4} + C = -\frac{1}{4} \cos^4 \theta + C.$$

7. Let $u = x^2$. Then $du = 2x\, dx$ and $x\, dx = \frac{1}{2}\, du$, so $\int x \sin(x^2)\, dx = \int \sin u (\frac{1}{2}\, du) = -\frac{1}{2} \cos u + C = -\frac{1}{2} \cos(x^2) + C$.

9. Let $u = 1 - 2x$. Then $du = -2\, dx$ and $dx = -\frac{1}{2}\, du$, so
$$\int (1 - 2x)^9\, dx = \int u^9 \left(-\tfrac{1}{2}\, du\right) = -\frac{1}{2} \cdot \frac{1}{10} u^{10} + C = -\frac{1}{20}(1 - 2x)^{10} + C.$$

11. Let $u = 2x + x^2$. Then $du = (2 + 2x)\, dx = 2(1 + x)\, dx$ and $(x + 1)\, dx = \frac{1}{2}\, du$, so
$$\int (x + 1) \sqrt{2x + x^2}\, dx = \int \sqrt{u}\, \left(\tfrac{1}{2}\, du\right) = \frac{1}{2} \frac{u^{3/2}}{3/2} + C = \frac{1}{3}(2x + x^2)^{3/2} + C.$$

  Or: Let $u = \sqrt{2x + x^2}$. Then $u^2 = 2x + x^2 \;\Rightarrow\; 2u\, du = (2 + 2x)\, dx \;\Rightarrow\; u\, du = (1 + x)\, dx$, so
$$\int (x + 1)\sqrt{2x + x^2}\, dx = \int u \cdot u\, du = \int u^2\, du = \frac{1}{3} u^3 + C = \frac{1}{3}(2x + x^2)^{3/2} + C.$$

13. Let $u = 3t$. Then $du = 3\, dt$ and $dt = \frac{1}{3}\, du$, so $\int \sec 3t \tan 3t\, dt = \int \sec u \tan u\, \left(\frac{1}{3}\, du\right) = \frac{1}{3} \sec u + C = \frac{1}{3} \sec 3t + C$.

15. Let $u = 3ax + bx^3$. Then $du = (3a + 3bx^2)\, dx = 3(a + bx^2)\, dx$, so
$$\int \frac{a + bx^2}{\sqrt{3ax + bx^3}}\, dx = \int \frac{\frac{1}{3}\, du}{u^{1/2}} = \frac{1}{3} \int u^{-1/2}\, du = \frac{1}{3} \cdot 2u^{1/2} + C = \frac{2}{3} \sqrt{3ax + bx^3} + C.$$

17. Let $u = \tan \theta$. Then $du = \sec^2 \theta\, d\theta$, so $\int \sec^2 \theta \tan^3 \theta\, d\theta = \int u^3\, du = \frac{1}{4} u^4 + C = \frac{1}{4} \tan^4 \theta + C$.

19. Let $u = x^3 + 3x$. Then $du = (3x^2 + 3)\, dx$ and $\frac{1}{3}\, du = (x^2 + 1)\, dx$, so
$$\int (x^2 + 1)(x^3 + 3x)^4\, dx = \int u^4 \left(\tfrac{1}{3}\, du\right) = \frac{1}{3} \cdot \frac{1}{5} u^5 + C = \frac{1}{15}(x^3 + 3x)^5 + C.$$

21. Let $u = \sin x$. Then $du = \cos x\, dx$, so $\int \frac{\cos x}{\sin^2 x}\, dx = \int \frac{1}{u^2}\, du = \int u^{-2}\, du = \frac{u^{-1}}{-1} + C = -\frac{1}{u} + C = -\frac{1}{\sin x} + C$
  [or $-\csc x + C$].

23. Let $u = \cot x$. Then $du = -\csc^2 x\, dx$ and $\csc^2 x\, dx = -du$, so
$$\int \sqrt{\cot x}\, \csc^2 x\, dx = \int \sqrt{u}\, (-du) = -\frac{u^{3/2}}{3/2} + C = -\frac{2}{3}(\cot x)^{3/2} + C.$$

25. Let $u = \sec x$. Then $du = \sec x \tan x\, dx$, so
$$\int \sec^3 x \tan x\, dx = \int \sec^2 x\, (\sec x \tan x)\, dx = \int u^2\, du = \frac{1}{3} u^3 + C = \frac{1}{3} \sec^3 x + C.$$

27. Let $u = 1 + z^3$. Then $du = 3z^2\,dz$ and $z^2\,dz = \frac{1}{3}\,du$, so

$$\int \frac{z^2}{\sqrt[3]{1+z^3}}\,dz = \int u^{-1/3}\left(\tfrac{1}{3}\,du\right) = \tfrac{1}{3}\cdot\tfrac{3}{2}u^{2/3} + C = \tfrac{1}{2}(1+z^3)^{2/3} + C.$$

29. Let $u = 2x + 5$. Then $du = 2\,dx$ and $x = \frac{1}{2}(u-5)$, so

$$\int x(2x+5)^8\,dx = \int \tfrac{1}{2}(u-5)u^8\left(\tfrac{1}{2}\,du\right) = \tfrac{1}{4}\int(u^9 - 5u^8)\,du$$
$$= \tfrac{1}{4}\left(\tfrac{1}{10}u^{10} - \tfrac{5}{9}u^9\right) + C = \tfrac{1}{40}(2x+5)^{10} - \tfrac{5}{36}(2x+5)^9 + C$$

31. Let $u = \frac{\pi}{2}t$, so $du = \frac{\pi}{2}\,dt$. When $t = 0$, $u = 0$; when $t = 1$, $u = \frac{\pi}{2}$. Thus,

$$\int_0^1 \cos(\pi t/2)\,dt = \int_0^{\pi/2} \cos u\left(\tfrac{2}{\pi}\,du\right) = \tfrac{2}{\pi}\left[\sin u\right]_0^{\pi/2} = \tfrac{2}{\pi}\left(\sin\tfrac{\pi}{2} - \sin 0\right) = \tfrac{2}{\pi}(1 - 0) = \tfrac{2}{\pi}$$

33. Let $u = 1 + 7x$, so $du = 7\,dx$. When $x = 0$, $u = 1$; when $x = 1$, $u = 8$. Thus,

$$\int_0^1 \sqrt[3]{1+7x}\,dx = \int_1^8 u^{1/3}\left(\tfrac{1}{7}\,du\right) = \tfrac{1}{7}\left[\tfrac{3}{4}u^{4/3}\right]_1^8 = \tfrac{3}{28}(8^{4/3} - 1^{4/3}) = \tfrac{3}{28}(16 - 1) = \tfrac{45}{28}$$

35. Let $u = t/4$, so $du = \frac{1}{4}\,dt$. When $t = 0$, $u = 0$; when $t = \pi$, $u = \pi/4$. Thus,

$$\int_0^\pi \sec^2(t/4)\,dt = \int_0^{\pi/4} \sec^2 u\,(4\,du) = 4\left[\tan u\right]_0^{\pi/4} = 4\left(\tan\tfrac{\pi}{4} - \tan 0\right) = 4(1 - 0) = 4.$$

37. Let $u = x^2 + a^2$, so $du = 2x\,dx$ and $x\,dx = \frac{1}{2}\,du$. When $x = 0$, $u = a^2$; when $x = a$, $u = 2a^2$. Thus,

$$\int_0^a x\sqrt{x^2+a^2}\,dx = \int_{a^2}^{2a^2} u^{1/2}\left(\tfrac{1}{2}\,du\right) = \tfrac{1}{2}\left[\tfrac{2}{3}u^{3/2}\right]_{a^2}^{2a^2} = \left[\tfrac{1}{3}u^{3/2}\right]_{a^2}^{2a^2} = \tfrac{1}{3}\left[(2a^2)^{3/2} - (a^2)^{3/2}\right] = \tfrac{1}{3}\left(2\sqrt{2} - 1\right)a^3$$

39. Let $u = x - 1$, so $u + 1 = x$ and $du = dx$. When $x = 1$, $u = 0$; when $x = 2$, $u = 1$. Thus,

$$\int_1^2 x\sqrt{x-1}\,dx = \int_0^1 (u+1)\sqrt{u}\,du = \int_0^1 (u^{3/2} + u^{1/2})\,du = \left[\tfrac{2}{5}u^{5/2} + \tfrac{2}{3}u^{3/2}\right]_0^1 = \tfrac{2}{5} + \tfrac{2}{3} = \tfrac{16}{15}.$$

41. Let $u = x^{-2}$, so $du = -2x^{-3}\,dx$. When $x = \frac{1}{2}$, $u = 4$; when $x = 1$, $u = 1$. Thus,

$$\int_{1/2}^1 \frac{\cos(x^{-2})}{x^3}\,dx = \int_4^1 \cos u\left(\frac{du}{-2}\right) = \frac{1}{2}\int_1^4 \cos u\,du = \frac{1}{2}\left[\sin u\right]_1^4 = \frac{1}{2}(\sin 4 - \sin 1).$$

43. $\int_{-\pi/4}^{\pi/4}(x^3 + x^4\tan x)\,dx = 0$ by Theorem 6(b), since $f(x) = x^3 + x^4\tan x$ is an odd function.

45. $f_{\text{ave}} = \frac{1}{b-a}\int_a^b f(x)\,dx = \frac{1}{4-0}\int_0^4 (4x - x^2)\,dx = \frac{1}{4}\left[2x^2 - \tfrac{1}{3}x^3\right]_0^4 = \frac{1}{4}\left[(32 - \tfrac{64}{3}) - 0\right] = \frac{1}{4}\left(\tfrac{32}{3}\right) = \frac{8}{3}$

47. $h_{\text{ave}} = \frac{1}{\pi - 0}\int_0^\pi \cos^4 x\,\sin x\,dx = \frac{1}{\pi}\int_1^{-1} u^4(-du)$ $[u = \cos x,\ du = -\sin x\,dx]$

$$= \frac{1}{\pi}\int_{-1}^1 u^4\,du = \frac{1}{\pi}\cdot 2\int_0^1 u^4\,du \quad\text{[by Theorem 6(a)]} = \frac{2}{\pi}\left[\tfrac{1}{5}u^5\right]_0^1 = \frac{2}{5\pi}$$

49. First write the integral as a sum of two integrals:

$I = \int_{-2}^2 (x+3)\sqrt{4 - x^2}\,dx = I_1 + I_2 = \int_{-2}^2 x\sqrt{4-x^2}\,dx + \int_{-2}^2 3\sqrt{4-x^2}\,dx$. $I_1 = 0$ by Theorem 6(b), since

$f(x) = x\sqrt{4 - x^2}$ is an odd function and we are integrating from $x = -2$ to $x = 2$. We interpret I_2 as three times the area of

a semicircle with radius 2, so $I = 0 + 3\cdot\frac{1}{2}\left(\pi\cdot 2^2\right) = 6\pi$.

51. The volume of inhaled air in the lungs at time t is

$$V(t) = \int_0^t f(u)\, du = \int_0^t \frac{1}{2}\sin\left(\frac{2\pi}{5}u\right) du = \int_0^{2\pi t/5} \frac{1}{2}\sin v\left(\frac{5}{2\pi}\,dv\right) \quad \left[\text{substitute } v = \frac{2\pi}{5}u,\; dv = \frac{2\pi}{5}\,du\right]$$

$$= \frac{5}{4\pi}\Big[-\cos v\Big]_0^{2\pi t/5} = \frac{5}{4\pi}\left[-\cos\left(\frac{2\pi}{5}t\right)+1\right] = \frac{5}{4\pi}\left[1-\cos\left(\frac{2\pi}{5}t\right)\right] \text{ liters}$$

53. Let $u = 2x$. Then $du = 2\,dx$, so $\int_0^2 f(2x)\,dx = \int_0^4 f(u)\left(\frac{1}{2}\,du\right) = \frac{1}{2}\int_0^4 f(u)\,du = \frac{1}{2}(10) = 5$.

55. Let $u = -x$. Then $du = -dx$, so

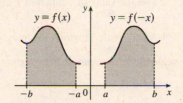

$$\int_a^b f(-x)\,dx = \int_{-a}^{-b} f(u)(-du) = \int_{-b}^{-a} f(u)\,du = \int_{-b}^{-a} f(x)\,dx.$$

From the diagram, we see that the equality follows from the fact that we are

reflecting the graph of f, and the limits of integration, about the y-axis.

57. Let $u = 1 - x$. Then $x = 1 - u$ and $dx = -du$, so

$$\int_0^1 x^a(1-x)^b\,dx = \int_1^0 (1-u)^a u^b(-du) = \int_0^1 u^b(1-u)^a\,du = \int_0^1 x^b(1-x)^a\,dx.$$

4 Review

CONCEPT CHECK

1. (a) $\sum_{i=1}^n f(x_i^*)\,\Delta x_i$ is an expression for a Riemann sum of a function f.

x_i^* is a point in the ith subinterval $[x_{i-1}, x_i]$ and $\Delta x_i = x_i - x_{i-1}$ is the length of the ith subinterval.

(b) See Figure 3 in Section 4.2.

(c) In Section 4.2, see Figure 5 and the paragraph before it.

2. (a) See the definition of the definite integral in Definition 4.2.2. If f is continuous and the subintervals have equal lengths, then the expression for the definite integral simplifies to the one in Theorem 4.2.4.

(b) It's the area under the curve $y = f(x)$ from a to b.

(c) In Section 4.2, see Figure 6 and the paragraph before it.

3. See the statement of the Midpoint Rule on page 217.

4. (a) See the Evaluation Theorem page 224.

(b) See the Net Change Theorem on page 228.

(c) $\int_{t_1}^{t_2} r(t)\,dt$ represents the change in the amount of water in the reservoir between time t_1 and time t_2.

5. (a) $\int f(x)\,dx$ is the family of functions $\{F \mid F' = f\}$. Any two such functions differ by a constant.

(b) The connection is given by the Net Change Theorem: $\int_a^b f(x)\,dx = \left[\int f(x)\,dx\right]_a^b$ if f is continuous.

6. See the Fundamental Theorem of Calculus on page 237.

7. (a) $\int_{60}^{120} v(t)\, dt$ represents the change in position of the particle from $t = 60$ to $t = 120$ seconds.

   (b) $\int_{60}^{120} |v(t)|\, dt$ represents the total distance traveled by the particle from $t = 60$ to 120 seconds.

   (c) $\int_{60}^{120} a(t)\, dt$ represents the change in the velocity of the particle from $t = 60$ to $t = 120$ seconds.

8. (a) The average value of a function f on an interval $[a, b]$ is $f_{\text{ave}} = \dfrac{1}{b - a} \displaystyle\int_a^b f(x)\, dx$.

   (b) The Mean Value Theorem for Integrals says that there is a number c at which the value of f is exactly equal to the average value of the function, that is, $f(c) = f_{\text{ave}}$. For a geometric interpretation of the Mean Value Theorem for Integrals, see Figure 10 in Section 4.4 and the discussion that accompanies it.

9. One process undoes what the other one does. The precise version of this statement is given by the Fundamental Theorem of Calculus. See the statement of this theorem and the paragraph that follows it on page 237.

10. See the Substitution Rule (4.5.4). This says that it is permissible to operate with the dx after an integral sign as if it were a differential.

TRUE-FALSE QUIZ

1. True by Property 2 of the Integral in Section 4.2.

3. True by Property 3 of the Integral in Section 4.2.

5. False.   For example, let $f(x) = x^2$. Then $\int_0^1 \sqrt{x^2}\, dx = \int_0^1 x\, dx = \frac{1}{2}$, but $\sqrt{\int_0^1 x^2\, dx} = \sqrt{\frac{1}{3}} = \frac{1}{\sqrt{3}}$.

7. True by Comparison Property 7 of the Integral in Section 4.2.

9. True.   The integrand is an odd function that is continuous on $[-1, 1]$.

11. False.   For example, the function $y = |x|$ is continuous on $\mathbb{R}$, but has no derivative at $x = 0$.

13. True by Property 5 of Integrals.

15. False. $\int_a^b f(x)\, dx$ is a constant, so $\dfrac{d}{dx}\left(\int_a^b f(x)\, dx\right) = 0$, not $f(x)$ [unless $f(x) = 0$]. Compare the given statement carefully with FTC1, in which the upper limit in the integral is x.

17. False.   The function $f(x) = 1/x^4$ is not bounded on the interval $[-2, 1]$. It has an infinite discontinuity at $x = 0$, so it is not integrable on the interval. (If the integral were to exist, a positive value would be expected, by Comparison Property 6 of Integrals.)

EXERCISES

1. (a)

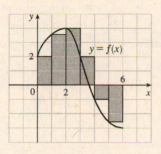

$$L_6 = \sum_{i=1}^{6} f(x_{i-1})\,\Delta x \quad [\Delta x = \tfrac{6-0}{6} = 1]$$

$$= f(x_0)\cdot 1 + f(x_1)\cdot 1 + f(x_2)\cdot 1 + f(x_3)\cdot 1 + f(x_4)\cdot 1 + f(x_5)\cdot 1$$

$$\approx 2 + 3.5 + 4 + 2 + (-1) + (-2.5) = 8$$

The Riemann sum represents the sum of the areas of the four rectangles above the x-axis minus the sum of the areas of the two rectangles below the x-axis.

(b)

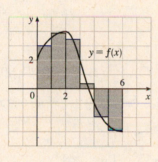

$$M_6 = \sum_{i=1}^{6} f(\overline{x}_i)\,\Delta x \quad [\Delta x = \tfrac{6-0}{6} = 1]$$

$$= f(\overline{x}_1)\cdot 1 + f(\overline{x}_2)\cdot 1 + f(\overline{x}_3)\cdot 1 + f(\overline{x}_4)\cdot 1 + f(\overline{x}_5)\cdot 1 + f(\overline{x}_6)\cdot 1$$

$$= f(0.5) + f(1.5) + f(2.5) + f(3.5) + f(4.5) + f(5.5)$$

$$\approx 3 + 3.9 + 3.4 + 0.3 + (-2) + (-2.9) = 5.7$$

The Riemann sum represents the sum of the areas of the four rectangles above the x-axis minus the sum of the areas of the two rectangles below the x-axis.

3. $\int_0^1 \left(x + \sqrt{1-x^2}\,\right) dx = \int_0^1 x\,dx + \int_0^1 \sqrt{1-x^2}\,dx = I_1 + I_2$.

I_1 can be interpreted as the area of the triangle shown in the figure and I_2 can be interpreted as the area of the quarter-circle.

Area $= \tfrac{1}{2}(1)(1) + \tfrac{1}{4}(\pi)(1)^2 = \tfrac{1}{2} + \tfrac{\pi}{4}$.

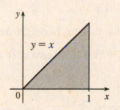
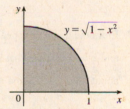

5. First note that either a or b must be the graph of $\int_0^x f(t)\,dt$, since $\int_0^0 f(t)\,dt = 0$, and $c(0) \neq 0$. Now notice that $b > 0$ when c is increasing, and that $c > 0$ when a is increasing. It follows that c is the graph of $f(x)$, b is the graph of $f'(x)$, and a is the graph of $\int_0^x f(t)\,dt$.

7. $\int_1^2 (8x^3 + 3x^2)\,dx = \left[8 \cdot \tfrac{1}{4}x^4 + 3 \cdot \tfrac{1}{3}x^3\right]_1^2 = \left[2x^4 + x^3\right]_1^2 = (2 \cdot 2^4 + 2^3) - (2 + 1) = 40 - 3 = 37$

9. $\int_0^1 (1 - x^9)\,dx = \left[x - \tfrac{1}{10}x^{10}\right]_0^1 = \left(1 - \tfrac{1}{10}\right) - 0 = \tfrac{9}{10}$

11. $\int_1^9 \dfrac{\sqrt{u} - 2u^2}{u}\,du = \int_1^9 (u^{-1/2} - 2u)\,du = \left[2u^{1/2} - u^2\right]_1^9 = (6 - 81) - (2 - 1) = -76$

13. Let $u = y^2 + 1$, so $du = 2y\,dy$ and $y\,dy = \tfrac{1}{2}\,du$. When $y = 0$, $u = 1$; when $y = 1$, $u = 2$. Thus,

$$\int_0^1 y(y^2 + 1)^5\,dy = \int_1^2 u^5\left(\tfrac{1}{2}\,du\right) = \tfrac{1}{2}\left[\tfrac{1}{6}u^6\right]_1^2 = \tfrac{1}{12}(64 - 1) = \tfrac{63}{12} = \tfrac{21}{4}.$$

15. Let $u = v^3$, so $du = 3v^2\,dv$. When $v = 0$, $u = 0$; when $v = 1$, $u = 1$. Thus,

$\int_0^1 v^2 \cos(v^3)\,dv = \int_0^1 \cos u \left(\frac{1}{3}\,du\right) = \frac{1}{3}\big[\sin u\big]_0^1 = \frac{1}{3}(\sin 1 - 0) = \frac{1}{3}\sin 1$.

17. $\displaystyle\int_{-\pi/4}^{\pi/4} \frac{t^4 \tan t}{2 + \cos t}\,dt = 0$ by Theorem 4.5.6(b), since $f(t) = \dfrac{t^4 \tan t}{2 + \cos t}$ is an odd function.

19. Let $u = 2\theta$. Then $du = 2\,d\theta$, so

$\int_0^{\pi/8} \sec 2\theta \tan 2\theta\,d\theta = \int_0^{\pi/4} \sec u \tan u \left(\frac{1}{2}\,du\right) = \frac{1}{2}\big[\sec u\big]_0^{\pi/4} = \frac{1}{2}\left(\sec\frac{\pi}{4} - \sec 0\right) = \frac{1}{2}(\sqrt{2} - 1) = \frac{1}{2}\sqrt{2} - \frac{1}{2}$.

21. Let $u = x^2 + 4x$. Then $du = (2x + 4)\,dx = 2(x + 2)\,dx$, so

$\int \dfrac{x + 2}{\sqrt{x^2 + 4x}}\,dx = \int u^{-1/2}\left(\frac{1}{2}\,du\right) = \frac{1}{2} \cdot 2u^{1/2} + C = \sqrt{u} + C = \sqrt{x^2 + 4x} + C$.

23. Let $u = \sin \pi t$. Then $du = \pi \cos \pi t\,dt$, so $\int \sin \pi t \cos \pi t\,dt = \int u\left(\frac{1}{\pi}\,du\right) = \frac{1}{\pi} \cdot \frac{1}{2}u^2 + C = \frac{1}{2\pi}(\sin \pi t)^2 + C$.

25. Since $x^2 - 4 < 0$ for $0 \le x < 2$ and $x^2 - 4 > 0$ for $2 < x \le 3$, we have $\left|x^2 - 4\right| = -(x^2 - 4) = 4 - x^2$ for $0 \le x < 2$ and

$\left|x^2 - 4\right| = x^2 - 4$ for $2 < x \le 3$. Thus,

$$\int_0^3 \left|x^2 - 4\right|\,dx = \int_0^2 (4 - x^2)\,dx + \int_2^3 (x^2 - 4)\,dx = \left[4x - \frac{x^3}{3}\right]_0^2 + \left[\frac{x^3}{3} - 4x\right]_2^3$$

$$= \left(8 - \frac{8}{3}\right) - 0 + (9 - 12) - \left(\frac{8}{3} - 8\right) = \frac{16}{3} - 3 + \frac{16}{3} = \frac{32}{3} - \frac{9}{3} = \frac{23}{3}$$

27. From the graph, it appears that the area under the curve $y = x\sqrt{x}$ between $x = 0$ and $x = 4$ is somewhat less than half the area of an 8×4 rectangle, so perhaps about 13 or 14. To find the exact value, we evaluate

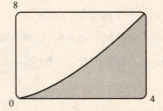

$\int_0^4 x\sqrt{x}\,dx = \int_0^4 x^{3/2}\,dx = \left[\frac{2}{5}x^{5/2}\right]_0^4 = \frac{2}{5}(4)^{5/2} = \frac{64}{5} = 12.8$.

29. By FTC1, $F(x) = \int_1^x \sqrt{1 + t^4}\,dt \implies F'(x) = \sqrt{1 + x^4}$.

31. $y = \displaystyle\int_{\sqrt{x}}^x \frac{\cos \theta}{\theta}\,d\theta = \int_1^x \frac{\cos \theta}{\theta}\,d\theta + \int_{\sqrt{x}}^1 \frac{\cos \theta}{\theta}\,d\theta = \int_1^x \frac{\cos \theta}{\theta}\,d\theta - \int_1^{\sqrt{x}} \frac{\cos \theta}{\theta}\,d\theta \implies$

$y' = \dfrac{\cos x}{x} - \dfrac{\cos \sqrt{x}}{\sqrt{x}}\dfrac{1}{2\sqrt{x}} = \dfrac{2\cos x - \cos\sqrt{x}}{2x}$

33. If $1 \le x \le 3$, then $\sqrt{1^2 + 3} \le \sqrt{x^2 + 3} \le \sqrt{3^2 + 3} \implies 2 \le \sqrt{x^2 + 3} \le 2\sqrt{3}$, so

$2(3 - 1) \le \int_1^3 \sqrt{x^2 + 3}\,dx \le 2\sqrt{3}(3 - 1)$; that is, $4 \le \int_1^3 \sqrt{x^2 + 3}\,dx \le 4\sqrt{3}$.

35. Let $f(x) = \sqrt{1 + x^3}$ on $[0, 1]$. The Midpoint Rule with $n = 5$ gives

$$\int_0^1 \sqrt{1 + x^3}\,dx \approx \frac{1}{5}[f(0.1) + f(0.3) + f(0.5) + f(0.7) + f(0.9)]$$

$$= \frac{1}{5}\left[\sqrt{1 + (0.1)^3} + \sqrt{1 + (0.3)^3} + \cdots + \sqrt{1 + (0.9)^3}\right] \approx 1.110$$

37. Note that $r(t) = b'(t)$, where $b(t) =$ the number of barrels of oil consumed up to time t. So, by the Net Change Theorem,

$\int_0^3 r(t)\, dt = b(3) - b(0)$ represents the number of barrels of oil consumed from Jan. 1, 2000, through Jan. 1, 2003.

39. We use the Midpoint Rule with $n = 6$ and $\Delta t = \frac{24-0}{6} = 4$. The increase in the bee population was

$$\int_0^{24} r(t)\, dt \approx M_6 = 4[r(2) + r(6) + r(10) + r(14) + r(18) + r(22)]$$
$$\approx 4[50 + 1000 + 7000 + 8550 + 1350 + 150] = 4(18{,}100) = 72{,}400$$

41. $\displaystyle \lim_{h \to 0} f_{\text{ave}} = \lim_{h \to 0} \frac{1}{(x+h) - x} \int_x^{x+h} f(t)\, dt = \lim_{h \to 0} \frac{F(x+h) - F(x)}{h}$, where $F(x) = \int_a^x f(t)\, dt$. But we recognize this

limit as being $F'(x)$ by the definition of a derivative. Therefore, $\displaystyle \lim_{h \to 0} f_{\text{ave}} = F'(x) = f(x)$ by FTC1.

43. Let $u = f(x)$ and $du = f'(x)\, dx$. So $2 \int_a^b f(x) f'(x)\, dx = 2 \int_{f(a)}^{f(b)} u\, du = \left[u^2\right]_{f(a)}^{f(b)} = [f(b)]^2 - [f(a)]^2$.

5 □ Inverse Functions: Exponential, Logarithmic, and Inverse Trigonometric Functions

5.1 Inverse Functions

1. (a) See Definition 1.

   (b) It must pass the Horizontal Line Test.

3. f is not one-to-one because $2 \neq 6$, but $f(2) = 2.0 = f(6)$.

5. We could draw a horizontal line that intersects the graph in more than one point. Thus, by the Horizontal Line Test, the function is not one-to-one.

7. No horizontal line intersects the graph more than once. Thus, by the Horizontal Line Test, the function is one-to-one.

9. The graph of $f(x) = x^2 - 2x$ is a parabola with axis of symmetry $x = -\dfrac{b}{2a} = -\dfrac{-2}{2(1)} = 1$. Pick any x-values equidistant from 1 to find two equal function values. For example, $f(0) = 0$ and $f(2) = 0$, so f is not one-to-one.

11. $g(x) = 1/x$. $x_1 \neq x_2$ $\Rightarrow$ $1/x_1 \neq 1/x_2$ $\Rightarrow$ $g(x_1) \neq g(x_2)$, so g is one-to-one.

    Geometric solution: The graph of g is the hyperbola shown in Figure 9 in Section 1.2. It passes the Horizontal Line Test, so g is one-to-one.

13. A football will attain every height h up to its maximum height twice: once on the way up, and again on the way down. Thus, even if t_1 does not equal t_2, $f(t_1)$ may equal $f(t_2)$, so f is not 1-1.

15. (a) Since f is 1-1, $f(6) = 17$ $\Leftrightarrow$ $f^{-1}(17) = 6$.

    (b) Since f is 1-1, $f^{-1}(3) = 2$ $\Leftrightarrow$ $f(2) = 3$.

17. $h(x) = x + \sqrt{x}$ $\Rightarrow$ $h'(x) = 1 + 1/(2\sqrt{x}) > 0$ on $(0, \infty)$. So h is increasing and hence, 1-1. By inspection, $h(4) = 4 + \sqrt{4} = 6$, so $h^{-1}(6) = 4$.

19. We solve $C = \frac{5}{9}(F - 32)$ for F: $\frac{9}{5}C = F - 32$ $\Rightarrow$ $F = \frac{9}{5}C + 32$. This gives us a formula for the inverse function, that is, the Fahrenheit temperature F as a function of the Celsius temperature C. $F \geq -459.67$ $\Rightarrow$ $\frac{9}{5}C + 32 \geq -459.67$ $\Rightarrow$ $\frac{9}{5}C \geq -491.67$ $\Rightarrow$ $C \geq -273.15$, the domain of the inverse function.

21. $y = f(x) = 3 - 2x$ $\Rightarrow$ $2x = 3 - y$ $\Rightarrow$ $x = \dfrac{3 - y}{2}$. Interchange x and y: $y = \dfrac{3 - x}{2}$. So $f^{-1}(x) = \dfrac{3 - x}{2}$.

23. $y = f(x) = 1 + \sqrt{2 + 3x}$ $(y \geq 1)$ $\Rightarrow$ $y - 1 = \sqrt{2 + 3x}$ $\Rightarrow$ $(y - 1)^2 = 2 + 3x$ $\Rightarrow$ $(y - 1)^2 - 2 = 3x$ $\Rightarrow$ $x = \frac{1}{3}(y - 1)^2 - \frac{2}{3}$. Interchange x and y: $y = \frac{1}{3}(x - 1)^2 - \frac{2}{3}$. So $f^{-1}(x) = \frac{1}{3}(x - 1)^2 - \frac{2}{3}$. Note that the domain of f^{-1} is $x \geq 1$.

25. For $f(x) = \dfrac{1 - \sqrt{x}}{1 + \sqrt{x}}$, the domain is $x \geq 0$. $f(0) = 1$ and as x increases, y decreases. As $x \to \infty$,

$$\frac{1 - \sqrt{x}}{1 + \sqrt{x}} \cdot \frac{1/\sqrt{x}}{1/\sqrt{x}} = \frac{1/\sqrt{x} - 1}{1/\sqrt{x} + 1} \to \frac{-1}{1} = -1, \text{ so the range of } f \text{ is } -1 < y \leq 1. \text{ Thus, the domain of } f^{-1} \text{ is } -1 < x \leq 1.$$

$$y = \frac{1 - \sqrt{x}}{1 + \sqrt{x}} \ \Rightarrow \ y(1 + \sqrt{x}) = 1 - \sqrt{x} \ \Rightarrow \ y + y\sqrt{x} = 1 - \sqrt{x} \ \Rightarrow \ \sqrt{x} + y\sqrt{x} = 1 - y \ \Rightarrow$$

$$\sqrt{x}(1 + y) = 1 - y \ \Rightarrow \ \sqrt{x} = \frac{1 - y}{1 + y} \ \Rightarrow \ x = \left(\frac{1 - y}{1 + y}\right)^2. \text{ Interchange } x \text{ and } y: \ y = \left(\frac{1 - x}{1 + x}\right)^2. \text{ So}$$

$$f^{-1}(x) = \left(\frac{1 - x}{1 + x}\right)^2 \text{ with } -1 < x \leq 1.$$

27. $y = f(x) = x^4 + 1 \ \Rightarrow \ y - 1 = x^4 \ \Rightarrow \ x = \sqrt[4]{y - 1}$ [not $\pm$ since

$x \geq 0$]. Interchange x and y: $y = \sqrt[4]{x - 1}$. So $f^{-1}(x) = \sqrt[4]{x - 1}$. The

graph of $y = \sqrt[4]{x - 1}$ is just the graph of $y = \sqrt[4]{x}$ shifted right one unit.

From the graph, we see that f and f^{-1} are reflections about the line $y = x$.

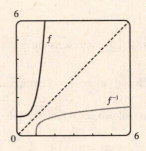

29. Reflect the graph of f about the line $y = x$. The points $(-1, -2)$, $(1, -1)$,

$(2, 2)$, and $(3, 3)$ on f are reflected to $(-2, -1)$, $(-1, 1)$, $(2, 2)$, and $(3, 3)$

on f^{-1}.

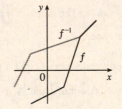

31. (a) $y = f(x) = \sqrt{1 - x^2}$ $(0 \leq x \leq 1$ and note that $y \geq 0)$ $\Rightarrow$

$y^2 = 1 - x^2 \ \Rightarrow \ x^2 = 1 - y^2 \ \Rightarrow \ x = \sqrt{1 - y^2}$. So

$f^{-1}(x) = \sqrt{1 - x^2}$, $0 \leq x \leq 1$. We see that f^{-1} and f are the same

function.

(b) The graph of f is the portion of the circle $x^2 + y^2 = 1$ with $0 \leq x \leq 1$ and

$0 \leq y \leq 1$ (quarter-circle in the first quadrant). The graph of f is symmetric

with respect to the line $y = x$, so its reflection about $y = x$ is itself, that is,

$f^{-1} = f$.

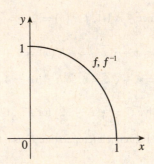

33. (a) $x_1 \neq x_2 \ \Rightarrow \ x_1^3 \neq x_2^3 \ \Rightarrow \ f(x_1) \neq f(x_2)$, so f is one-to-one.

(b) $f'(x) = 3x^2$ and $f(2) = 8 \ \Rightarrow \ f^{-1}(8) = 2$, so $(f^{-1})'(8) = 1/f'(f^{-1}(8)) = 1/f'(2) = \frac{1}{12}$.

(c) $y = x^3 \ \Rightarrow \ x = y^{1/3}$. Interchanging x and y gives $y = x^{1/3}$, so $f^{-1}(x) = x^{1/3}$. Domain$(f^{-1}) = $ range$(f) = \mathbb{R}$.

Range$(f^{-1}) = $ domain$(f) = \mathbb{R}$.

(d) $f^{-1}(x) = x^{1/3}$ $\Rightarrow$ $(f^{-1})'(x) = \frac{1}{3}x^{-2/3}$ $\Rightarrow$

$(f^{-1})'(8) = \frac{1}{3}\left(\frac{1}{4}\right) = \frac{1}{12}$ as in part (b).

(e)

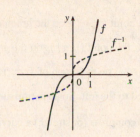

35. (a) Since $x \geq 0$, $x_1 \neq x_2$ $\Rightarrow$ $x_1^2 \neq x_2^2$ $\Rightarrow$ $9 - x_1^2 \neq 9 - x_2^2$ $\Rightarrow$ $f(x_1) \neq f(x_2)$, so f is 1-1.

(b) $f'(x) = -2x$ and $f(1) = 8$ $\Rightarrow$ $f^{-1}(8) = 1$, so $(f^{-1})'(8) = \dfrac{1}{f'(f^{-1}(8))} = \dfrac{1}{f'(1)} = \dfrac{1}{(-2)} = -\dfrac{1}{2}$.

(c) $y = 9 - x^2$ $\Rightarrow$ $x^2 = 9 - y$ $\Rightarrow$ $x = \sqrt{9 - y}$.

Interchange x and y: $y = \sqrt{9 - x}$, so $f^{-1}(x) = \sqrt{9 - x}$.

Domain(f^{-1}) = range (f) = $[0, 9]$.

Range(f^{-1}) = domain (f) = $[0, 3]$.

(e)

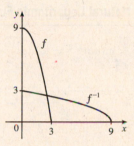

(d) $(f^{-1})'(x) = -1/(2\sqrt{9 - x})$ $\Rightarrow$ $(f^{-1})'(8) = -\frac{1}{2}$ as in part (b).

37. $f(0) = 4$ $\Rightarrow$ $f^{-1}(4) = 0$, and $f(x) = 2x^3 + 3x^2 + 7x + 4$ $\Rightarrow$ $f'(x) = 6x^2 + 6x + 7$ and $f'(0) = 7$.

Thus, $(f^{-1})'(4) = \dfrac{1}{f'(f^{-1}(4))} = \dfrac{1}{f'(0)} = \dfrac{1}{7}$.

39. $f(0) = 3$ $\Rightarrow$ $f^{-1}(3) = 0$, and $f(x) = 3 + x^2 + \tan(\pi x/2)$ $\Rightarrow$ $f'(x) = 2x + \frac{\pi}{2}\sec^2(\pi x/2)$ and

$f'(0) = \frac{\pi}{2} \cdot 1 = \frac{\pi}{2}$. Thus, $(f^{-1})'(3) = 1/f'(f^{-1}(3)) = 1/f'(0) = 2/\pi$.

41. $f(4) = 5$ $\Rightarrow$ $f^{-1}(5) = 4$. Thus, $(f^{-1})'(5) = \dfrac{1}{f'(f^{-1}(5))} = \dfrac{1}{f'(4)} = \dfrac{1}{2/3} = \dfrac{3}{2}$.

43. $f(x) = \int_3^x \sqrt{1 + t^3}\, dt$ $\Rightarrow$ $f'(x) = \sqrt{1 + x^3} > 0$, so f is an increasing function and it has an inverse. Since

$f(3) = \int_3^3 \sqrt{1 + t^3}\, dt = 0$, $f^{-1}(0) = 3$. Thus, $(f^{-1})'(0) = \dfrac{1}{f'(f^{-1}(0))} = \dfrac{1}{f'(3)} = \dfrac{1}{\sqrt{1 + 3^3}} = \dfrac{1}{\sqrt{28}}$.

45. We see that the graph of $y = f(x) = \sqrt{x^3 + x^2 + x + 1}$ is increasing, so

f is 1-1. Enter $x = \sqrt{y^3 + y^2 + y + 1}$ and use your CAS to solve the

equation for y. Using Derive, we get two (irrelevant) solutions involving

imaginary expressions, as well as one which can be simplified to the

following:

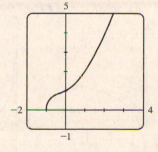

$$y = f^{-1}(x) = -\frac{\sqrt[3]{4}}{6}\left(\sqrt[3]{D - 27x^2 + 20} - \sqrt[3]{D + 27x^2 - 20} + \sqrt[3]{2}\right)$$

where $D = 3\sqrt{3}\sqrt{27x^4 - 40x^2 + 16}$. Maple and Mathematica each give two complex expressions and one real expression,

and the real expression is equivalent to that given by Derive. For example, Maple's expression simplifies to

$$\frac{1}{6}\frac{M^{2/3} - 8 - 2M^{1/3}}{2M^{1/3}}, \text{ where } M = 108x^2 + 12\sqrt{48 - 120x^2 + 81x^4} - 80.$$

47. (a) If the point (x, y) is on the graph of $y = f(x)$, then the point $(x - c, y)$ is that point shifted c units to the left. Since f is 1-1, the point (y, x) is on the graph of $y = f^{-1}(x)$ and the point corresponding to $(x - c, y)$ on the graph of f is $(y, x - c)$ on the graph of f^{-1}. Thus, the curve's reflection is shifted *down* the same number of units as the curve itself is shifted to the left. So an expression for the inverse function is $g^{-1}(x) = f^{-1}(x) - c$.

(b) If we compress (or stretch) a curve horizontally, the curve's reflection in the line $y = x$ is compressed (or stretched) *vertically* by the same factor. Using this geometric principle, we see that the inverse of $h(x) = f(cx)$ can be expressed as $h^{-1}(x) = (1/c) f^{-1}(x)$.

5.2 The Natural Logarithmic Function

1. $\ln \sqrt{ab} = \ln(ab)^{1/2} = \frac{1}{2}\ln(ab) = \frac{1}{2}(\ln a + \ln b) = \frac{1}{2}\ln a + \frac{1}{2}\ln b$ [assuming that the variables are positive]

3. $\ln \dfrac{x^2}{y^3 z^4} = \ln x^2 - \ln(y^3 z^4) = 2\ln x - (\ln y^3 + \ln z^4) = 2\ln x - 3\ln y - 4\ln z$

5. $\ln 5 + 5\ln 3 = \ln 5 + \ln 3^5$ [by Law 3]

$\qquad = \ln(5 \cdot 3^5)$ [by Law 1]

$\qquad = \ln 1215$

7. $\frac{1}{3}\ln(x + 2)^3 + \frac{1}{2}\left[\ln x - \ln(x^2 + 3x + 2)^2\right] = \ln[(x+2)^3]^{1/3} + \frac{1}{2}\ln \dfrac{x}{(x^2 + 3x + 2)^2}$ [by Laws 3, 2]

$\qquad = \ln(x + 2) + \ln \dfrac{\sqrt{x}}{x^2 + 3x + 2}$ [by Law 3]

$\qquad = \ln \dfrac{(x + 2)\sqrt{x}}{(x + 1)(x + 2)}$ [by Law 1]

$\qquad = \ln \dfrac{\sqrt{x}}{x + 1}$

Note that since $\ln x$ is defined for $x > 0$, we have $x + 1$, $x + 2$, and $x^2 + 3x + 2$ all positive, and hence their logarithms are defined.

9. Reflect the graph of $y = \ln x$ about the x-axis to obtain the graph of $y = -\ln x$.

11.

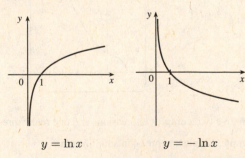

$y = \ln x$ $\qquad$ $y = -\ln x$

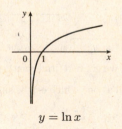

$y = \ln x$

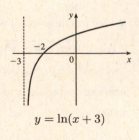

$y = \ln(x + 3)$

13. Let $t = x^2 - 9$. Then as $x \to 3^+$, $t \to 0^+$, and $\displaystyle\lim_{x \to 3^+} \ln(x^2 - 9) = \lim_{t \to 0^+} \ln t = -\infty$ by (4).

15. $f(x) = \sqrt{x}\ln x \;\Rightarrow\; f'(x) = \dfrac{1}{2\sqrt{x}}\ln x + \sqrt{x}\left(\dfrac{1}{x}\right) = \dfrac{\ln x + 2}{2\sqrt{x}}$

17. $f(x) = \sin(\ln x) \;\Rightarrow\; f'(x) = \cos(\ln x)\cdot\dfrac{d}{dx}\ln x = \cos(\ln x)\cdot\dfrac{1}{x} = \dfrac{\cos(\ln x)}{x}$

19. $f(x) = \ln\dfrac{1}{x} \;\Rightarrow\; f'(x) = \dfrac{1}{1/x}\dfrac{d}{dx}\left(\dfrac{1}{x}\right) = x\left(-\dfrac{1}{x^2}\right) = -\dfrac{1}{x}.$

Another solution: $f(x) = \ln\dfrac{1}{x} = \ln 1 - \ln x = -\ln x \;\Rightarrow\; f'(x) = -\dfrac{1}{x}.$

21. $g(x) = \ln\dfrac{a-x}{a+x} = \ln(a-x) - \ln(a+x) \;\Rightarrow\;$

$g'(x) = \dfrac{1}{a-x}(-1) - \dfrac{1}{a+x} = \dfrac{-(a+x)-(a-x)}{(a-x)(a+x)} = \dfrac{-2a}{a^2-x^2}$

23. $G(y) = \ln\dfrac{(2y+1)^5}{\sqrt{y^2+1}} = \ln(2y+1)^5 - \ln(y^2+1)^{1/2} = 5\ln(2y+1) - \tfrac{1}{2}\ln(y^2+1) \;\Rightarrow\;$

$G'(y) = 5\cdot\dfrac{1}{2y+1}\cdot 2 - \dfrac{1}{2}\cdot\dfrac{1}{y^2+1}\cdot 2y = \dfrac{10}{2y+1} - \dfrac{y}{y^2+1} \;\left[\text{or } \dfrac{8y^2-y+10}{(2y+1)(y^2+1)}\right]$

25. $g(x) = \ln\left(x\sqrt{x^2-1}\right) = \ln x + \ln(x^2-1)^{1/2} = \ln x + \tfrac{1}{2}\ln(x^2-1) \;\Rightarrow\;$

$g'(x) = \dfrac{1}{x} + \dfrac{1}{2}\cdot\dfrac{1}{x^2-1}\cdot 2x = \dfrac{1}{x} + \dfrac{x}{x^2-1} = \dfrac{x^2-1+x\cdot x}{x(x^2-1)} = \dfrac{2x^2-1}{x(x^2-1)}$

27. $f(u) = \dfrac{\ln u}{1+\ln(2u)} \;\Rightarrow\;$

$f'(u) = \dfrac{[1+\ln(2u)]\cdot\frac{1}{u} - \ln u\cdot\frac{1}{2u}\cdot 2}{[1+\ln(2u)]^2} = \dfrac{\frac{1}{u}[1+\ln(2u)-\ln u]}{[1+\ln(2u)]^2} = \dfrac{1+(\ln 2+\ln u)-\ln u}{u[1+\ln(2u)]^2} = \dfrac{1+\ln 2}{u[1+\ln(2u)]^2}$

29. $y = \ln\left|2-x-5x^2\right| \;\Rightarrow\; y' = \dfrac{1}{2-x-5x^2}\cdot(-1-10x) = \dfrac{-10x-1}{2-x-5x^2} \text{ or } \dfrac{10x+1}{5x^2+x-2}$

31. $y = \tan\left[\ln(ax+b)\right] \;\Rightarrow\; y' = \sec^2[\ln(ax+b)]\cdot\dfrac{1}{ax+b}\cdot a = \sec^2[\ln(ax+b)]\dfrac{a}{ax+b}$

33. $y = x^2\ln(2x) \;\Rightarrow\; y' = x^2\cdot\dfrac{1}{2x}\cdot 2 + \ln(2x)\cdot(2x) = x + 2x\ln(2x) \;\Rightarrow\;$

$y'' = 1 + 2x\cdot\dfrac{1}{2x}\cdot 2 + \ln(2x)\cdot 2 = 1 + 2 + 2\ln(2x) = 3 + 2\ln(2x)$

35. $f(x) = \dfrac{x}{1-\ln(x-1)} \;\Rightarrow\;$

$f'(x) = \dfrac{[1-\ln(x-1)]\cdot 1 - x\cdot\frac{-1}{x-1}}{[1-\ln(x-1)]^2} = \dfrac{\frac{(x-1)[1-\ln(x-1)]+x}{x-1}}{[1-\ln(x-1)]^2} = \dfrac{x-1-(x-1)\ln(x-1)+x}{(x-1)[1-\ln(x-1)]^2}$

$= \dfrac{2x-1-(x-1)\ln(x-1)}{(x-1)[1-\ln(x-1)]^2}$

$\text{Dom}(f) = \{x \mid x-1 > 0 \;\text{ and }\; 1-\ln(x-1) \neq 0\} = \{x \mid x > 1 \;\text{ and }\; \ln(x-1) \neq 1\}$

$\qquad = \{x \mid x > 1 \;\text{ and }\; x-1 \neq e^1\} = \{x \mid x > 1 \;\text{ and }\; x \neq 1+e\} = (1, 1+e)\cup(1+e, \infty)$

37. $f(x) = \dfrac{\ln x}{1 + x^2}$ $\Rightarrow$ $f'(x) = \dfrac{(1 + x^2)\left(\dfrac{1}{x}\right) - (\ln x)(2x)}{(1 + x^2)^2}$, so $f'(1) = \dfrac{2(1) - 0(2)}{2^2} = \dfrac{2}{4} = \dfrac{1}{2}$.

39. $y = \sin(2\ln x)$ $\Rightarrow$ $y' = \cos(2\ln x) \cdot \dfrac{2}{x}$. At $(1, 0)$, $y' = \cos 0 \cdot \dfrac{2}{1} = 2$, so an equation of the tangent line is

$y - 0 = 2 \cdot (x - 1)$, or $y = 2x - 2$.

41. $y = \ln(x^2 + y^2)$ $\Rightarrow$ $y' = \dfrac{1}{x^2 + y^2} \dfrac{d}{dx}(x^2 + y^2)$ $\Rightarrow$ $y' = \dfrac{2x + 2yy'}{x^2 + y^2}$ $\Rightarrow$ $x^2 y' + y^2 y' = 2x + 2yy'$ $\Rightarrow$

$x^2 y' + y^2 y' - 2yy' = 2x$ $\Rightarrow$ $(x^2 + y^2 - 2y)y' = 2x$ $\Rightarrow$ $y' = \dfrac{2x}{x^2 + y^2 - 2y}$

43. $f(x) = \ln(x - 1)$ $\Rightarrow$ $f'(x) = \dfrac{1}{(x - 1)} = (x - 1)^{-1}$ $\Rightarrow$ $f''(x) = -(x - 1)^{-2}$ $\Rightarrow$ $f'''(x) = 2(x - 1)^{-3}$ $\Rightarrow$

$f^{(4)}(x) = -2 \cdot 3(x - 1)^{-4}$ $\Rightarrow$ $\cdots$ $\Rightarrow$ $f^{(n)}(x) = (-1)^{n-1} \cdot 2 \cdot 3 \cdot 4 \cdot \cdots \cdot (n - 1)(x - 1)^{-n} = (-1)^{n-1} \dfrac{(n - 1)!}{(x - 1)^n}$

45. $y = f(x) = \ln(\sin x)$

A. $D = \{x \text{ in } \mathbb{R} \mid \sin x > 0\} = \displaystyle\bigcup_{n=-\infty}^{\infty} (2n\pi, (2n + 1)\pi) = \cdots \cup (-4\pi, -3\pi) \cup (-2\pi, -\pi) \cup (0, \pi) \cup (2\pi, 3\pi) \cup \cdots$

B. No y-intercept; x-intercepts: $f(x) = 0$ $\Leftrightarrow$ $\ln(\sin x) = 0$ $\Leftrightarrow$ $\sin x = e^0 = 1$ $\Leftrightarrow$ $x = 2n\pi + \dfrac{\pi}{2}$ for each

integer n. **C.** f is periodic with period 2π. **D.** $\displaystyle\lim_{x \to (2n\pi)^+} f(x) = -\infty$ and $\displaystyle\lim_{x \to [(2n+1)\pi]^-} f(x) = -\infty$, so the lines

$x = n\pi$ are VAs for all integers n. **E.** $f'(x) = \dfrac{\cos x}{\sin x} = \cot x$, so $f'(x) > 0$ when $2n\pi < x < 2n\pi + \dfrac{\pi}{2}$ for each

integer n, and $f'(x) < 0$ when $2n\pi + \dfrac{\pi}{2} < x < (2n + 1)\pi$. Thus, f is increasing on $\left(2n\pi, 2n\pi + \dfrac{\pi}{2}\right)$ and

decreasing on $\left(2n\pi + \dfrac{\pi}{2}, (2n + 1)\pi\right)$ for each integer n.   **H.**

F. Local maximum values $f\left(2n\pi + \dfrac{\pi}{2}\right) = 0$, no local minimum.

G. $f''(x) = -\csc^2 x < 0$, so f is CD on $(2n\pi, (2n + 1)\pi)$ for

each integer n. No IP

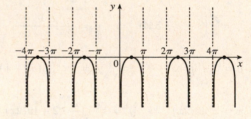

47. $y = f(x) = \ln(1 + x^2)$ **A.** $D = \mathbb{R}$ **B.** Both intercepts are 0.   **C.** $f(-x) = f(x)$, so the curve is symmetric about the

y-axis.   **D.** $\displaystyle\lim_{x \to \pm\infty} \ln(1 + x^2) = \infty$, no asymptotes.   **E.** $f'(x) = \dfrac{2x}{1 + x^2} > 0$ $\Leftrightarrow$

$x > 0$, so f is increasing on $(0, \infty)$ and decreasing on $(-\infty, 0)$.   **H.**

F. $f(0) = 0$ is a local and absolute minimum.

G. $f''(x) = \dfrac{2(1 + x^2) - 2x(2x)}{(1 + x^2)^2} = \dfrac{2(1 - x^2)}{(1 + x^2)^2} > 0$ $\Leftrightarrow$

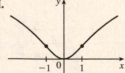

$|x| < 1$, so f is CU on $(-1, 1)$, CD on $(-\infty, -1)$ and $(1, \infty)$.

IP $(1, \ln 2)$ and $(-1, \ln 2)$.

49. We use the CAS to calculate $f'(x) = \dfrac{2 + \sin x + x \cos x}{2x + x \sin x}$ and

$f''(x) = \dfrac{2x^2 \sin x + 4 \sin x - \cos^2 x + x^2 + 5}{x^2 (\cos^2 x - 4 \sin x - 5)}$. From the graphs, it

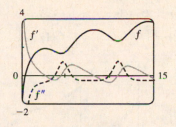

seems that $f' > 0$ (and so f is increasing) on approximately the intervals

$(0, 2.7)$, $(4.5, 8.2)$ and $(10.9, 14.3)$. It seems that f'' changes sign

(indicating inflection points) at $x \approx 3.8$, 5.7, 10.0 and 12.0.

Looking back at the graph of $f(x) = \ln(2x + x \sin x)$, this implies that the inflection points have approximate coordinates

$(3.8, 1.7)$, $(5.7, 2.1)$, $(10.0, 2.7)$, and $(12.0, 2.9)$.

51. $y = (x^2 + 2)^2 (x^4 + 4)^4 \;\Rightarrow\; \ln y = \ln[(x^2 + 2)^2 (x^4 + 4)^4] \;\Rightarrow\; \ln y = 2 \ln(x^2 + 2) + 4 \ln(x^4 + 4) \;\Rightarrow$

$\dfrac{1}{y} y' = 2 \cdot \dfrac{1}{x^2 + 2} \cdot 2x + 4 \cdot \dfrac{1}{x^4 + 4} \cdot 4x^3 \;\Rightarrow\; y' = y \left(\dfrac{4x}{x^2 + 2} + \dfrac{16x^3}{x^4 + 4} \right) \;\Rightarrow$

$y' = (x^2 + 2)^2 (x^4 + 4)^4 \left(\dfrac{4x}{x^2 + 2} + \dfrac{16x^3}{x^4 + 4} \right)$

53. $y = \sqrt{\dfrac{x - 1}{x^4 + 1}} \;\Rightarrow\; \ln y = \ln \left(\dfrac{x - 1}{x^4 + 1} \right)^{1/2} \;\Rightarrow\; \ln y = \dfrac{1}{2} \ln(x - 1) - \dfrac{1}{2} \ln(x^4 + 1) \;\Rightarrow$

$\dfrac{1}{y} y' = \dfrac{1}{2} \dfrac{1}{x - 1} - \dfrac{1}{2} \dfrac{1}{x^4 + 1} \cdot 4x^3 \;\Rightarrow\; y' = y \left(\dfrac{1}{2(x - 1)} - \dfrac{2x^3}{x^4 + 1} \right) \;\Rightarrow\; y' = \sqrt{\dfrac{x - 1}{x^4 + 1}} \left(\dfrac{1}{2x - 2} - \dfrac{2x^3}{x^4 + 1} \right)$

55. $\displaystyle\int_1^2 \dfrac{dt}{8 - 3t} = \left[-\dfrac{1}{3} \ln |8 - 3t| \right]_1^2 = -\dfrac{1}{3} \ln 2 - \left(-\dfrac{1}{3} \ln 5 \right) = \dfrac{1}{3}(\ln 5 - \ln 2) = \dfrac{1}{3} \ln \dfrac{5}{2}$

Or: Let $u = 8 - 3t$. Then $du = -3\, dt$, so

$\displaystyle\int_1^2 \dfrac{dt}{8 - 3t} = \int_5^2 \dfrac{-\frac{1}{3}\, du}{u} = \left[-\dfrac{1}{3} \ln |u| \right]_5^2 = -\dfrac{1}{3} \ln 2 - \left(-\dfrac{1}{3} \ln 5 \right) = \dfrac{1}{3}(\ln 5 - \ln 2) = \dfrac{1}{3} \ln \dfrac{5}{2}$.

57. $\displaystyle\int_1^e \dfrac{x^2 + x + 1}{x}\, dx = \int_1^e \left(x + 1 + \dfrac{1}{x} \right) dx = \left[\tfrac{1}{2} x^2 + x + \ln x \right]_1^e = \left(\tfrac{1}{2} e^2 + e + 1 \right) - \left(\tfrac{1}{2} + 1 + 0 \right)$

$= \tfrac{1}{2} e^2 + e - \tfrac{1}{2}$

59. Let $u = \ln x$. Then $du = \dfrac{dx}{x} \;\Rightarrow\; \displaystyle\int \dfrac{(\ln x)^2}{x}\, dx = \int u^2\, du = \dfrac{1}{3} u^3 + C = \dfrac{1}{3} (\ln x)^3 + C$.

61. $\displaystyle\int \dfrac{\sin 2x}{1 + \cos^2 x}\, dx = 2 \int \dfrac{\sin x \cos x}{1 + \cos^2 x}\, dx = 2I$. Let $u = \cos x$. Then $du = -\sin x\, dx$, so

$2I = -2 \displaystyle\int \dfrac{u\, du}{1 + u^2} = -2 \cdot \tfrac{1}{2} \ln(1 + u^2) + C = -\ln(1 + u^2) + C = -\ln(1 + \cos^2 x) + C$.

Or: Let $u = 1 + \cos^2 x$.

63. (a) $\dfrac{d}{dx} (\ln |\sin x| + C) = \dfrac{1}{\sin x} \cos x = \cot x$

(b) Let $u = \sin x$. Then $du = \cos x\, dx$, so $\displaystyle\int \cot x\, dx = \int \dfrac{\cos x}{\sin x}\, dx = \int \dfrac{du}{u} = \ln |u| + C = \ln |\sin x| + C$.

65. $f(x) = 2x + \ln x \Rightarrow f'(x) = 2 + 1/x$. If $g = f^{-1}$, then $f(1) = 2 \Rightarrow g(2) = 1$, so
$g'(2) = 1/f'(g(2)) = 1/f'(1) = \frac{1}{3}$.

67. (a)

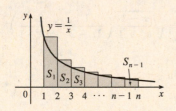

We interpret $\ln 1.5$ as the area under the curve $y = 1/x$ from $x = 1$ to $x = 1.5$. The area of the rectangle $BCDE$ is $\frac{1}{2} \cdot \frac{2}{3} = \frac{1}{3}$. The area of the trapezoid $ABCD$ is $\frac{1}{2} \cdot \frac{1}{2}\left(1 + \frac{2}{3}\right) = \frac{5}{12}$. Thus, by comparing areas, we observe that $\frac{1}{3} < \ln 1.5 < \frac{5}{12}$.

(b) With $f(t) = 1/t$, $n = 10$, and $\Delta t = 0.05$, we have

$$\ln 1.5 = \int_1^{1.5} (1/t)\, dt \approx (0.05)[f(1.025) + f(1.075) + \cdots + f(1.475)]$$

$$= (0.05)\left[\tfrac{1}{1.025} + \tfrac{1}{1.075} + \cdots + \tfrac{1}{1.475}\right] \approx 0.4054$$

69.

The area of R_i is $\dfrac{1}{i+1}$ and so $\dfrac{1}{2} + \dfrac{1}{3} + \cdots + \dfrac{1}{n} < \displaystyle\int_1^n \dfrac{1}{t}\, dt = \ln n$.

The area of S_i is $\dfrac{1}{i}$ and so $1 + \dfrac{1}{2} + \cdots + \dfrac{1}{n-1} > \displaystyle\int_1^n \dfrac{1}{t}\, dt = \ln n$.

71. The curve and the line will determine a region when they intersect at two or more points. So we solve the equation $x/(x^2 + 1) = mx \Rightarrow x = 0$ or

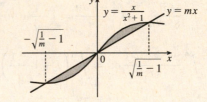

$mx^2 + m - 1 = 0 \Rightarrow x = 0$ or $x = \dfrac{\pm\sqrt{-4(m)(m-1)}}{2m} = \pm\sqrt{\dfrac{1}{m} - 1}$.

Note that if $m = 1$, this has only the solution $x = 0$, and no region is

determined. But if $1/m - 1 > 0 \Leftrightarrow 1/m > 1 \Leftrightarrow 0 < m < 1$, then there are two solutions. [Another way of seeing this

is to observe that the slope of the tangent to $y = x/(x^2 + 1)$ at the origin is $y' = 1$ and therefore we must have $0 < m < 1$.]

Note that we cannot just integrate between the positive and negative roots, since the curve and the line cross at the origin.

Since mx and $x/(x^2 + 1)$ are both odd functions, the total area is twice the area between the curves on the interval

$\left[0, \sqrt{1/m - 1}\right]$. So the total area enclosed is

$$2\int_0^{\sqrt{1/m-1}} \left[\frac{x}{x^2+1} - mx\right] dx = 2\left[\tfrac{1}{2}\ln(x^2+1) - \tfrac{1}{2}mx^2\right]_0^{\sqrt{1/m-1}}$$

$$= \left[\ln\left(\frac{1}{m} - 1 + 1\right) - m\left(\frac{1}{m} - 1\right)\right] - (\ln 1 - 0)$$

$$= \ln\left(\frac{1}{m}\right) + m - 1 = m - \ln m - 1$$

73. If $f(x) = \ln(1+x)$, then $f'(x) = \dfrac{1}{1+x}$, so $f'(0) = 1$.

Thus, $\displaystyle\lim_{x\to 0}\frac{\ln(1+x)}{x} = \lim_{x\to 0}\frac{f(x)}{x} = \lim_{x\to 0}\frac{f(x)-f(0)}{x-0} = f'(0) = 1$.

5.3   The Natural Exponential Function

1. (a) The number e is the value of a such that the slope of the tangent line at $x = 0$ on the graph of $y = a^x$ is exactly 1.

(b) $e \approx 2.71828$

(c)

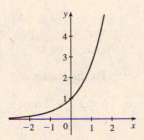

The function value at $x = 0$ is 1 and the slope at $x = 0$ is 1.

3. (a) $e^{-2\ln 5} = \left(e^{\ln 5}\right)^{-2} \overset{(4)}{=} 5^{-2} = \dfrac{1}{5^2} = \dfrac{1}{25}$ (b) $\ln\!\left(\ln e^{e^{10}}\right) \overset{(5)}{=} \ln(e^{10}) \overset{(5)}{=} 10$

5. (a) $e^{7-4x} = 6 \quad\Leftrightarrow\quad 7 - 4x = \ln 6 \quad\Leftrightarrow\quad 7 - \ln 6 = 4x \quad\Leftrightarrow\quad x = \frac{1}{4}(7 - \ln 6)$

(b) $\ln(3x - 10) = 2 \quad\Leftrightarrow\quad 3x - 10 = e^2 \quad\Leftrightarrow\quad 3x = e^2 + 10 \quad\Leftrightarrow\quad x = \frac{1}{3}(e^2 + 10)$

7. (a) $e^{3x+1} = k \quad\Leftrightarrow\quad 3x + 1 = \ln k \quad\Leftrightarrow\quad x = \frac{1}{3}(\ln k - 1)$

(b) $\ln x + \ln(x - 1) = \ln(x(x-1)) = 1 \quad\Leftrightarrow\quad x(x-1) = e^1 \quad\Leftrightarrow\quad x^2 - x - e = 0$. The quadratic formula (with $a = 1$,

$b = -1$, and $c = -e$) gives $x = \frac{1}{2}\left(1 \pm \sqrt{1 + 4e}\right)$, but we reject the negative root since the natural logarithm is not

defined for $x < 0$. So $x = \frac{1}{2}\left(1 + \sqrt{1 + 4e}\right)$.

9. (a) $\ln x < 0 \quad\Rightarrow\quad x < e^0 \quad\Rightarrow\quad x < 1$. Since the domain of $f(x) = \ln x$ is $x > 0$, the solution of the original inequality

is $0 < x < 1$.

(b) $e^x > 5 \quad\Rightarrow\quad \ln e^x > \ln 5 \quad\Rightarrow\quad x > \ln 5$

11.

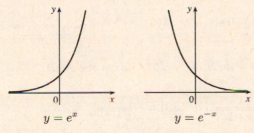

$y = e^x$ $y = e^{-x}$

13. We start with the graph of $y = e^x$ (Figure 2) and reflect about the y-axis to get the graph of $y = e^{-x}$. Then we compress

the graph vertically by a factor of 2 to obtain the graph of $y = \frac{1}{2}e^{-x}$ and then reflect about the x-axis to get the graph of

$y = -\frac{1}{2}e^{-x}$. Finally, we shift the graph upward one unit to get the graph of $y = 1 - \frac{1}{2}e^{-x}$.

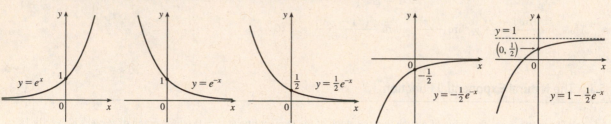

15. (a) For $f(x) = \sqrt{3 - e^{2x}}$, we must have $3 - e^{2x} \geq 0 \;\Rightarrow\; e^{2x} \leq 3 \;\Rightarrow\; 2x \leq \ln 3 \;\Rightarrow\; x \leq \frac{1}{2}\ln 3$.

Thus, the domain of f is $(-\infty, \frac{1}{2}\ln 3]$.

(b) $y = f(x) = \sqrt{3 - e^{2x}}$ [note that $y \geq 0$] $\;\Rightarrow\; y^2 = 3 - e^{2x} \;\Rightarrow\; e^{2x} = 3 - y^2 \;\Rightarrow\; 2x = \ln(3 - y^2) \;\Rightarrow\;$

$x = \frac{1}{2}\ln(3 - y^2)$. Interchange x and y: $y = \frac{1}{2}\ln(3 - x^2)$. So $f^{-1}(x) = \frac{1}{2}\ln(3 - x^2)$. For the domain of f^{-1},

we must have $3 - x^2 > 0 \;\Rightarrow\; x^2 < 3 \;\Rightarrow\; |x| < \sqrt{3} \;\Rightarrow\; -\sqrt{3} < x < \sqrt{3} \;\Rightarrow\; 0 \leq x < \sqrt{3}$ since $x \geq 0$. Note

that the domain of f^{-1}, $[0, \sqrt{3})$, equals the range of f.

17. Divide numerator and denominator by e^{3x}: $\displaystyle\lim_{x \to \infty} \frac{e^{3x} - e^{-3x}}{e^{3x} + e^{-3x}} = \lim_{x \to \infty} \frac{1 - e^{-6x}}{1 + e^{-6x}} = \frac{1 - 0}{1 + 0} = 1$

19. Let $t = 3/(2 - x)$. As $x \to 2^+$, $t \to -\infty$. So $\displaystyle\lim_{x \to 2^+} e^{3/(2-x)} = \lim_{t \to -\infty} e^t = 0$ by (6).

21. Since $-1 \leq \cos x \leq 1$ and $e^{-2x} > 0$, we have $-e^{-2x} \leq e^{-2x}\cos x \leq e^{-2x}$. We know that $\displaystyle\lim_{x \to \infty}(-e^{-2x}) = 0$ and

$\displaystyle\lim_{x \to \infty}(e^{-2x}) = 0$, so by the Squeeze Theorem, $\displaystyle\lim_{x \to \infty}(e^{-2x}\cos x) = 0$.

23. By the Product Rule, $f(x) = (x^3 + 2x)e^x \;\Rightarrow\;$

$$f'(x) = (x^3 + 2x)(e^x)' + e^x(x^3 + 2x)' = (x^3 + 2x)e^x + e^x(3x^2 + 2)$$
$$= e^x[(x^3 + 2x) + (3x^2 + 2)] = e^x(x^3 + 3x^2 + 2x + 2)$$

25. By (9), $y = e^{ax^3} \;\Rightarrow\; y' = e^{ax^3}\dfrac{d}{dx}(ax^3) = 3ax^2 e^{ax^3}$.

27. $f(u) = e^{1/u} \;\Rightarrow\; f'(u) = e^{1/u}\cdot\dfrac{d}{du}\left(\dfrac{1}{u}\right) = e^{1/u}\left(\dfrac{-1}{u^2}\right) = \left(\dfrac{-1}{u^2}\right)e^{1/u}$

29. $F(t) = e^{t\sin 2t} \;\Rightarrow\; F'(t) = e^{t\sin 2t}(t\sin 2t)' = e^{t\sin 2t}(t\cdot 2\cos 2t + \sin 2t\cdot 1) = e^{t\sin 2t}(2t\cos 2t + \sin 2t)$

31. $y = \sqrt{1 + 2e^{3x}} \;\Rightarrow\; y' = \dfrac{1}{2}(1 + 2e^{3x})^{-1/2}\dfrac{d}{dx}(1 + 2e^{3x}) = \dfrac{1}{2\sqrt{1 + 2e^{3x}}}(2e^{3x}\cdot 3) = \dfrac{3e^{3x}}{\sqrt{1 + 2e^{3x}}}$

33. $y = e^{e^x} \;\Rightarrow\; y' = e^{e^x}\cdot\dfrac{d}{dx}(e^x) = e^{e^x}\cdot e^x$ or $e^{e^x + x}$

35. By the Quotient Rule, $y = \dfrac{ae^x + b}{ce^x + d}$ $\Rightarrow$

$$y' = \frac{(ce^x + d)(ae^x) - (ae^x + b)(ce^x)}{(ce^x + d)^2} = \frac{(ace^x + ad - ace^x - bc)e^x}{(ce^x + d)^2} = \frac{(ad - bc)e^x}{(ce^x + d)^2}.$$

37. $y = e^{2x} \cos \pi x$ $\Rightarrow$ $y' = e^{2x}(-\pi \sin \pi x) + (\cos \pi x)(2e^{2x}) = e^{2x}(2 \cos \pi x - \pi \sin \pi x).$

At $(0, 1)$, $y' = 1(2 - 0) = 2$, so an equation of the tangent line is $y - 1 = 2(x - 0)$, or $y = 2x + 1.$

39. $\dfrac{d}{dx}(e^{x/y}) = \dfrac{d}{dx}(x - y)$ $\Rightarrow$ $e^{x/y} \cdot \dfrac{d}{dx}\left(\dfrac{x}{y}\right) = 1 - y'$ $\Rightarrow$

$$e^{x/y} \cdot \frac{y \cdot 1 - x \cdot y'}{y^2} = 1 - y' \ \Rightarrow \ e^{x/y} \cdot \frac{1}{y} - \frac{xe^{x/y}}{y^2} \cdot y' = 1 - y' \ \Rightarrow \ y' - \frac{xe^{x/y}}{y^2} \cdot y' = 1 - \frac{e^{x/y}}{y} \ \Rightarrow$$

$$y'\left(1 - \frac{xe^{x/y}}{y^2}\right) = \frac{y - e^{x/y}}{y} \ \Rightarrow \ y' = \frac{\dfrac{y - e^{x/y}}{y}}{\dfrac{y^2 - xe^{x/y}}{y^2}} = \frac{y(y - e^{x/y})}{y^2 - xe^{x/y}}$$

41. $y = e^{rx}$ $\Rightarrow$ $y' = re^{rx}$ $\Rightarrow$ $y'' = r^2 e^{rx}$, so if $y = e^{rx}$ satisfies the differential equation $y'' + 6y' + 8y = 0$,

then $r^2 e^{rx} + 6re^{rx} + 8e^{rx} = 0$; that is, $e^{rx}(r^2 + 6r + 8) = 0$. Since $e^{rx} > 0$ for all x, we must have $r^2 + 6r + 8 = 0$,

or $(r + 2)(r + 4) = 0$, so $r = -2$ or -4.

43. $f(x) = e^{2x}$ $\Rightarrow$ $f'(x) = 2e^{2x}$ $\Rightarrow$ $f''(x) = 2 \cdot 2e^{2x} = 2^2 e^{2x}$ $\Rightarrow$

$f'''(x) = 2^2 \cdot 2e^{2x} = 2^3 e^{2x}$ $\Rightarrow$ $\cdots$ $\Rightarrow$ $f^{(n)}(x) = 2^n e^{2x}$

45. (a) $f(x) = e^x + x$ is continuous on $\mathbb{R}$ and $f(-1) = e^{-1} - 1 < 0 < 1 = f(0)$, so by the Intermediate Value Theorem,

$e^x + x = 0$ has a root in $(-1, 0)$.

(b) $f(x) = e^x + x$ $\Rightarrow$ $f'(x) = e^x + 1$, so $x_{n+1} = x_n - \dfrac{e^{x_n} + x_n}{e^{x_n} + 1}$. Using $x_1 = -0.5$, we get $x_2 \approx -0.566311$,

$x_3 \approx -0.567143 \approx x_4$, so the root is -0.567143 to six decimal places.

47. (a) $\displaystyle\lim_{t \to \infty} p(t) = \lim_{t \to \infty} \frac{1}{1 + ae^{-kt}} = \frac{1}{1 + a \cdot 0} = 1$, since $k > 0$ $\Rightarrow$ $-kt \to -\infty$ $\Rightarrow$ $e^{-kt} \to 0.$

(b) $p(t) = (1 + ae^{-kt})^{-1}$ $\Rightarrow$ $\dfrac{dp}{dt} = -(1 + ae^{-kt})^{-2}(-kae^{-kt}) = \dfrac{kae^{-kt}}{(1 + ae^{-kt})^2}$

(c)

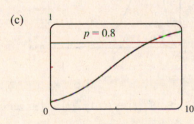

From the graph of $p(t) = (1 + 10e^{-0.5t})^{-1}$, it seems that $p(t) = 0.8$

(indicating that 80% of the population has heard the rumor) when

$t \approx 7.4$ hours.

49. $f(x) = x - e^x$ $\Rightarrow$ $f'(x) = 1 - e^x = 0$ $\Leftrightarrow$ $e^x = 1$ $\Leftrightarrow$ $x = 0$. Now $f'(x) > 0$ for all $x < 0$ and $f'(x) < 0$ for all

$x > 0$, so the absolute maximum value is $f(0) = 0 - 1 = -1.$

51. (a) $f(x) = (1-x)e^{-x}$ $\Rightarrow$ $f'(x) = (1-x)(-e^{-x}) + e^{-x}(-1) = e^{-x}(x-2) > 0$ $\Rightarrow$ $x > 2$, so f is increasing on $(2, \infty)$ and decreasing on $(-\infty, 2)$.

(b) $f''(x) = e^{-x}(1) + (x-2)(-e^{-x}) = e^{-x}(3-x) > 0$ $\Leftrightarrow$ $x < 3$, so f is CU on $(-\infty, 3)$ and CD on $(3, \infty)$.

(c) f'' changes sign at $x = 3$, so there is an IP at $(3, -2e^{-3})$.

53. $y = 1/(1 + e^{-x})$ **A.** $D = \mathbb{R}$ **B.** No x-intercept; y-intercept $= f(0) = \frac{1}{2}$. **C.** No symmetry

D. $\lim\limits_{x \to \infty} 1/(1 + e^{-x}) = \frac{1}{1+0} = 1$ and $\lim\limits_{x \to -\infty} 1/(1 + e^{-x}) = 0$ since $\lim\limits_{x \to -\infty} e^{-x} = \infty$, so f has horizontal asymptotes

$y = 0$ and $y = 1$. **E.** $f'(x) = -(1 + e^{-x})^{-2}(-e^{-x}) = e^{-x}/(1 + e^{-x})^2$. This is positive for all x, so f is increasing on $\mathbb{R}$.

F. No extreme values **G.** $f''(x) = \dfrac{(1 + e^{-x})^2(-e^{-x}) - e^{-x}(2)(1 + e^{-x})(-e^{-x})}{(1 + e^{-x})^4} = \dfrac{e^{-x}(e^{-x} - 1)}{(1 + e^{-x})^3}$

The second factor in the numerator is negative for $x > 0$ and positive for $x < 0$, **H.**

and the other factors are always positive, so f is CU on $(-\infty, 0)$ and CD

on $(0, \infty)$. IP at $\left(0, \frac{1}{2}\right)$

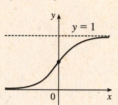

55. $y = f(x) = e^{-1/(x+1)}$ **A.** $D = \{x \mid x \ne -1\} = (-\infty, -1) \cup (-1, \infty)$ **B.** No x-intercept; y-intercept $= f(0) = e^{-1}$

C. No symmetry **D.** $\lim\limits_{x \to \pm\infty} e^{-1/(x+1)} = 1$ since $-1/(x+1) \to 0$, so $y = 1$ is a HA. $\lim\limits_{x \to -1^+} e^{-1/(x+1)} = 0$ since

$-1/(x+1) \to -\infty$, $\lim\limits_{x \to -1^-} e^{-1/(x+1)} = \infty$ since $-1/(x+1) \to \infty$, so $x = -1$ is a VA.

E. $f'(x) = e^{-1/(x+1)}/(x+1)^2$ $\Rightarrow$ $f'(x) > 0$ for all x except 1, so

f is increasing on $(-\infty, -1)$ and $(-1, \infty)$. **F.** No extreme values **H.**

G. $f''(x) = \dfrac{e^{-1/(x+1)}}{(x+1)^4} + \dfrac{e^{-1/(x+1)}(-2)}{(x+1)^3} = -\dfrac{e^{-1/(x+1)}(2x+1)}{(x+1)^4}$ $\Rightarrow$

$f''(x) > 0$ $\Leftrightarrow$ $2x + 1 < 0$ $\Leftrightarrow$ $x < -\frac{1}{2}$, so f is CU on $(-\infty, -1)$

and $\left(-1, -\frac{1}{2}\right)$, and CD on $\left(-\frac{1}{2}, \infty\right)$. f has an IP at $\left(-\frac{1}{2}, e^{-2}\right)$.

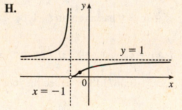

57. $S(t) = At^p e^{-kt}$ with $A = 0.01$, $p = 4$, and $k = 0.07$. We will find the

zeros of f'' for $f(t) = t^p e^{-kt}$.

$f'(t) = t^p(-ke^{-kt}) + e^{-kt}(pt^{p-1}) = e^{-kt}(-kt^p + pt^{p-1})$

$f''(t) = e^{-kt}(-kpt^{p-1} + p(p-1)t^{p-2}) + (-kt^p + pt^{p-1})(-ke^{-kt})$

$\quad = t^{p-2}e^{-kt}[-kpt + p(p-1) + k^2t^2 - kpt]$

$\quad = t^{p-2}e^{-kt}(k^2t^2 - 2kpt + p^2 - p)$

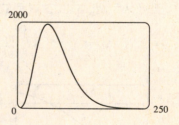

Using the given values of p and k gives us $f''(t) = t^2 e^{-0.07t}(0.0049t^2 - 0.56t + 12)$. So $S''(t) = 0.01f''(t)$ and its zeros

are $t = 0$ and the solutions of $0.0049t^2 - 0.56t + 12 = 0$, which are $t_1 = \frac{200}{7} \approx 28.57$ and $t_2 = \frac{600}{7} \approx 85.71$.

At t_1 minutes, the rate of increase of the level of medication in the bloodstream is at its greatest and at t_2 minutes, the rate of

decrease is the greatest.

59. $f(x) = e^{x^3-x} \to 0$ as $x \to -\infty$, and
$f(x) \to \infty$ as $x \to \infty$. From the graph,
it appears that f has a local minimum of
about $f(0.58) = 0.68$, and a local
maximum of about $f(-0.58) = 1.47$.

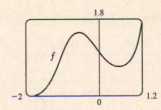 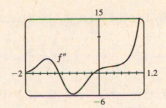

To find the exact values, we calculate

$f'(x) = (3x^2 - 1)e^{x^3-x}$, which is 0 when $3x^2 - 1 = 0$ $\Leftrightarrow$ $x = \pm\frac{1}{\sqrt{3}}$. The negative root corresponds to the local

maximum $f\left(-\frac{1}{\sqrt{3}}\right) = e^{(-1/\sqrt{3})^3 - (-1/\sqrt{3})} = e^{2\sqrt{3}/9}$, and the positive root corresponds to the local minimum

$f\left(\frac{1}{\sqrt{3}}\right) = e^{(1/\sqrt{3})^3 - (1/\sqrt{3})} = e^{-2\sqrt{3}/9}$. To estimate the inflection points, we calculate and graph

$f''(x) = \frac{d}{dx}\left[(3x^2 - 1)e^{x^3-x}\right] = (3x^2 - 1)e^{x^3-x}(3x^2 - 1) + e^{x^3-x}(6x) = e^{x^3-x}(9x^4 - 6x^2 + 6x + 1)$.

From the graph, it appears that $f''(x)$ changes sign (and thus f has inflection points) at $x \approx -0.15$ and $x \approx -1.09$. From the

graph of f, we see that these x-values correspond to inflection points at about $(-0.15, 1.15)$ and $(-1.09, 0.82)$.

61. $\displaystyle\int_0^2 \frac{dx}{e^{\pi x}} = \int_0^2 e^{-\pi x}\, dx = \left[-\frac{1}{\pi}e^{-\pi x}\right]_0^2 = -\frac{1}{\pi}e^{-2\pi} + \frac{1}{\pi}e^0 = \frac{1}{\pi}(1 - e^{-2\pi})$

63. Let $u = 1 + e^x$. Then $du = e^x\, dx$, so $\int e^x \sqrt{1 + e^x}\, dx = \int \sqrt{u}\, du = \frac{2}{3}u^{3/2} + C = \frac{2}{3}(1 + e^x)^{3/2} + C$.

65. Let $u = \tan x$. Then $du = \sec^2 x\, dx$, so $\int e^{\tan x} \sec^2 x\, dx = \int e^u\, du = e^u + C = e^{\tan x} + C$.

67. Let $u = 1/x$, so $du = -1/x^2\, dx$. When $x = 1$, $u = 1$; when $x = 2$, $u = \frac{1}{2}$. Thus,

$\displaystyle\int_1^2 \frac{e^{1/x}}{x^2}\, dx = \int_1^{1/2} e^u\, (-du) = -\left[e^u\right]_1^{1/2} = -(e^{1/2} - e) = e - \sqrt{e}$.

69. The rate is measured in liters per minute. Integrating from $t = 0$ minutes to $t = 60$ minutes will give us the total amount of oil

that leaks out (in liters) during the first hour.

$\int_0^{60} r(t)\, dt = \int_0^{60} 100 e^{-0.01t}\, dt \qquad [u = -0.01t, du = -0.01dt]$

$= 100 \int_0^{-0.6} e^u (-100\, du) = -10{,}000 \left[e^u\right]_0^{-0.6} = -10{,}000(e^{-0.6} - 1) \approx 4511.9 \approx 4512$ liters

71. We use Theorem 5.1.7. Note that $f(0) = 3 + 0 + e^0 = 4$, so $f^{-1}(4) = 0$. Also $f'(x) = 1 + e^x$. Therefore,

$\left(f^{-1}\right)'(4) = \dfrac{1}{f'(f^{-1}(4))} = \dfrac{1}{f'(0)} = \dfrac{1}{1 + e^0} = \dfrac{1}{2}$.

73. Using the second law of logarithms and Equation 5, we have $\ln(e^x/e^y) = \ln e^x - \ln e^y = x - y = \ln(e^{x-y})$. Since $\ln$ is a

one-to-one function, it follows that $e^x/e^y = e^{x-y}$.

75. (a) Let $f(x) = e^x - 1 - x$. Now $f(0) = e^0 - 1 = 0$, and for $x \geq 0$, we have $f'(x) = e^x - 1 \geq 0$. Now, since $f(0) = 0$ and

f is increasing on $[0, \infty)$, $f(x) \geq 0$ for $x \geq 0$ $\Rightarrow$ $e^x - 1 - x \geq 0$ $\Rightarrow$ $e^x \geq 1 + x$.

(b) For $0 \leq x \leq 1$, $x^2 \leq x$, so $e^{x^2} \leq e^x$ [since e^x is increasing]. Hence [from (a)] $1 + x^2 \leq e^{x^2} \leq e^x$.

So $\frac{4}{3} = \int_0^1 (1 + x^2)\, dx \leq \int_0^1 e^{x^2}\, dx \leq \int_0^1 e^x\, dx = e - 1 < e$ $\Rightarrow$ $\frac{4}{3} \leq \int_0^1 e^{x^2}\, dx \leq e$.

77. (a) By Exercise 75(a), the result holds for $n = 1$. Suppose that $e^x \geq 1 + x + \frac{x^2}{2!} + \cdots + \frac{x^k}{k!}$ for $x \geq 0$.

Let $f(x) = e^x - 1 - x - \frac{x^2}{2!} - \cdots - \frac{x^k}{k!} - \frac{x^{k+1}}{(k+1)!}$. Then $f'(x) = e^x - 1 - x - \cdots - \frac{x^k}{k!} \geq 0$ by assumption. Hence

$f(x)$ is increasing on $(0, \infty)$. So $0 \leq x$ implies that $0 = f(0) \leq f(x) = e^x - 1 - x - \cdots - \frac{x^k}{k!} - \frac{x^{k+1}}{(k+1)!}$, and hence

$e^x \geq 1 + x + \cdots + \frac{x^k}{k!} + \frac{x^{k+1}}{(k+1)!}$ for $x \geq 0$. Therefore, for $x \geq 0$, $e^x \geq 1 + x + \frac{x^2}{2!} + \cdots + \frac{x^n}{n!}$ for every positive

integer n, by mathematical induction.

(b) Taking $n = 4$ and $x = 1$ in (a), we have $e = e^1 \geq 1 + \frac{1}{2} + \frac{1}{6} + \frac{1}{24} = 2.708\overline{3} > 2.7$.

(c) $e^x \geq 1 + x + \cdots + \frac{x^k}{k!} + \frac{x^{k+1}}{(k+1)!} \quad \Rightarrow \quad \frac{e^x}{x^k} \geq \frac{1}{x^k} + \frac{1}{x^{k-1}} + \cdots + \frac{1}{k!} + \frac{x}{(k+1)!} \geq \frac{x}{(k+1)!}$.

But $\lim\limits_{x \to \infty} \frac{x}{(k+1)!} = \infty$, so $\lim\limits_{x \to \infty} \frac{e^x}{x^k} = \infty$.

5.4 General Logarithmic and Exponential Functions

1. (a) $a^x = e^{x \ln a}$

(b) The domain of $f(x) = a^x$ is $\mathbb{R}$.

(c) The range of $f(x) = a^x$ $[a \neq 1]$ is $(0, \infty)$.

(d) (i) See Figure 1.   (ii) See Figure 3.   (iii) See Figure 2.

3. Since $a^x = e^{x \ln a}$, $4^{-\pi} = e^{-\pi \ln 4}$.

5. Since $a^x = e^{x \ln a}$, $10^{x^2} = e^{x^2 \ln 10}$.

7. (a) $\log_5 125 = 3$ since $5^3 = 125$.   (b) $\log_3 \frac{1}{27} = -3$ since $3^{-3} = \frac{1}{3^3} = \frac{1}{27}$.

9. (a) $\log_2 6 - \log_2 15 + \log_2 20 = \log_2\left(\frac{6}{15}\right) + \log_2 20$ [by Law 2]

$= \log_2\left(\frac{6}{15} \cdot 20\right)$ [by Law 1]

$= \log_2 8$, and $\log_2 8 = 3$ since $2^3 = 8$.

(b) $\log_3 100 - \log_3 18 - \log_3 50 = \log_3\left(\frac{100}{18}\right) - \log_3 50 = \log_3\left(\frac{100}{18 \cdot 50}\right)$

$= \log_3\left(\frac{1}{9}\right)$, and $\log_3\left(\frac{1}{9}\right) = -2$ since $3^{-2} = \frac{1}{9}$.

11. All of these graphs approach 0 as $x \to -\infty$, all of them pass through the point

$(0, 1)$, and all of them are increasing and approach ∞ as $x \to \infty$. The larger the

base, the faster the function increases for $x > 0$, and the faster it approaches 0 as

$x \to -\infty$.

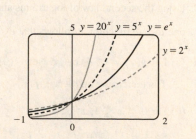

13. (a) $\log_{12} e = \dfrac{\ln e}{\ln 12} = \dfrac{1}{\ln 12} \approx 0.402430$ (b) $\log_6 13.54 = \dfrac{\ln 13.54}{\ln 6} \approx 1.454240$

(c) $\log_2 \pi = \dfrac{\ln \pi}{\ln 2} \approx 1.651496$

15. To graph these functions, we use $\log_{1.5} x = \dfrac{\ln x}{\ln 1.5}$ and $\log_{50} x = \dfrac{\ln x}{\ln 50}$.

These graphs all approach $-\infty$ as $x \to 0^+$, and they all pass through the

point $(1, 0)$. Also, they are all increasing, and all approach ∞ as $x \to \infty$.

The functions with larger bases increase extremely slowly, and the ones with

smaller bases do so somewhat more quickly. The functions with large bases

approach the y-axis more closely as $x \to 0^+$.

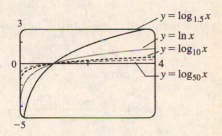

17. Use $y = Ca^x$ with the points $(1, 6)$ and $(3, 24)$. $6 = Ca^1$ $[C = \frac{6}{a}]$ and $24 = Ca^3$ $\Rightarrow$ $24 = \left(\dfrac{6}{a}\right)a^3$ $\Rightarrow$

$4 = a^2$ $\Rightarrow$ $a = 2$ [since $a > 0$] and $C = \frac{6}{2} = 3$. The function is $f(x) = 3 \cdot 2^x$.

19. (a) 2 ft = 24 in, $f(24) = 24^2$ in = 576 in = 48 ft. $g(24) = 2^{24}$ in = $2^{24}/(12 \cdot 5280)$ mi ≈ 265 mi

(b) 3 ft = 36 in, so we need x such that $\log_2 x = 36$ $\Leftrightarrow$ $x = 2^{36} = 68{,}719{,}476{,}736$. In miles, this is

$68{,}719{,}476{,}736 \text{ in} \cdot \dfrac{1 \text{ ft}}{12 \text{ in}} \cdot \dfrac{1 \text{ mi}}{5280 \text{ ft}} \approx 1{,}084{,}587.7 \text{ mi}.$

21. $\displaystyle\lim_{t \to \infty} 2^{-t^2} = \lim_{u \to -\infty} 2^u$ [where $u = -t^2$] $= 0$

23. $f(x) = x^5 + 5^x$ $\Rightarrow$ $f'(x) = 5x^4 + 5^x \ln 5$

25. $f(t) = 10^{\sqrt{t}}$ $\Rightarrow$ $f'(t) = 10^{\sqrt{t}} \ln 10 \dfrac{d}{dt}\left(\sqrt{t}\right) = \dfrac{10^{\sqrt{t}} \ln 10}{2\sqrt{t}}$

27. $L(v) = \tan\left(4^{v^2}\right)$ $\Rightarrow$ $L'(v) = \sec^2\left(4^{v^2}\right) \dfrac{d}{dv}\left(4^{v^2}\right) = \sec^2\left(4^{v^2}\right) \cdot 4^{v^2} \ln 4 \dfrac{d}{dv}\left(v^2\right) = 2v \ln 4 \sec^2\left(4^{v^2}\right) \cdot 4^{v^2}$

29. $y = 2x \log_{10} \sqrt{x} = 2x \log_{10} x^{1/2} = 2x \cdot \frac{1}{2} \log_{10} x = x \log_{10} x$ $\Rightarrow$ $y' = x \cdot \dfrac{1}{x \ln 10} + \log_{10} x \cdot 1 = \dfrac{1}{\ln 10} + \log_{10} x$

Note: $\dfrac{1}{\ln 10} = \dfrac{\ln e}{\ln 10} = \log_{10} e$, so the answer could be written as $\dfrac{1}{\ln 10} + \log_{10} x = \log_{10} e + \log_{10} x = \log_{10} ex$.

31. $y = x^x$ $\Rightarrow$ $\ln y = \ln x^x$ $\Rightarrow$ $\ln y = x \ln x$ $\Rightarrow$ $y'/y = x(1/x) + (\ln x) \cdot 1$ $\Rightarrow$ $y' = y(1 + \ln x)$ $\Rightarrow$

$y' = x^x(1 + \ln x)$

33. $y = x^{\sin x}$ $\Rightarrow$ $\ln y = \ln x^{\sin x}$ $\Rightarrow$ $\ln y = \sin x \ln x$ $\Rightarrow$ $\dfrac{y'}{y} = (\sin x) \cdot \dfrac{1}{x} + (\ln x)(\cos x)$ $\Rightarrow$

$y' = y\left(\dfrac{\sin x}{x} + \ln x \cos x\right)$ $\Rightarrow$ $y' = x^{\sin x}\left(\dfrac{\sin x}{x} + \ln x \cos x\right)$

35. $y = (\cos x)^x \;\Rightarrow\; \ln y = \ln(\cos x)^x \;\Rightarrow\; \ln y = x \ln \cos x \;\Rightarrow\; \dfrac{1}{y}\, y' = x \cdot \dfrac{1}{\cos x} \cdot (-\sin x) + \ln \cos x \cdot 1 \;\Rightarrow$

$y' = y\left(\ln \cos x - \dfrac{x \sin x}{\cos x}\right) \;\Rightarrow\; y' = (\cos x)^x (\ln \cos x - x \tan x)$

37. $y = (\tan x)^{1/x} \;\Rightarrow\; \ln y = \ln(\tan x)^{1/x} \;\Rightarrow\; \ln y = \dfrac{1}{x} \ln \tan x \;\Rightarrow$

$\dfrac{1}{y}\, y' = \dfrac{1}{x} \cdot \dfrac{1}{\tan x} \cdot \sec^2 x + \ln \tan x \cdot \left(-\dfrac{1}{x^2}\right) \;\Rightarrow\; y' = y\left(\dfrac{\sec^2 x}{x \tan x} - \dfrac{\ln \tan x}{x^2}\right) \;\Rightarrow$

$y' = (\tan x)^{1/x}\left(\dfrac{\sec^2 x}{x \tan x} - \dfrac{\ln \tan x}{x^2}\right) \quad \text{or} \quad y' = (\tan x)^{1/x} \cdot \dfrac{1}{x}\left(\csc x \sec x - \dfrac{\ln \tan x}{x}\right)$

39. $y = 10^x \;\Rightarrow\; y' = 10^x \ln 10$, so at $(1, 10)$, the slope of the tangent line is $10^1 \ln 10 = 10 \ln 10$, and its equation is

$y - 10 = 10 \ln 10(x - 1)$, or $y = (10 \ln 10)x + 10(1 - \ln 10)$.

41. $\displaystyle\int_1^2 10^t\, dt = \left[\dfrac{10^t}{\ln 10}\right]_1^2 = \dfrac{10^2}{\ln 10} - \dfrac{10^1}{\ln 10} = \dfrac{100 - 10}{\ln 10} = \dfrac{90}{\ln 10}$

43. $\displaystyle\int \dfrac{\log_{10} x}{x}\, dx = \int \dfrac{(\ln x)/(\ln 10)}{x}\, dx = \dfrac{1}{\ln 10} \int \dfrac{\ln x}{x}\, dx$. Now put $u = \ln x$, so $du = \dfrac{1}{x}\, dx$, and the expression becomes

$\dfrac{1}{\ln 10} \displaystyle\int u\, du = \dfrac{1}{\ln 10}\left(\tfrac{1}{2} u^2 + C_1\right) = \dfrac{1}{2 \ln 10}(\ln x)^2 + C.$

Or: The substitution $u = \log_{10} x$ gives $du = \dfrac{dx}{x \ln 10}$ and we get $\displaystyle\int \dfrac{\log_{10} x}{x}\, dx = \tfrac{1}{2} \ln 10 (\log_{10} x)^2 + C.$

45. Let $u = \sin\theta$. Then $du = \cos\theta\, d\theta$ and $\displaystyle\int 3^{\sin\theta} \cos\theta\, d\theta = \int 3^u\, du = \dfrac{3^u}{\ln 3} + C = \dfrac{1}{\ln 3} 3^{\sin\theta} + C.$

47. $y = \dfrac{10^x}{10^x + 1} \;\Leftrightarrow\; (10^x + 1)y = 10^x \;\Leftrightarrow\; 10^x \cdot y + y = 10^x \;\Leftrightarrow\; y = 10^x - 10^x y \;\Leftrightarrow$

$y = 10^x(1 - y) \;\Leftrightarrow\; 10^x = \dfrac{y}{1 - y} \;\Leftrightarrow\; \log_{10} 10^x = \log_{10}\left(\dfrac{y}{1 - y}\right) \;\Leftrightarrow\; x = \log_{10} y - \log_{10}(1 - y).$

Interchange x and y: $y = \log_{10} x - \log_{10}(1 - x)$ is the inverse function.

49. $\displaystyle\lim_{x \to 0^+} x^{-\ln x} = \lim_{x \to 0^+} \left(e^{\ln x}\right)^{-\ln x} = \lim_{x \to 0^+} e^{-(\ln x)^2} = 0$ since $-(\ln x)^2 \to -\infty$ as $x \to 0^+$.

51. Using Definition 1 and the second law of exponents for e^x, we have $a^{x-y} = e^{(x-y)\ln a} = e^{x \ln a - y \ln a} = \dfrac{e^{x \ln a}}{e^{y \ln a}} = \dfrac{a^x}{a^y}.$

53. Let $\log_a x = r$ and $\log_a y = s$. Then $a^r = x$ and $a^s = y$.

(a) $xy = a^r a^s = a^{r+s} \;\Rightarrow\; \log_a(xy) = r + s = \log_a x + \log_a y$

(b) $\dfrac{x}{y} = \dfrac{a^r}{a^s} = a^{r-s} \;\Rightarrow\; \log_a \dfrac{x}{y} = r - s = \log_a x - \log_a y$

(c) $x^y = (a^r)^y = a^{ry} \;\Rightarrow\; \log_a(x^y) = ry = y \log_a x$

5.5   Exponential Growth and Decay

1. The relative growth rate is $\frac{1}{P}\frac{dP}{dt} = 0.7944$, so $\frac{dP}{dt} = 0.7944P$ and, by Theorem 2, $P(t) = P(0)e^{0.7944t} = 2e^{0.7944t}$.

   Thus, $P(6) = 2e^{0.7944(6)} \approx 234.99$ or about 235 members.

3. (a) By Theorem 2, $P(t) = P(0)e^{kt} = 100e^{kt}$. Now $P(1) = 100e^{k(1)} = 420 \ \Rightarrow \ e^k = \frac{420}{100} \ \Rightarrow \ k = \ln 4.2$.

   So $P(t) = 100e^{(\ln 4.2)t} = 100(4.2)^t$.

   (b) $P(3) = 100(4.2)^3 = 7408.8 \approx 7409$ bacteria

   (c) $dP/dt = kP \ \Rightarrow \ P'(3) = k \cdot P(3) = (\ln 4.2)\big(100(4.2)^3\big)$ [from part (a)] $\approx 10{,}632$ bacteria/h

   (d) $P(t) = 100(4.2)^t = 10{,}000 \ \Rightarrow \ (4.2)^t = 100 \ \Rightarrow \ t = (\ln 100)/(\ln 4.2) \approx 3.2$ hours

5. (a) Let the population (in millions) in the year t be $P(t)$. Since the initial time is the year 1750, we substitute $t - 1750$ for t in

   Theorem 2, so the exponential model gives $P(t) = P(1750)e^{k(t-1750)}$. Then $P(1800) = 980 = 790e^{k(1800-1750)} \ \Rightarrow$

   $\frac{980}{790} = e^{k(50)} \ \Rightarrow \ \ln \frac{980}{790} = 50k \ \Rightarrow \ k = \frac{1}{50}\ln \frac{980}{790} \approx 0.0043104$. So with this model, we have

   $P(1900) = 790e^{k(1900-1750)} \approx 1508$ million, and $P(1950) = 790e^{k(1950-1750)} \approx 1871$ million. Both of these

   estimates are much too low.

   (b) In this case, the exponential model gives $P(t) = P(1850)e^{k(t-1850)} \ \Rightarrow \ P(1900) = 1650 = 1260e^{k(1900-1850)} \ \Rightarrow$

   $\ln \frac{1650}{1260} = k(50) \ \Rightarrow \ k = \frac{1}{50}\ln \frac{1650}{1260} \approx 0.005393$. So with this model, we estimate

   $P(1950) = 1260e^{k(1950-1850)} \approx 2161$ million. This is still too low, but closer than the estimate of $P(1950)$ in part (a).

   (c) The exponential model gives $P(t) = P(1900)e^{k(t-1900)} \ \Rightarrow \ P(1950) = 2560 = 1650e^{k(1950-1900)} \ \Rightarrow$

   $\ln \frac{2560}{1650} = k(50) \ \Rightarrow \ k = \frac{1}{50}\ln \frac{2560}{1650} \approx 0.008785$. With this model, we estimate

   $P(2000) = 1650e^{k(2000-1900)} \approx 3972$ million. This is much too low. The discrepancy is explained by the fact that the

   world birth rate (average yearly number of births per person) is about the same as always, whereas the mortality rate

   (especially the infant mortality rate) is much lower, owing mostly to advances in medical science and to the wars in the first

   part of the twentieth century. The exponential model assumes, among other things, that the birth and mortality rates will

   remain constant.

7. (a) If $y = [N_2O_5]$ then by Theorem 2, $\frac{dy}{dt} = -0.0005y \ \Rightarrow \ y(t) = y(0)e^{-0.0005t} = Ce^{-0.0005t}$.

   (b) $y(t) = Ce^{-0.0005t} = 0.9C \ \Rightarrow \ e^{-0.0005t} = 0.9 \ \Rightarrow \ -0.0005t = \ln 0.9 \ \Rightarrow \ t = -2000\ln 0.9 \approx 211$ s

9. (a) If $y(t)$ is the mass (in mg) remaining after t years, then $y(t) = y(0)e^{kt} = 100e^{kt}$.

   $y(30) = 100e^{30k} = \frac{1}{2}(100) \ \Rightarrow \ e^{30k} = \frac{1}{2} \ \Rightarrow \ k = -(\ln 2)/30 \ \Rightarrow \ y(t) = 100e^{-(\ln 2)t/30} = 100 \cdot 2^{-t/30}$ mg.

(b) $y(100) = 100 \cdot 2^{-100/30} \approx 9.92$ mg

(c) $100e^{-(\ln 2)t/30} = 1 \;\Rightarrow\; -(\ln 2)t/30 = \ln\frac{1}{100} \;\Rightarrow\; t = -30\,\frac{\ln 0.01}{\ln 2} \approx 199.3$ years

11. Let $y(t)$ be the level of radioactivity. Thus, $y(t) = y(0)e^{-kt}$ and k is determined by using the half-life:

$$y(5730) = \tfrac{1}{2}y(0) \;\Rightarrow\; y(0)e^{-k(5730)} = \tfrac{1}{2}y(0) \;\Rightarrow\; e^{-5730k} = \tfrac{1}{2} \;\Rightarrow\; -5730k = \ln\tfrac{1}{2} \;\Rightarrow\; k = -\frac{\ln\frac{1}{2}}{5730} = \frac{\ln 2}{5730}.$$

If 74% of the ^{14}C remains, then we know that $y(t) = 0.74y(0) \;\Rightarrow\; 0.74 = e^{-t(\ln 2)/5730} \;\Rightarrow\; \ln 0.74 = -\frac{t\ln 2}{5730} \;\Rightarrow$

$t = -\dfrac{5730(\ln 0.74)}{\ln 2} \approx 2489 \approx 2500$ years.

13. (a) Using Newton's Law of Cooling, $\dfrac{dT}{dt} = k(T - T_s)$, we have $\dfrac{dT}{dt} = k(T - 75)$. Now let $y = T - 75$, so

$y(0) = T(0) - 75 = 185 - 75 = 110$, so y is a solution of the initial-value problem $dy/dt = ky$ with $y(0) = 110$ and by

Theorem 2 we have $y(t) = y(0)e^{kt} = 110e^{kt}$.

$y(30) = 110e^{30k} = 150 - 75 \;\Rightarrow\; e^{30k} = \frac{75}{110} = \frac{15}{22} \;\Rightarrow\; k = \frac{1}{30}\ln\frac{15}{22}$, so $y(t) = 110e^{\frac{1}{30}t\ln\left(\frac{15}{22}\right)}$ and

$y(45) = 110e^{\frac{45}{30}\ln\left(\frac{15}{22}\right)} \approx 62°$F. Thus, $T(45) \approx 62 + 75 = 137°$F.

(b) $T(t) = 100 \;\Rightarrow\; y(t) = 25.$ $y(t) = 110e^{\frac{1}{30}t\ln\left(\frac{15}{22}\right)} = 25 \;\Rightarrow\; e^{\frac{1}{30}t\ln\left(\frac{15}{22}\right)} = \frac{25}{110} \;\Rightarrow\; \frac{1}{30}t\ln\frac{15}{22} = \ln\frac{25}{110} \;\Rightarrow$

$t = \dfrac{30\ln\frac{25}{110}}{\ln\frac{15}{22}} \approx 116$ min.

15. $\dfrac{dT}{dt} = k(T - 20)$. Letting $y = T - 20$, we get $\dfrac{dy}{dt} = ky$, so $y(t) = y(0)e^{kt}$. $y(0) = T(0) - 20 = 5 - 20 = -15$, so

$y(25) = y(0)e^{25k} = -15e^{25k}$, and $y(25) = T(25) - 20 = 10 - 20 = -10$, so $-15e^{25k} = -10 \;\Rightarrow\; e^{25k} = \frac{2}{3}$. Thus,

$25k = \ln\left(\frac{2}{3}\right)$ and $k = \frac{1}{25}\ln\left(\frac{2}{3}\right)$, so $y(t) = y(0)e^{kt} = -15e^{(1/25)\ln(2/3)t}$. More simply, $e^{25k} = \frac{2}{3} \;\Rightarrow\; e^{k} = \left(\frac{2}{3}\right)^{1/25} \;\Rightarrow$

$e^{kt} = \left(\frac{2}{3}\right)^{t/25} \;\Rightarrow\; y(t) = -15 \cdot \left(\frac{2}{3}\right)^{t/25}$.

(a) $T(50) = 20 + y(50) = 20 - 15 \cdot \left(\frac{2}{3}\right)^{50/25} = 20 - 15 \cdot \left(\frac{2}{3}\right)^{2} = 20 - \frac{20}{3} = 13.\overline{3}\,°$C

(b) $15 = T(t) = 20 + y(t) = 20 - 15 \cdot \left(\frac{2}{3}\right)^{t/25} \;\Rightarrow\; 15 \cdot \left(\frac{2}{3}\right)^{t/25} = 5 \;\Rightarrow\; \left(\frac{2}{3}\right)^{t/25} = \frac{1}{3} \;\Rightarrow$

$(t/25)\ln\left(\frac{2}{3}\right) = \ln\left(\frac{1}{3}\right) \;\Rightarrow\; t = 25\ln\left(\frac{1}{3}\right)/\ln\left(\frac{2}{3}\right) \approx 67.74$ min.

17. (a) Let $P(h)$ be the pressure at altitude h. Then $dP/dh = kP \;\Rightarrow\; P(h) = P(0)e^{kh} = 101.3e^{kh}$.

$P(1000) = 101.3e^{1000k} = 87.14 \;\Rightarrow\; 1000k = \ln\left(\frac{87.14}{101.3}\right) \;\Rightarrow\; k = \frac{1}{1000}\ln\left(\frac{87.14}{101.3}\right) \;\Rightarrow$

$P(h) = 101.3\,e^{\frac{1}{1000}h\ln\left(\frac{87.14}{101.3}\right)}$, so $P(3000) = 101.3e^{3\ln\left(\frac{87.14}{101.3}\right)} \approx 64.5$ kPa.

(b) $P(6187) = 101.3\,e^{\frac{6187}{1000}\ln\left(\frac{87.14}{101.3}\right)} \approx 39.9$ kPa

19. Using $A = A_0\left(1 + \dfrac{r}{n}\right)^{nt}$ with $A_0 = 3000$, $r = 0.05$, and $t = 5$, we have:

(a) Annually: $n = 1$; $\qquad A = 3000\left(1 + \frac{0.05}{1}\right)^{1 \cdot 5} = \$3828.84$

(b) Semiannually: $n = 2$; $\qquad A = 3000\left(1 + \frac{0.05}{2}\right)^{2 \cdot 5} = \$3840.25$

(c) Monthly: $n = 12$; $\qquad A = 3000\left(1 + \frac{0.05}{12}\right)^{12 \cdot 5} = \$3850.08$

(d) Weekly: $n = 52$; $\qquad A = 3000\left(1 + \frac{0.05}{52}\right)^{52 \cdot 5} = \$3851.61$

(e) Daily: $n = 365$; $\qquad A = 3000\left(1 + \frac{0.05}{365}\right)^{365 \cdot 5} = \$3852.01$

(f) Continuously: $\qquad A = 3000e^{(0.05)5} = \$3852.08$

5.6 Inverse Trigonometric Functions

1. (a) $\sin^{-1}\left(\frac{\sqrt{3}}{2}\right) = \frac{\pi}{3}$ since $\sin\frac{\pi}{3} = \frac{\sqrt{3}}{2}$ and $\frac{\pi}{3}$ is in $\left[-\frac{\pi}{2}, \frac{\pi}{2}\right]$.

(b) $\cos^{-1}(-1) = \pi$ since $\cos\pi = -1$ and π is in $[0, \pi]$.

3. (a) $\arctan 1 = \frac{\pi}{4}$ since $\tan\frac{\pi}{4} = 1$ and $\frac{\pi}{4}$ is in $\left(-\frac{\pi}{2}, \frac{\pi}{2}\right)$.

(b) $\sin^{-1}\left(\frac{1}{\sqrt{2}}\right) = \frac{\pi}{4}$ since $\sin\frac{\pi}{4} = \frac{1}{\sqrt{2}}$ and $\frac{\pi}{4}$ is in $\left[-\frac{\pi}{2}, \frac{\pi}{2}\right]$.

5. (a) In general, $\tan(\arctan x) = x$ for any real number x. Thus, $\tan(\arctan 10) = 10$.

(b) $\sin^{-1}\left(\sin\frac{7\pi}{3}\right) = \sin^{-1}\left(\sin\frac{\pi}{3}\right) = \sin^{-1}\frac{\sqrt{3}}{2} = \frac{\pi}{3}$ since $\sin\frac{\pi}{3} = \frac{\sqrt{3}}{2}$ and $\frac{\pi}{3}$ is in $\left[-\frac{\pi}{2}, \frac{\pi}{2}\right]$.

[Recall that $\frac{7\pi}{3} = \frac{\pi}{3} + 2\pi$ and the sine function is periodic with period 2π.]

7. Let $y = \sin^{-1} x$. Then $-\frac{\pi}{2} \le y \le \frac{\pi}{2}$ $\Rightarrow$ $\cos y \ge 0$, so $\cos(\sin^{-1} x) = \cos y = \sqrt{1 - \sin^2 y} = \sqrt{1 - x^2}$.

9. Let $y = \tan^{-1} x$. Then $\tan y = x$, so from the triangle we see that

$$\sin(\tan^{-1} x) = \sin y = \frac{x}{\sqrt{1 + x^2}}.$$

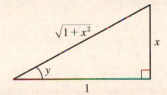

11. Let $y = \cos^{-1} x$. Then $\cos y = x$ and $0 \le y \le \pi$ $\Rightarrow$ $-\sin y \dfrac{dy}{dx} = 1$ $\Rightarrow$

$$\frac{dy}{dx} = -\frac{1}{\sin y} = -\frac{1}{\sqrt{1 - \cos^2 y}} = -\frac{1}{\sqrt{1 - x^2}}. \quad \text{[Note that } \sin y \ge 0 \text{ for } 0 \le y \le \pi.]$$

13. Let $y = \cot^{-1} x$. Then $\cot y = x$ $\Rightarrow$ $-\csc^2 y \dfrac{dy}{dx} = 1$ $\Rightarrow$ $\dfrac{dy}{dx} = -\dfrac{1}{\csc^2 y} = -\dfrac{1}{1 + \cot^2 y} = -\dfrac{1}{1 + x^2}$.

15. Let $y = \csc^{-1} x$. Then $\csc y = x$ $\Rightarrow$ $-\csc y \cot y \dfrac{dy}{dx} = 1$ $\Rightarrow$

$$\frac{dy}{dx} = -\frac{1}{\csc y \cot y} = -\frac{1}{\csc y \sqrt{\csc^2 y - 1}} = -\frac{1}{x\sqrt{x^2 - 1}}. \text{ Note that } \cot y \ge 0 \text{ on the domain of } \csc^{-1} x.$$

17. $y = (\tan^{-1} x)^2 \quad \Rightarrow \quad y' = 2(\tan^{-1} x)^1 \cdot \dfrac{d}{dx}(\tan^{-1} x) = 2\tan^{-1} x \cdot \dfrac{1}{1+x^2} = \dfrac{2\tan^{-1} x}{1+x^2}$

19. $y = \sin^{-1}(2x + 1) \quad \Rightarrow$

$$y' = \frac{1}{\sqrt{1 - (2x+1)^2}} \cdot \frac{d}{dx}(2x+1) = \frac{1}{\sqrt{1-(4x^2+4x+1)}} \cdot 2 = \frac{2}{\sqrt{-4x^2-4x}} = \frac{1}{\sqrt{-x^2-x}}$$

21. $G(x) = \sqrt{1-x^2}\arccos x \quad \Rightarrow \quad G'(x) = \sqrt{1-x^2} \cdot \dfrac{-1}{\sqrt{1-x^2}} + \arccos x \cdot \dfrac{1}{2}(1-x^2)^{-1/2}(-2x) = -1 - \dfrac{x\arccos x}{\sqrt{1-x^2}}$

23. $h(t) = \cot^{-1}(t) + \cot^{-1}(1/t) \quad \Rightarrow$

$$h'(t) = -\frac{1}{1+t^2} - \frac{1}{1+(1/t)^2} \cdot \frac{d}{dt}\frac{1}{t} = -\frac{1}{1+t^2} - \frac{t^2}{t^2+1} \cdot \left(-\frac{1}{t^2}\right) = -\frac{1}{1+t^2} + \frac{1}{t^2+1} = 0.$$

Note that this makes sense because $h(t) = \dfrac{\pi}{2}$ for $t > 0$ and $h(t) = \dfrac{3\pi}{2}$ for $t < 0$.

25. $y = \arctan(\cos\theta) \quad \Rightarrow \quad y' = \dfrac{1}{1+(\cos\theta)^2}(-\sin\theta) = -\dfrac{\sin\theta}{1+\cos^2\theta}$

27. $y = x\sin^{-1} x + \sqrt{1-x^2} \quad \Rightarrow$

$$y' = x \cdot \frac{1}{\sqrt{1-x^2}} + (\sin^{-1} x)(1) + \frac{1}{2}(1-x^2)^{-1/2}(-2x) = \frac{x}{\sqrt{1-x^2}} + \sin^{-1} x - \frac{x}{\sqrt{1-x^2}} = \sin^{-1} x$$

29. $y = \arccos\left(\dfrac{b + a\cos x}{a + b\cos x}\right) \quad \Rightarrow$

$$y' = -\frac{1}{\sqrt{1 - \left(\dfrac{b+a\cos x}{a+b\cos x}\right)^2}}\frac{(a+b\cos x)(-a\sin x) - (b+a\cos x)(-b\sin x)}{(a+b\cos x)^2}$$

$$= \frac{1}{\sqrt{a^2 + b^2\cos^2 x - b^2 - a^2\cos^2 x}}\frac{(a^2-b^2)\sin x}{|a+b\cos x|} = \frac{1}{\sqrt{a^2-b^2}\sqrt{1-\cos^2 x}}\frac{(a^2-b^2)\sin x}{|a+b\cos x|} = \frac{\sqrt{a^2-b^2}}{|a+b\cos x|}\frac{\sin x}{|\sin x|}$$

But $0 \le x \le \pi$, so $|\sin x| = \sin x$. Also $a > b > 0 \quad \Rightarrow \quad b\cos x \ge -b > -a$, so $a + b\cos x > 0$. Thus $y' = \dfrac{\sqrt{a^2-b^2}}{a+b\cos x}$.

31. $g(x) = x\sin^{-1}\left(\dfrac{x}{4}\right) + \sqrt{16-x^2} \quad \Rightarrow \quad g'(x) = \sin^{-1}\left(\dfrac{x}{4}\right) + \dfrac{x}{4\sqrt{1-(x/4)^2}} - \dfrac{x}{\sqrt{16-x^2}} = \sin^{-1}\left(\dfrac{x}{4}\right) \quad \Rightarrow$

$g'(2) = \sin^{-1}\left(\dfrac{1}{2}\right) = \dfrac{\pi}{6}$

33. Let $t = e^x$. As $x \to \infty$, $t \to \infty$. $\displaystyle\lim_{x\to\infty}\arctan(e^x) = \lim_{t\to\infty}\arctan t = \dfrac{\pi}{2}$.

35.

$\dfrac{dx}{dt} = 2$ ft/s, $\sin\theta = \dfrac{x}{10} \quad \Rightarrow \quad \theta = \sin^{-1}\left(\dfrac{x}{10}\right)$, $\dfrac{d\theta}{dx} = \dfrac{1/10}{\sqrt{1-(x/10)^2}}$,

$\dfrac{d\theta}{dt} = \dfrac{d\theta}{dx}\dfrac{dx}{dt} = \dfrac{1/10}{\sqrt{1-(x/10)^2}}(2)$ rad/s, $\dfrac{d\theta}{dt}\bigg]_{x=6} = \dfrac{2/10}{\sqrt{1-(6/10)^2}}$ rad/s $= \dfrac{1}{4}$ rad/s

37.

From the figure, $\tan \alpha = \dfrac{5}{x}$ and $\tan \beta = \dfrac{2}{3-x}$. Since

$$\alpha + \beta + \theta = 180° = \pi, \ \theta = \pi - \tan^{-1}\left(\frac{5}{x}\right) - \tan^{-1}\left(\frac{2}{3-x}\right) \ \Rightarrow$$

$$\frac{d\theta}{dx} = -\frac{1}{1+\left(\frac{5}{x}\right)^2}\left(-\frac{5}{x^2}\right) - \frac{1}{1+\left(\frac{2}{3-x}\right)^2}\left[\frac{2}{(3-x)^2}\right]$$

$$= \frac{x^2}{x^2+25}\cdot\frac{5}{x^2} - \frac{(3-x)^2}{(3-x)^2+4}\cdot\frac{2}{(3-x)^2}.$$

Now $\dfrac{d\theta}{dx} = 0 \ \Rightarrow \ \dfrac{5}{x^2+25} = \dfrac{2}{x^2-6x+13} \ \Rightarrow \ 2x^2 + 50 = 5x^2 - 30x + 65 \ \Rightarrow$

$3x^2 - 30x + 15 = 0 \ \Rightarrow \ x^2 - 10x + 5 = 0 \ \Rightarrow \ x = 5 \pm 2\sqrt{5}$. We reject the root with the $+$ sign, since it is

larger than 3. $d\theta/dx > 0$ for $x < 5 - 2\sqrt{5}$ and $d\theta/dx < 0$ for $x > 5 - 2\sqrt{5}$, so θ is maximized when

$|AP| = x = 5 - 2\sqrt{5} \approx 0.53$.

39. $\displaystyle\int_{1/\sqrt{3}}^{\sqrt{3}} \frac{8}{1+x^2}\,dx = \Big[8\arctan x\Big]_{1/\sqrt{3}}^{\sqrt{3}} = 8\left(\frac{\pi}{3} - \frac{\pi}{6}\right) = 8\left(\frac{\pi}{6}\right) = \frac{4\pi}{3}$

41. Let $u = \sin^{-1} x$, so $du = \dfrac{dx}{\sqrt{1-x^2}}$. When $x = 0$, $u = 0$; when $x = \frac{1}{2}$, $u = \frac{\pi}{6}$. Thus,

$$\int_0^{1/2} \frac{\sin^{-1} x}{\sqrt{1-x^2}}\,dx = \int_0^{\pi/6} u\,du = \left[\frac{u^2}{2}\right]_0^{\pi/6} = \frac{\pi^2}{72}.$$

43. Let $u = 1 + x^2$. Then $du = 2x\,dx$, so

$$\int \frac{1+x}{1+x^2}\,dx = \int \frac{1}{1+x^2}\,dx + \int \frac{x}{1+x^2}\,dx = \tan^{-1} x + \int \frac{\frac{1}{2}\,du}{u} = \tan^{-1} x + \frac{1}{2}\ln|u| + C$$

$$= \tan^{-1} x + \frac{1}{2}\ln\left|1+x^2\right| + C = \tan^{-1} x + \frac{1}{2}\ln(1+x^2) + C \quad [\text{since } 1 + x^2 > 0].$$

45. Let $u = t^3$. Then $du = 3t^2\,dt$ and $\displaystyle\int \frac{t^2}{\sqrt{1-t^6}}\,dt = \int \frac{\frac{1}{3}\,du}{\sqrt{1-u^2}} = \frac{1}{3}\sin^{-1} u + C = \frac{1}{3}\sin^{-1}(t^3) + C$.

47. Let $u = \sqrt{x}$. Then $du = \dfrac{dx}{2\sqrt{x}}$ and $\displaystyle\int \frac{dx}{\sqrt{x}\,(1+x)} = \int \frac{2\,du}{1+u^2} = 2\tan^{-1} u + C = 2\tan^{-1}\sqrt{x} + C$.

49. $y = \sec^{-1} x \ \Rightarrow \ \sec y = x \ \Rightarrow \ \sec y \tan y \dfrac{dy}{dx} = 1 \ \Rightarrow \ \dfrac{dy}{dx} = \dfrac{1}{\sec y \tan y}$. Now $\tan^2 y = \sec^2 y - 1 = x^2 - 1$, so

$\tan y = \pm\sqrt{x^2 - 1}$. For $y \in \left[0, \frac{\pi}{2}\right)$, $x \geq 1$, so $\sec y = x = |x|$ and $\tan y \geq 0 \ \Rightarrow \ \dfrac{dy}{dx} = \dfrac{1}{x\sqrt{x^2-1}} = \dfrac{1}{|x|\sqrt{x^2-1}}$.

For $y \in \left(\frac{\pi}{2}, \pi\right]$, $x \leq -1$, so $|x| = -x$ and $\tan y = -\sqrt{x^2-1} \ \Rightarrow$

$$\frac{dy}{dx} = \frac{1}{\sec y \tan y} = \frac{1}{x\left(-\sqrt{x^2-1}\right)} = \frac{1}{(-x)\sqrt{x^2-1}} = \frac{1}{|x|\sqrt{x^2-1}}$$

5.7 Hyperbolic Functions

1. (a) $\sinh 0 = \frac{1}{2}(e^0 - e^0) = 0$ (b) $\cosh 0 = \frac{1}{2}(e^0 + e^0) = \frac{1}{2}(1+1) = 1$

3. (a) $\sinh(\ln 2) = \dfrac{e^{\ln 2} - e^{-\ln 2}}{2} = \dfrac{e^{\ln 2} - (e^{\ln 2})^{-1}}{2} = \dfrac{2 - 2^{-1}}{2} = \dfrac{2 - \frac{1}{2}}{2} = \dfrac{3}{4}$

   (b) $\sinh 2 = \frac{1}{2}(e^2 - e^{-2}) \approx 3.62686$

5. (a) $\operatorname{sech} 0 = \dfrac{1}{\cosh 0} = \dfrac{1}{1} = 1$ (b) $\cosh^{-1} 1 = 0$ because $\cosh 0 = 1$.

7. $\sinh(-x) = \frac{1}{2}[e^{-x} - e^{-(-x)}] = \frac{1}{2}(e^{-x} - e^x) = -\frac{1}{2}(e^{-x} - e^x) = -\sinh x$

9. $\cosh x + \sinh x = \frac{1}{2}(e^x + e^{-x}) + \frac{1}{2}(e^x - e^{-x}) = \frac{1}{2}(2e^x) = e^x$

11. $\sinh x \cosh y + \cosh x \sinh y = \left[\frac{1}{2}(e^x - e^{-x})\right]\left[\frac{1}{2}(e^y + e^{-y})\right] + \left[\frac{1}{2}(e^x + e^{-x})\right]\left[\frac{1}{2}(e^y - e^{-y})\right]$

$$= \tfrac{1}{4}[(e^{x+y} + e^{x-y} - e^{-x+y} - e^{-x-y}) + (e^{x+y} - e^{x-y} + e^{-x+y} - e^{-x-y})]$$

$$= \tfrac{1}{4}(2e^{x+y} - 2e^{-x-y}) = \tfrac{1}{2}[e^{x+y} - e^{-(x+y)}] = \sinh(x+y)$$

13. Putting $y = x$ in the result from Exercise 11, we have

$$\sinh 2x = \sinh(x+x) = \sinh x \cosh x + \cosh x \sinh x = 2 \sinh x \cosh x.$$

15. By Exercise 9, $(\cosh x + \sinh x)^n = (e^x)^n = e^{nx} = \cosh nx + \sinh nx$.

17. $\operatorname{sech} x = \dfrac{1}{\cosh x} \;\Rightarrow\; \operatorname{sech} x = \dfrac{1}{5/3} = \dfrac{3}{5}$.

   $\cosh^2 x - \sinh^2 x = 1 \;\Rightarrow\; \sinh^2 x = \cosh^2 x - 1 = \left(\frac{5}{3}\right)^2 - 1 = \frac{16}{9} \;\Rightarrow\; \sinh x = \frac{4}{3}$ [because $x > 0$].

   $\operatorname{csch} x = \dfrac{1}{\sinh x} \;\Rightarrow\; \operatorname{csch} x = \dfrac{1}{4/3} = \dfrac{3}{4}$.

   $\tanh x = \dfrac{\sinh x}{\cosh x} \;\Rightarrow\; \tanh x = \dfrac{4/3}{5/3} = \dfrac{4}{5}$.

   $\coth x = \dfrac{1}{\tanh x} \;\Rightarrow\; \coth x = \dfrac{1}{4/5} = \dfrac{5}{4}$.

19. (a) $\displaystyle\lim_{x \to \infty} \tanh x = \lim_{x \to \infty} \frac{e^x - e^{-x}}{e^x + e^{-x}} \cdot \frac{e^{-x}}{e^{-x}} = \lim_{x \to \infty} \frac{1 - e^{-2x}}{1 + e^{-2x}} = \frac{1 - 0}{1 + 0} = 1$

   (b) $\displaystyle\lim_{x \to -\infty} \tanh x = \lim_{x \to -\infty} \frac{e^x - e^{-x}}{e^x + e^{-x}} \cdot \frac{e^x}{e^x} = \lim_{x \to -\infty} \frac{e^{2x} - 1}{e^{2x} + 1} = \frac{0 - 1}{0 + 1} = -1$

   (c) $\displaystyle\lim_{x \to \infty} \sinh x = \lim_{x \to \infty} \frac{e^x - e^{-x}}{2} = \infty$

   (d) $\displaystyle\lim_{x \to -\infty} \sinh x = \lim_{x \to -\infty} \frac{e^x - e^{-x}}{2} = -\infty$

   (e) $\displaystyle\lim_{x \to \infty} \operatorname{sech} x = \lim_{x \to \infty} \frac{2}{e^x + e^{-x}} = 0$

(f) $\lim\limits_{x\to\infty} \coth x = \lim\limits_{x\to\infty} \dfrac{e^x + e^{-x}}{e^x - e^{-x}} \cdot \dfrac{e^{-x}}{e^{-x}} = \lim\limits_{x\to\infty} \dfrac{1 + e^{-2x}}{1 - e^{-2x}} = \dfrac{1+0}{1-0} = 1$ [*Or:* Use part (a)]

(g) $\lim\limits_{x\to 0^+} \coth x = \lim\limits_{x\to 0^+} \dfrac{\cosh x}{\sinh x} = \infty$, since $\sinh x \to 0$ through positive values and $\cosh x \to 1$.

(h) $\lim\limits_{x\to 0^-} \coth x = \lim\limits_{x\to 0^-} \dfrac{\cosh x}{\sinh x} = -\infty$, since $\sinh x \to 0$ through negative values and $\cosh x \to 1$.

(i) $\lim\limits_{x\to -\infty} \operatorname{csch} x = \lim\limits_{x\to -\infty} \dfrac{2}{e^x - e^{-x}} = 0$

21. Let $y = \sinh^{-1} x$. Then $\sinh y = x$ and, by Example 1(a), $\cosh^2 y - \sinh^2 y = 1 \;\Rightarrow\;$ [with $\cosh y > 0$]

$\cosh y = \sqrt{1 + \sinh^2 y} = \sqrt{1 + x^2}$. So by Exercise 9, $e^y = \sinh y + \cosh y = x + \sqrt{1 + x^2} \;\Rightarrow\; y = \ln\!\left(x + \sqrt{1 + x^2}\,\right)$.

23. (a) Let $y = \tanh^{-1} x$. Then $x = \tanh y = \dfrac{\sinh y}{\cosh y} = \dfrac{(e^y - e^{-y})/2}{(e^y + e^{-y})/2} \cdot \dfrac{e^y}{e^y} = \dfrac{e^{2y} - 1}{e^{2y} + 1} \;\Rightarrow\; xe^{2y} + x = e^{2y} - 1 \;\Rightarrow\;$

$1 + x = e^{2y} - xe^{2y} \;\Rightarrow\; 1 + x = e^{2y}(1 - x) \;\Rightarrow\; e^{2y} = \dfrac{1+x}{1-x} \;\Rightarrow\; 2y = \ln\!\left(\dfrac{1+x}{1-x}\right) \;\Rightarrow\; y = \tfrac{1}{2}\ln\!\left(\dfrac{1+x}{1-x}\right)$.

(b) Let $y = \tanh^{-1} x$. Then $x = \tanh y$, so from Exercise 14 we have

$e^{2y} = \dfrac{1 + \tanh y}{1 - \tanh y} = \dfrac{1+x}{1-x} \;\Rightarrow\; 2y = \ln\!\left(\dfrac{1+x}{1-x}\right) \;\Rightarrow\; y = \tfrac{1}{2}\ln\!\left(\dfrac{1+x}{1-x}\right)$.

25. (a) Let $y = \cosh^{-1} x$. Then $\cosh y = x$ and $y \geq 0 \;\Rightarrow\; \sinh y \dfrac{dy}{dx} = 1 \;\Rightarrow\;$

$\dfrac{dy}{dx} = \dfrac{1}{\sinh y} = \dfrac{1}{\sqrt{\cosh^2 y - 1}} = \dfrac{1}{\sqrt{x^2 - 1}}$ [since $\sinh y \geq 0$ for $y \geq 0$].   *Or:* Use Formula 4.

(b) Let $y = \tanh^{-1} x$. Then $\tanh y = x \;\Rightarrow\; \operatorname{sech}^2 y \dfrac{dy}{dx} = 1 \;\Rightarrow\; \dfrac{dy}{dx} = \dfrac{1}{\operatorname{sech}^2 y} = \dfrac{1}{1 - \tanh^2 y} = \dfrac{1}{1 - x^2}$.

Or: Use Formula 5.

(c) Let $y = \operatorname{sech}^{-1} x$. Then $\operatorname{sech} y = x \;\Rightarrow\; -\operatorname{sech} y \tanh y \dfrac{dy}{dx} = 1 \;\Rightarrow\;$

$\dfrac{dy}{dx} = -\dfrac{1}{\operatorname{sech} y \tanh y} = -\dfrac{1}{\operatorname{sech} y \sqrt{1 - \operatorname{sech}^2 y}} = -\dfrac{1}{x\sqrt{1 - x^2}}$. [Note that $y > 0$ and so $\tanh y > 0$.]

27. $f(x) = x\sinh x - \cosh x \;\Rightarrow\; f'(x) = x\,(\sinh x)' + \sinh x \cdot 1 - \sinh x = x\cosh x$

29. $h(x) = \ln(\cosh x) \;\Rightarrow\; h'(x) = \dfrac{1}{\cosh x}(\cosh x)' = \dfrac{\sinh x}{\cosh x} = \tanh x$

31. $y = e^{\cosh 3x} \;\Rightarrow\; y' = e^{\cosh 3x} \cdot \sinh 3x \cdot 3 = 3e^{\cosh 3x} \sinh 3x$

33. $f(t) = \operatorname{sech}^2(e^t) = [\operatorname{sech}(e^t)]^2 \;\Rightarrow\;$

$f'(t) = 2[\operatorname{sech}(e^t)]\,[\operatorname{sech}(e^t)]' = 2\operatorname{sech}(e^t)\left[-\operatorname{sech}(e^t)\tanh(e^t) \cdot e^t\right] = -2e^t \operatorname{sech}^2(e^t)\tanh(e^t)$

35. $G(x) = \dfrac{1 - \cosh x}{1 + \cosh x} \;\;\Rightarrow$

$$G'(x) = \frac{(1 + \cosh x)(-\sinh x) - (1 - \cosh x)(\sinh x)}{(1 + \cosh x)^2} = \frac{-\sinh x - \sinh x \cosh x - \sinh x + \sinh x \cosh x}{(1 + \cosh x)^2}$$

$$= \frac{-2 \sinh x}{(1 + \cosh x)^2}$$

37. $y = \cosh^{-1}\sqrt{x} \;\;\Rightarrow\;\; y' = \dfrac{1}{\sqrt{(\sqrt{x})^2 - 1}} \dfrac{d}{dx}(\sqrt{x}) = \dfrac{1}{\sqrt{x-1}} \dfrac{1}{2\sqrt{x}} = \dfrac{1}{2\sqrt{x(x-1)}}$

39. $y = x \sinh^{-1}(x/3) - \sqrt{9 + x^2} \;\;\Rightarrow$

$$y' = \sinh^{-1}\left(\frac{x}{3}\right) + x\frac{1/3}{\sqrt{1 + (x/3)^2}} - \frac{2x}{2\sqrt{9 + x^2}} = \sinh^{-1}\left(\frac{x}{3}\right) + \frac{x}{\sqrt{9 + x^2}} - \frac{x}{\sqrt{9 + x^2}} = \sinh^{-1}\left(\frac{x}{3}\right)$$

41. $y = \coth^{-1}(\sec x) \;\;\Rightarrow$

$$y' = \frac{1}{1 - (\sec x)^2} \frac{d}{dx}(\sec x) = \frac{\sec x \tan x}{1 - \sec^2 x} = \frac{\sec x \tan x}{1 - (\tan^2 x + 1)} = \frac{\sec x \tan x}{-\tan^2 x}$$

$$= -\frac{\sec x}{\tan x} = -\frac{1/\cos x}{\sin x/\cos x} = -\frac{1}{\sin x} = -\csc x$$

43. $\dfrac{d}{dx} \arctan(\tanh x) = \dfrac{1}{1 + (\tanh x)^2} \dfrac{d}{dx}(\tanh x) = \dfrac{\operatorname{sech}^2 x}{1 + \tanh^2 x} = \dfrac{1/\cosh^2 x}{1 + (\sinh^2 x)/\cosh^2 x}$

$$= \frac{1}{\cosh^2 x + \sinh^2 x} = \frac{1}{\cosh 2x} \; [\text{by the identity for } \cosh(x+y)] \; = \operatorname{sech} 2x$$

45. As the depth d of the water gets large, the fraction $\dfrac{2\pi d}{L}$ gets large, and from Figure 3 or Exercise 19(a), $\tanh\left(\dfrac{2\pi d}{L}\right)$

approaches 1. Thus, $v = \sqrt{\dfrac{gL}{2\pi} \tanh\left(\dfrac{2\pi d}{L}\right)} \approx \sqrt{\dfrac{gL}{2\pi}(1)} = \sqrt{\dfrac{gL}{2\pi}}$.

47. (a) $y = 20\cosh(x/20) - 15 \;\;\Rightarrow\;\; y' = 20\sinh(x/20) \cdot \frac{1}{20} = \sinh(x/20)$. Since the right pole is positioned at $x = 7$,

we have $y'(7) = \sinh\frac{7}{20} \approx 0.3572$.

(b) If α is the angle between the tangent line and the x-axis, then $\tan \alpha = $ slope of the line $= \sinh\frac{7}{20}$, so

$\alpha = \tan^{-1}\left(\sinh\frac{7}{20}\right) \approx 0.343$ rad $\approx 19.66°$. Thus, the angle between the line and the pole is $\theta = 90° - \alpha \approx 70.34°$.

49. (a) From Exercise 48, the shape of the cable is given by $y = f(x) = \dfrac{T}{\rho g}\cosh\left(\dfrac{\rho g x}{T}\right)$. The shape is symmetric about the

y-axis, so the lowest point is $(0, f(0)) = \left(0, \dfrac{T}{\rho g}\right)$ and the poles are at $x = \pm 100$. We want to find T when the lowest

point is 60 m, so $\dfrac{T}{\rho g} = 60 \;\;\Rightarrow\;\; T = 60\rho g = (60 \text{ m})(2 \text{ kg/m})(9.8 \text{ m/s}^2) = 1176 \; \dfrac{\text{kg-m}}{\text{s}^2}$, or 1176 N (newtons).

The height of each pole is $f(100) = \dfrac{T}{\rho g}\cosh\left(\dfrac{\rho g \cdot 100}{T}\right) = 60\cosh\left(\dfrac{100}{60}\right) \approx 164.50$ m.

(b) If the tension is doubled from T to $2T$, then the low point is doubled since $\dfrac{T}{\rho g} = 60 \;\Rightarrow\; \dfrac{2T}{\rho g} = 120$. The new low point

is 120 m. The height of the poles is now $f(100) = \dfrac{2T}{\rho g}\cosh\left(\dfrac{\rho g \cdot 100}{2T}\right) = 120\cosh\left(\dfrac{100}{120}\right) \approx 164.13$ m, just a slight

decrease.

51. (a) $y = A\sinh mx + B\cosh mx \;\Rightarrow\; y' = mA\cosh mx + mB\sinh mx \;\Rightarrow$

$y'' = m^2 A\sinh mx + m^2 B\cosh mx = m^2(A\sinh mx + B\cosh mx) = m^2 y$

(b) From part (a), a solution of $y'' = 9y$ is $y(x) = A\sinh 3x + B\cosh 3x$. So $-4 = y(0) = A\sinh 0 + B\cosh 0 = B$, so

$B = -4$. Now $y'(x) = 3A\cosh 3x - 12\sinh 3x \;\Rightarrow\; 6 = y'(0) = 3A \;\Rightarrow\; A = 2$, so $y = 2\sinh 3x - 4\cosh 3x$.

53. Let $u = \cosh x$. Then $du = \sinh x\, dx$, so $\int \sinh x \cosh^2 x\, dx = \int u^2\, du = \tfrac{1}{3}u^3 + C = \tfrac{1}{3}\cosh^3 x + C$.

55. Let $u = \sqrt{x}$. Then $du = \dfrac{dx}{2\sqrt{x}}$ and $\displaystyle\int \dfrac{\sinh\sqrt{x}}{\sqrt{x}}\, dx = \int \sinh u \cdot 2\, du = 2\cosh u + C = 2\cosh\sqrt{x} + C$.

57. $\displaystyle\int \dfrac{\cosh x}{\cosh^2 x - 1}\, dx = \int \dfrac{\cosh x}{\sinh^2 x}\, dx = \int \dfrac{\cosh x}{\sinh x}\cdot\dfrac{1}{\sinh x}\, dx = \int \coth x \operatorname{csch} x\, dx = -\operatorname{csch} x + C$

59. Let $t = 3u$. Then $dt = 3\, du$ and

$\displaystyle\int_4^6 \dfrac{1}{\sqrt{t^2 - 9}}\, dt = \int_{4/3}^2 \dfrac{1}{\sqrt{9u^2 - 9}}\, 3\, du = \int_{4/3}^2 \dfrac{du}{\sqrt{u^2 - 1}} = \Big[\cosh^{-1} u\Big]_{4/3}^2 = \cosh^{-1} 2 - \cosh^{-1}\left(\tfrac{4}{3}\right)$ or

$\displaystyle = \Big[\cosh^{-1} u\Big]_{4/3}^2 = \Big[\ln\left(u + \sqrt{u^2 - 1}\right)\Big]_{4/3}^2 = \ln\left(2 + \sqrt{3}\right) - \ln\left(\dfrac{4 + \sqrt{7}}{3}\right) = \ln\left(\dfrac{6 + 3\sqrt{3}}{4 + \sqrt{7}}\right)$

61. Let $u = e^x$. Then $du = e^x\, dx$ and $\displaystyle\int \dfrac{e^x}{1 - e^{2x}}\, dx = \int \dfrac{du}{1 - u^2} = \tanh^{-1} u + C = \tanh^{-1}(e^x) + C$

$\left[\text{or } \dfrac{1}{2}\ln\left(\dfrac{1 + e^x}{1 - e^x}\right) + C\right]$.

63. The tangent to $y = \cosh x$ has slope 1 when $y' = \sinh x = 1 \;\Rightarrow\; x = \sinh^{-1} 1 = \ln\left(1 + \sqrt{2}\right)$, by Equation 3.

Since $\sinh x = 1$ and $y = \cosh x = \sqrt{1 + \sinh^2 x}$, we have $\cosh x = \sqrt{2}$. The point is $\left(\ln\left(1 + \sqrt{2}\right), \sqrt{2}\right)$.

5.8 Indeterminate Forms and l'Hospital's Rule

Note: The use of l'Hospital's Rule is indicated by an H above the equal sign: $\overset{\text{H}}{=}$

1. This limit has the form $\tfrac{0}{0}$. We can simply factor and simplify to evaluate the limit.

$\displaystyle\lim_{x\to 1} \dfrac{x^2 - 1}{x^2 - x} = \lim_{x\to 1} \dfrac{(x+1)(x-1)}{x(x-1)} = \lim_{x\to 1} \dfrac{x+1}{x} = \dfrac{1+1}{1} = 2$

3. This limit has the form $\tfrac{0}{0}$. $\displaystyle\lim_{x\to(\pi/2)^+} \dfrac{\cos x}{1 - \sin x} \overset{\text{H}}{=} \lim_{x\to(\pi/2)^+} \dfrac{-\sin x}{-\cos x} = \lim_{x\to(\pi/2)^+} \tan x = -\infty$.

5. This limit has the form $\tfrac{0}{0}$. $\displaystyle\lim_{t\to 0} \dfrac{e^{2t} - 1}{\sin t} \overset{\text{H}}{=} \lim_{t\to 0} \dfrac{2e^{2t}}{\cos t} = \dfrac{2(1)}{1} = 2$

7. This limit has the form $\frac{0}{0}$. $\displaystyle\lim_{\theta\to\pi/2}\frac{1-\sin\theta}{1+\cos 2\theta} \overset{\text{H}}{=} \lim_{\theta\to\pi/2}\frac{-\cos\theta}{-2\sin 2\theta} \overset{\text{H}}{=} \lim_{\theta\to\pi/2}\frac{\sin\theta}{-4\cos 2\theta} = \frac{1}{4}$

9. $\displaystyle\lim_{x\to 0^+}[(\ln x)/x] = -\infty$ since $\ln x \to -\infty$ as $x \to 0^+$ and dividing by small values of x just increases the magnitude of the

quotient $(\ln x)/x$. L'Hospital's Rule does not apply.

11. This limit has the form $\frac{0}{0}$. $\displaystyle\lim_{t\to 1}\frac{t^8-1}{t^5-1} \overset{\text{H}}{=} \lim_{t\to 1}\frac{8t^7}{5t^4} = \frac{8}{5}\lim_{t\to 1}t^3 = \frac{8}{5}(1) = \frac{8}{5}$

13. This limit has the form $\frac{0}{0}$. $\displaystyle\lim_{x\to 0}\frac{e^x-1-x}{x^2} \overset{\text{H}}{=} \lim_{x\to 0}\frac{e^x-1}{2x} \overset{\text{H}}{=} \lim_{x\to 0}\frac{e^x}{2} = \frac{1}{2}$

15. This limit has the form $\frac{0}{0}$. $\displaystyle\lim_{x\to 0}\frac{x3^x}{3^x-1} \overset{\text{H}}{=} \lim_{x\to 0}\frac{x3^x\ln 3+3^x}{3^x\ln 3} = \lim_{x\to 0}\frac{3^x(x\ln 3+1)}{3^x\ln 3} = \lim_{x\to 0}\frac{x\ln 3+1}{\ln 3} = \frac{1}{\ln 3}$

17. This limit has the form $\frac{0}{0}$. $\displaystyle\lim_{x\to 1}\frac{1-x+\ln x}{1+\cos\pi x} \overset{\text{H}}{=} \lim_{x\to 1}\frac{-1+1/x}{-\pi\sin\pi x} \overset{\text{H}}{=} \lim_{x\to 1}\frac{-1/x^2}{-\pi^2\cos\pi x} = \frac{-1}{-\pi^2(-1)} = -\frac{1}{\pi^2}$

19. This limit has the form $\frac{0}{0}$. $\displaystyle\lim_{x\to 1}\frac{x^a-ax+a-1}{(x-1)^2} \overset{\text{H}}{=} \lim_{x\to 1}\frac{ax^{a-1}-a}{2(x-1)} \overset{\text{H}}{=} \lim_{x\to 1}\frac{a(a-1)x^{a-2}}{2} = \frac{a(a-1)}{2}$

21. This limit has the form $\frac{0}{0}$. $\displaystyle\lim_{x\to 0}\frac{\cos x-1+\frac{1}{2}x^2}{x^4} \overset{\text{H}}{=} \lim_{x\to 0}\frac{-\sin x+x}{4x^3} \overset{\text{H}}{=} \lim_{x\to 0}\frac{-\cos x+1}{12x^2} \overset{\text{H}}{=} \lim_{x\to 0}\frac{\sin x}{24x} \overset{\text{H}}{=} \lim_{x\to 0}\frac{\cos x}{24} = \frac{1}{24}$

23. This limit has the form $\infty\cdot 0$. We'll change it to the form $\frac{0}{0}$.

$\displaystyle\lim_{x\to 0}\cot 2x\sin 6x = \lim_{x\to 0}\frac{\sin 6x}{\tan 2x} \overset{\text{H}}{=} \lim_{x\to 0}\frac{6\cos 6x}{2\sec^2 2x} = \frac{6(1)}{2(1)^2} = 3$

25. This limit has the form $\infty\cdot 0$. $\displaystyle\lim_{x\to\infty}x^3e^{-x^2} = \lim_{x\to\infty}\frac{x^3}{e^{x^2}} \overset{\text{H}}{=} \lim_{x\to\infty}\frac{3x^2}{2xe^{x^2}} = \lim_{x\to\infty}\frac{3x}{2e^{x^2}} \overset{\text{H}}{=} \lim_{x\to\infty}\frac{3}{4xe^{x^2}} = 0$

27. This limit has the form $0\cdot(-\infty)$.

$\displaystyle\lim_{x\to 1^+}\ln x\,\tan(\pi x/2) = \lim_{x\to 1^+}\frac{\ln x}{\cot(\pi x/2)} \overset{\text{H}}{=} \lim_{x\to 1^+}\frac{1/x}{(-\pi/2)\csc^2(\pi x/2)} = \frac{1}{(-\pi/2)(1)^2} = -\frac{2}{\pi}$

29. This limit has the form $\infty-\infty$.

$\displaystyle\lim_{x\to 0^+}\left(\frac{1}{x}-\frac{1}{e^x-1}\right) = \lim_{x\to 0^+}\frac{e^x-1-x}{x(e^x-1)} \overset{\text{H}}{=} \lim_{x\to 0^+}\frac{e^x-1}{xe^x+e^x-1} \overset{\text{H}}{=} \lim_{x\to 0^+}\frac{e^x}{xe^x+e^x+e^x} = \frac{1}{0+1+1} = \frac{1}{2}$

31. The limit has the form $\infty-\infty$ and we will change the form to a product by factoring out x.

$\displaystyle\lim_{x\to\infty}(x-\ln x) = \lim_{x\to\infty}x\left(1-\frac{\ln x}{x}\right) = \infty$ since $\displaystyle\lim_{x\to\infty}\frac{\ln x}{x} \overset{\text{H}}{=} \lim_{x\to\infty}\frac{1/x}{1} = 0$.

33. $y = x^{\sqrt{x}} \;\Rightarrow\; \ln y = \sqrt{x}\,\ln x$, so

$$\lim_{x\to 0^+}\ln y = \lim_{x\to 0^+}\sqrt{x}\,\ln x = \lim_{x\to 0^+}\frac{\ln x}{x^{-1/2}} \overset{\text{H}}{=} \lim_{x\to 0^+}\frac{1/x}{-\frac{1}{2}x^{-3/2}} = -2\lim_{x\to 0^+}\sqrt{x} = 0 \;\Rightarrow$$

$$\lim_{x\to 0^+}x^{\sqrt{x}} = \lim_{x\to 0^+}e^{\ln y} = e^0 = 1.$$

35. $y = (1-2x)^{1/x} \;\Rightarrow\; \ln y = \frac{1}{x}\ln(1-2x)$, so $\displaystyle\lim_{x\to 0}\ln y = \lim_{x\to 0}\frac{\ln(1-2x)}{x} \overset{\text{H}}{=} \lim_{x\to 0}\frac{-2/(1-2x)}{1} = -2 \;\Rightarrow$

$$\lim_{x\to 0}(1-2x)^{1/x} = \lim_{x\to 0}e^{\ln y} = e^{-2}.$$

37. $y = x^{1/(1-x)} \;\Rightarrow\; \ln y = \frac{1}{1-x}\ln x$, so $\displaystyle\lim_{x\to 1^+}\ln y = \lim_{x\to 1^+}\frac{1}{1-x}\ln x = \lim_{x\to 1^+}\frac{\ln x}{1-x} \overset{\text{H}}{=} \lim_{x\to 1^+}\frac{1/x}{-1} = -1 \;\Rightarrow$

$$\lim_{x\to 1^+}x^{1/(1-x)} = \lim_{x\to 1^+}e^{\ln y} = e^{-1} = \frac{1}{e}.$$

39.

From the graph, if $x = 500$, $y \approx 7.36$. The limit has the form 1^∞.

Now $y = \left(1 + \dfrac{2}{x}\right)^x \;\Rightarrow\; \ln y = x\ln\left(1 + \dfrac{2}{x}\right) \;\Rightarrow$

$$\lim_{x\to\infty}\ln y = \lim_{x\to\infty}\frac{\ln(1+2/x)}{1/x} \overset{\text{H}}{=} \lim_{x\to\infty}\frac{\dfrac{1}{1+2/x}\left(-\dfrac{2}{x^2}\right)}{-1/x^2}$$

$$= 2\lim_{x\to\infty}\frac{1}{1+2/x} = 2(1) = 2 \;\Rightarrow$$

$$\lim_{x\to\infty}\left(1+\frac{2}{x}\right)^x = \lim_{x\to\infty}e^{\ln y} = e^2 \;[\approx 7.39]$$

41. $\displaystyle\lim_{x\to\infty}\frac{e^x}{x^n} \overset{\text{H}}{=} \lim_{x\to\infty}\frac{e^x}{nx^{n-1}} \overset{\text{H}}{=} \lim_{x\to\infty}\frac{e^x}{n(n-1)x^{n-2}} \overset{\text{H}}{=} \cdots \overset{\text{H}}{=} \lim_{x\to\infty}\frac{e^x}{n!} = \infty$

43. $\displaystyle\lim_{x\to\infty}\frac{x}{\sqrt{x^2+1}} \overset{\text{H}}{=} \lim_{x\to\infty}\frac{1}{\frac{1}{2}(x^2+1)^{-1/2}(2x)} = \lim_{x\to\infty}\frac{\sqrt{x^2+1}}{x}$. Repeated applications of l'Hospital's Rule result in the

original limit or the limit of the reciprocal of the function. Another method is to try dividing the numerator and denominator

by x: $\displaystyle\lim_{x\to\infty}\frac{x}{\sqrt{x^2+1}} = \lim_{x\to\infty}\frac{x/x}{\sqrt{x^2/x^2+1/x^2}} = \lim_{x\to\infty}\frac{1}{\sqrt{1+1/x^2}} = \frac{1}{1} = 1$

45. First we will find $\displaystyle\lim_{n\to\infty}\left(1+\frac{r}{n}\right)^{nt}$, which is of the form 1^∞. $y = \left(1+\dfrac{r}{n}\right)^{nt} \;\Rightarrow\; \ln y = nt\ln\left(1+\dfrac{r}{n}\right)$, so

$$\lim_{n\to\infty}\ln y = \lim_{n\to\infty}nt\ln\left(1+\frac{r}{n}\right) = t\lim_{n\to\infty}\frac{\ln(1+r/n)}{1/n} \overset{\text{H}}{=} t\lim_{n\to\infty}\frac{(-r/n^2)}{(1+r/n)(-1/n^2)} = t\lim_{n\to\infty}\frac{r}{1+i/n} = tr \;\Rightarrow$$

$\displaystyle\lim_{n\to\infty}y = e^{rt}$. Thus, as $n\to\infty$, $A = A_0\left(1+\dfrac{r}{n}\right)^{nt} \to A_0 e^{rt}$.

47. $\displaystyle\lim_{E\to 0^+} P(E) = \lim_{E\to 0^+} \left(\frac{e^E + e^{-E}}{e^E - e^{-E}} - \frac{1}{E} \right)$

$\displaystyle = \lim_{E\to 0^+} \frac{E(e^E + e^{-E}) - 1(e^E - e^{-E})}{(e^E - e^{-E})\,E} = \lim_{E\to 0^+} \frac{Ee^E + Ee^{-E} - e^E + e^{-E}}{Ee^E - Ee^{-E}}$ $\qquad$ [form is $\frac{0}{0}$]

$\displaystyle \overset{\text{H}}{=} \lim_{E\to 0^+} \frac{Ee^E + e^E\cdot 1 + E\left(-e^{-E}\right) + e^{-E}\cdot 1 - e^E + \left(-e^{-E}\right)}{Ee^E + e^E\cdot 1 - \left[E(-e^{-E}) + e^{-E}\cdot 1\right]}$

$\displaystyle = \lim_{E\to 0^+} \frac{Ee^E - Ee^{-E}}{Ee^E + e^E + Ee^{-E} - e^{-E}} = \lim_{E\to 0^+} \frac{e^E - e^{-E}}{e^E + \dfrac{e^E}{E} + e^{-E} - \dfrac{e^{-E}}{E}}$ $\qquad$ [divide by E]

$\displaystyle = \frac{0}{2+L}, \quad \text{where } L = \lim_{E\to 0^+} \frac{e^E - e^{-E}}{E} \quad \text{[form is } \tfrac{0}{0}\text{]} \quad \overset{\text{H}}{=} \lim_{E\to 0^+} \frac{e^E + e^{-E}}{1} = \frac{1+1}{1} = 2$

Thus, $\displaystyle\lim_{E\to 0^+} P(E) = \frac{0}{2+2} = 0$.

49. We see that both numerator and denominator approach 0, so we can use l'Hospital's Rule:

$$\lim_{x\to a} \frac{\sqrt{2a^3 x - x^4} - a\sqrt[3]{aax}}{a - \sqrt[4]{ax^3}} \overset{\text{H}}{=} \lim_{x\to a} \frac{\frac{1}{2}(2a^3 x - x^4)^{-1/2}(2a^3 - 4x^3) - a\left(\frac{1}{3}\right)(aax)^{-2/3}a^2}{-\frac{1}{4}(ax^3)^{-3/4}(3ax^2)}$$

$$= \frac{\frac{1}{2}(2a^3 a - a^4)^{-1/2}(2a^3 - 4a^3) - \frac{1}{3}a^3(a^2 a)^{-2/3}}{-\frac{1}{4}(aa^3)^{-3/4}(3aa^2)}$$

$$= \frac{(a^4)^{-1/2}(-a^3) - \frac{1}{3}a^3(a^3)^{-2/3}}{-\frac{3}{4}a^3(a^4)^{-3/4}} = \frac{-a - \frac{1}{3}a}{-\frac{3}{4}} = \frac{4}{3}\left(\frac{4}{3}a\right) = \frac{16}{9}a$$

51. The limit, $L = \displaystyle\lim_{x\to\infty} \left[x - x^2 \ln\left(\frac{1+x}{x}\right) \right] = \lim_{x\to\infty} \left[x - x^2 \ln\left(\frac{1}{x} + 1\right) \right]$. Let $t = 1/x$, so as $x\to\infty$, $t\to 0^+$.

$$L = \lim_{t\to 0^+} \left[\frac{1}{t} - \frac{1}{t^2}\ln(t+1) \right] = \lim_{t\to 0^+} \frac{t - \ln(t+1)}{t^2} \overset{\text{H}}{=} \lim_{t\to 0^+} \frac{1 - \dfrac{1}{t+1}}{2t} = \lim_{t\to 0^+} \frac{t/(t+1)}{2t} = \lim_{t\to 0^+} \frac{1}{2(t+1)} = \frac{1}{2}$$

Note: Starting the solution by factoring out x or x^2 leads to a more complicated solution.

53. Since $f(2) = 0$, the given limit has the form $\frac{0}{0}$.

$$\lim_{x\to 0} \frac{f(2+3x) + f(2+5x)}{x} \overset{\text{H}}{=} \lim_{x\to 0} \frac{f'(2+3x)\cdot 3 + f'(2+5x)\cdot 5}{1} = f'(2)\cdot 3 + f'(2)\cdot 5 = 8f'(2) = 8\cdot 7 = 56$$

55. Since $\displaystyle\lim_{h\to 0}[f(x+h) - f(x-h)] = f(x) - f(x) = 0$ (f is differentiable and hence continuous) and $\displaystyle\lim_{h\to 0} 2h = 0$, we use

l'Hospital's Rule:

$$\lim_{h\to 0} \frac{f(x+h) - f(x-h)}{2h} \overset{\text{H}}{=} \lim_{h\to 0} \frac{f'(x+h)(1) - f'(x-h)(-1)}{2} = \frac{f'(x) + f'(x)}{2} = \frac{2f'(x)}{2} = f'(x)$$

$\dfrac{f(x+h) - f(x-h)}{2h}$ is the slope of the secant line between

$(x - h, f(x - h))$ and $(x + h, f(x + h))$. As $h \to 0$, this line gets closer

to the tangent line and its slope approaches $f'(x)$.

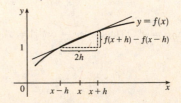

57. (a) We show that $\lim\limits_{x \to 0} \dfrac{f(x)}{x^n} = 0$ for every integer $n \geq 0$. Let $y = \dfrac{1}{x^2}$. Then

$$\lim_{x \to 0} \frac{f(x)}{x^{2n}} = \lim_{x \to 0} \frac{e^{-1/x^2}}{(x^2)^n} = \lim_{y \to \infty} \frac{y^n}{e^y} \overset{\text{H}}{=} \lim_{y \to \infty} \frac{ny^{n-1}}{e^y} \overset{\text{H}}{=} \cdots \overset{\text{H}}{=} \lim_{y \to \infty} \frac{n!}{e^y} = 0 \;\Rightarrow$$

$$\lim_{x \to 0} \frac{f(x)}{x^n} = \lim_{x \to 0} x^n \frac{f(x)}{x^{2n}} = \lim_{x \to 0} x^n \lim_{x \to 0} \frac{f(x)}{x^{2n}} = 0. \text{ Thus, } f'(0) = \lim_{x \to 0} \frac{f(x) - f(0)}{x - 0} = \lim_{x \to 0} \frac{f(x)}{x} = 0.$$

(b) Using the Chain Rule and the Quotient Rule we see that $f^{(n)}(x)$ exists for $x \neq 0$. In fact, we prove by induction that for each $n \geq 0$, there is a polynomial p_n and a non-negative integer k_n with $f^{(n)}(x) = p_n(x)f(x)/x^{k_n}$ for $x \neq 0$. This is true for $n = 0$; suppose it is true for the nth derivative. Then $f'(x) = f(x)(2/x^3)$, so

$$f^{(n+1)}(x) = \left[x^{k_n} \left[p_n'(x)\, f(x) + p_n(x)\, f'(x) \right] - k_n x^{k_n-1} p_n(x)\, f(x) \right] x^{-2k_n}$$

$$= \left[x^{k_n} p_n'(x) + p_n(x)(2/x^3) - k_n x^{k_n-1} p_n(x) \right] f(x) x^{-2k_n}$$

$$= \left[x^{k_n+3} p_n'(x) + 2p_n(x) - k_n x^{k_n+2} p_n(x) \right] f(x) x^{-(2k_n+3)}$$

which has the desired form.

Now we show by induction that $f^{(n)}(0) = 0$ for all n. By part (a), $f'(0) = 0$. Suppose that $f^{(n)}(0) = 0$. Then

$$f^{(n+1)}(0) = \lim_{x \to 0} \frac{f^{(n)}(x) - f^{(n)}(0)}{x - 0} = \lim_{x \to 0} \frac{f^{(n)}(x)}{x} = \lim_{x \to 0} \frac{p_n(x)\, f(x)/x^{k_n}}{x} = \lim_{x \to 0} \frac{p_n(x)\, f(x)}{x^{k_n+1}}$$

$$= \lim_{x \to 0} p_n(x) \lim_{x \to 0} \frac{f(x)}{x^{k_n+1}} = p_n(0) \cdot 0 = 0$$

5   Review

CONCEPT CHECK

1. (a) A function f is called a *one-to-one function* if it never takes on the same value twice; that is, if $f(x_1) \neq f(x_2)$ whenever $x_1 \neq x_2$. (Or, f is 1-1 if each output corresponds to only one input.)

Use the Horizontal Line Test: A function is one-to-one if and only if no horizontal line intersects its graph more than once.

(b) If f is a one-to-one function with domain A and range B, then its *inverse function* f^{-1} has domain B and range A and is defined by

$$f^{-1}(y) = x \;\Leftrightarrow\; f(x) = y$$

for any y in B. The graph of f^{-1} is obtained by reflecting the graph of f about the line $y = x$.

(c) $(f^{-1})'(a) = \dfrac{1}{f'(f^{-1}(a))}$

2. (a) $e = \lim\limits_{x \to 0} (1 + x)^{1/x}$

(b) $e \approx 2.71828$

(c) The differentiation formula for $y = a^x$ $[y' = a^x \ln a]$ is simplest when $a = e$ because $\ln e = 1$.

(d) The differentiation formula for $y = \log_a x$ $[y' = 1/(x \ln a)]$ is simplest when $a = e$ because $\ln e = 1$.

3. (a) The function $f(x) = e^x$ has domain $\mathbb{R}$ and range $(0, \infty)$.

(b) The function $f(x) = \ln x$ has domain $(0, \infty)$ and range $\mathbb{R}$.

(c) The graphs are reflections of one another about the line $y = x$. See Figure 5.3.1.

(d) $\log_a x = \dfrac{\ln x}{\ln a}$

4. (a) The inverse sine function $f(x) = \sin^{-1} x$ is defined as follows:

$$\sin^{-1} x = y \quad \Leftrightarrow \quad \sin y = x \quad \text{and} \quad -\frac{\pi}{2} \le y \le \frac{\pi}{2}$$

Its domain is $-1 \le x \le 1$ and its range is $-\dfrac{\pi}{2} \le y \le \dfrac{\pi}{2}$.

(b) The inverse cosine function $f(x) = \cos^{-1} x$ is defined as follows:

$$\cos^{-1} x = y \quad \Leftrightarrow \quad \cos y = x \quad \text{and} \quad 0 \le y \le \pi$$

Its domain is $-1 \le x \le 1$ and its range is $0 \le y \le \pi$.

(c) See Definition 5.6.7. Domain $= \mathbb{R}$, Range $= \left(-\frac{\pi}{2}, \frac{\pi}{2}\right)$. See Figure 10 in Section 5.6.

5. $\sinh x = \dfrac{e^x - e^{-x}}{2}$, $\cosh x = \dfrac{e^x + e^{-x}}{2}$, $\tanh x = \dfrac{\sinh x}{\cosh x} = \dfrac{e^x - e^{-x}}{e^x + e^{-x}}$

6. (a) $y = e^x \;\Rightarrow\; y' = e^x$ $\qquad$ (b) $y = a^x \;\Rightarrow\; y' = a^x \ln a$

(c) $y = \ln x \;\Rightarrow\; y' = 1/x$ $\qquad$ (d) $y = \log_a x \;\Rightarrow\; y' = 1/(x \ln a)$

(e) $y = \sin^{-1} x \;\Rightarrow\; y' = 1/\sqrt{1 - x^2}$ $\qquad$ (f) $y = \cos^{-1} x \;\Rightarrow\; y' = -1/\sqrt{1 - x^2}$

(g) $y = \tan^{-1} x \;\Rightarrow\; y' = 1/(1 + x^2)$ $\qquad$ (h) $y = \sinh x \;\Rightarrow\; y' = \cosh x$

(i) $y = \cosh x \;\Rightarrow\; y' = \sinh x$ $\qquad$ (j) $y = \tanh x \;\Rightarrow\; y' = \operatorname{sech}^2 x$

(k) $y = \sinh^{-1} x \;\Rightarrow\; y' = 1/\sqrt{1 + x^2}$ $\qquad$ (l) $y = \cosh^{-1} x \;\Rightarrow\; y' = 1/\sqrt{x^2 - 1}$

(m) $y = \tanh^{-1} x \;\Rightarrow\; y' = 1/(1 - x^2)$

7. (a) $\dfrac{dy}{dt} = ky$; the relative growth rate, $\dfrac{1}{y}\dfrac{dy}{dt}$, is constant.

(b) The equation in part (a) is an appropriate model for population growth, assuming that there is enough room and nutrition to support the growth.

(c) If $y(0) = y_0$, then the solution is $y(t) = y_0 e^{kt}$.

8. (a) See l'Hospital's Rule and the three notes that follow it in Section 5.8.

(b) Write fg as $\dfrac{f}{1/g}$ or $\dfrac{g}{1/f}$.

(c) Convert the difference into a quotient using a common denominator, rationalizing, factoring, or some other method.

(d) Convert the power to a product by taking the natural logarithm of both sides of $y = f^g$ or by writing f^g as $e^{g \ln f}$.

TRUE-FALSE QUIZ

1. True. If f is one-to-one, with domain $\mathbb{R}$, then $f^{-1}(f(6)) = 6$ by the first cancellation equation (5.1.4).

3. False. For example, $\cos\frac{\pi}{2} = \cos\left(-\frac{\pi}{2}\right)$, so $\cos x$ is not 1-1.

5. True, since $\ln x$ is an increasing function on $(0, \infty)$.

7. True. We can divide by e^x since $e^x \neq 0$ for every x.

9. False. Let $x = e$. Then $(\ln x)^6 = (\ln e)^6 = 1^6 = 1$, but $6\ln x = 6\ln e = 6 \cdot 1 = 6 \neq 1 = (\ln x)^6$. What *is* true, however, is that $\ln(x^6) = 6\ln x$ for $x > 0$.

11. False. $\ln 10$ is a constant, so its derivative, $\dfrac{d}{dx}(\ln 10)$, is 0, not $\frac{1}{10}$.

13. False. The "-1" is not an exponent; it is an indication of an inverse function.

15. True. See Figure 5.7.2.

EXERCISES

1. No. f is not 1-1 because the graph of f fails the Horizontal Line Test.

3. (a) $f^{-1}(3) = 7$ since $f(7) = 3$.
  (b) $(f^{-1})'(3) = \dfrac{1}{f'(f^{-1}(3))} = \dfrac{1}{f'(7)} = \dfrac{1}{8}$

5.
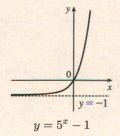
$$y = 5^x - 1$$

7.
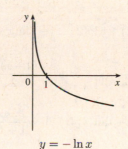
$$y = -\ln x$$

9.
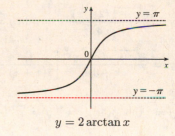
$$y = 2\arctan x$$

11. (a) $e^{2\ln 3} = (e^{\ln 3})^2 = 3^2 = 9$

  (b) $\log_{10} 25 + \log_{10} 4 = \log_{10}(25 \cdot 4) = \log_{10} 100 = \log_{10} 10^2 = 2$

13. (a) $e^x = 5 \;\Rightarrow\; x = \ln 5$

  (b) $\ln x = 2 \;\Rightarrow\; x = e^2$

15. (a) $\ln(x+1) + \ln(x-1) = 1 \;\Rightarrow\; \ln[(x+1)(x-1)] = 1 \;\Rightarrow\; \ln(x^2 - 1) = \ln e \;\Rightarrow\; x^2 - 1 = e \;\Rightarrow$
  $x^2 = e + 1 \;\Rightarrow\; x = \sqrt{e+1}$ since $\ln(x-1)$ is defined only when $x > 1$.

  (b) $\log_5(c^x) = d \;\Rightarrow\; x\log_5 c = d \;\Rightarrow\; x = \dfrac{d}{\log_5 c}$.

  Or: $\log_5(c^x) = d \;\Rightarrow\; 5^d = c^x \;\Rightarrow\; \ln 5^d = \ln c^x \;\Rightarrow\; d\ln 5 = x\ln c \;\Rightarrow\; x = \dfrac{d\ln 5}{\ln c}$.

17. $y = \ln(x \ln x) \implies y' = \dfrac{1}{x \ln x}(x \ln x)' = \dfrac{1}{x \ln x}\left(x \cdot \dfrac{1}{x} + \ln x \cdot 1\right) = \dfrac{1 + \ln x}{x \ln x}$

Another method: $y = \ln(x \ln x) = \ln x + \ln \ln x \implies y' = \dfrac{1}{x} + \dfrac{1}{\ln x} \cdot \dfrac{1}{x} = \dfrac{\ln x + 1}{x \ln x}$

19. $y = \dfrac{e^{1/x}}{x^2} \implies y' = \dfrac{x^2(e^{1/x})' - e^{1/x}(x^2)'}{(x^2)^2} = \dfrac{x^2(e^{1/x})(-1/x^2) - e^{1/x}(2x)}{x^4} = \dfrac{-e^{1/x}(1 + 2x)}{x^4}$

21. $y = \sqrt{\arctan x} \implies y' = \dfrac{1}{2}(\arctan x)^{-1/2}\dfrac{d}{dx}(\arctan x) = \dfrac{1}{2\sqrt{\arctan x}\,(1 + x^2)}$

23. $f(t) = t^2 \ln t \implies f'(t) = t^2 \cdot \dfrac{1}{t} + (\ln t)(2t) = t + 2t \ln t$ or $t(1 + 2 \ln t)$

25. $y = 3^{x \ln x} \implies y' = 3^{x \ln x}(\ln 3)\dfrac{d}{dx}(x \ln x) = 3^{x \ln x}(\ln 3)\left(x \cdot \dfrac{1}{x} + \ln x \cdot 1\right) = 3^{x \ln x}(\ln 3)(1 + \ln x)$

27. $y = x \sinh(x^2) \implies y' = x \cosh(x^2) \cdot 2x + \sinh(x^2) \cdot 1 = 2x^2 \cosh(x^2) + \sinh(x^2)$

29. $h(\theta) = e^{\tan 2\theta} \implies h'(\theta) = e^{\tan 2\theta} \cdot \sec^2 2\theta \cdot 2 = 2 \sec^2(2\theta)\, e^{\tan 2\theta}$

31. $y = \ln \sin x - \frac{1}{2}\sin^2 x \implies y' = \dfrac{1}{\sin x} \cdot \cos x - \frac{1}{2} \cdot 2 \sin x \cdot \cos x = \cot x - \sin x \cos x$

33. $y = \log_5(1 + 2x) \implies y' = \dfrac{1}{(1 + 2x)\ln 5}\dfrac{d}{dx}(1 + 2x) = \dfrac{2}{(1 + 2x)\ln 5}$

35. $y = \dfrac{\sqrt{x+1}\,(2-x)^5}{(x+3)^7} \implies \ln y = \frac{1}{2}\ln(x+1) + 5\ln(2-x) - 7\ln(x+3) \implies \dfrac{y'}{y} = \dfrac{1}{2(x+1)} + \dfrac{-5}{2-x} - \dfrac{7}{x+3} \implies$

$y' = \dfrac{\sqrt{x+1}\,(2-x)^5}{(x+3)^7}\left[\dfrac{1}{2(x+1)} - \dfrac{5}{2-x} - \dfrac{7}{x+3}\right]$ or $y' = \dfrac{(2-x)^4(3x^2 - 55x - 52)}{2\sqrt{x+1}\,(x+3)^8}$.

37. $y = x \tan^{-1}(4x) \implies y' = x \cdot \dfrac{1}{1 + (4x)^2} \cdot 4 + \tan^{-1}(4x) \cdot 1 = \dfrac{4x}{1 + 16x^2} + \tan^{-1}(4x)$

39. $y = \ln(\cosh 3x) \implies y' = (1/\cosh 3x)(\sinh 3x)(3) = 3 \tanh 3x$

41. $y = \cosh^{-1}(\sinh x) \implies y' = (\cosh x)/\sqrt{\sinh^2 x - 1}$

43. $y = \cos\left(e^{\sqrt{\tan 3x}}\right) \implies$

$y' = -\sin\left(e^{\sqrt{\tan 3x}}\right) \cdot \left(e^{\sqrt{\tan 3x}}\right)' = -\sin\left(e^{\sqrt{\tan 3x}}\right) e^{\sqrt{\tan 3x}} \cdot \frac{1}{2}(\tan 3x)^{-1/2} \cdot \sec^2(3x) \cdot 3$

$= \dfrac{-3\sin\left(e^{\sqrt{\tan 3x}}\right) e^{\sqrt{\tan 3x}} \sec^2(3x)}{2\sqrt{\tan 3x}}$

45. $f(x) = e^{g(x)} \implies f'(x) = e^{g(x)}g'(x)$

47. $f(x) = \ln|g(x)| \Rightarrow f'(x) = \dfrac{1}{g(x)} g'(x) = \dfrac{g'(x)}{g(x)}$

49. $f(x) = 2^x \Rightarrow f'(x) = 2^x \ln 2 \Rightarrow f''(x) = 2^x (\ln 2)^2 \Rightarrow \cdots \Rightarrow f^{(n)}(x) = 2^x (\ln 2)^n$

51. We first show it is true for $n = 1$: $f'(x) = e^x + xe^x = (x+1)e^x$. We now assume it is true for $n = k$:

$f^{(k)}(x) = (x+k)e^x$. With this assumption, we must show it is true for $n = k+1$:

$$f^{(k+1)}(x) = \frac{d}{dx}\left[f^{(k)}(x)\right] = \frac{d}{dx}\left[(x+k)e^x\right] = e^x + (x+k)e^x = \left[x + (k+1)\right]e^x.$$

Therefore, $f^{(n)}(x) = (x+n)e^x$ by mathematical induction.

53. $y = [\ln(x+4)]^2 \Rightarrow y' = 2[\ln(x+4)]^1 \cdot \dfrac{1}{x+4} \cdot 1 = 2\,\dfrac{\ln(x+4)}{x+4}$ and $y' = 0 \Leftrightarrow \ln(x+4) = 0 \Leftrightarrow$

$x + 4 = e^0 \Rightarrow x + 4 = 1 \Leftrightarrow x = -3$, so the tangent is horizontal at the point $(-3, 0)$.

55. (a) The line $x - 4y = 1$ has slope $\frac{1}{4}$. A tangent to $y = e^x$ has slope $\frac{1}{4}$ when $y' = e^x = \frac{1}{4} \Rightarrow x = \ln\frac{1}{4} = -\ln 4$.

Since $y = e^x$, the y-coordinate is $\frac{1}{4}$ and the point of tangency is $\left(-\ln 4, \frac{1}{4}\right)$. Thus, an equation of the tangent line

is $y - \frac{1}{4} = \frac{1}{4}(x + \ln 4)$ or $y = \frac{1}{4}x + \frac{1}{4}(\ln 4 + 1)$.

(b) The slope of the tangent at the point (a, e^a) is $\dfrac{d}{dx}\,e^x\bigg|_{x=a} = e^a$. Thus, an equation of the tangent line is

$y - e^a = e^a(x - a)$. We substitute $x = 0$, $y = 0$ into this equation, since we want the line to pass through the origin:

$0 - e^a = e^a(0 - a) \Leftrightarrow -e^a = e^a(-a) \Leftrightarrow a = 1$. So an equation of the tangent line at the point $(a, e^a) = (1, e)$

is $y - e = e(x - 1)$ or $y = ex$.

57. (a) $y(t) = y(0)e^{kt} = 200e^{kt} \Rightarrow y(0.5) = 200e^{0.5k} = 360 \Rightarrow e^{0.5k} = 1.8 \Rightarrow 0.5k = \ln 1.8 \Rightarrow$

$k = 2\ln 1.8 = \ln(1.8)^2 = \ln 3.24 \Rightarrow y(t) = 200e^{(\ln 3.24)t} = 200(3.24)^t$

(b) $y(4) = 200(3.24)^4 \approx 22{,}040$ bacteria

(c) $y'(t) = 200(3.24)^t \cdot \ln 3.24$, so $y'(4) = 200(3.24)^4 \cdot \ln 3.24 \approx 25{,}910$ bacteria per hour

(d) $200(3.24)^t = 10{,}000 \Rightarrow (3.24)^t = 50 \Rightarrow t\ln 3.24 = \ln 50 \Rightarrow t = \ln 50 / \ln 3.24 \approx 3.33$ hours

59. (a) $C'(t) = -kC(t) \Rightarrow C(t) = C(0)e^{-kt}$ by Theorem 5.5.2. But $C(0) = C_0$, so $C(t) = C_0 e^{-kt}$.

(b) $C(30) = \frac{1}{2}C_0$ since the concentration is reduced by half. Thus, $\frac{1}{2}C_0 = C_0 e^{-30k} \Rightarrow \ln\frac{1}{2} = -30k \Rightarrow$

$k = -\frac{1}{30}\ln\frac{1}{2} = \frac{1}{30}\ln 2$. Since 10% of the original concentration remains if 90% is eliminated, we want the value of t

such that $C(t) = \frac{1}{10}C_0$. Therefore, $\frac{1}{10}C_0 = C_0 e^{-t(\ln 2)/30} \Rightarrow \ln 0.1 = -t(\ln 2)/30 \Rightarrow t = -\frac{30}{\ln 2}\ln 0.1 \approx 100$ h.

61. Let $t = 1/x$. Then as $x \to 0^+$, $t \to \infty$, and $\displaystyle\lim_{x \to 0^+} \tan^{-1}(1/x) = \lim_{t \to \infty} \tan^{-1} t = \frac{\pi}{2}$.

63. Let $t = 2/(x-3)$. As $x \to 3^-$, $t \to -\infty$. $\displaystyle\lim_{x \to 3^-} e^{2/(x-3)} = \lim_{t \to -\infty} e^t = 0$

65. Let $t = \sinh x$. As $x \to 0^+$, $t \to 0^+$. $\displaystyle\lim_{x \to 0^+} \ln(\sinh x) = \lim_{t \to 0^+} \ln t = -\infty$

67. $\displaystyle\lim_{x \to \infty} \frac{(1+2^x)/2^x}{(1-2^x)/2^x} = \lim_{x \to \infty} \frac{1/2^x + 1}{1/2^x - 1} = \frac{0+1}{0-1} = -1$

69. This limit has the form $\frac{0}{0}$. $\displaystyle\lim_{x \to 0} \frac{e^x - 1}{\tan x} \overset{\text{H}}{=} \lim_{x \to 0} \frac{e^x}{\sec^2 x} = \frac{1}{1} = 1$

71. This limit has the form $\frac{0}{0}$. $\displaystyle\lim_{x \to 0} \frac{e^{4x} - 1 - 4x}{x^2} \overset{\text{H}}{=} \lim_{x \to 0} \frac{4e^{4x} - 4}{2x} \overset{\text{H}}{=} \lim_{x \to 0} \frac{16e^{4x}}{2} = \lim_{x \to 0} 8e^{4x} = 8 \cdot 1 = 8$

73. This limit has the form $\infty \cdot 0$.

$$\lim_{x \to -\infty} (x^2 - x^3)e^{2x} = \lim_{x \to -\infty} \frac{x^2 - x^3}{e^{-2x}} \quad \left[\tfrac{\infty}{\infty} \text{ form}\right] \overset{\text{H}}{=} \lim_{x \to -\infty} \frac{2x - 3x^2}{-2e^{-2x}} \quad \left[\tfrac{\infty}{\infty} \text{ form}\right]$$

$$\overset{\text{H}}{=} \lim_{x \to -\infty} \frac{2 - 6x}{4e^{-2x}} \quad \left[\tfrac{\infty}{\infty} \text{ form}\right] \overset{\text{H}}{=} \lim_{x \to -\infty} \frac{-6}{-8e^{-2x}} = 0$$

75. This limit has the form $\infty - \infty$.

$$\lim_{x \to 1^+} \left(\frac{x}{x-1} - \frac{1}{\ln x} \right) = \lim_{x \to 1^+} \left(\frac{x \ln x - x + 1}{(x-1)\ln x} \right) \overset{\text{H}}{=} \lim_{x \to 1^+} \frac{x \cdot (1/x) + \ln x - 1}{(x-1) \cdot (1/x) + \ln x} = \lim_{x \to 1^+} \frac{\ln x}{1 - 1/x + \ln x}$$

$$\overset{\text{H}}{=} \lim_{x \to 1^+} \frac{1/x}{1/x^2 + 1/x} = \frac{1}{1+1} = \frac{1}{2}$$

77. $\displaystyle\int \left(\frac{1-x}{x} \right)^2 dx = \int \left(\frac{1}{x} - 1 \right)^2 dx = \int \left(\frac{1}{x^2} - \frac{2}{x} + 1 \right) dx = -\frac{1}{x} - 2\ln|x| + x + C$

79. Let $u = -2y^2$. Then $du = -4y\,dy$ and $\int_0^1 ye^{-2y^2}\,dy = \int_0^{-2} e^u \left(-\frac{1}{4}\,du\right) = -\frac{1}{4}\left[e^u\right]_0^{-2} = -\frac{1}{4}(e^{-2} - 1) = \frac{1}{4}(1 - e^{-2})$.

81. Let $u = e^x$, so $du = e^x\,dx$. When $x = 0$, $u = 1$; when $x = 1$, $u = e$. Thus,

$$\int_0^1 \frac{e^x}{1 + e^{2x}}\,dx = \int_1^e \frac{1}{1 + u^2}\,du = \left[\arctan u\right]_1^e = \arctan e - \arctan 1 = \arctan e - \frac{\pi}{4}.$$

83. Let $u = \sqrt{x}$. Then $du = \dfrac{dx}{2\sqrt{x}} \;\Rightarrow\; \displaystyle\int \frac{e^{\sqrt{x}}}{\sqrt{x}}\,dx = 2\int e^u\,du = 2e^u + C = 2e^{\sqrt{x}} + C$.

85. Let $u = x^2 + 2x$. Then $du = (2x + 2)\,dx = 2(x + 1)\,dx$ and

$$\int \frac{x+1}{x^2 + 2x}\,dx = \int \frac{\frac{1}{2}\,du}{u} = \frac{1}{2}\ln|u| + C = \frac{1}{2}\ln\left|x^2 + 2x\right| + C.$$

87. Let $u = \ln(\cos x)$. Then $du = \dfrac{-\sin x}{\cos x}\,dx = -\tan x\,dx \;\Rightarrow$

$\int \tan x \ln(\cos x)\,dx = -\int u\,du = -\frac{1}{2}u^2 + C = -\frac{1}{2}[\ln(\cos x)]^2 + C.$

89. $f(x) = \ln x + \tan^{-1} x \;\Rightarrow\; f(1) = \ln 1 + \tan^{-1} 1 = \frac{\pi}{4} \;\Rightarrow\; f^{-1}\!\left(\frac{\pi}{4}\right) = 1.$

$f'(x) = \dfrac{1}{x} + \dfrac{1}{1+x^2}$, so $(f^{-1})'\!\left(\frac{\pi}{4}\right) = \dfrac{1}{f'(1)} = \dfrac{1}{3/2} = \dfrac{2}{3}.$

91.

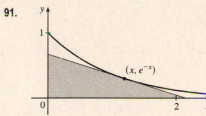

We find the equation of a tangent to the curve $y = e^{-x}$, so that we can find the x- and y-intercepts of this tangent, and then we can find the area of the triangle.

The slope of the tangent at the point $\left(a, e^{-a}\right)$ is given by $\dfrac{d}{dx}e^{-x}\bigg]_{x=a} = -e^{-a},$

and so the equation of the tangent is $y - e^{-a} = -e^{-a}(x-a) \;\Leftrightarrow\;$

$y = e^{-a}(a - x + 1).$

The y-intercept of this line is $y = e^{-a}(a - 0 + 1) = e^{-a}(a+1)$. To find the x-intercept we set $y = 0 \;\Rightarrow\;$

$e^{-a}(a - x + 1) = 0 \;\Rightarrow\; x = a + 1$. So the area of the triangle is $A(a) = \frac{1}{2}\left[e^{-a}(a+1)\right](a+1) = \frac{1}{2}e^{-a}(a+1)^2$. We

differentiate this with respect to a: $A'(a) = \frac{1}{2}\left[e^{-a}(2)(a+1) + (a+1)^2 e^{-a}(-1)\right] = \frac{1}{2}e^{-a}(1 - a^2)$. This is 0

at $a = \pm 1$, and the root $a = 1$ gives a maximum, by the First Derivative Test. So the maximum area of the triangle is

$A(1) = \frac{1}{2}e^{-1}(1+1)^2 = 2e^{-1} = 2/e.$

6 ☐ TECHNIQUES OF INTEGRATION

6.1 Integration by Parts

1. Let $u = \ln x$, $dv = x^2\,dx$ $\Rightarrow$ $du = \frac{1}{x}\,dx$, $v = \frac{1}{3}x^3$. Then by Equation 2,

$$\int x^2 \ln x\,dx = (\ln x)\left(\tfrac{1}{3}x^3\right) - \int \left(\tfrac{1}{3}x^3\right)\left(\tfrac{1}{x}\right)\,dx = \tfrac{1}{3}x^3 \ln x - \tfrac{1}{3}\int x^2\,dx = \tfrac{1}{3}x^3 \ln x - \tfrac{1}{3}\left(\tfrac{1}{3}x^3\right) + C$$

$$= \tfrac{1}{3}x^3 \ln x - \tfrac{1}{9}x^3 + C \quad \left[\text{or } \tfrac{1}{3}x^3\left(\ln x - \tfrac{1}{3}\right) + C\right]$$

3. Let $u = x$, $dv = \cos 5x\,dx$ $\Rightarrow$ $du = dx$, $v = \frac{1}{5}\sin 5x$. Then by Equation 2,

$$\int x \cos 5x\,dx = \tfrac{1}{5}x \sin 5x - \int \tfrac{1}{5}\sin 5x\,dx = \tfrac{1}{5}x \sin 5x + \tfrac{1}{25}\cos 5x + C.$$

5. Let $u = t$, $dv = e^{-3t}\,dt$ $\Rightarrow$ $du = dt$, $v = -\frac{1}{3}e^{-3t}$. Then by Equation 2,

$$\int t e^{-3t}\,dt = -\tfrac{1}{3}t e^{-3t} - \int -\tfrac{1}{3}e^{-3t}\,dt = -\tfrac{1}{3}t e^{-3t} + \tfrac{1}{3}\int e^{-3t}\,dt = -\tfrac{1}{3}t e^{-3t} - \tfrac{1}{9}e^{-3t} + C.$$

7. First let $u = x^2 + 2x$, $dv = \cos x\,dx$ $\Rightarrow$ $du = (2x + 2)\,dx$, $v = \sin x$. Then by Equation 2,

$I = \int (x^2 + 2x)\cos x\,dx = (x^2 + 2x)\sin x - \int (2x + 2)\sin x\,dx$. Next let $U = 2x + 2$, $dV = \sin x\,dx$ $\Rightarrow$ $dU = 2\,dx$,

$V = -\cos x$, so $\int (2x + 2)\sin x\,dx = -(2x + 2)\cos x - \int -2\cos x\,dx = -(2x + 2)\cos x + 2\sin x$. Thus,

$I = (x^2 + 2x)\sin x + (2x + 2)\cos x - 2\sin x + C.$

9. Let $u = \ln(2x + 1)$, $dv = dx$ $\Rightarrow$ $du = \dfrac{2}{2x + 1}\,dx$, $v = x$. Then

$$\int \ln(2x + 1)\,dx = x \ln(2x + 1) - \int \frac{2x}{2x + 1}\,dx = x \ln(2x + 1) - \int \frac{(2x + 1) - 1}{2x + 1}\,dx$$

$$= x \ln(2x + 1) - \int \left(1 - \frac{1}{2x + 1}\right)\,dx = x \ln(2x + 1) - x + \tfrac{1}{2}\ln(2x + 1) + C$$

$$= \tfrac{1}{2}(2x + 1)\ln(2x + 1) - x + C$$

11. Let $u = \arctan 4t$, $dv = dt$ $\Rightarrow$ $du = \dfrac{4}{1 + (4t)^2}\,dt = \dfrac{4}{1 + 16t^2}\,dt$, $v = t$. Then

$$\int \arctan 4t\,dt = t \arctan 4t - \int \frac{4t}{1 + 16t^2}\,dt = t \arctan 4t - \frac{1}{8}\int \frac{32t}{1 + 16t^2}\,dt = t \arctan 4t - \tfrac{1}{8}\ln(1 + 16t^2) + C.$$

13. First let $u = \sin 3\theta$, $dv = e^{2\theta}\,d\theta$ $\Rightarrow$ $du = 3\cos 3\theta\,d\theta$, $v = \frac{1}{2}e^{2\theta}$. Then

$I = \int e^{2\theta}\sin 3\theta\,d\theta = \frac{1}{2}e^{2\theta}\sin 3\theta - \frac{3}{2}\int e^{2\theta}\cos 3\theta\,d\theta$. Next let $U = \cos 3\theta$, $dV = e^{2\theta}\,d\theta$ $\Rightarrow$ $dU = -3\sin 3\theta\,d\theta$,

$V = \frac{1}{2}e^{2\theta}$ to get $\int e^{2\theta}\cos 3\theta\,d\theta = \frac{1}{2}e^{2\theta}\cos 3\theta + \frac{3}{2}\int e^{2\theta}\sin 3\theta\,d\theta$. Substituting in the previous formula gives

$I = \frac{1}{2}e^{2\theta}\sin 3\theta - \frac{3}{4}e^{2\theta}\cos 3\theta - \frac{9}{4}\int e^{2\theta}\sin 3\theta\,d\theta = \frac{1}{2}e^{2\theta}\sin 3\theta - \frac{3}{4}e^{2\theta}\cos 3\theta - \frac{9}{4}I$ $\Rightarrow$

$\frac{13}{4}I = \frac{1}{2}e^{2\theta}\sin 3\theta - \frac{3}{4}e^{2\theta}\cos 3\theta + C_1$. Hence, $I = \frac{1}{13}e^{2\theta}(2\sin 3\theta - 3\cos 3\theta) + C$, where $C = \frac{4}{13}C_1$.

15. Let $u = xe^{2x}$, $dv = \dfrac{1}{(1+2x)^2}\,dx$ $\Rightarrow$ $du = (x \cdot 2e^{2x} + e^{2x} \cdot 1)\,dx = e^{2x}(2x+1)\,dx$, $v = -\dfrac{1}{2(1+2x)}$.

Then by Equation 2,

$$\int \frac{xe^{2x}}{(1+2x)^2}\,dx = -\frac{xe^{2x}}{2(1+2x)} + \frac{1}{2}\int \frac{e^{2x}(2x+1)}{1+2x}\,dx = -\frac{xe^{2x}}{2(1+2x)} + \frac{1}{2}\int e^{2x}\,dx = -\frac{xe^{2x}}{2(1+2x)} + \frac{1}{4}e^{2x} + C$$

The answer could be written as $\dfrac{e^{2x}}{4(2x+1)} + C$.

17. Let $u = x$, $dv = \cos \pi x\,dx$ $\Rightarrow$ $du = dx$, $v = \frac{1}{\pi}\sin \pi x$. Then

$$\int_0^{1/2} x \cos \pi x\,dx = \left[\frac{1}{\pi}x \sin \pi x\right]_0^{1/2} - \int_0^{1/2} \frac{1}{\pi}\sin \pi x\,dx = \frac{1}{2\pi} - 0 - \frac{1}{\pi}\left[-\frac{1}{\pi}\cos \pi x\right]_0^{1/2}$$

$$= \frac{1}{2\pi} + \frac{1}{\pi^2}(0-1) = \frac{1}{2\pi} - \frac{1}{\pi^2} \text{ or } \frac{\pi - 2}{2\pi^2}$$

19. Let $u = \ln r$, $dv = r^3\,dr$ $\Rightarrow$ $du = \frac{1}{r}\,dr$, $v = \frac{1}{4}r^4$. Then

$$\int_1^3 r^3 \ln r\,dr = \left[\tfrac{1}{4}r^4 \ln r\right]_1^3 - \int_1^3 \tfrac{1}{4}r^3\,dr = \tfrac{81}{4}\ln 3 - 0 - \tfrac{1}{4}\left[\tfrac{1}{4}r^4\right]_1^3 = \tfrac{81}{4}\ln 3 - \tfrac{1}{16}(81-1) = \tfrac{81}{4}\ln 3 - 5.$$

21. Let $u = t$, $dv = \cosh t\,dt$ $\Rightarrow$ $du = dt$, $v = \sinh t$. Then

$$\int_0^1 t \cosh t\,dt = \left[t \sinh t\right]_0^1 - \int_0^1 \sinh t\,dt = (\sinh 1 - \sinh 0) - \left[\cosh t\right]_0^1 = \sinh 1 - (\cosh 1 - \cosh 0)$$

$$= \sinh 1 - \cosh 1 + 1.$$

We can use the definitions of sinh and cosh to write the answer in terms of e:

$$\sinh 1 - \cosh 1 + 1 = \tfrac{1}{2}(e^1 - e^{-1}) - \tfrac{1}{2}(e^1 + e^{-1}) + 1 = -e^{-1} + 1 = 1 - 1/e.$$

23. Let $u = \cos^{-1} x$, $dv = dx$ $\Rightarrow$ $du = -\dfrac{dx}{\sqrt{1-x^2}}$, $v = x$. Then

$$I = \int_0^{1/2} \cos^{-1} x\,dx = \left[x \cos^{-1} x\right]_0^{1/2} + \int_0^{1/2} \frac{x\,dx}{\sqrt{1-x^2}} = \frac{1}{2} \cdot \frac{\pi}{3} + \int_1^{3/4} t^{-1/2}\left[-\tfrac{1}{2}dt\right], \text{ where } t = 1 - x^2 \Rightarrow$$

$dt = -2x\,dx$. Thus, $I = \frac{\pi}{6} + \frac{1}{2}\int_{3/4}^1 t^{-1/2}\,dt = \frac{\pi}{6} + \left[\sqrt{t}\right]_{3/4}^1 = \frac{\pi}{6} + 1 - \frac{\sqrt{3}}{2} = \frac{1}{6}\left(\pi + 6 - 3\sqrt{3}\right)$.

25. Let $u = (\ln x)^2$, $dv = dx$ $\Rightarrow$ $du = \dfrac{2}{x}\ln x\,dx$, $v = x$. By (6), $I = \int_1^2 (\ln x)^2\,dx = \left[x(\ln x)^2\right]_1^2 - 2\int_1^2 \ln x\,dx$.

To evaluate the last integral, let $U = \ln x$, $dV = dx$ $\Rightarrow$ $dU = \dfrac{1}{x}\,dx$, $V = x$. Thus,

$$I = \left[x(\ln x)^2\right]_1^2 - 2\left(\left[x \ln x\right]_1^2 - \int_1^2 dx\right) = \left[x(\ln x)^2 - 2x \ln x + 2x\right]_1^2$$

$$= \left(2(\ln 2)^2 - 4 \ln 2 + 4\right) - (0 - 0 + 2) = 2(\ln 2)^2 - 4 \ln 2 + 2$$

27. Let $y = \sqrt{x}$, so that $dy = \frac{1}{2}x^{-1/2}\,dx = \dfrac{1}{2\sqrt{x}}\,dx = \dfrac{1}{2y}\,dx$. Thus, $\int \cos \sqrt{x}\,dx = \int \cos y\,(2y\,dy) = 2\int y \cos y\,dy$. Now

use parts with $u = y$, $dv = \cos y\,dy$, $du = dy$, $v = \sin y$ to get $\int y \cos y\,dy = y \sin y - \int \sin y\,dy = y \sin y + \cos y + C_1$,

so $\int \cos \sqrt{x}\,dx = 2y \sin y + 2 \cos y + C = 2\sqrt{x} \sin \sqrt{x} + 2 \cos \sqrt{x} + C$.

29. Let $x = \theta^2$, so that $dx = 2\theta\, d\theta$. Thus, $\displaystyle\int_{\sqrt{\pi/2}}^{\sqrt{\pi}} \theta^3 \cos(\theta^2)\, d\theta = \int_{\sqrt{\pi/2}}^{\sqrt{\pi}} \theta^2 \cos(\theta^2) \cdot \tfrac{1}{2}(2\theta\, d\theta) = \tfrac{1}{2}\int_{\pi/2}^{\pi} x \cos x\, dx$. Now use

parts with $u = x$, $dv = \cos x\, dx$, $du = dx$, $v = \sin x$ to get

$$\tfrac{1}{2}\int_{\pi/2}^{\pi} x \cos x\, dx = \tfrac{1}{2}\left(\big[x \sin x\big]_{\pi/2}^{\pi} - \int_{\pi/2}^{\pi} \sin x\, dx \right) = \tfrac{1}{2}\big[x \sin x + \cos x\big]_{\pi/2}^{\pi}$$

$$= \tfrac{1}{2}(\pi \sin \pi + \cos \pi) - \tfrac{1}{2}\left(\tfrac{\pi}{2}\sin\tfrac{\pi}{2} + \cos\tfrac{\pi}{2}\right) = \tfrac{1}{2}(\pi \cdot 0 - 1) - \tfrac{1}{2}\left(\tfrac{\pi}{2}\cdot 1 + 0\right) = -\tfrac{1}{2} - \tfrac{\pi}{4}$$

31. (a) Take $n = 2$ in Example 6 to get $\displaystyle\int \sin^2 x\, dx = -\tfrac{1}{2}\cos x \sin x + \tfrac{1}{2}\int 1\, dx = \dfrac{x}{2} - \dfrac{\sin 2x}{4} + C$.

(b) $\displaystyle\int \sin^4 x\, dx = -\tfrac{1}{4}\cos x \sin^3 x + \tfrac{3}{4}\int \sin^2 x\, dx = -\tfrac{1}{4}\cos x \sin^3 x + \tfrac{3}{8}x - \tfrac{3}{16}\sin 2x + C$.

33. (a) From Example 6, $\displaystyle\int \sin^n x\, dx = -\dfrac{1}{n}\cos x \sin^{n-1} x + \dfrac{n-1}{n}\int \sin^{n-2} x\, dx$. Using (6),

$$\int_0^{\pi/2} \sin^n x\, dx = \left[-\dfrac{\cos x \sin^{n-1} x}{n}\right]_0^{\pi/2} + \dfrac{n-1}{n}\int_0^{\pi/2} \sin^{n-2} x\, dx$$

$$= (0 - 0) + \dfrac{n-1}{n}\int_0^{\pi/2} \sin^{n-2} x\, dx = \dfrac{n-1}{n}\int_0^{\pi/2} \sin^{n-2} x\, dx$$

(b) Using $n = 3$ in part (a), we have $\int_0^{\pi/2} \sin^3 x\, dx = \tfrac{2}{3}\int_0^{\pi/2} \sin x\, dx = \left[-\tfrac{2}{3}\cos x\right]_0^{\pi/2} = \tfrac{2}{3}$.

Using $n = 5$ in part (a), we have $\int_0^{\pi/2} \sin^5 x\, dx = \tfrac{4}{5}\int_0^{\pi/2} \sin^3 x\, dx = \tfrac{4}{5}\cdot\tfrac{2}{3} = \tfrac{8}{15}$.

(c) The formula holds for $n = 1$ (that is, $2n + 1 = 3$) by (b). Assume it holds for some $k \geq 1$. Then

$$\int_0^{\pi/2} \sin^{2k+1} x\, dx = \dfrac{2 \cdot 4 \cdot 6 \cdots (2k)}{3 \cdot 5 \cdot 7 \cdots (2k+1)}. \text{ By Example 6,}$$

$$\int_0^{\pi/2} \sin^{2k+3} x\, dx = \dfrac{2k+2}{2k+3}\int_0^{\pi/2} \sin^{2k+1} x\, dx = \dfrac{2k+2}{2k+3} \cdot \dfrac{2 \cdot 4 \cdot 6 \cdots (2k)}{3 \cdot 5 \cdot 7 \cdots (2k+1)}$$

$$= \dfrac{2 \cdot 4 \cdot 6 \cdots (2k)[2(k+1)]}{3 \cdot 5 \cdot 7 \cdots (2k+1)[2(k+1)+1]},$$

so the formula holds for $n = k + 1$. By induction, the formula holds for all $n \geq 1$.

35. Let $u = (\ln x)^n$, $dv = dx$ $\Rightarrow$ $du = n(\ln x)^{n-1}(dx/x)$, $v = x$. By Equation 2,

$\int (\ln x)^n\, dx = x(\ln x)^n - \int nx(\ln x)^{n-1}(dx/x) = x(\ln x)^n - n\int (\ln x)^{n-1}\, dx$.

37. $\int \tan^n x\, dx = \int \tan^{n-2} x \tan^2 x\, dx = \int \tan^{n-2} x\, (\sec^2 x - 1)\, dx = \int \tan^{n-2} x \sec^2 x\, dx - \int \tan^{n-2} x\, dx$

$= I - \int \tan^{n-2} x\, dx.$

Let $u = \tan^{n-2} x$, $dv = \sec^2 x\, dx$ $\Rightarrow$ $du = (n-2)\tan^{n-3} x \sec^2 x\, dx$, $v = \tan x$. Then, by Equation 2,

$$I = \tan^{n-1} x - (n-2)\int \tan^{n-2} x \sec^2 x\, dx$$

$$1I = \tan^{n-1} x - (n-2)I$$

$$(n-1)I = \tan^{n-1} x$$

$$I = \dfrac{\tan^{n-1} x}{n-1}$$

Returning to the original integral, $\int \tan^n x\, dx = \dfrac{\tan^{n-1} x}{n-1} - \int \tan^{n-2} x\, dx$.

39. Take $n = 3$ in Exercise 35 to get $\int (\ln x)^3 \, dx = x \, (\ln x)^3 - 3 \int (\ln x)^2 \, dx = x(\ln x)^3 - 3x(\ln x)^2 + 6x \ln x - 6x + C$

[by Exercise 25].

Or: Instead of using Exercise 25, apply Exercise 35 again with $n = 2$.

41. $f_{\text{ave}} = \dfrac{1}{b-a} \displaystyle\int_a^b f(x) \, dx = \dfrac{1}{\pi/4 - 0} \int_0^{\pi/4} x \sec^2 x \, dx \qquad \begin{bmatrix} u = x, & dv = \sec^2 x \, dx \\ du = dx, & v = \tan x \end{bmatrix}$

$\qquad = \dfrac{4}{\pi} \left\{ \Big[x \tan x \Big]_0^{\pi/4} - \int_0^{\pi/4} \tan x \, dx \right\} = \dfrac{4}{\pi} \left\{ \dfrac{\pi}{4} - \Big[\ln |\sec x| \Big]_0^{\pi/4} \right\} = \dfrac{4}{\pi} \left(\dfrac{\pi}{4} - \ln \sqrt{2} \right)$

$\qquad = 1 - \dfrac{4}{\pi} \ln \sqrt{2} \ \text{ or } \ 1 - \dfrac{2}{\pi} \ln 2$

43. Since $v(t) > 0$ for all t, the desired distance is $s(t) = \int_0^t v(w) \, dw = \int_0^t w^2 e^{-w} \, dw$.

First let $u = w^2$, $dv = e^{-w} \, dw \ \Rightarrow \ du = 2w \, dw$, $v = -e^{-w}$. Then $s(t) = \big[-w^2 e^{-w} \big]_0^t + 2 \int_0^t w e^{-w} \, dw$.

Next let $U = w$, $dV = e^{-w} \, dw \ \Rightarrow \ dU = dw$, $V = -e^{-w}$. Then

$$s(t) = -t^2 e^{-t} + 2\left(\big[-w e^{-w} \big]_0^t + \int_0^t e^{-w} \, dw \right) = -t^2 e^{-t} + 2\left(-t e^{-t} + 0 + \big[-e^{-w} \big]_0^t \right)$$

$$= -t^2 e^{-t} + 2(-t e^{-t} - e^{-t} + 1) = -t^2 e^{-t} - 2t e^{-t} - 2e^{-t} + 2 = 2 - e^{-t}(t^2 + 2t + 2) \text{ meters}$$

45. For $I = \int_1^4 x f''(x) \, dx$, let $u = x$, $dv = f''(x) \, dx \ \Rightarrow \ du = dx$, $v = f'(x)$. Then

$I = \big[x f'(x) \big]_1^4 - \int_1^4 f'(x) \, dx = 4 f'(4) - 1 \cdot f'(1) - [f(4) - f(1)] = 4 \cdot 3 - 1 \cdot 5 - (7 - 2) = 12 - 5 - 5 = 2.$

We used the fact that f'' is continuous to guarantee that I exists.

6.2 Trigonometric Integrals and Substitutions

The symbols $\overset{s}{=}$ and $\overset{c}{=}$ indicate the use of the substitutions $\{u = \sin x, du = \cos x \, dx\}$ and $\{u = \cos x, du = -\sin x \, dx\}$, respectively.

1. $\int \sin^2 x \cos^3 x \, dx = \int \sin^2 x \cos^2 x \cos x \, dx = \int \sin^2 x \, (1 - \sin^2 x) \cos x \, dx$

$\qquad \overset{s}{=} \int u^2 (1 - u^2) \, du = \int (u^2 - u^4) \, du = \dfrac{1}{3} u^3 - \dfrac{1}{5} u^5 + C = \dfrac{1}{3} \sin^3 x - \dfrac{1}{5} \sin^5 x + C$

3. $\int_0^{\pi/2} \sin^7 \theta \cos^5 \theta \, d\theta = \int_0^{\pi/2} \sin^7 \theta \cos^4 \theta \cos \theta \, d\theta = \int_0^{\pi/2} \sin^7 \theta \, (1 - \sin^2 \theta)^2 \cos \theta \, d\theta$

$\qquad \overset{s}{=} \int_0^1 u^7 (1 - u^2)^2 \, du = \int_0^1 u^7 (1 - 2u^2 + u^4) \, du = \int_0^1 (u^7 - 2u^9 + u^{11}) \, du$

$\qquad = \left[\dfrac{1}{8} u^8 - \dfrac{1}{5} u^{10} + \dfrac{1}{12} u^{12} \right]_0^1 = \left(\dfrac{1}{8} - \dfrac{1}{5} + \dfrac{1}{12} \right) - 0 = \dfrac{15 - 24 + 10}{120} = \dfrac{1}{120}$

5. $\int_0^{\pi/2} \cos^2 \theta \, d\theta = \int_0^{\pi/2} \dfrac{1}{2}(1 + \cos 2\theta) \, d\theta \qquad$ [half-angle identity]

$\qquad = \dfrac{1}{2} \Big[\theta + \dfrac{1}{2} \sin 2\theta \Big]_0^{\pi/2} = \dfrac{1}{2} \Big[\big(\dfrac{\pi}{2} + 0 \big) - (0 + 0) \Big] = \dfrac{\pi}{4}$

7. $\int_0^{\pi} \cos^4 (2t) \, dt = \int_0^{\pi} [\cos^2 (2t)]^2 \, dt = \int_0^{\pi} \Big[\dfrac{1}{2}(1 + \cos(2 \cdot 2t)) \Big]^2 \, dt \qquad$ [half-angle identity]

$\qquad = \dfrac{1}{4} \int_0^{\pi} [1 + 2 \cos 4t + \cos^2 (4t)] \, dt = \dfrac{1}{4} \int_0^{\pi} [1 + 2 \cos 4t + \dfrac{1}{2}(1 + \cos 8t)] \, dt$

$\qquad = \dfrac{1}{4} \int_0^{\pi} \big(\dfrac{3}{2} + 2 \cos 4t + \dfrac{1}{2} \cos 8t \big) \, dt = \dfrac{1}{4} \Big[\dfrac{3}{2} t + \dfrac{1}{2} \sin 4t + \dfrac{1}{16} \sin 8t \Big]_0^{\pi} = \dfrac{1}{4} \Big[\big(\dfrac{3}{2} \pi + 0 + 0 \big) - 0 \Big] = \dfrac{3}{8} \pi$

9. $\int_0^{\pi/2} \sin^2 x \cos^2 x \, dx = \int_0^{\pi/2} \frac{1}{4}(4 \sin^2 x \cos^2 x) \, dx = \int_0^{\pi/2} \frac{1}{4}(2 \sin x \cos x)^2 dx = \frac{1}{4} \int_0^{\pi/2} \sin^2 2x \, dx$

$= \frac{1}{4} \int_0^{\pi/2} \frac{1}{2}(1 - \cos 4x) \, dx = \frac{1}{8} \int_0^{\pi/2} (1 - \cos 4x) \, dx = \frac{1}{8} \left[x - \frac{1}{4} \sin 4x \right]_0^{\pi/2} = \frac{1}{8}\left(\frac{\pi}{2}\right) = \frac{\pi}{16}$

11. $\int t \sin^2 t \, dt = \int t \left[\frac{1}{2}(1 - \cos 2t) \right] \, dt = \frac{1}{2} \int (t - t \cos 2t) \, dt = \frac{1}{2} \int t \, dt - \frac{1}{2} \int t \cos 2t \, dt$

$= \frac{1}{2}\left(\frac{1}{2}t^2\right) - \frac{1}{2}\left(\frac{1}{2}t \sin 2t - \int \frac{1}{2} \sin 2t \, dt\right) \qquad \begin{bmatrix} u = t, & dv = \cos 2t \, dt \\ du = dt, & v = \frac{1}{2} \sin 2t \end{bmatrix}$

$= \frac{1}{4}t^2 - \frac{1}{4}t \sin 2t + \frac{1}{2}\left(-\frac{1}{4} \cos 2t\right) + C = \frac{1}{4}t^2 - \frac{1}{4}t \sin 2t - \frac{1}{8} \cos 2t + C$

13. $\int \cos^2 x \tan^3 x \, dx = \int \frac{\sin^3 x}{\cos x} \, dx \overset{c}{=} \int \frac{(1 - u^2)(-du)}{u} = \int \left[\frac{-1}{u} + u \right] du$

$= -\ln|u| + \frac{1}{2}u^2 + C = \frac{1}{2} \cos^2 x - \ln|\cos x| + C$

15. $\int \frac{1 - \sin x}{\cos x} \, dx = \int (\sec x - \tan x) \, dx = \ln|\sec x + \tan x| - \ln|\sec x| + C \qquad \begin{bmatrix} \text{by (1) and the boxed} \\ \text{formula above it} \end{bmatrix}$

$= \ln|(\sec x + \tan x) \cos x| + C = \ln|1 + \sin x| + C$

$= \ln(1 + \sin x) + C \quad \text{since } 1 + \sin x \geq 0$

Or: $\int \frac{1 - \sin x}{\cos x} \, dx = \int \frac{1 - \sin x}{\cos x} \cdot \frac{1 + \sin x}{1 + \sin x} \, dx = \int \frac{(1 - \sin^2 x) dx}{\cos x \, (1 + \sin x)} = \int \frac{\cos x \, dx}{1 + \sin x}$

$= \int \frac{dw}{w} \qquad [\text{where } w = 1 + \sin x, \, dw = \cos x \, dx]$

$= \ln|w| + C = \ln|1 + \sin x| + C = \ln(1 + \sin x) + C$

17. $\int \tan x \sec^3 x \, dx = \int \tan x \sec x \sec^2 x \, dx = \int u^2 \, du \qquad [u = \sec x, \, du = \sec x \tan x \, dx]$

$= \frac{1}{3}u^3 + C = \frac{1}{3} \sec^3 x + C$

19. $\int \tan^2 x \, dx = \int (\sec^2 x - 1) \, dx = \tan x - x + C$

21. Let $u = \tan x$. Then $du = \sec^2 x \, dx$, so

$\int \tan^4 x \sec^6 x \, dx = \int \tan^4 x \sec^4 x \, (\sec^2 x \, dx) = \int \tan^4 x (1 + \tan^2 x)^2 \, (\sec^2 x \, dx)$

$= \int u^4 (1 + u^2)^2 \, du = \int (u^8 + 2u^6 + u^4) \, du$

$= \frac{1}{9}u^9 + \frac{2}{7}u^7 + \frac{1}{5}u^5 + C = \frac{1}{9} \tan^9 x + \frac{2}{7} \tan^7 x + \frac{1}{5} \tan^5 x + C$

23. $\int_0^{\pi/3} \tan^5 x \sec^4 x \, dx = \int_0^{\pi/3} \tan^5 x \, (\tan^2 x + 1) \sec^2 x \, dx = \int_0^{\sqrt{3}} u^5 (u^2 + 1) \, du \qquad [u = \tan x, \, du = \sec^2 x \, dx]$

$= \int_0^{\sqrt{3}} (u^7 + u^5) \, du = \left[\frac{1}{8}u^8 + \frac{1}{6}u^6 \right]_0^{\sqrt{3}} = \frac{81}{8} + \frac{27}{6} = \frac{81}{8} + \frac{9}{2} = \frac{81}{8} + \frac{36}{8} = \frac{117}{8}$

Alternate solution:

$\int_0^{\pi/3} \tan^5 x \sec^4 x \, dx = \int_0^{\pi/3} \tan^4 x \sec^3 x \sec x \tan x \, dx = \int_0^{\pi/3} (\sec^2 x - 1)^2 \sec^3 x \sec x \tan x \, dx$

$= \int_1^2 (u^2 - 1)^2 u^3 \, du \quad [u = \sec x, \, du = \sec x \tan x \, dx] \quad = \int_1^2 (u^4 - 2u^2 + 1) u^3 \, du$

$= \int_1^2 (u^7 - 2u^5 + u^3) \, du = \left[\frac{1}{8}u^8 - \frac{1}{3}u^6 + \frac{1}{4}u^4 \right]_1^2 = \left(32 - \frac{64}{3} + 4\right) - \left(\frac{1}{8} - \frac{1}{3} + \frac{1}{4}\right) = \frac{117}{8}$

25. $\int \tan^3 x \sec x \, dx = \int \tan^2 x \sec x \tan x \, dx = \int (\sec^2 x - 1) \sec x \tan x \, dx$

$$= \int (u^2 - 1) \, du \quad [u = \sec x, du = \sec x \tan x \, dx] \quad = \tfrac{1}{3} u^3 - u + C = \tfrac{1}{3} \sec^3 x - \sec x + C$$

27. $\int \tan^5 x \, dx = \int (\sec^2 x - 1)^2 \tan x \, dx = \int \sec^4 x \tan x \, dx - 2 \int \sec^2 x \tan x \, dx + \int \tan x \, dx$

$$= \int \sec^3 x \sec x \tan x \, dx - 2 \int \tan x \sec^2 x \, dx + \int \tan x \, dx$$

$$= \tfrac{1}{4} \sec^4 x - \tan^2 x + \ln|\sec x| + C \quad [\text{or } \tfrac{1}{4}\sec^4 x - \sec^2 x + \ln|\sec x| + C \,]$$

29. $\int_{\pi/6}^{\pi/2} \cot^2 x \, dx = \int_{\pi/6}^{\pi/2} (\csc^2 x - 1) \, dx = \left[-\cot x - x \right]_{\pi/6}^{\pi/2} = \left(0 - \tfrac{\pi}{2} \right) - \left(-\sqrt{3} - \tfrac{\pi}{6} \right) = \sqrt{3} - \tfrac{\pi}{3}$

31. $\int_{\pi/4}^{\pi/2} \cot^5 \phi \csc^3 \phi \, d\phi = \int_{\pi/4}^{\pi/2} \cot^4 \phi \csc^2 \phi \csc \phi \cot \phi \, d\phi = \int_{\pi/4}^{\pi/2} (\csc^2 \phi - 1)^2 \csc^2 \phi \csc \phi \cot \phi \, d\phi$

$$= \int_{\sqrt{2}}^{1} (u^2 - 1)^2 u^2 \, (-du) \quad [u = \csc \phi, du = -\csc \phi \cot \phi \, d\phi]$$

$$= \int_{1}^{\sqrt{2}} (u^6 - 2u^4 + u^2) \, du = \left[\tfrac{1}{7} u^7 - \tfrac{2}{5} u^5 + \tfrac{1}{3} u^3 \right]_{1}^{\sqrt{2}} = \left(\tfrac{8}{7}\sqrt{2} - \tfrac{8}{5}\sqrt{2} + \tfrac{2}{3}\sqrt{2} \right) - \left(\tfrac{1}{7} - \tfrac{2}{5} + \tfrac{1}{3} \right)$$

$$= \frac{120 - 168 + 70}{105} \sqrt{2} - \frac{15 - 42 + 35}{105} = \frac{22}{105} \sqrt{2} - \frac{8}{105}$$

33. $I = \int \csc x \, dx = \int \frac{\csc x \, (\csc x - \cot x)}{\csc x - \cot x} \, dx = \int \frac{-\csc x \cot x + \csc^2 x}{\csc x - \cot x} \, dx.$ Let $u = \csc x - \cot x \quad \Rightarrow$

$du = (-\csc x \cot x + \csc^2 x) \, dx.$ Then $I = \int du/u = \ln|u| = \ln|\csc x - \cot x| + C.$

35. $\int_0^{\pi/6} \sqrt{1 + \cos 2x} \, dx = \int_0^{\pi/6} \sqrt{1 + (2\cos^2 x - 1)} \, dx = \int_0^{\pi/6} \sqrt{2\cos^2 x} \, dx = \sqrt{2} \int_0^{\pi/6} \sqrt{\cos^2 x} \, dx$

$$= \sqrt{2} \int_0^{\pi/6} |\cos x| \, dx = \sqrt{2} \int_0^{\pi/6} \cos x \, dx \quad [\text{since } \cos x > 0 \text{ for } 0 \le x \le \pi/6]$$

$$= \sqrt{2} \left[\sin x \right]_0^{\pi/6} = \sqrt{2} \left(\tfrac{1}{2} - 0 \right) = \tfrac{1}{2}\sqrt{2}$$

37. (a) $\tfrac{1}{2}[\cos(A - B) - \cos(A + B)] = \tfrac{1}{2}[(\cos A \cos B + \sin A \sin B) - (\cos A \cos B - \sin A \sin B)]$

$$= \tfrac{1}{2}(2 \sin A \sin B) = \sin A \sin B$$

(b) By part (a): $\int \sin 5x \sin 2x \, dx = \int \tfrac{1}{2}[\cos(5x - 2x) - \cos(5x + 2x)] \, dx$

$$= \tfrac{1}{2} \int (\cos 3x - \cos 7x) \, dx = \tfrac{1}{6} \sin 3x - \tfrac{1}{14} \sin 7x + C$$

39. Let $x = 2 \sin \theta$, where $-\pi/2 \le \theta \le \pi/2$. Then $dx = 2 \cos \theta \, d\theta$ and

$\sqrt{4 - x^2} = \sqrt{4 - 4\sin^2\theta} = \sqrt{4\cos^2\theta} = 2 |\cos \theta| = 2 \cos \theta.$

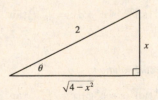

Thus, $\int \frac{dx}{x^2 \sqrt{4 - x^2}} = \int \frac{2 \cos \theta}{4 \sin^2\theta (2 \cos \theta)} \, d\theta = \frac{1}{4} \int \csc^2\theta \, d\theta$

$$= -\frac{1}{4} \cot \theta + C = -\frac{\sqrt{4 - x^2}}{4x} + C \quad [\text{see figure}]$$

41. Let $x = 2 \sec \theta$, where $0 \le \theta < \frac{\pi}{2}$ or $\pi \le \theta < \frac{3\pi}{2}$.   Then $dx = 2 \sec \theta \tan \theta \, d\theta$ and

$$\sqrt{x^2 - 4} = \sqrt{4 \sec^2 \theta - 4} = \sqrt{4(\sec^2 \theta - 1)}$$

$$= \sqrt{4 \tan^2 \theta} = 2 |\tan \theta| = 2 \tan \theta \quad \text{for the relevant values of } \theta$$

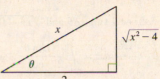

$$\int \frac{\sqrt{x^2 - 4}}{x} \, dx = \int \frac{2 \tan \theta}{2 \sec \theta} 2 \sec \theta \tan \theta \, d\theta = 2 \int \tan^2 \theta \, d\theta$$

$$= 2 \int (\sec^2 \theta - 1) \, d\theta = 2 (\tan \theta - \theta) + C = 2 \left[\frac{\sqrt{x^2 - 4}}{2} - \sec^{-1}\left(\frac{x}{2}\right) \right] + C$$

$$= \sqrt{x^2 - 4} - 2 \sec^{-1}\left(\frac{x}{2}\right) + C$$

43. Let $t = \sec \theta$, so $dt = \sec \theta \tan \theta \, d\theta$, $t = \sqrt{2} \;\Rightarrow\; \theta = \frac{\pi}{4}$, and $t = 2 \;\Rightarrow\; \theta = \frac{\pi}{3}$. Then

$$\int_{\sqrt{2}}^{2} \frac{1}{t^3 \sqrt{t^2 - 1}} \, dt = \int_{\pi/4}^{\pi/3} \frac{1}{\sec^3 \theta \tan \theta} \sec \theta \tan \theta \, d\theta = \int_{\pi/4}^{\pi/3} \frac{1}{\sec^2 \theta} \, d\theta = \int_{\pi/4}^{\pi/3} \cos^2 \theta \, d\theta$$

$$= \int_{\pi/4}^{\pi/3} \tfrac{1}{2} (1 + \cos 2\theta) \, d\theta = \tfrac{1}{2} \left[\theta + \tfrac{1}{2} \sin 2\theta \right]_{\pi/4}^{\pi/3}$$

$$= \tfrac{1}{2} \left[\left(\tfrac{\pi}{3} + \tfrac{1}{2} \tfrac{\sqrt{3}}{2} \right) - \left(\tfrac{\pi}{4} + \tfrac{1}{2} \cdot 1 \right) \right] = \tfrac{1}{2} \left(\tfrac{\pi}{12} + \tfrac{\sqrt{3}}{4} - \tfrac{1}{2} \right) = \tfrac{\pi}{24} + \tfrac{\sqrt{3}}{8} - \tfrac{1}{4}$$

45. Let $x = a \tan \theta$, where $a > 0$ and $-\frac{\pi}{2} < \theta < \frac{\pi}{2}$. Then $dx = a \sec^2 \theta \, d\theta$, $x = 0 \;\Rightarrow\; \theta = 0$, and $x = a \;\Rightarrow\; \theta = \frac{\pi}{4}$. Thus,

$$\int_0^a \frac{dx}{(a^2 + x^2)^{3/2}} = \int_0^{\pi/4} \frac{a \sec^2 \theta \, d\theta}{[a^2(1 + \tan^2 \theta)]^{3/2}} = \int_0^{\pi/4} \frac{a \sec^2 \theta \, d\theta}{a^3 \sec^3 \theta} = \frac{1}{a^2} \int_0^{\pi/4} \cos \theta \, d\theta = \frac{1}{a^2} \Big[\sin \theta \Big]_0^{\pi/4}$$

$$= \frac{1}{a^2} \left(\frac{\sqrt{2}}{2} - 0 \right) = \frac{1}{\sqrt{2} \, a^2}.$$

47. Let $x = 4 \tan \theta$, where $-\frac{\pi}{2} < \theta < \frac{\pi}{2}$. Then $dx = 4 \sec^2 \theta \, d\theta$ and

$$\sqrt{x^2 + 16} = \sqrt{16 \tan^2 \theta + 16} = \sqrt{16(\tan^2 \theta + 1)}$$

$$= \sqrt{16 \sec^2 \theta} = 4 |\sec \theta|$$

$$= 4 \sec \theta \quad \text{for the relevant values of } \theta.$$

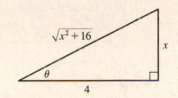

$$\int \frac{dx}{\sqrt{x^2 + 16}} = \int \frac{4 \sec^2 \theta \, d\theta}{4 \sec \theta} = \int \sec \theta \, d\theta = \ln|\sec \theta + \tan \theta| + C_1 = \ln \left| \frac{\sqrt{x^2 + 16}}{4} + \frac{x}{4} \right| + C_1$$

$$= \ln \left| \sqrt{x^2 + 16} + x \right| - \ln|4| + C_1 = \ln\left(\sqrt{x^2 + 16} + x\right) + C, \quad \text{where } C = C_1 - \ln 4.$$

(Since $\sqrt{x^2 + 16} + x > 0$, we don't need the absolute value.)

49. Let $2x = \sin \theta$, where $-\frac{\pi}{2} \le \theta \le \frac{\pi}{2}$. Then $x = \tfrac{1}{2} \sin \theta$, $dx = \tfrac{1}{2} \cos \theta \, d\theta$, and $\sqrt{1 - 4x^2} = \sqrt{1 - (2x)^2} = \cos \theta$.

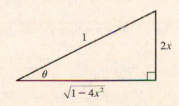

$$\int \sqrt{1 - 4x^2} \, dx = \int \cos \theta \left(\tfrac{1}{2} \cos \theta \right) d\theta = \tfrac{1}{4} \int (1 + \cos 2\theta) \, d\theta$$

$$= \tfrac{1}{4} \left(\theta + \tfrac{1}{2} \sin 2\theta \right) + C = \tfrac{1}{4} (\theta + \sin \theta \cos \theta) + C$$

$$= \tfrac{1}{4} \left[\sin^{-1}(2x) + 2x \sqrt{1 - 4x^2} \right] + C$$

51. Let $x = 3\sec\theta$, where $0 \le \theta < \frac{\pi}{2}$ or $\pi \le \theta < \frac{3\pi}{2}$. Then

$dx = 3\sec\theta\,\tan\theta\,d\theta$ and $\sqrt{x^2 - 9} = 3\tan\theta$, so

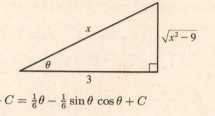

$$\int \frac{\sqrt{x^2-9}}{x^3}\,dx = \int \frac{3\tan\theta}{27\sec^3\theta}\,3\sec\theta\,\tan\theta\,d\theta = \frac{1}{3}\int \frac{\tan^2\theta}{\sec^2\theta}\,d\theta$$

$$= \frac{1}{3}\int \sin^2\theta\,d\theta = \frac{1}{3}\int \frac{1}{2}(1 - \cos 2\theta)\,d\theta = \frac{1}{6}\theta - \frac{1}{12}\sin 2\theta + C = \frac{1}{6}\theta - \frac{1}{6}\sin\theta\,\cos\theta + C$$

$$= \frac{1}{6}\sec^{-1}\!\left(\frac{x}{3}\right) - \frac{1}{6}\frac{\sqrt{x^2-9}}{x}\frac{3}{x} + C = \frac{1}{6}\sec^{-1}\!\left(\frac{x}{3}\right) - \frac{\sqrt{x^2-9}}{2x^2} + C$$

53. Let $x = \frac{3}{5}\sin\theta$, so $dx = \frac{3}{5}\cos\theta\,d\theta$, $x = 0 \Rightarrow \theta = 0$, and $x = 0.6 \Rightarrow \theta = \frac{\pi}{2}$. Then

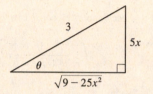

$$\int_0^{0.6} \frac{x^2}{\sqrt{9 - 25x^2}}\,dx = \int_0^{\pi/2} \frac{\left(\frac{3}{5}\right)^2\sin^2\theta}{3\cos\theta}\left(\frac{3}{5}\cos\theta\,d\theta\right) = \frac{9}{125}\int_0^{\pi/2}\sin^2\theta\,d\theta$$

$$= \frac{9}{125}\int_0^{\pi/2}\frac{1}{2}(1 - \cos 2\theta)\,d\theta = \frac{9}{250}\left[\theta - \frac{1}{2}\sin 2\theta\right]_0^{\pi/2}$$

$$= \frac{9}{250}\left[\left(\frac{\pi}{2} - 0\right) - 0\right] = \frac{9}{500}\pi$$

55. Let $u = x^2 - 7$, so $du = 2x\,dx$. Then $\displaystyle\int \frac{x}{\sqrt{x^2-7}}\,dx = \frac{1}{2}\int \frac{1}{\sqrt{u}}\,du = \frac{1}{2}\cdot 2\sqrt{u} + C = \sqrt{x^2 - 7} + C$.

57. Let $x = \tan\theta$, where $-\frac{\pi}{2} < \theta < \frac{\pi}{2}$. Then $dx = \sec^2\theta\,d\theta$

and $\sqrt{1 + x^2} = \sec\theta$, so

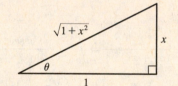

$$\int \frac{\sqrt{1+x^2}}{x}\,dx = \int \frac{\sec\theta}{\tan\theta}\sec^2\theta\,d\theta = \int \frac{\sec\theta}{\tan\theta}(1 + \tan^2\theta)\,d\theta$$

$$= \int (\csc\theta + \sec\theta\,\tan\theta)\,d\theta$$

$$= \ln|\csc\theta - \cot\theta| + \sec\theta + C \qquad \text{[by Exercise 33]}$$

$$= \ln\left|\frac{\sqrt{1+x^2}}{x} - \frac{1}{x}\right| + \frac{\sqrt{1+x^2}}{1} + C = \ln\left|\frac{\sqrt{1+x^2}-1}{x}\right| + \sqrt{1+x^2} + C$$

59. Let $u = x^2$, $du = 2x\,dx$. Then

$$\int x\sqrt{1-x^4}\,dx = \int \sqrt{1-u^2}\left(\frac{1}{2}\,du\right) = \frac{1}{2}\int \cos\theta\cdot\cos\theta\,d\theta \qquad \begin{bmatrix} \text{where } u = \sin\theta,\, du = \cos\theta\,d\theta, \\ \text{and } \sqrt{1-u^2} = \cos\theta \end{bmatrix}$$

$$= \frac{1}{2}\int \frac{1}{2}(1 + \cos 2\theta)\,d\theta = \frac{1}{4}\theta + \frac{1}{8}\sin 2\theta + C = \frac{1}{4}\theta + \frac{1}{4}\sin\theta\,\cos\theta + C$$

$$= \frac{1}{4}\sin^{-1}u + \frac{1}{4}u\sqrt{1-u^2} + C = \frac{1}{4}\sin^{-1}(x^2) + \frac{1}{4}x^2\sqrt{1-x^4} + C$$

61. $9x^2 + 6x - 8 = (3x + 1)^2 - 9$, so let $u = 3x + 1$, $du = 3\,dx$. Then $\displaystyle\int \frac{dx}{\sqrt{9x^2 + 6x - 8}} = \int \frac{\frac{1}{3}\,du}{\sqrt{u^2 - 9}}$.

Now let $u = 3\sec\theta$, where $0 \le \theta < \frac{\pi}{2}$ or $\pi \le \theta < \frac{3\pi}{2}$. Then $du = 3\sec\theta\,\tan\theta\,d\theta$ and $\sqrt{u^2 - 9} = 3\tan\theta$, so

$$\int \frac{\frac{1}{3}\,du}{\sqrt{u^2-9}} = \int \frac{\sec\theta\,\tan\theta\,d\theta}{3\tan\theta} = \frac{1}{3}\int \sec\theta\,d\theta = \frac{1}{3}\ln|\sec\theta + \tan\theta| + C_1 = \frac{1}{3}\ln\left|\frac{u + \sqrt{u^2-9}}{3}\right| + C_1$$

$$= \frac{1}{3}\ln\left|u + \sqrt{u^2-9}\right| + C = \frac{1}{3}\ln\left|3x + 1 + \sqrt{9x^2 + 6x - 8}\right| + C$$

63. $5 + 4x - x^2 = -(x^2 - 4x + 4) + 9 = -(x-2)^2 + 9$. Let

$x - 2 = 3\sin\theta$, $-\frac{\pi}{2} \le \theta \le \frac{\pi}{2}$, so $dx = 3\cos\theta\, d\theta$. Then

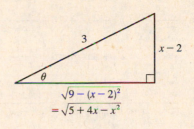

$$\int \sqrt{5 + 4x - x^2}\, dx = \int \sqrt{9 - (x-2)^2}\, dx = \int \sqrt{9 - 9\sin^2\theta}\, 3\cos\theta\, d\theta$$

$$= \int \sqrt{9\cos^2\theta}\, 3\cos\theta\, d\theta = \int 9\cos^2\theta\, d\theta$$

$$= \frac{9}{2}\int (1 + \cos 2\theta)\, d\theta = \frac{9}{2}\left(\theta + \frac{1}{2}\sin 2\theta\right) + C$$

$$= \frac{9}{2}\theta + \frac{9}{4}\sin 2\theta + C = \frac{9}{2}\theta + \frac{9}{4}(2\sin\theta\cos\theta) + C$$

$$= \frac{9}{2}\sin^{-1}\left(\frac{x-2}{3}\right) + \frac{9}{2}\cdot\frac{x-2}{3}\cdot\frac{\sqrt{5 + 4x - x^2}}{3} + C$$

$$= \frac{9}{2}\sin^{-1}\left(\frac{x-2}{3}\right) + \frac{1}{2}(x-2)\sqrt{5 + 4x - x^2} + C$$

65. $s = f(t) = \int_0^t \sin\omega u \cos^2\omega u\, du$. Let $y = \cos\omega u \;\Rightarrow\; dy = -\omega\sin\omega u\, du$. Then

$$s = -\frac{1}{\omega}\int_1^{\cos\omega t} y^2\, dy = -\frac{1}{\omega}\left[\frac{1}{3}y^3\right]_1^{\cos\omega t} = \frac{1}{3\omega}(1 - \cos^3\omega t).$$

67. The average value of $f(x) = \sqrt{x^2 - 1}/x$ on the interval $[1, 7]$ is

$$\frac{1}{7 - 1}\int_1^7 \frac{\sqrt{x^2 - 1}}{x}\, dx = \frac{1}{6}\int_0^\alpha \frac{\tan\theta}{\sec\theta}\cdot\sec\theta\tan\theta\, d\theta \qquad \left[\begin{array}{l}\text{where } x = \sec\theta,\, dx = \sec\theta\tan\theta\, d\theta, \\ \sqrt{x^2 - 1} = \tan\theta,\text{ and } \alpha = \sec^{-1}7\end{array}\right]$$

$$= \frac{1}{6}\int_0^\alpha \tan^2\theta\, d\theta = \frac{1}{6}\int_0^\alpha (\sec^2\theta - 1)\, d\theta = \frac{1}{6}\left[\tan\theta - \theta\right]_0^\alpha$$

$$= \frac{1}{6}(\tan\alpha - \alpha) = \frac{1}{6}\left(\sqrt{48} - \sec^{-1}7\right)$$

69. Area of $\triangle POQ = \frac{1}{2}(r\cos\theta)(r\sin\theta) = \frac{1}{2}r^2\sin\theta\cos\theta$. Area of region $PQR = \int_{r\cos\theta}^r \sqrt{r^2 - x^2}\, dx$.

Let $x = r\cos u \;\Rightarrow\; dx = -r\sin u\, du$ for $\theta \le u \le \frac{\pi}{2}$. Then we obtain

$$\int \sqrt{r^2 - x^2}\, dx = \int r\sin u\, (-r\sin u)\, du = -r^2\int \sin^2 u\, du = -\frac{1}{2}r^2(u - \sin u\cos u) + C$$

$$= -\frac{1}{2}r^2\cos^{-1}(x/r) + \frac{1}{2}x\sqrt{r^2 - x^2} + C$$

so   area of region $PQR = \frac{1}{2}\left[-r^2\cos^{-1}(x/r) + x\sqrt{r^2 - x^2}\right]_{r\cos\theta}^r$

$$= \frac{1}{2}\left[0 - (-r^2\theta + r\cos\theta\, r\sin\theta)\right] = \frac{1}{2}r^2\theta - \frac{1}{2}r^2\sin\theta\cos\theta$$

and thus, (area of sector POR) = (area of $\triangle POQ$) + (area of region PQR) = $\frac{1}{2}r^2\theta$.

6.3 Partial Fractions

1. (a) $\dfrac{1 + 6x}{(4x - 3)(2x + 5)} = \dfrac{A}{4x - 3} + \dfrac{B}{2x + 5}$

(b) $\dfrac{10}{5x^2 - 2x^3} = \dfrac{10}{x^2(5 - 2x)} = \dfrac{A}{x} + \dfrac{B}{x^2} + \dfrac{C}{5 - 2x}$

3. (a) $\dfrac{x^4 + 1}{x^5 + 4x^3} = \dfrac{x^4 + 1}{x^3(x^2 + 4)} = \dfrac{A}{x} + \dfrac{B}{x^2} + \dfrac{C}{x^3} + \dfrac{Dx + E}{x^2 + 4}$

(b) $\dfrac{1}{(x^2 - 9)^2} = \dfrac{1}{[(x + 3)(x - 3)]^2} = \dfrac{1}{(x + 3)^2(x - 3)^2} = \dfrac{A}{x + 3} + \dfrac{B}{(x + 3)^2} + \dfrac{C}{x - 3} + \dfrac{D}{(x - 3)^2}$

5. (a) $\dfrac{x^6}{x^2 - 4} = x^4 + 4x^2 + 16 + \dfrac{64}{(x+2)(x-2)}$ [by long division]

$$= x^4 + 4x^2 + 16 + \dfrac{A}{x+2} + \dfrac{B}{x-2}$$

(b) $\dfrac{x^4}{(x^2 - x + 1)(x^2 + 2)^2} = \dfrac{Ax + B}{x^2 - x + 1} + \dfrac{Cx + D}{x^2 + 2} + \dfrac{Ex + F}{(x^2 + 2)^2}$

7. $\displaystyle\int \dfrac{x^4}{x - 1}\, dx = \int \left(x^3 + x^2 + x + 1 + \dfrac{1}{x - 1} \right) dx$ [by division] $= \dfrac{1}{4}x^4 + \dfrac{1}{3}x^3 + \dfrac{1}{2}x^2 + x + \ln|x - 1| + C$

9. $\dfrac{5x + 1}{(2x + 1)(x - 1)} = \dfrac{A}{2x + 1} + \dfrac{B}{x - 1}$. Multiply both sides by $(2x + 1)(x - 1)$ to get $5x + 1 = A(x - 1) + B(2x + 1)$ $\Rightarrow$

$5x + 1 = Ax - A + 2Bx + B$ $\Rightarrow$ $5x + 1 = (A + 2B)x + (-A + B)$.

The coefficients of x must be equal and the constant terms are also equal, so $A + 2B = 5$ and

$-A + B = 1$. Adding these equations gives us $3B = 6$ $\Leftrightarrow$ $B = 2$, and hence, $A = 1$. Thus,

$$\int \dfrac{5x + 1}{(2x + 1)(x - 1)}\, dx = \int \left(\dfrac{1}{2x + 1} + \dfrac{2}{x - 1} \right) dx = \tfrac{1}{2}\ln|2x + 1| + 2\ln|x - 1| + C.$$

Another method: Substituting 1 for x in the equation $5x + 1 = A(x - 1) + B(2x + 1)$ gives $6 = 3B$ $\Leftrightarrow$ $B = 2$.

Substituting $-\tfrac{1}{2}$ for x gives $-\tfrac{3}{2} = -\tfrac{3}{2}A$ $\Leftrightarrow$ $A = 1$.

11. $\dfrac{2}{2x^2 + 3x + 1} = \dfrac{2}{(2x + 1)(x + 1)} = \dfrac{A}{2x + 1} + \dfrac{B}{x + 1}$.   Multiply both sides by $(2x + 1)(x + 1)$ to get

$2 = A(x + 1) + B(2x + 1)$. The coefficients of x must be equal and the constant terms are also equal, so $A + 2B = 0$ and

$A + B = 2$. Subtracting the second equation from the first gives $B = -2$, and hence, $A = 4$. Thus,

$$\int_0^1 \dfrac{2}{2x^2 + 3x + 1}\, dx = \int_0^1 \left(\dfrac{4}{2x + 1} - \dfrac{2}{x + 1} \right) dx = \left[\dfrac{4}{2}\ln|2x + 1| - 2\ln|x + 1| \right]_0^1 = (2\ln 3 - 2\ln 2) - 0 = 2\ln\dfrac{3}{2}.$$

Another method: Substituting -1 for x in the equation $2 = A(x + 1) + B(2x + 1)$ gives $2 = -B$ $\Leftrightarrow$ $B = -2$.

Substituting $-\tfrac{1}{2}$ for x gives $2 = \tfrac{1}{2}A$ $\Leftrightarrow$ $A = 4$.

13. $\displaystyle\int \dfrac{ax}{x^2 - bx}\, dx = \int \dfrac{ax}{x(x - b)}\, dx = \int \dfrac{a}{x - b}\, dx = a\ln|x - b| + C$

15. $\dfrac{2x + 3}{(x + 1)^2} = \dfrac{A}{x + 1} + \dfrac{B}{(x + 1)^2}$ $\Rightarrow$ $2x + 3 = A(x + 1) + B$. Take $x = -1$ to get $B = 1$, and equate coefficients

of x to get $A = 2$. Now

$$\int_0^1 \dfrac{2x + 3}{(x + 1)^2}\, dx = \int_0^1 \left[\dfrac{2}{x + 1} + \dfrac{1}{(x + 1)^2} \right] dx = \left[2\ln(x + 1) - \dfrac{1}{x + 1} \right]_0^1$$

$$= 2\ln 2 - \tfrac{1}{2} - (2\ln 1 - 1) = 2\ln 2 + \tfrac{1}{2}$$

17. $\dfrac{4y^2 - 7y - 12}{y(y+2)(y-3)} = \dfrac{A}{y} + \dfrac{B}{y+2} + \dfrac{C}{y-3}$ ⇒ $4y^2 - 7y - 12 = A(y+2)(y-3) + By(y-3) + Cy(y+2)$. Setting

$y = 0$ gives $-12 = -6A$, so $A = 2$. Setting $y = -2$ gives $18 = 10B$, so $B = \frac{9}{5}$. Setting $y = 3$ gives $3 = 15C$, so $C = \frac{1}{5}$.

Now

$$\int_1^2 \frac{4y^2 - 7y - 12}{y(y+2)(y-3)}\, dy = \int_1^2 \left(\frac{2}{y} + \frac{9/5}{y+2} + \frac{1/5}{y-3} \right) dy = \left[2\ln|y| + \tfrac{9}{5}\ln|y+2| + \tfrac{1}{5}\ln|y-3| \right]_1^2$$

$$= 2\ln 2 + \tfrac{9}{5}\ln 4 + \tfrac{1}{5}\ln 1 - 2\ln 1 - \tfrac{9}{5}\ln 3 - \tfrac{1}{5}\ln 2$$

$$= 2\ln 2 + \tfrac{18}{5}\ln 2 - \tfrac{1}{5}\ln 2 - \tfrac{9}{5}\ln 3 = \tfrac{27}{5}\ln 2 - \tfrac{9}{5}\ln 3 = \tfrac{9}{5}(3\ln 2 - \ln 3) = \tfrac{9}{5}\ln\tfrac{8}{3}$$

19. $\dfrac{x^2+1}{(x-3)(x-2)^2} = \dfrac{A}{x-3} + \dfrac{B}{x-2} + \dfrac{C}{(x-2)^2}$. Multiply both sides by $(x-3)(x-2)^2$ to get

$x^2 + 1 = A(x-2)^2 + B(x-3)(x-2) + C(x-3)$. Setting $x = 2$ gives $5 = -C$ ⇔ $C = -5$. Setting $x = 3$

gives $10 = A$. Equating coefficients of x^2 gives $1 = A + B$, so $B = -9$. Thus,

$$\int \frac{x^2+1}{(x-3)(x-2)^2}\, dx = \int \left(\frac{10}{x-3} - \frac{9}{x-2} - \frac{5}{(x-2)^2} \right) dx = 10\ln|x-3| - 9\ln|x-2| + \frac{5}{x-2} + C$$

21.

$$x^2 + 4 \overline{\smash{\big)}\ \begin{array}{r} x \phantom{{}+4} \\ x^3 + 0x^2 + 0x + 4 \\ \underline{x^3 \phantom{{}+0x^2} + 4x} \\ -4x + 4 \end{array}}$$

By long division, $\dfrac{x^3+4}{x^2+4} = x + \dfrac{-4x+4}{x^2+4}$. Thus,

$$\int \frac{x^3+4}{x^2+4}\, dx = \int \left(x + \frac{-4x+4}{x^2+4} \right) dx = \int \left(x - \frac{4x}{x^2+4} + \frac{4}{x^2+2^2} \right) dx$$

$$= \tfrac{1}{2}x^2 - 4 \cdot \tfrac{1}{2}\ln|x^2+4| + 4 \cdot \tfrac{1}{2}\tan^{-1}\left(\frac{x}{2}\right) + C = \tfrac{1}{2}x^2 - 2\ln(x^2+4) + 2\tan^{-1}\left(\frac{x}{2}\right) + C$$

23. $\dfrac{10}{(x-1)(x^2+9)} = \dfrac{A}{x-1} + \dfrac{Bx+C}{x^2+9}$. Multiply both sides by $(x-1)(x^2+9)$ to get

$10 = A(x^2+9) + (Bx+C)(x-1)$ $(\star)$. Substituting 1 for x gives $10 = 10A$ ⇔ $A = 1$. Substituting 0 for x gives

$10 = 9A - C$ ⇒ $C = 9(1) - 10 = -1$. The coefficients of the x^2-terms in $(\star)$ must be equal, so $0 = A + B$ ⇒

$B = -1$. Thus,

$$\int \frac{10}{(x-1)(x^2+9)}\, dx = \int \left(\frac{1}{x-1} + \frac{-x-1}{x^2+9} \right) dx = \int \left(\frac{1}{x-1} - \frac{x}{x^2+9} - \frac{1}{x^2+9} \right) dx$$

$$= \ln|x-1| - \tfrac{1}{2}\ln(x^2+9) - \tfrac{1}{3}\tan^{-1}\left(\frac{x}{3}\right) + C$$

In the second term we used the substitution $u = x^2 + 9$ and in the last term we used Formula 9.

25. $\dfrac{x^3 + x^2 + 2x + 1}{(x^2 + 1)(x^2 + 2)} = \dfrac{Ax + B}{x^2 + 1} + \dfrac{Cx + D}{x^2 + 2}$. Multiply both sides by $(x^2 + 1)(x^2 + 2)$ to get

$$x^3 + x^2 + 2x + 1 = (Ax + B)(x^2 + 2) + (Cx + D)(x^2 + 1) \quad \Leftrightarrow$$

$$x^3 + x^2 + 2x + 1 = (Ax^3 + Bx^2 + 2Ax + 2B) + (Cx^3 + Dx^2 + Cx + D) \quad \Leftrightarrow$$

$x^3 + x^2 + 2x + 1 = (A + C)x^3 + (B + D)x^2 + (2A + C)x + (2B + D)$. Comparing coefficients gives us the following

system of equations:

$$A + C = 1 \quad \textbf{(1)} \qquad B + D = 1 \quad \textbf{(2)}$$
$$2A + C = 2 \quad \textbf{(3)} \qquad 2B + D = 1 \quad \textbf{(4)}$$

Subtracting equation **(1)** from equation **(3)** gives us $A = 1$, so $C = 0$. Subtracting equation **(2)** from equation **(4)** gives us

$B = 0$, so $D = 1$. Thus, $I = \displaystyle\int \dfrac{x^3 + x^2 + 2x + 1}{(x^2 + 1)(x^2 + 2)}\,dx = \int \left(\dfrac{x}{x^2 + 1} + \dfrac{1}{x^2 + 2} \right) dx$. For $\displaystyle\int \dfrac{x}{x^2 + 1}\,dx$, let $u = x^2 + 1$

so $du = 2x\,dx$ and then $\displaystyle\int \dfrac{x}{x^2 + 1}\,dx = \dfrac{1}{2}\int \dfrac{1}{u}\,du = \dfrac{1}{2}\ln|u| + C = \dfrac{1}{2}\ln(x^2 + 1) + C$. For $\displaystyle\int \dfrac{1}{x^2 + 2}\,dx$, use

Formula 10 with $a = \sqrt{2}$. So $\displaystyle\int \dfrac{1}{x^2 + 2}\,dx = \int \dfrac{1}{x^2 + (\sqrt{2})^2}\,dx = \dfrac{1}{\sqrt{2}}\tan^{-1}\dfrac{x}{\sqrt{2}} + C$.

Thus, $I = \dfrac{1}{2}\ln(x^2 + 1) + \dfrac{1}{\sqrt{2}}\tan^{-1}\dfrac{x}{\sqrt{2}} + C$.

27. $\displaystyle\int \dfrac{x + 4}{x^2 + 2x + 5}\,dx = \int \dfrac{x + 1}{x^2 + 2x + 5}\,dx + \int \dfrac{3}{x^2 + 2x + 5}\,dx = \dfrac{1}{2}\int \dfrac{(2x + 2)\,dx}{x^2 + 2x + 5} + \int \dfrac{3\,dx}{(x + 1)^2 + 4}$

$\qquad = \dfrac{1}{2}\ln|x^2 + 2x + 5| + 3\displaystyle\int \dfrac{2\,du}{4(u^2 + 1)} \qquad \begin{bmatrix} \text{where } x + 1 = 2u, \\ \text{and } dx = 2\,du \end{bmatrix}$

$\qquad = \dfrac{1}{2}\ln(x^2 + 2x + 5) + \dfrac{3}{2}\tan^{-1}u + C = \dfrac{1}{2}\ln(x^2 + 2x + 5) + \dfrac{3}{2}\tan^{-1}\left(\dfrac{x + 1}{2} \right) + C$

29. $\dfrac{1}{x^3 - 1} = \dfrac{1}{(x - 1)(x^2 + x + 1)} = \dfrac{A}{x - 1} + \dfrac{Bx + C}{x^2 + x + 1} \quad \Rightarrow \quad 1 = A(x^2 + x + 1) + (Bx + C)(x - 1)$.

Take $x = 1$ to get $A = \frac{1}{3}$. Equating coefficients of x^2 and then comparing the constant terms, we get $0 = \frac{1}{3} + B$, $1 = \frac{1}{3} - C$,

so $B = -\frac{1}{3}$, $C = -\frac{2}{3} \quad \Rightarrow$

$$\int \dfrac{1}{x^3 - 1}\,dx = \int \dfrac{\frac{1}{3}}{x - 1}\,dx + \int \dfrac{-\frac{1}{3}x - \frac{2}{3}}{x^2 + x + 1}\,dx = \frac{1}{3}\ln|x - 1| - \frac{1}{3}\int \dfrac{x + 2}{x^2 + x + 1}\,dx$$

$$= \frac{1}{3}\ln|x - 1| - \frac{1}{3}\int \dfrac{x + 1/2}{x^2 + x + 1}\,dx - \frac{1}{3}\int \dfrac{(3/2)\,dx}{(x + 1/2)^2 + 3/4}$$

$$= \frac{1}{3}\ln|x - 1| - \frac{1}{6}\ln(x^2 + x + 1) - \frac{1}{2}\left(\dfrac{2}{\sqrt{3}} \right)\tan^{-1}\left(\dfrac{x + \frac{1}{2}}{\sqrt{3}/2} \right) + K$$

$$= \frac{1}{3}\ln|x - 1| - \frac{1}{6}\ln(x^2 + x + 1) - \frac{1}{\sqrt{3}}\tan^{-1}\left(\frac{1}{\sqrt{3}}(2x + 1) \right) + K$$

31. $\dfrac{1}{x(x^2+4)^2} = \dfrac{A}{x} + \dfrac{Bx+C}{x^2+4} + \dfrac{Dx+E}{(x^2+4)^2}$ ⇒ $1 = A(x^2+4)^2 + (Bx+C)x(x^2+4) + (Dx+E)x$. Setting $x = 0$

gives $1 = 16A$, so $A = \frac{1}{16}$. Now compare coefficients.

$$1 = \tfrac{1}{16}(x^4 + 8x^2 + 16) + (Bx^2 + Cx)(x^2 + 4) + Dx^2 + Ex$$

$$1 = \tfrac{1}{16}x^4 + \tfrac{1}{2}x^2 + 1 + Bx^4 + Cx^3 + 4Bx^2 + 4Cx + Dx^2 + Ex$$

$$1 = \left(\tfrac{1}{16} + B\right)x^4 + Cx^3 + \left(\tfrac{1}{2} + 4B + D\right)x^2 + (4C + E)x + 1$$

So $B + \frac{1}{16} = 0$ ⇒ $B = -\frac{1}{16}, C = 0, \frac{1}{2} + 4B + D = 0$ ⇒ $D = -\frac{1}{4}$, and $4C + E = 0$ ⇒ $E = 0$. Thus,

$$\int \frac{dx}{x(x^2+4)^2} = \int \left(\frac{\frac{1}{16}}{x} + \frac{-\frac{1}{16}x}{x^2+4} + \frac{-\frac{1}{4}x}{(x^2+4)^2}\right) dx = \frac{1}{16}\ln|x| - \frac{1}{16}\cdot\frac{1}{2}\ln|x^2+4| - \frac{1}{4}\left(-\frac{1}{2}\right)\frac{1}{x^2+4} + C$$

$$= \frac{1}{16}\ln|x| - \frac{1}{32}\ln(x^2+4) + \frac{1}{8(x^2+4)} + C$$

33. $\displaystyle\int \frac{x-3}{(x^2+2x+4)^2}\,dx = \int \frac{x-3}{(x^2+2x+4)^2}\,dx = \int \frac{x-3}{[(x+1)^2+3]^2}\,dx = \int \frac{u-4}{(u^2+3)^2}\,du$ [with $u = x+1$]

$$= \int \frac{u\,du}{(u^2+3)^2} - 4\int \frac{du}{(u^2+3)^2} = \frac{1}{2}\int \frac{dv}{v^2} - 4\int \frac{\sqrt{3}\sec^2\theta\,d\theta}{9\sec^4\theta} \quad \begin{bmatrix} v = u^2 + 3 \text{ in the first integral;} \\ u = \sqrt{3}\tan\theta \text{ in the second} \end{bmatrix}$$

$$= \frac{-1}{(2v)} - \frac{4\sqrt{3}}{9}\int \cos^2\theta\,d\theta = \frac{-1}{2(u^2+3)} - \frac{2\sqrt{3}}{9}(\theta + \sin\theta\cos\theta) + C$$

$$= \frac{-1}{2(x^2+2x+4)} - \frac{2\sqrt{3}}{9}\left[\tan^{-1}\left(\frac{x+1}{\sqrt{3}}\right) + \frac{\sqrt{3}(x+1)}{x^2+2x+4}\right] + C$$

$$= \frac{-1}{2(x^2+2x+4)} - \frac{2\sqrt{3}}{9}\tan^{-1}\left(\frac{x+1}{\sqrt{3}}\right) - \frac{2(x+1)}{3(x^2+2x+4)} + C$$

35. Let $u = \sqrt{x}$, so $u^2 = x$ and $dx = 2u\,du$. Thus,

$$\int_9^{16} \frac{\sqrt{x}}{x-4}\,dx = \int_3^4 \frac{u}{u^2-4}\,2u\,du = 2\int_3^4 \frac{u^2}{u^2-4}\,du = 2\int_3^4 \left(1 + \frac{4}{u^2-4}\right)du \quad \text{[by long division]}$$

$$= 2 + 8\int_3^4 \frac{du}{(u+2)(u-2)} \quad (\star)$$

Multiply $\dfrac{1}{(u+2)(u-2)} = \dfrac{A}{u+2} + \dfrac{B}{u-2}$ by $(u+2)(u-2)$ to get $1 = A(u-2) + B(u+2)$. Equating coefficients we

get $A + B = 0$ and $-2A + 2B = 1$. Solving gives us $B = \frac{1}{4}$ and $A = -\frac{1}{4}$, so $\dfrac{1}{(u+2)(u-2)} = \dfrac{-1/4}{u+2} + \dfrac{1/4}{u-2}$ and $(\star)$ is

$$2 + 8\int_3^4 \left(\frac{-1/4}{u+2} + \frac{1/4}{u-2}\right)du = 2 + 8\left[-\tfrac{1}{4}\ln|u+2| + \tfrac{1}{4}\ln|u-2|\right]_3^4 = 2 + \left[2\ln|u-2| - 2\ln|u+2|\right]_3^4$$

$$= 2 + 2\left[\ln\left|\frac{u-2}{u+2}\right|\right]_3^4 = 2 + 2\left(\ln\tfrac{2}{6} - \ln\tfrac{1}{5}\right) = 2 + 2\ln\frac{2/6}{1/5}$$

$$= 2 + 2\ln\tfrac{5}{3} \text{ or } 2 + \ln\left(\tfrac{5}{3}\right)^2 = 2 + \ln\tfrac{25}{9}$$

37. Let $u = \sqrt[3]{x^2 + 1}$. Then $x^2 = u^3 - 1$, $2x\,dx = 3u^2\,du$ $\Rightarrow$

$$\int \frac{x^3\,dx}{\sqrt[3]{x^2 + 1}} = \int \frac{(u^3 - 1)\frac{3}{2}u^2\,du}{u} = \frac{3}{2}\int (u^4 - u)\,du$$

$$= \frac{3}{10}u^5 - \frac{3}{4}u^2 + C = \frac{3}{10}(x^2 + 1)^{5/3} - \frac{3}{4}(x^2 + 1)^{2/3} + C$$

39. Let $u = e^x$. Then $x = \ln u$, $dx = \dfrac{du}{u}$ $\Rightarrow$

$$\int \frac{e^{2x}\,dx}{e^{2x} + 3e^x + 2} = \int \frac{u^2\,(du/u)}{u^2 + 3u + 2} = \int \frac{u\,du}{(u + 1)(u + 2)} = \int \left[\frac{-1}{u + 1} + \frac{2}{u + 2}\right] du$$

$$= 2\ln|u + 2| - \ln|u + 1| + C = \ln \frac{(e^x + 2)^2}{e^x + 1} + C$$

41. Let $u = \ln(x^2 - x + 2)$, $dv = dx$. Then $du = \dfrac{2x - 1}{x^2 - x + 2}\,dx$, $v = x$, and (by integration by parts)

$$\int \ln(x^2 - x + 2)\,dx = x\ln(x^2 - x + 2) - \int \frac{2x^2 - x}{x^2 - x + 2}\,dx = x\ln(x^2 - x + 2) - \int \left(2 + \frac{x - 4}{x^2 - x + 2}\right)dx$$

$$= x\ln(x^2 - x + 2) - 2x - \int \frac{\frac{1}{2}(2x - 1)}{x^2 - x + 2}\,dx + \frac{7}{2}\int \frac{dx}{(x - \frac{1}{2})^2 + \frac{7}{4}}$$

$$= x\ln(x^2 - x + 2) - 2x - \frac{1}{2}\ln(x^2 - x + 2) + \frac{7}{2}\int \frac{\frac{\sqrt{7}}{2}\,du}{\frac{7}{4}(u^2 + 1)} \qquad \begin{bmatrix} \text{where } x - \frac{1}{2} = \frac{\sqrt{7}}{2}u, \\ dx = \frac{\sqrt{7}}{2}\,du, \\ (x - \frac{1}{2})^2 + \frac{7}{4} = \frac{7}{4}(u^2 + 1) \end{bmatrix}$$

$$= (x - \tfrac{1}{2})\ln(x^2 - x + 2) - 2x + \sqrt{7}\tan^{-1}u + C$$

$$= (x - \tfrac{1}{2})\ln(x^2 - x + 2) - 2x + \sqrt{7}\tan^{-1}\frac{2x - 1}{\sqrt{7}} + C$$

43. $\dfrac{P + S}{P[(r - 1)P - S]} = \dfrac{A}{P} + \dfrac{B}{(r - 1)P - S}$ $\Rightarrow$ $P + S = A[(r - 1)P - S] + BP = [(r - 1)A + B]P - AS$ $\Rightarrow$

$(r - 1)A + B = 1$, $-A = 1$ $\Rightarrow$ $A = -1$, $B = r$. Now

$$t = \int \frac{P + S}{P[(r - 1)P - S]}\,dP = \int \left[\frac{-1}{P} + \frac{r}{(r - 1)P - S}\right]dP = -\int \frac{dP}{P} + \frac{r}{r - 1}\int \frac{r - 1}{(r - 1)P - S}\,dP$$

so $t = -\ln P + \dfrac{r}{r - 1}\ln|(r - 1)P - S| + C$. Here $r = 0.10$ and $S = 900$, so

$t = -\ln P + \dfrac{0.1}{-0.9}\ln|-0.9P - 900| + C = -\ln P - \tfrac{1}{9}\ln(|-1|\,|0.9P + 900|) = -\ln P - \tfrac{1}{9}\ln(0.9P + 900) + C$.

When $t = 0$, $P = 10{,}000$, so $0 = -\ln 10{,}000 - \tfrac{1}{9}\ln(9900) + C$. Thus, $C = \ln 10{,}000 + \tfrac{1}{9}\ln 9900$ [≈ 10.2326], so our equation becomes

$$t = \ln 10{,}000 - \ln P + \tfrac{1}{9}\ln 9900 - \tfrac{1}{9}\ln(0.9P + 900) = \ln \frac{10{,}000}{P} + \frac{1}{9}\ln \frac{9900}{0.9P + 900}$$

$$= \ln \frac{10{,}000}{P} + \frac{1}{9}\ln \frac{1100}{0.1P + 100} = \ln \frac{10{,}000}{P} + \frac{1}{9}\ln \frac{11{,}000}{P + 1000}$$

45. There are only finitely many values of x where $Q(x) = 0$ (assuming that Q is not the zero polynomial). At all other values of x, $F(x)/Q(x) = G(x)/Q(x)$, so $F(x) = G(x)$. In other words, the values of F and G agree at all except perhaps finitely many values of x. By continuity of F and G, the polynomials F and G must agree at those values of x too.

More explicitly: if a is a value of x such that $Q(a) = 0$, then $Q(x) \neq 0$ for all x sufficiently close to a. Thus,

$$F(a) = \lim_{x \to a} F(x) \qquad \text{[by continuity of } F\text{]}$$

$$= \lim_{x \to a} G(x) \qquad \text{[whenever } Q(x) \neq 0\text{]}$$

$$= G(a) \qquad \text{[by continuity of } G\text{]}$$

47. If $a \neq 0$ and n is a positive integer, then $f(x) = \dfrac{1}{x^n(x-a)} = \dfrac{A_1}{x} + \dfrac{A_2}{x^2} + \cdots + \dfrac{A_n}{x^n} + \dfrac{B}{x-a}$. Multiply both sides by

$x^n(x-a)$ to get $1 = A_1 x^{n-1}(x-a) + A_2 x^{n-2}(x-a) + \cdots + A_n(x-a) + Bx^n$. Let $x = a$ in the last equation to get

$1 = Ba^n \Rightarrow B = 1/a^n$. So

$$f(x) - \frac{B}{x-a} = \frac{1}{x^n(x-a)} - \frac{1}{a^n(x-a)} = \frac{a^n - x^n}{x^n a^n(x-a)} = -\frac{x^n - a^n}{a^n x^n(x-a)}$$

$$= -\frac{(x-a)(x^{n-1} + x^{n-2}a + x^{n-3}a^2 + \cdots + xa^{n-2} + a^{n-1})}{a^n x^n(x-a)}$$

$$= -\left(\frac{x^{n-1}}{a^n x^n} + \frac{x^{n-2}a}{a^n x^n} + \frac{x^{n-3}a^2}{a^n x^n} + \cdots + \frac{xa^{n-2}}{a^n x^n} + \frac{a^{n-1}}{a^n x^n} \right)$$

$$= -\frac{1}{a^n x} - \frac{1}{a^{n-1}x^2} - \frac{1}{a^{n-2}x^3} - \cdots - \frac{1}{a^2 x^{n-1}} - \frac{1}{ax^n}$$

Thus, $f(x) = \dfrac{1}{x^n(x-a)} = -\dfrac{1}{a^n x} - \dfrac{1}{a^{n-1}x^2} - \cdots - \dfrac{1}{ax^n} + \dfrac{1}{a^n(x-a)}$.

6.4 Integration with Tables and Computer Algebra Systems

Keep in mind that there are several ways to approach many of these exercises, and different methods can lead to different forms of the answer.

1. $\displaystyle\int_0^{\pi/8} \arctan 2x \, dx = \frac{1}{2}\int_0^{\pi/4} \arctan u \, du \qquad [u = 2x, \ du = 2\,dx]$

$$\overset{89}{=} \frac{1}{2}\left[u \arctan u - \frac{1}{2}\ln(1 + u^2) \right]_0^{\pi/4} = \frac{1}{2}\left\{ \left[\frac{\pi}{4}\arctan\frac{\pi}{4} - \frac{1}{2}\ln\left(1 + \frac{\pi^2}{16}\right) \right] - 0 \right\}$$

$$= \frac{\pi}{8}\arctan\frac{\pi}{4} - \frac{1}{4}\ln\left(1 + \frac{\pi^2}{16}\right)$$

3. $\displaystyle\int \frac{\cos x}{\sin^2 x - 9}\,dx = \int \frac{1}{u^2 - 9}\,du \quad \begin{bmatrix} u = \sin x, \\ du = \cos x\,dx \end{bmatrix} \overset{20}{=} \frac{1}{2(3)}\ln\left|\frac{u-3}{u+3}\right| + C = \frac{1}{6}\ln\left|\frac{\sin x - 3}{\sin x + 3}\right| + C$

5. Let $u = 2x$ and $a = 3$. Then $du = 2\,dx$ and

$$\int \frac{dx}{x^2\sqrt{4x^2 + 9}} = \int \frac{\frac{1}{2}\,du}{\dfrac{u^2}{4}\sqrt{u^2 + a^2}} = 2\int \frac{du}{u^2\sqrt{a^2 + u^2}} \overset{28}{=} -2\frac{\sqrt{a^2 + u^2}}{a^2 u} + C$$

$$= -2\frac{\sqrt{4x^2 + 9}}{9 \cdot 2x} + C = -\frac{\sqrt{4x^2 + 9}}{9x} + C$$

7. $\int x^3 \sin x \, dx \overset{84}{=} -x^3 \cos x + 3 \int x^2 \cos x \, dx$, $\int x^2 \cos x \, dx \overset{85}{=} x^2 \sin x - 2 \int x \sin x \, dx$, and

$\int x \sin x \, dx \overset{82}{=} \sin x - x \cos x + C$. Substituting, we get

$\int x^3 \sin x \, dx = -x^3 \cos x + 3\left[x^2 \sin x - 2(\sin x - x \cos x)\right] + C = -x^3 \cos x + 3x^2 \sin x - 6 \sin x + 6x \cos x + C$.

So $\int_0^\pi x^3 \sin x \, dx = \left[-x^3 \cos x + 3x^2 \sin x - 6 \sin x + 6x \cos x\right]_0^\pi = \left(-\pi^3 \cdot -1 + 6\pi \cdot -1\right) - (0) = \pi^3 - 6\pi$.

9. $\displaystyle\int \frac{\tan^3(1/z)}{z^2} \, dz \quad \begin{bmatrix} u = 1/z, \\ du = -dz/z^2 \end{bmatrix} = -\int \tan^3 u \, du \overset{69}{=} -\tfrac{1}{2}\tan^2 u - \ln|\cos u| + C$

$\qquad\qquad\qquad\qquad\qquad\qquad = -\tfrac{1}{2}\tan^2\left(\dfrac{1}{z}\right) - \ln\left|\cos\left(\dfrac{1}{z}\right)\right| + C$

11. Let $z = 6 + 4y - 4y^2 = 6 - (4y^2 - 4y + 1) + 1 = 7 - (2y - 1)^2$, $u = 2y - 1$, and $a = \sqrt{7}$. Then $z = a^2 - u^2$, $du = 2\,dy$,

and

$\int y\sqrt{6 + 4y - 4y^2} \, dy = \int y\sqrt{z} \, dy = \int \tfrac{1}{2}(u+1)\sqrt{a^2 - u^2}\,\tfrac{1}{2}\,du = \tfrac{1}{4}\int u\sqrt{a^2 - u^2}\,du + \tfrac{1}{4}\int \sqrt{a^2 - u^2}\,du$

$\qquad\qquad\qquad\qquad = \tfrac{1}{4}\int \sqrt{a^2 - u^2}\,du - \tfrac{1}{8}\int (-2u)\sqrt{a^2 - u^2}\,du$

$\qquad\qquad\qquad\qquad \overset{30}{=} \dfrac{u}{8}\sqrt{a^2 - u^2} + \dfrac{a^2}{8}\sin^{-1}\left(\dfrac{u}{a}\right) - \dfrac{1}{8}\int \sqrt{w}\,dw \quad \begin{bmatrix} w = a^2 - u^2, \\ dw = -2u\,du \end{bmatrix}$

$\qquad\qquad\qquad\qquad = \dfrac{2y-1}{8}\sqrt{6 + 4y - 4y^2} + \dfrac{7}{8}\sin^{-1}\dfrac{2y-1}{\sqrt{7}} - \dfrac{1}{8}\cdot\dfrac{2}{3}w^{3/2} + C$

$\qquad\qquad\qquad\qquad = \dfrac{2y-1}{8}\sqrt{6 + 4y - 4y^2} + \dfrac{7}{8}\sin^{-1}\dfrac{2y-1}{\sqrt{7}} - \dfrac{1}{12}(6 + 4y - 4y^2)^{3/2} + C$

This can be rewritten as

$\sqrt{6 + 4y - 4y^2}\left[\dfrac{1}{8}(2y - 1) - \dfrac{1}{12}(6 + 4y - 4y^2)\right] + \dfrac{7}{8}\sin^{-1}\dfrac{2y-1}{\sqrt{7}} + C$

$\qquad\qquad = \left(\dfrac{1}{3}y^2 - \dfrac{1}{12}y - \dfrac{5}{8}\right)\sqrt{6 + 4y - 4y^2} + \dfrac{7}{8}\sin^{-1}\left(\dfrac{2y-1}{\sqrt{7}}\right) + C$

$\qquad\qquad = \dfrac{1}{24}(8y^2 - 2y - 15)\sqrt{6 + 4y - 4y^2} + \dfrac{7}{8}\sin^{-1}\left(\dfrac{2y-1}{\sqrt{7}}\right) + C$

13. Let $u = \sin x$. Then $du = \cos x \, dx$, so

$\int \sin^2 x \cos x \ln(\sin x) \, dx = \int u^2 \ln u \, du \overset{101}{=} \dfrac{u^{2+1}}{(2+1)^2}\left[(2+1)\ln u - 1\right] + C = \tfrac{1}{9}u^3(3\ln u - 1) + C$

$\qquad\qquad\qquad\qquad\qquad\qquad = \tfrac{1}{9}\sin^3 x\left[3\ln(\sin x) - 1\right] + C$

15. Let $u = e^x$ and $a = \sqrt{3}$. Then $du = e^x \, dx$ and

$\displaystyle\int \frac{e^x}{3 - e^{2x}} \, dx = \int \frac{du}{a^2 - u^2} \overset{19}{=} \frac{1}{2a}\ln\left|\frac{u+a}{u-a}\right| + C = \frac{1}{2\sqrt{3}}\ln\left|\frac{e^x + \sqrt{3}}{e^x - \sqrt{3}}\right| + C$.

17. $\displaystyle\int \frac{x^4 \, dx}{\sqrt{x^{10} - 2}} = \int \frac{x^4 \, dx}{\sqrt{(x^5)^2 - 2}} = \frac{1}{5}\int \frac{du}{\sqrt{u^2 - 2}} \quad \begin{bmatrix} u = x^5, \\ du = 5x^4 \, dx \end{bmatrix}$

$\qquad\qquad \overset{43}{=} \tfrac{1}{5}\ln\left|u + \sqrt{u^2 - 2}\right| + C = \tfrac{1}{5}\ln\left|x^5 + \sqrt{x^{10} - 2}\right| + C$

19. Let $u = \ln x$ and $a = 2$. Then $du = dx/x$ and

$$\int \frac{\sqrt{4 + (\ln x)^2}}{x}\, dx = \int \sqrt{a^2 + u^2}\, du \overset{21}{=} \frac{u}{2}\sqrt{a^2 + u^2} + \frac{a^2}{2}\ln\left(u + \sqrt{a^2 + u^2}\right) + C$$

$$= \tfrac{1}{2}(\ln x)\sqrt{4 + (\ln x)^2} + 2\ln\left[\ln x + \sqrt{4 + (\ln x)^2}\right] + C$$

21. Let $u = e^x$. Then $x = \ln u$, $dx = du/u$, so

$$\int \sqrt{e^{2x} - 1}\, dx = \int \frac{\sqrt{u^2 - 1}}{u}\, du \overset{41}{=} \sqrt{u^2 - 1} - \cos^{-1}(1/u) + C = \sqrt{e^{2x} - 1} - \cos^{-1}(e^{-x}) + C.$$

23. (a) $\dfrac{d}{du}\left[\dfrac{1}{b^3}\left(a + bu - \dfrac{a^2}{a + bu} - 2a\ln|a + bu|\right) + C\right] = \dfrac{1}{b^3}\left[b + \dfrac{ba^2}{(a + bu)^2} - \dfrac{2ab}{(a + bu)}\right]$

$$= \frac{1}{b^3}\left[\frac{b(a + bu)^2 + ba^2 - (a + bu)2ab}{(a + bu)^2}\right]$$

$$= \frac{1}{b^3}\left[\frac{b^3 u^2}{(a + bu)^2}\right] = \frac{u^2}{(a + bu)^2}$$

(b) Let $t = a + bu \;\Rightarrow\; dt = b\,du$. Note that $u = \dfrac{t - a}{b}$ and $du = \dfrac{1}{b}\,dt$.

$$\int \frac{u^2\, du}{(a + bu)^2} = \frac{1}{b^3}\int \frac{(t - a)^2}{t^2}\, dt = \frac{1}{b^3}\int \frac{t^2 - 2at + a^2}{t^2}\, dt = \frac{1}{b^3}\int\left(1 - \frac{2a}{t} + \frac{a^2}{t^2}\right) dt$$

$$= \frac{1}{b^3}\left(t - 2a\ln|t| - \frac{a^2}{t}\right) + C = \frac{1}{b^3}\left(a + bu - \frac{a^2}{a + bu} - 2a\ln|a + bu|\right) + C$$

25. Maple and Mathematica both give $\int \sec^4 x\, dx = \frac{2}{3}\tan x + \frac{1}{3}\tan x \sec^2 x$, while Derive gives the second

term as $\dfrac{\sin x}{3\cos^3 x} = \dfrac{1}{3}\dfrac{\sin x}{\cos x}\dfrac{1}{\cos^2 x} = \dfrac{1}{3}\tan x \sec^2 x$. Using Formula 77, we get

$\int \sec^4 x\, dx = \frac{1}{3}\tan x \sec^2 x + \frac{2}{3}\int \sec^2 x\, dx = \frac{1}{3}\tan x \sec^2 x + \frac{2}{3}\tan x + C$.

27. Derive gives $\int x^2 \sqrt{x^2 + 4}\, dx = \frac{1}{4}x(x^2 + 2)\sqrt{x^2 + 4} - 2\ln\left(\sqrt{x^2 + 4} + x\right)$. Maple gives

$\frac{1}{4}x(x^2 + 4)^{3/2} - \frac{1}{2}x\sqrt{x^2 + 4} - 2\,\mathrm{arcsinh}\left(\frac{1}{2}x\right)$. Applying the command `convert(%,ln);` yields

$$\tfrac{1}{4}x(x^2 + 4)^{3/2} - \tfrac{1}{2}x\sqrt{x^2 + 4} - 2\ln\left(\tfrac{1}{2}x + \tfrac{1}{2}\sqrt{x^2 + 4}\right) = \tfrac{1}{4}x(x^2 + 4)^{1/2}\left[(x^2 + 4) - 2\right] - 2\ln\left[(x + \sqrt{x^2 + 4})/2\right]$$

$$= \tfrac{1}{4}x(x^2 + 2)\sqrt{x^2 + 4} - 2\ln\left(\sqrt{x^2 + 4} + x\right) + 2\ln 2$$

Mathematica gives $\frac{1}{4}x(2 + x^2)\sqrt{3 + x^2} - 2\,\mathrm{arcsinh}(x/2)$. Applying the `TrigToExp` and `Simplify` commands gives

$\frac{1}{4}\left[x(2 + x^2)\sqrt{4 + x^2} - 8\log\left(\frac{1}{2}(x + \sqrt{4 + x^2})\right)\right] = \frac{1}{4}x(x^2 + 2)\sqrt{x^2 + 4} - 2\ln\left(x + \sqrt{4 + x^2}\right) + 2\ln 2$, so all are

equivalent (without constant).

Now use Formula 22 to get

$$\int x^2 \sqrt{2^2 + x^2}\, dx = \frac{x}{8}(2^2 + 2x^2)\sqrt{2^2 + x^2} - \frac{2^4}{8}\ln\left(x + \sqrt{2^2 + x^2}\right) + C$$

$$= \frac{x}{8}(2)(2 + x^2)\sqrt{4 + x^2} - 2\ln\left(x + \sqrt{4 + x^2}\right) + C$$

$$= \tfrac{1}{4}x(x^2 + 2)\sqrt{x^2 + 4} - 2\ln\left(\sqrt{x^2 + 4} + x\right) + C$$

29. Derive and Maple give $\displaystyle\int \cos^4 x\,dx = \frac{\sin x \cos^3 x}{4} + \frac{3\sin x \cos x}{8} + \frac{3x}{8}$, while Mathematica gives

$$\frac{3x}{8} + \frac{1}{4}\sin(2x) + \frac{1}{32}\sin(4x) = \frac{3x}{8} + \frac{1}{4}(2\sin x \cos x) + \frac{1}{32}(2\sin 2x \cos 2x)$$

$$= \frac{3x}{8} + \frac{1}{2}\sin x \cos x + \frac{1}{16}[2\sin x \cos x\,(2\cos^2 x - 1)]$$

$$= \frac{3x}{8} + \frac{1}{2}\sin x \cos x + \frac{1}{4}\sin x \cos^3 x - \frac{1}{8}\sin x \cos x,$$

so all are equivalent.

Using tables,

$$\int \cos^4 x\,dx \overset{74}{=} \tfrac{1}{4}\cos^3 x \sin x + \tfrac{3}{4}\int \cos^2 x\,dx \overset{64}{=} \tfrac{1}{4}\cos^3 x \sin x + \tfrac{3}{4}\left(\tfrac{1}{2}x + \tfrac{1}{4}\sin 2x\right) + C$$

$$= \tfrac{1}{4}\cos^3 x \sin x + \tfrac{3}{8}x + \tfrac{3}{16}(2\sin x \cos x) + C = \tfrac{1}{4}\cos^3 x \sin x + \tfrac{3}{8}x + \tfrac{3}{8}\sin x \cos x + C$$

31. Maple gives $\int \tan^5 x\,dx = \tfrac{1}{4}\tan^4 x - \tfrac{1}{2}\tan^2 x + \tfrac{1}{2}\ln(1 + \tan^2 x)$, Mathematica gives

$\int \tan^5 x\,dx = \tfrac{1}{4}[-1 - 2\cos(2x)]\sec^4 x - \ln(\cos x)$, and Derive gives $\int \tan^5 x\,dx = \tfrac{1}{4}\tan^4 x - \tfrac{1}{2}\tan^2 x - \ln(\cos x)$.

These expressions are equivalent, and none includes absolute value bars or a constant of integration. Note that Mathematica's

and Derive's expressions suggest that the integral is undefined where $\cos x < 0$, which is not the case. Using Formula 75,

$\int \tan^5 x\,dx = \frac{1}{5-1}\tan^{5-1} x - \int \tan^{5-2} x\,dx = \tfrac{1}{4}\tan^4 x - \int \tan^3 x\,dx$. Using Formula 69,

$\int \tan^3 x\,dx = \tfrac{1}{2}\tan^2 x + \ln|\cos x| + C$, so $\int \tan^5 x\,dx = \tfrac{1}{4}\tan^4 x - \tfrac{1}{2}\tan^2 x - \ln|\cos x| + C$.

33. (a) $\displaystyle F(x) = \int f(x)\,dx = \int \frac{1}{x\sqrt{1-x^2}}\,dx \overset{35}{=} -\frac{1}{1}\ln\left|\frac{1 + \sqrt{1-x^2}}{x}\right| + C = -\ln\left|\frac{1 + \sqrt{1-x^2}}{x}\right| + C.$

f has domain $\{x \mid x \neq 0, 1 - x^2 > 0\} = \{x \mid x \neq 0, |x| < 1\} = (-1, 0) \cup (0, 1)$. F has the same domain.

(b) Derive gives $F(x) = \ln\left(\sqrt{1-x^2} - 1\right) - \ln x$ and Mathematica gives $F(x) = \ln x - \ln\left(1 + \sqrt{1-x^2}\right)$.

Both are correct if you take absolute values of the logarithm arguments, and both would then have the

same domain. Maple gives $F(x) = -\operatorname{arctanh}\left(1/\sqrt{1-x^2}\right)$. This function has domain

$\left\{x \mid |x| < 1, -1 < 1/\sqrt{1-x^2} < 1\right\} = \left\{x \mid |x| < 1, 1/\sqrt{1-x^2} < 1\right\} = \left\{x \mid |x| < 1, \sqrt{1-x^2} > 1\right\} = \emptyset,$

the empty set! If we apply the command `convert(%,ln);` to Maple's answer, we get

$-\dfrac{1}{2}\ln\left(\dfrac{1}{\sqrt{1-x^2}} + 1\right) + \dfrac{1}{2}\ln\left(1 - \dfrac{1}{\sqrt{1-x^2}}\right)$, which has the same domain, $\emptyset$.

6.5 Approximate Integration

1. (a) $\Delta x = (b - a)/n = (4 - 0)/2 = 2$

$$L_2 = \sum_{i=1}^{2} f(x_{i-1})\,\Delta x = f(x_0) \cdot 2 + f(x_1) \cdot 2 = 2\,[f(0) + f(2)] = 2(0.5 + 2.5) = 6$$

$$R_2 = \sum_{i=1}^{2} f(x_i) \, \Delta x = f(x_1) \cdot 2 + f(x_2) \cdot 2 = 2\,[f(2) + f(4)] = 2(2.5 + 3.5) = 12$$

$$M_2 = \sum_{i=1}^{2} f(\overline{x}_i)\Delta x = f(\overline{x}_1) \cdot 2 + f(\overline{x}_2) \cdot 2 = 2\,[f(1) + f(3)] \approx 2(1.6 + 3.2) = 9.6$$

(b)

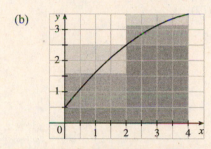

L_2 is an underestimate, since the area under the small rectangles is less than the area under the curve, and R_2 is an overestimate, since the area under the large rectangles is greater than the area under the curve. It appears that M_2 is an overestimate, though it is fairly close to I. See the solution to Exercise 39 for a proof of the fact that if f is concave down on $[a, b]$, then the Midpoint Rule is an overestimate of $\int_a^b f(x)\,dx$.

(c) $T_2 = \left(\frac{1}{2}\,\Delta x\right)[f(x_0) + 2f(x_1) + f(x_2)] = \frac{2}{2}[f(0) + 2f(2) + f(4)] = 0.5 + 2(2.5) + 3.5 = 9.$

This approximation is an underestimate, since the graph is concave down. Thus, $T_2 = 9 < I$. See the solution to Exercise 39 for a general proof of this conclusion.

(d) For any n, we will have $L_n < T_n < I < M_n < R_n$.

3. $f(x) = \cos(x^2)$, $\Delta x = \frac{1-0}{4} = \frac{1}{4}$

(a) $T_4 = \frac{1}{4 \cdot 2}\left[f(0) + 2f\left(\frac{1}{4}\right) + 2f\left(\frac{2}{4}\right) + 2f\left(\frac{3}{4}\right) + f(1)\right] \approx 0.895759$

(b) $M_4 = \frac{1}{4}\left[f\left(\frac{1}{8}\right) + f\left(\frac{3}{8}\right) + f\left(\frac{5}{8}\right) + f\left(\frac{7}{8}\right)\right] \approx 0.908907$

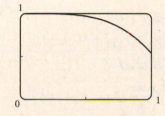

The graph shows that f is concave down on $[0, 1]$. So T_4 is an underestimate and M_4 is an overestimate. We can conclude that

$0.895759 < \int_0^1 \cos(x^2)\,dx < 0.908907.$

5. (a) $f(x) = \dfrac{x}{1 + x^2}$, $\Delta x = \dfrac{b - a}{n} = \dfrac{2 - 0}{10} = \dfrac{1}{5}$

$M_{10} = \frac{1}{5}\left[f\left(\frac{1}{10}\right) + f\left(\frac{3}{10}\right) + f\left(\frac{5}{10}\right) + \cdots + f\left(\frac{19}{10}\right)\right] \approx 0.806598$

(b) $S_{10} = \frac{1}{5 \cdot 3}\left[f(0) + 4f\left(\frac{1}{5}\right) + 2f\left(\frac{2}{5}\right) + 4f\left(\frac{3}{5}\right) + 2f\left(\frac{4}{5}\right) + \cdots + 4f\left(\frac{9}{5}\right) + f(2)\right] \approx 0.804779$

Actual: $I = \displaystyle\int_0^2 \frac{x}{1 + x^2}\,dx = \left[\frac{1}{2}\ln\left|1 + x^2\right|\right]_0^2$ $[u = 1 + x^2,\, du = 2x\,dx]$

$= \frac{1}{2}\ln 5 - \frac{1}{2}\ln 1 = \frac{1}{2}\ln 5 \approx 0.804719$

Errors: $E_M = $ actual $- M_{10} = I - M_{10} \approx -0.001879$

$E_S = $ actual $- S_{10} = I - S_{10} \approx -0.000060$

7. $f(x) = \sqrt{x^3 - 1}$, $\Delta x = \dfrac{b - a}{n} = \dfrac{2 - 1}{10} = \dfrac{1}{10}$

(a) $T_{10} = \frac{1}{10 \cdot 2}[f(1) + 2f(1.1) + 2f(1.2) + 2f(1.3) + 2f(1.4) + 2f(1.5)$

$\qquad\qquad + 2f(1.6) + 2f(1.7) + 2f(1.8) + 2f(1.9) + f(2)]$

≈ 1.506361

(b) $M_{10} = \frac{1}{10}[f(1.05) + f(1.15) + f(1.25) + f(1.35) + f(1.45) + f(1.55) + f(1.65) + f(1.75) + f(1.85) + f(1.95)]$

≈ 1.518362

(c) $S_{10} = \frac{1}{10 \cdot 3}[f(1) + 4f(1.1) + 2f(1.2) + 4f(1.3) + 2f(1.4)$

$+ 4f(1.5) + 2f(1.6) + 4f(1.7) + 2f(1.8) + 4f(1.9) + f(2)]$

≈ 1.511519

9. $f(x) = \frac{e^x}{1 + x^2}, \Delta x = \frac{b - a}{n} = \frac{2 - 0}{10} = \frac{1}{5}$

(a) $T_{10} = \frac{1}{5 \cdot 2}[f(0) + 2f(0.2) + 2f(0.4) + 2f(0.6) + 2f(0.8) + 2f(1)$

$+ 2f(1.2) + 2f(1.4) + 2f(1.6) + 2f(1.8) + f(2)]$

≈ 2.660833

(b) $M_{10} = \frac{1}{5}[f(0.1) + f(0.3) + f(0.5) + f(0.7) + f(0.9) + f(1.1) + f(1.3) + f(1.5) + f(1.7) + f(1.9)]$

≈ 2.664377

(c) $S_{10} = \frac{1}{5 \cdot 3}[f(0) + 4f(0.2) + 2f(0.4) + 4f(0.6) + 2f(0.8)$

$+ 4f(1) + 2f(1.2) + 4f(1.4) + 2f(1.6) + 4f(1.8) + f(2)]$

≈ 2.663244

11. $f(t) = e^{\sqrt{t}} \sin t, \Delta t = \frac{4 - 0}{8} = \frac{1}{2}$

(a) $T_8 = \frac{1}{2 \cdot 2}\left[f(0) + 2f\left(\frac{1}{2}\right) + 2f(1) + 2f\left(\frac{3}{2}\right) + 2f(2) + 2f\left(\frac{5}{2}\right) + 2f(3) + 2f\left(\frac{7}{2}\right) + f(4)\right] \approx 4.513618$

(b) $M_8 = \frac{1}{2}\left[f\left(\frac{1}{4}\right) + f\left(\frac{3}{4}\right) + f\left(\frac{5}{4}\right) + f\left(\frac{7}{4}\right) + f\left(\frac{9}{4}\right) + f\left(\frac{11}{4}\right) + f\left(\frac{13}{4}\right) + f\left(\frac{15}{4}\right)\right] \approx 4.748256$

(c) $S_8 = \frac{1}{2 \cdot 3}\left[f(0) + 4f\left(\frac{1}{2}\right) + 2f(1) + 4f\left(\frac{3}{2}\right) + 2f(2) + 4f\left(\frac{5}{2}\right) + 2f(3) + 4f\left(\frac{7}{2}\right) + f(4)\right] \approx 4.675111$

13. $f(x) = \frac{\cos x}{x}, \Delta x = \frac{5 - 1}{8} = \frac{1}{2}$

(a) $T_8 = \frac{1}{2 \cdot 2}\left[f(1) + 2f\left(\frac{3}{2}\right) + 2f(2) + \cdots + 2f(4) + 2f\left(\frac{9}{2}\right) + f(5)\right] \approx -0.495333$

(b) $M_8 = \frac{1}{2}\left[f\left(\frac{5}{4}\right) + f\left(\frac{7}{4}\right) + f\left(\frac{9}{4}\right) + f\left(\frac{11}{4}\right) + f\left(\frac{13}{4}\right) + f\left(\frac{15}{4}\right) + f\left(\frac{17}{4}\right) + f\left(\frac{19}{4}\right)\right] \approx -0.543321$

(c) $S_8 = \frac{1}{2 \cdot 3}\left[f(1) + 4f\left(\frac{3}{2}\right) + 2f(2) + 4f\left(\frac{5}{2}\right) + 2f(3) + 4f\left(\frac{7}{2}\right) + 2f(4) + 4f\left(\frac{9}{2}\right) + f(5)\right] \approx -0.526123$

15. $f(y) = \frac{1}{1 + y^5}, \Delta y = \frac{3 - 0}{6} = \frac{1}{2}$

(a) $T_6 = \frac{1}{2 \cdot 2}\left[f(0) + 2f\left(\frac{1}{2}\right) + 2f\left(\frac{2}{2}\right) + 2f\left(\frac{3}{2}\right) + 2f\left(\frac{4}{2}\right) + 2f\left(\frac{5}{2}\right) + f(3)\right] \approx 1.064275$

(b) $M_6 = \frac{1}{2}\left[f\left(\frac{1}{4}\right) + f\left(\frac{3}{4}\right) + f\left(\frac{5}{4}\right) + f\left(\frac{7}{4}\right) + f\left(\frac{9}{4}\right) + f\left(\frac{11}{4}\right)\right] \approx 1.067416$

(c) $S_6 = \frac{1}{2 \cdot 3}\left[f(0) + 4f\left(\frac{1}{2}\right) + 2f\left(\frac{2}{2}\right) + 4f\left(\frac{3}{2}\right) + 2f\left(\frac{4}{2}\right) + 4f\left(\frac{5}{2}\right) + f(3)\right] \approx 1.074915$

17. $f(x) = \cos(x^2), \Delta x = \frac{1 - 0}{8} = \frac{1}{8}$

(a) $T_8 = \frac{1}{8 \cdot 2}\left\{f(0) + 2\left[f\left(\frac{1}{8}\right) + f\left(\frac{2}{8}\right) + \cdots + f\left(\frac{7}{8}\right)\right] + f(1)\right\} \approx 0.902333$

$M_8 = \frac{1}{8}\left[f\left(\frac{1}{16}\right) + f\left(\frac{3}{16}\right) + f\left(\frac{5}{16}\right) + \cdots + f\left(\frac{15}{16}\right)\right] = 0.905620$

(b) $f(x) = \cos(x^2), f'(x) = -2x \sin(x^2), f''(x) = -2 \sin(x^2) - 4x^2 \cos(x^2)$. For $0 \le x \le 1$, sin and cos are positive,

so $|f''(x)| = 2 \sin(x^2) + 4x^2 \cos(x^2) \le 2 \cdot 1 + 4 \cdot 1 \cdot 1 = 6$ since $\sin(x^2) \le 1$ and $\cos(x^2) \le 1$ for all x, and $x^2 \leq 1$ for $0 \leq x \leq 1$. So for $n = 8$, we take $K = 6$, $a = 0$, and $b = 1$ in Theorem 3, to get

$|E_T| \leq 6 \cdot 1^3/(12 \cdot 8^2) = \frac{1}{128} = 0.0078125$ and $|E_M| \leq \frac{1}{256} = 0.00390625$. [A better estimate is obtained by noting

from a graph of f'' that $|f''(x)| \leq 4$ for $0 \leq x \leq 1$.]

(c) Take $K = 6$ [as in part (b)] in Theorem 3. $\quad |E_T| \leq \dfrac{K(b-a)^3}{12n^2} \leq 0.0001 \quad \Leftrightarrow \quad \dfrac{6(1-0)^3}{12n^2} \leq 10^{-4} \quad \Leftrightarrow$

$\dfrac{1}{2n^2} \leq \dfrac{1}{10^4} \quad \Leftrightarrow \quad 2n^2 \geq 10^4 \quad \Leftrightarrow \quad n^2 \geq 5000 \quad \Leftrightarrow \quad n \geq 71$. Take $n = 71$ for T_n. For E_M, again take $K = 6$ in

Theorem 3 to get $|E_M| \leq 10^{-4} \quad \Leftrightarrow \quad 4n^2 \geq 10^4 \quad \Leftrightarrow \quad n^2 \geq 2500 \quad \Leftrightarrow \quad n \geq 50$. Take $n = 50$ for M_n.

19. $f(x) = \sin x$, $\Delta x = \dfrac{\pi - 0}{10} = \dfrac{\pi}{10}$

(a) $T_{10} = \dfrac{\pi}{10 \cdot 2}\left[f(0) + 2f\left(\dfrac{\pi}{10}\right) + 2f\left(\dfrac{2\pi}{10}\right) + \cdots + 2f\left(\dfrac{9\pi}{10}\right) + f(\pi)\right] \approx 1.983524$

$M_{10} = \dfrac{\pi}{10}\left[f\left(\dfrac{\pi}{20}\right) + f\left(\dfrac{3\pi}{20}\right) + f\left(\dfrac{5\pi}{20}\right) + \cdots + f\left(\dfrac{19\pi}{20}\right)\right] \approx 2.008248$

$S_{10} = \dfrac{\pi}{10 \cdot 3}\left[f(0) + 4f\left(\dfrac{\pi}{10}\right) + 2f\left(\dfrac{2\pi}{10}\right) + 4f\left(\dfrac{3\pi}{10}\right) + \cdots + 4f\left(\dfrac{9\pi}{10}\right) + f(\pi)\right] \approx 2.000110$

Since $I = \int_0^{\pi} \sin x \, dx = \left[-\cos x\right]_0^{\pi} = 1 - (-1) = 2$, $E_T = I - T_{10} \approx 0.016476$, $E_M = I - M_{10} \approx -0.008248$,

and $E_S = I - S_{10} \approx -0.000110$.

(b) $f(x) = \sin x \quad \Rightarrow \quad \left|f^{(n)}(x)\right| \leq 1$, so take $K = 1$ for all error estimates.

$|E_T| \leq \dfrac{K(b-a)^3}{12n^2} = \dfrac{1(\pi - 0)^3}{12(10)^2} = \dfrac{\pi^3}{1200} \approx 0.025839. \quad |E_M| \leq \dfrac{|E_T|}{2} = \dfrac{\pi^3}{2400} \approx 0.012919.$

$|E_S| \leq \dfrac{K(b-a)^5}{180n^4} = \dfrac{1(\pi - 0)^5}{180(10)^4} = \dfrac{\pi^5}{1,800,000} \approx 0.000170.$

The actual error is about 64% of the error estimate in all three cases.

(c) $|E_T| \leq 0.00001 \quad \Leftrightarrow \quad \dfrac{\pi^3}{12n^2} \leq \dfrac{1}{10^5} \quad \Leftrightarrow \quad n^2 \geq \dfrac{10^5 \pi^3}{12} \quad \Rightarrow \quad n \geq 508.3$. Take $n = 509$ for T_n.

$|E_M| \leq 0.00001 \quad \Leftrightarrow \quad \dfrac{\pi^3}{24n^2} \leq \dfrac{1}{10^5} \quad \Leftrightarrow \quad n^2 \geq \dfrac{10^5 \pi^3}{24} \quad \Rightarrow \quad n \geq 359.4$. Take $n = 360$ for M_n.

$|E_S| \leq 0.00001 \quad \Leftrightarrow \quad \dfrac{\pi^5}{180n^4} \leq \dfrac{1}{10^5} \quad \Leftrightarrow \quad n^4 \geq \dfrac{10^5 \pi^5}{180} \quad \Rightarrow \quad n \geq 20.3$.

Take $n = 22$ for S_n (since n must be even).

21. (a) Using a CAS, we differentiate $f(x) = e^{\cos x}$ twice, and find that

$f''(x) = e^{\cos x}(\sin^2 x - \cos x)$. From the graph, we see that the maximum

value of $|f''(x)|$ occurs at the endpoints of the interval $[0, 2\pi]$.

Since $f''(0) = -e$, we can use $K = e$ or $K = 2.8$.

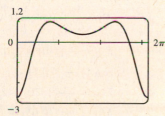

(b) A CAS gives $M_{10} \approx 7.954926518$. (In Maple, use

`Student[Calculus1][RiemannSum]` or `Student[Calculus1][ApproximateInt]`.)

(c) Using Theorem 3 for the Midpoint Rule, with $K = e$, we get $|E_M| \leq \dfrac{e(2\pi - 0)^3}{24 \cdot 10^2} \approx 0.280945995.$

With $K = 2.8$, we get $|E_M| \leq \dfrac{2.8(2\pi - 0)^3}{24 \cdot 10^2} = 0.289391916.$

(d) A CAS gives $I \approx 7.954926521$.

(e) The actual error is only about 3×10^{-9}, much less than the estimate in part (c).

(f) We use the CAS to differentiate twice more, and then graph

$$f^{(4)}(x) = e^{\cos x}(\sin^4 x - 6\sin^2 x \cos x + 3 - 7\sin^2 x + \cos x).$$

From the graph, we see that the maximum value of $\left|f^{(4)}(x)\right|$ occurs at the

endpoints of the interval $[0, 2\pi]$. Since $f^{(4)}(0) = 4e$, we can use $K = 4e$

or $K = 10.9$.

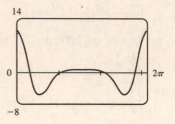

(g) A CAS gives $S_{10} \approx 7.953789422$. (In Maple, use `Student[Calculus1][ApproximateInt]`.)

(h) Using Theorem 4 with $K = 4e$, we get $|E_S| \leq \dfrac{4e(2\pi - 0)^5}{180 \cdot 10^4} \approx 0.059153618$.

With $K = 10.9$, we get $|E_S| \leq \dfrac{10.9(2\pi - 0)^5}{180 \cdot 10^4} \approx 0.059299814$.

(i) The actual error is about $7.954926521 - 7.953789422 \approx 0.00114$. This is quite a bit smaller than the estimate in part (h), though the difference is not nearly as great as it was in the case of the Midpoint Rule.

(j) To ensure that $|E_S| \leq 0.0001$, we use Theorem 4: $|E_S| \leq \dfrac{4e(2\pi)^5}{180 \cdot n^4} \leq 0.0001 \quad \Rightarrow \quad \dfrac{4e(2\pi)^5}{180 \cdot 0.0001} \leq n^4 \quad \Rightarrow$

$n^4 \geq 5{,}915{,}362 \quad \Leftrightarrow \quad n \geq 49.3$. So we must take $n \geq 50$ to ensure that $|I - S_n| \leq 0.0001$.

($K = 10.9$ leads to the same value of n.)

23. $I = \int_0^1 xe^x\,dx = \left[(x-1)e^x\right]_0^1$ [parts or Formula 96] $= 0 - (-1) = 1$, $f(x) = xe^x$, $\Delta x = 1/n$

$n = 5$: $\quad L_5 = \frac{1}{5}[f(0) + f(0.2) + f(0.4) + f(0.6) + f(0.8)] \approx 0.742943$

$\qquad R_5 = \frac{1}{5}[f(0.2) + f(0.4) + f(0.6) + f(0.8) + f(1)] \approx 1.286599$

$\qquad T_5 = \frac{1}{5 \cdot 2}[f(0) + 2f(0.2) + 2f(0.4) + 2f(0.6) + 2f(0.8) + f(1)] \approx 1.014771$

$\qquad M_5 = \frac{1}{5}[f(0.1) + f(0.3) + f(0.5) + f(0.7) + f(0.9)] \approx 0.992621$

$\qquad E_L = I - L_5 \approx 1 - 0.742943 = 0.257057$

$\qquad E_R \approx 1 - 1.286599 = -0.286599$

$\qquad E_T \approx 1 - 1.014771 = -0.014771$

$\qquad E_M \approx 1 - 0.992621 = 0.007379$

$n = 10$: $\quad L_{10} = \frac{1}{10}[f(0) + f(0.1) + f(0.2) + \cdots + f(0.9)] \approx 0.867782$

$\qquad R_{10} = \frac{1}{10}[f(0.1) + f(0.2) + \cdots + f(0.9) + f(1)] \approx 1.139610$

$\qquad T_{10} = \frac{1}{10 \cdot 2}\{f(0) + 2[f(0.1) + f(0.2) + \cdots + f(0.9)] + f(1)\} \approx 1.003696$

$\qquad M_{10} = \frac{1}{10}[f(0.05) + f(0.15) + \cdots + f(0.85) + f(0.95)] \approx 0.998152$

$\qquad E_L = I - L_{10} \approx 1 - 0.867782 = 0.132218$

$\qquad E_R \approx 1 - 1.139610 = -0.139610$

$\qquad E_T \approx 1 - 1.003696 = -0.003696$

$\qquad E_M \approx 1 - 0.998152 = 0.001848$

$n = 20$: $L_{20} = \frac{1}{20}[f(0) + f(0.05) + f(0.10) + \cdots + f(0.95)] \approx 0.932967$

$R_{20} = \frac{1}{20}[f(0.05) + f(0.10) + \cdots + f(0.95) + f(1)] \approx 1.068881$

$T_{20} = \frac{1}{20 \cdot 2}\{f(0) + 2[f(0.05) + f(0.10) + \cdots + f(0.95)] + f(1)\} \approx 1.000924$

$M_{20} = \frac{1}{20}[f(0.025) + f(0.075) + f(0.125) + \cdots + f(0.975)] \approx 0.999538$

$E_L = I - L_{20} \approx 1 - 0.932967 = 0.067033$

$E_R \approx 1 - 1.068881 = -0.068881$

$E_T \approx 1 - 1.000924 = -0.000924$

$E_M \approx 1 - 0.999538 = 0.000462$

n	L_n	R_n	T_n	M_n
5	0.742943	1.286599	1.014771	0.992621
10	0.867782	1.139610	1.003696	0.998152
20	0.932967	1.068881	1.000924	0.999538

n	E_L	E_R	E_T	E_M
5	0.257057	−0.286599	−0.014771	0.007379
10	0.132218	−0.139610	−0.003696	0.001848
20	0.067033	−0.068881	−0.000924	0.000462

Observations:

1. E_L and E_R are always opposite in sign, as are E_T and E_M.

2. As n is doubled, E_L and E_R are decreased by about a factor of 2, and E_T and E_M are decreased by a factor of about 4.

3. The Midpoint approximation is about twice as accurate as the Trapezoidal approximation.

4. All the approximations become more accurate as the value of n increases.

5. The Midpoint and Trapezoidal approximations are much more accurate than the endpoint approximations.

25. $\Delta x = (b-a)/n = (6-0)/6 = 1$

(a) $T_6 = \frac{\Delta x}{2}[f(0) + 2f(1) + 2f(2) + 2f(3) + 2f(4) + 2f(5) + f(6)]$

$\approx \frac{1}{2}[3 + 2(5) + 2(4) + 2(2) + 2(2.8) + 2(4) + 1]$

$= \frac{1}{2}(39.6) = 19.8$

(b) $M_6 = \Delta x[f(0.5) + f(1.5) + f(2.5) + f(3.5) + f(4.5) + f(5.5)]$

$\approx 1[4.5 + 4.7 + 2.6 + 2.2 + 3.4 + 3.2]$

$= 20.6$

(c) $S_6 = \frac{\Delta x}{3}[f(0) + 4f(1) + 2f(2) + 4f(3) + 2f(4) + 4f(5) + f(6)]$

$\approx \frac{1}{3}[3 + 4(5) + 2(4) + 4(2) + 2(2.8) + 4(4) + 1]$

$= \frac{1}{3}(61.6) = 20.5\overline{3}$

27. By the Net Change Theorem, the increase in velocity is equal to $\int_0^6 a(t)\,dt$. We use Simpson's Rule with $n = 6$ and

$\Delta t = (6-0)/6 = 1$ to estimate this integral:

$\int_0^6 a(t)\,dt \approx S_6 = \frac{1}{3}[a(0) + 4a(1) + 2a(2) + 4a(3) + 2a(4) + 4a(5) + a(6)]$

$\approx \frac{1}{3}[0 + 4(0.5) + 2(4.1) + 4(9.8) + 2(12.9) + 4(9.5) + 0] = \frac{1}{3}(113.2) = 37.7\overline{3}$ ft/s

29. $T_{\text{ave}} = \frac{1}{24-0} \int_0^{24} T(t)\, dt \approx \frac{1}{24} S_{12} = \frac{1}{24} \frac{24-0}{3(12)} [T(0) + 4T(2) + 2T(4) + 4T(6) + 2T(8) + 4T(10) + 2T(12)$

$$+ 4T(14) + 2T(16) + 4T(18) + 2T(20) + 4T(22) + T(24)]$$

$$\approx \tfrac{1}{36}[67 + 4(65) + 2(62) + 4(58) + 2(56) + 4(61) + 2(63) + 4(68)$$

$$+ 2(71) + 4(69) + 2(67) + 4(66) + 64]$$

$$= \tfrac{1}{36}(2317) = 64.36\overline{1}°\text{F}.$$

The average temperature was about $64.4°\text{F}$.

31. (a) $\int_1^5 f(x)\, dx \approx M_4 = \frac{5-1}{4}[f(1.5) + f(2.5) + f(3.5) + f(4.5)] = 1(2.9 + 3.6 + 4.0 + 3.9) = 14.4$

(b) $-2 \le f''(x) \le 3 \;\Rightarrow\; |f''(x)| \le 3 \;\Rightarrow\; K = 3$, since $|f''(x)| \le K$. The error estimate for the Midpoint Rule is

$$|E_M| \le \frac{K(b-a)^3}{24n^2} = \frac{3(5-1)^3}{24(4)^2} = \frac{1}{2}.$$

33. By the Net Change Theorem, the energy used is equal to $\int_0^6 P(t)\, dt$. We use Simpson's Rule with $n = 12$ and

$\Delta t = \frac{6-0}{12} = \frac{1}{2}$ to estimate this integral:

$$\int_0^6 P(t)\, dt \approx S_{12} = \frac{1/2}{3}[P(0) + 4P(0.5) + 2P(1) + 4P(1.5) + 2P(2) + 4P(2.5) + 2P(3)$$

$$+ 4P(3.5) + 2P(4) + 4P(4.5) + 2P(5) + 4P(5.5) + P(6)]$$

$$= \tfrac{1}{6}[1814 + 4(1735) + 2(1686) + 4(1646) + 2(1637) + 4(1609) + 2(1604)$$

$$+ 4(1611) + 2(1621) + 4(1666) + 2(1745) + 4(1886) + 2052]$$

$$= \tfrac{1}{6}(61{,}064) = 10{,}177.\overline{3} \text{ megawatt-hours}$$

35. $I(\theta) = \dfrac{N^2 \sin^2 k}{k^2}$, where $k = \dfrac{\pi N d \sin\theta}{\lambda}$, $N = 10{,}000$, $d = 10^{-4}$, and $\lambda = 632.8 \times 10^{-9}$. So $I(\theta) = \dfrac{(10^4)^2 \sin^2 k}{k^2}$,

where $k = \dfrac{\pi (10^4)(10^{-4}) \sin\theta}{632.8 \times 10^{-9}}$. Now $n = 10$ and $\Delta\theta = \dfrac{10^{-6} - (-10^{-6})}{10} = 2 \times 10^{-7}$, so

$M_{10} = 2 \times 10^{-7}[I(-0.0000009) + I(-0.0000007) + \cdots + I(0.0000009)] \approx 59.4.$

37. Consider the function f whose graph is shown. The area $\int_0^2 f(x)\, dx$

is close to 2. The Trapezoidal Rule gives

$T_2 = \frac{2-0}{2 \cdot 2}[f(0) + 2f(1) + f(2)] = \frac{1}{2}[1 + 2 \cdot 1 + 1] = 2.$

The Midpoint Rule gives $M_2 = \frac{2-0}{2}[f(0.5) + f(1.5)] = 1[0 + 0] = 0,$

so the Trapezoidal Rule is more accurate.

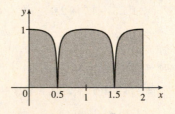

39. Since the Trapezoidal and Midpoint approximations on the interval $[a, b]$ are the sums of the Trapezoidal and Midpoint

approximations on the subintervals $[x_{i-1}, x_i]$, $i = 1, 2, \ldots, n$, we can focus our attention on one such interval. The condition

$f''(x) < 0$ for $a \le x \le b$ means that the graph of f is concave down as in Figure 5. In that figure, T_n is the area of the

trapezoid $AQRD$, $\int_a^b f(x)\, dx$ is the area of the region $AQPRD$, and M_n is the area of the trapezoid $ABCD$, so

$T_n < \int_a^b f(x)\, dx < M_n$. In general, the condition $f'' < 0$ implies that the graph of f on $[a, b]$ lies above the chord joining the points $(a, f(a))$ and $(b, f(b))$. Thus, $\int_a^b f(x)\,dx > T_n$. Since M_n is the area under a tangent to the graph, and since $f'' < 0$

implies that the tangent lies above the graph, we also have $M_n > \int_a^b f(x)\,dx$. Thus, $T_n < \int_a^b f(x)\,dx < M_n$.

41. $T_n = \frac{1}{2}\Delta x\,[f(x_0) + 2f(x_1) + \cdots + 2f(x_{n-1}) + f(x_n)]$ and

$M_n = \Delta x\,[f(\overline{x}_1) + f(\overline{x}_2) + \cdots + f(\overline{x}_{n-1}) + f(\overline{x}_n)]$, where $\overline{x}_i = \frac{1}{2}(x_{i-1} + x_i)$. Now

$T_{2n} = \frac{1}{2}\left(\frac{1}{2}\Delta x\right)[f(x_0) + 2f(\overline{x}_1) + 2f(x_1) + 2f(\overline{x}_2) + 2f(x_2) + \cdots + 2f(\overline{x}_{n-1}) + 2f(x_{n-1}) + 2f(\overline{x}_n) + f(x_n)]$ so

$\frac{1}{2}(T_n + M_n) = \frac{1}{2}T_n + \frac{1}{2}M_n$

$\qquad = \frac{1}{4}\Delta x[f(x_0) + 2f(x_1) + \cdots + 2f(x_{n-1}) + f(x_n)] + \frac{1}{4}\Delta x[2f(\overline{x}_1) + 2f(\overline{x}_2) + \cdots + 2f(\overline{x}_{n-1}) + 2f(\overline{x}_n)]$

$\qquad = T_{2n}$

6.6   Improper Integrals

1. (a) Since $y = \dfrac{x}{x - 1}$ has an infinite discontinuity at $x = 1$, $\displaystyle\int_1^2 \frac{x}{x-1}\,dx$ is a Type 2 improper integral.

(b) Since $\displaystyle\int_0^\infty \frac{1}{1 + x^3}\,dx$ has an infinite interval of integration, it is an improper integral of Type 1.

(c) Since $\displaystyle\int_{-\infty}^\infty x^2 e^{-x^2}\,dx$ has an infinite interval of integration, it is an improper integral of Type 1.

(d) Since $y = \cot x$ has an infinite discontinuity at $x = 0$, $\int_0^{\pi/4} \cot x\,dx$ is a Type 2 improper integral.

3. The area under the graph of $y = 1/x^3 = x^{-3}$ between $x = 1$ and $x = t$ is

$A(t) = \int_1^t x^{-3}\,dx = \left[-\frac{1}{2}x^{-2}\right]_1^t = -\frac{1}{2}t^{-2} - \left(-\frac{1}{2}\right) = \frac{1}{2} - 1/(2t^2)$. So the area for $1 \le x \le 10$ is

$A(10) = 0.5 - 0.005 = 0.495$, the area for $1 \le x \le 100$ is $A(100) = 0.5 - 0.00005 = 0.49995$, and the area for

$1 \le x \le 1000$ is $A(1000) = 0.5 - 0.0000005 = 0.4999995$. The total area under the curve for $x \ge 1$ is

$\displaystyle\lim_{t\to\infty} A(t) = \lim_{t\to\infty}\left[\frac{1}{2} - 1/(2t^2)\right] = \frac{1}{2}$.

5. $\displaystyle\int_3^\infty \frac{1}{(x-2)^{3/2}}\,dx = \lim_{t\to\infty}\int_3^t (x-2)^{-3/2}\,dx = \lim_{t\to\infty}\left[-2\,(x-2)^{-1/2}\right]_3^t \qquad [u = x - 2,\, du = dx]$

$\qquad\qquad = \displaystyle\lim_{t\to\infty}\left(\frac{-2}{\sqrt{t-2}} + \frac{2}{\sqrt{1}}\right) = 0 + 2 = 2.$ Convergent

7. $\displaystyle\int_{-\infty}^0 \frac{1}{3 - 4x}\,dx = \lim_{t\to-\infty}\int_t^0 \frac{1}{3-4x}\,dx = \lim_{t\to-\infty}\left[-\frac{1}{4}\ln|3 - 4x|\right]_t^0 = \lim_{t\to-\infty}\left[-\frac{1}{4}\ln 3 + \frac{1}{4}\ln|3 - 4t|\right] = \infty.$

Divergent

9. $\displaystyle\int_2^\infty e^{-5p}\,dp = \lim_{t\to\infty}\int_2^t e^{-5p}\,dp = \lim_{t\to\infty}\left[-\frac{1}{5}e^{-5p}\right]_2^t = \lim_{t\to\infty}\left(-\frac{1}{5}e^{-5t} + \frac{1}{5}e^{-10}\right) = 0 + \frac{1}{5}e^{-10} = \frac{1}{5}e^{-10}.$ Convergent

11. $\displaystyle\int_0^\infty \frac{x^2}{\sqrt{1 + x^3}}\,dx = \lim_{t\to\infty}\int_0^t \frac{x^2}{\sqrt{1+x^3}}\,dx = \lim_{t\to\infty}\left[\frac{2}{3}\sqrt{1+x^3}\right]_0^t = \lim_{t\to\infty}\left(\frac{2}{3}\sqrt{1+t^3} - \frac{2}{3}\right) = \infty.$ Divergent

13. $\int_{-\infty}^{\infty} xe^{-x^2}\,dx = \int_{-\infty}^{0} xe^{-x^2}\,dx + \int_{0}^{\infty} xe^{-x^2}\,dx.$

$\int_{-\infty}^{0} xe^{-x^2}\,dx = \lim_{t \to -\infty} \left(-\frac{1}{2}\right)\left[e^{-x^2}\right]_{t}^{0} = \lim_{t \to -\infty} \left(-\frac{1}{2}\right)\left(1 - e^{-t^2}\right) = -\frac{1}{2} \cdot 1 = -\frac{1}{2}$, and

$\int_{0}^{\infty} xe^{-x^2}\,dx = \lim_{t \to \infty} \left(-\frac{1}{2}\right)\left[e^{-x^2}\right]_{0}^{t} = \lim_{t \to \infty} \left(-\frac{1}{2}\right)\left(e^{-t^2} - 1\right) = -\frac{1}{2} \cdot (-1) = \frac{1}{2}.$

Therefore, $\int_{-\infty}^{\infty} xe^{-x^2}\,dx = -\frac{1}{2} + \frac{1}{2} = 0.$ Convergent

15. $\int_{-\infty}^{0} ze^{2z}\,dz = \lim_{t \to -\infty} \int_{t}^{0} ze^{2z}\,dz = \lim_{t \to -\infty} \left[\frac{1}{2}ze^{2z} - \frac{1}{4}e^{2z}\right]_{t}^{0} \quad \begin{bmatrix} \text{integration by parts with} \\ u = z,\, dv = e^{2z}\,dz \end{bmatrix}$

$= \lim_{t \to -\infty} \left[\left(0 - \frac{1}{4}\right) - \left(\frac{1}{2}te^{2t} - \frac{1}{4}e^{2t}\right)\right] = -\frac{1}{4} - 0 + 0 \quad \text{[by l'Hospital's Rule]} = -\frac{1}{4}.$ Convergent

17. $\int_{1}^{\infty} \frac{\ln x}{x}\,dx = \lim_{t \to \infty} \left[\frac{(\ln x)^2}{2}\right]_{1}^{t} \quad \begin{bmatrix} \text{by substitution with} \\ u = \ln x,\, du = dx/x \end{bmatrix} = \lim_{t \to \infty} \frac{(\ln t)^2}{2} = \infty.$ Divergent

19. $\int_{1}^{\infty} \frac{1}{x^2 + x}\,dx = \lim_{t \to \infty} \int_{1}^{t} \frac{1}{x(x+1)}\,dx = \lim_{t \to \infty} \int_{1}^{t} \left(\frac{1}{x} - \frac{1}{x+1}\right)dx \quad \text{[partial fractions]}$

$= \lim_{t \to \infty} \left[\ln|x| - \ln|x+1|\right]_{1}^{t} = \lim_{t \to \infty} \left[\ln\left|\frac{x}{x+1}\right|\right]_{1}^{t} = \lim_{t \to \infty} \left(\ln\frac{t}{t+1} - \ln\frac{1}{2}\right) = 0 - \ln\frac{1}{2} = \ln 2.$

Convergent

21. $\int_{-\infty}^{\infty} \frac{x^2}{9 + x^6}\,dx = \int_{-\infty}^{0} \frac{x^2}{9 + x^6}\,dx + \int_{0}^{\infty} \frac{x^2}{9 + x^6}\,dx = 2\int_{0}^{\infty} \frac{x^2}{9 + x^6}\,dx \quad \text{[since the integrand is even]}.$

Now $\int \frac{x^2\,dx}{9 + x^6} \quad \begin{bmatrix} u = x^3 \\ du = 3x^2\,dx \end{bmatrix} = \int \frac{\frac{1}{3}\,du}{9 + u^2} \quad \begin{bmatrix} u = 3v \\ du = 3\,dv \end{bmatrix} = \int \frac{\frac{1}{3}(3\,dv)}{9 + 9v^2} = \frac{1}{9}\int \frac{dv}{1 + v^2}$

$= \frac{1}{9}\tan^{-1} v + C = \frac{1}{9}\tan^{-1}\left(\frac{u}{3}\right) + C = \frac{1}{9}\tan^{-1}\left(\frac{x^3}{3}\right) + C,$

so $2\int_{0}^{\infty} \frac{x^2}{9 + x^6}\,dx = 2\lim_{t \to \infty} \int_{0}^{t} \frac{x^2}{9 + x^6}\,dx = 2\lim_{t \to \infty} \left[\frac{1}{9}\tan^{-1}\left(\frac{x^3}{3}\right)\right]_{0}^{t} = 2\lim_{t \to \infty} \frac{1}{9}\tan^{-1}\left(\frac{t^3}{3}\right) = \frac{2}{9} \cdot \frac{\pi}{2} = \frac{\pi}{9}.$

Convergent

23. $\int_{0}^{1} \frac{3}{x^5}\,dx = \lim_{t \to 0^+} \int_{t}^{1} 3x^{-5}\,dx = \lim_{t \to 0^+} \left[-\frac{3}{4x^4}\right]_{t}^{1} = -\frac{3}{4}\lim_{t \to 0^+}\left(1 - \frac{1}{t^4}\right) = \infty.$ Divergent

25. $\int_{-2}^{14} \frac{dx}{\sqrt[4]{x + 2}} = \lim_{t \to -2^+} \int_{t}^{14} (x+2)^{-1/4}\,dx = \lim_{t \to -2^+} \left[\frac{4}{3}(x+2)^{3/4}\right]_{t}^{14} = \frac{4}{3}\lim_{t \to -2^+}\left[16^{3/4} - (t+2)^{3/4}\right]$

$= \frac{4}{3}(8 - 0) = \frac{32}{3}.$ Convergent

27. There is an infinite discontinuity at $x = 1$. $\int_{0}^{9} \frac{1}{\sqrt[3]{x - 1}}\,dx = \int_{0}^{1} (x-1)^{-1/3}\,dx + \int_{1}^{9} (x-1)^{-1/3}\,dx.$

Here $\int_{0}^{1} (x-1)^{-1/3}\,dx = \lim_{t \to 1^-} \int_{0}^{t} (x-1)^{-1/3}\,dx = \lim_{t \to 1^-} \left[\frac{3}{2}(x-1)^{2/3}\right]_{0}^{t} = \lim_{t \to 1^-} \left[\frac{3}{2}(t-1)^{2/3} - \frac{3}{2}\right] = -\frac{3}{2}$

and $\int_1^9 (x-1)^{-1/3}\,dx = \lim\limits_{t\to 1^+} \int_t^9 (x-1)^{-1/3}\,dx = \lim\limits_{t\to 1^+}\left[\frac{3}{2}(x-1)^{2/3}\right]_t^9 = \lim\limits_{t\to 1^+}\left[6 - \frac{3}{2}(t-1)^{2/3}\right] = 6$. Thus,

$\int_0^9 \dfrac{1}{\sqrt[3]{x-1}}\,dx = -\dfrac{3}{2} + 6 = \dfrac{9}{2}$.   **Convergent**

29. There is an infinite discontinuity at $x = 0$. $\int_{-1}^1 \dfrac{e^x}{e^x - 1}\,dx = \int_{-1}^0 \dfrac{e^x}{e^x - 1}\,dx + \int_0^1 \dfrac{e^x}{e^x - 1}\,dx$.

$\int_{-1}^0 \dfrac{e^x}{e^x - 1}\,dx = \lim\limits_{t\to 0^-} \int_{-1}^t \dfrac{e^x}{e^x - 1}\,dx = \lim\limits_{t\to 0^-}\left[\ln|e^x - 1|\right]_{-1}^t = \lim\limits_{t\to 0^-}\left[\ln|e^t - 1| - \ln|e^{-1} - 1|\right] = -\infty$,

so $\int_{-1}^1 \dfrac{e^x}{e^x - 1}\,dx$ is divergent. The integral $\int_0^1 \dfrac{e^x}{e^x - 1}\,dx$ also diverges since

$\int_0^1 \dfrac{e^x}{e^x - 1}\,dx = \lim\limits_{t\to 0^+} \int_t^1 \dfrac{e^x}{e^x - 1}\,dx = \lim\limits_{t\to 0^+}\left[\ln|e^x - 1|\right]_t^1 = \lim\limits_{t\to 0^+}\left[\ln|e - 1| - \ln|e^t - 1|\right] = \infty$.   **Divergent**

31. $I = \int_0^2 z^2 \ln z\,dz = \lim\limits_{t\to 0^+} \int_t^2 z^2 \ln z\,dz = \lim\limits_{t\to 0^+}\left[\dfrac{z^3}{3^2}(3\ln z - 1)\right]_t^2$ $\begin{bmatrix}\text{integrate by parts}\\ \text{or use Formula 101}\end{bmatrix}$

$= \lim\limits_{t\to 0^+}\left[\dfrac{8}{9}(3\ln 2 - 1) - \dfrac{1}{9}t^3(3\ln t - 1)\right] = \dfrac{8}{3}\ln 2 - \dfrac{8}{9} - \dfrac{1}{9}\lim\limits_{t\to 0^+}\left[t^3(3\ln t - 1)\right] = \dfrac{8}{3}\ln 2 - \dfrac{8}{9} - \dfrac{1}{9}L$.

Now $L = \lim\limits_{t\to 0^+}\left[t^3(3\ln t - 1)\right] = \lim\limits_{t\to 0^+} \dfrac{3\ln t - 1}{t^{-3}} \overset{\text{H}}{=} \lim\limits_{t\to 0^+} \dfrac{3/t}{-3/t^4} = \lim\limits_{t\to 0^+}(-t^3) = 0$.

Thus, $L = 0$ and $I = \dfrac{8}{3}\ln 2 - \dfrac{8}{9}$.   **Convergent**

33.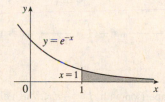

Area $= \int_1^\infty e^{-x}\,dx = \lim\limits_{t\to\infty} \int_1^t e^{-x}\,dx = \lim\limits_{t\to\infty}\left[-e^{-x}\right]_1^t$

$= \lim\limits_{t\to\infty}(-e^{-t} + e^{-1}) = 0 + e^{-1} = 1/e$

35.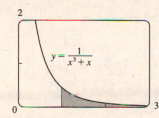

Area $= \int_1^\infty \dfrac{1}{x^3 + x}\,dx = \lim\limits_{t\to\infty} \int_1^t \dfrac{1}{x(x^2 + 1)}\,dx$

$= \lim\limits_{t\to\infty} \int_1^t \left(\dfrac{1}{x} - \dfrac{x}{x^2 + 1}\right)dx$ [partial fractions]

$= \lim\limits_{t\to\infty}\left[\ln|x| - \dfrac{1}{2}\ln|x^2 + 1|\right]_1^t = \lim\limits_{t\to\infty}\left[\ln \dfrac{x}{\sqrt{x^2 + 1}}\right]_1^t$

$= \lim\limits_{t\to\infty}\left(\ln \dfrac{t}{\sqrt{t^2 + 1}} - \ln \dfrac{1}{\sqrt{2}}\right) = \ln 1 - \ln 2^{-1/2} = \dfrac{1}{2}\ln 2$

37.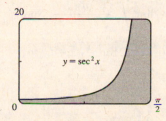

Area $= \int_0^{\pi/2} \sec^2 x\,dx = \lim\limits_{t\to(\pi/2)^-} \int_0^t \sec^2 x\,dx = \lim\limits_{t\to(\pi/2)^-}\left[\tan x\right]_0^t$

$= \lim\limits_{t\to(\pi/2)^-}(\tan t - 0) = \infty$

Infinite area

39. (a)

t	$\int_1^t g(x)\,dx$
2	0.447453
5	0.577101
10	0.621306
100	0.668479
1000	0.672957
10,000	0.673407

$$g(x) = \frac{\sin^2 x}{x^2}.$$

It appears that the integral is convergent.

(b) $-1 \le \sin x \le 1 \;\Rightarrow\; 0 \le \sin^2 x \le 1 \;\Rightarrow\; 0 \le \dfrac{\sin^2 x}{x^2} \le \dfrac{1}{x^2}$. Since $\displaystyle\int_1^\infty \frac{1}{x^2}\,dx$ is convergent

[Equation 2 with $p = 2 > 1$], $\displaystyle\int_1^\infty \frac{\sin^2 x}{x^2}\,dx$ is convergent by the Comparison Theorem.

(c)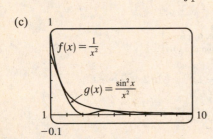

Since $\int_1^\infty f(x)\,dx$ is finite and the area under $g(x)$ is less than the area under $f(x)$ on any interval $[1, t]$, $\int_1^\infty g(x)\,dx$ must be finite; that is, the integral is convergent.

41. For $x > 0$, $\dfrac{x}{x^3 + 1} < \dfrac{x}{x^3} = \dfrac{1}{x^2}$. $\displaystyle\int_1^\infty \frac{1}{x^2}\,dx$ is convergent by Theorem 2 with $p = 2 > 1$, so $\displaystyle\int_1^\infty \frac{x}{x^3 + 1}\,dx$ is convergent

by the Comparison Theorem. $\displaystyle\int_0^1 \frac{x}{x^3 + 1}\,dx$ is a constant, so $\displaystyle\int_0^\infty \frac{x}{x^3 + 1}\,dx = \int_0^1 \frac{x}{x^3 + 1}\,dx + \int_1^\infty \frac{x}{x^3 + 1}\,dx$ is also

convergent.

43. For $x > 1$, $f(x) = \dfrac{x + 1}{\sqrt{x^4 - x}} > \dfrac{x + 1}{\sqrt{x^4}} > \dfrac{x}{x^2} = \dfrac{1}{x}$, so $\displaystyle\int_2^\infty f(x)\,dx$ diverges by comparison with $\displaystyle\int_2^\infty \frac{1}{x}\,dx$, which diverges

by Theorem 2 with $p = 1 \le 1$. Thus, $\int_1^\infty f(x)\,dx = \int_1^2 f(x)\,dx + \int_2^\infty f(x)\,dx$ also diverges.

45. For $0 < x \le 1$, $\dfrac{\sec^2 x}{x\sqrt{x}} > \dfrac{1}{x^{3/2}}$. Now

$$I = \int_0^1 x^{-3/2}\,dx = \lim_{t \to 0^+} \int_t^1 x^{-3/2}\,dx = \lim_{t \to 0^+} \left[-2x^{-1/2} \right]_t^1 = \lim_{t \to 0^+} \left(-2 + \frac{2}{\sqrt{t}} \right) = \infty, \text{ so } I \text{ is divergent, and by}$$

comparison, $\displaystyle\int_0^1 \frac{\sec^2 x}{x\sqrt{x}}$ is divergent.

47. $\displaystyle\int_0^\infty \frac{dx}{\sqrt{x}\,(1 + x)} = \int_0^1 \frac{dx}{\sqrt{x}\,(1 + x)} + \int_1^\infty \frac{dx}{\sqrt{x}\,(1 + x)} = \lim_{t \to 0^+} \int_t^1 \frac{dx}{\sqrt{x}\,(1 + x)} + \lim_{t \to \infty} \int_1^t \frac{dx}{\sqrt{x}\,(1 + x)}$. Now

$$\int \frac{dx}{\sqrt{x}\,(1 + x)} = \int \frac{2u\,du}{u(1 + u^2)} \quad \begin{bmatrix} u = \sqrt{x},\, x = u^2, \\ dx = 2u\,du \end{bmatrix} = 2 \int \frac{du}{1 + u^2} = 2\tan^{-1} u + C = 2\tan^{-1}\sqrt{x} + C, \text{ so}$$

$$\int_0^\infty \frac{dx}{\sqrt{x}\,(1 + x)} = \lim_{t \to 0^+} \left[2\tan^{-1}\sqrt{x} \right]_t^1 + \lim_{t \to \infty} \left[2\tan^{-1}\sqrt{x} \right]_1^t$$

$$= \lim_{t \to 0^+} \left[2\left(\tfrac{\pi}{4}\right) - 2\tan^{-1}\sqrt{t} \right] + \lim_{t \to \infty} \left[2\tan^{-1}\sqrt{t} - 2\left(\tfrac{\pi}{4}\right) \right] = \tfrac{\pi}{2} - 0 + 2\left(\tfrac{\pi}{2}\right) - \tfrac{\pi}{2} = \pi.$$

49. If $p = 1$, then $\int_0^1 \dfrac{dx}{x^p} = \lim\limits_{t \to 0^+} \int_t^1 \dfrac{dx}{x} = \lim\limits_{t \to 0^+} [\ln x]_t^1 = \infty.$ Divergent

If $p \neq 1$, then $\int_0^1 \dfrac{dx}{x^p} = \lim\limits_{t \to 0^+} \int_t^1 \dfrac{dx}{x^p}$ [note that the integral is not improper if $p < 0$]

$$= \lim\limits_{t \to 0^+} \left[\frac{x^{-p+1}}{-p+1}\right]_t^1 = \lim\limits_{t \to 0^+} \frac{1}{1-p}\left[1 - \frac{1}{t^{p-1}}\right]$$

If $p > 1$, then $p - 1 > 0$, so $\dfrac{1}{t^{p-1}} \to \infty$ as $t \to 0^+$, and the integral diverges.

If $p < 1$, then $p - 1 < 0$, so $\dfrac{1}{t^{p-1}} \to 0$ as $t \to 0^+$ and $\int_0^1 \dfrac{dx}{x^p} = \dfrac{1}{1-p}\left[\lim\limits_{t \to 0^+}\left(1 - t^{1-p}\right)\right] = \dfrac{1}{1-p}.$

Thus, the integral converges if and only if $p < 1$, and in that case its value is $\dfrac{1}{1-p}$.

51. (a) $I = \int_{-\infty}^{\infty} x\,dx = \int_{-\infty}^{0} x\,dx + \int_0^{\infty} x\,dx$, and $\int_0^{\infty} x\,dx = \lim\limits_{t \to \infty}\int_0^t x\,dx = \lim\limits_{t \to \infty}\left[\frac{1}{2}x^2\right]_0^t = \lim\limits_{t \to \infty}\left[\frac{1}{2}t^2 - 0\right] = \infty,$

so I is divergent.

(b) $\int_{-t}^{t} x\,dx = \left[\frac{1}{2}x^2\right]_{-t}^{t} = \frac{1}{2}t^2 - \frac{1}{2}t^2 = 0$, so $\lim\limits_{t \to \infty}\int_{-t}^{t} x\,dx = 0$. Therefore, $\int_{-\infty}^{\infty} x\,dx \neq \lim\limits_{t \to \infty}\int_{-t}^{t} x\,dx.$

53. We would expect a small percentage of bulbs to burn out in the first few hundred hours, most of the bulbs to burn out after close to 700 hours, and a few overachievers to burn on and on.

(a)

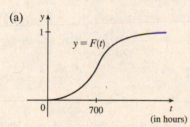

(b) $r(t) = F'(t)$ is the rate at which the fraction $F(t)$ of burnt-out bulbs increases as t increases. This could be interpreted as a fractional burnout rate.

(c) $\int_0^{\infty} r(t)\,dt = \lim\limits_{x \to \infty} F(x) = 1$, since all of the bulbs will eventually burn out.

55. $I = \int_0^{\infty} te^{kt}\,dt = \lim\limits_{s \to \infty}\left[\dfrac{1}{k^2}(kt-1)e^{kt}\right]_0^s$ [Formula 96, or parts] $= \lim\limits_{s \to \infty}\left[\left(\dfrac{1}{k}se^{ks} - \dfrac{1}{k^2}e^{ks}\right) - \left(-\dfrac{1}{k^2}\right)\right].$

Since $k < 0$ the first two terms approach 0 (you can verify that the first term does so with l'Hospital's Rule), so the limit is equal to $1/k^2$. Thus, $M = -kI = -k\left(1/k^2\right) = -1/k = -1/(-0.000121) \approx 8264.5$ years.

57. $I = \int_a^{\infty} \dfrac{1}{x^2+1}\,dx = \lim\limits_{t \to \infty}\int_a^t \dfrac{1}{x^2+1}\,dx = \lim\limits_{t \to \infty}\left[\tan^{-1}x\right]_a^t = \lim\limits_{t \to \infty}\left(\tan^{-1}t - \tan^{-1}a\right) = \dfrac{\pi}{2} - \tan^{-1}a.$

$I < 0.001 \;\Rightarrow\; \dfrac{\pi}{2} - \tan^{-1}a < 0.001 \;\Rightarrow\; \tan^{-1}a > \dfrac{\pi}{2} - 0.001 \;\Rightarrow\; a > \tan\left(\dfrac{\pi}{2} - 0.001\right) \approx 1000.$

59. We use integration by parts: let $u = x$, $dv = xe^{-x^2}\,dx \;\Rightarrow\; du = dx$, $v = -\frac{1}{2}e^{-x^2}$. So

$$\int_0^{\infty} x^2 e^{-x^2}\,dx = \lim\limits_{t \to \infty}\left[-\frac{1}{2}xe^{-x^2}\right]_0^t + \frac{1}{2}\int_0^{\infty} e^{-x^2}\,dx = \lim\limits_{t \to \infty}\left[-\frac{t}{2e^{t^2}}\right] + \frac{1}{2}\int_0^{\infty} e^{-x^2}\,dx = \frac{1}{2}\int_0^{\infty} e^{-x^2}\,dx$$

(The limit is 0 by l'Hospital's Rule.)

61. For the first part of the integral, let $x = 2\tan\theta \implies dx = 2\sec^2\theta\, d\theta$.

$$\int \frac{1}{\sqrt{x^2+4}}\, dx = \int \frac{2\sec^2\theta}{2\sec\theta}\, d\theta = \int \sec\theta\, d\theta = \ln|\sec\theta + \tan\theta|.$$

From the figure, $\tan\theta = \dfrac{x}{2}$, and $\sec\theta = \dfrac{\sqrt{x^2+4}}{2}$. So

$$I = \int_0^\infty \left(\frac{1}{\sqrt{x^2+4}} - \frac{C}{x+2} \right) dx = \lim_{t\to\infty} \left[\ln\left| \frac{\sqrt{x^2+4}}{2} + \frac{x}{2} \right| - C\ln|x+2| \right]_0^t$$

$$= \lim_{t\to\infty} \left[\ln\frac{\sqrt{t^2+4}+t}{2} - C\ln(t+2) - (\ln 1 - C\ln 2) \right]$$

$$= \lim_{t\to\infty} \left[\ln\left(\frac{\sqrt{t^2+4}+t}{2(t+2)^C} \right) + \ln 2^C \right] = \ln\left(\lim_{t\to\infty} \frac{t+\sqrt{t^2+4}}{(t+2)^C} \right) + \ln 2^{C-1}$$

Now $L = \lim\limits_{t\to\infty} \dfrac{t+\sqrt{t^2+4}}{(t+2)^C} \overset{\text{H}}{=} \lim\limits_{t\to\infty} \dfrac{1+t/\sqrt{t^2+4}}{C(t+2)^{C-1}} = \dfrac{2}{C \lim\limits_{t\to\infty}(t+2)^{C-1}}.$

If $C < 1$, $L = \infty$ and I diverges.

If $C = 1$, $L = 2$ and I converges to $\ln 2 + \ln 2^0 = \ln 2$.

If $C > 1$, $L = 0$ and I diverges to $-\infty$.

63. No, $I = \int_0^\infty f(x)\, dx$ must be *divergent*. Since $\lim\limits_{x\to\infty} f(x) = 1$, there must exist an N such that if $x \geq N$, then $f(x) \geq \frac{1}{2}$.

Thus, $I = I_1 + I_2 = \int_0^N f(x)\, dx + \int_N^\infty f(x)\, dx$, where I_1 is an ordinary definite integral that has a finite value, and I_2 is improper and diverges by comparison with the divergent integral $\int_N^\infty \frac{1}{2}\, dx$.

6 Review

CONCEPT CHECK

1. See Formula 6.1.1 or 6.1.2. We try to choose $u = f(x)$ to be a function that becomes simpler when differentiated (or at least not more complicated) as long as $dv = g'(x)\, dx$ can be readily integrated to give v.

2. See the margin note on page 325.

3. If $\sqrt{a^2-x^2}$ occurs, try $x = a\sin\theta$; if $\sqrt{a^2+x^2}$ occurs, try $x = a\tan\theta$, and if $\sqrt{x^2-a^2}$ occurs, try $x = a\sec\theta$. See the Table of Trigonometric Substitutions on page 328.

4. See Equation 2 and Expressions 6, 8, and 10 in Section 6.3.

5. See the Midpoint Rule, the Trapezoidal Rule, and Simpson's Rule, as well as their associated error bounds, all in Section 6.6. We would expect the best estimate to be given by Simpson's Rule.

6. See Definitions 1(a), (b), and (c) in Section 6.6.

7. See Definitions 3(b), (a), and (c) in Section 6.6.

8. See the Comparison Theorem on page 365.

TRUE-FALSE QUIZ

1. False.   Since the numerator has a higher degree than the denominator, $\dfrac{x\left(x^2+4\right)}{x^2-4} = x + \dfrac{8x}{x^2-4} = x + \dfrac{A}{x+2} + \dfrac{B}{x-2}$.

3. False.   It can be put in the form $\dfrac{A}{x} + \dfrac{B}{x^2} + \dfrac{C}{x-4}$.

5. False.   This is an improper integral, since the denominator vanishes at $x=1$.

$$\int_0^4 \frac{x}{x^2-1}\,dx = \int_0^1 \frac{x}{x^2-1}\,dx + \int_1^4 \frac{x}{x^2-1}\,dx \text{ and}$$

$$\int_0^1 \frac{x}{x^2-1}\,dx = \lim_{t \to 1^-}\int_0^t \frac{x}{x^2-1}\,dx = \lim_{t \to 1^-}\left[\tfrac{1}{2}\ln\left|x^2-1\right|\right]_0^t = \lim_{t \to 1^-}\tfrac{1}{2}\ln\left|t^2-1\right| = \infty$$

So the integral diverges.

7. False.   See Exercise 51 in Section 6.6.

9. (a) True.   See the end of Section 6.4.

   (b) False.   Examples include the functions $f(x) = e^{x^2}$, $g(x) = \sin(x^2)$, and $h(x) = \dfrac{\sin x}{x}$.

11. False.   If $f(x) = 1/x$, then f is continuous and decreasing on $[1, \infty)$ with $\lim\limits_{x \to \infty} f(x) = 0$, but $\int_1^\infty f(x)\,dx$ is divergent.

13. False.   Take $f(x) = 1$ for all x and $g(x) = -1$ for all x. Then $\int_a^\infty f(x)\,dx = \infty$ [divergent]

   and $\int_a^\infty g(x)\,dx = -\infty$ [divergent],   but $\int_a^\infty [f(x)+g(x)]\,dx = 0$ [convergent].

EXERCISES

1. $\displaystyle\int_1^2 \frac{(x+1)^2}{x}\,dx = \int_1^2 \frac{x^2+2x+1}{x}\,dx = \int_1^2 \left(x + 2 + \frac{1}{x}\right)dx = \left[\frac{1}{2}x^2 + 2x + \ln|x|\right]_1^2$

$\qquad = (2 + 4 + \ln 2) - \left(\tfrac{1}{2} + 2 + 0\right) = \tfrac{7}{2} + \ln 2$

3. $\displaystyle\int_0^{\pi/2} \sin\theta\, e^{\cos\theta}\,d\theta = \int_1^0 e^u\,(-du)$ $\begin{bmatrix} u = \cos\theta, \\ du = -\sin\theta\,d\theta \end{bmatrix}$

$\qquad = \displaystyle\int_0^1 e^u\,du = \left[e^u\right]_0^1 = e^1 - e^0 = e - 1$

5. $\displaystyle\int \frac{dt}{2t^2+3t+1} = \int \frac{1}{(2t+1)(t+1)}\,dt = \int \left(\frac{2}{2t+1} - \frac{1}{t+1}\right)dt$ [partial fractions] $= \ln|2t+1| - \ln|t+1| + C$

7. Let $u = \ln t$, $du = dt/t$. Then $\displaystyle\int \frac{\sin(\ln t)}{t}\,dt = \int \sin u\,du = -\cos u + C = -\cos(\ln t) + C$.

9. $\displaystyle\int_1^4 x^{3/2}\ln x\,dx$ $\begin{bmatrix} u = \ln x, & dv = x^{3/2}\,dx, \\ du = dx/x & v = \tfrac{2}{5}x^{5/2} \end{bmatrix}$ $= \dfrac{2}{5}\left[x^{5/2}\ln x\right]_1^4 - \dfrac{2}{5}\displaystyle\int_1^4 x^{3/2}\,dx = \tfrac{2}{5}(32\ln 4 - \ln 1) - \tfrac{2}{5}\left[\tfrac{2}{5}x^{5/2}\right]_1^4$

$\qquad = \tfrac{2}{5}(64\ln 2) - \tfrac{4}{25}(32-1) = \tfrac{128}{5}\ln 2 - \tfrac{124}{25}$ $\left[\text{or } \tfrac{64}{5}\ln 4 - \tfrac{124}{25}\right]$

11. Let $x = \sec\theta$. Then

$$\int_1^2 \frac{\sqrt{x^2 - 1}}{x}\,dx = \int_0^{\pi/3} \frac{\tan\theta}{\sec\theta}\,\sec\theta\,\tan\theta\,d\theta = \int_0^{\pi/3} \tan^2\theta\,d\theta = \int_0^{\pi/3} (\sec^2\theta - 1)\,d\theta = \big[\tan\theta - \theta\big]_0^{\pi/3} = \sqrt{3} - \frac{\pi}{3}.$$

13. $\displaystyle\int \frac{dx}{x^3 + x} = \int \left(\frac{1}{x} - \frac{x}{x^2 + 1}\right) dx = \ln|x| - \frac{1}{2}\ln(x^2 + 1) + C$

15. $\displaystyle\int_0^{\pi/2} \sin^3\theta\,\cos^2\theta\,d\theta = \int_0^{\pi/2}(1 - \cos^2\theta)\cos^2\theta\,\sin\theta\,d\theta = \int_1^0 (1 - u^2)u^2\,(-du)$ $\qquad \begin{bmatrix} u = \cos\theta, \\ du = -\sin\theta\,d\theta \end{bmatrix}$

$$= \int_0^1 (u^2 - u^4)\,du = \big[\tfrac{1}{3}u^3 - \tfrac{1}{5}u^5\big]_0^1 = \big(\tfrac{1}{3} - \tfrac{1}{5}\big) - 0 = \tfrac{2}{15}$$

17. Integrate by parts with $u = x$, $dv = \sec x\,\tan x\,dx \;\Rightarrow\; du = dx$, $v = \sec x$:

$$\int x\sec x\,\tan x\,dx = x\sec x - \int \sec x\,dx \overset{14}{=} x\sec x - \ln|\sec x + \tan x| + C.$$

19. $\displaystyle\int \frac{x + 1}{9x^2 + 6x + 5}\,dx = \int \frac{x + 1}{(9x^2 + 6x + 1) + 4}\,dx = \int \frac{x + 1}{(3x + 1)^2 + 4}\,dx$ $\qquad \begin{bmatrix} u = 3x + 1, \\ du = 3\,dx \end{bmatrix}$

$$= \int \frac{\big[\tfrac{1}{3}(u - 1)\big] + 1}{u^2 + 4}\left(\tfrac{1}{3}\,du\right) = \frac{1}{3}\cdot\frac{1}{3}\int \frac{(u - 1) + 3}{u^2 + 4}\,du$$

$$= \frac{1}{9}\int \frac{u}{u^2 + 4}\,du + \frac{1}{9}\int \frac{2}{u^2 + 2^2}\,du = \frac{1}{9}\cdot\frac{1}{2}\ln(u^2 + 4) + \frac{2}{9}\cdot\frac{1}{2}\tan^{-1}\!\left(\frac{1}{2}u\right) + C$$

$$= \frac{1}{18}\ln(9x^2 + 6x + 5) + \frac{1}{9}\tan^{-1}\!\big[\tfrac{1}{2}(3x + 1)\big] + C$$

21. $\displaystyle\int \frac{dx}{\sqrt{x^2 - 4x}} = \int \frac{dx}{\sqrt{(x^2 - 4x + 4) - 4}} = \int \frac{dx}{\sqrt{(x - 2)^2 - 2^2}}$

$$= \int \frac{2\sec\theta\,\tan\theta\,d\theta}{2\tan\theta} \qquad \begin{bmatrix} x - 2 = 2\sec\theta, \\ dx = 2\sec\theta\,\tan\theta\,d\theta \end{bmatrix}$$

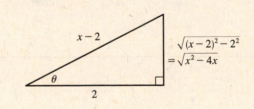

$$= \int \sec\theta\,d\theta = \ln|\sec\theta + \tan\theta| + C_1$$

$$= \ln\left|\frac{x - 2}{2} + \frac{\sqrt{x^2 - 4x}}{2}\right| + C_1$$

$$= \ln\left|x - 2 + \sqrt{x^2 - 4x}\,\right| + C, \text{ where } C = C_1 - \ln 2$$

23. Let $u = \cot 4x$. Then $du = -4\csc^2 4x\,dx \;\Rightarrow\;$

$$\int \csc^4 4x\,dx = \int (\cot^2 4x + 1)\csc^2 4x\,dx = \int (u^2 + 1)\big(-\tfrac{1}{4}\,du\big) = -\tfrac{1}{4}\big(\tfrac{1}{3}u^3 + u\big) + C = -\tfrac{1}{12}(\cot^3 4x + 3\cot 4x) + C.$$

25. $\displaystyle\frac{3x^3 - x^2 + 6x - 4}{(x^2 + 1)(x^2 + 2)} = \frac{Ax + B}{x^2 + 1} + \frac{Cx + D}{x^2 + 2} \;\Rightarrow\; 3x^3 - x^2 + 6x - 4 = (Ax + B)(x^2 + 2) + (Cx + D)(x^2 + 1).$

Equating the coefficients gives $A + C = 3$, $B + D = -1$, $2A + C = 6$, and $2B + D = -4 \;\Rightarrow\;$

$A = 3$, $C = 0$, $B = -3$, and $D = 2$. Now

$$\int \frac{3x^3 - x^2 + 6x - 4}{(x^2 + 1)(x^2 + 2)}\,dx = 3\int \frac{x - 1}{x^2 + 1}\,dx + 2\int \frac{dx}{x^2 + 2} = \frac{3}{2}\ln(x^2 + 1) - 3\tan^{-1}x + \sqrt{2}\tan^{-1}\!\left(\frac{x}{\sqrt{2}}\right) + C.$$

27. $\displaystyle\int_0^{\pi/2} \cos^3 x\,\sin 2x\,dx = \int_0^{\pi/2} \cos^3 x\,(2\sin x\,\cos x)\,dx = \int_0^{\pi/2} 2\cos^4 x\,\sin x\,dx = \big[-\tfrac{2}{5}\cos^5 x\big]_0^{\pi/2} = \tfrac{2}{5}$

29. The integrand is an odd function, so $\int_{-3}^{3} \dfrac{x}{1+|x|}\, dx = 0$ [by 4.5.6(b)].

31. Let $u = \sqrt{e^x - 1}$. Then $u^2 = e^x - 1$ and $2u\, du = e^x\, dx$. Also, $e^x + 8 = u^2 + 9$. Thus,

$$\int_0^{\ln 10} \frac{e^x \sqrt{e^x - 1}}{e^x + 8}\, dx = \int_0^3 \frac{u \cdot 2u\, du}{u^2 + 9} = 2\int_0^3 \frac{u^2}{u^2 + 9}\, du = 2\int_0^3 \left(1 - \frac{9}{u^2 + 9}\right) du$$

$$= 2\left[u - \frac{9}{3} \tan^{-1}\!\left(\frac{u}{3}\right)\right]_0^3 = 2\big[(3 - 3\tan^{-1} 1) - 0\big] = 2\left(3 - 3 \cdot \frac{\pi}{4}\right) = 6 - \frac{3\pi}{2}$$

33. Let $x = 2\sin\theta \;\Rightarrow\; (4 - x^2)^{3/2} = (2\cos\theta)^3$, $dx = 2\cos\theta\, d\theta$, so

$$\int \frac{x^2}{(4 - x^2)^{3/2}}\, dx = \int \frac{4\sin^2\theta}{8\cos^3\theta} 2\cos\theta\, d\theta = \int \tan^2\theta\, d\theta = \int (\sec^2\theta - 1)\, d\theta$$

$$= \tan\theta - \theta + C = \frac{x}{\sqrt{4 - x^2}} - \sin^{-1}\!\left(\frac{x}{2}\right) + C$$

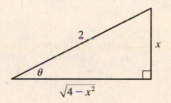

35. $\displaystyle \int \frac{1}{\sqrt{x} + x^{3/2}}\, dx = \int \frac{dx}{\sqrt{x\,(1 + \sqrt{x})}} = \int \frac{dx}{\sqrt{x}\sqrt{1 + \sqrt{x}}} \quad \begin{bmatrix} u = 1 + \sqrt{x}, \\ du = \dfrac{dx}{2\sqrt{x}} \end{bmatrix} = \int \frac{2\, du}{\sqrt{u}} = \int 2u^{-1/2}\, du$

$$= 4\sqrt{u} + C = 4\sqrt{1 + \sqrt{x}} + C$$

37. $\int (\cos x + \sin x)^2 \cos 2x\, dx = \int (\cos^2 x + 2\sin x \cos x + \sin^2 x) \cos 2x\, dx = \int (1 + \sin 2x) \cos 2x\, dx$

$$= \int \cos 2x\, dx + \tfrac{1}{2}\int \sin 4x\, dx = \tfrac{1}{2}\sin 2x - \tfrac{1}{8}\cos 4x + C$$

Or: $\int (\cos x + \sin x)^2 \cos 2x\, dx = \int (\cos x + \sin x)^2 (\cos^2 x - \sin^2 x)\, dx$

$$= \int (\cos x + \sin x)^3 (\cos x - \sin x)\, dx = \tfrac{1}{4}(\cos x + \sin x)^4 + C_1$$

39. We'll integrate $I = \displaystyle\int \frac{xe^{2x}}{(1 + 2x)^2}\, dx$ by parts with $u = xe^{2x}$ and $dv = \dfrac{dx}{(1 + 2x)^2}$. Then $du = (x \cdot 2e^{2x} + e^{2x} \cdot 1)\, dx$

and $v = -\dfrac{1}{2} \cdot \dfrac{1}{1 + 2x}$, so

$$I = -\frac{1}{2} \cdot \frac{xe^{2x}}{1 + 2x} - \int \left[-\frac{1}{2} \cdot \frac{e^{2x}(2x + 1)}{1 + 2x}\right] dx = -\frac{xe^{2x}}{4x + 2} + \frac{1}{2} \cdot \frac{1}{2} e^{2x} + C = e^{2x}\left(\frac{1}{4} - \frac{x}{4x + 2}\right) + C$$

Thus, $\displaystyle\int_0^{1/2} \frac{xe^{2x}}{(1 + 2x)^2}\, dx = \left[e^{2x}\left(\frac{1}{4} - \frac{x}{4x + 2}\right)\right]_0^{1/2} = e\left(\frac{1}{4} - \frac{1}{8}\right) - 1\left(\frac{1}{4} - 0\right) = \frac{1}{8}e - \frac{1}{4}$.

41. $\displaystyle\int_1^{\infty} \frac{1}{(2x + 1)^3}\, dx = \lim_{t\to\infty} \int_1^t \frac{1}{(2x + 1)^3}\, dx = \lim_{t\to\infty} \int_1^t \tfrac{1}{2}(2x + 1)^{-3} 2\, dx = \lim_{t\to\infty} \left[-\frac{1}{4(2x + 1)^2}\right]_1^t$

$$= -\frac{1}{4} \lim_{t\to\infty} \left[\frac{1}{(2t + 1)^2} - \frac{1}{9}\right] = -\frac{1}{4}\left(0 - \frac{1}{9}\right) = \frac{1}{36}$$

43. $\displaystyle\int \frac{dx}{x\ln x} \quad \begin{bmatrix} u = \ln x, \\ du = dx/x \end{bmatrix} = \int \frac{du}{u} = \ln|u| + C = \ln|\ln x| + C$, so

$$\int_2^{\infty} \frac{dx}{x\ln x} = \lim_{t\to\infty} \int_2^t \frac{dx}{x\ln x} = \lim_{t\to\infty} \Big[\ln|\ln x|\Big]_2^t = \lim_{t\to\infty} [\ln(\ln t) - \ln(\ln 2)] = \infty, \text{ so the integral is divergent.}$$

45. $\int_0^4 \dfrac{\ln x}{\sqrt{x}}\, dx = \lim\limits_{t \to 0^+} \int_t^4 \dfrac{\ln x}{\sqrt{x}}\, dx \overset{\star}{=} \lim\limits_{t \to 0^+} \left[2\sqrt{x}\,\ln x - 4\sqrt{x}\right]_t^4$

$\qquad = \lim\limits_{t \to 0^+} \left[(2 \cdot 2\ln 4 - 4 \cdot 2) - (2\sqrt{t}\,\ln t - 4\sqrt{t}) \right] \overset{\star\star}{=} (4\ln 4 - 8) - (0 - 0) = 4\ln 4 - 8$

$(\star)$ $\qquad$ Let $u = \ln x,\, dv = \dfrac{1}{\sqrt{x}}\, dx \;\Rightarrow\; du = \dfrac{1}{x}\, dx,\, v = 2\sqrt{x}$. Then

$\qquad\qquad\qquad\qquad \int \dfrac{\ln x}{\sqrt{x}}\, dx = 2\sqrt{x}\,\ln x - 2\int \dfrac{dx}{\sqrt{x}} = 2\sqrt{x}\,\ln x - 4\sqrt{x} + C$

$(\star\star)$ $\qquad\qquad \lim\limits_{t \to 0^+} \left(2\sqrt{t}\,\ln t \right) = \lim\limits_{t \to 0^+} \dfrac{2\ln t}{t^{-1/2}} \overset{\text{H}}{=} \lim\limits_{t \to 0^+} \dfrac{2/t}{-\frac{1}{2}t^{-3/2}} = \lim\limits_{t \to 0^+} \left(-4\sqrt{t} \right) = 0$

47. $\int_0^1 \dfrac{x-1}{\sqrt{x}}\, dx = \lim\limits_{t \to 0^+} \int_t^1 \left(\dfrac{x}{\sqrt{x}} - \dfrac{1}{\sqrt{x}} \right) dx = \lim\limits_{t \to 0^+} \int_t^1 \left(x^{1/2} - x^{-1/2} \right) dx = \lim\limits_{t \to 0^+} \left[\tfrac{2}{3}x^{3/2} - 2x^{1/2} \right]_t^1$

$\qquad = \lim\limits_{t \to 0^+} \left[\left(\tfrac{2}{3} - 2 \right) - \left(\tfrac{2}{3}t^{3/2} - 2t^{1/2} \right) \right] = -\tfrac{4}{3} - 0 = -\tfrac{4}{3}$

49. Let $u = 2x + 1$. Then

$\int_{-\infty}^{\infty} \dfrac{dx}{4x^2 + 4x + 5} = \int_{-\infty}^{\infty} \dfrac{\frac{1}{2}\, du}{u^2 + 4} = \tfrac{1}{2} \int_{-\infty}^0 \dfrac{du}{u^2 + 4} + \tfrac{1}{2} \int_0^{\infty} \dfrac{du}{u^2 + 4}$

$\qquad = \tfrac{1}{2} \lim\limits_{t \to -\infty} \left[\tfrac{1}{2}\tan^{-1}\left(\tfrac{1}{2}u\right) \right]_t^0 + \tfrac{1}{2} \lim\limits_{t \to \infty} \left[\tfrac{1}{2}\tan^{-1}\left(\tfrac{1}{2}u\right) \right]_0^t = \tfrac{1}{4}\left[0 - \left(-\tfrac{\pi}{2} \right) \right] + \tfrac{1}{4}\left[\tfrac{\pi}{2} - 0 \right] = \tfrac{\pi}{4}$.

51. $\int \sqrt{4x^2 - 4x - 3}\, dx = \int \sqrt{(2x-1)^2 - 4}\, dx \quad \begin{bmatrix} u = 2x - 1, \\ du = 2\, dx \end{bmatrix} = \int \sqrt{u^2 - 2^2} \left(\tfrac{1}{2}\, du \right)$

$\qquad \overset{39}{=} \tfrac{1}{2}\left(\dfrac{u}{2}\sqrt{u^2 - 2^2} - \dfrac{2^2}{2}\ln\left| u + \sqrt{u^2 - 2^2} \right| \right) + C = \tfrac{1}{4}u\sqrt{u^2 - 4} - \ln\left| u + \sqrt{u^2 - 4} \right| + C$

$\qquad = \tfrac{1}{4}(2x-1)\sqrt{4x^2 - 4x - 3} - \ln\left| 2x - 1 + \sqrt{4x^2 - 4x - 3} \right| + C$

53. Let $u = \sin x$, so that $du = \cos x\, dx$. Then

$\int \cos x\sqrt{4 + \sin^2 x}\, dx = \int \sqrt{2^2 + u^2}\, du \overset{21}{=} \dfrac{u}{2}\sqrt{2^2 + u^2} + \dfrac{2^2}{2}\ln\left(u + \sqrt{2^2 + u^2} \right) + C$

$\qquad\qquad\qquad = \tfrac{1}{2}\sin x\sqrt{4 + \sin^2 x} + 2\ln\left(\sin x + \sqrt{4 + \sin^2 x} \right) + C$

55. For $n \geq 0$, $\int_0^{\infty} x^n\, dx = \lim\limits_{t \to \infty} \left[x^{n+1}/(n+1) \right]_0^t = \infty$. For $n < 0$, $\int_0^{\infty} x^n\, dx = \int_0^1 x^n\, dx + \int_1^{\infty} x^n\, dx$. Both integrals are

improper. By (6.6.2), the second integral diverges if $-1 \leq n < 0$. By Exercise 6.6.49, the first integral diverges if $n \leq -1$.

Thus, $\int_0^{\infty} x^n\, dx$ is divergent for all values of n.

57. $f(x) = \dfrac{1}{\ln x}$, $\Delta x = \dfrac{b-a}{n} = \dfrac{4-2}{10} = \dfrac{1}{5}$

(a) $T_{10} = \dfrac{1}{5 \cdot 2}\{ f(2) + 2[f(2.2) + f(2.4) + \cdots + f(3.8)] + f(4) \} \approx 1.925444$

(b) $M_{10} = \dfrac{1}{5}[f(2.1) + f(2.3) + f(2.5) + \cdots + f(3.9)] \approx 1.920915$

(c) $S_{10} = \dfrac{1}{5 \cdot 3}[f(2) + 4f(2.2) + 2f(2.4) + \cdots + 2f(3.6) + 4f(3.8) + f(4)] \approx 1.922470$

59. $f(x) = \dfrac{1}{\ln x}$ $\Rightarrow$ $f'(x) = -\dfrac{1}{x(\ln x)^2}$ $\Rightarrow$ $f''(x) = \dfrac{2 + \ln x}{x^2(\ln x)^3} = \dfrac{2}{x^2(\ln x)^3} + \dfrac{1}{x^2(\ln x)^2}$. Note that each term of

$f''(x)$ decreases on $[2, 4]$, so we'll take $K = f''(2) \approx 2.022$. $|E_T| \le \dfrac{K(b-a)^3}{12n^2} \approx \dfrac{2.022(4-2)^3}{12(10)^2} = 0.01348$ and

$|E_M| \le \dfrac{K(b-a)^3}{24n^2} = 0.00674$. $|E_T| \le 0.00001$ $\Leftrightarrow$ $\dfrac{2.022(8)}{12n^2} \le \dfrac{1}{10^5}$ $\Leftrightarrow$ $n^2 \ge \dfrac{10^5(2.022)(8)}{12}$ $\Rightarrow$ $n \ge 367.2$.

Take $n = 368$ for T_n. $|E_M| \le 0.00001$ $\Leftrightarrow$ $n^2 \ge \dfrac{10^5(2.022)(8)}{24}$ $\Rightarrow$ $n \ge 259.6$. Take $n = 260$ for M_n.

61. $\Delta t = \left(\tfrac{10}{60} - 0\right) / 10 = \tfrac{1}{60}$.

Distance traveled $= \int_0^{10} v \, dt \approx S_{10}$

$\qquad = \tfrac{1}{60 \cdot 3}[40 + 4(42) + 2(45) + 4(49) + 2(52) + 4(54) + 2(56) + 4(57) + 2(57) + 4(55) + 56]$

$\qquad = \tfrac{1}{180}(1544) = 8.5\overline{7}$ mi

63. (a) $f(x) = \sin(\sin x)$. A CAS gives

$\qquad f^{(4)}(x) = \sin(\sin x)[\cos^4 x + 7\cos^2 x - 3]$

$\qquad\qquad + \cos(\sin x)[6\cos^2 x \sin x + \sin x]$

From the graph, we see that $\left|f^{(4)}(x)\right| < 3.8$ for $x \in [0, \pi]$.

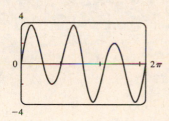

(b) We use Simpson's Rule with $f(x) = \sin(\sin x)$ and $\Delta x = \tfrac{\pi}{10}$:

$\qquad \int_0^\pi f(x)\, dx \approx \tfrac{\pi}{10 \cdot 3}\left[f(0) + 4f\left(\tfrac{\pi}{10}\right) + 2f\left(\tfrac{2\pi}{10}\right) + \cdots + 4f\left(\tfrac{9\pi}{10}\right) + f(\pi)\right] \approx 1.786721$

From part (a), we know that $\left|f^{(4)}(x)\right| < 3.8$ on $[0, \pi]$, so we use Theorem 6.5.4 with $K = 3.8$, and estimate the error

as $|E_S| \le \dfrac{3.8(\pi - 0)^5}{180(10)^4} \approx 0.000646$.

(c) If we want the error to be less than 0.00001, we must have $|E_S| \le \dfrac{3.8\pi^5}{180n^4} \le 0.00001$,

so $n^4 \ge \dfrac{3.8\pi^5}{180(0.00001)} \approx 646{,}041.6$ $\Rightarrow$ $n \ge 28.35$. Since n must be even for Simpson's Rule, we must have $n \ge 30$

to ensure the desired accuracy.

65. By the Fundamental Theorem of Calculus,

$\int_0^\infty f'(x)\, dx = \lim\limits_{t \to \infty} \int_0^t f'(x)\, dx = \lim\limits_{t \to \infty} [f(t) - f(0)] = \lim\limits_{t \to \infty} f(t) - f(0) = 0 - f(0) = -f(0)$.

7 □ APPLICATIONS OF INTEGRATION

7.1 Areas Between Curves

1. $A = \int_{x=0}^{x=4} (y_T - y_B)\,dx = \int_0^4 \left[(5x - x^2) - x\right] dx = \int_0^4 (4x - x^2)\,dx = \left[2x^2 - \frac{1}{3}x^3\right]_0^4 = \left(32 - \frac{64}{3}\right) - (0) = \frac{32}{3}$

3. $A = \int_{y=-1}^{y=1} (x_R - x_L)\,dy = \int_{-1}^1 \left[e^y - (y^2 - 2)\right] dy = \int_{-1}^1 (e^y - y^2 + 2)\,dy$

$= \left[e^y - \frac{1}{3}y^3 + 2y\right]_{-1}^1 = (e^1 - \frac{1}{3} + 2) - (e^{-1} + \frac{1}{3} - 2) = e - \frac{1}{e} + \frac{10}{3}$

5. $A = \int_{-1}^1 \left[e^x - (x^2 - 1)\right] dx = \left[e^x - \frac{1}{3}x^3 + x\right]_{-1}^1$

$= (e - \frac{1}{3} + 1) - (e^{-1} + \frac{1}{3} - 1) = e - \frac{1}{e} + \frac{4}{3}$

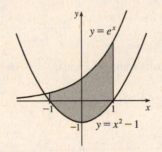
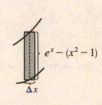

7. The curves intersect when $(x - 2)^2 = x \iff x^2 - 4x + 4 = x \iff x^2 - 5x + 4 = 0 \iff$

$(x - 1)(x - 4) = 0 \iff x = 1$ or 4.

$A = \int_1^4 \left[x - (x - 2)^2\right] dx = \int_1^4 (-x^2 + 5x - 4)\,dx$

$= \left[-\frac{1}{3}x^3 + \frac{5}{2}x^2 - 4x\right]_1^4$

$= \left(-\frac{64}{3} + 40 - 16\right) - \left(-\frac{1}{3} + \frac{5}{2} - 4\right)$

$= \frac{9}{2}$

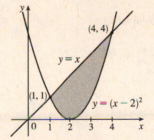
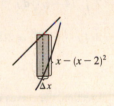

9. The curves intersect when $1 - y^2 = y^2 - 1 \iff 2 = 2y^2 \iff y^2 = 1 \iff y = \pm 1$.

$A = \int_{-1}^1 \left[(1 - y^2) - (y^2 - 1)\right] dy$

$= \int_{-1}^1 2(1 - y^2)\,dy$

$= 2 \cdot 2 \int_0^1 (1 - y^2)\,dy$

$= 4\left[y - \frac{1}{3}y^3\right]_0^1 = 4(1 - \frac{1}{3}) = \frac{8}{3}$

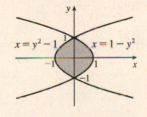
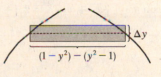

11. $12 - x^2 = x^2 - 6 \iff 2x^2 = 18 \iff$

$x^2 = 9 \iff x = \pm 3$, so

$$A = \int_{-3}^{3} \left[(12 - x^2) - (x^2 - 6)\right] dx$$

$$= 2 \int_{0}^{3} \left(18 - 2x^2\right) dx \qquad \text{[by symmetry]}$$

$$= 2\left[18x - \tfrac{2}{3}x^3\right]_0^3 = 2\left[(54 - 18) - 0\right]$$

$$= 2(36) = 72$$

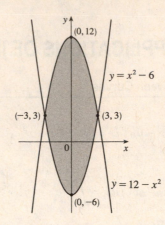

13. $e^x = xe^x \iff e^x - xe^x = 0 \iff e^x(1 - x) = 0 \iff x = 1$.

$$A = \int_{0}^{1} (e^x - xe^x) \, dx$$

$$= \left[e^x - (xe^x - e^x)\right]_0^1 \quad \text{[use parts with } u = x \text{ and } dv = e^x \, dx]$$

$$= \left[2e^x - xe^x\right]_0^1 = (2e - e) - (2 - 0) = e - 2$$

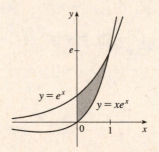

15. $2y^2 = 4 + y^2 \iff y^2 = 4 \iff y = \pm 2$, so

$$A = \int_{-2}^{2} \left[(4 + y^2) - 2y^2\right] dy$$

$$= 2 \int_{0}^{2} (4 - y^2) \, dy \qquad \text{[by symmetry]}$$

$$= 2\left[4y - \tfrac{1}{3}y^3\right]_0^2 = 2\left(8 - \tfrac{8}{3}\right) = \tfrac{32}{3}$$

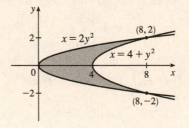

17. By inspection, the curves intersect at $x = \pm\tfrac{1}{2}$.

$$A = \int_{-1/2}^{1/2} \left[\cos \pi x - (4x^2 - 1)\right] dx$$

$$= 2 \int_{0}^{1/2} (\cos \pi x - 4x^2 + 1) \, dx \quad \text{[by symmetry]}$$

$$= 2\left[\tfrac{1}{\pi} \sin \pi x - \tfrac{4}{3}x^3 + x\right]_0^{1/2} = 2\left[\left(\tfrac{1}{\pi} - \tfrac{1}{6} + \tfrac{1}{2}\right) - 0\right]$$

$$= 2\left(\tfrac{1}{\pi} + \tfrac{1}{3}\right) = \tfrac{2}{\pi} + \tfrac{2}{3}$$

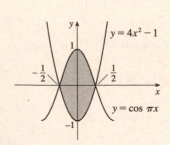

19. $1/x = x \iff 1 = x^2 \iff x = \pm 1$ and $1/x = \frac{1}{4}x \iff$

$4 = x^2 \iff x = \pm 2$, so for $x > 0$,

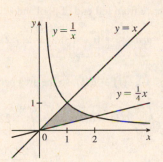

$$A = \int_0^1 \left(x - \frac{1}{4}x \right) dx + \int_1^2 \left(\frac{1}{x} - \frac{1}{4}x \right) dx$$

$$= \int_0^1 \left(\frac{3}{4}x \right) dx + \int_1^2 \left(\frac{1}{x} - \frac{1}{4}x \right) dx$$

$$= \left[\frac{3}{8}x^2 \right]_0^1 + \left[\ln|x| - \frac{1}{8}x^2 \right]_1^2$$

$$= \frac{3}{8} + \left(\ln 2 - \frac{1}{2} \right) - \left(0 - \frac{1}{8} \right) = \ln 2$$

21. $\cos x = \sin 2x = 2 \sin x \cos x \iff 2 \sin x \cos x - \cos x = 0 \iff \cos x (2 \sin x - 1) = 0 \iff$

$2 \sin x = 1$ or $\cos x = 0 \iff x = \frac{\pi}{6}$ or $\frac{\pi}{2}$.

$$A = \int_0^{\pi/6} (\cos x - \sin 2x)\, dx + \int_{\pi/6}^{\pi/2} (\sin 2x - \cos x)\, dx$$

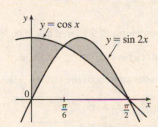

$$= \left[\sin x + \frac{1}{2} \cos 2x \right]_0^{\pi/6} + \left[-\frac{1}{2} \cos 2x - \sin x \right]_{\pi/6}^{\pi/2}$$

$$= \left(\frac{1}{2} + \frac{1}{2} \cdot \frac{1}{2} \right) - \left(0 + \frac{1}{2} \cdot 1 \right) + \left[-\frac{1}{2} \cdot (-1) - 1 \right] - \left(-\frac{1}{2} \cdot \frac{1}{2} - \frac{1}{2} \right)$$

$$= \frac{3}{4} - \frac{1}{2} - \frac{1}{2} + \frac{3}{4} = \frac{1}{2}$$

23.

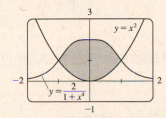

Graph `Y₁=2/(1+x^4)` and `Y₂=x^2`. We see that $Y_1 > Y_2$ on $(-1, 1)$, so the area is given by $\int_{-1}^1 \left(\dfrac{2}{1+x^4} - x^2 \right) dx$. Evaluate the integral with a command such as `fnInt(Y₁-Y₂,x,-1,1)` to get 2.80123 to five decimal places.

Another method: Graph $f(x) = $ `Y₁=2/(1+x^4)-x^2` and from the graph evaluate $\int f(x)\, dx$ from -1 to 1.

25.

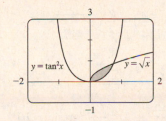

The curves intersect at $x = 0$ and $x = a \approx 0.749363$.

$$A = \int_0^a \left(\sqrt{x} - \tan^2 x \right) dx \approx 0.25142$$

27. As in Example 3, we approximate the distance between the two cars after ten seconds using Simpson's Rule

with $\Delta t = 1\text{ s} = \frac{1}{3600}$ h.

$$\text{distance}_{\text{Kelly}} - \text{distance}_{\text{Chris}} = \int_0^{10} v_K\, dt - \int_0^{10} v_C\, dt = \int_0^{10} (v_K - v_C)\, dt \approx S_{10}$$

$$= \frac{1}{3 \cdot 3600} [(0 - 0) + 4(22 - 20) + 2(37 - 32) + 4(52 - 46) + 2(61 - 54) + 4(71 - 62)$$

$$+ 2(80 - 69) + 4(86 - 75) + 2(93 - 81) + 4(98 - 86) + (102 - 90)]$$

$$= \frac{1}{10,800} (242) = \frac{121}{5400}\text{ mi}$$

So after 10 seconds, Kelly's car is about $\dfrac{121}{5400}\text{ mi} \left(5280\dfrac{\text{ft}}{\text{mi}} \right) \approx 118$ ft ahead of Chris's.

29. If x = distance from left end of pool and $w = w(x)$ = width at x, then Simpson's Rule with $n = 8$ and $\Delta x = 2$ gives

Area $= \int_0^{16} w \, dx \approx \frac{2}{3}[0 + 4(6.2) + 2(7.2) + 4(6.8) + 2(5.6) + 4(5.0) + 2(4.8) + 4(4.8) + 0] = \frac{2}{3}(126.4) \approx 84 \text{ m}^2$.

31. For $0 \le t \le 10$, $b(t) > d(t)$, so the area between the curves is given by

$$\int_0^{10} [b(t) - d(t)] \, dt = \int_0^{10} (2200e^{0.024t} - 1460e^{0.018t}) \, dt = \left[\frac{2200}{0.024} e^{0.024t} - \frac{1460}{0.018} e^{0.018t} \right]_0^{10}$$

$$= \left(\frac{275{,}000}{3} e^{0.24} - \frac{730{,}000}{9} e^{0.18} \right) - \left(\frac{275{,}000}{3} - \frac{730{,}000}{9} \right) \approx 8868 \text{ people}$$

This area A represents the increase in population over a 10-year period.

33. Let the equation of the large circle be $x^2 + y^2 = R^2$. Then the equation of

the small circle is $x^2 + (y - b)^2 = r^2$, where $b = \sqrt{R^2 - r^2}$ is the distance

between the centers of the circles. The desired area is

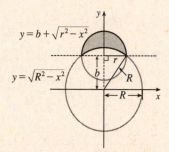

$A = \int_{-r}^{r} \left[(b + \sqrt{r^2 - x^2}) - \sqrt{R^2 - x^2} \right] dx$

$= 2 \int_0^r (b + \sqrt{r^2 - x^2} - \sqrt{R^2 - x^2}) \, dx$

$= 2 \int_0^r b \, dx + 2 \int_0^r \sqrt{r^2 - x^2} \, dx - 2 \int_0^r \sqrt{R^2 - x^2} \, dx$

The first integral is just $2br = 2r \sqrt{R^2 - r^2}$. The second integral represents the area of a quarter-circle of radius r, so its value

is $\frac{1}{4}\pi r^2$. To evaluate the other integral, note that

$$\int \sqrt{a^2 - x^2} \, dx = \int a^2 \cos^2 \theta \, d\theta \quad [x = a \sin \theta, \, dx = a \cos \theta \, d\theta] \quad = \left(\tfrac{1}{2} a^2 \right) \int (1 + \cos 2\theta) \, d\theta$$

$$= \tfrac{1}{2} a^2 \left(\theta + \tfrac{1}{2} \sin 2\theta \right) + C = \tfrac{1}{2} a^2 (\theta + \sin \theta \cos \theta) + C$$

$$= \frac{a^2}{2} \arcsin \left(\frac{x}{a} \right) + \frac{a^2}{2} \left(\frac{x}{a} \right) \frac{\sqrt{a^2 - x^2}}{a} + C = \frac{a^2}{2} \arcsin \left(\frac{x}{a} \right) + \frac{x}{2} \sqrt{a^2 - x^2} + C$$

Thus, the desired area is

$$A = 2r \sqrt{R^2 - r^2} + 2\left(\tfrac{1}{4}\pi r^2 \right) - \left[R^2 \arcsin(x/R) + x \sqrt{R^2 - x^2} \right]_0^r$$

$$= 2r \sqrt{R^2 - r^2} + \tfrac{1}{2}\pi r^2 - \left[R^2 \arcsin(r/R) + r \sqrt{R^2 - r^2} \right] = r \sqrt{R^2 - r^2} + \tfrac{\pi}{2} r^2 - R^2 \arcsin(r/R)$$

35. We first assume that $c > 0$, since c can be replaced by $-c$ in both equations without changing the graphs, and if $c = 0$ the

curves do not enclose a region. We see from the graph that the enclosed area A lies between $x = -c$ and $x = c$, and by

symmetry, it is equal to four times the area in the first quadrant. The enclosed area is

$A = 4 \int_0^c (c^2 - x^2) \, dx = 4 \left[c^2 x - \tfrac{1}{3} x^3 \right]_0^c = 4 \left(c^3 - \tfrac{1}{3} c^3 \right) = 4 \left(\tfrac{2}{3} c^3 \right) = \tfrac{8}{3} c^3$

So $A = 576 \iff \tfrac{8}{3} c^3 = 576 \iff c^3 = 216 \iff c = \sqrt[3]{216} = 6$.

Note that $c = -6$ is another solution, since the graphs are the same.

37.

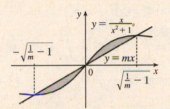

By the symmetry of the problem, we consider only the first quadrant, where

$y = x^2 \;\Rightarrow\; x = \sqrt{y}$. We are looking for a number b such that

$$\int_0^b \sqrt{y}\,dy = \int_b^4 \sqrt{y}\,dy \;\Rightarrow\; \tfrac{2}{3}\Big[y^{3/2}\Big]_0^b = \tfrac{2}{3}\Big[y^{3/2}\Big]_b^4 \;\Rightarrow$$

$$b^{3/2} = 4^{3/2} - b^{3/2} \;\Rightarrow\; 2b^{3/2} = 8 \;\Rightarrow\; b^{3/2} = 4 \;\Rightarrow\; b = 4^{2/3} \approx 2.52.$$

39. The area under the graph of f from 0 to t is equal to $\int_0^t f(x)\,dx$, so the requirement is that $\int_0^t f(x)\,dx = t^3$ for all t. We

differentiate both sides of this equation with respect to t (with the help of FTC1) to get $f(t) = 3t^2$. This function is positive

and continuous, as required.

41. The curve and the line will determine a region when they intersect at two or

more points. So we solve the equation $x/(x^2 + 1) = mx \;\Rightarrow$

$x = x(mx^2 + m) \;\Rightarrow\; x(mx^2 + m) - x = 0 \;\Rightarrow$

$x(mx^2 + m - 1) = 0 \;\Rightarrow\; x = 0$ or $mx^2 + m - 1 = 0 \;\Rightarrow$

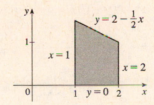

$x = 0$ or $x^2 = \dfrac{1-m}{m} \;\Rightarrow\; x = 0$ or $x = \pm\sqrt{\dfrac{1}{m} - 1}$. Note that if $m = 1$, this has only the solution $x = 0$, and no region

is determined. But if $1/m - 1 > 0 \;\Leftrightarrow\; 1/m > 1 \;\Leftrightarrow\; 0 < m < 1$, then there are two solutions. [Another way of seeing

this is to observe that the slope of the tangent to $y = x/(x^2 + 1)$ at the origin is $y'(0) = 1$ and therefore we must have

$0 < m < 1$.] Note that we cannot just integrate between the positive and negative roots, since the curve and the line cross at

the origin. Since mx and $x/(x^2 + 1)$ are both odd functions, the total area is twice the area between the curves on the interval

$\left[0, \sqrt{1/m - 1}\right]$. So the total area enclosed is

$$2\int_0^{\sqrt{1/m-1}} \left[\frac{x}{x^2+1} - mx\right] dx = 2\Big[\tfrac{1}{2}\ln(x^2+1) - \tfrac{1}{2}mx^2\Big]_0^{\sqrt{1/m-1}} = [\ln(1/m - 1 + 1) - m(1/m - 1)] - (\ln 1 - 0)$$

$$= \ln(1/m) - 1 + m = m - \ln m - 1$$

7.2 Volumes

1. A cross-section is a disk with radius $2 - \tfrac{1}{2}x$, so its area is $A(x) = \pi\left(2 - \tfrac{1}{2}x\right)^2$.

$$V = \int_1^2 A(x)\,dx = \int_1^2 \pi\left(2 - \tfrac{1}{2}x\right)^2 dx$$

$$= \pi\int_1^2 \left(4 - 2x + \tfrac{1}{4}x^2\right) dx$$

$$= \pi\Big[4x - x^2 + \tfrac{1}{12}x^3\Big]_1^2$$

$$= \pi\Big[\big(8 - 4 + \tfrac{8}{12}\big) - \big(4 - 1 + \tfrac{1}{12}\big)\Big]$$

$$= \pi\Big(1 + \tfrac{7}{12}\Big) = \tfrac{19}{12}\pi$$

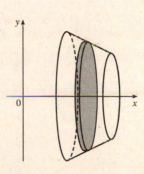

3. A cross-section is a disk with radius $2\sqrt{y}$, so its area is

$$A(y) = \pi\left(2\sqrt{y}\right)^2.$$

$$V = \int_0^9 A(y)\,dy = \int_0^9 \pi\left(2\sqrt{y}\right)^2 dy = 4\pi \int_0^9 y\,dy$$

$$= 4\pi\left[\tfrac{1}{2}y^2\right]_0^9 = 2\pi(81) = 162\pi$$

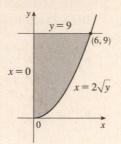 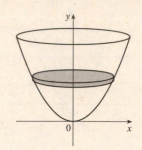

5. A cross-section is a washer (annulus) with inner radius x^3 and outer radius x, so its area is

$$A(x) = \pi(x)^2 - \pi(x^3)^2 = \pi(x^2 - x^6).$$

$$V = \int_0^1 A(x)\,dx = \int_0^1 \pi(x^2 - x^6)\,dx$$

$$= \pi\left[\tfrac{1}{3}x^3 - \tfrac{1}{7}x^7\right]_0^1 = \pi\left(\tfrac{1}{3} - \tfrac{1}{7}\right) = \tfrac{4}{21}\pi$$

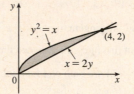 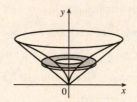

7. A cross-section is a washer with inner radius y^2 and outer radius $2y$, so its area is

$$A(y) = \pi(2y)^2 - \pi(y^2)^2 = \pi(4y^2 - y^4).$$

$$V = \int_0^2 A(y)\,dy = \pi \int_0^2 (4y^2 - y^4)\,dy$$

$$= \pi\left[\tfrac{4}{3}y^3 - \tfrac{1}{5}y^5\right]_0^2 = \pi\left(\tfrac{32}{3} - \tfrac{32}{5}\right) = \tfrac{64}{15}\pi$$

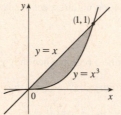 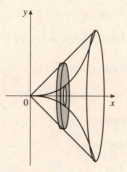

9. A cross-section is a washer with inner radius $1 - \sqrt{x}$ and outer radius $1 - x$, so its area is

$$A(x) = \pi(1 - x)^2 - \pi\left(1 - \sqrt{x}\right)^2$$

$$= \pi\left[(1 - 2x + x^2) - \left(1 - 2\sqrt{x} + x\right)\right]$$

$$= \pi\left(-3x + x^2 + 2\sqrt{x}\right).$$

$$V = \int_0^1 A(x)\,dx = \pi \int_0^1 \left(-3x + x^2 + 2\sqrt{x}\right)dx$$

$$= \pi\left[-\tfrac{3}{2}x^2 + \tfrac{1}{3}x^3 + \tfrac{4}{3}x^{3/2}\right]_0^1 = \pi\left(-\tfrac{3}{2} + \tfrac{5}{3}\right) = \tfrac{\pi}{6}$$

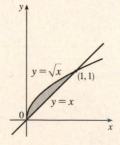 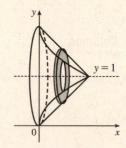

11. A cross-section is a washer with inner radius $(1 + \sec x) - 1 = \sec x$ and outer radius $3 - 1 = 2$, so its area is

$$A(x) = \pi\left[2^2 - (\sec x)^2\right] = \pi(4 - \sec^2 x).$$

$$V = \int_{-\pi/3}^{\pi/3} A(x)\,dx = \int_{-\pi/3}^{\pi/3} \pi(4 - \sec^2 x)\,dx$$

$$= 2\pi \int_{0}^{\pi/3} (4 - \sec^2 x)\,dx \qquad \text{[by symmetry]}$$

$$= 2\pi\Big[4x - \tan x\Big]_{0}^{\pi/3} = 2\pi\left[\left(\tfrac{4\pi}{3} - \sqrt{3}\right) - 0\right]$$

$$= 2\pi\left(\tfrac{4\pi}{3} - \sqrt{3}\right)$$

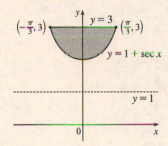
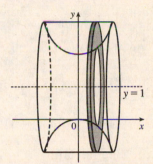

13. A cross-section is a disk with radius $1/x$, so its area is $A(x) = \pi(1/x)^2$.

$$V = \int_{1}^{2} A(x)\,dx = \int_{1}^{2} \pi\left(\frac{1}{x}\right)^2 dx$$

$$= \pi \int_{1}^{2} \frac{1}{x^2}\,dx = \pi\left[-\frac{1}{x}\right]_{1}^{2}$$

$$= \pi\left[-\tfrac{1}{2} - (-1)\right] = \tfrac{\pi}{2}$$

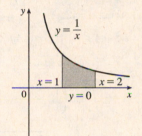

15. The curves $x - y = 1$ and $y = x^2 - 4x + 3$ intersect when

$$x - 1 = x^2 - 4x + 3 \quad \Leftrightarrow \quad 0 = x^2 - 5x + 4 \quad \Leftrightarrow$$

$$0 = (x - 1)(x - 4) \quad \Leftrightarrow \quad x = 1 \text{ or } 4.$$ A cross-section is a washer with

inner radius $3 - (x - 1)$ and outer radius $3 - (x^2 - 4x + 3)$, so its area is

$$A(x) = \pi[3 - (x^2 - 4x + 3)]^2 - \pi[3 - (x - 1)]^2.$$

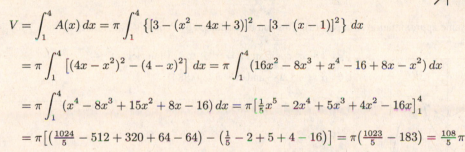

$$V = \int_{1}^{4} A(x)\,dx = \pi \int_{1}^{4} \left\{[3 - (x^2 - 4x + 3)]^2 - [3 - (x - 1)]^2\right\} dx$$

$$= \pi \int_{1}^{4} \left[(4x - x^2)^2 - (4 - x)^2\right] dx = \pi \int_{1}^{4} (16x^2 - 8x^3 + x^4 - 16 + 8x - x^2)\,dx$$

$$= \pi \int_{1}^{4} (x^4 - 8x^3 + 15x^2 + 8x - 16)\,dx = \pi\left[\tfrac{1}{5}x^5 - 2x^4 + 5x^3 + 4x^2 - 16x\right]_{1}^{4}$$

$$= \pi\left[\left(\tfrac{1024}{5} - 512 + 320 + 64 - 64\right) - \left(\tfrac{1}{5} - 2 + 5 + 4 - 16\right)\right] = \pi\left(\tfrac{1023}{5} - 183\right) = \tfrac{108}{5}\pi$$

17. $y = \sqrt{x} \;\Rightarrow\; x = y^2$ and $y = x^3 \;\Rightarrow\; x = \sqrt[3]{y}$. A cross-section is a

washer with inner radius $1 - \sqrt[3]{y}$ and outer radius $1 - y^2$, so its area is

$$A(y) = \pi(1 - y^2)^2 - \pi\left(1 - \sqrt[3]{y}\right)^2.$$

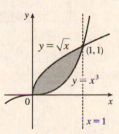

$$V = \int_0^1 A(y)\,dy = \int_0^1 \left[\pi(1 - y^2)^2 - \pi\left(1 - \sqrt[3]{y}\right)^2\right] dy$$

$$= \pi \int_0^1 \left[(1 - 2y^2 + y^4) - (1 - 2y^{1/3} + y^{2/3})\right] dy$$

$$= \pi \int_0^1 (-2y^2 + y^4 + 2y^{1/3} - y^{2/3})\,dy = \pi\left[-\tfrac{2}{3}y^3 + \tfrac{1}{5}y^5 + \tfrac{3}{2}y^{4/3} - \tfrac{3}{5}y^{5/3}\right]_0^1 = \pi\left(-\tfrac{2}{3} + \tfrac{1}{5} + \tfrac{3}{2} - \tfrac{3}{5}\right) = \tfrac{13\pi}{30}$$

19. (a) About the x-axis:

$$V = \int_{-1}^{1} \pi(e^{-x^2})^2 \, dx = 2\pi \int_{0}^{1} e^{-2x^2} \, dx \quad \text{[by symmetry]}$$

$$\approx 3.75825$$

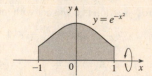

(b) About $y = -1$:

$$V = \int_{-1}^{1} \pi \left\{ [e^{-x^2} - (-1)]^2 - [0 - (-1)]^2 \right\} dx$$

$$= 2\pi \int_{0}^{1} [(e^{-x^2} + 1)^2 - 1] \, dx = 2\pi \int_{0}^{1} (e^{-2x^2} + 2e^{-x^2}) \, dx$$

$$\approx 13.14312$$

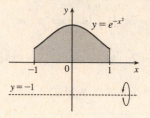

21. (a) About $y = 2$:

$$x^2 + 4y^2 = 4 \quad \Rightarrow \quad 4y^2 = 4 - x^2 \quad \Rightarrow \quad y^2 = 1 - x^2/4 \quad \Rightarrow$$

$$y = \pm\sqrt{1 - x^2/4}$$

$$V = \int_{-2}^{2} \pi \left\{ \left[2 - \left(-\sqrt{1 - x^2/4} \right) \right]^2 - \left(2 - \sqrt{1 - x^2/4} \right)^2 \right\} dx$$

$$= 2\pi \int_{0}^{2} 8\sqrt{1 - x^2/4} \, dx \approx 78.95684$$

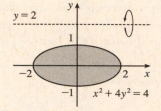

(b) About $x = 2$:

$$x^2 + 4y^2 = 4 \quad \Rightarrow \quad x^2 = 4 - 4y^2 \quad \Rightarrow \quad x = \pm\sqrt{4 - 4y^2}$$

$$V = \int_{-1}^{1} \pi \left\{ \left[2 - \left(-\sqrt{4 - 4y^2} \right) \right]^2 - \left(2 - \sqrt{4 - 4y^2} \right)^2 \right\} dy$$

$$= 2\pi \int_{0}^{1} 8\sqrt{4 - 4y^2} \, dy \approx 78.95684$$

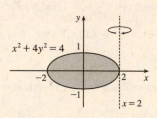

[Notice that this is the same approximation as in part (a). This can be explained by Pappus's Theorem in Section 7.6.]

23.

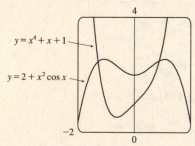

$y = 2 + x^2 \cos x$ and $y = x^4 + x + 1$ intersect at

$x = a \approx -1.288$ and $x = b \approx 0.884$.

$$V = \pi \int_{a}^{b} [(2 + x^2 \cos x)^2 - (x^4 + x + 1)^2] \, dx \approx 23.780$$

25. $V = \pi \displaystyle\int_{0}^{\pi} \left\{ [\sin^2 x - (-1)]^2 - [0 - (-1)]^2 \right\} dx$

$\overset{\text{CAS}}{=} \frac{11}{8}\pi^2$

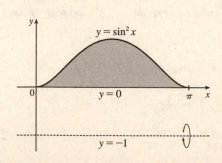

27. (a) $\pi \int_0^{\pi/2} \cos^2 x \, dx$ describes the volume of the solid obtained by rotating the region

$\mathcal{R} = \{(x, y) \mid 0 \le x \le \frac{\pi}{2}, 0 \le y \le \cos x\}$ of the xy-plane about the x-axis.

(b) $\pi \int_0^1 (y^4 - y^8) \, dy = \pi \int_0^1 \left[(y^2)^2 - (y^4)^2\right] dy$ describes the volume of the solid obtained by rotating the region

$\mathcal{R} = \{(x, y) \mid 0 \le y \le 1, y^4 \le x \le y^2\}$ of the xy-plane about the y-axis.

29. There are 10 subintervals over the 15-cm length, so we'll use $n = 10/2 = 5$ for the Midpoint Rule.

$$V = \int_0^{15} A(x) \, dx \approx M_5 = \frac{15-0}{5}[A(1.5) + A(4.5) + A(7.5) + A(10.5) + A(13.5)]$$

$$= 3(18 + 79 + 106 + 128 + 39) = 3 \cdot 370 = 1110 \text{ cm}^3$$

31. We'll form a right circular cone with height h and base radius r by

revolving the line $y = \frac{r}{h}x$ about the x-axis.

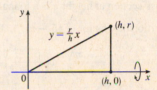

$$V = \pi \int_0^h \left(\frac{r}{h}x\right)^2 dx = \pi \int_0^h \frac{r^2}{h^2} x^2 dx = \pi \frac{r^2}{h^2}\left[\frac{1}{3}x^3\right]_0^h$$

$$= \pi \frac{r^2}{h^2}\left(\frac{1}{3}h^3\right) = \frac{1}{3}\pi r^2 h$$

Another solution: Revolve $x = -\frac{r}{h}y + r$ about the y-axis.

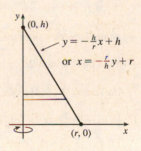

$$V = \pi \int_0^h \left(-\frac{r}{h}y + r\right)^2 dy \overset{*}{=} \pi \int_0^h \left[\frac{r^2}{h^2}y^2 - \frac{2r^2}{h}y + r^2\right] dy$$

$$= \pi \left[\frac{r^2}{3h^2}y^3 - \frac{r^2}{h}y^2 + r^2 y\right]_0^h = \pi\left(\frac{1}{3}r^2 h - r^2 h + r^2 h\right) = \frac{1}{3}\pi r^2 h$$

$*$ Or use substitution with $u = r - \frac{r}{h}y$ and $du = -\frac{r}{h}dy$ to get

$$\pi \int_r^0 u^2 \left(-\frac{h}{r} du\right) = -\pi \frac{h}{r}\left[\frac{1}{3}u^3\right]_r^0 = -\pi \frac{h}{r}\left(-\frac{1}{3}r^3\right) = \frac{1}{3}\pi r^2 h.$$

33. $x^2 + y^2 = r^2 \iff x^2 = r^2 - y^2$

$$V = \pi \int_{r-h}^r (r^2 - y^2) \, dy = \pi\left[r^2 y - \frac{y^3}{3}\right]_{r-h}^r = \pi\left\{\left[r^3 - \frac{r^3}{3}\right] - \left[r^2(r-h) - \frac{(r-h)^3}{3}\right]\right\}$$

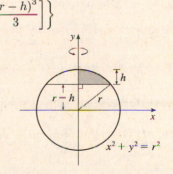

$$= \pi\left\{\tfrac{2}{3}r^3 - \tfrac{1}{3}(r-h)\left[3r^2 - (r-h)^2\right]\right\}$$

$$= \tfrac{1}{3}\pi\left\{2r^3 - (r-h)\left[3r^2 - (r^2 - 2rh + h^2)\right]\right\}$$

$$= \tfrac{1}{3}\pi\left\{2r^3 - (r-h)\left[2r^2 + 2rh - h^2\right]\right\}$$

$$= \tfrac{1}{3}\pi\left(2r^3 - 2r^3 - 2r^2 h + rh^2 + 2r^2 h + 2rh^2 - h^3\right)$$

$$= \tfrac{1}{3}\pi\left(3rh^2 - h^3\right) = \tfrac{1}{3}\pi h^2(3r - h), \text{ or, equivalently, } \pi h^2\left(r - \frac{h}{3}\right)$$

35. For a cross-section at height y, we see from similar triangles that $\dfrac{\alpha/2}{b/2} = \dfrac{h-y}{h}$, so $\alpha = b\left(1 - \dfrac{y}{h}\right)$.

Similarly, for cross-sections having $2b$ as their base and β replacing α, $\beta = 2b\left(1 - \dfrac{y}{h}\right)$. So

$$V = \int_0^h A(y)\,dy = \int_0^h \left[b\left(1 - \frac{y}{h}\right)\right]\left[2b\left(1 - \frac{y}{h}\right)\right] dy$$

$$= \int_0^h 2b^2\left(1 - \frac{y}{h}\right)^2 dy = 2b^2 \int_0^h \left(1 - \frac{2y}{h} + \frac{y^2}{h^2}\right) dy$$

$$= 2b^2\left[y - \frac{y^2}{h} + \frac{y^3}{3h^2}\right]_0^h = 2b^2\left[h - h + \tfrac{1}{3}h\right]$$

$$= \tfrac{2}{3}b^2 h \quad \left[\,= \tfrac{1}{3}Bh \text{ where } B \text{ is the area of the base, as with any pyramid.}\right]$$

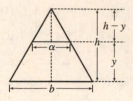

37. A cross-section at height z is a triangle similar to the base, so we'll multiply the legs of the base triangle, 3 and 4, by a proportionality factor of $(5-z)/5$. Thus, the triangle at height z has area

$$A(z) = \frac{1}{2} \cdot 3\left(\frac{5-z}{5}\right) \cdot 4\left(\frac{5-z}{5}\right) = 6\left(1 - \frac{z}{5}\right)^2, \text{ so}$$

$$V = \int_0^5 A(z)\,dz = 6\int_0^5 \left(1 - \frac{z}{5}\right)^2 dz = 6\int_1^0 u^2(-5\,du) \qquad \begin{bmatrix} u = 1 - z/5, \\ du = -\tfrac{1}{5}\,dz \end{bmatrix}$$

$$= -30\left[\tfrac{1}{3}u^3\right]_1^0 = -30\left(-\tfrac{1}{3}\right) = 10 \text{ cm}^3$$

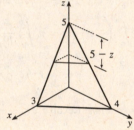

39. If l is a leg of the isosceles right triangle and $2y$ is the hypotenuse,

then $l^2 + l^2 = (2y)^2 \;\Rightarrow\; 2l^2 = 4y^2 \;\Rightarrow\; l^2 = 2y^2.$

$$V = \int_{-2}^2 A(x)\,dx = 2\int_0^2 A(x)\,dx = 2\int_0^2 \tfrac{1}{2}(l)(l)\,dx = 2\int_0^2 y^2\,dx$$

$$= 2\int_0^2 \tfrac{1}{4}(36 - 9x^2)\,dx = \tfrac{9}{2}\int_0^2 (4 - x^2)\,dx$$

$$= \tfrac{9}{2}\left[4x - \tfrac{1}{3}x^3\right]_0^2 = \tfrac{9}{2}\left(8 - \tfrac{8}{3}\right) = 24$$

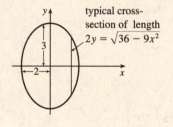

typical cross-section of length $2y = \sqrt{36 - 9x^2}$

41. The cross-section of the base corresponding to the coordinate x has length

$y = 1 - x$. The corresponding square with side s has area

$A(x) = s^2 = (1-x)^2 = 1 - 2x + x^2$. Therefore,

$$V = \int_0^1 A(x)\,dx = \int_0^1 (1 - 2x + x^2)\,dx$$

$$= \left[x - x^2 + \tfrac{1}{3}x^3\right]_0^1 = \left(1 - 1 + \tfrac{1}{3}\right) - 0 = \tfrac{1}{3}$$

Or: $\displaystyle\int_0^1 (1-x)^2\,dx = \int_1^0 u^2(-du) \quad [u = 1 - x] = \left[\tfrac{1}{3}u^3\right]_0^1 = \tfrac{1}{3}$

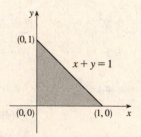

43. The cross-section of the base b corresponding to the coordinate x has length $1 - x^2$. The height h also has length $1 - x^2$, so the corresponding isosceles triangle has area $A(x) = \frac{1}{2}bh = \frac{1}{2}(1 - x^2)^2$. Therefore,

$$V = \int_{-1}^{1} \frac{1}{2}(1 - x^2)^2 \, dx$$

$$= 2 \cdot \frac{1}{2} \int_{0}^{1} (1 - 2x^2 + x^4) \, dx \quad \text{[by symmetry]}$$

$$= \left[x - \frac{2}{3}x^3 + \frac{1}{5}x^5 \right]_0^1 = \left(1 - \frac{2}{3} + \frac{1}{5}\right) - 0 = \frac{8}{15}$$

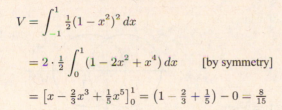

45. (a) The radius of the barrel is the same at each end by symmetry, since the function $y = R - cx^2$ is even. Since the barrel is obtained by rotating the graph of the function y about the x-axis, this radius is equal to the value of y at $x = \frac{1}{2}h$, which is $R - c\left(\frac{1}{2}h\right)^2 = R - d = r$.

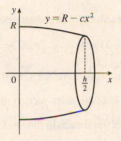

(b) The barrel is symmetric about the y-axis, so its volume is twice the volume of that part of the barrel for $x > 0$. Also, the barrel is a volume of rotation, so

$$V = 2 \int_0^{h/2} \pi y^2 \, dx = 2\pi \int_0^{h/2} \left(R - cx^2\right)^2 dx = 2\pi \left[R^2 x - \frac{2}{3}Rcx^3 + \frac{1}{5}c^2 x^5 \right]_0^{h/2}$$

$$= 2\pi \left(\frac{1}{2}R^2 h - \frac{1}{12}Rch^3 + \frac{1}{160}c^2 h^5 \right)$$

Trying to make this look more like the expression we want, we rewrite it as $V = \frac{1}{3}\pi h \left[2R^2 + \left(R^2 - \frac{1}{2}Rch^2 + \frac{3}{80}c^2 h^4 \right) \right]$.

But $R^2 - \frac{1}{2}Rch^2 + \frac{3}{80}c^2 h^4 = \left(R - \frac{1}{4}ch^2 \right)^2 - \frac{1}{40}c^2 h^4 = (R - d)^2 - \frac{2}{5}\left(\frac{1}{4}ch^2\right)^2 = r^2 - \frac{2}{5}d^2$.

Substituting this back into V, we see that $V = \frac{1}{3}\pi h\left(2R^2 + r^2 - \frac{2}{5}d^2\right)$, as required.

47. (a) The torus is obtained by rotating the circle $(x - R)^2 + y^2 = r^2$ about the y-axis. Solving for x, we see that the right half of the circle is given by $x = R + \sqrt{r^2 - y^2} = f(y)$ and the left half by $x = R - \sqrt{r^2 - y^2} = g(y)$.

So

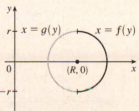

$$V = \pi \int_{-r}^{r} \left\{ [f(y)]^2 - [g(y)]^2 \right\} dy$$

$$= 2\pi \int_0^r \left[\left(R^2 + 2R\sqrt{r^2 - y^2} + r^2 - y^2 \right) - \left(R^2 - 2R\sqrt{r^2 - y^2} + r^2 - y^2 \right) \right] dy$$

$$= 2\pi \int_0^r 4R\sqrt{r^2 - y^2} \, dy = 8\pi R \int_0^r \sqrt{r^2 - y^2} \, dy$$

(b) Observe that the integral represents a quarter of the area of a circle with radius r, so

$$8\pi R \int_0^r \sqrt{r^2 - y^2} \, dy = 8\pi R \cdot \frac{1}{4}\pi r^2 = 2\pi^2 r^2 R.$$

49. (a) Volume$(S_1) = \int_0^h A(z) \, dz =$ Volume(S_2) since the cross-sectional area $A(z)$ at height z is the same for both solids.

(b) By Cavalieri's Principle, the volume of the cylinder in the figure is the same as that of a right circular cylinder with radius r and height h, that is, $\pi r^2 h$.

51. The volume is obtained by rotating the area common to two circles of radius r, as shown. The volume of the right half is

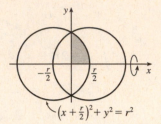

$$V_{\text{right}} = \pi \int_0^{r/2} y^2\, dx = \pi \int_0^{r/2} \left[r^2 - \left(\tfrac{1}{2}r + x \right)^2 \right] dx$$

$$= \pi \left[r^2 x - \tfrac{1}{3}\left(\tfrac{1}{2}r + x \right)^3 \right]_0^{r/2} = \pi \left[\left(\tfrac{1}{2}r^3 - \tfrac{1}{3}r^3 \right) - \left(0 - \tfrac{1}{24}r^3 \right) \right] = \tfrac{5}{24}\pi r^3$$

So by symmetry, the total volume is twice this, or $\frac{5}{12}\pi r^3$.

Another solution: We observe that the volume is the twice the volume of a cap of a sphere, so we can use the formula from Exercise 33 with $h = \tfrac{1}{2}r$: $V = 2 \cdot \tfrac{1}{3}\pi h^2(3r - h) = \tfrac{2}{3}\pi\left(\tfrac{1}{2}r \right)^2\left(3r - \tfrac{1}{2}r \right) = \tfrac{5}{12}\pi r^3$.

53. Take the x-axis to be the axis of the cylindrical hole of radius r. A quarter of the cross-section through y, perpendicular to the y-axis, is the rectangle shown. Using the Pythagorean Theorem twice, we see that the dimensions of this rectangle are $x = \sqrt{R^2 - y^2}$ and $z = \sqrt{r^2 - y^2}$, so

$$\tfrac{1}{4}A(y) = xz = \sqrt{r^2 - y^2}\,\sqrt{R^2 - y^2}, \text{ and}$$

$$V = \int_{-r}^{r} A(y)\, dy = \int_{-r}^{r} 4\sqrt{r^2 - y^2}\,\sqrt{R^2 - y^2}\, dy = 8\int_0^r \sqrt{r^2 - y^2}\,\sqrt{R^2 - y^2}\, dy$$

7.3 Volumes by Cylindrical Shells

1.

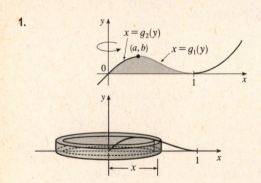

If we were to use the "washer" method, we would first have to locate the local maximum point (a, b) of $y = x(x - 1)^2$ using the methods of Chapter 3. Then we would have to solve the equation $y = x(x - 1)^2$ for x in terms of y to obtain the functions $x = g_1(y)$ and $x = g_2(y)$ shown in the first figure. This step would be difficult because it involves the cubic formula. Finally we would find the volume using

$$V = \pi \int_0^b \left\{ [g_1(y)]^2 - [g_2(y)]^2 \right\} dy.$$

Using shells, we find that a typical approximating shell has radius x, so its circumference is $2\pi x$. Its height is y, that is, $x(x - 1)^2$. So the total volume is

$$V = \int_0^1 2\pi x \left[x(x - 1)^2 \right] dx = 2\pi \int_0^1 \left(x^4 - 2x^3 + x^2 \right) dx = 2\pi \left[\frac{x^5}{5} - 2\frac{x^4}{4} + \frac{x^3}{3} \right]_0^1 = \frac{\pi}{15}$$

3. $V = \int_0^1 2\pi x \sqrt[3]{x}\, dx = 2\pi \int_0^1 x^{4/3}\, dx$

$= 2\pi \left[\frac{3}{7} x^{7/3}\right]_0^1 = 2\pi \left(\frac{3}{7}\right) = \frac{6}{7}\pi$

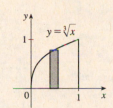

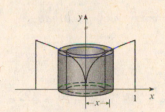

5. $V = \int_0^1 2\pi x e^{-x^2}\, dx$. Let $u = x^2$.

Thus, $du = 2x\, dx$, so

$V = \pi \int_0^1 e^{-u}\, du = \pi \left[-e^{-u}\right]_0^1 = \pi(1 - 1/e)$.

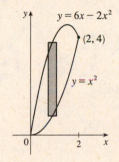

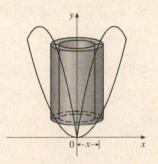

7. $x^2 = 6x - 2x^2 \quad\Leftrightarrow\quad 3x^2 - 6x = 0 \quad\Leftrightarrow\quad 3x(x - 2) = 0 \quad\Leftrightarrow\quad x = 0$ or 2.

$V = \int_0^2 2\pi x[(6x - 2x^2) - x^2]\, dx$

$= 2\pi \int_0^2 (-3x^3 + 6x^2)\, dx$

$= 2\pi \left[-\frac{3}{4} x^4 + 2x^3\right]_0^2$

$= 2\pi(-12 + 16) = 8\pi$

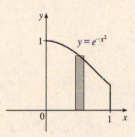

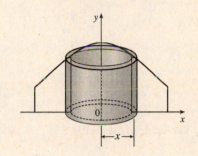

9. $xy = 1 \quad\Rightarrow\quad x = \frac{1}{y}$, so

$V = 2\pi \int_1^3 y\left(\frac{1}{y}\right) dy$

$= 2\pi \int_1^3 dy = 2\pi \left[y\right]_1^3$

$= 2\pi(3 - 1) = 4\pi$

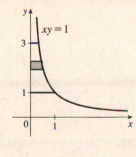

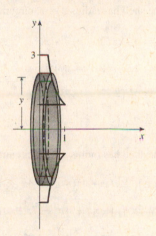

11. $V = 2\pi \int_0^8 \left[y(\sqrt[3]{y} - 0) \right] dy$

$= 2\pi \int_0^8 y^{4/3} \, dy = 2\pi \left[\frac{3}{7} y^{7/3} \right]_0^8$

$= \frac{6\pi}{7}(8^{7/3}) = \frac{6\pi}{7}(2^7) = \frac{768}{7}\pi$

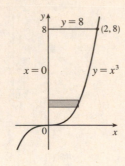
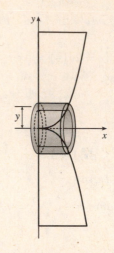

13. The height of the shell is $2 - \left[1 + (y - 2)^2 \right] = 1 - (y - 2)^2 = 1 - \left(y^2 - 4y + 4 \right) = -y^2 + 4y - 3.$

$V = 2\pi \int_1^3 y(-y^2 + 4y - 3) \, dy$

$= 2\pi \int_1^3 (-y^3 + 4y^2 - 3y) \, dy$

$= 2\pi \left[-\frac{1}{4} y^4 + \frac{4}{3} y^3 - \frac{3}{2} y^2 \right]_1^3$

$= 2\pi \left[\left(-\frac{81}{4} + 36 - \frac{27}{2} \right) - \left(-\frac{1}{4} + \frac{4}{3} - \frac{3}{2} \right) \right]$

$= 2\pi \left(\frac{8}{3} \right) = \frac{16}{3}\pi$

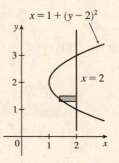
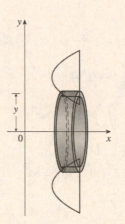

15. The shell has radius $2 - x$, circumference $2\pi(2 - x)$, and height x^4.

$V = \int_0^1 2\pi(2 - x)x^4 \, dx$

$= 2\pi \int_0^1 (2x^4 - x^5) \, dx$

$= 2\pi \left[\frac{2}{5} x^5 - \frac{1}{6} x^6 \right]_0^1$

$= 2\pi \left[\left(\frac{2}{5} - \frac{1}{6} \right) - 0 \right] = 2\pi \left(\frac{7}{30} \right) = \frac{7}{15}\pi$

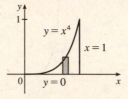
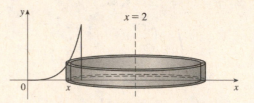

17. The shell has radius $x - 1$, circumference $2\pi(x - 1)$, and height $(4x - x^2) - 3 = -x^2 + 4x - 3$.

$V = \int_1^3 2\pi(x - 1)(-x^2 + 4x - 3) \, dx$

$= 2\pi \int_1^3 (-x^3 + 5x^2 - 7x + 3) \, dx$

$= 2\pi \left[-\frac{1}{4} x^4 + \frac{5}{3} x^3 - \frac{7}{2} x^2 + 3x \right]_1^3$

$= 2\pi \left[\left(-\frac{81}{4} + 45 - \frac{63}{2} + 9 \right) - \left(-\frac{1}{4} + \frac{5}{3} - \frac{7}{2} + 3 \right) \right]$

$= 2\pi \left(\frac{4}{3} \right) = \frac{8}{3}\pi$

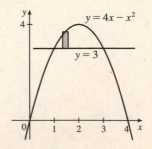
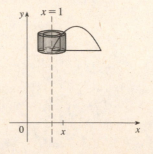

19. The shell has radius $1 - y$, circumference $2\pi(1 - y)$, and height $1 - \sqrt[3]{y}$ $\left[y = x^3 \iff x = \sqrt[3]{y} \right]$.

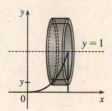
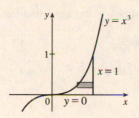

$$V = \int_0^1 2\pi(1 - y)(1 - y^{1/3}) \, dy$$
$$= 2\pi \int_0^1 (1 - y - y^{1/3} + y^{4/3}) \, dy$$
$$= 2\pi \left[y - \tfrac{1}{2}y^2 - \tfrac{3}{4}y^{4/3} + \tfrac{3}{7}y^{7/3} \right]_0^1$$
$$= 2\pi \left[\left(1 - \tfrac{1}{2} - \tfrac{3}{4} + \tfrac{3}{7} \right) - 0 \right]$$
$$= 2\pi \left(\tfrac{5}{28} \right) = \tfrac{5}{14}\pi$$

21. (a) $V = 2\pi \displaystyle\int_0^2 x(xe^{-x}) \, dx = 2\pi \int_0^2 x^2 e^{-x} \, dx$

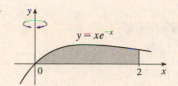

(b) $V \approx 4.06300$

23. (a) $V = 2\pi \displaystyle\int_{-\pi/2}^{\pi/2} (\pi - x)[\cos^4 x - (-\cos^4 x)] \, dx$

$$= 4\pi \int_{-\pi/2}^{\pi/2} (\pi - x) \cos^4 x \, dx$$

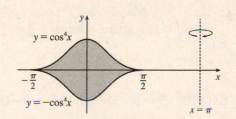

(b) $V \approx 46.50942$

25. (a) $V = \int_0^\pi 2\pi(4 - y) \sqrt{\sin y} \, dy$ **(b)** $V \approx 36.57476$

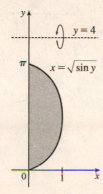

27. $\Delta x = \dfrac{\pi/4 - 0}{4} = \dfrac{\pi}{16}$.

$$V = \int_0^{\pi/4} 2\pi x \tan x \, dx \approx 2\pi \cdot \tfrac{\pi}{16} \left(\tfrac{\pi}{32} \tan \tfrac{\pi}{32} + \tfrac{3\pi}{32} \tan \tfrac{3\pi}{32} + \tfrac{5\pi}{32} \tan \tfrac{5\pi}{32} + \tfrac{7\pi}{32} \tan \tfrac{7\pi}{32} \right) \approx 1.142$$

29. $\int_0^3 2\pi x^5 \, dx = 2\pi \int_0^3 x(x^4) \, dx$. The solid is obtained by rotating the region $0 \le y \le x^4$, $0 \le x \le 3$ about the y-axis using cylindrical shells.

31. $\int_0^1 2\pi(3-y)(1-y^2)\,dy$. The solid is obtained by rotating the region bounded by (i) $x = 1 - y^2$, $x = 0$, and $y = 0$ or

(ii) $x = y^2$, $x = 1$, and $y = 0$ about the line $y = 3$ using cylindrical shells.

33. Use shells:

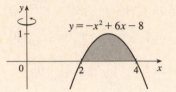

$$V = \int_2^4 2\pi x(-x^2 + 6x - 8)\,dx = 2\pi \int_2^4 (-x^3 + 6x^2 - 8x)\,dx$$

$$= 2\pi\left[-\tfrac{1}{4}x^4 + 2x^3 - 4x^2\right]_2^4$$

$$= 2\pi[(-64 + 128 - 64) - (-4 + 16 - 16)]$$

$$= 2\pi(4) = 8\pi$$

35. Use washers: $y^2 - x^2 = 1 \;\Rightarrow\; y = \pm\sqrt{x^2 \pm 1}$

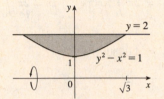

$$V = \int_{-\sqrt{3}}^{\sqrt{3}} \pi\left[(2-0)^2 - \left(\sqrt{x^2+1} - 0\right)^2\right]dx$$

$$= 2\pi \int_0^{\sqrt{3}} [4 - (x^2 + 1)]\,dx \qquad \text{[by symmetry]}$$

$$= 2\pi \int_0^{\sqrt{3}} (3 - x^2)\,dx = 2\pi\left[3x - \tfrac{1}{3}x^3\right]_0^{\sqrt{3}}$$

$$= 2\pi(3\sqrt{3} - \sqrt{3}) = 4\sqrt{3}\,\pi$$

37. Use disks: $x^2 + (y-1)^2 = 1 \;\Leftrightarrow\; x = \pm\sqrt{1 - (y-1)^2}$

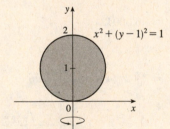

$$V = \pi \int_0^2 \left[\sqrt{1-(y-1)^2}\right]^2 dy = \pi \int_0^2 (2y - y^2)\,dy$$

$$= \pi\left[y^2 - \tfrac{1}{3}y^3\right]_0^2 = \pi\left(4 - \tfrac{8}{3}\right) = \tfrac{4}{3}\pi$$

39. Use shells:

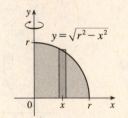

$$V = 2\int_0^r 2\pi x \sqrt{r^2 - x^2}\,dx = -2\pi \int_0^r (r^2 - x^2)^{1/2}(-2x)\,dx$$

$$= \left[-2\pi \cdot \tfrac{2}{3}(r^2 - x^2)^{3/2}\right]_0^r = -\tfrac{4}{3}\pi(0 - r^3) = \tfrac{4}{3}\pi r^3$$

41. $V = 2\pi \int_0^r x\left(-\dfrac{h}{r}x + h\right)dx = 2\pi h \int_0^r \left(-\dfrac{x^2}{r} + x\right)dx$

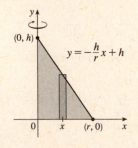

$$= 2\pi h\left[-\dfrac{x^3}{3r} + \dfrac{x^2}{2}\right]_0^r = 2\pi h\,\dfrac{r^2}{6} = \dfrac{\pi r^2 h}{3}$$

43. Using the formula for volumes of rotation and the figure, we see that

Volume $= \int_0^d \pi b^2 \, dy - \int_0^c \pi a^2 \, dy - \int_c^d \pi \left[f^{-1}(y) \right]^2 \, dy = \pi b^2 d - \pi a^2 c - \int_c^d \pi \left[f^{-1}(y) \right]^2 \, dy.$ Let $y = f(x)$,

which gives $dy = f'(x) \, dx$ and $f^{-1}(y) = x$, so that $V = \pi b^2 d - \pi a^2 c - \pi \int_a^b x^2 f'(x) \, dx.$

Now integrate by parts with $u = x^2$, and $dv = f'(x) \, dx$ $\Rightarrow$ $du = 2x \, dx$, $v = f(x)$, and

$\int_a^b x^2 f'(x) \, dx = \left[x^2 f(x) \right]_a^b - \int_a^b 2x f(x) \, dx = b^2 f(b) - a^2 f(a) - \int_a^b 2x f(x) \, dx$, but $f(a) = c$ and $f(b) = d$ $\Rightarrow$

$V = \pi b^2 d - \pi a^2 c - \pi \left[b^2 d - a^2 c - \int_a^b 2x f(x) \, dx \right] = \int_a^b 2\pi x f(x) \, dx.$

7.4 Arc Length

1. $y = 2x - 5$ $\Rightarrow$ $L = \int_{-1}^3 \sqrt{1 + (dy/dx)^2} \, dx = \int_{-1}^3 \sqrt{1 + (2)^2} \, dx = \sqrt{5} \, [3 - (-1)] = 4\sqrt{5}.$

The arc length can be calculated using the distance formula, since the curve is a line segment, so

$L = [\text{distance from } (-1, -7) \text{ to } (3, 1)] = \sqrt{[3 - (-1)]^2 + [1 - (-7)]^2} = \sqrt{80} = 4\sqrt{5}$

3. $y = \sin x$ $\Rightarrow$ $dy/dx = \cos x$ $\Rightarrow$ $1 + (dy/dx)^2 = 1 + \cos^2 x.$ So $L = \int_0^\pi \sqrt{1 + \cos^2 x} \, dx \approx 3.8202.$

5. $x = \sqrt{y} - y$ $\Rightarrow$ $dx/dy = 1/(2\sqrt{y}) - 1$ $\Rightarrow$ $1 + (dx/dy)^2 = \left(\dfrac{1}{2\sqrt{y}} - 1 \right)^2.$

So $L = \int_1^4 \sqrt{1 + \left(\dfrac{1}{2\sqrt{y}} - 1 \right)^2} \, dy \approx 3.6095.$

7. $y = 1 + 6x^{3/2}$ $\Rightarrow$ $dy/dx = 9x^{1/2}$ $\Rightarrow$ $1 + (dy/dx)^2 = 1 + 81x.$ So

$L = \int_0^1 \sqrt{1 + 81x} \, dx = \int_1^{82} u^{1/2} \left(\dfrac{1}{81} \, du \right) \quad \begin{bmatrix} u = 1 + 81x, \\ du = 81 \, dx \end{bmatrix} = \dfrac{1}{81} \cdot \dfrac{2}{3} \left[u^{3/2} \right]_1^{82} = \dfrac{2}{243} \left(82\sqrt{82} - 1 \right)$

9.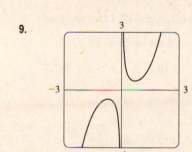

$y = \dfrac{x^3}{3} + \dfrac{1}{4x}$ $\Rightarrow$ $y' = x^2 - \dfrac{1}{4x^2}$ $\Rightarrow$

$1 + (y')^2 = 1 + \left(x^4 - \dfrac{1}{2} + \dfrac{1}{16x^4} \right) = x^4 + \dfrac{1}{2} + \dfrac{1}{16x^4} = \left(x^2 + \dfrac{1}{4x^2} \right)^2.$ So

$L = \int_1^2 \sqrt{1 + (y')^2} \, dx = \int_1^2 \left| x^2 + \dfrac{1}{4x^2} \right| \, dx = \int_1^2 \left(x^2 + \dfrac{1}{4x^2} \right) \, dx$

$= \left[\dfrac{1}{3} x^3 - \dfrac{1}{4x} \right]_1^2 = \left(\dfrac{8}{3} - \dfrac{1}{8} \right) - \left(\dfrac{1}{3} - \dfrac{1}{4} \right) = \dfrac{7}{3} + \dfrac{1}{8} = \dfrac{59}{24}$

11. $x = \dfrac{1}{3} \sqrt{y} \, (y - 3) = \dfrac{1}{3} y^{3/2} - y^{1/2}$ $\Rightarrow$ $dx/dy = \dfrac{1}{2} y^{1/2} - \dfrac{1}{2} y^{-1/2}$ $\Rightarrow$

$1 + (dx/dy)^2 = 1 + \dfrac{1}{4} y - \dfrac{1}{2} + \dfrac{1}{4} y^{-1} = \dfrac{1}{4} y + \dfrac{1}{2} + \dfrac{1}{4} y^{-1} = \left(\dfrac{1}{2} y^{1/2} + \dfrac{1}{2} y^{-1/2} \right)^2.$ So

$L = \int_1^9 \left(\dfrac{1}{2} y^{1/2} + \dfrac{1}{2} y^{-1/2} \right) dy = \dfrac{1}{2} \left[\dfrac{2}{3} y^{3/2} + 2y^{1/2} \right]_1^9 = \dfrac{1}{2} \left[\left(\dfrac{2}{3} \cdot 27 + 2 \cdot 3 \right) - \left(\dfrac{2}{3} \cdot 1 + 2 \cdot 1 \right) \right]$

$= \dfrac{1}{2} \left(24 - \dfrac{8}{3} \right) = \dfrac{1}{2} \left(\dfrac{64}{3} \right) = \dfrac{32}{3}.$

13. $y = \ln(\sec x) \;\Rightarrow\; \dfrac{dy}{dx} = \dfrac{\sec x \tan x}{\sec x} = \tan x \;\Rightarrow\; 1 + \left(\dfrac{dy}{dx}\right)^2 = 1 + \tan^2 x = \sec^2 x$, so

$$L = \int_0^{\pi/4} \sqrt{\sec^2 x}\, dx = \int_0^{\pi/4} |\sec x|\, dx = \int_0^{\pi/4} \sec x\, dx = \Big[\ln(\sec x + \tan x)\Big]_0^{\pi/4}$$

$$= \ln(\sqrt{2} + 1) - \ln(1 + 0) = \ln(\sqrt{2} + 1)$$

15. $y = \dfrac{1}{4}x^2 - \dfrac{1}{2}\ln x \;\Rightarrow\; y' = \dfrac{1}{2}x - \dfrac{1}{2x} \;\Rightarrow\; 1 + (y')^2 = 1 + \left(\dfrac{1}{4}x^2 - \dfrac{1}{2} + \dfrac{1}{4x^2}\right) = \dfrac{1}{4}x^2 + \dfrac{1}{2} + \dfrac{1}{4x^2} = \left(\dfrac{1}{2}x + \dfrac{1}{2x}\right)^2.$

So

$$L = \int_1^2 \sqrt{1 + (y')^2}\, dx = \int_1^2 \left|\dfrac{1}{2}x + \dfrac{1}{2x}\right| dx = \int_1^2 \left(\dfrac{1}{2}x + \dfrac{1}{2x}\right) dx$$

$$= \left[\dfrac{1}{4}x^2 + \dfrac{1}{2}\ln|x|\right]_1^2 = \left(1 + \dfrac{1}{2}\ln 2\right) - \left(\dfrac{1}{4} + 0\right) = \dfrac{3}{4} + \dfrac{1}{2}\ln 2$$

17. $y = \ln(1 - x^2) \;\Rightarrow\; y' = \dfrac{1}{1 - x^2}\cdot(-2x) \;\Rightarrow$

$$1 + \left(\dfrac{dy}{dx}\right)^2 = 1 + \dfrac{4x^2}{(1 - x^2)^2} = \dfrac{1 - 2x^2 + x^4 + 4x^2}{(1 - x^2)^2} = \dfrac{1 + 2x^2 + x^4}{(1 - x^2)^2} = \dfrac{(1 + x^2)^2}{(1 - x^2)^2} \;\Rightarrow$$

$$\sqrt{1 + \left(\dfrac{dy}{dx}\right)^2} = \sqrt{\left(\dfrac{1 + x^2}{1 - x^2}\right)^2} = \dfrac{1 + x^2}{1 - x^2} = -1 + \dfrac{2}{1 - x^2} \quad \text{[by division]} \quad = -1 + \dfrac{1}{1 + x} + \dfrac{1}{1 - x} \quad \text{[partial fractions]}.$$

So $L = \displaystyle\int_0^{1/2} \left(-1 + \dfrac{1}{1 + x} + \dfrac{1}{1 - x}\right) dx = \Big[-x + \ln|1 + x| - \ln|1 - x|\Big]_0^{1/2} = \left(-\dfrac{1}{2} + \ln\dfrac{3}{2} - \ln\dfrac{1}{2}\right) - 0 = \ln 3 - \dfrac{1}{2}.$

19.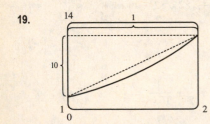

From the figure, the length of the curve is slightly larger than the hypotenuse of the triangle formed by the points $(1, 2)$, $(1, 12)$, and $(2, 12)$. This length is about $\sqrt{10^2 + 1^2} \approx 10$, so we might estimate the length to be 10.

$y = x^2 + x^3 \;\Rightarrow\; y' = 2x + 3x^2 \;\Rightarrow\; 1 + (y')^2 = 1 + (2x + 3x^2)^2.$

So $L = \int_1^2 \sqrt{1 + (2x + 3x^2)^2}\, dx \approx 10.0556.$

21. $y = x \sin x \;\Rightarrow\; dy/dx = x \cos x + (\sin x)(1) \;\Rightarrow\; 1 + (dy/dx)^2 = 1 + (x \cos x + \sin x)^2.$ Let

$f(x) = \sqrt{1 + (dy/dx)^2} = \sqrt{1 + (x \cos x + \sin x)^2}.$ Then $L = \int_0^{2\pi} f(x)\, dx$. Since $n = 10$, $\Delta x = \dfrac{2\pi - 0}{10} = \dfrac{\pi}{5}$. Now

$L \approx S_{10}$

$$= \dfrac{\pi/5}{3}\left[f(0) + 4f\!\left(\dfrac{\pi}{5}\right) + 2f\!\left(\dfrac{2\pi}{5}\right) + 4f\!\left(\dfrac{3\pi}{5}\right) + 2f\!\left(\dfrac{4\pi}{5}\right) + 4f\!\left(\dfrac{5\pi}{5}\right) + 2f\!\left(\dfrac{6\pi}{5}\right) + 4f\!\left(\dfrac{7\pi}{5}\right) + 2f\!\left(\dfrac{8\pi}{5}\right) + 4f\!\left(\dfrac{9\pi}{5}\right) + f(2\pi)\right]$$

≈ 15.498085

The value of the integral produced by a calculator is 15.374568 (to six decimal places).

23. $y = \ln(1 + x^3)$ $\Rightarrow$ $dy/dx = \dfrac{1}{1 + x^3} \cdot 3x^2$ $\Rightarrow$ $L = \int_0^5 f(x)\,dx$, where $f(x) = \sqrt{1 + 9x^4/(1+x^3)^2}$. Since $n = 10$,

$\Delta x = \frac{5-0}{10} = \frac{1}{2}$. Now

$L \approx S_{10}$

$= \frac{1/2}{3}[f(0) + 4f(0.5) + 2f(1) + 4f(1.5) + 2f(2) + 4f(2.5) + 2f(3) + 4f(3.5) + 2f(4) + 4f(4.5) + f(5)]$

≈ 7.094570

The value of the integral produced by a calculator is 7.118819 (to six decimal places).

25. $x = \ln\left(1 - y^2\right)$ $\Rightarrow$ $\dfrac{dx}{dy} = \dfrac{-2y}{1 - y^2}$ $\Rightarrow$ $1 + \left(\dfrac{dx}{dy}\right)^2 = 1 + \dfrac{4y^2}{(1-y^2)^2} = \dfrac{(1+y^2)^2}{(1-y^2)^2}$. So

$$L = \int_0^{1/2} \sqrt{\frac{(1+y^2)^2}{(1-y^2)^2}}\,dy = \int_0^{1/2} \frac{1+y^2}{1-y^2}\,dy = \ln 3 - \frac{1}{2} \quad \text{[from a CAS]}$$

27. $y^{2/3} = 1 - x^{2/3}$ $\Rightarrow$ $y = (1 - x^{2/3})^{3/2}$ $\Rightarrow$

$\dfrac{dy}{dx} = \frac{3}{2}(1 - x^{2/3})^{1/2}\left(-\frac{2}{3}x^{-1/3}\right) = -x^{-1/3}(1 - x^{2/3})^{1/2}$ $\Rightarrow$

$\left(\dfrac{dy}{dx}\right)^2 = x^{-2/3}(1 - x^{2/3}) = x^{-2/3} - 1$. Thus

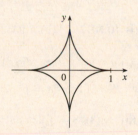

$L = 4\int_0^1 \sqrt{1 + (x^{-2/3} - 1)}\,dx = 4\int_0^1 x^{-1/3}\,dx = 4\lim_{t\to 0^+}\left[\frac{3}{2}x^{2/3}\right]_t^1 = 6.$

29. $y = 2x^{3/2}$ $\Rightarrow$ $y' = 3x^{1/2}$ $\Rightarrow$ $1 + (y')^2 = 1 + 9x$. The arc length function with starting point $P_0(1, 2)$ is

$s(x) = \int_1^x \sqrt{1 + 9t}\,dt = \left[\frac{2}{27}(1 + 9t)^{3/2}\right]_1^x = \frac{2}{27}\left[(1 + 9x)^{3/2} - 10\sqrt{10}\right].$

31. $y = \sin^{-1}x + \sqrt{1 - x^2}$ $\Rightarrow$ $y' = \dfrac{1}{\sqrt{1-x^2}} - \dfrac{x}{\sqrt{1-x^2}} = \dfrac{1-x}{\sqrt{1-x^2}}$ $\Rightarrow$

$1 + (y')^2 = 1 + \dfrac{(1-x)^2}{1-x^2} = \dfrac{1-x^2+1-2x+x^2}{1-x^2} = \dfrac{2-2x}{1-x^2} = \dfrac{2(1-x)}{(1+x)(1-x)} = \dfrac{2}{1+x}$ $\Rightarrow$

$\sqrt{1 + (y')^2} = \sqrt{\dfrac{2}{1+x}}$. Thus, the arc length function with starting point $(0, 1)$ is given by

$s(x) = \int_0^x \sqrt{1 + [f'(t)]^2}\,dt = \int_0^x \sqrt{\dfrac{2}{1+t}}\,dt = \sqrt{2}\left[2\sqrt{1+t}\right]_0^x = 2\sqrt{2}\left(\sqrt{1+x} - 1\right).$

33. The prey hits the ground when $y = 0$ $\Leftrightarrow$ $180 - \frac{1}{45}x^2 = 0$ $\Leftrightarrow$ $x^2 = 45 \cdot 180$ $\Rightarrow$ $x = \sqrt{8100} = 90$,

since x must be positive. $y' = -\frac{2}{45}x$ $\Rightarrow$ $1 + (y')^2 = 1 + \frac{4}{45^2}x^2$, so the distance traveled by the prey is

$L = \int_0^{90} \sqrt{1 + \dfrac{4}{45^2}x^2}\,dx = \int_0^4 \sqrt{1 + u^2}\left(\frac{45}{2}\,du\right) \qquad \begin{bmatrix} u = \frac{2}{45}x, \\ du = \frac{2}{45}\,dx \end{bmatrix}$

$\overset{21}{=} \frac{45}{2}\left[\frac{1}{2}u\sqrt{1+u^2} + \frac{1}{2}\ln\left(u + \sqrt{1+u^2}\right)\right]_0^4 = \frac{45}{2}\left[2\sqrt{17} + \frac{1}{2}\ln\left(4 + \sqrt{17}\right)\right] = 45\sqrt{17} + \frac{45}{4}\ln\left(4 + \sqrt{17}\right) \approx 209.1\text{ m}$

35. The sine wave has amplitude 1 and period 14, since it goes through two periods in a distance of 28 in., so its equation is

$y = 1 \sin\left(\frac{2\pi}{14}x\right) = \sin\left(\frac{\pi}{7}x\right)$. The width w of the flat metal sheet needed to make the panel is the arc length of the sine curve

from $x = 0$ to $x = 28$. We set up the integral to evaluate w using the arc length formula with $\frac{dy}{dx} = \frac{\pi}{7}\cos\left(\frac{\pi}{7}x\right)$:

$L = \int_0^{28} \sqrt{1 + \left[\frac{\pi}{7}\cos\left(\frac{\pi}{7}x\right)\right]^2}\, dx = 2\int_0^{14} \sqrt{1 + \left[\frac{\pi}{7}\cos\left(\frac{\pi}{7}x\right)\right]^2}\, dx$. This integral would be very difficult to evaluate exactly,

so we use a CAS, and find that $L \approx 29.36$ inches.

7.5 Area of a Surface of Revolution

1. (a) (i) $y = \tan x \;\Rightarrow\; dy/dx = \sec^2 x \;\Rightarrow\; ds = \sqrt{1 + (dy/dx)^2}\, dx = \sqrt{1 + \sec^4 x}\, dx$. By (7), an integral for the

area of the surface obtained by rotating the curve about the x-axis is $S = \int 2\pi y\, ds = \int_0^{\pi/3} 2\pi \tan x \sqrt{1 + \sec^4 x}\, dx$.

(ii) By (8), an integral for the area of the surface obtained by rotating the curve about the y-axis is

$S = \int 2\pi x\, ds = \int_0^{\pi/3} 2\pi x \sqrt{1 + \sec^4 x}\, dx$.

(b) (i) 10.5017   (ii) 7.9353

3. (a) (i) $y = e^{-x^2} \;\Rightarrow\; dy/dx = e^{-x^2}\cdot(-2x) \;\Rightarrow\; ds = \sqrt{1 + (dy/dx)^2}\, dx = \sqrt{1 + 4x^2 e^{-2x^2}}\, dx$. By (7),

$S = \int 2\pi y\, ds = \int_{-1}^{1} 2\pi e^{-x^2}\sqrt{1 + 4x^2 e^{-2x^2}}\, dx$.

(ii) By (8), $S = \int 2\pi x\, ds = \int_0^1 2\pi x \sqrt{1 + 4x^2 e^{-2x^2}}\, dx$ [symmetric about the y-axis]

(b) (i) 11.0753   (ii) 3.9603

5. $y = x^3 \;\Rightarrow\; y' = 3x^2$. So

$$S = \int_0^2 2\pi y \sqrt{1 + (y')^2}\, dx = 2\pi \int_0^2 x^3 \sqrt{1 + 9x^4}\, dx \qquad [u = 1 + 9x^4,\, du = 36x^3\, dx]$$

$$= \frac{2\pi}{36}\int_1^{145} \sqrt{u}\, du = \frac{\pi}{18}\left[\frac{2}{3}u^{3/2}\right]_1^{145} = \frac{\pi}{27}\left(145\sqrt{145} - 1\right)$$

7. $y = \sqrt{1 + 4x} \;\Rightarrow\; y' = \frac{1}{2}(1 + 4x)^{-1/2}(4) = \dfrac{2}{\sqrt{1 + 4x}} \;\Rightarrow\; \sqrt{1 + (y')^2} = \sqrt{1 + \dfrac{4}{1 + 4x}} = \sqrt{\dfrac{5 + 4x}{1 + 4x}}$. So

$$S = \int_1^5 2\pi y \sqrt{1 + (y')^2}\, dx = 2\pi \int_1^5 \sqrt{1 + 4x}\, \sqrt{\frac{5 + 4x}{1 + 4x}}\, dx = 2\pi \int_1^5 \sqrt{4x + 5}\, dx$$

$$= 2\pi \int_9^{25} \sqrt{u}\,\left(\tfrac{1}{4}\, du\right) \quad \begin{bmatrix} u = 4x + 5, \\ du = 4\, dx \end{bmatrix} = \frac{2\pi}{4}\left[\frac{2}{3}u^{3/2}\right]_9^{25} = \frac{\pi}{3}\left(25^{3/2} - 9^{3/2}\right) = \frac{\pi}{3}(125 - 27) = \frac{98}{3}\pi$$

9. $y = \sin \pi x \;\Rightarrow\; y' = \pi \cos \pi x \;\Rightarrow\; 1 + (y')^2 = 1 + \pi^2 \cos^2(\pi x)$. So

$$S = \int_0^1 2\pi y \sqrt{1 + (y')^2}\, dx = 2\pi \int_0^1 \sin \pi x \sqrt{1 + \pi^2 \cos^2(\pi x)}\, dx \qquad \begin{bmatrix} u = \pi \cos \pi x, \\ du = -\pi^2 \sin \pi x\, dx \end{bmatrix}$$

$$= 2\pi \int_{\pi}^{-\pi} \sqrt{1 + u^2}\,\left(-\frac{1}{\pi^2}\, du\right) = \frac{2}{\pi}\int_{-\pi}^{\pi} \sqrt{1 + u^2}\, du$$

$$= \frac{4}{\pi}\int_0^{\pi} \sqrt{1 + u^2}\, du \overset{21}{=} \frac{4}{\pi}\left[\frac{u}{2}\sqrt{1 + u^2} + \frac{1}{2}\ln\left(u + \sqrt{1 + u^2}\right)\right]_0^{\pi}$$

$$= \frac{4}{\pi}\left[\left(\frac{\pi}{2}\sqrt{1 + \pi^2} + \frac{1}{2}\ln\left(\pi + \sqrt{1 + \pi^2}\right)\right) - 0\right] = 2\sqrt{1 + \pi^2} + \frac{2}{\pi}\ln\left(\pi + \sqrt{1 + \pi^2}\right)$$

11. $x = \frac{1}{3}(y^2 + 2)^{3/2}$ $\Rightarrow$ $dx/dy = \frac{1}{2}(y^2 + 2)^{1/2}(2y) = y\sqrt{y^2 + 2}$ $\Rightarrow$ $1 + (dx/dy)^2 = 1 + y^2(y^2 + 2) = (y^2 + 1)^2$.

So $S = 2\pi \int_1^2 y(y^2 + 1)\, dy = 2\pi \left[\frac{1}{4}y^4 + \frac{1}{2}y^2\right]_1^2 = 2\pi \left(4 + 2 - \frac{1}{4} - \frac{1}{2}\right) = \frac{21\pi}{2}$.

13. $y = \sqrt[3]{x}$ $\Rightarrow$ $x = y^3$ $\Rightarrow$ $1 + (dx/dy)^2 = 1 + 9y^4$. So

$$S = 2\pi \int_1^2 x\sqrt{1 + (dx/dy)^2}\, dy = 2\pi \int_1^2 y^3 \sqrt{1 + 9y^4}\, dy = \frac{2\pi}{36} \int_1^2 \sqrt{1 + 9y^4}\, 36y^3\, dy = \frac{\pi}{18}\left[\frac{2}{3}(1 + 9y^4)^{3/2}\right]_1^2$$

$$= \frac{\pi}{27}\left(145\sqrt{145} - 10\sqrt{10}\right)$$

15. $x = \sqrt{a^2 - y^2}$ $\Rightarrow$ $dx/dy = \frac{1}{2}(a^2 - y^2)^{-1/2}(-2y) = -y/\sqrt{a^2 - y^2}$ $\Rightarrow$

$$1 + \left(\frac{dx}{dy}\right)^2 = 1 + \frac{y^2}{a^2 - y^2} = \frac{a^2 - y^2}{a^2 - y^2} + \frac{y^2}{a^2 - y^2} = \frac{a^2}{a^2 - y^2} \quad \Rightarrow$$

$$S = \int_0^{a/2} 2\pi\sqrt{a^2 - y^2}\,\frac{a}{\sqrt{a^2 - y^2}}\, dy = 2\pi \int_0^{a/2} a\, dy = 2\pi a\big[y\big]_0^{a/2} = 2\pi a\left(\frac{a}{2} - 0\right) = \pi a^2.$$

Note that this is $\frac{1}{4}$ the surface area of a sphere of radius a, and the length of the interval $y = 0$ to $y = a/2$ is $\frac{1}{4}$ the length of the interval $y = -a$ to $y = a$.

17. $y = x^3$ and $0 \le y \le 1$ $\Rightarrow$ $y' = 3x^2$ and $0 \le x \le 1$.

$$S = \int_0^1 2\pi x\sqrt{1 + (3x^2)^2}\, dx = 2\pi \int_0^3 \sqrt{1 + u^2}\,\frac{1}{6}\, du \quad \begin{bmatrix} u = 3x^2, \\ du = 6x\, dx \end{bmatrix} = \frac{\pi}{3}\int_0^3 \sqrt{1 + u^2}\, du$$

$$\overset{21}{=} \text{[or use CAS]} \ \frac{\pi}{3}\left[\frac{1}{2}u\sqrt{1 + u^2} + \frac{1}{2}\ln\left(u + \sqrt{1 + u^2}\right)\right]_0^3 = \frac{\pi}{3}\left[\frac{3}{2}\sqrt{10} + \frac{1}{2}\ln\left(3 + \sqrt{10}\right)\right] = \frac{\pi}{6}\left[3\sqrt{10} + \ln\left(3 + \sqrt{10}\right)\right]$$

19. Since $a > 0$, the curve $3ay^2 = x(a - x)^2$ only has points with $x \ge 0$.

$[3ay^2 \ge 0 \ \Rightarrow \ x(a - x)^2 \ge 0 \ \Rightarrow \ x \ge 0.]$

The curve is symmetric about the x-axis (since the equation is unchanged

when y is replaced by $-y$). $y = 0$ when $x = 0$ or a, so the curve's loop

extends from $x = 0$ to $x = a$.

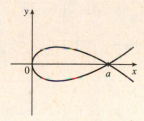

$$\frac{d}{dx}(3ay^2) = \frac{d}{dx}[x(a - x)^2] \ \Rightarrow \ 6ay\frac{dy}{dx} = x \cdot 2(a - x)(-1) + (a - x)^2 \ \Rightarrow \ \frac{dy}{dx} = \frac{(a - x)[-2x + a - x]}{6ay} \ \Rightarrow$$

$$\left(\frac{dy}{dx}\right)^2 = \frac{(a - x)^2(a - 3x)^2}{36a^2y^2} = \frac{(a - x)^2(a - 3x)^2}{36a^2} \cdot \frac{3a}{x(a - x)^2} \quad \begin{bmatrix} \text{the last fraction} \\ \text{is } 1/y^2 \end{bmatrix} = \frac{(a - 3x)^2}{12ax} \ \Rightarrow$$

$$1 + \left(\frac{dy}{dx}\right)^2 = 1 + \frac{a^2 - 6ax + 9x^2}{12ax} = \frac{12ax}{12ax} + \frac{a^2 - 6ax + 9x^2}{12ax} = \frac{a^2 + 6ax + 9x^2}{12ax} = \frac{(a + 3x)^2}{12ax} \quad \text{for } x \ne 0.$$

(a) $\displaystyle S = \int_{x=0}^a 2\pi y\, ds = 2\pi \int_0^a \frac{\sqrt{x}\,(a - x)}{\sqrt{3a}} \cdot \frac{a + 3x}{\sqrt{12ax}}\, dx = 2\pi \int_0^a \frac{(a - x)(a + 3x)}{6a}\, dx$

$$= \frac{\pi}{3a}\int_0^a (a^2 + 2ax - 3x^2)\, dx = \frac{\pi}{3a}\left[a^2 x + ax^2 - x^3\right]_0^a = \frac{\pi}{3a}(a^3 + a^3 - a^3) = \frac{\pi}{3a} \cdot a^3 = \frac{\pi a^2}{3}.$$

Note that we have rotated the top half of the loop about the x-axis. This generates the full surface.

(b) We must rotate the full loop about the y-axis, so we get double the area obtained by rotating the top half of the loop:

$$S = 2 \cdot 2\pi \int_{x=0}^{a} x\,ds = 4\pi \int_0^a x\,\frac{a+3x}{\sqrt{12ax}}\,dx = \frac{4\pi}{2\sqrt{3a}} \int_0^a x^{1/2}(a+3x)\,dx = \frac{2\pi}{\sqrt{3a}} \int_0^a (ax^{1/2} + 3x^{3/2})\,dx$$

$$= \frac{2\pi}{\sqrt{3a}} \left[\frac{2}{3}ax^{3/2} + \frac{6}{5}x^{5/2} \right]_0^a = \frac{2\pi\sqrt{3}}{3\sqrt{a}} \left(\frac{2}{3}a^{5/2} + \frac{6}{5}a^{5/2} \right) = \frac{2\pi\sqrt{3}}{3} \left(\frac{2}{3} + \frac{6}{5} \right)a^2 = \frac{2\pi\sqrt{3}}{3}\left(\frac{28}{15}\right)a^2$$

$$= \frac{56\pi\sqrt{3}\,a^2}{45}$$

21. (a) $\dfrac{x^2}{a^2} + \dfrac{y^2}{b^2} = 1 \;\Rightarrow\; \dfrac{y\,(dy/dx)}{b^2} = -\dfrac{x}{a^2} \;\Rightarrow\; \dfrac{dy}{dx} = -\dfrac{b^2 x}{a^2 y} \;\Rightarrow$

$$1 + \left(\frac{dy}{dx}\right)^2 = 1 + \frac{b^4 x^2}{a^4 y^2} = \frac{b^4 x^2 + a^4 y^2}{a^4 y^2} = \frac{b^4 x^2 + a^4 b^2\left(1 - x^2/a^2\right)}{a^4 b^2\left(1 - x^2/a^2\right)} = \frac{a^4 b^2 + b^4 x^2 - a^2 b^2 x^2}{a^4 b^2 - a^2 b^2 x^2}$$

$$= \frac{a^4 + b^2 x^2 - a^2 x^2}{a^4 - a^2 x^2} = \frac{a^4 - (a^2 - b^2)x^2}{a^2(a^2 - x^2)}$$

The ellipsoid's surface area is twice the area generated by rotating the first-quadrant portion of the ellipse about the x-axis. Thus,

$$S = 2 \int_0^a 2\pi y \sqrt{1 + \left(\frac{dy}{dx}\right)^2}\,dx = 4\pi \int_0^a \frac{b}{a}\sqrt{a^2 - x^2}\,\frac{\sqrt{a^4 - (a^2-b^2)x^2}}{a\sqrt{a^2-x^2}}\,dx = \frac{4\pi b}{a^2} \int_0^a \sqrt{a^4 - (a^2-b^2)x^2}\,dx$$

$$= \frac{4\pi b}{a^2} \int_0^{a\sqrt{a^2-b^2}} \sqrt{a^4 - u^2}\,\frac{du}{\sqrt{a^2 - b^2}} \quad \left[u = \sqrt{a^2-b^2}\,x\right] \stackrel{30}{=} \frac{4\pi b}{a^2\sqrt{a^2-b^2}} \left[\frac{u}{2}\sqrt{a^4 - u^2} + \frac{a^4}{2}\sin^{-1}\left(\frac{u}{a^2}\right) \right]_0^{a\sqrt{a^2-b^2}}$$

$$= \frac{4\pi b}{a^2\sqrt{a^2-b^2}} \left[\frac{a\sqrt{a^2-b^2}}{2}\sqrt{a^4 - a^2(a^2-b^2)} + \frac{a^4}{2}\sin^{-1}\frac{\sqrt{a^2-b^2}}{a} \right] = 2\pi \left[b^2 + \frac{a^2 b\sin^{-1}\dfrac{\sqrt{a^2-b^2}}{a}}{\sqrt{a^2-b^2}} \right]$$

(b) $\dfrac{x^2}{a^2} + \dfrac{y^2}{b^2} = 1 \;\Rightarrow\; \dfrac{x\,(dx/dy)}{a^2} = -\dfrac{y}{b^2} \;\Rightarrow\; \dfrac{dx}{dy} = -\dfrac{a^2 y}{b^2 x} \;\Rightarrow$

$$1 + \left(\frac{dx}{dy}\right)^2 = 1 + \frac{a^4 y^2}{b^4 x^2} = \frac{b^4 x^2 + a^4 y^2}{b^4 x^2} = \frac{b^4 a^2(1 - y^2/b^2) + a^4 y^2}{b^4 a^2(1 - y^2/b^2)} = \frac{a^2 b^4 - a^2 b^2 y^2 + a^4 y^2}{a^2 b^4 - a^2 b^2 y^2}$$

$$= \frac{b^4 - b^2 y^2 + a^2 y^2}{b^4 - b^2 y^2} = \frac{b^4 - (b^2 - a^2)y^2}{b^2(b^2 - y^2)}$$

The oblate spheroid's surface area is twice the area generated by rotating the first-quadrant portion of the ellipse about the y-axis. Thus,

$$S = 2 \int_0^b 2\pi x \sqrt{1 + \left(\frac{dx}{dy}\right)^2}\,dy = 4\pi \int_0^b \frac{a}{b}\sqrt{b^2 - y^2}\,\frac{\sqrt{b^4 - (b^2-a^2)y^2}}{b\sqrt{b^2-y^2}}\,dy$$

$$= \frac{4\pi a}{b^2} \int_0^b \sqrt{b^4 - (b^2-a^2)\,y^2}\,dy = \frac{4\pi a}{b^2} \int_0^b \sqrt{b^4 + (a^2-b^2)\,y^2}\,dy \qquad \left[\text{since } a > b\right]$$

$$= \frac{4\pi a}{b^2} \int_0^{b\sqrt{a^2-b^2}} \sqrt{b^4 + u^2}\,\frac{du}{\sqrt{a^2-b^2}} \qquad \left[u = \sqrt{a^2-b^2}\,y\right]$$

$$\stackrel{21}{=} \frac{4\pi a}{b^2\sqrt{a^2-b^2}} \left[\frac{u}{2}\sqrt{b^4 + u^2} + \frac{b^4}{2}\ln\left(u + \sqrt{b^4 + u^2}\right) \right]_0^{b\sqrt{a^2-b^2}}$$

$$= \frac{4\pi a}{b^2\sqrt{a^2-b^2}} \left\{ \left[\frac{b\sqrt{a^2-b^2}}{2}(ab) + \frac{b^4}{2}\ln\left(b\sqrt{a^2-b^2} + ab\right) \right] - \left[0 + \frac{b^4}{2}\ln(b^2) \right] \right\}$$

$$= \frac{4\pi a}{b^2\sqrt{a^2-b^2}} \left[\frac{ab^2\sqrt{a^2-b^2}}{2} + \frac{b^4}{2}\ln\frac{b\sqrt{a^2-b^2} + ab}{b^2} \right] = 2\pi a^2 + \frac{2\pi ab^2}{\sqrt{a^2-b^2}}\ln\frac{\sqrt{a^2-b^2} + a}{b}$$

23. The analogue of $f(x_i^*)$ in the derivation of (4) is now $c - f(x_i^*)$, so

$$S = \lim_{n \to \infty} \sum_{i=1}^{n} 2\pi[c - f(x_i^*)] \sqrt{1 + [f'(x_i^*)]^2} \, \Delta x = \int_a^b 2\pi[c - f(x)] \sqrt{1 + [f'(x)]^2} \, dx.$$

25. For the upper semicircle, $f(x) = \sqrt{r^2 - x^2}$, $f'(x) = -x/\sqrt{r^2 - x^2}$. The surface area generated is

$$S_1 = \int_{-r}^{r} 2\pi\left(r - \sqrt{r^2 - x^2}\right)\sqrt{1 + \frac{x^2}{r^2 - x^2}} \, dx = 4\pi \int_0^r \left(r - \sqrt{r^2 - x^2}\right) \frac{r}{\sqrt{r^2 - x^2}} \, dx$$

$$= 4\pi \int_0^r \left(\frac{r^2}{\sqrt{r^2 - x^2}} - r\right) dx$$

For the lower semicircle, $f(x) = -\sqrt{r^2 - x^2}$ and $f'(x) = \dfrac{x}{\sqrt{r^2 - x^2}}$, so $S_2 = 4\pi \int_0^r \left(\dfrac{r^2}{\sqrt{r^2 - x^2}} + r\right) dx.$

Thus, the total area is $S = S_1 + S_2 = 8\pi \int_0^r \left(\dfrac{r^2}{\sqrt{r^2 - x^2}}\right) dx = 8\pi\left[r^2 \sin^{-1}\left(\dfrac{x}{r}\right)\right]_0^r = 8\pi r^2 \left(\dfrac{\pi}{2}\right) = 4\pi^2 r^2.$

7.6 Applications to Physics and Engineering

1. $W = \int_a^b f(x) \, dx = \int_1^{10} 5x^{-2} \, dx = 5\left[-x^{-1}\right]_1^{10} = 5\left(-\frac{1}{10} + 1\right) = 4.5$ ft-lb

3. The force function is given by $F(x)$ (in newtons) and the work (in joules) is the area under the curve, given by

$\int_0^8 F(x) \, dx = \int_0^4 F(x) \, dx + \int_4^8 F(x) \, dx = \frac{1}{2}(4)(30) + (4)(30) = 180$ J.

5. According to Hooke's Law, the force required to maintain a spring stretched x units beyond its natural length is proportional

to x, that is, $f(x) = kx$. Here, the amount stretched is 4 in. $= \frac{1}{3}$ ft and the force is 10 lb. Thus, $10 = k\left(\frac{1}{3}\right)$ $\Rightarrow$

$k = 30$ lb/ft, and $f(x) = 30x$. The work done in stretching the spring from its natural length to 6 in. $= \frac{1}{2}$ ft beyond its natural

length is $W = \int_0^{1/2} 30x \, dx = \left[15x^2\right]_0^{1/2} = \frac{15}{4}$ ft-lb.

7. (a) If $\int_0^{0.12} kx \, dx = 2$ J, then $2 = \left[\frac{1}{2}kx^2\right]_0^{0.12} = \frac{1}{2}k(0.0144) = 0.0072k$ and $k = \frac{2}{0.0072} = \frac{2500}{9} \approx 277.78$ N/m.

Thus, the work needed to stretch the spring from 35 cm to 40 cm is

$\int_{0.05}^{0.10} \frac{2500}{9} x \, dx = \left[\frac{1250}{9}x^2\right]_{1/20}^{1/10} = \frac{1250}{9}\left(\frac{1}{100} - \frac{1}{400}\right) = \frac{25}{24} \approx 1.04$ J.

(b) $f(x) = kx$, so $30 = \frac{2500}{9}x$ and $x = \frac{270}{2500}$ m $= 10.8$ cm

In Exercises 9–16, n is the number of subintervals of length Δx, and x_i^* is a sample point in the ith subinterval $[x_{i-1}, x_i]$.

9. (a) The portion of the rope from x ft to $(x + \Delta x)$ ft below the top of the building weighs $\frac{1}{2}\Delta x$ lb and must be lifted x_i^* ft,

so its contribution to the total work is $\frac{1}{2}x_i^* \Delta x$ ft-lb. The total work is

$$W = \lim_{n \to \infty} \sum_{i=1}^{n} \frac{1}{2}x_i^* \Delta x = \int_0^{50} \frac{1}{2}x \, dx = \left[\frac{1}{4}x^2\right]_0^{50} = \frac{2500}{4} = 625 \text{ ft-lb}$$

Notice that the exact height of the building does not matter (as long as it is more than 50 ft).

(b) When half the rope is pulled to the top of the building, the work to lift the top half of the rope is

$W_1 = \int_0^{25} \frac{1}{2}x\,dx = \left[\frac{1}{4}x^2\right]_0^{25} = \frac{625}{4}$ ft-lb. The bottom half of the rope is lifted 25 ft and the work needed to accomplish

that is $W_2 = \int_{25}^{50} \frac{1}{2} \cdot 25\,dx = \frac{25}{2}\left[x\right]_{25}^{50} = \frac{625}{2}$ ft-lb. The total work done in pulling half the rope to the top of the building

is $W = W_1 + W_2 = \frac{625}{2} + \frac{625}{4} = \frac{3}{4} \cdot 625 = \frac{1875}{4}$ ft-lb.

11. The work needed to lift the cable is $\lim\limits_{n\to\infty} \sum_{i=1}^{n} 2x_i^* \,\Delta x = \int_0^{500} 2x\,dx = \left[x^2\right]_0^{500} = 250{,}000$ ft-lb. The work needed to lift

the coal is $800\text{ lb} \cdot 500\text{ ft} = 400{,}000$ ft-lb. Thus, the total work required is $250{,}000 + 400{,}000 = 650{,}000$ ft-lb.

13. At a height of x meters $(0 \le x \le 12)$, the mass of the rope is $(0.8\text{ kg/m})(12 - x\text{ m}) = (9.6 - 0.8x)$ kg and the mass of the

water is $\left(\frac{36}{12}\text{ kg/m}\right)(12 - x\text{ m}) = (36 - 3x)$ kg. The mass of the bucket is 10 kg, so the total mass is

$(9.6 - 0.8x) + (36 - 3x) + 10 = (55.6 - 3.8x)$ kg, and hence, the total force is $9.8(55.6 - 3.8x)$ N. The work needed to lift

the bucket Δx m through the ith subinterval of $[0, 12]$ is $9.8(55.6 - 3.8x_i^*)\Delta x$, so the total work is

$$W = \lim_{n\to\infty} \sum_{i=1}^{n} 9.8(55.6 - 3.8x_i^*)\,\Delta x = \int_0^{12} (9.8)(55.6 - 3.8x)\,dx = 9.8\left[55.6x - 1.9x^2\right]_0^{12} = 9.8(393.6) \approx 3857 \text{ J}$$

15. A "slice" of water Δx m thick and lying at a depth of x_i^* m (where $0 \le x_i^* \le \frac{1}{2}$) has volume $(2 \times 1 \times \Delta x)$ m^3, a mass of

$2000\,\Delta x$ kg, weighs about $(9.8)(2000\,\Delta x) = 19{,}600\,\Delta x$ N, and thus requires about $19{,}600x_i^*\,\Delta x$ J of work for its removal.

So $W = \lim\limits_{n\to\infty} \sum_{i=1}^{n} 19{,}600x_i^*\,\Delta x = \int_0^{1/2} 19{,}600x\,dx = \left[9800x^2\right]_0^{1/2} = 2450$ J.

17. (a) A rectangular "slice" of water Δx m thick and lying x m above the bottom has width x m and volume $8x\,\Delta x$ m^3.

It weighs about $(9.8 \times 1000)(8x\,\Delta x)$ N, and must be lifted $(5 - x)$ m by the pump, so the work needed is about

$(9.8 \times 10^3)(5 - x)(8x\,\Delta x)$ J. The total work required is

$$W \approx \int_0^3 (9.8 \times 10^3)(5 - x)8x\,dx = (9.8 \times 10^3)\int_0^3 (40x - 8x^2)\,dx = (9.8 \times 10^3)\left[20x^2 - \tfrac{8}{3}x^3\right]_0^3$$
$$= (9.8 \times 10^3)(180 - 72) = (9.8 \times 10^3)(108) = 1058.4 \times 10^3 \approx 1.06 \times 10^6 \text{ J}$$

(b) If only 4.7×10^5 J of work is done, then only the water above a certain level (call

it h) will be pumped out. So we use the same formula as in part (a), except that the

work is fixed, and we are trying to find the lower limit of integration:

$4.7 \times 10^5 \approx \int_h^3 (9.8 \times 10^3)(5 - x)8x\,dx = (9.8 \times 10^3)\left[20x^2 - \tfrac{8}{3}x^3\right]_h^3 \Leftrightarrow$

$\frac{4.7}{9.8} \times 10^2 \approx 48 = \left(20 \cdot 3^2 - \tfrac{8}{3} \cdot 3^3\right) - \left(20h^2 - \tfrac{8}{3}h^3\right) \Leftrightarrow$

$2h^3 - 15h^2 + 45 = 0$. To find the solution of this equation, we plot $2h^3 - 15h^2 + 45$ between $h = 0$ and $h = 3$.

We see that the equation is satisfied for $h \approx 2.0$. So the depth of water remaining in the tank is about 2.0 m.

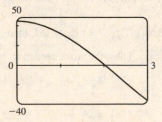

19. $V = \pi r^2 x$, so V is a function of x and P can also be regarded as a function of x. If $V_1 = \pi r^2 x_1$ and $V_2 = \pi r^2 x_2$, then

$$W = \int_{x_1}^{x_2} F(x)\,dx = \int_{x_1}^{x_2} \pi r^2 P(V(x))\,dx = \int_{x_1}^{x_2} P(V(x))\,dV(x) \qquad [\text{Let } V(x) = \pi r^2 x, \text{ so } dV(x) = \pi r^2\,dx.]$$
$$= \int_{V_1}^{V_2} P(V)\,dV \quad \text{by the Substitution Rule.}$$

21. (a) $W = \int_a^b F(r)\,dr = \int_a^b G\frac{m_1 m_2}{r^2}\,dr = Gm_1 m_2 \left[\frac{-1}{r}\right]_a^b = Gm_1 m_2 \left(\frac{1}{a} - \frac{1}{b}\right)$

(b) By part (a), $W = GMm \left(\dfrac{1}{R} - \dfrac{1}{R + 1{,}000{,}000}\right)$ where M = mass of the earth in kg, R = radius of the earth in m,

and m = mass of satellite in kg. (Note that 1000 km = 1,000,000 m.) Thus,

$$W = (6.67 \times 10^{-11})(5.98 \times 10^{24})(1000) \times \left(\frac{1}{6.37 \times 10^6} - \frac{1}{7.37 \times 10^6}\right) \approx 8.50 \times 10^9 \text{ J}$$

23. From Exercise 22, $W = \dfrac{GMm}{R}$. The initial kinetic energy supplies the needed work, so $\dfrac{1}{2}mv_0^2 = \dfrac{GMm}{R}$ $\Rightarrow$

$v_0 = \sqrt{2GM/R}$.

25. Set up a vertical x-axis as shown, with $x = 0$ at the water's surface and x increasing in the

downward direction. Then the area of the ith rectangular strip is $6\,\Delta x$ and the pressure on

the strip is δx_i^* (where $\delta \approx 62.5$ lb/ft^3). Thus, the hydrostatic force on the strip is

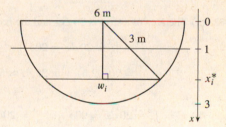

$\delta x_i^* \cdot 6\,\Delta x$ and the total hydrostatic force $\approx \sum_{i=1}^n \delta x_i^* \cdot 6\,\Delta x$. The total force

$$F = \lim_{n \to \infty} \sum_{i=1}^n \delta x_i^* \cdot 6\,\Delta x = \int_2^6 \delta x \cdot 6\,dx = 6\delta \int_2^6 x\,dx = 6\delta \left[\tfrac{1}{2}x^2\right]_2^6 = 6\delta(18 - 2) = 96\delta \approx 6000 \text{ lb}$$

27. Set up a vertical x-axis as shown. The base of the triangle shown in the figure

has length $\sqrt{3^2 - (x_i^*)^2}$, so $w_i = 2\sqrt{9 - (x_i^*)^2}$, and the area of the ith

rectangular strip is $2\sqrt{9 - (x_i^*)^2}\,\Delta x$. The ith rectangular strip is $(x_i^* - 1)$ m

below the surface level of the water, so the pressure on the strip is $\rho g(x_i^* - 1)$.

The hydrostatic force on the strip is $\rho g(x_i^* - 1) \cdot 2\sqrt{9 - (x_i^*)^2}\,\Delta x$ and the total

force on the plate $\approx \sum_{i=1}^n \rho g(x_i^* - 1) \cdot 2\sqrt{9 - (x_i^*)^2}\,\Delta x$. The total force

$$F = \lim \sum_{i=1}^n \rho g(x_i^* - 1) \cdot 2\sqrt{9 - (x_i^*)^2}\,\Delta x = 2\rho g \int_1^3 (x - 1)\sqrt{9 - x^2}\,dx$$

$$= 2\rho g \int_1^3 x\sqrt{9 - x^2}\,dx - 2\rho g \int_1^3 \sqrt{9 - x^2}\,dx \overset{30}{=} 2\rho g \left[-\tfrac{1}{3}(9 - x^2)^{3/2}\right]_1^3 - 2\rho g \left[\frac{x}{2}\sqrt{9 - x^2} + \frac{9}{2}\sin^{-1}\left(\frac{x}{3}\right)\right]_1^3$$

$$= 2\rho g \left[0 + \tfrac{1}{3}(8\sqrt{8})\right] - 2\rho g \left[(0 + \tfrac{9}{2} \cdot \tfrac{\pi}{2}) - (\tfrac{1}{2}\sqrt{8} + \tfrac{9}{2}\sin^{-1}(\tfrac{1}{3}))\right]$$

$$= \tfrac{32}{3}\sqrt{2}\,\rho g - \tfrac{9\pi}{2}\rho g + 2\sqrt{2}\,\rho g + 9\left[\sin^{-1}(\tfrac{1}{3})\right]\rho g = \left(\tfrac{38}{3}\sqrt{2} - \tfrac{9\pi}{2} + 9\sin^{-1}(\tfrac{1}{3})\right)\rho g$$

$$\approx 6.835 \cdot 1000 \cdot 9.8 \approx 6.7 \times 10^4 \text{ N}$$

Note: If you set up a typical coordinate system with the water level at $y = -1$, then $F = \int_{-3}^{-1} \rho g(-1 - y)2\sqrt{9 - y^2}\,dy$.

29. Set up a vertical x-axis as shown. Then the area of the ith rectangular strip is

$$\left(2 - \frac{2}{\sqrt{3}} x_i^*\right) \Delta x. \quad \left[\text{By similar triangles, } \frac{w_i}{2} = \frac{\sqrt{3} - x_i^*}{\sqrt{3}}, \text{ so } w_i = 2 - \frac{2}{\sqrt{3}} x_i^*.\right]$$

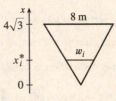

The pressure on the strip is $\rho g x_i^*$, so the hydrostatic force on the strip is

$\rho g x_i^* \left(2 - \frac{2}{\sqrt{3}} x_i^*\right) \Delta x$ and the hydrostatic force on the plate $\approx \sum\limits_{i=1}^{n} \rho g x_i^* \left(2 - \frac{2}{\sqrt{3}} x_i^*\right) \Delta x$.

The total force

$$F = \lim_{n \to \infty} \sum_{i=1}^{n} \rho g x_i^* \left(2 - \frac{2}{\sqrt{3}} x_i^*\right) \Delta x = \int_0^{\sqrt{3}} \rho g x \left(2 - \frac{2}{\sqrt{3}} x\right) dx = \rho g \int_0^{\sqrt{3}} \left(2x - \frac{2}{\sqrt{3}} x^2\right) dx$$

$$= \rho g \left[x^2 - \frac{2}{3\sqrt{3}} x^3\right]_0^{\sqrt{3}} = \rho g \left[(3 - 2) - 0\right] = \rho g \approx 1000 \cdot 9.8 = 9.8 \times 10^3 \text{ N}$$

31. By similar triangles, $\dfrac{8}{4\sqrt{3}} = \dfrac{w_i}{x_i^*} \;\Rightarrow\; w_i = \dfrac{2x_i^*}{\sqrt{3}}$. The area of the ith

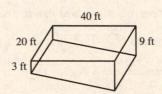

rectangular strip is $\dfrac{2x_i^*}{\sqrt{3}} \Delta x$ and the pressure on it is $\rho g \left(4\sqrt{3} - x_i^*\right)$.

$$F = \int_0^{4\sqrt{3}} \rho g \left(4\sqrt{3} - x\right) \frac{2x}{\sqrt{3}} dx = 8\rho g \int_0^{4\sqrt{3}} x\, dx - \frac{2\rho g}{\sqrt{3}} \int_0^{4\sqrt{3}} x^2\, dx$$

$$= 4\rho g \left[x^2\right]_0^{4\sqrt{3}} - \frac{2\rho g}{3\sqrt{3}} \left[x^3\right]_0^{4\sqrt{3}} = 192\rho g - \frac{2\rho g}{3\sqrt{3}} 64 \cdot 3\sqrt{3} = 192\rho g - 128\rho g = 64\rho g$$

$$\approx 64(840)(9.8) \approx 5.27 \times 10^5 \text{ N}$$

33. (a) The area of a strip is $20 \Delta x$ and the pressure on it is δx_i.

$$F = \int_0^3 \delta x 20\, dx = 20\delta \left[\tfrac{1}{2} x^2\right]_0^3 = 20\delta \cdot \tfrac{9}{2} = 90\delta$$

$$= 90(62.5) = 5625 \text{ lb} \approx 5.63 \times 10^3 \text{ lb}$$

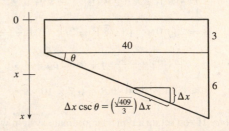

(b) $F = \int_0^9 \delta x 20\, dx = 20\delta \left[\tfrac{1}{2} x^2\right]_0^9 = 20\delta \cdot \tfrac{81}{2} = 810\delta = 810(62.5) = 50{,}625 \text{ lb} \approx 5.06 \times 10^4 \text{ lb}.$

(c) For the first 3 ft, the length of the side is constant at 40 ft. For $3 < x \le 9$, we can use similar triangles to find the length a:

$\dfrac{a}{40} = \dfrac{9 - x}{6} \;\Rightarrow\; a = 40 \cdot \dfrac{9 - x}{6}.$

$$F = \int_0^3 \delta x 40\, dx + \int_3^9 \delta x (40) \frac{9 - x}{6} dx = 40\delta \left[\tfrac{1}{2} x^2\right]_0^3 + \tfrac{20}{3}\delta \int_3^9 (9x - x^2)\, dx = 180\delta + \tfrac{20}{3}\delta \left[\tfrac{9}{2} x^2 - \tfrac{1}{3} x^3\right]_3^9$$

$$= 180\delta + \tfrac{20}{3}\delta \left[\left(\tfrac{729}{2} - 243\right) - \left(\tfrac{81}{2} - 9\right)\right] = 180\delta + 600\delta = 780\delta = 780(62.5) = 48{,}750 \text{ lb} \approx 4.88 \times 10^4 \text{ lb}$$

(d) For any right triangle with hypotenuse on the bottom,

$\sin \theta = \dfrac{\Delta x}{\text{hypotenuse}} \;\Rightarrow\;$

hypotenuse $= \Delta x \csc \theta = \Delta x \dfrac{\sqrt{40^2 + 6^2}}{6} = \dfrac{\sqrt{409}}{3} \Delta x.$

$$F = \int_3^9 \delta x 20 \frac{\sqrt{409}}{3} dx = \tfrac{1}{3}\left(20\sqrt{409}\right)\delta \left[\tfrac{1}{2} x^2\right]_3^9$$

$$= \tfrac{1}{3} \cdot 10\sqrt{409}\,\delta (81 - 9) \approx 303{,}356 \text{ lb} \approx 3.03 \times 10^5 \text{ lb}$$

35. $F = \int_{7.0}^{9.4} 64 x w(x)\,dx$, where $w(x)$ is the width of the plate at depth x. From the table, we see that $\Delta x = 0.4$, so using

Simpson's Rule to estimate F, we get

$$F \approx 64\,\tfrac{0.4}{3}\left[7.0 w(7.0) + 4(7.4) w(7.4) + 2(7.8) w(7.8) + 4(8.2) w(8.2) + 2(8.6) w(8.6) + 4(9.0) w(9.0) + 9.4 w(9.4)\right]$$

$$= \tfrac{25.6}{3}\left[7(1.2) + 29.6(1.8) + 15.6(2.9) + 32.8(3.8) + 17.2(3.6) + 36(4.2) + 9.4(4.4)\right]$$

$$= \tfrac{25.6}{3}(486.04) \approx 4148 \text{ lb}$$

37. The mass is $m = \sum_{i=1}^{3} m_i = 4 + 2 + 4 = 10$. The moment about the x-axis is $M_x = \sum_{i=1}^{3} m_i y_i = 4(-3) + 2(1) + 4(5) = 10$.

The moment about the y-axis is $M_y = \sum_{i=1}^{3} m_i x_i = 4(2) + 2(-3) + 4(3) = 14$. The center of mass is

$$(\overline{x}, \overline{y}) = \left(\frac{M_y}{m}, \frac{M_x}{m}\right) = \left(\frac{14}{10}, \frac{10}{10}\right) = (1.4, 1).$$

39. The region in the figure is "right-heavy" and "bottom-heavy," so we know that

$\overline{x} > 0.5$ and $\overline{y} < 1$, and we might guess that $\overline{x} = 0.7$ and $\overline{y} = 0.7$.

$A = \int_0^1 2x\,dx = \left[x^2\right]_0^1 = 1 - 0 = 1.$

$\overline{x} = \frac{1}{A} \int_0^1 x(2x)\,dx = \frac{1}{1}\left[\frac{2}{3}x^3\right]_0^1 = \frac{2}{3}.$

$\overline{y} = \frac{1}{A} \int_0^1 \frac{1}{2}(2x)^2\,dx = \frac{1}{1}\int_0^1 2x^2\,dx = \left[\frac{2}{3}x^3\right]_0^1 = \frac{2}{3}.$

Thus, the centroid is $(\overline{x}, \overline{y}) = \left(\frac{2}{3}, \frac{2}{3}\right).$

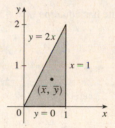

41. The region in the figure is "right-heavy" and "bottom-heavy," so we know

$\overline{x} > 0.5$ and $\overline{y} < 1$, and we might guess that $\overline{x} = 0.6$ and $\overline{y} = 0.9$.

$A = \int_0^1 e^x\,dx = \left[e^x\right]_0^1 = e - 1.$

$\overline{x} = \frac{1}{A}\int_0^1 x e^x\,dx = \frac{1}{e-1}\left[x e^x - e^x\right]_0^1$ [by parts]

$\qquad = \frac{1}{e-1}\left[0 - (-1)\right] = \frac{1}{e-1}.$

$\overline{y} = \frac{1}{A}\int_0^1 \frac{1}{2}(e^x)^2\,dx = \frac{1}{e-1} \cdot \frac{1}{4}\left[e^{2x}\right]_0^1 = \frac{1}{4(e-1)}(e^2 - 1) = \frac{e+1}{4}.$

Thus, the centroid is $(\overline{x}, \overline{y}) = \left(\frac{1}{e-1}, \frac{e+1}{4}\right) \approx (0.58, 0.93).$

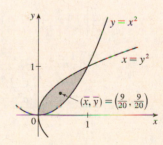

43. $A = \int_0^1 (x^{1/2} - x^2)\,dx = \left[\frac{2}{3}x^{3/2} - \frac{1}{3}x^3\right]_0^1 = \left(\frac{2}{3} - \frac{1}{3}\right) - 0 = \frac{1}{3}.$

$\overline{x} = \frac{1}{A}\int_0^1 x(x^{1/2} - x^2)\,dx = 3\int_0^1 (x^{3/2} - x^3)\,dx$

$\qquad = 3\left[\frac{2}{5}x^{5/2} - \frac{1}{4}x^4\right]_0^1 = 3\left(\frac{2}{5} - \frac{1}{4}\right) = 3\left(\frac{3}{20}\right) = \frac{9}{20}.$

$\overline{y} = \frac{1}{A}\int_0^1 \frac{1}{2}\left[(x^{1/2})^2 - (x^2)^2\right]\,dx = 3\left(\frac{1}{2}\right)\int_0^1 (x - x^4)\,dx$

$\qquad = \frac{3}{2}\left[\frac{1}{2}x^2 - \frac{1}{5}x^5\right]_0^1 = \frac{3}{2}\left(\frac{1}{2} - \frac{1}{5}\right) = \frac{3}{2}\left(\frac{3}{10}\right) = \frac{9}{20}.$

Thus, the centroid is $(\overline{x}, \overline{y}) = \left(\frac{9}{20}, \frac{9}{20}\right).$

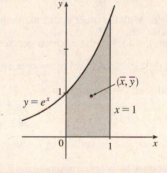

45. $A = \int_0^{\pi/4} (\cos x - \sin x)\, dx = \big[\sin x + \cos x\big]_0^{\pi/4} = \sqrt{2} - 1.$

$\overline{x} = A^{-1} \int_0^{\pi/4} x(\cos x - \sin x)\, dx$

$\quad = A^{-1} \big[x(\sin x + \cos x) + \cos x - \sin x\big]_0^{\pi/4}$　　[integration by parts]

$\quad = A^{-1}\left(\frac{\pi}{4}\sqrt{2} - 1\right) = \dfrac{\frac{1}{4}\pi\sqrt{2} - 1}{\sqrt{2} - 1}.$

$\overline{y} = A^{-1} \int_0^{\pi/4} \frac{1}{2}(\cos^2 x - \sin^2 x)\, dx = \frac{1}{2A} \int_0^{\pi/4} \cos 2x\, dx = \frac{1}{4A}\big[\sin 2x\big]_0^{\pi/4} = \dfrac{1}{4A} = \dfrac{1}{4\left(\sqrt{2} - 1\right)}.$

Thus, the centroid is $(\overline{x}, \overline{y}) = \left(\dfrac{\pi\sqrt{2} - 4}{4\left(\sqrt{2} - 1\right)}, \dfrac{1}{4\left(\sqrt{2} - 1\right)}\right) \approx (0.27, 0.60).$

[Figure: region bounded by $y = \sin x$ and $y = \cos x$ with $(\overline{x}, \overline{y}) \approx (0.27, 0.60)$]

47. The line has equation $y = \frac{3}{4}x.$　$A = \frac{1}{2}(4)(3) = 6,$ so $m = \rho A = 10(6) = 60.$

$$M_x = \rho \int_0^4 \frac{1}{2}\left(\frac{3}{4}x\right)^2 dx = 10 \int_0^4 \frac{9}{32}x^2\, dx = \frac{45}{16}\left[\frac{1}{3}x^3\right]_0^4 = \frac{45}{16}\left(\frac{64}{3}\right) = 60$$

$$M_y = \rho \int_0^4 x\left(\frac{3}{4}x\right) dx = \frac{15}{2} \int_0^4 x^2\, dx = \frac{15}{2}\left[\frac{1}{3}x^3\right]_0^4 = \frac{15}{2}\left(\frac{64}{3}\right) = 160$$

$\overline{x} = \dfrac{M_y}{m} = \dfrac{160}{60} = \dfrac{8}{3}$ and $\overline{y} = \dfrac{M_x}{m} = \dfrac{60}{60} = 1.$ Thus, the centroid is $(\overline{x}, \overline{y}) = \left(\frac{8}{3}, 1\right).$

49. Choose x- and y-axes so that the base (one side of the triangle) lies along the x-axis with the other vertex along the positive y-axis as shown. From geometry, we know the medians intersect at a point $\frac{2}{3}$ of the way from each vertex (along the median) to the opposite side. The median from B goes to the midpoint $\left(\frac{1}{2}(a + c), 0\right)$ of side AC, so the point of intersection of the medians is $\left(\frac{2}{3} \cdot \frac{1}{2}(a + c), \frac{1}{3}b\right) = \left(\frac{1}{3}(a + c), \frac{1}{3}b\right).$

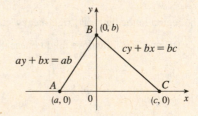

This can also be verified by finding the equations of two medians, and solving them simultaneously to find their point of intersection. Now let us compute the location of the centroid of the triangle. The area is $A = \frac{1}{2}(c - a)b.$

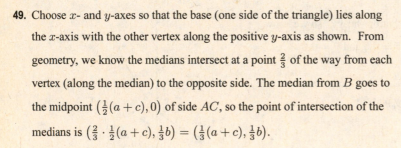

$$\overline{x} = \frac{1}{A}\left[\int_a^0 x \cdot \frac{b}{a}(a - x)\, dx + \int_0^c x \cdot \frac{b}{c}(c - x)\, dx\right] = \frac{1}{A}\left[\frac{b}{a}\int_a^0 (ax - x^2)\, dx + \frac{b}{c}\int_0^c (cx - x^2)\, dx\right]$$

$$= \frac{b}{Aa}\left[\frac{1}{2}ax^2 - \frac{1}{3}x^3\right]_a^0 + \frac{b}{Ac}\left[\frac{1}{2}cx^2 - \frac{1}{3}x^3\right]_0^c = \frac{b}{Aa}\left[-\frac{1}{2}a^3 + \frac{1}{3}a^3\right] + \frac{b}{Ac}\left[\frac{1}{2}c^3 - \frac{1}{3}c^3\right]$$

$$= \frac{2}{a(c - a)} \cdot \frac{-a^3}{6} + \frac{2}{c(c - a)} \cdot \frac{c^3}{6} = \frac{1}{3(c - a)}(c^2 - a^2) = \frac{a + c}{3}$$

[continued]

and $\bar{y} = \dfrac{1}{A}\left[\displaystyle\int_a^0 \dfrac{1}{2}\left(\dfrac{b}{a}(a-x)\right)^2 dx + \int_0^c \dfrac{1}{2}\left(\dfrac{b}{c}(c-x)\right)^2 dx\right]$

$\qquad = \dfrac{1}{A}\left[\dfrac{b^2}{2a^2}\displaystyle\int_a^0 (a^2 - 2ax + x^2)\,dx + \dfrac{b^2}{2c^2}\int_0^c (c^2 - 2cx + x^2)\,dx\right]$

$\qquad = \dfrac{1}{A}\left[\dfrac{b^2}{2a^2}\left[a^2 x - ax^2 + \tfrac{1}{3}x^3\right]_a^0 + \dfrac{b^2}{2c^2}\left[c^2 x - cx^2 + \tfrac{1}{3}x^3\right]_0^c\right]$

$\qquad = \dfrac{1}{A}\left[\dfrac{b^2}{2a^2}(-a^3 + a^3 - \tfrac{1}{3}a^3) + \dfrac{b^2}{2c^2}(c^3 - c^3 + \tfrac{1}{3}c^3)\right] = \dfrac{1}{A}\left[\dfrac{b^2}{6}(-a + c)\right] = \dfrac{2}{(c-a)\,b}\cdot\dfrac{(c-a)b^2}{6} = \dfrac{b}{3}$

Thus, the centroid is $(\bar{x}, \bar{y}) = \left(\dfrac{a+c}{3}, \dfrac{b}{3}\right)$, as claimed.

Remarks: Actually the computation of $\bar{y}$ is all that is needed. By considering each side of the triangle in turn to be the base, we see that the centroid is $\frac{1}{3}$ of the way from each side to the opposite vertex and must therefore be the intersection of the medians.

The computation of $\bar{y}$ in this problem (and many others) can be simplified by using horizontal rather than vertical approximating rectangles. If the length of a thin rectangle at coordinate y is $\ell(y)$, then its area is $\ell(y)\,\Delta y$, its mass is $\rho\ell(y)\,\Delta y$, and its moment about the x-axis is $\Delta M_x = \rho y \ell(y)\,\Delta y$. Thus,

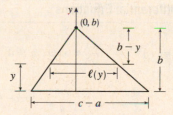

$$M_x = \int \rho y \ell(y)\,dy \qquad \text{and} \qquad \bar{y} = \dfrac{\int \rho y \ell(y)\,dy}{\rho A} = \dfrac{1}{A}\int y\ell(y)\,dy$$

In this problem, $\ell(y) = \dfrac{c-a}{b}(b-y)$ by similar triangles, so

$$\bar{y} = \dfrac{1}{A}\int_0^b \dfrac{c-a}{b}\,y(b-y)\,dy = \dfrac{2}{b^2}\int_0^b (by - y^2)\,dy = \dfrac{2}{b^2}\left[\tfrac{1}{2}by^2 - \tfrac{1}{3}y^3\right]_0^b = \dfrac{2}{b^2}\cdot\dfrac{b^3}{6} = \dfrac{b}{3}$$

Notice that only one integral is needed when this method is used.

51. Divide the lamina into two triangles and one rectangle with respective masses of 2, 2 and 4, so that the total mass is 8. Using the result of Exercise 49, the triangles have centroids $\left(-1, \tfrac{2}{3}\right)$ and $\left(1, \tfrac{2}{3}\right)$. The centroid of the rectangle (its center) is $\left(0, -\tfrac{1}{2}\right)$.

So, using Formulas 9 and 11, we have $\bar{y} = \dfrac{M_x}{m} = \dfrac{1}{m}\displaystyle\sum_{i=1}^{3} m_i\,y_i = \tfrac{1}{8}\left[2\left(\tfrac{2}{3}\right) + 2\left(\tfrac{2}{3}\right) + 4\left(-\tfrac{1}{2}\right)\right] = \tfrac{1}{8}\left(\tfrac{2}{3}\right) = \tfrac{1}{12}$, and $\bar{x} = 0$,

since the lamina is symmetric about the line $x = 0$. Thus, the centroid is $(\bar{x}, \bar{y}) = \left(0, \tfrac{1}{12}\right)$.

53. A cone of height h and radius r can be generated by rotating a right triangle about one of its legs as shown. By Exercise 49, $\bar{x} = \tfrac{1}{3}r$, so by the Theorem of Pappus, the volume of the cone is

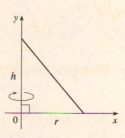

$$V = Ad = \left(\tfrac{1}{2}\cdot \text{base}\cdot\text{height}\right)\cdot(2\pi\bar{x}) = \tfrac{1}{2}rh\cdot 2\pi\left(\tfrac{1}{3}r\right) = \tfrac{1}{3}\pi r^2 h.$$

55. Suppose the region lies between two curves $y = f(x)$ and $y = g(x)$ where $f(x) \geq g(x)$, as illustrated in Figure 13.

Choose points x_i with $a = x_0 < x_1 < \cdots < x_n = b$ and choose x_i^* to be the midpoint of the ith subinterval; that is,

$x_i^* = \overline{x}_i = \frac{1}{2}(x_{i-1} + x_i)$. Then the centroid of the ith approximating rectangle R_i is its center $C_i = \left(\overline{x}_i, \frac{1}{2}[f(\overline{x}_i) + g(\overline{x}_i)]\right)$.

Its area is $[f(\overline{x}_i) - g(\overline{x}_i)]\,\Delta x$, so its mass is

$\rho[f(\overline{x}_i) - g(\overline{x}_i)]\,\Delta x$. Thus, $M_y(R_i) = \rho[f(\overline{x}_i) - g(\overline{x}_i)]\,\Delta x \cdot \overline{x}_i = \rho\overline{x}_i\,[f(\overline{x}_i) - g(\overline{x}_i)]\,\Delta x$ and

$M_x(R_i) = \rho[f(\overline{x}_i) - g(\overline{x}_i)]\,\Delta x \cdot \frac{1}{2}[f(\overline{x}_i) + g(\overline{x}_i)] = \rho \cdot \frac{1}{2}\left[f(\overline{x}_i)^2 - g(\overline{x}_i)^2\right]\Delta x$. Summing over i and taking the limit

as $n \to \infty$, we get $M_y = \lim\limits_{n\to\infty} \sum_i \rho\overline{x}_i\,[f(\overline{x}_i) - g(\overline{x}_i)]\,\Delta x = \rho \int_a^b x[f(x) - g(x)]\,dx$ and

$M_x = \lim\limits_{n\to\infty} \sum_i \rho \cdot \frac{1}{2}\left[f(\overline{x}_i)^2 - g(\overline{x}_i)^2\right]\Delta x = \rho \int_a^b \frac{1}{2}\left[f(x)^2 - g(x)^2\right]dx$.

Thus, $\overline{x} = \dfrac{M_y}{m} = \dfrac{M_y}{\rho A} = \dfrac{1}{A}\displaystyle\int_a^b x[f(x) - g(x)]\,dx$ and $\overline{y} = \dfrac{M_x}{m} = \dfrac{M_x}{\rho A} = \dfrac{1}{A}\displaystyle\int_a^b \frac{1}{2}\left[f(x)^2 - g(x)^2\right]dx$.

7.7 Differential Equations

1. $\dfrac{dy}{dx} = xy^2 \;\Rightarrow\; \dfrac{dy}{y^2} = x\,dx \;[y \neq 0] \;\Rightarrow\; \displaystyle\int y^{-2}\,dy = \int x\,dx \;\Rightarrow\; -y^{-1} = \frac{1}{2}x^2 + C \;\Rightarrow$

$\dfrac{1}{y} = -\frac{1}{2}x^2 - C \;\Rightarrow\; y = \dfrac{1}{-\frac{1}{2}x^2 - C} = \dfrac{2}{K - x^2}$, where $K = -2C$. $y = 0$ is also a solution.

3. $xy^2y' = x + 1 \;\Rightarrow\; y^2\dfrac{dy}{dx} = \dfrac{x + 1}{x} \;\Rightarrow\; y^2\,dy = \left(1 + \dfrac{1}{x}\right)dx \;\Rightarrow\; \displaystyle\int y^2\,dy = \int\left(1 + \dfrac{1}{x}\right)dx \;\Rightarrow$

$\frac{1}{3}y^3 = x + \ln|x| + C \;\Rightarrow\; y^3 = 3x + 3\ln|x| + 3C \;\Rightarrow\; y = \sqrt[3]{3x + 3\ln|x| + K}$, where $K = 3C$.

5. $(y + \sin y)\,y' = x + x^3 \;\Rightarrow\; (y + \sin y)\dfrac{dy}{dx} = x + x^3 \;\Rightarrow\; \displaystyle\int(y + \sin y)\,dy = \int(x + x^3)\,dx \;\Rightarrow$

$\frac{1}{2}y^2 - \cos y = \frac{1}{2}x^2 + \frac{1}{4}x^4 + C$. We cannot solve explicitly for y.

7. $\dfrac{dp}{dt} = t^2p - p + t^2 - 1 = p(t^2 - 1) + 1(t^2 - 1) = (p + 1)(t^2 - 1) \;\Rightarrow\; \dfrac{1}{p + 1}\,dp = (t^2 - 1)\,dt \;\Rightarrow$

$\displaystyle\int \dfrac{1}{p + 1}\,dp = \int(t^2 - 1)\,dt \;\Rightarrow\; \ln|p + 1| = \frac{1}{3}t^3 - t + C \;\Rightarrow\; |p + 1| = e^{t^3/3 - t + C} \;\Rightarrow\; p + 1 = \pm e^C e^{t^3/3 - t} \;\Rightarrow$

$p = Ke^{t^3/3 - t} - 1$, where $K = \pm e^C$. Since $p = -1$ is also a solution, K can equal 0, and hence, K can be any real number.

9. $\dfrac{dy}{dx} = \dfrac{x}{y} \;\Rightarrow\; y\,dy = x\,dx \;\Rightarrow\; \displaystyle\int y\,dy = \int x\,dx \;\Rightarrow\; \frac{1}{2}y^2 = \frac{1}{2}x^2 + C$. $\;y(0) = -3 \;\Rightarrow$

$\frac{1}{2}(-3)^2 = \frac{1}{2}(0)^2 + C \;\Rightarrow\; C = \frac{9}{2}$, so $\frac{1}{2}y^2 = \frac{1}{2}x^2 + \frac{9}{2} \;\Rightarrow\; y^2 = x^2 + 9 \;\Rightarrow\; y = -\sqrt{x^2 + 9}$ since $y(0) = -3 < 0$.

11. $\dfrac{du}{dt} = \dfrac{2t + \sec^2 t}{2u}$, $u(0) = -5$. $\;\displaystyle\int 2u\,du = \int(2t + \sec^2 t)\,dt \;\Rightarrow\; u^2 = t^2 + \tan t + C$,

where $[u(0)]^2 = 0^2 + \tan 0 + C \;\Rightarrow\; C = (-5)^2 = 25$. Therefore, $u^2 = t^2 + \tan t + 25$, so $u = \pm\sqrt{t^2 + \tan t + 25}$.

Since $u(0) = -5$, we must have $u = -\sqrt{t^2 + \tan t + 25}$.

13. $y' \tan x = a + y$, $0 < x < \pi/2$ $\Rightarrow$ $\dfrac{dy}{dx} = \dfrac{a+y}{\tan x}$ $\Rightarrow$ $\dfrac{dy}{a+y} = \cot x \, dx$ $[a + y \neq 0]$ $\Rightarrow$

$\displaystyle\int \dfrac{dy}{a+y} = \int \dfrac{\cos x}{\sin x} \, dx$ $\Rightarrow$ $\ln|a+y| = \ln|\sin x| + C$ $\Rightarrow$ $|a+y| = e^{\ln|\sin x|+C} = e^{\ln|\sin x|} \cdot e^{C} = e^{C} |\sin x|$ $\Rightarrow$

$a + y = K \sin x$, where $K = \pm e^{C}$. (In our derivation, K was nonzero, but we can restore the excluded case

$y = -a$ by allowing K to be zero.) $y(\pi/3) = a$ $\Rightarrow$ $a + a = K \sin\left(\dfrac{\pi}{3}\right)$ $\Rightarrow$ $2a = K \dfrac{\sqrt{3}}{2}$ $\Rightarrow$ $K = \dfrac{4a}{\sqrt{3}}$.

Thus, $a + y = \dfrac{4a}{\sqrt{3}} \sin x$ and so $y = \dfrac{4a}{\sqrt{3}} \sin x - a$.

15. If the slope at the point (x, y) is xy, then we have $\dfrac{dy}{dx} = xy$ $\Rightarrow$ $\dfrac{dy}{y} = x \, dx$ $[y \neq 0]$ $\Rightarrow$ $\displaystyle\int \dfrac{dy}{y} = \int x \, dx$ $\Rightarrow$

$\ln|y| = \tfrac{1}{2}x^2 + C$. $y(0) = 1$ $\Rightarrow$ $\ln 1 = 0 + C$ $\Rightarrow$ $C = 0$. Thus, $|y| = e^{x^2/2}$ $\Rightarrow$ $y = \pm e^{x^2/2}$, so $y = e^{x^2/2}$

since $y(0) = 1 > 0$. Note that $y = 0$ is not a solution because it doesn't satisfy the initial condition $y(0) = 1$.

17. (a) $y' = 2x \sqrt{1 - y^2}$ $\Rightarrow$ $\dfrac{dy}{dx} = 2x \sqrt{1 - y^2}$ $\Rightarrow$ $\dfrac{dy}{\sqrt{1 - y^2}} = 2x \, dx$ $\Rightarrow$ $\displaystyle\int \dfrac{dy}{\sqrt{1 - y^2}} = \int 2x \, dx$ $\Rightarrow$

$\sin^{-1} y = x^2 + C$ for $-\dfrac{\pi}{2} \leq x^2 + C \leq \dfrac{\pi}{2}$.

(b) $y(0) = 0$ $\Rightarrow$ $\sin^{-1} 0 = 0^2 + C$ $\Rightarrow$ $C = 0$,

so $\sin^{-1} y = x^2$ and $y = \sin(x^2)$ for $-\sqrt{\pi/2} \leq x \leq \sqrt{\pi/2}$.

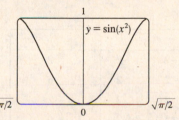

(c) For $\sqrt{1 - y^2}$ to be a real number, we must have $-1 \leq y \leq 1$; that is, $-1 \leq y(0) \leq 1$. Thus, the initial-value problem

$y' = 2x \sqrt{1 - y^2}$, $y(0) = 2$ does *not* have a solution.

19. $\dfrac{dy}{dx} = \dfrac{\sin x}{\sin y}$, $y(0) = \dfrac{\pi}{2}$.   So $\int \sin y \, dy = \int \sin x \, dx$ $\Leftrightarrow$

$-\cos y = -\cos x + C$ $\Leftrightarrow$ $\cos y = \cos x - C$. From the initial condition,

we need $\cos \dfrac{\pi}{2} = \cos 0 - C$ $\Rightarrow$ $0 = 1 - C$ $\Rightarrow$ $C = 1$, so the solution is

$\cos y = \cos x - 1$. Note that we cannot take $\cos^{-1}$ of both sides, since that would

unnecessarily restrict the solution to the case where $-1 \leq \cos x - 1$ $\Leftrightarrow$ $0 \leq \cos x$,

as $\cos^{-1}$ is defined only on $[-1, 1]$. Instead we plot the graph using Maple's

`plots[implicitplot]` or Mathematica's `Plot[Evaluate[···]]`.

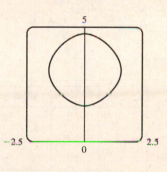

21. $y' = 2 - y$. The slopes at each point are independent of x, so the slopes are the same along each line parallel to the x-axis.

Thus, III is the direction field for this equation. Note that for $y = 2$, $y' = 0$.

23. $y' = x + y - 1 = 0$ on the line $y = -x + 1$. Direction field IV satisfies this condition. Notice also that on the line $y = -x$ we

have $y' = -1$, which is true in IV.

25. (a) $y(0) = 1$

(b) $y(0) = 2$

(c) $y(0) = -1$

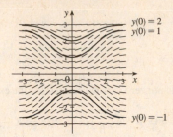

27.

x	y	$y' = \frac{1}{2}y$
0	0	0
0	1	0.5
0	2	1
0	-3	-1.5
0	-2	-1

Note that for $y = 0$, $y' = 0$. The three solution curves sketched go through $(0, 0)$, $(0, 1)$, and $(0, -1)$.

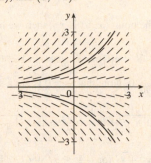

29.

x	y	$y' = y - 2x$
-2	-2	2
-2	2	6
2	2	-2
2	-2	-6

Note that $y' = 0$ for any point on the line $y = 2x$. The slopes are positive to the left of the line and negative to the right of the line. The solution curve in the graph passes through $(1, 0)$.

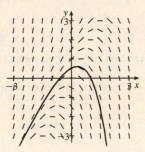

31.

x	y	$y' = y + xy$
0	± 2	± 2
1	± 2	± 4
-3	± 2	∓ 4

Note that $y' = y(x + 1) = 0$ for any point on $y = 0$ or on $x = -1$. The slopes are positive when the factors y and $x + 1$ have the same sign and negative when they have opposite signs. The solution curve in the graph passes through $(0, 1)$.

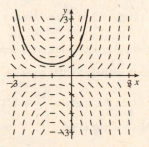

33. (a) $\dfrac{dP}{dt} = k(M - P)$ is always positive, so the level of performance P is increasing. As P gets close to M, dP/dt gets close to 0; that is, the performance levels off.

(b) $\dfrac{dP}{dt} = k(M - P) \;\Leftrightarrow\; \displaystyle\int \dfrac{dP}{P - M} = \int (-k)\,dt \;\Leftrightarrow\; \ln|P - M| = -kt + C \;\Leftrightarrow\; |P - M| = e^{-kt+C} \;\Leftrightarrow\;$

$P - M = Ae^{-kt} \;\left[A = \pm e^C\right] \;\Leftrightarrow\; P = M + Ae^{-kt}$. If we assume that performance is at level 0 when $t = 0$, then

$P(0) = 0 \;\Leftrightarrow\; 0 = M + A \;\Leftrightarrow\; A = -M \;\Leftrightarrow\; P(t) = M - Me^{-kt}.$ $\displaystyle\lim_{t \to \infty} P(t) = M - M \cdot 0 = M.$

35. (a) $\dfrac{dC}{dt} = r - kC \;\Rightarrow\; \dfrac{dC}{dt} = -(kC - r) \;\Rightarrow\; \displaystyle\int \dfrac{dC}{kC - r} = \int -dt \;\Rightarrow\; (1/k)\ln|kC - r| = -t + M_1 \;\Rightarrow\;$

$\ln|kC - r| = -kt + M_2 \;\Rightarrow\; |kC - r| = e^{-kt+M_2} \;\Rightarrow\; kC - r = M_3 e^{-kt} \;\Rightarrow\; kC = M_3 e^{-kt} + r \;\Rightarrow\;$

$C(t) = M_4 e^{-kt} + r/k.$ $C(0) = C_0 \;\Rightarrow\; C_0 = M_4 + r/k \;\Rightarrow\; M_4 = C_0 - r/k \;\Rightarrow\;$

$C(t) = (C_0 - r/k)e^{-kt} + r/k.$

(b) If $C_0 < r/k$, then $C_0 - r/k < 0$ and the formula for $C(t)$ shows that $C(t)$ increases and $\displaystyle\lim_{t \to \infty} C(t) = r/k.$

As t increases, the formula for $C(t)$ shows how the role of C_0 steadily diminishes as that of r/k increases.

37. The differential equation is a logistic equation with $k = 0.00008$, carrying capacity $M = 1000$, and initial population $y_0 = P(0) = 100$. So Equation 8 gives the population at time t as

$$P(t) = \frac{100 \cdot 1000}{100 + (1000 - 100)e^{-0.08t}} = \frac{100{,}000}{100 + 900e^{-0.08t}} = \frac{1000}{1 + 9e^{-0.08t}}$$

So the population sizes when $t = 40$ and 80 are

$$P(40) = \frac{1000}{1 + 9e^{-3.2}} \approx 731.6 \qquad P(80) = \frac{1000}{1 + 9e^{-6.4}} \approx 985.3$$

The population reaches 900 when $\dfrac{1000}{1 + 9e^{-0.08t}} = 900.$ Solving this equation for t, we get $1 + 9e^{-0.08t} = \dfrac{10}{9} \;\Rightarrow\;$

$e^{-0.08t} = \dfrac{1}{81} \;\Rightarrow\; -0.08t = \ln\dfrac{1}{81} = -\ln 81 \;\Rightarrow\; t = \dfrac{\ln 81}{0.08} \approx 54.9.$ So the population reaches 900 when t is approximately 55.

39. (a) Our assumption is that $\dfrac{dy}{dt} = ky(1 - y)$, where y is the fraction of the population that has heard the rumor.

(b) The equation in part (a) is the logistic differential equation (7) with $M = 1$, so the solution is given by (8):

$$y = \frac{y_0}{y_0 + (1 - y_0)e^{-kt}}.$$

(c) Let t be the number of hours since 8 AM. Then $y_0 = y(0) = \dfrac{80}{1000} = 0.08$ and $y(4) = \dfrac{1}{2}$, so

$\dfrac{1}{2} = y(4) = \dfrac{0.08}{0.08 + 0.92e^{-4k}}.$ Thus, $0.08 + 0.92e^{-4k} = 0.16$, $e^{-4k} = \dfrac{0.08}{0.92} = \dfrac{2}{23}$, and $e^{-k} = \left(\dfrac{2}{23}\right)^{1/4}$,

so $y = \dfrac{0.08}{0.08 + 0.92(2/23)^{t/4}} = \dfrac{2}{2 + 23(2/23)^{t/4}}.$ Solving this equation for t, we get

$2y + 23y\left(\dfrac{2}{23}\right)^{t/4} = 2 \;\Rightarrow\; \left(\dfrac{2}{23}\right)^{t/4} = \dfrac{2 - 2y}{23y} \;\Rightarrow\; \left(\dfrac{2}{23}\right)^{t/4} = \dfrac{2}{23} \cdot \dfrac{1 - y}{y} \;\Rightarrow\; \left(\dfrac{2}{23}\right)^{t/4-1} = \dfrac{1 - y}{y}.$

[continued]

It follows that $\dfrac{t}{4} - 1 = \dfrac{\ln[(1-y)/y]}{\ln\frac{2}{23}}$, so $t = 4\left[1 + \dfrac{\ln((1-y)/y)}{\ln\frac{2}{23}}\right]$.

When $y = 0.9$, $\dfrac{1-y}{y} = \dfrac{1}{9}$, so $t = 4\left(1 - \dfrac{\ln 9}{\ln\frac{2}{23}}\right) \approx 7.6$ h or 7 h 36 min. Thus, 90% of the population will have heard

the rumor by 3:36 PM.

41. (a) $\dfrac{dy}{dt} = ky(M - y)$ $\Rightarrow$

$$\dfrac{d^2y}{dt^2} = ky\left(-\dfrac{dy}{dt}\right) + k(M-y)\dfrac{dy}{dt} = k\dfrac{dy}{dt}(M - 2y) = k[ky(M-y)](M-2y) = k^2y(M-y)(M-2y)$$

(b) y grows fastest when y' has a maximum, that is, when $y'' = 0$. From part (a), $y'' = 0$ $\Leftrightarrow$ $y = 0, y = M$, or $y = M/2$.

Since $0 < y < M$, we see that $y'' = 0$ $\Leftrightarrow$ $y = M/2$.

43. (a) Let $y(t)$ be the amount of salt (in kg) after t minutes. Then $y(0) = 15$. The amount of liquid in the tank is 1000 L at all

times, so the concentration at time t (in minutes) is $y(t)/1000$ kg/L and $\dfrac{dy}{dt} = -\left[\dfrac{y(t)}{1000}\dfrac{\text{kg}}{\text{L}}\right]\left(10\dfrac{\text{L}}{\text{min}}\right) = -\dfrac{y(t)}{100}\dfrac{\text{kg}}{\text{min}}$.

$\displaystyle\int\dfrac{dy}{y} = -\dfrac{1}{100}\int dt$ $\Rightarrow$ $\ln y = -\dfrac{t}{100} + C$, and $y(0) = 15$ $\Rightarrow$ $\ln 15 = C$, so $\ln y = \ln 15 - \dfrac{t}{100}$.

It follows that $\ln\left(\dfrac{y}{15}\right) = -\dfrac{t}{100}$ and $\dfrac{y}{15} = e^{-t/100}$, so $y = 15e^{-t/100}$ kg.

(b) After 20 minutes, $y = 15e^{-20/100} = 15e^{-0.2} \approx 12.3$ kg.

45. Let $y(t)$ be the amount of alcohol in the vat after t minutes. Then $y(0) = 0.04(500) = 20$ gal. The amount of beer in the vat

is 500 gallons at all times, so the percentage at time t (in minutes) is $y(t)/500 \times 100$, and the change in the amount of alcohol

with respect to time t is $\dfrac{dy}{dt} = \text{rate in} - \text{rate out} = 0.06\left(5\dfrac{\text{gal}}{\text{min}}\right) - \dfrac{y(t)}{500}\left(5\dfrac{\text{gal}}{\text{min}}\right) = 0.3 - \dfrac{y}{100} = \dfrac{30-y}{100}\dfrac{\text{gal}}{\text{min}}$.

Hence, $\displaystyle\int\dfrac{dy}{30-y} = \int\dfrac{dt}{100}$ and $-\ln|30-y| = \frac{1}{100}t + C$. Because $y(0) = 20$, we have $-\ln 10 = C$, so

$-\ln|30-y| = \frac{1}{100}t - \ln 10$ $\Rightarrow$ $\ln|30-y| = -t/100 + \ln 10$ $\Rightarrow$ $\ln|30-y| = \ln e^{-t/100} + \ln 10$ $\Rightarrow$

$\ln|30-y| = \ln(10e^{-t/100})$ $\Rightarrow$ $|30-y| = 10e^{-t/100}$. Since y is continuous, $y(0) = 20$, and the right-hand side is

never zero, we deduce that $30 - y$ is always positive. Thus, $30 - y = 10e^{-t/100}$ $\Rightarrow$ $y = 30 - 10e^{-t/100}$. The

percentage of alcohol is $p(t) = y(t)/500 \times 100 = y(t)/5 = 6 - 2e^{-t/100}$. The percentage of alcohol after one hour is

$p(60) = 6 - 2e^{-60/100} \approx 4.9$.

47. Assume that the raindrop begins at rest, so that $v(0) = 0$. $dm/dt = km$ and $(mv)' = gm$ $\Rightarrow$ $mv' + vm' = gm$ $\Rightarrow$

$mv' + v(km) = gm$ $\Rightarrow$ $v' + vk = g$ $\Rightarrow$ $\dfrac{dv}{dt} = g - kv$ $\Rightarrow$ $\displaystyle\int\dfrac{dv}{g-kv} = \int dt$ $\Rightarrow$

$-(1/k)\ln|g - kv| = t + C$ $\Rightarrow$ $\ln|g - kv| = -kt - kC$ $\Rightarrow$ $g - kv = Ae^{-kt}$. $v(0) = 0$ $\Rightarrow$ $A = g$.

So $kv = g - ge^{-kt}$ $\Rightarrow$ $v = (g/k)(1 - e^{-kt})$. Since $k > 0$, as $t \to \infty$, $e^{-kt} \to 0$ and therefore, $\displaystyle\lim_{t\to\infty}v(t) = g/k$.

49. (a) The rate of growth of the area is jointly proportional to $\sqrt{A(t)}$ and $M - A(t)$; that is, the rate is proportional to the product of those two quantities. So for some constant k, $dA/dt = k\sqrt{A}\,(M - A)$. We are interested in the maximum of the function dA/dt (when the tissue grows the fastest), so we differentiate, using the Chain Rule and then substituting for dA/dt from the differential equation:

$$\frac{d}{dt}\left(\frac{dA}{dt}\right) = k\left[\sqrt{A}\,(-1)\frac{dA}{dt} + (M - A)\cdot\tfrac{1}{2}A^{-1/2}\frac{dA}{dt}\right] = \tfrac{1}{2}kA^{-1/2}\frac{dA}{dt}\left[-2A + (M - A)\right]$$

$$= \tfrac{1}{2}kA^{-1/2}\left[k\sqrt{A}(M - A)\right][M - 3A] = \tfrac{1}{2}k^2(M - A)(M - 3A)$$

This is 0 when $M - A = 0$ [this situation never actually occurs, since the graph of $A(t)$ is asymptotic to the line $y = M$, as in the logistic model] and when $M - 3A = 0 \;\Leftrightarrow\; A(t) = M/3$. This represents a maximum by the First Derivative Test, since $\dfrac{d}{dt}\left(\dfrac{dA}{dt}\right)$ goes from positive to negative when $A(t) = M/3$.

(b) From the CAS, we get $A(t) = M\left(\dfrac{Ce^{\sqrt{M}kt} - 1}{Ce^{\sqrt{M}kt} + 1}\right)^2$. To get C in terms of the initial area A_0 and the maximum area M,

we substitute $t = 0$ and $A = A_0 = A(0)$: $A_0 = M\left(\dfrac{C - 1}{C + 1}\right)^2 \;\Leftrightarrow\; (C + 1)\sqrt{A_0} = (C - 1)\sqrt{M} \;\Leftrightarrow$

$C\sqrt{A_0} + \sqrt{A_0} = C\sqrt{M} - \sqrt{M} \;\Leftrightarrow\; \sqrt{M} + \sqrt{A_0} = C\sqrt{M} - C\sqrt{A_0} \;\Leftrightarrow$

$\sqrt{M} + \sqrt{A_0} = C\left(\sqrt{M} - \sqrt{A_0}\right) \;\Leftrightarrow\; C = \dfrac{\sqrt{M} + \sqrt{A_0}}{\sqrt{M} - \sqrt{A_0}}$. [Notice that if $A_0 = 0$, then $C = 1$.]

7 Review

CONCEPT CHECK

1. (a) See Section 7.1, Figure 2 and Equations 7.1.1 and 7.1.2.

(b) Instead of using "top minus bottom" and integrating from left to right, we use "right minus left" and integrate from bottom to top. See Figures 8 and 9 in Section 7.1.

2. The numerical value of the area represents the number of meters by which Sue is ahead of Kathy after 1 minute.

3. (a) See the discussion on pages 377–78.

(b) See the discussion between Examples 5 and 6 in Section 7.2. If the cross-section is a disk, find the radius in terms of x or y and use $A = \pi(\text{radius})^2$. If the cross-section is a washer, find the inner radius r_{in} and outer radius r_{out} and use $A = \pi\left(r_{\text{out}}^2\right) - \pi\left(r_{\text{in}}^2\right)$.

4. (a) $V = 2\pi rh\,\Delta r = (\text{circumference})(\text{height})(\text{thickness})$

(b) For a typical shell, find the circumference and height in terms of x or y and calculate $V = \int_a^b (\text{circumference})(\text{height})(dx \text{ or } dy)$, where a and b are the limits on x or y.

(c) Sometimes slicing produces washers or disks whose radii are difficult (or impossible) to find explicitly. On other occasions, the cylindrical shell method leads to an easier integral than slicing does.

5. (a) The length of a curve is defined to be the limit of the lengths of the inscribed polygons, as described near Figure 3 in Section 7.4.

   (b) See Equation 7.4.2.

   (c) See Equation 7.4.4.

6. (a) $S = \int_a^b 2\pi f(x) \sqrt{1 + [f'(x)]^2}\, dx$

   (b) If $x = g(y), c \le y \le d$, then $S = \int_c^d 2\pi y \sqrt{1 + [g'(y)]^2}\, dy$.

   (c) $S = \int_a^b 2\pi x \sqrt{1 + [f'(x)]^2}\, dx$ or $S = \int_c^d 2\pi g(y) \sqrt{1 + [g'(y)]^2}\, dy$

7. $\int_0^6 f(x)\, dx$ represents the amount of work done. Its units are newton-meters, or joules.

8. Let $c(x)$ be the cross-sectional length of the wall (measured parallel to the surface of the fluid) at depth x. Then the hydrostatic force against the wall is given by $F = \int_a^b \delta x c(x)\, dx$, where a and b are the lower and upper limits for x at points of the wall and δ is the weight density of the fluid.

9. (a) The center of mass is the point at which the plate balances horizontally.

   (b) See Equations 7.6.12.

10. If a plane region $\mathcal{R}$ that lies entirely on one side of a line ℓ in its plane is rotated about ℓ, then the volume of the resulting solid is the product of the area of $\mathcal{R}$ and the distance traveled by the centroid of $\mathcal{R}$.

11. (a) A differential equation is an equation that contains an unknown function and one or more of its derivatives.

    (b) The order of a differential equation is the order of the highest derivative that occurs in the equation.

    (c) An initial condition is a condition of the form $y(t_0) = y_0$.

12. See the paragraph preceding Example 6 in Section 7.7.

13. A separable equation is a first-order differential equation in which the expression for dy/dx can be factored as a function of x times a function of y, that is, $dy/dx = g(x)f(y)$. We can solve the equation by integrating both sides of the equation $dy/f(y) = g(x)dx$ and solving for y.

EXERCISES

1. The curves intersect when $x^2 = 4x - x^2 \iff 2x^2 - 4x = 0 \iff$
   $2x(x - 2) = 0 \iff x = 0$ or 2.
   $A = \int_0^2 \left[(4x - x^2) - x^2\right] dx = \int_0^2 (4x - 2x^2)\, dx$
   $= \left[2x^2 - \frac{2}{3}x^3\right]_0^2 = \left[(8 - \frac{16}{3}) - 0\right] = \frac{8}{3}$

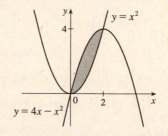

3. If $x \geq 0$, then $|x| = x$, and the graphs intersect when $x = 1 - 2x^2 \iff 2x^2 + x - 1 = 0 \iff (2x - 1)(x + 1) = 0 \iff$

$x = \frac{1}{2}$ or -1, but $-1 < 0$. By symmetry, we can double the area from $x = 0$ to $x = \frac{1}{2}$.

$A = 2 \int_0^{1/2} \left[(1 - 2x^2) - x \right] dx = 2 \int_0^{1/2} (-2x^2 - x + 1) \, dx$

$= 2\left[-\frac{2}{3}x^3 - \frac{1}{2}x^2 + x \right]_0^{1/2} = 2\left[\left(-\frac{1}{12} - \frac{1}{8} + \frac{1}{2} \right) - 0 \right]$

$= 2\left(\frac{7}{24} \right) = \frac{7}{12}$

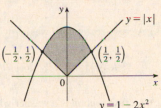

5. Using washers with inner radius x^2 and outer radius $2x$, we have

$V = \pi \int_0^2 \left[(2x)^2 - (x^2)^2 \right] dx = \pi \int_0^2 (4x^2 - x^4) \, dx$

$= \pi \left[\frac{4}{3}x^3 - \frac{1}{5}x^5 \right]_0^2 = \pi \left(\frac{32}{3} - \frac{32}{5} \right)$

$= 32\pi \cdot \frac{2}{15} = \frac{64}{15}\pi$

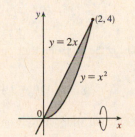

7. $V = \pi \int_{-3}^3 \left\{ \left[(9 - y^2) - (-1) \right]^2 - \left[0 - (-1) \right]^2 \right\} dy$

$= 2\pi \int_0^3 \left[(10 - y^2)^2 - 1 \right] dy = 2\pi \int_0^3 (100 - 20y^2 + y^4 - 1) \, dy$

$= 2\pi \int_0^3 (99 - 20y^2 + y^4) \, dy = 2\pi \left[99y - \frac{20}{3}y^3 + \frac{1}{5}y^5 \right]_0^3$

$= 2\pi \left(297 - 180 + \frac{243}{5} \right) = \frac{1656}{5}\pi$

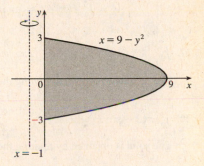

9. The graph of $x^2 - y^2 = a^2$ is a hyperbola with right and left branches.

Solving for y gives us $y^2 = x^2 - a^2 \implies y = \pm\sqrt{x^2 - a^2}$.

We'll use shells and the height of each shell is

$\sqrt{x^2 - a^2} - \left(-\sqrt{x^2 - a^2} \right) = 2\sqrt{x^2 - a^2}$.

The volume is $V = \int_a^{a+h} 2\pi x \cdot 2\sqrt{x^2 - a^2} \, dx$. To evaluate, let $u = x^2 - a^2$,

so $du = 2x \, dx$ and $x \, dx = \frac{1}{2} du$. When $x = a$, $u = 0$, and when $x = a + h$,

$u = (a + h)^2 - a^2 = a^2 + 2ah + h^2 - a^2 = 2ah + h^2$.

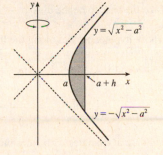

Thus, $V = 4\pi \int_0^{2ah+h^2} \sqrt{u} \left(\frac{1}{2} du \right) = 2\pi \left[\frac{2}{3}u^{3/2} \right]_0^{2ah+h^2} = \frac{4}{3}\pi \left(2ah + h^2 \right)^{3/2}$.

11. A shell has radius $\frac{\pi}{2} - x$, circumference $2\pi\left(\frac{\pi}{2} - x \right)$, and height $\cos^2 x - \frac{1}{4}$.

$y = \cos^2 x$ intersects $y = \frac{1}{4}$ when $\cos^2 x = \frac{1}{4} \iff$

$\cos x = \pm\frac{1}{2}$ $[|x| \leq \pi/2] \iff x = \pm\frac{\pi}{3}$.

$V = \int_{-\pi/3}^{\pi/3} 2\pi\left(\frac{\pi}{2} - x \right)\left(\cos^2 x - \frac{1}{4} \right) dx$

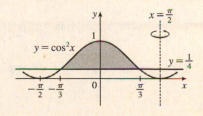

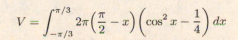

13. (a) A cross-section is a washer with inner radius x^2 and outer radius x.

$$V = \int_0^1 \pi\left[(x)^2 - (x^2)^2\right] dx = \int_0^1 \pi(x^2 - x^4)\, dx = \pi\left[\tfrac{1}{3}x^3 - \tfrac{1}{5}x^5\right]_0^1 = \pi\left[\tfrac{1}{3} - \tfrac{1}{5}\right] = \tfrac{2}{15}\pi$$

(b) A cross-section is a washer with inner radius y and outer radius $\sqrt{y}$.

$$V = \int_0^1 \pi\left[\left(\sqrt{y}\right)^2 - y^2\right] dy = \int_0^1 \pi(y - y^2)\, dy = \pi\left[\tfrac{1}{2}y^2 - \tfrac{1}{3}y^3\right]_0^1 = \pi\left[\tfrac{1}{2} - \tfrac{1}{3}\right] = \tfrac{\pi}{6}$$

(c) A cross-section is a washer with inner radius $2 - x$ and outer radius $2 - x^2$.

$$V = \int_0^1 \pi\left[(2 - x^2)^2 - (2 - x)^2\right] dx = \int_0^1 \pi(x^4 - 5x^2 + 4x)\, dx = \pi\left[\tfrac{1}{5}x^5 - \tfrac{5}{3}x^3 + 2x^2\right]_0^1 = \pi\left[\tfrac{1}{5} - \tfrac{5}{3} + 2\right] = \tfrac{8}{15}\pi$$

15. (a) Using the Midpoint Rule on $[0, 1]$ with $f(x) = \tan(x^2)$ and $n = 4$, we estimate

$$A = \int_0^1 \tan(x^2)\, dx \approx \tfrac{1}{4}\left[\tan\left(\left(\tfrac{1}{8}\right)^2\right) + \tan\left(\left(\tfrac{3}{8}\right)^2\right) + \tan\left(\left(\tfrac{5}{8}\right)^2\right) + \tan\left(\left(\tfrac{7}{8}\right)^2\right)\right] \approx \tfrac{1}{4}(1.53) \approx 0.38$$

(b) Using the Midpoint Rule on $[0, 1]$ with $f(x) = \pi \tan^2(x^2)$ (for disks) and $n = 4$, we estimate

$$V = \int_0^1 f(x)\, dx \approx \tfrac{1}{4}\pi\left[\tan^2\left(\left(\tfrac{1}{8}\right)^2\right) + \tan^2\left(\left(\tfrac{3}{8}\right)^2\right) + \tan^2\left(\left(\tfrac{5}{8}\right)^2\right) + \tan^2\left(\left(\tfrac{7}{8}\right)^2\right)\right] \approx \tfrac{\pi}{4}(1.114) \approx 0.87$$

17. $\int_0^{\pi/2} 2\pi x \cos x\, dx = \int_0^{\pi/2} (2\pi x) \cos x\, dx$

The solid is obtained by rotating the region $\mathcal{R} = \left\{(x, y) \mid 0 \le x \le \tfrac{\pi}{2}, 0 \le y \le \cos x\right\}$ about the y-axis.

19. $\int_0^\pi \pi(2 - \sin x)^2\, dx$

The solid is obtained by rotating the region $\mathcal{R} = \{(x, y) \mid 0 \le x \le \pi, 0 \le y \le 2 - \sin x\}$ about the x-axis.

21. Take the base to be the disk $x^2 + y^2 \le 9$. Then $V = \int_{-3}^3 A(x)\, dx$, where $A(x_0)$ is the area of the isosceles right triangle whose hypotenuse lies along the line $x = x_0$ in the xy-plane. The length of the hypotenuse is $2\sqrt{9 - x^2}$ and the length of each leg is $\sqrt{2}\sqrt{9 - x^2}$. $A(x) = \tfrac{1}{2}\left(\sqrt{2}\sqrt{9 - x^2}\right)^2 = 9 - x^2$, so

$$V = 2\int_0^3 A(x)\, dx = 2\int_0^3 (9 - x^2)\, dx = 2\left[9x - \tfrac{1}{3}x^3\right]_0^3 = 2(27 - 9) = 36$$

23. Equilateral triangles with sides measuring $\tfrac{1}{4}x$ meters have height $\tfrac{1}{4}x \sin 60° = \tfrac{\sqrt{3}}{8}x$. Therefore,

$$A(x) = \tfrac{1}{2} \cdot \tfrac{1}{4}x \cdot \tfrac{\sqrt{3}}{8}x = \tfrac{\sqrt{3}}{64}x^2. \quad V = \int_0^{20} A(x)\, dx = \tfrac{\sqrt{3}}{64}\int_0^{20} x^2\, dx = \tfrac{\sqrt{3}}{64}\left[\tfrac{1}{3}x^3\right]_0^{20} = \tfrac{8000\sqrt{3}}{64 \cdot 3} = \tfrac{125\sqrt{3}}{3}\ \text{m}^3.$$

25. $y = \tfrac{1}{6}(x^2 + 4)^{3/2} \ \Rightarrow \ dy/dx = \tfrac{1}{4}(x^2 + 4)^{1/2}(2x) \ \Rightarrow$

$$1 + (dy/dx)^2 = 1 + \left[\tfrac{1}{2}x(x^2 + 4)^{1/2}\right]^2 = 1 + \tfrac{1}{4}x^2(x^2 + 4) = \tfrac{1}{4}x^4 + x^2 + 1 = \left(\tfrac{1}{2}x^2 + 1\right)^2.$$

Thus, $L = \int_0^3 \sqrt{\left(\tfrac{1}{2}x^2 + 1\right)^2}\, dx = \int_0^3 \left(\tfrac{1}{2}x^2 + 1\right) dx = \left[\tfrac{1}{6}x^3 + x\right]_0^3 = \tfrac{15}{2}.$

27. (a) $y = \dfrac{x^4}{16} + \dfrac{1}{2x^2} = \dfrac{1}{16}x^4 + \dfrac{1}{2}x^{-2}$ $\Rightarrow$ $\dfrac{dy}{dx} = \dfrac{1}{4}x^3 - x^{-3}$ $\Rightarrow$

$1 + (dy/dx)^2 = 1 + \left(\frac{1}{4}x^3 - x^{-3}\right)^2 = 1 + \frac{1}{16}x^6 - \frac{1}{2} + x^{-6} = \frac{1}{16}x^6 + \frac{1}{2} + x^{-6} = \left(\frac{1}{4}x^3 + x^{-3}\right)^2$.

Thus, $L = \int_1^2 \left(\frac{1}{4}x^3 + x^{-3}\right) dx = \left[\frac{1}{16}x^4 - \frac{1}{2}x^{-2}\right]_1^2 = \left(1 - \frac{1}{8}\right) - \left(\frac{1}{16} - \frac{1}{2}\right) = \frac{21}{16}$.

(b) $S = \int_1^2 2\pi x\left(\frac{1}{4}x^3 + x^{-3}\right) dx = 2\pi \int_1^2 \left(\frac{1}{4}x^4 + x^{-2}\right) dx = 2\pi\left[\frac{1}{20}x^5 - \frac{1}{x}\right]_1^2$

$= 2\pi\left[\left(\frac{32}{20} - \frac{1}{2}\right) - \left(\frac{1}{20} - 1\right)\right] = 2\pi\left(\frac{8}{5} - \frac{1}{2} - \frac{1}{20} + 1\right) = 2\pi\left(\frac{41}{20}\right) = \frac{41}{10}\pi$

29. $y = \sin x$ $\Rightarrow$ $y' = \cos x$ $\Rightarrow$ $1 + (y')^2 = 1 + \cos^2 x$. Let $f(x) = \sqrt{1 + \cos^2 x}$. Then

$L = \int_0^\pi f(x)\,dx \approx S_{10}$

$= \dfrac{(\pi - 0)/10}{3}\left[f(0) + 4f\left(\frac{\pi}{10}\right) + 2f\left(\frac{2\pi}{10}\right) + 4f\left(\frac{3\pi}{10}\right) + 2f\left(\frac{4\pi}{10}\right)\right.$

$\left. + 4f\left(\frac{5\pi}{10}\right) + 2f\left(\frac{6\pi}{10}\right) + 4f\left(\frac{7\pi}{10}\right) + 2f\left(\frac{8\pi}{10}\right) + 4f\left(\frac{9\pi}{10}\right) + f(\pi)\right]$

≈ 3.820188

31. $y = \int_1^x \sqrt{\sqrt{t} - 1}\,dt$ $\Rightarrow$ $dy/dx = \sqrt{\sqrt{x} - 1}$ $\Rightarrow$ $1 + (dy/dx)^2 = 1 + \left(\sqrt{x} - 1\right) = \sqrt{x}$.

Thus, $L = \int_1^{16} \sqrt{\sqrt{x}}\,dx = \int_1^{16} x^{1/4}\,dx = \frac{4}{5}\left[x^{5/4}\right]_1^{16} = \frac{4}{5}(32 - 1) = \frac{124}{5}$.

33. $f(x) = kx$ $\Rightarrow$ $30\,\text{N} = k(15 - 12)\,\text{cm}$ $\Rightarrow$ $k = 10\,\text{N/cm} = 1000\,\text{N/m}$. $20\,\text{cm} - 12\,\text{cm} = 0.08\,\text{m}$ $\Rightarrow$

$W = \int_0^{0.08} kx\,dx = 1000\int_0^{0.08} x\,dx = 500\left[x^2\right]_0^{0.08} = 500(0.08)^2 = 3.2\,\text{N·m} = 3.2\,\text{J}$.

35. (a) The parabola has equation $y = ax^2$ with vertex at the origin and passing through

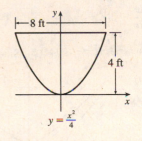

$(4, 4)$. $4 = a \cdot 4^2$ $\Rightarrow$ $a = \frac{1}{4}$ $\Rightarrow$ $y = \frac{1}{4}x^2$ $\Rightarrow$ $x^2 = 4y$ $\Rightarrow$

$x = 2\sqrt{y}$. Each circular disk has radius $2\sqrt{y}$ and is moved $4 - y$ ft.

$W = \int_0^4 \pi\left(2\sqrt{y}\right)^2 62.5(4 - y)\,dy = 250\pi \int_0^4 y(4 - y)\,dy$

$= 250\pi\left[2y^2 - \frac{1}{3}y^3\right]_0^4 = 250\pi\left(32 - \frac{64}{3}\right) = \frac{8000\pi}{3} \approx 8378\,\text{ft-lb}$

(b) In part (a) we knew the final water level (0) but not the amount of work done. Here

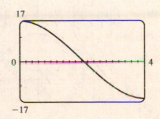

we use the same equation, except with the work fixed, and the lower limit of

integration (that is, the final water level—call it h) unknown: $W = 4000$ $\Leftrightarrow$

$250\pi\left[2y^2 - \frac{1}{3}y^3\right]_h^4 = 4000$ $\Leftrightarrow$ $\frac{16}{\pi} = \left[\left(32 - \frac{64}{3}\right) - \left(2h^2 - \frac{1}{3}h^3\right)\right]$ $\Leftrightarrow$

$h^3 - 6h^2 + 32 - \frac{48}{\pi} = 0$. We graph the function $f(h) = h^3 - 6h^2 + 32 - \frac{48}{\pi}$

on the interval $[0, 4]$ to see where it is 0. From the graph, $f(h) = 0$ for $h \approx 2.1$.

So the depth of water remaining is about 2.1 ft.

37. As in Example 5 of Section 7.5, $\dfrac{a}{2-x} = \dfrac{1}{2}$ $\Rightarrow$ $2a = 2 - x$ and $w = 2(1.5 + a) = 3 + 2a = 3 + 2 - x = 5 - x$.

Thus, $F = \int_0^2 \rho g x(5-x)\,dx = \rho g\left[\frac{5}{2}x^2 - \frac{1}{3}x^3\right]_0^2 = \rho g\left(10 - \frac{8}{3}\right) = \frac{22}{3}\delta$ $[\rho g = \delta]$ $\approx \frac{22}{3} \cdot 62.5 \approx 458$ lb.

39. $A = \int_0^4 \left(\sqrt{x} - \frac{1}{2}x\right)dx = \left[\frac{2}{3}x^{3/2} - \frac{1}{4}x^2\right]_0^4 = \frac{16}{3} - 4 = \frac{4}{3}$

$\bar{x} = \frac{1}{A}\int_0^4 x\left(\sqrt{x} - \frac{1}{2}x\right)dx = \frac{3}{4}\int_0^4 \left(x^{3/2} - \frac{1}{2}x^2\right)dx$

$= \frac{3}{4}\left[\frac{2}{5}x^{5/2} - \frac{1}{6}x^3\right]_0^4 = \frac{3}{4}\left(\frac{64}{5} - \frac{64}{6}\right) = \frac{3}{4}\left(\frac{64}{30}\right) = \frac{8}{5}$

$\bar{y} = \frac{1}{A}\int_0^4 \frac{1}{2}\left[\left(\sqrt{x}\right)^2 - \left(\frac{1}{2}x\right)^2\right]dx = \frac{3}{4}\int_0^4 \frac{1}{2}\left(x - \frac{1}{4}x^2\right)dx = \frac{3}{8}\left[\frac{1}{2}x^2 - \frac{1}{12}x^3\right]_0^4 = \frac{3}{8}\left(8 - \frac{16}{3}\right) = \frac{3}{8}\left(\frac{8}{3}\right) = 1$

Thus, the centroid is $(\bar{x}, \bar{y}) = \left(\frac{8}{5}, 1\right)$.

41. The centroid of this circle, $(1, 0)$, travels a distance $2\pi(1)$ when the lamina is rotated about the y-axis. The area of the circle is $\pi(1)^2$. So by the Theorem of Pappus, $V = A(2\pi\bar{x}) = \pi(1)^2 2\pi(1) = 2\pi^2$.

43. $2ye^{y^2}y' = 2x + 3\sqrt{x}$ $\Rightarrow$ $2ye^{y^2}\dfrac{dy}{dx} = 2x + 3\sqrt{x}$ $\Rightarrow$ $2ye^{y^2}\,dy = \left(2x + 3\sqrt{x}\right)dx$ $\Rightarrow$

$\int 2ye^{y^2}\,dy = \int\left(2x + 3\sqrt{x}\right)dx$ $\Rightarrow$ $e^{y^2} = x^2 + 2x^{3/2} + C$ $\Rightarrow$ $y^2 = \ln(x^2 + 2x^{3/2} + C)$ $\Rightarrow$

$y = \pm\sqrt{\ln(x^2 + 2x^{3/2} + C)}$

45. $\dfrac{dr}{dt} + 2tr = r$ $\Rightarrow$ $\dfrac{dr}{dt} = r - 2tr = r(1 - 2t)$ $\Rightarrow$ $\displaystyle\int \dfrac{dr}{r} = \int (1 - 2t)\,dt$ $\Rightarrow$ $\ln|r| = t - t^2 + C$ $\Rightarrow$

$|r| = e^{t-t^2+C} = ke^{t-t^2}$. Since $r(0) = 5$, $5 = ke^0 = k$. Thus, $r(t) = 5e^{t-t^2}$.

47. (a)

We sketch the direction field and four solution curves, as shown. Note that the slope $y' = x/y$ is not defined on the line $y = 0$.

(b) $y' = x/y$ $\Leftrightarrow$ $y\,dy = x\,dx$ $\Leftrightarrow$ $y^2 = x^2 + C$. For $C = 0$, this is the pair of lines $y = \pm x$. For $C \neq 0$, it is the hyperbola $x^2 - y^2 = -C$.

8 □ SERIES

8.1 Sequences

1. (a) A sequence is an ordered list of numbers. It can also be defined as a function whose domain is the set of positive integers.

   (b) The terms a_n approach 8 as n becomes large. In fact, we can make a_n as close to 8 as we like by taking n sufficiently large.

   (c) The terms a_n become large as n becomes large. In fact, we can make a_n as large as we like by taking n sufficiently large.

3. The first six terms of $a_n = \dfrac{n}{2n+1}$ are $\dfrac{1}{3}, \dfrac{2}{5}, \dfrac{3}{7}, \dfrac{4}{9}, \dfrac{5}{11}, \dfrac{6}{13}$. It appears that the sequence is approaching $\dfrac{1}{2}$.

   $$\lim_{n\to\infty} \frac{n}{2n+1} = \lim_{n\to\infty} \frac{1}{2+1/n} = \frac{1}{2}$$

5. $\left\{-3, 2, -\dfrac{4}{3}, \dfrac{8}{9}, -\dfrac{16}{27}, \ldots\right\}$. The first term is -3 and each term is $-\dfrac{2}{3}$ times the preceding one, so $a_n = -3\left(-\dfrac{2}{3}\right)^{n-1}$.

7. $\left\{\dfrac{1}{2}, -\dfrac{4}{3}, \dfrac{9}{4}, -\dfrac{16}{5}, \dfrac{25}{6}, \ldots\right\}$. The numerator of the nth term is n^2 and its denominator is $n+1$. Including the alternating signs, we get $a_n = (-1)^{n+1}\dfrac{n^2}{n+1}$.

9. $a_n = 1 - (0.2)^n$, so $\lim\limits_{n\to\infty} a_n = 1 - 0 = 1$ by (8). **Converges**

11. $a_n = \dfrac{3+5n^2}{n+n^2} = \dfrac{(3+5n^2)/n^2}{(n+n^2)/n^2} = \dfrac{5+3/n^2}{1+1/n}$, so $a_n \to \dfrac{5+0}{1+0} = 5$ as $n \to \infty$. **Converges**

13. If $b_n = \dfrac{2n\pi}{1+8n}$, then $\lim\limits_{n\to\infty} b_n = \lim\limits_{n\to\infty} \dfrac{(2n\pi)/n}{(1+8n)/n} = \lim\limits_{n\to\infty} \dfrac{2\pi}{1/n+8} = \dfrac{2\pi}{8} = \dfrac{\pi}{4}$. Since $\tan$ is continuous at $\frac{\pi}{4}$, by the

   Continuity and Convergence Theorem, $\lim\limits_{n\to\infty} \tan\left(\dfrac{2n\pi}{1+8n}\right) = \tan\left(\lim\limits_{n\to\infty} \dfrac{2n\pi}{1+8n}\right) = \tan\dfrac{\pi}{4} = 1$. **Converges**

15. $a_n = \dfrac{n^2}{\sqrt{n^3+4n}} = \dfrac{n^2/\sqrt{n^3}}{\sqrt{n^3+4n}/\sqrt{n^3}} = \dfrac{\sqrt{n}}{\sqrt{1+4/n^2}}$, so $a_n \to \infty$ as $n \to \infty$ since $\lim\limits_{n\to\infty} \sqrt{n} = \infty$ and

   $\lim\limits_{n\to\infty} \sqrt{1+4/n^2} = 1$. **Diverges**

17. $\lim\limits_{n\to\infty} |a_n| = \lim\limits_{n\to\infty} \left|\dfrac{(-1)^n}{2\sqrt{n}}\right| = \dfrac{1}{2}\lim\limits_{n\to\infty}\dfrac{1}{n^{1/2}} = \dfrac{1}{2}(0) = 0$, so $\lim\limits_{n\to\infty} a_n = 0$ by (6). **Converges**

19. $a_n = \cos(n/2)$. This sequence diverges since the terms don't approach any particular real number as $n \to \infty$. The terms take on values between -1 and 1.

21. $a_n = \dfrac{(2n-1)!}{(2n+1)!} = \dfrac{(2n-1)!}{(2n+1)(2n)(2n-1)!} = \dfrac{1}{(2n+1)(2n)} \to 0$ as $n \to \infty$. **Converges**

23. $a_n = n^2 e^{-n} = \dfrac{n^2}{e^n}$. Since $\lim\limits_{x\to\infty} \dfrac{x^2}{e^x} \overset{\text{H}}{=} \lim\limits_{x\to\infty} \dfrac{2x}{e^x} \overset{\text{H}}{=} \lim\limits_{x\to\infty} \dfrac{2}{e^x} = 0$, it follows from Theorem 3 that $\lim\limits_{n\to\infty} a_n = 0$. **Converges**

25. $0 \le \dfrac{\cos^2 n}{2^n} \le \dfrac{1}{2^n}$ [since $0 \le \cos^2 n \le 1$], so since $\displaystyle\lim_{n \to \infty} \dfrac{1}{2^n} = 0$, $\left\{ \dfrac{\cos^2 n}{2^n} \right\}$ converges to 0 by the Squeeze Theorem.

27. $y = \left(1 + \dfrac{2}{x}\right)^x \implies \ln y = x \ln\left(1 + \dfrac{2}{x}\right)$, so

$$\lim_{x \to \infty} \ln y = \lim_{x \to \infty} \frac{\ln(1 + 2/x)}{1/x} \overset{\text{H}}{=} \lim_{x \to \infty} \frac{\left(\dfrac{1}{1 + 2/x}\right)\left(-\dfrac{2}{x^2}\right)}{-1/x^2} = \lim_{x \to \infty} \frac{2}{1 + 2/x} = 2 \implies$$

$\displaystyle\lim_{x \to \infty} \left(1 + \dfrac{2}{x}\right)^x = \lim_{x \to \infty} e^{\ln y} = e^2$, so by Theorem 3, $\displaystyle\lim_{n \to \infty}\left(1 + \dfrac{2}{n}\right)^n = e^2$.   Converges

29. $\{0, 1, 0, 0, 1, 0, 0, 0, 1, \ldots\}$ diverges since the sequence takes on only two values, 0 and 1, and never stays arbitrarily close to

either one (or any other value) for n sufficiently large.

31. $a_n = \ln(2n^2 + 1) - \ln(n^2 + 1) = \ln\left(\dfrac{2n^2 + 1}{n^2 + 1}\right) = \ln\left(\dfrac{2 + 1/n^2}{1 + 1/n^2}\right) \to \ln 2$ as $n \to \infty$.   Converges

33. (a) $a_n = 1000(1.06)^n \implies a_1 = 1060$, $a_2 = 1123.60$, $a_3 = 1191.02$, $a_4 = 1262.48$, and $a_5 = 1338.23$.

(b) $\displaystyle\lim_{n \to \infty} a_n = 1000 \lim_{n \to \infty} (1.06)^n$, so the sequence diverges by (8) with $r = 1.06 > 1$.

35. Since $\{a_n\}$ is a decreasing sequence, $a_n > a_{n+1}$ for all $n \ge 1$. Because all of its terms lie between 5 and 8, $\{a_n\}$ is a

bounded sequence. By the Monotonic Sequence Theorem, $\{a_n\}$ is convergent; that is, $\{a_n\}$ has a limit L. L must be less than

8 since $\{a_n\}$ is decreasing, so $5 \le L < 8$.

37. $a_n = \dfrac{1}{2n + 3}$ is decreasing since $a_{n+1} = \dfrac{1}{2(n + 1) + 3} = \dfrac{1}{2n + 5} < \dfrac{1}{2n + 3} = a_n$ for each $n \ge 1$. The sequence is

bounded since $0 < a_n \le \frac{1}{5}$ for all $n \ge 1$. Note that $a_1 = \frac{1}{5}$.

39. The terms of $a_n = n(-1)^n$ alternate in sign, so the sequence is not monotonic. The first five terms are $-1, 2, -3, 4$, and -5.

Since $\displaystyle\lim_{n \to \infty} |a_n| = \lim_{n \to \infty} n = \infty$, the sequence is not bounded.

41. For $\left\{\sqrt{2}, \sqrt{2\sqrt{2}}, \sqrt{2\sqrt{2\sqrt{2}}}, \ldots\right\}$, $a_1 = 2^{1/2}$, $a_2 = 2^{3/4}$, $a_3 = 2^{7/8}, \ldots$, so $a_n = 2^{(2^n - 1)/2^n} = 2^{1 - (1/2^n)}$.

$\displaystyle\lim_{n \to \infty} a_n = \lim_{n \to \infty} 2^{1 - (1/2^n)} = 2^1 = 2.$

Alternate solution: Let $L = \displaystyle\lim_{n \to \infty} a_n$. (We could show the limit exists by showing that $\{a_n\}$ is bounded and increasing.)

Then L must satisfy $L = \sqrt{2 \cdot L} \implies L^2 = 2L \implies L(L - 2) = 0$. $L \ne 0$ since the sequence increases, so $L = 2$.

43. We show by induction that $\{a_n\}$ is increasing and bounded above by 3. Let P_n be the proposition that $a_{n+1} > a_n$ and

$0 < a_n < 3$. Clearly P_1 is true. Assume that P_n is true.

Then $a_{n+1} > a_n \implies \dfrac{1}{a_{n+1}} < \dfrac{1}{a_n} \implies -\dfrac{1}{a_{n+1}} > -\dfrac{1}{a_n}$. Now $a_{n+2} = 3 - \dfrac{1}{a_{n+1}} > 3 - \dfrac{1}{a_n} = a_{n+1} \iff P_{n+1}$.

This proves that $\{a_n\}$ is increasing and bounded above by 3, so $1 = a_1 < a_n < 3$, that is, $\{a_n\}$ is bounded, and hence convergent by the Monotonic Sequence Theorem. If $L = \lim\limits_{n\to\infty} a_n$, then $\lim\limits_{n\to\infty} a_{n+1} = L$ also, so L must satisfy

$$L = 3 - 1/L \quad\Rightarrow\quad L^2 - 3L + 1 = 0 \quad\Rightarrow\quad L = \tfrac{3 \pm \sqrt{5}}{2}. \text{ But } L > 1, \text{ so } L = \tfrac{3 + \sqrt{5}}{2}.$$

45. (a) Let a_n be the number of rabbit pairs in the nth month. Clearly $a_1 = 1 = a_2$. In the nth month, each pair that is 2 or more months old (that is, a_{n-2} pairs) will produce a new pair to add to the a_{n-1} pairs already present. Thus, $a_n = a_{n-1} + a_{n-2}$, so that $\{a_n\} = \{f_n\}$, the Fibonacci sequence.

(b) $a_n = \dfrac{f_{n+1}}{f_n} \;\Rightarrow\; a_{n-1} = \dfrac{f_n}{f_{n-1}} = \dfrac{f_{n-1} + f_{n-2}}{f_{n-1}} = 1 + \dfrac{f_{n-2}}{f_{n-1}} = 1 + \dfrac{1}{f_{n-1}/f_{n-2}} = 1 + \dfrac{1}{a_{n-2}}.$ If $L = \lim\limits_{n\to\infty} a_n$,

then $L = \lim\limits_{n\to\infty} a_{n-1}$ and $L = \lim\limits_{n\to\infty} a_{n-2}$, so L must satisfy $L = 1 + \dfrac{1}{L} \;\Rightarrow\; L^2 - L - 1 = 0 \;\Rightarrow\; L = \tfrac{1+\sqrt{5}}{2}$

[since L must be positive].

47. $(0.8)^n < 0.000001 \;\Rightarrow\; \ln(0.8)^n < \ln(0.000001) \;\Rightarrow\; n\ln(0.8) < \ln(0.000001) \;\Rightarrow\; n > \dfrac{\ln(0.000001)}{\ln(0.8)} \;\Rightarrow$

$n > 61.9$, so n must be at least 62 to satisfy the given inequality.

49. Theorem 6: If $\lim\limits_{n\to\infty} |a_n| = 0$ then $\lim\limits_{n\to\infty} -|a_n| = 0$, and since $-|a_n| \le a_n \le |a_n|$, we have that $\lim\limits_{n\to\infty} a_n = 0$ by the Squeeze Theorem.

51. To Prove: If $\lim\limits_{n\to\infty} a_n = 0$ and $\{b_n\}$ is bounded, then $\lim\limits_{n\to\infty} (a_n b_n) = 0$.

Proof: Since $\{b_n\}$ is bounded, there is a positive number M such that $|b_n| \le M$ and hence, $|a_n|\,|b_n| \le |a_n|\,M$ for all $n \ge 1$. Let $\varepsilon > 0$ be given. Since $\lim\limits_{n\to\infty} a_n = 0$, there is an integer N such that $|a_n - 0| < \dfrac{\varepsilon}{M}$ if $n > N$. Then

$|a_n b_n - 0| = |a_n b_n| = |a_n|\,|b_n| \le |a_n|\,M = |a_n - 0|\,M < \dfrac{\varepsilon}{M} \cdot M = \varepsilon$ for all $n > N$. Since ε was arbitrary,

$\lim\limits_{n\to\infty} (a_n b_n) = 0.$

53. (a) Suppose $\{p_n\}$ converges to p. Then $p_{n+1} = \dfrac{bp_n}{a + p_n} \;\Rightarrow\; \lim\limits_{n\to\infty} p_{n+1} = \dfrac{b \lim\limits_{n\to\infty} p_n}{a + \lim\limits_{n\to\infty} p_n} \;\Rightarrow\; p = \dfrac{bp}{a + p} \;\Rightarrow$

$p^2 + ap = bp \;\Rightarrow\; p(p + a - b) = 0 \;\Rightarrow\; p = 0 \text{ or } p = b - a.$

(b) $p_{n+1} = \dfrac{bp_n}{a + p_n} = \dfrac{\left(\dfrac{b}{a}\right)p_n}{1 + \dfrac{p_n}{a}} < \left(\dfrac{b}{a}\right)p_n$ since $1 + \dfrac{p_n}{a} > 1.$

(c) By part (b), $p_1 < \left(\dfrac{b}{a}\right)p_0,\ p_2 < \left(\dfrac{b}{a}\right)p_1 < \left(\dfrac{b}{a}\right)^2 p_0,\ p_3 < \left(\dfrac{b}{a}\right)p_2 < \left(\dfrac{b}{a}\right)^3 p_0,$ etc. In general, $p_n < \left(\dfrac{b}{a}\right)^n p_0,$

so $\lim\limits_{n\to\infty} p_n \le \lim\limits_{n\to\infty} \left(\dfrac{b}{a}\right)^n \cdot p_0 = 0$ since $b < a$. $\left[\text{By (8), } \lim\limits_{n\to\infty} r^n = 0 \text{ if } -1 < r < 1. \text{ Here } r = \dfrac{b}{a} \in (0, 1).\right]$

(d) Let $a < b$. We first show, by induction, that if $p_0 < b - a$, then $p_n < b - a$ and $p_{n+1} > p_n$.

For $n = 0$, we have $p_1 - p_0 = \dfrac{bp_0}{a + p_0} - p_0 = \dfrac{p_0(b - a - p_0)}{a + p_0} > 0$ since $p_0 < b - a$. So $p_1 > p_0.$

Now we suppose the assertion is true for $n = k$, that is, $p_k < b - a$ and $p_{k+1} > p_k$. Then

$$b - a - p_{k+1} = b - a - \frac{bp_k}{a + p_k} = \frac{a(b - a) + bp_k - ap_k - bp_k}{a + p_k} = \frac{a(b - a - p_k)}{a + p_k} > 0 \text{ because } p_k < b - a. \text{ So}$$

$$p_{k+1} < b - a. \text{ And } p_{k+2} - p_{k+1} = \frac{bp_{k+1}}{a + p_{k+1}} - p_{k+1} = \frac{p_{k+1}(b - a - p_{k+1})}{a + p_{k+1}} > 0 \text{ since } p_{k+1} < b - a. \text{ Therefore,}$$

$p_{k+2} > p_{k+1}$. Thus, the assertion is true for $n = k + 1$. It is therefore true for all n by mathematical induction.

A similar proof by induction shows that if $p_0 > b - a$, then $p_n > b - a$ and $\{p_n\}$ is decreasing.

In either case the sequence $\{p_n\}$ is bounded and monotonic, so it is convergent by the Monotonic Sequence Theorem. It then follows from part (a) that $\lim_{n \to \infty} p_n = b - a$.

8.2 Series

1. (a) A sequence is an ordered list of numbers whereas a series is the *sum* of a list of numbers.

   (b) A series is convergent if the sequence of partial sums is a convergent sequence. A series is divergent if it is not convergent.

3. For $\sum_{n=1}^{\infty} \frac{1}{n^3}$, $a_n = \frac{1}{n^3}$. $s_1 = a_1 = \frac{1}{1^3} = 1$, $s_2 = s_1 + a_2 = 1 + \frac{1}{2^3} = 1.125$, $s_3 = s_2 + a_3 \approx 1.1620$,

   $s_4 = s_3 + a_4 \approx 1.1777$, $s_5 = s_4 + a_5 \approx 1.1857$, $s_6 = s_5 + a_6 \approx 1.1903$, $s_7 = s_6 + a_7 \approx 1.1932$, and

   $s_8 = s_7 + a_8 \approx 1.1952$. It appears that the series is convergent.

5. For $\sum_{n=1}^{\infty} \frac{n}{1 + \sqrt{n}}$, $a_n = \frac{n}{1 + \sqrt{n}}$. $s_1 = a_1 = \frac{1}{1 + \sqrt{1}} = 0.5$, $s_2 = s_1 + a_2 = 0.5 + \frac{2}{1 + \sqrt{2}} \approx 1.3284$,

   $s_3 = s_2 + a_3 \approx 2.4265$, $s_4 = s_3 + a_4 \approx 3.7598$, $s_5 = s_4 + a_5 \approx 5.3049$, $s_6 = s_5 + a_6 \approx 7.0443$,

   $s_7 = s_6 + a_7 \approx 8.9644$, $s_8 = s_7 + a_8 \approx 11.0540$. It appears that the series is divergent.

7. $10 - 2 + 0.4 - 0.08 + \cdots$ is a geometric series with ratio $-\frac{2}{10} = -\frac{1}{5}$. Since $|r| = \frac{1}{5} < 1$, the series converges to

   $$\frac{a}{1 - r} = \frac{10}{1 - (-1/5)} = \frac{10}{6/5} = \frac{50}{6} = \frac{25}{3}.$$

9. $\sum_{n=1}^{\infty} \frac{(-3)^{n-1}}{4^n} = \frac{1}{4} \sum_{n=1}^{\infty} \left(-\frac{3}{4}\right)^{n-1}$. The latter series is geometric with $a = 1$ and ratio $r = -\frac{3}{4}$. Since $|r| = \frac{3}{4} < 1$, it

   converges to $\frac{1}{1 - (-3/4)} = \frac{4}{7}$. Thus, the given series converges to $\left(\frac{1}{4}\right)\left(\frac{4}{7}\right) = \frac{1}{7}$.

11. $\sum_{n=0}^{\infty} \frac{\pi^n}{3^{n+1}} = \frac{1}{3} \sum_{n=0}^{\infty} \left(\frac{\pi}{3}\right)^n$ is a geometric series with ratio $r = \frac{\pi}{3}$. Since $|r| > 1$, the series diverges.

13. $\sum_{n=1}^{\infty} \frac{3^n}{e^{n-1}} = \sum_{n=1}^{\infty} \frac{3 \cdot 3^{n-1}}{e^{n-1}} = \sum_{n=1}^{\infty} 3 \left(\frac{3}{e}\right)^{n-1}$. This is a geometric series with $r = 3/e$ and $e < 3$, so $r > 1$ and the series is

   divergent.

15. $\sum_{n=1}^{\infty} \frac{n - 1}{3n - 1}$ diverges by the Test for Divergence since $\lim_{n \to \infty} a_n = \lim_{n \to \infty} \frac{n - 1}{3n - 1} = \frac{1}{3} \neq 0$.

17. Converges.

$$\sum_{n=1}^{\infty} \frac{1+2^n}{3^n} = \sum_{n=1}^{\infty} \left(\frac{1}{3^n} + \frac{2^n}{3^n} \right) = \sum_{n=1}^{\infty} \left[\left(\frac{1}{3} \right)^n + \left(\frac{2}{3} \right)^n \right] \qquad \text{[sum of two convergent geometric series]}$$

$$= \frac{1/3}{1-1/3} + \frac{2/3}{1-2/3} = \frac{1}{2} + 2 = \frac{5}{2}$$

19. $\displaystyle\sum_{n=1}^{\infty} \sqrt[n]{2} = 2 + \sqrt{2} + \sqrt[3]{2} + \sqrt[4]{2} + \cdots$ diverges by the Test for Divergence since

$$\lim_{n\to\infty} a_n = \lim_{n\to\infty} \sqrt[n]{2} = \lim_{n\to\infty} 2^{1/n} = 2^0 = 1 \neq 0.$$

21. $\displaystyle\sum_{n=1}^{\infty} \arctan n$ diverges by the Test for Divergence since $\displaystyle\lim_{n\to\infty} a_n = \lim_{n\to\infty} \arctan n = \frac{\pi}{2} \neq 0.$

23. $\dfrac{1}{3} + \dfrac{1}{6} + \dfrac{1}{9} + \dfrac{1}{12} + \dfrac{1}{15} + \cdots = \displaystyle\sum_{n=1}^{\infty} \frac{1}{3n} = \frac{1}{3} \sum_{n=1}^{\infty} \frac{1}{n}$. This is a constant multiple of the divergent harmonic series, so

it diverges.

25. Using partial fractions, the partial sums of the series $\displaystyle\sum_{n=2}^{\infty} \frac{2}{n^2-1}$ are

$$s_n = \sum_{i=2}^{n} \frac{2}{(i-1)(i+1)} = \sum_{i=2}^{n} \left(\frac{1}{i-1} - \frac{1}{i+1} \right)$$

$$= \left(1 - \frac{1}{3} \right) + \left(\frac{1}{2} - \frac{1}{4} \right) + \left(\frac{1}{3} - \frac{1}{5} \right) + \cdots + \left(\frac{1}{n-3} - \frac{1}{n-1} \right) + \left(\frac{1}{n-2} - \frac{1}{n} \right)$$

This sum is a telescoping series and $s_n = 1 + \dfrac{1}{2} - \dfrac{1}{n-1} - \dfrac{1}{n}$.

Thus, $\displaystyle\sum_{n=2}^{\infty} \frac{2}{n^2-1} = \lim_{n\to\infty} s_n = \lim_{n\to\infty} \left(1 + \frac{1}{2} - \frac{1}{n-1} - \frac{1}{n} \right) = \frac{3}{2}.$

27. For the series $\displaystyle\sum_{n=1}^{\infty} \frac{3}{n(n+3)}$, $s_n = \displaystyle\sum_{i=1}^{n} \frac{3}{i(i+3)} = \sum_{i=1}^{n} \left(\frac{1}{i} - \frac{1}{i+3} \right)$ [using partial fractions]. The latter sum is

$$\left(1 - \tfrac{1}{4} \right) + \left(\tfrac{1}{2} - \tfrac{1}{5} \right) + \left(\tfrac{1}{3} - \tfrac{1}{6} \right) + \left(\tfrac{1}{4} - \tfrac{1}{7} \right) + \cdots + \left(\tfrac{1}{n-3} - \tfrac{1}{n} \right) + \left(\tfrac{1}{n-2} - \tfrac{1}{n+1} \right) + \left(\tfrac{1}{n-1} - \tfrac{1}{n+2} \right) + \left(\tfrac{1}{n} - \tfrac{1}{n+3} \right)$$

$$= 1 + \tfrac{1}{2} + \tfrac{1}{3} - \tfrac{1}{n+1} - \tfrac{1}{n+2} - \tfrac{1}{n+3} \qquad \text{[telescoping series]}$$

Thus, $\displaystyle\sum_{n=1}^{\infty} \frac{3}{n(n+3)} = \lim_{n\to\infty} s_n = \lim_{n\to\infty} \left(1 + \tfrac{1}{2} + \tfrac{1}{3} - \tfrac{1}{n+1} - \tfrac{1}{n+2} - \tfrac{1}{n+3} \right) = 1 + \tfrac{1}{2} + \tfrac{1}{3} = \frac{11}{6}.$ Converges

29. (a) Many people would guess that $x < 1$, but note that x consists of an infinite number of 9s.

(b) $x = 0.99999\ldots = \dfrac{9}{10} + \dfrac{9}{100} + \dfrac{9}{1000} + \dfrac{9}{10,000} + \cdots = \displaystyle\sum_{n=1}^{\infty} \frac{9}{10^n}$, which is a geometric series with $a_1 = 0.9$ and

$r = 0.1$. Its sum is $\dfrac{0.9}{1-0.1} = \dfrac{0.9}{0.9} = 1$, that is, $x = 1$.

(c) The number 1 has two decimal representations, $1.00000\ldots$ and $0.99999\ldots.$

(d) Except for 0, all rational numbers that have a terminating decimal representation can be written in more than one way. For

example, 0.5 can be written as $0.49999\ldots$ as well as $0.50000\ldots.$

31. $0.\overline{8} = \dfrac{8}{10} + \dfrac{8}{10^2} + \cdots$ is a geometric series with $a = \dfrac{8}{10}$ and $r = \dfrac{1}{10}$. It converges to $\dfrac{a}{1-r} = \dfrac{8/10}{1 - 1/10} = \dfrac{8}{9}$.

33. $2.\overline{516} = 2 + \dfrac{516}{10^3} + \dfrac{516}{10^6} + \cdots$. Now $\dfrac{516}{10^3} + \dfrac{516}{10^6} + \cdots$ is a geometric series with $a = \dfrac{516}{10^3}$ and $r = \dfrac{1}{10^3}$. It converges to

$\dfrac{a}{1-r} = \dfrac{516/10^3}{1 - 1/10^3} = \dfrac{516/10^3}{999/10^3} = \dfrac{516}{999}$. Thus, $2.\overline{516} = 2 + \dfrac{516}{999} = \dfrac{2514}{999} = \dfrac{838}{333}$.

35. $\displaystyle\sum_{n=1}^{\infty} (-5)^n x^n = \sum_{n=1}^{\infty} (-5x)^n$ is a geometric series with $r = -5x$, so the series converges $\Leftrightarrow$ $|r| < 1$ $\Leftrightarrow$

$|-5x| < 1$ $\Leftrightarrow$ $|x| < \dfrac{1}{5}$, that is, $-\dfrac{1}{5} < x < \dfrac{1}{5}$. In that case, the sum of the series is $\dfrac{a}{1-r} = \dfrac{-5x}{1 - (-5x)} = \dfrac{-5x}{1 + 5x}$.

37. $\displaystyle\sum_{n=0}^{\infty} \dfrac{(x-2)^n}{3^n} = \sum_{n=0}^{\infty} \left(\dfrac{x-2}{3}\right)^n$ is a geometric series with $r = \dfrac{x-2}{3}$, so the series converges $\Leftrightarrow$ $|r| < 1$ $\Leftrightarrow$

$\left|\dfrac{x-2}{3}\right| < 1$ $\Leftrightarrow$ $-1 < \dfrac{x-2}{3} < 1$ $\Leftrightarrow$ $-3 < x - 2 < 3$ $\Leftrightarrow$ $-1 < x < 5$. In that case, the sum of the series is

$\dfrac{a}{1-r} = \dfrac{1}{1 - \dfrac{x-2}{3}} = \dfrac{1}{\dfrac{3 - (x-2)}{3}} = \dfrac{3}{5 - x}$.

39. For $n = 1$, $a_1 = 0$ since $s_1 = 0$. For $n > 1$,

$$a_n = s_n - s_{n-1} = \dfrac{n-1}{n+1} - \dfrac{(n-1)-1}{(n-1)+1} = \dfrac{(n-1)n - (n+1)(n-2)}{(n+1)n} = \dfrac{2}{n(n+1)}$$

Also, $\displaystyle\sum_{n=1}^{\infty} a_n = \lim_{n\to\infty} s_n = \lim_{n\to\infty} \dfrac{1 - 1/n}{1 + 1/n} = 1$.

41. (a) The quantity of the drug in the body after the first tablet is 150 mg. After the second tablet, there is 150 mg plus 5%

of the first 150-mg tablet, that is, $[150 + 150(0.05)]$ mg. After the third tablet, the quantity is

$[150 + 150(0.05) + 150(0.05)^2] = 157.875$ mg. After n tablets, the quantity (in mg) is

$150 + 150(0.05) + \cdots + 150(0.05)^{n-1}$. We can use Formula 3 to write this as $\dfrac{150(1 - 0.05^n)}{1 - 0.05} = \dfrac{3000}{19}(1 - 0.05^n)$.

(b) The number of milligrams remaining in the body in the long run is $\displaystyle\lim_{n\to\infty}\left[\dfrac{3000}{19}(1 - 0.05^n)\right] = \dfrac{3000}{19}(1 - 0) \approx 157.895$,

only 0.02 mg more than the amount after 3 tablets.

43. (a) The first step in the chain occurs when the local government spends D dollars. The people who receive it spend a

fraction c of those D dollars, that is, Dc dollars. Those who receive the Dc dollars spend a fraction c of it, that is,

Dc^2 dollars. Continuing in this way, we see that the total spending after n transactions is

$$S_n = D + Dc + Dc^2 + \cdots + Dc^{n-1} = \dfrac{D(1 - c^n)}{1 - c} \text{ by (3)}.$$

(b) $\displaystyle\lim_{n\to\infty} S_n = \lim_{n\to\infty} \dfrac{D(1 - c^n)}{1 - c} = \dfrac{D}{1 - c} \lim_{n\to\infty} (1 - c^n) = \dfrac{D}{1 - c}$ $\left[\text{since } 0 < c < 1 \Rightarrow \displaystyle\lim_{n\to\infty} c^n = 0\right]$

$= \dfrac{D}{s}$ $[\text{since } c + s = 1] = kD$ $[\text{since } k = 1/s]$

If $c = 0.8$, then $s = 1 - c = 0.2$ and the multiplier is $k = 1/s = 5$.

45. $\sum\limits_{n=2}^{\infty} (1+c)^{-n}$ is a geometric series with $a = (1+c)^{-2}$ and $r = (1+c)^{-1}$, so the series converges when

$\left|(1+c)^{-1}\right| < 1 \;\Leftrightarrow\; |1+c| > 1 \;\Leftrightarrow\; 1+c > 1 \text{ or } 1+c < -1 \;\Leftrightarrow\; c > 0 \text{ or } c < -2.$ We calculate the sum of the

series and set it equal to 2: $\dfrac{(1+c)^{-2}}{1-(1+c)^{-1}} = 2 \;\Leftrightarrow\; \left(\dfrac{1}{1+c}\right)^2 = 2 - 2\left(\dfrac{1}{1+c}\right) \;\Leftrightarrow\; 1 = 2(1+c)^2 - 2(1+c) \;\Leftrightarrow$

$2c^2 + 2c - 1 = 0 \;\Leftrightarrow\; c = \dfrac{-2 \pm \sqrt{12}}{4} = \dfrac{\pm\sqrt{3}-1}{2}.$ However, the negative root is inadmissible because $-2 < \dfrac{-\sqrt{3}-1}{2} < 0.$

So $c = \dfrac{\sqrt{3}-1}{2}.$

47. Let d_n be the diameter of C_n. We draw lines from the centers of the C_i to

the center of D (or C), and using the Pythagorean Theorem, we can write

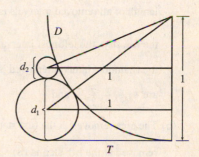

$1^2 + \left(1 - \tfrac{1}{2}d_1\right)^2 = \left(1 + \tfrac{1}{2}d_1\right)^2 \;\Leftrightarrow$

$1 = \left(1 + \tfrac{1}{2}d_1\right)^2 - \left(1 - \tfrac{1}{2}d_1\right)^2 = 2d_1$ [difference of squares] $\Rightarrow\; d_1 = \tfrac{1}{2}.$

Similarly,

$1 = \left(1 + \tfrac{1}{2}d_2\right)^2 - \left(1 - d_1 - \tfrac{1}{2}d_2\right)^2 = 2d_2 + 2d_1 - d_1^2 - d_1 d_2$

$= (2 - d_1)(d_1 + d_2) \;\Leftrightarrow$

$d_2 = \dfrac{1}{2-d_1} - d_1 = \dfrac{(1-d_1)^2}{2-d_1},\; 1 = \left(1 + \tfrac{1}{2}d_3\right)^2 - \left(1 - d_1 - d_2 - \tfrac{1}{2}d_3\right)^2 \;\Leftrightarrow\; d_3 = \dfrac{[1 - (d_1 + d_2)]^2}{2 - (d_1 + d_2)},$ and in general,

$d_{n+1} = \dfrac{\left(1 - \sum_{i=1}^{n} d_i\right)^2}{2 - \sum_{i=1}^{n} d_i}.$ If we actually calculate d_2 and d_3 from the formulas above, we find that they are $\dfrac{1}{6} = \dfrac{1}{2 \cdot 3}$ and

$\dfrac{1}{12} = \dfrac{1}{3 \cdot 4}$ respectively, so we suspect that in general, $d_n = \dfrac{1}{n(n+1)}.$ To prove this, we use induction: Assume that for all

$k \leq n,\; d_k = \dfrac{1}{k(k+1)} = \dfrac{1}{k} - \dfrac{1}{k+1}.$ Then $\sum\limits_{i=1}^{n} d_i = 1 - \dfrac{1}{n+1} = \dfrac{n}{n+1}$ [telescoping sum]. Substituting this into our

formula for d_{n+1}, we get $d_{n+1} = \dfrac{\left[1 - \dfrac{n}{n+1}\right]^2}{2 - \left(\dfrac{n}{n+1}\right)} = \dfrac{\dfrac{1}{(n+1)^2}}{\dfrac{n+2}{n+1}} = \dfrac{1}{(n+1)(n+2)},$ and the induction is complete.

Now, we observe that the partial sums $\sum_{i=1}^{n} d_i$ of the diameters of the circles approach 1 as $n \to \infty$; that is,

$\sum\limits_{n=1}^{\infty} a_n = \sum\limits_{n=1}^{\infty} \dfrac{1}{n(n+1)} = 1,$ which is what we wanted to prove.

49. The series $1 - 1 + 1 - 1 + 1 - 1 + \cdots$ diverges (geometric series with $r = -1$) so we cannot say that

$0 = 1 - 1 + 1 - 1 + 1 - 1 + \cdots.$

51. $\sum_{n=1}^{\infty} ca_n = \lim\limits_{n \to \infty} \sum_{i=1}^{n} ca_i = \lim\limits_{n \to \infty} c \sum_{i=1}^{n} a_i = c \lim\limits_{n \to \infty} \sum_{i=1}^{n} a_i = c \sum_{n=1}^{\infty} a_n,$ which exists by hypothesis.

53. Suppose on the contrary that $\sum(a_n + b_n)$ converges. Then $\sum(a_n + b_n)$ and $\sum a_n$ are convergent series. So by

Theorem 8(iii), $\sum[(a_n + b_n) - a_n]$ would also be convergent. But $\sum[(a_n + b_n) - a_n] = \sum b_n,$ a contradiction, since

$\sum b_n$ is given to be divergent.

55. The partial sums $\{s_n\}$ form an increasing sequence, since $s_n - s_{n-1} = a_n > 0$ for all n. Also, the sequence $\{s_n\}$ is bounded since $s_n \le 1000$ for all n. So by the Monotonic Sequence Theorem, the sequence of partial sums converges, that is, the series $\sum a_n$ is convergent.

57. (a) At the first step, only the interval $\left(\frac{1}{3}, \frac{2}{3}\right)$ (length $\frac{1}{3}$) is removed. At the second step, we remove the intervals $\left(\frac{1}{9}, \frac{2}{9}\right)$ and $\left(\frac{7}{9}, \frac{8}{9}\right)$, which have a total length of $2 \cdot \left(\frac{1}{3}\right)^2$. At the third step, we remove 2^2 intervals, each of length $\left(\frac{1}{3}\right)^3$. In general, at the nth step we remove 2^{n-1} intervals, each of length $\left(\frac{1}{3}\right)^n$, for a length of $2^{n-1} \cdot \left(\frac{1}{3}\right)^n = \frac{1}{3}\left(\frac{2}{3}\right)^{n-1}$. Thus, the total length of all removed intervals is $\sum_{n=1}^{\infty} \frac{1}{3}\left(\frac{2}{3}\right)^{n-1} = \frac{1/3}{1 - 2/3} = 1$ [geometric series with $a = \frac{1}{3}$ and $r = \frac{2}{3}$]. Notice that at the nth step, the leftmost interval that is removed is $\left(\left(\frac{1}{3}\right)^n, \left(\frac{2}{3}\right)^n\right)$, so we never remove 0, and 0 is in the Cantor set. Also, the rightmost interval removed is $\left(1 - \left(\frac{2}{3}\right)^n, 1 - \left(\frac{1}{3}\right)^n\right)$, so 1 is never removed. Some other numbers in the Cantor set are $\frac{1}{3}, \frac{2}{3}, \frac{1}{9}, \frac{2}{9}, \frac{7}{9}$, and $\frac{8}{9}$.

(b) The area removed at the first step is $\frac{1}{9}$; at the second step, $8 \cdot \left(\frac{1}{9}\right)^2$; at the third step, $(8)^2 \cdot \left(\frac{1}{9}\right)^3$. In general, the area removed at the nth step is $(8)^{n-1}\left(\frac{1}{9}\right)^n = \frac{1}{9}\left(\frac{8}{9}\right)^{n-1}$, so the total area of all removed squares is

$$\sum_{n=1}^{\infty} \frac{1}{9}\left(\frac{8}{9}\right)^{n-1} = \frac{1/9}{1 - 8/9} = 1.$$

59. (a) For $\sum_{n=1}^{\infty} \frac{n}{(n+1)!}$, $s_1 = \frac{1}{1 \cdot 2} = \frac{1}{2}$, $s_2 = \frac{1}{2} + \frac{2}{1 \cdot 2 \cdot 3} = \frac{5}{6}$, $s_3 = \frac{5}{6} + \frac{3}{1 \cdot 2 \cdot 3 \cdot 4} = \frac{23}{24}$,

$s_4 = \frac{23}{24} + \frac{4}{1 \cdot 2 \cdot 3 \cdot 4 \cdot 5} = \frac{119}{120}$. The denominators are $(n+1)!$, so a guess would be $s_n = \frac{(n+1)! - 1}{(n+1)!}$.

(b) For $n = 1$, $s_1 = \frac{1}{2} = \frac{2! - 1}{2!}$, so the formula holds for $n = 1$. Assume $s_k = \frac{(k+1)! - 1}{(k+1)!}$. Then

$$s_{k+1} = \frac{(k+1)! - 1}{(k+1)!} + \frac{k+1}{(k+2)!} = \frac{(k+1)! - 1}{(k+1)!} + \frac{k+1}{(k+1)!(k+2)} = \frac{(k+2)! - (k+2) + k + 1}{(k+2)!}$$

$$= \frac{(k+2)! - 1}{(k+2)!}$$

Thus, the formula is true for $n = k + 1$. So by induction, the guess is correct.

(c) $\lim_{n \to \infty} s_n = \lim_{n \to \infty} \frac{(n+1)! - 1}{(n+1)!} = \lim_{n \to \infty} \left[1 - \frac{1}{(n+1)!}\right] = 1$ and so $\sum_{n=1}^{\infty} \frac{n}{(n+1)!} = 1$.

8.3 The Integral and Comparison Tests

1. The picture shows that $a_2 = \frac{1}{2^{1.3}} < \int_1^2 \frac{1}{x^{1.3}}\,dx$,

$a_3 = \frac{1}{3^{1.3}} < \int_2^3 \frac{1}{x^{1.3}}\,dx$, and so on, so $\sum_{n=2}^{\infty} \frac{1}{n^{1.3}} < \int_1^{\infty} \frac{1}{x^{1.3}}\,dx$. The

integral converges by (6.6.2) with $p = 1.3 > 1$, so the series converges.

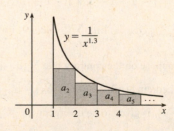

3. (a) We cannot say anything about $\sum a_n$. If $a_n > b_n$ for all n and $\sum b_n$ is convergent, then $\sum a_n$ could be convergent or

divergent. (See the note after Example 4.)

(b) If $a_n < b_n$ for all n, then $\sum a_n$ is convergent. [This is part (i) of the Comparison Test.]

5. $\sum\limits_{n=1}^{\infty} n^b$ is a p-series with $p = -b$. $\sum\limits_{n=1}^{\infty} b^n$ is a geometric series. By (1), the p-series is convergent if $p > 1$. In this case,

$\sum\limits_{n=1}^{\infty} n^b = \sum\limits_{n=1}^{\infty} \left(1/n^{-b}\right)$, so $-b > 1 \iff b < -1$ are the values for which the series converge. A geometric series

$\sum\limits_{n=1}^{\infty} ar^{n-1}$ converges if $|r| < 1$, so $\sum\limits_{n=1}^{\infty} b^n$ converges if $|b| < 1 \iff -1 < b < 1$.

7. The function $f(x) = 1/\sqrt[5]{x} = x^{-1/5}$ is continuous, positive, and decreasing on $[1, \infty)$, so the Integral Test applies.

$\int_1^{\infty} x^{-1/5}\,dx = \lim\limits_{t\to\infty} \int_1^t x^{-1/5}\,dx = \lim\limits_{t\to\infty} \left[\frac{5}{4}x^{4/5}\right]_1^t = \lim\limits_{t\to\infty}\left(\frac{5}{4}t^{4/5} - \frac{5}{4}\right) = \infty$, so $\sum\limits_{n=1}^{\infty} 1/\sqrt[5]{n}$ diverges.

9. $\dfrac{n}{2n^3 + 1} < \dfrac{n}{2n^3} = \dfrac{1}{2n^2} < \dfrac{1}{n^2}$ for all $n \ge 1$, so $\sum\limits_{n=1}^{\infty} \dfrac{n}{2n^3 + 1}$ converges by comparison with $\sum\limits_{n=1}^{\infty} \dfrac{1}{n^2}$, which converges

because it is a p-series with $p = 2 > 1$.

11. The series $\sum\limits_{n=1}^{\infty} \dfrac{1}{n^{0.85}}$ is a p-series with $p = 0.85 \le 1$, so it diverges by (1). Therefore, the series $\sum\limits_{n=1}^{\infty} \dfrac{2}{n^{0.85}}$ must also diverge,

for if it converged, then $\sum\limits_{n=1}^{\infty} \dfrac{1}{n^{0.85}}$ would have to converge [by Theorem 8(i) in Section 8.2].

13. $1 + \dfrac{1}{8} + \dfrac{1}{27} + \dfrac{1}{64} + \dfrac{1}{125} + \cdots = \sum\limits_{n=1}^{\infty} \dfrac{1}{n^3}$. This is a p-series with $p = 3 > 1$, so it converges by (1).

15. $1 + \dfrac{1}{3} + \dfrac{1}{5} + \dfrac{1}{7} + \dfrac{1}{9} + \cdots = \sum\limits_{n=1}^{\infty} \dfrac{1}{2n - 1}$. The function $f(x) = \dfrac{1}{2x - 1}$ is continuous, positive, and decreasing on $[1, \infty)$,

so the Integral Test applies.

$\int_1^{\infty} \dfrac{1}{2x - 1}\,dx = \lim\limits_{t\to\infty} \int_1^t \dfrac{1}{2x - 1}\,dx = \lim\limits_{t\to\infty}\left[\frac{1}{2}\ln|2x - 1|\right]_1^t = \frac{1}{2}\lim\limits_{t\to\infty}\left(\ln(2t - 1) - 0\right) = \infty$, so the series $\sum\limits_{n=1}^{\infty} \dfrac{1}{2n - 1}$

diverges.

17. $f(x) = xe^{-x}$ is continuous and positive on $[1, \infty)$. $f'(x) = -xe^{-x} + e^{-x} = e^{-x}(1 - x) < 0$ for $x > 1$, so f is decreasing

on $[1, \infty)$. Thus, the Integral Test applies.

$\int_1^{\infty} xe^{-x}\,dx = \lim\limits_{b\to\infty} \int_1^b xe^{-x}\,dx = \lim\limits_{b\to\infty}\left[-xe^{-x} - e^{-x}\right]_1^b$ [by parts] $= \lim\limits_{b\to\infty}\left[-be^{-b} - e^{-b} + e^{-1} + e^{-1}\right] = 2/e$

since $\lim\limits_{b\to\infty} be^{-b} = \lim\limits_{b\to\infty} (b/e^b) \overset{\text{H}}{=} \lim\limits_{b\to\infty}(1/e^b) = 0$ and $\lim\limits_{b\to\infty} e^{-b} = 0$. Thus, $\sum\limits_{n=1}^{\infty} ne^{-n}$ converges.

19. $f(x) = \dfrac{1}{x \ln x}$ is continuous and positive on $[2, \infty)$, and also decreasing since $f'(x) = -\dfrac{1 + \ln x}{x^2 (\ln x)^2} < 0$ for $x > 2$, so we can

use the Integral Test. $\displaystyle\int_2^\infty \dfrac{1}{x \ln x}\, dx = \lim_{t \to \infty} \left[\ln(\ln x)\right]_2^t = \lim_{t \to \infty} \left[\ln(\ln t) - \ln(\ln 2)\right] = \infty$, so the series $\displaystyle\sum_{n=2}^\infty \dfrac{1}{n \ln n}$ diverges.

21. $\dfrac{\cos^2 n}{n^2 + 1} \le \dfrac{1}{n^2 + 1} < \dfrac{1}{n^2}$, so the series $\displaystyle\sum_{n=1}^\infty \dfrac{\cos^2 n}{n^2 + 1}$ converges by comparison with the p-series $\displaystyle\sum_{n=1}^\infty \dfrac{1}{n^2}$ $[p = 2 > 1]$.

23. $\dfrac{n-1}{n\, 4^n}$ is positive for $n > 1$ and $\dfrac{n-1}{n\, 4^n} < \dfrac{n}{n\, 4^n} = \dfrac{1}{4^n} = \left(\dfrac{1}{4}\right)^n$, so $\displaystyle\sum_{n=1}^\infty \dfrac{n-1}{n\, 4^n}$ converges by comparison with the convergent

geometric series $\displaystyle\sum_{n=1}^\infty \left(\dfrac{1}{4}\right)^n$.

25. Use the Limit Comparison Test with $a_n = \dfrac{1 + 4^n}{1 + 3^n}$ and $b_n = \dfrac{4^n}{3^n}$:

$$\lim_{n \to \infty} \frac{a_n}{b_n} = \lim_{n \to \infty} \frac{\dfrac{1 + 4^n}{1 + 3^n}}{\dfrac{4^n}{3^n}} = \lim_{n \to \infty} \frac{1 + 4^n}{1 + 3^n} \cdot \frac{3^n}{4^n} = \lim_{n \to \infty} \frac{1 + 4^n}{4^n} \cdot \frac{3^n}{1 + 3^n} = \lim_{n \to \infty} \left(\frac{1}{4^n} + 1\right) \cdot \frac{1}{\dfrac{1}{3^n} + 1} = 1 > 0$$

Since the geometric series $\sum b_n = \sum \left(\tfrac{4}{3}\right)^n$ diverges, so does $\displaystyle\sum_{n=1}^\infty \dfrac{1 + 4^n}{1 + 3^n}$. Alternatively, use the Comparison Test with

$\dfrac{1 + 4^n}{1 + 3^n} > \dfrac{1 + 4^n}{3^n + 3^n} > \dfrac{4^n}{2(3^n)} = \dfrac{1}{2}\left(\dfrac{4}{3}\right)^n$ or use the Test for Divergence.

27. $\dfrac{2 + (-1)^n}{n\sqrt{n}} \le \dfrac{3}{n\sqrt{n}}$, and $\displaystyle\sum_{n=1}^\infty \dfrac{3}{n\sqrt{n}}$ converges because it is a constant multiple of the convergent p-series $\displaystyle\sum_{n=1}^\infty \dfrac{1}{n\sqrt{n}}$

$\left[p = \tfrac{3}{2} > 1\right]$, so the given series converges by the Comparison Test.

29. Use the Limit Comparison Test with $a_n = \sin\left(\dfrac{1}{n}\right)$ and $b_n = \dfrac{1}{n}$. Then $\sum a_n$ and $\sum b_n$ are series with positive terms and

$$\lim_{n \to \infty} \frac{a_n}{b_n} = \lim_{n \to \infty} \frac{\sin(1/n)}{1/n} = \lim_{\theta \to 0} \frac{\sin\theta}{\theta} = 1 > 0. \text{ Since } \sum_{n=1}^\infty b_n \text{ is the divergent harmonic series,}$$

$\displaystyle\sum_{n=1}^\infty \sin(1/n)$ also diverges. [Note that we could also use l'Hospital's Rule to evaluate the limit:

$$\lim_{x \to \infty} \frac{\sin(1/x)}{1/x} \overset{\mathrm{H}}{=} \lim_{x \to \infty} \frac{\cos(1/x) \cdot (-1/x^2)}{-1/x^2} = \lim_{x \to \infty} \cos\frac{1}{x} = \cos 0 = 1.]$$

31. We have already shown (in Exercise 19) that when $p = 1$ the series $\displaystyle\sum_{n=2}^\infty \dfrac{1}{n(\ln n)^p}$ diverges, so assume that $p \ne 1$.

$f(x) = \dfrac{1}{x(\ln x)^p}$ is continuous and positive on $[2, \infty)$, and $f'(x) = -\dfrac{p + \ln x}{x^2(\ln x)^{p+1}} < 0$ if $x > e^{-p}$, so that f is eventually

decreasing and we can use the Integral Test.

$$\int_2^\infty \frac{1}{x(\ln x)^p}\, dx = \lim_{t \to \infty} \left[\frac{(\ln x)^{1-p}}{1-p}\right]_2^t \quad [\text{for } p \ne 1] = \lim_{t \to \infty} \left[\frac{(\ln t)^{1-p}}{1-p} - \frac{(\ln 2)^{1-p}}{1-p}\right]$$

This limit exists whenever $1 - p < 0 \iff p > 1$, so the series converges for $p > 1$.

33. (a)

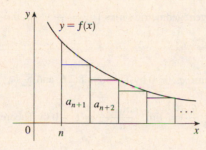 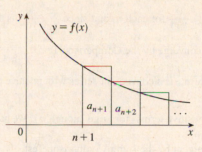

We use the same notation and ideas as in the Integral Test, assuming that f is decreasing on $[n, \infty)$. Comparing the areas of the rectangles with the area under $y = f(x)$ for $x > n$ in the first figure, we see that

$$R_n = a_{n+1} + a_{n+2} + \cdots \le \int_n^\infty f(x)\,dx$$

Similarly, we see from the second figure that

$$R_n = a_{n+1} + a_{n+2} + \cdots \ge \int_{n+1}^\infty f(x)\,dx$$

So we have proved that $\int_{n+1}^\infty f(x)\,dx \le R_n \le \int_n^\infty f(x)\,dx$.

(b) If we add s_n to each side of the inequalities in part (a), we get $s_n + \int_{n+1}^\infty f(x)\,dx \le s \le s_n + \int_n^\infty f(x)\,dx$ because $s_n + R_n = s$.

35. (a) $f(x) = \dfrac{1}{x^2}$ is positive and continuous and $f'(x) = -\dfrac{2}{x^3}$ is negative for $x > 0$, and so the Integral Test applies.

$$\sum_{n=1}^\infty \frac{1}{n^2} \approx s_{10} = \frac{1}{1^2} + \frac{1}{2^2} + \frac{1}{3^2} + \cdots + \frac{1}{10^2} \approx 1.549768. \text{ From Exercise 33(a) we have}$$

$$R_{10} \le \int_{10}^\infty \frac{1}{x^2}\,dx = \lim_{t \to \infty} \left[\frac{-1}{x}\right]_{10}^t = \lim_{t \to \infty} \left(-\frac{1}{t} + \frac{1}{10}\right) = \frac{1}{10}, \text{ so the error is at most } 0.1.$$

(b) $s_{10} + \displaystyle\int_{11}^\infty \frac{1}{x^2}\,dx \le s \le s_{10} + \int_{10}^\infty \frac{1}{x^2}\,dx \quad \Rightarrow \quad s_{10} + \frac{1}{11} \le s \le s_{10} + \frac{1}{10} \quad \Rightarrow$

$1.549768 + 0.090909 = 1.640677 \le s \le 1.549768 + 0.1 = 1.649768$, so we get $s \approx 1.64522$ (the average of 1.640677 and 1.649768) with error ≤ 0.005 (the maximum of $1.649768 - 1.64522$ and $1.64522 - 1.640677$, rounded up).

(c) $R_n \le \displaystyle\int_n^\infty \frac{1}{x^2}\,dx = \frac{1}{n}$. So $R_n < 0.001$ if $\dfrac{1}{n} < \dfrac{1}{1000} \quad \Leftrightarrow \quad n > 1000$.

37. (a) From the figure, $a_2 + a_3 + \cdots + a_n \le \int_1^n f(x)\,dx$, so with

$$f(x) = \frac{1}{x}, \ \frac{1}{2} + \frac{1}{3} + \frac{1}{4} + \cdots + \frac{1}{n} \le \int_1^n \frac{1}{x}\,dx = \ln n.$$

Thus, $s_n = 1 + \dfrac{1}{2} + \dfrac{1}{3} + \dfrac{1}{4} + \cdots + \dfrac{1}{n} \le 1 + \ln n.$

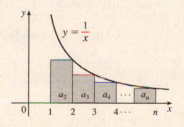

(b) By part (a), $s_{10^6} \le 1 + \ln 10^6 \approx 14.82 < 15$ and

$s_{10^9} \le 1 + \ln 10^9 \approx 21.72 < 22.$

39. Since $\dfrac{d_n}{10^n} \leq \dfrac{9}{10^n}$ for each n, and since $\displaystyle\sum_{n=1}^{\infty} \dfrac{9}{10^n}$ is a convergent geometric series $\left(|r| = \frac{1}{10} < 1\right)$, $0.d_1 d_2 d_3 \ldots = \displaystyle\sum_{n=1}^{\infty} \dfrac{d_n}{10^n}$

will always converge by the Comparison Test.

41. Yes. Since $\sum a_n$ is a convergent series with positive terms, $\displaystyle\lim_{n\to\infty} a_n = 0$ by Theorem 8.2.6, and $\sum b_n = \sum \sin(a_n)$ is a

series with positive terms (for large enough n). We have $\displaystyle\lim_{n\to\infty} \dfrac{b_n}{a_n} = \lim_{n\to\infty} \dfrac{\sin(a_n)}{a_n} = 1 > 0$ by Theorem 1.4.6. Thus, $\sum b_n$

is also convergent by the Limit Comparison Test.

43. (a) Since $\displaystyle\lim_{n\to\infty} \dfrac{a_n}{b_n} = \infty$, there is an integer N such that $\dfrac{a_n}{b_n} > 1$ whenever $n > N$. (Take $M = 1$ in Definition 8.1.5.)

   Then $a_n > b_n$ whenever $n > N$ and since $\sum b_n$ is divergent, $\sum a_n$ is also divergent by the Comparison Test.

   (b) (i) If $a_n = \dfrac{1}{\ln n}$ and $b_n = \dfrac{1}{n}$ for $n \geq 2$, then $\displaystyle\lim_{n\to\infty} \dfrac{a_n}{b_n} = \lim_{n\to\infty} \dfrac{n}{\ln n} = \lim_{x\to\infty} \dfrac{x}{\ln x} \overset{\text{H}}{=} \lim_{x\to\infty} \dfrac{1}{1/x} = \lim_{x\to\infty} x = \infty$,

   so by part (a), $\displaystyle\sum_{n=2}^{\infty} \dfrac{1}{\ln n}$ is divergent.

   (ii) If $a_n = \dfrac{\ln n}{n}$ and $b_n = \dfrac{1}{n}$, then $\displaystyle\sum_{n=1}^{\infty} b_n$ is the divergent harmonic series and $\displaystyle\lim_{n\to\infty} \dfrac{a_n}{b_n} = \lim_{n\to\infty} \ln n = \lim_{x\to\infty} \ln x = \infty$,

   so $\displaystyle\sum_{n=1}^{\infty} a_n$ diverges by part (a).

45. Since $\sum a_n$ converges, $\displaystyle\lim_{n\to\infty} a_n = 0$, so there exists N such that $|a_n - 0| < 1$ for all $n > N$ $\Rightarrow$ $0 \leq a_n < 1$ for

all $n > N$ $\Rightarrow$ $0 \leq a_n^2 \leq a_n$. Since $\sum a_n$ converges, so does $\sum a_n^2$ by the Comparison Test.

47. $\displaystyle\lim_{n\to\infty} n a_n = \lim_{n\to\infty} \dfrac{a_n}{1/n}$, so we apply the Limit Comparison Test with $b_n = \dfrac{1}{n}$. Since $\displaystyle\lim_{n\to\infty} n a_n > 0$ we know that either both

series converge or both series diverge, and we also know that $\displaystyle\sum_{n=1}^{\infty} \dfrac{1}{n}$ diverges [p-series with $p = 1$]. Therefore, $\sum a_n$ must be

divergent.

8.4 Other Convergence Tests

1. (a) An alternating series is a series whose terms are alternately positive and negative.

   (b) An alternating series $\displaystyle\sum_{n=1}^{\infty} a_n = \sum_{n=1}^{\infty} (-1)^{n-1} b_n$, where $b_n = |a_n|$, converges if $0 < b_{n+1} \leq b_n$ for all n and $\displaystyle\lim_{n\to\infty} b_n = 0$.

   (This is the Alternating Series Test.)

   (c) The error involved in using the partial sum s_n as an approximation to the total sum s is the remainder $R_n = s - s_n$ and the

   size of the error is smaller than b_{n+1}; that is, $|R_n| \leq b_{n+1}$. (This is the Alternating Series Estimation Theorem.)

3. $\dfrac{4}{7} - \dfrac{4}{8} + \dfrac{4}{9} - \dfrac{4}{10} + \dfrac{4}{11} - \cdots = \displaystyle\sum_{n=1}^{\infty} (-1)^{n-1} \dfrac{4}{n+6}$. Now $b_n = \dfrac{4}{n+6} > 0$, $\{b_n\}$ is decreasing, and $\displaystyle\lim_{n\to\infty} b_n = 0$, so the

series converges by the Alternating Series Test.

5. $\displaystyle\sum_{n=1}^{\infty} a_n = \sum_{n=1}^{\infty} (-1)^{n-1} \frac{1}{2n+1} = \sum_{n=1}^{\infty} (-1)^{n-1} b_n$. Now $b_n = \dfrac{1}{2n+1} > 0$, $\{b_n\}$ is decreasing, and $\displaystyle\lim_{n\to\infty} b_n = 0$, so the

series converges by the Alternating Series Test.

7. $\displaystyle\sum_{n=1}^{\infty} a_n = \sum_{n=1}^{\infty} (-1)^n \frac{3n-1}{2n+1} = \sum_{n=1}^{\infty} (-1)^n b_n$. Now $\displaystyle\lim_{n\to\infty} b_n = \lim_{n\to\infty} \frac{3-1/n}{2+1/n} = \frac{3}{2} \neq 0$. Since $\displaystyle\lim_{n\to\infty} a_n \neq 0$

(in fact the limit does not exist), the series diverges by the Test for Divergence.

9. The series $\displaystyle\sum_{n=1}^{\infty} \frac{(-1)^{n+1}}{n^6}$ satisfies (i) of the Alternating Series Test because $\dfrac{1}{(n+1)^6} < \dfrac{1}{n^6}$ and (ii) $\displaystyle\lim_{n\to\infty} \frac{1}{n^6} = 0$, so the

series is convergent. Now $b_5 = \dfrac{1}{5^6} = 0.000064 > 0.00005$ and $b_6 = \dfrac{1}{6^6} \approx 0.00002 < 0.00005$, so by the Alternating Series

Estimation Theorem, $n = 5$. (That is, since the 6th term is less than the desired error, we need to add the first 5 terms to get the

sum to the desired accuracy.)

11. The series $\displaystyle\sum_{n=0}^{\infty} \frac{(-1)^n}{10^n\,n!}$ satisfies (i) of the Alternating Series Test because $\dfrac{1}{10^{n+1}(n+1)!} < \dfrac{1}{10^n\,n!}$ and (ii) $\displaystyle\lim_{n\to\infty} \frac{1}{10^n\,n!} = 0$,

so the series is convergent. Now $b_3 = \dfrac{1}{10^3\,3!} \approx 0.000\,167 > 0.000\,005$ and $b_4 = \dfrac{1}{10^4\,4!} = 0.000\,004 < 0.000\,005$, so by

the Alternating Series Estimation Theorem, $n = 4$ (since the series starts with $n = 0$, not $n = 1$). (That is, since the 5th term

is less than the desired error, we need to add the first 4 terms to get the sum to the desired accuracy.)

13. $b_4 = \dfrac{1}{8!} = \dfrac{1}{40{,}320} \approx 0.000\,025$, so

$$\sum_{n=1}^{\infty} \frac{(-1)^n}{(2n)!} \approx s_3 = \sum_{n=1}^{3} \frac{(-1)^n}{(2n)!} = -\frac{1}{2} + \frac{1}{24} - \frac{1}{720} \approx -0.459\,722$$

Adding b_4 to s_3 does not change the fourth decimal place of s_3, so by the Alternating Series Estimation Theorem, the sum of

the series, correct to four decimal places, is -0.4597.

15. $b_7 = \dfrac{7^2}{10^7} = 0.000\,004\,9$, so

$$\sum_{n=1}^{\infty} \frac{(-1)^{n-1}n^2}{10^n} \approx s_6 = \sum_{n=1}^{6} \frac{(-1)^{n-1}n^2}{10^n} = \frac{1}{10} - \frac{4}{100} + \frac{9}{1000} - \frac{16}{10{,}000} + \frac{25}{100{,}000} - \frac{36}{1{,}000{,}000} = 0.067\,614$$

Adding b_7 to s_6 does not change the fourth decimal place of s_6, so by the Alternating Series Estimation Theorem, the sum of

the series, correct to four decimal places, is 0.0676.

17. $\displaystyle\sum_{n=1}^{\infty} \frac{(-1)^{n-1}}{n} = 1 - \frac{1}{2} + \frac{1}{3} - \frac{1}{4} + \cdots + \frac{1}{49} - \frac{1}{50} + \frac{1}{51} - \frac{1}{52} + \cdots$. The 50th partial sum of this series is an

underestimate, since $\displaystyle\sum_{n=1}^{\infty} \frac{(-1)^{n-1}}{n} = s_{50} + \left(\frac{1}{51} - \frac{1}{52}\right) + \left(\frac{1}{53} - \frac{1}{54}\right) + \cdots$, and the terms in parentheses are all positive.

The result can be seen geometrically in Figure 1.

19. $\displaystyle\lim_{n\to\infty} \left| \frac{a_{n+1}}{a_n} \right| = \lim_{n\to\infty} \left| \frac{n+1}{5^{n+1}} \cdot \frac{5^n}{n} \right| = \lim_{n\to\infty} \left| \frac{1}{5} \cdot \frac{n+1}{n} \right| = \frac{1}{5} \lim_{n\to\infty} \frac{1+1/n}{1} = \frac{1}{5}(1) = \frac{1}{5} < 1$, so the series $\displaystyle\sum_{n=1}^{\infty} \frac{n}{5^n}$ is

absolutely convergent by the Ratio Test.

21. Using the Ratio Test, $\lim\limits_{n\to\infty}\left|\dfrac{a_{n+1}}{a_n}\right| = \lim\limits_{n\to\infty}\left|\dfrac{(-10)^{n+1}}{(n+1)!}\cdot\dfrac{n!}{(-10)^n}\right| = \lim\limits_{n\to\infty}\left|\dfrac{-10}{n+1}\right| = 0 < 1$, so the series $\sum\limits_{n=0}^{\infty}\dfrac{(-10)^n}{n!}$ is absolutely convergent.

23. $b_n = \dfrac{1}{5n+1} > 0$ for $n \geq 0$, $\{b_n\}$ is decreasing for $n \geq 0$, and $\lim\limits_{n\to\infty} b_n = 0$, so $\sum\limits_{n=0}^{\infty}\dfrac{(-1)^n}{5n+1}$ converges by the Alternating

Series Test. To determine absolute convergence, choose $a_n = \dfrac{1}{n}$ to get

$\lim\limits_{n\to\infty}\dfrac{a_n}{b_n} = \lim\limits_{n\to\infty}\dfrac{1/n}{1/(5n+1)} = \lim\limits_{n\to\infty}\dfrac{5n+1}{n} = 5 > 0$, so $\sum\limits_{n=1}^{\infty}\dfrac{1}{5n+1}$ diverges by the Limit Comparison Test with the

harmonic series. Thus, the series $\sum\limits_{n=0}^{\infty}\dfrac{(-1)^n}{5n+1}$ is conditionally convergent.

25. $\lim\limits_{k\to\infty}\left|\dfrac{a_{k+1}}{a_k}\right| = \lim\limits_{k\to\infty}\left[\dfrac{(k+1)\left(\frac{2}{3}\right)^{k+1}}{k\left(\frac{2}{3}\right)^k}\right] = \lim\limits_{k\to\infty}\dfrac{k+1}{k}\left(\dfrac{2}{3}\right)^1 = \dfrac{2}{3}\lim\limits_{k\to\infty}\left(1+\dfrac{1}{k}\right) = \tfrac{2}{3}(1) = \tfrac{2}{3} < 1$, so the series

$\sum\limits_{n=1}^{\infty} k\left(\dfrac{2}{3}\right)^k$ is absolutely convergent by the Ratio Test. Since the terms of this series are positive, absolute convergence is the

same as convergence.

27. $\lim\limits_{n\to\infty}\left|\dfrac{a_{n+1}}{a_n}\right| = \lim\limits_{n\to\infty}\left[\dfrac{10^{n+1}}{(n+2)4^{2(n+1)+1}}\cdot\dfrac{(n+1)4^{2n+1}}{10^n}\right] = \lim\limits_{n\to\infty}\left[\dfrac{10^{n+1}}{(n+2)4^{2n+3}}\cdot\dfrac{(n+1)4^{2n+1}}{10^n}\right]$

$\qquad = \lim\limits_{n\to\infty}\left(\dfrac{10}{4^2}\cdot\dfrac{n+1}{n+2}\right) = \dfrac{5}{8} < 1$,

so the series $\sum\limits_{n=1}^{\infty}\dfrac{10^n}{(n+1)4^{2n+1}}$ is absolutely convergent by the Ratio Test. Since the terms of this series are positive, absolute

convergence is the same as convergence.

29. $\dfrac{|\cos(n\pi/3)|}{n!} \leq \dfrac{1}{n!}$ and $\sum\limits_{n=1}^{\infty}\dfrac{1}{n!}$ converges (use the Ratio Test), so the series $\sum\limits_{n=1}^{\infty}\dfrac{\cos(n\pi/3)}{n!}$ converges absolutely by the

Comparison Test.

31. $\left|\dfrac{(-1)^n\arctan n}{n^2}\right| < \dfrac{\pi/2}{n^2}$, so since $\sum\limits_{n=1}^{\infty}\dfrac{\pi/2}{n^2} = \dfrac{\pi}{2}\sum\limits_{n=1}^{\infty}\dfrac{1}{n^2}$ converges ($p = 2 > 1$), the given series $\sum\limits_{n=1}^{\infty}\dfrac{(-1)^n\arctan n}{n^2}$

converges absolutely by the Comparison Test.

33. $\lim\limits_{n\to\infty}\sqrt[n]{|a_n|} = \lim\limits_{n\to\infty}\dfrac{n^2+1}{2n^2+1} = \lim\limits_{n\to\infty}\dfrac{1+1/n^2}{2+1/n^2} = \dfrac{1}{2} < 1$, so the series $\sum\limits_{n=1}^{\infty}\left(\dfrac{n^2+1}{2n^2+1}\right)^n$ is absolutely convergent by the

Root Test.

35. $\lim\limits_{n\to\infty}\sqrt[n]{|a_n|} = \lim\limits_{n\to\infty}\sqrt[n]{\left(1+\dfrac{1}{n}\right)^{n^2}} = \lim\limits_{n\to\infty}\left(1+\dfrac{1}{n}\right)^n = e > 1$ [see the margin note on page 468], so the series

$\sum\limits_{n=1}^{\infty}\left(1+\dfrac{1}{n}\right)^{n^2}$ diverges by the Root Test.

37. Use the Ratio Test with the series

$$1 - \frac{1 \cdot 3}{3!} + \frac{1 \cdot 3 \cdot 5}{5!} - \frac{1 \cdot 3 \cdot 5 \cdot 7}{7!} + \cdots + (-1)^{n-1}\frac{1 \cdot 3 \cdot 5 \cdots (2n-1)}{(2n-1)!} + \cdots = \sum_{n=1}^{\infty}(-1)^{n-1}\frac{1 \cdot 3 \cdot 5 \cdots (2n-1)}{(2n-1)!}.$$

$$\lim_{n \to \infty}\left|\frac{a_{n+1}}{a_n}\right| = \lim_{n \to \infty}\left|\frac{(-1)^n \cdot 1 \cdot 3 \cdot 5 \cdots (2n-1)[2(n+1)-1]}{[2(n+1)-1]!} \cdot \frac{(2n-1)!}{(-1)^{n-1} \cdot 1 \cdot 3 \cdot 5 \cdots (2n-1)}\right|$$

$$= \lim_{n \to \infty}\left|\frac{(-1)(2n+1)(2n-1)!}{(2n+1)(2n)(2n-1)!}\right| = \lim_{n \to \infty}\frac{1}{2n} = 0 < 1,$$

so the given series is absolutely convergent and therefore convergent.

39. $\displaystyle\sum_{n=1}^{\infty}\frac{2 \cdot 4 \cdot 6 \cdots (2n)}{n!} = \sum_{n=1}^{\infty}\frac{(2 \cdot 1) \cdot (2 \cdot 2) \cdot (2 \cdot 3) \cdots (2 \cdot n)}{n!} = \sum_{n=1}^{\infty}\frac{2^n n!}{n!} = \sum_{n=1}^{\infty}2^n$, which diverges by the Test for

Divergence since $\displaystyle\lim_{n \to \infty}2^n = \infty$.

41. The series $\displaystyle\sum_{n=1}^{\infty}\frac{b_n^n \cos n\pi}{n} = \sum_{n=1}^{\infty}(-1)^n\frac{b_n^n}{n}$, where $b_n > 0$ for $n \geq 1$ and $\displaystyle\lim_{n \to \infty}b_n = \frac{1}{2}$.

$$\lim_{n \to \infty}\left|\frac{a_{n+1}}{a_n}\right| = \lim_{n \to \infty}\left|\frac{(-1)^{n+1}b_n^{n+1}}{n+1} \cdot \frac{n}{(-1)^n b_n^n}\right| = \lim_{n \to \infty}b_n\frac{n}{n+1} = \frac{1}{2}(1) = \frac{1}{2} < 1,$$ so the series $\displaystyle\sum_{n=1}^{\infty}\frac{b_n^n \cos n\pi}{n}$ is

absolutely convergent by the Ratio Test.

43. (a) $\displaystyle\lim_{n \to \infty}\left|\frac{1/(n+1)^3}{1/n^3}\right| = \lim_{n \to \infty}\frac{n^3}{(n+1)^3} = \lim_{n \to \infty}\frac{1}{(1+1/n)^3} = 1$. Inconclusive

(b) $\displaystyle\lim_{n \to \infty}\left|\frac{(n+1)}{2^{n+1}} \cdot \frac{2^n}{n}\right| = \lim_{n \to \infty}\frac{n+1}{2n} = \lim_{n \to \infty}\left(\frac{1}{2} + \frac{1}{2n}\right) = \frac{1}{2}$. Conclusive (convergent)

(c) $\displaystyle\lim_{n \to \infty}\left|\frac{(-3)^n}{\sqrt{n+1}} \cdot \frac{\sqrt{n}}{(-3)^{n-1}}\right| = 3\lim_{n \to \infty}\sqrt{\frac{n}{n+1}} = 3\lim_{n \to \infty}\sqrt{\frac{1}{1+1/n}} = 3$. Conclusive (divergent)

(d) $\displaystyle\lim_{n \to \infty}\left|\frac{\sqrt{n+1}}{1+(n+1)^2} \cdot \frac{1+n^2}{\sqrt{n}}\right| = \lim_{n \to \infty}\left[\sqrt{1 + \frac{1}{n}} \cdot \frac{1/n^2 + 1}{1/n^2 + (1+1/n)^2}\right] = 1$. Inconclusive

45. (a) $\displaystyle\lim_{n \to \infty}\left|\frac{a_{n+1}}{a_n}\right| = \lim_{n \to \infty}\left|\frac{x^{n+1}}{(n+1)!} \cdot \frac{n!}{x^n}\right| = \lim_{n \to \infty}\left|\frac{x}{n+1}\right| = |x|\lim_{n \to \infty}\frac{1}{n+1} = |x| \cdot 0 = 0 < 1$, so by the Ratio Test the

series $\displaystyle\sum_{n=0}^{\infty}\frac{x^n}{n!}$ converges for all x.

(b) Since the series of part (a) always converges, we must have $\displaystyle\lim_{n \to \infty}\frac{x^n}{n!} = 0$ by Theorem 8.2.6.

47. (i) Following the hint, we get that $|a_n| < r^n$ for $n \geq N$, and so since the geometric series $\sum_{n=1}^{\infty}r^n$ converges $[0 < r < 1]$,

the series $\sum_{n=N}^{\infty}|a_n|$ converges as well by the Comparison Test, and hence so does $\sum_{n=1}^{\infty}|a_n|$, so $\sum_{n=1}^{\infty}a_n$ is absolutely

convergent.

(ii) If $\displaystyle\lim_{n \to \infty}\sqrt[n]{|a_n|} = L > 1$, then there is an integer N such that $\sqrt[n]{|a_n|} > 1$ for all $n \geq N$, so $|a_n| > 1$ for $n \geq N$. Thus,

$\displaystyle\lim_{n \to \infty}a_n \neq 0$, so $\sum_{n=1}^{\infty}a_n$ diverges by the Test for Divergence.

(iii) Consider $\displaystyle\sum_{n=1}^{\infty}\frac{1}{n}$ [diverges] and $\displaystyle\sum_{n=1}^{\infty}\frac{1}{n^2}$ [converges]. For each sum, $\displaystyle\lim_{n \to \infty}\sqrt[n]{|a_n|} = 1$, so the Root Test is inconclusive.

8.5 Power Series

1. A power series is a series of the form $\sum_{n=0}^{\infty} c_n x^n = c_0 + c_1 x + c_2 x^2 + c_3 x^3 + \cdots$, where x is a variable and the c_n's are constants called the coefficients of the series.

   More generally, a series of the form $\sum_{n=0}^{\infty} c_n (x-a)^n = c_0 + c_1 (x-a) + c_2 (x-a)^2 + \cdots$ is called a power series in $(x-a)$ or a power series centered at a or a power series about a, where a is a constant.

3. If $a_n = (-1)^n n x^n$, then

$$\lim_{n \to \infty} \left| \frac{a_{n+1}}{a_n} \right| = \lim_{n \to \infty} \left| \frac{(-1)^{n+1}(n+1)x^{n+1}}{(-1)^n \, nx^n} \right| = \lim_{n \to \infty} \left| (-1) \frac{n+1}{n} x \right| = \lim_{n \to \infty} \left[\left(1 + \frac{1}{n} \right) |x| \right] = |x|. \text{ By the Ratio Test, the}$$

series $\sum_{n=1}^{\infty} (-1)^n n x^n$ converges when $|x| < 1$, so the radius of convergence $R = 1$. Now we'll check the endpoints, that is,

$x = \pm 1$. Both series $\sum_{n=1}^{\infty} (-1)^n n(\pm 1)^n = \sum_{n=1}^{\infty} (\mp 1)^n n$ diverge by the Test for Divergence since $\lim_{n \to \infty} |(\mp 1)^n n| = \infty$. Thus,

the interval of convergence is $I = (-1, 1)$.

5. If $a_n = \dfrac{x^n}{2n-1}$, then $\lim_{n \to \infty} \left| \dfrac{a_{n+1}}{a_n} \right| = \lim_{n \to \infty} \left| \dfrac{x^{n+1}}{2n+1} \cdot \dfrac{2n-1}{x^n} \right| = \lim_{n \to \infty} \left(\dfrac{2n-1}{2n+1} |x| \right) = \lim_{n \to \infty} \left(\dfrac{2 - 1/n}{2 + 1/n} |x| \right) = |x|.$ By

the Ratio Test, the series $\sum_{n=1}^{\infty} \dfrac{x^n}{2n-1}$ converges when $|x| < 1$, so $R = 1$. When $x = 1$, the series $\sum_{n=1}^{\infty} \dfrac{1}{2n-1}$ diverges by

comparison with $\sum_{n=1}^{\infty} \dfrac{1}{2n}$ since $\dfrac{1}{2n-1} > \dfrac{1}{2n}$ and $\dfrac{1}{2} \sum_{n=1}^{\infty} \dfrac{1}{n}$ diverges since it is a constant multiple of the harmonic series.

When $x = -1$, the series $\sum_{n=1}^{\infty} \dfrac{(-1)^n}{2n-1}$ converges by the Alternating Series Test. Thus, the interval of convergence is $[-1, 1)$.

7. If $a_n = \dfrac{x^n}{n!}$, then $\lim_{n \to \infty} \left| \dfrac{a_{n+1}}{a_n} \right| = \lim_{n \to \infty} \left| \dfrac{x^{n+1}}{(n+1)!} \cdot \dfrac{n!}{x^n} \right| = \lim_{n \to \infty} \left| \dfrac{x}{n+1} \right| = |x| \lim_{n \to \infty} \dfrac{1}{n+1} = |x| \cdot 0 = 0 < 1$ for *all* real x.

   So, by the Ratio Test, $R = \infty$ and $I = (-\infty, \infty)$.

9. If $a_n = (-1)^n \dfrac{n^2 x^n}{2^n}$, then

$$\lim_{n \to \infty} \left| \frac{a_{n+1}}{a_n} \right| = \lim_{n \to \infty} \left| \frac{(n+1)^2 x^{n+1}}{2^{n+1}} \cdot \frac{2^n}{n^2 x^n} \right| = \lim_{n \to \infty} \left| \frac{x(n+1)^2}{2n^2} \right| = \lim_{n \to \infty} \left[\frac{|x|}{2} \left(1 + \frac{1}{n} \right)^2 \right] = \frac{|x|}{2} (1)^2 = \tfrac{1}{2} |x|. \text{ By the}$$

Ratio Test, the series $\sum_{n=1}^{\infty} (-1)^n \dfrac{n^2 x^n}{2^n}$ converges when $\tfrac{1}{2} |x| < 1 \iff |x| < 2$, so the radius of convergence is $R = 2$.

When $x = \pm 2$, both series $\sum_{n=1}^{\infty} (-1)^n \dfrac{n^2(\pm 2)^n}{2^n} = \sum_{n=1}^{\infty} (\mp 1)^n n^2$ diverge by the Test for Divergence since

$\lim_{n \to \infty} |(\mp 1)^n n^2| = \infty$. Thus, the interval of convergence is $I = (-2, 2)$.

11. If $a_n = (-1)^n \dfrac{x^n}{4^n \ln n}$, then $\lim_{n \to \infty} \left| \dfrac{a_{n+1}}{a_n} \right| = \lim_{n \to \infty} \left| \dfrac{x^{n+1}}{4^{n+1} \ln(n+1)} \cdot \dfrac{4^n \ln n}{x^n} \right| = \dfrac{|x|}{4} \lim_{n \to \infty} \dfrac{\ln n}{\ln(n+1)} = \dfrac{|x|}{4} \cdot 1$

    [by l'Hospital's Rule] $= \dfrac{|x|}{4}$. By the Ratio Test, the series converges when $\dfrac{|x|}{4} < 1 \iff |x| < 4$, so $R = 4$. When

$x = -4$, $\sum_{n=2}^{\infty} (-1)^n \dfrac{x^n}{4^n \ln n} = \sum_{n=2}^{\infty} \dfrac{[(-1)(-4)]^n}{4^n \ln n} = \sum_{n=2}^{\infty} \dfrac{1}{\ln n}$. Since $\ln n < n$ for $n \geq 2$, $\dfrac{1}{\ln n} > \dfrac{1}{n}$ and $\sum_{n=2}^{\infty} \dfrac{1}{n}$ is the

divergent harmonic series (without the $n = 1$ term), $\sum_{n=2}^{\infty} \dfrac{1}{\ln n}$ is divergent by the Comparison Test. When $x = 4$,

$\sum_{n=2}^{\infty} (-1)^n \dfrac{x^n}{4^n \ln n} = \sum_{n=2}^{\infty} (-1)^n \dfrac{1}{\ln n}$, which converges by the Alternating Series Test. Thus, $I = (-4, 4]$.

13. If $a_n = \dfrac{(-3)^n x^n}{n^{3/2}}$, then

$$\lim_{n \to \infty} \left| \dfrac{a_{n+1}}{a_n} \right| = \lim_{n \to \infty} \left| \dfrac{(-3)^{n+1} x^{n+1}}{(n+1)^{3/2}} \cdot \dfrac{n^{3/2}}{(-3)^n x^n} \right| = \lim_{n \to \infty} \left| -3x \left(\dfrac{n}{n+1} \right)^{3/2} \right| = 3 |x| \lim_{n \to \infty} \left(\dfrac{1}{1 + 1/n} \right)^{3/2}$$

$$= 3 |x| (1) = 3 |x|$$

By the Ratio Test, the series $\sum_{n=1}^{\infty} \dfrac{(-3)^n}{n \sqrt{n}} x^n$ converges when $3 |x| < 1 \iff |x| < \frac{1}{3}$, so $R = \frac{1}{3}$. When $x = \frac{1}{3}$, the series

$\sum_{n=1}^{\infty} \dfrac{(-1)^n}{n^{3/2}}$ converges by the Alternating Series Test. When $x = -\frac{1}{3}$, the series $\sum_{n=1}^{\infty} \dfrac{1}{n^{3/2}}$ is a convergent p-series

$\left(p = \frac{3}{2} > 1 \right)$. Thus, the interval of convergence is $\left[-\frac{1}{3}, \frac{1}{3} \right]$.

15. If $a_n = \dfrac{(x-2)^n}{n^2 + 1}$, then $\lim_{n \to \infty} \left| \dfrac{a_{n+1}}{a_n} \right| = \lim_{n \to \infty} \left| \dfrac{(x-2)^{n+1}}{(n+1)^2 + 1} \cdot \dfrac{n^2 + 1}{(x-2)^n} \right| = |x - 2| \lim_{n \to \infty} \dfrac{n^2 + 1}{(n+1)^2 + 1} = |x - 2|$. By the

Ratio Test, the series $\sum_{n=0}^{\infty} \dfrac{(x-2)^n}{n^2 + 1}$ converges when $|x - 2| < 1$ $[R = 1]$ $\iff$ $-1 < x - 2 < 1$ $\iff$ $1 < x < 3$. When

$x = 1$, the series $\sum_{n=0}^{\infty} (-1)^n \dfrac{1}{n^2 + 1}$ converges by the Alternating Series Test; when $x = 3$, the series $\sum_{n=0}^{\infty} \dfrac{1}{n^2 + 1}$ converges by

comparison with the p-series $\sum_{n=1}^{\infty} \dfrac{1}{n^2}$ $[p = 2 > 1]$. Thus, the interval of convergence is $I = [1, 3]$.

17. $a_n = \dfrac{n}{b^n} (x - a)^n$, where $b > 0$.

$$\lim_{n \to \infty} \left| \dfrac{a_{n+1}}{a_n} \right| = \lim_{n \to \infty} \dfrac{(n+1) |x - a|^{n+1}}{b^{n+1}} \cdot \dfrac{b^n}{n |x - a|^n} = \lim_{n \to \infty} \left(1 + \dfrac{1}{n} \right) \dfrac{|x - a|}{b} = \dfrac{|x - a|}{b}.$$

By the Ratio Test, the series converges when $\dfrac{|x - a|}{b} < 1$ $\iff$ $|x - a| < b$ $[\text{so } R = b]$ $\iff$ $-b < x - a < b$ $\iff$

$a - b < x < a + b$. When $|x - a| = b$, $\lim_{n \to \infty} |a_n| = \lim_{n \to \infty} n = \infty$, so the series diverges. Thus, $I = (a - b, a + b)$.

19. If $a_n = n! (2x - 1)^n$, then $\lim_{n \to \infty} \left| \dfrac{a_{n+1}}{a_n} \right| = \lim_{n \to \infty} \left| \dfrac{(n+1)! (2x - 1)^{n+1}}{n! (2x - 1)^n} \right| = \lim_{n \to \infty} (n + 1) |2x - 1| \to \infty$ as $n \to \infty$

for all $x \neq \frac{1}{2}$. Since the series diverges for all $x \neq \frac{1}{2}$, $R = 0$ and $I = \{ \frac{1}{2} \}$.

21. If $a_n = \dfrac{x^n}{1 \cdot 3 \cdot 5 \cdot \cdots \cdot (2n - 1)}$, then

$$\lim_{n \to \infty} \left| \dfrac{a_{n+1}}{a_n} \right| = \lim_{n \to \infty} \left| \dfrac{x^{n+1}}{1 \cdot 3 \cdot 5 \cdot \cdots \cdot (2n - 1)(2n + 1)} \cdot \dfrac{1 \cdot 3 \cdot 5 \cdot \cdots \cdot (2n - 1)}{x^n} \right| = \lim_{n \to \infty} \dfrac{|x|}{2n + 1} = 0 < 1. \text{ Thus, by}$$

the Ratio Test, the series $\sum_{n=1}^{\infty} \dfrac{x^n}{1 \cdot 3 \cdot 5 \cdot \cdots \cdot (2n - 1)}$ converges for *all* real x and we have $R = \infty$ and $I = (-\infty, \infty)$.

23. (a) We are given that the power series $\sum_{n=0}^{\infty} c_n x^n$ is convergent for $x = 4$. So by Theorem 3, it must converge for at least

$-4 < x \le 4$. In particular, it converges when $x = -2$; that is, $\sum_{n=0}^{\infty} c_n(-2)^n$ is convergent.

(b) It does not follow that $\sum_{n=0}^{\infty} c_n(-4)^n$ is necessarily convergent. [See the comments after Theorem 3 about convergence at

the endpoint of an interval. An example is $c_n = (-1)^n/(n4^n)$.]

25. If $a_n = \dfrac{(n!)^k}{(kn)!} x^n$, then

$$\lim_{n \to \infty} \left| \frac{a_{n+1}}{a_n} \right| = \lim_{n \to \infty} \frac{[(n+1)!]^k (kn)!}{(n!)^k [k(n+1)]!} |x| = \lim_{n \to \infty} \frac{(n+1)^k}{(kn+k)(kn+k-1) \cdots (kn+2)(kn+1)} |x|$$

$$= \lim_{n \to \infty} \left[\frac{(n+1)}{(kn+1)} \frac{(n+1)}{(kn+2)} \cdots \frac{(n+1)}{(kn+k)} \right] |x|$$

$$= \lim_{n \to \infty} \left[\frac{n+1}{kn+1} \right] \lim_{n \to \infty} \left[\frac{n+1}{kn+2} \right] \cdots \lim_{n \to \infty} \left[\frac{n+1}{kn+k} \right] |x|$$

$$= \left(\frac{1}{k} \right)^k |x| < 1 \quad \Leftrightarrow \quad |x| < k^k \text{ for convergence, and the radius of convergence is } R = k^k.$$

27. No. If a power series is centered at a, its interval of convergence is symmetric about a. If a power series has an infinite radius

of convergence, then its interval of convergence must be $(-\infty, \infty)$, not $[0, \infty)$.

29. (a) If $a_n = \dfrac{(-1)^n x^{2n+1}}{n!(n+1)! 2^{2n+1}}$, then

$$\lim_{n \to \infty} \left| \frac{a_{n+1}}{a_n} \right| = \lim_{n \to \infty} \left| \frac{x^{2n+3}}{(n+1)!(n+2)! 2^{2n+3}} \cdot \frac{n!(n+1)! 2^{2n+1}}{x^{2n+1}} \right| = \left(\frac{x}{2} \right)^2 \lim_{n \to \infty} \frac{1}{(n+1)(n+2)} = 0 \text{ for all } x.$$

So $J_1(x)$ converges for all x and its domain is $(-\infty, \infty)$.

(b), (c) The initial terms of $J_1(x)$ up to $n = 5$ are $a_0 = \dfrac{x}{2}$,

$a_1 = -\dfrac{x^3}{16}, a_2 = \dfrac{x^5}{384}, a_3 = -\dfrac{x^7}{18,432}, a_4 = \dfrac{x^9}{1,474,560}$,

and $a_5 = -\dfrac{x^{11}}{176,947,200}$. The partial sums seem to

approximate $J_1(x)$ well near the origin, but as $|x|$ increases,

we need to take a large number of terms to get a good

approximation.

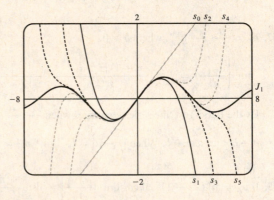

31. $s_{2n-1} = 1 + 2x + x^2 + 2x^3 + x^4 + 2x^5 + \cdots + x^{2n-2} + 2x^{2n-1}$

$= 1(1 + 2x) + x^2(1 + 2x) + x^4(1 + 2x) + \cdots + x^{2n-2}(1 + 2x) = (1 + 2x)(1 + x^2 + x^4 + \cdots + x^{2n-2})$

$= (1 + 2x)\dfrac{1 - x^{2n}}{1 - x^2}$ [by (8.2.3) with $r = x^2$] $\to \dfrac{1 + 2x}{1 - x^2}$ as $n \to \infty$ by (8.2.4), when $|x| < 1$.

Also $s_{2n} = s_{2n-1} + x^{2n} \to \dfrac{1 + 2x}{1 - x^2}$ since $x^{2n} \to 0$ for $|x| < 1$. Therefore, $s_n \to \dfrac{1 + 2x}{1 - x^2}$ since s_{2n} and s_{2n-1} both

approach $\dfrac{1 + 2x}{1 - x^2}$ as $n \to \infty$. Thus, the interval of convergence is $(-1, 1)$ and $f(x) = \dfrac{1 + 2x}{1 - x^2}$.

33. We use the Root Test on the series $\sum c_n x^n$. We need $\lim\limits_{n\to\infty} \sqrt[n]{|c_n x^n|} = |x| \lim\limits_{n\to\infty} \sqrt[n]{|c_n|} = c\,|x| < 1$ for convergence, or $|x| < 1/c$, so $R = 1/c$.

35. For $2 < x < 3$, $\sum c_n x^n$ diverges and $\sum d_n x^n$ converges. By Exercise 8.2.53, $\sum(c_n + d_n)\,x^n$ diverges. Since both series converge for $|x| < 2$, the radius of convergence of $\sum(c_n + d_n)\,x^n$ is 2.

8.6   Representing Functions as Power Series

1. If $f(x) = \sum\limits_{n=0}^{\infty} c_n x^n$ has radius of convergence 10, then $f'(x) = \sum\limits_{n=1}^{\infty} n c_n x^{n-1}$ also has radius of convergence 10 by Theorem 2.

3. Our goal is to write the function in the form $\dfrac{1}{1-r}$, and then use Equation 1 to represent the function as a sum of a power series. $f(x) = \dfrac{1}{1+x} = \dfrac{1}{1-(-x)} = \sum\limits_{n=0}^{\infty}(-x)^n = \sum\limits_{n=0}^{\infty}(-1)^n x^n$ with $|-x| < 1 \;\Leftrightarrow\; |x| < 1$, so $R = 1$ and $I = (-1,1)$.

5. $f(x) = \dfrac{2}{3-x} = \dfrac{2}{3}\left(\dfrac{1}{1-x/3}\right) = \dfrac{2}{3}\sum\limits_{n=0}^{\infty}\left(\dfrac{x}{3}\right)^n$ or, equivalently, $2\sum\limits_{n=0}^{\infty}\dfrac{1}{3^{n+1}}x^n$. The series converges when $\left|\dfrac{x}{3}\right| < 1$, that is, when $|x| < 3$, so $R = 3$ and $I = (-3,3)$.

7. $f(x) = \dfrac{x}{9+x^2} = \dfrac{x}{9}\left[\dfrac{1}{1+(x/3)^2}\right] = \dfrac{x}{9}\left[\dfrac{1}{1-\{-(x/3)^2\}}\right] = \dfrac{x}{9}\sum\limits_{n=0}^{\infty}\left[-\left(\dfrac{x}{3}\right)^2\right]^n = \dfrac{x}{9}\sum\limits_{n=0}^{\infty}(-1)^n\dfrac{x^{2n}}{9^n} = \sum\limits_{n=0}^{\infty}(-1)^n\dfrac{x^{2n+1}}{9^{n+1}}$

The geometric series $\sum\limits_{n=0}^{\infty}\left[-\left(\dfrac{x}{3}\right)^2\right]^n$ converges when $\left|-\left(\dfrac{x}{3}\right)^2\right| < 1 \;\Leftrightarrow\; \dfrac{|x^2|}{9} < 1 \;\Leftrightarrow\; |x|^2 < 9 \;\Leftrightarrow\; |x| < 3$, so $R = 3$ and $I = (-3,3)$.

9. $f(x) = \dfrac{1+x}{1-x} = (1+x)\left(\dfrac{1}{1-x}\right) = (1+x)\sum\limits_{n=0}^{\infty}x^n = \sum\limits_{n=0}^{\infty}x^n + \sum\limits_{n=0}^{\infty}x^{n+1} = 1 + \sum\limits_{n=1}^{\infty}x^n + \sum\limits_{n=1}^{\infty}x^n = 1 + 2\sum\limits_{n=1}^{\infty}x^n$.

The series converges when $|x| < 1$, so $R = 1$ and $I = (-1,1)$.

A second approach: $f(x) = \dfrac{1+x}{1-x} = \dfrac{-(1-x)+2}{1-x} = -1 + 2\left(\dfrac{1}{1-x}\right) = -1 + 2\sum\limits_{n=0}^{\infty}x^n = 1 + 2\sum\limits_{n=1}^{\infty}x^n$.

A third approach:

$f(x) = \dfrac{1+x}{1-x} = (1+x)\left(\dfrac{1}{1-x}\right) = (1+x)(1+x+x^2+x^3+\cdots)$

$\qquad = (1+x+x^2+x^3+\cdots) + (x+x^2+x^3+x^4+\cdots) = 1 + 2x + 2x^2 + 2x^3 + \cdots = 1 + 2\sum\limits_{n=1}^{\infty}x^n$.

11. $f(x) = \dfrac{3}{x^2-x-2} = \dfrac{3}{(x-2)(x+1)} = \dfrac{A}{x-2} + \dfrac{B}{x+1} \;\Rightarrow\; 3 = A(x+1) + B(x-2)$. Let $x = 2$ to get $A = 1$ and $x = -1$ to get $B = -1$. Thus

$\dfrac{3}{x^2-x-2} = \dfrac{1}{x-2} - \dfrac{1}{x+1} = \dfrac{1}{-2}\left(\dfrac{1}{1-(x/2)}\right) - \dfrac{1}{1-(-x)} = -\dfrac{1}{2}\sum\limits_{n=0}^{\infty}\left(\dfrac{x}{2}\right)^n - \sum\limits_{n=0}^{\infty}(-x)^n$

$\qquad\qquad = \sum\limits_{n=0}^{\infty}\left[-\dfrac{1}{2}\left(\dfrac{1}{2}\right)^n - 1(-1)^n\right]x^n = \sum\limits_{n=0}^{\infty}\left[(-1)^{n+1} - \dfrac{1}{2^{n+1}}\right]x^n$

[continued]

We represented f as the sum of two geometric series; the first converges for $x \in (-2, 2)$ and the second converges for $(-1, 1)$.

Thus, the sum converges for $x \in (-1, 1) = I$.

13. (a) $f(x) = \dfrac{1}{(1+x)^2} = \dfrac{d}{dx}\left(\dfrac{-1}{1+x}\right) = -\dfrac{d}{dx}\left[\displaystyle\sum_{n=0}^{\infty}(-1)^n x^n\right]$ [from Exercise 3]

$= \displaystyle\sum_{n=1}^{\infty}(-1)^{n+1}nx^{n-1}$ [from Theorem 2(i)] $= \displaystyle\sum_{n=0}^{\infty}(-1)^n(n+1)x^n$ with $R = 1$.

In the last step, note that we *decreased* the initial value of the summation variable n by 1, and then *increased* each

occurrence of n in the term by 1 [also note that $(-1)^{n+2} = (-1)^n$].

(b) $f(x) = \dfrac{1}{(1+x)^3} = -\dfrac{1}{2}\dfrac{d}{dx}\left[\dfrac{1}{(1+x)^2}\right] = -\dfrac{1}{2}\dfrac{d}{dx}\left[\displaystyle\sum_{n=0}^{\infty}(-1)^n(n+1)x^n\right]$ [from part (a)]

$= -\dfrac{1}{2}\displaystyle\sum_{n=1}^{\infty}(-1)^n(n+1)nx^{n-1} = \dfrac{1}{2}\displaystyle\sum_{n=0}^{\infty}(-1)^n(n+2)(n+1)x^n$ with $R = 1$.

(c) $f(x) = \dfrac{x^2}{(1+x)^3} = x^2 \cdot \dfrac{1}{(1+x)^3} = x^2 \cdot \dfrac{1}{2}\displaystyle\sum_{n=0}^{\infty}(-1)^n(n+2)(n+1)x^n$ [from part (b)]

$= \dfrac{1}{2}\displaystyle\sum_{n=0}^{\infty}(-1)^n(n+2)(n+1)x^{n+2}$

To write the power series with x^n rather than x^{n+2}, we will *decrease* each occurrence of n in the term by 2 and *increase*

the initial value of the summation variable by 2. This gives us $\dfrac{1}{2}\displaystyle\sum_{n=2}^{\infty}(-1)^n(n)(n-1)x^n$ with $R = 1$.

15. $f(x) = \ln(5 - x) = -\displaystyle\int \dfrac{dx}{5-x} = -\dfrac{1}{5}\int \dfrac{dx}{1-x/5} = -\dfrac{1}{5}\int\left[\displaystyle\sum_{n=0}^{\infty}\left(\dfrac{x}{5}\right)^n\right]dx = C - \dfrac{1}{5}\displaystyle\sum_{n=0}^{\infty}\dfrac{x^{n+1}}{5^n(n+1)} = C - \displaystyle\sum_{n=1}^{\infty}\dfrac{x^n}{n\,5^n}$

Putting $x = 0$, we get $C = \ln 5$. The series converges for $|x/5| < 1 \iff |x| < 5$, so $R = 5$.

17. We know that $\dfrac{1}{1+4x} = \dfrac{1}{1-(-4x)} = \displaystyle\sum_{n=0}^{\infty}(-4x)^n$. Differentiating, we get

$\dfrac{-4}{(1+4x)^2} = \displaystyle\sum_{n=1}^{\infty}(-4)^n nx^{n-1} = \displaystyle\sum_{n=0}^{\infty}(-4)^{n+1}(n+1)x^n$, so

$f(x) = \dfrac{x}{(1+4x)^2} = \dfrac{-x}{4} \cdot \dfrac{-4}{(1+4x)^2} = \dfrac{-x}{4}\displaystyle\sum_{n=0}^{\infty}(-4)^{n+1}(n+1)x^n = \displaystyle\sum_{n=0}^{\infty}(-1)^n 4^n(n+1)x^{n+1}$

for $|-4x| < 1 \iff |x| < \dfrac{1}{4}$, so $R = \dfrac{1}{4}$.

19. By Example 5, $\dfrac{1}{(1-x)^2} = \displaystyle\sum_{n=0}^{\infty}(n+1)x^n$. Thus,

$$f(x) = \dfrac{1+x}{(1-x)^2} = \dfrac{1}{(1-x)^2} + \dfrac{x}{(1-x)^2} = \displaystyle\sum_{n=0}^{\infty}(n+1)x^n + \displaystyle\sum_{n=0}^{\infty}(n+1)x^{n+1}$$

$$= \displaystyle\sum_{n=0}^{\infty}(n+1)x^n + \displaystyle\sum_{n=1}^{\infty}nx^n \qquad \text{[make the starting values equal]}$$

$$= 1 + \displaystyle\sum_{n=1}^{\infty}[(n+1)+n]x^n = 1 + \displaystyle\sum_{n=1}^{\infty}(2n+1)x^n = \displaystyle\sum_{n=0}^{\infty}(2n+1)x^n \text{ with } R = 1.$$

21. $f(x) = \dfrac{x}{x^2+16} = \dfrac{x}{16}\left(\dfrac{1}{1-(-x^2/16)}\right) = \dfrac{x}{16}\sum\limits_{n=0}^{\infty}\left(-\dfrac{x^2}{16}\right)^n = \dfrac{x}{16}\sum\limits_{n=0}^{\infty}(-1)^n\dfrac{1}{16^n}x^{2n} = \sum\limits_{n=0}^{\infty}(-1)^n\dfrac{1}{16^{n+1}}x^{2n+1}$.

The series converges when $\left|-x^2/16\right| < 1 \;\Leftrightarrow\; x^2 < 16 \;\Leftrightarrow\; |x| < 4$, so $R = 4$. The partial sums are $s_1 = \dfrac{x}{16}$,

$s_2 = s_1 - \dfrac{x^3}{16^2},\; s_3 = s_2 + \dfrac{x^5}{16^3},\; s_4 = s_3 - \dfrac{x^7}{16^4},\; s_5 = s_4 + \dfrac{x^9}{16^5},\;\ldots$. Note that s_1 corresponds to the first term of the infinite

sum, regardless of the value of the summation variable and the value of the exponent.

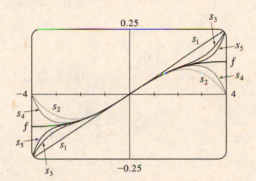

As n increases, $s_n(x)$ approximates f better on the interval of convergence, which is $(-4, 4)$.

23. $f(x) = \ln\left(\dfrac{1+x}{1-x}\right) = \ln(1+x) - \ln(1-x) = \displaystyle\int\dfrac{dx}{1+x} + \int\dfrac{dx}{1-x} = \int\dfrac{dx}{1-(-x)} + \int\dfrac{dx}{1-x}$

$= \displaystyle\int\left[\sum_{n=0}^{\infty}(-1)^n x^n + \sum_{n=0}^{\infty}x^n\right]dx = \int\left[(1 - x + x^2 - x^3 + x^4 - \cdots) + (1 + x + x^2 + x^3 + x^4 + \cdots)\right]dx$

$= \displaystyle\int(2 + 2x^2 + 2x^4 + \cdots)\,dx = \int\sum_{n=0}^{\infty}2x^{2n}\,dx = C + \sum_{n=0}^{\infty}\dfrac{2x^{2n+1}}{2n+1}$

But $f(0) = \ln\frac{1}{1} = 0$, so $C = 0$ and we have $f(x) = \displaystyle\sum_{n=0}^{\infty}\dfrac{2x^{2n+1}}{2n+1}$ with $R = 1$. If $x = \pm 1$, then $f(x) = \pm 2\displaystyle\sum_{n=0}^{\infty}\dfrac{1}{2n+1}$,

which both diverge by the Limit Comparison Test with $b_n = \dfrac{1}{n}$.

The partial sums are $s_1 = \dfrac{2x}{1},\; s_2 = s_1 + \dfrac{2x^3}{3},\; s_3 = s_2 + \dfrac{2x^5}{5},\;\ldots$.

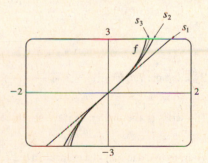

As n increases, $s_n(x)$ approximates f better on the interval of convergence,

which is $(-1, 1)$.

25. $\dfrac{t}{1-t^8} = t\cdot\dfrac{1}{1-t^8} = t\displaystyle\sum_{n=0}^{\infty}(t^8)^n = \sum_{n=0}^{\infty}t^{8n+1} \;\Rightarrow\; \int\dfrac{t}{1-t^8}\,dt = C + \sum_{n=0}^{\infty}\dfrac{t^{8n+2}}{8n+2}$. The series for $\dfrac{1}{1-t^8}$ converges

when $\left|t^8\right| < 1 \;\Leftrightarrow\; |t| < 1$, so $R = 1$ for that series and also the series for $t/(1-t^8)$. By Theorem 2, the series for

$\displaystyle\int\dfrac{t}{1-t^8}\,dt$ also has $R = 1$.

27. From Example 6, $\ln(1 + x) = \sum_{n=1}^{\infty} (-1)^{n-1} \dfrac{x^n}{n}$ for $|x| < 1$, so $x^2 \ln(1 + x) = \sum_{n=1}^{\infty} (-1)^{n-1} \dfrac{x^{n+2}}{n}$ and

$$\int x^2 \ln(1 + x)\,dx = C + \sum_{n=1}^{\infty} (-1)^n \dfrac{x^{n+3}}{n(n+3)}.$$ $R = 1$ for the series for $\ln(1 + x)$, so $R = 1$ for the series representing

$x^2 \ln(1 + x)$ as well. By Theorem 2, the series for $\displaystyle\int x^2 \ln(1 + x)\,dx$ also has $R = 1$.

29. $\dfrac{1}{1 + x^5} = \dfrac{1}{1 - (-x^5)} = \sum_{n=0}^{\infty} \left(-x^5\right)^n = \sum_{n=0}^{\infty} (-1)^n x^{5n}$ $\Rightarrow$

$$\int \dfrac{1}{1 + x^5}\,dx = \int \sum_{n=0}^{\infty} (-1)^n x^{5n}\,dx = C + \sum_{n=0}^{\infty} (-1)^n \dfrac{x^{5n+1}}{5n+1}.$$ Thus,

$$I = \int_0^{0.2} \dfrac{1}{1 + x^5}\,dx = \left[x - \dfrac{x^6}{6} + \dfrac{x^{11}}{11} - \cdots \right]_0^{0.2} = 0.2 - \dfrac{(0.2)^6}{6} + \dfrac{(0.2)^{11}}{11} - \cdots.$$ The series is alternating, so if we use

the first two terms, the error is at most $(0.2)^{11}/11 \approx 1.9 \times 10^{-9}$. So $I \approx 0.2 - (0.2)^6/6 \approx 0.199\,989$ to six decimal places.

31. We substitute $3x$ for x in Example 7, and find that

$$\int x \arctan(3x)\,dx = \int x \sum_{n=0}^{\infty} (-1)^n \dfrac{(3x)^{2n+1}}{2n+1}\,dx = \int \sum_{n=0}^{\infty} (-1)^n \dfrac{3^{2n+1} x^{2n+2}}{2n+1}\,dx = C + \sum_{n=0}^{\infty} (-1)^n \dfrac{3^{2n+1} x^{2n+3}}{(2n+1)(2n+3)}$$

So
$$\int_0^{0.1} x \arctan(3x)\,dx = \left[\dfrac{3x^3}{1 \cdot 3} - \dfrac{3^3 x^5}{3 \cdot 5} + \dfrac{3^5 x^7}{5 \cdot 7} - \dfrac{3^7 x^9}{7 \cdot 9} + \cdots \right]_0^{0.1}$$

$$= \dfrac{1}{10^3} - \dfrac{9}{5 \times 10^5} + \dfrac{243}{35 \times 10^7} - \dfrac{2187}{63 \times 10^9} + \cdots.$$

The series is alternating, so if we use three terms, the error is at most $\dfrac{2187}{63 \times 10^9} \approx 3.5 \times 10^{-8}$. So

$$\int_0^{0.1} x \arctan(3x)\,dx \approx \dfrac{1}{10^3} - \dfrac{9}{5 \times 10^5} + \dfrac{243}{35 \times 10^7} \approx 0.000\,983$$ to six decimal places.

33. By Example 7, $\arctan x = x - \dfrac{x^3}{3} + \dfrac{x^5}{5} - \dfrac{x^7}{7} + \cdots$, so $\arctan 0.2 = 0.2 - \dfrac{(0.2)^3}{3} + \dfrac{(0.2)^5}{5} - \dfrac{(0.2)^7}{7} + \cdots.$

The series is alternating, so if we use three terms, the error is at most $\dfrac{(0.2)^7}{7} \approx 0.000\,002.$

Thus, to five decimal places, $\arctan 0.2 \approx 0.2 - \dfrac{(0.2)^3}{3} + \dfrac{(0.2)^5}{5} \approx 0.197\,40.$

35. (a) $J_0(x) = \sum_{n=0}^{\infty} \frac{(-1)^n x^{2n}}{2^{2n}(n!)^2}$, $J_0'(x) = \sum_{n=1}^{\infty} \frac{(-1)^n 2nx^{2n-1}}{2^{2n}(n!)^2}$, and $J_0''(x) = \sum_{n=1}^{\infty} \frac{(-1)^n 2n(2n-1)x^{2n-2}}{2^{2n}(n!)^2}$, so

$$x^2 J_0''(x) + xJ_0'(x) + x^2 J_0(x) = \sum_{n=1}^{\infty} \frac{(-1)^n 2n(2n-1)x^{2n}}{2^{2n}(n!)^2} + \sum_{n=1}^{\infty} \frac{(-1)^n 2nx^{2n}}{2^{2n}(n!)^2} + \sum_{n=0}^{\infty} \frac{(-1)^n x^{2n+2}}{2^{2n}(n!)^2}$$

$$= \sum_{n=1}^{\infty} \frac{(-1)^n 2n(2n-1)x^{2n}}{2^{2n}(n!)^2} + \sum_{n=1}^{\infty} \frac{(-1)^n 2nx^{2n}}{2^{2n}(n!)^2} + \sum_{n=1}^{\infty} \frac{(-1)^{n-1} x^{2n}}{2^{2n-2}[(n-1)!]^2}$$

$$= \sum_{n=1}^{\infty} \frac{(-1)^n 2n(2n-1)x^{2n}}{2^{2n}(n!)^2} + \sum_{n=1}^{\infty} \frac{(-1)^n 2nx^{2n}}{2^{2n}(n!)^2} + \sum_{n=1}^{\infty} \frac{(-1)^n (-1)^{-1} 2^2 n^2 x^{2n}}{2^{2n}(n!)^2}$$

$$= \sum_{n=1}^{\infty} (-1)^n \left[\frac{2n(2n-1) + 2n - 2^2 n^2}{2^{2n}(n!)^2}\right] x^{2n}$$

$$= \sum_{n=1}^{\infty} (-1)^n \left[\frac{4n^2 - 2n + 2n - 4n^2}{2^{2n}(n!)^2}\right] x^{2n} = 0$$

(b) $\int_0^1 J_0(x)\, dx = \int_0^1 \left[\sum_{n=0}^{\infty} \frac{(-1)^n x^{2n}}{2^{2n}(n!)^2}\right] dx = \int_0^1 \left(1 - \frac{x^2}{4} + \frac{x^4}{64} - \frac{x^6}{2304} + \cdots\right) dx$

$$= \left[x - \frac{x^3}{3\cdot 4} + \frac{x^5}{5\cdot 64} - \frac{x^7}{7\cdot 2304} + \cdots\right]_0^1 = 1 - \frac{1}{12} + \frac{1}{320} - \frac{1}{16{,}128} + \cdots$$

Since $\frac{1}{16{,}128} \approx 0.000062$, it follows from The Alternating Series Estimation Theorem that, correct to three decimal places,

$\int_0^1 J_0(x)\, dx \approx 1 - \frac{1}{12} + \frac{1}{320} \approx 0.920$.

37. (a) $f(x) = \sum_{n=0}^{\infty} \frac{x^n}{n!} \Rightarrow f'(x) = \sum_{n=1}^{\infty} \frac{nx^{n-1}}{n!} = \sum_{n=1}^{\infty} \frac{x^{n-1}}{(n-1)!} = \sum_{n=0}^{\infty} \frac{x^n}{n!} = f(x)$

(b) By Theorem 5.5.2, the only solution to the differential equation $df(x)/dx = f(x)$ is $f(x) = Ke^x$, but $f(0) = 1$, so $K = 1$ and $f(x) = e^x$.

Or: We could solve the equation $df(x)/dx = f(x)$ as a separable differential equation.

39. If $a_n = \dfrac{x^n}{n^2}$, then by the Ratio Test, $\lim_{n\to\infty}\left|\dfrac{a_{n+1}}{a_n}\right| = \lim_{n\to\infty}\left|\dfrac{x^{n+1}}{(n+1)^2}\cdot\dfrac{n^2}{x^n}\right| = |x|\lim_{n\to\infty}\left(\dfrac{n}{n+1}\right)^2 = |x| < 1$ for

convergence, so $R = 1$. When $x = \pm 1$, $\sum_{n=1}^{\infty}\left|\dfrac{x^n}{n^2}\right| = \sum_{n=1}^{\infty}\dfrac{1}{n^2}$ which is a convergent p-series ($p = 2 > 1$), so the interval of

convergence for f is $[-1, 1]$. By Theorem 2, the radii of convergence of f' and f'' are both 1, so we need only check the

endpoints. $f(x) = \sum_{n=1}^{\infty}\dfrac{x^n}{n^2} \Rightarrow f'(x) = \sum_{n=1}^{\infty}\dfrac{nx^{n-1}}{n^2} = \sum_{n=0}^{\infty}\dfrac{x^n}{n+1}$, and this series diverges for $x = 1$ (harmonic series)

and converges for $x = -1$ (Alternating Series Test), so the interval of convergence is $[-1, 1)$. $f''(x) = \sum_{n=1}^{\infty}\dfrac{nx^{n-1}}{n+1}$ diverges

at both 1 and -1 (Test for Divergence) since $\lim_{n\to\infty}\dfrac{n}{n+1} = 1 \neq 0$, so its interval of convergence is $(-1, 1)$.

41. By Example 7, $\tan^{-1} x = \sum_{n=0}^{\infty} (-1)^n \dfrac{x^{2n+1}}{2n+1}$ for $|x| < 1$. In particular, for $x = \dfrac{1}{\sqrt{3}}$, we

have $\dfrac{\pi}{6} = \tan^{-1}\left(\dfrac{1}{\sqrt{3}}\right) = \sum_{n=0}^{\infty} (-1)^n \dfrac{(1/\sqrt{3})^{2n+1}}{2n+1} = \sum_{n=0}^{\infty} (-1)^n \left(\dfrac{1}{3}\right)^n \dfrac{1}{\sqrt{3}} \dfrac{1}{2n+1}$, so

$\pi = \dfrac{6}{\sqrt{3}} \sum_{n=0}^{\infty} \dfrac{(-1)^n}{(2n+1)3^n} = 2\sqrt{3} \sum_{n=0}^{\infty} \dfrac{(-1)^n}{(2n+1)3^n}$.

8.7 Taylor and Maclaurin Series

1. Using Theorem 5 with $\sum_{n=0}^{\infty} b_n(x-5)^n$, $b_n = \dfrac{f^{(n)}(a)}{n!}$, so $b_8 = \dfrac{f^{(8)}(5)}{8!}$.

3. Since $f^{(n)}(0) = (n+1)!$, Equation 7 gives the Maclaurin series

$\sum_{n=0}^{\infty} \dfrac{f^{(n)}(0)}{n!} x^n = \sum_{n=0}^{\infty} \dfrac{(n+1)!}{n!} x^n = \sum_{n=0}^{\infty} (n+1)x^n$. Applying the Ratio Test with $a_n = (n+1)x^n$ gives us

$\lim_{n\to\infty} \left| \dfrac{a_{n+1}}{a_n} \right| = \lim_{n\to\infty} \left| \dfrac{(n+2)x^{n+1}}{(n+1)x^n} \right| = |x| \lim_{n\to\infty} \dfrac{n+2}{n+1} = |x| \cdot 1 = |x|$. For convergence, we must have $|x| < 1$, so the

radius of convergence $R = 1$.

5.

n	$f^{(n)}(x)$	$f^{(n)}(0)$
0	$(1-x)^{-2}$	1
1	$2(1-x)^{-3}$	2
2	$6(1-x)^{-4}$	6
3	$24(1-x)^{-5}$	24
4	$120(1-x)^{-6}$	120
⋮	⋮	⋮

$(1-x)^{-2} = f(0) + f'(0)x + \dfrac{f''(0)}{2!}x^2 + \dfrac{f'''(0)}{3!}x^3 + \dfrac{f^{(4)}(0)}{4!}x^4 + \cdots$

$= 1 + 2x + \dfrac{6}{2}x^2 + \dfrac{24}{6}x^3 + \dfrac{120}{24}x^4 + \cdots$

$= 1 + 2x + 3x^2 + 4x^3 + 5x^4 + \cdots = \sum_{n=0}^{\infty} (n+1)x^n$

$\lim_{n\to\infty} \left| \dfrac{a_{n+1}}{a_n} \right| = \lim_{n\to\infty} \left| \dfrac{(n+2)x^{n+1}}{(n+1)x^n} \right| = |x| \lim_{n\to\infty} \dfrac{n+2}{n+1} = |x|\,(1) = |x| < 1$

for convergence, so $R = 1$.

7.

n	$f^{(n)}(x)$	$f^{(n)}(0)$
0	$\sin \pi x$	0
1	$\pi \cos \pi x$	π
2	$-\pi^2 \sin \pi x$	0
3	$-\pi^3 \cos \pi x$	$-\pi^3$
4	$\pi^4 \sin \pi x$	0
5	$\pi^5 \cos \pi x$	π^5
⋮	⋮	⋮

$\sin \pi x = f(0) + f'(0)x + \dfrac{f''(0)}{2!}x^2 + \dfrac{f'''(0)}{3!}x^3$

$\qquad + \dfrac{f^{(4)}(0)}{4!}x^4 + \dfrac{f^{(5)}(0)}{5!}x^5 + \cdots$

$= 0 + \pi x + 0 - \dfrac{\pi^3}{3!}x^3 + 0 + \dfrac{\pi^5}{5!}x^5 + \cdots$

$= \pi x - \dfrac{\pi^3}{3!}x^3 + \dfrac{\pi^5}{5!}x^5 - \dfrac{\pi^7}{7!}x^7 + \cdots$

$= \sum_{n=0}^{\infty} (-1)^n \dfrac{\pi^{2n+1}}{(2n+1)!} x^{2n+1}$

$\lim_{n\to\infty} \left| \dfrac{a_{n+1}}{a_n} \right| = \lim_{n\to\infty} \left| \dfrac{\pi^{2n+3} x^{2n+3}}{(2n+3)!} \cdot \dfrac{(2n+1)!}{\pi^{2n+1} x^{2n+1}} \right| = \lim_{n\to\infty} \dfrac{\pi^2 x^2}{(2n+3)(2n+2)} = 0 < 1$ for all x, so $R = \infty$.

9.

n	$f^{(n)}(x)$	$f^{(n)}(0)$
0	$\sinh x$	0
1	$\cosh x$	1
2	$\sinh x$	0
3	$\cosh x$	1
4	$\sinh x$	0
⋮	⋮	⋮

$$f^{(n)}(0) = \begin{cases} 0 & \text{if } n \text{ is even} \\ 1 & \text{if } n \text{ is odd} \end{cases} \qquad \text{so } \sinh x = \sum_{n=0}^{\infty} \frac{x^{2n+1}}{(2n+1)!}.$$

Use the Ratio Test to find R. If $a_n = \dfrac{x^{2n+1}}{(2n+1)!}$, then

$$\lim_{n\to\infty} \left| \frac{a_{n+1}}{a_n} \right| = \lim_{n\to\infty} \left| \frac{x^{2n+3}}{(2n+3)!} \cdot \frac{(2n+1)!}{x^{2n+1}} \right| = x^2 \cdot \lim_{n\to\infty} \frac{1}{(2n+3)(2n+2)}$$

$$= 0 < 1 \quad \text{for all } x, \text{ so } R = \infty.$$

11.

n	$f^{(n)}(x)$	$f^{(n)}(1)$
0	$x^4 - 3x^2 + 1$	-1
1	$4x^3 - 6x$	-2
2	$12x^2 - 6$	6
3	$24x$	24
4	24	24
5	0	0
6	0	0
⋮	⋮	⋮

$f^{(n)}(x) = 0$ for $n \geq 5$, so f has a finite series expansion about $a = 1$.

$$f(x) = x^4 - 3x^2 + 1 = \sum_{n=0}^{4} \frac{f^{(n)}(1)}{n!}(x-1)^n$$

$$= \frac{-1}{0!}(x-1)^0 + \frac{-2}{1!}(x-1)^1 + \frac{6}{2!}(x-1)^2$$

$$+ \frac{24}{3!}(x-1)^3 + \frac{24}{4!}(x-1)^4$$

$$= -1 - 2(x-1) + 3(x-1)^2 + 4(x-1)^3 + (x-1)^4$$

A finite series converges for all x, so $R = \infty$.

13.

n	$f^{(n)}(x)$	$f^{(n)}(2)$
0	$\ln x$	$\ln 2$
1	$1/x$	$1/2$
2	$-1/x^2$	$-1/2^2$
3	$2/x^3$	$2/2^3$
4	$-6/x^4$	$-6/2^4$
5	$24/x^5$	$24/2^5$
⋮	⋮	⋮

$$f(x) = \ln x = \sum_{n=0}^{\infty} \frac{f^{(n)}(2)}{n!}(x-2)^n$$

$$= \frac{\ln 2}{0!}(x-2)^0 + \frac{1}{1!\,2^1}(x-2)^1 + \frac{-1}{2!\,2^2}(x-2)^2 + \frac{2}{3!\,2^3}(x-2)^3$$

$$+ \frac{-6}{4!\,2^4}(x-2)^4 + \frac{24}{5!\,2^5}(x-2)^5 + \cdots$$

$$= \ln 2 + \sum_{n=1}^{\infty} (-1)^{n+1} \frac{(n-1)!}{n!\,2^n}(x-2)^n$$

$$= \ln 2 + \sum_{n=1}^{\infty} (-1)^{n+1} \frac{1}{n\,2^n}(x-2)^n$$

$$\lim_{n\to\infty} \left| \frac{a_{n+1}}{a_n} \right| = \lim_{n\to\infty} \left| \frac{(-1)^{n+2}(x-2)^{n+1}}{(n+1)\,2^{n+1}} \cdot \frac{n\,2^n}{(-1)^{n+1}(x-2)^n} \right| = \lim_{n\to\infty} \left| \frac{(-1)(x-2)n}{(n+1)2} \right| = \lim_{n\to\infty} \left(\frac{n}{n+1} \right) \frac{|x-2|}{2}$$

$$= \frac{|x-2|}{2} < 1 \quad \text{for convergence, so } |x-2| < 2 \text{ and } R = 2.$$

15.

n	$f^{(n)}(x)$	$f^{(n)}(3)$
0	e^{2x}	e^6
1	$2e^{2x}$	$2e^6$
2	$2^2 e^{2x}$	$4e^6$
3	$2^3 e^{2x}$	$8e^6$
4	$2^4 e^{2x}$	$16e^6$
⋮	⋮	⋮

$$f(x) = e^{2x} = \sum_{n=0}^{\infty} \frac{f^{(n)}(3)}{n!}(x-3)^n$$

$$= \frac{e^6}{0!}(x-3)^0 + \frac{2e^6}{1!}(x-3)^1 + \frac{4e^6}{2!}(x-3)^2$$

$$+ \frac{8e^6}{3!}(x-3)^3 + \frac{16e^6}{4!}(x-3)^4 + \cdots$$

$$= \sum_{n=0}^{\infty} \frac{2^n e^6}{n!}(x-3)^n$$

$$\lim_{n\to\infty}\left|\frac{a_{n+1}}{a_n}\right| = \lim_{n\to\infty}\left|\frac{2^{n+1}e^6(x-3)^{n+1}}{(n+1)!} \cdot \frac{n!}{2^n e^6 (x-3)^n}\right| = \lim_{n\to\infty}\frac{2|x-3|}{n+1} = 0 < 1 \quad \text{for all } x, \text{ so } R = \infty.$$

17.

n	$f^{(n)}(x)$	$f^{(n)}(\pi)$
0	$\cos x$	-1
1	$-\sin x$	0
2	$-\cos x$	1
3	$\sin x$	0
4	$\cos x$	-1
⋮	⋮	⋮

$$f(x) = \cos x = \sum_{k=0}^{\infty} \frac{f^{(k)}(\pi)}{k!}(x-\pi)^k$$

$$= -1 + \frac{(x-\pi)^2}{2!} - \frac{(x-\pi)^4}{4!} + \frac{(x-\pi)^6}{6!} - \cdots$$

$$= \sum_{n=0}^{\infty} (-1)^{n+1}\frac{(x-\pi)^{2n}}{(2n)!}$$

$$\lim_{n\to\infty}\left|\frac{a_{n+1}}{a_n}\right| = \lim_{n\to\infty}\left[\frac{|x-\pi|^{2n+2}}{(2n+2)!} \cdot \frac{(2n)!}{|x-\pi|^{2n}}\right]$$

$$= \lim_{n\to\infty}\frac{|x-\pi|^2}{(2n+2)(2n+1)} = 0 < 1 \quad \text{for all } x, \text{ so } R = \infty.$$

19. If $f(x) = \sin \pi x$, then $f^{(n+1)}(x) = \pm\pi^{n+1}\sin \pi x$ or $\pm\pi^{n+1}\cos \pi x$. In each case, $\left|f^{(n+1)}(x)\right| \le \pi^{n+1}$, so by Formula 9

with $a = 0$ and $M = \pi^{n+1}$, $|R_n(x)| \le \dfrac{\pi^{n+1}}{(n+1)!}|x|^{n+1} = \dfrac{|\pi x|^{n+1}}{(n+1)!}$. Thus, $|R_n(x)| \to 0$ as $n \to \infty$ by Equation 11.

So $\displaystyle\lim_{n\to\infty} R_n(x) = 0$ and, by Theorem 8, the series in Exercise 7 represents $\sin \pi x$ for all x.

21. If $f(x) = \sinh x$, then by Taylor's Formula $R_n(x) = \dfrac{f^{(n+1)}(z)}{(n+1)!}x^{n+1}$, where $0 < |z| < |x|$. But for all n,

$\left|f^{(n+1)}(z)\right| \le \cosh z \le \cosh x$ (because all derivatives are either sinh or cosh, $|\sinh z| < |\cosh z|$ for all z, and

$|z| < |x| \Rightarrow \cosh z < \cosh x$). So $|R_n(x)| \le \dfrac{\cosh x}{(n+1)!}x^{n+1} \to 0$ as $n \to \infty$ by Equation 11. So by Theorem 8, the series

represents $\sinh x$ for all x.

23. $\sqrt[4]{1-x} = [1+(-x)]^{1/4} = \displaystyle\sum_{n=0}^{\infty}\binom{1/4}{n}(-x)^n = 1 + \frac{1}{4}(-x) + \frac{\frac{1}{4}\left(-\frac{3}{4}\right)}{2!}(-x)^2 + \frac{\frac{1}{4}\left(-\frac{3}{4}\right)\left(-\frac{7}{4}\right)}{3!}(-x)^3 + \cdots$

$= 1 - \dfrac{1}{4}x + \displaystyle\sum_{n=2}^{\infty}\frac{(-1)^{n-1}(-1)^n \cdot [3 \cdot 7 \cdot \cdots \cdot (4n-5)]}{4^n \cdot n!}x^n = 1 - \dfrac{1}{4}x - \displaystyle\sum_{n=2}^{\infty}\frac{3 \cdot 7 \cdot \cdots \cdot (4n-5)}{4^n \cdot n!}x^n$

and $|-x| < 1 \quad \Leftrightarrow \quad |x| < 1$, so $R = 1$.

25. $\dfrac{1}{(2+x)^3} = \dfrac{1}{[2(1+x/2)]^3} = \dfrac{1}{8}\left(1+\dfrac{x}{2}\right)^{-3} = \dfrac{1}{8}\displaystyle\sum_{n=0}^{\infty}\binom{-3}{n}\left(\dfrac{x}{2}\right)^n$. The binomial coefficient is

$$\binom{-3}{n} = \dfrac{(-3)(-4)(-5)\cdots\cdots(-3-n+1)}{n!} = \dfrac{(-3)(-4)(-5)\cdots\cdots[-(n+2)]}{n!}$$

$$= \dfrac{(-1)^n \cdot 2 \cdot 3 \cdot 4 \cdot 5 \cdots\cdots (n+1)(n+2)}{2 \cdot n!} = \dfrac{(-1)^n (n+1)(n+2)}{2}$$

Thus, $\dfrac{1}{(2+x)^3} = \dfrac{1}{8}\displaystyle\sum_{n=0}^{\infty}\dfrac{(-1)^n(n+1)(n+2)}{2}\dfrac{x^n}{2^n} = \sum_{n=0}^{\infty}\dfrac{(-1)^n(n+1)(n+2)x^n}{2^{n+4}}$ for $\left|\dfrac{x}{2}\right| < 1 \;\Leftrightarrow\; |x| < 2$, so $R = 2$.

27. $\sin x = \displaystyle\sum_{n=0}^{\infty}(-1)^n\dfrac{x^{2n+1}}{(2n+1)!} \;\Rightarrow\; f(x) = \sin(\pi x) = \sum_{n=0}^{\infty}(-1)^n\dfrac{(\pi x)^{2n+1}}{(2n+1)!} = \sum_{n=0}^{\infty}(-1)^n\dfrac{\pi^{2n+1}}{(2n+1)!}x^{2n+1}, R = \infty$.

29. $e^x = \displaystyle\sum_{n=0}^{\infty}\dfrac{x^n}{n!} \;\Rightarrow\; e^{2x} = \sum_{n=0}^{\infty}\dfrac{(2x)^n}{n!} = \sum_{n=0}^{\infty}\dfrac{2^n x^n}{n!}$, so $f(x) = e^x + e^{2x} = \sum_{n=0}^{\infty}\dfrac{1}{n!}x^n + \sum_{n=0}^{\infty}\dfrac{2^n}{n!}x^n = \sum_{n=0}^{\infty}\dfrac{2^n+1}{n!}x^n$,

$R = \infty$.

31. $\cos x = \displaystyle\sum_{n=0}^{\infty}(-1)^n\dfrac{x^{2n}}{(2n)!} \;\Rightarrow\; \cos\left(\tfrac{1}{2}x^2\right) = \sum_{n=0}^{\infty}(-1)^n\dfrac{\left(\tfrac{1}{2}x^2\right)^{2n}}{(2n)!} = \sum_{n=0}^{\infty}(-1)^n\dfrac{x^{4n}}{2^{2n}(2n)!}$, so

$$f(x) = x\cos\left(\tfrac{1}{2}x^2\right) = \sum_{n=0}^{\infty}(-1)^n\dfrac{1}{2^{2n}(2n)!}x^{4n+1}, R = \infty.$$

33. We must write the binomial in the form (1+ expression), so we'll factor out a 4.

$$\dfrac{x}{\sqrt{4+x^2}} = \dfrac{x}{\sqrt{4(1+x^2/4)}} = \dfrac{x}{2\sqrt{1+x^2/4}} = \dfrac{x}{2}\left(1+\dfrac{x^2}{4}\right)^{-1/2} = \dfrac{x}{2}\sum_{n=0}^{\infty}\binom{-\tfrac{1}{2}}{n}\left(\dfrac{x^2}{4}\right)^n$$

$$= \dfrac{x}{2}\left[1+\left(-\tfrac{1}{2}\right)\dfrac{x^2}{4} + \dfrac{\left(-\tfrac{1}{2}\right)\left(-\tfrac{3}{2}\right)}{2!}\left(\dfrac{x^2}{4}\right)^2 + \dfrac{\left(-\tfrac{1}{2}\right)\left(-\tfrac{3}{2}\right)\left(-\tfrac{5}{2}\right)}{3!}\left(\dfrac{x^2}{4}\right)^3 + \cdots\right]$$

$$= \dfrac{x}{2} + \dfrac{x}{2}\sum_{n=1}^{\infty}(-1)^n\dfrac{1\cdot3\cdot5\cdots\cdots(2n-1)}{2^n\cdot4^n\cdot n!}x^{2n}$$

$$= \dfrac{x}{2} + \sum_{n=1}^{\infty}(-1)^n\dfrac{1\cdot3\cdot5\cdots\cdots(2n-1)}{n!\,2^{3n+1}}x^{2n+1} \text{ and } \dfrac{x^2}{4} < 1 \;\Leftrightarrow\; \dfrac{|x|}{2} < 1 \;\Leftrightarrow\; |x| < 2, \text{ so } R = 2.$$

35. $\sin^2 x = \dfrac{1}{2}(1-\cos 2x) = \dfrac{1}{2}\left[1 - \displaystyle\sum_{n=0}^{\infty}\dfrac{(-1)^n(2x)^{2n}}{(2n)!}\right] = \dfrac{1}{2}\left[1 - 1 - \sum_{n=1}^{\infty}\dfrac{(-1)^n(2x)^{2n}}{(2n)!}\right] = \sum_{n=1}^{\infty}\dfrac{(-1)^{n+1}2^{2n-1}x^{2n}}{(2n)!}$,

$R = \infty$

37. $\cos x = \displaystyle\sum_{n=0}^{\infty}(-1)^n\dfrac{x^{2n}}{(2n)!} \;\Rightarrow$

$$f(x) = \cos(x^2) = \sum_{n=0}^{\infty}\dfrac{(-1)^n(x^2)^{2n}}{(2n)!} = \sum_{n=0}^{\infty}\dfrac{(-1)^n x^{4n}}{(2n)!}$$

$$= 1 - \tfrac{1}{2}x^4 + \tfrac{1}{24}x^8 - \tfrac{1}{720}x^{12} + \cdots$$

The series for $\cos x$ converges for all x, so the same is true of the series for

$f(x)$, that is, $R = \infty$. Notice that, as n increases, $T_n(x)$ becomes a better

approximation to $f(x)$.

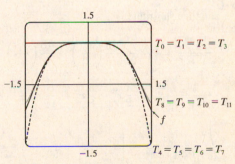

39. $5° = 5° \left(\dfrac{\pi}{180°}\right) = \dfrac{\pi}{36}$ radians and $\cos x = \displaystyle\sum_{n=0}^{\infty} (-1)^n \dfrac{x^{2n}}{(2n)!} = 1 - \dfrac{x^2}{2!} + \dfrac{x^4}{4!} - \dfrac{x^6}{6!} + \cdots$, so

$\cos \dfrac{\pi}{36} = 1 - \dfrac{(\pi/36)^2}{2!} + \dfrac{(\pi/36)^4}{4!} - \dfrac{(\pi/36)^6}{6!} + \cdots$. Now $1 - \dfrac{(\pi/36)^2}{2!} \approx 0.99619$ and adding $\dfrac{(\pi/36)^4}{4!} \approx 2.4 \times 10^{-6}$

does not affect the fifth decimal place, so $\cos 5° \approx 0.99619$ by the Alternating Series Estimation Theorem.

41. (a) $1/\sqrt{1-x^2} = \left[1 + (-x^2)\right]^{-1/2} = 1 + \left(-\tfrac{1}{2}\right)(-x^2) + \dfrac{\left(-\frac{1}{2}\right)\left(-\frac{3}{2}\right)}{2!}(-x^2)^2 + \dfrac{\left(-\frac{1}{2}\right)\left(-\frac{3}{2}\right)\left(-\frac{5}{2}\right)}{3!}(-x^2)^3 + \cdots$

$\qquad = 1 + \displaystyle\sum_{n=1}^{\infty} \dfrac{1 \cdot 3 \cdot 5 \cdot \cdots \cdot (2n-1)}{2^n \cdot n!} x^{2n}$

(b) $\sin^{-1} x = \displaystyle\int \dfrac{1}{\sqrt{1-x^2}}\, dx = C + x + \sum_{n=1}^{\infty} \dfrac{1 \cdot 3 \cdot 5 \cdot \cdots \cdot (2n-1)}{(2n+1)2^n \cdot n!} x^{2n+1}$

$\qquad = x + \displaystyle\sum_{n=1}^{\infty} \dfrac{1 \cdot 3 \cdot 5 \cdot \cdots \cdot (2n-1)}{(2n+1)2^n \cdot n!} x^{2n+1}$ since $0 = \sin^{-1} 0 = C$.

43. $\cos x = \displaystyle\sum_{n=0}^{\infty} (-1)^n \dfrac{x^{2n}}{(2n)!} \;\Rightarrow\; \cos(x^3) = \sum_{n=0}^{\infty} (-1)^n \dfrac{(x^3)^{2n}}{(2n)!} = \sum_{n=0}^{\infty} (-1)^n \dfrac{x^{6n}}{(2n)!} \;\Rightarrow$

$x\cos(x^3) = \displaystyle\sum_{n=0}^{\infty} (-1)^n \dfrac{x^{6n+1}}{(2n)!} \;\Rightarrow\; \int x\cos(x^3)\, dx = C + \sum_{n=0}^{\infty} (-1)^n \dfrac{x^{6n+2}}{(6n+2)(2n)!}$, with $R = \infty$.

45. $\cos x = \displaystyle\sum_{n=0}^{\infty} (-1)^n \dfrac{x^{2n}}{(2n)!} \;\Rightarrow\; \cos x - 1 = \sum_{n=1}^{\infty} (-1)^n \dfrac{x^{2n}}{(2n)!} \;\Rightarrow\; \dfrac{\cos x - 1}{x} = \sum_{n=1}^{\infty} (-1)^n \dfrac{x^{2n-1}}{(2n)!} \;\Rightarrow$

$\displaystyle\int \dfrac{\cos x - 1}{x}\, dx = C + \sum_{n=1}^{\infty} (-1)^n \dfrac{x^{2n}}{2n \cdot (2n)!}$, with $R = \infty$.

47. By Exercise 43, $\displaystyle\int x\cos(x^3)\, dx = C + \sum_{n=0}^{\infty} (-1)^n \dfrac{x^{6n+2}}{(6n+2)(2n)!}$, so

$\displaystyle\int_0^1 x\cos(x^3)\, dx = \left[\sum_{n=0}^{\infty} (-1)^n \dfrac{x^{6n+2}}{(6n+2)(2n)!}\right]_0^1 = \sum_{n=0}^{\infty} \dfrac{(-1)^n}{(6n+2)(2n)!} = \dfrac{1}{2} - \dfrac{1}{8 \cdot 2!} + \dfrac{1}{14 \cdot 4!} - \dfrac{1}{20 \cdot 6!} + \cdots$, but

$\dfrac{1}{20 \cdot 6!} = \dfrac{1}{14{,}400} \approx 0.000\,069$, so $\displaystyle\int_0^1 x\cos(x^3)\, dx \approx \dfrac{1}{2} - \dfrac{1}{16} + \dfrac{1}{336} \approx 0.440$ (correct to three decimal places) by the

Alternating Series Estimation Theorem.

49. We first find a series representation for $f(x) = (1+x)^{-1/2}$,

and then substitute.

$\dfrac{1}{\sqrt{1+x}} = 1 - \dfrac{x}{2} + \dfrac{3}{4}\left(\dfrac{x^2}{2!}\right) - \dfrac{15}{8}\left(\dfrac{x^3}{3!}\right) + \cdots \;\Rightarrow$

$\dfrac{1}{\sqrt{1+x^3}} = 1 - \dfrac{1}{2}x^3 + \dfrac{3}{8}x^6 - \dfrac{5}{16}x^9 + \cdots \;\Rightarrow$

n	$f^{(n)}(x)$	$f^{(n)}(0)$
0	$(1+x)^{-1/2}$	1
1	$-\frac{1}{2}(1+x)^{-3/2}$	$-\frac{1}{2}$
2	$\frac{3}{4}(1+x)^{-5/2}$	$\frac{3}{4}$
3	$-\frac{15}{8}(1+x)^{-7/2}$	$-\frac{15}{8}$
⋮	⋮	⋮

$\displaystyle\int_0^{0.1} \dfrac{dx}{\sqrt{1+x^3}} = \left[x - \dfrac{1}{8}x^4 + \dfrac{3}{56}x^7 - \dfrac{1}{32}x^{10} + \cdots\right]_0^{0.1} \approx (0.1) - \dfrac{1}{8}(0.1)^4$, by the Alternating Series Estimation

Theorem, since $\frac{3}{56}(0.1)^7 \approx 0.000\,000\,005\,4 < 10^{-8}$, which is the maximum desired error. Therefore,

$$\int_0^{0.1} \frac{dx}{\sqrt{1+x^3}} \approx 0.099\,987\,50.$$

51. $\displaystyle\lim_{x \to 0} \frac{x - \ln(1+x)}{x^2} = \lim_{x \to 0} \frac{x - \left(x - \frac{1}{2}x^2 + \frac{1}{3}x^3 - \frac{1}{4}x^4 + \frac{1}{5}x^5 - \cdots\right)}{x^2} = \lim_{x \to 0} \frac{\frac{1}{2}x^2 - \frac{1}{3}x^3 + \frac{1}{4}x^4 - \frac{1}{5}x^5 + \cdots}{x^2}$

$\displaystyle = \lim_{x \to 0}\left(\frac{1}{2} - \frac{1}{3}x + \frac{1}{4}x^2 - \frac{1}{5}x^3 + \cdots\right) = \frac{1}{2}$

since power series are continuous functions.

53. $\displaystyle\lim_{x \to 0} \frac{\sin x - x + \frac{1}{6}x^3}{x^5} = \lim_{x \to 0} \frac{\left(x - \frac{1}{3!}x^3 + \frac{1}{5!}x^5 - \frac{1}{7!}x^7 + \cdots\right) - x + \frac{1}{6}x^3}{x^5}$

$\displaystyle = \lim_{x \to 0} \frac{\frac{1}{5!}x^5 - \frac{1}{7!}x^7 + \cdots}{x^5} = \lim_{x \to 0}\left(\frac{1}{5!} - \frac{x^2}{7!} + \frac{x^4}{9!} - \cdots\right) = \frac{1}{5!} = \frac{1}{120}$

since power series are continuous functions.

55. As in Example 9(a), we have $e^{-x^2} = 1 - \frac{x^2}{1!} + \frac{x^4}{2!} - \frac{x^6}{3!} + \cdots$ and we know that $\cos x = 1 - \frac{x^2}{2!} + \frac{x^4}{4!} - \cdots$ from

Equation 17. Therefore, $e^{-x^2}\cos x = \left(1 - x^2 + \frac{1}{2}x^4 - \cdots\right)\left(1 - \frac{1}{2}x^2 + \frac{1}{24}x^4 - \cdots\right)$. Writing only the terms with

degree ≤ 4, we get $e^{-x^2}\cos x = 1 - \frac{1}{2}x^2 + \frac{1}{24}x^4 - x^2 + \frac{1}{2}x^4 + \frac{1}{2}x^4 + \cdots = 1 - \frac{3}{2}x^2 + \frac{25}{24}x^4 + \cdots$.

57. $\dfrac{x}{\sin x} = \dfrac{x}{x - \frac{1}{6}x^3 + \frac{1}{120}x^5 - \cdots}$.

$$
\begin{array}{r}
1 + \frac{1}{6}x^2 + \frac{7}{360}x^4 + \cdots \\[4pt]
x - \frac{1}{6}x^3 + \frac{1}{120}x^5 - \cdots \,\overline{\big)\ x} \\[4pt]
\underline{x - \frac{1}{6}x^3 + \frac{1}{120}x^5 - \cdots} \\[4pt]
\frac{1}{6}x^3 - \frac{1}{120}x^5 + \cdots \\[4pt]
\underline{\frac{1}{6}x^3 - \frac{1}{36}x^5 + \cdots} \\[4pt]
\frac{7}{360}x^5 + \cdots \\[4pt]
\underline{\frac{7}{360}x^5 + \cdots} \\[4pt]
\cdots
\end{array}
$$

From the long division above, $\dfrac{x}{\sin x} = 1 + \frac{1}{6}x^2 + \frac{7}{360}x^4 + \cdots$.

59. $\displaystyle\sum_{n=0}^{\infty} (-1)^n \frac{x^{4n}}{n!} = \sum_{n=0}^{\infty} \frac{\left(-x^4\right)^n}{n!} = e^{-x^4}$

61. $\displaystyle\sum_{n=0}^{\infty} \frac{(-1)^n \pi^{2n+1}}{4^{2n+1}(2n+1)!} = \sum_{n=0}^{\infty} \frac{(-1)^n \left(\frac{\pi}{4}\right)^{2n+1}}{(2n+1)!} = \sin\frac{\pi}{4} = \frac{1}{\sqrt{2}}$

63. $3 + \dfrac{9}{2!} + \dfrac{27}{3!} + \dfrac{81}{4!} + \cdots = \dfrac{3^1}{1!} + \dfrac{3^2}{2!} + \dfrac{3^3}{3!} + \dfrac{3^4}{4!} + \cdots = \displaystyle\sum_{n=1}^{\infty} \frac{3^n}{n!} = \sum_{n=0}^{\infty} \frac{3^n}{n!} - 1 = e^3 - 1$

65. (a) $[1 + (-x)]^{-2} = 1 + (-2)(-x) + \dfrac{(-2)(-3)}{2!}(-x)^2 + \dfrac{(-2)(-3)(-4)}{3!}(-x)^3 + \cdots$

$$= 1 + 2x + 3x^2 + 4x^3 + \cdots = \sum_{n=0}^{\infty} (n+1)x^n,$$

so $\dfrac{x}{(1-x)^2} = x \sum_{n=0}^{\infty} (n+1)x^n = \sum_{n=0}^{\infty} (n+1)x^{n+1} = \sum_{n=1}^{\infty} nx^n.$

(b) With $x = \frac{1}{2}$ in part (a), we have $\sum_{n=1}^{\infty} n\left(\frac{1}{2}\right)^n = \sum_{n=1}^{\infty} \dfrac{n}{2^n} = \dfrac{\frac{1}{2}}{\left(1 - \frac{1}{2}\right)^2} = \dfrac{\frac{1}{2}}{\frac{1}{4}} = 2.$

67. If p is an nth-degree polynomial, then $p^{(i)}(x) = 0$ for $i > n$, so its Taylor series at a is $p(x) = \sum_{i=0}^{n} \dfrac{p^{(i)}(a)}{i!}(x-a)^i.$

Put $x - a = 1$, so that $x = a + 1$. Then $p(a + 1) = \sum_{i=0}^{n} \dfrac{p^{(i)}(a)}{i!}.$

This is true for any a, so replace a by x: $p(x + 1) = \sum_{i=0}^{n} \dfrac{p^{(i)}(x)}{i!}$

69. (a) $g(x) = \sum_{n=0}^{\infty} \binom{k}{n} x^n \;\; \Rightarrow \;\; g'(x) = \sum_{n=1}^{\infty} \binom{k}{n} nx^{n-1}$, so

$(1 + x)g'(x) = (1 + x) \sum_{n=1}^{\infty} \binom{k}{n} nx^{n-1} = \sum_{n=1}^{\infty} \binom{k}{n} nx^{n-1} + \sum_{n=1}^{\infty} \binom{k}{n} nx^n$

$$= \sum_{n=0}^{\infty} \binom{k}{n+1}(n+1)x^n + \sum_{n=0}^{\infty} \binom{k}{n} nx^n \qquad \begin{bmatrix} \text{Replace } n \text{ with } n+1 \\ \text{ in the first series} \end{bmatrix}$$

$$= \sum_{n=0}^{\infty} (n+1)\dfrac{k(k-1)(k-2)\cdots(k-n+1)(k-n)}{(n+1)!}x^n + \sum_{n=0}^{\infty} \left[(n)\dfrac{k(k-1)(k-2)\cdots(k-n+1)}{n!}\right]x^n$$

$$= \sum_{n=0}^{\infty} \dfrac{(n+1)k(k-1)(k-2)\cdots(k-n+1)}{(n+1)!}\left[(k-n)+n\right]x^n$$

$$= k \sum_{n=0}^{\infty} \dfrac{k(k-1)(k-2)\cdots(k-n+1)}{n!}x^n = k \sum_{n=0}^{\infty} \binom{k}{n} x^n = kg(x)$$

Thus, $g'(x) = \dfrac{kg(x)}{1+x}$.

(b) $h(x) = (1+x)^{-k}g(x) \;\; \Rightarrow$

$$h'(x) = -k(1+x)^{-k-1}g(x) + (1+x)^{-k}g'(x) \qquad \text{[Product Rule]}$$

$$= -k(1+x)^{-k-1}g(x) + (1+x)^{-k}\dfrac{kg(x)}{1+x} \qquad \text{[from part (a)]}$$

$$= -k(1+x)^{-k-1}g(x) + k(1+x)^{-k-1}g(x) = 0$$

(c) From part (b) we see that $h(x)$ must be constant for $x \in (-1, 1)$, so $h(x) = h(0) = 1$ for $x \in (-1, 1)$.

Thus, $h(x) = 1 = (1+x)^{-k}g(x) \;\; \Leftrightarrow \;\; g(x) = (1+x)^k$ for $x \in (-1, 1)$.

8.8   Applications of Taylor Polynomials

1. (a)

n	$f^{(n)}(x)$	$f^{(n)}(0)$	$T_n(x)$
0	$\cos x$	1	1
1	$-\sin x$	0	1
2	$-\cos x$	-1	$1 - \frac{1}{2}x^2$
3	$\sin x$	0	$1 - \frac{1}{2}x^2$
4	$\cos x$	1	$1 - \frac{1}{2}x^2 + \frac{1}{24}x^4$
5	$-\sin x$	0	$1 - \frac{1}{2}x^2 + \frac{1}{24}x^4$
6	$-\cos x$	-1	$1 - \frac{1}{2}x^2 + \frac{1}{24}x^4 - \frac{1}{720}x^6$

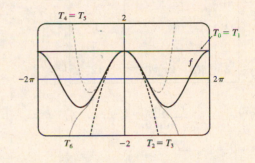

(b)

x	f	$T_0 = T_1$	$T_2 = T_3$	$T_4 = T_5$	T_6
$\frac{\pi}{4}$	0.7071	1	0.6916	0.7074	0.7071
$\frac{\pi}{2}$	0	1	-0.2337	0.0200	-0.0009
π	-1	1	-3.9348	0.1239	-1.2114

(c) As n increases, $T_n(x)$ is a good approximation to $f(x)$ on a larger and larger interval.

3.

n	$f^{(n)}(x)$	$f^{(n)}(2)$
0	$1/x$	$\frac{1}{2}$
1	$-1/x^2$	$-\frac{1}{4}$
2	$2/x^3$	$\frac{1}{4}$
3	$-6/x^4$	$-\frac{3}{8}$

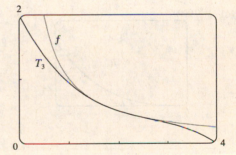

$$T_3(x) = \sum_{n=0}^{3} \frac{f^{(n)}(2)}{n!}(x-2)^n$$

$$= \frac{\frac{1}{2}}{0!} - \frac{\frac{1}{4}}{1!}(x-2) + \frac{\frac{1}{4}}{2!}(x-2)^2 - \frac{\frac{3}{8}}{3!}(x-2)^3$$

$$= \frac{1}{2} - \frac{1}{4}(x-2) + \frac{1}{8}(x-2)^2 - \frac{1}{16}(x-2)^3$$

5.

n	$f^{(n)}(x)$	$f^{(n)}(\pi/2)$
0	$\cos x$	0
1	$-\sin x$	-1
2	$-\cos x$	0
3	$\sin x$	1

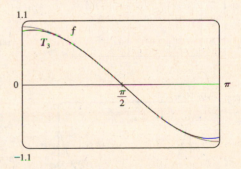

$$T_3(x) = \sum_{n=0}^{3} \frac{f^{(n)}(\pi/2)}{n!}\left(x - \tfrac{\pi}{2}\right)^n$$

$$= -\left(x - \tfrac{\pi}{2}\right) + \tfrac{1}{6}\left(x - \tfrac{\pi}{2}\right)^3$$

7.

n	$f^{(n)}(x)$	$f^{(n)}(0)$
0	xe^{-2x}	0
1	$(1-2x)e^{-2x}$	1
2	$4(x-1)e^{-2x}$	-4
3	$4(3-2x)e^{-2x}$	12

$$T_3(x) = \sum_{n=0}^{3} \frac{f^{(n)}(0)}{n!}x^n = \frac{0}{1}\cdot 1 + \frac{1}{1}x^1 + \frac{-4}{2}x^2 + \frac{12}{6}x^3 = x - 2x^2 + 2x^3$$

9.

n	$f^{(n)}(x)$	$f^{(n)}(4)$
0	$\sqrt{x}$	2
1	$\frac{1}{2}x^{-1/2}$	$\frac{1}{4}$
2	$-\frac{1}{4}x^{-3/2}$	$-\frac{1}{32}$
3	$\frac{3}{8}x^{-5/2}$	

(a) $f(x) = \sqrt{x} \approx T_2(x) = 2 + \frac{1}{4}(x-4) - \frac{1/32}{2!}(x-4)^2$

$$= 2 + \frac{1}{4}(x-4) - \frac{1}{64}(x-4)^2$$

(b) $|R_2(x)| \le \dfrac{M}{3!}|x-4|^3$, where $|f'''(x)| \le M$. Now $4 \le x \le 4.2 \Rightarrow$

$|x-4| \le 0.2 \Rightarrow |x-4|^3 \le 0.008$. Since $f'''(x)$ is decreasing

on $[4, 4.2]$, we can take $M = |f'''(4)| = \frac{3}{8}4^{-5/2} = \frac{3}{256}$, so

$$|R_2(x)| \le \frac{3/256}{6}(0.008) = \frac{0.008}{512} = 0.000\,015\,625.$$

(c)

From the graph of $|R_2(x)| = |\sqrt{x} - T_2(x)|$, it seems that the

error is less than 1.52×10^{-5} on $[4, 4.2]$.

11.

n	$f^{(n)}(x)$	$f^{(n)}(1)$
0	$x^{2/3}$	1
1	$\frac{2}{3}x^{-1/3}$	$\frac{2}{3}$
2	$-\frac{2}{9}x^{-4/3}$	$-\frac{2}{9}$
3	$\frac{8}{27}x^{-7/3}$	$\frac{8}{27}$
4	$-\frac{56}{81}x^{-10/3}$	

(a) $f(x) = x^{2/3} \approx T_3(x) = 1 + \frac{2}{3}(x-1) - \frac{2/9}{2!}(x-1)^2 + \frac{8/27}{3!}(x-1)^3$

$$= 1 + \frac{2}{3}(x-1) - \frac{1}{9}(x-1)^2 + \frac{4}{81}(x-1)^3$$

(b) $|R_3(x)| \le \dfrac{M}{4!}|x-1|^4$, where $\left|f^{(4)}(x)\right| \le M$. Now $0.8 \le x \le 1.2 \Rightarrow$

$|x-1| \le 0.2 \Rightarrow |x-1|^4 \le 0.0016$. Since $\left|f^{(4)}(x)\right|$ is decreasing

on $[0.8, 1.2]$, we can take $M = \left|f^{(4)}(0.8)\right| = \frac{56}{81}(0.8)^{-10/3}$, so

$$|R_3(x)| \le \frac{\frac{56}{81}(0.8)^{-10/3}}{24}(0.0016) \approx 0.000\,096\,97.$$

(c)

From the graph of $|R_3(x)| = \left|x^{2/3} - T_3(x)\right|$, it seems that the

error is less than $0.000\,053\,3$ on $[0.8, 1.2]$.

13.

n	$f^{(n)}(x)$	$f^{(n)}(0)$
0	e^{x^2}	1
1	$e^{x^2}(2x)$	0
2	$e^{x^2}(2+4x^2)$	2
3	$e^{x^2}(12x+8x^3)$	0
4	$e^{x^2}(12+48x^2+16x^4)$	

(a) $f(x) = e^{x^2} \approx T_3(x) = 1 + \dfrac{2}{2!}x^2 = 1 + x^2$

(b) $|R_3(x)| \le \dfrac{M}{4!}|x|^4$, where $\left|f^{(4)}(x)\right| \le M$. Now $0 \le x \le 0.1 \Rightarrow$

$x^4 \le (0.1)^4$, and letting $x = 0.1$ gives

$|R_3(x)| \le \dfrac{e^{0.01}(12 + 0.48 + 0.0016)}{24}(0.1)^4 \approx 0.00006.$

(c)

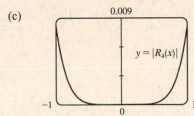

0.00008

$y = |R_3(x)|$

0   0.1

From the graph of $|R_3(x)| = \left|e^{x^2} - T_3(x)\right|$, it appears that the

error is less than $0.000\,051$ on $[0, 0.1]$.

15.

n	$f^{(n)}(x)$	$f^{(n)}(0)$
0	$x\sin x$	0
1	$\sin x + x\cos x$	0
2	$2\cos x - x\sin x$	2
3	$-3\sin x - x\cos x$	0
4	$-4\cos x + x\sin x$	-4
5	$5\sin x + x\cos x$	

(a) $f(x) = x\sin x \approx T_4(x) = \dfrac{2}{2!}(x-0)^2 + \dfrac{-4}{4!}(x-0)^4 = x^2 - \dfrac{1}{6}x^4$

(b) $|R_4(x)| \le \dfrac{M}{5!}|x|^5$, where $\left|f^{(5)}(x)\right| \le M$. Now $-1 \le x \le 1 \Rightarrow$

$|x| \le 1$, and a graph of $f^{(5)}(x)$ shows that $\left|f^{(5)}(x)\right| \le 5$ for $-1 \le x \le 1$.

Thus, we can take $M = 5$ and get $|R_4(x)| \le \dfrac{5}{5!} \cdot 1^5 = \dfrac{1}{24} = 0.041\overline{6}.$

(c)

0.009

$y = |R_4(x)|$

-1 0   1

From the graph of $|R_4(x)| = |x\sin x - T_4(x)|$, it seems that the

error is less than 0.0082 on $[-1, 1]$.

17. From Exercise 5, $\cos x = -\left(x - \frac{\pi}{2}\right) + \frac{1}{6}\left(x - \frac{\pi}{2}\right)^3 + R_3(x)$, where $|R_3(x)| \le \dfrac{M}{4!}\left|x - \frac{\pi}{2}\right|^4$ with

$\left|f^{(4)}(x)\right| = |\cos x| \le M = 1$. Now $x = 80° = (90° - 10°) = \left(\frac{\pi}{2} - \frac{\pi}{18}\right) = \frac{4\pi}{9}$ radians, so the error is

$\left|R_3\left(\frac{4\pi}{9}\right)\right| \le \frac{1}{24}\left(\frac{\pi}{18}\right)^4 \approx 0.000\,039$, which means our estimate would *not* be accurate to five decimal places. However,

$T_3 = T_4$, so we can use $\left|R_4\left(\frac{4\pi}{9}\right)\right| \le \frac{1}{120}\left(\frac{\pi}{18}\right)^5 \approx 0.000\,001$. Therefore, to five decimal places,

$\cos 80° \approx -\left(-\frac{\pi}{18}\right) + \frac{1}{6}\left(-\frac{\pi}{18}\right)^3 \approx 0.17365.$

19. All derivatives of e^x are e^x, so the remainder term is $R_n(x) = \dfrac{e^z}{(n+1)!}x^{n+1}$, where $0 < z < 0.1$. So we want

$R_n(0.1) \le \dfrac{e^{0.1}}{(n+1)!}(0.1)^{n+1} < 0.00001$, and we find that $n = 3$ satisfies this inequality. [In fact $R_3(0.1) < 0.0000046$.]

So four terms are required ($n = 0, 1, 2, 3$).

21. $\sin x = x - \frac{1}{3!}x^3 + \frac{1}{5!}x^5 - \cdots$. By the Alternating Series

Estimation Theorem, the error in the approximation

$\sin x = x - \frac{1}{3!}x^3$ is less than $\left|\frac{1}{5!}x^5\right| < 0.01$ ⇔

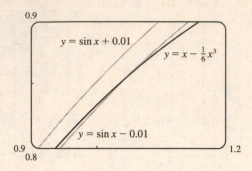

$\left|x^5\right| < 120(0.01)$ ⇔ $|x| < (1.2)^{1/5} \approx 1.037$. The curves

$y = x - \frac{1}{6}x^3$ and $y = \sin x - 0.01$ intersect at $x \approx 1.043$, so

the graph confirms our estimate. Since both the sine function

and the given approximation are odd functions, we need to check the estimate only for $x > 0$. Thus, the desired range of

values for x is $-1.037 < x < 1.037$.

23. Let $s(t)$ be the position function of the car, and for convenience set $s(0) = 0$. The velocity of the car is $v(t) = s'(t)$ and the

acceleration is $a(t) = s''(t)$, so the second degree Taylor polynomial is $T_2(t) = s(0) + v(0)t + \frac{a(0)}{2}t^2 = 20t + t^2$. We

estimate the distance traveled during the next second to be $s(1) \approx T_2(1) = 20 + 1 = 21$ m. The function $T_2(t)$ would not be

accurate over a full minute, since the car could not possibly maintain an acceleration of 2 m/s² for that long (if it did, its final

speed would be 140 m/s $\approx$ 313 mi/h!).

25. $E = \dfrac{q}{D^2} - \dfrac{q}{(D+d)^2} = \dfrac{q}{D^2} - \dfrac{q}{D^2(1+d/D)^2} = \dfrac{q}{D^2}\left[1 - \left(1 + \dfrac{d}{D}\right)^{-2}\right].$

We use the Binomial Series to expand $(1 + d/D)^{-2}$:

$$E = \frac{q}{D^2}\left[1 - \left(1 - 2\left(\frac{d}{D}\right) + \frac{2\cdot 3}{2!}\left(\frac{d}{D}\right)^2 - \frac{2\cdot 3\cdot 4}{3!}\left(\frac{d}{D}\right)^3 + \cdots\right)\right] = \frac{q}{D^2}\left[2\left(\frac{d}{D}\right) - 3\left(\frac{d}{D}\right)^2 + 4\left(\frac{d}{D}\right)^3 - \cdots\right]$$

$$\approx \frac{q}{D^2}\cdot 2\left(\frac{d}{D}\right) = 2qd\cdot\frac{1}{D^3}$$

when D is much larger than d; that is, when P is far away from the dipole.

27. (a) L is the length of the arc subtended by the angle θ, so $L = R\theta$ ⇒

$\theta = L/R$. Now $\sec\theta = (R+C)/R$ ⇒ $R\sec\theta = R + C$ ⇒

$C = R\sec\theta - R = R\sec(L/R) - R.$

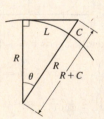

(b) First we'll find a Taylor polynomial $T_4(x)$ for $f(x) = \sec x$ at $x = 0$.

n	$f^{(n)}(x)$	$f^{(n)}(0)$
0	$\sec x$	1
1	$\sec x \tan x$	0
2	$\sec x(2\tan^2 x + 1)$	1
3	$\sec x \tan x(6\tan^2 x + 5)$	0
4	$\sec x(24\tan^4 x + 28\tan^2 x + 5)$	5

Thus, $f(x) = \sec x \approx T_4(x) = 1 + \frac{1}{2!}(x-0)^2 + \frac{5}{4!}(x-0)^4 = 1 + \frac{1}{2}x^2 + \frac{5}{24}x^4$. By part (a),

$$C \approx R\left[1 + \frac{1}{2}\left(\frac{L}{R}\right)^2 + \frac{5}{24}\left(\frac{L}{R}\right)^4\right] - R = R + \frac{1}{2}R \cdot \frac{L^2}{R^2} + \frac{5}{24}R \cdot \frac{L^4}{R^4} - R = \frac{L^2}{2R} + \frac{5L^4}{24R^3}.$$

(c) Taking $L = 100$ km and $R = 6370$ km, the formula in part (a) says that

$$C = R\sec(L/R) - R = 6370\sec(100/6370) - 6370 \approx 0.785\,009\,965\,44 \text{ km}.$$

The formula in part (b) says that $C \approx \dfrac{L^2}{2R} + \dfrac{5L^4}{24R^3} = \dfrac{100^2}{2 \cdot 6370} + \dfrac{5 \cdot 100^4}{24 \cdot 6370^3} \approx 0.785\,009\,957\,36$ km.

The difference between these two results is only $0.000\,000\,008\,08$ km, or $0.000\,008\,08$ m!

29. Using Taylor's Formula with $n = 1$, $a = x_n$, $x = r$, we get $f(r) = f(x_n) + f'(x_n)(r - x_n) + R_1(x)$, where

$R_1(x) = \frac{1}{2}f''(z)(r - x_n)^2$ and z lies betwen x_n and r. But r is a root, so $f(r) = 0$ and Taylor's Formula becomes

$0 = f(x_n) + f'(x_n)(r - x_n) + \frac{1}{2}f''(z)(r - x_n)^2$. Taking the first two terms to the left side and dividing by $f'(x_n)$,

we have $x_n - r - \dfrac{f(x_n)}{f'(x_n)} = \dfrac{1}{2}\dfrac{f''(z)}{f'(x_n)}|x_n - r|^2$. By the formula for Newton's Method, we have

$$|x_{n+1} - r| = \left|x_n - \frac{f(x_n)}{f'(x_n)} - r\right| = \frac{1}{2}\frac{|f''(z)|}{|f'(x_n)|}|x_n - r|^2 \leq \frac{M}{2K}|x_n - r|^2 \text{ since } |f''(z)| \leq M \text{ and } |f'(x_n)| \geq K.$$

8   Review

CONCEPT CHECK

1. (a) See Definition 8.1.1.

   (b) See Definition 8.2.2.

   (c) The terms of the sequence $\{a_n\}$ approach 3 as n becomes large.

   (d) By adding sufficiently many terms of the series, we can make the partial sums as close to 3 as we like.

2. (a) A sequence $\{a_n\}$ is bounded if there are numbers m and M such that $m \leq a_n \leq M$ for all $n \geq 1$.

   (b) A sequence is monotonic if it is either increasing or decreasing.

   (c) By Theorem 8.1.11, every bounded, monotonic sequence is convergent.

3. (a) See (4) in Section 8.2.

   (b) The p-series $\displaystyle\sum_{n=1}^{\infty} \frac{1}{n^p}$ is convergent if $p > 1$.

4. If $\sum a_n = 3$, then $\displaystyle\lim_{n \to \infty} a_n = 0$ and $\displaystyle\lim_{n \to \infty} s_n = 3$.

5. (a) *Test for Divergence:* If $\displaystyle\lim_{n \to \infty} a_n$ does not exist or if $\displaystyle\lim_{n \to \infty} a_n \neq 0$, then the series $\sum_{n=1}^{\infty} a_n$ is divergent.

(b) *Integral Test:* Suppose f is a continuous, positive, decreasing function on $[1, \infty)$ and let $a_n = f(n)$. Then the series $\sum_{n=1}^{\infty} a_n$ is convergent if and only if the improper integral $\int_1^{\infty} f(x)\, dx$ is convergent. In other words:

   (i) If $\int_1^{\infty} f(x)\, dx$ is convergent, then $\sum_{n=1}^{\infty} a_n$ is convergent.

   (ii) If $\int_1^{\infty} f(x)\, dx$ is divergent, then $\sum_{n=1}^{\infty} a_n$ is divergent.

(c) *Comparison Test:* Suppose that $\sum a_n$ and $\sum b_n$ are series with positive terms.

   (i) If $\sum b_n$ is convergent and $a_n \le b_n$ for all n, then $\sum a_n$ is also convergent.

   (ii) If $\sum b_n$ is divergent and $a_n \ge b_n$ for all n, then $\sum a_n$ is also divergent.

(d) *Limit Comparison Test:* Suppose that $\sum a_n$ and $\sum b_n$ are series with positive terms. If $\lim_{n \to \infty} (a_n / b_n) = c$, where c is a finite number and $c > 0$, then either both series converge or both diverge.

(e) *Alternating Series Test:* If the alternating series $\sum_{n=1}^{\infty} (-1)^{n-1} b_n = b_1 - b_2 + b_3 - b_4 + b_5 - b_6 + \cdots$ $[b_n > 0]$ satisfies (i) $b_{n+1} \le b_n$ for all n and (ii) $\lim_{n \to \infty} b_n = 0$, then the series is convergent.

(f) *Ratio Test:*

   (i) If $\lim_{n \to \infty} \left| \dfrac{a_{n+1}}{a_n} \right| = L < 1$, then the series $\sum_{n=1}^{\infty} a_n$ is absolutely convergent (and therefore convergent).

   (ii) If $\lim_{n \to \infty} \left| \dfrac{a_{n+1}}{a_n} \right| = L > 1$ or $\lim_{n \to \infty} \left| \dfrac{a_{n+1}}{a_n} \right| = \infty$, then the series $\sum_{n=1}^{\infty} a_n$ is divergent.

   (iii) If $\lim_{n \to \infty} \left| \dfrac{a_{n+1}}{a_n} \right| = 1$, the Ratio Test is inconclusive; that is, no conclusion can be drawn about the convergence or divergence of $\sum a_n$.

(g) *Root Test:*

   (i) If $\lim_{n \to \infty} \sqrt[n]{|a_n|} = L < 1$, then the series $\sum_{n=1}^{\infty} a_n$ is absolutely convergent (and therefore convergent).

   (ii) If $\lim_{n \to \infty} \sqrt[n]{|a_n|} = L > 1$ or $\lim_{n \to \infty} \sqrt[n]{|a_n|} = \infty$, then the series $\sum_{n=1}^{\infty} a_n$ is divergent.

   (iii) If $\lim_{n \to \infty} \sqrt[n]{|a_n|} = 1$, the Root Test is inconclusive.

6. (a) A series $\sum a_n$ is called *absolutely convergent* if the series of absolute values $\sum |a_n|$ is convergent.

   (b) If a series $\sum a_n$ is absolutely convergent, then it is convergent.

   (c) A series $\sum a_n$ is called *conditionally convergent* if it is convergent but not absolutely convergent.

7. By adding terms until you reach the desired accuracy given by the Alternating Series Estimation Theorem.

8. (a) $\sum_{n=0}^{\infty} c_n (x - a)^n$

   (b) Given the power series $\sum_{n=0}^{\infty} c_n (x - a)^n$, the radius of convergence is:

   (i) 0 if the series converges only when $x = a$

   (ii) ∞ if the series converges for all x, or

   (iii) a positive number R such that the series converges if $|x - a| < R$ and diverges if $|x - a| > R$.

(c) The interval of convergence of a power series is the interval that consists of all values of x for which the series converges. Corresponding to the cases in part (b), the interval of convergence is: (i) the single point $\{a\}$, (ii) all real numbers, that is, the real number line $(-\infty, \infty)$, or (iii) an interval with endpoints $a - R$ and $a + R$ which can contain neither, either, or both of the endpoints. In this case, we must test the series for convergence at each endpoint to determine the interval of convergence.

9. (a), (b) See Theorem 8.6.2.

10. (a) $T_n(x) = \sum\limits_{i=0}^{n} \dfrac{f^{(i)}(a)}{i!} (x - a)^i$ (b) $\sum\limits_{n=0}^{\infty} \dfrac{f^{(n)}(a)}{n!} (x - a)^n$

(c) $\sum\limits_{n=0}^{\infty} \dfrac{f^{(n)}(0)}{n!} x^n$ $[a = 0$ in part (b)] (d) See Theorem 8.7.8.

(e) See Taylor's Formula (8.7.9).

11. (a)–(f) See Table 1 on page 490.

12. See the binomial series (8.7.18) for the expansion. The radius of convergence for the binomial series is 1.

TRUE-FALSE QUIZ

1. False. See Note 2 on page 447.

3. True. If $\lim\limits_{n \to \infty} a_n = L$, then as $n \to \infty$, $2n + 1 \to \infty$, so $a_{2n+1} \to L$.

5. False. For example, take $c_n = (-1)^n/(n6^n)$.

7. False, since $\lim\limits_{n \to \infty} \left| \dfrac{a_{n+1}}{a_n} \right| = \lim\limits_{n \to \infty} \left| \dfrac{1}{(n + 1)^3} \cdot \dfrac{n^3}{1} \right| = \lim\limits_{n \to \infty} \left| \dfrac{n^3}{(n + 1)^3} \cdot \dfrac{1/n^3}{1/n^3} \right| = \lim\limits_{n \to \infty} \dfrac{1}{(1 + 1/n)^3} = 1.$ x

9. False. See the note after Example 4 in Section 8.3.

11. True. See (8) in Section 8.1.

13. True. By Theorem 8.7.5 the coefficient of x^3 is $\dfrac{f'''(0)}{3!} = \dfrac{1}{3} \Rightarrow f'''(0) = 2.$

Or: Use Theorem 8.6.2 to differentiate f three times.

15. False. For example, let $a_n = b_n = (-1)^n$. Then $\{a_n\}$ and $\{b_n\}$ are divergent, but $a_n b_n = 1$, so $\{a_n b_n\}$ is convergent.

17. True by Theorem 8.4.1. $\left[\sum (-1)^n a_n \text{ is absolutely convergent and hence convergent.} \right]$

19. True. $0.99999\ldots = 0.9 + 0.9(0.1)^1 + 0.9(0.1)^2 + 0.9(0.1)^3 + \cdots = \sum\limits_{n=1}^{\infty} (0.9)(0.1)^{n-1} = \dfrac{0.9}{1 - 0.1} = 1$ by the formula for the sum of a geometric series $[S = a_1/(1 - r)]$ with ratio r satisfying $|r| < 1$.

21. True. A finite number of terms doesn't affect convergence or divergence of a series.

EXERCISES

1. $\left\{\dfrac{2+n^3}{1+2n^3}\right\}$ converges since $\displaystyle\lim_{n\to\infty}\dfrac{2+n^3}{1+2n^3} = \lim_{n\to\infty}\dfrac{2/n^3+1}{1/n^3+2} = \dfrac{1}{2}$.

3. $\displaystyle\lim_{n\to\infty} a_n = \lim_{n\to\infty}\dfrac{n^3}{1+n^2} = \lim_{n\to\infty}\dfrac{n}{1/n^2+1} = \infty$, so the sequence diverges.

5. $|a_n| = \left|\dfrac{n\sin n}{n^2+1}\right| \le \dfrac{n}{n^2+1} < \dfrac{1}{n}$, so $|a_n| \to 0$ as $n \to \infty$. Thus, $\displaystyle\lim_{n\to\infty} a_n = 0$. The sequence $\{a_n\}$ is convergent.

7. $\left\{\left(1+\dfrac{3}{n}\right)^{4n}\right\}$ is convergent. Let $y = \left(1+\dfrac{3}{x}\right)^{4x}$. Then

$$\lim_{x\to\infty}\ln y = \lim_{x\to\infty} 4x\ln(1+3/x) = \lim_{x\to\infty}\dfrac{\ln(1+3/x)}{1/(4x)} \overset{\text{H}}{=} \lim_{x\to\infty}\dfrac{\dfrac{1}{1+3/x}\left(-\dfrac{3}{x^2}\right)}{-1/(4x^2)} = \lim_{x\to\infty}\dfrac{12}{1+3/x} = 12, \text{ so}$$

$$\lim_{x\to\infty} y = \lim_{n\to\infty}\left(1+\dfrac{3}{n}\right)^{4n} = e^{12}.$$

9. $\dfrac{n}{n^3+1} < \dfrac{n}{n^3} = \dfrac{1}{n^2}$, so $\displaystyle\sum_{n=1}^{\infty}\dfrac{n}{n^3+1}$ converges by the Comparison Test with the convergent p-series $\displaystyle\sum_{n=1}^{\infty}\dfrac{1}{n^2}$ $[\,p=2>1\,]$.

11. $\displaystyle\lim_{n\to\infty}\left|\dfrac{a_{n+1}}{a_n}\right| = \lim_{n\to\infty}\left[\dfrac{(n+1)^3}{5^{n+1}}\cdot\dfrac{5^n}{n^3}\right] = \lim_{n\to\infty}\left(1+\dfrac{1}{n}\right)^3\cdot\dfrac{1}{5} = \dfrac{1}{5} < 1$, so $\displaystyle\sum_{n=1}^{\infty}\dfrac{n^3}{5^n}$ converges by the Ratio Test.

13. Let $f(x) = \dfrac{1}{x\sqrt{\ln x}}$. Then f is continuous, positive, and decreasing on $[2,\infty)$, so the Integral Test applies.

$$\int_2^{\infty} f(x)\,dx = \lim_{t\to\infty}\int_2^t \dfrac{1}{x\sqrt{\ln x}}\,dx \quad\left[u = \ln x,\, du = \dfrac{1}{x}\,dx\right] = \lim_{t\to\infty}\int_{\ln 2}^{\ln t} u^{-1/2}\,du = \lim_{t\to\infty}\left[2\sqrt{u}\,\right]_{\ln 2}^{\ln t}$$

$$= \lim_{t\to\infty}\left(2\sqrt{\ln t} - 2\sqrt{\ln 2}\right) = \infty,$$

so the series $\displaystyle\sum_{n=2}^{\infty}\dfrac{1}{n\sqrt{\ln n}}$ diverges.

15. $|a_n| = \left|\dfrac{\cos 3n}{1+(1.2)^n}\right| \le \dfrac{1}{1+(1.2)^n} < \dfrac{1}{(1.2)^n} = \left(\dfrac{5}{6}\right)^n$, so $\displaystyle\sum_{n=1}^{\infty} |a_n|$ converges by comparison with the convergent geometric

series $\displaystyle\sum_{n=1}^{\infty}\left(\dfrac{5}{6}\right)^n$ $\left[r = \dfrac{5}{6} < 1\right]$. It follows that $\displaystyle\sum_{n=1}^{\infty} a_n$ converges (by Theorem 1 in Section 8.4).

17. $\displaystyle\lim_{n\to\infty}\left|\dfrac{a_{n+1}}{a_n}\right| = \lim_{n\to\infty}\dfrac{1\cdot3\cdot5\cdot\cdots\cdot(2n-1)(2n+1)}{5^{n+1}(n+1)!}\cdot\dfrac{5^n\,n!}{1\cdot3\cdot5\cdot\cdots\cdot(2n-1)} = \lim_{n\to\infty}\dfrac{2n+1}{5(n+1)} = \dfrac{2}{5} < 1$, so the series

converges by the Ratio Test.

19. $b_n = \dfrac{\sqrt{n}}{n+1} > 0$, $\{b_n\}$ is decreasing, and $\displaystyle\lim_{n\to\infty} b_n = 0$, so the series $\displaystyle\sum_{n=1}^{\infty}(-1)^{n-1}\dfrac{\sqrt{n}}{n+1}$ converges by the Alternating

Series Test.

21. Consider the series of absolute values: $\sum\limits_{n=1}^{\infty} n^{-1/3}$ is a p-series with $p = \frac{1}{3} \leq 1$ and is therefore divergent. But if we apply the

Alternating Series Test, we see that $b_n = \dfrac{1}{\sqrt[3]{n}} > 0$, $\{b_n\}$ is decreasing, and $\lim\limits_{n \to \infty} b_n = 0$, so the series $\sum\limits_{n=1}^{\infty} (-1)^{n-1} n^{-1/3}$

converges. Thus, $\sum\limits_{n=1}^{\infty} (-1)^{n-1} n^{-1/3}$ is conditionally convergent.

23. $\left| \dfrac{a_{n+1}}{a_n} \right| = \left| \dfrac{(-1)^{n+1}(n+2)3^{n+1}}{2^{2n+3}} \cdot \dfrac{2^{2n+1}}{(-1)^n(n+1)3^n} \right| = \dfrac{n+2}{n+1} \cdot \dfrac{3}{4} = \dfrac{1+(2/n)}{1+(1/n)} \cdot \dfrac{3}{4} \to \dfrac{3}{4} < 1$ as $n \to \infty$, so by the Ratio

Test, $\sum\limits_{n=1}^{\infty} \dfrac{(-1)^n(n+1)3^n}{2^{2n+1}}$ is absolutely convergent.

25. $\dfrac{2^{2n+1}}{5^n} = \dfrac{2^{2n} \cdot 2^1}{5^n} = \dfrac{(2^2)^n \cdot 2}{5^n} = 2\left(\dfrac{4}{5}\right)^n$, so $\sum\limits_{n=1}^{\infty} \dfrac{2^{2n+1}}{5^n} = 2 \sum\limits_{n=1}^{\infty} \left(\dfrac{4}{5}\right)^n$ is a geometric series with $a = \dfrac{8}{5}$ and $r = \dfrac{4}{5}$.

Since $|r| = \dfrac{4}{5} < 1$, the series converges to $\dfrac{a}{1-r} = \dfrac{8/5}{1-4/5} = \dfrac{8/5}{1/5} = 8$.

27. $\sum\limits_{n=1}^{\infty} [\tan^{-1}(n+1) - \tan^{-1} n] = \lim\limits_{n \to \infty} s_n$

$$= \lim\limits_{n \to \infty} [(\tan^{-1} 2 - \tan^{-1} 1) + (\tan^{-1} 3 - \tan^{-1} 2) + \cdots + (\tan^{-1}(n+1) - \tan^{-1} n)]$$

$$= \lim\limits_{n \to \infty} [\tan^{-1}(n+1) - \tan^{-1} 1] = \tfrac{\pi}{2} - \tfrac{\pi}{4} = \tfrac{\pi}{4}$$

29. $1 - e + \dfrac{e^2}{2!} - \dfrac{e^3}{3!} + \dfrac{e^4}{4!} - \cdots = \sum\limits_{n=0}^{\infty} (-1)^n \dfrac{e^n}{n!} = \sum\limits_{n=0}^{\infty} \dfrac{(-e)^n}{n!} = e^{-e}$ since $e^x = \sum\limits_{n=0}^{\infty} \dfrac{x^n}{n!}$ for all x.

31. $\cosh x = \dfrac{1}{2}(e^x + e^{-x}) = \dfrac{1}{2}\left(\sum\limits_{n=0}^{\infty} \dfrac{x^n}{n!} + \sum\limits_{n=0}^{\infty} \dfrac{(-x)^n}{n!} \right)$

$$= \dfrac{1}{2}\left[\left(1 + x + \dfrac{x^2}{2!} + \dfrac{x^3}{3!} + \dfrac{x^4}{4!} + \cdots\right) + \left(1 - x + \dfrac{x^2}{2!} - \dfrac{x^3}{3!} + \dfrac{x^4}{4!} - \cdots\right) \right]$$

$$= \dfrac{1}{2}\left(2 + 2 \cdot \dfrac{x^2}{2!} + 2 \cdot \dfrac{x^4}{4!} + \cdots\right) = 1 + \dfrac{1}{2}x^2 + \sum\limits_{n=2}^{\infty} \dfrac{x^{2n}}{(2n)!} \geq 1 + \dfrac{1}{2}x^2 \quad \text{for all } x$$

33. $\sum\limits_{n=1}^{\infty} \dfrac{(-1)^{n+1}}{n^5} = 1 - \dfrac{1}{32} + \dfrac{1}{243} - \dfrac{1}{1024} + \dfrac{1}{3125} - \dfrac{1}{7776} + \dfrac{1}{16{,}807} - \dfrac{1}{32{,}768} + \cdots.$

Since $b_8 = \dfrac{1}{8^5} = \dfrac{1}{32{,}768} < 0.000031$, $\sum\limits_{n=1}^{\infty} \dfrac{(-1)^{n+1}}{n^5} \approx \sum\limits_{n=1}^{7} \dfrac{(-1)^{n+1}}{n^5} \approx 0.9721$.

35. Use the Limit Comparison Test. $\lim\limits_{n \to \infty} \left| \dfrac{\left(\frac{n+1}{n}\right)a_n}{a_n} \right| = \lim\limits_{n \to \infty} \dfrac{n+1}{n} = \lim\limits_{n \to \infty} \left(1 + \dfrac{1}{n}\right) = 1 > 0.$

Since $\sum |a_n|$ is convergent, so is $\sum \left| \left(\dfrac{n+1}{n}\right)a_n \right|$, by the Limit Comparison Test.

37. $\lim\limits_{n \to \infty} \left| \dfrac{a_{n+1}}{a_n} \right| = \lim\limits_{n \to \infty} \left[\dfrac{|x+2|^{n+1}}{(n+1)4^{n+1}} \cdot \dfrac{n \, 4^n}{|x+2|^n} \right] = \lim\limits_{n \to \infty} \left[\dfrac{n}{n+1} \dfrac{|x+2|}{4} \right] = \dfrac{|x+2|}{4} < 1 \quad \Leftrightarrow \quad |x+2| < 4$, so $R = 4$.

$|x+2| < 4 \quad \Leftrightarrow \quad -4 < x + 2 < 4 \quad \Leftrightarrow \quad -6 < x < 2$. If $x = -6$, then the series $\sum\limits_{n=1}^{\infty} \dfrac{(x+2)^n}{n \, 4^n}$ becomes

$\displaystyle\sum_{n=1}^{\infty} \frac{(-4)^n}{n4^n} = \sum_{n=1}^{\infty} \frac{(-1)^n}{n}$, the alternating harmonic series, which converges by the Alternating Series Test. When $x = 2$, the

series becomes the harmonic series $\displaystyle\sum_{n=1}^{\infty} \frac{1}{n}$, which diverges. Thus, $I = [-6, 2)$.

39. $\displaystyle\lim_{n\to\infty} \left| \frac{a_{n+1}}{a_n} \right| = \lim_{n\to\infty} \left| \frac{2^{n+1}(x-3)^{n+1}}{\sqrt{n+4}} \cdot \frac{\sqrt{n+3}}{2^n(x-3)^n} \right| = 2\,|x-3| \lim_{n\to\infty} \sqrt{\frac{n+3}{n+4}} = 2\,|x-3| < 1 \iff |x-3| < \frac{1}{2}$,

so $R = \frac{1}{2}$. $|x-3| < \frac{1}{2} \iff -\frac{1}{2} < x-3 < \frac{1}{2} \iff \frac{5}{2} < x < \frac{7}{2}$. For $x = \frac{7}{2}$, the series $\displaystyle\sum_{n=1}^{\infty} \frac{2^n(x-3)^n}{\sqrt{n+3}}$ becomes

$\displaystyle\sum_{n=0}^{\infty} \frac{1}{\sqrt{n+3}} = \sum_{n=3}^{\infty} \frac{1}{n^{1/2}}$, which diverges $\left[p = \frac{1}{2} \le 1\right]$, but for $x = \frac{5}{2}$, we get $\displaystyle\sum_{n=0}^{\infty} \frac{(-1)^n}{\sqrt{n+3}}$, which is a convergent

alternating series, so $I = \left[\frac{5}{2}, \frac{7}{2}\right)$.

41.

n	$f^{(n)}(x)$	$f^{(n)}\left(\frac{\pi}{6}\right)$
0	$\sin x$	$\frac{1}{2}$
1	$\cos x$	$\frac{\sqrt{3}}{2}$
2	$-\sin x$	$-\frac{1}{2}$
3	$-\cos x$	$-\frac{\sqrt{3}}{2}$
4	$\sin x$	$\frac{1}{2}$
⋮	⋮	⋮

$\sin x = f\left(\frac{\pi}{6}\right) + f'\left(\frac{\pi}{6}\right)\left(x-\frac{\pi}{6}\right) + \frac{f''\left(\frac{\pi}{6}\right)}{2!}\left(x-\frac{\pi}{6}\right)^2 + \frac{f^{(3)}\left(\frac{\pi}{6}\right)}{3!}\left(x-\frac{\pi}{6}\right)^3 + \frac{f^{(4)}\left(\frac{\pi}{6}\right)}{4!}\left(x-\frac{\pi}{6}\right)^4 + \cdots$

$= \frac{1}{2}\left[1 - \frac{1}{2!}\left(x-\frac{\pi}{6}\right)^2 + \frac{1}{4!}\left(x-\frac{\pi}{6}\right)^4 - \cdots\right] + \frac{\sqrt{3}}{2}\left[\left(x-\frac{\pi}{6}\right) - \frac{1}{3!}\left(x-\frac{\pi}{6}\right)^3 + \cdots\right]$

$= \frac{1}{2}\displaystyle\sum_{n=0}^{\infty} (-1)^n \frac{1}{(2n)!}\left(x-\frac{\pi}{6}\right)^{2n} + \frac{\sqrt{3}}{2}\sum_{n=0}^{\infty} (-1)^n \frac{1}{(2n+1)!}\left(x-\frac{\pi}{6}\right)^{2n+1}$

43. $\dfrac{1}{1+x} = \dfrac{1}{1-(-x)} = \displaystyle\sum_{n=0}^{\infty} (-x)^n = \sum_{n=0}^{\infty} (-1)^n x^n$ for $|x| < 1 \implies \dfrac{x^2}{1+x} = \displaystyle\sum_{n=0}^{\infty} (-1)^n x^{n+2}$ with $R = 1$.

45. $\displaystyle\int \frac{1}{4-x}\,dx = -\ln(4-x) + C$ and

$\displaystyle\int \frac{1}{4-x}\,dx = \frac{1}{4}\int \frac{1}{1-x/4}\,dx = \frac{1}{4}\int \sum_{n=0}^{\infty} \left(\frac{x}{4}\right)^n dx = \frac{1}{4}\int \sum_{n=0}^{\infty} \frac{x^n}{4^n}\,dx = \frac{1}{4}\sum_{n=0}^{\infty} \frac{x^{n+1}}{4^n(n+1)} + C$. So

$\ln(4-x) = -\dfrac{1}{4}\displaystyle\sum_{n=0}^{\infty} \frac{x^{n+1}}{4^n(n+1)} + C = -\sum_{n=0}^{\infty} \frac{x^{n+1}}{4^{n+1}(n+1)} + C = -\sum_{n=1}^{\infty} \frac{x^n}{n4^n} + C$. Putting $x = 0$, we get $C = \ln 4$.

Thus, $f(x) = \ln(4-x) = \ln 4 - \displaystyle\sum_{n=1}^{\infty} \frac{x^n}{n4^n}$. The series converges for $|x/4| < 1 \iff |x| < 4$, so $R = 4$.

Another solution:

$\ln(4-x) = \ln[4(1-x/4)] = \ln 4 + \ln(1-x/4) = \ln 4 + \ln[1 + (-x/4)]$

$\qquad = \ln 4 + \displaystyle\sum_{n=1}^{\infty} (-1)^{n+1} \frac{(-x/4)^n}{n}$ [from Table 1] $= \ln 4 + \sum_{n=1}^{\infty} (-1)^{2n+1} \frac{x^n}{n4^n} = \ln 4 - \sum_{n=1}^{\infty} \frac{x^n}{n4^n}$.

47. $\sin x = \sum\limits_{n=0}^{\infty} \dfrac{(-1)^n x^{2n+1}}{(2n+1)!} \;\Rightarrow\; \sin(x^4) = \sum\limits_{n=0}^{\infty} \dfrac{(-1)^n (x^4)^{2n+1}}{(2n+1)!} = \sum\limits_{n=0}^{\infty} \dfrac{(-1)^n x^{8n+4}}{(2n+1)!}$ for all x, so the radius of

convergence is ∞.

49. $f(x) = \dfrac{1}{\sqrt[4]{16-x}} = \dfrac{1}{\sqrt[4]{16(1-x/16)}} = \dfrac{1}{\sqrt[4]{16}\left(1-\frac{1}{16}x\right)^{1/4}} = \frac{1}{2}\left(1-\frac{1}{16}x\right)^{-1/4}$

$\quad = \dfrac{1}{2}\left[1 + \left(-\frac{1}{4}\right)\left(-\frac{x}{16}\right) + \dfrac{\left(-\frac{1}{4}\right)\left(-\frac{5}{4}\right)}{2!}\left(-\frac{x}{16}\right)^2 + \dfrac{\left(-\frac{1}{4}\right)\left(-\frac{5}{4}\right)\left(-\frac{9}{4}\right)}{3!}\left(-\frac{x}{16}\right)^3 + \cdots\right]$

$\quad = \dfrac{1}{2} + \sum\limits_{n=1}^{\infty} \dfrac{1\cdot 5\cdot 9\cdot\;\cdots\;\cdot(4n-3)}{2\cdot 4^n\cdot n!\cdot 16^n}\,x^n = \dfrac{1}{2} + \sum\limits_{n=1}^{\infty} \dfrac{1\cdot 5\cdot 9\cdot\;\cdots\;\cdot(4n-3)}{2^{6n+1}\,n!}\,x^n$

for $\left|-\dfrac{x}{16}\right| < 1 \;\Leftrightarrow\; |x| < 16$, so $R = 16$.

51. $e^x = \sum\limits_{n=0}^{\infty} \dfrac{x^n}{n!}$, so $\dfrac{e^x}{x} = \dfrac{1}{x}\sum\limits_{n=0}^{\infty} \dfrac{x^n}{n!} = \sum\limits_{n=0}^{\infty} \dfrac{x^{n-1}}{n!} = x^{-1} + \sum\limits_{n=1}^{\infty} \dfrac{x^{n-1}}{n!} = \dfrac{1}{x} + \sum\limits_{n=1}^{\infty} \dfrac{x^{n-1}}{n!}$ and

$\displaystyle\int \dfrac{e^x}{x}\,dx = C + \ln|x| + \sum\limits_{n=1}^{\infty} \dfrac{x^n}{n\cdot n!}$.

53. (a)

n	$f^{(n)}(x)$	$f^{(n)}(1)$
0	$x^{1/2}$	1
1	$\frac{1}{2}x^{-1/2}$	$\frac{1}{2}$
2	$-\frac{1}{4}x^{-3/2}$	$-\frac{1}{4}$
3	$\frac{3}{8}x^{-5/2}$	$\frac{3}{8}$
4	$-\frac{15}{16}x^{-7/2}$	$-\frac{15}{16}$
⋮	⋮	⋮

$\sqrt{x} \approx T_3(x) = 1 + \dfrac{1/2}{1!}(x-1) - \dfrac{1/4}{2!}(x-1)^2 + \dfrac{3/8}{3!}(x-1)^3$

$\qquad = 1 + \frac{1}{2}(x-1) - \frac{1}{8}(x-1)^2 + \frac{1}{16}(x-1)^3$

(c) By Taylor's Formula, $R_3(x) = \dfrac{f^{(4)}(z)}{4!}(x-1)^4 = -\dfrac{5(x-1)^4}{128\,z^{7/2}}$,

with z between x and 1. If $0.9 \le x \le 1.1$, then $0 \le |x-1| \le 0.1$

and $z^{7/2} > (0.9)^{7/2}$ so $|R_3(x)| < \dfrac{5(0.1)^4}{128(0.9)^{7/2}} < 0.000006$.

(b)

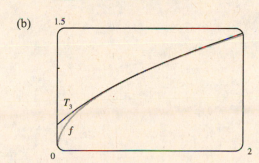

(d)

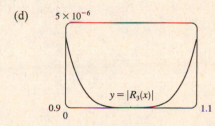

From the graph of $|R_3(x)| = |\sqrt{x} - T_3(x)|$, it appears that

the error is less than 5×10^{-6} on $[0.9, 1.1]$.

55. $\sin x = \sum\limits_{n=0}^{\infty} (-1)^n \dfrac{x^{2n+1}}{(2n+1)!} = x - \dfrac{x^3}{3!} + \dfrac{x^5}{5!} - \dfrac{x^7}{7!} + \cdots$, so $\sin x - x = -\dfrac{x^3}{3!} + \dfrac{x^5}{5!} - \dfrac{x^7}{7!} + \cdots$ and

$\dfrac{\sin x - x}{x^3} = -\dfrac{1}{3!} + \dfrac{x^2}{5!} - \dfrac{x^4}{7!} + \cdots$. Thus, $\lim\limits_{x \to 0} \dfrac{\sin x - x}{x^3} = \lim\limits_{x \to 0} \left(-\dfrac{1}{6} + \dfrac{x^2}{120} - \dfrac{x^4}{5040} + \cdots \right) = -\dfrac{1}{6}$.

57. We use induction, hypothesizing that $a_{n-1} < a_n < 2$. Note first that $1 < a_2 = \frac{1}{3}(1+4) = \frac{5}{3} < 2$, so the hypothesis holds

for $n = 2$. Now assume that $a_{k-1} < a_k < 2$. Then $a_k = \frac{1}{3}(a_{k-1} + 4) < \frac{1}{3}(a_k + 4) < \frac{1}{3}(2+4) = 2$. So $a_k < a_{k+1} < 2$,

and the induction is complete. To find the limit of the sequence, we note that $L = \lim\limits_{n \to \infty} a_n = \lim\limits_{n \to \infty} a_{n+1} \implies$

$L = \frac{1}{3}(L+4) \implies L = 2$.

59. $f(x) = \sum\limits_{n=0}^{\infty} c_n x^n \implies f(-x) = \sum\limits_{n=0}^{\infty} c_n (-x)^n = \sum\limits_{n=0}^{\infty} (-1)^n c_n x^n$

(a) If f is an odd function, then $f(-x) = -f(x) \implies \sum\limits_{n=0}^{\infty} (-1)^n c_n x^n = \sum\limits_{n=0}^{\infty} -c_n x^n$. The coefficients of any power series

are uniquely determined (by Theorem 8.7.5), so $(-1)^n c_n = -c_n$.

If n is even, then $(-1)^n = 1$, so $c_n = -c_n \implies 2c_n = 0 \implies c_n = 0$. Thus, all even coefficients are 0, that is,
$c_0 = c_2 = c_4 = \cdots = 0$.

(b) If f is even, then $f(-x) = f(x) \implies \sum\limits_{n=0}^{\infty} (-1)^n c_n x^n = \sum\limits_{n=0}^{\infty} c_n x^n \implies (-1)^n c_n = c_n$.

If n is odd, then $(-1)^n = -1$, so $-c_n = c_n \implies 2c_n = 0 \implies c_n = 0$. Thus, all odd coefficients are 0,

that is, $c_1 = c_3 = c_5 = \cdots = 0$.

9 □ PARAMETRIC EQUATIONS AND POLAR COORDINATES

9.1 Parametric Curves

1. $x = t^2 + t$, $y = t^2 - t$, $-2 \le t \le 2$

t	-2	-1	0	1	2
x	2	0	0	2	6
y	6	2	0	0	2

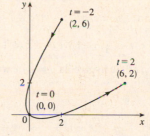

3. $x = \cos^2 t$, $y = 1 - \sin t$, $0 \le t \le \pi/2$

t	0	$\pi/6$	$\pi/3$	$\pi/2$
x	1	3/4	1/4	0
y	1	1/2	$1 - \frac{\sqrt{3}}{2} \approx 0.13$	0

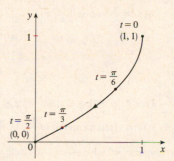

5. $x = 3 - 4t$, $y = 2 - 3t$

(a)

t	-1	0	1	2
x	7	3	-1	-5
y	5	2	-1	-4

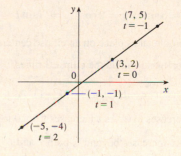

(b) $x = 3 - 4t \;\Rightarrow\; 4t = -x + 3 \;\Rightarrow\; t = -\frac{1}{4}x + \frac{3}{4}$, so

$$y = 2 - 3t = 2 - 3\left(-\tfrac{1}{4}x + \tfrac{3}{4}\right) = 2 + \tfrac{3}{4}x - \tfrac{9}{4} \;\Rightarrow\; y = \tfrac{3}{4}x - \tfrac{1}{4}$$

7. $x = \sqrt{t}$, $y = 1 - t$

(a)

t	0	1	2	3	4
x	0	1	1.414	1.732	2
y	1	0	-1	-2	-3

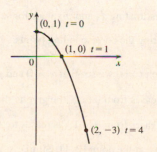

(b) $x = \sqrt{t} \;\Rightarrow\; t = x^2 \;\Rightarrow\; y = 1 - t = 1 - x^2$. Since $t \ge 0$, $x \ge 0$.

So the curve is the right half of the parabola $y = 1 - x^2$.

9. (a) $x = \sin \frac{1}{2}\theta$, $y = \cos \frac{1}{2}\theta$, $-\pi \le \theta \le \pi$. **(b)**

$x^2 + y^2 = \sin^2 \frac{1}{2}\theta + \cos^2 \frac{1}{2}\theta = 1$. For $-\pi \le \theta \le 0$, we have

$-1 \le x \le 0$ and $0 \le y \le 1$. For $0 < \theta \le \pi$, we have $0 < x \le 1$

and $1 > y \ge 0$. The graph is a semicircle.

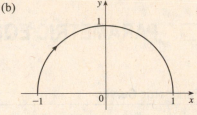

11. (a) $x = \sin t$, $y = \csc t$, $0 < t < \frac{\pi}{2}$. $\quad y = \csc t = \dfrac{1}{\sin t} = \dfrac{1}{x}$. **(b)**

For $0 < t < \frac{\pi}{2}$, we have $0 < x < 1$ and $y > 1$. Thus, the curve is the

portion of the hyperbola $y = 1/x$ with $y > 1$.

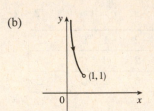

13. (a) $x = e^{2t} \quad \Rightarrow \quad 2t = \ln x \quad \Rightarrow \quad t = \frac{1}{2}\ln x$. **(b)**

$y = t + 1 = \frac{1}{2}\ln x + 1$.

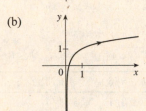

15. $x = 3 + 2\cos t$, $y = 1 + 2\sin t$, $\pi/2 \le t \le 3\pi/2$. By Example 4 with $r = 2$, $h = 3$, and $k = 1$, the motion of the particle

takes place on a circle centered at $(3, 1)$ with a radius of 2. As t goes from $\frac{\pi}{2}$ to $\frac{3\pi}{2}$, the particle starts at the point $(3, 3)$ and

moves counterclockwise along the circle $(x - 3)^2 + (y - 1)^2 = 4$ to $(3, -1)$ [one-half of a circle].

17. $x = 5\sin t$, $y = 2\cos t \quad \Rightarrow \quad \sin t = \dfrac{x}{5}$, $\cos t = \dfrac{y}{2}$. $\quad \sin^2 t + \cos^2 t = 1 \quad \Rightarrow \quad \left(\dfrac{x}{5}\right)^2 + \left(\dfrac{y}{2}\right)^2 = 1$. The motion of the

particle takes place on an ellipse centered at $(0, 0)$. As t goes from $-\pi$ to 5π, the particle starts at the point $(0, -2)$ and moves

clockwise around the ellipse 3 times.

19. When $t = -1$, $(x, y) = (0, -1)$. As t increases to 0, x decreases to -1 and y

increases to 0. As t increases from 0 to 1, x increases to 0 and y increases to 1.

As t increases beyond 1, both x and y increase. For $t < -1$, x is positive and

decreasing and y is negative and increasing. We could achieve greater accuracy

by estimating x- and y-values for selected values of t from the given graphs and

plotting the corresponding points.

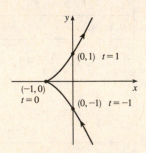

21. When $t = 0$ we see that $x = 0$ and $y = 0$, so the curve starts at the origin. As t

increases from 0 to $\frac{1}{2}$, the graphs show that y increases from 0 to 1 while x

increases from 0 to 1, decreases to 0 and to -1, then increases back to 0, so we

arrive at the point $(0, 1)$. Similarly, as t increases from $\frac{1}{2}$ to 1, y decreases from 1

to 0 while x repeats its pattern, and we arrive back at the origin. We could achieve greater accuracy by estimating x- and

y-values for selected values of t from the given graphs and plotting the corresponding points.

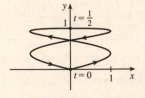

23. Use $y = t$ and $x = t - 2\sin \pi t$ with a t-interval of $[-\pi, \pi]$.

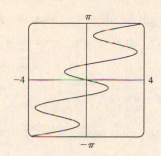

25. (a) $x = x_1 + (x_2 - x_1)t$, $y = y_1 + (y_2 - y_1)t$, $0 \le t \le 1$. Clearly the curve passes through $P_1(x_1, y_1)$ when $t = 0$ and

through $P_2(x_2, y_2)$ when $t = 1$. For $0 < t < 1$, x is strictly between x_1 and x_2 and y is strictly between y_1 and y_2. For

every value of t, x and y satisfy the relation $y - y_1 = \dfrac{y_2 - y_1}{x_2 - x_1}(x - x_1)$, which is the equation of the line through

$P_1(x_1, y_1)$ and $P_2(x_2, y_2)$.

Finally, any point (x, y) on that line satisfies $\dfrac{y - y_1}{y_2 - y_1} = \dfrac{x - x_1}{x_2 - x_1}$; if we call that common value t, then the given

parametric equations yield the point (x, y); and any (x, y) on the line between $P_1(x_1, y_1)$ and $P_2(x_2, y_2)$ yields a value of

t in $[0, 1]$. So the given parametric equations exactly specify the line segment from $P_1(x_1, y_1)$ to $P_2(x_2, y_2)$.

(b) $x = -2 + [3 - (-2)]t = -2 + 5t$ and $y = 7 + (-1 - 7)t = 7 - 8t$ for $0 \le t \le 1$.

27. The circle $x^2 + (y - 1)^2 = 4$ has center $(0, 1)$ and radius 2, so by Example 4 it can be represented by $x = 2\cos t$,

$y = 1 + 2\sin t$, $0 \le t \le 2\pi$. This representation gives us the circle with a counterclockwise orientation starting at $(2, 1)$.

(a) To get a clockwise orientation, we could change the equations to $x = 2\cos t$, $y = 1 - 2\sin t$, $0 \le t \le 2\pi$.

(b) To get three times around in the counterclockwise direction, we use the original equations $x = 2\cos t$, $y = 1 + 2\sin t$ with

the domain expanded to $0 \le t \le 6\pi$.

(c) To start at $(0, 3)$ using the original equations, we must have $x_1 = 0$; that is, $2\cos t = 0$. Hence, $t = \frac{\pi}{2}$. So we use

$x = 2\cos t$, $y = 1 + 2\sin t$, $\frac{\pi}{2} \le t \le \frac{3\pi}{2}$.

Alternatively, if we want t to start at 0, we could change the equations of the curve. For example, we could use

$x = -2\sin t$, $y = 1 + 2\cos t$, $0 \le t \le \pi$.

29. *Big circle:* It's centered at $(2, 2)$ with a radius of 2, so by Example 4, parametric equations are

$$x = 2 + 2\cos t, \qquad y = 2 + 2\sin t, \qquad 0 \le t \le 2\pi$$

Small circles: They are centered at $(1, 3)$ and $(3, 3)$ with a radius of 0.1. By Example 4, parametric equations are

(left)	$x = 1 + 0.1\cos t$,	$y = 3 + 0.1\sin t$,	$0 \le t \le 2\pi$

and

(right)	$x = 3 + 0.1\cos t$,	$y = 3 + 0.1\sin t$,	$0 \le t \le 2\pi$

Semicircle: It's the lower half of a circle centered at $(2, 2)$ with radius 1. By Example 4, parametric equations are

$$x = 2 + 1\cos t, \qquad y = 2 + 1\sin t, \qquad \pi \le t \le 2\pi$$

To get all four graphs on the same screen with a typical graphing calculator, we need to change the last t-interval to $[0, 2\pi]$ in order to match the others. We can do this by changing t to $0.5t$. This change gives us the upper half. There are several ways to get the lower half—one is to change the "+" to a "−" in the y-assignment, giving us

$$x = 2 + 1\cos(0.5t), \qquad y = 2 - 1\sin(0.5t), \qquad 0 \le t \le 2\pi$$

31. (a) $x = t^3 \;\Rightarrow\; t = x^{1/3}$, so $y = t^2 = x^{2/3}$.

We get the entire curve $y = x^{2/3}$ traversed in a left to right direction.

(b) $x = t^6 \;\Rightarrow\; t = x^{1/6}$, so $y = t^4 = x^{4/6} = x^{2/3}$.

Since $x = t^6 \ge 0$, we only get the right half of the curve $y = x^{2/3}$.

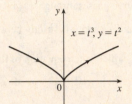

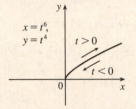

(c) $x = e^{-3t} = (e^{-t})^3$ $\;[$so $e^{-t} = x^{1/3}]$,

$y = e^{-2t} = (e^{-t})^2 = (x^{1/3})^2 = x^{2/3}$.

If $t < 0$, then x and y are both larger than 1. If $t > 0$, then x and y are between 0 and 1. Since $x > 0$ and $y > 0$, the curve never quite reaches the origin.

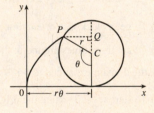

33. The case $\frac{\pi}{2} < \theta < \pi$ is illustrated. C has coordinates $(r\theta, r)$ as in Example 7, and Q has coordinates $(r\theta, r + r\cos(\pi - \theta)) = (r\theta, r(1 - \cos\theta))$ [since $\cos(\pi - \alpha) = \cos\pi\cos\alpha + \sin\pi\sin\alpha = -\cos\alpha$], so P has coordinates $(r\theta - r\sin(\pi - \theta), r(1 - \cos\theta)) = (r(\theta - \sin\theta), r(1 - \cos\theta))$ [since $\sin(\pi - \alpha) = \sin\pi\cos\alpha - \cos\pi\sin\alpha = \sin\alpha$]. Again we have the parametric equations $x = r(\theta - \sin\theta)$, $y = r(1 - \cos\theta)$.

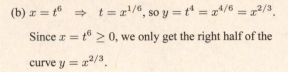

35. It is apparent that $x = |OQ|$ and $y = |QP| = |ST|$. From the diagram, $x = |OQ| = a\cos\theta$ and $y = |ST| = b\sin\theta$. Thus, the parametric equations are $x = a\cos\theta$ and $y = b\sin\theta$. To eliminate θ we rearrange: $\sin\theta = y/b \;\Rightarrow\; \sin^2\theta = (y/b)^2$ and $\cos\theta = x/a \;\Rightarrow\; \cos^2\theta = (x/a)^2$. Adding the two equations: $\sin^2\theta + \cos^2\theta = 1 = x^2/a^2 + y^2/b^2$. Thus, we have an ellipse.

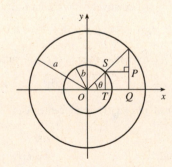

37. (a)

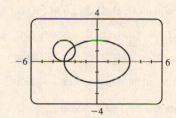

There are 2 points of intersection:

$(-3, 0)$ and approximately $(-2.1, 1.4)$.

(b) A collision point occurs when $x_1 = x_2$ and $y_1 = y_2$ for the same t. So solve the equations:

$$3 \sin t = -3 + \cos t \quad \textbf{(1)}$$

$$2 \cos t = 1 + \sin t \quad \textbf{(2)}$$

From **(2)**, $\sin t = 2 \cos t - 1$. Substituting into **(1)**, we get $3(2 \cos t - 1) = -3 + \cos t \;\Rightarrow\; 5 \cos t = 0 \;\textbf{(⋆)}\;\Rightarrow$ $\cos t = 0 \;\Rightarrow\; t = \frac{\pi}{2}$ or $\frac{3\pi}{2}$. We check that $t = \frac{3\pi}{2}$ satisfies **(1)** and **(2)** but $t = \frac{\pi}{2}$ does not. So the only collision point occurs when $t = \frac{3\pi}{2}$, and this gives the point $(-3, 0)$. [We could check our work by graphing x_1 and x_2 together as functions of t and, on another plot, y_1 and y_2 as functions of t. If we do so, we see that the only value of t for which *both* pairs of graphs intersect is $t = \frac{3\pi}{2}$.]

(c) The circle is centered at $(3, 1)$ instead of $(-3, 1)$. There are still 2 intersection points: $(3, 0)$ and $(2.1, 1.4)$, but there are no collision points, since **(⋆)** in part (b) becomes $5 \cos t = 6 \;\Rightarrow\; \cos t = \frac{6}{5} > 1$.

39. $x = t^2$, $y = t^3 - ct$.   We use a graphing device to produce the graphs for various values of c with $-\pi \le t \le \pi$. Note that all the members of the family are symmetric about the x-axis. For $c < 0$, the graph does not cross itself, but for $c = 0$ it has a cusp at $(0, 0)$ and for $c > 0$ the graph crosses itself at $x = c$, so the loop grows larger as c increases.

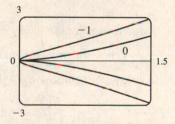

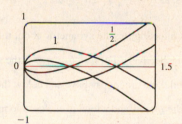

41. $x = t + a \cos t$, $y = t + a \sin t$, $a > 0$.   From the first figure, we see that curves roughly follow the line $y = x$, and they start having loops when a is between 1.4 and 1.6. The loops increase in size as a increases.

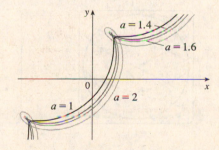

While not required, the following is a solution to determine the *exact* values for which the curve has a loop, that is, we seek the values of a for which there exist parameter values t and u such that $t < u$ and

$(t + a \cos t, t + a \sin t) = (u + a \cos u, u + a \sin u)$.   [continued]

In the diagram at the left, T denotes the point (t, t), U the point (u, u), and P the point $(t + a\cos t, t + a\sin t) = (u + a\cos u, u + a\sin u)$.

Since $\overline{PT} = \overline{PU} = a$, the triangle PTU is isosceles. Therefore its base angles, $\alpha = \angle PTU$ and $\beta = \angle PUT$ are equal. Since $\alpha = t - \frac{\pi}{4}$ and $\beta = 2\pi - \frac{3\pi}{4} - u = \frac{5\pi}{4} - u$, the relation $\alpha = \beta$ implies that

$$u + t = \frac{3\pi}{2} \quad \textbf{(1)}.$$

Since $\overline{TU} = \text{distance}((t, t), (u, u)) = \sqrt{2(u - t)^2} = \sqrt{2}\,(u - t)$, we see that

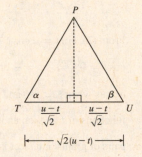

$$\cos\alpha = \frac{\frac{1}{2}\overline{TU}}{\overline{PT}} = \frac{(u - t)/\sqrt{2}}{a}, \text{ so } u - t = \sqrt{2}\,a\cos\alpha, \text{ that is,}$$

$u - t = \sqrt{2}\,a\cos\left(t - \frac{\pi}{4}\right)$ **(2)**. Now $\cos\left(t - \frac{\pi}{4}\right) = \sin\left[\frac{\pi}{2} - \left(t - \frac{\pi}{4}\right)\right] = \sin\left(\frac{3\pi}{4} - t\right)$,

so we can rewrite **(2)** as $u - t = \sqrt{2}\,a\sin\left(\frac{3\pi}{4} - t\right)$ **(2′)**. Subtracting **(2′)** from **(1)** and

dividing by 2, we obtain $t = \frac{3\pi}{4} - \frac{\sqrt{2}}{2}a\sin\left(\frac{3\pi}{4} - t\right)$, or $\frac{3\pi}{4} - t = \frac{a}{\sqrt{2}}\sin\left(\frac{3\pi}{4} - t\right)$ **(3)**.

Since $a > 0$ and $t < u$, it follows from **(2′)** that $\sin\left(\frac{3\pi}{4} - t\right) > 0$. Thus from **(3)** we see that $t < \frac{3\pi}{4}$. [We have

implicitly assumed that $0 < t < \pi$ by the way we drew our diagram, but we lost no generality by doing so since replacing t

by $t + 2\pi$ merely increases x and y by 2π. The curve's basic shape repeats every time we change t by 2π.] Solving for a in

(3), we get $a = \dfrac{\sqrt{2}\left(\frac{3\pi}{4} - t\right)}{\sin\left(\frac{3\pi}{4} - t\right)}$. Write $z = \frac{3\pi}{4} - t$. Then $a = \dfrac{\sqrt{2}\,z}{\sin z}$, where $z > 0$. Now $\sin z < z$ for $z > 0$, so $a > \sqrt{2}$.

$\left[\text{As } z \to 0^{+}, \text{ that is, as } t \to \left(\frac{3\pi}{4}\right)^{-}, a \to \sqrt{2}\right].$

43. Note that all the Lissajous figures are symmetric about the x-axis. The parameters a and b simply stretch the graph in the

x- and y-directions respectively. For $a = b = n = 1$ the graph is simply a circle with radius 1. For $n = 2$ the graph crosses

itself at the origin and there are loops above and below the x-axis. In general, the figures have $n - 1$ points of intersection,

all of which are on the y-axis, and a total of n closed loops.

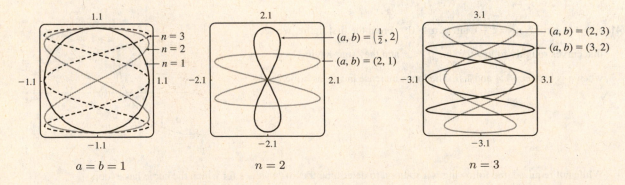

9.2 Calculus with Parametric Curves

1. $x = t\sin t$, $y = t^2 + t$ $\Rightarrow$ $\dfrac{dy}{dt} = 2t + 1$, $\dfrac{dx}{dt} = t\cos t + \sin t$, and $\dfrac{dy}{dx} = \dfrac{dy/dt}{dx/dt} = \dfrac{2t + 1}{t\cos t + \sin t}$.

3. $x = 1 + 4t - t^2$, $y = 2 - t^3$; $t = 1$. $\dfrac{dy}{dt} = -3t^2$, $\dfrac{dx}{dt} = 4 - 2t$, and $\dfrac{dy}{dx} = \dfrac{dy/dt}{dx/dt} = \dfrac{-3t^2}{4 - 2t}$. When $t = 1$,

   $(x, y) = (4, 1)$ and $dy/dx = -\frac{3}{2}$, so an equation of the tangent to the curve at the point corresponding to $t = 1$ is

   $y - 1 = -\frac{3}{2}(x - 4)$, or $y = -\frac{3}{2}x + 7$.

5. $x = t\cos t$, $y = t\sin t$; $t = \pi$. $\dfrac{dy}{dt} = t\cos t + \sin t$, $\dfrac{dx}{dt} = t(-\sin t) + \cos t$, and $\dfrac{dy}{dx} = \dfrac{dy/dt}{dx/dt} = \dfrac{t\cos t + \sin t}{-t\sin t + \cos t}$.

   When $t = \pi$, $(x, y) = (-\pi, 0)$ and $dy/dx = -\pi/(-1) = \pi$, so an equation of the tangent to the curve at the point

   corresponding to $t = \pi$ is $y - 0 = \pi[x - (-\pi)]$, or $y = \pi x + \pi^2$.

7. (a) $x = 1 + \ln t$, $y = t^2 + 2$; $(1, 3)$. $\dfrac{dy}{dt} = 2t$, $\dfrac{dx}{dt} = \dfrac{1}{t}$, and $\dfrac{dy}{dx} = \dfrac{dy/dt}{dx/dt} = \dfrac{2t}{1/t} = 2t^2$. At $(1, 3)$,

   $x = 1 + \ln t = 1$ $\Rightarrow$ $\ln t = 0$ $\Rightarrow$ $t = 1$ and $\dfrac{dy}{dx} = 2$, so an equation of the tangent is $y - 3 = 2(x - 1)$,

   or $y = 2x + 1$.

   (b) $x = 1 + \ln t$ $\Rightarrow$ $\ln t = x - 1$ $\Rightarrow$ $t = e^{x-1}$, so $y = t^2 + 2 = (e^{x-1})^2 + 2 = e^{2x-2} + 2$, and $y' = e^{2x-2} \cdot 2$.

   At $(1, 3)$, $y' = e^{2(1)-2} \cdot 2 = 2$, so an equation of the tangent is $y - 3 = 2(x - 1)$, or $y = 2x + 1$.

9. $x = t^2 + 1$, $y = t^2 + t$ $\Rightarrow$ $\dfrac{dy}{dx} = \dfrac{dy/dt}{dx/dt} = \dfrac{2t + 1}{2t} = 1 + \dfrac{1}{2t}$ $\Rightarrow$ $\dfrac{d^2y}{dx^2} = \dfrac{\frac{d}{dt}\left(\frac{dy}{dx}\right)}{dx/dt} = \dfrac{-1/(2t^2)}{2t} = -\dfrac{1}{4t^3}$.

   The curve is CU when $\dfrac{d^2y}{dx^2} > 0$, that is, when $t < 0$.

11. $x = e^t$, $y = te^{-t}$ $\Rightarrow$ $\dfrac{dy}{dx} = \dfrac{dy/dt}{dx/dt} = \dfrac{-te^{-t} + e^{-t}}{e^t} = \dfrac{e^{-t}(1 - t)}{e^t} = e^{-2t}(1 - t)$ $\Rightarrow$

   $\dfrac{d^2y}{dx^2} = \dfrac{\frac{d}{dt}\left(\frac{dy}{dx}\right)}{dx/dt} = \dfrac{e^{-2t}(-1) + (1 - t)(-2e^{-2t})}{e^t} = \dfrac{e^{-2t}(-1 - 2 + 2t)}{e^t} = e^{-3t}(2t - 3)$. The curve is CU when

   $\dfrac{d^2y}{dx^2} > 0$, that is, when $t > \frac{3}{2}$.

13. $x = t^3 - 3t$, $y = t^2 - 3$. $\dfrac{dy}{dt} = 2t$, so $\dfrac{dy}{dt} = 0$ $\Leftrightarrow$ $t = 0$ $\Leftrightarrow$

   $(x, y) = (0, -3)$. $\dfrac{dx}{dt} = 3t^2 - 3 = 3(t + 1)(t - 1)$, so $\dfrac{dx}{dt} = 0$ $\Leftrightarrow$

   $t = -1$ or 1 $\Leftrightarrow$ $(x, y) = (2, -2)$ or $(-2, -2)$. The curve has a horizontal

   tangent at $(0, -3)$ and vertical tangents at $(2, -2)$ and $(-2, -2)$.

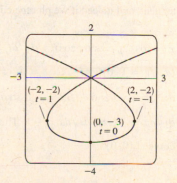

15. $x = 2\cos\theta$, $y = \sin 2\theta$.

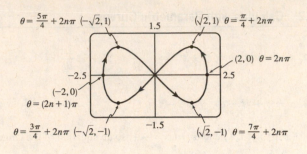

$dy/d\theta = 2\cos 2\theta$, so $dy/d\theta = 0 \Leftrightarrow 2\theta = \frac{\pi}{2} + n\pi$

(n an integer) $\Leftrightarrow \theta = \frac{\pi}{4} + \frac{\pi}{2}n \Leftrightarrow$

$(x, y) = (\pm\sqrt{2}, \pm 1)$. Also, $dx/d\theta = -2\sin\theta$, so

$dx/d\theta = 0 \Leftrightarrow \theta = n\pi \Leftrightarrow (x, y) = (\pm 2, 0)$. The

curve has horizontal tangents at $(\pm\sqrt{2}, \pm 1)$ (four points),

and vertical tangents at $(\pm 2, 0)$.

17. From the graph, it appears that the rightmost point on the curve $x = t - t^6$, $y = e^t$

is about $(0.6, 2)$. To find the exact coordinates, we find the value of t for which the

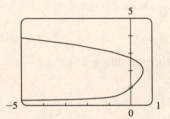

graph has a vertical tangent, that is, $0 = dx/dt = 1 - 6t^5 \Leftrightarrow t = 1/\sqrt[5]{6}$.

Hence, the rightmost point is

$$\left(1/\sqrt[5]{6} - 1/\left(6\sqrt[5]{6}\right), e^{1/\sqrt[5]{6}}\right) = \left(5 \cdot 6^{-6/5}, e^{6^{-1/5}}\right) \approx (0.58, 2.01).$$

19. We graph the curve $x = t^4 - 2t^3 - 2t^2$, $y = t^3 - t$ in the viewing rectangle $[-2, 1.1]$ by $[-0.5, 0.5]$. This rectangle

corresponds approximately to $t \in [-1, 0.8]$.

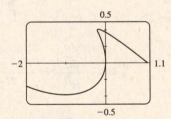
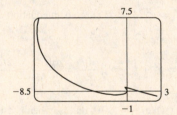

We estimate that the curve has horizontal tangents at about $(-1, -0.4)$ and $(-0.17, 0.39)$ and vertical tangents at

about $(0, 0)$ and $(-0.19, 0.37)$. We calculate $\dfrac{dy}{dx} = \dfrac{dy/dt}{dx/dt} = \dfrac{3t^2 - 1}{4t^3 - 6t^2 - 4t}$. The horizontal tangents occur when

$dy/dt = 3t^2 - 1 = 0 \Leftrightarrow t = \pm\frac{1}{\sqrt{3}}$, so both horizontal tangents are shown in our graph. The vertical tangents occur when

$dx/dt = 2t(2t^2 - 3t - 2) = 0 \Leftrightarrow 2t(2t + 1)(t - 2) = 0 \Leftrightarrow t = 0, -\frac{1}{2}$ or 2. It seems that we have missed one vertical

tangent, and indeed if we plot the curve on the t-interval $[-1.2, 2.2]$ we see that there is another vertical tangent at $(-8, 6)$.

21. $x = \cos t$, $y = \sin t \cos t$. $dx/dt = -\sin t$, $dy/dt = -\sin^2 t + \cos^2 t = \cos 2t$.

$(x, y) = (0, 0) \Leftrightarrow \cos t = 0 \Leftrightarrow t$ is an odd multiple of $\frac{\pi}{2}$. When $t = \frac{\pi}{2}$,

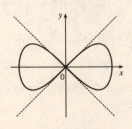

$dx/dt = -1$ and $dy/dt = -1$, so $dy/dx = 1$. When $t = \frac{3\pi}{2}$, $dx/dt = 1$ and

$dy/dt = -1$. So $dy/dx = -1$. Thus, $y = x$ and $y = -x$ are both tangent to the

curve at $(0, 0)$.

23. $x = r\theta - d\sin\theta$, $y = r - d\cos\theta$.

(a) $\dfrac{dx}{d\theta} = r - d\cos\theta$, $\dfrac{dy}{d\theta} = d\sin\theta$, so $\dfrac{dy}{dx} = \dfrac{d\sin\theta}{r - d\cos\theta}$.

(b) If $0 < d < r$, then $|d\cos\theta| \le d < r$, so $r - d\cos\theta \ge r - d > 0$. This shows that $dx/d\theta$ never vanishes,

so the trochoid can have no vertical tangent if $d < r$.

25. $x = 2t^3$, $y = 1 + 4t - t^2$ $\Rightarrow$ $\dfrac{dy}{dx} = \dfrac{dy/dt}{dx/dt} = \dfrac{4 - 2t}{6t^2}$. Now solve $\dfrac{dy}{dx} = 1$ $\Leftrightarrow$ $\dfrac{4 - 2t}{6t^2} = 1$ $\Leftrightarrow$

$6t^2 + 2t - 4 = 0$ $\Leftrightarrow$ $2(3t - 2)(t + 1) = 0$ $\Leftrightarrow$ $t = \frac{2}{3}$ or $t = -1$. If $t = \frac{2}{3}$, the point is $\left(\frac{16}{27}, \frac{29}{9}\right)$, and if $t = -1$,

the point is $(-2, -4)$.

27. By symmetry of the ellipse about the x- and y-axes,

$$A = 4 \int_0^a y\,dx = 4 \int_{\pi/2}^0 b\sin\theta\,(-a\sin\theta)\,d\theta = 4ab \int_0^{\pi/2} \sin^2\theta\,d\theta = 4ab \int_0^{\pi/2} \tfrac{1}{2}(1 - \cos 2\theta)\,d\theta$$

$$= 2ab\big[\theta - \tfrac{1}{2}\sin 2\theta\big]_0^{\pi/2} = 2ab\left(\tfrac{\pi}{2}\right) = \pi ab$$

29. The curve $x = 1 + e^t$, $y = t - t^2 = t(1 - t)$ intersects the x-axis when $y = 0$,

that is, when $t = 0$ and $t = 1$. The corresponding values of x are 2 and $1 + e$.

The shaded area is given by

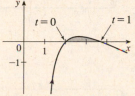

$$\int_{x=2}^{x=1+e} (y_T - y_B)\,dx = \int_{t=0}^{t=1} [y(t) - 0]\,x'(t)\,dt = \int_0^1 (t - t^2)e^t\,dt$$

$$= \int_0^1 te^t\,dt - \int_0^1 t^2 e^t\,dt = \int_0^1 te^t\,dt - \big[t^2 e^t\big]_0^1 + 2\int_0^1 te^t\,dt \qquad \text{[Formula 97 or parts]}$$

$$= 3\int_0^1 te^t\,dt - (e - 0) = 3\big[(t - 1)e^t\big]_0^1 - e \qquad \text{[Formula 96 or parts]}$$

$$= 3[0 - (-1)] - e = 3 - e$$

31. $x = r\theta - d\sin\theta$, $y = r - d\cos\theta$.

$$A = \int_0^{2\pi r} y\,dx = \int_0^{2\pi} (r - d\cos\theta)(r - d\cos\theta)\,d\theta = \int_0^{2\pi} (r^2 - 2dr\cos\theta + d^2\cos^2\theta)\,d\theta$$

$$= \big[r^2\theta - 2dr\sin\theta + \tfrac{1}{2}d^2\left(\theta + \tfrac{1}{2}\sin 2\theta\right)\big]_0^{2\pi} = 2\pi r^2 + \pi d^2$$

33. $x = t + e^{-t}$, $y = t - e^{-t}$, $0 \le t \le 2$. $dx/dt = 1 - e^{-t}$ and $dy/dt = 1 + e^{-t}$, so

$$(dx/dt)^2 + (dy/dt)^2 = (1 - e^{-t})^2 + (1 + e^{-t})^2 = 1 - 2e^{-t} + e^{-2t} + 1 + 2e^{-t} + e^{-2t} = 2 + 2e^{-2t}.$$

Thus, $L = \int_a^b \sqrt{(dx/dt)^2 + (dy/dt)^2}\,dt = \int_0^2 \sqrt{2 + 2e^{-2t}}\,dt \approx 3.1416$.

35. $x = t - 2\sin t$, $y = 1 - 2\cos t$, $0 \le t \le 4\pi$. $dx/dt = 1 - 2\cos t$ and $dy/dt = 2\sin t$, so

$$(dx/dt)^2 + (dy/dt)^2 = (1 - 2\cos t)^2 + (2\sin t)^2 = 1 - 4\cos t + 4\cos^2 t + 4\sin^2 t = 5 - 4\cos t.$$

Thus, $L = \int_a^b \sqrt{(dx/dt)^2 + (dy/dt)^2}\,dt = \int_0^{4\pi} \sqrt{5 - 4\cos t}\,dt \approx 26.7298$.

37. $x = 1 + 3t^2$, $y = 4 + 2t^3$, $0 \le t \le 1$. $dx/dt = 6t$ and $dy/dt = 6t^2$, so $(dx/dt)^2 + (dy/dt)^2 = 36t^2 + 36t^4$

Thus, $L = \int_0^1 \sqrt{36t^2 + 36t^4}\, dt = \int_0^1 6t\sqrt{1 + t^2}\, dt = 6\int_1^2 \sqrt{u}\, \left(\frac{1}{2}\, du\right)$ $\quad [u = 1 + t^2, du = 2t\, dt]$

$\qquad = 3\left[\frac{2}{3} u^{3/2}\right]_1^2 = 2(2^{3/2} - 1) = 2(2\sqrt{2} - 1)$

39. $x = t\sin t$, $y = t\cos t$, $0 \le t \le 1$. $\dfrac{dx}{dt} = t\cos t + \sin t$ and $\dfrac{dy}{dt} = -t\sin t + \cos t$, so

$\left(\dfrac{dx}{dt}\right)^2 + \left(\dfrac{dy}{dt}\right)^2 = t^2\cos^2 t + 2t\sin t\cos t + \sin^2 t + t^2\sin^2 t - 2t\sin t\cos t + \cos^2 t$

$\qquad = t^2(\cos^2 t + \sin^2 t) + \sin^2 t + \cos^2 t = t^2 + 1.$

Thus, $L = \int_0^1 \sqrt{t^2 + 1}\, dt \overset{21}{=} \left[\frac{1}{2} t\sqrt{t^2 + 1} + \frac{1}{2}\ln\left(t + \sqrt{t^2 + 1}\right)\right]_0^1 = \frac{1}{2}\sqrt{2} + \frac{1}{2}\ln\left(1 + \sqrt{2}\right)$.

41.

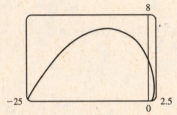

$x = e^t\cos t$, $y = e^t\sin t$, $0 \le t \le \pi$.

$\left(\dfrac{dx}{dt}\right)^2 + \left(\dfrac{dy}{dt}\right)^2 = [e^t(\cos t - \sin t)]^2 + [e^t(\sin t + \cos t)]^2$

$\qquad = (e^t)^2(\cos^2 t - 2\cos t\sin t + \sin^2 t)$

$\qquad\qquad + (e^t)^2(\sin^2 t + 2\sin t\cos t + \cos^2 t)$

$\qquad = e^{2t}(2\cos^2 t + 2\sin^2 t) = 2e^{2t}$

Thus, $L = \int_0^\pi \sqrt{2e^{2t}}\, dt = \int_0^\pi \sqrt{2}\, e^t\, dt = \sqrt{2}\left[e^t\right]_0^\pi = \sqrt{2}\,(e^\pi - 1)$.

43.

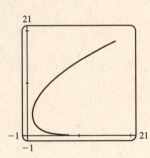

$x = e^t - t$, $y = 4e^{t/2}$, $-8 \le t \le 3$.

$(dx/dt)^2 + (dy/dt)^2 = (e^t - 1)^2 + (2e^{t/2})^2 = e^{2t} - 2e^t + 1 + 4e^t$

$\qquad = e^{2t} + 2e^t + 1 = (e^t + 1)^2$

$L = \int_{-8}^3 \sqrt{(e^t + 1)^2}\, dt = \int_{-8}^3 (e^t + 1)\, dt = \left[e^t + t\right]_{-8}^{3t}$

$\qquad = (e^3 + 3) - (e^{-8} - 8) = e^3 - e^{-8} + 11$

45. $x = t - e^t$, $y = t + e^t$, $-6 \le t \le 6$.

$\left(\dfrac{dx}{dt}\right)^2 + \left(\dfrac{dy}{dt}\right)^2 = (1 - e^t)^2 + (1 + e^t)^2 = (1 - 2e^t + e^{2t}) + (1 + 2e^t + e^{2t}) = 2 + 2e^{2t}$, so $L = \int_{-6}^6 \sqrt{2 + 2e^{2t}}\, dt$.

Set $f(t) = \sqrt{2 + 2e^{2t}}$. Then by Simpson's Rule with $n = 6$ and $\Delta t = \frac{6 - (-6)}{6} = 2$, we get

$L \approx \frac{2}{3}[f(-6) + 4f(-4) + 2f(-2) + 4f(0) + 2f(2) + 4f(4) + f(6)] \approx 612.3053$.

47. $x = \sin^2 t$, $y = \cos^2 t$, $0 \le t \le 3\pi$.

$(dx/dt)^2 + (dy/dt)^2 = (2\sin t\cos t)^2 + (-2\cos t\sin t)^2 = 8\sin^2 t\cos^2 t = 2\sin^2 2t \quad \Rightarrow$

Distance $= \int_0^{3\pi} \sqrt{2}\, |\sin 2t|\, dt = 6\sqrt{2}\int_0^{\pi/2} \sin 2t\, dt$ [by symmetry] $= -3\sqrt{2}\left[\cos 2t\right]_0^{\pi/2} = -3\sqrt{2}\,(-1 - 1) = 6\sqrt{2}$.

The full curve is traversed as t goes from 0 to $\frac{\pi}{2}$, because the curve is the segment of $x + y = 1$ that lies in the first quadrant

(since $x, y \ge 0$), and this segment is completely traversed as t goes from 0 to $\frac{\pi}{2}$. Thus, $L = \int_0^{\pi/2} \sin 2t\, dt = \sqrt{2}$, as above.

49. $x = a \sin \theta$, $y = b \cos \theta$, $0 \le \theta \le 2\pi$.

$$\left(\frac{dx}{dt}\right)^2 + \left(\frac{dy}{dt}\right)^2 = (a \cos \theta)^2 + (-b \sin \theta)^2 = a^2 \cos^2 \theta + b^2 \sin^2 \theta = a^2(1 - \sin^2 \theta) + b^2 \sin^2 \theta$$

$$= a^2 - (a^2 - b^2) \sin^2 \theta = a^2 - c^2 \sin^2 \theta = a^2 \left(1 - \frac{c^2}{a^2} \sin^2 \theta\right) = a^2(1 - e^2 \sin^2 \theta)$$

So $L = 4 \int_0^{\pi/2} \sqrt{a^2 \left(1 - e^2 \sin^2 \theta\right)} \, d\theta$ [by symmetry] $= 4a \int_0^{\pi/2} \sqrt{1 - e^2 \sin^2 \theta} \, d\theta$.

51. (a) $x = 11 \cos t - 4 \cos(11t/2)$, $y = 11 \sin t - 4 \sin(11t/2)$.

Notice that $0 \le t \le 2\pi$ does not give the complete curve because

$x(0) \ne x(2\pi)$. In fact, we must take $t \in [0, 4\pi]$ in order to obtain the

complete curve, since the first term in each of the parametric equations has

period 2π and the second has period $\frac{2\pi}{11/2} = \frac{4\pi}{11}$, and the least common

integer multiple of these two numbers is 4π.

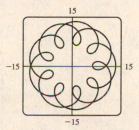

(b) We use the CAS to find the derivatives dx/dt and dy/dt, and then use Theorem 5 to find the arc length. Recent versions

of Maple express the integral $\int_0^{4\pi} \sqrt{(dx/dt)^2 + (dy/dt)^2} \, dt$ as $88E(2\sqrt{2}\,i)$, where $E(x)$ is the elliptic integral

$\int_0^1 \dfrac{\sqrt{1 - x^2 t^2}}{\sqrt{1 - t^2}} \, dt$ and i is the imaginary number $\sqrt{-1}$.

Some earlier versions of Maple (as well as Mathematica) cannot do the integral exactly, so we use the command

`evalf(Int(sqrt(diff(x,t)^2+diff(y,t)^2),t=0..4*Pi));` to estimate the length, and find that the arc

length is approximately 294.03. Derive's `Para_arc_length` function in the utility file `Int_apps` simplifies the

integral to $11 \int_0^{4\pi} \sqrt{-4 \cos t \, \cos\left(\frac{11t}{2}\right) - 4 \sin t \, \sin\left(\frac{11t}{2}\right) + 5} \, dt$.

53. The coordinates of T are $(r \cos \theta, r \sin \theta)$. Since TP was unwound from

arc TA, TP has length $r\theta$. Also $\angle PTQ = \angle PTR - \angle QTR = \frac{1}{2}\pi - \theta$,

so P has coordinates $x = r \cos \theta + r\theta \cos\left(\frac{1}{2}\pi - \theta\right) = r(\cos \theta + \theta \sin \theta)$,

$y = r \sin \theta - r\theta \sin\left(\frac{1}{2}\pi - \theta\right) = r(\sin \theta - \theta \cos \theta)$.

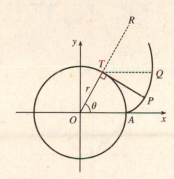

9.3 Polar Coordinates

1. (a) $\left(2, \frac{\pi}{3}\right)$

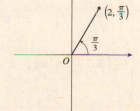

By adding 2π to $\frac{\pi}{3}$, we obtain the point $\left(2, \frac{7\pi}{3}\right)$. The direction

opposite $\frac{\pi}{3}$ is $\frac{4\pi}{3}$, so $\left(-2, \frac{4\pi}{3}\right)$ is a point that satisfies the $r < 0$

requirement.

(b) $\left(1, -\frac{3\pi}{4}\right)$

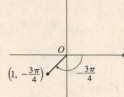

$r > 0$: $\left(1, -\frac{3\pi}{4} + 2\pi\right) = \left(1, \frac{5\pi}{4}\right)$

$r < 0$: $\left(-1, -\frac{3\pi}{4} + \pi\right) = \left(-1, \frac{\pi}{4}\right)$

(c) $\left(-1, \frac{\pi}{2}\right)$

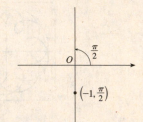

$r > 0$: $\left(-(-1), \frac{\pi}{2} + \pi\right) = \left(1, \frac{3\pi}{2}\right)$

$r < 0$: $\left(-1, \frac{\pi}{2} + 2\pi\right) = \left(-1, \frac{5\pi}{2}\right)$

3. (a)

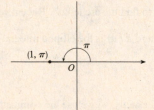

$x = 1\cos\pi = 1(-1) = -1$ and

$y = 1\sin\pi = 1(0) = 0$ give us

the Cartesian coordinates $(-1, 0)$.

(b)

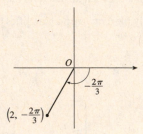

$x = 2\cos\left(-\frac{2\pi}{3}\right) = 2\left(-\frac{1}{2}\right) = -1$ and

$y = 2\sin\left(-\frac{2\pi}{3}\right) = 2\left(-\frac{\sqrt{3}}{2}\right) = -\sqrt{3}$

give us $\left(-1, -\sqrt{3}\right)$.

(c)

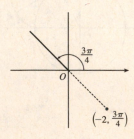

$x = -2\cos\frac{3\pi}{4} = -2\left(-\frac{\sqrt{2}}{2}\right) = \sqrt{2}$ and

$y = -2\sin\frac{3\pi}{4} = -2\left(\frac{\sqrt{2}}{2}\right) = -\sqrt{2}$

gives us $\left(\sqrt{2}, -\sqrt{2}\right)$.

5. (a) $x = 2$ and $y = -2$ $\Rightarrow$ $r = \sqrt{2^2 + (-2)^2} = 2\sqrt{2}$ and $\theta = \tan^{-1}\left(\frac{-2}{2}\right) = -\frac{\pi}{4}$. Since $(2, -2)$ is in the fourth

quadrant, the polar coordinates are (i) $\left(2\sqrt{2}, \frac{7\pi}{4}\right)$ and (ii) $\left(-2\sqrt{2}, \frac{3\pi}{4}\right)$.

(b) $x = -1$ and $y = \sqrt{3}$ $\Rightarrow$ $r = \sqrt{(-1)^2 + \left(\sqrt{3}\right)^2} = 2$ and $\theta = \tan^{-1}\left(\frac{\sqrt{3}}{-1}\right) = \frac{2\pi}{3}$. Since $\left(-1, \sqrt{3}\right)$ is in the second

quadrant, the polar coordinates are (i) $\left(2, \frac{2\pi}{3}\right)$ and (ii) $\left(-2, \frac{5\pi}{3}\right)$.

7. The curves $r = 1$ and $r = 2$ represent circles with center O and radii 1 and 2. The region in the plane satisfying $1 \le r \le 2$ consists of both circles and the shaded region between them in the figure.

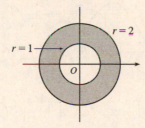

9. $r \ge 0$, $\pi/4 \le \theta \le 3\pi/4$.

$\theta = k$ represents a line through O.

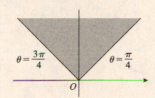

11. $2 < r < 3$, $\frac{5\pi}{3} \le \theta \le \frac{7\pi}{3}$

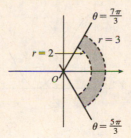

13. $r = 2\cos\theta \Rightarrow r^2 = 2r\cos\theta \Leftrightarrow x^2 + y^2 = 2x \Leftrightarrow x^2 - 2x + 1 + y^2 = 1 \Leftrightarrow (x-1)^2 + y^2 = 1$, a circle of radius 1 centered at $(1, 0)$. The first two equations are actually equivalent since $r^2 = 2r\cos\theta \Rightarrow r(r - 2\cos\theta) = 0 \Rightarrow r = 0$ or $r = 2\cos\theta$. But $r = 2\cos\theta$ gives the point $r = 0$ (the pole) when $\theta = 0$. Thus, the equation $r = 2\cos\theta$ is equivalent to the compound condition ($r = 0$ or $r = 2\cos\theta$).

15. $r^2 \cos 2\theta = 1 \Leftrightarrow r^2(\cos^2\theta - \sin^2\theta) = 1 \Leftrightarrow (r\cos\theta)^2 - (r\sin\theta)^2 = 1 \Leftrightarrow x^2 - y^2 = 1$, a hyperbola centered at the origin with foci on the x-axis.

17. $y = 1 + 3x \Leftrightarrow r\sin\theta = 1 + 3r\cos\theta \Leftrightarrow r\sin\theta - 3r\cos\theta = 1 \Leftrightarrow r(\sin\theta - 3\cos\theta) = 1 \Leftrightarrow$

$r = \dfrac{1}{\sin\theta - 3\cos\theta}$

19. $x^2 + y^2 = 2cx \Leftrightarrow r^2 = 2cr\cos\theta \Leftrightarrow r^2 - 2cr\cos\theta = 0 \Leftrightarrow r(r - 2c\cos\theta) = 0 \Leftrightarrow r = 0$ or $r = 2c\cos\theta$. $r = 0$ is included in $r = 2c\cos\theta$ when $\theta = \frac{\pi}{2} + n\pi$, so the curve is represented by the single equation $r = 2c\cos\theta$.

21. (a) The description leads immediately to the polar equation $\theta = \frac{\pi}{6}$, and the Cartesian equation $y = \tan\left(\frac{\pi}{6}\right)x = \frac{1}{\sqrt{3}}x$ is slightly more difficult to derive.

(b) The easier description here is the Cartesian equation $x = 3$.

23. $r = -2\sin\theta$

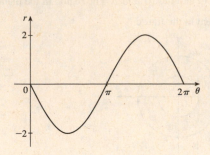 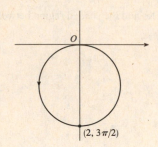

25. $r = 2(1 + \cos\theta)$

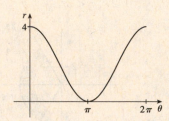 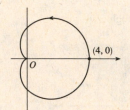

27. $r = \theta, \quad \theta \geq 0$

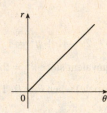 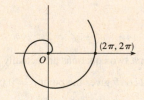

29. $r = 4\sin 3\theta$

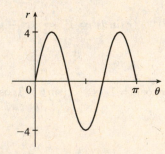 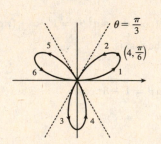

31. $r = 2\cos 4\theta$

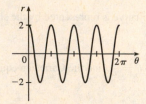 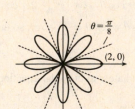

33. $r = 1 - 2\sin\theta$

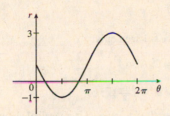

35. $r^2 = 9\sin 2\theta$

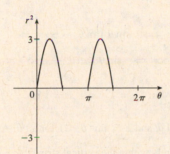

37. $r = 2 + \sin 3\theta$

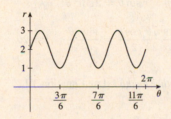

39. $r = 1 + 2\cos 2\theta$

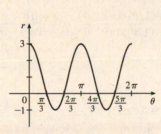

41. For $\theta = 0$, π, and 2π, r has its minimum value of about 0.5. For $\theta = \frac{\pi}{2}$ and $\frac{3\pi}{2}$, r attains its maximum value of 2.

We see that the graph has a similar shape for $0 \le \theta \le \pi$ and $\pi \le \theta \le 2\pi$.

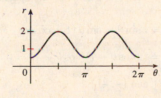

43. $x = r\cos\theta = (4 + 2\sec\theta)\cos\theta = 4\cos\theta + 2$. Now, $r \to \infty \ \Rightarrow$

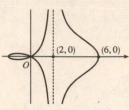

$(4 + 2\sec\theta) \to \infty \ \Rightarrow \ \theta \to \left(\frac{\pi}{2}\right)^{-}$ or $\theta \to \left(\frac{3\pi}{2}\right)^{+}$ [since we need only

consider $0 \le \theta < 2\pi$], so $\lim\limits_{r\to\infty} x = \lim\limits_{\theta\to\pi/2^{-}} (4\cos\theta + 2) = 2$. Also,

$r \to -\infty \ \Rightarrow \ (4 + 2\sec\theta) \to -\infty \ \Rightarrow \ \theta \to \left(\frac{\pi}{2}\right)^{+}$ or $\theta \to \left(\frac{3\pi}{2}\right)^{-}$, so

$\lim\limits_{r\to-\infty} x = \lim\limits_{\theta\to\pi/2^{+}} (4\cos\theta + 2) = 2$. Therefore, $\lim\limits_{r\to\pm\infty} x = 2 \ \Rightarrow \ x = 2$ is a vertical asymptote.

45. To show that $x = 1$ is an asymptote we must prove $\lim\limits_{r\to\pm\infty} x = 1$.

$x = (r)\cos\theta = (\sin\theta\tan\theta)\cos\theta = \sin^2\theta$. Now, $r \to \infty \ \Rightarrow \ \sin\theta\tan\theta \to \infty \ \Rightarrow$

$\theta \to \left(\frac{\pi}{2}\right)^{-}$, so $\lim\limits_{r\to\infty} x = \lim\limits_{\theta\to\pi/2^{-}} \sin^2\theta = 1$. Also, $r \to -\infty \ \Rightarrow \ \sin\theta\tan\theta \to -\infty \ \Rightarrow$

$\theta \to \left(\frac{\pi}{2}\right)^{+}$, so $\lim\limits_{r\to-\infty} x = \lim\limits_{\theta\to\pi/2^{+}} \sin^2\theta = 1$. Therefore, $\lim\limits_{r\to\pm\infty} x = 1 \ \Rightarrow \ x = 1$ is

a vertical asymptote. Also notice that $x = \sin^2\theta \ge 0$ for all θ, and $x = \sin^2\theta \le 1$ for all θ. And $x \ne 1$, since the curve is not

defined at odd multiples of $\frac{\pi}{2}$. Therefore, the curve lies entirely within the vertical strip $0 \le x < 1$.

47. $r = 2\sin\theta \ \Rightarrow \ x = r\cos\theta = 2\sin\theta\cos\theta = \sin 2\theta$, $y = r\sin\theta = 2\sin^2\theta \ \Rightarrow$

$$\frac{dy}{dx} = \frac{dy/d\theta}{dx/d\theta} = \frac{2\cdot 2\sin\theta\cos\theta}{\cos 2\theta\cdot 2} = \frac{\sin 2\theta}{\cos 2\theta} = \tan 2\theta$$

When $\theta = \frac{\pi}{6}$, $\frac{dy}{dx} = \tan\left(2\cdot\frac{\pi}{6}\right) = \tan\frac{\pi}{3} = \sqrt{3}$. [*Another method:* Use Equation 3.]

49. $r = 1/\theta \ \Rightarrow \ x = r\cos\theta = (\cos\theta)/\theta$, $y = r\sin\theta = (\sin\theta)/\theta \ \Rightarrow$

$$\frac{dy}{dx} = \frac{dy/d\theta}{dx/d\theta} = \frac{\sin\theta(-1/\theta^2) + (1/\theta)\cos\theta}{\cos\theta(-1/\theta^2) - (1/\theta)\sin\theta} \cdot \frac{\theta^2}{\theta^2} = \frac{-\sin\theta + \theta\cos\theta}{-\cos\theta - \theta\sin\theta}$$

When $\theta = \pi$, $\frac{dy}{dx} = \frac{-0 + \pi(-1)}{-(-1) - \pi(0)} = \frac{-\pi}{1} = -\pi$.

51. $r = 3\cos\theta \ \Rightarrow \ x = r\cos\theta = 3\cos\theta\cos\theta$, $y = r\sin\theta = 3\cos\theta\sin\theta \ \Rightarrow$

$\frac{dy}{d\theta} = -3\sin^2\theta + 3\cos^2\theta = 3\cos 2\theta = 0 \ \Rightarrow \ 2\theta = \frac{\pi}{2}$ or $\frac{3\pi}{2} \ \Leftrightarrow \ \theta = \frac{\pi}{4}$ or $\frac{3\pi}{4}$.

So the tangent is horizontal at $\left(\frac{3}{\sqrt{2}}, \frac{\pi}{4}\right)$ and $\left(-\frac{3}{\sqrt{2}}, \frac{3\pi}{4}\right)$ $\left[\text{same as }\left(\frac{3}{\sqrt{2}}, -\frac{\pi}{4}\right)\right]$.

$\frac{dx}{d\theta} = -6\sin\theta\cos\theta = -3\sin 2\theta = 0 \ \Rightarrow \ 2\theta = 0$ or $\pi \ \Leftrightarrow \ \theta = 0$ or $\frac{\pi}{2}$. So the tangent is vertical at $(3, 0)$ and $\left(0, \frac{\pi}{2}\right)$.

53. $r = 1 + \cos\theta \ \Rightarrow \ x = r\cos\theta = \cos\theta(1 + \cos\theta)$, $y = r\sin\theta = \sin\theta(1 + \cos\theta) \ \Rightarrow$

$\frac{dy}{d\theta} = (1 + \cos\theta)\cos\theta - \sin^2\theta = 2\cos^2\theta + \cos\theta - 1 = (2\cos\theta - 1)(\cos\theta + 1) = 0 \ \Rightarrow \ \cos\theta = \frac{1}{2}$ or $-1 \ \Rightarrow$

$\theta = \frac{\pi}{3}, \pi$, or $\frac{5\pi}{3} \ \Rightarrow$ horizontal tangent at $\left(\frac{3}{2}, \frac{\pi}{3}\right)$, $(0, \pi)$, and $\left(\frac{3}{2}, \frac{5\pi}{3}\right)$.

$\frac{dx}{d\theta} = -(1 + \cos\theta)\sin\theta - \cos\theta\sin\theta = -\sin\theta(1 + 2\cos\theta) = 0 \ \Rightarrow \ \sin\theta = 0$ or $\cos\theta = -\frac{1}{2} \ \Rightarrow$

$\theta = 0, \pi, \frac{2\pi}{3},$ or $\frac{4\pi}{3}$ $\Rightarrow$ vertical tangent at $(2, 0)$, $\left(\frac{1}{2}, \frac{2\pi}{3}\right)$, and $\left(\frac{1}{2}, \frac{4\pi}{3}\right)$.

Note that the tangent is horizontal, not vertical when $\theta = \pi$, since $\displaystyle\lim_{\theta \to \pi} \frac{dy/d\theta}{dx/d\theta} = 0$.

55. $r = a \sin \theta + b \cos \theta \Rightarrow r^2 = ar \sin \theta + br \cos \theta \Rightarrow x^2 + y^2 = ay + bx \Rightarrow$

$x^2 - bx + \left(\frac{1}{2}b\right)^2 + y^2 - ay + \left(\frac{1}{2}a\right)^2 = \left(\frac{1}{2}b\right)^2 + \left(\frac{1}{2}a\right)^2 \Rightarrow \left(x - \frac{1}{2}b\right)^2 + \left(y - \frac{1}{2}a\right)^2 = \frac{1}{4}(a^2 + b^2)$, and this is a circle

with center $\left(\frac{1}{2}b, \frac{1}{2}a\right)$ and radius $\frac{1}{2}\sqrt{a^2 + b^2}$.

57. $r = e^{\sin \theta} - 2\cos(4\theta)$.

The parameter interval is $[0, 2\pi]$.

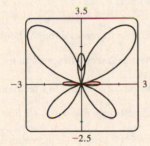

59. $r = 1 + \cos^{999} \theta$. The parameter interval is $[0, 2\pi]$.

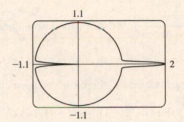

61. It appears that the graph of $r = 1 + \sin\left(\theta - \frac{\pi}{6}\right)$ is the same shape as the graph of $r = 1 + \sin \theta$, but rotated counterclockwise about the origin by $\frac{\pi}{6}$. Similarly, the graph of $r = 1 + \sin\left(\theta - \frac{\pi}{3}\right)$ is rotated by $\frac{\pi}{3}$. In general, the graph of $r = f(\theta - \alpha)$ is the same shape as that of $r = f(\theta)$, but rotated counterclockwise through α about the origin. That is, for any point (r_0, θ_0) on the curve $r = f(\theta)$, the point $(r_0, \theta_0 + \alpha)$ is on the curve $r = f(\theta - \alpha)$, since $r_0 = f(\theta_0) = f((\theta_0 + \alpha) - \alpha)$.

63. (a) $r = \sin n\theta$.

$n = 2$

$n = 3$

$n = 4$

$n = 5$

From the graphs, it seems that when n is even, the number of loops in the curve (called a rose) is $2n$, and when n is odd, the number of loops is simply n. This is because in the case of n odd, every point on the graph is traversed twice, due to the fact that

$$r(\theta + \pi) = \sin[n(\theta + \pi)] = \sin n\theta \cos n\pi + \cos n\theta \sin n\pi = \begin{cases} \sin n\theta & \text{if } n \text{ is even} \\ -\sin n\theta & \text{if } n \text{ is odd} \end{cases}$$

(b) The graph of $r = |\sin n\theta|$ has $2n$ loops whether n is odd or even, since $r(\theta + \pi) = r(\theta)$.

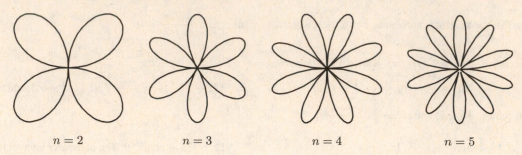

$$n = 2 \qquad n = 3 \qquad n = 4 \qquad n = 5$$

65. $r = \dfrac{1 - a\cos\theta}{1 + a\cos\theta}$. We start with $a = 0$, since in this case the curve is simply the circle $r = 1$.

As a increases, the graph moves to the left, and its right side becomes flattened. As a increases through about 0.4, the right

side seems to grow a dimple, which upon closer investigation (with narrower θ-ranges) seems to appear at $a \approx 0.42$ [the

actual value is $\sqrt{2} - 1$]. As $a \to 1$, this dimple becomes more pronounced, and the curve begins to stretch out horizontally,

until at $a = 1$ the denominator vanishes at $\theta = \pi$, and the dimple becomes an actual cusp. For $a > 1$ we must choose our

parameter interval carefully, since $r \to \infty$ as $1 + a\cos\theta \to 0 \iff \theta \to \pm\cos^{-1}(-1/a)$. As a increases from 1, the curve

splits into two parts. The left part has a loop, which grows larger as a increases, and the right part grows broader vertically,

and its left tip develops a dimple when $a \approx 2.42$ [actually, $\sqrt{2} + 1$]. As a increases, the dimple grows more and more

pronounced. If $a < 0$, we get the same graph as we do for the corresponding positive a-value, but with a rotation through π

about the pole, as happened when c was replaced with $-c$ in Exercise 64.

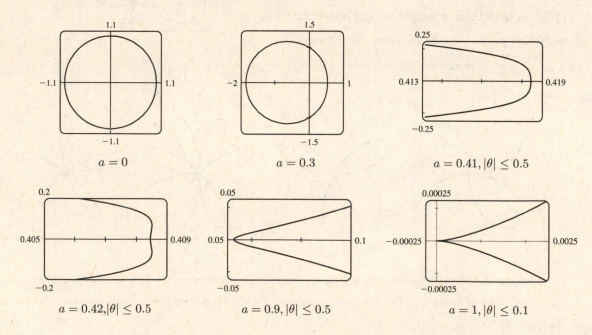

$$a = 0 \qquad\qquad a = 0.3 \qquad\qquad a = 0.41, |\theta| \le 0.5$$

$$a = 0.42, |\theta| \le 0.5 \qquad\qquad a = 0.9, |\theta| \le 0.5 \qquad\qquad a = 1, |\theta| \le 0.1$$

[continued]

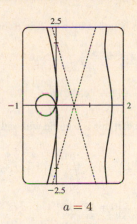

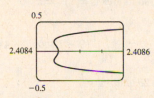

$a = 2.41, |\theta - \pi| \leq 0.2$

$a = 2$

$a = 2.42, |\theta - \pi| \leq 0.2$

$a = 4$

67. $\tan\psi = \tan(\phi - \theta) = \dfrac{\tan\phi - \tan\theta}{1 + \tan\phi\,\tan\theta} = \dfrac{\dfrac{dy}{dx} - \tan\theta}{1 + \dfrac{dy}{dx}\tan\theta} = \dfrac{\dfrac{dy/d\theta}{dx/d\theta} - \tan\theta}{1 + \dfrac{dy/d\theta}{dx/d\theta}\tan\theta}$

$= \dfrac{\dfrac{dy}{d\theta} - \dfrac{dx}{d\theta}\tan\theta}{\dfrac{dx}{d\theta} + \dfrac{dy}{d\theta}\tan\theta} = \dfrac{\left(\dfrac{dr}{d\theta}\sin\theta + r\cos\theta\right) - \tan\theta\left(\dfrac{dr}{d\theta}\cos\theta - r\sin\theta\right)}{\left(\dfrac{dr}{d\theta}\cos\theta - r\sin\theta\right) + \tan\theta\left(\dfrac{dr}{d\theta}\sin\theta + r\cos\theta\right)} = \dfrac{r\cos\theta + r\cdot\dfrac{\sin^2\theta}{\cos\theta}}{\dfrac{dr}{d\theta}\cos\theta + \dfrac{dr}{d\theta}\cdot\dfrac{\sin^2\theta}{\cos\theta}}$

$= \dfrac{r\cos^2\theta + r\sin^2\theta}{\dfrac{dr}{d\theta}\cos^2\theta + \dfrac{dr}{d\theta}\sin^2\theta} = \dfrac{r}{dr/d\theta}$

9.4 Areas and Lengths in Polar Coordinates

1. $r = e^{-\theta/4}$, $\pi/2 \leq \theta \leq \pi$.

$A = \displaystyle\int_{\pi/2}^{\pi} \tfrac{1}{2}r^2\,d\theta = \int_{\pi/2}^{\pi} \tfrac{1}{2}(e^{-\theta/4})^2\,d\theta = \int_{\pi/2}^{\pi} \tfrac{1}{2}e^{-\theta/2}\,d\theta = \tfrac{1}{2}\Big[-2e^{-\theta/2}\Big]_{\pi/2}^{\pi} = -1(e^{-\pi/2} - e^{-\pi/4}) = e^{-\pi/4} - e^{-\pi/2}$

3. $r^2 = 9\sin 2\theta$, $r \geq 0$, $0 \leq \theta \leq \pi/2$.

$A = \displaystyle\int_0^{\pi/2} \tfrac{1}{2}r^2\,d\theta = \int_0^{\pi/2} \tfrac{1}{2}(9\sin 2\theta)\,d\theta = \tfrac{9}{2}\Big[-\tfrac{1}{2}\cos 2\theta\Big]_0^{\pi/2} = -\tfrac{9}{4}(-1 - 1) = \tfrac{9}{2}$

5. $r = \sqrt{\theta}$, $0 \leq \theta \leq 2\pi$. $A = \displaystyle\int_0^{2\pi} \tfrac{1}{2}r^2\,d\theta = \int_0^{2\pi} \tfrac{1}{2}\left(\sqrt{\theta}\right)^2\,d\theta = \int_0^{2\pi} \tfrac{1}{2}\theta\,d\theta = \Big[\tfrac{1}{4}\theta^2\Big]_0^{2\pi} = \pi^2$

7. $r = 4 + 3\sin\theta$, $-\dfrac{\pi}{2} \leq \theta \leq \dfrac{\pi}{2}$.

$A = \displaystyle\int_{-\pi/2}^{\pi/2} \tfrac{1}{2}((4 + 3\sin\theta)^2\,d\theta = \tfrac{1}{2}\int_{-\pi/2}^{\pi/2}(16 + 24\sin\theta + 9\sin^2\theta)\,d\theta$

$= \tfrac{1}{2}\displaystyle\int_{-\pi/2}^{\pi/2}(16 + 9\sin^2\theta)\,d\theta$ [odd function]

$= \tfrac{1}{2}\cdot 2\displaystyle\int_0^{\pi/2}\Big[16 + 9\cdot\tfrac{1}{2}(1 - \cos 2\theta)\Big]\,d\theta$ [even function]

$= \displaystyle\int_0^{\pi/2}\Big(\tfrac{41}{2} - \tfrac{9}{2}\cos 2\theta\Big)\,d\theta = \Big[\tfrac{41}{2}\theta - \tfrac{9}{4}\sin 2\theta\Big]_0^{\pi/2} = \Big(\tfrac{41\pi}{4} - 0\Big) - (0 - 0) = \tfrac{41\pi}{4}$

9. The area is bounded by $r = 2\sin\theta$ for $\theta = 0$ to $\theta = \pi$.

$$A = \int_0^\pi \tfrac{1}{2} r^2 \, d\theta = \tfrac{1}{2} \int_0^\pi (2\sin\theta)^2 \, d\theta = \tfrac{1}{2} \int_0^\pi 4\sin^2\theta \, d\theta$$

$$= 2 \int_0^\pi \tfrac{1}{2}(1 - \cos 2\theta) d\theta = \left[\theta - \tfrac{1}{2}\sin 2\theta\right]_0^\pi = \pi$$

Also, note that this is a circle with radius 1, so its area is $\pi(1)^2 = \pi$.

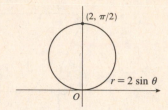

11. $A = \displaystyle\int_0^{2\pi} \tfrac{1}{2} r^2 \, d\theta = \int_0^{2\pi} \tfrac{1}{2}(3 + 2\cos\theta)^2 \, d\theta = \tfrac{1}{2}\int_0^{2\pi} (9 + 12\cos\theta + 4\cos^2\theta) \, d\theta$

$$= \tfrac{1}{2} \int_0^{2\pi} \left[9 + 12\cos\theta + 4 \cdot \tfrac{1}{2}(1 + \cos 2\theta)\right] d\theta$$

$$= \tfrac{1}{2} \int_0^{2\pi} (11 + 12\cos\theta + 2\cos 2\theta) \, d\theta = \tfrac{1}{2}\left[11\theta + 12\sin\theta + \sin 2\theta\right]_0^{2\pi}$$

$$= \tfrac{1}{2}(22\pi) = 11\pi$$

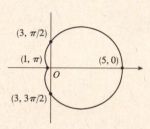

13. $A = \displaystyle\int_0^{2\pi} \tfrac{1}{2} r^2 \, d\theta = \int_0^{2\pi} \tfrac{1}{2}\left(\sqrt{1 + \cos^2(5\theta)}\right)^2 d\theta$

$$= \tfrac{1}{2} \int_0^{2\pi} (1 + \cos^2(5\theta)) \, d\theta = \tfrac{1}{2} \int_0^{2\pi} \left[1 + \tfrac{1}{2}(1 + \cos 10\theta)\right] d\theta$$

$$= \tfrac{1}{2}\left[\tfrac{3}{2}\theta + \tfrac{1}{20}\sin 10\theta\right]_0^{2\pi} = \tfrac{1}{2}(3\pi) = \tfrac{3}{2}\pi$$

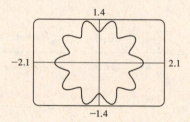

15. The curve passes through the pole when $r = 0 \;\Rightarrow\; 4\cos 3\theta = 0 \;\Rightarrow\; \cos 3\theta = 0 \;\Rightarrow\; 3\theta = \tfrac{\pi}{2} + \pi n \;\Rightarrow$

$\theta = \tfrac{\pi}{6} + \tfrac{\pi}{3} n$. The part of the shaded loop above the polar axis is traced out for

$\theta = 0$ to $\theta = \pi/6$, so we'll use $-\pi/6$ and $\pi/6$ as our limits of integration.

$$A = \int_{-\pi/6}^{\pi/6} \tfrac{1}{2}(4\cos 3\theta)^2 \, d\theta = 2 \int_0^{\pi/6} \tfrac{1}{2}(16\cos^2 3\theta) \, d\theta$$

$$= 16 \int_0^{\pi/6} \tfrac{1}{2}(1 + \cos 6\theta) \, d\theta = 8\left[\theta + \tfrac{1}{6}\sin 6\theta\right]_0^{\pi/6} = 8\left(\tfrac{\pi}{6}\right) = \tfrac{4}{3}\pi$$

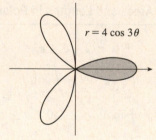

17.

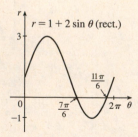

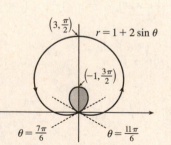

This is a limaçon, with inner loop traced out between $\theta = \tfrac{7\pi}{6}$ and $\tfrac{11\pi}{6}$ [found by solving $r = 0$].

$$A = 2 \int_{7\pi/6}^{3\pi/2} \tfrac{1}{2}(1 + 2\sin\theta)^2 \, d\theta = \int_{7\pi/6}^{3\pi/2} (1 + 4\sin\theta + 4\sin^2\theta) \, d\theta = \int_{7\pi/6}^{3\pi/2} \left[1 + 4\sin\theta + 4 \cdot \tfrac{1}{2}(1 - \cos 2\theta)\right] d\theta$$

$$= \left[\theta - 4\cos\theta + 2\theta - \sin 2\theta\right]_{7\pi/6}^{3\pi/2} = \left(\tfrac{9\pi}{2}\right) - \left(\tfrac{7\pi}{2} + 2\sqrt{3} - \tfrac{\sqrt{3}}{2}\right) = \pi - \tfrac{3\sqrt{3}}{2}$$

19. $2\cos\theta = 1 \Rightarrow \cos\theta = \frac{1}{2} \Rightarrow \theta = \frac{\pi}{3}$ or $\frac{5\pi}{3}$.

$$A = 2\int_0^{\pi/3}\frac{1}{2}[(2\cos\theta)^2 - 1^2]\,d\theta = \int_0^{\pi/3}(4\cos^2\theta - 1)\,d\theta$$

$$= \int_0^{\pi/3}\left\{4\left[\frac{1}{2}(1+\cos 2\theta)\right] - 1\right\}d\theta = \int_0^{\pi/3}(1 + 2\cos 2\theta)\,d\theta$$

$$= \left[\theta + \sin 2\theta\right]_0^{\pi/3} = \frac{\pi}{3} + \frac{\sqrt{3}}{2}$$

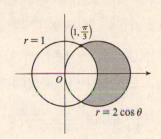

21. $3\cos\theta = 1 + \cos\theta \Leftrightarrow \cos\theta = \frac{1}{2} \Rightarrow \theta = \frac{\pi}{3}$ or $-\frac{\pi}{3}$.

$$A = 2\int_0^{\pi/3}\frac{1}{2}[(3\cos\theta)^2 - (1+\cos\theta)^2]\,d\theta$$

$$= \int_0^{\pi/3}(8\cos^2\theta - 2\cos\theta - 1)\,d\theta = \int_0^{\pi/3}[4(1+\cos 2\theta) - 2\cos\theta - 1]\,d\theta$$

$$= \int_0^{\pi/3}(3 + 4\cos 2\theta - 2\cos\theta)\,d\theta = \left[3\theta + 2\sin 2\theta - 2\sin\theta\right]_0^{\pi/3}$$

$$= \pi + \sqrt{3} - \sqrt{3} = \pi$$

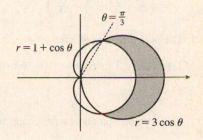

23. $\sqrt{3}\cos\theta = \sin\theta \Rightarrow \sqrt{3} = \dfrac{\sin\theta}{\cos\theta} \Rightarrow \tan\theta = \sqrt{3} \Rightarrow \theta = \frac{\pi}{3}$.

$$A = \int_0^{\pi/3}\frac{1}{2}(\sin\theta)^2\,d\theta + \int_{\pi/3}^{\pi/2}\frac{1}{2}\left(\sqrt{3}\cos\theta\right)^2 d\theta$$

$$= \int_0^{\pi/3}\frac{1}{2}\cdot\frac{1}{2}(1-\cos 2\theta)\,d\theta + \int_{\pi/3}^{\pi/2}\frac{1}{2}\cdot 3\cdot\frac{1}{2}(1+\cos 2\theta)\,d\theta$$

$$= \frac{1}{4}\left[\theta - \frac{1}{2}\sin 2\theta\right]_0^{\pi/3} + \frac{3}{4}\left[\theta + \frac{1}{2}\sin 2\theta\right]_{\pi/3}^{\pi/2}$$

$$= \frac{1}{4}\left[\left(\frac{\pi}{3} - \frac{\sqrt{3}}{4}\right) - 0\right] + \frac{3}{4}\left[\left(\frac{\pi}{2} + 0\right) - \left(\frac{\pi}{3} + \frac{\sqrt{3}}{4}\right)\right]$$

$$= \frac{\pi}{12} - \frac{\sqrt{3}}{16} + \frac{\pi}{8} - \frac{3\sqrt{3}}{16} = \frac{5\pi}{24} - \frac{\sqrt{3}}{4}$$

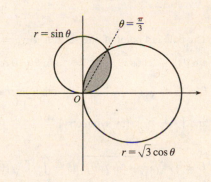

25. $\sin 2\theta = \cos 2\theta \Rightarrow \dfrac{\sin 2\theta}{\cos 2\theta} = 1 \Rightarrow \tan 2\theta = 1 \Rightarrow 2\theta = \frac{\pi}{4} \Rightarrow$

$\theta = \frac{\pi}{8} \Rightarrow$

$$A = 8\cdot 2\int_0^{\pi/8}\frac{1}{2}\sin^2 2\theta\,d\theta = 8\int_0^{\pi/8}\frac{1}{2}(1-\cos 4\theta)\,d\theta$$

$$= 4\left[\theta - \frac{1}{4}\sin 4\theta\right]_0^{\pi/8} = 4\left(\frac{\pi}{8} - \frac{1}{4}\cdot 1\right) = \frac{\pi}{2} - 1$$

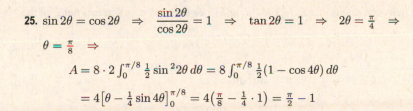

27. The darker shaded region (from $\theta = 0$ to $\theta = 2\pi/3$) represents $\frac{1}{2}$ of the desired area plus $\frac{1}{2}$ of the area of the inner loop. From this area, we'll subtract $\frac{1}{2}$ of the area of the inner loop (the lighter shaded region from $\theta = 2\pi/3$ to $\theta = \pi$), and then double that difference to obtain the desired area.

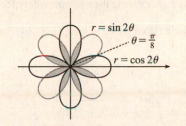
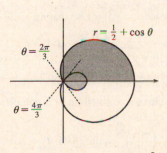

[continued]

$$A = 2\left[\int_0^{2\pi/3} \tfrac{1}{2}\left(\tfrac{1}{2} + \cos\theta\right)^2 d\theta - \int_{2\pi/3}^{\pi} \tfrac{1}{2}\left(\tfrac{1}{2} + \cos\theta\right)^2 d\theta\right]$$

$$= \int_0^{2\pi/3}\left(\tfrac{1}{4} + \cos\theta + \cos^2\theta\right) d\theta - \int_{2\pi/3}^{\pi}\left(\tfrac{1}{4} + \cos\theta + \cos^2\theta\right) d\theta$$

$$= \int_0^{2\pi/3}\left[\tfrac{1}{4} + \cos\theta + \tfrac{1}{2}(1 + \cos 2\theta)\right] d\theta - \int_{2\pi/3}^{\pi}\left[\tfrac{1}{4} + \cos\theta + \tfrac{1}{2}(1 + \cos 2\theta)\right] d\theta$$

$$= \left[\frac{\theta}{4} + \sin\theta + \frac{\theta}{2} + \frac{\sin 2\theta}{4}\right]_0^{2\pi/3} - \left[\frac{\theta}{4} + \sin\theta + \frac{\theta}{2} + \frac{\sin 2\theta}{4}\right]_{2\pi/3}^{\pi}$$

$$= \left(\frac{\pi}{6} + \frac{\sqrt{3}}{2} + \frac{\pi}{3} - \frac{\sqrt{3}}{8}\right) - \left(\frac{\pi}{4} + \frac{\pi}{2}\right) + \left(\frac{\pi}{6} + \frac{\sqrt{3}}{2} + \frac{\pi}{3} - \frac{\sqrt{3}}{8}\right) = \frac{\pi}{4} + \frac{3}{4}\sqrt{3} = \tfrac{1}{4}\left(\pi + 3\sqrt{3}\right)$$

29. The pole is a point of intersection.

$$1 + \sin\theta = 3\sin\theta \quad \Rightarrow \quad 1 = 2\sin\theta \quad \Rightarrow \quad \sin\theta = \tfrac{1}{2} \quad \Rightarrow$$

$\theta = \frac{\pi}{6}$ or $\frac{5\pi}{6}$.

The other two points of intersection are $\left(\frac{3}{2}, \frac{\pi}{6}\right)$ and $\left(\frac{3}{2}, \frac{5\pi}{6}\right)$.

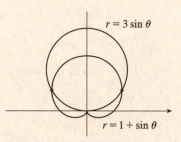

31. The pole is a point of intersection. $\sin\theta = \sin 2\theta = 2\sin\theta\cos\theta \quad \Leftrightarrow$

$\sin\theta\,(1 - 2\cos\theta) = 0 \quad \Leftrightarrow \quad \sin\theta = 0$ or $\cos\theta = \tfrac{1}{2} \quad \Rightarrow$

$\theta = 0, \pi, \frac{\pi}{3},$ or $-\frac{\pi}{3} \quad \Rightarrow \quad$ the other intersection points are $\left(\frac{\sqrt{3}}{2}, \frac{\pi}{3}\right)$

and $\left(\frac{\sqrt{3}}{2}, \frac{2\pi}{3}\right)$ [by symmetry].

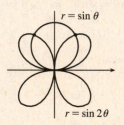

33. $L = \int_a^b \sqrt{r^2 + (dr/d\theta)^2}\, d\theta = \int_0^{\pi/3} \sqrt{(3\sin\theta)^2 + (3\cos\theta)^2}\, d\theta = \int_0^{\pi/3} \sqrt{9(\sin^2\theta + \cos^2\theta)}\, d\theta$

$\quad = 3\int_0^{\pi/3} d\theta = 3[\theta]_0^{\pi/3} = 3\left(\frac{\pi}{3}\right) = \pi.$

As a check, note that the circumference of a circle with radius $\frac{3}{2}$ is $2\pi\left(\frac{3}{2}\right) = 3\pi$, and since $\theta = 0$ to $\pi = \frac{\pi}{3}$ traces out $\frac{1}{3}$ of the

circle (from $\theta = 0$ to $\theta = \pi$), $\frac{1}{3}(3\pi) = \pi.$

35. $L = \int_a^b \sqrt{r^2 + (dr/d\theta)^2}\, d\theta = \int_0^{2\pi} \sqrt{(\theta^2)^2 + (2\theta)^2}\, d\theta = \int_0^{2\pi} \sqrt{\theta^4 + 4\theta^2}\, d\theta$

$\quad = \int_0^{2\pi} \sqrt{\theta^2(\theta^2 + 4)}\, d\theta = \int_0^{2\pi} \theta\sqrt{\theta^2 + 4}\, d\theta$

Now let $u = \theta^2 + 4$, so that $du = 2\theta\, d\theta \quad \left[\theta\, d\theta = \tfrac{1}{2}\, du\right]$ and

$$\int_0^{2\pi} \theta\sqrt{\theta^2 + 4}\, d\theta = \int_4^{4\pi^2 + 4} \tfrac{1}{2}\sqrt{u}\, du = \tfrac{1}{2} \cdot \tfrac{2}{3}\left[u^{3/2}\right]_4^{4(\pi^2 + 1)} = \tfrac{1}{3}\left[4^{3/2}(\pi^2 + 1)^{3/2} - 4^{3/2}\right] = \tfrac{8}{3}\left[(\pi^2 + 1)^{3/2} - 1\right]$$

37. The curve $r = \sin(6\sin\theta)$ is completely traced with $0 \leq \theta \leq \pi$. $\quad r = \sin(6\sin\theta) \quad \Rightarrow \quad \dfrac{dr}{d\theta} = \cos(6\sin\theta) \cdot 6\cos\theta$, so

$$r^2 + \left(\frac{dr}{d\theta}\right)^2 = \sin^2(6\sin\theta) + 36\cos^2\theta\,\cos^2(6\sin\theta) \quad \Rightarrow \quad L \int_0^{\pi} \sqrt{\sin^2(6\sin\theta) + 36\cos^2\theta\,\cos^2(6\sin\theta)}\, d\theta \approx 8.0091.$$

9.5   Conic Sections in Polar Coordinates

1. The directrix $x = 4$ is to the right of the focus at the origin, so we use the form with "$+ e\cos\theta$" in the denominator.

(See Theorem 8 and Figure 8.) An equation is $r = \dfrac{ed}{1 + e\cos\theta} = \dfrac{\frac{1}{2}\cdot 4}{1 + \frac{1}{2}\cos\theta} = \dfrac{4}{2 + \cos\theta}$.

3. The directrix $y = 2$ is above the focus at the origin, so we use the form with "$+ e\sin\theta$" in the denominator. An equation is

$$r = \frac{ed}{1 + e\sin\theta} = \frac{1.5(2)}{1 + 1.5\sin\theta} = \frac{6}{2 + 3\sin\theta}.$$

5. The vertex $(4, 3\pi/2)$ is 4 units below the focus at the origin, so the directrix is 8 units below the focus ($d = 8$), and we use the

form with "$-e\sin\theta$" in the denominator.

$e = 1$ for a parabola, so an equation is $r = \dfrac{ed}{1 - e\sin\theta} = \dfrac{1(8)}{1 - 1\sin\theta} = \dfrac{8}{1 - \sin\theta}$.

7. The directrix $r = 4\sec\theta$ (equivalent to $r\cos\theta = 4$ or $x = 4$) is to the right of the focus at the origin, so we will use the form

with "$+ e\cos\theta$" in the denominator. The distance from the focus to the directrix is $d = 4$, so an equation is

$$r = \frac{ed}{1 + e\cos\theta} = \frac{\frac{1}{2}(4)}{1 + \frac{1}{2}\cos\theta} \cdot \frac{2}{2} = \frac{4}{2 + \cos\theta}.$$

9. $r = \dfrac{4}{5 - 4\sin\theta} \cdot \dfrac{1/5}{1/5} = \dfrac{4/5}{1 - \frac{4}{5}\sin\theta}$, where $e = \frac{4}{5}$ and $ed = \frac{4}{5} \;\Rightarrow\; d = 1$.

(a) Eccentricity $= e = \frac{4}{5}$

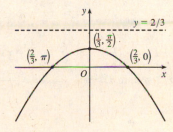

(b) Since $e = \frac{4}{5} < 1$, the conic is an ellipse.

(c) Since "$- e\sin\theta$" appears in the denominator, the directrix is below the focus

   at the origin, $d = |Fl| = 1$, so an equation of the directrix is $y = -1$.

(d) The vertices are $\left(4, \frac{\pi}{2}\right)$ and $\left(\frac{4}{9}, \frac{3\pi}{2}\right)$.

11. $r = \dfrac{2}{3 + 3\sin\theta} \cdot \dfrac{1/3}{1/3} = \dfrac{2/3}{1 + 1\sin\theta}$, where $e = 1$ and $ed = \frac{2}{3} \;\Rightarrow\; d = \frac{2}{3}$.

(a) Eccentricity $= e = 1$

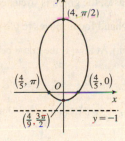

(b) Since $e = 1$, the conic is a parabola.

(c) Since "$+ e\sin\theta$" appears in the denominator, the directrix is above the focus

   at the origin. $d = |Fl| = \frac{2}{3}$, so an equation of the directrix is $y = \frac{2}{3}$.

(d) The vertex is at $\left(\frac{1}{3}, \frac{\pi}{2}\right)$, midway between the focus and directrix.

13. $r = \dfrac{9}{6 + 2\cos\theta} \cdot \dfrac{1/6}{1/6} = \dfrac{3/2}{1 + \frac{1}{3}\cos\theta}$, where $e = \frac{1}{3}$ and $ed = \frac{3}{2} \;\Rightarrow\; d = \frac{9}{2}$.

(a) Eccentricity $= e = \frac{1}{3}$

(b) Since $e = \frac{1}{3} < 1$, the conic is an ellipse.

(c) Since "$+e \cos \theta$" appears in the denominator, the directrix is to the right of

the focus at the origin. $d = |Fl| = \frac{9}{2}$, so an equation of the directrix is

$x = \frac{9}{2}$.

(d) The vertices are $\left(\frac{9}{8}, 0\right)$ and $\left(\frac{9}{4}, \pi\right)$, so the center is midway between them,

that is, $\left(\frac{9}{16}, \pi\right)$.

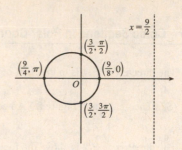

15. $r = \dfrac{3}{4 - 8\cos\theta} \cdot \dfrac{1/4}{1/4} = \dfrac{3/4}{1 - 2\cos\theta}$, where $e = 2$ and $ed = \frac{3}{4} \Rightarrow d = \frac{3}{8}$.

(a) Eccentricity $= e = 2$

(b) Since $e = 2 > 1$, the conic is a hyperbola.

(c) Since "$-e \cos \theta$" appears in the denominator, the directrix is to the left of

the focus at the origin. $d = |Fl| = \frac{3}{8}$, so an equation of the directrix is

$x = -\frac{3}{8}$.

(d) The vertices are $\left(-\frac{3}{4}, 0\right)$ and $\left(\frac{1}{4}, \pi\right)$, so the center is midway between them,

that is, $\left(\frac{1}{2}, \pi\right)$.

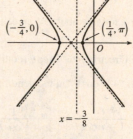

17. For $e < 1$ the curve is an ellipse. It is nearly circular when e is close to 0. As e

increases, the graph is stretched out to the right, and grows larger (that is, its

right-hand focus moves to the right while its left-hand focus remains at the

origin.) At $e = 1$, the curve becomes a parabola with focus at the origin.

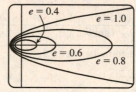

19. $|PF| = e\,|Pl| \Rightarrow r = e[d - r\cos(\pi - \theta)] = e(d + r\cos\theta) \Rightarrow$

$r(1 - e\cos\theta) = ed \Rightarrow r = \dfrac{ed}{1 - e\cos\theta}$

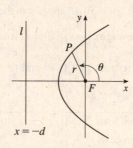

21. $|PF| = e\,|Pl| \Rightarrow r = e[d - r\sin(\theta - \pi)] = e(d + r\sin\theta) \Rightarrow$

$r(1 - e\sin\theta) = ed \Rightarrow r = \dfrac{ed}{1 - e\sin\theta}$

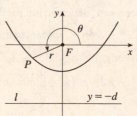

23. (a) If the directrix is $x = -d$, then $r = \dfrac{ed}{1 - e\cos\theta}$ [see Figure 8(b)], and, from (6), $a^2 = \dfrac{e^2 d^2}{(1 - e^2)^2} \Rightarrow$

$ed = a(1 - e^2)$. Therefore, $r = \dfrac{a(1 - e^2)}{1 - e\cos\theta}$.

(b) $e = 0.017$ and the major axis $= 2a = 2.99 \times 10^8 \Rightarrow a = 1.495 \times 10^8$.

Therefore $r = \dfrac{1.495 \times 10^8 \left[1 - (0.017)^2\right]}{1 - 0.017 \cos \theta} \approx \dfrac{1.49 \times 10^8}{1 - 0.017 \cos \theta}$.

25. Here $2a = $ length of major axis $= 36.18$ AU $\Rightarrow a = 18.09$ AU and $e = 0.97$. By Exercise 23(a), the equation of the orbit

is $r = \dfrac{18.09[1 - (0.97)^2]}{1 + 0.97 \cos \theta} \approx \dfrac{1.07}{1 + 0.97 \cos \theta}$. By Exercise 24(a), the maximum distance from the comet to the sun is

$18.09(1 + 0.97) \approx 35.64$ AU or about 3.314 billion miles.

27. The minimum distance is at perihelion, where $4.6 \times 10^7 = r = a(1 - e) = a(1 - 0.206) = a(0.794) \Rightarrow$

$a = 4.6 \times 10^7/0.794$. So the maximum distance, which is at aphelion, is

$r = a(1 + e) = (4.6 \times 10^7/0.794)(1.206) \approx 7.0 \times 10^7$ km.

29. From Exercise 27, we have $e = 0.206$ and $a(1 - e) = 4.6 \times 10^7$ km. Thus, $a = 4.6 \times 10^7/0.794$. From Exercise 23, we can

write the equation of Mercury's orbit as $r = a\dfrac{1 - e^2}{1 - e \cos \theta}$. So since $\dfrac{dr}{d\theta} = \dfrac{-a(1 - e^2)e \sin \theta}{(1 - e \cos \theta)^2} \Rightarrow$

$r^2 + \left(\dfrac{dr}{d\theta}\right)^2 = \dfrac{a^2(1 - e^2)^2}{(1 - e \cos \theta)^2} + \dfrac{a^2(1 - e^2)^2 e^2 \sin^2 \theta}{(1 - e \cos \theta)^4} = \dfrac{a^2(1 - e^2)^2}{(1 - e \cos \theta)^4}(1 - 2e \cos \theta + e^2)$

the length of the orbit is $L = \displaystyle\int_0^{2\pi} \sqrt{r^2 + (dr/d\theta)^2}\, d\theta = a(1 - e^2)\int_0^{2\pi} \dfrac{\sqrt{1 + e^2 - 2e \cos \theta}}{(1 - e \cos \theta)^2}\, d\theta \approx 3.6 \times 10^8$ km.

This seems reasonable, since Mercury's orbit is nearly circular, and the circumference of a circle of radius a

is $2\pi a \approx 3.6 \times 10^8$ km.

9 Review

CONCEPT CHECK

1. (a) A parametric curve is a set of points of the form $(x, y) = (f(t), g(t))$, where f and g are continuous functions of a variable t.

(b) Sketching a parametric curve, like sketching the graph of a function, is difficult to do in general. We can plot points on the curve by finding $f(t)$ and $g(t)$ for various values of t, either by hand or with a calculator or computer. Sometimes, when f and g are given by formulas, we can eliminate t from the equations $x = f(t)$ and $y = g(t)$ to get a Cartesian equation relating x and y. It may be easier to graph that equation than to work with the original formulas for x and y in terms of t.

2. (a) You can find $\dfrac{dy}{dx}$ as a function of t by calculating $\dfrac{dy}{dx} = \dfrac{dy/dt}{dx/dt}$ [if $dx/dt \neq 0$].

(b) Calculate the area as $\int_a^b y\, dx = \int_\alpha^\beta g(t) f'(t) dt$ [or $\int_\beta^\alpha g(t) f'(t) dt$ if the leftmost point is $(f(\beta), g(\beta))$ rather than $(f(\alpha), g(\alpha))$].

3. $L = \int_\alpha^\beta \sqrt{(dx/dt)^2 + (dy/dt)^2}\, dt = \int_\alpha^\beta \sqrt{[f'(t)]^2 + [g'(t)]^2}\, dt$

4. (a) See Figure 5 in Section 9.3.

(b) $x = r\cos\theta$, $y = r\sin\theta$

(c) To find a polar representation (r, θ) with $r \geq 0$ and $0 \leq \theta < 2\pi$, first calculate $r = \sqrt{x^2 + y^2}$. Then θ is specified by $\cos\theta = x/r$ and $\sin\theta = y/r$.

5. (a) Calculate $\dfrac{dy}{dx} = \dfrac{\dfrac{dy}{d\theta}}{\dfrac{dx}{d\theta}} = \dfrac{\dfrac{d}{d\theta}(y)}{\dfrac{d}{d\theta}(x)} = \dfrac{\dfrac{d}{d\theta}(r\sin\theta)}{\dfrac{d}{d\theta}(r\cos\theta)} = \dfrac{\left(\dfrac{dr}{d\theta}\right)\sin\theta + r\cos\theta}{\left(\dfrac{dr}{d\theta}\right)\cos\theta - r\sin\theta}$, where $r = f(\theta)$.

(b) Calculate $A = \int_a^b \frac{1}{2} r^2 \, d\theta = \int_a^b \frac{1}{2}[f(\theta)]^2 \, d\theta$

(c) $L = \int_a^b \sqrt{(dx/d\theta)^2 + (dy/d\theta)^2} \, d\theta = \int_a^b \sqrt{r^2 + (dr/d\theta)^2} \, d\theta = \int_a^b \sqrt{[f(\theta)]^2 + [f'(\theta)]^2} \, d\theta$

6. (a) If a conic section has focus F and corresponding directrix l, then the eccentricity e is the fixed ratio $|PF| / |Pl|$ for points P of the conic section.

(b) $e < 1$ for an ellipse; $e > 1$ for a hyperbola; $e = 1$ for a parabola.

(c) $x = d$: $r = \dfrac{ed}{1 + e\cos\theta}$. $x = -d$: $r = \dfrac{ed}{1 - e\cos\theta}$. $y = d$: $r = \dfrac{ed}{1 + e\sin\theta}$. $y = -d$: $r = \dfrac{ed}{1 - e\sin\theta}$.

TRUE-FALSE QUIZ

1. False. Consider the curve defined by $x = f(t) = (t - 1)^3$ and $y = g(t) = (t - 1)^2$. Then $g'(t) = 2(t - 1)$, so $g'(1) = 0$, but its graph has a *vertical* tangent when $t = 1$. *Note:* The statement is true if $f'(1) \neq 0$ when $g'(1) = 0$.

3. False. For example, if $f(t) = \cos t$ and $g(t) = \sin t$ for $0 \leq t \leq 4\pi$, then the curve is a circle of radius 1, hence its length is 2π, but $\int_0^{4\pi} \sqrt{[f'(t)]^2 + [g'(t)]^2} \, dt = \int_0^{4\pi} \sqrt{(-\sin t)^2 + (\cos t)^2} \, dt = \int_0^{4\pi} 1 \, dt = 4\pi$, since as t increases from 0 to 4π, the circle is traversed twice.

5. True. The curve $r = 1 - \sin 2\theta$ is unchanged if we rotate it through $180°$ about O because $1 - \sin 2(\theta + \pi) = 1 - \sin(2\theta + 2\pi) = 1 - \sin 2\theta$. So it's unchanged if we replace r by $-r$. In other words, it's the same curve as $r = -(1 - \sin 2\theta) = \sin 2\theta - 1$.

7. False. The first pair of equations gives the portion of the parabola $y = x^2$ with $x \geq 0$, whereas the second pair of equations traces out the whole parabola $y = x^2$.

EXERCISES

1. $x = t^2 + 4t$, $y = 2 - t$, $-4 \leq t \leq 1$. $t = 2 - y$, so

$x = (2 - y)^2 + 4(2 - y) = 4 - 4y + y^2 + 8 - 4y = y^2 - 8y + 12$ $\Leftrightarrow$

$x + 4 = y^2 - 8y + 16 = (y - 4)^2$. This is part of a parabola with vertex

$(-4, 4)$, opening to the right.

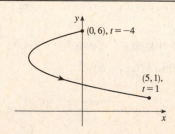

3. $y = \sec\theta = \dfrac{1}{\cos\theta} = \dfrac{1}{x}$.   Since $0 \le \theta \le \pi/2, 0 < x \le 1$ and $y \ge 1$.

This is part of the hyperbola $y = 1/x$.

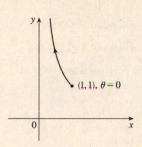

5. Three different sets of parametric equations for the curve $y = \sqrt{x}$ are

(i) $x = t,\ \ y = \sqrt{t}$

(ii) $x = t^4,\ \ y = t^2$

(iii) $x = \tan^2 t,\ \ y = \tan t,\ \ 0 \le t < \pi/2$

There are many other sets of equations that also give this curve.

7. (a)
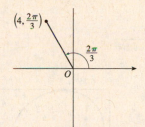
The Cartesian coordinates are $x = 4\cos\dfrac{2\pi}{3} = 4\left(-\dfrac{1}{2}\right) = -2$ and

$y = 4\sin\dfrac{2\pi}{3} = 4\left(\dfrac{\sqrt{3}}{2}\right) = 2\sqrt{3}$, that is, the point $\left(-2, 2\sqrt{3}\right)$.

(b) Given $x = -3$ and $y = 3$, we have $r = \sqrt{(-3)^2 + 3^2} = \sqrt{18} = 3\sqrt{2}$. Also, $\tan\theta = \dfrac{y}{x} \ \Rightarrow\ \tan\theta = \dfrac{3}{-3}$, and since

$(-3, 3)$ is in the second quadrant, $\theta = \dfrac{3\pi}{4}$. Thus, one set of polar coordinates for $(-3, 3)$ is $\left(3\sqrt{2}, \dfrac{3\pi}{4}\right)$, and two others are

$\left(3\sqrt{2}, \dfrac{11\pi}{4}\right)$ and $\left(-3\sqrt{2}, \dfrac{7\pi}{4}\right)$.

9. $r = 1 - \cos\theta$. This cardioid is
symmetric about the polar axis.

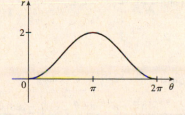
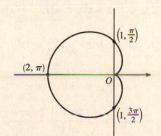

11. $r = \cos 3\theta$. This is a
three-leaved rose. The curve is
traced twice.

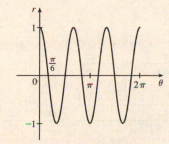
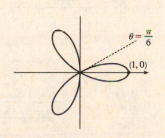

13. $r = 1 + \cos 2\theta$. The curve is symmetric about the pole and both the horizontal and vertical axes.

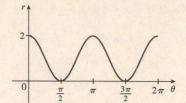

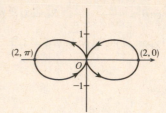

15. $r = \dfrac{3}{1 + 2\sin\theta}$ $\Rightarrow$ $e = 2 > 1$, so the conic is a hyperbola. $de = 3$ $\Rightarrow$ $d = \frac{3}{2}$ and the form "$+2\sin\theta$" imply that the directrix is above the focus at the origin and has equation $y = \frac{3}{2}$. The vertices are $\left(1, \frac{\pi}{2}\right)$ and $\left(-3, \frac{3\pi}{2}\right)$.

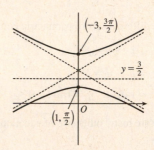

17. $x + y = 2$ $\Leftrightarrow$ $r\cos\theta + r\sin\theta = 2$ $\Leftrightarrow$ $r(\cos\theta + \sin\theta) = 2$ $\Leftrightarrow$ $r = \dfrac{2}{\cos\theta + \sin\theta}$

19. $r = (\sin\theta)/\theta$. As $\theta \to \pm\infty$, $r \to 0$. As $\theta \to 0$, $r \to 1$. In the first figure, there are an infinite number of x-intercepts at $x = \pi n$, n a nonzero integer. These correspond to pole points in the second figure.

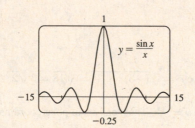

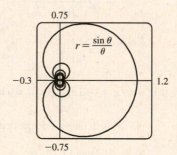

21. $x = \ln t$, $y = 1 + t^2$; $t = 1$. $\dfrac{dy}{dt} = 2t$ and $\dfrac{dx}{dt} = \dfrac{1}{t}$, so $\dfrac{dy}{dx} = \dfrac{dy/dt}{dx/dt} = \dfrac{2t}{1/t} = 2t^2$.

When $t = 1$, $(x, y) = (0, 2)$ and $dy/dx = 2$.

23. $r = e^{-\theta}$ $\Rightarrow$ $y = r\sin\theta = e^{-\theta}\sin\theta$ and $x = r\cos\theta = e^{-\theta}\cos\theta$ $\Rightarrow$

$$\frac{dy}{dx} = \frac{dy/d\theta}{dx/d\theta} = \frac{\frac{dr}{d\theta}\sin\theta + r\cos\theta}{\frac{dr}{d\theta}\cos\theta - r\sin\theta} = \frac{-e^{-\theta}\sin\theta + e^{-\theta}\cos\theta}{-e^{-\theta}\cos\theta - e^{-\theta}\sin\theta} \cdot \frac{-e^{\theta}}{-e^{\theta}} = \frac{\sin\theta - \cos\theta}{\cos\theta + \sin\theta}.$$

When $\theta = \pi$, $\dfrac{dy}{dx} = \dfrac{0 - (-1)}{-1 + 0} = \dfrac{1}{-1} = -1$.

25. $x = t + \sin t$, $y = t - \cos t$ $\Rightarrow$ $\dfrac{dy}{dx} = \dfrac{dy/dt}{dx/dt} = \dfrac{1 + \sin t}{1 + \cos t}$ $\Rightarrow$

$$\frac{d^2y}{dx^2} = \frac{\dfrac{d}{dt}\left(\dfrac{dy}{dx}\right)}{dx/dt} = \frac{\dfrac{(1 + \cos t)\cos t - (1 + \sin t)(-\sin t)}{(1 + \cos t)^2}}{1 + \cos t} = \frac{\cos t + \cos^2 t + \sin t + \sin^2 t}{(1 + \cos t)^3} = \frac{1 + \cos t + \sin t}{(1 + \cos t)^3}$$

27. We graph the curve $x = t^3 - 3t$, $y = t^2 + t + 1$ for $-2.2 \le t \le 1.2$.

By zooming in or using a cursor, we find that the lowest point is about

$(1.4, 0.75)$. To find the exact values, we find the t-value at which

$$dy/dt = 2t + 1 = 0 \iff t = -\tfrac{1}{2} \iff (x, y) = \left(\tfrac{11}{8}, \tfrac{3}{4}\right).$$

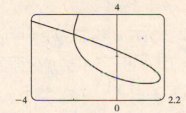

29. $x = 2a \cos t - a \cos 2t \implies \dfrac{dx}{dt} = -2a \sin t + 2a \sin 2t = 2a \sin t (2 \cos t - 1) = 0 \iff$

$\sin t = 0$ or $\cos t = \tfrac{1}{2} \implies t = 0, \tfrac{\pi}{3}, \pi$, or $\tfrac{5\pi}{3}$.

$y = 2a \sin t - a \sin 2t \implies \dfrac{dy}{dt} = 2a \cos t - 2a \cos 2t = 2a(1 + \cos t - 2\cos^2 t) = 2a(1 - \cos t)(1 + 2\cos t) = 0 \implies$

$t = 0, \tfrac{2\pi}{3}$, or $\tfrac{4\pi}{3}$.

Thus the graph has vertical tangents where $t = \tfrac{\pi}{3}, \pi$, and $\tfrac{5\pi}{3}$, and horizontal tangents where $t = \tfrac{2\pi}{3}$ and $\tfrac{4\pi}{3}$. To determine

what the slope is where $t = 0$, we use l'Hospital's Rule to evaluate $\displaystyle\lim_{t \to 0} \dfrac{dy/dt}{dx/dt} = 0$, so there is a horizontal tangent there.

t	x	y
0	a	0
$\tfrac{\pi}{3}$	$\tfrac{3}{2}a$	$\tfrac{\sqrt{3}}{2}a$
$\tfrac{2\pi}{3}$	$-\tfrac{1}{2}a$	$\tfrac{3\sqrt{3}}{2}a$
π	$-3a$	0
$\tfrac{4\pi}{3}$	$-\tfrac{1}{2}a$	$-\tfrac{3\sqrt{3}}{2}a$
$\tfrac{5\pi}{3}$	$\tfrac{3}{2}a$	$-\tfrac{\sqrt{3}}{2}a$

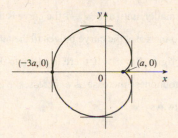

31. The curve $r^2 = 9 \cos 5\theta$ has 10 "petals." For instance, for $-\tfrac{\pi}{10} \le \theta \le \tfrac{\pi}{10}$, there are two petals, one with $r > 0$ and one

with $r < 0$.

$$A = 10 \int_{-\pi/10}^{\pi/10} \tfrac{1}{2} r^2 \, d\theta = 5 \int_{-\pi/10}^{\pi/10} 9 \cos 5\theta \, d\theta = 5 \cdot 9 \cdot 2 \int_{0}^{\pi/10} \cos 5\theta \, d\theta = 18 \big[\sin 5\theta \big]_{0}^{\pi/10} = 18$$

33. The curves intersect when $4 \cos \theta = 2 \implies \cos \theta = \tfrac{1}{2} \implies \theta = \pm \tfrac{\pi}{3}$

for $-\pi \le \theta \le \pi$. The points of intersection are $\left(2, \tfrac{\pi}{3}\right)$ and $\left(2, -\tfrac{\pi}{3}\right)$.

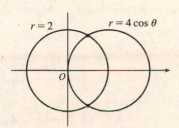

35. The curves intersect where $2 \sin \theta = \sin \theta + \cos \theta \implies$

$\sin \theta = \cos \theta \implies \theta = \tfrac{\pi}{4}$, and also at the origin (at which $\theta = \tfrac{3\pi}{4}$

on the second curve).

$$A = \int_{0}^{\pi/4} \tfrac{1}{2}(2 \sin \theta)^2 \, d\theta + \int_{\pi/4}^{3\pi/4} \tfrac{1}{2}(\sin \theta + \cos \theta)^2 \, d\theta$$

$$= \int_{0}^{\pi/4} (1 - \cos 2\theta) \, d\theta + \tfrac{1}{2} \int_{\pi/4}^{3\pi/4} (1 + \sin 2\theta) \, d\theta$$

$$= \big[\theta - \tfrac{1}{2} \sin 2\theta \big]_{0}^{\pi/4} + \big[\tfrac{1}{2}\theta - \tfrac{1}{4} \cos 2\theta \big]_{\pi/4}^{3\pi/4} = \tfrac{1}{2}(\pi - 1)$$

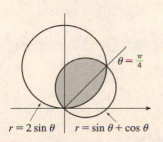

37. $x = 3t^2$, $y = 2t^3$.

$$L = \int_0^2 \sqrt{(dx/dt)^2 + (dy/dt)^2}\, dt = \int_0^2 \sqrt{(6t)^2 + (6t^2)^2}\, dt = \int_0^2 \sqrt{36t^2 + 36t^4}\, dt = \int_0^2 \sqrt{36t^2}\,\sqrt{1 + t^2}\, dt$$

$$= \int_0^2 6\,|t|\,\sqrt{1 + t^2}\, dt = 6\int_0^2 t\,\sqrt{1 + t^2}\, dt = 6\int_1^5 u^{1/2}\left(\tfrac{1}{2} du\right) \qquad \left[u = 1 + t^2,\, du = 2t\, dt\right]$$

$$= 6 \cdot \tfrac{1}{2} \cdot \tfrac{2}{3}\left[u^{3/2}\right]_1^5 = 2(5^{3/2} - 1) = 2\left(5\sqrt{5} - 1\right)$$

39. $L = \int_\pi^{2\pi} \sqrt{r^2 + (dr/d\theta)^2}\, d\theta = \int_\pi^{2\pi} \sqrt{(1/\theta)^2 + (-1/\theta^2)^2}\, d\theta = \int_\pi^{2\pi} \dfrac{\sqrt{\theta^2 + 1}}{\theta^2}\, d\theta$

$$\overset{24}{=} \left[-\dfrac{\sqrt{\theta^2 + 1}}{\theta} + \ln\left(\theta + \sqrt{\theta^2 + 1}\right)\right]_\pi^{2\pi} = \dfrac{\sqrt{\pi^2 + 1}}{\pi} - \dfrac{\sqrt{4\pi^2 + 1}}{2\pi} + \ln\left(\dfrac{2\pi + \sqrt{4\pi^2 + 1}}{\pi + \sqrt{\pi^2 + 1}}\right)$$

$$= \dfrac{2\sqrt{\pi^2 + 1} - \sqrt{4\pi^2 + 1}}{2\pi} + \ln\left(\dfrac{2\pi + \sqrt{4\pi^2 + 1}}{\pi + \sqrt{\pi^2 + 1}}\right)$$

41. For all c except -1, the curve is asymptotic to the line $x = 1$. For $c < -1$, the curve bulges to the right near $y = 0$. As c increases, the bulge becomes smaller, until at $c = -1$ the curve is the straight line $x = 1$. As c continues to increase, the curve bulges to the left, until at $c = 0$ there is a cusp at the origin. For $c > 0$, there is a loop to the left of the origin, whose size and roundness increase as c increases. Note that the x-intercept of the curve is always $-c$.

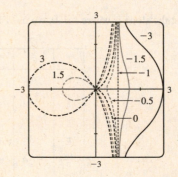

43. Directrix $x = 4 \Rightarrow d = 4$, so $e = \frac{1}{3} \Rightarrow r = \dfrac{ed}{1 + e\cos\theta} = \dfrac{4}{3 + \cos\theta}$.

45. In polar coordinates, an equation for the circle is $r = 2a\sin\theta$. Thus, the coordinates of Q are $x = r\cos\theta = 2a\sin\theta\,\cos\theta$ and $y = r\sin\theta = 2a\sin^2\theta$. The coordinates of R are $x = 2a\cot\theta$ and $y = 2a$. Since P is the midpoint of QR, we use the midpoint formula to get $x = a(\sin\theta\,\cos\theta + \cot\theta)$ and $y = a(1 + \sin^2\theta)$.

10 ☐ VECTORS AND THE GEOMETRY OF SPACE

10.1 Three-Dimensional Coordinate Systems

1. We start at the origin, which has coordinates $(0, 0, 0)$. First we move 4 units along the positive x-axis, affecting only the x-coordinate, bringing us to the point $(4, 0, 0)$. We then move 3 units straight downward, in the negative z-direction. Thus only the z-coordinate is affected, and we arrive at $(4, 0, -3)$.

3. The distance from a point to the yz-plane is the absolute value of the x-coordinate of the point. $C(2, 4, 6)$ has the x-coordinate with the smallest absolute value, so C is the point closest to the yz-plane. $A(-4, 0, -1)$ must lie in the xz-plane since the distance from A to the xz-plane, given by the y-coordinate of A, is 0.

5. The equation $x + y = 2$ represents the set of all points in $\mathbb{R}^3$ whose x- and y-coordinates have a sum of 2, or equivalently where $y = 2 - x$. This is the set $\{(x, 2 - x, z) \mid x \in \mathbb{R}, z \in \mathbb{R}\}$ which is a vertical plane that intersects the xy-plane in the line $y = 2 - x$, $z = 0$.

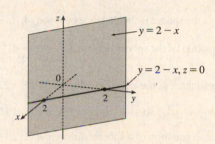

7. (a) We can find the lengths of the sides of the triangle by using the distance formula between pairs of vertices:

$$|PQ| = \sqrt{(7 - 3)^2 + [0 - (-2)]^2 + [1 - (-3)]^2} = \sqrt{16 + 4 + 16} = 6$$

$$|QR| = \sqrt{(1 - 7)^2 + (2 - 0)^2 + (1 - 1)^2} = \sqrt{36 + 4 + 0} = \sqrt{40} = 2\sqrt{10}$$

$$|RP| = \sqrt{(3 - 1)^2 + (-2 - 2)^2 + (-3 - 1)^2} = \sqrt{4 + 16 + 16} = 6$$

The longest side is QR, but the Pythagorean Theorem is not satisfied: $|PQ|^2 + |RP|^2 \neq |QR|^2$. Thus PQR is not a right triangle. PQR is isosceles, as two sides have the same length.

(b) Compute the lengths of the sides of the triangle by using the distance formula between pairs of vertices:

$$|PQ| = \sqrt{(4 - 2)^2 + [1 - (-1)]^2 + (1 - 0)^2} = \sqrt{4 + 4 + 1} = 3$$

$$|QR| = \sqrt{(4 - 4)^2 + (-5 - 1)^2 + (4 - 1)^2} = \sqrt{0 + 36 + 9} = \sqrt{45} = 3\sqrt{5}$$

$$|RP| = \sqrt{(2 - 4)^2 + [-1 - (-5)]^2 + (0 - 4)^2} = \sqrt{4 + 16 + 16} = 6$$

Since the Pythagorean Theorem is satisfied by $|PQ|^2 + |RP|^2 = |QR|^2$, PQR is a right triangle. PQR is not isosceles, as no two sides have the same length.

9. (a) First we find the distances between points:

$$|AB| = \sqrt{(3-2)^2 + (7-4)^2 + (-2-2)^2} = \sqrt{26}$$

$$|BC| = \sqrt{(1-3)^2 + (3-7)^2 + [3-(-2)]^2} = \sqrt{45} = 3\sqrt{5}$$

$$|AC| = \sqrt{(1-2)^2 + (3-4)^2 + (3-2)^2} = \sqrt{3}$$

In order for the points to lie on a straight line, the sum of the two shortest distances must be equal to the longest distance.
Since $\sqrt{26} + \sqrt{3} \neq 3\sqrt{5}$, the three points do not lie on a straight line.

(b) First we find the distances between points:

$$|DE| = \sqrt{(1-0)^2 + [-2-(-5)]^2 + (4-5)^2} = \sqrt{11}$$

$$|EF| = \sqrt{(3-1)^2 + [4-(-2)]^2 + (2-4)^2} = \sqrt{44} = 2\sqrt{11}$$

$$|DF| = \sqrt{(3-0)^2 + [4-(-5)]^2 + (2-5)^2} = \sqrt{99} = 3\sqrt{11}$$

Since $|DE| + |EF| = |DF|$, the three points lie on a straight line.

11. The radius of the sphere is the distance between $(4, 3, -1)$ and $(3, 8, 1)$: $r = \sqrt{(3-4)^2 + (8-3)^2 + [1-(-1)]^2} = \sqrt{30}$.
Thus, an equation of the sphere is $(x-3)^2 + (y-8)^2 + (z-1)^2 = 30$.

13. Completing squares in the equation $x^2 + y^2 + z^2 - 2x - 4y + 8z = 15$ gives

$(x^2 - 2x + 1) + (y^2 - 4y + 4) + (z^2 + 8z + 16) = 15 + 1 + 4 + 16 \quad \Rightarrow \quad (x-1)^2 + (y-2)^2 + (z+4)^2 = 36$, which we

recognize as an equation of a sphere with center $(1, 2, -4)$ and radius 6.

15. Completing squares in the equation $2x^2 - 8x + 2y^2 + 2z^2 + 24z = 1$ gives

$2(x^2 - 4x + 4) + 2y^2 + 2(z^2 + 12z + 36) = 1 + 8 + 72 \quad \Rightarrow \quad 2(x-2)^2 + 2y^2 + 2(z+6)^2 = 81 \quad \Rightarrow$

$(x-2)^2 + y^2 + (z+6)^2 = \frac{81}{2}$, which we recognize as an equation of a sphere with center $(2, 0, -6)$ and

radius $\sqrt{\frac{81}{2}} = 9/\sqrt{2}$.

17. (a) If the midpoint of the line segment from $P_1(x_1, y_1, z_1)$ to $P_2(x_2, y_2, z_2)$ is $Q = \left(\dfrac{x_1 + x_2}{2}, \dfrac{y_1 + y_2}{2}, \dfrac{z_1 + z_2}{2} \right)$,

then the distances $|P_1 Q|$ and $|Q P_2|$ are equal, and each is half of $|P_1 P_2|$. We verify that this is the case:

$$|P_1 P_2| = \sqrt{(x_2 - x_1)^2 + (y_2 - y_1)^2 + (z_2 - z_1)^2}$$

$$|P_1 Q| = \sqrt{\left[\tfrac{1}{2}(x_1 + x_2) - x_1 \right]^2 + \left[\tfrac{1}{2}(y_1 + y_2) - y_1 \right]^2 + \left[\tfrac{1}{2}(z_1 + z_2) - z_1 \right]^2}$$

$$= \sqrt{\left(\tfrac{1}{2}x_2 - \tfrac{1}{2}x_1 \right)^2 + \left(\tfrac{1}{2}y_2 - \tfrac{1}{2}y_1 \right)^2 + \left(\tfrac{1}{2}z_2 - \tfrac{1}{2}z_1 \right)^2}$$

$$= \sqrt{\left(\tfrac{1}{2} \right)^2 \left[(x_2 - x_1)^2 + (y_2 - y_1)^2 + (z_2 - z_1)^2 \right]} = \tfrac{1}{2}\sqrt{(x_2 - x_1)^2 + (y_2 - y_1)^2 + (z_2 - z_1)^2}$$

$$= \tfrac{1}{2}|P_1 P_2|$$

$$|QP_2| = \sqrt{\left[x_2 - \tfrac{1}{2}(x_1 + x_2)\right]^2 + \left[y_2 - \tfrac{1}{2}(y_1 + y_2)\right]^2 + \left[z_2 - \tfrac{1}{2}(z_1 + z_2)\right]^2}$$

$$= \sqrt{\left(\tfrac{1}{2}x_2 - \tfrac{1}{2}x_1\right)^2 + \left(\tfrac{1}{2}y_2 - \tfrac{1}{2}y_1\right)^2 + \left(\tfrac{1}{2}z_2 - \tfrac{1}{2}z_1\right)^2} = \sqrt{\left(\tfrac{1}{2}\right)^2\left[(x_2 - x_1)^2 + (y_2 - y_1)^2 + (z_2 - z_1)^2\right]}$$

$$= \tfrac{1}{2}\sqrt{(x_2 - x_1)^2 + (y_2 - y_1)^2 + (z_2 - z_1)^2} = \tfrac{1}{2}|P_1 P_2|$$

So Q is indeed the midpoint of $P_1 P_2$.

(b) By part (a), the midpoints of sides AB, BC and CA are $P_1\left(-\tfrac{1}{2}, 1, 4\right)$, $P_2\left(1, \tfrac{1}{2}, 5\right)$ and $P_3\left(\tfrac{5}{2}, \tfrac{3}{2}, 4\right)$. (Recall that a median of a triangle is a line segment from a vertex to the midpoint of the opposite side.) Then the lengths of the medians are:

$$|AP_2| = \sqrt{0^2 + \left(\tfrac{1}{2} - 2\right)^2 + (5 - 3)^2} = \sqrt{\tfrac{9}{4} + 4} = \sqrt{\tfrac{25}{4}} = \tfrac{5}{2}$$

$$|BP_3| = \sqrt{\left(\tfrac{5}{2} + 2\right)^2 + \left(\tfrac{3}{2}\right)^2 + (4 - 5)^2} = \sqrt{\tfrac{81}{4} + \tfrac{9}{4} + 1} = \sqrt{\tfrac{94}{4}} = \tfrac{1}{2}\sqrt{94}$$

$$|CP_1| = \sqrt{\left(-\tfrac{1}{2} - 4\right)^2 + (1 - 1)^2 + (4 - 5)^2} = \sqrt{\tfrac{81}{4} + 1} = \tfrac{1}{2}\sqrt{85}$$

19. (a) Since the sphere touches the xy-plane, its radius is the distance from its center, $(2, -3, 6)$, to the xy-plane, namely 6.

Therefore $r = 6$ and an equation of the sphere is $(x - 2)^2 + (y + 3)^2 + (z - 6)^2 = 6^2 = 36$.

(b) The radius of this sphere is the distance from its center $(2, -3, 6)$ to the yz-plane, which is 2. Therefore, an equation is

$$(x - 2)^2 + (y + 3)^2 + (z - 6)^2 = 4.$$

(c) Here the radius is the distance from the center $(2, -3, 6)$ to the xz-plane, which is 3. Therefore, an equation is

$$(x - 2)^2 + (y + 3)^2 + (z - 6)^2 = 9.$$

21. The equation $x = 5$ represents a plane parallel to the yz-plane and 5 units in front of it.

23. The inequality $y < 8$ represents a half-space consisting of all points to the left of the plane $y = 8$.

25. The inequality $0 \le z \le 6$ represents all points on or between the horizontal planes $z = 0$ (the xy-plane) and $z = 6$.

27. The inequality $x^2 + y^2 + z^2 \le 3$ is equivalent to $\sqrt{x^2 + y^2 + z^2} \le \sqrt{3}$, so the region consists of those points whose distance from the origin is at most $\sqrt{3}$. This is the set of all points on or inside the sphere with radius $\sqrt{3}$ and center $(0, 0, 0)$.

29. Here $x^2 + z^2 \le 9$ or equivalently $\sqrt{x^2 + z^2} \le 3$ which describes the set of all points in $\mathbb{R}^3$ whose distance from the y-axis is at most 3. Thus, the inequality represents the region consisting of all points on or inside a circular cylinder of radius 3 with axis the y-axis.

31. This describes all points whose x-coordinate is between 0 and 5, that is, $0 < x < 5$.

33. This describes a region all of whose points have a distance to the origin which is greater than r, but smaller than R. So inequalities describing the region are $r < \sqrt{x^2 + y^2 + z^2} < R$, or $r^2 < x^2 + y^2 + z^2 < R^2$.

35. We need to find a set of points $\{P(x, y, z) \mid |AP| = |BP|\}$.

$$\sqrt{(x + 1)^2 + (y - 5)^2 + (z - 3)^2} = \sqrt{(x - 6)^2 + (y - 2)^2 + (z + 2)^2} \quad \Rightarrow$$

$$(x + 1)^2 + (y - 5) + (z - 3)^2 = (x - 6)^2 + (y - 2)^2 + (z + 2)^2 \quad \Rightarrow$$

$x^2 + 2x + 1 + y^2 - 10y + 25 + z^2 - 6z + 9 = x^2 - 12x + 36 + y^2 - 4y + 4 + z^2 + 4z + 4 \Rightarrow 14x - 6y - 10z = 9$.

Thus the set of points is a plane perpendicular to the line segment joining A and B (since this plane must contain the

perpendicular bisector of the line segment AB).

37. The sphere $x^2 + y^2 + z^2 = 4$ has center $(0, 0, 0)$ and radius 2. Completing squares in $x^2 - 4x + y^2 - 4y + z^2 - 4z = -11$

gives $(x^2 - 4x + 4) + (y^2 - 4y + 4) + (z^2 - 4z + 4) = -11 + 4 + 4 + 4 \Rightarrow (x-2)^2 + (y-2)^2 + (z-2)^2 = 1$,

so this is the sphere with center $(2, 2, 2)$ and radius 1. The (shortest) distance between the spheres is measured along

the line segment connecting their centers. The distance between $(0, 0, 0)$ and $(2, 2, 2)$ is

$\sqrt{(2-0)^2 + (2-0)^2 + (2-0)^2} = \sqrt{12} = 2\sqrt{3}$, and subtracting the radius of each circle, the distance between the

spheres is $2\sqrt{3} - 2 - 1 = 2\sqrt{3} - 3$.

10.2   Vectors

1. Vectors are equal when they share the same length and direction (but not necessarily location). Using the symmetry of the

parallelogram as a guide, we see that $\overrightarrow{AB} = \overrightarrow{DC}$, $\overrightarrow{DA} = \overrightarrow{CB}$, $\overrightarrow{DE} = \overrightarrow{EB}$, and $\overrightarrow{EA} = \overrightarrow{CE}$.

3. (a)   (b)   (c)

(d)   (e)   (f)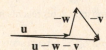

5. $\mathbf{a} = \langle 3 - (-1), 2 - 1 \rangle = \langle 4, 1 \rangle$

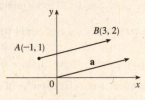

7. $\mathbf{a} = \langle 2 - 0, 3 - 3, -1 - 1 \rangle = \langle 2, 0, -2 \rangle$

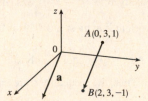

9. $\langle -1, 4 \rangle + \langle 6, -2 \rangle = \langle -1 + 6, 4 + (-2) \rangle = \langle 5, 2 \rangle$

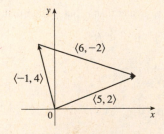

11. $\langle 3, 0, 1 \rangle + \langle 0, 8, 0 \rangle = \langle 3 + 0, 0 + 8, 1 + 0 \rangle$

$= \langle 3, 8, 1 \rangle$

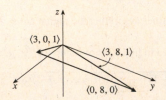

13. $\mathbf{a} + \mathbf{b} = \langle 5 + (-3), -12 + (-6) \rangle = \langle 2, -18 \rangle$

$2\mathbf{a} + 3\mathbf{b} = \langle 10, -24 \rangle + \langle -9, -18 \rangle = \langle 1, -42 \rangle$

$|\mathbf{a}| = \sqrt{5^2 + (-12)^2} = \sqrt{169} = 13$

$|\mathbf{a} - \mathbf{b}| = |\langle 5 - (-3), -12 - (-6) \rangle| = |\langle 8, -6 \rangle| = \sqrt{8^2 + (-6)^2} = \sqrt{100} = 10$

15. $\mathbf{a} + \mathbf{b} = (\mathbf{i} + 2\mathbf{j} - 3\mathbf{k}) + (-2\mathbf{i} - \mathbf{j} + 5\mathbf{k}) = -\mathbf{i} + \mathbf{j} + 2\mathbf{k}$

$2\mathbf{a} + 3\mathbf{b} = 2(\mathbf{i} + 2\mathbf{j} - 3\mathbf{k}) + 3(-2\mathbf{i} - \mathbf{j} + 5\mathbf{k}) = 2\mathbf{i} + 4\mathbf{j} - 6\mathbf{k} - 6\mathbf{i} - 3\mathbf{j} + 15\mathbf{k} = -4\mathbf{i} + \mathbf{j} + 9\mathbf{k}$

$|\mathbf{a}| = \sqrt{1^2 + 2^2 + (-3)^2} = \sqrt{14}$

$|\mathbf{a} - \mathbf{b}| = |(\mathbf{i} + 2\mathbf{j} - 3\mathbf{k}) - (-2\mathbf{i} - \mathbf{j} + 5\mathbf{k})| = |3\mathbf{i} + 3\mathbf{j} - 8\mathbf{k}| = \sqrt{3^2 + 3^2 + (-8)^2} = \sqrt{82}$

17. The vector $8\mathbf{i} - \mathbf{j} + 4\mathbf{k}$ has length $|8\mathbf{i} - \mathbf{j} + 4\mathbf{k}| = \sqrt{8^2 + (-1)^2 + 4^2} = \sqrt{81} = 9$, so by the unit vector with the same

direction is $\frac{1}{9}(8\mathbf{i} - \mathbf{j} + 4\mathbf{k}) = \frac{8}{9}\mathbf{i} - \frac{1}{9}\mathbf{j} + \frac{4}{9}\mathbf{k}$.

19.

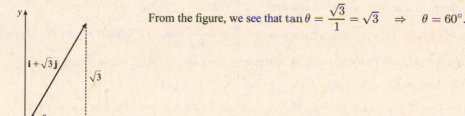

From the figure, we see that $\tan\theta = \frac{\sqrt{3}}{1} = \sqrt{3} \quad \Rightarrow \quad \theta = 60°$.

21. From the figure, we see that the x-component of $\mathbf{v}$ is

$v_1 = |\mathbf{v}| \cos(\pi/3) = 4 \cdot \frac{1}{2} = 2$ and the y-component is

$v_2 = |\mathbf{v}| \sin(\pi/3) = 4 \cdot \frac{\sqrt{3}}{2} = 2\sqrt{3}$. Thus

$\mathbf{v} = \langle v_1, v_2 \rangle = \langle 2, 2\sqrt{3} \rangle$.

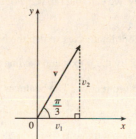

23. The velocity vector $\mathbf{v}$ makes an angle of $40°$ with the horizontal and

has magnitude equal to the speed at which the football was thrown.

From the figure, we see that the horizontal component of $\mathbf{v}$ is

$|\mathbf{v}| \cos 40° = 60 \cos 40° \approx 45.96$ ft/s and the vertical component

is $|\mathbf{v}| \sin 40° = 60 \sin 40° \approx 38.57$ ft/s.

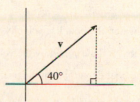

25. The given force vectors can be expressed in terms of their horizontal and vertical components as $-300\,\mathbf{i}$ and

$200 \cos 60°\,\mathbf{i} + 200 \sin 60°\,\mathbf{j} = 200\left(\frac{1}{2}\right)\mathbf{i} + 200\left(\frac{\sqrt{3}}{2}\right)\mathbf{j} = 100\mathbf{i} + 100\sqrt{3}\,\mathbf{j}$. The resultant force $\mathbf{F}$ is the sum of

these two vectors: $\mathbf{F} = (-300 + 100)\mathbf{i} + (0 + 100\sqrt{3})\mathbf{j} = -200\mathbf{i} + 100\sqrt{3}\,\mathbf{j}$. Then we have

$|\mathbf{F}| \approx \sqrt{(-200)^2 + (100\sqrt{3})^2} = \sqrt{70{,}000} = 100\sqrt{7} \approx 264.6$ N. Let θ be the angle $\mathbf{F}$ makes with the positive x-axis. Then $\tan\theta = \dfrac{100\sqrt{3}}{-200} = -\dfrac{\sqrt{3}}{2}$ and the terminal point of $\mathbf{F}$ lies in the second quadrant, so

$$\theta = \tan^{-1}\left(-\frac{\sqrt{3}}{2}\right) + 180° \approx -40.9° + 180° = 139.1°.$$

27. With respect to the water's surface, the woman's velocity is the vector sum of the velocity of the ship with respect

to the water, and the woman's velocity with respect to the ship. If we let north be the positive y-direction, then

$\mathbf{v} = \langle 0, 22 \rangle + \langle -3, 0 \rangle = \langle -3, 22 \rangle$. The woman's speed is $|\mathbf{v}| = \sqrt{9 + 484} \approx 22.2$ mi/h. The vector $\mathbf{v}$ makes an angle θ

with the east, where $\theta = \tan^{-1}\left(\dfrac{22}{-3}\right) \approx 98°$. Therefore, the woman's direction is about $\text{N}(98 - 90)°\text{W} = \text{N}8°\text{W}$.

29. Let $\mathbf{T}_1$ and $\mathbf{T}_2$ represent the tension vectors in each side of the

clothesline as shown in the figure. $\mathbf{T}_1$ and $\mathbf{T}_2$ have equal vertical

components and opposite horizontal components, so we can write

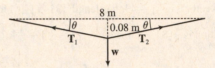

$\mathbf{T}_1 = -a\,\mathbf{i} + b\,\mathbf{j}$ and $\mathbf{T}_2 = a\,\mathbf{i} + b\,\mathbf{j}$ $[a, b > 0]$. By similar triangles, $\dfrac{b}{a} = \dfrac{0.08}{4}$ $\Rightarrow$ $a = 50b$. The force due to gravity

acting on the shirt has magnitude $0.8g \approx (0.8)(9.8) = 7.84$ N, hence we have $\mathbf{w} = -7.84\,\mathbf{j}$. The resultant $\mathbf{T}_1 + \mathbf{T}_2$

of the tensile forces counterbalances $\mathbf{w}$, so $\mathbf{T}_1 + \mathbf{T}_2 = -\mathbf{w}$ $\Rightarrow$ $(-a\,\mathbf{i} + b\,\mathbf{j}) + (a\,\mathbf{i} + b\,\mathbf{j}) = 7.84\,\mathbf{j}$ $\Rightarrow$

$(-50b\,\mathbf{i} + b\,\mathbf{j}) + (50b\,\mathbf{i} + b\,\mathbf{j}) = 2b\,\mathbf{j} = 7.84\,\mathbf{j}$ $\Rightarrow$ $b = \dfrac{7.84}{2} = 3.92$ and $a = 50b = 196$. Thus the tensions are

$\mathbf{T}_1 = -a\,\mathbf{i} + b\,\mathbf{j} = -196\,\mathbf{i} + 3.92\,\mathbf{j}$ and $\mathbf{T}_2 = a\,\mathbf{i} + b\,\mathbf{j} = 196\,\mathbf{i} + 3.92\,\mathbf{j}$.

Alternatively, we can find the value of θ and proceed as in Example 7.

31. (a) Set up coordinate axes so that the boatman is at the origin, the canal is

bordered by the y-axis and the line $x = 3$, and the current flows in the

negative y-direction. The boatman wants to reach the point $(3, 2)$. Let θ be

the angle, measured from the positive y-axis, in the direction he should

steer. (See the figure.)

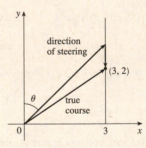

In still water, the boat has velocity $\mathbf{v}_b = \langle 13\sin\theta, 13\cos\theta \rangle$ and the velocity of the current is $\mathbf{v}_c \langle 0, -3.5 \rangle$, so the true path

of the boat is determined by the velocity vector $\mathbf{v} = \mathbf{v}_b + \mathbf{v}_c = \langle 13\sin\theta, 13\cos\theta - 3.5 \rangle$. Let t be the time (in hours)

after the boat departs; then the position of the boat at time t is given by $t\mathbf{v}$ and the boat crosses the canal when

$t\mathbf{v} = \langle 13\sin\theta, 13\cos\theta - 3.5 \rangle t = \langle 3, 2 \rangle$. Thus $13(\sin\theta)t = 3$ $\Rightarrow$ $t = \dfrac{3}{13\sin\theta}$ and $(13\cos\theta - 3.5)t = 2$.

Substituting gives $(13\cos\theta - 3.5)\,\dfrac{3}{13\sin\theta} = 2$ $\Rightarrow$ $39\cos\theta - 10.5 = 26\sin\theta$ **(1)**. Squaring both sides, we have

$$1521\cos^2\theta - 819\cos\theta + 110.25 = 676\sin^2\theta = 676\left(1 - \cos^2\theta\right)$$

$$2197\cos^2\theta - 819\cos\theta - 565.75 = 0$$

The quadratic formula gives

$$\cos\theta = \frac{819 \pm \sqrt{(-819)^2 - 4(2197)(-565.75)}}{2(2197)}$$

$$= \frac{819 \pm \sqrt{5{,}642{,}572}}{4394} \approx 0.72699 \text{ or } -0.35421$$

The acute value for θ is approximately $\cos^{-1}(0.72699) \approx 43.4°$. Thus the boatman should steer in the direction that is 43.4° from the bank, toward upstream.

Alternate solution: We could solve (1) graphically by plotting $y = 39\cos\theta - 10.5$ and $y = 26\sin\theta$ on a graphing device and finding the appoximate intersection point $(0.757, 17.85)$. Thus $\theta \approx 0.757$ radians or equivalently 43.4°.

(b) From part (a) we know the trip is completed when $t = \dfrac{3}{13\sin\theta}$. But $\theta \approx 43.4°$, so the time required is approximately

$$\frac{3}{13\sin 43.4°} \approx 0.336 \text{ hours or } 20.2 \text{ minutes.}$$

33. The slope of the tangent line to the graph of $y = x^2$ at the point $(2, 4)$ is

$$\left.\frac{dy}{dx}\right|_{x=2} = 2x\Big|_{x=2} = 4$$

and a parallel vector is $\mathbf{i} + 4\mathbf{j}$ which has length $|\mathbf{i} + 4\mathbf{j}| = \sqrt{1^2 + 4^2} = \sqrt{17}$, so unit vectors parallel to the tangent line are $\pm\frac{1}{\sqrt{17}}(\mathbf{i} + 4\mathbf{j})$.

35. (a), (b)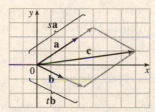

(c) From the sketch, we estimate that $s \approx 1.3$ and $t \approx 1.6$.

(d) $\mathbf{c} = s\mathbf{a} + t\mathbf{b} \Leftrightarrow 7 = 3s + 2t$ and $1 = 2s - t$.

Solving these equations gives $s = \frac{9}{7}$ and $t = \frac{11}{7}$.

37. $|\mathbf{r} - \mathbf{r}_0|$ is the distance between the points (x, y, z) and (x_0, y_0, z_0), so the set of points is a sphere with radius 1 and center (x_0, y_0, z_0).

Alternate method: $|\mathbf{r} - \mathbf{r}_0| = 1 \Leftrightarrow \sqrt{(x - x_0)^2 + (y - y_0)^2 + (z - z_0)^2} = 1 \Leftrightarrow$
$(x - x_0)^2 + (y - y_0)^2 + (z - z_0)^2 = 1$, which is the equation of a sphere with radius 1 and center (x_0, y_0, z_0).

39. $\mathbf{a} + (\mathbf{b} + \mathbf{c}) = \langle a_1, a_2 \rangle + (\langle b_1, b_2 \rangle + \langle c_1, c_2 \rangle) = \langle a_1, a_2 \rangle + \langle b_1 + c_1, b_2 + c_2 \rangle$

$\qquad = \langle a_1 + b_1 + c_1, a_2 + b_2 + c_2 \rangle = \langle (a_1 + b_1) + c_1, (a_2 + b_2) + c_2 \rangle$

$\qquad = \langle a_1 + b_1, a_2 + b_2 \rangle + \langle c_1, c_2 \rangle = (\langle a_1, a_2 \rangle + \langle b_1, b_2 \rangle) + \langle c_1, c_2 \rangle$

$\qquad = (\mathbf{a} + \mathbf{b}) + \mathbf{c}$

41. Consider triangle ABC, where D and E are the midpoints of AB and BC. We know that $\overrightarrow{AB} + \overrightarrow{BC} = \overrightarrow{AC}$ **(1)** and
$\overrightarrow{DB} + \overrightarrow{BE} = \overrightarrow{DE}$ **(2)**. However, $\overrightarrow{DB} = \frac{1}{2}\overrightarrow{AB}$, and $\overrightarrow{BE} = \frac{1}{2}\overrightarrow{BC}$. Substituting these expressions for $\overrightarrow{DB}$ and $\overrightarrow{BE}$ into
(2) gives $\frac{1}{2}\overrightarrow{AB} + \frac{1}{2}\overrightarrow{BC} = \overrightarrow{DE}$. Comparing this with **(1)** gives $\overrightarrow{DE} = \frac{1}{2}\overrightarrow{AC}$. Therefore $\overrightarrow{AC}$ and $\overrightarrow{DE}$ are parallel and
$\left|\overrightarrow{DE}\right| = \frac{1}{2}\left|\overrightarrow{AC}\right|$.

10.3 The Dot Product

1. (a) $\mathbf{a} \cdot \mathbf{b}$ is a scalar, and the dot product is defined only for vectors, so $(\mathbf{a} \cdot \mathbf{b}) \cdot \mathbf{c}$ has no meaning.

   (b) $(\mathbf{a} \cdot \mathbf{b}) \mathbf{c}$ is a scalar multiple of a vector, so it does have meaning.

   (c) Both $|\mathbf{a}|$ and $\mathbf{b} \cdot \mathbf{c}$ are scalars, so $|\mathbf{a}| (\mathbf{b} \cdot \mathbf{c})$ is an ordinary product of real numbers, and has meaning.

   (d) Both $\mathbf{a}$ and $\mathbf{b} + \mathbf{c}$ are vectors, so the dot product $\mathbf{a} \cdot (\mathbf{b} + \mathbf{c})$ has meaning.

   (e) $\mathbf{a} \cdot \mathbf{b}$ is a scalar, but $\mathbf{c}$ is a vector, and so the two quantities cannot be added and $\mathbf{a} \cdot \mathbf{b} + \mathbf{c}$ has no meaning.

   (f) $|\mathbf{a}|$ is a scalar, and the dot product is defined only for vectors, so $|\mathbf{a}| \cdot (\mathbf{b} + \mathbf{c})$ has no meaning.

3. $\mathbf{a} \cdot \mathbf{b} = \left\langle -2, \frac{1}{3} \right\rangle \cdot \langle -5, 12 \rangle = (-2)(-5) + \left(\frac{1}{3} \right)(12) = 10 + 4 = 14$

5. $\mathbf{a} \cdot \mathbf{b} = \left\langle 4, 1, \frac{1}{4} \right\rangle \cdot \langle 6, -3, -8 \rangle = (4)(6) + (1)(-3) + \left(\frac{1}{4} \right)(-8) = 19$

7. $\mathbf{a} \cdot \mathbf{b} = (2\,\mathbf{i} + \mathbf{j}) \cdot (\mathbf{i} - \mathbf{j} + \mathbf{k}) = (2)(1) + (1)(-1) + (0)(1) = 1$

9. By Theorem 3, $\mathbf{a} \cdot \mathbf{b} = |\mathbf{a}| \, |\mathbf{b}| \cos \theta = (6)(5) \cos \frac{2\pi}{3} = 30 \left(-\frac{1}{2} \right) = -15$.

11. $\mathbf{u}$, $\mathbf{v}$, and $\mathbf{w}$ are all unit vectors, so the triangle is an equilateral triangle. Thus the angle between $\mathbf{u}$ and $\mathbf{v}$ is $60°$ and

    $\mathbf{u} \cdot \mathbf{v} = |\mathbf{u}| \, |\mathbf{v}| \cos 60° = (1)(1)\left(\frac{1}{2} \right) = \frac{1}{2}$. If $\mathbf{w}$ is moved so it has the same initial point as $\mathbf{u}$, we can see that the angle

    between them is $120°$ and we have $\mathbf{u} \cdot \mathbf{w} = |\mathbf{u}| \, |\mathbf{w}| \cos 120° = (1)(1)\left(-\frac{1}{2} \right) = -\frac{1}{2}$.

13. (a) $\mathbf{i} \cdot \mathbf{j} = \langle 1, 0, 0 \rangle \cdot \langle 0, 1, 0 \rangle = (1)(0) + (0)(1) + (0)(0) = 0$. Similarly, $\mathbf{j} \cdot \mathbf{k} = (0)(0) + (1)(0) + (0)(1) = 0$ and

    $\mathbf{k} \cdot \mathbf{i} = (0)(1) + (0)(0) + (1)(0) = 0$.

    Another method: Because $\mathbf{i}$, $\mathbf{j}$, and $\mathbf{k}$ are mutually perpendicular, the cosine factor in each dot product is $\cos \frac{\pi}{2} = 0$.

    (b) By Property 1 of the dot product, $\mathbf{i} \cdot \mathbf{i} = |\mathbf{i}|^2 = 1^2 = 1$ since $\mathbf{i}$ is a unit vector. Similarly, $\mathbf{j} \cdot \mathbf{j} = |\mathbf{j}|^2 = 1$ and

    $\mathbf{k} \cdot \mathbf{k} = |\mathbf{k}|^2 = 1$.

15. $|\mathbf{a}| = \sqrt{4^2 + 3^2} = 5$, $|\mathbf{b}| = \sqrt{2^2 + (-1)^2} = \sqrt{5}$, and $\mathbf{a} \cdot \mathbf{b} = (4)(2) + (3)(-1) = 5$. From the definition of the dot product,

    we have $\cos \theta = \dfrac{\mathbf{a} \cdot \mathbf{b}}{|\mathbf{a}| \, |\mathbf{b}|} = \dfrac{5}{5 \cdot \sqrt{5}} = \dfrac{1}{\sqrt{5}}$. So the angle between $\mathbf{a}$ and $\mathbf{b}$ is $\theta = \cos^{-1} \left(\dfrac{1}{\sqrt{5}} \right) \approx 63°$.

17. $|\mathbf{a}| = \sqrt{4^2 + (-3)^2 + 1^2} = \sqrt{26}$, $|\mathbf{b}| = \sqrt{2^2 + 0^2 + (-1)^2} = \sqrt{5}$, and $\mathbf{a} \cdot \mathbf{b} = (4)(2) + (-3)(0) + (1)(-1) = 7$.

    Then $\cos \theta = \dfrac{\mathbf{a} \cdot \mathbf{b}}{|\mathbf{a}| \, |\mathbf{b}|} = \dfrac{7}{\sqrt{26} \cdot \sqrt{5}} = \dfrac{7}{\sqrt{130}}$ and $\theta = \cos^{-1} \left(\dfrac{7}{\sqrt{130}} \right) \approx 52°$.

19. (a) $\mathbf{a} \cdot \mathbf{b} = (-5)(6) + (3)(-8) + (7)(2) = -40 \neq 0$, so $\mathbf{a}$ and $\mathbf{b}$ are not orthogonal. Also, since $\mathbf{a}$ is not a scalar multiple

    of $\mathbf{b}$, $\mathbf{a}$ and $\mathbf{b}$ are not parallel.

    (b) $\mathbf{a} \cdot \mathbf{b} = (4)(-3) + (6)(2) = 0$, so $\mathbf{a}$ and $\mathbf{b}$ are orthogonal (and not parallel).

    (c) $\mathbf{a} \cdot \mathbf{b} = (-1)(3) + (2)(4) + (5)(-1) = 0$, so $\mathbf{a}$ and $\mathbf{b}$ are orthogonal (and not parallel).

    (d) Because $\mathbf{a} = -\frac{2}{3}\,\mathbf{b}$, $\mathbf{a}$ and $\mathbf{b}$ are parallel.

21. $\overrightarrow{QP} = \langle -1, -3, 2 \rangle$, $\overrightarrow{QR} = \langle 4, -2, -1 \rangle$, and $\overrightarrow{QP} \cdot \overrightarrow{QR} = -4 + 6 - 2 = 0$. Thus $\overrightarrow{QP}$ and $\overrightarrow{QR}$ are orthogonal, so the angle of the triangle at vertex Q is a right angle.

23. Let $\mathbf{a} = a_1\mathbf{i} + a_2\mathbf{j} + a_3\mathbf{k}$ be a vector orthogonal to both $\mathbf{i}+\mathbf{j}$ and $\mathbf{i}+\mathbf{k}$. Then $\mathbf{a} \cdot (\mathbf{i}+\mathbf{j}) = 0 \Leftrightarrow a_1 + a_2 = 0$ and $\mathbf{a} \cdot (\mathbf{i}+\mathbf{k}) = 0 \Leftrightarrow a_1 + a_3 = 0$, so $a_1 = -a_2 = -a_3$. Furthermore $\mathbf{a}$ is to be a unit vector, so $1 = a_1^2 + a_2^2 + a_3^2 = 3a_1^2$ implies $a_1 = \pm\frac{1}{\sqrt{3}}$. Thus $\mathbf{a} = \frac{1}{\sqrt{3}}\mathbf{i} - \frac{1}{\sqrt{3}}\mathbf{j} - \frac{1}{\sqrt{3}}\mathbf{k}$ and $\mathbf{a} = -\frac{1}{\sqrt{3}}\mathbf{i} + \frac{1}{\sqrt{3}}\mathbf{j} + \frac{1}{\sqrt{3}}\mathbf{k}$ are two such unit vectors.

25. The line $2x - y = 3 \Leftrightarrow y = 2x - 3$ has slope 2, so a vector parallel to the line is $\mathbf{a} = \langle 1, 2 \rangle$. The line $3x + y = 7 \Leftrightarrow y = -3x + 7$ has slope -3, so a vector parallel to the line is $\mathbf{b} = \langle 1, -3 \rangle$. The angle between the lines is the same as the angle θ between the vectors. Here we have $\mathbf{a} \cdot \mathbf{b} = (1)(1) + (2)(-3) = -5$, $|\mathbf{a}| = \sqrt{1^2 + 2^2} = \sqrt{5}$, and $|\mathbf{b}| = \sqrt{1^2 + (-3)^2} = \sqrt{10}$, so $\cos\theta = \frac{\mathbf{a} \cdot \mathbf{b}}{|\mathbf{a}||\mathbf{b}|} = \frac{-5}{\sqrt{5}\cdot\sqrt{10}} = \frac{-5}{5\sqrt{2}} = -\frac{1}{\sqrt{2}}$ or $-\frac{\sqrt{2}}{2}$. Thus $\theta = 135°$, and the acute angle between the lines is $180° - 135° = 45°$.

27. The curves $y = x^2$ and $y = x^3$ meet when $x^2 = x^3 \Leftrightarrow x^3 - x^2 = 0 \Leftrightarrow x^2(x-1) = 0 \Leftrightarrow x = 0, x = 1$. We have $\frac{d}{dx}x^2 = 2x$ and $\frac{d}{dx}x^3 = 3x^2$, so the tangent lines of both curves have slope 0 at $x = 0$. Thus the angle between the curves is $0°$ at the point $(0,0)$. For $x = 1$, $\frac{d}{dx}x^2\Big|_{x=1} = 2$ and $\frac{d}{dx}x^3\Big|_{x=1} = 3$ so the tangent lines at the point $(1,1)$ have slopes 2 and 3. Vectors parallel to the tangent lines are $\langle 1, 2 \rangle$ and $\langle 1, 3 \rangle$, and the angle θ between them is given by

$$\cos\theta = \frac{\langle 1, 2 \rangle \cdot \langle 1, 3 \rangle}{|\langle 1, 2 \rangle| \, |\langle 1, 3 \rangle|} = \frac{1+6}{\sqrt{5}\sqrt{10}} = \frac{7}{5\sqrt{2}}$$

Thus $\theta = \cos^{-1}\left(\frac{7}{5\sqrt{2}}\right) \approx 8.1°$.

29. $|\mathbf{a}| = \sqrt{(-5)^2 + 12^2} = \sqrt{169} = 13$. The scalar projection of $\mathbf{b}$ onto $\mathbf{a}$ is $\text{comp}_\mathbf{a}\,\mathbf{b} = \frac{\mathbf{a}\cdot\mathbf{b}}{|\mathbf{a}|} = \frac{-5\cdot 4 + 12\cdot 6}{13} = 4$ and the vector projection of $\mathbf{b}$ onto $\mathbf{a}$ is $\text{proj}_\mathbf{a}\,\mathbf{b} = \left(\frac{\mathbf{a}\cdot\mathbf{b}}{|\mathbf{a}|}\right)\frac{\mathbf{a}}{|\mathbf{a}|} = 4\cdot\frac{1}{13}\langle -5, 12 \rangle = \langle -\frac{20}{13}, -\frac{48}{13} \rangle$.

31. $|\mathbf{a}| = \sqrt{9+36+4} = 7$ so the scalar projection of $\mathbf{b}$ onto $\mathbf{a}$ is $\text{comp}_\mathbf{a}\mathbf{b} = \frac{\mathbf{a}\cdot\mathbf{b}}{|\mathbf{a}|} = \frac{1}{7}(3+12-6) = \frac{9}{7}$. The vector projection of $\mathbf{b}$ onto $\mathbf{a}$ is $\text{proj}_\mathbf{a}\mathbf{b} = \frac{9}{7}\frac{\mathbf{a}}{|\mathbf{a}|} = \frac{9}{7}\cdot\frac{1}{7}\langle 3, 6, -2 \rangle = \frac{9}{49}\langle 3, 6, -2 \rangle = \langle \frac{27}{49}, \frac{54}{49}, -\frac{18}{49} \rangle$.

33. $(\text{orth}_\mathbf{a}\,\mathbf{b})\cdot\mathbf{a} = (\mathbf{b} - \text{proj}_\mathbf{a}\,\mathbf{b})\cdot\mathbf{a} = \mathbf{b}\cdot\mathbf{a} - (\text{proj}_\mathbf{a}\,\mathbf{b})\cdot\mathbf{a} = \mathbf{b}\cdot\mathbf{a} - \frac{\mathbf{a}\cdot\mathbf{b}}{|\mathbf{a}|^2}\mathbf{a}\cdot\mathbf{a} = \mathbf{b}\cdot\mathbf{a} - \frac{\mathbf{a}\cdot\mathbf{b}}{|\mathbf{a}|^2}|\mathbf{a}|^2 = \mathbf{b}\cdot\mathbf{a} - \mathbf{a}\cdot\mathbf{b} = 0$.

So they are orthogonal by (7).

35. $\text{comp}_\mathbf{a}\,\mathbf{b} = \frac{\mathbf{a}\cdot\mathbf{b}}{|\mathbf{a}|} = 2 \Leftrightarrow \mathbf{a}\cdot\mathbf{b} = 2|\mathbf{a}| = 2\sqrt{10}$. If $\mathbf{b} = \langle b_1, b_2, b_3 \rangle$, then we need $3b_1 + 0b_2 - 1b_3 = 2\sqrt{10}$. One possible solution is obtained by taking $b_1 = 0, b_2 = 0, b_3 = -2\sqrt{10}$. In general, $\mathbf{b} = \langle s, t, 3s - 2\sqrt{10} \rangle$, $s, t \in \mathbb{R}$.

37. The displacement vector is $\mathbf{D} = (6-0)\,\mathbf{i} + (12-10)\,\mathbf{j} + (20-8)\,\mathbf{k} = 6\,\mathbf{i} + 2\,\mathbf{j} + 12\,\mathbf{k}$ so, , the work done is

$$W = \mathbf{F} \cdot \mathbf{D} = (8\,\mathbf{i} - 6\,\mathbf{j} + 9\,\mathbf{k}) \cdot (6\,\mathbf{i} + 2\,\mathbf{j} + 12\,\mathbf{k}) = 48 - 12 + 108 = 144 \text{ joules.}$$

39. Here $|\mathbf{D}| = 80$ ft, $|\mathbf{F}| = 30$ lb, and $\theta = 40°$. Thus

$$W = \mathbf{F} \cdot \mathbf{D} = |\mathbf{F}|\,|\mathbf{D}| \cos \theta = (30)(80) \cos 40° = 2400 \cos 40° \approx 1839 \text{ ft-lb.}$$

41. First note that $\mathbf{n} = \langle a, b \rangle$ is perpendicular to the line, because if $Q_1 = (a_1, b_1)$ and $Q_2 = (a_2, b_2)$ lie on the line, then

$\mathbf{n} \cdot \overrightarrow{Q_1 Q_2} = aa_2 - aa_1 + bb_2 - bb_1 = 0$, since $aa_2 + bb_2 = -c = aa_1 + bb_1$ from the equation of the line.

Let $P_2 = (x_2, y_2)$ lie on the line. Then the distance from P_1 to the line is the absolute value of the scalar projection

of $\overrightarrow{P_1 P_2}$ onto $\mathbf{n}$. $\quad \text{comp}_{\mathbf{n}}\left(\overrightarrow{P_1 P_2}\right) = \dfrac{|\mathbf{n} \cdot \langle x_2 - x_1, y_2 - y_1 \rangle|}{|\mathbf{n}|} = \dfrac{|ax_2 - ax_1 + by_2 - by_1|}{\sqrt{a^2 + b^2}} = \dfrac{|ax_1 + by_1 + c|}{\sqrt{a^2 + b^2}}$

since $ax_2 + by_2 = -c$. The required distance is $\dfrac{|(3)(-2) + (-4)(3) + 5|}{\sqrt{3^2 + (-4)^2}} = \dfrac{13}{5}$.

43. For convenience, consider the unit cube positioned so that its back left corner is at the origin, and its edges lie along the

coordinate axes. The diagonal of the cube that begins at the origin and ends at $(1, 1, 1)$ has vector representation $\langle 1, 1, 1 \rangle$.

The angle θ between this vector and the vector of the edge which also begins at the origin and runs along the x-axis [that is,

$\langle 1, 0, 0 \rangle$] is given by $\cos \theta = \dfrac{\langle 1, 1, 1 \rangle \cdot \langle 1, 0, 0 \rangle}{|\langle 1, 1, 1 \rangle|\,|\langle 1, 0, 0 \rangle|} = \dfrac{1}{\sqrt{3}} \quad \Rightarrow \quad \theta = \cos^{-1}\left(\dfrac{1}{\sqrt{3}}\right) \approx 55°$.

45. Consider the H—C—H combination consisting of the sole carbon atom and the two hydrogen atoms that are at $(1, 0, 0)$ and

$(0, 1, 0)$ (or any H—C—H combination, for that matter). Vector representations of the line segments emanating from the

carbon atom and extending to these two hydrogen atoms are $\left\langle 1 - \frac{1}{2}, 0 - \frac{1}{2}, 0 - \frac{1}{2} \right\rangle = \left\langle \frac{1}{2}, -\frac{1}{2}, -\frac{1}{2} \right\rangle$ and

$\left\langle 0 - \frac{1}{2}, 1 - \frac{1}{2}, 0 - \frac{1}{2} \right\rangle = \left\langle -\frac{1}{2}, \frac{1}{2}, -\frac{1}{2} \right\rangle$. The bond angle, θ, is therefore given by

$$\cos \theta = \dfrac{\left\langle \frac{1}{2}, -\frac{1}{2}, -\frac{1}{2} \right\rangle \cdot \left\langle -\frac{1}{2}, \frac{1}{2}, -\frac{1}{2} \right\rangle}{\left|\left\langle \frac{1}{2}, -\frac{1}{2}, -\frac{1}{2} \right\rangle\right|\,\left|\left\langle -\frac{1}{2}, \frac{1}{2}, -\frac{1}{2} \right\rangle\right|} = \dfrac{-\frac{1}{4} - \frac{1}{4} + \frac{1}{4}}{\sqrt{\frac{3}{4}}\sqrt{\frac{3}{4}}} = -\dfrac{1}{3} \quad \Rightarrow \quad \theta = \cos^{-1}\left(-\tfrac{1}{3}\right) \approx 109.5°.$$

47. Let $\mathbf{a} = \langle a_1, a_2, a_3 \rangle$ and $= \langle b_1, b_2, b_3 \rangle$.

Property 2: $\quad \mathbf{a} \cdot \mathbf{b} = \langle a_1, a_2, a_3 \rangle \cdot \langle b_1, b_2, b_3 \rangle = a_1 b_1 + a_2 b_2 + a_3 b_3$

$$= b_1 a_1 + b_2 a_2 + b_3 a_3 = \langle b_1, b_2, b_3 \rangle \cdot \langle a_1, a_2, a_3 \rangle = \mathbf{b} \cdot \mathbf{a}$$

Property 4: $\quad (c\,\mathbf{a}) \cdot \mathbf{b} = \langle ca_1, ca_2, ca_3 \rangle \cdot \langle b_1, b_2, b_3 \rangle = (ca_1)b_1 + (ca_2)b_2 + (ca_3)b_3$

$$= c\,(a_1 b_1 + a_2 b_2 + a_3 b_3) = c\,(\mathbf{a} \cdot \mathbf{b}) = a_1(cb_1) + a_2(cb_2) + a_3(cb_3)$$

$$= \langle a_1, a_2, a_3 \rangle \cdot \langle cb_1, cb_2, cb_3 \rangle = \mathbf{a} \cdot (c\,\mathbf{b})$$

Property 5: $\quad \mathbf{0} \cdot \mathbf{a} = \langle 0, 0, 0 \rangle \cdot \langle a_1, a_2, a_3 \rangle = (0)(a_1) + (0)(a_2) + (0)(a_3) = 0$

49. $|\mathbf{a} \cdot \mathbf{b}| = |\,|\mathbf{a}|\,|\mathbf{b}| \cos \theta| = |\mathbf{a}|\,|\mathbf{b}|\,|\cos \theta|$. Since $|\cos \theta| \leq 1$, $|\mathbf{a} \cdot \mathbf{b}| = |\mathbf{a}|\,|\mathbf{b}|\,|\cos \theta| \leq |\mathbf{a}|\,|\mathbf{b}|$.

Note: We have equality in the case of $\cos \theta = \pm 1$, so $\theta = 0$ or $\theta = \pi$, thus equality when $\mathbf{a}$ and $\mathbf{b}$ are parallel.

51. (a)

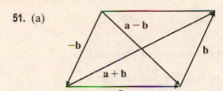

The Parallelogram Law states that the sum of the squares of the lengths of the diagonals of a parallelogram equals the sum of the squares of its (four) sides.

(b) $|\mathbf{a} + \mathbf{b}|^2 = (\mathbf{a} + \mathbf{b}) \cdot (\mathbf{a} + \mathbf{b}) = |\mathbf{a}|^2 + 2(\mathbf{a} \cdot \mathbf{b}) + |\mathbf{b}|^2$ and $|\mathbf{a} - \mathbf{b}|^2 = (\mathbf{a} - \mathbf{b}) \cdot (\mathbf{a} - \mathbf{b}) = |\mathbf{a}|^2 - 2(\mathbf{a} \cdot \mathbf{b}) + |\mathbf{b}|^2$.

Adding these two equations gives $|\mathbf{a} + \mathbf{b}|^2 + |\mathbf{a} - \mathbf{b}|^2 = 2|\mathbf{a}|^2 + 2|\mathbf{b}|^2$.

10.4   The Cross Product

1. $\mathbf{a} \times \mathbf{b} = \begin{vmatrix} \mathbf{i} & \mathbf{j} & \mathbf{k} \\ 6 & 0 & -2 \\ 0 & 8 & 0 \end{vmatrix} = \begin{vmatrix} 0 & -2 \\ 8 & 0 \end{vmatrix} \mathbf{i} - \begin{vmatrix} 6 & -2 \\ 0 & 0 \end{vmatrix} \mathbf{j} + \begin{vmatrix} 6 & 0 \\ 0 & 8 \end{vmatrix} \mathbf{k}$

$= [0 - (-16)]\,\mathbf{i} - (0 - 0)\,\mathbf{j} + (48 - 0)\,\mathbf{k} = 16\,\mathbf{i} + 48\,\mathbf{k}$

Now $(\mathbf{a} \times \mathbf{b}) \cdot \mathbf{a} = \langle 16, 0, 48 \rangle \cdot \langle 6, 0, -2 \rangle = 96 + 0 - 96 = 0$ and $(\mathbf{a} \times \mathbf{b}) \cdot \mathbf{b} = \langle 16, 0, 48 \rangle \cdot \langle 0, 8, 0 \rangle = 0 + 0 + 0 = 0$, so $\mathbf{a} \times \mathbf{b}$ is orthogonal to both $\mathbf{a}$ and $\mathbf{b}$.

3. $\mathbf{a} \times \mathbf{b} = \begin{vmatrix} \mathbf{i} & \mathbf{j} & \mathbf{k} \\ 1 & 3 & -2 \\ -1 & 0 & 5 \end{vmatrix} = \begin{vmatrix} 3 & -2 \\ 0 & 5 \end{vmatrix} \mathbf{i} - \begin{vmatrix} 1 & -2 \\ -1 & 5 \end{vmatrix} \mathbf{j} + \begin{vmatrix} 1 & 3 \\ -1 & 0 \end{vmatrix} \mathbf{k}$

$= (15 - 0)\,\mathbf{i} - (5 - 2)\,\mathbf{j} + [0 - (-3)]\,\mathbf{k} = 15\,\mathbf{i} - 3\,\mathbf{j} + 3\,\mathbf{k}$

Since $(\mathbf{a} \times \mathbf{b}) \cdot \mathbf{a} = (15\,\mathbf{i} - 3\,\mathbf{j} + 3\,\mathbf{k}) \cdot (\mathbf{i} + 3\,\mathbf{j} - 2\,\mathbf{k}) = 15 - 9 - 6 = 0$, $\mathbf{a} \times \mathbf{b}$ is orthogonal to $\mathbf{a}$.

Since $(\mathbf{a} \times \mathbf{b}) \cdot \mathbf{b} = (15\,\mathbf{i} - 3\,\mathbf{j} + 3\,\mathbf{k}) \cdot (-\mathbf{i} + 5\,\mathbf{k}) = -15 + 0 + 15 = 0$, $\mathbf{a} \times \mathbf{b}$ is orthogonal to $\mathbf{b}$.

5. $\mathbf{a} \times \mathbf{b} = \begin{vmatrix} \mathbf{i} & \mathbf{j} & \mathbf{k} \\ 1 & -1 & -1 \\ \frac{1}{2} & 1 & \frac{1}{2} \end{vmatrix} = \begin{vmatrix} -1 & -1 \\ 1 & \frac{1}{2} \end{vmatrix} \mathbf{i} - \begin{vmatrix} 1 & -1 \\ \frac{1}{2} & \frac{1}{2} \end{vmatrix} \mathbf{j} + \begin{vmatrix} 1 & -1 \\ \frac{1}{2} & 1 \end{vmatrix} \mathbf{k}$

$= \left[-\frac{1}{2} - (-1) \right] \mathbf{i} - \left[\frac{1}{2} - \left(-\frac{1}{2} \right) \right] \mathbf{j} + \left[1 - \left(-\frac{1}{2} \right) \right] \mathbf{k} = \frac{1}{2}\,\mathbf{i} - \mathbf{j} + \frac{3}{2}\,\mathbf{k}$

Now $(\mathbf{a} \times \mathbf{b}) \cdot \mathbf{a} = \left(\frac{1}{2}\,\mathbf{i} - \mathbf{j} + \frac{3}{2}\,\mathbf{k} \right) \cdot (\mathbf{i} - \mathbf{j} - \mathbf{k}) = \frac{1}{2} + 1 - \frac{3}{2} = 0$ and

$(\mathbf{a} \times \mathbf{b}) \cdot \mathbf{b} = \left(\frac{1}{2}\,\mathbf{i} - \mathbf{j} + \frac{3}{2}\,\mathbf{k} \right) \cdot \left(\frac{1}{2}\,\mathbf{i} + \mathbf{j} + \frac{1}{2}\,\mathbf{k} \right) = \frac{1}{4} - 1 + \frac{3}{4} = 0$, so $\mathbf{a} \times \mathbf{b}$ is orthogonal to both $\mathbf{a}$ and $\mathbf{b}$.

7. $\mathbf{a} \times \mathbf{b} = \begin{vmatrix} \mathbf{i} & \mathbf{j} & \mathbf{k} \\ t & 1 & 1/t \\ t^2 & t^2 & 1 \end{vmatrix} = \begin{vmatrix} 1 & 1/t \\ t^2 & 1 \end{vmatrix} \mathbf{i} - \begin{vmatrix} t & 1/t \\ t^2 & 1 \end{vmatrix} \mathbf{j} + \begin{vmatrix} t & 1 \\ t^2 & t^2 \end{vmatrix} \mathbf{k}$

$= (1 - t)\,\mathbf{i} - (t - t)\,\mathbf{j} + (t^3 - t^2)\,\mathbf{k} = (1 - t)\,\mathbf{i} + (t^3 - t^2)\,\mathbf{k}$

Since $(\mathbf{a} \times \mathbf{b}) \cdot \mathbf{a} = \langle 1 - t, 0, t^3 - t^2 \rangle \cdot \langle t, 1, 1/t \rangle = t - t^2 + 0 + t^2 - t = 0$, $\mathbf{a} \times \mathbf{b}$ is orthogonal to $\mathbf{a}$.

Since $(\mathbf{a} \times \mathbf{b}) \cdot \mathbf{b} = \langle 1 - t, 0, t^3 - t^2 \rangle \cdot \langle t^2, t^2, 1 \rangle = t^2 - t^3 + 0 + t^3 - t^2 = 0$, $\mathbf{a} \times \mathbf{b}$ is orthogonal to $\mathbf{b}$.

9. According to the discussion preceding Theorem 8, $\mathbf{i} \times \mathbf{j} = \mathbf{k}$, so $(\mathbf{i} \times \mathbf{j}) \times \mathbf{k} = \mathbf{k} \times \mathbf{k} = \mathbf{0}$ [by Example 2].

11. $(\mathbf{j} - \mathbf{k}) \times (\mathbf{k} - \mathbf{i}) = (\mathbf{j} - \mathbf{k}) \times \mathbf{k} + (\mathbf{j} - \mathbf{k}) \times (-\mathbf{i})$ by Property 3 of Theorem 8

$$= \mathbf{j} \times \mathbf{k} + (-\mathbf{k}) \times \mathbf{k} + \mathbf{j} \times (-\mathbf{i}) + (-\mathbf{k}) \times (-\mathbf{i}) \quad \text{by Property 4 of Theorem 8}$$

$$= (\mathbf{j} \times \mathbf{k}) + (-1)(\mathbf{k} \times \mathbf{k}) + (-1)(\mathbf{j} \times \mathbf{i}) + (-1)^2 (\mathbf{k} \times \mathbf{i}) \quad \text{by Property 2 of Theorem 8}$$

$$= \mathbf{i} + (-1)\,\mathbf{0} + (-1)(-\mathbf{k}) + \mathbf{j} = \mathbf{i} + \mathbf{j} + \mathbf{k} \quad \text{by Example 2 and}$$

the discussion preceding Theorem 8

13. (a) Since $\mathbf{b} \times \mathbf{c}$ is a vector, the dot product $\mathbf{a} \cdot (\mathbf{b} \times \mathbf{c})$ is meaningful and is a scalar.

(b) $\mathbf{b} \cdot \mathbf{c}$ is a scalar, so $\mathbf{a} \times (\mathbf{b} \cdot \mathbf{c})$ is meaningless, as the cross product is defined only for two *vectors*.

(c) Since $\mathbf{b} \times \mathbf{c}$ is a vector, the cross product $\mathbf{a} \times (\mathbf{b} \times \mathbf{c})$ is meaningful and results in another vector.

(d) $\mathbf{b} \cdot \mathbf{c}$ is a scalar, so the dot product $\mathbf{a} \cdot (\mathbf{b} \cdot \mathbf{c})$ is meaningless, as the dot product is defined only for two vectors.

(e) Since $(\mathbf{a} \cdot \mathbf{b})$ and $(\mathbf{c} \cdot \mathbf{d})$ are both scalars, the cross product $(\mathbf{a} \cdot \mathbf{b}) \times (\mathbf{c} \cdot \mathbf{d})$ is meaningless.

(f) $\mathbf{a} \times \mathbf{b}$ and $\mathbf{c} \times \mathbf{d}$ are both vectors, so the dot product $(\mathbf{a} \times \mathbf{b}) \cdot (\mathbf{c} \times \mathbf{d})$ is meaningful and is a scalar.

15. If we sketch $\mathbf{u}$ and $\mathbf{v}$ starting from the same initial point, we see that the

angle between them is $60°$. Using Theorem 6, we have

$$|\mathbf{u} \times \mathbf{v}| = |\mathbf{u}|\,|\mathbf{v}| \sin \theta = (12)(16) \sin 60° = 192 \cdot \frac{\sqrt{3}}{2} = 96\sqrt{3}.$$

By the right-hand rule, $\mathbf{u} \times \mathbf{v}$ is directed into the page.

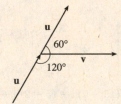

17. $\mathbf{a} \times \mathbf{b} = \begin{vmatrix} \mathbf{i} & \mathbf{j} & \mathbf{k} \\ 2 & -1 & 3 \\ 4 & 2 & 1 \end{vmatrix} = \begin{vmatrix} -1 & 3 \\ 2 & 1 \end{vmatrix} \mathbf{i} - \begin{vmatrix} 2 & 3 \\ 4 & 1 \end{vmatrix} \mathbf{j} + \begin{vmatrix} 2 & -1 \\ 4 & 2 \end{vmatrix} \mathbf{k} = (-1-6)\,\mathbf{i} - (2-12)\,\mathbf{j} + [4-(-4)]\,\mathbf{k} = -7\,\mathbf{i} + 10\,\mathbf{j} + 8\,\mathbf{k}$

$\mathbf{b} \times \mathbf{a} = \begin{vmatrix} \mathbf{i} & \mathbf{j} & \mathbf{k} \\ 4 & 2 & 1 \\ 2 & -1 & 3 \end{vmatrix} = \begin{vmatrix} 2 & 1 \\ -1 & 3 \end{vmatrix} \mathbf{i} - \begin{vmatrix} 4 & 1 \\ 2 & 3 \end{vmatrix} \mathbf{j} + \begin{vmatrix} 4 & 2 \\ 2 & -1 \end{vmatrix} \mathbf{k} = [6-(-1)]\,\mathbf{i} - (12-2)\,\mathbf{j} + (-4-4)\,\mathbf{k} = 7\,\mathbf{i} - 10\,\mathbf{j} - 8\,\mathbf{k}$

Notice $\mathbf{a} \times \mathbf{b} = -\mathbf{b} \times \mathbf{a}$ here, as we know is always true by Property 1 of Theorem 8.

19. By Theorem 5, the cross product of two vectors is orthogonal to both vectors. So we calculate

$$\langle 3, 2, 1 \rangle \times \langle -1, 1, 0 \rangle = \begin{vmatrix} \mathbf{i} & \mathbf{j} & \mathbf{k} \\ 3 & 2 & 1 \\ -1 & 1 & 0 \end{vmatrix} = \begin{vmatrix} 2 & 1 \\ 1 & 0 \end{vmatrix} \mathbf{i} - \begin{vmatrix} 3 & 1 \\ -1 & 0 \end{vmatrix} \mathbf{j} + \begin{vmatrix} 3 & 2 \\ -1 & 1 \end{vmatrix} \mathbf{k} = -\mathbf{i} - \mathbf{j} + 5\,\mathbf{k}.$$

So two unit vectors orthogonal to both are $\pm \dfrac{\langle -1, -1, 5 \rangle}{\sqrt{1+1+25}} = \pm \dfrac{\langle -1, -1, 5 \rangle}{3\sqrt{3}}$, that is, $\left\langle -\dfrac{1}{3\sqrt{3}}, -\dfrac{1}{3\sqrt{3}}, \dfrac{5}{3\sqrt{3}} \right\rangle$

and $\left\langle \dfrac{1}{3\sqrt{3}}, \dfrac{1}{3\sqrt{3}}, -\dfrac{5}{3\sqrt{3}} \right\rangle$.

21. Let $\mathbf{a} = \langle a_1, a_2, a_3 \rangle$. Then

$$\mathbf{0} \times \mathbf{a} = \begin{vmatrix} \mathbf{i} & \mathbf{j} & \mathbf{k} \\ 0 & 0 & 0 \\ a_1 & a_2 & a_3 \end{vmatrix} = \begin{vmatrix} 0 & 0 \\ a_2 & a_3 \end{vmatrix} \mathbf{i} - \begin{vmatrix} 0 & 0 \\ a_1 & a_3 \end{vmatrix} \mathbf{j} + \begin{vmatrix} 0 & 0 \\ a_1 & a_2 \end{vmatrix} \mathbf{k} = \mathbf{0},$$

$$\mathbf{a} \times \mathbf{0} = \begin{vmatrix} \mathbf{i} & \mathbf{j} & \mathbf{k} \\ a_1 & a_2 & a_3 \\ 0 & 0 & 0 \end{vmatrix} = \begin{vmatrix} a_2 & a_3 \\ 0 & 0 \end{vmatrix} \mathbf{i} - \begin{vmatrix} a_1 & a_3 \\ 0 & 0 \end{vmatrix} \mathbf{j} + \begin{vmatrix} a_1 & a_2 \\ 0 & 0 \end{vmatrix} \mathbf{k} = \mathbf{0}.$$

23. $\mathbf{a} \times \mathbf{b} = \langle a_2 b_3 - a_3 b_2, a_3 b_1 - a_1 b_3, a_1 b_2 - a_2 b_1 \rangle$

$\qquad = \langle (-1)(b_2 a_3 - b_3 a_2), (-1)(b_3 a_1 - b_1 a_3), (-1)(b_1 a_2 - b_2 a_1) \rangle$

$\qquad = -\langle b_2 a_3 - b_3 a_2, b_3 a_1 - b_1 a_3, b_1 a_2 - b_2 a_1 \rangle = -\mathbf{b} \times \mathbf{a}$

25. $\mathbf{a} \times (\mathbf{b} + \mathbf{c}) = \mathbf{a} \times \langle b_1 + c_1, b_2 + c_2, b_3 + c_3 \rangle$

$\qquad = \langle a_2(b_3 + c_3) - a_3(b_2 + c_2), a_3(b_1 + c_1) - a_1(b_3 + c_3), a_1(b_2 + c_2) - a_2(b_1 + c_1) \rangle$

$\qquad = \langle a_2 b_3 + a_2 c_3 - a_3 b_2 - a_3 c_2, a_3 b_1 + a_3 c_1 - a_1 b_3 - a_1 c_3, a_1 b_2 + a_1 c_2 - a_2 b_1 - a_2 c_1 \rangle$

$\qquad = \langle (a_2 b_3 - a_3 b_2) + (a_2 c_3 - a_3 c_2), (a_3 b_1 - a_1 b_3) + (a_3 c_1 - a_1 c_3), (a_1 b_2 - a_2 b_1) + (a_1 c_2 - a_2 c_1) \rangle$

$\qquad = \langle a_2 b_3 - a_3 b_2, a_3 b_1 - a_1 b_3, a_1 b_2 - a_2 b_1 \rangle + \langle a_2 c_3 - a_3 c_2, a_3 c_1 - a_1 c_3, a_1 c_2 - a_2 c_1 \rangle$

$\qquad = (\mathbf{a} \times \mathbf{b}) + (\mathbf{a} \times \mathbf{c})$

27. By plotting the vertices, we can see that the parallelogram is determined by the

vectors $\overrightarrow{AB} = \langle 2, 3 \rangle$ and $\overrightarrow{AD} = \langle 4, -2 \rangle$. We know that the area of the parallelogram

determined by two vectors is equal to the length of the cross product of these vectors.

In order to compute the cross product, we consider the vector $\overrightarrow{AB}$ as the three-

dimensional vector $\langle 2, 3, 0 \rangle$ (and similarly for $\overrightarrow{AD}$), and then the area of

parallelogram $ABCD$ is

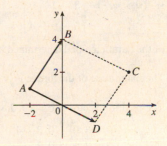

$$\left| \overrightarrow{AB} \times \overrightarrow{AD} \right| = \left\| \begin{matrix} \mathbf{i} & \mathbf{j} & \mathbf{k} \\ 2 & 3 & 0 \\ 4 & -2 & 0 \end{matrix} \right\| = |(0)\mathbf{i} - (0)\mathbf{j} + (-4 - 12)\mathbf{k}| = |-16\,\mathbf{k}| = 16$$

29. (a) Because the plane through P, Q, and R contains the vectors $\overrightarrow{PQ}$ and $\overrightarrow{PR}$, a vector orthogonal to both of these vectors

(such as their cross product) is also orthogonal to the plane. Here $\overrightarrow{PQ} = \langle -3, 1, 2 \rangle$ and $\overrightarrow{PR} = \langle 3, 2, 4 \rangle$, so

$$\overrightarrow{PQ} \times \overrightarrow{PR} = \langle (1)(4) - (2)(2), (2)(3) - (-3)(4), (-3)(2) - (1)(3) \rangle = \langle 0, 18, -9 \rangle$$

Therefore, $\langle 0, 18, -9 \rangle$ (or any nonzero scalar multiple thereof, such as $\langle 0, 2, -1 \rangle$) is orthogonal to the plane through P, Q,

and R.

(b) Note that the area of the triangle determined by P, Q, and R is equal to half of the area of the

parallelogram determined by the three points. From part (a), the area of the parallelogram is

$$\left| \overrightarrow{PQ} \times \overrightarrow{PR} \right| = |\langle 0, 18, -9 \rangle| = \sqrt{0 + 324 + 81} = \sqrt{405} = 9\sqrt{5}, \text{ so the area of the triangle is } \tfrac{1}{2} \cdot 9\sqrt{5} = \tfrac{9}{2}\sqrt{5}.$$

31. (a) $\overrightarrow{PQ} = \langle 4, 3, -2 \rangle$ and $\overrightarrow{PR} = \langle 5, 5, 1 \rangle$, so a vector orthogonal to the plane through P, Q, and R is

$\overrightarrow{PQ} \times \overrightarrow{PR} = \langle (3)(1) - (-2)(5), (-2)(5) - (4)(1), (4)(5) - (3)(5) \rangle = \langle 13, -14, 5 \rangle$ [or any scalar mutiple thereof].

(b) The area of the parallelogram determined by $\overrightarrow{PQ}$ and $\overrightarrow{PR}$ is

$\left| \overrightarrow{PQ} \times \overrightarrow{PR} \right| = |\langle 13, -14, 5 \rangle| = \sqrt{13^2 + (-14)^2 + 5^2} = \sqrt{390}$, so the area of triangle PQR is $\frac{1}{2}\sqrt{390}$.

33. By Equation 11, the volume of the parallelepiped determined by $\mathbf{a}$, $\mathbf{b}$, and $\mathbf{c}$ is the magnitude of their scalar triple product,

which is $\mathbf{a} \cdot (\mathbf{b} \times \mathbf{c}) = \begin{vmatrix} 1 & 2 & 3 \\ -1 & 1 & 2 \\ 2 & 1 & 4 \end{vmatrix} = 1 \begin{vmatrix} 1 & 2 \\ 1 & 4 \end{vmatrix} - 2 \begin{vmatrix} -1 & 2 \\ 2 & 4 \end{vmatrix} + 3 \begin{vmatrix} -1 & 1 \\ 2 & 1 \end{vmatrix} = 1(4 - 2) - 2(-4 - 4) + 3(-1 - 2) = 9$.

Thus the volume of the parallelepiped is 9 cubic units.

35. $\mathbf{a} = \overrightarrow{PQ} = \langle 4, 2, 2 \rangle$, $\mathbf{b} = \overrightarrow{PR} = \langle 3, 3, -1 \rangle$, and $\mathbf{c} = \overrightarrow{PS} = \langle 5, 5, 1 \rangle$.

$\mathbf{a} \cdot (\mathbf{b} \times \mathbf{c}) = \begin{vmatrix} 4 & 2 & 2 \\ 3 & 3 & -1 \\ 5 & 5 & 1 \end{vmatrix} = 4 \begin{vmatrix} 3 & -1 \\ 5 & 1 \end{vmatrix} - 2 \begin{vmatrix} 3 & -1 \\ 5 & 1 \end{vmatrix} + 2 \begin{vmatrix} 3 & 3 \\ 5 & 5 \end{vmatrix} = 32 - 16 + 0 = 16$,

so the volume of the parallelepiped is 16 cubic units.

37. $\mathbf{u} \cdot (\mathbf{v} \times \mathbf{w}) = \begin{vmatrix} 1 & 5 & -2 \\ 3 & -1 & 0 \\ 5 & 9 & -4 \end{vmatrix} = 1 \begin{vmatrix} -1 & 0 \\ 9 & -4 \end{vmatrix} - 5 \begin{vmatrix} 3 & 0 \\ 5 & -4 \end{vmatrix} + (-2) \begin{vmatrix} 3 & -1 \\ 5 & 9 \end{vmatrix} = 4 + 60 - 64 = 0$, which says that the volume

of the parallelepiped determined by $\mathbf{u}$, $\mathbf{v}$ and $\mathbf{w}$ is 0, and thus these three vectors are coplanar.

39. The magnitude of the torque is $|\boldsymbol{\tau}| = |\mathbf{r} \times \mathbf{F}| = |\mathbf{r}|\,|\mathbf{F}| \sin\theta = (0.18 \text{ m})(60 \text{ N}) \sin(70 + 10)° = 10.8 \sin 80° \approx 10.6 \text{ N·m}$.

41. The position vector is $\mathbf{r} = \langle 0, 0.3, 0 \rangle$ and $\mathbf{F}$ has direction $\langle 0, 3, -4 \rangle$. The angle θ between them can be determined by

$\cos\theta = \dfrac{\langle 0, 0.3, 0 \rangle \cdot \langle 0, 3, -4 \rangle}{|\langle 0, 0.3, 0 \rangle|\,|\langle 0, 3, -4 \rangle|} \quad \Rightarrow \quad \cos\theta = \dfrac{0.9}{(0.3)(5)} \quad \Rightarrow \quad \cos\theta = 0.6 \quad \Rightarrow \quad \theta \approx 53.1°$. Then $|\boldsymbol{\tau}| = |\mathbf{r}|\,|\mathbf{F}|\sin\theta \quad \Rightarrow$

$100 = 0.3\,|\mathbf{F}| \sin 53.1° \quad \Rightarrow \quad |\mathbf{F}| \approx 417 \text{ N}$.

43. From Theorem 6 we have $|\mathbf{a} \times \mathbf{b}| = |\mathbf{a}|\,|\mathbf{b}| \sin\theta$, where θ is the angle between $\mathbf{a}$ and $\mathbf{b}$, and from Theorem 10.3.3 we have

$\mathbf{a} \cdot \mathbf{b} = |\mathbf{a}|\,|\mathbf{b}| \cos\theta \quad \Rightarrow \quad |\mathbf{a}|\,|\mathbf{b}| = \dfrac{\mathbf{a} \cdot \mathbf{b}}{\cos\theta}$. Substituting the second equation into the first gives $|\mathbf{a} \times \mathbf{b}| = \dfrac{\mathbf{a} \cdot \mathbf{b}}{\cos\theta} \sin\theta$, so

$\dfrac{|\mathbf{a} \times \mathbf{b}|}{\mathbf{a} \cdot \mathbf{b}} = \tan\theta$. Here $|\mathbf{a} \times \mathbf{b}| = |\langle 1, 2, 2 \rangle| = \sqrt{1 + 4 + 4} = 3$, so $\tan\theta = \dfrac{|\mathbf{a} \times \mathbf{b}|}{\mathbf{a} \cdot \mathbf{b}} = \dfrac{3}{\sqrt{3}} = \sqrt{3} \quad \Rightarrow \quad \theta = 60°$.

45. (a)

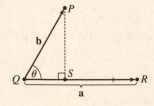

The distance between a point and a line is the length of the perpendicular

from the point to the line, here $\left| \overrightarrow{PS} \right| = d$. But referring to triangle PQS,

$d = \left| \overrightarrow{PS} \right| = \left| \overrightarrow{QP} \right| \sin\theta = |\mathbf{b}| \sin\theta$. But θ is the angle between $\overrightarrow{QP} = \mathbf{b}$

and $\overrightarrow{QR} = \mathbf{a}$. Thus by the definition of the cross product, $\sin\theta = \dfrac{|\mathbf{a} \times \mathbf{b}|}{|\mathbf{a}|\,|\mathbf{b}|}$

and so $d = |\mathbf{b}| \sin\theta = \dfrac{|\mathbf{b}|\,|\mathbf{a} \times \mathbf{b}|}{|\mathbf{a}|\,|\mathbf{b}|} = \dfrac{|\mathbf{a} \times \mathbf{b}|}{|\mathbf{a}|}$.

(b) $\mathbf{a} = \overrightarrow{QR} = \langle -1, -2, -1 \rangle$ and $\mathbf{b} = \overrightarrow{QP} = \langle 1, -5, -7 \rangle$. Then

$$\mathbf{a} \times \mathbf{b} = \langle (-2)(-7) - (-1)(-5), (-1)(1) - (-1)(-7), (-1)(-5) - (-2)(1) \rangle = \langle 9, -8, 7 \rangle.$$

Thus the distance is $d = \dfrac{|\mathbf{a} \times \mathbf{b}|}{|\mathbf{a}|} = \frac{1}{\sqrt{6}} \sqrt{81 + 64 + 49} = \sqrt{\frac{194}{6}} = \sqrt{\frac{97}{3}}$.

47. From Theorem 6 we have $|\mathbf{a} \times \mathbf{b}| = |\mathbf{a}|\,|\mathbf{b}| \sin \theta$ so

$$|\mathbf{a} \times \mathbf{b}|^2 = |\mathbf{a}|^2 |\mathbf{b}|^2 \sin^2 \theta = |\mathbf{a}|^2 |\mathbf{b}|^2 \left(1 - \cos^2 \theta \right)$$
$$= |\mathbf{a}|^2 |\mathbf{b}|^2 - (|\mathbf{a}|\,|\mathbf{b}| \cos \theta)^2 = |\mathbf{a}|^2 |\mathbf{b}|^2 - (\mathbf{a} \cdot \mathbf{b})^2$$

by Theorem 10.3.3.

49. $(\mathbf{a} - \mathbf{b}) \times (\mathbf{a} + \mathbf{b}) = (\mathbf{a} - \mathbf{b}) \times \mathbf{a} + (\mathbf{a} - \mathbf{b}) \times \mathbf{b}$ ⟶ Property 3 of Theorem 8

$\qquad\qquad = \mathbf{a} \times \mathbf{a} + (-\mathbf{b}) \times \mathbf{a} + \mathbf{a} \times \mathbf{b} + (-\mathbf{b}) \times \mathbf{b}$ ⟶ Property 4 of Theorem 8

$\qquad\qquad = (\mathbf{a} \times \mathbf{a}) - (\mathbf{b} \times \mathbf{a}) + (\mathbf{a} \times \mathbf{b}) - (\mathbf{b} \times \mathbf{b})$ ⟶ Property 2 of Theorem 8 (with $c = -1$)

$\qquad\qquad = \mathbf{0} - (\mathbf{b} \times \mathbf{a}) + (\mathbf{a} \times \mathbf{b}) - \mathbf{0}$ ⟶ by Example 2

$\qquad\qquad = (\mathbf{a} \times \mathbf{b}) + (\mathbf{a} \times \mathbf{b})$ ⟶ Property 1 of Theorem 8

$\qquad\qquad = 2(\mathbf{a} \times \mathbf{b})$

51. $\mathbf{a} \times (\mathbf{b} \times \mathbf{c}) + \mathbf{b} \times (\mathbf{c} \times \mathbf{a}) + \mathbf{c} \times (\mathbf{a} \times \mathbf{b})$

$\qquad = [(\mathbf{a} \cdot \mathbf{c})\mathbf{b} - (\mathbf{a} \cdot \mathbf{b})\mathbf{c}] + [(\mathbf{b} \cdot \mathbf{a})\mathbf{c} - (\mathbf{b} \cdot \mathbf{c})\mathbf{a}] + [(\mathbf{c} \cdot \mathbf{b})\mathbf{a} - (\mathbf{c} \cdot \mathbf{a})\mathbf{b}]$ by Exercise 50

$\qquad = (\mathbf{a} \cdot \mathbf{c})\mathbf{b} - (\mathbf{a} \cdot \mathbf{b})\mathbf{c} + (\mathbf{a} \cdot \mathbf{b})\mathbf{c} - (\mathbf{b} \cdot \mathbf{c})\mathbf{a} + (\mathbf{b} \cdot \mathbf{c})\mathbf{a} - (\mathbf{a} \cdot \mathbf{c})\mathbf{b} = \mathbf{0}$

53. (a) No. If $\mathbf{a} \cdot \mathbf{b} = \mathbf{a} \cdot \mathbf{c}$, then $\mathbf{a} \cdot (\mathbf{b} - \mathbf{c}) = 0$, so $\mathbf{a}$ is perpendicular to $\mathbf{b} - \mathbf{c}$, which can happen if $\mathbf{b} \neq \mathbf{c}$. For example, let $\mathbf{a} = \langle 1, 1, 1 \rangle$, $\mathbf{b} = \langle 1, 0, 0 \rangle$ and $\mathbf{c} = \langle 0, 1, 0 \rangle$.

(b) No. If $\mathbf{a} \times \mathbf{b} = \mathbf{a} \times \mathbf{c}$ then $\mathbf{a} \times (\mathbf{b} - \mathbf{c}) = \mathbf{0}$, which implies that $\mathbf{a}$ is parallel to $\mathbf{b} - \mathbf{c}$, which of course can happen if $\mathbf{b} \neq \mathbf{c}$.

(c) Yes. Since $\mathbf{a} \cdot \mathbf{c} = \mathbf{a} \cdot \mathbf{b}$, $\mathbf{a}$ is perpendicular to $\mathbf{b} - \mathbf{c}$, by part (a). From part (b), $\mathbf{a}$ is also parallel to $\mathbf{b} - \mathbf{c}$. Thus since $\mathbf{a} \neq \mathbf{0}$ but is both parallel and perpendicular to $\mathbf{b} - \mathbf{c}$, we have $\mathbf{b} - \mathbf{c} = \mathbf{0}$, so $\mathbf{b} = \mathbf{c}$.

10.5   Equations of Lines and Planes

1. (a) True; each of the first two lines has a direction vector parallel to the direction vector of the third line, so these vectors are each scalar multiples of the third direction vector. Then the first two direction vectors are also scalar multiples of each other, so these vectors, and hence the two lines, are parallel.

(b) False; for example, the x- and y-axes are both perpendicular to the z-axis, yet the x- and y-axes are not parallel.

(c) True; each of the first two planes has a normal vector parallel to the normal vector of the third plane, so these two normal vectors are parallel to each other and the planes are parallel.

(d) False; for example, the xy- and yz-planes are not parallel, yet they are both perpendicular to the xz-plane.

(e) False; the x- and y-axes are not parallel, yet they are both parallel to the plane $z = 1$.

(f) True; if each line is perpendicular to a plane, then the lines' direction vectors are both parallel to a normal vector for the plane. Thus, the direction vectors are parallel to each other and the lines are parallel.

(g) False; the planes $y = 1$ and $z = 1$ are not parallel, yet they are both parallel to the x-axis.

(h) True; if each plane is perpendicular to a line, then any normal vector for each plane is parallel to a direction vector for the line. Thus, the normal vectors are parallel to each other and the planes are parallel.

(i) True; see Figure 9 and the accompanying discussion.

(j) False; they can be skew, as in Example 3.

(k) True. Consider any normal vector for the plane and any direction vector for the line. If the normal vector is perpendicular to the direction vector, the line and plane are parallel. Otherwise, the vectors meet at an angle θ, $0° \leq \theta < 90°$, and the line will intersect the plane at an angle $90° - \theta$.

3. For this line, we have $\mathbf{r}_0 = 2\mathbf{i} + 2.4\mathbf{j} + 3.5\mathbf{k}$ and $\mathbf{v} = 3\mathbf{i} + 2\mathbf{j} - \mathbf{k}$, so a vector equation is
$\mathbf{r} = \mathbf{r}_0 + t\mathbf{v} = (2\mathbf{i} + 2.4\mathbf{j} + 3.5\mathbf{k}) + t(3\mathbf{i} + 2\mathbf{j} - \mathbf{k}) = (2 + 3t)\mathbf{i} + (2.4 + 2t)\mathbf{j} + (3.5 - t)\mathbf{k}$ and parametric equations are $x = 2 + 3t$, $y = 2.4 + 2t$, $z = 3.5 - t$.

5. A line perpendicular to the given plane has the same direction as a normal vector to the plane, such as $\mathbf{n} = \langle 1, 3, 1 \rangle$. So $\mathbf{r}_0 = \mathbf{i} + 6\mathbf{k}$, and we can take $\mathbf{v} = \mathbf{i} + 3\mathbf{j} + \mathbf{k}$. Then a vector equation is $\mathbf{r} = (\mathbf{i} + 6\mathbf{k}) + t(\mathbf{i} + 3\mathbf{j} + \mathbf{k}) = (1 + t)\mathbf{i} + 3t\mathbf{j} + (6 + t)\mathbf{k}$, and parametric equations are $x = 1 + t$, $y = 3t$, $z = 6 + t$.

7. The vector $\mathbf{v} = \langle 2 - 0, 1 - \frac{1}{2}, -3 - 1 \rangle = \langle 2, \frac{1}{2}, -4 \rangle$ is parallel to the line. Letting $P_0 = (2, 1, -3)$, parametric equations are $x = 2 + 2t$, $y = 1 + \frac{1}{2}t$, $z = -3 - 4t$, while symmetric equations are $\dfrac{x - 2}{2} = \dfrac{y - 1}{1/2} = \dfrac{z + 3}{-4}$ or

$\dfrac{x - 2}{2} = 2y - 2 = \dfrac{z + 3}{-4}$.

9. The line has direction $\mathbf{v} = \langle 1, 2, 1 \rangle$. Letting $P_0 = (1, -1, 1)$, parametric equations are $x = 1 + t$, $y = -1 + 2t$, $z = 1 + t$ and symmetric equations are $x - 1 = \dfrac{y + 1}{2} = z - 1$.

11. Direction vectors of the lines are $\mathbf{v}_1 = \langle -2 - (-4), 0 - (-6), -3 - 1 \rangle = \langle 2, 6, -4 \rangle$ and $\mathbf{v}_2 = \langle 5 - 10, 3 - 18, 14 - 4 \rangle = \langle -5, -15, 10 \rangle$, and since $\mathbf{v}_2 = -\frac{5}{2}\mathbf{v}_1$, the direction vectors and thus the lines are parallel.

13. (a) The line passes through the point $(1, -5, 6)$ and a direction vector for the line is $\langle -1, 2, -3 \rangle$, so symmetric equations for the line are $\dfrac{x - 1}{-1} = \dfrac{y + 5}{2} = \dfrac{z - 6}{-3}$.

(b) The line intersects the xy-plane when $z = 0$, so we need $\dfrac{x - 1}{-1} = \dfrac{y + 5}{2} = \dfrac{0 - 6}{-3}$ or $\dfrac{x - 1}{-1} = 2 \implies x = -1$,

$\dfrac{y + 5}{2} = 2 \implies y = -1$. Thus the point of intersection with the xy-plane is $(-1, -1, 0)$. Similarly for the yz-plane, we need $x = 0 \implies 1 = \dfrac{y + 5}{2} = \dfrac{z - 6}{-3} \implies y = -3$, $z = 3$. Thus the line intersects the yz-plane at $(0, -3, 3)$. For the xz-plane, we need $y = 0 \implies \dfrac{x - 1}{-1} = \dfrac{5}{2} = \dfrac{z - 6}{-3} \implies x = -\frac{3}{2}$, $z = -\frac{3}{2}$. So the line intersects the xz-plane at $\left(-\frac{3}{2}, 0, -\frac{3}{2} \right)$.

15. From , the line segment from $\mathbf{r}_0 = 2\,\mathbf{i} - \mathbf{j} + 4\,\mathbf{k}$ to $\mathbf{r}_1 = 4\,\mathbf{i} + 6\,\mathbf{j} + \mathbf{k}$ is

$$\mathbf{r}(t) = (1-t)\,\mathbf{r}_0 + t\,\mathbf{r}_1 = (1-t)(2\,\mathbf{i} - \mathbf{j} + 4\,\mathbf{k}) + t(4\,\mathbf{i} + 6\,\mathbf{j} + \mathbf{k}) = (2\,\mathbf{i} - \mathbf{j} + 4\,\mathbf{k}) + t(2\,\mathbf{i} + 7\,\mathbf{j} - 3\,\mathbf{k}), 0 \le t \le 1.$$

17. Since the direction vectors $\langle 2, -1, 3 \rangle$ and $\langle 4, -2, 5 \rangle$ are not scalar multiples of each other, the lines aren't parallel. For the lines to intersect, we must be able to find one value of t and one value of s that produce the same point from the respective parametric equations. Thus we need to satisfy the following three equations: $3 + 2t = 1 + 4s$, $4 - t = 3 - 2s$, $1 + 3t = 4 + 5s$. Solving the last two equations we get $t = 1$, $s = 0$ and checking, we see that these values don't satisfy the first equation. Thus the lines aren't parallel and don't intersect, so they must be skew lines.

19. Since the direction vectors $\langle 1, -2, -3 \rangle$ and $\langle 1, 3, -7 \rangle$ aren't scalar multiples of each other, the lines aren't parallel. Parametric equations of the lines are $L_1: x = 2 + t, y = 3 - 2t, z = 1 - 3t$ and $L_2: x = 3 + s, y = -4 + 3s, z = 2 - 7s$. Thus, for the lines to intersect, the three equations $2 + t = 3 + s$, $3 - 2t = -4 + 3s$, and $1 - 3t = 2 - 7s$ must be satisfied simultaneously. Solving the first two equations gives $t = 2$, $s = 1$ and checking, we see that these values do satisfy the third equation, so the lines intersect when $t = 2$ and $s = 1$, that is, at the point $(4, -1, -5)$.

21. $\mathbf{i} + 4\,\mathbf{j} + \mathbf{k} = \langle 1, 4, 1 \rangle$ is a normal vector to the plane and $\left(-1, \tfrac{1}{2}, 3\right)$ is a point on the plane, so setting $a = 1$, $b = 4$, $c = 1$, $x_0 = -1$, $y_0 = \tfrac{1}{2}$, $z_0 = 3$ in Equation 7 gives $1[x - (-1)] + 4\left(y - \tfrac{1}{2}\right) + 1(z - 3) = 0$ or $x + 4y + z = 4$ as an equation of the plane.

23. Since the two planes are parallel, they will have the same normal vectors. So we can take $\mathbf{n} = \langle 5, -1, -1 \rangle$, and an equation of the plane is $5(x - 1) - 1[y - (-1)] - 1[z - (-1)] = 0$ or $5x - y - z = 7$.

25. Here the vectors $\mathbf{a} = \langle 1 - 0, 0 - 1, 1 - 1 \rangle = \langle 1, -1, 0 \rangle$ and $\mathbf{b} = \langle 1 - 0, 1 - 1, 0 - 1 \rangle = \langle 1, 0, -1 \rangle$ lie in the plane, so $\mathbf{a} \times \mathbf{b}$ is a normal vector to the plane. Thus, we can take $\mathbf{n} = \mathbf{a} \times \mathbf{b} = \langle 1 - 0, 0 + 1, 0 + 1 \rangle = \langle 1, 1, 1 \rangle$. If P_0 is the point $(0, 1, 1)$, an equation of the plane is $1(x - 0) + 1(y - 1) + 1(z - 1) = 0$ or $x + y + z = 2$.

27. If we first find two nonparallel vectors in the plane, their cross product will be a normal vector to the plane. Since the given line lies in the plane, its direction vector $\mathbf{a} = \langle -2, 5, 4 \rangle$ is one vector in the plane. We can verify that the given point $(6, 0, -2)$ does not lie on this line, so to find another nonparallel vector $\mathbf{b}$ which lies in the plane, we can pick any point on the line and find a vector connecting the points. If we put $t = 0$, we see that $(4, 3, 7)$ is on the line, so $\mathbf{b} = \langle 6 - 4, 0 - 3, -2 - 7 \rangle = \langle 2, -3, -9 \rangle$ and $\mathbf{n} = \mathbf{a} \times \mathbf{b} = \langle -45 + 12, 8 - 18, 6 - 10 \rangle = \langle -33, -10, -4 \rangle$. Thus, an equation of the plane is $-33(x - 6) - 10(y - 0) - 4[z - (-2)] = 0$ or $33x + 10y + 4z = 190$.

29. A direction vector for the line of intersection is $\mathbf{a} = \mathbf{n}_1 \times \mathbf{n}_2 = \langle 1, 1, -1 \rangle \times \langle 2, -1, 3 \rangle = \langle 2, -5, -3 \rangle$, and $\mathbf{a}$ is parallel to the desired plane. Another vector parallel to the plane is the vector connecting any point on the line of intersection to the given point $(-1, 2, 1)$ in the plane. Setting $x = 0$, the equations of the planes reduce to $y - z = 2$ and $-y + 3z = 1$ with simultaneous solution $y = \tfrac{7}{2}$ and $z = \tfrac{3}{2}$. So a point on the line is $\left(0, \tfrac{7}{2}, \tfrac{3}{2}\right)$ and another vector parallel to the plane is $\left\langle -1, -\tfrac{3}{2}, -\tfrac{1}{2} \right\rangle$. Then a normal vector to the plane is $\mathbf{n} = \langle 2, -5, -3 \rangle \times \left\langle -1, -\tfrac{3}{2}, -\tfrac{1}{2} \right\rangle = \langle -2, 4, -8 \rangle$ and an equation of the plane is $-2(x + 1) + 4(y - 2) - 8(z - 1) = 0$ or $x - 2y + 4z = -1$.

31. If a plane is perpendicular to two other planes, its normal vector is perpendicular to the normal vectors of the other two planes. Thus $\langle 2, 1, -2 \rangle \times \langle 1, 0, 3 \rangle = \langle 3 - 0, -2 - 6, 0 - 1 \rangle = \langle 3, -8, -1 \rangle$ is a normal vector to the desired plane. The point $(1, 5, 1)$ lies on the plane, so an equation is $3(x - 1) - 8(y - 5) - (z - 1) = 0$ or $3x - 8y - z = -38$.

33. Substitute the parametric equations of the line into the equation of the plane: $(3 - t) - (2 + t) + 2(5t) = 9 \Rightarrow$ $8t = 8 \Rightarrow t = 1$. Therefore, the point of intersection of the line and the plane is given by $x = 3 - 1 = 2$, $y = 2 + 1 = 3$, and $z = 5(1) = 5$, that is, the point $(2, 3, 5)$.

35. Normal vectors for the planes are $\mathbf{n}_1 = \langle 1, 1, 1 \rangle$ and $\mathbf{n}_2 = \langle 1, -1, 1 \rangle$. The normals are not parallel, so neither are the planes. Furthermore, $\mathbf{n}_1 \cdot \mathbf{n}_2 = 1 - 1 + 1 = 1 \neq 0$, so the planes aren't perpendicular. The angle between them is given by

$$\cos\theta = \frac{\mathbf{n}_1 \cdot \mathbf{n}_2}{|\mathbf{n}_1|\,|\mathbf{n}_2|} = \frac{1}{\sqrt{3}\,\sqrt{3}} = \frac{1}{3} \Rightarrow \theta = \cos^{-1}\left(\tfrac{1}{3}\right) \approx 70.5°.$$

37. The normals are $\mathbf{n}_1 = \langle 1, -4, 2 \rangle$ and $\mathbf{n}_2 = \langle 2, -8, 4 \rangle$. Since $\mathbf{n}_2 = 2\mathbf{n}_1$, the normals (and thus the planes) are parallel.

39. (a) To find a point on the line of intersection, set one of the variables equal to a constant, say $z = 0$. (This will fail if the line of intersection does not cross the xy-plane; in that case, try setting x or y equal to 0.) The equations of the two planes reduce to $x + y = 1$ and $x + 2y = 1$. Solving these two equations gives $x = 1$, $y = 0$. Thus a point on the line is $(1, 0, 0)$. A vector $\mathbf{v}$ in the direction of this intersecting line is perpendicular to the normal vectors of both planes, so we can take $\mathbf{v} = \mathbf{n}_1 \times \mathbf{n}_2 = \langle 1, 1, 1 \rangle \times \langle 1, 2, 2 \rangle = \langle 2 - 2, 1 - 2, 2 - 1 \rangle = \langle 0, -1, 1 \rangle$. By Equations 2, parametric equations for the line are $x = 1$, $y = -t$, $z = t$.

(b) The angle between the planes satisfies $\cos\theta = \dfrac{\mathbf{n}_1 \cdot \mathbf{n}_2}{|\mathbf{n}_1|\,|\mathbf{n}_2|} = \dfrac{1 + 2 + 2}{\sqrt{3}\,\sqrt{9}} = \dfrac{5}{3\sqrt{3}}$. Therefore $\theta = \cos^{-1}\left(\dfrac{5}{3\sqrt{3}}\right) \approx 15.8°$.

41. The plane contains the points $(a, 0, 0)$, $(0, b, 0)$ and $(0, 0, c)$. Thus the vectors $\mathbf{a} = \langle -a, b, 0 \rangle$ and $\mathbf{b} = \langle -a, 0, c \rangle$ lie in the plane, and $\mathbf{n} = \mathbf{a} \times \mathbf{b} = \langle bc - 0, 0 + ac, 0 + ab \rangle = \langle bc, ac, ab \rangle$ is a normal vector to the plane. The equation of the plane is therefore $bcx + acy + abz = abc + 0 + 0$ or $bcx + acy + abz = abc$. Notice that if $a \neq 0$, $b \neq 0$ and $c \neq 0$ then we can rewrite the equation as $\dfrac{x}{a} + \dfrac{y}{b} + \dfrac{z}{c} = 1$. This is a good equation to remember!

43. Two vectors which are perpendicular to the required line are the normal of the given plane, $\langle 1, 1, 1 \rangle$, and a direction vector for the given line, $\langle 1, -1, 2 \rangle$. So a direction vector for the required line is $\langle 1, 1, 1 \rangle \times \langle 1, -1, 2 \rangle = \langle 3, -1, -2 \rangle$. Thus L is given by $\langle x, y, z \rangle = \langle 0, 1, 2 \rangle + t \langle 3, -1, -2 \rangle$, or in parametric form, $x = 3t$, $y = 1 - t$, $z = 2 - 2t$.

45. Let P_i have normal vector $\mathbf{n}_i$. Then $\mathbf{n}_1 = \langle 3, 6, -3 \rangle$, $\mathbf{n}_2 = \langle 4, -12, 8 \rangle$, $\mathbf{n}_3 = \langle 3, -9, 6 \rangle$, $\mathbf{n}_4 = \langle 1, 2, -1 \rangle$. Now $\mathbf{n}_1 = 3\mathbf{n}_4$, so $\mathbf{n}_1$ and $\mathbf{n}_4$ are parallel, and hence P_1 and P_4 are parallel; similarly P_2 and P_3 are parallel because $\mathbf{n}_2 = \frac{4}{3}\mathbf{n}_3$. However, $\mathbf{n}_1$ and $\mathbf{n}_2$ are not parallel (so not all four planes are parallel). Notice that the point $(2, 0, 0)$ lies on both P_1 and P_4, so these two planes are identical. The point $\left(\frac{5}{4}, 0, 0\right)$ lies on P_2 but not on P_3, so these are different planes.

47. Let $Q = (1, 3, 4)$ and $R = (2, 1, 1)$, points on the line corresponding to $t = 0$ and $t = 1$. Let

$P = (4, 1, -2)$. Then $\mathbf{a} = \overrightarrow{QR} = \langle 1, -2, -3 \rangle$, $\mathbf{b} = \overrightarrow{QP} = \langle 3, -2, -6 \rangle$. The distance is

$$d = \frac{|\mathbf{a} \times \mathbf{b}|}{|\mathbf{a}|} = \frac{|\langle 1, -2, -3 \rangle \times \langle 3, -2, -6 \rangle|}{|\langle 1, -2, -3 \rangle|} = \frac{|\langle 6, -3, 4 \rangle|}{|\langle 1, -2, -3 \rangle|} = \frac{\sqrt{6^2 + (-3)^2 + 4^2}}{\sqrt{1^2 + (-2)^2 + (-3)^2}} = \frac{\sqrt{61}}{\sqrt{14}} = \sqrt{\frac{61}{14}}.$$

49. By Equation 9, the distance is $D = \dfrac{|ax_1 + by_1 + cz_1 + d|}{\sqrt{a^2 + b^2 + c^2}} = \dfrac{|3(1) + 2(-2) + 6(4) - 5|}{\sqrt{3^2 + 2^2 + 6^2}} = \dfrac{|18|}{\sqrt{49}} = \dfrac{18}{7}.$

51. Put $y = z = 0$ in the equation of the first plane to get the point $(2, 0, 0)$ on the plane. Because the planes are parallel, the

distance D between them is the distance from $(2, 0, 0)$ to the second plane. By Equation 9,

$$D = \frac{|4(2) - 6(0) + 2(0) - 3|}{\sqrt{4^2 + (-6)^2 + (2)^2}} = \frac{5}{\sqrt{56}} = \frac{5}{2\sqrt{14}} \text{ or } \frac{5\sqrt{14}}{28}.$$

53. The distance between two parallel planes is the same as the distance between a point on one of the planes and the other plane.

Let $P_0 = (x_0, y_0, z_0)$ be a point on the plane given by $ax + by + cz + d_1 = 0$. Then $ax_0 + by_0 + cz_0 + d_1 = 0$ and the

distance between P_0 and the plane given by $ax + by + cz + d_2 = 0$ is, from Equation 9,

$$D = \frac{|ax_0 + by_0 + cz_0 + d_2|}{\sqrt{a^2 + b^2 + c^2}} = \frac{|-d_1 + d_2|}{\sqrt{a^2 + b^2 + c^2}} = \frac{|d_1 - d_2|}{\sqrt{a^2 + b^2 + c^2}}.$$

55. L_1: $x = y = z$ $\Rightarrow$ $x = y$ **(1)**. L_2: $x + 1 = y/2 = z/3$ $\Rightarrow$ $x + 1 = y/2$ **(2)**. The solution of **(1)** and **(2)** is

$x = y = -2$. However, when $x = -2$, $x = z$ $\Rightarrow$ $z = -2$, but $x + 1 = z/3$ $\Rightarrow$ $z = -3$, a contradiction. Hence the

lines do not intersect. For L_1, $\mathbf{v}_1 = \langle 1, 1, 1 \rangle$, and for L_2, $\mathbf{v}_2 = \langle 1, 2, 3 \rangle$, so the lines are not parallel. Thus the lines are skew

lines. If two lines are skew, they can be viewed as lying in two parallel planes and so the distance between the skew lines

would be the same as the distance between these parallel planes. The common normal vector to the planes must be

perpendicular to both $\langle 1, 1, 1 \rangle$ and $\langle 1, 2, 3 \rangle$, the direction vectors of the two lines. So set

$\mathbf{n} = \langle 1, 1, 1 \rangle \times \langle 1, 2, 3 \rangle = \langle 3 - 2, -3 + 1, 2 - 1 \rangle = \langle 1, -2, 1 \rangle$. From above, we know that $(-2, -2, -2)$ and $(-2, -2, -3)$

are points of L_1 and L_2 respectively. So in the notation of Equation 8, $1(-2) - 2(-2) + 1(-2) + d_1 = 0$ $\Rightarrow$ $d_1 = 0$ and

$1(-2) - 2(-2) + 1(-3) + d_2 = 0$ $\Rightarrow$ $d_2 = 1$.

By Exercise 53, the distance between these two skew lines is $D = \dfrac{|0 - 1|}{\sqrt{1 + 4 + 1}} = \dfrac{1}{\sqrt{6}}$.

Alternate solution (without reference to planes): A vector which is perpendicular to both of the lines is

$\mathbf{n} = \langle 1, 1, 1 \rangle \times \langle 1, 2, 3 \rangle = \langle 1, -2, 1 \rangle$. Pick any point on each of the lines, say $(-2, -2, -2)$ and $(-2, -2, -3)$, and form the

vector $\mathbf{b} = \langle 0, 0, 1 \rangle$ connecting the two points. The distance between the two skew lines is the absolute value of the scalar

projection of $\mathbf{b}$ along $\mathbf{n}$, that is, $D = \dfrac{|\mathbf{n} \cdot \mathbf{b}|}{|\mathbf{n}|} = \dfrac{|1 \cdot 0 - 2 \cdot 0 + 1 \cdot 1|}{\sqrt{1 + 4 + 1}} = \dfrac{1}{\sqrt{6}}.$

57. A direction vector for L_1 is $\mathbf{v}_1 = \langle 2, 0, -1 \rangle$ and a direction vector for L_2 is $\mathbf{v}_2 = \langle 3, 2, 2 \rangle$. These vectors are not parallel so neither are the lines. Parametric equations for the lines are L_1: $x = 2t$, $y = 0$, $z = -t$, and L_2: $x = 1 + 3s$, $y = -1 + 2s$, $z = 1 + 2s$. No values of t and s satisfy these equations simultaneously, so the lines don't intersect and hence are skew. We can view the lines as lying in two parallel planes; a common normal vector to the planes is $\mathbf{n} = \mathbf{v}_1 \times \mathbf{v}_2 = \langle 2, -7, 4 \rangle$. Line L_1 passes through the origin, so $(0, 0, 0)$ lies on one of the planes, and $(1, -1, 1)$ is a point on L_2 and therefore on the other plane. Equations of the planes then are $2x - 7y + 4z = 0$ and $2x - 7y + 4z - 13 = 0$, and by Exercise 53, the distance between the two skew lines is $D = \dfrac{|0 - (-13)|}{\sqrt{4 + 49 + 16}} = \dfrac{13}{\sqrt{69}}$.

Alternate solution (without reference to planes): Direction vectors of the two lines are $\mathbf{v}_1 = \langle 2, 0, -1 \rangle$ and $\mathbf{v}_2 = \langle 3, 2, 2 \rangle$. Then $\mathbf{n} = \mathbf{v}_1 \times \mathbf{v}_2 = \langle 2, -7, 4 \rangle$ is perpendicular to both lines. Pick any point on each of the lines, say $(0, 0, 0)$ and $(1, -1, 1)$, and form the vector $\mathbf{b} = \langle 1, -1, 1 \rangle$ connecting the two points. Then the distance between the two skew lines is the absolute value of the scalar projection of $\mathbf{b}$ along $\mathbf{n}$, that is, $D = \dfrac{|\mathbf{n} \cdot \mathbf{b}|}{|\mathbf{n}|} = \dfrac{|2 + 7 + 4|}{\sqrt{4 + 49 + 16}} = \dfrac{13}{\sqrt{69}}$.

59. If $a \neq 0$, then $ax + by + cz + d = 0 \implies a(x + d/a) + b(y - 0) + c(z - 0) = 0$ which by (7) is the scalar equation of the plane through the point $(-d/a, 0, 0)$ with normal vector $\langle a, b, c \rangle$. Similarly, if $b \neq 0$ (or if $c \neq 0$) the equation of the plane can be rewritten as $a(x - 0) + b(y + d/b) + c(z - 0) = 0$ [or as $a(x - 0) + b(y - 0) + c(z + d/c) = 0$] which by (7) is the scalar equation of a plane through the point $(0, -d/b, 0)$ [or the point $(0, 0, -d/c)$] with normal vector $\langle a, b, c \rangle$.

10.6 Cylinders and Quadric Surfaces

1. (a) In $\mathbb{R}^2$, the equation $y = x^2$ represents a parabola.

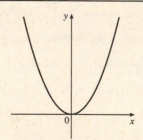

(b) In $\mathbb{R}^3$, the equation $y = x^2$ doesn't involve z, so any horizontal plane with equation $z = k$ intersects the graph in a curve with equation $y = x^2$. Thus, the surface is a parabolic cylinder, made up of infinitely many shifted copies of the same parabola. The rulings are parallel to the z-axis.

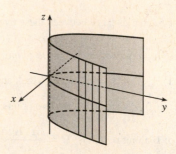

(c) In $\mathbb{R}^3$, the equation $z = y^2$ also represents a parabolic

cylinder. Since x doesn't appear, the graph is formed by

moving the parabola $z = y^2$ in the direction of the x-axis.

Thus, the rulings of the cylinder are parallel to the x-axis.

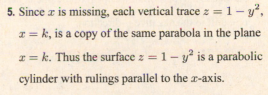

3. Since y is missing from the equation, the vertical traces

$x^2 + z^2 = 1$, $y = k$, are copies of the same circle in

the plane $y = k$. Thus the surface $x^2 + z^2 = 1$ is a

circular cylinder with rulings parallel to the y-axis.

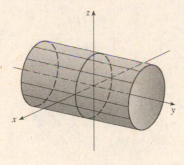

5. Since x is missing, each vertical trace $z = 1 - y^2$,

$x = k$, is a copy of the same parabola in the plane

$x = k$. Thus the surface $z = 1 - y^2$ is a parabolic

cylinder with rulings parallel to the x-axis.

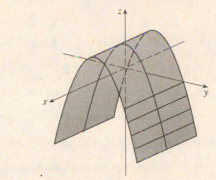

7. Since z is missing, each horizontal trace $xy = 1$,

$z = k$, is a copy of the same hyperbola in the plane

$z = k$. Thus the surface $xy = 1$ is a hyperbolic

cylinder with rulings parallel to the z-axis.

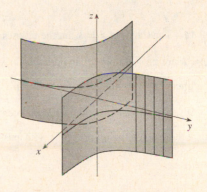

9. (a) The traces of $x^2 + y^2 - z^2 = 1$ in $x = k$ are $y^2 - z^2 = 1 - k^2$, a family of hyperbolas. (Note that the hyperbolas are

oriented differently for $-1 < k < 1$ than for $k < -1$ or $k > 1$.) The traces in $y = k$ are $x^2 - z^2 = 1 - k^2$, a similar

family of hyperbolas. The traces in $z = k$ are $x^2 + y^2 = 1 + k^2$, a family of circles. For $k = 0$, the trace in the

xy-plane, the circle is of radius 1. As $|k|$ increases, so does the radius of the circle. This behavior, combined with the

hyperbolic vertical traces, gives the graph of the hyperboloid of one sheet in Table 1.

(b) The shape of the surface is unchanged, but the hyperboloid is rotated so that its axis is the y-axis. Traces in $y = k$ are circles, while traces in $x = k$ and $z = k$ are hyperbolas.

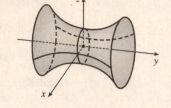

(c) Completing the square in y gives $x^2 + (y+1)^2 - z^2 = 1$. The surface is a hyperboloid identical to the one in part (a) but shifted one unit in the negative y-direction.

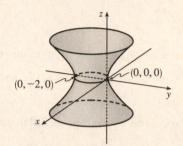

11. For $x = y^2 + 4z^2$, the traces in $x = k$ are $y^2 + 4z^2 = k$. When $k > 0$ we have a family of ellipses. When $k = 0$ we have just a point at the origin, and the trace is empty for $k < 0$. The traces in $y = k$ are $x = 4z^2 + k^2$, a family of parabolas opening in the positive x-direction. Similarly, the traces in $z = k$ are $x = y^2 + 4k^2$, a family of parabolas opening in the positive x-direction. We recognize the graph as an elliptic paraboloid with axis the x-axis and vertex the origin.

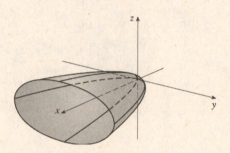

13. $x^2 = y^2 + 4z^2$. The traces in $x = k$ are the ellipses $y^2 + 4z^2 = k^2$. The traces in $y = k$ are $x^2 - 4z^2 = k^2$, hyperbolas for $k \neq 0$ and two intersecting lines if $k = 0$. Similarly, the traces in $z = k$ are $x^2 - y^2 = 4k^2$, hyperbolas for $k \neq 0$ and two intersecting lines if $k = 0$. We recognize the graph as an elliptic cone with axis the x-axis and vertex the origin.

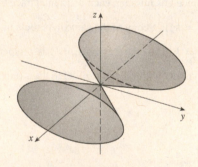

15. $-x^2 + 4y^2 - z^2 = 4$. The traces in $x = k$ are the hyperbolas $4y^2 - z^2 = 4 + k^2$. The traces in $y = k$ are $x^2 + z^2 = 4k^2 - 4$, a family of circles for $|k| > 1$, and the traces in $z = k$ are $4y^2 - x^2 = 4 + k^2$, a family of hyperbolas. Thus the surface is a hyperboloid of two sheets with axis the y-axis.

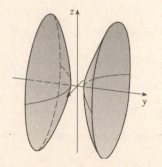

17. $36x^2 + y^2 + 36z^2 = 36$. The traces in $x = k$ are $y^2 + 36z^2 = 36(1 - k^2)$,

a family of ellipses for $|k| < 1$. (The traces are a single point for $|k| = 1$

and are empty for $|k| > 1$.) The traces in $y = k$ are the circles

$36x^2 + 36z^2 = 36 - k^2 \quad \Leftrightarrow \quad x^2 + z^2 = 1 - \frac{1}{36}k^2, |k| < 6$, and the

traces in $z = k$ are the ellipses $36x^2 + y^2 = 36(1 - k^2), |k| < 1$. The

graph is an ellipsoid centered at the origin with intercepts $x = \pm 1$, $y = \pm 6$,

$z = \pm 1$.

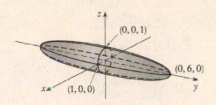

19. $y = z^2 - x^2$. The traces in $x = k$ are the parabolas $y = z^2 - k^2$;

the traces in $y = k$ are $k = z^2 - x^2$, which are hyperbolas (note the hyperbolas

are oriented differently for $k > 0$ than for $k < 0$); and the traces in $z = k$ are

the parabolas $y = k^2 - x^2$. Thus, $\frac{y}{1} = \frac{z^2}{1^2} - \frac{x^2}{1^2}$ is a hyperbolic paraboloid.

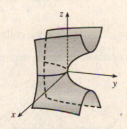

21. $y^2 = x^2 + \frac{1}{9}z^2$ or $y^2 = x^2 + \frac{z^2}{9}$ represents an elliptic

cone with vertex $(0, 0, 0)$ and axis the y-axis.

23. $x^2 + 2y - 2z^2 = 0$ or $2y = 2z^2 - x^2$ or $y = z^2 - \frac{x^2}{2}$

represents a hyperbolic paraboloid with center $(0, 0, 0)$.

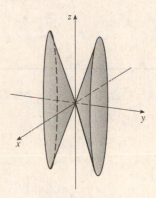

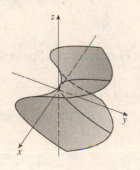

25. Completing squares in y and z gives

$4x^2 + (y - 2)^2 + 4(z - 3)^2 = 4$ or

$x^2 + \frac{(y - 2)^2}{4} + (z - 3)^2 = 1$, an ellipsoid with

center $(0, 2, 3)$.

27. Completing squares in all three variables gives

$(x - 2)^2 - (y + 1)^2 + (z - 1)^2 = 0$ or

$(y + 1)^2 = (x - 2)^2 + (z - 1)^2$, a circular cone with

center $(2, -1, 1)$ and axis the horizontal line $x = 2$,

$z = 1$.

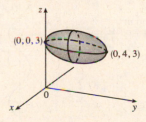

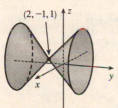

29.

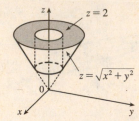

31. Let $P = (x, y, z)$ be an arbitrary point equidistant from $(-1, 0, 0)$ and the plane $x = 1$. Then the distance from P to

$(-1, 0, 0)$ is $\sqrt{(x+1)^2 + y^2 + z^2}$ and the distance from P to the plane $x = 1$ is $|x - 1|$. So

$|x - 1| = \sqrt{(x+1)^2 + y^2 + z^2} \iff (x-1)^2 = (x+1)^2 + y^2 + z^2 \iff x^2 - 2x + 1 = x^2 + 2x + 1 + y^2 + z^2 \iff$

$-4x = y^2 + z^2$. Thus the collection of all such points P is a circular paraboloid with vertex at the origin, axis the x-axis,

which opens in the negative direction.

33.

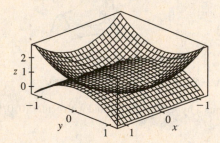

The curve of intersection looks like a bent ellipse. The projection

of this curve onto the xy-plane is the set of points $(x, y, 0)$ which

satisfy $x^2 + y^2 = 1 - y^2 \iff x^2 + 2y^2 = 1 \iff$

$x^2 + \dfrac{y^2}{\left(1/\sqrt{2}\right)^2} = 1$. This is an equation of an ellipse.

10.7 Vector Functions and Space Curves

1. The component functions $\sqrt{4 - t^2}$, e^{-3t}, and $\ln(t + 1)$ are all defined when $4 - t^2 \geq 0 \implies -2 \leq t \leq 2$ and

$t + 1 > 0 \implies t > -1$, so the domain of $\mathbf{r}$ is $(-1, 2]$.

3. $\displaystyle\lim_{t \to 0} e^{-3t} = e^0 = 1$, $\displaystyle\lim_{t \to 0} \frac{t^2}{\sin^2 t} = \lim_{t \to 0} \frac{1}{\dfrac{\sin^2 t}{t^2}} = \frac{1}{\displaystyle\lim_{t \to 0} \frac{\sin^2 t}{t^2}} = \frac{1}{\left(\displaystyle\lim_{t \to 0} \frac{\sin t}{t}\right)^2} = \frac{1}{1^2} = 1$,

and $\displaystyle\lim_{t \to 0} \cos 2t = \cos 0 = 1$. Thus

$\displaystyle\lim_{t \to 0} \left(e^{-3t}\,\mathbf{i} + \frac{t^2}{\sin^2 t}\,\mathbf{j} + \cos 2t\,\mathbf{k} \right) = \left[\lim_{t \to 0} e^{-3t}\right]\mathbf{i} + \left[\lim_{t \to 0} \frac{t^2}{\sin^2 t}\right]\mathbf{j} + \left[\lim_{t \to 0} \cos 2t\right]\mathbf{k} = \mathbf{i} + \mathbf{j} + \mathbf{k}$.

5. The corresponding parametric equations for this curve are $x = \sin t$, $y = t$.

We can make a table of values, or we can eliminate the parameter: $t = y \implies$

$x = \sin y$, with $y \in \mathbb{R}$. By comparing different values of t, we find the direction in

which t increases as indicated in the graph.

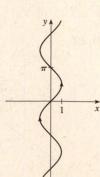

7. The corresponding parametric equations are $x = t$, $y = 2 - t$, $z = 2t$, which are

parametric equations of a line through the point $(0, 2, 0)$ and with direction vector

$\langle 1, -1, 2 \rangle$.

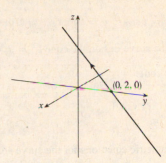

9. The corresponding parametric equations are $x = 1$, $y = \cos t$, $z = 2 \sin t$.

Eliminating the parameter in y and z gives $y^2 + (z/2)^2 = \cos^2 t + \sin^2 t = 1$

or $y^2 + z^2/4 = 1$. Since $x = 1$, the curve is an ellipse centered at $(1, 0, 0)$ in

the plane $x = 1$.

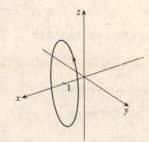

11. The parametric equations are $x = t^2$, $y = t^4$, $z = t^6$. These are positive

for $t \neq 0$ and 0 when $t = 0$. So the curve lies entirely in the first octant.

The projection of the graph onto the xy-plane is $y = x^2$, $y > 0$, a half parabola.

Onto the xz-plane $z = x^3$, $z > 0$, a half cubic, and the yz-plane, $y^3 = z^2$.

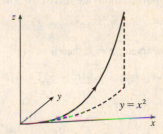

13. Taking $\mathbf{r}_0 = \langle 2, 0, 0 \rangle$ and $\mathbf{r}_1 = \langle 6, 2, -2 \rangle$, we have from Equation 10.5.4

$\mathbf{r}(t) = (1 - t)\mathbf{r}_0 + t\mathbf{r}_1 = (1 - t)\langle 2, 0, 0 \rangle + t\langle 6, 2, -2 \rangle$, $0 \le t \le 1$ or $\mathbf{r}(t) = \langle 2 + 4t, 2t, -2t \rangle$, $0 \le t \le 1$.

Parametric equations are $x = 2 + 4t$, $y = 2t$, $z = -2t$, $0 \le t \le 1$.

15. Taking $\mathbf{r}_0 = \langle 0, -1, 1 \rangle$ and $\mathbf{r}_1 = \langle \frac{1}{2}, \frac{1}{3}, \frac{1}{4} \rangle$, we have

$\mathbf{r}(t) = (1 - t)\mathbf{r}_0 + t\mathbf{r}_1 = (1 - t)\langle 0, -1, 1 \rangle + t\langle \frac{1}{2}, \frac{1}{3}, \frac{1}{4} \rangle$, $0 \le t \le 1$ or $\mathbf{r}(t) = \langle \frac{1}{2}t, -1 + \frac{4}{3}t, 1 - \frac{3}{4}t \rangle$, $0 \le t \le 1$.

Parametric equations are $x = \frac{1}{2}t$, $y = -1 + \frac{4}{3}t$, $z = 1 - \frac{3}{4}t$, $0 \le t \le 1$.

17. $x = t \cos t$, $y = t$, $z = t \sin t$, $t \ge 0$. At any point (x, y, z) on the curve, $x^2 + z^2 = t^2 \cos^2 t + t^2 \sin^2 t = t^2 = y^2$ so the

curve lies on the circular cone $x^2 + z^2 = y^2$ with axis the y-axis. Also notice that $y \ge 0$; the graph is II.

19. $x = t$, $y = 1/(1 + t^2)$, $z = t^2$. At any point on the curve we have $z = x^2$, so the curve lies on a parabolic cylinder parallel

to the y-axis. Notice that $0 < y \le 1$ and $z \ge 0$. Also the curve passes through $(0, 1, 0)$ when $t = 0$ and $y \to 0$, $z \to \infty$ as

$t \to \pm\infty$, so the graph must be V.

21. $x = \cos 8t$, $y = \sin 8t$, $z = e^{0.8t}$, $t \ge 0$. $x^2 + y^2 = \cos^2 8t + \sin^2 8t = 1$, so the curve lies on a circular cylinder with

axis the z-axis. A point (x, y, z) on the curve lies directly above the point $(x, y, 0)$, which moves counterclockwise around the

unit circle in the xy-plane as t increases. The curve starts at $(1, 0, 1)$, when $t = 0$, and $z \to \infty$ (at an increasing rate) as

$t \to \infty$, so the graph is IV.

23. If $x = t \cos t$, $y = t \sin t$, $z = t$, then $x^2 + y^2 = t^2 \cos^2 t + t^2 \sin^2 t = t^2 = z^2$, so the curve lies on the cone $z^2 = x^2 + y^2$. Since $z = t$, the curve is a spiral on this cone.

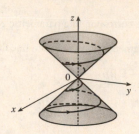

25. Parametric equations for the curve are $x = t$, $y = 0$, $z = 2t - t^2$. Substituting into the equation of the paraboloid gives $2t - t^2 = t^2 \Rightarrow 2t = 2t^2 \Rightarrow t = 0, 1$. Since $\mathbf{r}(0) = \mathbf{0}$ and $\mathbf{r}(1) = \mathbf{i} + \mathbf{k}$, the points of intersection are $(0, 0, 0)$ and $(1, 0, 1)$.

27. If $t = -1$, then $x = 1$, $y = 4$, $z = 0$, so the curve passes through the point $(1, 4, 0)$. If $t = 3$, then $x = 9$, $y = -8$, $z = 28$, so the curve passes through the point $(9, -8, 28)$. For the point $(4, 7, -6)$ to be on the curve, we require $y = 1 - 3t = 7 \Rightarrow t = -2$. But then $z = 1 + (-2)^3 = -7 \neq -6$, so $(4, 7, -6)$ is not on the curve.

29. Both equations are solved for z, so we can substitute to eliminate z: $\sqrt{x^2 + y^2} = 1 + y \Rightarrow x^2 + y^2 = 1 + 2y + y^2 \Rightarrow x^2 = 1 + 2y \Rightarrow y = \frac{1}{2}(x^2 - 1)$. We can form parametric equations for the curve C of intersection by choosing a parameter $x = t$, then $y = \frac{1}{2}(t^2 - 1)$ and $z = 1 + y = 1 + \frac{1}{2}(t^2 - 1) = \frac{1}{2}(t^2 + 1)$. Thus a vector function representing C is $\mathbf{r}(t) = t\,\mathbf{i} + \frac{1}{2}(t^2 - 1)\,\mathbf{j} + \frac{1}{2}(t^2 + 1)\,\mathbf{k}$.

31.

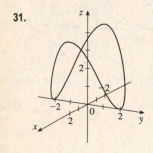
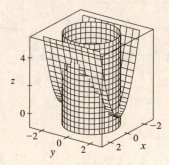

The projection of the curve C of intersection onto the xy-plane is the circle $x^2 + y^2 = 4, z = 0$. Then we can write $x = 2 \cos t$, $y = 2 \sin t$, $0 \le t \le 2\pi$. Since C also lies on the surface $z = x^2$, we have $z = x^2 = (2 \cos t)^2 = 4 \cos^2 t$. Then parametric equations for C are $x = 2 \cos t$, $y = 2 \sin t$, $z = 4 \cos^2 t$, $0 \le t \le 2\pi$.

33. Since $(x + 2)^2 = t^2 = y - 1 \Rightarrow$ (a), (c)

$y = (x + 2)^2 + 1$, the curve is a

parabola.

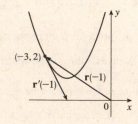

(b) $\mathbf{r}'(t) = \langle 1, 2t \rangle$,

$\mathbf{r}'(-1) = \langle 1, -2 \rangle$

35. $x = \sin t$, $y = 2 \cos t$ so   (a), (c)

$x^2 + (y/2)^2 = 1$ and the curve is

an ellipse.

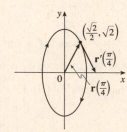

(b) $\mathbf{r}'(t) = \cos t\,\mathbf{i} - 2 \sin t\,\mathbf{j}$,

$\mathbf{r}'\left(\frac{\pi}{4}\right) = \frac{\sqrt{2}}{2}\mathbf{i} - \sqrt{2}\,\mathbf{j}$

37. Since $x = e^{2t} = (e^t)^2 = y^2$, the curve is part of a parabola. Note that here $x > 0$, $y > 0$.

(a), (c)

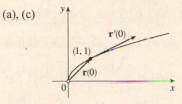

(b) $\mathbf{r}'(t) = 2e^{2t}\,\mathbf{i} + e^t\,\mathbf{j}$,

$\mathbf{r}'(0) = 2\,\mathbf{i} + \mathbf{j}$

39. $\mathbf{r}'(t) = \left\langle \dfrac{d}{dt}\,[t\sin t],\ \dfrac{d}{dt}\,[t^2],\ \dfrac{d}{dt}\,[t\cos 2t] \right\rangle = \langle t\cos t + \sin t,\ 2t,\ t(-\sin 2t)\cdot 2 + \cos 2t \rangle$

$= \langle t\cos t + \sin t,\ 2t,\ \cos 2t - 2t\sin 2t \rangle$

41. $\mathbf{r}(t) = e^{t^2}\,\mathbf{i} - \mathbf{j} + \ln(1 + 3t)\,\mathbf{k} \ \Rightarrow \ \mathbf{r}'(t) = 2te^{t^2}\,\mathbf{i} + \dfrac{3}{1 + 3t}\,\mathbf{k}$

43. $\mathbf{r}'(t) = \mathbf{0} + \mathbf{b} + 2t\,\mathbf{c} = \mathbf{b} + 2t\,\mathbf{c}$ by Formulas 1 and 3 of Theorem 5.

45. $\mathbf{r}'(t) = -\sin t\,\mathbf{i} + 3\,\mathbf{j} + 4\cos 2t\,\mathbf{k} \ \Rightarrow \ \mathbf{r}'(0) = 3\,\mathbf{j} + 4\,\mathbf{k}$. Thus

$\mathbf{T}(0) = \dfrac{\mathbf{r}'(0)}{|\mathbf{r}'(0)|} = \dfrac{1}{\sqrt{0^2 + 3^2 + 4^2}}\,(3\,\mathbf{j} + 4\,\mathbf{k}) = \tfrac{1}{5}(3\,\mathbf{j} + 4\,\mathbf{k}) = \tfrac{3}{5}\,\mathbf{j} + \tfrac{4}{5}\,\mathbf{k}$.

47. $\mathbf{r}(t) = \langle t, t^2, t^3 \rangle \ \Rightarrow \ \mathbf{r}'(t) = \langle 1, 2t, 3t^2 \rangle$. Then $\mathbf{r}'(1) = \langle 1, 2, 3 \rangle$ and $|\mathbf{r}'(1)| = \sqrt{1^2 + 2^2 + 3^2} = \sqrt{14}$, so

$\mathbf{T}(1) = \dfrac{\mathbf{r}'(1)}{|\mathbf{r}'(1)|} = \dfrac{1}{\sqrt{14}}\,\langle 1, 2, 3 \rangle = \left\langle \dfrac{1}{\sqrt{14}}, \dfrac{2}{\sqrt{14}}, \dfrac{3}{\sqrt{14}} \right\rangle$. $\mathbf{r}''(t) = \langle 0, 2, 6t \rangle$, so

$$\mathbf{r}'(t) \times \mathbf{r}''(t) = \begin{vmatrix} \mathbf{i} & \mathbf{j} & \mathbf{k} \\ 1 & 2t & 3t^2 \\ 0 & 2 & 6t \end{vmatrix} = \begin{vmatrix} 2t & 3t^2 \\ 2 & 6t \end{vmatrix}\mathbf{i} - \begin{vmatrix} 1 & 3t^2 \\ 0 & 6t \end{vmatrix}\mathbf{j} + \begin{vmatrix} 1 & 2t \\ 0 & 2 \end{vmatrix}\mathbf{k}$$

$$= (12t^2 - 6t^2)\,\mathbf{i} - (6t - 0)\,\mathbf{j} + (2 - 0)\,\mathbf{k} = \langle 6t^2, -6t, 2 \rangle$$

49. The vector equation for the curve is $\mathbf{r}(t) = \langle 1 + 2\sqrt{t}, t^3 - t, t^3 + t \rangle$, so $\mathbf{r}'(t) = \langle 1/\sqrt{t}, 3t^2 - 1, 3t^2 + 1 \rangle$. The point $(3, 0, 2)$ corresponds to $t = 1$, so the tangent vector there is $\mathbf{r}'(1) = \langle 1, 2, 4 \rangle$. Thus, the tangent line goes through the point $(3, 0, 2)$ and is parallel to the vector $\langle 1, 2, 4 \rangle$. Parametric equations are $x = 3 + t$, $y = 2t$, $z = 2 + 4t$.

51. The vector equation for the curve is $\mathbf{r}(t) = \langle e^{-t}\cos t, e^{-t}\sin t, e^{-t} \rangle$, so

$$\mathbf{r}'(t) = \langle e^{-t}(-\sin t) + (\cos t)(-e^{-t}), e^{-t}\cos t + (\sin t)(-e^{-t}), (-e^{-t}) \rangle$$

$$= \langle -e^{-t}(\cos t + \sin t), e^{-t}(\cos t - \sin t), -e^{-t} \rangle$$

The point $(1, 0, 1)$ corresponds to $t = 0$, so the tangent vector there is

$\mathbf{r}'(0) = \langle -e^0(\cos 0 + \sin 0), e^0(\cos 0 - \sin 0), -e^0 \rangle = \langle -1, 1, -1 \rangle$. Thus, the tangent line is parallel to the vector $\langle -1, 1, -1 \rangle$ and parametric equations are $x = 1 + (-1)t = 1 - t$, $y = 0 + 1 \cdot t = t$, $z = 1 + (-1)t = 1 - t$.

53. First we parametrize the curve C of intersection. The projection of C onto the xy-plane is contained in the circle $x^2 + y^2 = 25$, $z = 0$, so we can write $x = 5\cos t$, $y = 5\sin t$. C also lies on the cylinder $y^2 + z^2 = 20$, and $z \geq 0$ near the point $(3, 4, 2)$, so we can write $z = \sqrt{20 - y^2} = \sqrt{20 - 25\sin^2 t}$. A vector equation then for C is

$\mathbf{r}(t) = \left\langle 5\cos t, 5\sin t, \sqrt{20 - 25\sin^2 t} \right\rangle \quad \Rightarrow \quad \mathbf{r}'(t) = \left\langle -5\sin t, 5\cos t, \frac{1}{2}(20 - 25\sin^2 t)^{-1/2}(-50\sin t\cos t) \right\rangle.$

The point $(3, 4, 2)$ corresponds to $t = \cos^{-1}\left(\frac{3}{5}\right)$, so the tangent vector there is

$\mathbf{r}'\left(\cos^{-1}\left(\frac{3}{5}\right)\right) = \left\langle -5\left(\frac{4}{5}\right), 5\left(\frac{3}{5}\right), \frac{1}{2}\left(20 - 25\left(\frac{4}{5}\right)^2\right)^{-1/2}\left(-50\left(\frac{4}{5}\right)\left(\frac{3}{5}\right)\right) \right\rangle = \langle -4, 3, -6 \rangle.$

The tangent line is parallel to this vector and passes through $(3, 4, 2)$, so a vector equation for the line

is $\mathbf{r}(t) = (3 - 4t)\mathbf{i} + (4 + 3t)\mathbf{j} + (2 - 6t)\mathbf{k}.$

55. $\mathbf{r}(t) = \langle t\cos t, t, t\sin t \rangle \quad \Rightarrow \quad \mathbf{r}'(t) = \langle \cos t - t\sin t, 1, t\cos t + \sin t \rangle.$

At $(-\pi, \pi, 0)$, $t = \pi$ and $\mathbf{r}'(\pi) = \langle -1, 1, -\pi \rangle.$ Thus, parametric equations

of the tangent line are $x = -\pi - t, \; y = \pi + t, \; z = -\pi t.$

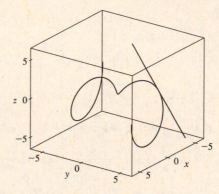

57. The angle of intersection of the two curves is the angle between the two tangent vectors to the curves at the point of

intersection. Since $\mathbf{r}'_1(t) = \langle 1, 2t, 3t^2 \rangle$ and $t = 0$ at $(0, 0, 0)$, $\mathbf{r}'_1(0) = \langle 1, 0, 0 \rangle$ is a tangent vector to $\mathbf{r}_1$ at $(0, 0, 0)$. Similarly,

$\mathbf{r}'_2(t) = \langle \cos t, 2\cos 2t, 1 \rangle$ and since $\mathbf{r}_2(0) = \langle 0, 0, 0 \rangle$, $\mathbf{r}'_2(0) = \langle 1, 2, 1 \rangle$ is a tangent vector to $\mathbf{r}_2$ at $(0, 0, 0)$. If θ is the angle

between these two tangent vectors, then $\cos\theta = \frac{1}{\sqrt{1}\sqrt{6}}\langle 1, 0, 0 \rangle \cdot \langle 1, 2, 1 \rangle = \frac{1}{\sqrt{6}}$ and $\theta = \cos^{-1}\left(\frac{1}{\sqrt{6}}\right) \approx 66°.$

59. $\int_0^2 (t\,\mathbf{i} - t^3\,\mathbf{j} + 3t^5\,\mathbf{k})\,dt = \left(\int_0^2 t\,dt\right)\mathbf{i} - \left(\int_0^2 t^3\,dt\right)\mathbf{j} + \left(\int_0^2 3t^5\,dt\right)\mathbf{k}$

$= \left[\frac{1}{2}t^2\right]_0^2 \mathbf{i} - \left[\frac{1}{4}t^4\right]_0^2 \mathbf{j} + \left[\frac{1}{2}t^6\right]_0^2 \mathbf{k}$

$= \frac{1}{2}(4 - 0)\mathbf{i} - \frac{1}{4}(16 - 0)\mathbf{j} + \frac{1}{2}(64 - 0)\mathbf{k} = 2\mathbf{i} - 4\mathbf{j} + 32\mathbf{k}$

61. $\int_0^{\pi/2} (3\sin^2 t\cos t\,\mathbf{i} + 3\sin t\cos^2 t\,\mathbf{j} + 2\sin t\cos t\,\mathbf{k})\,dt$

$= \left(\int_0^{\pi/2} 3\sin^2 t\cos t\,dt\right)\mathbf{i} + \left(\int_0^{\pi/2} 3\sin t\cos^2 t\,dt\right)\mathbf{j} + \left(\int_0^{\pi/2} 2\sin t\cos t\,dt\right)\mathbf{k}$

$= \left[\sin^3 t\right]_0^{\pi/2}\mathbf{i} + \left[-\cos^3 t\right]_0^{\pi/2}\mathbf{j} + \left[\sin^2 t\right]_0^{\pi/2}\mathbf{k} = (1 - 0)\mathbf{i} + (0 + 1)\mathbf{j} + (1 - 0)\mathbf{k} = \mathbf{i} + \mathbf{j} + \mathbf{k}$

63. $\int (\sec^2 t\,\mathbf{i} + t(t^2 + 1)^3\,\mathbf{j} + t^2\ln t\,\mathbf{k})\,dt = \left(\int \sec^2 t\,dt\right)\mathbf{i} + \left(\int t(t^2 + 1)^3\,dt\right)\mathbf{j} + \left(\int t^2\ln t\,dt\right)\mathbf{k}$

$= \tan t\,\mathbf{i} + \frac{1}{8}(t^2 + 1)^4\,\mathbf{j} + \left(\frac{1}{3}t^3\ln t - \frac{1}{9}t^3\right)\mathbf{k} + \mathbf{C},$

where $\mathbf{C}$ is a vector constant of integration. [For the z-component, integrate by parts with $u = \ln t, \; dv = t^2\,dt.$]

65. $\mathbf{r}'(t) = 2t\,\mathbf{i} + 3t^2\,\mathbf{j} + \sqrt{t}\,\mathbf{k} \quad \Rightarrow \quad \mathbf{r}(t) = t^2\,\mathbf{i} + t^3\,\mathbf{j} + \frac{2}{3}t^{3/2}\,\mathbf{k} + \mathbf{C},$ where $\mathbf{C}$ is a constant vector.

But $\mathbf{i} + \mathbf{j} = \mathbf{r}(1) = \mathbf{i} + \mathbf{j} + \frac{2}{3}\mathbf{k} + \mathbf{C}.$ Thus $\mathbf{C} = -\frac{2}{3}\mathbf{k}$ and $\mathbf{r}(t) = t^2\,\mathbf{i} + t^3\,\mathbf{j} + \left(\frac{2}{3}t^{3/2} - \frac{2}{3}\right)\mathbf{k}.$

67. For the particles to collide, we require $\mathbf{r}_1(t) = \mathbf{r}_2(t) \quad \Leftrightarrow \quad \langle t^2, 7t - 12, t^2 \rangle = \langle 4t - 3, t^2, 5t - 6 \rangle.$ Equating components

gives $t^2 = 4t - 3, \; 7t - 12 = t^2,$ and $t^2 = 5t - 6.$ From the first equation, $t^2 - 4t + 3 = 0 \quad \Leftrightarrow \quad (t - 3)(t - 1) = 0$ so $t = 1$

or $t = 3$. $t = 1$ does not satisfy the other two equations, but $t = 3$ does. The particles collide when $t = 3$, at the point $(9, 9, 9)$.

69. Let $\mathbf{u}(t) = \langle u_1(t), u_2(t), u_3(t) \rangle$ and $\mathbf{v}(t) = \langle v_1(t), v_2(t), v_3(t) \rangle$. In each part of this problem the basic procedure is to use Equation 1 and then analyze the individual component functions using the limit properties we have already developed for real-valued functions.

(a) $\displaystyle\lim_{t \to a} \mathbf{u}(t) + \lim_{t \to a} \mathbf{v}(t) = \left\langle \lim_{t \to a} u_1(t), \lim_{t \to a} u_2(t), \lim_{t \to a} u_3(t) \right\rangle + \left\langle \lim_{t \to a} v_1(t), \lim_{t \to a} v_2(t), \lim_{t \to a} v_3(t) \right\rangle$ and the limits of these

component functions must each exist since the vector functions both possess limits as $t \to a$. Then adding the two vectors and using the addition property of limits for real-valued functions, we have that

$$\lim_{t \to a} \mathbf{u}(t) + \lim_{t \to a} \mathbf{v}(t) = \left\langle \lim_{t \to a} u_1(t) + \lim_{t \to a} v_1(t), \lim_{t \to a} u_2(t) + \lim_{t \to a} v_2(t), \lim_{t \to a} u_3(t) + \lim_{t \to a} v_3(t) \right\rangle$$

$$= \left\langle \lim_{t \to a} [u_1(t) + v_1(t)], \lim_{t \to a} [u_2(t) + v_2(t)], \lim_{t \to a} [u_3(t) + v_3(t)] \right\rangle$$

$$= \lim_{t \to a} \langle u_1(t) + v_1(t), u_2(t) + v_2(t), u_3(t) + v_3(t) \rangle \qquad \text{[using (1) backward]}$$

$$= \lim_{t \to a} [\mathbf{u}(t) + \mathbf{v}(t)]$$

(b) $\displaystyle\lim_{t \to a} c\mathbf{u}(t) = \lim_{t \to a} \langle cu_1(t), cu_2(t), cu_3(t) \rangle = \left\langle \lim_{t \to a} cu_1(t), \lim_{t \to a} cu_2(t), \lim_{t \to a} cu_3(t) \right\rangle$

$$= \left\langle c \lim_{t \to a} u_1(t), c \lim_{t \to a} u_2(t), c \lim_{t \to a} u_3(t) \right\rangle = c \left\langle \lim_{t \to a} u_1(t), \lim_{t \to a} u_2(t), \lim_{t \to a} u_3(t) \right\rangle$$

$$= c \lim_{t \to a} \langle u_1(t), u_2(t), u_3(t) \rangle = c \lim_{t \to a} \mathbf{u}(t)$$

(c) $\displaystyle\lim_{t \to a} \mathbf{u}(t) \cdot \lim_{t \to a} \mathbf{v}(t) = \left\langle \lim_{t \to a} u_1(t), \lim_{t \to a} u_2(t), \lim_{t \to a} u_3(t) \right\rangle \cdot \left\langle \lim_{t \to a} v_1(t), \lim_{t \to a} v_2(t), \lim_{t \to a} v_3(t) \right\rangle$

$$= \left[\lim_{t \to a} u_1(t) \right] \left[\lim_{t \to a} v_1(t) \right] + \left[\lim_{t \to a} u_2(t) \right] \left[\lim_{t \to a} v_2(t) \right] + \left[\lim_{t \to a} u_3(t) \right] \left[\lim_{t \to a} v_3(t) \right]$$

$$= \lim_{t \to a} u_1(t)v_1(t) + \lim_{t \to a} u_2(t)v_2(t) + \lim_{t \to a} u_3(t)v_3(t)$$

$$= \lim_{t \to a} [u_1(t)v_1(t) + u_2(t)v_2(t) + u_3(t)v_3(t)] = \lim_{t \to a} [\mathbf{u}(t) \cdot \mathbf{v}(t)]$$

(d) $\displaystyle\lim_{t \to a} \mathbf{u}(t) \times \lim_{t \to a} \mathbf{v}(t) = \left\langle \lim_{t \to a} u_1(t), \lim_{t \to a} u_2(t), \lim_{t \to a} u_3(t) \right\rangle \times \left\langle \lim_{t \to a} v_1(t), \lim_{t \to a} v_2(t), \lim_{t \to a} v_3(t) \right\rangle$

$$= \left\langle \left[\lim_{t \to a} u_2(t) \right] \left[\lim_{t \to a} v_3(t) \right] - \left[\lim_{t \to a} u_3(t) \right] \left[\lim_{t \to a} v_2(t) \right], \right.$$

$$\left[\lim_{t \to a} u_3(t) \right] \left[\lim_{t \to a} v_1(t) \right] - \left[\lim_{t \to a} u_1(t) \right] \left[\lim_{t \to a} v_3(t) \right],$$

$$\left. \left[\lim_{t \to a} u_1(t) \right] \left[\lim_{t \to a} v_2(t) \right] - \left[\lim_{t \to a} u_2(t) \right] \left[\lim_{t \to a} v_1(t) \right] \right\rangle$$

$$= \left\langle \lim_{t \to a} [u_2(t)v_3(t) - u_3(t)v_2(t)], \lim_{t \to a} [u_3(t)v_1(t) - u_1(t)v_3(t)], \right.$$

$$\left. \lim_{t \to a} [u_1(t)v_2(t) - u_2(t)v_1(t)] \right\rangle$$

$$= \lim_{t \to a} \langle u_2(t)v_3(t) - u_3(t)v_2(t), u_3(t)v_1(t) - u_1(t)v_3(t), u_1(t)v_2(t) - u_2(t)v_1(t) \rangle$$

$$= \lim_{t \to a} [\mathbf{u}(t) \times \mathbf{v}(t)]$$

71. $\dfrac{d}{dt}\left[\mathbf{u}(t)+\mathbf{v}(t)\right]=\dfrac{d}{dt}\left\langle u_1(t)+v_1(t),u_2(t)+v_2(t),u_3(t)+v_3(t)\right\rangle$

$$=\left\langle \dfrac{d}{dt}\left[u_1(t)+v_1(t)\right],\dfrac{d}{dt}\left[u_2(t)+v_2(t)\right],\dfrac{d}{dt}\left[u_3(t)+v_3(t)\right]\right\rangle$$

$$=\left\langle u_1'(t)+v_1'(t),u_2'(t)+v_2'(t),u_3'(t)+v_3'(t)\right\rangle$$

$$=\left\langle u_1'(t),u_2'(t),u_3'(t)\right\rangle+\left\langle v_1'(t),v_2'(t),v_3'(t)\right\rangle=\mathbf{u}'(t)+\mathbf{v}'(t)$$

73. $\dfrac{d}{dt}\left[\mathbf{u}(t)\times\mathbf{v}(t)\right]=\dfrac{d}{dt}\left\langle u_2(t)v_3(t)-u_3(t)v_2(t),u_3(t)v_1(t)-u_1(t)v_3(t),u_1(t)v_2(t)-u_2(t)v_1(t)\right\rangle$

$$=\left\langle u_2'v_3(t)+u_2(t)v_3'(t)-u_3'(t)v_2(t)-u_3(t)v_2'(t),\right.$$

$$u_3'(t)v_1(t)+u_3(t)v_1'(t)-u_1'(t)v_3(t)-u_1(t)v_3'(t),$$

$$\left.u_1'(t)v_2(t)+u_1(t)v_2'(t)-u_2'(t)v_1(t)-u_2(t)v_1'(t)\right\rangle$$

$$=\left\langle u_2'(t)v_3(t)-u_3'(t)v_2(t),u_3'(t)v_1(t)-u_1'(t)v_3(t),u_1'(t)v_2(t)-u_2'(t)v_1(t)\right\rangle$$

$$+\left\langle u_2(t)v_3'(t)-u_3(t)v_2'(t),u_3(t)v_1'(t)-u_1(t)v_3'(t),u_1(t)v_2'(t)-u_2(t)v_1'(t)\right\rangle$$

$$=\mathbf{u}'(t)\times\mathbf{v}(t)+\mathbf{u}(t)\times\mathbf{v}'(t)$$

Alternate solution: Let $\mathbf{r}(t)=\mathbf{u}(t)\times\mathbf{v}(t)$. Then

$$\mathbf{r}(t+h)-\mathbf{r}(t)=\left[\mathbf{u}(t+h)\times\mathbf{v}(t+h)\right]-\left[\mathbf{u}(t)\times\mathbf{v}(t)\right]$$

$$=\left[\mathbf{u}(t+h)\times\mathbf{v}(t+h)\right]-\left[\mathbf{u}(t)\times\mathbf{v}(t)\right]+\left[\mathbf{u}(t+h)\times\mathbf{v}(t)\right]-\left[\mathbf{u}(t+h)\times\mathbf{v}(t)\right]$$

$$=\mathbf{u}(t+h)\times\left[\mathbf{v}(t+h)-\mathbf{v}(t)\right]+\left[\mathbf{u}(t+h)-\mathbf{u}(t)\right]\times\mathbf{v}(t)$$

(Be careful of the order of the cross product.) Dividing through by h and taking the limit as $h\to0$ we have

$$\mathbf{r}'(t)=\lim_{h\to0}\dfrac{\mathbf{u}(t+h)\times\left[\mathbf{v}(t+h)-\mathbf{v}(t)\right]}{h}+\lim_{h\to0}\dfrac{\left[\mathbf{u}(t+h)-\mathbf{u}(t)\right]\times\mathbf{v}(t)}{h}=\mathbf{u}(t)\times\mathbf{v}'(t)+\mathbf{u}'(t)\times\mathbf{v}(t)$$

by Exercise 69(a) and Definition 3.

75. $\dfrac{d}{dt}\left[\mathbf{u}(t)\cdot\mathbf{v}(t)\right]=\mathbf{u}'(t)\cdot\mathbf{v}(t)+\mathbf{u}(t)\cdot\mathbf{v}'(t)$ [by Formula 4 of Theorem 5]

$$=\left\langle\cos t,-\sin t,1\right\rangle\cdot\left\langle t,\cos t,\sin t\right\rangle+\left\langle\sin t,\cos t,t\right\rangle\cdot\left\langle1,-\sin t,\cos t\right\rangle$$

$$=t\cos t-\cos t\sin t+\sin t+\sin t-\cos t\sin t+t\cos t$$

$$=2t\cos t+2\sin t-2\cos t\sin t$$

77. By Formula 4 of Theorem 5, $f'(t)=\mathbf{u}'(t)\cdot\mathbf{v}(t)+\mathbf{u}(t)\cdot\mathbf{v}'(t)$, and $\mathbf{v}'(t)=\left\langle1,2t,3t^2\right\rangle$, so

$$f'(2)=\mathbf{u}'(2)\cdot\mathbf{v}(2)+\mathbf{u}(2)\cdot\mathbf{v}'(2)=\left\langle3,0,4\right\rangle\cdot\left\langle2,4,8\right\rangle+\left\langle1,2,-1\right\rangle\cdot\left\langle1,4,12\right\rangle=6+0+32+1+8-12=35.$$

79. $\dfrac{d}{dt}\left[\mathbf{r}(t)\times\mathbf{r}'(t)\right]=\mathbf{r}'(t)\times\mathbf{r}'(t)+\mathbf{r}(t)\times\mathbf{r}''(t)$ by Formula 5 of Theorem 5. But $\mathbf{r}'(t)\times\mathbf{r}'(t)=\mathbf{0}$ (by Example 2 in

Section 10.4). Thus, $\dfrac{d}{dt}\left[\mathbf{r}(t)\times\mathbf{r}'(t)\right]=\mathbf{r}(t)\times\mathbf{r}''(t)$.

81. $\dfrac{d}{dt}\left|\mathbf{r}(t)\right|=\dfrac{d}{dt}\left[\mathbf{r}(t)\cdot\mathbf{r}(t)\right]^{1/2}=\tfrac{1}{2}\left[\mathbf{r}(t)\cdot\mathbf{r}(t)\right]^{-1/2}\left[2\mathbf{r}(t)\cdot\mathbf{r}'(t)\right]=\dfrac{1}{\left|\mathbf{r}(t)\right|}\mathbf{r}(t)\cdot\mathbf{r}'(t)$

83. Since $\mathbf{u}(t) = \mathbf{r}(t) \cdot [\mathbf{r}'(t) \times \mathbf{r}''(t)]$,

$$\mathbf{u}'(t) = \mathbf{r}'(t) \cdot [\mathbf{r}'(t) \times \mathbf{r}''(t)] + \mathbf{r}(t) \cdot \frac{d}{dt}[\mathbf{r}'(t) \times \mathbf{r}''(t)]$$

$$= 0 + \mathbf{r}(t) \cdot [\mathbf{r}''(t) \times \mathbf{r}''(t) + \mathbf{r}'(t) \times \mathbf{r}'''(t)] \qquad \text{[since } \mathbf{r}'(t) \perp \mathbf{r}'(t) \times \mathbf{r}''(t)]$$

$$= \mathbf{r}(t) \cdot [\mathbf{r}'(t) \times \mathbf{r}'''(t)] \qquad \text{[since } \mathbf{r}''(t) \times \mathbf{r}''(t) = \mathbf{0}]$$

10.8   Arc Length and Curvature

1. $\mathbf{r}(t) = \langle t, 3\cos t, 3\sin t \rangle \;\Rightarrow\; \mathbf{r}'(t) = \langle 1, -3\sin t, 3\cos t \rangle \;\Rightarrow\;$

$|\mathbf{r}'(t)| = \sqrt{1^2 + (-3\sin t)^2 + (3\cos t)^2} = \sqrt{1 + 9(\sin^2 t + \cos^2 t)} = \sqrt{10}$.

Then using Formula 3, we have $L = \int_{-5}^{5} |\mathbf{r}'(t)|\, dt = \int_{-5}^{5} \sqrt{10}\, dt = \sqrt{10}\, t\Big]_{-5}^{5} = 10\sqrt{10}$.

3. $\mathbf{r}(t) = \mathbf{i} + t^2\,\mathbf{j} + t^3\,\mathbf{k} \;\Rightarrow\; \mathbf{r}'(t) = 2t\,\mathbf{j} + 3t^2\,\mathbf{k} \;\Rightarrow\; |\mathbf{r}'(t)| = \sqrt{4t^2 + 9t^4} = t\sqrt{4 + 9t^2}$ [since $t \geq 0$].

Then $L = \int_0^1 |\mathbf{r}'(t)|\, dt = \int_0^1 t\sqrt{4 + 9t^2}\, dt = \frac{1}{18} \cdot \frac{2}{3}(4 + 9t^2)^{3/2}\Big]_0^1 = \frac{1}{27}(13^{3/2} - 4^{3/2}) = \frac{1}{27}(13^{3/2} - 8)$.

5. $\mathbf{r}(t) = \langle t^2, t^3, t^4 \rangle \;\Rightarrow\; \mathbf{r}'(t) = \langle 2t, 3t^2, 4t^3 \rangle \;\Rightarrow\; |\mathbf{r}'(t)| = \sqrt{(2t)^2 + (3t^2)^2 + (4t^3)^2} = \sqrt{4t^2 + 9t^4 + 16t^6}$, so

$L = \int_0^2 |\mathbf{r}'(t)|\, dt = \int_0^2 \sqrt{4t^2 + 9t^4 + 16t^6}\, dt \approx 18.6833$.

7. The projection of the curve C onto the xy-plane is the curve $x^2 = 2y$ or $y = \frac{1}{2}x^2$, $z = 0$. Then we can choose the parameter

$x = t \;\Rightarrow\; y = \frac{1}{2}t^2$. Since C also lies on the surface $3z = xy$, we have $z = \frac{1}{3}xy = \frac{1}{3}(t)(\frac{1}{2}t^2) = \frac{1}{6}t^3$. Then parametric

equations for C are $x = t$, $y = \frac{1}{2}t^2$, $z = \frac{1}{6}t^3$ and the corresponding vector equation is $\mathbf{r}(t) = \langle t, \frac{1}{2}t^2, \frac{1}{6}t^3 \rangle$. The origin

corresponds to $t = 0$ and the point $(6, 18, 36)$ corresponds to $t = 6$, so

$$L = \int_0^6 |\mathbf{r}'(t)|\, dt = \int_0^6 \left|\langle 1, t, \tfrac{1}{2}t^2 \rangle\right| dt = \int_0^6 \sqrt{1^2 + t^2 + \left(\tfrac{1}{2}t^2\right)^2}\, dt = \int_0^6 \sqrt{1 + t^2 + \tfrac{1}{4}t^4}\, dt$$

$$= \int_0^6 \sqrt{(1 + \tfrac{1}{2}t^2)^2}\, dt = \int_0^6 (1 + \tfrac{1}{2}t^2)\, dt = \left[t + \tfrac{1}{6}t^3\right]_0^6 = 6 + 36 = 42$$

9. $\mathbf{r}(t) = 2t\,\mathbf{i} + (1 - 3t)\,\mathbf{j} + (5 + 4t)\,\mathbf{k} \;\Rightarrow\; \mathbf{r}'(t) = 2\,\mathbf{i} - 3\,\mathbf{j} + 4\,\mathbf{k}$ and $\frac{ds}{dt} = |\mathbf{r}'(t)| = \sqrt{4 + 9 + 16} = \sqrt{29}$. Then

$s = s(t) = \int_0^t |\mathbf{r}'(u)|\, du = \int_0^t \sqrt{29}\, du = \sqrt{29}\, t$. Therefore, $t = \frac{1}{\sqrt{29}}\, s$, and substituting for t in the original equation, we

have $\mathbf{r}(t(s)) = \frac{2}{\sqrt{29}}\, s\,\mathbf{i} + \left(1 - \frac{3}{\sqrt{29}}\, s\right)\mathbf{j} + \left(5 + \frac{4}{\sqrt{29}}\, s\right)\mathbf{k}$.

11. Here $\mathbf{r}(t) = \langle 3\sin t, 4t, 3\cos t \rangle$, so $\mathbf{r}'(t) = \langle 3\cos t, 4, -3\sin t \rangle$ and $|\mathbf{r}'(t)| = \sqrt{9\cos^2 t + 16 + 9\sin^2 t} = \sqrt{25} = 5$.

The point $(0, 0, 3)$ corresponds to $t = 0$, so the arc length function beginning at $(0, 0, 3)$ and measuring in the positive

direction is given by $s(t) = \int_0^t |\mathbf{r}'(u)|\, du = \int_0^t 5\, du = 5t$. $s(t) = 5 \;\Rightarrow\; 5t = 5 \;\Rightarrow\; t = 1$, thus your location after

moving 5 units along the curve is $(3\sin 1, 4, 3\cos 1)$.

13. (a) $\mathbf{r}(t) = \langle t, 3\cos t, 3\sin t \rangle$ $\Rightarrow$ $\mathbf{r}'(t) = \langle 1, -3\sin t, 3\cos t \rangle$ $\Rightarrow$ $|\mathbf{r}'(t)| = \sqrt{1 + 9\sin^2 t + 9\cos^2 t} = \sqrt{10}$.

Then $\mathbf{T}(t) = \dfrac{\mathbf{r}'(t)}{|\mathbf{r}'(t)|} = \dfrac{1}{\sqrt{10}}\langle 1, -3\sin t, 3\cos t \rangle$ or $\left\langle \dfrac{1}{\sqrt{10}}, -\dfrac{3}{\sqrt{10}}\sin t, \dfrac{3}{\sqrt{10}}\cos t \right\rangle$.

$\mathbf{T}'(t) = \dfrac{1}{\sqrt{10}}\langle 0, -3\cos t, -3\sin t \rangle$ $\Rightarrow$ $|\mathbf{T}'(t)| = \dfrac{1}{\sqrt{10}}\sqrt{0 + 9\cos^2 t + 9\sin^2 t} = \dfrac{3}{\sqrt{10}}$. Thus

$\mathbf{N}(t) = \dfrac{\mathbf{T}'(t)}{|\mathbf{T}'(t)|} = \dfrac{1/\sqrt{10}}{3/\sqrt{10}}\langle 0, -3\cos t, -3\sin t \rangle = \langle 0, -\cos t, -\sin t \rangle$.

(b) $\kappa(t) = \dfrac{|\mathbf{T}'(t)|}{|\mathbf{r}'(t)|} = \dfrac{3/\sqrt{10}}{\sqrt{10}} = \dfrac{3}{10}$

15. (a) $\mathbf{r}(t) = \langle \sqrt{2}\,t, e^t, e^{-t} \rangle$ $\Rightarrow$ $\mathbf{r}'(t) = \langle \sqrt{2}, e^t, -e^{-t} \rangle$ $\Rightarrow$ $|\mathbf{r}'(t)| = \sqrt{2 + e^{2t} + e^{-2t}} = \sqrt{(e^t + e^{-t})^2} = e^t + e^{-t}$.

Then

$\mathbf{T}(t) = \dfrac{\mathbf{r}'(t)}{|\mathbf{r}'(t)|} = \dfrac{1}{e^t + e^{-t}}\langle \sqrt{2}, e^t, -e^{-t} \rangle = \dfrac{1}{e^{2t} + 1}\langle \sqrt{2}\,e^t, e^{2t}, -1 \rangle$ $\left[\text{after multiplying by }\dfrac{e^t}{e^t}\right]$ and

$\mathbf{T}'(t) = \dfrac{1}{e^{2t} + 1}\langle \sqrt{2}\,e^t, 2e^{2t}, 0 \rangle - \dfrac{2e^{2t}}{(e^{2t} + 1)^2}\langle \sqrt{2}\,e^t, e^{2t}, -1 \rangle$

$= \dfrac{1}{(e^{2t} + 1)^2}\left[(e^{2t} + 1)\langle \sqrt{2}\,e^t, 2e^{2t}, 0 \rangle - 2e^{2t}\langle \sqrt{2}\,e^t, e^{2t}, -1 \rangle \right] = \dfrac{1}{(e^{2t} + 1)^2}\langle \sqrt{2}\,e^t(1 - e^{2t}), 2e^{2t}, 2e^{2t} \rangle$

Then

$|\mathbf{T}'(t)| = \dfrac{1}{(e^{2t} + 1)^2}\sqrt{2e^{2t}(1 - 2e^{2t} + e^{4t}) + 4e^{4t} + 4e^{4t}} = \dfrac{1}{(e^{2t} + 1)^2}\sqrt{2e^{2t}(1 + 2e^{2t} + e^{4t})}$

$= \dfrac{1}{(e^{2t} + 1)^2}\sqrt{2e^{2t}(1 + e^{2t})^2} = \dfrac{\sqrt{2}\,e^t(1 + e^{2t})}{(e^{2t} + 1)^2} = \dfrac{\sqrt{2}\,e^t}{e^{2t} + 1}$

Therefore

$\mathbf{N}(t) = \dfrac{\mathbf{T}'(t)}{|\mathbf{T}'(t)|} = \dfrac{e^{2t} + 1}{\sqrt{2}\,e^t}\dfrac{1}{(e^{2t} + 1)^2}\langle \sqrt{2}\,e^t(1 - e^{2t}), 2e^{2t}, 2e^{2t} \rangle$

$= \dfrac{1}{\sqrt{2}\,e^t(e^{2t} + 1)}\langle \sqrt{2}\,e^t(1 - e^{2t}), 2e^{2t}, 2e^{2t} \rangle = \dfrac{1}{e^{2t} + 1}\langle 1 - e^{2t}, \sqrt{2}\,e^t, \sqrt{2}\,e^t \rangle$

(b) $\kappa(t) = \dfrac{|\mathbf{T}'(t)|}{|\mathbf{r}'(t)|} = \dfrac{\sqrt{2}\,e^t}{e^{2t} + 1} \cdot \dfrac{1}{e^t + e^{-t}} = \dfrac{\sqrt{2}\,e^t}{e^{3t} + 2e^t + e^{-t}} = \dfrac{\sqrt{2}\,e^{2t}}{e^{4t} + 2e^{2t} + 1} = \dfrac{\sqrt{2}\,e^{2t}}{(e^{2t} + 1)^2}$

17. $\mathbf{r}(t) = t^3\,\mathbf{j} + t^2\,\mathbf{k}$ $\Rightarrow$ $\mathbf{r}'(t) = 3t^2\,\mathbf{j} + 2t\,\mathbf{k}$, $\mathbf{r}''(t) = 6t\,\mathbf{j} + 2\,\mathbf{k}$, $|\mathbf{r}'(t)| = \sqrt{0^2 + (3t^2)^2 + (2t)^2} = \sqrt{9t^4 + 4t^2}$,

$\mathbf{r}'(t) \times \mathbf{r}''(t) = -6t^2\,\mathbf{i}$, $|\mathbf{r}'(t) \times \mathbf{r}''(t)| = 6t^2$. Then $\kappa(t) = \dfrac{|\mathbf{r}'(t) \times \mathbf{r}''(t)|}{|\mathbf{r}'(t)|^3} = \dfrac{6t^2}{\left(\sqrt{9t^4 + 4t^2}\right)^3} = \dfrac{6t^2}{(9t^4 + 4t^2)^{3/2}}$.

19. $\mathbf{r}(t) = 3t\,\mathbf{i} + 4\sin t\,\mathbf{j} + 4\cos t\,\mathbf{k}$ $\Rightarrow$ $\mathbf{r}'(t) = 3\,\mathbf{i} + 4\cos t\,\mathbf{j} - 4\sin t\,\mathbf{k}$, $\mathbf{r}''(t) = -4\sin t\,\mathbf{j} - 4\cos t\,\mathbf{k}$,

$|\mathbf{r}'(t)| = \sqrt{9 + 16\cos^2 t + 16\sin^2 t} = \sqrt{9 + 16} = 5$, $\mathbf{r}'(t) \times \mathbf{r}''(t) = -16\,\mathbf{i} + 12\cos t\,\mathbf{j} - 12\sin t\,\mathbf{k}$,

$|\mathbf{r}'(t) \times \mathbf{r}''(t)| = \sqrt{256 + 144\cos^2 t + 144\sin^2 t} = \sqrt{400} = 20$. Then $\kappa(t) = \dfrac{|\mathbf{r}'(t) \times \mathbf{r}''(t)|}{|\mathbf{r}'(t)|^3} = \dfrac{20}{5^3} = \dfrac{4}{25}$.

21. $\mathbf{r}(t) = \langle t, t^2, t^3 \rangle$ $\Rightarrow$ $\mathbf{r}'(t) = \langle 1, 2t, 3t^2 \rangle$.   The point $(1, 1, 1)$ corresponds to $t = 1$, and $\mathbf{r}'(1) = \langle 1, 2, 3 \rangle$ $\Rightarrow$

$|\mathbf{r}'(1)| = \sqrt{1 + 4 + 9} = \sqrt{14}$. $\mathbf{r}''(t) = \langle 0, 2, 6t \rangle$ $\Rightarrow$ $\mathbf{r}''(1) = \langle 0, 2, 6 \rangle$. $\mathbf{r}'(1) \times \mathbf{r}''(1) = \langle 6, -6, 2 \rangle$, so

$|\mathbf{r}'(1) \times \mathbf{r}''(1)| = \sqrt{36 + 36 + 4} = \sqrt{76}$. Then $\kappa(1) = \dfrac{|\mathbf{r}'(1) \times \mathbf{r}''(1)|}{|\mathbf{r}'(1)|^3} = \dfrac{\sqrt{76}}{\sqrt{14}^3} = \dfrac{1}{7}\sqrt{\dfrac{19}{14}}$.

23. $f(x) = x^4$, $f'(x) = 4x^3$, $f''(x) = 12x^2$, $\kappa(x) = \dfrac{|f''(x)|}{[1 + (f'(x))^2]^{3/2}} = \dfrac{|12x^2|}{[1 + (4x^3)^2]^{3/2}} = \dfrac{12x^2}{(1 + 16x^6)^{3/2}}$

25. $f(x) = xe^x$, $f'(x) = xe^x + e^x$, $f''(x) = xe^x + 2e^x$,

$\kappa(x) = \dfrac{|f''(x)|}{[1 + (f'(x))^2]^{3/2}} = \dfrac{|xe^x + 2e^x|}{[1 + (xe^x + e^x)^2]^{3/2}} = \dfrac{|x + 2|\, e^x}{[1 + (xe^x + e^x)^2]^{3/2}}$

27. Since $y' = y'' = e^x$, the curvature is $\kappa(x) = \dfrac{|y''(x)|}{[1 + (y'(x))^2]^{3/2}} = \dfrac{e^x}{(1 + e^{2x})^{3/2}} = e^x(1 + e^{2x})^{-3/2}$.

To find the maximum curvature, we first find the critical numbers of $\kappa(x)$:

$\kappa'(x) = e^x(1 + e^{2x})^{-3/2} + e^x\left(-\frac{3}{2}\right)(1 + e^{2x})^{-5/2}(2e^{2x}) = e^x\dfrac{1 + e^{2x} - 3e^{2x}}{(1 + e^{2x})^{5/2}} = e^x\dfrac{1 - 2e^{2x}}{(1 + e^{2x})^{5/2}}$.

$\kappa'(x) = 0$ when $1 - 2e^{2x} = 0$, so $e^{2x} = \frac{1}{2}$ or $x = -\frac{1}{2}\ln 2$. And since $1 - 2e^{2x} > 0$ for $x < -\frac{1}{2}\ln 2$ and $1 - 2e^{2x} < 0$

for $x > -\frac{1}{2}\ln 2$, the maximum curvature is attained at the point $\left(-\frac{1}{2}\ln 2, e^{(-\ln 2)/2}\right) = \left(-\frac{1}{2}\ln 2, \frac{1}{\sqrt{2}}\right)$.

Since $\lim\limits_{x \to \infty} e^x(1 + e^{2x})^{-3/2} = 0$, $\kappa(x)$ approaches 0 as $x \to \infty$.

29. (a) C appears to be changing direction more quickly at P than Q, so we would expect the curvature to be greater at P.

(b) First we sketch approximate osculating circles at P and Q. Using the

axes scale as a guide, we measure the radius of the osculating circle

at P to be approximately 0.8 units, thus $\rho = \dfrac{1}{\kappa}$ $\Rightarrow$

$\kappa = \dfrac{1}{\rho} \approx \dfrac{1}{0.8} \approx 1.3$. Similarly, we estimate the radius of the

osculating circle at Q to be 1.4 units, so $\kappa = \dfrac{1}{\rho} \approx \dfrac{1}{1.4} \approx 0.7$.

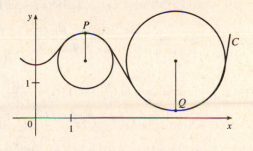

31. $y = x^{-2}$ $\Rightarrow$ $y' = -2x^{-3}$, $y'' = 6x^{-4}$, and

$\kappa(x) = \dfrac{|y''|}{[1 + (y')^2]^{3/2}} = \dfrac{|6x^{-4}|}{[1 + (-2x^{-3})^2]^{3/2}} = \dfrac{6}{x^4(1 + 4x^{-6})^{3/2}}$.

The appearance of the two humps in this graph is perhaps a little surprising, but it is

explained by the fact that $y = x^{-2}$ increases asymptotically at the origin from both

directions, and so its graph has very little bend there. [Note that $\kappa(0)$ is undefined.]

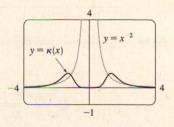

33. $\mathbf{r}(t) = \langle te^t, e^{-t}, \sqrt{2}\, t \rangle \;\Rightarrow\; \mathbf{r}'(t) = \langle (t+1)e^t, -e^{-t}, \sqrt{2} \rangle, \quad \mathbf{r}''(t) = \langle (t+2)e^t, e^{-t}, 0 \rangle.$ Then

$$\mathbf{r}'(t) \times \mathbf{r}''(t) = \langle -\sqrt{2}\, e^{-t}, \sqrt{2}(t+2)e^t, 2t+3 \rangle, \quad |\mathbf{r}'(t) \times \mathbf{r}''(t)| = \sqrt{2e^{-2t} + 2(t+2)^2 e^{2t} + (2t+3)^2},$$

$$|\mathbf{r}'(t)| = \sqrt{(t+1)^2 e^{2t} + e^{-2t} + 2}, \quad \text{and} \quad \kappa(t) = \frac{|\mathbf{r}'(t) \times \mathbf{r}''(t)|}{|\mathbf{r}'(t)|^3} = \frac{\sqrt{2e^{-2t} + 2(t+2)^2 e^{2t} + (2t+3)^2}}{[(t+1)^2 e^{2t} + e^{-2t} + 2]^{3/2}}.$$

We plot the space curve and its curvature function for $-5 \le t \le 5$ below.

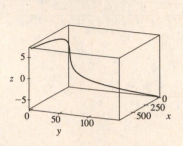
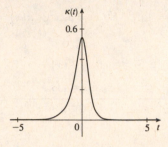

From the graph of $\kappa(t)$ we see that curvature is maximized for $t = 0$, so the curve bends most sharply at the point $(0, 1, 0)$. The curve bends more gradually as we move away from this point, becoming almost linear. This is reflected in the curvature graph, where $\kappa(t)$ becomes nearly 0 as $|t|$ increases.

35. Notice that the curve b has two inflection points at which the graph appears almost straight. We would expect the curvature to be 0 or nearly 0 at these values, but the curve a isn't near 0 there. Thus, a must be the graph of $y = f(x)$ rather than the graph of curvature, and b is the graph of $y = \kappa(x)$.

37. $x = e^t \cos t \;\Rightarrow\; \dot{x} = e^t(\cos t - \sin t) \;\Rightarrow\; \ddot{x} = e^t(-\sin t - \cos t) + e^t(\cos t - \sin t) = -2e^t \sin t,$

$y = e^t \sin t \;\Rightarrow\; \dot{y} = e^t(\cos t + \sin t) \;\Rightarrow\; \ddot{y} = e^t(-\sin t + \cos t) + e^t(\cos t + \sin t) = 2e^t \cos t.$ Then

$$\kappa(t) = \frac{|\dot{x}\ddot{y} - \dot{y}\ddot{x}|}{[\dot{x}^2 + \dot{y}^2]^{3/2}} = \frac{\left| e^t(\cos t - \sin t)(2e^t \cos t) - e^t(\cos t + \sin t)(-2e^t \sin t) \right|}{\left([e^t(\cos t - \sin t)]^2 + [e^t(\cos t + \sin t)]^2 \right)^{3/2}}$$

$$= \frac{\left| 2e^{2t}(\cos^2 t - \sin t \cos t + \sin t \cos t + \sin^2 t) \right|}{\left[e^{2t}(\cos^2 t - 2\cos t \sin t + \sin^2 t + \cos^2 t + 2\cos t \sin t + \sin^2 t) \right]^{3/2}} = \frac{\left| 2e^{2t}(1) \right|}{\left[e^{2t}(1+1) \right]^{3/2}} = \frac{2e^{2t}}{e^{3t}(2)^{3/2}} = \frac{1}{\sqrt{2}\, e^t}$$

39. $\left(1, \frac{2}{3}, 1 \right)$ corresponds to $t = 1$. $\mathbf{T}(t) = \dfrac{\mathbf{r}'(t)}{|\mathbf{r}'(t)|} = \dfrac{\langle 2t, 2t^2, 1 \rangle}{\sqrt{4t^2 + 4t^4 + 1}} = \dfrac{\langle 2t, 2t^2, 1 \rangle}{2t^2 + 1}$, so $\mathbf{T}(1) = \left\langle \frac{2}{3}, \frac{2}{3}, \frac{1}{3} \right\rangle$.

$\mathbf{T}'(t) = -4t(2t^2+1)^{-2}\langle 2t, 2t^2, 1 \rangle + (2t^2+1)^{-1}\langle 2, 4t, 0 \rangle$ [by Formula 3 of Theorem 10.7.5]

$$= (2t^2+1)^{-2}\langle -8t^2 + 4t^2 + 2, -8t^3 + 8t^3 + 4t, -4t \rangle = 2(2t^2+1)^{-2}\langle 1 - 2t^2, 2t, -2t \rangle$$

$$\mathbf{N}(t) = \frac{\mathbf{T}'(t)}{|\mathbf{T}'(t)|} = \frac{2(2t^2+1)^{-2}\langle 1 - 2t^2, 2t, -2t \rangle}{2(2t^2+1)^{-2}\sqrt{(1-2t^2)^2 + (2t)^2 + (-2t)^2}} = \frac{\langle 1 - 2t^2, 2t, -2t \rangle}{\sqrt{1 - 4t^2 + 4t^4 + 8t^2}} = \frac{\langle 1 - 2t^2, 2t, -2t \rangle}{1 + 2t^2}$$

$\mathbf{N}(1) = \left\langle -\frac{1}{3}, \frac{2}{3}, -\frac{2}{3} \right\rangle$ and $\mathbf{B}(1) = \mathbf{T}(1) \times \mathbf{N}(1) = \left\langle -\frac{4}{9} - \frac{2}{9}, -\left(-\frac{4}{9} + \frac{1}{9} \right), \frac{4}{9} + \frac{2}{9} \right\rangle = \left\langle -\frac{2}{3}, \frac{1}{3}, \frac{2}{3} \right\rangle.$

41. $(0, \pi, -2)$ corresponds to $t = \pi$. $\mathbf{r}(t) = \langle 2\sin 3t, t, 2\cos 3t \rangle \;\Rightarrow$

$$\mathbf{T}(t) = \frac{\mathbf{r}'(t)}{|\mathbf{r}'(t)|} = \frac{\langle 6\cos 3t, 1, -6\sin 3t \rangle}{\sqrt{36\cos^2 3t + 1 + 36\sin^2 3t}} = \frac{1}{\sqrt{37}}\langle 6\cos 3t, 1, -6\sin 3t \rangle.$$

$\mathbf{T}(\pi) = \frac{1}{\sqrt{37}}\langle -6, 1, 0 \rangle$ is a normal vector for the normal plane, and so $\langle -6, 1, 0 \rangle$ is also normal. Thus an equation for the plane is $-6(x - 0) + 1(y - \pi) + 0(z + 2) = 0$ or $y - 6x = \pi$.

$$\mathbf{T}'(t) = \frac{1}{\sqrt{37}} \langle -18 \sin 3t, 0, -18 \cos 3t \rangle \quad \Rightarrow \quad |\mathbf{T}'(t)| = \frac{\sqrt{18^2 \sin^2 3t + 18^2 \cos^2 3t}}{\sqrt{37}} = \frac{18}{\sqrt{37}} \quad \Rightarrow$$

$$\mathbf{N}(t) = \frac{\mathbf{T}'(t)}{|\mathbf{T}'(t)|} = \langle -\sin 3t, 0, -\cos 3t \rangle. \text{ So } \mathbf{N}(\pi) = \langle 0, 0, 1 \rangle \text{ and } \mathbf{B}(\pi) = \frac{1}{\sqrt{37}} \langle -6, 1, 0 \rangle \times \langle 0, 0, 1 \rangle = \frac{1}{\sqrt{37}} \langle 1, 6, 0 \rangle.$$

Since $\mathbf{B}(\pi)$ is a normal to the osculating plane, so is $\langle 1, 6, 0 \rangle$.

An equation for the plane is $1(x - 0) + 6(y - \pi) + 0(z + 2) = 0$ or $x + 6y = 6\pi$.

43. The ellipse is given by the parametric equations $x = 2 \cos t$, $y = 3 \sin t$, so using the result from Exercise 36,

$$\kappa(t) = \frac{|\dot{x}\ddot{y} - \ddot{x}\dot{y}|}{[\dot{x}^2 + \dot{y}^2]^{3/2}} = \frac{|(-2 \sin t)(-3 \sin t) - (3 \cos t)(-2 \cos t)|}{(4 \sin^2 t + 9 \cos^2 t)^{3/2}} = \frac{6}{(4 \sin^2 t + 9 \cos^2 t)^{3/2}}.$$

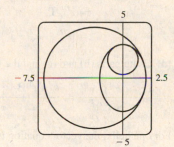

At $(2, 0)$, $t = 0$. Now $\kappa(0) = \frac{6}{27} = \frac{2}{9}$, so the radius of the osculating circle is

$1/\kappa(0) = \frac{9}{2}$ and its center is $\left(-\frac{5}{2}, 0\right)$. Its equation is therefore $\left(x + \frac{5}{2}\right)^2 + y^2 = \frac{81}{4}$.

At $(0, 3)$, $t = \frac{\pi}{2}$, and $\kappa\left(\frac{\pi}{2}\right) = \frac{6}{8} = \frac{3}{4}$. So the radius of the osculating circle is $\frac{4}{3}$ and

its center is $\left(0, \frac{5}{3}\right)$. Hence its equation is $x^2 + \left(y - \frac{5}{3}\right)^2 = \frac{16}{9}$.

45. The tangent vector is normal to the normal plane, and the vector $\langle 6, 6, -8 \rangle$ is normal to the given plane.

But $\mathbf{T}(t) \parallel \mathbf{r}'(t)$ and $\langle 6, 6, -8 \rangle \parallel \langle 3, 3, -4 \rangle$, so we need to find t such that $\mathbf{r}'(t) \parallel \langle 3, 3, -4 \rangle$.

$\mathbf{r}(t) = \langle t^3, 3t, t^4 \rangle \quad \Rightarrow \quad \mathbf{r}'(t) = \langle 3t^2, 3, 4t^3 \rangle \parallel \langle 3, 3, -4 \rangle$ when $t = -1$. So the planes are parallel at the point $(-1, -3, 1)$.

47. $\kappa = \left| \dfrac{d\mathbf{T}}{ds} \right| = \left| \dfrac{d\mathbf{T}/dt}{ds/dt} \right| = \dfrac{|d\mathbf{T}/dt|}{ds/dt}$ and $\mathbf{N} = \dfrac{d\mathbf{T}/dt}{|d\mathbf{T}/dt|}$, so $\kappa\mathbf{N} = \dfrac{\left|\dfrac{d\mathbf{T}}{dt}\right| \dfrac{d\mathbf{T}}{dt}}{\left|\dfrac{d\mathbf{T}}{dt}\right| \dfrac{ds}{dt}} = \dfrac{d\mathbf{T}/dt}{ds/dt} = \dfrac{d\mathbf{T}}{ds}$ by the Chain Rule.

49. (a) $|\mathbf{B}| = 1 \quad \Rightarrow \quad \mathbf{B} \cdot \mathbf{B} = 1 \quad \Rightarrow \quad \dfrac{d}{ds}(\mathbf{B} \cdot \mathbf{B}) = 0 \quad \Rightarrow \quad 2\dfrac{d\mathbf{B}}{ds} \cdot \mathbf{B} = 0 \quad \Rightarrow \quad \dfrac{d\mathbf{B}}{ds} \perp \mathbf{B}$

(b) $\mathbf{B} = \mathbf{T} \times \mathbf{N} \quad \Rightarrow$

$$\frac{d\mathbf{B}}{ds} = \frac{d}{ds}(\mathbf{T} \times \mathbf{N}) = \frac{d}{dt}(\mathbf{T} \times \mathbf{N}) \frac{1}{ds/dt} = \frac{d}{dt}(\mathbf{T} \times \mathbf{N}) \frac{1}{|\mathbf{r}'(t)|} = [(\mathbf{T}' \times \mathbf{N}) + (\mathbf{T} \times \mathbf{N}')] \frac{1}{|\mathbf{r}'(t)|}$$

$$= \left[\left(\mathbf{T} \times \frac{\mathbf{T}'}{|\mathbf{T}'|} \right) + (\mathbf{T} \times \mathbf{N}') \right] \frac{1}{|\mathbf{r}'(t)|} = \frac{\mathbf{T} \times \mathbf{N}'}{|\mathbf{r}'(t)|} \quad \Rightarrow \quad \frac{d\mathbf{B}}{ds} \perp \mathbf{T}$$

(c) $\mathbf{B} = \mathbf{T} \times \mathbf{N} \quad \Rightarrow \quad \mathbf{T} \perp \mathbf{N}, \mathbf{B} \perp \mathbf{T}$ and $\mathbf{B} \perp \mathbf{N}$. So $\mathbf{B}, \mathbf{T}$ and $\mathbf{N}$ form an orthogonal set of vectors in the three-

dimensional space $\mathbb{R}^3$. From parts (a) and (b), $d\mathbf{B}/ds$ is perpendicular to both $\mathbf{B}$ and $\mathbf{T}$, so $d\mathbf{B}/ds$ is parallel to $\mathbf{N}$.

Therefore, $d\mathbf{B}/ds = -\tau(s)\mathbf{N}$, where $\tau(s)$ is a scalar.

(d) Since $\mathbf{B} = \mathbf{T} \times \mathbf{N}$, $\mathbf{T} \perp \mathbf{N}$ and both $\mathbf{T}$ and $\mathbf{N}$ are unit vectors, $\mathbf{B}$ is a unit vector mutually perpendicular to both $\mathbf{T}$ and

$\mathbf{N}$. For a plane curve, $\mathbf{T}$ and $\mathbf{N}$ always lie in the plane of the curve, so that $\mathbf{B}$ is a constant unit vector always

perpendicular to the plane. Thus $d\mathbf{B}/ds = \mathbf{0}$, but $d\mathbf{B}/ds = -\tau(s)\mathbf{N}$ and $\mathbf{N} \neq \mathbf{0}$, so $\tau(s) = 0$.

51. (a) $\mathbf{r}' = s'\,\mathbf{T} \;\Rightarrow\; \mathbf{r}'' = s''\,\mathbf{T} + s'\,\mathbf{T}' = s''\,\mathbf{T} + s'\,\dfrac{d\mathbf{T}}{ds}\,s' = s''\,\mathbf{T} + \kappa(s')^2\,\mathbf{N}$ by the first Serret-Frenet formula.

(b) Using part (a), we have

$$\mathbf{r}' \times \mathbf{r}'' = (s'\,\mathbf{T}) \times [s''\,\mathbf{T} + \kappa(s')^2\,\mathbf{N}]$$
$$= [(s'\,\mathbf{T}) \times (s''\,\mathbf{T})] + [(s'\,\mathbf{T}) \times (\kappa(s')^2\,\mathbf{N})] \qquad \text{[by Property 3 of Theorem 10.4.8]}$$
$$= (s's'')(\mathbf{T} \times \mathbf{T}) + \kappa(s')^3(\mathbf{T} \times \mathbf{N}) = \mathbf{0} + \kappa(s')^3\,\mathbf{B} = \kappa(s')^3\,\mathbf{B}$$

(c) Using part (a), we have

$$\mathbf{r}''' = [s''\,\mathbf{T} + \kappa(s')^2\,\mathbf{N}]' = s'''\,\mathbf{T} + s''\,\mathbf{T}' + \kappa'(s')^2\,\mathbf{N} + 2\kappa s' s''\,\mathbf{N} + \kappa(s')^2\,\mathbf{N}'$$
$$= s'''\,\mathbf{T} + s''\,\dfrac{d\mathbf{T}}{ds}\,s' + \kappa'(s')^2\,\mathbf{N} + 2\kappa s' s''\,\mathbf{N} + \kappa(s')^2\,\dfrac{d\mathbf{N}}{ds}\,s'$$
$$= s'''\,\mathbf{T} + s''\,s'\,\kappa\,\mathbf{N} + \kappa'(s')^2\,\mathbf{N} + 2\kappa s' s''\,\mathbf{N} + \kappa(s')^3(-\kappa\,\mathbf{T} + \tau\,\mathbf{B}) \qquad \text{[by the second formula]}$$
$$= [s''' - \kappa^2(s')^3]\,\mathbf{T} + [3\kappa s' s'' + \kappa'(s')^2]\,\mathbf{N} + \kappa\tau(s')^3\,\mathbf{B}$$

(d) Using parts (b) and (c) and the facts that $\mathbf{B} \cdot \mathbf{T} = 0$, $\mathbf{B} \cdot \mathbf{N} = 0$, and $\mathbf{B} \cdot \mathbf{B} = 1$, we get

$$\dfrac{(\mathbf{r}' \times \mathbf{r}'') \cdot \mathbf{r}'''}{|\mathbf{r}' \times \mathbf{r}''|^2} = \dfrac{\kappa(s')^3\,\mathbf{B} \cdot \left\{[s''' - \kappa^2(s')^3]\,\mathbf{T} + [3\kappa s' s'' + \kappa'(s')^2]\,\mathbf{N} + \kappa\tau(s')^3\,\mathbf{B}\right\}}{|\kappa(s')^3\,\mathbf{B}|^2} = \dfrac{\kappa(s')^3\,\kappa\tau(s')^3}{[\kappa(s')^3]^2} = \tau.$$

53. For one helix, the vector equation is $\mathbf{r}(t) = \langle 10\cos t, 10\sin t, 34t/(2\pi) \rangle$ (measuring in angstroms), because the radius of each

helix is 10 angstroms, and z increases by 34 angstroms for each increase of 2π in t. Using the arc length formula, letting t go

from 0 to $2.9 \times 10^8 \times 2\pi$, we find the approximate length of each helix to be

$$L = \int_0^{2.9\times10^8\times2\pi} |\mathbf{r}'(t)|\,dt = \int_0^{2.9\times10^8\times2\pi} \sqrt{(-10\sin t)^2 + (10\cos t)^2 + \left(\tfrac{34}{2\pi}\right)^2}\,dt = \sqrt{100 + \left(\tfrac{34}{2\pi}\right)^2}\,t\,\Bigg]_0^{2.9\times10^8\times2\pi}$$

$$= 2.9 \times 10^8 \times 2\pi \sqrt{100 + \left(\tfrac{34}{2\pi}\right)^2} \approx 2.07 \times 10^{10}\ \text{Å} \text{ — more than two meters!}$$

10.9 Motion in Space: Velocity and Acceleration

1. $\mathbf{r}(t) = \left\langle -\tfrac{1}{2}t^2, t \right\rangle \;\Rightarrow\;$ At $t = 2$:

$\mathbf{v}(t) = \mathbf{r}'(t) = \langle -t, 1 \rangle$ $\mathbf{v}(2) = \langle -2, 1 \rangle$

$\mathbf{a}(t) = \mathbf{r}''(t) = \langle -1, 0 \rangle$ $\mathbf{a}(2) = \langle -1, 0 \rangle$

$|\mathbf{v}(t)| = \sqrt{t^2 + 1}$

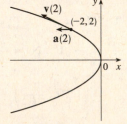

3. $\mathbf{r}(t) = 3\cos t\,\mathbf{i} + 2\sin t\,\mathbf{j} \;\Rightarrow\;$ At $t = \pi/3$:

$\mathbf{v}(t) = -3\sin t\,\mathbf{i} + 2\cos t\,\mathbf{j}$ $\mathbf{v}\left(\tfrac{\pi}{3}\right) = -\tfrac{3\sqrt{3}}{2}\,\mathbf{i} + \mathbf{j}$

$\mathbf{a}(t) = -3\cos t\,\mathbf{i} - 2\sin t\,\mathbf{j}$ $\mathbf{a}\left(\tfrac{\pi}{3}\right) = -\tfrac{3}{2}\,\mathbf{i} - \sqrt{3}\,\mathbf{j}$

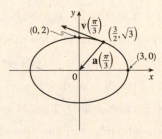

$|\mathbf{v}(t)| = \sqrt{9\sin^2 t + 4\cos^2 t} = \sqrt{4 + 5\sin^2 t}$

Notice that $x^2/9 + y^2/4 = \sin^2 t + \cos^2 t = 1$, so the path is an ellipse.

5. $\mathbf{r}(t) = t\,\mathbf{i} + t^2\,\mathbf{j} + 2\,\mathbf{k}$ $\Rightarrow$ At $t = 1$:

$\mathbf{v}(t) = \mathbf{i} + 2t\,\mathbf{j}$ $\mathbf{v}(1) = \mathbf{i} + 2\,\mathbf{j}$

$\mathbf{a}(t) = 2\,\mathbf{j}$ $\mathbf{a}(1) = 2\,\mathbf{j}$

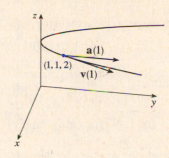

$|\mathbf{v}(t)| = \sqrt{1 + 4t^2}$

Here $x = t$, $y = t^2$ $\Rightarrow$ $y = x^2$ and $z = 2$, so the path of the particle is a

parabola in the plane $z = 2$.

7. $\mathbf{r}(t) = \langle t^2 + t, t^2 - t, t^3 \rangle$ $\Rightarrow$ $\mathbf{v}(t) = \mathbf{r}'(t) = \langle 2t + 1, 2t - 1, 3t^2 \rangle$, $\mathbf{a}(t) = \mathbf{v}'(t) = \langle 2, 2, 6t \rangle$,

$|\mathbf{v}(t)| = \sqrt{(2t+1)^2 + (2t-1)^2 + (3t^2)^2} = \sqrt{9t^4 + 8t^2 + 2}$.

9. $\mathbf{r}(t) = \sqrt{2}\,t\,\mathbf{i} + e^t\,\mathbf{j} + e^{-t}\,\mathbf{k}$ $\Rightarrow$ $\mathbf{v}(t) = \mathbf{r}'(t) = \sqrt{2}\,\mathbf{i} + e^t\,\mathbf{j} - e^{-t}\,\mathbf{k}$, $\mathbf{a}(t) = \mathbf{v}'(t) = e^t\,\mathbf{j} + e^{-t}\,\mathbf{k}$,

$|\mathbf{v}(t)| = \sqrt{2 + e^{2t} + e^{-2t}} = \sqrt{(e^t + e^{-t})^2} = e^t + e^{-t}$.

11. $\mathbf{a}(t) = \mathbf{i} + 2\,\mathbf{j}$ $\Rightarrow$ $\mathbf{v}(t) = \int \mathbf{a}(t)\,dt = \int (\mathbf{i} + 2\,\mathbf{j})\,dt = t\,\mathbf{i} + 2t\,\mathbf{j} + \mathbf{C}$ and $\mathbf{k} = \mathbf{v}(0) = \mathbf{C}$,

so $\mathbf{C} = \mathbf{k}$ and $\mathbf{v}(t) = t\,\mathbf{i} + 2t\,\mathbf{j} + \mathbf{k}$. $\mathbf{r}(t) = \int \mathbf{v}(t)\,dt = \int (t\,\mathbf{i} + 2t\,\mathbf{j} + \mathbf{k})\,dt = \frac{1}{2}t^2\,\mathbf{i} + t^2\,\mathbf{j} + t\,\mathbf{k} + \mathbf{D}$.

But $\mathbf{i} = \mathbf{r}(0) = \mathbf{D}$, so $\mathbf{D} = \mathbf{i}$ and $\mathbf{r}(t) = \left(\frac{1}{2}t^2 + 1\right)\mathbf{i} + t^2\,\mathbf{j} + t\,\mathbf{k}$.

13. (a) $\mathbf{a}(t) = 2t\,\mathbf{i} + \sin t\,\mathbf{j} + \cos 2t\,\mathbf{k}$ $\Rightarrow$ (b)

$\mathbf{v}(t) = \int (2t\,\mathbf{i} + \sin t\,\mathbf{j} + \cos 2t\,\mathbf{k})\,dt = t^2\,\mathbf{i} - \cos t\,\mathbf{j} + \frac{1}{2}\sin 2t\,\mathbf{k} + \mathbf{C}$

and $\mathbf{i} = \mathbf{v}(0) = -\mathbf{j} + \mathbf{C}$, so $\mathbf{C} = \mathbf{i} + \mathbf{j}$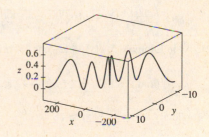

and $\mathbf{v}(t) = (t^2 + 1)\,\mathbf{i} + (1 - \cos t)\,\mathbf{j} + \frac{1}{2}\sin 2t\,\mathbf{k}$.

$\mathbf{r}(t) = \int [(t^2 + 1)\,\mathbf{i} + (1 - \cos t)\,\mathbf{j} + \frac{1}{2}\sin 2t\,\mathbf{k}]\,dt$

$= \left(\frac{1}{3}t^3 + t\right)\mathbf{i} + (t - \sin t)\,\mathbf{j} - \frac{1}{4}\cos 2t\,\mathbf{k} + \mathbf{D}$

But $\mathbf{j} = \mathbf{r}(0) = -\frac{1}{4}\mathbf{k} + \mathbf{D}$, so $\mathbf{D} = \mathbf{j} + \frac{1}{4}\mathbf{k}$ and $\mathbf{r}(t) = \left(\frac{1}{3}t^3 + t\right)\mathbf{i} + (t - \sin t + 1)\,\mathbf{j} + \left(\frac{1}{4} - \frac{1}{4}\cos 2t\right)\mathbf{k}$.

15. $\mathbf{r}(t) = \langle t^2, 5t, t^2 - 16t \rangle$ $\Rightarrow$ $\mathbf{v}(t) = \langle 2t, 5, 2t - 16 \rangle$, $|\mathbf{v}(t)| = \sqrt{4t^2 + 25 + 4t^2 - 64t + 256} = \sqrt{8t^2 - 64t + 281}$

and $\dfrac{d}{dt}|\mathbf{v}(t)| = \frac{1}{2}(8t^2 - 64t + 281)^{-1/2}(16t - 64)$. This is zero if and only if the numerator is zero, that is,

$16t - 64 = 0$ or $t = 4$. Since $\dfrac{d}{dt}|\mathbf{v}(t)| < 0$ for $t < 4$ and $\dfrac{d}{dt}|\mathbf{v}(t)| > 0$ for $t > 4$, the minimum speed of $\sqrt{153}$ is attained

at $t = 4$ units of time.

17. $|\mathbf{F}(t)| = 20$ N in the direction of the positive z-axis, so $\mathbf{F}(t) = 20\,\mathbf{k}$. Also $m = 4$ kg, $\mathbf{r}(0) = \mathbf{0}$ and $\mathbf{v}(0) = \mathbf{i} - \mathbf{j}$.

Since $20\,\mathbf{k} = \mathbf{F}(t) = 4\,\mathbf{a}(t)$, $\mathbf{a}(t) = 5\,\mathbf{k}$. Then $\mathbf{v}(t) = 5t\,\mathbf{k} + \mathbf{c}_1$ where $\mathbf{c}_1 = \mathbf{i} - \mathbf{j}$ so $\mathbf{v}(t) = \mathbf{i} - \mathbf{j} + 5t\,\mathbf{k}$ and the

speed is $|\mathbf{v}(t)| = \sqrt{1 + 1 + 25t^2} = \sqrt{25t^2 + 2}$. Also $\mathbf{r}(t) = t\,\mathbf{i} - t\,\mathbf{j} + \frac{5}{2}t^2\,\mathbf{k} + \mathbf{c}_2$ and $\mathbf{0} = \mathbf{r}(0)$, so $\mathbf{c}_2 = \mathbf{0}$

and $\mathbf{r}(t) = t\,\mathbf{i} - t\,\mathbf{j} + \frac{5}{2}t^2\,\mathbf{k}$.

19. $|\mathbf{v}(0)| = 200$ m/s and, since the angle of elevation is $60°$, a unit vector in the direction of the velocity is

$(\cos 60°)\mathbf{i} + (\sin 60°)\mathbf{j} = \frac{1}{2}\mathbf{i} + \frac{\sqrt{3}}{2}\mathbf{j}$. Thus $\mathbf{v}(0) = 200\left(\frac{1}{2}\mathbf{i} + \frac{\sqrt{3}}{2}\mathbf{j}\right) = 100\,\mathbf{i} + 100\sqrt{3}\,\mathbf{j}$ and if we set up the axes so that the projectile starts at the origin, then $\mathbf{r}(0) = \mathbf{0}$. Ignoring air resistance, the only force is that due to gravity, so

$\mathbf{F}(t) = m\mathbf{a}(t) = -mg\,\mathbf{j}$ where $g \approx 9.8$ m/s^2. Thus $\mathbf{a}(t) = -9.8\,\mathbf{j}$ and, integrating, we have $\mathbf{v}(t) = -9.8t\,\mathbf{j} + \mathbf{C}$. But

$100\,\mathbf{i} + 100\,\sqrt{3}\,\mathbf{j} = \mathbf{v}(0) = \mathbf{C}$, so $\mathbf{v}(t) = 100\,\mathbf{i} + \left(100\,\sqrt{3} - 9.8t\right)\mathbf{j}$ and then (integrating again)

$\mathbf{r}(t) = 100\,t\,\mathbf{i} + \left(100\,\sqrt{3}\,t - 4.9t^2\right)\mathbf{j} + \mathbf{D}$ where $\mathbf{0} = \mathbf{r}(0) = \mathbf{D}$. Thus the position function of the projectile is

$\mathbf{r}(t) = 100\,t\,\mathbf{i} + \left(100\,\sqrt{3}\,t - 4.9t^2\right)\mathbf{j}$.

(a) Parametric equations for the projectile are $x(t) = 100t$, $y(t) = 100\,\sqrt{3}\,t - 4.9t^2$. The projectile reaches the ground when

$y(t) = 0$ (and $t > 0$) $\;\Rightarrow\;$ $100\,\sqrt{3}\,t - 4.9t^2 = t\left(100\,\sqrt{3} - 4.9t\right) = 0$ $\;\Rightarrow\;$ $t = \frac{100\sqrt{3}}{4.9} \approx 35.3$ s. So the range is

$x\left(\frac{100\sqrt{3}}{4.9}\right) = 100\left(\frac{100\sqrt{3}}{4.9}\right) \approx 3535$ m.

(b) The maximum height is reached when $y(t)$ has a critical number (or equivalently, when the vertical component

of velocity is 0): $y'(t) = 0$ $\;\Rightarrow\;$ $100\,\sqrt{3} - 9.8t = 0$ $\;\Rightarrow\;$ $t = \frac{100\sqrt{3}}{9.8} \approx 17.7$ s. Thus the maximum height is

$y\left(\frac{100\sqrt{3}}{9.8}\right) = 100\,\sqrt{3}\left(\frac{100\sqrt{3}}{9.8}\right) - 4.9\left(\frac{100\sqrt{3}}{9.8}\right)^2 \approx 1531$ m.

(c) From part (a), impact occurs at $t = \frac{100\sqrt{3}}{4.9}$ s. Thus, the velocity at impact is

$\mathbf{v}\left(\frac{100\sqrt{3}}{4.9}\right) = 100\,\mathbf{i} + \left[100\,\sqrt{3} - 9.8\left(\frac{100\sqrt{3}}{4.9}\right)\right]\mathbf{j} = 100\,\mathbf{i} - 100\,\sqrt{3}\,\mathbf{j}$ and the speed is

$\left|\mathbf{v}\left(\frac{100\sqrt{3}}{4.9}\right)\right| = \sqrt{10{,}000 + 30{,}000} = 200$ m/s.

21. As in Example 5, $\mathbf{r}(t) = (v_0 \cos 45°)t\,\mathbf{i} + \left[(v_0 \sin 45°)t - \frac{1}{2}gt^2\right]\mathbf{j} = \frac{1}{2}\left[v_0\sqrt{2}\,t\,\mathbf{i} + \left(v_0\sqrt{2}\,t - gt^2\right)\mathbf{j}\right]$. The ball lands when

$y = 0$ (and $t > 0$) $\;\Rightarrow\;$ $t = \dfrac{v_0\sqrt{2}}{g}$ s. Now since it lands 90 m away, $90 = x = \frac{1}{2}v_0\sqrt{2}\,\dfrac{v_0\sqrt{2}}{g}$ or $v_0^2 = 90g$ and the initial

velocity is $v_0 = \sqrt{90g} \approx 30$ m/s.

23. Let α be the angle of elevation. Then $v_0 = 150$ m/s and from Example 5, the horizontal distance traveled by the projectile is

$d = \dfrac{v_0^2 \sin 2\alpha}{g}$. Thus $\dfrac{150^2 \sin 2\alpha}{g} = 800$ $\;\Rightarrow\;$ $\sin 2\alpha = \dfrac{800g}{150^2} \approx 0.3484$ $\;\Rightarrow\;$ $2\alpha \approx 20.4°$ or $180 - 20.4 = 159.6°$.

Two angles of elevation then are $\alpha \approx 10.2°$ and $\alpha \approx 79.8°$.

25. Place the catapult at the origin and assume the catapult is 100 meters from the city, so the city lies between $(100, 0)$

and $(600, 0)$. The initial speed is $v_0 = 80$ m/s and let θ be the angle the catapult is set at. As in Example 5, the trajectory of

the catapulted rock is given by $\mathbf{r}(t) = (80 \cos \theta)t\,\mathbf{i} + \left[(80 \sin \theta)t - 4.9t^2\right]\mathbf{j}$. The top of the near city wall is at $(100, 15)$,

which the rock will hit when $(80 \cos \theta)\,t = 100$ $\;\Rightarrow\;$ $t = \dfrac{5}{4 \cos \theta}$ and $(80 \sin \theta)t - 4.9t^2 = 15$ $\;\Rightarrow\;$

$80 \sin \theta \cdot \dfrac{5}{4 \cos \theta} - 4.9\left(\dfrac{5}{4 \cos \theta}\right)^2 = 15$ $\;\Rightarrow\;$ $100 \tan \theta - 7.65625 \sec^2 \theta = 15$. Replacing $\sec^2 \theta$ with $\tan^2 \theta + 1$ gives

$7.65625 \tan^2 \theta - 100 \tan \theta + 22.65625 = 0$. Using the quadratic formula, we have $\tan \theta \approx 0.230635$, 12.8306 $\;\Rightarrow\;$

$\theta \approx 13.0°$, $85.5°$. So for $13.0° < \theta < 85.5°$, the rock will land beyond the near city wall. The base of the far wall is located at $(600, 0)$ which the rock hits if $(80 \cos \theta)t = 600 \Rightarrow t = \dfrac{15}{2 \cos \theta}$ and $(80 \sin \theta)t - 4.9t^2 = 0 \Rightarrow$

$80 \sin \theta \cdot \dfrac{15}{2 \cos \theta} - 4.9 \left(\dfrac{15}{2 \cos \theta} \right)^2 = 0 \Rightarrow 600 \tan \theta - 275.625 \sec^2 \theta = 0 \Rightarrow$

$275.625 \tan^2 \theta - 600 \tan \theta + 275.625 = 0$. Solutions are $\tan \theta \approx 0.658678, 1.51819 \Rightarrow \theta \approx 33.4°, 56.6°$. Thus the

rock lands beyond the enclosed city ground for $33.4° < \theta < 56.6°$, and the angles that allow the rock to land on city ground

are $13.0° < \theta < 33.4°$, $56.6° < \theta < 85.5°$. If you consider that the rock can hit the far wall and bounce back into the city, we

calculate the angles that cause the rock to hit the top of the wall at $(600, 15)$: $(80 \cos \theta)t = 600 \Rightarrow t = \dfrac{15}{2 \cos \theta}$ and

$(80 \sin \theta)t - 4.9t^2 = 15 \Rightarrow 600 \tan \theta - 275.625 \sec^2 \theta = 15 \Rightarrow 275.625 \tan^2 \theta - 600 \tan \theta + 290.625 = 0$.

Solutions are $\tan \theta \approx 0.727506, 1.44936 \Rightarrow \theta \approx 36.0°, 55.4°$, so the catapult should be set with angle θ where

$13.0° < \theta < 36.0°$, $55.4° < \theta < 85.5°$.

27. Here $\mathbf{a}(t) = -4\mathbf{j} - 32\mathbf{k}$ so $\mathbf{v}(t) = -4t\mathbf{j} - 32t\mathbf{k} + \mathbf{v}_0 = -4t\mathbf{j} - 32t\mathbf{k} + 50\mathbf{i} + 80\mathbf{k} = 50\mathbf{i} - 4t\mathbf{j} + (80 - 32t)\mathbf{k}$ and

$\mathbf{r}(t) = 50t\mathbf{i} - 2t^2\mathbf{j} + (80t - 16t^2)\mathbf{k}$ (note that $\mathbf{r}_0 = \mathbf{0}$). The ball lands when the z-component of $\mathbf{r}(t)$ is zero

and $t > 0$: $80t - 16t^2 = 16t(5 - t) = 0 \Rightarrow t = 5$. The position of the ball then is

$\mathbf{r}(5) = 50(5)\mathbf{i} - 2(5)^2\mathbf{j} + [80(5) - 16(5)^2]\mathbf{k} = 250\mathbf{i} - 50\mathbf{j}$ or equivalently the point $(250, -50, 0)$. This is a distance of

$\sqrt{250^2 + (-50)^2 + 0^2} = \sqrt{65{,}000} \approx 255$ ft from the origin at an angle of $\tan^{-1}\left(\dfrac{50}{250} \right) \approx 11.3°$ from the eastern direction

toward the south. The speed of the ball is $|\mathbf{v}(5)| = |50\mathbf{i} - 20\mathbf{j} - 80\mathbf{k}| = \sqrt{50^2 + (-20)^2 + (-80)^2} = \sqrt{9300} \approx 96.4$ ft/s.

29. (a) After t seconds, the boat will be $5t$ meters west of point A. The velocity

of the water at that location is $\frac{3}{400}(5t)(40 - 5t)\mathbf{j}$. The velocity of the

boat in still water is $5\mathbf{i}$, so the resultant velocity of the boat is

$\mathbf{v}(t) = 5\mathbf{i} + \frac{3}{400}(5t)(40 - 5t)\mathbf{j} = 5\mathbf{i} + \left(\frac{3}{2}t - \frac{3}{16}t^2 \right)\mathbf{j}$. Integrating, we obtain

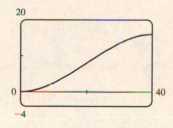

$\mathbf{r}(t) = 5t\mathbf{i} + \left(\frac{3}{4}t^2 - \frac{1}{16}t^3 \right)\mathbf{j} + \mathbf{C}$. If we place the origin at A (and consider $\mathbf{j}$

to coincide with the northern direction) then $\mathbf{r}(0) = \mathbf{0} \Rightarrow \mathbf{C} = \mathbf{0}$ and we have $\mathbf{r}(t) = 5t\mathbf{i} + \left(\frac{3}{4}t^2 - \frac{1}{16}t^3 \right)\mathbf{j}$. The boat

reaches the east bank after 8 s, and it is located at $\mathbf{r}(8) = 5(8)\mathbf{i} + \left(\frac{3}{4}(8)^2 - \frac{1}{16}(8)^3 \right)\mathbf{j} = 40\mathbf{i} + 16\mathbf{j}$. Thus the boat is 16 m

downstream.

(b) Let α be the angle north of east that the boat heads. Then the velocity of the boat in still water is given by

$5(\cos \alpha)\mathbf{i} + 5(\sin \alpha)\mathbf{j}$. At t seconds, the boat is $5(\cos \alpha)t$ meters from the west bank, at which point the velocity

of the water is $\frac{3}{400}[5(\cos \alpha)t][40 - 5(\cos \alpha)t]\mathbf{j}$. The resultant velocity of the boat is given by

$\mathbf{v}(t) = 5(\cos \alpha)\mathbf{i} + \left[5 \sin \alpha + \frac{3}{400}(5t \cos \alpha)(40 - 5t \cos \alpha) \right]\mathbf{j} = (5 \cos \alpha)\mathbf{i} + \left(5 \sin \alpha + \frac{3}{2}t \cos \alpha - \frac{3}{16}t^2 \cos^2 \alpha \right)\mathbf{j}$.

Integrating, $\mathbf{r}(t) = (5t \cos \alpha)\mathbf{i} + \left(5t \sin \alpha + \frac{3}{4}t^2 \cos \alpha - \frac{1}{16}t^3 \cos^2 \alpha \right)\mathbf{j}$ (where we have again placed

the origin at A). The boat will reach the east bank when $5t \cos \alpha = 40 \Rightarrow t = \dfrac{40}{5 \cos \alpha} = \dfrac{8}{\cos \alpha}$.

In order to land at point $B(40, 0)$ we need $5t \sin \alpha + \frac{3}{4}t^2 \cos \alpha - \frac{1}{16}t^3 \cos^2 \alpha = 0 \Rightarrow$

$$5\left(\frac{8}{\cos\alpha}\right)\sin\alpha + \frac{3}{4}\left(\frac{8}{\cos\alpha}\right)^2\cos\alpha - \frac{1}{16}\left(\frac{8}{\cos\alpha}\right)^3\cos^2\alpha = 0 \quad \Rightarrow \quad \frac{1}{\cos\alpha}(40\sin\alpha + 48 - 32) = 0 \quad \Rightarrow$$

$40\sin\alpha + 16 = 0 \quad \Rightarrow \quad \sin\alpha = -\frac{2}{5}$. Thus $\alpha = \sin^{-1}\left(-\frac{2}{5}\right) \approx -23.6°$, so the boat should head 23.6° south of

east (upstream). The path does seem realistic. The boat initially heads

upstream to counteract the effect of the current. Near the center of the river,

the current is stronger and the boat is pushed downstream. When the boat

nears the eastern bank, the current is slower and the boat is able to progress

upstream to arrive at point B.

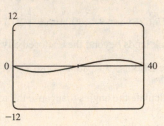

31. $\mathbf{r}(t) = \cos t\,\mathbf{i} + \sin t\,\mathbf{j} + t\,\mathbf{k} \quad \Rightarrow \quad \mathbf{r}'(t) = -\sin t\,\mathbf{i} + \cos t\,\mathbf{j} + \mathbf{k}, \quad |\mathbf{r}'(t)| = \sqrt{\sin^2 t + \cos^2 t + 1} = \sqrt{2},$

$\mathbf{r}''(t) = -\cos t\,\mathbf{i} - \sin t\,\mathbf{j}, \quad \mathbf{r}'(t) \times \mathbf{r}''(t) = \sin t\,\mathbf{i} - \cos t\,\mathbf{j} + \mathbf{k}.$

Then $a_T = \dfrac{\mathbf{r}'(t)\cdot\mathbf{r}''(t)}{|\mathbf{r}'(t)|} = \dfrac{\sin t\,\cos t - \sin t\,\cos t}{\sqrt{2}} = 0$ and $a_N = \dfrac{|\mathbf{r}'(t)\times\mathbf{r}''(t)|}{|\mathbf{r}'(t)|} = \dfrac{\sqrt{\sin^2 t + \cos^2 t + 1}}{\sqrt{2}} = \dfrac{\sqrt{2}}{\sqrt{2}} = 1.$

33. $\mathbf{r}(t) = (3t - t^3)\,\mathbf{i} + 3t^2\,\mathbf{j} \quad \Rightarrow \quad \mathbf{r}'(t) = (3 - 3t^2)\,\mathbf{i} + 6t\,\mathbf{j},$

$|\mathbf{r}'(t)| = \sqrt{(3-3t^2)^2 + (6t)^2} = \sqrt{9 + 18t^2 + 9t^4} = \sqrt{(3-3t^2)^2} = 3 + 3t^2, \mathbf{r}''(t) = -6t\,\mathbf{i} + 6\,\mathbf{j},$

$\mathbf{r}'(t) \times \mathbf{r}''(t) = (18 + 18t^2)\,\mathbf{k}.$ Then Equation 9 gives

$a_T = \dfrac{\mathbf{r}'(t)\cdot\mathbf{r}''(t)}{|\mathbf{r}'(t)|} = \dfrac{(3-3t^2)(-6t) + (6t)(6)}{3+3t^2} = \dfrac{18t + 18t^3}{3+3t^2} = \dfrac{18t(1+t^2)}{3(1+t^2)} = 6t \quad \Big[\text{or by Equation 8,}$

$a_T = v' = \dfrac{d}{dt}\big[3 + 3t^2\big] = 6t\Big]$ and Equation 10 gives $a_N = \dfrac{|\mathbf{r}'(t)\times\mathbf{r}''(t)|}{|\mathbf{r}'(t)|} = \dfrac{18 + 18t^2}{3+3t^2} = \dfrac{18(1+t^2)}{3(1+t^2)} = 6.$

35. If the engines are turned off at time t, then the spacecraft will continue to travel in the direction of $\mathbf{v}(t)$, so we need a t such

that for some scalar $s > 0$, $\mathbf{r}(t) + s\,\mathbf{v}(t) = \langle 6, 4, 9\rangle.$ $\quad \mathbf{v}(t) = \mathbf{r}'(t) = \mathbf{i} + \dfrac{1}{t}\,\mathbf{j} + \dfrac{8t}{(t^2+1)^2}\,\mathbf{k} \quad \Rightarrow$

$\mathbf{r}(t) + s\,\mathbf{v}(t) = \left\langle 3 + t + s,\, 2 + \ln t + \dfrac{s}{t},\, 7 - \dfrac{4}{t^2+1} + \dfrac{8st}{(t^2+1)^2}\right\rangle \quad \Rightarrow \quad 3 + t + s = 6 \quad \Rightarrow \quad s = 3 - t,$

so $7 - \dfrac{4}{t^2+1} + \dfrac{8(3-t)t}{(t^2+1)^2} = 9 \quad \Leftrightarrow \quad \dfrac{24t - 12t^2 - 4}{(t^2+1)^2} = 2 \quad \Leftrightarrow \quad t^4 + 8t^2 - 12t + 3 = 0.$

It is easily seen that $t = 1$ is a root of this polynomial. Also $2 + \ln 1 + \dfrac{3-1}{1} = 4$, so $t = 1$ is the desired solution.

10   Review

CONCEPT CHECK

---

1. A scalar is a real number, while a vector is a quantity that has both a real-valued magnitude and a direction.

2. To add two vectors geometrically, we can use either the Triangle Law or the Parallelogram Law, as illustrated in Figures 3

and 4 in Section 10.2. (See also the definition of vector addition on page 549.) Algebraically, we add the corresponding

components of the vectors.

3. For $c > 0$, $c\mathbf{a}$ is a vector with the same direction as $\mathbf{a}$ and length c times the length of $\mathbf{a}$. If $c < 0$, $c\mathbf{a}$ points in the opposite direction as $\mathbf{a}$ and has length $|c|$ times the length of $\mathbf{a}$. (See Figures 7 and 15 in Section 10.2.) Algebraically, to find $c\mathbf{a}$ we multiply each component of $\mathbf{a}$ by c.

4. See (1) in Section 10.2.

5. See Theorem 10.3.3 and Definition 10.3.1.

6. The dot product can be used to determine the work done moving an object given the force and displacement vectors. The dot product can also be used to find the angle between two vectors and the scalar projection of one vector onto another. In particular, the dot product can determine if two vectors are orthogonal.

7. See the boxed equations as well as Figures 3 and 4 and the accompanying discussion on page 560.

8. See Theorem 10.4.6 and the preceding discussion; use either (4) or (7) in Section 10.4.

9. The cross product can be used to determine torque if the force and position vectors are known. In addition, the cross product can be used to create a vector orthogonal to two given vectors as well as to determine if two vectors are parallel. The cross product can also be used to find the area of a parallelogram determined by two vectors.

10. (a) The area of the parallelogram determined by $\mathbf{a}$ and $\mathbf{b}$ is the length of the cross product: $|\mathbf{a} \times \mathbf{b}|$.

    (b) The volume of the parallelepiped determined by $\mathbf{a}$, $\mathbf{b}$, and $\mathbf{c}$ is the magnitude of their scalar triple product: $|\mathbf{a} \cdot (\mathbf{b} \times \mathbf{c})|$.

11. If an equation of the plane is known, it can be written as $ax + by + cz + d = 0$. A normal vector, which is perpendicular to the plane, is $\langle a, b, c \rangle$ (or any scalar multiple of $\langle a, b, c \rangle$). If an equation is not known, we can use points on the plane to find two non-parallel vectors which lie in the plane. The cross product of these vectors is a vector perpendicular to the plane.

12. The angle between two intersecting planes is defined as the acute angle between their normal vectors. We can find this angle using Corollary 10.3.6.

13. See (1), (2), and (3) in Section 10.5.

14. See (5), (6), and (7) in Section 10.5.

15. (a) Two (nonzero) vectors are parallel if and only if one is a scalar multiple of the other. In addition, two nonzero vectors are parallel if and only if their cross product is $\mathbf{0}$.

    (b) Two vectors are perpendicular if and only if their dot product is 0.

    (c) Two planes are parallel if and only if their normal vectors are parallel.

16. (a) Determine the vectors $\overrightarrow{PQ} = \langle a_1, a_2, a_3 \rangle$ and $\overrightarrow{PR} = \langle b_1, b_2, b_3 \rangle$. If there is a scalar t such that $\langle a_1, a_2, a_3 \rangle = t \langle b_1, b_2, b_3 \rangle$, then the vectors are parallel and the points must all lie on the same line.

    Alternatively, if $\overrightarrow{PQ} \times \overrightarrow{PR} = \mathbf{0}$, then $\overrightarrow{PQ}$ and $\overrightarrow{PR}$ are parallel, so P, Q, and R are collinear.

    Thirdly, an algebraic method is to determine an equation of the line joining two of the points, and then check whether or not the third point satisfies this equation.

(b) Find the vectors $\overrightarrow{PQ} = \mathbf{a}$, $\overrightarrow{PR} = \mathbf{b}$, $\overrightarrow{PS} = \mathbf{c}$. $\mathbf{a} \times \mathbf{b}$ is normal to the plane formed by P, Q and R, and so S lies on this plane if $\mathbf{a} \times \mathbf{b}$ and $\mathbf{c}$ are orthogonal, that is, if $(\mathbf{a} \times \mathbf{b}) \cdot \mathbf{c} = 0$. (Or use the reasoning in Example 5 in Section 10.4.) Alternatively, find an equation for the plane determined by three of the points and check whether or not the fourth point satisfies this equation.

17. (a) See Exercise 10.4.45.

(b) See Example 7 in Section 10.5.

18. The traces of a surface are the curves of intersection of the surface with planes parallel to the coordinate planes. We can find the trace in the plane $x = k$ (parallel to the yz-plane) by setting $x = k$ and determining the curve represented by the resulting equation. Traces in the planes $y = k$ (parallel to the xz-plane) and $z = k$ (parallel to the xy-plane) are found similarly.

19. See Table 1 in Section 10.6.

20. A vector function is a function whose domain is a set of real numbers and whose range is a set of vectors. To find the derivative or integral, we can differentiate or integrate each component of the vector function.

21. The tip of the moving vector $\mathbf{r}(t)$ of a continuous vector function traces out a space curve.

22. The tangent vector to a smooth curve at a point P with position vector $\mathbf{r}(t)$ is the vector $\mathbf{r}'(t)$. The tangent line at P is the line through P parallel to the tangent vector $\mathbf{r}'(t)$. The unit tangent vector is $\mathbf{T}(t) = \dfrac{\mathbf{r}'(t)}{|\mathbf{r}'(t)|}$.

23. (a)–(f) See Theorem 10.7.5.

24. Use Formula 10.8.2, or equivalently, 10.8.3.

25. (a) The curvature of a curve is $\kappa = \left| \dfrac{d\mathbf{T}}{ds} \right|$ where $\mathbf{T}$ is the unit tangent vector.

(b) $\kappa(t) = \left| \dfrac{\mathbf{T}'(t)}{\mathbf{r}'(t)} \right|$ (c) $\kappa(t) = \dfrac{|\mathbf{r}'(t) \times \mathbf{r}''(t)|}{|\mathbf{r}'(t)|^3}$ (d) $\kappa(x) = \dfrac{|f''(x)|}{[1 + (f'(x))^2]^{3/2}}$

26. The unit normal vector: $\mathbf{N}(t) = \dfrac{\mathbf{T}'(t)}{|\mathbf{T}'(t)|}$.   The binormal vector: $\mathbf{B}(t) = \mathbf{T}(t) \times \mathbf{N}(t)$.

27. (a) If $\mathbf{r}(t)$ is the position vector of the particle on the space curve, the velocity $\mathbf{v}(t) = \mathbf{r}'(t)$, the speed is given by $|\mathbf{v}(t)|$, and the acceleration $\mathbf{a}(t) = \mathbf{v}'(t) = \mathbf{r}''(t)$.

(b) $\mathbf{a} = a_T \mathbf{T} + a_N \mathbf{N}$ where $a_T = v'$ and $a_N = \kappa v^2$.

28. See the statement of Kepler's Laws on page 612.

TRUE-FALSE QUIZ

1. This is false, as the dot product of two vectors is a scalar, not a vector.

3. False. For example, if $\mathbf{u} = \mathbf{i}$ and $\mathbf{v} = \mathbf{j}$ then $|\mathbf{u} \cdot \mathbf{v}| = |0| = 0$ but $|\mathbf{u}|\,|\mathbf{v}| = 1 \cdot 1 = 1$. In fact, by Theorem 10.3.3, $|\mathbf{u} \cdot \mathbf{v}| = \big|\,|\mathbf{u}|\,|\mathbf{v}| \cos\theta\,\big|$.

5. True, by Property 2 of the dot product. (See page 557.)

7. True. If θ is the angle between $\mathbf{u}$ and $\mathbf{v}$, then by Theorem 10.4.6, $|\mathbf{u} \times \mathbf{v}| = |\mathbf{u}|\,|\mathbf{v}| \sin \theta = |\mathbf{v}|\,|\mathbf{u}| \sin \theta = |\mathbf{v} \times \mathbf{u}|$.

(Or, by Theorem 10.4.8, $|\mathbf{u} \times \mathbf{v}| = |-\mathbf{v} \times \mathbf{u}| = |-1|\,|\mathbf{v} \times \mathbf{u}| = |\mathbf{v} \times \mathbf{u}|$.)

9. Property 2 of the cross product tells us that this is true.

11. This is true by Theorem 10.4.8.

13. This is true because $\mathbf{u} \times \mathbf{v}$ is orthogonal to $\mathbf{u}$ (see Theorem 10.4.5), and the dot product of two orthogonal vectors is 0.

15. This is false. A normal vector to the plane is $\mathbf{n} = \langle 6, -2, 4 \rangle$. Because $\langle 3, -1, 2 \rangle = \frac{1}{2}\mathbf{n}$, the vector is parallel to $\mathbf{n}$ and hence perpendicular to the plane.

17. This is false. In $\mathbb{R}^2$, $x^2 + y^2 = 1$ represents a circle, but $\{(x, y, z) \mid x^2 + y^2 = 1\}$ represents a *three-dimensional surface*, namely, a circular cylinder with axis the z-axis.

19. False. For example, $\mathbf{i} \cdot \mathbf{j} = 0$ but $\mathbf{i} \neq \mathbf{0}$ and $\mathbf{j} \neq \mathbf{0}$.

21. This is true. If $\mathbf{u}$ and $\mathbf{v}$ are both nonzero, then by in , $\mathbf{u} \cdot \mathbf{v} = 0$ implies that $\mathbf{u}$ and $\mathbf{v}$ are orthogonal. But $\mathbf{u} \times \mathbf{v} = \mathbf{0}$ implies that $\mathbf{u}$ and $\mathbf{v}$ are parallel (see). Two nonzero vectors can't be both parallel and orthogonal, so at least one of $\mathbf{u}$, $\mathbf{v}$ must be $\mathbf{0}$.

23. True. If we reparametrize the curve by replacing $u = t^3$, we have $\mathbf{r}(u) = u\,\mathbf{i} + 2u\,\mathbf{j} + 3u\,\mathbf{k}$, which is a line through the origin with direction vector $\mathbf{i} + 2\,\mathbf{j} + 3\,\mathbf{k}$.

25. False. The vector function represents a line, but the line does not pass through the origin; the x-component is 0 only for $t = 0$ which corresponds to the point $(0, 3, 0)$ not $(0, 0, 0)$.

27. False. By Formula 5 of Theorem 10.7.5, $\dfrac{d}{dt}\,[\mathbf{u}(t) \times \mathbf{v}(t)] = \mathbf{u}'(t) \times \mathbf{v}(t) + \mathbf{u}(t) \times \mathbf{v}'(t)$.

29. False. κ is the magnitude of the rate of change of the unit tangent vector $\mathbf{T}$ with respect to arc length s, not with respect to t.

31. True. At an inflection point where f is twice continuously differentiable we must have $f''(x) = 0$, and by Equation 10.8.11, the curvature is 0 there.

33. False. If $\mathbf{r}(t)$ is the position of a moving particle at time t and $|\mathbf{r}(t)| = 1$ then the particle lies on the unit circle or the unit sphere, but this does not mean that the speed $|\mathbf{r}'(t)|$ must be constant. As a counterexample, let $\mathbf{r}(t) = \langle t, \sqrt{1 - t^2} \rangle$, then $\mathbf{r}'(t) = \langle 1, -t/\sqrt{1 - t^2} \rangle$ and $|\mathbf{r}(t)| = \sqrt{t^2 + 1 - t^2} = 1$ but $|\mathbf{r}'(t)| = \sqrt{1 + t^2/(1 - t^2)} = 1/\sqrt{1 - t^2}$ which is not constant.

EXERCISES

1. (a) The radius of the sphere is the distance between the points $(-1, 2, 1)$ and $(6, -2, 3)$, namely,

$\sqrt{[6 - (-1)]^2 + (-2 - 2)^2 + (3 - 1)^2} = \sqrt{69}$. By the formula for an equation of a sphere (see page 546), an equation of the sphere with center $(-1, 2, 1)$ and radius $\sqrt{69}$ is $(x + 1)^2 + (y - 2)^2 + (z - 1)^2 = 69$.

(b) The intersection of this sphere with the yz-plane is the set of points on the sphere whose x-coordinate is 0. Putting $x = 0$ into the equation, we have $(y - 2)^2 + (z - 1)^2 = 68, x = 0$ which represents a circle in the yz-plane with center $(0, 2, 1)$ and radius $\sqrt{68}$.

(c) Completing squares gives $(x - 4)^2 + (y + 1)^2 + (z + 3)^2 = -1 + 16 + 1 + 9 = 25$. Thus the sphere is centered at $(4, -1, -3)$ and has radius 5.

3. $\mathbf{u} \cdot \mathbf{v} = |\mathbf{u}|\,|\mathbf{v}|\cos 45° = (2)(3)\frac{\sqrt{2}}{2} = 3\sqrt{2}$. $\quad |\mathbf{u} \times \mathbf{v}| = |\mathbf{u}|\,|\mathbf{v}|\sin 45° = (2)(3)\frac{\sqrt{2}}{2} = 3\sqrt{2}$.

By the right-hand rule, $\mathbf{u} \times \mathbf{v}$ is directed out of the page.

5. For the two vectors to be orthogonal, we need $\langle 3, 2, x \rangle \cdot \langle 2x, 4, x \rangle = 0 \quad \Leftrightarrow \quad (3)(2x) + (2)(4) + (x)(x) = 0 \quad \Leftrightarrow$

$x^2 + 6x + 8 = 0 \quad \Leftrightarrow \quad (x + 2)(x + 4) = 0 \quad \Leftrightarrow \quad x = -2$ or $x = -4$.

7. (a) $(\mathbf{u} \times \mathbf{v}) \cdot \mathbf{w} = \mathbf{u} \cdot (\mathbf{v} \times \mathbf{w}) = 2$

(b) $\mathbf{u} \cdot (\mathbf{w} \times \mathbf{v}) = \mathbf{u} \cdot [-(\mathbf{v} \times \mathbf{w})] = -\mathbf{u} \cdot (\mathbf{v} \times \mathbf{w}) = -2$

(c) $\mathbf{v} \cdot (\mathbf{u} \times \mathbf{w}) = (\mathbf{v} \times \mathbf{u}) \cdot \mathbf{w} = -(\mathbf{u} \times \mathbf{v}) \cdot \mathbf{w} = -2$

(d) $(\mathbf{u} \times \mathbf{v}) \cdot \mathbf{v} = \mathbf{u} \cdot (\mathbf{v} \times \mathbf{v}) = \mathbf{u} \cdot \mathbf{0} = 0$

9. For simplicity, consider a unit cube positioned with its back left corner at the origin. Vector representations of the diagonals joining the points $(0, 0, 0)$ to $(1, 1, 1)$ and $(1, 0, 0)$ to $(0, 1, 1)$ are $\langle 1, 1, 1 \rangle$ and $\langle -1, 1, 1 \rangle$. Let θ be the angle between these two vectors. $\quad \langle 1, 1, 1 \rangle \cdot \langle -1, 1, 1 \rangle = -1 + 1 + 1 = 1 = |\langle 1, 1, 1 \rangle|\,|\langle -1, 1, 1 \rangle|\cos\theta = 3\cos\theta \quad \Rightarrow \quad \cos\theta = \frac{1}{3} \quad \Rightarrow$

$\theta = \cos^{-1}\left(\frac{1}{3}\right) \approx 71°$.

11. $\overrightarrow{AB} = \langle 1, 0, -1 \rangle$, $\overrightarrow{AC} = \langle 0, 4, 3 \rangle$, so

(a) a vector perpendicular to the plane is $\overrightarrow{AB} \times \overrightarrow{AC} = \langle 0 + 4, -(3 + 0), 4 - 0 \rangle = \langle 4, -3, 4 \rangle$.

(b) $\frac{1}{2}\left|\overrightarrow{AB} \times \overrightarrow{AC}\right| = \frac{1}{2}\sqrt{16 + 9 + 16} = \frac{\sqrt{41}}{2}$.

13. Let F_1 be the magnitude of the force directed $20°$ away from the direction of shore, and let F_2 be the magnitude of the other force. Separating these forces into components parallel to the direction of the resultant force and perpendicular to it gives

$F_1\cos 20° + F_2\cos 30° = 255$ **(1)**, and $F_1\sin 20° - F_2\sin 30° = 0 \quad \Rightarrow \quad F_1 = F_2\dfrac{\sin 30°}{\sin 20°}$ **(2)**. Substituting **(2)**

into **(1)** gives $F_2(\sin 30° \cot 20° + \cos 30°) = 255 \quad \Rightarrow \quad F_2 \approx 114$ N. Substituting this into **(2)** gives $F_1 \approx 166$ N.

15. The line has direction $\mathbf{v} = \langle -3, 2, 3 \rangle$. Letting $P_0 = (4, -1, 2)$, parametric equations are

$x = 4 - 3t, \; y = -1 + 2t, \; z = 2 + 3t$.

17. A direction vector for the line is a normal vector for the plane, $\mathbf{n} = \langle 2, -1, 5 \rangle$, and parametric equations for the line are

$x = -2 + 2t, \; y = 2 - t, \; z = 4 + 5t$.

19. Here the vectors $\mathbf{a} = \langle 4 - 3, 0 - (-1), 2 - 1 \rangle = \langle 1, 1, 1 \rangle$ and $\mathbf{b} = \langle 6 - 3, 3 - (-1), 1 - 1 \rangle = \langle 3, 4, 0 \rangle$ lie in the plane,

so $\mathbf{n} = \mathbf{a} \times \mathbf{b} = \langle -4, 3, 1 \rangle$ is a normal vector to the plane and an equation of the plane is

$-4(x - 3) + 3(y - (-1)) + 1(z - 1) = 0$ or $-4x + 3y + z = -14$.

21. Substitution of the parametric equations into the equation of the plane gives $2x - y + z = 2(2 - t) - (1 + 3t) + 4t = 2$ ⟹

$-t + 3 = 2$ ⟹ $t = 1$. When $t = 1$, the parametric equations give $x = 2 - 1 = 1$, $y = 1 + 3 = 4$ and $z = 4$. Therefore,

the point of intersection is $(1, 4, 4)$.

23. Since the direction vectors $\langle 2, 3, 4 \rangle$ and $\langle 6, -1, 2 \rangle$ aren't parallel, neither are the lines. For the lines to intersect, the three

equations $1 + 2t = -1 + 6s$, $2 + 3t = 3 - s$, $3 + 4t = -5 + 2s$ must be satisfied simultaneously. Solving the first two

equations gives $t = \frac{1}{5}$, $s = \frac{2}{5}$ and checking we see these values don't satisfy the third equation. Thus the lines aren't parallel

and they don't intersect, so they must be skew.

25. $\mathbf{n_1} = \langle 1, 0, -1 \rangle$ and $\mathbf{n_2} = \langle 0, 1, 2 \rangle$. Setting $z = 0$, it is easy to see that $(1, 3, 0)$ is a point on the line of intersection of

$x - z = 1$ and $y + 2z = 3$. The direction of this line is $\mathbf{v_1} = \mathbf{n_1} \times \mathbf{n_2} = \langle 1, -2, 1 \rangle$. A second vector parallel to the desired

plane is $\mathbf{v_2} = \langle 1, 1, -2 \rangle$, since it is perpendicular to $x + y - 2z = 1$. Therefore, the normal of the plane in question is

$\mathbf{n} = \mathbf{v_1} \times \mathbf{v_2} = \langle 4 - 1, 1 + 2, 1 + 2 \rangle = 3 \langle 1, 1, 1 \rangle$. Taking $(x_0, y_0, z_0) = (1, 3, 0)$, the equation we are looking for is

$(x - 1) + (y - 3) + z = 0$ ⟺ $x + y + z = 4$.

27. By Exercise 10.5.53, $D = \dfrac{|-2 - (-24)|}{\sqrt{3^2 + 1^2 + (-4)^2}} = \dfrac{22}{\sqrt{26}}$.

29. The equation $x = z$ represents a plane perpendicular to

the xz-plane and intersecting the xz-plane in the line

$x = z$, $y = 0$.

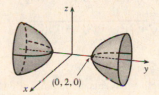

31. The equation $x^2 = y^2 + 4z^2$ represents a (right elliptical)

cone with vertex at the origin and axis the x-axis.

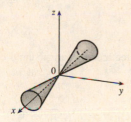

33. An equivalent equation is $-x^2 + \dfrac{y^2}{4} - z^2 = 1$, a

hyperboloid of two sheets with axis the y-axis. For

$|y| > 2$, traces parallel to the xz-plane are circles.

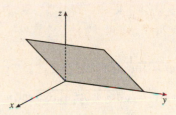

35. Completing the square in y gives

$$4x^2 + 4(y - 1)^2 + z^2 = 4 \text{ or } x^2 + (y - 1)^2 + \dfrac{z^2}{4} = 1,$$

an ellipsoid centered at $(0, 1, 0)$.

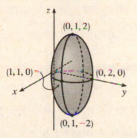

37. $4x^2 + y^2 = 16 \quad \Leftrightarrow \quad \dfrac{x^2}{4} + \dfrac{y^2}{16} = 1$. The equation of the ellipsoid is $\dfrac{x^2}{4} + \dfrac{y^2}{16} + \dfrac{z^2}{c^2} = 1$, since the horizontal trace in the

plane $z = 0$ must be the original ellipse. The traces of the ellipsoid in the yz-plane must be circles since the surface is obtained

by rotation about the x-axis. Therefore, $c^2 = 16$ and the equation of the ellipsoid is $\dfrac{x^2}{4} + \dfrac{y^2}{16} + \dfrac{z^2}{16} = 1 \quad \Leftrightarrow$

$4x^2 + y^2 + z^2 = 16.$

39. (a) The corresponding parametric equations for the curve are $x = t$,

$y = \cos \pi t$, $z = \sin \pi t$. Since $y^2 + z^2 = 1$, the curve is contained in a

circular cylinder with axis the x-axis. Since $x = t$, the curve is a helix.

(b) $\mathbf{r}(t) = t\,\mathbf{i} + \cos \pi t\,\mathbf{j} + \sin \pi t\,\mathbf{k} \quad \Rightarrow$

$\mathbf{r}'(t) = \mathbf{i} - \pi \sin \pi t\,\mathbf{j} + \pi \cos \pi t\,\mathbf{k} \quad \Rightarrow$

$\mathbf{r}''(t) = -\pi^2 \cos \pi t\,\mathbf{j} - \pi^2 \sin \pi t\,\mathbf{k}$

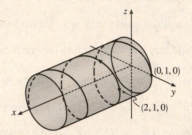

41. The projection of the curve C of intersection onto the xy-plane is the circle $x^2 + y^2 = 16, z = 0$. So we can write

$x = 4\cos t$, $y = 4\sin t$, $0 \le t \le 2\pi$. From the equation of the plane, we have $z = 5 - x = 5 - 4\cos t$, so parametric

equations for C are $x = 4\cos t$, $y = 4\sin t$, $z = 5 - 4\cos t, 0 \le t \le 2\pi$, and the corresponding vector function is

$\mathbf{r}(t) = 4\cos t\,\mathbf{i} + 4\sin t\,\mathbf{j} + (5 - 4\cos t)\,\mathbf{k}, 0 \le t \le 2\pi.$

43. $\int_0^1 (t^2\,\mathbf{i} + t\cos \pi t\,\mathbf{j} + \sin \pi t\,\mathbf{k})\,dt = \left(\int_0^1 t^2\,dt\right)\mathbf{i} + \left(\int_0^1 t\cos \pi t\,dt\right)\mathbf{j} + \left(\int_0^1 \sin \pi t\,dt\right)\mathbf{k}$

$= \left[\tfrac{1}{3}t^3\right]_0^1 \mathbf{i} + \left(\left[\tfrac{t}{\pi}\sin \pi t\right]_0^1 - \int_0^1 \tfrac{1}{\pi}\sin \pi t\,dt\right)\mathbf{j} + \left[-\tfrac{1}{\pi}\cos \pi t\right]_0^1 \mathbf{k}$

$= \tfrac{1}{3}\mathbf{i} + \left[\tfrac{1}{\pi^2}\cos \pi t\right]_0^1 \mathbf{j} + \tfrac{2}{\pi}\mathbf{k} = \tfrac{1}{3}\mathbf{i} - \tfrac{2}{\pi^2}\mathbf{j} + \tfrac{2}{\pi}\mathbf{k}$

where we integrated by parts in the y-component.

45. $\mathbf{r}(t) = \langle t^2, t^3, t^4 \rangle \quad \Rightarrow \quad \mathbf{r}'(t) = \langle 2t, 3t^2, 4t^3 \rangle \quad \Rightarrow \quad |\mathbf{r}'(t)| = \sqrt{4t^2 + 9t^4 + 16t^6}$ and

$L = \int_0^3 |\mathbf{r}'(t)|\,dt = \int_0^3 \sqrt{4t^2 + 9t^4 + 16t^6}\,dt$. Using Simpson's Rule with $f(t) = \sqrt{4t^2 + 9t^4 + 16t^6}$ and $n = 6$ we

have $\Delta t = \dfrac{3-0}{6} = \tfrac{1}{2}$ and

$L \approx \dfrac{\Delta t}{3}\left[f(0) + 4f\!\left(\tfrac{1}{2}\right) + 2f(1) + 4f\!\left(\tfrac{3}{2}\right) + 2f(2) + 4f\!\left(\tfrac{5}{2}\right) + f(3)\right]$

$= \tfrac{1}{6}\Bigg[\sqrt{0+0+0} + 4\cdot\sqrt{4\left(\tfrac{1}{2}\right)^2 + 9\left(\tfrac{1}{2}\right)^4 + 16\left(\tfrac{1}{2}\right)^6} + 2\cdot\sqrt{4(1)^2 + 9(1)^4 + 16(1)^6}$

$+ 4\cdot\sqrt{4\left(\tfrac{3}{2}\right)^2 + 9\left(\tfrac{3}{2}\right)^4 + 16\left(\tfrac{3}{2}\right)^6} + 2\cdot\sqrt{4(2)^2 + 9(2)^4 + 16(2)^6}$

$+ 4\cdot\sqrt{4\left(\tfrac{5}{2}\right)^2 + 9\left(\tfrac{5}{2}\right)^4 + 16\left(\tfrac{5}{2}\right)^6} + \sqrt{4(3)^2 + 9(3)^4 + 16(3)^6}\Bigg]$

≈ 86.631

47. The angle of intersection of the two curves, θ, is the angle between their respective tangents at the point of intersection.

For both curves the point $(1, 0, 0)$ occurs when $t = 0$.

$\mathbf{r}_1'(t) = -\sin t\,\mathbf{i} + \cos t\,\mathbf{j} + \mathbf{k} \quad \Rightarrow \quad \mathbf{r}_1'(0) = \mathbf{j} + \mathbf{k}$ and $\mathbf{r}_2'(t) = \mathbf{i} + 2t\,\mathbf{j} + 3t^2\,\mathbf{k} \quad \Rightarrow \quad \mathbf{r}_2'(0) = \mathbf{i}$.

$\mathbf{r}_1'(0) \cdot \mathbf{r}_2'(0) = (\mathbf{j} + \mathbf{k}) \cdot \mathbf{i} = 0$. Therefore, the curves intersect in a right angle, that is, $\theta = \tfrac{\pi}{2}$.

49. (a) $\mathbf{T}(t) = \dfrac{\mathbf{r}'(t)}{|\mathbf{r}'(t)|} = \dfrac{\langle t^2, t, 1 \rangle}{|\langle t^2, t, 1 \rangle|} = \dfrac{\langle t^2, t, 1 \rangle}{\sqrt{t^4 + t^2 + 1}}$

(b) $\mathbf{T}'(t) = -\frac{1}{2}(t^4 + t^2 + 1)^{-3/2}(4t^3 + 2t)\langle t^2, t, 1 \rangle + (t^4 + t^2 + 1)^{-1/2}\langle 2t, 1, 0 \rangle$

$\quad = \dfrac{-2t^3 - t}{(t^4 + t^2 + 1)^{3/2}} \langle t^2, t, 1 \rangle + \dfrac{1}{(t^4 + t^2 + 1)^{1/2}} \langle 2t, 1, 0 \rangle$

$\quad = \dfrac{\langle -2t^5 - t^3, -2t^4 - t^2, -2t^3 - t \rangle + \langle 2t^5 + 2t^3 + 2t, t^4 + t^2 + 1, 0 \rangle}{(t^4 + t^2 + 1)^{3/2}} = \dfrac{\langle t^3 + 2t, -t^4 + 1, -2t^3 - t \rangle}{(t^4 + t^2 + 1)^{3/2}}$

$|\mathbf{T}'(t)| = \dfrac{\sqrt{t^6 + 4t^4 + 4t^2 + t^8 - 2t^4 + 1 + 4t^6 + 4t^4 + t^2}}{(t^4 + t^2 + 1)^{3/2}} = \dfrac{\sqrt{t^8 + 5t^6 + 6t^4 + 5t^2 + 1}}{(t^4 + t^2 + 1)^{3/2}}$ and

$\mathbf{N}(t) = \dfrac{\langle t^3 + 2t, 1 - t^4, -2t^3 - t \rangle}{\sqrt{t^8 + 5t^6 + 6t^4 + 5t^2 + 1}}.$

(c) $\kappa(t) = \dfrac{|\mathbf{T}'(t)|}{|\mathbf{r}'(t)|} = \dfrac{\sqrt{t^8 + 5t^6 + 6t^4 + 5t^2 + 1}}{(t^4 + t^2 + 1)^2}$ or $\dfrac{\sqrt{t^4 + 4t^2 + 1}}{(t^4 + t^2 + 1)^{3/2}}$

51. $y' = 4x^3$, $y'' = 12x^2$ and $\kappa(x) = \dfrac{|y''|}{[1 + (y')^2]^{3/2}} = \dfrac{|12x^2|}{(1 + 16x^6)^{3/2}}$, so $\kappa(1) = \dfrac{12}{17^{3/2}}$.

53. $\mathbf{r}(t) = t \ln t \, \mathbf{i} + t \, \mathbf{j} + e^{-t}\,\mathbf{k}$, $\mathbf{v}(t) = \mathbf{r}'(t) = (1 + \ln t)\mathbf{i} + \mathbf{j} - e^{-t}\,\mathbf{k}$,

$|\mathbf{v}(t)| = \sqrt{(1 + \ln t)^2 + 1^2 + (-e^{-t})^2} = \sqrt{2 + 2\ln t + (\ln t)^2 + e^{-2t}}$, $\mathbf{a}(t) = \mathbf{v}'(t) = \frac{1}{t}\mathbf{i} + e^{-t}\,\mathbf{k}$

55. We set up the axes so that the shot leaves the athlete's hand 7 ft above the origin. Then we are given $\mathbf{r}(0) = 7\mathbf{j}$,

$|\mathbf{v}(0)| = 43$ ft/s, and $\mathbf{v}(0)$ has direction given by a $45°$ angle of elevation. Then a unit vector in the direction of $\mathbf{v}(0)$ is

$\frac{1}{\sqrt{2}}(\mathbf{i} + \mathbf{j})$ $\Rightarrow$ $\mathbf{v}(0) = \frac{43}{\sqrt{2}}(\mathbf{i} + \mathbf{j})$. Assuming air resistance is negligible, the only external force is due to gravity, so as in

Example 10.9.5 we have $\mathbf{a} = -g\,\mathbf{j}$ where here $g \approx 32$ ft/s^2. Since $\mathbf{v}'(t) = \mathbf{a}(t)$, we integrate, giving $\mathbf{v}(t) = -gt\,\mathbf{j} + \mathbf{C}$

where $\mathbf{C} = \mathbf{v}(0) = \frac{43}{\sqrt{2}}(\mathbf{i} + \mathbf{j})$ $\Rightarrow$ $\mathbf{v}(t) = \frac{43}{\sqrt{2}}\mathbf{i} + \left(\frac{43}{\sqrt{2}} - gt\right)\mathbf{j}$. Since $\mathbf{r}'(t) = \mathbf{v}(t)$ we integrate again, so

$\mathbf{r}(t) = \frac{43}{\sqrt{2}}t\,\mathbf{i} + \left(\frac{43}{\sqrt{2}}t - \frac{1}{2}gt^2\right)\mathbf{j} + \mathbf{D}$. But $\mathbf{D} = \mathbf{r}(0) = 7\mathbf{j}$ $\Rightarrow$ $\mathbf{r}(t) = \frac{43}{\sqrt{2}}t\,\mathbf{i} + \left(\frac{43}{\sqrt{2}}t - \frac{1}{2}gt^2 + 7\right)\mathbf{j}$.

(a) At 2 seconds, the shot is at $\mathbf{r}(2) = \frac{43}{\sqrt{2}}(2)\mathbf{i} + \left(\frac{43}{\sqrt{2}}(2) - \frac{1}{2}g(2)^2 + 7\right)\mathbf{j} \approx 60.8\,\mathbf{i} + 3.8\,\mathbf{j}$, so the shot is about 3.8 ft above

the ground, at a horizontal distance of 60.8 ft from the athlete.

(b) The shot reaches its maximum height when the vertical component of velocity is 0: $\frac{43}{\sqrt{2}} - gt = 0$ $\Rightarrow$

$t = \dfrac{43}{\sqrt{2}\,g} \approx 0.95$ s. Then $\mathbf{r}(0.95) \approx 28.9\,\mathbf{i} + 21.4\,\mathbf{j}$, so the maximum height is approximately 21.4 ft.

(c) The shot hits the ground when the vertical component of $\mathbf{r}(t)$ is 0, so $\frac{43}{\sqrt{2}}t - \frac{1}{2}gt^2 + 7 = 0$ $\Rightarrow$

$-16t^2 + \frac{43}{\sqrt{2}}t + 7 = 0$ $\Rightarrow$ $t \approx 2.11$ s. $\mathbf{r}(2.11) \approx 64.2\,\mathbf{i} - 0.08\,\mathbf{j}$, thus the shot lands approximately 64.2 ft from the

athlete.

57. By the Fundamental Theorem of Calculus, $\mathbf{r}'(t) = \langle \sin(\frac{1}{2}\pi t^2), \cos(\frac{1}{2}\pi t^2) \rangle$, $|\mathbf{r}'(t)| = 1$ and so $\mathbf{T}(t) = \mathbf{r}'(t)$.

Thus $\mathbf{T}'(t) = \pi t \langle \cos(\frac{1}{2}\pi t^2), -\sin(\frac{1}{2}\pi t^2) \rangle$ and the curvature is $\kappa = |\mathbf{T}'(t)| = \sqrt{(\pi t)^2(1)} = \pi\,|t|$.

11 □ PARTIAL DERIVATIVES

11.1 Functions of Several Variables

1. (a) $g(2, -1) = \cos(2 + 2(-1)) = \cos(0) = 1$

   (b) $x + 2y$ is defined for all choices of values for x and y and the cosine function is defined for all input values, so the domain of g is $\mathbb{R}^2$.

   (c) The range of the cosine function is $[-1, 1]$ and $x + 2y$ generates all possible input values for the cosine function, so the range of $\cos(x + 2y)$ is $[-1, 1]$.

3. (a) $f(1, 1, 1) = \sqrt{1} + \sqrt{1} + \sqrt{1} + \ln(4 - 1^2 - 1^2 - 1^2) = 3 + \ln 1 = 3$

   (b) $\sqrt{x}$, $\sqrt{y}$, $\sqrt{z}$ are defined only when $x \geq 0$, $y \geq 0$, $z \geq 0$, and $\ln(4 - x^2 - y^2 - z^2)$ is defined when

   $4 - x^2 - y^2 - z^2 > 0 \quad \Leftrightarrow \quad x^2 + y^2 + z^2 < 4$, thus the domain is

   $\{(x, y, z) \mid x^2 + y^2 + z^2 < 4, \ x \geq 0, \ y \geq 0, \ z \geq 0\}$, the portion of the interior of a sphere of radius 2, centered at the origin, that is in the first octant.

5. $\sqrt{2x - y}$ is defined only when $2x - y \geq 0$, or $y \leq 2x$.

   So the domain of f is $\{(x, y) \mid y \leq 2x\}$.

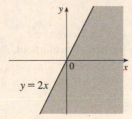

7. $\ln(9 - x^2 - 9y^2)$ is defined only when

   $9 - x^2 - 9y^2 > 0$, or $\frac{1}{9}x^2 + y^2 < 1$. So the domain of f is $\{(x, y) \mid \frac{1}{9}x^2 + y^2 < 1\}$, the interior of an ellipse.

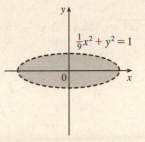

9. $\sqrt{y - x^2}$ is defined only when $y - x^2 \geq 0$, or $y \geq x^2$.

   In addition, f is not defined if $1 - x^2 = 0 \quad \Leftrightarrow$

   $x = \pm 1$. Thus the domain of f is

   $\{(x, y) \mid y \geq x^2, \ x \neq \pm 1\}$.

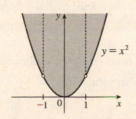

11. We need $1 - x^2 - y^2 - z^2 \geq 0$ or $x^2 + y^2 + z^2 \leq 1$,

    so $D = \{(x, y, z) \mid x^2 + y^2 + z^2 \leq 1\}$ (the points inside or on the sphere of radius 1, center the origin).

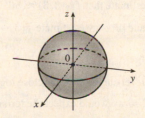

13. $z = 10 - 4x - 5y$ or $4x + 5y + z = 10$, a plane with intercepts 2.5, 2, and 10.

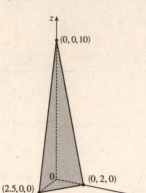

15. $z = y^2 + 1$, a parabolic cylinder

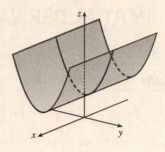

17. $z = 9 - x^2 - 9y^2$, an elliptic paraboloid opening downward with vertex at $(0, 0, 9)$.

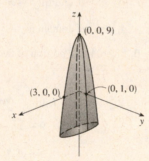

19. $z = \sqrt{4 - 4x^2 - y^2}$ so $4x^2 + y^2 + z^2 = 4$ or $x^2 + \dfrac{y^2}{4} + \dfrac{z^2}{4} = 1$ and $z \geq 0$, the top half of an ellipsoid.

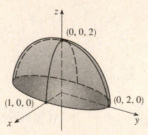

21. The point $(-3, 3)$ lies between the level curves with z-values 50 and 60. Since the point is a little closer to the level curve with $z = 60$, we estimate that $f(-3, 3) \approx 56$. The point $(3, -2)$ appears to be just about halfway between the level curves with z-values 30 and 40, so we estimate $f(3, -2) \approx 35$. The graph rises as we approach the origin, gradually from above, steeply from below.

23. Near A, the level curves are very close together, indicating that the terrain is quite steep. At B, the level curves are much farther apart, so we would expect the terrain to be much less steep than near A, perhaps almost flat.

25. The level curves are $(y - 2x)^2 = k$ or $y = 2x \pm \sqrt{k}$, $k \geq 0$, a family of pairs of parallel lines.

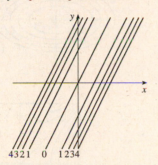

27. The level curves are $\sqrt{x} + y = k$ or $y = -\sqrt{x} + k$, a family of vertical translations of the graph of the root function $y = -\sqrt{x}$.

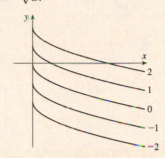

29. The level curves are $ye^x = k$ or $y = ke^{-x}$, a family of exponential curves.

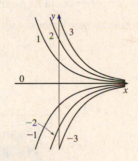

31. The level curves are $\sqrt{y^2 - x^2} = k$ or $y^2 - x^2 = k^2$, $k \geq 0$. When $k = 0$ the level curve is the pair of lines $y = \pm x$. For $k > 0$, the level curves are hyperbolas with axis the y-axis.

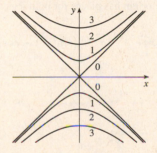

33. The contour map consists of the level curves $k = x^2 + 9y^2$, a family of ellipses with major axis the x-axis. (Or, if $k = 0$, the origin.)

The graph of $f(x, y)$ is the surface $z = x^2 + 9y^2$, an elliptic paraboloid.

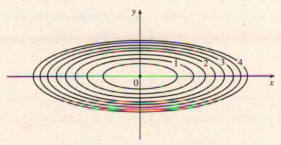

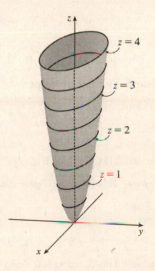

If we visualize lifting each ellipse $k = x^2 + 9y^2$ of the contour map to the plane $z = k$, we have horizontal traces that indicate the shape of the graph of f.

35. The isothermals are given by $k = 100/(1 + x^2 + 2y^2)$ or $x^2 + 2y^2 = (100 - k)/k$ $[0 < k \le 100]$, a family of ellipses.

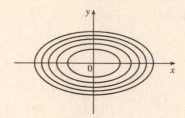

37. $f(x, y) = xy^2 - x^3$

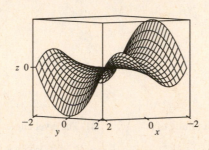

The traces parallel to the yz-plane (such as the left-front trace in the graph above) are parabolas; those parallel to the xz-plane (such as the right-front trace) are cubic curves. The surface is called a monkey saddle because a monkey sitting on the surface near the origin has places for both legs and tail to rest.

39. $f(x, y) = e^{-(x^2+y^2)/3} \left(\sin(x^2) + \cos(y^2) \right)$

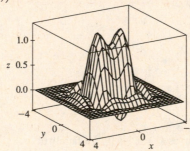
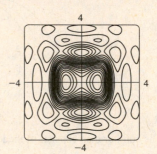

41. $z = \sin(xy)$ (a) C   (b) II

Reasons: This function is periodic in both x and y, and the function is the same when x is interchanged with y, so its graph is symmetric about the plane $y = x$. In addition, the function is 0 along the x- and y-axes. These conditions are satisfied only by C and II.

43. $z = \sin(x - y)$ (a) F   (b) I

Reasons: This function is periodic in both x and y but is constant along the lines $y = x + k$, a condition satisfied only by F and I.

45. $z = (1 - x^2)(1 - y^2)$ (a) B   (b) VI

Reasons: This function is 0 along the lines $x = \pm 1$ and $y = \pm 1$. The only contour map in which this could occur is VI. Also note that the trace in the xz-plane is the parabola $z = 1 - x^2$ and the trace in the yz-plane is the parabola $z = 1 - y^2$, so the graph is B.

47. $k = x + 3y + 5z$ is a family of parallel planes with normal vector $\langle 1, 3, 5 \rangle$.

49. Equations for the level surfaces are $k = y^2 + z^2$. For $k > 0$, we have a family of circular cylinders with axis the x-axis and radius $\sqrt{k}$. When $k = 0$ the level surface is the x-axis. (There are no level surfaces for $k < 0$.)

51. (a) The graph of g is the graph of f shifted upward 2 units.

(b) The graph of g is the graph of f stretched vertically by a factor of 2.

(c) The graph of g is the graph of f reflected about the xy-plane.

(d) The graph of $g(x, y) = -f(x, y) + 2$ is the graph of f reflected about the xy-plane and then shifted upward 2 units.

53.

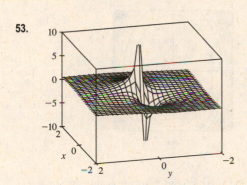

$f(x, y) = \dfrac{x + y}{x^2 + y^2}$. As both x and y become large, the function values appear to approach 0, regardless of which direction is considered. As (x, y) approaches the origin, the graph exhibits asymptotic behavior. From some directions, $f(x, y) \to \infty$, while in others $f(x, y) \to -\infty$. (These are the vertical spikes visible in the graph.) If the graph is examined carefully, however, one can see that $f(x, y)$ approaches 0 along the line $y = -x$.

55. $f(x, y) = e^{cx^2 + y^2}$. First, if $c = 0$, the graph is the cylindrical surface $z = e^{y^2}$ (whose level curves are parallel lines). When $c > 0$, the vertical trace above the y-axis remains fixed while the sides of the surface in the x-direction "curl" upward, giving the graph a shape resembling an elliptic paraboloid. The level curves of the surface are ellipses centered at the origin.

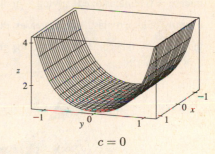

$c = 0$

For $0 < c < 1$, the ellipses have major axis the x-axis and the eccentricity increases as $c \to 0$.

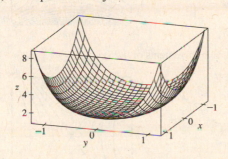

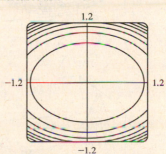

$c = 0.5$ (level curves in increments of 1)

[continued]

For $c = 1$ the level curves are circles centered at the origin.

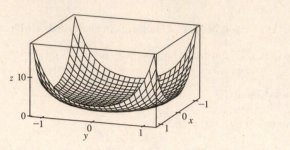
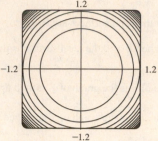

$c = 1$ (level curves in increments of 1)

When $c > 1$, the level curves are ellipses with major axis the y-axis, and the eccentricity increases as c increases.

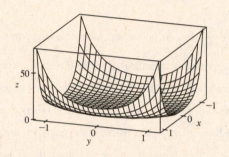
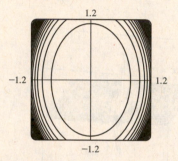

$c = 2$ (level curves in increments of 4)

For values of $c < 0$, the sides of the surface in the x-direction curl downward and approach the xy-plane (while the vertical trace $x = 0$ remains fixed), giving a saddle-shaped appearance to the graph near the point $(0, 0, 1)$. The level curves consist of a family of hyperbolas. As c decreases, the surface becomes flatter in the x-direction and the surface's approach to the curve in the trace $x = 0$ becomes steeper, as the graphs demonstrate.

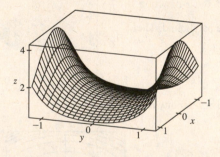
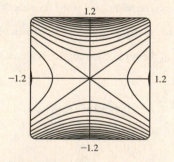

$c = -0.5$ (level curves in increments of 0.25)

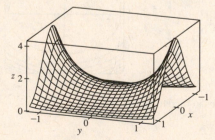
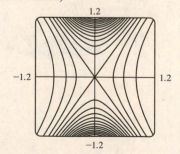

$c = -2$ (level curves in increments of 0.25)

11.2 Limits and Continuity

1. In general, we can't say anything about $f(3,1)$! $\lim\limits_{(x,y)\to(3,1)} f(x,y) = 6$ means that the values of $f(x,y)$ approach 6 as

   (x,y) approaches, but is not equal to, $(3,1)$. If f is continuous, we know that $\lim\limits_{(x,y)\to(a,b)} f(x,y) = f(a,b)$, so

   $\lim\limits_{(x,y)\to(3,1)} f(x,y) = f(3,1) = 6$.

3. $f(x,y) = 5x^3 - x^2y^2$ is a polynomial, and hence continuous, so $\lim\limits_{(x,y)\to(1,2)} f(x,y) = f(1,2) = 5(1)^3 - (1)^2(2)^2 = 1$.

5. $f(x,y) = (x^4 - 4y^2)/(x^2 + 2y^2)$. First approach $(0,0)$ along the x-axis. Then $f(x,0) = x^4/x^2 = x^2$ for $x \neq 0$, so

   $f(x,y) \to 0$. Now approach $(0,0)$ along the y-axis. For $y \neq 0$, $f(0,y) = -4y^2/2y^2 = -2$, so $f(x,y) \to -2$. Since f has

   two different limits along two different lines, the limit does not exist.

7. $f(x,y) = (y^2 \sin^2 x)/(x^4 + y^4)$. On the x-axis, $f(x,0) = 0$ for $x \neq 0$, so $f(x,y) \to 0$ as $(x,y) \to (0,0)$ along the

   x-axis. Approaching $(0,0)$ along the line $y = x$, $f(x,x) = \dfrac{x^2 \sin^2 x}{x^4 + x^4} = \dfrac{\sin^2 x}{2x^2} = \dfrac{1}{2}\left(\dfrac{\sin x}{x}\right)^2$ for $x \neq 0$ and

   $\lim\limits_{x\to 0} \dfrac{\sin x}{x} = 1$, so $f(x,y) \to \frac{1}{2}$. Since f has two different limits along two different lines, the limit does not exist.

9. $f(x,y) = \dfrac{xy}{\sqrt{x^2 + y^2}}$. We can see that the limit along any line through $(0,0)$ is 0, as well as along other paths through

   $(0,0)$ such as $x = y^2$ and $y = x^2$. So we suspect that the limit exists and equals 0; we use the Squeeze Theorem to prove our

   assertion. $0 \le \left|\dfrac{xy}{\sqrt{x^2 + y^2}}\right| \le |x|$ since $|y| \le \sqrt{x^2 + y^2}$, and $|x| \to 0$ as $(x,y) \to (0,0)$. So $\lim\limits_{(x,y)\to(0,0)} f(x,y) = 0$.

11. Let $f(x,y) = \dfrac{x^2 y e^y}{x^4 + 4y^2}$. Then $f(x,0) = 0$ for $x \neq 0$, so $f(x,y) \to 0$ as $(x,y) \to (0,0)$ along the x-axis. Approaching

    $(0,0)$ along the y-axis or the line $y = x$ also gives a limit of 0. But $f(x,x^2) = \dfrac{x^2 x^2 e^{x^2}}{x^4 + 4(x^2)^2} = \dfrac{x^4 e^{x^2}}{5x^4} = \dfrac{e^{x^2}}{5}$ for $x \neq 0$, so

    $f(x,y) \to e^0/5 = \frac{1}{5}$ as $(x,y) \to (0,0)$ along the parabola $y = x^2$. Thus the limit doesn't exist.

13. $\lim\limits_{(x,y)\to(0,0)} \dfrac{x^2 + y^2}{\sqrt{x^2 + y^2 + 1} - 1} = \lim\limits_{(x,y)\to(0,0)} \dfrac{x^2 + y^2}{\sqrt{x^2 + y^2 + 1} - 1} \cdot \dfrac{\sqrt{x^2 + y^2 + 1} + 1}{\sqrt{x^2 + y^2 + 1} + 1}$

    $= \lim\limits_{(x,y)\to(0,0)} \dfrac{(x^2 + y^2)\left(\sqrt{x^2 + y^2 + 1} + 1\right)}{x^2 + y^2} = \lim\limits_{(x,y)\to(0,0)} \left(\sqrt{x^2 + y^2 + 1} + 1\right) = 2$

15. $f(x,y,z) = \dfrac{xy + yz^2 + xz^2}{x^2 + y^2 + z^4}$. Then $f(x,0,0) = 0/x^2 = 0$ for $x \neq 0$, so as $(x,y,z) \to (0,0,0)$ along the x-axis,

    $f(x,y,z) \to 0$. But $f(x,x,0) = x^2/(2x^2) = \frac{1}{2}$ for $x \neq 0$, so as $(x,y,z) \to (0,0,0)$ along the line $y = x$, $z = 0$,

    $f(x,y,z) \to \frac{1}{2}$. Thus the limit doesn't exist.

17.

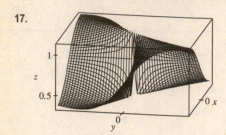

From the ridges on the graph, we see that as $(x, y) \to (0, 0)$ along the lines under the two ridges, $f(x, y)$ approaches different values. So the limit does not exist.

19. $h(x, y) = g(f(x, y)) = (2x + 3y - 6)^2 + \sqrt{2x + 3y - 6}$. Since f is a polynomial, it is continuous on $\mathbb{R}^2$ and g is

continuous on its domain $\{t \mid t \geq 0\}$. Thus h is continuous on its domain.

$D = \{(x, y) \mid 2x + 3y - 6 \geq 0\} = \{(x, y) \mid y \geq -\frac{2}{3}x + 2\}$, which consists of all points on or above the line $y = -\frac{2}{3}x + 2$.

21. $F(x, y) = \dfrac{1 + x^2 + y^2}{1 - x^2 - y^2}$ is a rational function and thus is continuous on its domain

$\{(x, y) \mid 1 - x^2 - y^2 \neq 0\} = \{(x, y) \mid x^2 + y^2 \neq 1\}$.

23. $G(x, y) = \ln(x^2 + y^2 - 4) = g(f(x, y))$ where $f(x, y) = x^2 + y^2 - 4$, continuous on $\mathbb{R}^2$, and $g(t) = \ln t$, continuous on its

domain $\{t \mid t > 0\}$. Thus G is continuous on its domain $\{(x, y) \mid x^2 + y^2 - 4 > 0\} = \{(x, y) \mid x^2 + y^2 > 4\}$, the exterior

of the circle $x^2 + y^2 = 4$.

25. $f(x, y, z) = h(g(x, y, z))$ where $g(x, y, z) = x^2 + y^2 + z^2$, a polynomial that is continuous

everywhere, and $h(t) = \arcsin t$, continuous on $[-1, 1]$. Thus f is continuous on its domain

$\{(x, y, z) \mid -1 \leq x^2 + y^2 + z^2 \leq 1\} = \{(x, y, z) \mid x^2 + y^2 + z^2 \leq 1\}$, so f is continuous on the unit ball.

27. $f(x, y) = \begin{cases} \dfrac{x^2 y^3}{2x^2 + y^2} & \text{if } (x, y) \neq (0, 0) \\ 1 & \text{if } (x, y) = (0, 0) \end{cases}$ The first piece of f is a rational function defined everywhere except at the

origin, so f is continuous on $\mathbb{R}^2$ except possibly at the origin. Since $x^2 \leq 2x^2 + y^2$, we have $\left| x^2 y^3 / (2x^2 + y^2) \right| \leq \left| y^3 \right|$. We

know that $\left| y^3 \right| \to 0$ as $(x, y) \to (0, 0)$. So, by the Squeeze Theorem, $\displaystyle \lim_{(x,y) \to (0,0)} f(x, y) = \lim_{(x,y) \to (0,0)} \frac{x^2 y^3}{2x^2 + y^2} = 0$.

But $f(0, 0) = 1$, so f is discontinuous at $(0, 0)$. Therefore, f is continuous on the set $\{(x, y) \mid (x, y) \neq (0, 0)\}$.

29. $\displaystyle \lim_{(x,y) \to (0,0)} \frac{x^3 + y^3}{x^2 + y^2} = \lim_{r \to 0^+} \frac{(r \cos \theta)^3 + (r \sin \theta)^3}{r^2} = \lim_{r \to 0^+} (r \cos^3 \theta + r \sin^3 \theta) = 0$

31. $\displaystyle \lim_{(x,y) \to (0,0)} \frac{e^{-x^2 - y^2} - 1}{x^2 + y^2} = \lim_{r \to 0^+} \frac{e^{-r^2} - 1}{r^2} = \lim_{r \to 0^+} \frac{e^{-r^2}(-2r)}{2r}$ [using l'Hospital's Rule]

$$= \lim_{r \to 0^+} -e^{-r^2} = -e^0 = -1$$

33. Since $|\mathbf{x} - \mathbf{a}|^2 = |\mathbf{x}|^2 + |\mathbf{a}|^2 - 2 |\mathbf{x}| |\mathbf{a}| \cos \theta \geq |\mathbf{x}|^2 + |\mathbf{a}|^2 - 2 |\mathbf{x}| |\mathbf{a}| = (|\mathbf{x}| - |\mathbf{a}|)^2$, we have $\left| |\mathbf{x}| - |\mathbf{a}| \right| \leq |\mathbf{x} - \mathbf{a}|$. Let

$\epsilon > 0$ be given and set $\delta = \epsilon$. Then if $0 < |\mathbf{x} - \mathbf{a}| < \delta$, $\left| |\mathbf{x}| - |\mathbf{a}| \right| \leq |\mathbf{x} - \mathbf{a}| < \delta = \epsilon$. Hence $\lim_{\mathbf{x} \to \mathbf{a}} |\mathbf{x}| = |\mathbf{a}|$ and

$f(\mathbf{x}) = |\mathbf{x}|$ is continuous on $\mathbb{R}^n$.

11.3 Partial Derivatives

1. (a) $\partial T/\partial x$ represents the rate of change of T when we fix y and t and consider T as a function of the single variable x, which

describes how quickly the temperature changes when longitude changes but latitude and time are constant. $\partial T/\partial y$

represents the rate of change of T when we fix x and t and consider T as a function of y, which describes how quickly the

temperature changes when latitude changes but longitude and time are constant. $\partial T/\partial t$ represents the rate of change of T

when we fix x and y and consider T as a function of t, which describes how quickly the temperature changes over time for

a constant longitude and latitude.

(b) $f_x(158, 21, 9)$ represents the rate of change of temperature at longitude 158°W, latitude 21°N at 9:00 AM when only

longitude varies. Since the air is warmer to the west than to the east, increasing longitude results in an increased air

temperature, so we would expect $f_x(158, 21, 9)$ to be positive. $f_y(158, 21, 9)$ represents the rate of change of temperature

at the same time and location when only latitude varies. Since the air is warmer to the south and cooler to the north,

increasing latitude results in a decreased air temperature, so we would expect $f_y(158, 21, 9)$ to be negative. $f_t(158, 21, 9)$

represents the rate of change of temperature at the same time and location when only time varies. Since typically air

temperature increases from the morning to the afternoon as the sun warms it, we would expect $f_t(158, 21, 9)$ to be

positive.

3. (a) If we start at $(1, 2)$ and move in the positive x-direction, the graph of f increases. Thus $f_x(1, 2)$ is positive.

(b) If we start at $(1, 2)$ and move in the positive y-direction, the graph of f decreases. Thus $f_y(1, 2)$ is negative.

5. $f(x, y) = 16 - 4x^2 - y^2 \ \Rightarrow \ f_x(x, y) = -8x$ and $f_y(x, y) = -2y \ \Rightarrow \ f_x(1, 2) = -8$ and $f_y(1, 2) = -4$. The graph

of f is the paraboloid $z = 16 - 4x^2 - y^2$ and the vertical plane $y = 2$ intersects it in the parabola $z = 12 - 4x^2$, $y = 2$

(the curve C_1 in the first figure). The slope of the tangent line
to this parabola at $(1, 2, 8)$ is $f_x(1, 2) = -8$. Similarly the
plane $x = 1$ intersects the paraboloid in the parabola
$z = 12 - y^2$, $x = 1$ (the curve C_2 in the second figure) and
the slope of the tangent line at $(1, 2, 8)$ is $f_y(1, 2) = -4$.

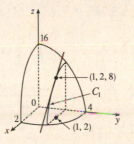
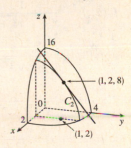

7. $f(x, y) = y^5 - 3xy \ \Rightarrow \ f_x(x, y) = 0 - 3y = -3y, \ f_y(x, y) = 5y^4 - 3x$

9. $f(x, t) = e^{-t} \cos \pi x \ \Rightarrow \ f_x(x, t) = e^{-t}(-\sin \pi x)(\pi) = -\pi e^{-t} \sin \pi x, \ f_t(x, t) = e^{-t}(-1) \cos \pi x = -e^{-t} \cos \pi x$

11. $f(x, y) = x/y = xy^{-1} \ \Rightarrow \ f_x(x, y) = y^{-1} = 1/y, \ f_y(x, y) = -xy^{-2} = -x/y^2$

13. $f(x, y) = \dfrac{ax + by}{cx + dy} \ \Rightarrow \ f_x(x, y) = \dfrac{(cx + dy)(a) - (ax + by)(c)}{(cx + dy)^2} = \dfrac{(ad - bc)y}{(cx + dy)^2},$

$f_y(x, y) = \dfrac{(cx + dy)(b) - (ax + by)(d)}{(cx + dy)^2} = \dfrac{(bc - ad)x}{(cx + dy)^2}$

15. $g(u,v) = (u^2v - v^3)^5 \;\Rightarrow\; g_u(u,v) = 5(u^2v - v^3)^4 \cdot 2uv = 10uv(u^2v - v^3)^4,$

$g_v(u,v) = 5(u^2v - v^3)^4(u^2 - 3v^2) = 5(u^2 - 3v^2)(u^2v - v^3)^4$

17. $R(p,q) = \tan^{-1}(pq^2) \;\Rightarrow\; R_p(p,q) = \dfrac{1}{1 + (pq^2)^2} \cdot q^2 = \dfrac{q^2}{1 + p^2q^4},\; R_q(p,q) = \dfrac{1}{1 + (pq^2)^2} \cdot 2pq = \dfrac{2pq}{1 + p^2q^4}$

19. $F(x,y) = \displaystyle\int_y^x \cos(e^t)\, dt \;\Rightarrow\; F_x(x,y) = \dfrac{\partial}{\partial x} \int_y^x \cos(e^t)\, dt = \cos(e^x)$ by the Fundamental Theorem of Calculus, Part 1;

$F_y(x,y) = \dfrac{\partial}{\partial y} \displaystyle\int_y^x \cos(e^t)\, dt = \dfrac{\partial}{\partial y}\left[-\int_x^y \cos(e^t)\, dt \right] = -\dfrac{\partial}{\partial y}\int_x^y \cos(e^t)\, dt = -\cos(e^y).$

21. $f(x,y,z) = xz - 5x^2y^3z^4 \;\Rightarrow\; f_x(x,y,z) = z - 10xy^3z^4,\; f_y(x,y,z) = -15x^2y^2z^4,\; f_z(x,y,z) = x - 20x^2y^3z^3$

23. $w = \ln(x + 2y + 3z) \;\Rightarrow\; \dfrac{\partial w}{\partial x} = \dfrac{1}{x + 2y + 3z},\; \dfrac{\partial w}{\partial y} = \dfrac{2}{x + 2y + 3z},\; \dfrac{\partial w}{\partial z} = \dfrac{3}{x + 2y + 3z}$

25. $u = xy\sin^{-1}(yz) \;\Rightarrow\; \dfrac{\partial u}{\partial x} = y\sin^{-1}(yz),\; \dfrac{\partial u}{\partial y} = xy \cdot \dfrac{1}{\sqrt{1 - (yz)^2}}\,(z) + \sin^{-1}(yz) \cdot x = \dfrac{xyz}{\sqrt{1 - y^2z^2}} + x\sin^{-1}(yz),$

$\dfrac{\partial u}{\partial z} = xy \cdot \dfrac{1}{\sqrt{1 - (yz)^2}}\,(y) = \dfrac{xy^2}{\sqrt{1 - y^2z^2}}$

27. $h(x,y,z,t) = x^2y\cos(z/t) \;\Rightarrow\; h_x(x,y,z,t) = 2xy\cos(z/t),\; h_y(x,y,z,t) = x^2\cos(z/t),$

$h_z(x,y,z,t) = -x^2y\sin(z/t)(1/t) = (-x^2y/t)\sin(z/t),\; h_t(x,y,z,t) = -x^2y\sin(z/t)(-zt^{-2}) = (x^2yz/t^2)\sin(z/t)$

29. $u = \sqrt{x_1^2 + x_2^2 + \cdots + x_n^2}.$ For each $i = 1, \ldots, n,\; u_{x_i} = \tfrac{1}{2}\left(x_1^2 + x_2^2 + \cdots + x_n^2\right)^{-1/2}(2x_i) = \dfrac{x_i}{\sqrt{x_1^2 + x_2^2 + \cdots + x_n^2}}.$

31. $f(x,y) = \ln\left(x + \sqrt{x^2 + y^2}\right) \;\Rightarrow$

$f_x(x,y) = \dfrac{1}{x + \sqrt{x^2 + y^2}}\left[1 + \tfrac{1}{2}(x^2 + y^2)^{-1/2}(2x) \right] = \dfrac{1}{x + \sqrt{x^2 + y^2}}\left(1 + \dfrac{x}{\sqrt{x^2 + y^2}} \right),$

so $f_x(3,4) = \dfrac{1}{3 + \sqrt{3^2 + 4^2}}\left(1 + \dfrac{3}{\sqrt{3^2 + 4^2}} \right) = \tfrac{1}{8}\left(1 + \tfrac{3}{5} \right) = \tfrac{1}{5}.$

33. $f(x,y,z) = \dfrac{y}{x + y + z} \;\Rightarrow\; f_y(x,y,z) = \dfrac{1(x + y + z) - y(1)}{(x + y + z)^2} = \dfrac{x + z}{(x + y + z)^2},$

so $f_y(2,1,-1) = \dfrac{2 + (-1)}{(2 + 1 + (-1))^2} = \dfrac{1}{4}.$

35. $f(x,y) = xy^2 - x^3y \;\Rightarrow$

$\begin{aligned} f_x(x,y) &= \lim_{h \to 0} \dfrac{f(x+h,y) - f(x,y)}{h} = \lim_{h \to 0} \dfrac{(x+h)y^2 - (x+h)^3y - (xy^2 - x^3y)}{h} \\[2mm] &= \lim_{h \to 0} \dfrac{h(y^2 - 3x^2y - 3xyh - yh^2)}{h} = \lim_{h \to 0}(y^2 - 3x^2y - 3xyh - yh^2) = y^2 - 3x^2y \end{aligned}$

$\begin{aligned} f_y(x,y) &= \lim_{h \to 0} \dfrac{f(x,y+h) - f(x,y)}{h} = \lim_{h \to 0} \dfrac{x(y+h)^2 - x^3(y+h) - (xy^2 - x^3y)}{h} = \lim_{h \to 0} \dfrac{h(2xy + xh - x^3)}{h} \\[2mm] &= \lim_{h \to 0}(2xy + xh - x^3) = 2xy - x^3 \end{aligned}$

37. $f(x, y) = x^2 y^3 \quad \Rightarrow \quad f_x = 2xy^3, \quad f_y = 3x^2 y^2$

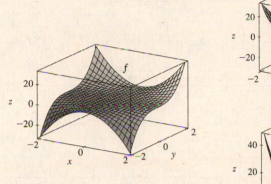

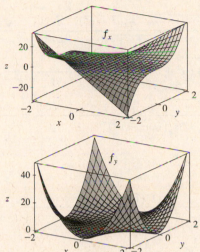

Note that traces of f in planes parallel to the xz-plane are parabolas which open downward for $y < 0$ and upward for $y > 0$, and the traces of f_x in these planes are straight lines, which have negative slopes for $y < 0$ and positive slopes for $y > 0$. The traces of f in planes parallel to the yz-plane are cubic curves, and the traces of f_y in these planes are parabolas.

39. $x^2 + 2y^2 + 3z^2 = 1 \quad \Rightarrow \quad \dfrac{\partial}{\partial x}(x^2 + 2y^2 + 3z^2) = \dfrac{\partial}{\partial x}(1) \quad \Rightarrow \quad 2x + 0 + 6z\dfrac{\partial z}{\partial x} = 0 \quad \Rightarrow \quad 6z\dfrac{\partial z}{\partial x} = -2x \quad \Rightarrow$

$\dfrac{\partial z}{\partial x} = \dfrac{-2x}{6z} = -\dfrac{x}{3z}$, and $\dfrac{\partial}{\partial y}(x^2 + 2y^2 + 3z^2) = \dfrac{\partial}{\partial y}(1) \quad \Rightarrow \quad 0 + 4y + 6z\dfrac{\partial z}{\partial y} = 0 \quad \Rightarrow \quad 6z\dfrac{\partial z}{\partial y} = -4y \quad \Rightarrow$

$\dfrac{\partial z}{\partial y} = \dfrac{-4y}{6z} = -\dfrac{2y}{3z}$.

41. $e^z = xyz \quad \Rightarrow \quad \dfrac{\partial}{\partial x}(e^z) = \dfrac{\partial}{\partial x}(xyz) \quad \Rightarrow \quad e^z\dfrac{\partial z}{\partial x} = y\left(x\dfrac{\partial z}{\partial x} + z \cdot 1\right) \quad \Rightarrow \quad e^z\dfrac{\partial z}{\partial x} - xy\dfrac{\partial z}{\partial x} = yz \quad \Rightarrow$

$(e^z - xy)\dfrac{\partial z}{\partial x} = yz$, so $\dfrac{\partial z}{\partial x} = \dfrac{yz}{e^z - xy}$.

$\dfrac{\partial}{\partial y}(e^z) = \dfrac{\partial}{\partial y}(xyz) \quad \Rightarrow \quad e^z\dfrac{\partial z}{\partial y} = x\left(y\dfrac{\partial z}{\partial x} + z \cdot 1\right) \quad \Rightarrow \quad e^z\dfrac{\partial z}{\partial y} - xy\dfrac{\partial z}{\partial y} = xz \quad \Rightarrow \quad (e^z - xy)\dfrac{\partial z}{\partial y} = xz$, so

$\dfrac{\partial z}{\partial y} = \dfrac{xz}{e^z - xy}$.

43. (a) $z = f(x) + g(y) \quad \Rightarrow \quad \dfrac{\partial z}{\partial x} = f'(x), \quad \dfrac{\partial z}{\partial y} = g'(y)$

(b) $z = f(x + y)$. Let $u = x + y$. Then $\dfrac{\partial z}{\partial x} = \dfrac{df}{du}\dfrac{\partial u}{\partial x} = \dfrac{df}{du}(1) = f'(u) = f'(x + y)$,

$\dfrac{\partial z}{\partial y} = \dfrac{df}{du}\dfrac{\partial u}{\partial y} = \dfrac{df}{du}(1) = f'(u) = f'(x + y)$.

45. $f(x, y) = x^3 y^5 + 2x^4 y \quad \Rightarrow \quad f_x(x, y) = 3x^2 y^5 + 8x^3 y, \ f_y(x, y) = 5x^3 y^4 + 2x^4$. Then $f_{xx}(x, y) = 6xy^5 + 24x^2 y$,

$f_{xy}(x, y) = 15x^2 y^4 + 8x^3, \ f_{yx}(x, y) = 15x^2 y^4 + 8x^3$, and $f_{yy}(x, y) = 20x^3 y^3$.

47. $w = \sqrt{u^2 + v^2}$ $\Rightarrow$ $w_u = \frac{1}{2}(u^2 + v^2)^{-1/2} \cdot 2u = \frac{u}{\sqrt{u^2 + v^2}}$, $w_v = \frac{1}{2}(u^2 + v^2)^{-1/2} \cdot 2v = \frac{v}{\sqrt{u^2 + v^2}}$. Then

$$w_{uu} = \frac{1 \cdot \sqrt{u^2 + v^2} - u \cdot \frac{1}{2}(u^2 + v^2)^{-1/2}(2u)}{\left(\sqrt{u^2 + v^2}\right)^2} = \frac{\sqrt{u^2 + v^2} - u^2/\sqrt{u^2 + v^2}}{u^2 + v^2} = \frac{u^2 + v^2 - u^2}{(u^2 + v^2)^{3/2}} = \frac{v^2}{(u^2 + v^2)^{3/2}},$$

$$w_{uv} = u\left(-\tfrac{1}{2}\right)\left(u^2 + v^2\right)^{-3/2}(2v) = -\frac{uv}{(u^2 + v^2)^{3/2}}, \quad w_{vu} = v\left(-\tfrac{1}{2}\right)\left(u^2 + v^2\right)^{-3/2}(2u) = -\frac{uv}{(u^2 + v^2)^{3/2}},$$

$$w_{vv} = \frac{1 \cdot \sqrt{u^2 + v^2} - v \cdot \frac{1}{2}(u^2 + v^2)^{-1/2}(2v)}{\left(\sqrt{u^2 + v^2}\right)^2} = \frac{\sqrt{u^2 + v^2} - v^2/\sqrt{u^2 + v^2}}{u^2 + v^2} = \frac{u^2 + v^2 - v^2}{(u^2 + v^2)^{3/2}} = \frac{u^2}{(u^2 + v^2)^{3/2}}.$$

49. $z = \arctan \dfrac{x + y}{1 - xy}$ $\Rightarrow$

$$z_x = \frac{1}{1 + \left(\frac{x+y}{1-xy}\right)^2} \cdot \frac{(1)(1 - xy) - (x + y)(-y)}{(1 - xy)^2} = \frac{1 + y^2}{(1 - xy)^2 + (x + y)^2} = \frac{1 + y^2}{1 + x^2 + y^2 + x^2 y^2}$$

$$= \frac{1 + y^2}{(1 + x^2)(1 + y^2)} = \frac{1}{1 + x^2},$$

$$z_y = \frac{1}{1 + \left(\frac{x+y}{1-xy}\right)^2} \cdot \frac{(1)(1 - xy) - (x + y)(-x)}{(1 - xy)^2} = \frac{1 + x^2}{(1 - xy)^2 + (x + y)^2} = \frac{1 + x^2}{(1 + x^2)(1 + y^2)} = \frac{1}{1 + y^2}.$$

Then $z_{xx} = -(1 + x^2)^{-2} \cdot 2x = -\dfrac{2x}{(1 + x^2)^2}$, $z_{xy} = 0$, $z_{yx} = 0$, $z_{yy} = -(1 + y^2)^{-2} \cdot 2y = -\dfrac{2y}{(1 + y^2)^2}$.

51. $u = x^4 y^3 - y^4$ $\Rightarrow$ $u_x = 4x^3 y^3$, $u_{xy} = 12x^3 y^2$ and $u_y = 3x^4 y^2 - 4y^3$, $u_{yx} = 12x^3 y^2$.

Thus $u_{xy} = u_{yx}$.

53. $f(x, y) = x^4 y^2 - x^3 y$ $\Rightarrow$ $f_x = 4x^3 y^2 - 3x^2 y$, $f_{xx} = 12x^2 y^2 - 6xy$, $f_{xxx} = 24xy^2 - 6y$ and

$f_{xy} = 8x^3 y - 3x^2$, $f_{xyx} = 24x^2 y - 6x$.

55. $f(x, y, z) = e^{xyz^2}$ $\Rightarrow$ $f_x = e^{xyz^2} \cdot yz^2 = yz^2 e^{xyz^2}$, $f_{xy} = yz^2 \cdot e^{xyz^2}(xz^2) + e^{xyz^2} \cdot z^2 = (xyz^4 + z^2)e^{xyz^2}$,

$f_{xyz} = (xyz^4 + z^2) \cdot e^{xyz^2}(2xyz) + e^{xyz^2} \cdot (4xyz^3 + 2z) = (2x^2 y^2 z^5 + 6xyz^3 + 2z)e^{xyz^2}$.

57. $u = e^{r\theta} \sin\theta$ $\Rightarrow$ $\dfrac{\partial u}{\partial \theta} = e^{r\theta} \cos\theta + \sin\theta \cdot e^{r\theta}(r) = e^{r\theta}(\cos\theta + r\sin\theta)$,

$$\frac{\partial^2 u}{\partial r \, \partial \theta} = e^{r\theta}(\sin\theta) + (\cos\theta + r\sin\theta) e^{r\theta}(\theta) = e^{r\theta}(\sin\theta + \theta\cos\theta + r\theta\sin\theta),$$

$$\frac{\partial^3 u}{\partial r^2 \, \partial \theta} = e^{r\theta}(\theta\sin\theta) + (\sin\theta + \theta\cos\theta + r\theta\sin\theta) \cdot e^{r\theta}(\theta) = \theta e^{r\theta}(2\sin\theta + \theta\cos\theta + r\theta\sin\theta).$$

59. Assuming that the third partial derivatives of f are continuous (easily verified), we can write $f_{xzy} = f_{yxz}$. Then

$f(x, y, z) = xy^2 z^3 + \arcsin\left(x\sqrt{z}\right)$ $\Rightarrow$ $f_y = 2xyz^3 + 0$, $f_{yx} = 2yz^3$, and $f_{yxz} = 6yz^2 = f_{xzy}$.

61. $u = e^{-\alpha^2 k^2 t} \sin kx$ $\Rightarrow$ $u_x = ke^{-\alpha^2 k^2 t} \cos kx$, $u_{xx} = -k^2 e^{-\alpha^2 k^2 t} \sin kx$, and $u_t = -\alpha^2 k^2 e^{-\alpha^2 k^2 t} \sin kx$.

Thus $\alpha^2 u_{xx} = u_t$.

63. $u = \dfrac{1}{\sqrt{x^2 + y^2 + z^2}}$ $\Rightarrow$ $u_x = \left(-\frac{1}{2}\right)(x^2 + y^2 + z^2)^{-3/2}(2x) = -x(x^2 + y^2 + z^2)^{-3/2}$ and

$$u_{xx} = -(x^2 + y^2 + z^2)^{-3/2} - x\left(-\frac{3}{2}\right)(x^2 + y^2 + z^2)^{-5/2}(2x) = \frac{2x^2 - y^2 - z^2}{(x^2 + y^2 + z^2)^{5/2}}.$$

By symmetry, $u_{yy} = \dfrac{2y^2 - x^2 - z^2}{(x^2 + y^2 + z^2)^{5/2}}$ and $u_{zz} = \dfrac{2z^2 - x^2 - y^2}{(x^2 + y^2 + z^2)^{5/2}}.$

Thus $u_{xx} + u_{yy} + u_{zz} = \dfrac{2x^2 - y^2 - z^2 + 2y^2 - x^2 - z^2 + 2z^2 - x^2 - y^2}{(x^2 + y^2 + z^2)^{5/2}} = 0.$

65. Let $v = x + at,\ w = x - at.$ Then $u_t = \dfrac{\partial[f(v) + g(w)]}{\partial t} = \dfrac{df(v)}{dv}\dfrac{\partial v}{\partial t} + \dfrac{dg(w)}{dw}\dfrac{\partial w}{\partial t} = af'(v) - ag'(w)$ and

$$u_{tt} = \frac{\partial[af'(v) - ag'(w)]}{\partial t} = a[af''(v) + ag''(w)] = a^2[f''(v) + g''(w)].$$ Similarly, by using the Chain Rule we have

$u_x = f'(v) + g'(w)$ and $u_{xx} = f''(v) + g''(w).$ Thus $u_{tt} = a^2 u_{xx}.$

67. $z_x = e^y + ye^x,\ z_{xx} = ye^x,\ \partial^3 z/\partial x^3 = ye^x.$ By symmetry $z_y = xe^y + e^x,\ z_{yy} = xe^y,\ \partial^3 z/\partial y^3 = xe^y.$

Then $\partial^3 z/\partial x \partial y^2 = e^y$ and $\partial^3 z/\partial x^2 \partial y = e^x.$ Thus $z = xe^y + ye^x$ satisfies the given partial differential equation.

69. By the Chain Rule, taking the partial derivative of both sides with respect to R_1 gives

$$\frac{\partial R^{-1}}{\partial R}\frac{\partial R}{\partial R_1} = \frac{\partial\left[(1/R_1) + (1/R_2) + (1/R_3)\right]}{\partial R_1} \quad \text{or} \quad -R^{-2}\frac{\partial R}{\partial R_1} = -R_1^{-2}. \text{ Thus } \frac{\partial R}{\partial R_1} = \frac{R^2}{R_1^2}.$$

71. $\left(P + \dfrac{n^2 a}{V^2}\right)(V - nb) = nRT$ $\Rightarrow$ $T = \dfrac{1}{nR}\left(P + \dfrac{n^2 a}{V^2}\right)(V - nb),$ so $\dfrac{\partial T}{\partial P} = \dfrac{1}{nR}(1)(V - nb) = \dfrac{V - nb}{nR}.$

We can also write $P + \dfrac{n^2 a}{V^2} = \dfrac{nRT}{V - nb}$ $\Rightarrow$ $P = \dfrac{nRT}{V - nb} - \dfrac{n^2 a}{V^2} = nRT(V - nb)^{-1} - n^2 aV^{-2},$ so

$$\frac{\partial P}{\partial V} = -nRT(V - nb)^{-2}(1) + 2n^2 aV^{-3} = \frac{2n^2 a}{V^3} - \frac{nRT}{(V - nb)^2}.$$

73. $\dfrac{\partial K}{\partial m} = \frac{1}{2}v^2,\ \dfrac{\partial K}{\partial v} = mv,\ \dfrac{\partial^2 K}{\partial v^2} = m.$ Thus $\dfrac{\partial K}{\partial m} \cdot \dfrac{\partial^2 K}{\partial v^2} = \frac{1}{2}v^2 m = K.$

75. $f_x(x, y) = x + 4y$ $\Rightarrow$ $f_{xy}(x, y) = 4$ and $f_y(x, y) = 3x - y$ $\Rightarrow$ $f_{yx}(x, y) = 3.$ Since f_{xy} and f_{yx} are continuous

everywhere but $f_{xy}(x, y) \neq f_{yx}(x, y),$ Clairaut's Theorem implies that such a function $f(x, y)$ does not exist.

77. By the geometry of partial derivatives, the slope of the tangent line is $f_x(1, 2).$ By implicit differentiation of

$4x^2 + 2y^2 + z^2 = 16,$ we get $8x + 2z\,(\partial z/\partial x) = 0$ $\Rightarrow$ $\partial z/\partial x = -4x/z,$ so when $x = 1$ and $z = 2$ we have

$\partial z/\partial x = -2.$ So the slope is $f_x(1, 2) = -2.$ Thus the tangent line is given by $z - 2 = -2(x - 1),\ y = 2.$ Taking the

parameter to be $t = x - 1,$ we can write parametric equations for this line: $x = 1 + t,\ y = 2,\ z = 2 - 2t.$

79. By Clairaut's Theorem, $f_{xyy} = (f_{xy})_y = (f_{yx})_y = f_{yxy} = (f_y)_{xy} = (f_y)_{yx} = f_{yyx}.$

81. Let $g(x) = f(x, 0) = x(x^2)^{-3/2}e^0 = x\,|x|^{-3}.$ But we are using the point $(1, 0),$ so near $(1, 0),\ g(x) = x^{-2}.$ Then

$g'(x) = -2x^{-3}$ and $g'(1) = -2,$ so using (1) we have $f_x(1, 0) = g'(1) = -2.$

83. (a)

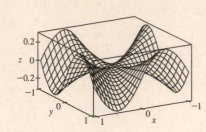

(b) For $(x, y) \neq (0, 0)$,

$$f_x(x, y) = \frac{(3x^2 y - y^3)(x^2 + y^2) - (x^3 y - xy^3)(2x)}{(x^2 + y^2)^2}$$

$$= \frac{x^4 y + 4x^2 y^3 - y^5}{(x^2 + y^2)^2}$$

and by symmetry $f_y(x, y) = \dfrac{x^5 - 4x^3 y^2 - xy^4}{(x^2 + y^2)^2}$.

(c) $f_x(0, 0) = \lim\limits_{h \to 0} \dfrac{f(h, 0) - f(0, 0)}{h} = \lim\limits_{h \to 0} \dfrac{(0/h^2) - 0}{h} = 0$ and $f_y(0, 0) = \lim\limits_{h \to 0} \dfrac{f(0, h) - f(0, 0)}{h} = 0$.

(d) By (3), $f_{xy}(0, 0) = \dfrac{\partial f_x}{\partial y} = \lim\limits_{h \to 0} \dfrac{f_x(0, h) - f_x(0, 0)}{h} = \lim\limits_{h \to 0} \dfrac{(-h^5 - 0)/h^4}{h} = -1$ while by (2),

$$f_{yx}(0, 0) = \frac{\partial f_y}{\partial x} = \lim_{h \to 0} \frac{f_y(h, 0) - f_y(0, 0)}{h} = \lim_{h \to 0} \frac{h^5/h^4}{h} = 1.$$

(e) For $(x, y) \neq (0, 0)$, we use a CAS to compute

$$f_{xy}(x, y) = \frac{x^6 + 9x^4 y^2 - 9x^2 y^4 - y^6}{(x^2 + y^2)^3}$$

Now as $(x, y) \to (0, 0)$ along the x-axis, $f_{xy}(x, y) \to 1$ while as

$(x, y) \to (0, 0)$ along the y-axis, $f_{xy}(x, y) \to -1$. Thus f_{xy} isn't

continuous at $(0, 0)$ and Clairaut's Theorem doesn't apply, so there is

no contradiction. The graphs of f_{xy} and f_{yx} are identical except at the

origin, where we observe the discontinuity.

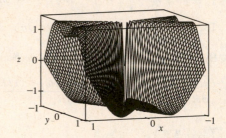

11.4 Tangent Planes and Linear Approximations

1. $z = f(x, y) = 3y^2 - 2x^2 + x \ \Rightarrow \ f_x(x, y) = -4x + 1$, $f_y(x, y) = 6y$, so $f_x(2, -1) = -7$, $f_y(2, -1) = -6$.

By Equation 2, an equation of the tangent plane is $z - (-3) = f_x(2, -1)(x - 2) + f_y(2, -1)[y - (-1)] \ \Rightarrow$

$z + 3 = -7(x - 2) - 6(y + 1)$ or $z = -7x - 6y + 5$.

3. $z = f(x, y) = \sqrt{xy} \ \Rightarrow \ f_x(x, y) = \frac{1}{2}(xy)^{-1/2} \cdot y = \frac{1}{2}\sqrt{y/x}$, $f_y(x, y) = \frac{1}{2}(xy)^{-1/2} \cdot x = \frac{1}{2}\sqrt{x/y}$, so $f_x(1, 1) = \frac{1}{2}$

and $f_y(1, 1) = \frac{1}{2}$. Thus an equation of the tangent plane is $z - 1 = f_x(1, 1)(x - 1) + f_y(1, 1)(y - 1) \ \Rightarrow$

$z - 1 = \frac{1}{2}(x - 1) + \frac{1}{2}(y - 1)$ or $x + y - 2z = 0$.

5. $z = f(x, y) = x\sin(x + y) \ \Rightarrow \ f_x(x, y) = x \cdot \cos(x + y) + \sin(x + y) \cdot 1 = x\cos(x + y) + \sin(x + y)$,

$f_y(x, y) = x\cos(x + y)$, so $f_x(-1, 1) = (-1)\cos 0 + \sin 0 = -1$, $f_y(-1, 1) = (-1)\cos 0 = -1$ and an equation of the

tangent plane is $z - 0 = (-1)(x + 1) + (-1)(y - 1)$ or $x + y + z = 0$.

7. $z = f(x, y) = x^2 + xy + 3y^2$, so $f_x(x, y) = 2x + y$ $\Rightarrow$ $f_x(1, 1) = 3$, $f_y(x, y) = x + 6y$ $\Rightarrow$ $f_y(1, 1) = 7$ and an

equation of the tangent plane is $z - 5 = 3(x - 1) + 7(y - 1)$ or $z = 3x + 7y - 5$. After zooming in, the surface and the

tangent plane become almost indistinguishable. (Here, the tangent plane is below the surface.) If we zoom in farther, the

surface and the tangent plane will appear to coincide.

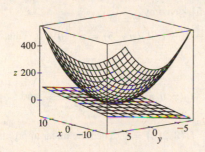
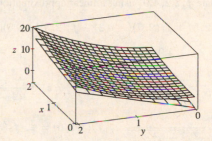

9. $f(x, y) = \dfrac{xy \sin (x - y)}{1 + x^2 + y^2}$.   A CAS gives $f_x(x, y) = \dfrac{y \sin (x - y) + xy \cos (x - y)}{1 + x^2 + y^2} - \dfrac{2x^2 y \sin (x - y)}{(1 + x^2 + y^2)^2}$ and

$f_y(x, y) = \dfrac{x \sin (x - y) - xy \cos (x - y)}{1 + x^2 + y^2} - \dfrac{2xy^2 \sin (x - y)}{(1 + x^2 + y^2)^2}$. We use the CAS to evaluate these at $(1, 1)$, and then

substitute the results into Equation 2 to compute an equation of the tangent plane: $z = \frac{1}{3}x - \frac{1}{3}y$. The surface and tangent

plane are shown in the first graph below. After zooming in, the surface and the tangent plane become almost indistinguishable,

as shown in the second graph. (Here, the tangent plane is shown with fewer traces than the surface.) If we zoom in farther, the

surface and the tangent plane will appear to coincide.

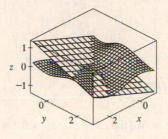
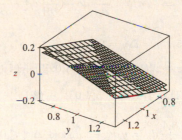

11. $f(x, y) = 1 + x \ln(xy - 5)$.   The partial derivatives are $f_x(x, y) = x \cdot \dfrac{1}{xy - 5} (y) + \ln(xy - 5) \cdot 1 = \dfrac{xy}{xy - 5} + \ln(xy - 5)$

and $f_y(x, y) = x \cdot \dfrac{1}{xy - 5} (x) = \dfrac{x^2}{xy - 5}$, so $f_x(2, 3) = 6$ and $f_y(2, 3) = 4$. Both f_x and f_y are continuous functions for

$xy > 5$, so by Theorem 8, f is differentiable at $(2, 3)$. By Equation 3, the linearization of f at $(2, 3)$ is given by

$L(x, y) = f(2, 3) + f_x(2, 3)(x - 2) + f_y(2, 3)(y - 3) = 1 + 6(x - 2) + 4(y - 3) = 6x + 4y - 23$.

13. $f(x, y) = e^{-xy} \cos y$.   The partial derivatives are $f_x(x, y) = e^{-xy}(-y) \cos y = -ye^{-xy} \cos y$ and

$f_y(x, y) = e^{-xy}(-\sin y) + (\cos y)e^{-xy}(-x) = -e^{-xy}(\sin y + x \cos y)$, so $f_x(\pi, 0) = 0$ and $f_y(\pi, 0) = -\pi$.

Both f_x and f_y are continuous functions, so f is differentiable at $(\pi, 0)$, and the linearization of f at $(\pi, 0)$ is

$L(x, y) = f(\pi, 0) + f_x(\pi, 0)(x - \pi) + f_y(\pi, 0)(y - 0) = 1 + 0(x - \pi) - \pi(y - 0) = 1 - \pi y$.

15. Let $f(x, y) = \dfrac{2x + 3}{4y + 1}$. Then $f_x(x, y) = \dfrac{2}{4y + 1}$ and $f_y(x, y) = (2x + 3)(-1)(4y + 1)^{-2}(4) = \dfrac{-8x - 12}{(4y + 1)^2}$. Both f_x and f_y

are continuous functions for $y \neq -\frac{1}{4}$, so by Theorem 8, f is differentiable at $(0, 0)$. We have $f_x(0, 0) = 2$, $f_y(0, 0) = -12$

and the linear approximation of f at $(0, 0)$ is $f(x, y) \approx f(0, 0) + f_x(0, 0)(x - 0) + f_y(0, 0)(y - 0) = 3 + 2x - 12y$.

17. We can estimate $f(2.2, 4.9)$ using a linear approximation of f at $(2, 5)$, given by

$f(x, y) \approx f(2, 5) + f_x(2, 5)(x - 2) + f_y(2, 5)(y - 5) = 6 + 1(x - 2) + (-1)(y - 5) = x - y + 9$. Thus

$f(2.2, 4.9) \approx 2.2 - 4.9 + 9 = 6.3$.

19. $f(x, y, z) = \sqrt{x^2 + y^2 + z^2}$ $\Rightarrow$ $f_x(x, y, z) = \dfrac{x}{\sqrt{x^2 + y^2 + z^2}}$, $f_y(x, y, z) = \dfrac{y}{\sqrt{x^2 + y^2 + z^2}}$, and

$f_z(x, y, z) = \dfrac{z}{\sqrt{x^2 + y^2 + z^2}}$, so $f_x(3, 2, 6) = \frac{3}{7}$, $f_y(3, 2, 6) = \frac{2}{7}$, $f_z(3, 2, 6) = \frac{6}{7}$. Then the linear approximation of f

at $(3, 2, 6)$ is given by

$$f(x, y, z) \approx f(3, 2, 6) + f_x(3, 2, 6)(x - 3) + f_y(3, 2, 6)(y - 2) + f_z(3, 2, 6)(z - 6)$$

$$= 7 + \tfrac{3}{7}(x - 3) + \tfrac{2}{7}(y - 2) + \tfrac{6}{7}(z - 6) = \tfrac{3}{7}x + \tfrac{2}{7}y + \tfrac{6}{7}z$$

Thus $\sqrt{(3.02)^2 + (1.97)^2 + (5.99)^2} = f(3.02, 1.97, 5.99) \approx \frac{3}{7}(3.02) + \frac{2}{7}(1.97) + \frac{6}{7}(5.99) \approx 6.9914$.

21. $m = p^5 q^3$ $\Rightarrow$ $dm = \dfrac{\partial m}{\partial p}\, dp + \dfrac{\partial m}{\partial q}\, dq = 5p^4 q^3\, dp + 3p^5 q^2\, dq$

23. $R = \alpha\beta^2 \cos\gamma$ $\Rightarrow$ $dR = \dfrac{\partial R}{\partial \alpha}\, d\alpha + \dfrac{\partial R}{\partial \beta}\, d\beta + \dfrac{\partial R}{\partial \gamma}\, d\gamma = \beta^2 \cos\gamma\, d\alpha + 2\alpha\beta \cos\gamma\, d\beta - \alpha\beta^2 \sin\gamma\, d\gamma$

25. $dx = \Delta x = 0.05$, $dy = \Delta y = 0.1$, $z = 5x^2 + y^2$, $z_x = 10x$, $z_y = 2y$. Thus when $x = 1$ and $y = 2$,

$dz = z_x(1, 2)\, dx + z_y(1, 2)\, dy = (10)(0.05) + (4)(0.1) = 0.9$ while

$\Delta z = f(1.05, 2.1) - f(1, 2) = 5(1.05)^2 + (2.1)^2 - 5 - 4 = 0.9225$.

27. $dA = \dfrac{\partial A}{\partial x}\, dx + \dfrac{\partial A}{\partial y}\, dy = y\, dx + x\, dy$ and $|\Delta x| \leq 0.1$, $|\Delta y| \leq 0.1$. We use $dx = 0.1$, $dy = 0.1$ with $x = 30$, $y = 24$; then

the maximum error in the area is about $dA = 24(0.1) + 30(0.1) = 5.4$ cm^2.

29. The volume of a can is $V = \pi r^2 h$ and $\Delta V \approx dV$ is an estimate of the amount of tin. Here $dV = 2\pi r h\, dr + \pi r^2\, dh$, so put

$dr = 0.04$, $dh = 0.08$ (0.04 on top, 0.04 on bottom) and then $\Delta V \approx dV = 2\pi(48)(0.04) + \pi(16)(0.08) \approx 16.08$ cm^3.

Thus the amount of tin is about 16 cm^3.

31. The errors in measurement are at most 2%, so $\left|\dfrac{\Delta w}{w}\right| \leq 0.02$ and $\left|\dfrac{\Delta h}{h}\right| \leq 0.02$. The relative error in the calculated surface

area is

$$\frac{\Delta S}{S} \approx \frac{dS}{S} = \frac{0.1091(0.425w^{0.425 - 1})h^{0.725}\, dw + 0.1091w^{0.425}(0.725h^{0.725 - 1})\, dh}{0.1091w^{0.425}h^{0.725}} = 0.425\frac{dw}{w} + 0.725\frac{dh}{h}$$

[continued]

To estimate the maximum relative error, we use $\dfrac{dw}{w} = \left|\dfrac{\Delta w}{w}\right| = 0.02$ and $\dfrac{dh}{h} = \left|\dfrac{\Delta h}{h}\right| = 0.02$ $\Rightarrow$

$\dfrac{dS}{S} = 0.425\,(0.02) + 0.725\,(0.02) = 0.023$. Thus the maximum percentage error is approximately 2.3%.

33. First we find $\dfrac{\partial R}{\partial R_1}$ implicitly by taking partial derivatives of both sides with respect to R_1:

$$\dfrac{\partial}{\partial R_1}\left(\dfrac{1}{R}\right) = \dfrac{\partial\,[(1/R_1) + (1/R_2) + (1/R_3)]}{\partial R_1} \quad\Rightarrow\quad -R^{-2}\dfrac{\partial R}{\partial R_1} = -R_1^{-2} \quad\Rightarrow\quad \dfrac{\partial R}{\partial R_1} = \dfrac{R^2}{R_1^2}.$$ Then by symmetry,

$\dfrac{\partial R}{\partial R_2} = \dfrac{R^2}{R_2^2}$, $\dfrac{\partial R}{\partial R_3} = \dfrac{R^2}{R_3^2}$. When $R_1 = 25$, $R_2 = 40$ and $R_3 = 50$, $\dfrac{1}{R} = \dfrac{17}{200}$ $\Leftrightarrow$ $R = \dfrac{200}{17}$ Ω. Since the possible error

for each R_i is 0.5%, the maximum error of R is attained by setting $\Delta R_i = 0.005 R_i$. So

$$\Delta R \approx dR = \dfrac{\partial R}{\partial R_1}\Delta R_1 + \dfrac{\partial R}{\partial R_2}\Delta R_2 + \dfrac{\partial R}{\partial R_3}\Delta R_3 = (0.005)R^2\left(\dfrac{1}{R_1} + \dfrac{1}{R_2} + \dfrac{1}{R_3}\right) = (0.005)R = \dfrac{1}{17} \approx 0.059\ \Omega.$$

35. $\Delta z = f(a + \Delta x, b + \Delta y) - f(a, b) = (a + \Delta x)^2 + (b + \Delta y)^2 - (a^2 + b^2)$

$\quad = a^2 + 2a\,\Delta x + (\Delta x)^2 + b^2 + 2b\,\Delta y + (\Delta y)^2 - a^2 - b^2 = 2a\,\Delta x + (\Delta x)^2 + 2b\,\Delta y + (\Delta y)^2$

But $f_x(a, b) = 2a$ and $f_y(a, b) = 2b$ and so $\Delta z = f_x(a, b)\,\Delta x + f_y(a, b)\,\Delta y + \Delta x\,\Delta x + \Delta y\,\Delta y$, which is Definition 7

with $\varepsilon_1 = \Delta x$ and $\varepsilon_2 = \Delta y$. Hence f is differentiable.

37. To show that f is continuous at (a, b) we need to show that $\displaystyle\lim_{(x,y)\to(a,b)} f(x, y) = f(a, b)$ or

equivalently $\displaystyle\lim_{(\Delta x, \Delta y)\to(0,0)} f(a + \Delta x, b + \Delta y) = f(a, b)$. Since f is differentiable at (a, b),

$f(a + \Delta x, b + \Delta y) - f(a, b) = \Delta z = f_x(a, b)\,\Delta x + f_y(a, b)\,\Delta y + \varepsilon_1\,\Delta x + \varepsilon_2\,\Delta y$, where ε_1 and $\varepsilon_2 \to 0$ as

$(\Delta x, \Delta y) \to (0, 0)$. Thus $f(a + \Delta x, b + \Delta y) = f(a, b) + f_x(a, b)\,\Delta x + f_y(a, b)\,\Delta y + \varepsilon_1\,\Delta x + \varepsilon_2\,\Delta y$. Taking the limit of

both sides as $(\Delta x, \Delta y) \to (0, 0)$ gives $\displaystyle\lim_{(\Delta x, \Delta y)\to(0,0)} f(a + \Delta x, b + \Delta y) = f(a, b)$. Thus f is continuous at (a, b).

11.5   The Chain Rule

1. $z = x^2 + y^2 + xy$, $x = \sin t$, $y = e^t$ $\Rightarrow$ $\dfrac{dz}{dt} = \dfrac{\partial z}{\partial x}\dfrac{dx}{dt} + \dfrac{\partial z}{\partial y}\dfrac{dy}{dt} = (2x + y)\cos t + (2y + x)e^t$

3. $w = xe^{y/z}$, $x = t^2$, $y = 1 - t$, $z = 1 + 2t$ $\Rightarrow$

$\dfrac{dw}{dt} = \dfrac{\partial w}{\partial x}\dfrac{dx}{dt} + \dfrac{\partial w}{\partial y}\dfrac{dy}{dt} + \dfrac{\partial w}{\partial z}\dfrac{dz}{dt} = e^{y/z}\cdot 2t + xe^{y/z}\left(\dfrac{1}{z}\right)\cdot(-1) + xe^{y/z}\left(-\dfrac{y}{z^2}\right)\cdot 2 = e^{y/z}\left(2t - \dfrac{x}{z} - \dfrac{2xy}{z^2}\right)$

5. $z = x^2 y^3$, $x = s\cos t$, $y = s\sin t$ $\Rightarrow$

$\quad \dfrac{\partial z}{\partial s} = \dfrac{\partial z}{\partial x}\dfrac{\partial x}{\partial s} + \dfrac{\partial z}{\partial y}\dfrac{\partial y}{\partial s} = 2xy^3\cos t + 3x^2 y^2\sin t$

$\quad \dfrac{\partial z}{\partial t} = \dfrac{\partial z}{\partial x}\dfrac{\partial x}{\partial t} + \dfrac{\partial z}{\partial y}\dfrac{\partial y}{\partial t} = (2xy^3)(-s\sin t) + (3x^2 y^2)(s\cos t) = -2sxy^3\sin t + 3sx^2 y^2\cos t$

7. $z = e^r \cos\theta, \; r = st, \; \theta = \sqrt{s^2 + t^2} \;\Rightarrow$

$$\frac{\partial z}{\partial s} = \frac{\partial z}{\partial r}\frac{\partial r}{\partial s} + \frac{\partial z}{\partial\theta}\frac{\partial\theta}{\partial s} = e^r \cos\theta \cdot t + e^r(-\sin\theta)\cdot\tfrac{1}{2}(s^2+t^2)^{-1/2}(2s) = te^r\cos\theta - e^r\sin\theta\cdot\frac{s}{\sqrt{s^2+t^2}}$$

$$= e^r\left(t\cos\theta - \frac{s}{\sqrt{s^2+t^2}}\sin\theta\right)$$

$$\frac{\partial z}{\partial t} = \frac{\partial z}{\partial r}\frac{\partial r}{\partial t} + \frac{\partial z}{\partial\theta}\frac{\partial\theta}{\partial t} = e^r\cos\theta\cdot s + e^r(-\sin\theta)\cdot\tfrac{1}{2}(s^2+t^2)^{-1/2}(2t) = se^r\cos\theta - e^r\sin\theta\cdot\frac{t}{\sqrt{s^2+t^2}}$$

$$= e^r\left(s\cos\theta - \frac{t}{\sqrt{s^2+t^2}}\sin\theta\right)$$

9. When $t = 3$, $x = g(3) = 2$ and $y = h(3) = 7$. By the Chain Rule (2),

$$\frac{dz}{dt} = \frac{\partial f}{\partial x}\frac{dx}{dt} + \frac{\partial f}{\partial y}\frac{dy}{dt} = f_x(2,7)g'(3) + f_y(2,7)\,h'(3) = (6)(5) + (-8)(-4) = 62.$$

11. $g(u,v) = f(x(u,v), y(u,v))$ where $x = e^u + \sin v$, $y = e^u + \cos v \;\Rightarrow$

$\dfrac{\partial x}{\partial u} = e^u$, $\dfrac{\partial x}{\partial v} = \cos v$, $\dfrac{\partial y}{\partial u} = e^u$, $\dfrac{\partial y}{\partial v} = -\sin v$. By the Chain Rule (3), $\dfrac{\partial g}{\partial u} = \dfrac{\partial f}{\partial x}\dfrac{\partial x}{\partial u} + \dfrac{\partial f}{\partial y}\dfrac{\partial y}{\partial u}$. Then

$$g_u(0,0) = f_x(x(0,0), y(0,0))\,x_u(0,0) + f_y(x(0,0), y(0,0))\,y_u(0,0) = f_x(1,2)(e^0) + f_y(1,2)(e^0) = 2(1) + 5(1) = 7.$$

Similarly, $\dfrac{\partial g}{\partial v} = \dfrac{\partial f}{\partial x}\dfrac{\partial x}{\partial v} + \dfrac{\partial f}{\partial y}\dfrac{\partial y}{\partial v}$. Then

$$g_v(0,0) = f_x(x(0,0), y(0,0))\,x_v(0,0) + f_y(x(0,0), y(0,0))\,y_v(0,0) = f_x(1,2)(\cos 0) + f_y(1,2)(-\sin 0)$$

$$= 2(1) + 5(0) = 2$$

13.

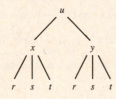

$u = f(x,y), \; x = x(r,s,t), \; y = y(r,s,t) \;\Rightarrow$

$$\frac{\partial u}{\partial r} = \frac{\partial u}{\partial x}\frac{\partial x}{\partial r} + \frac{\partial u}{\partial y}\frac{\partial y}{\partial r}, \quad \frac{\partial u}{\partial s} = \frac{\partial u}{\partial x}\frac{\partial x}{\partial s} + \frac{\partial u}{\partial y}\frac{\partial y}{\partial s},$$

$$\frac{\partial u}{\partial t} = \frac{\partial u}{\partial x}\frac{\partial x}{\partial t} + \frac{\partial u}{\partial y}\frac{\partial y}{\partial t}$$

15.

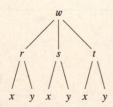

$w = f(r,s,t), \; r = r(x,y), \; s = s(x,y), \; t = t(x,y) \;\Rightarrow$

$$\frac{\partial w}{\partial x} = \frac{\partial w}{\partial r}\frac{\partial r}{\partial x} + \frac{\partial w}{\partial s}\frac{\partial s}{\partial x} + \frac{\partial w}{\partial t}\frac{\partial t}{\partial x}, \quad \frac{\partial w}{\partial y} = \frac{\partial w}{\partial r}\frac{\partial r}{\partial y} + \frac{\partial w}{\partial s}\frac{\partial s}{\partial y} + \frac{\partial w}{\partial t}\frac{\partial t}{\partial y}$$

17. $z = x^4 + x^2 y, \; x = s + 2t - u, \; y = stu^2 \;\Rightarrow$

$$\frac{\partial z}{\partial s} = \frac{\partial z}{\partial x}\frac{\partial x}{\partial s} + \frac{\partial z}{\partial y}\frac{\partial y}{\partial s} = (4x^3 + 2xy)(1) + (x^2)(tu^2),$$

$$\frac{\partial z}{\partial t} = \frac{\partial z}{\partial x}\frac{\partial x}{\partial t} + \frac{\partial z}{\partial y}\frac{\partial y}{\partial t} = (4x^3 + 2xy)(2) + (x^2)(su^2),$$

$$\frac{\partial z}{\partial u} = \frac{\partial z}{\partial x}\frac{\partial x}{\partial u} + \frac{\partial z}{\partial y}\frac{\partial y}{\partial u} = (4x^3 + 2xy)(-1) + (x^2)(2stu).$$

When $s = 4$, $t = 2$, and $u = 1$ we have $x = 7$ and $y = 8$,

so $\dfrac{\partial z}{\partial s} = (1484)(1) + (49)(2) = 1582$, $\dfrac{\partial z}{\partial t} = (1484)(2) + (49)(4) = 3164$, $\dfrac{\partial z}{\partial u} = (1484)(-1) + (49)(16) = -700$.

19. $w = xy + yz + zx$, $x = r\cos\theta$, $y = r\sin\theta$, $z = r\theta$ $\Rightarrow$

$$\frac{\partial w}{\partial r} = \frac{\partial w}{\partial x}\frac{\partial x}{\partial r} + \frac{\partial w}{\partial y}\frac{\partial y}{\partial r} + \frac{\partial w}{\partial z}\frac{\partial z}{\partial r} = (y+z)(\cos\theta) + (x+z)(\sin\theta) + (y+x)(\theta),$$

$$\frac{\partial w}{\partial \theta} = \frac{\partial w}{\partial x}\frac{\partial x}{\partial \theta} + \frac{\partial w}{\partial y}\frac{\partial y}{\partial \theta} + \frac{\partial w}{\partial z}\frac{\partial z}{\partial \theta} = (y+z)(-r\sin\theta) + (x+z)(r\cos\theta) + (y+x)(r).$$

When $r = 2$ and $\theta = \pi/2$ we have $x = 0$, $y = 2$, and $z = \pi$, so $\dfrac{\partial w}{\partial r} = (2 + \pi)(0) + (0 + \pi)(1) + (2 + 0)(\pi/2) = 2\pi$ and

$$\frac{\partial w}{\partial \theta} = (2 + \pi)(-2) + (0 + \pi)(0) + (2 + 0)(2) = -2\pi.$$

21. $N = \dfrac{p+q}{p+r}$, $p = u + vw$, $q = v + uw$, $r = w + uv$ $\Rightarrow$

$$\frac{\partial N}{\partial u} = \frac{\partial N}{\partial p}\frac{\partial p}{\partial u} + \frac{\partial N}{\partial q}\frac{\partial q}{\partial u} + \frac{\partial N}{\partial r}\frac{\partial r}{\partial u}$$

$$= \frac{(p+r)(1) - (p+q)(1)}{(p+r)^2}(1) + \frac{(p+r)(1) - (p+q)(0)}{(p+r)^2}(w) + \frac{(p+r)(0) - (p+q)(1)}{(p+r)^2}(v)$$

$$= \frac{(r-q) + (p+r)w - (p+q)v}{(p+r)^2},$$

$$\frac{\partial N}{\partial v} = \frac{\partial N}{\partial p}\frac{\partial p}{\partial v} + \frac{\partial N}{\partial q}\frac{\partial q}{\partial v} + \frac{\partial N}{\partial r}\frac{\partial r}{\partial v} = \frac{r-q}{(p+r)^2}(w) + \frac{p+r}{(p+r)^2}(1) + \frac{-(p+q)}{(p+r)^2}(u) = \frac{(r-q)w + (p+r) - (p+q)u}{(p+r)^2},$$

$$\frac{\partial N}{\partial w} = \frac{\partial N}{\partial p}\frac{\partial p}{\partial w} + \frac{\partial N}{\partial q}\frac{\partial q}{\partial w} + \frac{\partial N}{\partial r}\frac{\partial r}{\partial w} = \frac{r-q}{(p+r)^2}(v) + \frac{p+r}{(p+r)^2}(u) + \frac{-(p+q)}{(p+r)^2}(1) = \frac{(r-q)v + (p+r)u - (p+q)}{(p+r)^2}.$$

When $u = 2$, $v = 3$, and $w = 4$ we have $p = 14$, $q = 11$, and $r = 10$, so $\dfrac{\partial N}{\partial u} = \dfrac{-1 + (24)(4) - (25)(3)}{(24)^2} = \dfrac{20}{576} = \dfrac{5}{144}$,

$\dfrac{\partial N}{\partial v} = \dfrac{(-1)(4) + 24 - (25)(2)}{(24)^2} = \dfrac{-30}{576} = -\dfrac{5}{96}$, and $\dfrac{\partial N}{\partial w} = \dfrac{(-1)(3) + (24)(2) - 25}{(24)^2} = \dfrac{20}{576} = \dfrac{5}{144}$.

23. $\tan^{-1}(x^2 y) = x + xy^2$, so let $F(x, y) = \tan^{-1}(x^2 y) - x - xy^2 = 0$. Then

$$F_x(x, y) = \frac{1}{1 + (x^2 y)^2}(2xy) - 1 - y^2 = \frac{2xy}{1 + x^4 y^2} - 1 - y^2 = \frac{2xy - (1 + y^2)(1 + x^4 y^2)}{1 + x^4 y^2},$$

$$F_y(x, y) = \frac{1}{1 + (x^2 y)^2}(x^2) - 2xy = \frac{x^2}{1 + x^4 y^2} - 2xy = \frac{x^2 - 2xy(1 + x^4 y^2)}{1 + x^4 y^2}$$

and $$\frac{dy}{dx} = -\frac{F_x}{F_y} = -\frac{[2xy - (1 + y^2)(1 + x^4 y^2)]/(1 + x^4 y^2)}{[x^2 - 2xy(1 + x^4 y^2)]/(1 + x^4 y^2)} = \frac{(1 + y^2)(1 + x^4 y^2) - 2xy}{x^2 - 2xy(1 + x^4 y^2)}$$

$$= \frac{1 + x^4 y^2 + y^2 + x^4 y^4 - 2xy}{x^2 - 2xy - 2x^5 y^3}$$

25. $x^2 + 2y^2 + 3z^2 = 1$, so let $F(x, y, z) = x^2 + 2y^2 + 3z^2 - 1 = 0$. Then by Equations 7

$$\frac{\partial z}{\partial x} = -\frac{F_x}{F_z} = -\frac{2x}{6z} = -\frac{x}{3z} \quad \text{and} \quad \frac{\partial z}{\partial y} = -\frac{F_y}{F_z} = -\frac{4y}{6z} = -\frac{2y}{3z}.$$

27. $e^z = xyz$, so let $F(x, y, z) = e^z - xyz = 0$. Then $\dfrac{\partial z}{\partial x} = -\dfrac{F_x}{F_z} = -\dfrac{-yz}{e^z - xy} = \dfrac{yz}{e^z - xy}$ and

$$\frac{\partial z}{\partial y} = -\frac{F_y}{F_z} = -\frac{-xz}{e^z - xy} = \frac{xz}{e^z - xy}.$$

29. Since x and y are each functions of t, $T(x, y)$ is a function of t, so by the Chain Rule, $\dfrac{dT}{dt} = \dfrac{\partial T}{\partial x} \dfrac{dx}{dt} + \dfrac{\partial T}{\partial y} \dfrac{dy}{dt}$. After

3 seconds, $x = \sqrt{1+t} = \sqrt{1+3} = 2$, $y = 2 + \frac{1}{3}t = 2 + \frac{1}{3}(3) = 3$, $\dfrac{dx}{dt} = \dfrac{1}{2\sqrt{1+t}} = \dfrac{1}{2\sqrt{1+3}} = \dfrac{1}{4}$, and $\dfrac{dy}{dt} = \dfrac{1}{3}$.

Then $\dfrac{dT}{dt} = T_x(2, 3)\dfrac{dx}{dt} + T_y(2, 3)\dfrac{dy}{dt} = 4\left(\frac{1}{4}\right) + 3\left(\frac{1}{3}\right) = 2$. Thus the temperature is rising at a rate of $2°$C/s.

31. $C = 1449.2 + 4.6T - 0.055T^2 + 0.00029T^3 + 0.016D$, so $\dfrac{\partial C}{\partial T} = 4.6 - 0.11T + 0.00087T^2$ and $\dfrac{\partial C}{\partial D} = 0.016$.

According to the graph, the diver is experiencing a temperature of approximately $12.5°$C at $t = 20$ minutes, so

$\dfrac{\partial C}{\partial T} = 4.6 - 0.11(12.5) + 0.00087(12.5)^2 \approx 3.36$. By sketching tangent lines at $t = 20$ to the graphs given, we estimate

$\dfrac{dD}{dt} \approx \dfrac{1}{2}$ and $\dfrac{dT}{dt} \approx -\dfrac{1}{10}$. Then, by the Chain Rule, $\dfrac{dC}{dt} = \dfrac{\partial C}{\partial T}\dfrac{dT}{dt} + \dfrac{\partial C}{\partial D}\dfrac{dD}{dt} \approx (3.36)\left(-\frac{1}{10}\right) + (0.016)\left(\frac{1}{2}\right) \approx -0.33$.

Thus the speed of sound experienced by the diver is decreasing at a rate of approximately 0.33 m/s per minute.

33. (a) $V = \ell w h$, so by the Chain Rule,

$$\frac{dV}{dt} = \frac{\partial V}{\partial \ell}\frac{d\ell}{dt} + \frac{\partial V}{\partial w}\frac{dw}{dt} + \frac{\partial V}{\partial h}\frac{dh}{dt} = wh\frac{d\ell}{dt} + \ell h\frac{dw}{dt} + \ell w\frac{dh}{dt} = 2 \cdot 2 \cdot 2 + 1 \cdot 2 \cdot 2 + 1 \cdot 2 \cdot (-3) = 6 \text{ m}^3/\text{s}.$$

(b) $S = 2(\ell w + \ell h + wh)$, so by the Chain Rule,

$$\frac{dS}{dt} = \frac{\partial S}{\partial \ell}\frac{d\ell}{dt} + \frac{\partial S}{\partial w}\frac{dw}{dt} + \frac{\partial S}{\partial h}\frac{dh}{dt} = 2(w + h)\frac{d\ell}{dt} + 2(\ell + h)\frac{dw}{dt} + 2(\ell + w)\frac{dh}{dt}$$

$$= 2(2 + 2)2 + 2(1 + 2)2 + 2(1 + 2)(-3) = 10 \text{ m}^2/\text{s}$$

(c) $L^2 = \ell^2 + w^2 + h^2 \ \Rightarrow \ 2L\dfrac{dL}{dt} = 2\ell\dfrac{d\ell}{dt} + 2w\dfrac{dw}{dt} + 2h\dfrac{dh}{dt} = 2(1)(2) + 2(2)(2) + 2(2)(-3) = 0 \ \Rightarrow$

$dL/dt = 0$ m/s.

35. $\dfrac{dP}{dt} = 0.05$, $\dfrac{dT}{dt} = 0.15$, $V = 8.31\dfrac{T}{P}$ and $\dfrac{dV}{dt} = \dfrac{8.31}{P}\dfrac{dT}{dt} - 8.31\dfrac{T}{P^2}\dfrac{dP}{dt}$. Thus when $P = 20$ and $T = 320$,

$$\frac{dV}{dt} = 8.31\left[\frac{0.15}{20} - \frac{(0.05)(320)}{400}\right] \approx -0.27 \text{ L/s}.$$

37. (a) By the Chain Rule, $\dfrac{\partial z}{\partial r} = \dfrac{\partial z}{\partial x}\cos\theta + \dfrac{\partial z}{\partial y}\sin\theta$, $\dfrac{\partial z}{\partial \theta} = \dfrac{\partial z}{\partial x}(-r\sin\theta) + \dfrac{\partial z}{\partial y}r\cos\theta$.

(b) $\left(\dfrac{\partial z}{\partial r}\right)^2 = \left(\dfrac{\partial z}{\partial x}\right)^2 \cos^2\theta + 2\dfrac{\partial z}{\partial x}\dfrac{\partial z}{\partial y}\cos\theta\sin\theta + \left(\dfrac{\partial z}{\partial y}\right)^2 \sin^2\theta$,

$$\left(\frac{\partial z}{\partial \theta}\right)^2 = \left(\frac{\partial z}{\partial x}\right)^2 r^2 \sin^2 \theta - 2\frac{\partial z}{\partial x}\frac{\partial z}{\partial y} r^2 \cos \theta \sin \theta + \left(\frac{\partial z}{\partial y}\right)^2 r^2 \cos^2 \theta. \text{ Thus}$$

$$\left(\frac{\partial z}{\partial r}\right)^2 + \frac{1}{r^2}\left(\frac{\partial z}{\partial \theta}\right)^2 = \left[\left(\frac{\partial z}{\partial x}\right)^2 + \left(\frac{\partial z}{\partial y}\right)^2\right](\cos^2 \theta + \sin^2 \theta) = \left(\frac{\partial z}{\partial x}\right)^2 + \left(\frac{\partial z}{\partial y}\right)^2.$$

39. Let $u = x - y$. Then $\dfrac{\partial z}{\partial x} = \dfrac{dz}{du}\dfrac{\partial u}{\partial x} = \dfrac{dz}{du}$ and $\dfrac{\partial z}{\partial y} = \dfrac{dz}{du}(-1)$. Thus $\dfrac{\partial z}{\partial x} + \dfrac{\partial z}{\partial y} = 0$.

41. Let $u = x + at$, $v = x - at$. Then $z = f(u) + g(v)$, so $\partial z/\partial u = f'(u)$ and $\partial z/\partial v = g'(v)$.

Thus $\dfrac{\partial z}{\partial t} = \dfrac{\partial z}{\partial u}\dfrac{\partial u}{\partial t} + \dfrac{\partial z}{\partial v}\dfrac{\partial v}{\partial t} = af'(u) - ag'(v)$ and

$$\frac{\partial^2 z}{\partial t^2} = a\frac{\partial}{\partial t}[f'(u) - g'(v)] = a\left(\frac{df'(u)}{du}\frac{\partial u}{\partial t} - \frac{dg'(v)}{dv}\frac{\partial v}{\partial t}\right) = a^2 f''(u) + a^2 g''(v).$$

Similarly $\dfrac{\partial z}{\partial x} = f'(u) + g'(v)$ and $\dfrac{\partial^2 z}{\partial x^2} = f''(u) + g''(v)$. Thus $\dfrac{\partial^2 z}{\partial t^2} = a^2\dfrac{\partial^2 z}{\partial x^2}$.

43. $\dfrac{\partial z}{\partial s} = \dfrac{\partial z}{\partial x}2s + \dfrac{\partial z}{\partial y}2r$. Then

$$\frac{\partial^2 z}{\partial r \partial s} = \frac{\partial}{\partial r}\left(\frac{\partial z}{\partial x}2s\right) + \frac{\partial}{\partial r}\left(\frac{\partial z}{\partial y}2r\right)$$

$$= \frac{\partial^2 z}{\partial x^2}\frac{\partial x}{\partial r}2s + \frac{\partial}{\partial y}\left(\frac{\partial z}{\partial x}\right)\frac{\partial y}{\partial r}2s + \frac{\partial z}{\partial x}\frac{\partial}{\partial r}2s + \frac{\partial^2 z}{\partial y^2}\frac{\partial y}{\partial r}2r + \frac{\partial}{\partial x}\left(\frac{\partial z}{\partial y}\right)\frac{\partial x}{\partial r}2r + \frac{\partial z}{\partial y}2$$

$$= 4rs\frac{\partial^2 z}{\partial x^2} + \frac{\partial^2 z}{\partial y \partial x}4s^2 + 0 + 4rs\frac{\partial^2 z}{\partial y^2} + \frac{\partial^2 z}{\partial x \partial y}4r^2 + 2\frac{\partial z}{\partial y}$$

By the continuity of the partials, $\dfrac{\partial^2 z}{\partial r \partial s} = 4rs\dfrac{\partial^2 z}{\partial x^2} + 4rs\dfrac{\partial^2 z}{\partial y^2} + (4r^2 + 4s^2)\dfrac{\partial^2 z}{\partial x \partial y} + 2\dfrac{\partial z}{\partial y}$.

45. $\dfrac{\partial z}{\partial r} = \dfrac{\partial z}{\partial x}\cos \theta + \dfrac{\partial z}{\partial y}\sin \theta$ and $\dfrac{\partial z}{\partial \theta} = -\dfrac{\partial z}{\partial x}r\sin \theta + \dfrac{\partial z}{\partial y}r\cos \theta$. Then

$$\frac{\partial^2 z}{\partial r^2} = \cos \theta\left(\frac{\partial^2 z}{\partial x^2}\cos \theta + \frac{\partial^2 z}{\partial y \partial x}\sin \theta\right) + \sin \theta\left(\frac{\partial^2 z}{\partial y^2}\sin \theta + \frac{\partial^2 z}{\partial x \partial y}\cos \theta\right)$$

$$= \cos^2 \theta\frac{\partial^2 z}{\partial x^2} + 2\cos \theta \sin \theta\frac{\partial^2 z}{\partial x \partial y} + \sin^2 \theta\frac{\partial^2 z}{\partial y^2}$$

and

$$\frac{\partial^2 z}{\partial \theta^2} = -r\cos \theta\frac{\partial z}{\partial x} + (-r\sin \theta)\left(\frac{\partial^2 z}{\partial x^2}(-r\sin \theta) + \frac{\partial^2 z}{\partial y \partial x}r\cos \theta\right)$$

$$-r\sin \theta\frac{\partial z}{\partial y} + r\cos \theta\left(\frac{\partial^2 z}{\partial y^2}r\cos \theta + \frac{\partial^2 z}{\partial x \partial y}(-r\sin \theta)\right)$$

$$= -r\cos \theta\frac{\partial z}{\partial x} - r\sin \theta\frac{\partial z}{\partial y} + r^2\sin^2 \theta\frac{\partial^2 z}{\partial x^2} - 2r^2\cos \theta \sin \theta\frac{\partial^2 z}{\partial x \partial y} + r^2\cos^2 \theta\frac{\partial^2 z}{\partial y^2}$$

Thus

$$\frac{\partial^2 z}{\partial r^2} + \frac{1}{r^2}\frac{\partial^2 z}{\partial \theta^2} + \frac{1}{r}\frac{\partial z}{\partial r} = (\cos^2 \theta + \sin^2 \theta)\frac{\partial^2 z}{\partial x^2} + (\sin^2 \theta + \cos^2 \theta)\frac{\partial^2 z}{\partial y^2}$$

$$-\frac{1}{r}\cos \theta\frac{\partial z}{\partial x} - \frac{1}{r}\sin \theta\frac{\partial z}{\partial y} + \frac{1}{r}\left(\cos \theta\frac{\partial z}{\partial x} + \sin \theta\frac{\partial z}{\partial y}\right)$$

$$= \frac{\partial^2 z}{\partial x^2} + \frac{\partial^2 z}{\partial y^2} \text{ as desired.}$$

47. $F(x, y, z) = 0$ is assumed to define z as a function of x and y, that is, $z = f(x, y)$. So by (7), $\dfrac{\partial z}{\partial x} = -\dfrac{F_x}{F_z}$ since $F_z \neq 0$.

Similarly, it is assumed that $F(x, y, z) = 0$ defines x as a function of y and z, that is $x = h(x, z)$. Then $F(h(y, z), y, z) = 0$

and by the Chain Rule, $F_x \dfrac{\partial x}{\partial y} + F_y \dfrac{\partial y}{\partial y} + F_z \dfrac{\partial z}{\partial y} = 0$. But $\dfrac{\partial z}{\partial y} = 0$ and $\dfrac{\partial y}{\partial y} = 1$, so $F_x \dfrac{\partial x}{\partial y} + F_y = 0 \;\Rightarrow\; \dfrac{\partial x}{\partial y} = -\dfrac{F_y}{F_x}$.

A similar calculation shows that $\dfrac{\partial y}{\partial z} = -\dfrac{F_z}{F_y}$. Thus $\dfrac{\partial z}{\partial x}\dfrac{\partial x}{\partial y}\dfrac{\partial y}{\partial z} = \left(-\dfrac{F_x}{F_z}\right)\left(-\dfrac{F_y}{F_x}\right)\left(-\dfrac{F_z}{F_y}\right) = -1$.

11.6 Directional Derivatives and the Gradient Vector

1. $f(x, y) = ye^{-x} \;\Rightarrow\; f_x(x, y) = -ye^{-x}$ and $f_y(x, y) = e^{-x}$. If $\mathbf{u}$ is a unit vector in the direction of $\theta = 2\pi/3$, then

from Equation 6, $D_\mathbf{u} f(0, 4) = f_x(0, 4) \cos\left(\frac{2\pi}{3}\right) + f_y(0, 4) \sin\left(\frac{2\pi}{3}\right) = -4 \cdot \left(-\frac{1}{2}\right) + 1 \cdot \frac{\sqrt{3}}{2} = 2 + \frac{\sqrt{3}}{2}$.

3. $f(x, y) = \sin(2x + 3y)$

(a) $\nabla f(x, y) = \dfrac{\partial f}{\partial x} \mathbf{i} + \dfrac{\partial f}{\partial y} \mathbf{j} = [\cos(2x + 3y) \cdot 2]\, \mathbf{i} + [\cos(2x + 3y) \cdot 3]\, \mathbf{j} = 2\cos(2x + 3y)\, \mathbf{i} + 3\cos(2x + 3y)\, \mathbf{j}$

(b) $\nabla f(-6, 4) = (2\cos 0)\, \mathbf{i} + (3\cos 0)\, \mathbf{j} = 2\, \mathbf{i} + 3\, \mathbf{j}$

(c) By Equation 9, $D_\mathbf{u} f(-6, 4) = \nabla f(-6, 4) \cdot \mathbf{u} = (2\, \mathbf{i} + 3\, \mathbf{j}) \cdot \frac{1}{2}\left(\sqrt{3}\, \mathbf{i} - \mathbf{j}\right) = \frac{1}{2}\left(2\sqrt{3} - 3\right) = \sqrt{3} - \frac{3}{2}$.

5. $f(x, y, z) = x^2 yz - xyz^3$

(a) $\nabla f(x, y, z) = \langle f_x(x, y, z), f_y(x, y, z), f_z(x, y, z) \rangle = \langle 2xyz - yz^3,\, x^2 z - xz^3,\, x^2 y - 3xyz^2 \rangle$

(b) $\nabla f(2, -1, 1) = \langle -4 + 1, 4 - 2, -4 + 6 \rangle = \langle -3, 2, 2 \rangle$

(c) By Equation 14, $D_\mathbf{u} f(2, -1, 1) = \nabla f(2, -1, 1) \cdot \mathbf{u} = \langle -3, 2, 2 \rangle \cdot \langle 0, \frac{4}{5}, -\frac{3}{5} \rangle = 0 + \frac{8}{5} - \frac{6}{5} = \frac{2}{5}$.

7. $f(x, y) = e^x \sin y \;\Rightarrow\; \nabla f(x, y) = \langle e^x \sin y, e^x \cos y \rangle, \nabla f(0, \pi/3) = \left\langle \frac{\sqrt{3}}{2}, \frac{1}{2} \right\rangle$, and a

unit vector in the direction of $\mathbf{v}$ is $\mathbf{u} = \dfrac{1}{\sqrt{(-6)^2 + 8^2}} \langle -6, 8 \rangle = \frac{1}{10} \langle -6, 8 \rangle = \langle -\frac{3}{5}, \frac{4}{5} \rangle$, so

$D_\mathbf{u} f(0, \pi/3) = \nabla f(0, \pi/3) \cdot \mathbf{u} = \left\langle \frac{\sqrt{3}}{2}, \frac{1}{2} \right\rangle \cdot \langle -\frac{3}{5}, \frac{4}{5} \rangle = -\frac{3\sqrt{3}}{10} + \frac{4}{10} = \frac{4 - 3\sqrt{3}}{10}$.

9. $g(p, q) = p^4 - p^2 q^3 \;\Rightarrow\; \nabla g(p, q) = \left(4p^3 - 2pq^3\right) \mathbf{i} + \left(-3p^2 q^2\right) \mathbf{j}, \nabla g(2, 1) = 28\, \mathbf{i} - 12\, \mathbf{j}$, and a unit

vector in the direction of $\mathbf{v}$ is $\mathbf{u} = \dfrac{1}{\sqrt{1^2 + 3^2}} (\mathbf{i} + 3\, \mathbf{j}) = \frac{1}{\sqrt{10}} (\mathbf{i} + 3\, \mathbf{j})$, so

$D_\mathbf{u} g(2, 1) = \nabla g(2, 1) \cdot \mathbf{u} = (28\, \mathbf{i} - 12\, \mathbf{j}) \cdot \frac{1}{\sqrt{10}} (\mathbf{i} + 3\, \mathbf{j}) = \frac{1}{\sqrt{10}} (28 - 36) = -\frac{8}{\sqrt{10}}$ or $-\frac{4\sqrt{10}}{5}$.

11. $f(x, y, z) = xe^y + ye^z + ze^x \;\Rightarrow\; \nabla f(x, y, z) = \langle e^y + ze^x, xe^y + e^z, ye^z + e^x \rangle, \nabla f(0, 0, 0) = \langle 1, 1, 1 \rangle$, and a unit

vector in the direction of $\mathbf{v}$ is $\mathbf{u} = \dfrac{1}{\sqrt{25 + 1 + 4}} \langle 5, 1, -2 \rangle = \frac{1}{\sqrt{30}} \langle 5, 1, -2 \rangle$, so

$D_\mathbf{u} f(0, 0, 0) = \nabla f(0, 0, 0) \cdot \mathbf{u} = \langle 1, 1, 1 \rangle \cdot \frac{1}{\sqrt{30}} \langle 5, 1, -2 \rangle = \frac{4}{\sqrt{30}}$.

13. $f(x,y) = \sqrt{xy} \;\Rightarrow\; \nabla f(x,y) = \left\langle \frac{1}{2}(xy)^{-1/2}(y), \frac{1}{2}(xy)^{-1/2}(x) \right\rangle = \left\langle \dfrac{y}{2\sqrt{xy}}, \dfrac{x}{2\sqrt{xy}} \right\rangle$, so $\nabla f(2,8) = \left\langle 1, \frac{1}{4} \right\rangle$.

The unit vector in the direction of $\overrightarrow{PQ} = \langle 5-2, 4-8 \rangle = \langle 3, -4 \rangle$ is $\mathbf{u} = \left\langle \frac{3}{5}, -\frac{4}{5} \right\rangle$, so

$D_{\mathbf{u}} f(2,8) = \nabla f(2,8) \cdot \mathbf{u} = \left\langle 1, \frac{1}{4} \right\rangle \cdot \left\langle \frac{3}{5}, -\frac{4}{5} \right\rangle = \frac{2}{5}$.

15. $f(x,y) = \sin(xy) \;\Rightarrow\; \nabla f(x,y) = \langle y\cos(xy), x\cos(xy) \rangle$, $\nabla f(1,0) = \langle 0, 1 \rangle$. Thus the maximum rate of change is

$|\nabla f(1,0)| = 1$ in the direction $\langle 0, 1 \rangle$.

17. $f(x,y,z) = \sqrt{x^2 + y^2 + z^2} \;\Rightarrow$

$\nabla f(x,y,z) = \left\langle \frac{1}{2}(x^2+y^2+z^2)^{-1/2} \cdot 2x, \frac{1}{2}(x^2+y^2+z^2)^{-1/2} \cdot 2y, \frac{1}{2}(x^2+y^2+z^2)^{-1/2} \cdot 2z \right\rangle$

$= \left\langle \dfrac{x}{\sqrt{x^2+y^2+z^2}}, \dfrac{y}{\sqrt{x^2+y^2+z^2}}, \dfrac{z}{\sqrt{x^2+y^2+z^2}} \right\rangle,$

$\nabla f(3,6,-2) = \left\langle \frac{3}{\sqrt{49}}, \frac{6}{\sqrt{49}}, \frac{-2}{\sqrt{49}} \right\rangle = \left\langle \frac{3}{7}, \frac{6}{7}, -\frac{2}{7} \right\rangle$. Thus the maximum rate of change is

$|\nabla f(3,6,-2)| = \sqrt{\left(\frac{3}{7}\right)^2 + \left(\frac{6}{7}\right)^2 + \left(-\frac{2}{7}\right)^2} = \sqrt{\frac{9+36+4}{49}} = 1$ in the direction $\left\langle \frac{3}{7}, \frac{6}{7}, -\frac{2}{7} \right\rangle$ or equivalently $\langle 3, 6, -2 \rangle$.

19. (a) As in the proof of Theorem 15, $D_{\mathbf{u}} f = |\nabla f| \cos\theta$. Since the minimum value of $\cos\theta$ is -1 occurring when $\theta = \pi$, the

minimum value of $D_{\mathbf{u}} f$ is $-|\nabla f|$ occurring when $\theta = \pi$, that is when $\mathbf{u}$ is in the opposite direction of ∇f

(assuming $\nabla f \neq \mathbf{0}$).

(b) $f(x,y) = x^4 y - x^2 y^3 \;\Rightarrow\; \nabla f(x,y) = \langle 4x^3 y - 2xy^3, x^4 - 3x^2 y^2 \rangle$, so f decreases fastest at the point $(2,-3)$ in the

direction $-\nabla f(2,-3) = -\langle 12, -92 \rangle = \langle -12, 92 \rangle$.

21. The direction of fastest change is $\nabla f(x,y) = (2x-2)\mathbf{i} + (2y-4)\mathbf{j}$, so we need to find all points (x,y) where $\nabla f(x,y)$ is

parallel to $\mathbf{i} + \mathbf{j} \;\Leftrightarrow\; (2x-2)\mathbf{i} + (2y-4)\mathbf{j} = k(\mathbf{i}+\mathbf{j}) \;\Leftrightarrow\; k = 2x-2$ and $k = 2y - 4$. Then $2x - 2 = 2y - 4 \;\Rightarrow$

$y = x + 1$, so the direction of fastest change is $\mathbf{i} + \mathbf{j}$ at all points on the line $y = x + 1$.

23. $T = \dfrac{k}{\sqrt{x^2 + y^2 + z^2}}$ and $120 = T(1,2,2) = \dfrac{k}{3}$ so $k = 360$.

(a) $\mathbf{u} = \dfrac{\langle 1, -1, 1 \rangle}{\sqrt{3}}$,

$D_{\mathbf{u}} T(1,2,2) = \nabla T(1,2,2) \cdot \mathbf{u} = \left[-360(x^2+y^2+z^2)^{-3/2} \langle x, y, z \rangle \right]_{(1,2,2)} \cdot \mathbf{u} = -\frac{40}{3} \langle 1, 2, 2 \rangle \cdot \frac{1}{\sqrt{3}} \langle 1, -1, 1 \rangle = -\frac{40}{3\sqrt{3}}$

(b) From (a), $\nabla T = -360(x^2+y^2+z^2)^{-3/2} \langle x, y, z \rangle$, and since $\langle x, y, z \rangle$ is the position vector of the point (x,y,z), the

vector $-\langle x, y, z \rangle$, and thus ∇T, always points toward the origin.

25. $\nabla V(x,y,z) = \langle 10x - 3y + yz, xz - 3x, xy \rangle$, $\nabla V(3,4,5) = \langle 38, 6, 12 \rangle$

(a) $D_{\mathbf{u}} V(3,4,5) = \langle 38, 6, 12 \rangle \cdot \frac{1}{\sqrt{3}} \langle 1, 1, -1 \rangle = \frac{32}{\sqrt{3}}$

(b) $\nabla V(3,4,5) = \langle 38, 6, 12 \rangle$, or equivalently, $\langle 19, 3, 6 \rangle$.

(c) $|\nabla V(3,4,5)| = \sqrt{38^2 + 6^2 + 12^2} = \sqrt{1624} = 2\sqrt{406}$

27. A unit vector in the direction of $\overrightarrow{AB}$ is **i** and a unit vector in the direction of $\overrightarrow{AC}$ is **j**. Thus $D_{\overrightarrow{AB}} f(1,3) = f_x(1,3) = 3$ and

$D_{\overrightarrow{AC}} f(1,3) = f_y(1,3) = 26$. Therefore $\nabla f(1,3) = \langle f_x(1,3), f_y(1,3) \rangle = \langle 3, 26 \rangle$, and by definition,

$D_{\overrightarrow{AD}} f(1,3) = \nabla f \cdot \mathbf{u}$ where **u** is a unit vector in the direction of $\overrightarrow{AD}$, which is $\left\langle \frac{5}{13}, \frac{12}{13} \right\rangle$. Therefore,

$D_{\overrightarrow{AD}} f(1,3) = \langle 3, 26 \rangle \cdot \left\langle \frac{5}{13}, \frac{12}{13} \right\rangle = 3 \cdot \frac{5}{13} + 26 \cdot \frac{12}{13} = \frac{327}{13}$.

29. (a) $\nabla(au + bv) = \left\langle \dfrac{\partial(au+bv)}{\partial x}, \dfrac{\partial(au+bv)}{\partial y} \right\rangle = \left\langle a\dfrac{\partial u}{\partial x} + b\dfrac{\partial v}{\partial x}, a\dfrac{\partial u}{\partial y} + b\dfrac{\partial v}{\partial y} \right\rangle = a\left\langle \dfrac{\partial u}{\partial x}, \dfrac{\partial u}{\partial y} \right\rangle + b\left\langle \dfrac{\partial v}{\partial x}, \dfrac{\partial v}{\partial y} \right\rangle$

$\qquad = a\nabla u + b\nabla v$

(b) $\nabla(uv) = \left\langle v\dfrac{\partial u}{\partial x} + u\dfrac{\partial v}{\partial x}, v\dfrac{\partial u}{\partial y} + u\dfrac{\partial v}{\partial y} \right\rangle = v\left\langle \dfrac{\partial u}{\partial x}, \dfrac{\partial u}{\partial y} \right\rangle + u\left\langle \dfrac{\partial v}{\partial x}, \dfrac{\partial v}{\partial y} \right\rangle = v\nabla u + u\nabla v$

(c) $\nabla\left(\dfrac{u}{v}\right) = \left\langle \dfrac{v\dfrac{\partial u}{\partial x} - u\dfrac{\partial v}{\partial x}}{v^2}, \dfrac{v\dfrac{\partial u}{\partial y} - u\dfrac{\partial v}{\partial y}}{v^2} \right\rangle = \dfrac{v\left\langle \dfrac{\partial u}{\partial x}, \dfrac{\partial u}{\partial y} \right\rangle - u\left\langle \dfrac{\partial v}{\partial x}, \dfrac{\partial v}{\partial y} \right\rangle}{v^2} = \dfrac{v\nabla u - u\nabla v}{v^2}$

(d) $\nabla u^n = \left\langle \dfrac{\partial(u^n)}{\partial x}, \dfrac{\partial(u^n)}{\partial y} \right\rangle = \left\langle nu^{n-1}\dfrac{\partial u}{\partial x}, nu^{n-1}\dfrac{\partial u}{\partial y} \right\rangle = nu^{n-1}\nabla u$

31. Let $F(x,y,z) = 2(x-2)^2 + (y-1)^2 + (z-3)^2$. Then $2(x-2)^2 + (y-1)^2 + (z-3)^2 = 10$ is a level surface of F.

$F_x(x,y,z) = 4(x-2) \implies F_x(3,3,5) = 4$, $F_y(x,y,z) = 2(y-1) \implies F_y(3,3,5) = 4$, and

$F_z(x,y,z) = 2(z-3) \implies F_z(3,3,5) = 4$.

(a) Equation 19 gives an equation of the tangent plane at $(3,3,5)$ as $4(x-3) + 4(y-3) + 4(z-5) = 0 \iff$

$4x + 4y + 4z = 44$ or equivalently $x + y + z = 11$.

(b) By Equation 20, the normal line has symmetric equations $\dfrac{x-3}{4} = \dfrac{y-3}{4} = \dfrac{z-5}{4}$ or equivalently

$x - 3 = y - 3 = z - 5$. Corresponding parametric equations are $x = 3 + t$, $y = 3 + t$, $z = 5 + t$.

33. Let $F(x,y,z) = xyz^2$. Then $xyz^2 = 6$ is a level surface of F and $\nabla F(x,y,z) = \langle yz^2, xz^2, 2xyz \rangle$.

(a) $\nabla F(3,2,1) = \langle 2, 3, 12 \rangle$ is a normal vector for the tangent plane at $(3,2,1)$, so an equation of the tangent plane

is $2(x-3) + 3(y-2) + 12(z-1) = 0$ or $2x + 3y + 12z = 24$.

(b) The normal line has direction $\langle 2, 3, 12 \rangle$, so parametric equations are $x = 3 + 2t$, $y = 2 + 3t$, $z = 1 + 12t$, and

symmetric equations are $\dfrac{x-3}{2} = \dfrac{y-2}{3} = \dfrac{z-1}{12}$.

35. Let $F(x, y, z) = x + y + z - e^{xyz}$. Then $x + y + z = e^{xyz}$ is the level surface $F(x, y, z) = 0$,

and $\nabla F(x, y, z) = \langle 1 - yze^{xyz}, 1 - xze^{xyz}, 1 - xye^{xyz} \rangle$.

(a) $\nabla F(0, 0, 1) = \langle 1, 1, 1 \rangle$ is a normal vector for the tangent plane at $(0, 0, 1)$, so an equation of the tangent plane

is $1(x - 0) + 1(y - 0) + 1(z - 1) = 0$ or $x + y + z = 1$.

(b) The normal line has direction $\langle 1, 1, 1 \rangle$, so parametric equations are $x = t$, $y = t$, $z = 1 + t$, and symmetric equations are

$x = y = z - 1$.

37. $F(x, y, z) = xy + yz + zx$, $\nabla F(x, y, z) = \langle y + z, x + z, y + x \rangle$,

$\nabla F(1, 1, 1) = \langle 2, 2, 2 \rangle$, so an equation of the tangent plane is

$2x + 2y + 2z = 6$ or $x + y + z = 3$, and the normal line is given by

$x - 1 = y - 1 = z - 1$ or $x = y = z$. To graph the surface we solve for z:

$$z = \frac{3 - xy}{x + y}.$$

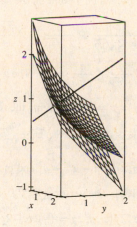

39. $f(x, y) = xy \implies \nabla f(x, y) = \langle y, x \rangle$, $\nabla f(3, 2) = \langle 2, 3 \rangle$. $\nabla f(3, 2)$

is perpendicular to the tangent line, so the tangent line has equation

$\nabla f(3, 2) \cdot \langle x - 3, y - 2 \rangle = 0 \implies \langle 2, 3 \rangle \cdot \langle x - 3, x - 2 \rangle = 0 \implies$

$2(x - 3) + 3(y - 2) = 0$ or $2x + 3y = 12$.

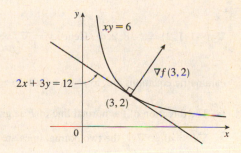

41. $\nabla F(x_0, y_0, z_0) = \left\langle \dfrac{2x_0}{a^2}, \dfrac{2y_0}{b^2}, \dfrac{2z_0}{c^2} \right\rangle$. Thus an equation of the tangent plane at (x_0, y_0, z_0) is

$\dfrac{2x_0}{a^2} x + \dfrac{2y_0}{b^2} y + \dfrac{2z_0}{c^2} z = 2\left(\dfrac{x_0^2}{a^2} + \dfrac{y_0^2}{b^2} + \dfrac{z_0^2}{c^2} \right) = 2(1) = 2$ since (x_0, y_0, z_0) is a point on the ellipsoid. Hence

$\dfrac{x_0}{a^2} x + \dfrac{y_0}{b^2} y + \dfrac{z_0}{c^2} z = 1$ is an equation of the tangent plane.

43. The hyperboloid $x^2 - y^2 - z^2 = 1$ is a level surface of $F(x, y, z) = x^2 - y^2 - z^2$ and $\nabla F(x, y, z) = \langle 2x, -2y, -2z \rangle$ is a

normal vector to the surface and hence a normal vector for the tangent plane at (x, y, z). The tangent plane is parallel to the

plane $z = x + y$ or $x + y - z = 0$ if and only if the corresponding normal vectors are parallel, so we need a point (x_0, y_0, z_0)

on the hyperboloid where $\langle 2x_0, -2y_0, -2z_0 \rangle = c\langle 1, 1, -1 \rangle$ or equivalently $\langle x_0, -y_0, -z_0 \rangle = k\langle 1, 1, -1 \rangle$ for some $k \neq 0$.

Then we must have $x_0 = k$, $y_0 = -k$, $z_0 = k$ and substituting into the equation of the hyperboloid gives

$k^2 - (-k)^2 - k^2 = 1 \iff -k^2 = 1$, an impossibility. Thus there is no such point on the hyperboloid.

45. Let $F(x, y, z) = x^2 + y^2 - z$. Then the paraboloid is the level surface $F(x, y, z) = 0$ and $\nabla F(x, y, z) = \langle 2x, 2y, -1 \rangle$, so

$\nabla F(1, 1, 2) = \langle 2, 2, -1 \rangle$ is a normal vector to the surface. Thus the normal line at $(1, 1, 2)$ is given by $x = 1 + 2t$,

$y = 1 + 2t$, $z = 2 - t$. Substitution into the equation of the paraboloid $z = x^2 + y^2$ gives $2 - t = (1 + 2t)^2 + (1 + 2t)^2 \Leftrightarrow$

$2 - t = 2 + 8t + 8t^2 \Leftrightarrow 8t^2 + 9t = 0 \Leftrightarrow t(8t + 9) = 0$. Thus the line intersects the paraboloid when $t = 0$,

corresponding to the given point $(1, 1, 2)$, or when $t = -\frac{9}{8}$, corresponding to the point $\left(-\frac{5}{4}, -\frac{5}{4}, \frac{25}{8}\right)$.

47. Let (x_0, y_0, z_0) be a point on the surface. Then an equation of the tangent plane at the point is

$$\frac{x}{2\sqrt{x_0}} + \frac{y}{2\sqrt{y_0}} + \frac{z}{2\sqrt{z_0}} = \frac{\sqrt{x_0} + \sqrt{y_0} + \sqrt{z_0}}{2}. \text{ But } \sqrt{x_0} + \sqrt{y_0} + \sqrt{z_0} = \sqrt{c}, \text{ so the equation is}$$

$\frac{x}{\sqrt{x_0}} + \frac{y}{\sqrt{y_0}} + \frac{z}{\sqrt{z_0}} = \sqrt{c}$. The x-, y-, and z-intercepts are $\sqrt{cx_0}$, $\sqrt{cy_0}$ and $\sqrt{cz_0}$ respectively. (The x-intercept is found by

setting $y = z = 0$ and solving the resulting equation for x, and the y- and z-intercepts are found similarly.) So the sum of the

intercepts is $\sqrt{c}\left(\sqrt{x_0} + \sqrt{y_0} + \sqrt{z_0}\right) = c$, a constant.

49. If $f(x, y, z) = z - x^2 - y^2$ and $g(x, y, z) = 4x^2 + y^2 + z^2$, then the tangent line is perpendicular to both ∇f and ∇g

at $(-1, 1, 2)$. The vector $\mathbf{v} = \nabla f \times \nabla g$ will therefore be parallel to the tangent line.

We have $\nabla f(x, y, z) = \langle -2x, -2y, 1 \rangle \Rightarrow \nabla f(-1, 1, 2) = \langle 2, -2, 1 \rangle$, and $\nabla g(x, y, z) = \langle 8x, 2y, 2z \rangle \Rightarrow$

$$\nabla g(-1, 1, 2) = \langle -8, 2, 4 \rangle. \text{ Hence } \mathbf{v} = \nabla f \times \nabla g = \begin{vmatrix} \mathbf{i} & \mathbf{j} & \mathbf{k} \\ 2 & -2 & 1 \\ -8 & 2 & 4 \end{vmatrix} = -10\mathbf{i} - 16\mathbf{j} - 12\mathbf{k}.$$

Parametric equations are: $x = -1 - 10t$, $y = 1 - 16t$, $z = 2 - 12t$.

51. (a) The direction of the normal line of F is given by ∇F, and that of G by ∇G. Assuming that

$\nabla F \neq 0 \neq \nabla G$, the two normal lines are perpendicular at P if $\nabla F \cdot \nabla G = 0$ at $P \Leftrightarrow$

$\langle \partial F/\partial x, \partial F/\partial y, \partial F/\partial z \rangle \cdot \langle \partial G/\partial x, \partial G/\partial y, \partial G/\partial z \rangle = 0$ at $P \Leftrightarrow F_x G_x + F_y G_y + F_z G_z = 0$ at P.

(b) Here $F = x^2 + y^2 - z^2$ and $G = x^2 + y^2 + z^2 - r^2$, so

$\nabla F \cdot \nabla G = \langle 2x, 2y, -2z \rangle \cdot \langle 2x, 2y, 2z \rangle = 4x^2 + 4y^2 - 4z^2 = 4F = 0$, since the point (x, y, z) lies on the graph of

$F = 0$. To see that this is true without using calculus, note that $G = 0$ is the equation of a sphere centered at the origin and

$F = 0$ is the equation of a right circular cone with vertex at the origin (which is generated by lines through the origin). At

any point of intersection, the sphere's normal line (which passes through the origin) lies on the cone, and thus is

perpendicular to the cone's normal line. So the surfaces with equations $F = 0$ and $G = 0$ are everywhere orthogonal.

53. Let $\mathbf{u} = \langle a, b \rangle$ and $\mathbf{v} = \langle c, d \rangle$. Then we know that at the given point, $D_{\mathbf{u}} f = \nabla f \cdot \mathbf{u} = af_x + bf_y$ and

$D_{\mathbf{v}} f = \nabla f \cdot \mathbf{v} = cf_x + df_y$. But these are just two linear equations in the two unknowns f_x and f_y, and since $\mathbf{u}$ and $\mathbf{v}$ are

not parallel, we can solve the equations to find $\nabla f = \langle f_x, f_y \rangle$ at the given point. In fact,

$$\nabla f = \left\langle \frac{d\, D_{\mathbf{u}} f - b\, D_{\mathbf{v}} f}{ad - bc}, \frac{a\, D_{\mathbf{v}} f - c\, D_{\mathbf{u}} f}{ad - bc} \right\rangle.$$

11.7   Maximum and Minimum Values

1. (a) First we compute $D(1,1) = f_{xx}(1,1)\, f_{yy}(1,1) - [f_{xy}(1,1)]^2 = (4)(2) - (1)^2 = 7$. Since $D(1,1) > 0$ and

   $f_{xx}(1,1) > 0$, f has a local minimum at $(1,1)$ by the Second Derivatives Test.

   (b) $D(1,1) = f_{xx}(1,1)\, f_{yy}(1,1) - [f_{xy}(1,1)]^2 = (4)(2) - (3)^2 = -1$. Since $D(1,1) < 0$, f has a saddle point at $(1,1)$ by

   the Second Derivatives Test.

3. $f(x,y) = x^2 + xy + y^2 + y$ $\Rightarrow$ $f_x = 2x + y$, $f_y = x + 2y + 1$, $f_{xx} = 2$, $f_{xy} = 1$, $f_{yy} = 2$. Then $f_x = 0$ implies

   $y = -2x$, and substitution into $f_y = x + 2y + 1 = 0$ gives $x + 2(-2x) + 1 = 0$ $\Rightarrow$ $-3x = -1$ $\Rightarrow$ $x = \frac{1}{3}$.

   Then $y = -\frac{2}{3}$ and the only critical point is $\left(\frac{1}{3}, -\frac{2}{3}\right)$.

   $D(x,y) = f_{xx}f_{yy} - (f_{xy})^2 = (2)(2) - (1)^2 = 3$, and since

   $D\left(\frac{1}{3}, -\frac{2}{3}\right) = 3 > 0$ and $f_{xx}\left(\frac{1}{3}, -\frac{2}{3}\right) = 2 > 0$, $f\left(\frac{1}{3}, -\frac{2}{3}\right) = -\frac{1}{3}$ is a local

   minimum by the Second Derivatives Test.

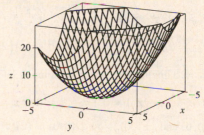

5. $f(x,y) = y^3 + 3x^2 y - 6x^2 - 6y^2 + 2$ $\Rightarrow$ $f_x = 6xy - 12x$, $f_y = 3y^2 + 3x^2 - 12y$, $f_{xx} = 6y - 12$, $f_{xy} = 6x$,

   $f_{yy} = 6y - 12$. Then $f_x = 0$ implies $6x(y - 2) = 0$, so $x = 0$ or $y = 2$. If $x = 0$ then substitution into $f_y = 0$ gives

   $3y^2 - 12y = 0$ $\Rightarrow$ $3y(y - 4) = 0$ $\Rightarrow$ $y = 0$ or $y = 4$, so we have critical points $(0,0)$ and $(0,4)$. If $y = 2$,

   substitution into $f_y = 0$ gives $12 + 3x^2 - 24 = 0$ $\Rightarrow$ $x^2 = 4$ $\Rightarrow$

   $x = \pm 2$, so we have critical points $(\pm 2, 2)$.

   $D(0,0) = (-12)(-12) - 0^2 = 144 > 0$ and $f_{xx}(0,0) = -12 < 0$, so

   $f(0,0) = 2$ is a local maximum. $D(0,4) = (12)(12) - 0^2 = 144 > 0$

   and $f_{xx}(0,4) = 12 > 0$, so $f(0,4) = -30$ is a local minimum.

   $D(\pm 2, 2) = (0)(0) - (\pm 12)^2 = -144 < 0$, so $(\pm 2, 2)$ are saddle points.

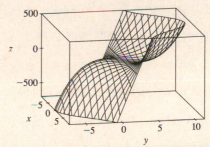

7. $f(x,y) = x^3 - 12xy + 8y^3$ $\Rightarrow$ $f_x = 3x^2 - 12y$, $f_y = -12x + 24y^2$, $f_{xx} = 6x$, $f_{xy} = -12$, $f_{yy} = 48y$. Then $f_x = 0$

   implies $x^2 = 4y$ and $f_y = 0$ implies $x = 2y^2$. Substituting the second equation into the first gives $(2y^2)^2 = 4y$ $\Rightarrow$

   $4y^4 = 4y$ $\Rightarrow$ $4y(y^3 - 1) = 0$ $\Rightarrow$ $y = 0$ or $y = 1$. If $y = 0$ then

   $x = 0$ and if $y = 1$ then $x = 2$, so the critical points are $(0,0)$ and $(2,1)$.

   $D(0,0) = (0)(0) - (-12)^2 = -144 < 0$, so $(0,0)$ is a saddle point.

   $D(2,1) = (12)(48) - (-12)^2 = 432 > 0$ and $f_{xx}(2,1) = 12 > 0$ so

   $f(2,1) = -8$ is a local minimum.

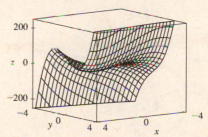

9. $f(x, y) = e^x \cos y \;\Rightarrow\; f_x = e^x \cos y,\; f_y = -e^x \sin y.$

Now $f_x = 0$ implies $\cos y = 0$ or $y = \frac{\pi}{2} + n\pi$ for n an integer.

But $\sin\left(\frac{\pi}{2} + n\pi\right) \neq 0$, so there are no critical points.

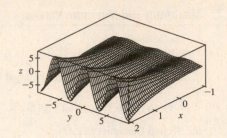

11. $f(x, y) = (x^2 + y^2)e^{y^2 - x^2} \;\Rightarrow$

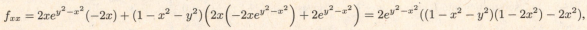

$f_x = (x^2 + y^2)e^{y^2 - x^2}(-2x) + 2xe^{y^2 - x^2} = 2xe^{y^2 - x^2}(1 - x^2 - y^2),$

$f_y = (x^2 + y^2)e^{y^2 - x^2}(2y) + 2ye^{y^2 - x^2} = 2ye^{y^2 - x^2}(1 + x^2 + y^2),$

$f_{xx} = 2xe^{y^2 - x^2}(-2x) + (1 - x^2 - y^2)\left(2x\left(-2xe^{y^2 - x^2}\right) + 2e^{y^2 - x^2}\right) = 2e^{y^2 - x^2}((1 - x^2 - y^2)(1 - 2x^2) - 2x^2),$

$f_{xy} = 2xe^{y^2 - x^2}(-2y) + 2x(2y)e^{y^2 - x^2}(1 - x^2 - y^2) = -4xye^{y^2 - x^2}(x^2 + y^2),$

$f_{yy} = 2ye^{y^2 - x^2}(2y) + (1 + x^2 + y^2)\left(2y\left(2ye^{y^2 - x^2}\right) + 2e^{y^2 - x^2}\right) = 2e^{y^2 - x^2}((1 + x^2 + y^2)(1 + 2y^2) + 2y^2).$

$f_y = 0$ implies $y = 0$, and substituting into $f_x = 0$ gives

$2xe^{-x^2}(1 - x^2) = 0 \;\Rightarrow\; x = 0$ or $x = \pm 1$. Thus the critical points are

$(0, 0)$ and $(\pm 1, 0)$. Now $D(0, 0) = (2)(2) - 0 > 0$ and $f_{xx}(0, 0) = 2 > 0$,

so $f(0, 0) = 0$ is a local minimum. $D(\pm 1, 0) = (-4e^{-1})(4e^{-1}) - 0 < 0$

so $(\pm 1, 0)$ are saddle points.

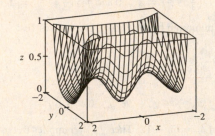

13. $f(x, y) = y^2 - 2y \cos x \;\Rightarrow\; f_x = 2y \sin x,\; f_y = 2y - 2\cos x,$

$f_{xx} = 2y \cos x,\; f_{xy} = 2\sin x,\; f_{yy} = 2.$ Then $f_x = 0$ implies $y = 0$ or

$\sin x = 0 \;\Rightarrow\; x = 0, \pi,$ or 2π for $-1 \leq x \leq 7$. Substituting $y = 0$ into

$f_y = 0$ gives $\cos x = 0 \;\Rightarrow\; x = \frac{\pi}{2}$ or $\frac{3\pi}{2}$, substituting $x = 0$ or $x = 2\pi$

into $f_y = 0$ gives $y = 1$, and substituting $x = \pi$ into $f_y = 0$ gives $y = -1$.

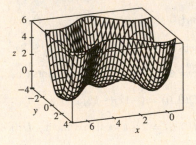

Thus the critical points are $(0, 1)$, $\left(\frac{\pi}{2}, 0\right)$, $(\pi, -1)$, $\left(\frac{3\pi}{2}, 0\right)$, and $(2\pi, 1)$.

$D\left(\frac{\pi}{2}, 0\right) = D\left(\frac{3\pi}{2}, 0\right) = -4 < 0$ so $\left(\frac{\pi}{2}, 0\right)$ and $\left(\frac{3\pi}{2}, 0\right)$ are saddle points. $D(0, 1) = D(\pi, -1) = D(2\pi, 1) = 4 > 0$ and

$f_{xx}(0, 1) = f_{xx}(\pi, -1) = f_{xx}(2\pi, 1) = 2 > 0$, so $f(0, 1) = f(\pi, -1) = f(2\pi, 1) = -1$ are local minima.

15. $f(x, y) = 3x^2 y + y^3 - 3x^2 - 3y^2 + 2$

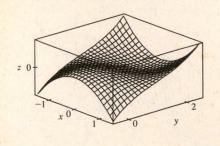

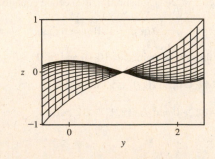

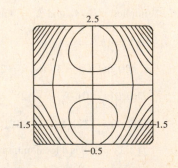

[continued]

From the graphs, it appears that f has a local maximum $f(0,0) \approx 2$ and a local minimum $f(0,2) \approx -2$. There appear to be saddle points near $(\pm 1, 1)$.

$f_x = 6xy - 6x$, $f_y = 3x^2 + 3y^2 - 6y$. Then $f_x = 0$ implies $x = 0$ or $y = 1$ and when $x = 0$, $f_y = 0$ implies $y = 0$ or $y = 2$; when $y = 1$, $f_y = 0$ implies $x^2 = 1$ or $x = \pm 1$. Thus the critical points are $(0,0)$, $(0,2)$, $(\pm 1, 1)$. Now $f_{xx} = 6y - 6$, $f_{yy} = 6y - 6$, $f_{xy} = 6x$, so $D(0,0) = D(0,2) = 36 > 0$ while $D(\pm 1, 1) = -36 < 0$ and $f_{xx}(0,0) = -6$, $f_{xx}(0,2) = 6$. Hence $(\pm 1, 1)$ are saddle points while $f(0,0) = 2$ is a local maximum and $f(0,2) = -2$ is a local minimum.

17. $f(x,y) = \sin x + \sin y + \sin(x+y)$, $0 \le x \le 2\pi$, $0 \le y \le 2\pi$

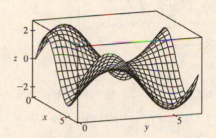 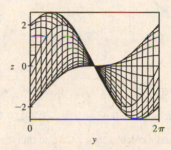 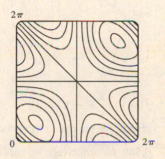

From the graphs it appears that f has a local maximum at about $(1, 1)$ with value approximately 2.6, a local minimum at about $(5, 5)$ with value approximately -2.6, and a saddle point at about $(3, 3)$.

$f_x = \cos x + \cos(x+y)$, $f_y = \cos y + \cos(x+y)$, $f_{xx} = -\sin x - \sin(x+y)$, $f_{yy} = -\sin y - \sin(x+y)$, $f_{xy} = -\sin(x+y)$. Setting $f_x = 0$ and $f_y = 0$ and subtracting gives $\cos x - \cos y = 0$ or $\cos x = \cos y$. Thus $x = y$ or $x = 2\pi - y$. If $x = y$, $f_x = 0$ becomes $\cos x + \cos 2x = 0$ or $2 \cos^2 x + \cos x - 1 = 0$, a quadratic in $\cos x$. Thus $\cos x = -1$ or $\frac{1}{2}$ and $x = \pi$, $\frac{\pi}{3}$, or $\frac{5\pi}{3}$, giving the critical points (π, π), $\left(\frac{\pi}{3}, \frac{\pi}{3}\right)$ and $\left(\frac{5\pi}{3}, \frac{5\pi}{3}\right)$. Similarly if $x = 2\pi - y$, $f_x = 0$ becomes $(\cos x) + 1 = 0$ and the resulting critical point is (π, π). Now $D(x,y) = \sin x \sin y + \sin x \sin(x+y) + \sin y \sin(x+y)$. So $D(\pi, \pi) = 0$ and the Second Derivatives Test doesn't apply. However, along the line $y = x$ we have $f(x,x) = 2 \sin x + \sin 2x = 2 \sin x + 2 \sin x \cos x = 2 \sin x (1 + \cos x)$, and $f(x,x) > 0$ for $0 < x < \pi$ while $f(x,x) < 0$ for $\pi < x < 2\pi$. Thus every disk with center (π, π) contains points where f is positive as well as points where f is negative, so the graph crosses its tangent plane $(z = 0)$ there and (π, π) is a saddle point. $D\left(\frac{\pi}{3}, \frac{\pi}{3}\right) = \frac{9}{4} > 0$ and $f_{xx}\left(\frac{\pi}{3}, \frac{\pi}{3}\right) < 0$ so $f\left(\frac{\pi}{3}, \frac{\pi}{3}\right) = \frac{3\sqrt{3}}{2}$ is a local maximum while $D\left(\frac{5\pi}{3}, \frac{5\pi}{3}\right) = \frac{9}{4} > 0$ and $f_{xx}\left(\frac{5\pi}{3}, \frac{5\pi}{3}\right) > 0$, so $f\left(\frac{5\pi}{3}, \frac{5\pi}{3}\right) = -\frac{3\sqrt{3}}{2}$ is a local minimum.

19. $f(x,y) = x^4 + y^4 - 4x^2 y + 2y$ $\Rightarrow$ $f_x(x,y) = 4x^3 - 8xy$ and $f_y(x,y) = 4y^3 - 4x^2 + 2$. $f_x = 0$ $\Rightarrow$ $4x(x^2 - 2y) = 0$, so $x = 0$ or $x^2 = 2y$. If $x = 0$ then substitution into $f_y = 0$ gives $4y^3 = -2$ $\Rightarrow$ $y = -\frac{1}{\sqrt[3]{2}}$, so $\left(0, -\frac{1}{\sqrt[3]{2}}\right)$ is a critical point. Substituting $x^2 = 2y$ into $f_y = 0$ gives $4y^3 - 8y + 2 = 0$. Using a graph, solutions are approximately $y = -1.526$, 0.259, and 1.267. (Alternatively, we could have used a calculator or a CAS to find these roots.) We have $x^2 = 2y$ $\Rightarrow$ $x = \pm\sqrt{2y}$, so $y = -1.526$ gives no real-valued solution for x, but $y = 0.259$ $\Rightarrow$ $x \approx \pm 0.720$ and $y = 1.267$ $\Rightarrow$ $x \approx \pm 1.592$. Thus to three decimal places, the critical points are $\left(0, -\frac{1}{\sqrt[3]{2}}\right) \approx (0, -0.794)$, $(\pm 0.720, 0.259)$, and $(\pm 1.592, 1.267)$. Now since $f_{xx} = 12x^2 - 8y$, $f_{xy} = -8x$, $f_{yy} = 12y^2$, and

$D = (12x^2 - 8y)(12y^2) - 64x^2$, we have $D(0, -0.794) > 0$, $f_{xx}(0, -0.794) > 0$, $D(\pm 0.720, 0.259) < 0$,

$D(\pm 1.592, 1.267) > 0$, and $f_{xx}(\pm 1.592, 1.267) > 0$. Therefore $f(0, -0.794) \approx -1.191$ and $f(\pm 1.592, 1.267) \approx -1.310$

are local minima, and $(\pm 0.720, 0.259)$ are saddle points. There is no highest point on the graph, but the lowest points are

approximately $(\pm 1.592, 1.267, -1.310)$.

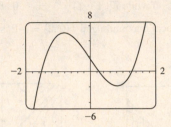

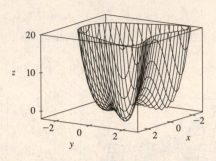

21. $f(x, y) = x^4 + y^3 - 3x^2 + y^2 + x - 2y + 1 \quad \Rightarrow \quad f_x(x, y) = 4x^3 - 6x + 1$ and $f_y(x, y) = 3y^2 + 2y - 2$. From the

graphs, we see that to three decimal places, $f_x = 0$ when $x \approx -1.301$, 0.170, or 1.131, and $f_y = 0$ when $y \approx -1.215$ or

0.549. (Alternatively, we could have used a calculator or a CAS to find these roots. We could also use the quadratic formula to

find the solutions of $f_y = 0$.) So, to three decimal places, f has critical points at $(-1.301, -1.215)$, $(-1.301, 0.549)$,

$(0.170, -1.215)$, $(0.170, 0.549)$, $(1.131, -1.215)$, and $(1.131, 0.549)$. Now since $f_{xx} = 12x^2 - 6$, $f_{xy} = 0$, $f_{yy} = 6y + 2$,

and $D = (12x^2 - 6)(6y + 2)$, we have $D(-1.301, -1.215) < 0$, $D(-1.301, 0.549) > 0$, $f_{xx}(-1.301, 0.549) > 0$,

$D(0.170, -1.215) > 0$, $f_{xx}(0.170, -1.215) < 0$, $D(0.170, 0.549) < 0$, $D(1.131, -1.215) < 0$, $D(1.131, 0.549) > 0$, and

$f_{xx}(1.131, 0.549) > 0$. Therefore, to three decimal places, $f(-1.301, 0.549) \approx -3.145$ and $f(1.131, 0.549) \approx -0.701$ are

local minima, $f(0.170, -1.215) \approx 3.197$ is a local maximum, and $(-1.301, -1.215)$, $(0.170, 0.549)$, and $(1.131, -1.215)$

are saddle points. There is no highest or lowest point on the graph.

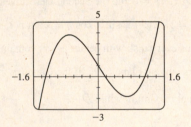

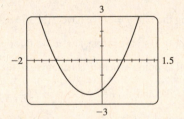

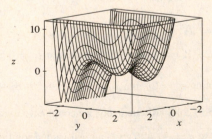

23. Since f is a polynomial it is continuous on D, so an absolute maximum and minimum exist. Here $f_x = 2x - 2$, $f_y = 2y$, and

setting $f_x = f_y = 0$ gives $(1, 0)$ as the only critical point (which is inside D), where $f(1, 0) = -1$. Along L_1: $x = 0$ and

$f(0, y) = y^2$ for $-2 \leq y \leq 2$, a quadratic function which attains its minimum at $y = 0$, where $f(0, 0) = 0$, and its maximum

at $y = \pm 2$, where $f(0, \pm 2) = 4$. Along L_2: $y = x - 2$ for $0 \leq x \leq 2$, and $f(x, x - 2) = 2x^2 - 6x + 4 = 2\left(x - \frac{3}{2}\right)^2 - \frac{1}{2}$,

a quadratic which attains its minimum at $x = \frac{3}{2}$, where $f\left(\frac{3}{2}, -\frac{1}{2}\right) = -\frac{1}{2}$, and its maximum at $x = 0$, where $f(0, -2) = 4$.

Along L_3: $y = 2 - x$ for $0 \leq x \leq 2$, and

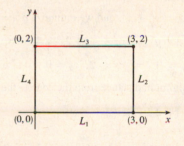

$f(x, 2 - x) = 2x^2 - 6x + 4 = 2\left(x - \frac{3}{2}\right)^2 - \frac{1}{2}$, a quadratic which attains

its minimum at $x = \frac{3}{2}$, where $f\left(\frac{3}{2}, \frac{1}{2}\right) = -\frac{1}{2}$, and its maximum at $x = 0$,

where $f(0, 2) = 4$. Thus the absolute maximum of f on D is $f(0, \pm 2) = 4$

and the absolute minimum is $f(1, 0) = -1$.

25. $f_x(x, y) = 2x + 2xy$, $f_y(x, y) = 2y + x^2$, and setting $f_x = f_y = 0$

gives $(0, 0)$ as the only critical point in D, with $f(0, 0) = 4$.

On L_1: $y = -1$, $f(x, -1) = 5$, a constant.

On L_2: $x = 1$, $f(1, y) = y^2 + y + 5$, a quadratic in y which attains its

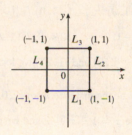

maximum at $(1, 1)$, $f(1, 1) = 7$ and its minimum at $\left(1, -\frac{1}{2}\right)$, $f\left(1, -\frac{1}{2}\right) = \frac{19}{4}$.

On L_3: $f(x, 1) = 2x^2 + 5$ which attains its maximum at $(-1, 1)$ and $(1, 1)$

with $f(\pm 1, 1) = 7$ and its minimum at $(0, 1)$, $f(0, 1) = 5$.

On L_4: $f(-1, y) = y^2 + y + 5$ with maximum at $(-1, 1)$, $f(-1, 1) = 7$ and minimum at $\left(-1, -\frac{1}{2}\right)$, $f\left(-1, -\frac{1}{2}\right) = \frac{19}{4}$.

Thus the absolute maximum is attained at both $(\pm 1, 1)$ with $f(\pm 1, 1) = 7$ and the absolute minimum on D is attained at

$(0, 0)$ with $f(0, 0) = 4$.

27. $f(x, y) = x^4 + y^4 - 4xy + 2$ is a polynomial and hence continuous on D, so

it has an absolute maximum and minimum on D. $f_x(x, y) = 4x^3 - 4y$ and

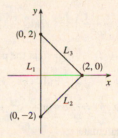

$f_y(x, y) = 4y^3 - 4x$; then $f_x = 0$ implies $y = x^3$, and substitution into

$f_y = 0 \Rightarrow x = y^3$ gives $x^9 - x = 0 \Rightarrow x(x^8 - 1) = 0 \Rightarrow x = 0$

or $x = \pm 1$. Thus the critical points are $(0, 0)$, $(1, 1)$, and $(-1, -1)$, but only

$(1, 1)$ with $f(1, 1) = 0$ is inside D. On L_1: $y = 0$, $f(x, 0) = x^4 + 2$,

$0 \leq x \leq 3$, a polynomial in x which attains its maximum at $x = 3$, $f(3, 0) = 83$, and its minimum at $x = 0$, $f(0, 0) = 2$.

On L_2: $x = 3$, $f(3, y) = y^4 - 12y + 83$, $0 \leq y \leq 2$, a polynomial in y which attains its minimum at $y = \sqrt[3]{3}$,

$f(3, \sqrt[3]{3}) = 83 - 9\sqrt[3]{3} \approx 70.0$, and its maximum at $y = 0$, $f(3, 0) = 83$.

On L_3: $y = 2$, $f(x, 2) = x^4 - 8x + 18$, $0 \leq x \leq 3$, a polynomial in x which attains its minimum at $x = \sqrt[3]{2}$,

$f(\sqrt[3]{2}, 2) = 18 - 6\sqrt[3]{2} \approx 10.4$, and its maximum at $x = 3$, $f(3, 2) = 75$. On L_4: $x = 0$, $f(0, y) = y^4 + 2$, $0 \leq y \leq 2$, a

polynomial in y which attains its maximum at $y = 2$, $f(0, 2) = 18$, and its minimum at $y = 0$, $f(0, 0) = 2$. Thus the absolute

maximum of f on D is $f(3, 0) = 83$ and the absolute minimum is $f(1, 1) = 0$.

29. $f(x, y) = -(x^2 - 1)^2 - (x^2 y - x - 1)^2 \;\Rightarrow\; f_x(x, y) = -2(x^2 - 1)(2x) - 2(x^2 y - x - 1)(2xy - 1)$ and

$f_y(x, y) = -2(x^2 y - x - 1)x^2$. Setting $f_y(x, y) = 0$ gives either $x = 0$ or $x^2 y - x - 1 = 0$.

There are no critical points for $x = 0$, since $f_x(0, y) = -2$, so we set $x^2 y - x - 1 = 0 \;\Leftrightarrow\; y = \dfrac{x + 1}{x^2}$ $[x \neq 0]$,

so $f_x\!\left(x, \dfrac{x + 1}{x^2}\right) = -2(x^2 - 1)(2x) - 2\left(x^2 \dfrac{x + 1}{x^2} - x - 1\right)\left(2x\dfrac{x + 1}{x^2} - 1\right) = -4x(x^2 - 1)$. Therefore

$f_x(x, y) = f_y(x, y) = 0$ at the points $(1, 2)$ and $(-1, 0)$. To classify these critical points, we calculate

$f_{xx}(x, y) = -12x^2 - 12x^2 y^2 + 12xy + 4y + 2, \; f_{yy}(x, y) = -2x^4,$

and $f_{xy}(x, y) = -8x^3 y + 6x^2 + 4x$. In order to use the Second Derivatives

Test we calculate

$D(-1, 0) = f_{xx}(-1, 0)\, f_{yy}(-1, 0) - [f_{xy}(-1, 0)]^2 = 16 > 0,$

$f_{xx}(-1, 0) = -10 < 0, \; D(1, 2) = 16 > 0$, and $f_{xx}(1, 2) = -26 < 0$, so

both $(-1, 0)$ and $(1, 2)$ give local maxima.

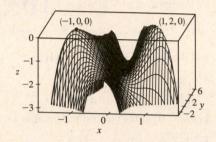

31. Let d be the distance from $(2, 0, -3)$ to any point (x, y, z) on the plane $x + y + z = 1$, so $d = \sqrt{(x - 2)^2 + y^2 + (z + 3)^2}$

where $z = 1 - x - y$, and we minimize $d^2 = f(x, y) = (x - 2)^2 + y^2 + (4 - x - y)^2$. Then

$f_x(x, y) = 2(x - 2) + 2(4 - x - y)(-1) = 4x + 2y - 12, \; f_y(x, y) = 2y + 2(4 - x - y)(-1) = 2x + 4y - 8$. Solving

$4x + 2y - 12 = 0$ and $2x + 4y - 8 = 0$ simultaneously gives $x = \frac{8}{3}, y = \frac{2}{3}$, so the only critical point is $\left(\frac{8}{3}, \frac{2}{3}\right)$. An absolute

minimum exists (since there is a minimum distance from the point to the plane) and it must occur at a critical point, so the

shortest distance occurs for $x = \frac{8}{3}, y = \frac{2}{3}$ for which $d = \sqrt{\left(\frac{8}{3} - 2\right)^2 + \left(\frac{2}{3}\right)^2 + \left(4 - \frac{8}{3} - \frac{2}{3}\right)^2} = \sqrt{\frac{4}{3}} = \frac{2}{\sqrt{3}}$.

33. Let d be the distance from the point $(4, 2, 0)$ to any point (x, y, z) on the cone, so $d = \sqrt{(x - 4)^2 + (y - 2)^2 + z^2}$ where

$z^2 = x^2 + y^2$, and we minimize $d^2 = (x - 4)^2 + (y - 2)^2 + x^2 + y^2 = f(x, y)$. Then

$f_x(x, y) = 2(x - 4) + 2x = 4x - 8, \; f_y(x, y) = 2(y - 2) + 2y = 4y - 4$, and the critical points occur when

$f_x = 0 \;\Rightarrow\; x = 2, \; f_y = 0 \;\Rightarrow\; y = 1$. Thus the only critical point is $(2, 1)$. An absolute minimum exists (since there is a

minimum distance from the cone to the point) which must occur at a critical point, so the points on the cone closest

to $(4, 2, 0)$ are $\left(2, 1, \pm\sqrt{5}\right)$.

35. $x + y + z = 100$, so maximize $f(x, y) = xy(100 - x - y)$. $\;f_x = 100y - 2xy - y^2, \; f_y = 100x - x^2 - 2xy,$

$f_{xx} = -2y, \; f_{yy} = -2x, \; f_{xy} = 100 - 2x - 2y$. Then $f_x = 0$ implies $y = 0$ or $y = 100 - 2x$. Substituting $y = 0$ into

$f_y = 0$ gives $x = 0$ or $x = 100$ and substituting $y = 100 - 2x$ into $f_y = 0$ gives $3x^2 - 100x = 0$ so $x = 0$ or $\frac{100}{3}$.

Thus the critical points are $(0, 0)$, $(100, 0)$, $(0, 100)$ and $\left(\frac{100}{3}, \frac{100}{3}\right)$.

$D(0, 0) = D(100, 0) = D(0, 100) = -10{,}000$ while $D\left(\frac{100}{3}, \frac{100}{3}\right) = \frac{10{,}000}{3}$ and $f_{xx}\left(\frac{100}{3}, \frac{100}{3}\right) = -\frac{200}{3} < 0$. Thus $(0, 0)$,

$(100, 0)$ and $(0, 100)$ are saddle points whereas $f\left(\frac{100}{3}, \frac{100}{3}\right)$ is a local maximum. Thus the numbers are $x = y = z = \frac{100}{3}$.

37. Center the sphere at the origin so that its equation is $x^2 + y^2 + z^2 = r^2$, and orient the inscribed rectangular box so that its

edges are parallel to the coordinate axes. Any vertex of the box satisfies $x^2 + y^2 + z^2 = r^2$, so take (x, y, z) to be the vertex in the first octant. Then the box has length $2x$, width $2y$, and height $2z = 2\sqrt{r^2 - x^2 - y^2}$ with volume given by

$$V(x, y) = (2x)(2y)\left(2\sqrt{r^2 - x^2 - y^2}\right) = 8xy\sqrt{r^2 - x^2 - y^2} \text{ for } 0 < x < r, 0 < y < r. \text{ Then}$$

$$V_x = (8xy) \cdot \tfrac{1}{2}(r^2 - x^2 - y^2)^{-1/2}(-2x) + \sqrt{r^2 - x^2 - y^2} \cdot 8y = \frac{8y(r^2 - 2x^2 - y^2)}{\sqrt{r^2 - x^2 - y^2}} \text{ and } V_y = \frac{8x(r^2 - x^2 - 2y^2)}{\sqrt{r^2 - x^2 - y^2}}.$$

Setting $V_x = 0$ gives $y = 0$ or $2x^2 + y^2 = r^2$, but $y > 0$ so only the latter solution applies. Similarly, $V_y = 0$ with $x > 0$

implies $x^2 + 2y^2 = r^2$. Substituting, we have $2x^2 + y^2 = x^2 + 2y^2 \;\Rightarrow\; x^2 = y^2 \;\Rightarrow\; y = x$. Then $x^2 + 2y^2 = r^2 \;\Rightarrow$

$3x^2 = r^2 \;\Rightarrow\; x = \sqrt{r^2/3} = r/\sqrt{3} = y$. Thus the only critical point is $(r/\sqrt{3}, r/\sqrt{3})$. There must be a maximum

volume and here it must occur at a critical point, so the maximum volume occurs when $x = y = r/\sqrt{3}$ and the maximum

volume is $V\left(\frac{r}{\sqrt{3}}, \frac{r}{\sqrt{3}}\right) = 8\left(\frac{r}{\sqrt{3}}\right)\left(\frac{r}{\sqrt{3}}\right)\sqrt{r^2 - \left(\frac{r}{\sqrt{3}}\right)^2 - \left(\frac{r}{\sqrt{3}}\right)^2} = \frac{8}{3\sqrt{3}}r^3$.

39. Maximize $f(x, y) = \dfrac{xy}{3}(6 - x - 2y)$, then the maximum volume is $V = xyz$.

$f_x = \tfrac{1}{3}(6y - 2xy - y^2) = \tfrac{1}{3}y(6 - 2x - 2y)$ and $f_y = \tfrac{1}{3}x(6 - x - 4y)$. Setting $f_x = 0$ and $f_y = 0$ gives the critical point

$(2, 1)$ which geometrically must give a maximum. Thus the volume of the largest such box is $V = (2)(1)\left(\tfrac{2}{3}\right) = \tfrac{4}{3}$.

41. Let the dimensions be x, y, and z; then $4x + 4y + 4z = c$ and the volume is

$V = xyz = xy\left(\tfrac{1}{4}c - x - y\right) = \tfrac{1}{4}cxy - x^2y - xy^2$, $x > 0$, $y > 0$. Then $V_x = \tfrac{1}{4}cy - 2xy - y^2$ and $V_y = \tfrac{1}{4}cx - x^2 - 2xy$,

so $V_x = 0 = V_y$ when $2x + y = \tfrac{1}{4}c$ and $x + 2y = \tfrac{1}{4}c$. Solving, we get $x = \tfrac{1}{12}c$, $y = \tfrac{1}{12}c$ and $z = \tfrac{1}{4}c - x - y = \tfrac{1}{12}c$. From

the geometrical nature of the problem, this critical point must give an absolute maximum. Thus the box is a cube with edge

length $\tfrac{1}{12}c$.

43. Let the dimensions be x, y and z, then minimize $xy + 2(xz + yz)$ if $xyz = 32{,}000$ cm³. Then

$f(x, y) = xy + [64{,}000(x + y)/xy] = xy + 64{,}000(x^{-1} + y^{-1})$, $f_x = y - 64{,}000x^{-2}$, $f_y = x - 64{,}000y^{-2}$.

And $f_x = 0$ implies $y = 64{,}000/x^2$; substituting into $f_y = 0$ implies $x^3 = 64{,}000$ or $x = 40$ and then $y = 40$. Now

$D(x, y) = [(2)(64{,}000)]^2 x^{-3}y^{-3} - 1 > 0$ for $(40, 40)$ and $f_{xx}(40, 40) > 0$ so this is indeed a minimum. Thus the

dimensions of the box are $x = y = 40$ cm, $z = 20$ cm.

45. Let x, y, z be the dimensions of the rectangular box. Then the volume of the box is xyz and

$L = \sqrt{x^2 + y^2 + z^2} \;\Rightarrow\; L^2 = x^2 + y^2 + z^2 \;\Rightarrow\; z = \sqrt{L^2 - x^2 - y^2}$.

Substituting, we have volume $V(x, y) = xy\sqrt{L^2 - x^2 - y^2}$ $(x, y > 0)$.

$V_x = xy \cdot \tfrac{1}{2}(L^2 - x^2 - y^2)^{-1/2}(-2x) + y\sqrt{L^2 - x^2 - y^2} = y\sqrt{L^2 - x^2 - y^2} - \dfrac{x^2y}{\sqrt{L^2 - x^2 - y^2}}$,

$V_y = x\sqrt{L^2 - x^2 - y^2} - \dfrac{xy^2}{\sqrt{L^2 - x^2 - y^2}}$. $V_x = 0$ implies $y(L^2 - x^2 - y^2) = x^2y \;\Rightarrow\; y(L^2 - 2x^2 - y^2) = 0 \;\Rightarrow$

$2x^2 + y^2 = L^2$ (since $y > 0$), and $V_y = 0$ implies $x(L^2 - x^2 - y^2) = xy^2 \;\Rightarrow\; x(L^2 - x^2 - 2y^2) = 0 \;\Rightarrow$

$x^2 + 2y^2 = L^2$ (since $x > 0$). Substituting $y^2 = L^2 - 2x^2$ into $x^2 + 2y^2 = L^2$ gives $x^2 + 2L^2 - 4x^2 = L^2 \;\Rightarrow$

$3x^2 = L^2 \implies x = L/\sqrt{3}$ (since $x > 0$) and then $y = \sqrt{L^2 - 2\left(L/\sqrt{3}\right)^2} = L/\sqrt{3}$.

So the only critical point is $\left(L/\sqrt{3}, L/\sqrt{3}\right)$ which, from the geometrical nature of the problem, must give an absolute

maximum. Thus the maximum volume is $V\left(L/\sqrt{3}, L/\sqrt{3}\right) = \left(L/\sqrt{3}\right)^2 \sqrt{L^2 - \left(L/\sqrt{3}\right)^2 - \left(L/\sqrt{3}\right)^2} = L^3/\left(3\sqrt{3}\right)$

cubic units.

47. Note that here the variables are m and b, and $f(m, b) = \sum\limits_{i=1}^{n} [y_i - (mx_i + b)]^2$. Then $f_m = \sum\limits_{i=1}^{n} -2x_i[y_i - (mx_i + b)] = 0$

implies $\sum\limits_{i=1}^{n} \left(x_i y_i - mx_i^2 - bx_i\right) = 0$ or $\sum\limits_{i=1}^{n} x_i y_i = m \sum\limits_{i=1}^{n} x_i^2 + b \sum\limits_{i=1}^{n} x_i$ and $f_b = \sum\limits_{i=1}^{n} -2[y_i - (mx_i + b)] = 0$ implies

$\sum\limits_{i=1}^{n} y_i = m \sum\limits_{i=1}^{n} x_i + \sum\limits_{i=1}^{n} b = m\left(\sum\limits_{i=1}^{n} x_i\right) + nb$. Thus we have the two desired equations.

Now $f_{mm} = \sum\limits_{i=1}^{n} 2x_i^2$, $f_{bb} = \sum\limits_{i=1}^{n} 2 = 2n$ and $f_{mb} = \sum\limits_{i=1}^{n} 2x_i$. And $f_{mm}(m, b) > 0$ always and

$D(m, b) = 4n\left(\sum\limits_{i=1}^{n} x_i^2\right) - 4\left(\sum\limits_{i=1}^{n} x_i\right)^2 = 4\left[n\left(\sum\limits_{i=1}^{n} x_i^2\right) - \left(\sum\limits_{i=1}^{n} x_i\right)^2\right] > 0$ always so the solutions of these two

equations do indeed minimize $\sum\limits_{i=1}^{n} d_i^2$.

11.8 Lagrange Multipliers

1. $f(x, y) = x^2 + y^2$, $g(x, y) = xy = 1$, and $\nabla f = \lambda \nabla g \implies \langle 2x, 2y \rangle = \langle \lambda y, \lambda x \rangle$, so $2x = \lambda y$, $2y = \lambda x$, and $xy = 1$.

From the last equation, $x \neq 0$ and $y \neq 0$, so $2x = \lambda y \implies \lambda = 2x/y$. Substituting, we have $2y = (2x/y)\,x \implies$

$y^2 = x^2 \implies y = \pm x$. But $xy = 1$, so $x = y = \pm 1$ and the possible points for the extreme values of f are $(1, 1)$ and

$(-1, -1)$. Here there is no maximum value, since the constraint $xy = 1$ allows x or y to become arbitrarily large, and hence

$f(x, y) = x^2 + y^2$ can be made arbitrarily large. The minimum value is $f(1, 1) = f(-1, -1) = 2$.

3. $f(x, y) = y^2 - x^2$, $g(x, y) = \frac{1}{4}x^2 + y^2 = 1$, and $\nabla f = \lambda \nabla g \implies \langle -2x, 2y \rangle = \langle \frac{1}{2}\lambda x, 2\lambda y \rangle$, so $-2x = \frac{1}{2}\lambda x$, $2y = 2\lambda y$,

and $\frac{1}{4}x^2 + y^2 = 1$. From the first equation we have $x(4 + \lambda) = 0 \implies x = 0$ or $\lambda = -4$. If $x = 0$ then the third equation

gives $y = \pm 1$. If $\lambda = -4$ then the second equation gives $2y = -8y \implies y = 0$, and substituting into the third equation,

we have $x = \pm 2$. Thus the possible extreme values of f occur at the points $(0, \pm 1)$ and $(\pm 2, 0)$. Evaluating f at these points,

we see that the maximum value is $f(0, \pm 1) = 1$ and the minimum is $f(\pm 2, 0) = -4$.

5. $f(x, y, z) = 2x + 2y + z$, $g(x, y, z) = x^2 + y^2 + z^2 = 9$, and $\nabla f = \lambda \nabla g \implies \langle 2, 2, 1 \rangle = \langle 2\lambda x, 2\lambda y, 2\lambda z \rangle$, so $2\lambda x = 2$,

$2\lambda y = 2$, $2\lambda z = 1$, and $x^2 + y^2 + z^2 = 9$. The first three equations imply $x = \dfrac{1}{\lambda}$, $y = \dfrac{1}{\lambda}$, and $z = \dfrac{1}{2\lambda}$. But substitution into

the fourth equation gives $\left(\dfrac{1}{\lambda}\right)^2 + \left(\dfrac{1}{\lambda}\right)^2 + \left(\dfrac{1}{2\lambda}\right)^2 = 9 \implies \dfrac{9}{4\lambda^2} = 9 \implies \lambda = \pm\frac{1}{2}$, so f has possible extreme values at

the points $(2, 2, 1)$ and $(-2, -2, -1)$. The maximum value of f on $x^2 + y^2 + z^2 = 9$ is $f(2, 2, 1) = 9$, and the minimum is

$f(-2, -2, -1) = -9$.

7. $f(x, y, z) = xyz$, $g(x, y, z) = x^2 + 2y^2 + 3z^2 = 6$. $\nabla f = \lambda \nabla g \implies \langle yz, xz, xy \rangle = \lambda \langle 2x, 4y, 6z \rangle$. If any of x, y, or z is

zero then $x = y = z = 0$ which contradicts $x^2 + 2y^2 + 3z^2 = 6$. Then $\lambda = (yz)/(2x) = (xz)/(4y) = (xy)/(6z)$ or

$x^2 = 2y^2$ and $z^2 = \frac{2}{3}y^2$. Thus $x^2 + 2y^2 + 3z^2 = 6$ implies $6y^2 = 6$ or $y = \pm 1$. Then the possible points are

$\left(\sqrt{2}, \pm 1, \sqrt{\frac{2}{3}}\right)$, $\left(\sqrt{2}, \pm 1, -\sqrt{\frac{2}{3}}\right)$, $\left(-\sqrt{2}, \pm 1, \sqrt{\frac{2}{3}}\right)$, $\left(-\sqrt{2}, \pm 1, -\sqrt{\frac{2}{3}}\right)$. The maximum value of f on the ellipsoid is

$\frac{2}{\sqrt{3}}$, occurring when all coordinates are positive or exactly two are negative and the minimum is $-\frac{2}{\sqrt{3}}$ occurring when 1 or 3 of

the coordinates are negative.

9. $f(x, y, z) = x^2 + y^2 + z^2$, $g(x, y, z) = x^4 + y^4 + z^4 = 1 \implies \nabla f = \langle 2x, 2y, 2z \rangle$, $\lambda \nabla g = \langle 4\lambda x^3, 4\lambda y^3, 4\lambda z^3 \rangle$.

Case 1: If $x \neq 0$, $y \neq 0$ and $z \neq 0$, then $\nabla f = \lambda \nabla g$ implies $\lambda = 1/(2x^2) = 1/(2y^2) = 1/(2z^2)$ or $x^2 = y^2 = z^2$ and

$3x^4 = 1$ or $x = \pm \frac{1}{\sqrt[4]{3}}$ giving the points $\left(\pm \frac{1}{\sqrt[4]{3}}, \frac{1}{\sqrt[4]{3}}, \frac{1}{\sqrt[4]{3}}\right)$, $\left(\pm \frac{1}{\sqrt[4]{3}}, -\frac{1}{\sqrt[4]{3}}, \frac{1}{\sqrt[4]{3}}\right)$, $\left(\pm \frac{1}{\sqrt[4]{3}}, \frac{1}{\sqrt[4]{3}}, -\frac{1}{\sqrt[4]{3}}\right)$, $\left(\pm \frac{1}{\sqrt[4]{3}}, -\frac{1}{\sqrt[4]{3}}, -\frac{1}{\sqrt[4]{3}}\right)$

all with an f-value of $\sqrt{3}$.

Case 2: If one of the variables equals zero and the other two are not zero, then the squares of the two nonzero coordinates are

equal with common value $\frac{1}{\sqrt{2}}$ and corresponding f value of $\sqrt{2}$.

Case 3: If exactly two of the variables are zero, then the third variable has value ± 1 with the corresponding f value of 1. Thus

on $x^4 + y^4 + z^4 = 1$, the maximum value of f is $\sqrt{3}$ and the minimum value is 1.

11. $f(x, y, z, t) = x + y + z + t$, $g(x, y, z, t) = x^2 + y^2 + z^2 + t^2 = 1 \implies \langle 1, 1, 1, 1 \rangle = \langle 2\lambda x, 2\lambda y, 2\lambda z, 2\lambda t \rangle$, so

$\lambda = 1/(2x) = 1/(2y) = 1/(2z) = 1/(2t)$ and $x = y = z = t$. But $x^2 + y^2 + z^2 + t^2 = 1$, so the possible points are

$\left(\pm \frac{1}{2}, \pm \frac{1}{2}, \pm \frac{1}{2}, \pm \frac{1}{2}\right)$. Thus the maximum value of f is $f\left(\frac{1}{2}, \frac{1}{2}, \frac{1}{2}, \frac{1}{2}\right) = 2$ and the minimum value is

$f\left(-\frac{1}{2}, -\frac{1}{2}, -\frac{1}{2}, -\frac{1}{2}\right) = -2$.

13. $f(x, y, z) = x + 2y$, $g(x, y, z) = x + y + z = 1$, $h(x, y, z) = y^2 + z^2 = 4 \implies \nabla f = \langle 1, 2, 0 \rangle$, $\lambda \nabla g = \langle \lambda, \lambda, \lambda \rangle$

and $\mu \nabla h = \langle 0, 2\mu y, 2\mu z \rangle$. Then $1 = \lambda$, $2 = \lambda + 2\mu y$ and $0 = \lambda + 2\mu z$ so $\mu y = \frac{1}{2} = -\mu z$ or $y = 1/(2\mu)$, $z = -1/(2\mu)$.

Thus $x + y + z = 1$ implies $x = 1$ and $y^2 + z^2 = 4$ implies $\mu = \pm \frac{1}{2\sqrt{2}}$. Then the possible points are $\left(1, \pm \sqrt{2}, \mp \sqrt{2}\right)$

and the maximum value is $f\left(1, \sqrt{2}, -\sqrt{2}\right) = 1 + 2\sqrt{2}$ and the minimum value is $f\left(1, -\sqrt{2}, \sqrt{2}\right) = 1 - 2\sqrt{2}$.

15. $f(x, y, z) = yz + xy$, $g(x, y, z) = xy = 1$, $h(x, y, z) = y^2 + z^2 = 1 \implies \nabla f = \langle y, x + z, y \rangle$, $\lambda \nabla g = \langle \lambda y, \lambda x, 0 \rangle$,

$\mu \nabla h = \langle 0, 2\mu y, 2\mu z \rangle$. Then $y = \lambda y$ implies $\lambda = 1$ [$y \neq 0$ since $g(x, y, z) = 1$], $x + z = \lambda x + 2\mu y$ and $y = 2\mu z$. Thus

$\mu = z/(2y) = y/(2y)$ or $y^2 = z^2$, and so $y^2 + z^2 = 1$ implies $y = \pm \frac{1}{\sqrt{2}}$, $z = \pm \frac{1}{\sqrt{2}}$. Then $xy = 1$ implies $x = \pm \sqrt{2}$ and

the possible points are $\left(\pm \sqrt{2}, \pm \frac{1}{\sqrt{2}}, \frac{1}{\sqrt{2}}\right)$, $\left(\pm \sqrt{2}, \pm \frac{1}{\sqrt{2}}, -\frac{1}{\sqrt{2}}\right)$. Hence the maximum of f subject to the constraints is

$f\left(\pm \sqrt{2}, \pm \frac{1}{\sqrt{2}}, \pm \frac{1}{\sqrt{2}}\right) = \frac{3}{2}$ and the minimum is $f\left(\pm \sqrt{2}, \pm \frac{1}{\sqrt{2}}, \mp \frac{1}{\sqrt{2}}\right) = \frac{1}{2}$.

Note: Since $xy = 1$ is one of the constraints we could have solved the problem by solving $f(y, z) = yz + 1$ subject to

$y^2 + z^2 = 1$.

17. $f(x, y) = x^2 + y^2 + 4x - 4y$. For the interior of the region, we find the critical points: $f_x = 2x + 4$, $f_y = 2y - 4$, so the

only critical point is $(-2, 2)$ (which is inside the region) and $f(-2, 2) = -8$. For the boundary, we use Lagrange multipliers.

$g(x, y) = x^2 + y^2 = 9$, so $\nabla f = \lambda \nabla g \Rightarrow \langle 2x + 4, 2y - 4 \rangle = \langle 2\lambda x, 2\lambda y \rangle$. Thus $2x + 4 = 2\lambda x$ and $2y - 4 = 2\lambda y$.

Adding the two equations gives $2x + 2y = 2\lambda x + 2\lambda y \Rightarrow x + y = \lambda(x + y) \Rightarrow (x + y)(\lambda - 1) = 0$, so

$x + y = 0 \Rightarrow y = -x$ or $\lambda - 1 = 0 \Rightarrow \lambda = 1$. But $\lambda = 1$ leads to a contradition in $2x + 4 = 2\lambda x$, so $y = -x$ and

$x^2 + y^2 = 9$ implies $2y^2 = 9 \Rightarrow y = \pm \frac{3}{\sqrt{2}}$. We have $f\left(\frac{3}{\sqrt{2}}, -\frac{3}{\sqrt{2}}\right) = 9 + 12\sqrt{2} \approx 25.97$ and

$f\left(-\frac{3}{\sqrt{2}}, \frac{3}{\sqrt{2}}\right) = 9 - 12\sqrt{2} \approx -7.97$, so the maximum value of f on the disk $x^2 + y^2 \leq 9$ is $f\left(\frac{3}{\sqrt{2}}, -\frac{3}{\sqrt{2}}\right) = 9 + 12\sqrt{2}$

and the minimum is $f(-2, 2) = -8$.

19. $f(x, y) = e^{-xy}$. For the interior of the region, we find the critical points: $f_x = -ye^{-xy}$, $f_y = -xe^{-xy}$, so the only

critical point is $(0, 0)$, and $f(0, 0) = 1$. For the boundary, we use Lagrange multipliers. $g(x, y) = x^2 + 4y^2 = 1 \Rightarrow$

$\lambda \nabla g = \langle 2\lambda x, 8\lambda y \rangle$, so setting $\nabla f = \lambda \nabla g$ we get $-ye^{-xy} = 2\lambda x$ and $-xe^{-xy} = 8\lambda y$. The first of these gives

$e^{-xy} = -2\lambda x/y$, and then the second gives $-x(-2\lambda x/y) = 8\lambda y \Rightarrow x^2 = 4y^2$. Solving this last equation with the

constraint $x^2 + 4y^2 = 1$ gives $x = \pm \frac{1}{\sqrt{2}}$ and $y = \pm \frac{1}{2\sqrt{2}}$. Now $f\left(\pm \frac{1}{\sqrt{2}}, \mp \frac{1}{2\sqrt{2}}\right) = e^{1/4} \approx 1.284$ and

$f\left(\pm \frac{1}{\sqrt{2}}, \pm \frac{1}{2\sqrt{2}}\right) = e^{-1/4} \approx 0.779$. The former are the maxima on the region and the latter are the minima.

21. At the extreme values of f, the level curves of f just touch the curve $g(x, y) = 8$ with a common tangent line. (See Figure 1

and the accompanying discussion.) We can observe several such occurrences on the contour map, but the level curve

$f(x, y) = c$ with the largest value of c which still intersects the curve $g(x, y) = 8$ is approximately $c = 59$, and the smallest

value of c corresponding to a level curve which intersects $g(x, y) = 8$ appears to be $c = 30$. Thus we estimate the maximum

value of f subject to the constraint $g(x, y) = 8$ to be about 59 and the minimum to be 30.

23. (a) $f(x, y) = x$, $g(x, y) = y^2 + x^4 - x^3 = 0 \Rightarrow \nabla f = \langle 1, 0 \rangle = \lambda \nabla g = \lambda \langle 4x^3 - 3x^2, 2y \rangle$. Then

$1 = \lambda(4x^3 - 3x^2)$ **(1)** and $0 = 2\lambda y$ **(2)**. We have $\lambda \neq 0$ from **(1)**, so **(2)** gives $y = 0$. Then, from the constraint equation,

$x^4 - x^3 = 0 \Rightarrow x^3(x - 1) = 0 \Rightarrow x = 0$ or $x = 1$. But $x = 0$ contradicts **(1)**, so the only possible extreme value

subject to the constraint is $f(1, 0) = 1$. (The question remains whether this is indeed the minimum of f.)

(b) The constraint is $y^2 + x^4 - x^3 = 0 \Leftrightarrow y^2 = x^3 - x^4$. The left side is non-negative, so we must have $x^3 - x^4 \geq 0$

which is true only for $0 \leq x \leq 1$. Therefore the minimum possible value for $f(x, y) = x$ is 0 which occurs for $x = y = 0$.

However, $\lambda \nabla g(0, 0) = \lambda \langle 0 - 0, 0 \rangle = \langle 0, 0 \rangle$ and $\nabla f(0, 0) = \langle 1, 0 \rangle$, so $\nabla f(0, 0) \neq \lambda \nabla g(0, 0)$ for all values of λ.

(c) Here $\nabla g(0, 0) = \mathbf{0}$ but the method of Lagrange multipliers requires that $\nabla g \neq \mathbf{0}$ everywhere on the constraint curve.

25. $P(L, K) = bL^\alpha K^{1-\alpha}$, $g(L, K) = mL + nK = p \Rightarrow \nabla P = \langle \alpha b L^{\alpha-1} K^{1-\alpha}, (1 - \alpha) b L^\alpha K^{-\alpha} \rangle$, $\lambda \nabla g = \langle \lambda m, \lambda n \rangle$.

Then $\alpha b(K/L)^{1-\alpha} = \lambda m$ and $(1 - \alpha) b(L/K)^\alpha = \lambda n$ and $mL + nK = p$, so $\alpha b(K/L)^{1-\alpha}/m = (1 - \alpha) b(L/K)^\alpha/n$ or

$n\alpha/[m(1 - \alpha)] = (L/K)^\alpha (L/K)^{1-\alpha}$ or $L = Kn\alpha/[m(1 - \alpha)]$. Substituting into $mL + nK = p$ gives $K = (1 - \alpha)p/n$

and $L = \alpha p/m$ for the maximum production.

27. Let the sides of the rectangle be x and y. Then $f(x, y) = xy$, $g(x, y) = 2x + 2y = p$ $\Rightarrow$ $\nabla f(x, y) = \langle y, x \rangle$, $\lambda \nabla g = \langle 2\lambda, 2\lambda \rangle$. Then $\lambda = \frac{1}{2}y = \frac{1}{2}x$ implies $x = y$ and the rectangle with maximum area is a square with side length $\frac{1}{4}p$.

29. Let $f(x, y, z) = d^2 = (x - 2)^2 + (y - 1)^2 + (z + 1)^2$, then we want to minimize f subject to the constraint $g(x, y, z) = x + y - z = 1$. $\nabla f = \lambda \nabla g$ $\Rightarrow$ $\langle 2(x - 2), 2(y - 1), 2(z + 1) \rangle = \lambda \langle 1, 1, -1 \rangle$, so $x = (\lambda + 4)/2$, $y = (\lambda + 2)/2$, $z = -(\lambda + 2)/2$. Substituting into the constraint equation gives $\frac{\lambda + 4}{2} + \frac{\lambda + 2}{2} + \frac{\lambda + 2}{2} = 1$ $\Rightarrow$ $3\lambda + 8 = 2$ $\Rightarrow$ $\lambda = -2$, so $x = 1$, $y = 0$, and $z = 0$. This must correspond to a minimum, so the shortest distance is $d = \sqrt{(1 - 2)^2 + (0 - 1)^2 + (0 + 1)^2} = \sqrt{3}$.

31. Let $f(x, y, z) = d^2 = (x - 4)^2 + (y - 2)^2 + z^2$. Then we want to minimize f subject to the constraint $g(x, y, z) = x^2 + y^2 - z^2 = 0$. $\nabla f = \lambda \nabla g$ $\Rightarrow$ $\langle 2(x - 4), 2(y - 2), 2z \rangle = \langle 2\lambda x, 2\lambda y, -2\lambda z \rangle$, so $x - 4 = \lambda x$, $y - 2 = \lambda y$, and $z = -\lambda z$. From the last equation we have $z + \lambda z = 0$ $\Rightarrow$ $z(1 + \lambda) = 0$, so either $z = 0$ or $\lambda = -1$. But from the constraint equation we have $z = 0$ $\Rightarrow$ $x^2 + y^2 = 0$ $\Rightarrow$ $x = y = 0$ which is not possible from the first two equations. So $\lambda = -1$ and $x - 4 = \lambda x$ $\Rightarrow$ $x = 2$, $y - 2 = \lambda y$ $\Rightarrow$ $y = 1$, and $x^2 + y^2 - z^2 = 0$ $\Rightarrow$ $4 + 1 - z^2 = 0$ $\Rightarrow$ $z = \pm\sqrt{5}$. This must correspond to a minimum, so the points on the cone closest to $(4, 2, 0)$ are $\left(2, 1, \pm\sqrt{5}\right)$.

33. $f(x, y, z) = xyz$, $g(x, y, z) = x + y + z = 100$ $\Rightarrow$ $\nabla f = \langle yz, xz, xy \rangle = \lambda \nabla g = \langle \lambda, \lambda, \lambda \rangle$. Then $\lambda = yz = xz = xy$ implies $x = y = z = \frac{100}{3}$.

35. If the dimensions are $2x$, $2y$, and $2z$, then maximize $f(x, y, z) = (2x)(2y)(2z) = 8xyz$ subject to $g(x, y, z) = x^2 + y^2 + z^2 = r^2$ ($x > 0$, $y > 0$, $z > 0$). Then $\nabla f = \lambda \nabla g$ $\Rightarrow$ $\langle 8yz, 8xz, 8xy \rangle = \lambda \langle 2x, 2y, 2z \rangle$ $\Rightarrow$ $8yz = 2\lambda x$, $8xz = 2\lambda y$, and $8xy = 2\lambda z$, so $\lambda = \frac{4yz}{x} = \frac{4xz}{y} = \frac{4xy}{z}$. This gives $x^2 z = y^2 z$ $\Rightarrow$ $x^2 = y^2$ (since $z \neq 0$) and $xy^2 = xz^2$ $\Rightarrow$ $z^2 = y^2$, so $x^2 = y^2 = z^2$ $\Rightarrow$ $x = y = z$, and substituting into the constraint equation gives $3x^2 = r^2$ $\Rightarrow$ $x = r/\sqrt{3} = y = z$. Thus the largest volume of such a box is $f\left(\frac{r}{\sqrt{3}}, \frac{r}{\sqrt{3}}, \frac{r}{\sqrt{3}}\right) = 8\left(\frac{r}{\sqrt{3}}\right)\left(\frac{r}{\sqrt{3}}\right)\left(\frac{r}{\sqrt{3}}\right) = \frac{8}{3\sqrt{3}}r^3$.

37. $f(x, y, z) = xyz$, $g(x, y, z) = x + 2y + 3z = 6$ $\Rightarrow$ $\nabla f = \langle yz, xz, xy \rangle = \lambda \nabla g = \langle \lambda, 2\lambda, 3\lambda \rangle$. Then $\lambda = yz = \frac{1}{2}xz = \frac{1}{3}xy$ implies $x = 2y$, $z = \frac{2}{3}y$. But $2y + 2y + 2y = 6$ so $y = 1$, $x = 2$, $z = \frac{2}{3}$ and the volume is $V = \frac{4}{3}$.

39. $f(x, y, z) = xyz$, $g(x, y, z) = 4(x + y + z) = c$ $\Rightarrow$ $\nabla f = \langle yz, xz, xy \rangle$, $\lambda \nabla g = \langle 4\lambda, 4\lambda, 4\lambda \rangle$. Thus $4\lambda = yz = xz = xy$ or $x = y = z = \frac{1}{12}c$ are the dimensions giving the maximum volume.

41. If the dimensions of the box are given by x, y, and z, then we need to find the maximum value of $f(x, y, z) = xyz$ $[x, y, z > 0]$ subject to the constraint $L = \sqrt{x^2 + y^2 + z^2}$ or $g(x, y, z) = x^2 + y^2 + z^2 = L^2$. $\nabla f = \lambda \nabla g$ $\Rightarrow$ $\langle yz, xz, xy \rangle = \lambda \langle 2x, 2y, 2z \rangle$, so $yz = 2\lambda x$ $\Rightarrow$ $\lambda = \frac{yz}{2x}$, $xz = 2\lambda y$ $\Rightarrow$ $\lambda = \frac{xz}{2y}$, and $xy = 2\lambda z$ $\Rightarrow$ $\lambda = \frac{xy}{2z}$.

[continued]

Thus $\lambda = \dfrac{yz}{2x} = \dfrac{xz}{2y}$ $\Rightarrow$ $x^2 = y^2$ [since $z \neq 0$] $\Rightarrow$ $x = y$ and $\lambda = \dfrac{yz}{2x} = \dfrac{xy}{2z}$ $\Rightarrow$ $x = z$ [since $y \neq 0$].

Substituting into the constraint equation gives $x^2 + x^2 + x^2 = L^2$ $\Rightarrow$ $x^2 = L^2/3$ $\Rightarrow$ $x = L/\sqrt{3} = y = z$ and the

maximum volume is $\left(L/\sqrt{3}\right)^3 = L^3/\left(3\sqrt{3}\right)$.

43. We need to find the extreme values of $f(x, y, z) = x^2 + y^2 + z^2$ subject to the two constraints $g(x, y, z) = x + y + 2z = 2$

and $h(x, y, z) = x^2 + y^2 - z = 0$. $\nabla f = \langle 2x, 2y, 2z \rangle$, $\lambda \nabla g = \langle \lambda, \lambda, 2\lambda \rangle$ and $\mu \nabla h = \langle 2\mu x, 2\mu y, -\mu \rangle$. Thus we need

$2x = \lambda + 2\mu x$ **(1)**, $2y = \lambda + 2\mu y$ **(2)**, $2z = 2\lambda - \mu$ **(3)**, $x + y + 2z = 2$ **(4)**, and $x^2 + y^2 - z = 0$ **(5)**.

From **(1)** and **(2)**, $2(x - y) = 2\mu(x - y)$, so if $x \neq y$, $\mu = 1$. Putting this in **(3)** gives $2z = 2\lambda - 1$ or $\lambda = z + \frac{1}{2}$, but putting

$\mu = 1$ into **(1)** says $\lambda = 0$. Hence $z + \frac{1}{2} = 0$ or $z = -\frac{1}{2}$. Then **(4)** and **(5)** become $x + y - 3 = 0$ and $x^2 + y^2 + \frac{1}{2} = 0$. The

last equation cannot be true, so this case gives no solution. So we must have $x = y$. Then **(4)** and **(5)** become $2x + 2z = 2$ and

$2x^2 - z = 0$ which imply $z = 1 - x$ and $z = 2x^2$. Thus $2x^2 = 1 - x$ or $2x^2 + x - 1 = (2x - 1)(x + 1) = 0$ so $x = \frac{1}{2}$ or

$x = -1$. The two points to check are $\left(\frac{1}{2}, \frac{1}{2}, \frac{1}{2}\right)$ and $(-1, -1, 2)$: $f\left(\frac{1}{2}, \frac{1}{2}, \frac{1}{2}\right) = \frac{3}{4}$ and $f(-1, -1, 2) = 6$. Thus $\left(\frac{1}{2}, \frac{1}{2}, \frac{1}{2}\right)$ is

the point on the ellipse nearest the origin and $(-1, -1, 2)$ is the one farthest from the origin.

45. $f(x, y, z) = ye^{x-z}$, $g(x, y, z) = 9x^2 + 4y^2 + 36z^2 = 36$, $h(x, y, z) = xy + yz = 1$. $\nabla f = \lambda \nabla g + \mu \nabla h$ $\Rightarrow$

$\left\langle ye^{x-z}, e^{x-z}, -ye^{x-z} \right\rangle = \lambda \langle 18x, 8y, 72z \rangle + \mu \langle y, x + z, y \rangle$, so $ye^{x-z} = 18\lambda x + \mu y$, $e^{x-z} = 8\lambda y + \mu(x + z)$,

$-ye^{x-z} = 72\lambda z + \mu y$, $9x^2 + 4y^2 + 36z^2 = 36$, $xy + yz = 1$. Using a CAS to solve these 5 equations simultaneously for x,

y, z, λ, and μ (in Maple, use the `allvalues` command), we get 4 real-valued solutions:

$$x \approx 0.222444, \quad y \approx -2.157012, \quad z \approx -0.686049, \quad \lambda \approx -0.200401, \quad \mu \approx 2.108584$$

$$x \approx -1.951921, \quad y \approx -0.545867, \quad z \approx 0.119973, \quad \lambda \approx 0.003141, \quad \mu \approx -0.076238$$

$$x \approx 0.155142, \quad y \approx 0.904622, \quad z \approx 0.950293, \quad \lambda \approx -0.012447, \quad \mu \approx 0.489938$$

$$x \approx 1.138731, \quad y \approx 1.768057, \quad z \approx -0.573138, \quad \lambda \approx 0.317141, \quad \mu \approx 1.862675$$

Substituting these values into f gives $f(0.222444, -2.157012, -0.686049) \approx -5.3506$,

$f(-1.951921, -0.545867, 0.119973) \approx -0.0688$, $f(0.155142, 0.904622, 0.950293) \approx 0.4084$,

$f(1.138731, 1.768057, -0.573138) \approx 9.7938$. Thus the maximum is approximately 9.7938, and the minimum is

approximately -5.3506.

47. (a) We wish to maximize $f(x_1, x_2, \ldots, x_n) = \sqrt[n]{x_1 x_2 \cdots x_n}$ subject to

$g(x_1, x_2, \ldots, x_n) = x_1 + x_2 + \cdots + x_n = c$ and $x_i > 0$.

$\nabla f = \left\langle \frac{1}{n}(x_1 x_2 \cdots x_n)^{\frac{1}{n} - 1}(x_2 \cdots x_n), \frac{1}{n}(x_1 x_2 \cdots x_n)^{\frac{1}{n} - 1}(x_1 x_3 \cdots x_n), \ldots, \frac{1}{n}(x_1 x_2 \cdots x_n)^{\frac{1}{n} - 1}(x_1 \cdots x_{n-1}) \right\rangle$

and $\lambda \nabla g = \langle \lambda, \lambda, \ldots, \lambda \rangle$, so we need to solve the system of equations

$$\frac{1}{n}(x_1 x_2 \cdots x_n)^{\frac{1}{n} - 1}(x_2 \cdots x_n) = \lambda \quad \Rightarrow \quad x_1^{1/n} x_2^{1/n} \cdots x_n^{1/n} = n\lambda x_1$$

$$\frac{1}{n}(x_1 x_2 \cdots x_n)^{\frac{1}{n} - 1}(x_1 x_3 \cdots x_n) = \lambda \quad \Rightarrow \quad x_1^{1/n} x_2^{1/n} \cdots x_n^{1/n} = n\lambda x_2$$

$$\vdots$$

$$\frac{1}{n}(x_1 x_2 \cdots x_n)^{\frac{1}{n} - 1}(x_1 \cdots x_{n-1}) = \lambda \quad \Rightarrow \quad x_1^{1/n} x_2^{1/n} \cdots x_n^{1/n} = n\lambda x_n$$

This implies $n\lambda x_1 = n\lambda x_2 = \cdots = n\lambda x_n$. Note $\lambda \neq 0$, otherwise we can't have all $x_i > 0$. Thus $x_1 = x_2 = \cdots = x_n$.

But $x_1 + x_2 + \cdots + x_n = c \implies nx_1 = c \implies x_1 = \dfrac{c}{n} = x_2 = x_3 = \cdots = x_n$. Then the only point where f can

have an extreme value is $\left(\dfrac{c}{n}, \dfrac{c}{n}, \ldots, \dfrac{c}{n}\right)$. Since we can choose values for $(x_1, x_2, \ldots, x_n)$ that make f as close to

zero (but not equal) as we like, f has no minimum value. Thus the maximum value is

$$f\left(\frac{c}{n}, \frac{c}{n}, \ldots, \frac{c}{n}\right) = \sqrt[n]{\frac{c}{n} \cdot \frac{c}{n} \cdots \frac{c}{n}} = \frac{c}{n}.$$

(b) From part (a), $\dfrac{c}{n}$ is the maximum value of f. Thus $f(x_1, x_2, \ldots, x_n) = \sqrt[n]{x_1 x_2 \cdots x_n} \leq \dfrac{c}{n}$. But

$x_1 + x_2 + \cdots + x_n = c$, so $\sqrt[n]{x_1 x_2 \cdots x_n} \leq \dfrac{x_1 + x_2 + \cdots + x_n}{n}$. These two means are equal when f attains its

maximum value $\dfrac{c}{n}$, but this can occur only at the point $\left(\dfrac{c}{n}, \dfrac{c}{n}, \ldots, \dfrac{c}{n}\right)$ we found in part (a). So the means are equal only

when $x_1 = x_2 = x_3 = \cdots = x_n = \dfrac{c}{n}$.

11   Review

CONCEPT CHECK

1. (a) A function f of two variables is a rule that assigns to each ordered pair (x, y) of real numbers in its domain a unique real

   number denoted by $f(x, y)$.

   (b) One way to visualize a function of two variables is by graphing it, resulting in the surface $z = f(x, y)$. Another method for

   visualizing a function of two variables is a contour map. The contour map consists of level curves of the function which are

   horizontal traces of the graph of the function projected onto the xy-plane. Also, we can use an arrow diagram such as

   Figure 1 in Section 11.1.

2. A function f of three variables is a rule that assigns to each ordered triple (x, y, z) in its domain a unique real number

   $f(x, y, z)$. We can visualize a function of three variables by examining its level surfaces $f(x, y, z) = k$, where k is a constant.

3. $\displaystyle \lim_{(x,y) \to (a,b)} f(x, y) = L$ means the values of $f(x, y)$ approach the number L as the point (x, y) approaches the point (a, b)

   along any path that is within the domain of f. We can show that a limit at a point does not exist by finding two different paths

   approaching the point along which $f(x, y)$ has different limits.

4. (a) See Definition 11.2.4.

   (b) If f is continuous on $\mathbb{R}^2$, its graph will appear as a surface without holes or breaks.

5. (a) See (2) and (3) in Section 11.3.

   (b) See "Interpretations of Partial Derivatives" on page 640.

   (c) To find f_x, regard y as a constant and differentiate $f(x, y)$ with respect to x. To find f_y, regard x as a constant and

   differentiate $f(x, y)$ with respect to y.

6. See the statement of Clairaut's Theorem on page 643.

7. (a) See (2) in Section 11.4.

(b) See (19) and the preceding discussion in Section 11.6.

8. See (3) and (4) and the accompanying discussion in Section 11.4. We can interpret the linearization of f at (a, b) geometrically as the linear function whose graph is the tangent plane to the graph of f at (a, b). Thus it is the linear function which best approximates f near (a, b).

9. (a) See Definition 11.4.7.

(b) Use Theorem 11.4.8.

10. See (10) and the associated discussion in Section 11.4.

11. See (2) and (3) in Section 11.5.

12. See (7) and the preceding discussion in Section 11.5.

13. (a) See Definition 11.6.2. We can interpret it as the rate of change of f at (x_0, y_0) in the direction of $\mathbf{u}$. Geometrically, if P is the point $(x_0, y_0, f(x_0, y_0))$ on the graph of f and C is the curve of intersection of the graph of f with the vertical plane that passes through P in the direction $\mathbf{u}$, the directional derivative of f at (x_0, y_0) in the direction of $\mathbf{u}$ is the slope of the tangent line to C at P. (See Figure 2 in Section 11.6 .)

(b) See Theorem 11.6.3.

14. (a) See (8) and (13) in Section 11.6.

(b) $D_{\mathbf{u}} f(x, y) = \nabla f(x, y) \cdot \mathbf{u}$ or $D_{\mathbf{u}} f(x, y, z) = \nabla f(x, y, z) \cdot \mathbf{u}$

(c) The gradient vector of a function points in the direction of maximum rate of increase of the function. On a graph of the function, the gradient points in the direction of steepest ascent.

15. (a) f has a local maximum at (a, b) if $f(x, y) \le f(a, b)$ when (x, y) is near (a, b).

(b) f has an absolute maximum at (a, b) if $f(x, y) \le f(a, b)$ for all points (x, y) in the domain of f.

(c) f has a local minimum at (a, b) if $f(x, y) \ge f(a, b)$ when (x, y) is near (a, b).

(d) f has an absolute minimum at (a, b) if $f(x, y) \ge f(a, b)$ for all points (x, y) in the domain of f.

(e) f has a saddle point at (a, b) if $f(a, b)$ is a local maximum in one direction but a local minimum in another.

16. (a) By Theorem 11.7.2, if f has a local maximum at (a, b) and the first-order partial derivatives of f exist there, then $f_x(a, b) = 0$ and $f_y(a, b) = 0$.

(b) A critical point of f is a point (a, b) such that $f_x(a, b) = 0$ and $f_y(a, b) = 0$ or one of these partial derivatives does not exist.

17. See (3) in Section 11.7.

18. (a) See Figure 7 and the accompanying discussion in Section 11.7.

(b) See Theorem 11.7.4.

(c) See the procedure outlined in (5) in Section 11.7.

19. See the discussion beginning on page 684; see "Two Constraints" on page 687.

TRUE-FALSE QUIZ

1. True. $f_y(a, b) = \lim\limits_{h \to 0} \dfrac{f(a, b + h) - f(a, b)}{h}$ from Equation 11.3.3. Let $h = y - b$. As $h \to 0$, $y \to b$. Then by substituting,

we get $f_y(a, b) = \lim\limits_{y \to b} \dfrac{f(a, y) - f(a, b)}{y - b}$.

3. False. $f_{xy} = \dfrac{\partial^2 f}{\partial y \, \partial x}$.

5. False. See Example 11.2.3.

7. True. If f has a local minimum and f is differentiable at (a, b) then by Theorem 11.7.2, $f_x(a, b) = 0$ and $f_y(a, b) = 0$, so
$\nabla f(a, b) = \langle f_x(a, b), f_y(a, b) \rangle = \langle 0, 0 \rangle = \mathbf{0}$.

9. False. $\nabla f(x, y) = \langle 0, 1/y \rangle$.

11. True. $\nabla f = \langle \cos x, \cos y \rangle$, so $|\nabla f| = \sqrt{\cos^2 x + \cos^2 y}$. But $|\cos \theta| \le 1$, so $|\nabla f| \le \sqrt{2}$. Now
$D_\mathbf{u} f(x, y) = \nabla f \cdot \mathbf{u} = |\nabla f| \, |\mathbf{u}| \cos \theta$, but $\mathbf{u}$ is a unit vector, so $|D_\mathbf{u} f(x, y)| \le \sqrt{2} \cdot 1 \cdot 1 = \sqrt{2}$.

EXERCISES

1. $\ln(x + y + 1)$ is defined only when $x + y + 1 > 0 \quad \Leftrightarrow \quad y > -x - 1$,
so the domain of f is $\{(x, y) \mid y > -x - 1\}$, all those points above the
line $y = -x - 1$.

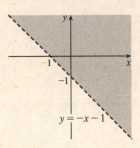

3. $z = f(x, y) = 1 - y^2$, a parabolic
cylinder

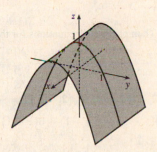

5. The level curves are $\sqrt{4x^2 + y^2} = k$ or $4x^2 + y^2 = k^2$, $k \ge 0$,
a family of ellipses.

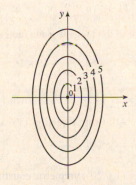

7.

9. f is a rational function, so it is continuous on its domain. Since f is defined at $(1, 1)$, we use direct substitution to evaluate the limit:

$$\lim_{(x,y)\to(1,1)} \frac{2xy}{x^2 + 2y^2} = \frac{2(1)(1)}{1^2 + 2(1)^2} = \frac{2}{3}.$$

11. $f(x,y) = (5y^3 + 2x^2y)^8 \quad\Rightarrow\quad f_x = 8(5y^3 + 2x^2y)^7(4xy) = 32xy(5y^3 + 2x^2y)^7,$

$f_y = 8(5y^3 + 2x^2y)^7(15y^2 + 2x^2) = (16x^2 + 120y^2)(5y^3 + 2x^2y)^7$

13. $F(\alpha,\beta) = \alpha^2 \ln(\alpha^2 + \beta^2) \quad\Rightarrow\quad F_\alpha = \alpha^2 \cdot \dfrac{1}{\alpha^2 + \beta^2}\,(2\alpha) + \ln(\alpha^2 + \beta^2) \cdot 2\alpha = \dfrac{2\alpha^3}{\alpha^2 + \beta^2} + 2\alpha \ln(\alpha^2 + \beta^2),$

$F_\beta = \alpha^2 \cdot \dfrac{1}{\alpha^2 + \beta^2}\,(2\beta) = \dfrac{2\alpha^2\beta}{\alpha^2 + \beta^2}$

15. $S(u,v,w) = u\arctan(v\sqrt{w}) \quad\Rightarrow\quad S_u = \arctan(v\sqrt{w}),\ S_v = u \cdot \dfrac{1}{1 + (v\sqrt{w})^2}\,(\sqrt{w}) = \dfrac{u\sqrt{w}}{1 + v^2w},$

$S_w = u \cdot \dfrac{1}{1 + (v\sqrt{w})^2}\left(v \cdot \tfrac{1}{2}w^{-1/2}\right) = \dfrac{uv}{2\sqrt{w}\,(1 + v^2w)}$

17. $f(x,y) = 4x^3 - xy^2 \quad\Rightarrow\quad f_x = 12x^2 - y^2,\ f_y = -2xy,\ f_{xx} = 24x,\ f_{yy} = -2x,\ f_{xy} = f_{yx} = -2y$

19. $f(x,y,z) = x^k y^l z^m \quad\Rightarrow\quad f_x = kx^{k-1}y^l z^m,\ f_y = lx^k y^{l-1} z^m,\ f_z = mx^k y^l z^{m-1},\ f_{xx} = k(k-1)x^{k-2}y^l z^m,$

$f_{yy} = l(l-1)x^k y^{l-2} z^m,\ f_{zz} = m(m-1)x^k y^l z^{m-2},\ f_{xy} = f_{yx} = klx^{k-1}y^{l-1}z^m,\ f_{xz} = f_{zx} = kmx^{k-1}y^l z^{m-1},$

$f_{yz} = f_{zy} = lmx^k y^{l-1} z^{m-1}$

21. $z = xy + xe^{y/x} \quad\Rightarrow\quad \dfrac{\partial z}{\partial x} = y - \dfrac{y}{x}e^{y/x} + e^{y/x},\ \dfrac{\partial z}{\partial y} = x + e^{y/x}$ and

$x\dfrac{\partial z}{\partial x} + y\dfrac{\partial z}{\partial y} = x\left(y - \dfrac{y}{x}e^{y/x} + e^{y/x}\right) + y\left(x + e^{y/x}\right) = xy - ye^{y/x} + xe^{y/x} + xy + ye^{y/x} = xy + xy + xe^{y/x} = xy + z.$

23. (a) $z_x = 6x + 2 \quad\Rightarrow\quad z_x(1, -2) = 8$ and $z_y = -2y \quad\Rightarrow\quad z_y(1, -2) = 4$, so an equation of the tangent plane is

$z - 1 = 8(x - 1) + 4(y + 2)$ or $z = 8x + 4y + 1$.

(b) A normal vector to the tangent plane (and the surface) at $(1, -2, 1)$ is $\langle 8, 4, -1\rangle$. Then parametric equations for the normal

line there are $x = 1 + 8t$, $y = -2 + 4t$, $z = 1 - t$, and symmetric equations are $\dfrac{x-1}{8} = \dfrac{y+2}{4} = \dfrac{z-1}{-1}$.

25. (a) Let $F(x,y,z) = x^2 + 2y^2 - 3z^2$. Then $F_x = 2x$, $F_y = 4y$, $F_z = -6z$, so $F_x(2, -1, 1) = 4$, $F_y(2, -1, 1) = -4$,

$F_z(2, -1, 1) = -6$. From Equation 11.6.19, an equation of the tangent plane is $4(x - 2) - 4(y + 1) - 6(z - 1) = 0$

or, equivalently, $2x - 2y - 3z = 3$.

(b) From Equations 11.6.20, symmetric equations for the normal line are $\dfrac{x-2}{4} = \dfrac{y+1}{-4} = \dfrac{z-1}{-6}$.

27. (a) Let $F(x, y, z) = x + 2y + 3z - \sin(xyz)$. Then $F_x = 1 - yz\cos(xyz)$, $F_y = 2 - xz\cos(xyz)$, $F_z = 3 - xy\cos(xyz)$,

so $F_x(2, -1, 0) = 1$, $F_y(2, -1, 0) = 2$, $F_z(2, -1, 0) = 5$. From Equation 11.6.19, an equation of the tangent plane is

$1(x - 2) + 2(y + 1) + 5(z - 0) = 0$ or $x + 2y + 5z = 0$.

(b) From Equations 11.6.20, symmetric equations for the normal line are $\dfrac{x - 2}{1} = \dfrac{y + 1}{2} = \dfrac{z}{5}$.

29. The hyperboloid is a level surface of the function $F(x, y, z) = x^2 + 4y^2 - z^2$, so a normal vector to the surface at (x_0, y_0, z_0)

is $\nabla F(x_0, y_0, z_0) = \langle 2x_0, 8y_0, -2z_0 \rangle$. A normal vector for the plane $2x + 2y + z = 5$ is $\langle 2, 2, 1 \rangle$. For the planes to be

parallel, we need the normal vectors to be parallel, so $\langle 2x_0, 8y_0, -2z_0 \rangle = k\langle 2, 2, 1 \rangle$, or $x_0 = k$, $y_0 = \frac{1}{4}k$, and $z_0 = -\frac{1}{2}k$.

But $x_0^2 + 4y_0^2 - z_0^2 = 4 \Rightarrow k^2 + \frac{1}{4}k^2 - \frac{1}{4}k^2 = 4 \Rightarrow k^2 = 4 \Rightarrow k = \pm 2$. So there are two such points:

$\left(2, \frac{1}{2}, -1\right)$ and $\left(-2, -\frac{1}{2}, 1\right)$.

31. $f(x, y, z) = x^3\sqrt{y^2 + z^2} \Rightarrow f_x(x, y, z) = 3x^2\sqrt{y^2 + z^2}$, $f_y(x, y, z) = \dfrac{yx^3}{\sqrt{y^2 + z^2}}$, $f_z(x, y, z) = \dfrac{zx^3}{\sqrt{y^2 + z^2}}$,

so $f(2, 3, 4) = 8(5) = 40$, $f_x(2, 3, 4) = 3(4)\sqrt{25} = 60$, $f_y(2, 3, 4) = \frac{3(8)}{\sqrt{25}} = \frac{24}{5}$, and $f_z(2, 3, 4) = \frac{4(8)}{\sqrt{25}} = \frac{32}{5}$. Then the

linear approximation of f at $(2, 3, 4)$ is

$$f(x, y, z) \approx f(2, 3, 4) + f_x(2, 3, 4)(x - 2) + f_y(2, 3, 4)(y - 3) + f_z(2, 3, 4)(z - 4)$$

$$= 40 + 60(x - 2) + \tfrac{24}{5}(y - 3) + \tfrac{32}{5}(z - 4) = 60x + \tfrac{24}{5}y + \tfrac{32}{5}z - 120$$

Then $(1.98)^3\sqrt{(3.01)^2 + (3.97)^2} = f(1.98, 3.01, 3.97) \approx 60(1.98) + \frac{24}{5}(3.01) + \frac{32}{5}(3.97) - 120 = 38.656$.

33. $\dfrac{du}{dp} = \dfrac{\partial u}{\partial x}\dfrac{dx}{dp} + \dfrac{\partial u}{\partial y}\dfrac{dy}{dp} + \dfrac{\partial u}{\partial z}\dfrac{dz}{dp} = 2xy^3(1 + 6p) + 3x^2y^2(pe^p + e^p) + 4z^3(p\cos p + \sin p)$

35. By the Chain Rule, $\dfrac{\partial z}{\partial s} = \dfrac{\partial z}{\partial x}\dfrac{\partial x}{\partial s} + \dfrac{\partial z}{\partial y}\dfrac{\partial y}{\partial s}$. When $s = 1$ and $t = 2$, $x = g(1, 2) = 3$ and $y = h(1, 2) = 6$, so

$\dfrac{\partial z}{\partial s} = f_x(3, 6)g_s(1, 2) + f_y(3, 6)h_s(1, 2) = (7)(-1) + (8)(-5) = -47$. Similarly, $\dfrac{\partial z}{\partial t} = \dfrac{\partial z}{\partial x}\dfrac{\partial x}{\partial t} + \dfrac{\partial z}{\partial y}\dfrac{\partial y}{\partial t}$, so

$\dfrac{\partial z}{\partial t} = f_x(3, 6)g_t(1, 2) + f_y(3, 6)h_t(1, 2) = (7)(4) + (8)(10) = 108$.

37. $\dfrac{\partial z}{\partial x} = 2xf'(x^2 - y^2)$, $\dfrac{\partial z}{\partial y} = 1 - 2yf'(x^2 - y^2)$ $\left[\text{where } f' = \dfrac{df}{d(x^2 - y^2)}\right]$. Then

$y\dfrac{\partial z}{\partial x} + x\dfrac{\partial z}{\partial y} = 2xyf'(x^2 - y^2) + x - 2xyf'(x^2 - y^2) = x$.

39. $\dfrac{\partial z}{\partial x} = \dfrac{\partial z}{\partial u}y + \dfrac{\partial z}{\partial v}\dfrac{-y}{x^2}$ and

$$\dfrac{\partial^2 z}{\partial x^2} = y\dfrac{\partial}{\partial x}\left(\dfrac{\partial z}{\partial u}\right) + \dfrac{2y}{x^3}\dfrac{\partial z}{\partial v} + \dfrac{-y}{x^2}\dfrac{\partial}{\partial x}\left(\dfrac{\partial z}{\partial v}\right) = \dfrac{2y}{x^3}\dfrac{\partial z}{\partial v} + y\left(\dfrac{\partial^2 z}{\partial u^2}y + \dfrac{\partial^2 z}{\partial v\, \partial u}\dfrac{-y}{x^2}\right) + \dfrac{-y}{x^2}\left(\dfrac{\partial^2 z}{\partial v^2}\dfrac{-y}{x^2} + \dfrac{\partial^2 z}{\partial u\, \partial v}y\right)$$

$$= \dfrac{2y}{x^3}\dfrac{\partial z}{\partial v} + y^2\dfrac{\partial^2 z}{\partial u^2} - \dfrac{2y^2}{x^2}\dfrac{\partial^2 z}{\partial u\, \partial v} + \dfrac{y^2}{x^4}\dfrac{\partial^2 z}{\partial v^2}$$

Also $\dfrac{\partial z}{\partial y} = x\dfrac{\partial z}{\partial u} + \dfrac{1}{x}\dfrac{\partial z}{\partial v}$ and

$$\frac{\partial^2 z}{\partial y^2} = x \frac{\partial}{\partial y}\left(\frac{\partial z}{\partial u}\right) + \frac{1}{x}\frac{\partial}{\partial y}\left(\frac{\partial z}{\partial v}\right) = x\left(\frac{\partial^2 z}{\partial u^2}x + \frac{\partial^2 z}{\partial v\,\partial u}\frac{1}{x}\right) + \frac{1}{x}\left(\frac{\partial^2 z}{\partial v^2}\frac{1}{x} + \frac{\partial^2 z}{\partial u\,\partial v}x\right) = x^2\frac{\partial^2 z}{\partial u^2} + 2\frac{\partial^2 z}{\partial u\,\partial v} + \frac{1}{x^2}\frac{\partial^2 z}{\partial v^2}$$

Thus

$$x^2\frac{\partial^2 z}{\partial x^2} - y^2\frac{\partial^2 z}{\partial y^2} = \frac{2y}{x}\frac{\partial z}{\partial v} + x^2 y^2\frac{\partial^2 z}{\partial u^2} - 2y^2\frac{\partial^2 z}{\partial u\,\partial v} + \frac{y^2}{x^2}\frac{\partial^2 z}{\partial v^2} - x^2 y^2\frac{\partial^2 z}{\partial u^2} - 2y^2\frac{\partial^2 z}{\partial u\,\partial v} - \frac{y^2}{x^2}\frac{\partial^2 z}{\partial v^2}$$

$$= \frac{2y}{x}\frac{\partial z}{\partial v} - 4y^2\frac{\partial^2 z}{\partial u\,\partial v} = 2v\frac{\partial z}{\partial v} - 4uv\frac{\partial^2 z}{\partial u\,\partial v}$$

since $y = xv = \dfrac{uv}{y}$ or $y^2 = uv$.

41. $f(x, y, z) = x^2 e^{yz^2} \Rightarrow \nabla f = \langle f_x, f_y, f_z\rangle = \left\langle 2xe^{yz^2}, x^2 e^{yz^2}\cdot z^2, x^2 e^{yz^2}\cdot 2yz\right\rangle = \left\langle 2xe^{yz^2}, x^2 z^2 e^{yz^2}, 2x^2 yz e^{yz^2}\right\rangle$

43. $f(x, y) = x^2 e^{-y} \Rightarrow \nabla f = \langle 2xe^{-y}, -x^2 e^{-y}\rangle$, $\nabla f(-2, 0) = \langle -4, -4\rangle$. The direction is given by $\langle 4, -3\rangle$, so

$\mathbf{u} = \dfrac{1}{\sqrt{4^2+(-3)^2}}\langle 4, -3\rangle = \frac{1}{5}\langle 4, -3\rangle$ and $D_{\mathbf{u}} f(-2, 0) = \nabla f(-2, 0)\cdot\mathbf{u} = \langle -4, -4\rangle\cdot\frac{1}{5}\langle 4, -3\rangle = \frac{1}{5}(-16+12) = -\frac{4}{5}$.

45. $\nabla f = \left\langle 2xy, x^2 + 1/(2\sqrt{y})\right\rangle$, $|\nabla f(2, 1)| = \left|\langle 4, \frac{9}{2}\rangle\right|$. Thus the maximum rate of change of f at $(2, 1)$ is $\frac{\sqrt{145}}{2}$ in the

direction $\langle 4, \frac{9}{2}\rangle$.

47. $f(x, y) = x^2 - xy + y^2 + 9x - 6y + 10 \Rightarrow f_x = 2x - y + 9$,

$f_y = -x + 2y - 6$, $f_{xx} = 2 = f_{yy}$, $f_{xy} = -1$. Then $f_x = 0$ and $f_y = 0$ imply

$y = 1$, $x = -4$. Thus the only critical point is $(-4, 1)$ and $f_{xx}(-4, 1) > 0$,

$D(-4, 1) = 3 > 0$, so $f(-4, 1) = -11$ is a local minimum.

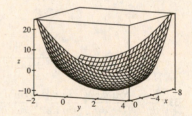

49. $f(x, y) = 3xy - x^2 y - xy^2 \Rightarrow f_x = 3y - 2xy - y^2$, $f_y = 3x - x^2 - 2xy$,

$f_{xx} = -2y$, $f_{yy} = -2x$, $f_{xy} = 3 - 2x - 2y$. Then $f_x = 0$ implies

$y(3 - 2x - y) = 0$ so $y = 0$ or $y = 3 - 2x$. Substituting into $f_y = 0$ implies

$x(3 - x) = 0$ or $3x(-1 + x) = 0$. Hence the critical points are $(0, 0)$, $(3, 0)$,

$(0, 3)$ and $(1, 1)$. $D(0, 0) = D(3, 0) = D(0, 3) = -9 < 0$ so $(0, 0)$, $(3, 0)$, and

$(0, 3)$ are saddle points. $D(1, 1) = 3 > 0$ and $f_{xx}(1, 1) = -2 < 0$, so

$f(1, 1) = 1$ is a local maximum.

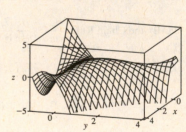

51. First solve inside D. Here $f_x = 4y^2 - 2xy^2 - y^3$, $f_y = 8xy - 2x^2 y - 3xy^2$.

Then $f_x = 0$ implies $y = 0$ or $y = 4 - 2x$, but $y = 0$ isn't inside D. Substituting

$y = 4 - 2x$ into $f_y = 0$ implies $x = 0$, $x = 2$ or $x = 1$, but $x = 0$ isn't inside D,

and when $x = 2$, $y = 0$ but $(2, 0)$ isn't inside D. Thus the only critical point inside

D is $(1, 2)$ and $f(1, 2) = 4$. Secondly we consider the boundary of D.

On L_1: $f(x, 0) = 0$ and so $f = 0$ on L_1. On L_2: $x = -y + 6$ and

$f(-y + 6, y) = y^2(6 - y)(-2) = -2(6y^2 - y^3)$ which has critical points

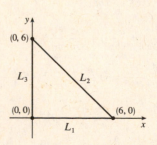

at $y = 0$ and $y = 4$. Then $f(6, 0) = 0$ while $f(2, 4) = -64$. On L_3: $f(0, y) = 0$, so $f = 0$ on L_3. Thus on D the absolute maximum of f is $f(1, 2) = 4$ while the absolute minimum is $f(2, 4) = -64$.

53. $f(x, y) = x^3 - 3x + y^4 - 2y^2$

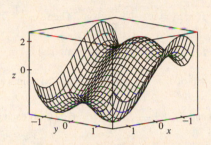
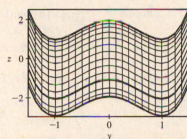
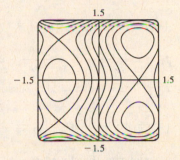

From the graphs, it appears that f has a local maximum $f(-1, 0) \approx 2$, local minima $f(1, \pm 1) \approx -3$, and saddle points at $(-1, \pm 1)$ and $(1, 0)$.

To find the exact quantities, we calculate $f_x = 3x^2 - 3 = 0 \iff x = \pm 1$ and $f_y = 4y^3 - 4y = 0 \iff y = 0, \pm 1$, giving the critical points estimated above. Also $f_{xx} = 6x$, $f_{xy} = 0$, $f_{yy} = 12y^2 - 4$, so using the Second Derivatives Test, $D(-1, 0) = 24 > 0$ and $f_{xx}(-1, 0) = -6 < 0$ indicating a local maximum $f(-1, 0) = 2$; $D(1, \pm 1) = 48 > 0$ and $f_{xx}(1, \pm 1) = 6 > 0$ indicating local minima $f(1, \pm 1) = -3$; and $D(-1, \pm 1) = -48$ and $D(1, 0) = -24$, indicating saddle points.

55. $f(x, y) = x^2 y$, $g(x, y) = x^2 + y^2 = 1 \implies \nabla f = \langle 2xy, x^2 \rangle = \lambda \nabla g = \langle 2\lambda x, 2\lambda y \rangle$. Then $2xy = 2\lambda x$ implies $x = 0$ or $y = \lambda$. If $x = 0$ then $x^2 + y^2 = 1$ gives $y = \pm 1$ and we have possible points $(0, \pm 1)$ where $f(0, \pm 1) = 0$. If $y = \lambda$ then $x^2 = 2\lambda y$ implies $x^2 = 2y^2$ and substitution into $x^2 + y^2 = 1$ gives $3y^2 = 1 \implies y = \pm \frac{1}{\sqrt{3}}$ and $x = \pm \sqrt{\frac{2}{3}}$. The corresponding possible points are $\left(\pm \sqrt{\frac{2}{3}}, \pm \frac{1}{\sqrt{3}} \right)$. The absolute maximum is $f\left(\pm \sqrt{\frac{2}{3}}, \frac{1}{\sqrt{3}} \right) = \frac{2}{3\sqrt{3}}$ while the absolute minimum is $f\left(\pm \sqrt{\frac{2}{3}}, -\frac{1}{\sqrt{3}} \right) = -\frac{2}{3\sqrt{3}}$.

57. $f(x, y, z) = xyz$, $g(x, y, z) = x^2 + y^2 + z^2 = 3$. $\nabla f = \lambda \nabla g \implies \langle yz, xz, xy \rangle = \lambda \langle 2x, 2y, 2z \rangle$. If any of x, y, or z is zero, then $x = y = z = 0$ which contradicts $x^2 + y^2 + z^2 = 3$. Then $\lambda = \frac{yz}{2x} = \frac{xz}{2y} = \frac{xy}{2z} \implies 2y^2 z = 2x^2 z \implies y^2 = x^2$, and similarly $2yz^2 = 2x^2 y \implies z^2 = x^2$. Substituting into the constraint equation gives $x^2 + x^2 + x^2 = 3 \implies x^2 = 1 = y^2 = z^2$. Thus the possible points are $(1, 1, \pm 1)$, $(1, -1, \pm 1)$, $(-1, 1, \pm 1)$, $(-1, -1, \pm 1)$. The absolute maximum is $f(1, 1, 1) = f(1, -1, -1) = f(-1, 1, -1) = f(-1, -1, 1) = 1$ and the absolute minimum is $f(1, 1, -1) = f(1, -1, 1) = f(-1, 1, 1) = f(-1, -1, -1) = -1$.

59. $f(x, y, z) = x^2 + y^2 + z^2$, $g(x, y, z) = xy^2 z^3 = 2 \implies \nabla f = \langle 2x, 2y, 2z \rangle = \lambda \nabla g = \langle \lambda y^2 z^3, 2\lambda xyz^3, 3\lambda xy^2 z^2 \rangle$. Since $xy^2 z^3 = 2$, $x \neq 0$, $y \neq 0$ and $z \neq 0$, so $2x = \lambda y^2 z^3$ **(1)**, $1 = \lambda xz^3$ **(2)**, $2 = 3\lambda xy^2 z$ **(3)**. Then **(2)** and **(3)** imply $\frac{1}{xz^3} = \frac{2}{3xy^2 z}$ or $y^2 = \frac{2}{3} z^2$ so $y = \pm z \sqrt{\frac{2}{3}}$. Similarly **(1)** and **(3)** imply $\frac{2x}{y^2 z^3} = \frac{2}{3xy^2 z}$ or $3x^2 = z^2$ so $x = \pm \frac{1}{\sqrt{3}} z$. But $xy^2 z^3 = 2$ so x and z must have the same sign, that is, $x = \frac{1}{\sqrt{3}} z$. Thus $g(x, y, z) = 2$ implies $\frac{1}{\sqrt{3}} z \left(\frac{2}{3} z^2 \right) z^3 = 2$ or

$z = \pm 3^{1/4}$ and the possible points are $(\pm 3^{-1/4}, 3^{-1/4}\sqrt{2}, \pm 3^{1/4})$, $(\pm 3^{-1/4}, -3^{-1/4}\sqrt{2}, \pm 3^{1/4})$. However at each of these points f takes on the same value, $2\sqrt{3}$. But $(2, 1, 1)$ also satisfies $g(x, y, z) = 2$ and $f(2, 1, 1) = 6 > 2\sqrt{3}$. Thus f has an absolute minimum value of $2\sqrt{3}$ and no absolute maximum subject to the constraint $xy^2z^3 = 2$.

Alternate solution: $g(x, y, z) = xy^2z^3 = 2$ implies $y^2 = \dfrac{2}{xz^3}$, so minimize $f(x, z) = x^2 + \dfrac{2}{xz^3} + z^2$. Then

$f_x = 2x - \dfrac{2}{x^2z^3}$, $f_z = -\dfrac{6}{xz^4} + 2z$, $f_{xx} = 2 + \dfrac{4}{x^3z^3}$, $f_{zz} = \dfrac{24}{xz^5} + 2$ and $f_{xz} = \dfrac{6}{x^2z^4}$. Now $f_x = 0$ implies

$2x^3z^3 - 2 = 0$ or $z = 1/x$. Substituting into $f_y = 0$ implies $-6x^3 + 2x^{-1} = 0$ or $x = \dfrac{1}{\sqrt[4]{3}}$, so the two critical points are

$\left(\pm\dfrac{1}{\sqrt[4]{3}}, \pm\sqrt[4]{3}\right)$. Then $D\left(\pm\dfrac{1}{\sqrt[4]{3}}, \pm\sqrt[4]{3}\right) = (2 + 4)\left(2 + \dfrac{24}{3}\right) - \left(\dfrac{6}{\sqrt{3}}\right)^2 > 0$ and $f_{xx}\left(\pm\dfrac{1}{\sqrt[4]{3}}, \pm\sqrt[4]{3}\right) = 6 > 0$, so each point

is a minimum. Finally, $y^2 = \dfrac{2}{xz^3}$, so the four points closest to the origin are $\left(\pm\dfrac{1}{\sqrt[4]{3}}, \dfrac{\sqrt{2}}{\sqrt[4]{3}}, \pm\sqrt[4]{3}\right)$, $\left(\pm\dfrac{1}{\sqrt[4]{3}}, -\dfrac{\sqrt{2}}{\sqrt[4]{3}}, \pm\sqrt[4]{3}\right)$.

61.

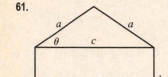

The area of the triangle is $\frac{1}{2}ca\sin\theta$ and the area of the rectangle is bc. Thus, the area of the whole object is $f(a, b, c) = \frac{1}{2}ca\sin\theta + bc$. The perimeter of the object is $g(a, b, c) = 2a + 2b + c = P$. To simplify $\sin\theta$ in terms of a, b, and c notice that $a^2\sin^2\theta + \left(\frac{1}{2}c\right)^2 = a^2 \Rightarrow \sin\theta = \dfrac{1}{2a}\sqrt{4a^2 - c^2}$.

Thus $f(a, b, c) = \dfrac{c}{4}\sqrt{4a^2 - c^2} + bc$. (Instead of using θ, we could just have

used the Pythagorean Theorem.) As a result, by Lagrange's method, we must find a, b, c, and λ by solving $\nabla f = \lambda\nabla g$ which gives the following equations: $ca(4a^2 - c^2)^{-1/2} = 2\lambda$ **(1)**, $c = 2\lambda$ **(2)**, $\frac{1}{4}(4a^2 - c^2)^{1/2} - \frac{1}{4}c^2(4a^2 - c^2)^{-1/2} + b = \lambda$ **(3)**, and $2a + 2b + c = P$ **(4)**. From **(2)**, $\lambda = \frac{1}{2}c$ and so **(1)** produces $ca(4a^2 - c^2)^{-1/2} = c \Rightarrow (4a^2 - c^2)^{1/2} = a \Rightarrow$

$4a^2 - c^2 = a^2 \Rightarrow c = \sqrt{3}\,a$ **(5)**. Similarly, since $(4a^2 - c^2)^{1/2} = a$ and $\lambda = \frac{1}{2}c$, **(3)** gives $\dfrac{a}{4} - \dfrac{c^2}{4a} + b = \dfrac{c}{2}$, so from

(5), $\dfrac{a}{4} - \dfrac{3a}{4} + b = \dfrac{\sqrt{3}\,a}{2} \Rightarrow -\dfrac{a}{2} - \dfrac{\sqrt{3}\,a}{2} = -b \Rightarrow b = \dfrac{a}{2}\left(1 + \sqrt{3}\right)$ **(6)**. Substituting **(5)** and **(6)** into **(4)** we get:

$2a + a\left(1 + \sqrt{3}\right) + \sqrt{3}\,a = P \Rightarrow 3a + 2\sqrt{3}\,a = P \Rightarrow a = \dfrac{P}{3 + 2\sqrt{3}} = \dfrac{2\sqrt{3} - 3}{3}P$ and thus

$b = \dfrac{\left(2\sqrt{3} - 3\right)\left(1 + \sqrt{3}\right)}{6}P = \dfrac{3 - \sqrt{3}}{6}P$ and $c = \left(2 - \sqrt{3}\right)P$.

12 □ MULTIPLE INTEGRALS

12.1 Double Integrals over Rectangles

1. (a) The subrectangles are shown in the figure.

The surface is the graph of $f(x, y) = xy$ and $\Delta A = 4$, so we estimate

$$V \approx \sum_{i=1}^{3} \sum_{j=1}^{2} f(x_i, y_j)\, \Delta A$$

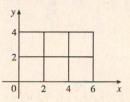

$$= f(2,2)\,\Delta A + f(2,4)\,\Delta A + f(4,2)\,\Delta A + f(4,4)\,\Delta A + f(6,2)\,\Delta A + f(6,4)\,\Delta A$$

$$= 4(4) + 8(4) + 8(4) + 16(4) + 12(4) + 24(4) = 288$$

(b) $V \approx \sum_{i=1}^{3} \sum_{j=1}^{2} f(\overline{x}_i, \overline{y}_j)\, \Delta A = f(1,1)\,\Delta A + f(1,3)\,\Delta A + f(3,1)\,\Delta A + f(3,3)\,\Delta A + f(5,1)\,\Delta A + f(5,3)\,\Delta A$

$$= 1(4) + 3(4) + 3(4) + 9(4) + 5(4) + 15(4) = 144$$

3. (a) The subrectangles are shown in the figure. Since $\Delta A = 1 \cdot \frac{1}{2} = \frac{1}{2}$, we estimate

$$\iint_R xe^{-xy}\, dA \approx \sum_{i=1}^{2} \sum_{j=1}^{2} f(x_{ij}^*, y_{ij}^*)\, \Delta A$$

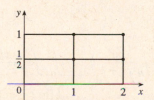

$$= f\left(1, \tfrac{1}{2}\right)\Delta A + f(1,1)\,\Delta A + f\left(2, \tfrac{1}{2}\right)\Delta A + f(2,1)\,\Delta A$$

$$= e^{-1/2}\left(\tfrac{1}{2}\right) + e^{-1}\left(\tfrac{1}{2}\right) + 2e^{-1}\left(\tfrac{1}{2}\right) + 2e^{-2}\left(\tfrac{1}{2}\right) \approx 0.990$$

(b) $\iint_R xe^{-xy}\, dA \approx \sum_{i=1}^{2} \sum_{j=1}^{2} f(\overline{x}_i, \overline{y}_j)\, \Delta A$

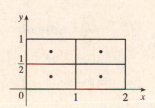

$$= f\left(\tfrac{1}{2}, \tfrac{1}{4}\right)\Delta A + f\left(\tfrac{1}{2}, \tfrac{3}{4}\right)\Delta A + f\left(\tfrac{3}{2}, \tfrac{1}{4}\right)\Delta A + f\left(\tfrac{3}{2}, \tfrac{3}{4}\right)\Delta A$$

$$= \tfrac{1}{2}e^{-1/8}\left(\tfrac{1}{2}\right) + \tfrac{1}{2}e^{-3/8}\left(\tfrac{1}{2}\right) + \tfrac{3}{2}e^{-3/8}\left(\tfrac{1}{2}\right) + \tfrac{3}{2}e^{-9/8}\left(\tfrac{1}{2}\right) \approx 1.151$$

5. With $m = n = 2$, we have $\Delta A = 4$. Using the contour map to estimate the value of f at the center of each subrectangle, we have

$$\iint_R f(x,y)\, dA \approx \sum_{i=1}^{2} \sum_{j=1}^{2} f(\overline{x}_i, \overline{y}_j)\, \Delta A = \Delta A[f(1,1) + f(1,3) + f(3,1) + f(3,3)] \approx 4(27 + 4 + 14 + 17) = 248$$

7. $z = 3 > 0$, so we can interpret the integral as the volume of the solid S that lies below the plane $z = 3$ and above the rectangle $[-2, 2] \times [1, 6]$. S is a rectangular solid, thus $\iint_R 3\, dA = 4 \cdot 5 \cdot 3 = 60$.

9. $z = f(x, y) = 4 - 2y \geq 0$ for $0 \leq y \leq 1$. Thus the integral represents the volume of that part of the rectangular solid $[0, 1] \times [0, 1] \times [0, 4]$ which lies below the plane $z = 4 - 2y$. So

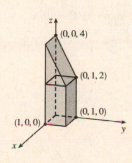

$$\iint_R (4 - 2y)\, dA = (1)(1)(2) + \tfrac{1}{2}(1)(1)(2) = 3$$

11. $\int_1^4 \int_0^2 (6x^2y - 2x)\, dy\, dx = \int_1^4 \left[3x^2y^2 - 2xy \right]_{y=0}^{y=2} dx = \int_1^4 (12x^2 - 4x)\, dx = \left[4x^3 - 2x^2 \right]_1^4 = (256 - 32) - (4 - 2) = 222$

13. $\int_0^2 \int_0^4 y^3 e^{2x}\, dy\, dx = \int_0^2 e^{2x}\, dx \int_0^4 y^3\, dy$ [as in Example 8] $= \left[\frac{1}{2} e^{2x} \right]_0^2 \left[\frac{1}{4} y^4 \right]_0^4 = \frac{1}{2}(e^4 - 1)(64 - 0) = 32(e^4 - 1)$

15. $\int_{-3}^3 \int_0^{\pi/2} (y + y^2 \cos x)\, dx\, dy = \int_{-3}^3 \left[xy + y^2 \sin x \right]_{x=0}^{x=\pi/2} dy$

$$= \int_{-3}^3 \left(\frac{\pi}{2} y + y^2 \right) dy = \left[\frac{\pi}{4} y^2 + \frac{1}{3} y^3 \right]_{-3}^3$$

$$= \left[\frac{9\pi}{4} + 9 - \left(\frac{9\pi}{4} - 9 \right) \right] = 18$$

17. $\int_1^4 \int_1^2 \left(\frac{x}{y} + \frac{y}{x} \right) dy\, dx = \int_1^4 \left[x \ln|y| + \frac{1}{x} \cdot \frac{1}{2} y^2 \right]_{y=1}^{y=2} dx = \int_1^4 \left(x \ln 2 + \frac{3}{2x} \right) dx = \left[\frac{1}{2} x^2 \ln 2 + \frac{3}{2} \ln|x| \right]_1^4$

$$= 8 \ln 2 + \frac{3}{2} \ln 4 - \frac{1}{2} \ln 2 = \frac{15}{2} \ln 2 + 3 \ln 4^{1/2} = \frac{21}{2} \ln 2$$

19. $\int_0^1 \int_0^1 v(u + v^2)^4\, du\, dv = \int_0^1 \left[\frac{1}{5} v(u + v^2)^5 \right]_{u=0}^{u=1} dv = \frac{1}{5} \int_0^1 v \left[(1 + v^2)^5 - (0 + v^2)^5 \right] dv$

$$= \frac{1}{5} \int_0^1 \left[v(1 + v^2)^5 - v^{11} \right] dv = \frac{1}{5} \left[\frac{1}{2} \cdot \frac{1}{6} (1 + v^2)^6 - \frac{1}{12} v^{12} \right]_0^1$$

[substitute $t = 1 + v^2 \;\Rightarrow\; dt = 2v\, dv$ in the first term]

$$= \frac{1}{60} \left[(2^6 - 1) - (1 - 0) \right] = \frac{1}{60}(63 - 1) = \frac{31}{30}$$

21. $\iint_R \frac{xy^2}{x^2 + 1}\, dA = \int_0^1 \int_{-3}^3 \frac{xy^2}{x^2 + 1}\, dy\, dx = \int_0^1 \frac{x}{x^2 + 1}\, dx \int_{-3}^3 y^2\, dy = \left[\frac{1}{2} \ln(x^2 + 1) \right]_0^1 \left[\frac{1}{3} y^3 \right]_{-3}^3$

$$= \frac{1}{2}(\ln 2 - \ln 1) \cdot \frac{1}{3}(27 + 27) = 9 \ln 2$$

23. $\int_0^{\pi/6} \int_0^{\pi/3} x \sin(x + y)\, dy\, dx$

$$= \int_0^{\pi/6} \left[-x \cos(x + y) \right]_{y=0}^{y=\pi/3} dx = \int_0^{\pi/6} \left[x \cos x - x \cos\left(x + \frac{\pi}{3} \right) \right] dx$$

$$= x \left[\sin x - \sin\left(x + \frac{\pi}{3} \right) \right]_0^{\pi/6} - \int_0^{\pi/6} \left[\sin x - \sin\left(x + \frac{\pi}{3} \right) \right] dx \qquad \text{[by integrating by parts separately for each term]}$$

$$= \frac{\pi}{6} \left[\frac{1}{2} - 1 \right] - \left[-\cos x + \cos\left(x + \frac{\pi}{3} \right) \right]_0^{\pi/6} = -\frac{\pi}{12} - \left[-\frac{\sqrt{3}}{2} + 0 - \left(-1 + \frac{1}{2} \right) \right] = \frac{\sqrt{3} - 1}{2} - \frac{\pi}{12}$$

25. $\iint_R y e^{-xy}\, dA = \int_0^3 \int_0^2 y e^{-xy}\, dx\, dy = \int_0^3 \left[-e^{-xy} \right]_{x=0}^{x=2} dy = \int_0^3 (-e^{-2y} + 1)\, dy = \left[\frac{1}{2} e^{-2y} + y \right]_0^3$

$$= \frac{1}{2} e^{-6} + 3 - \left(\frac{1}{2} + 0 \right) = \frac{1}{2} e^{-6} + \frac{5}{2}$$

27. $z = f(x, y) = 4 - x - 2y \geq 0$ for $0 \leq x \leq 1$ and $0 \leq y \leq 1$. So the solid

is the region in the first octant which lies below the plane $z = 4 - x - 2y$

and above $[0, 1] \times [0, 1]$.

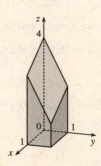

29. The solid lies under the plane $4x + 6y - 2z + 15 = 0$ or $z = 2x + 3y + \frac{15}{2}$ so

$$V = \iint_R \left(2x + 3y + \tfrac{15}{2}\right) dA = \int_{-1}^{1} \int_{-1}^{2} \left(2x + 3y + \tfrac{15}{2}\right) dx\, dy = \int_{-1}^{1} \left[x^2 + 3xy + \tfrac{15}{2}x\right]_{x=-1}^{x=2} dy$$

$$= \int_{-1}^{1} \left[(19 + 6y) - \left(-\tfrac{13}{2} - 3y\right)\right] dy = \int_{-1}^{1}\left(\tfrac{51}{2} + 9y\right) dy = \left[\tfrac{51}{2}y + \tfrac{9}{2}y^2\right]_{-1}^{1} = 30 - (-21) = 51$$

31. $V = \int_{-2}^{2} \int_{-1}^{1} \left(1 - \tfrac{1}{4}x^2 - \tfrac{1}{9}y^2\right) dx\, dy = 4 \int_{0}^{2} \int_{0}^{1} \left(1 - \tfrac{1}{4}x^2 - \tfrac{1}{9}y^2\right) dx\, dy$

$$= 4 \int_{0}^{2} \left[x - \tfrac{1}{12}x^3 - \tfrac{1}{9}y^2 x\right]_{x=0}^{x=1} dy = 4 \int_{0}^{2} \left(\tfrac{11}{12} - \tfrac{1}{9}y^2\right) dy = 4\left[\tfrac{11}{12}y - \tfrac{1}{27}y^3\right]_{0}^{2} = 4 \cdot \tfrac{83}{54} = \tfrac{166}{27}$$

33. Here we need the volume of the solid lying under the surface $z = x \sec^2 y$ and above the rectangle $R = [0, 2] \times [0, \pi/4]$ in the xy-plane.

$$V = \int_{0}^{2} \int_{0}^{\pi/4} x \sec^2 y\, dy\, dx = \int_{0}^{2} x\, dx \int_{0}^{\pi/4} \sec^2 y\, dy = \left[\tfrac{1}{2}x^2\right]_{0}^{2} \left[\tan y\right]_{0}^{\pi/4}$$

$$= (2 - 0)(\tan\tfrac{\pi}{4} - \tan 0) = 2(1 - 0) = 2$$

35. The solid lies below the surface $z = 2 + x^2 + (y - 2)^2$ and above the plane $z = 1$ for $-1 \le x \le 1$, $0 \le y \le 4$. The volume of the solid is the difference in volumes between the solid that lies under $z = 2 + x^2 + (y - 2)^2$ over the rectangle $R = [-1, 1] \times [0, 4]$ and the solid that lies under $z = 1$ over R.

$$V = \int_{0}^{4} \int_{-1}^{1} [2 + x^2 + (y - 2)^2]\, dx\, dy - \int_{0}^{4} \int_{-1}^{1} (1)\, dx\, dy$$

$$= \int_{0}^{4} \left[2x + \tfrac{1}{3}x^3 + x(y-2)^2\right]_{x=-1}^{x=1} dy - \int_{-1}^{1} dx \int_{0}^{4} dy$$

$$= \int_{0}^{4} \left[\left(2 + \tfrac{1}{3} + (y-2)^2\right) - \left(-2 - \tfrac{1}{3} - (y-2)^2\right)\right] dy - [x]_{-1}^{1}\, [y]_{0}^{4}$$

$$= \int_{0}^{4} \left[\tfrac{14}{3} + 2(y-2)^2\right] dy - [1 - (-1)][4 - 0] = \left[\tfrac{14}{3}y + \tfrac{2}{3}(y-2)^3\right]_{0}^{4} - (2)(4)$$

$$= \left[\left(\tfrac{56}{3} + \tfrac{16}{3}\right) - \left(0 - \tfrac{16}{3}\right)\right] - 8 = \tfrac{88}{3} - 8 = \tfrac{64}{3}$$

37. In Maple, we can calculate the integral by defining the integrand as `f` and then using the command `int(int(f,x=0..1),y=0..1);`. In Mathematica, we can use the command

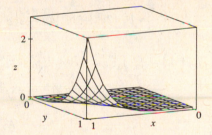

$$\text{Integrate[f,\{x,0,1\},\{y,0,1\}]}$$

We find that $\iint_R x^5 y^3 e^{xy}\, dA = 21e - 57 \approx 0.0839$. We can use `plot3d` (in Maple) or `Plot3D` (in Mathematica) to graph the function.

39. R is the rectangle $[-1, 1] \times [0, 5]$. Thus, $A(R) = 2 \cdot 5 = 10$ and

$$f_{ave} = \frac{1}{A(R)} \iint_R f(x, y)\, dA = \tfrac{1}{10} \int_{0}^{5} \int_{-1}^{1} x^2 y\, dx\, dy = \tfrac{1}{10} \int_{0}^{5} \left[\tfrac{1}{3}x^3 y\right]_{x=-1}^{x=1} dy = \tfrac{1}{10} \int_{0}^{5} \tfrac{2}{3}y\, dy = \tfrac{1}{10}\left[\tfrac{1}{3}y^2\right]_{0}^{5} = \tfrac{5}{6}.$$

41. If we divide R into mn subrectangles, $\iint_R k\, dA \approx \sum_{i=1}^{m} \sum_{j=1}^{n} f\left(x_{ij}^*, y_{ij}^*\right) \Delta A$ for any choice of sample points $\left(x_{ij}^*, y_{ij}^*\right)$.

But $f\left(x_{ij}^*, y_{ij}^*\right) = k$ always and $\sum_{i=1}^{m} \sum_{j=1}^{n} \Delta A = $ area of $R = (b - a)(d - c)$. Thus, no matter how we choose the sample points, $\sum\limits_{i=1}^{m} \sum\limits_{j=1}^{n} f(x_{ij}^*, y_{ij}^*)\, \Delta A = k \sum\limits_{i=1}^{m} \sum\limits_{j=1}^{n} \Delta A = k(b-a)(d-c)$ and so

$$\iint_R k\, dA = \lim_{m,n\to\infty} \sum_{i=1}^{m} \sum_{j=1}^{n} f(x_{ij}^*, y_{ij}^*)\, \Delta A = \lim_{m,n\to\infty} k \sum_{i=1}^{m} \sum_{j=1}^{n} \Delta A = \lim_{m,n\to\infty} k(b-a)(d-c) = k(b-a)(d-c).$$

43. $\displaystyle\iint_R \frac{xy}{1+x^4}\, dA = \int_{-1}^{1} \int_0^1 \frac{xy}{1+x^4}\, dy\, dx = \int_{-1}^{1} \frac{x}{1+x^4}\, dx \int_0^1 y\, dy$ [by Equation 11] but $f(x) = \dfrac{x}{1+x^4}$ is an odd

function so $\displaystyle\int_{-1}^{1} f(x)\, dx = 0$. Thus $\displaystyle\iint_R \frac{xy}{1+x^4}\, dA = 0 \cdot \int_0^1 y\, dy = 0$.

45. Let $f(x,y) = \dfrac{x-y}{(x+y)^3}$. Then a CAS gives $\int_0^1 \int_0^1 f(x,y)\, dy\, dx = \frac{1}{2}$ and $\int_0^1 \int_0^1 f(x,y)\, dx\, dy = -\frac{1}{2}$.

To explain the seeming violation of Fubini's Theorem, note that f has an infinite discontinuity at $(0,0)$ and thus does not

satisfy the conditions of Fubini's Theorem. In fact, both iterated integrals involve improper integrals which diverge at their

lower limits of integration.

12.2 Double Integrals over General Regions

1. $\int_0^4 \int_0^{\sqrt{y}} xy^2\, dx\, dy = \int_0^4 \left[\frac{1}{2}x^2 y^2\right]_{x=0}^{x=\sqrt{y}}\, dy = \int_0^4 \frac{1}{2} y^2 [(\sqrt{y})^2 - 0^2]\, dy = \frac{1}{2} \int_0^4 y^3\, dy = \frac{1}{2}\left[\frac{1}{4}y^4\right]_0^4 = \frac{1}{2}(64-0) = 32$

3. $\int_0^1 \int_{x^2}^{x} (1+2y)\, dy\, dx = \int_0^1 \left[y+y^2\right]_{y=x^2}^{y=x}\, dx = \int_0^1 \left[x + x^2 - x^2 - (x^2)^2\right]\, dx$

$$= \int_0^1 (x - x^4)\, dx = \left[\frac{1}{2}x^2 - \frac{1}{5}x^5\right]_0^1 = \frac{1}{2} - \frac{1}{5} - 0 + 0 = \frac{3}{10}$$

5. $\int_0^1 \int_0^{s^2} \cos(s^3)\, dt\, ds = \int_0^1 \left[t\cos(s^3)\right]_{t=0}^{t=s^2}\, ds = \int_0^1 s^2 \cos(s^3)\, ds = \frac{1}{3}\sin(s^3)\Big]_0^1 = \frac{1}{3}(\sin 1 - \sin 0) = \frac{1}{3}\sin 1$

7. $\iint_D y^2\, dA = \int_{-1}^{1} \int_{-y-2}^{y} y^2\, dx\, dy = \int_{-1}^{1} \left[xy^2\right]_{x=-y-2}^{x=y}\, dy = \int_{-1}^{1} y^2 \left[y - (-y-2)\right]\, dy$

$$= \int_{-1}^{1} (2y^3 + 2y^2)\, dy = \left[\frac{1}{2}y^4 + \frac{2}{3}y^3\right]_{-1}^{1} = \frac{1}{2} + \frac{2}{3} - \frac{1}{2} + \frac{2}{3} = \frac{4}{3}$$

9. $\iint_D x\, dA = \int_0^\pi \int_0^{\sin x} x\, dy\, dx = \int_0^\pi \left[xy\right]_{y=0}^{y=\sin x}\, dx = \int_0^\pi x\sin x\, dx$ $\left[\begin{array}{l}\text{integrate by parts} \\ \text{with } u=x,\, dv=\sin x\, dx\end{array}\right]$

$$= \left[-x\cos x + \sin x\right]_0^\pi = -\pi\cos\pi + \sin\pi + 0 - \sin 0 = \pi$$

11.

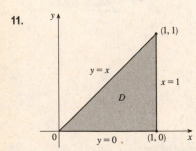

As a type I region, D lies between the lower boundary $y=0$ and the upper

boundary $y=x$ for $0 \le x \le 1$, so $D = \{(x,y) \mid 0 \le x \le 1,\, 0 \le y \le x\}$. If we

describe D as a type II region, D lies between the left boundary $x=y$ and the

right boundary $x=1$ for $0 \le y \le 1$, so $D = \{(x,y) \mid 0 \le y \le 1,\, y \le x \le 1\}$.

Thus $\iint_D x\, dA = \int_0^1 \int_0^x x\, dy\, dx = \int_0^1 \left[xy\right]_{y=0}^{y=x}\, dx = \int_0^1 x^2\, dx = \frac{1}{3}x^3\Big]_0^1 = \frac{1}{3}(1-0) = \frac{1}{3}$ or

$\iint_D x\, dA = \int_0^1 \int_y^1 x\, dx\, dy = \int_0^1 \left[\frac{1}{2}x^2\right]_{x=y}^{x=1}\, dy = \frac{1}{2}\int_0^1 (1-y^2)\, dy = \frac{1}{2}\left[y - \frac{1}{3}y^3\right]_0^1 = \frac{1}{2}\left[(1-\frac{1}{3}) - 0\right] = \frac{1}{3}$.

13.

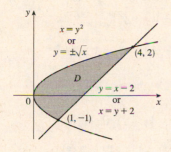

The curves $y = x - 2$ or $x = y + 2$ and $x = y^2$ intersect when $y + 2 = y^2$ $\Leftrightarrow$ $y^2 - y - 2 = 0$ $\Leftrightarrow$ $(y - 2)(y + 1) = 0$ $\Leftrightarrow$ $y = -1, y = 2$, so the points of intersection are $(1, -1)$ and $(4, 2)$. If we describe D as a type I region, the upper boundary curve is $y = \sqrt{x}$ but the lower boundary curve consists of two parts,

$$y = -\sqrt{x} \text{ for } 0 \le x \le 1 \text{ and } y = x - 2 \text{ for } 1 \le x \le 4.$$

Thus $D = \{(x, y) \mid 0 \le x \le 1, \ -\sqrt{x} \le y \le \sqrt{x}\} \cup \{(x, y) \mid 1 \le x \le 4, \ x - 2 \le y \le \sqrt{x}\}$ and

$\iint_D y \, dA = \int_0^1 \int_{-\sqrt{x}}^{\sqrt{x}} y \, dy \, dx + \int_1^4 \int_{x-2}^{\sqrt{x}} y \, dy \, dx$. If we describe D as a type II region, D is enclosed by the left boundary

$x = y^2$ and the right boundary $x = y + 2$ for $-1 \le y \le 2$, so $D = \{(x, y) \mid -1 \le y \le 2, \ y^2 \le x \le y + 2\}$ and

$\iint_D y \, dA = \int_{-1}^2 \int_{y^2}^{y+2} y \, dx \, dy$. In either case, the resulting iterated integrals are not difficult to evaluate but the region D is more simply described as a type II region, giving one iterated integral rather than a sum of two, so we evaluate the latter integral:

$$\iint_D y \, dA = \int_{-1}^2 \int_{y^2}^{y+2} y \, dx \, dy = \int_{-1}^2 \left[xy \right]_{x=y^2}^{x=y+2} dy = \int_{-1}^2 (y + 2 - y^2) y \, dy = \int_{-1}^2 (y^2 + 2y - y^3) \, dy$$

$$= \left[\tfrac{1}{3} y^3 + y^2 - \tfrac{1}{4} y^4 \right]_{-1}^2 = \left(\tfrac{8}{3} + 4 - 4 \right) - \left(-\tfrac{1}{3} + 1 - \tfrac{1}{4} \right) = \tfrac{9}{4}$$

15. $\int_0^1 \int_0^{x^2} x \cos y \, dy \, dx = \int_0^1 \left[x \sin y \right]_{y=0}^{y=x^2} dx = \int_0^1 x \sin x^2 \, dx = -\tfrac{1}{2} \cos x^2 \Big]_0^1 = \tfrac{1}{2} (1 - \cos 1)$

17.

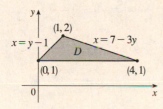

$$\iint_D y^2 \, dA = \int_1^2 \int_{y-1}^{7-3y} y^2 \, dx \, dy = \int_1^2 \left[xy^2 \right]_{x=y-1}^{x=7-3y} dy$$

$$= \int_1^2 \left[(7 - 3y) - (y - 1) \right] y^2 \, dy = \int_1^2 (8y^2 - 4y^3) \, dy$$

$$= \left[\tfrac{8}{3} y^3 - y^4 \right]_1^2 = \tfrac{64}{3} - 16 - \tfrac{8}{3} + 1 = \tfrac{11}{3}$$

19.

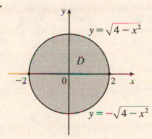

$$\int_{-2}^2 \int_{-\sqrt{4-x^2}}^{\sqrt{4-x^2}} (2x - y) \, dy \, dx$$

$$= \int_{-2}^2 \left[2xy - \tfrac{1}{2} y^2 \right]_{y=-\sqrt{4-x^2}}^{y=\sqrt{4-x^2}} dx$$

$$= \int_{-2}^2 \left[2x \sqrt{4-x^2} - \tfrac{1}{2} (4 - x^2) + 2x \sqrt{4-x^2} + \tfrac{1}{2} (4 - x^2) \right] dx$$

$$= \int_{-2}^2 4x \sqrt{4-x^2} \, dx = -\tfrac{4}{3} (4 - x^2)^{3/2} \Big]_{-2}^2 = 0$$

[Or, note that $4x \sqrt{4-x^2}$ is an odd function, so $\int_{-2}^2 4x \sqrt{4-x^2} \, dx = 0$.]

21.

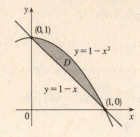

$$V = \int_0^1 \int_{1-x}^{1-x^2} (1 - x + 2y)\, dy\, dx = \int_0^1 \left[y - xy + y^2 \right]_{y=1-x}^{y=1-x^2} dx$$

$$= \int_0^1 \left[\left((1 - x^2) - x(1 - x^2) + (1 - x^2)^2 \right) \right.$$
$$\left. - \left((1 - x) - x(1 - x) + (1 - x)^2 \right) \right] dx$$

$$= \int_0^1 \left[(x^4 + x^3 - 3x^2 - x + 2) - (2x^2 - 4x + 2) \right] dx$$

$$= \int_0^1 (x^4 + x^3 - 5x^2 + 3x)\, dx = \left[\tfrac{1}{5}x^5 + \tfrac{1}{4}x^4 - \tfrac{5}{3}x^3 + \tfrac{3}{2}x^2 \right]_0^1$$

$$= \tfrac{1}{5} + \tfrac{1}{4} - \tfrac{5}{3} + \tfrac{3}{2} = \tfrac{17}{60}$$

23.

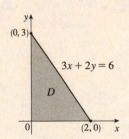

$$V = \int_1^2 \int_1^{7-3y} xy\, dx\, dy = \int_1^2 \left[\tfrac{1}{2}x^2 y \right]_{x=1}^{x=7-3y} dy$$

$$= \tfrac{1}{2} \int_1^2 (48y - 42y^2 + 9y^3)\, dy$$

$$= \tfrac{1}{2} \left[24y^2 - 14y^3 + \tfrac{9}{4}y^4 \right]_1^2 = \tfrac{31}{8}$$

25.

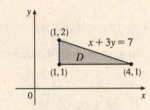

$$V = \int_0^2 \int_0^{3-\frac{3}{2}x} (6 - 3x - 2y)\, dy\, dx$$

$$= \int_0^2 \left[6y - 3xy - y^2 \right]_{y=0}^{y=3-\frac{3}{2}x} dx$$

$$= \int_0^2 \left[6(3 - \tfrac{3}{2}x) - 3x(3 - \tfrac{3}{2}x) - (3 - \tfrac{3}{2}x)^2 \right] dx$$

$$= \int_0^2 \left(\tfrac{9}{4}x^2 - 9x + 9 \right) dx = \left[\tfrac{3}{4}x^3 - \tfrac{9}{2}x^2 + 9x \right]_0^2 = 6 - 0 = 6$$

27.

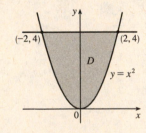

$$V = \int_{-2}^2 \int_{x^2}^4 x^2\, dy\, dx$$

$$= \int_{-2}^2 x^2 \left[y \right]_{y=x^2}^{y=4} dx = \int_{-2}^2 (4x^2 - x^4)\, dx$$

$$= \left[\tfrac{4}{3}x^3 - \tfrac{1}{5}x^5 \right]_{-2}^2 = \tfrac{32}{3} - \tfrac{32}{5} + \tfrac{32}{3} - \tfrac{32}{5} = \tfrac{128}{15}$$

29.

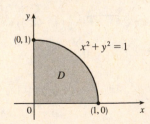

$$V = \int_0^1 \int_0^{\sqrt{1-x^2}} y\, dy\, dx = \int_0^1 \left[\frac{y^2}{2} \right]_{y=0}^{y=\sqrt{1-x^2}} dx$$

$$= \int_0^1 \frac{1 - x^2}{2}\, dx = \tfrac{1}{2} \left[x - \tfrac{1}{3}x^3 \right]_0^1 = \tfrac{1}{3}$$

31. The two bounding curves $y = 1 - x^2$ and $y = x^2 - 1$ intersect at $(\pm 1, 0)$ with $1 - x^2 \geq x^2 - 1$ on $[-1, 1]$. Within this region, the plane $z = 2x + 2y + 10$ is above the plane $z = 2 - x - y$, so

$$V = \int_{-1}^{1} \int_{x^2-1}^{1-x^2} (2x + 2y + 10) \, dy \, dx - \int_{-1}^{1} \int_{x^2-1}^{1-x^2} (2 - x - y) \, dy \, dx$$

$$= \int_{-1}^{1} \int_{x^2-1}^{1-x^2} (2x + 2y + 10 - (2 - x - y)) \, dy \, dx$$

$$= \int_{-1}^{1} \int_{x^2-1}^{1-x^2} (3x + 3y + 8) \, dy \, dx = \int_{-1}^{1} \left[3xy + \tfrac{3}{2}y^2 + 8y \right]_{y=x^2-1}^{y=1-x^2} dx$$

$$= \int_{-1}^{1} \left[3x(1 - x^2) + \tfrac{3}{2}(1 - x^2)^2 + 8(1 - x^2) - 3x(x^2 - 1) - \tfrac{3}{2}(x^2 - 1)^2 - 8(x^2 - 1) \right] dx$$

$$= \int_{-1}^{1} (-6x^3 - 16x^2 + 6x + 16) \, dx = \left[-\tfrac{3}{2}x^4 - \tfrac{16}{3}x^3 + 3x^2 + 16x \right]_{-1}^{1}$$

$$= -\tfrac{3}{2} - \tfrac{16}{3} + 3 + 16 + \tfrac{3}{2} - \tfrac{16}{3} - 3 + 16 = \tfrac{64}{3}$$

33. The solid lies below the plane $z = 1 - x - y$ or $x + y + z = 1$ and above the region $D = \{(x, y) \mid 0 \leq x \leq 1, 0 \leq y \leq 1 - x\}$ in the xy-plane. The solid is a tetrahedron.

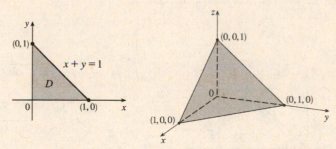

35. The two surfaces intersect in the circle $x^2 + y^2 = 1$, $z = 0$ and the region of integration is the disk D: $x^2 + y^2 \leq 1$.

Using a CAS, the volume is $\displaystyle\iint_D (1 - x^2 - y^2) \, dA = \int_{-1}^{1} \int_{-\sqrt{1-x^2}}^{\sqrt{1-x^2}} (1 - x^2 - y^2) \, dy \, dx = \frac{\pi}{2}$.

37.

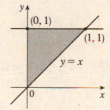

Because the region of integration is

$$D = \{(x, y) \mid 0 \leq x \leq y, 0 \leq y \leq 1\} = \{(x, y) \mid x \leq y \leq 1, 0 \leq x \leq 1\}$$

we have $\int_0^1 \int_0^y f(x, y) \, dx \, dy = \iint_D f(x, y) \, dA = \int_0^1 \int_x^1 f(x, y) \, dy \, dx$.

39.

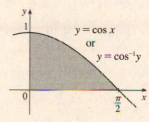

Because the region of integration is

$$D = \{(x, y) \mid 0 \leq y \leq \cos x, 0 \leq x \leq \pi/2\}$$
$$= \{(x, y) \mid 0 \leq x \leq \cos^{-1} y, 0 \leq y \leq 1\}$$

we have

$$\int_0^{\pi/2} \int_0^{\cos x} f(x, y) \, dy \, dx = \iint_D f(x, y) \, dA = \int_0^1 \int_0^{\cos^{-1} y} f(x, y) \, dx \, dy.$$

41.

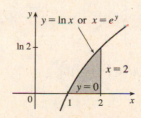

Because the region of integration is

$$D = \{(x, y) \mid 0 \leq y \leq \ln x, 1 \leq x \leq 2\} = \{(x, y) \mid e^y \leq x \leq 2, 0 \leq y \leq \ln 2\}$$

we have

$$\int_1^2 \int_0^{\ln x} f(x, y) \, dy \, dx = \iint_D f(x, y) \, dA = \int_0^{\ln 2} \int_{e^y}^2 f(x, y) \, dx \, dy$$

43.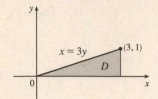

$$\int_0^1 \int_{3y}^3 e^{x^2}\, dx\, dy = \int_0^3 \int_0^{x/3} e^{x^2}\, dy\, dx = \int_0^3 \Big[e^{x^2} y \Big]_{y=0}^{y=x/3} dx$$

$$= \int_0^3 \left(\frac{x}{3}\right) e^{x^2}\, dx = \tfrac{1}{6} e^{x^2} \Big]_0^3 = \frac{e^9 - 1}{6}$$

45.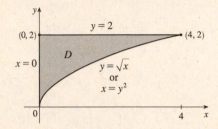

$$\int_0^4 \int_{\sqrt{x}}^2 \frac{1}{y^3 + 1}\, dy\, dx = \int_0^2 \int_0^{y^2} \frac{1}{y^3 + 1}\, dx\, dy$$

$$= \int_0^2 \frac{1}{y^3 + 1}\, \big[x \big]_{x=0}^{x=y^2}\, dy = \int_0^2 \frac{y^2}{y^3 + 1}\, dy$$

$$= \tfrac{1}{3} \ln \big| y^3 + 1 \big| \,\Big]_0^2 = \tfrac{1}{3}(\ln 9 - \ln 1) = \tfrac{1}{3} \ln 9$$

47.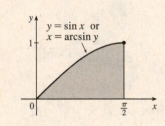

$$\int_0^1 \int_{\arcsin y}^{\pi/2} \cos x \sqrt{1 + \cos^2 x}\, dx\, dy$$

$$= \int_0^{\pi/2} \int_0^{\sin x} \cos x \sqrt{1 + \cos^2 x}\, dy\, dx$$

$$= \int_0^{\pi/2} \cos x \sqrt{1 + \cos^2 x}\, \big[y \big]_{y=0}^{y=\sin x}\, dx$$

$$= \int_0^{\pi/2} \cos x \sqrt{1 + \cos^2 x}\, \sin x\, dx \qquad \left[\begin{array}{l} \text{Let } u = \cos x,\ du = -\sin x\, dx, \\ \qquad\quad dx = du/(-\sin x) \end{array} \right]$$

$$= \int_1^0 -u\sqrt{1 + u^2}\, du = -\tfrac{1}{3}\big(1 + u^2\big)^{3/2} \Big]_1^0$$

$$= \tfrac{1}{3}\big(\sqrt{8} - 1\big) = \tfrac{1}{3}\big(2\sqrt{2} - 1\big)$$

49. $D = \{(x, y) \mid 0 \le x \le 1,\ -x + 1 \le y \le 1\} \cup \{(x, y) \mid -1 \le x \le 0,\ x + 1 \le y \le 1\}$

$\qquad \cup \{(x, y) \mid 0 \le x \le 1,\ -1 \le y \le x - 1\} \cup \{(x, y) \mid -1 \le x \le 0,\ -1 \le y \le -x - 1\}, \quad$ all type I.

$$\iint_D x^2\, dA = \int_0^1 \int_{1-x}^1 x^2\, dy\, dx + \int_{-1}^0 \int_{x+1}^1 x^2\, dy\, dx + \int_0^1 \int_{-1}^{x-1} x^2\, dy\, dx + \int_{-1}^0 \int_{-1}^{-x-1} x^2\, dy\, dx$$

$$= 4 \int_0^1 \int_{1-x}^1 x^2\, dy\, dx \qquad [\text{by symmetry of the regions and because } f(x, y) = x^2 \ge 0]$$

$$= 4 \int_0^1 x^3\, dx = 4\big[\tfrac{1}{4} x^4\big]_0^1 = 1$$

51. For $D = [0, 1] \times [0, 1]$, $0 \le \sqrt{x^3 + y^3} \le \sqrt{2}$ and $A(D) = 1$, so $0 \le \iint_D \sqrt{x^3 + y^3}\, dA \le \sqrt{2}$.

53. Since $m \le f(x, y) \le M$, $\iint_D m\, dA \le \iint_D f(x, y)\, dA \le \iint_D M\, dA$ by (8) $\Rightarrow$

$\quad m \iint_D 1\, dA \le \iint_D f(x, y)\, dA \le M \iint_D 1\, dA$ by (7) $\Rightarrow \quad mA(D) \le \iint_D f(x, y)\, dA \le MA(D)$ by (10).

55.

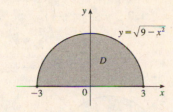

First we can write $\iint_D (x+2)\, dA = \iint_D x\, dA + \iint_D 2\, dA$. But $f(x,y) = x$ is

an odd function with respect to x [that is, $f(-x,y) = -f(x,y)$] and D is

symmetric with respect to x. Consequently, the volume above D and below the

graph of f is the same as the volume below D and above the graph of f, so

$\iint_D x\, dA = 0$. Also, $\iint_D 2\, dA = 2 \cdot A(D) = 2 \cdot \frac{1}{2}\pi(3)^2 = 9\pi$ since D is a half

disk of radius 3. Thus $\iint_D (x+2)\, dA = 0 + 9\pi = 9\pi$.

57. We can write $\iint_D (2x+3y)\, dA = \iint_D 2x\, dA + \iint_D 3y\, dA$. $\iint_D 2x\, dA$ represents the volume of the solid lying under the

plane $z = 2x$ and above the rectangle D. This solid region is a triangular cylinder with length b and whose cross-section is a

triangle with width a and height $2a$. (See the first figure.)

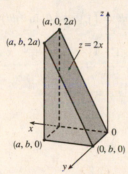

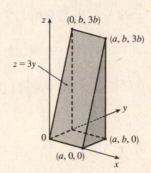

Thus its volume is $\frac{1}{2} \cdot a \cdot 2a \cdot b = a^2 b$. Similarly, $\iint_D 3y\, dA$ represents the volume of a triangular cylinder with length a,

triangular cross-section with width b and height $3b$, and volume $\frac{1}{2} \cdot b \cdot 3b \cdot a = \frac{3}{2}ab^2$. (See the second figure.) Thus

$$\iint_D (2x+3y)\, dA = a^2 b + \frac{3}{2}ab^2$$

59. $\iint_D \left(ax^3 + by^3 + \sqrt{a^2 - x^2}\right) dA = \iint_D ax^3\, dA + \iint_D by^3\, dA + \iint_D \sqrt{a^2 - x^2}\, dA$. Now ax^3 is odd with respect

to x and by^3 is odd with respect to y, and the region of integration is symmetric with respect to both x and y,

so $\iint_D ax^3\, dA = \iint_D by^3\, dA = 0$.

$\iint_D \sqrt{a^2 - x^2}\, dA$ represents the volume of the solid region under the

graph of $z = \sqrt{a^2 - x^2}$ and above the rectangle D, namely a half circular

cylinder with radius a and length $2b$ (see the figure) whose volume is

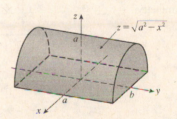

$\frac{1}{2} \cdot \pi r^2 h = \frac{1}{2}\pi a^2 (2b) = \pi a^2 b$. Thus

$\iint_D \left(ax^3 + by^3 + \sqrt{a^2 - x^2}\right) dA = 0 + 0 + \pi a^2 b = \pi a^2 b$.

12.3 Double Integrals in Polar Coordinates

1. The region R is more easily described by polar coordinates: $R = \left\{(r,\theta) \mid 0 \le r \le 4,\, 0 \le \theta \le \frac{3\pi}{2}\right\}$.

Thus $\iint_R f(x,y)\, dA = \int_0^{3\pi/2} \int_0^4 f(r\cos\theta, r\sin\theta)\, r\, dr\, d\theta$.

3. The region R is more easily described by rectangular coordinates: $R = \{(x, y) \mid -1 \leq x \leq 1, 0 \leq y \leq \frac{1}{2}x + \frac{1}{2}\}$.

Thus $\iint_R f(x, y)\, dA = \int_{-1}^{1} \int_0^{(x+1)/2} f(x, y)\, dy\, dx$.

5. The integral $\int_{\pi/4}^{3\pi/4} \int_1^2 r\, dr\, d\theta$ represents the area of the region

$R = \{(r, \theta) \mid 1 \leq r \leq 2, \pi/4 \leq \theta \leq 3\pi/4\}$, the top quarter portion of a

ring (annulus).

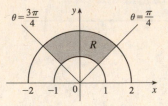

$\int_{\pi/4}^{3\pi/4} \int_1^2 r\, dr\, d\theta = \left(\int_{\pi/4}^{3\pi/4} d\theta \right)\left(\int_1^2 r\, dr \right)$

$= \left[\theta \right]_{\pi/4}^{3\pi/4} \left[\tfrac{1}{2}r^2 \right]_1^2 = \left(\tfrac{3\pi}{4} - \tfrac{\pi}{4} \right) \cdot \tfrac{1}{2}(4 - 1) = \tfrac{\pi}{2} \cdot \tfrac{3}{2} = \tfrac{3\pi}{4}$

7. The half disk D can be described in polar coordinates as $D = \{(r, \theta) \mid 0 \leq r \leq 5, 0 \leq \theta \leq \pi\}$. Then

$$\iint_D x^2 y\, dA = \int_0^\pi \int_0^5 (r\cos\theta)^2 (r\sin\theta)\, r\, dr\, d\theta = \left(\int_0^\pi \cos^2\theta \sin\theta\, d\theta \right)\left(\int_0^5 r^4\, dr \right)$$

$$= \left[-\tfrac{1}{3}\cos^3\theta \right]_0^\pi \left[\tfrac{1}{5}r^5 \right]_0^5 = -\tfrac{1}{3}(-1 - 1) \cdot 625 = \tfrac{1250}{3}$$

9. $\iint_R \sin(x^2 + y^2)\, dA = \int_0^{\pi/2} \int_1^3 \sin(r^2)\, r\, dr\, d\theta = \left(\int_0^{\pi/2} d\theta \right)\left(\int_1^3 r\sin(r^2)\, dr \right)$

$= \left[\theta \right]_0^{\pi/2} \left[-\tfrac{1}{2}\cos(r^2) \right]_1^3$

$= \left(\tfrac{\pi}{2} \right) \left[-\tfrac{1}{2}(\cos 9 - \cos 1) \right] = \tfrac{\pi}{4}(\cos 1 - \cos 9)$

11. R is the region shown in the figure, and can be described

by $R = \{(r, \theta) \mid 0 \leq \theta \leq \pi/4, 1 \leq r \leq 2\}$. Thus

$\iint_R \arctan(y/x)\, dA = \int_0^{\pi/4} \int_1^2 \arctan(\tan\theta)\, r\, dr\, d\theta$ since $y/x = \tan\theta$.

Also, $\arctan(\tan\theta) = \theta$ for $0 \leq \theta \leq \pi/4$, so the integral becomes

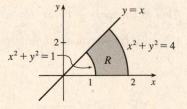

$\int_0^{\pi/4} \int_1^2 \theta\, r\, dr\, d\theta = \int_0^{\pi/4} \theta\, d\theta \int_1^2 r\, dr = \left[\tfrac{1}{2}\theta^2 \right]_0^{\pi/4} \left[\tfrac{1}{2}r^2 \right]_1^2 = \tfrac{\pi^2}{32} \cdot \tfrac{3}{2} = \tfrac{3}{64}\pi^2$.

13. $V = \iint_{x^2 + y^2 \leq 4} \sqrt{x^2 + y^2}\, dA = \int_0^{2\pi} \int_0^2 \sqrt{r^2}\, r\, dr\, d\theta = \int_0^{2\pi} d\theta \int_0^2 r^2\, dr = \left[\theta \right]_0^{2\pi} \left[\tfrac{1}{3}r^3 \right]_0^2 = 2\pi\left(\tfrac{8}{3} \right) = \tfrac{16}{3}\pi$

15. By symmetry,

$$V = 2 \iint_{x^2 + y^2 \leq a^2} \sqrt{a^2 - x^2 - y^2}\, dA = 2 \int_0^{2\pi} \int_0^a \sqrt{a^2 - r^2}\, r\, dr\, d\theta = 2 \int_0^{2\pi} d\theta \int_0^a r\sqrt{a^2 - r^2}\, dr$$

$$= 2\left[\theta \right]_0^{2\pi} \left[-\tfrac{1}{3}(a^2 - r^2)^{3/2} \right]_0^a = 2(2\pi)\left(0 + \tfrac{1}{3}a^3 \right) = \tfrac{4\pi}{3}a^3$$

17. The cone $z = \sqrt{x^2 + y^2}$ intersects the sphere $x^2 + y^2 + z^2 = 1$ when $x^2 + y^2 + \left(\sqrt{x^2 + y^2} \right)^2 = 1$ or $x^2 + y^2 = \tfrac{1}{2}$. So

$$V = \iint_{x^2 + y^2 \leq 1/2} \left(\sqrt{1 - x^2 - y^2} - \sqrt{x^2 + y^2} \right) dA = \int_0^{2\pi} \int_0^{1/\sqrt{2}} \left(\sqrt{1 - r^2} - r \right) r\, dr\, d\theta$$

$$= \int_0^{2\pi} d\theta \int_0^{1/\sqrt{2}} \left(r\sqrt{1 - r^2} - r^2 \right) dr = \left[\theta \right]_0^{2\pi} \left[-\tfrac{1}{3}(1 - r^2)^{3/2} - \tfrac{1}{3}r^3 \right]_0^{1/\sqrt{2}} = 2\pi\left(-\tfrac{1}{3} \right)\left(\tfrac{1}{\sqrt{2}} - 1 \right) = \tfrac{\pi}{3}\left(2 - \sqrt{2} \right)$$

19. The given solid is the region inside the cylinder $x^2 + y^2 = 4$ between the surfaces $z = \sqrt{64 - 4x^2 - 4y^2}$

and $z = -\sqrt{64 - 4x^2 - 4y^2}$. So

$$V = \iint\limits_{x^2+y^2 \leq 4} \left[\sqrt{64 - 4x^2 - 4y^2} - \left(-\sqrt{64 - 4x^2 - 4y^2}\right)\right] dA = \iint\limits_{x^2+y^2 \leq 4} 2\sqrt{64 - 4x^2 - 4y^2}\, dA$$

$$= 4\int_0^{2\pi}\int_0^2 \sqrt{16 - r^2}\, r\, dr\, d\theta = 4\int_0^{2\pi} d\theta \int_0^2 r\sqrt{16 - r^2}\, dr = 4\left[\theta\right]_0^{2\pi}\left[-\tfrac{1}{3}(16 - r^2)^{3/2}\right]_0^2$$

$$= 8\pi\left(-\tfrac{1}{3}\right)(12^{3/2} - 16^{2/3}) = \tfrac{8\pi}{3}\left(64 - 24\sqrt{3}\right)$$

21. One loop is given by the region

$D = \{(r, \theta)\,|-\pi/6 \leq \theta \leq \pi/6,\, 0 \leq r \leq \cos 3\theta\,\}$, so the area is

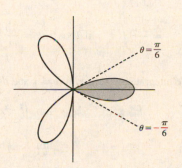

$$\iint_D dA = \int_{-\pi/6}^{\pi/6}\int_0^{\cos 3\theta} r\, dr\, d\theta = \int_{-\pi/6}^{\pi/6}\left[\tfrac{1}{2}r^2\right]_{r=0}^{r=\cos 3\theta} d\theta$$

$$= \int_{-\pi/6}^{\pi/6} \tfrac{1}{2}\cos^2 3\theta\, d\theta = 2\int_0^{\pi/6} \tfrac{1}{2}\left(\frac{1 + \cos 6\theta}{2}\right) d\theta$$

$$= \tfrac{1}{2}\left[\theta + \tfrac{1}{6}\sin 6\theta\right]_0^{\pi/6} = \frac{\pi}{12}$$

23.

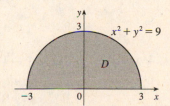

$$\int_{-3}^3\int_0^{\sqrt{9-x^2}} \sin(x^2 + y^2)\, dy\, dx = \int_0^\pi\int_0^3 \sin\left(r^2\right) r\, dr\, d\theta$$

$$= \int_0^\pi d\theta \int_0^3 r\sin\left(r^2\right) dr = \left[\theta\right]_0^\pi\left[-\tfrac{1}{2}\cos\left(r^2\right)\right]_0^3$$

$$= \pi\left(-\tfrac{1}{2}\right)(\cos 9 - 1) = \tfrac{\pi}{2}(1 - \cos 9)$$

25.

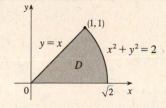

$$\int_0^{\pi/4}\int_0^{\sqrt{2}} (r\cos\theta + r\sin\theta)\, r\, dr\, d\theta = \int_0^{\pi/4}(\cos\theta + \sin\theta)\, d\theta \int_0^{\sqrt{2}} r^2\, dr$$

$$= \left[\sin\theta - \cos\theta\right]_0^{\pi/4}\left[\tfrac{1}{3}r^3\right]_0^{\sqrt{2}}$$

$$= \left[\tfrac{\sqrt{2}}{2} - \tfrac{\sqrt{2}}{2} - 0 + 1\right] \cdot \tfrac{1}{3}\left(2\sqrt{2} - 0\right) = \tfrac{2\sqrt{2}}{3}$$

27. The surface of the water in the pool is a circular disk D with radius 20 ft. If we place D on coordinate axes with the origin at

the center of D and define $f(x, y)$ to be the depth of the water at (x, y), then the volume of water in the pool is the volume of

the solid that lies above $D = \{(x, y)\,|\,x^2 + y^2 \leq 400\}$ and below the graph of $f(x, y)$. We can associate north with the

positive y-direction, so we are given that the depth is constant in the x-direction and the depth increases linearly in the

y-direction from $f(0, -20) = 2$ to $f(0, 20) = 7$. The trace in the yz-plane is a line segment from $(0, -20, 2)$ to $(0, 20, 7)$.

The slope of this line is $\frac{7-2}{20-(-20)} = \frac{1}{8}$, so an equation of the line is $z - 7 = \tfrac{1}{8}(y - 20) \Rightarrow z = \tfrac{1}{8}y + \tfrac{9}{2}$. Since $f(x, y)$ is

independent of x, $f(x, y) = \tfrac{1}{8}y + \tfrac{9}{2}$. Thus the volume is given by $\iint_D f(x, y)\, dA$, which is most conveniently evaluated

using polar coordinates. Then $D = \{(r, \theta)\,|\,0 \leq r \leq 20,\, 0 \leq \theta \leq 2\pi\}$ and substituting $x = r\cos\theta$, $y = r\sin\theta$ the integral becomes

$$\int_0^{2\pi}\int_0^{20} \left(\tfrac{1}{8}r\sin\theta + \tfrac{9}{2}\right)r\,dr\,d\theta = \int_0^{2\pi}\left[\tfrac{1}{24}r^3\sin\theta + \tfrac{9}{4}r^2\right]_{r=0}^{r=20}d\theta = \int_0^{2\pi}\left(\tfrac{1000}{3}\sin\theta + 900\right)d\theta$$

$$= \left[-\tfrac{1000}{3}\cos\theta + 900\theta\right]_0^{2\pi} = 1800\pi$$

Thus the pool contains $1800\pi \approx 5655$ ft^3 of water.

29. $\displaystyle\int_{1/\sqrt{2}}^{1}\int_{\sqrt{1-x^2}}^{x}xy\,dy\,dx + \int_1^{\sqrt{2}}\int_0^{x}xy\,dy\,dx + \int_{\sqrt{2}}^{2}\int_0^{\sqrt{4-x^2}}xy\,dy\,dx$

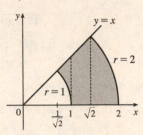

$$= \int_0^{\pi/4}\int_1^2 r^3\cos\theta\sin\theta\,dr\,d\theta = \int_0^{\pi/4}\left[\frac{r^4}{4}\cos\theta\sin\theta\right]_{r=1}^{r=2}d\theta$$

$$= \frac{15}{4}\int_0^{\pi/4}\sin\theta\cos\theta\,d\theta = \frac{15}{4}\left[\frac{\sin^2\theta}{2}\right]_0^{\pi/4} = \frac{15}{16}$$

31. (a) We integrate by parts with $u = x$ and $dv = xe^{-x^2}dx$. Then $du = dx$ and $v = -\tfrac{1}{2}e^{-x^2}$, so

$$\int_0^{\infty}x^2e^{-x^2}dx = \lim_{t\to\infty}\int_0^t x^2e^{-x^2}dx = \lim_{t\to\infty}\left(-\tfrac{1}{2}xe^{-x^2}\Big]_0^t + \int_0^t\tfrac{1}{2}e^{-x^2}dx\right)$$

$$= \lim_{t\to\infty}\left(-\tfrac{1}{2}te^{-t^2}\right) + \tfrac{1}{2}\int_0^{\infty}e^{-x^2}dx = 0 + \tfrac{1}{2}\int_0^{\infty}e^{-x^2}dx \qquad\text{[by l'Hospital's Rule]}$$

$$= \tfrac{1}{4}\int_{-\infty}^{\infty}e^{-x^2}dx \qquad\text{[since } e^{-x^2}\text{ is an even function]}$$

$$= \tfrac{1}{4}\sqrt{\pi} \qquad\text{[by Exercise 30(c)]}$$

(b) Let $u = \sqrt{x}$. Then $u^2 = x \;\Rightarrow\; dx = 2u\,du \;\Rightarrow$

$$\int_0^{\infty}\sqrt{x}\,e^{-x}dx = \lim_{t\to\infty}\int_0^t\sqrt{x}\,e^{-x}dx = \lim_{t\to\infty}\int_0^{\sqrt{t}}ue^{-u^2}2u\,du = 2\int_0^{\infty}u^2e^{-u^2}du = 2\left(\tfrac{1}{4}\sqrt{\pi}\right) \quad\text{[by part(a)]} = \tfrac{1}{2}\sqrt{\pi}.$$

12.4 Applications of Double Integrals

1. $Q = \iint_D \sigma(x,y)\,dA = \int_0^5\int_2^5 (2x+4y)\,dy\,dx = \int_0^5\left[2xy + 2y^2\right]_{y=2}^{y=5}dx$

$= \int_0^5 (10x + 50 - 4x - 8)\,dx = \int_0^5 (6x + 42)\,dx = \left[3x^2 + 42x\right]_0^5 = 75 + 210 = 285$ C

3. $m = \iint_D \rho(x,y)\,dA = \int_1^3\int_1^4 ky^2\,dy\,dx = k\int_1^3 dx\int_1^4 y^2\,dy = k\,[x]_1^3\left[\tfrac{1}{3}y^3\right]_1^4 = k(2)(21) = 42k,$

$\overline{x} = \frac{1}{m}\iint_D x\rho(x,y)\,dA = \frac{1}{42k}\int_1^3\int_1^4 kxy^2\,dy\,dx = \frac{1}{42}\int_1^3 x\,dx\int_1^4 y^2\,dy = \frac{1}{42}\left[\tfrac{1}{2}x^2\right]_1^3\left[\tfrac{1}{3}y^3\right]_1^4 = \frac{1}{42}(4)(21) = 2,$

$\overline{y} = \frac{1}{m}\iint_D y\rho(x,y)\,dA = \frac{1}{42k}\int_1^3\int_1^4 ky^3\,dy\,dx = \frac{1}{42}\int_1^3 dx\int_1^4 y^3\,dy = \frac{1}{42}\,[x]_1^3\left[\tfrac{1}{4}y^4\right]_1^4 = \frac{1}{42}(2)\left(\tfrac{255}{4}\right) = \frac{85}{28}$

Hence $m = 42k,\ (\overline{x}, \overline{y}) = \left(2, \tfrac{85}{28}\right).$

5. $m = \int_0^2\int_{x/2}^{3-x}(x+y)\,dy\,dx = \int_0^2\left[xy + \tfrac{1}{2}y^2\right]_{y=x/2}^{y=3-x}dx = \int_0^2\left[x\left(3 - \tfrac{3}{2}x\right) + \tfrac{1}{2}(3-x)^2 - \tfrac{1}{8}x^2\right]dx$

$= \int_0^2\left(-\tfrac{9}{8}x^2 + \tfrac{9}{2}\right)dx = \left[-\tfrac{9}{8}\left(\tfrac{1}{3}x^3\right) + \tfrac{9}{2}x\right]_0^2 = 6,$

$M_y = \int_0^2\int_{x/2}^{3-x}(x^2 + xy)\,dy\,dx = \int_0^2\left[x^2y + \tfrac{1}{2}xy^2\right]_{y=x/2}^{y=3-x}dx = \int_0^2\left(\tfrac{9}{2}x - \tfrac{9}{8}x^3\right)dx = \tfrac{9}{2},$

$M_x = \int_0^2\int_{x/2}^{3-y}(xy + y^2)\,dy\,dx = \int_0^2\left[\tfrac{1}{2}xy^2 + \tfrac{1}{3}y^3\right]_{y=x/2}^{y=3-x}dx = \int_0^2\left(9 - \tfrac{9}{2}x\right)dx = 9.$

Hence $m = 6,\ (\overline{x}, \overline{y}) = \left(\dfrac{M_y}{m}, \dfrac{M_x}{m}\right) = \left(\dfrac{3}{4}, \dfrac{3}{2}\right).$

7. $m = \int_{-1}^{1} \int_0^{1-x^2} ky \, dy \, dx = k \int_{-1}^{1} \left[\frac{1}{2}y^2\right]_{y=0}^{y=1-x^2} dx = \frac{1}{2}k \int_{-1}^{1} (1-x^2)^2 \, dx = \frac{1}{2}k \int_{-1}^{1} (1 - 2x^2 + x^4) \, dx$

$\qquad = \frac{1}{2}k\left[x - \frac{2}{3}x^3 + \frac{1}{5}x^5\right]_{-1}^1 = \frac{1}{2}k\left(1 - \frac{2}{3} + \frac{1}{5} + 1 - \frac{2}{3} + \frac{1}{5}\right) = \frac{8}{15}k,$

$M_y = \int_{-1}^1 \int_0^{1-x^2} kxy \, dy \, dx = k \int_{-1}^1 \left[\frac{1}{2}xy^2\right]_{y=0}^{y=1-x^2} dx = \frac{1}{2}k \int_{-1}^1 x(1-x^2)^2 \, dx = \frac{1}{2}k \int_{-1}^1 (x - 2x^3 + x^5) \, dx$

$\qquad = \frac{1}{2}k\left[\frac{1}{2}x^2 - \frac{1}{2}x^4 + \frac{1}{6}x^6\right]_{-1}^1 = \frac{1}{2}k\left(\frac{1}{2} - \frac{1}{2} + \frac{1}{6} - \frac{1}{2} + \frac{1}{2} - \frac{1}{6}\right) = 0,$

$M_x = \int_{-1}^1 \int_0^{1-x^2} ky^2 \, dy \, dx = k \int_{-1}^1 \left[\frac{1}{3}y^3\right]_{y=0}^{y=1-x^2} dx = \frac{1}{3}k \int_{-1}^1 (1-x^2)^3 \, dx = \frac{1}{3}k \int_{-1}^1 (1 - 3x^2 + 3x^4 - x^6) \, dx$

$\qquad = \frac{1}{3}k\left[x - x^3 + \frac{3}{5}x^5 - \frac{1}{7}x^7\right]_{-1}^1 = \frac{1}{3}k\left(1 - 1 + \frac{3}{5} - \frac{1}{7} + 1 - 1 + \frac{3}{5} - \frac{1}{7}\right) = \frac{32}{105}k.$

Hence $m = \frac{8}{15}k,\ (\overline{x}, \overline{y}) = \left(0, \frac{32k/105}{8k/15}\right) = \left(0, \frac{4}{7}\right).$

9. Note that $\sin(\pi x/L) \geq 0$ for $0 \leq x \leq L$.

$m = \int_0^L \int_0^{\sin(\pi x/L)} y \, dy \, dx = \int_0^L \frac{1}{2}\sin^2(\pi x/L) \, dx = \frac{1}{2}\left[\frac{1}{2}x - \frac{L}{4\pi}\sin(2\pi x/L)\right]_0^L = \frac{1}{4}L,$

$\qquad M_y = \int_0^L \int_0^{\sin(\pi x/L)} x \cdot y \, dy \, dx = \frac{1}{2}\int_0^L x\sin^2(\pi x/L) \, dx \quad \left[\begin{array}{l}\text{integrate by parts with} \\ u = x,\, dv = \sin^2(\pi x/L)\, dx\end{array}\right]$

$\qquad\qquad = \frac{1}{2}\cdot x\left(\frac{1}{2}x - \frac{L}{4\pi}\sin(2\pi x/L)\right)\big]_0^L - \frac{1}{2}\int_0^L \left[\frac{1}{2}x - \frac{L}{4\pi}\sin(2\pi x/L)\right] dx$

$\qquad\qquad = \frac{1}{4}L^2 - \frac{1}{2}\left[\frac{1}{4}x^2 + \frac{L^2}{4\pi^2}\cos(2\pi x/L)\right]_0^L = \frac{1}{4}L^2 - \frac{1}{2}\left(\frac{1}{4}L^2 + \frac{L^2}{4\pi^2} - \frac{L^2}{4\pi^2}\right) = \frac{1}{8}L^2,$

$\qquad M_x = \int_0^L \int_0^{\sin(\pi x/L)} y \cdot y \, dy \, dx = \int_0^L \frac{1}{3}\sin^3(\pi x/L) \, dx = \frac{1}{3}\int_0^L \left[1 - \cos^2(\pi x/L)\right]\sin(\pi x/L) \, dx$

$\qquad\qquad\qquad\quad [\text{substitute } u = \cos(\pi x/L)] \quad \Rightarrow \quad du = -\frac{\pi}{L}\sin(\pi x/L)]$

$\qquad\qquad = \frac{1}{3}\left(-\frac{L}{\pi}\right)\left[\cos(\pi x/L) - \frac{1}{3}\cos^3(\pi x/L)\right]_0^L = -\frac{L}{3\pi}\left(-1 + \frac{1}{3} - 1 + \frac{1}{3}\right) = \frac{4}{9\pi}L.$

Hence $m = \frac{L}{4},\ (\overline{x}, \overline{y}) = \left(\frac{L^2/8}{L/4}, \frac{4L/(9\pi)}{L/4}\right) = \left(\frac{L}{2}, \frac{16}{9\pi}\right).$

11. $\rho(x, y) = ky = kr\sin\theta,\ m = \int_0^{\pi/2} \int_0^1 kr^2 \sin\theta \, dr \, d\theta = \frac{1}{3}k \int_0^{\pi/2} \sin\theta \, d\theta = \frac{1}{3}k\left[-\cos\theta\right]_0^{\pi/2} = \frac{1}{3}k,$

$M_y = \int_0^{\pi/2} \int_0^1 kr^3 \sin\theta \cos\theta \, dr \, d\theta = \frac{1}{4}k \int_0^{\pi/2} \sin\theta \cos\theta \, d\theta = \frac{1}{8}k\left[-\cos 2\theta\right]_0^{\pi/2} = \frac{1}{8}k,$

$M_x = \int_0^{\pi/2} \int_0^1 kr^3 \sin^2\theta \, dr \, d\theta = \frac{1}{4}k \int_0^{\pi/2} \sin^2\theta \, d\theta = \frac{1}{8}k\left[\theta + \sin 2\theta\right]_0^{\pi/2} = \frac{\pi}{16}k.$

Hence $(\overline{x}, \overline{y}) = \left(\frac{3}{8}, \frac{3\pi}{16}\right).$

13.

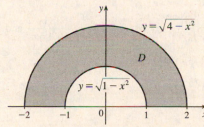

$\rho(x, y) = k\sqrt{x^2 + y^2} = kr,$

$m = \iint_D \rho(x, y) \, dA = \int_0^\pi \int_1^2 kr \cdot r \, dr \, d\theta$

$\qquad = k \int_0^\pi d\theta \int_1^2 r^2 \, dr = k(\pi)\left[\frac{1}{3}r^3\right]_1^2 = \frac{7}{3}\pi k,$

$M_y = \iint_D x\rho(x, y) \, dA = \int_0^\pi \int_1^2 (r\cos\theta)(kr)\, r \, dr \, d\theta = k \int_0^\pi \cos\theta \, d\theta \int_1^2 r^3 \, dr$

$\qquad = k\left[\sin\theta\right]_0^\pi \left[\frac{1}{4}r^4\right]_1^2 = k(0)\left(\frac{15}{4}\right) = 0$ \qquad [this is to be expected as the region and density function are symmetric about the y-axis]

[continued]

$$M_x = \iint_D y\rho(x,y)\,dA = \int_0^\pi \int_1^2 (r\sin\theta)(kr)\,r\,dr\,d\theta = k\int_0^\pi \sin\theta\,d\theta \int_1^2 r^3\,dr$$

$$= k\left[-\cos\theta\right]_0^\pi \left[\tfrac{1}{4}r^4\right]_1^2 = k(1+1)\left(\tfrac{15}{4}\right) = \tfrac{15}{2}k.$$

Hence $(\overline{x}, \overline{y}) = \left(0, \frac{15k/2}{7\pi k/3}\right) = \left(0, \frac{45}{14\pi}\right)$.

15. Placing the vertex opposite the hypotenuse at $(0,0)$, $\rho(x,y) = k(x^2+y^2)$. Then

$$m = \int_0^a \int_0^{a-x} k(x^2+y^2)\,dy\,dx = k\int_0^a \left[ax^2 - x^3 + \tfrac{1}{3}(a-x)^3\right]dx = k\left[\tfrac{1}{3}ax^3 - \tfrac{1}{4}x^4 - \tfrac{1}{12}(a-x)^4\right]_0^a = \tfrac{1}{6}ka^4.$$

By symmetry, $\quad M_y = M_x = \int_0^a \int_0^{a-x} ky(x^2+y^2)\,dy\,dx = k\int_0^a \left[\tfrac{1}{2}(a-x)^2 x^2 + \tfrac{1}{4}(a-x)^4\right]dx$

$$= k\left[\tfrac{1}{6}a^2x^3 - \tfrac{1}{4}ax^4 + \tfrac{1}{10}x^5 - \tfrac{1}{20}(a-x)^5\right]_0^a = \tfrac{1}{15}ka^5$$

Hence $(\overline{x}, \overline{y}) = \left(\tfrac{2}{5}a, \tfrac{2}{5}a\right)$.

17. $I_x = \iint_D y^2\rho(x,y)\,dA = \int_{-1}^1 \int_0^{1-x^2} y^2 \cdot ky\,dy\,dx = k\int_{-1}^1 \left[\tfrac{1}{4}y^4\right]_{y=0}^{y=1-x^2} dx = \tfrac{1}{4}k\int_{-1}^1 (1-x^2)^4\,dx$

$$= \tfrac{1}{4}k\int_{-1}^1 (x^8 - 4x^6 + 6x^4 - 4x^2 + 1)\,dx = \tfrac{1}{4}k\left[\tfrac{1}{9}x^9 - \tfrac{4}{7}x^7 + \tfrac{6}{5}x^5 - \tfrac{4}{3}x^3 + x\right]_{-1}^1 = \tfrac{64}{315}k,$$

$I_y = \iint_D x^2\rho(x,y)\,dA = \int_{-1}^1 \int_0^{1-x^2} kx^2 y\,dy\,dx = k\int_{-1}^1 \left[\tfrac{1}{2}x^2 y^2\right]_{y=0}^{y=1-x^2} dx = \tfrac{1}{2}k\int_{-1}^1 x^2(1-x^2)^2\,dx$

$$= \tfrac{1}{2}k\int_{-1}^1 (x^2 - 2x^4 + x^6)\,dx = \tfrac{1}{2}k\left[\tfrac{1}{3}x^3 - \tfrac{2}{5}x^5 + \tfrac{1}{7}x^7\right]_{-1}^1 = \tfrac{8}{105}k,$$

and $I_0 = I_x + I_y = \tfrac{64}{315}k + \tfrac{8}{105}k = \tfrac{88}{315}k$.

19. As in Exercise 15, we place the vertex opposite the hypotenuse at $(0,0)$ and the equal sides along the positive axes.

$I_x = \int_0^a \int_0^{a-x} y^2 k(x^2+y^2)\,dy\,dx = k\int_0^a \int_0^{a-x} (x^2y^2 + y^4)\,dy\,dx = k\int_0^a \left[\tfrac{1}{3}x^2y^3 + \tfrac{1}{5}y^5\right]_{y=0}^{y=a-x} dx$

$$= k\int_0^a \left[\tfrac{1}{3}x^2(a-x)^3 + \tfrac{1}{5}(a-x)^5\right]dx = k\left[\tfrac{1}{3}\left(\tfrac{1}{3}a^3x^3 - \tfrac{3}{4}a^2x^4 + \tfrac{3}{5}ax^5 - \tfrac{1}{6}x^6\right) - \tfrac{1}{30}(a-x)^6\right]_0^a = \tfrac{7}{180}ka^6,$$

$I_y = \int_0^a \int_0^{a-x} x^2 k(x^2+y^2)\,dy\,dx = k\int_0^a \int_0^{a-x}(x^4 + x^2y^2)\,dy\,dx = k\int_0^a \left[x^4 y + \tfrac{1}{3}x^2y^3\right]_{y=0}^{y=a-x} dx$

$$= k\int_0^a \left[x^4(a-x) + \tfrac{1}{3}x^2(a-x)^3\right]dx = k\left[\tfrac{1}{5}ax^5 - \tfrac{1}{6}x^6 + \tfrac{1}{3}\left(\tfrac{1}{3}a^3x^3 - \tfrac{3}{4}a^2x^4 + \tfrac{3}{5}ax^5 - \tfrac{1}{6}x^6\right)\right]_0^a = \tfrac{7}{180}ka^6,$$

and $I_0 = I_x + I_y = \tfrac{7}{90}ka^6$.

21. The right loop of the curve is given by $D = \{(r,\theta) \mid 0 \le r \le \cos 2\theta,\ -\pi/4 \le \theta \le \pi/4\}$. Using a CAS, we

find $m = \iint_D \rho(x,y)\,dA = \iint_D (x^2+y^2)\,dA = \int_{-\pi/4}^{\pi/4} \int_0^{\cos 2\theta} r^2\,r\,dr\,d\theta = \dfrac{3\pi}{64}$. Then

$$\overline{x} = \frac{1}{m}\iint_D x\rho(x,y)\,dA = \frac{64}{3\pi}\int_{-\pi/4}^{\pi/4}\int_0^{\cos 2\theta}(r\cos\theta)r^2\,r\,dr\,d\theta = \frac{64}{3\pi}\int_{-\pi/4}^{\pi/4}\int_0^{\cos 2\theta} r^4\cos\theta\,dr\,d\theta = \frac{16384\sqrt{2}}{10395\pi}\ \text{and}$$

$$\overline{y} = \frac{1}{m}\iint_D y\rho(x,y)\,dA = \frac{64}{3\pi}\int_{-\pi/4}^{\pi/4}\int_0^{\cos 2\theta}(r\sin\theta)r^2\,r\,dr\,d\theta = \frac{64}{3\pi}\int_{-\pi/4}^{\pi/4}\int_0^{\cos 2\theta} r^4\sin\theta\,dr\,d\theta = 0,\ \text{so}$$

$(\overline{x}, \overline{y}) = \left(\dfrac{16384\sqrt{2}}{10395\pi}, 0\right)$.

The moments of inertia are

$$I_x = \iint_D y^2\rho(x,y)\,dA = \int_{-\pi/4}^{\pi/4}\int_0^{\cos 2\theta}(r\sin\theta)^2 r^2\,r\,dr\,d\theta = \int_{-\pi/4}^{\pi/4}\int_0^{\cos 2\theta} r^5\sin^2\theta\,dr\,d\theta = \frac{5\pi}{384} - \frac{4}{105},$$

$$I_y = \iint_D x^2 \rho(x,y)\, dA = \int_{-\pi/4}^{\pi/4} \int_0^{\cos 2\theta} (r\cos\theta)^2\, r^2\, r\, dr\, d\theta = \int_{-\pi/4}^{\pi/4} \int_0^{\cos 2\theta} r^5 \cos^2\theta\, dr\, d\theta = \frac{5\pi}{384} + \frac{4}{105}, \text{ and}$$

$$I_0 = I_x + I_y = \frac{5\pi}{192}.$$

23. $I_x = \iint_D y^2 \rho(x,y)\, dA = \int_0^h \int_0^b \rho y^2\, dx\, dy = \rho \int_0^b dx \int_0^h y^2\, dy = \rho\, [x]_0^b \left[\frac{1}{3} y^3\right]_0^h = \rho b \left(\frac{1}{3} h^3\right) = \frac{1}{3}\rho b h^3,$

$I_y = \iint_D x^2 \rho(x,y)\, dA = \int_0^h \int_0^b \rho x^2\, dx\, dy = \rho \int_0^b x^2\, dx \int_0^h dy = \rho \left[\frac{1}{3} x^3\right]_0^b [y]_0^h = \frac{1}{3}\rho b^3 h,$

and $m = \rho\,(\text{area of rectangle}) = \rho b h$ since the lamina is homogeneous. Hence $\overline{\overline{x}}^2 = \dfrac{I_y}{m} = \dfrac{\frac{1}{3}\rho b^3 h}{\rho b h} = \dfrac{b^2}{3} \quad\Rightarrow\quad \overline{\overline{x}} = \dfrac{b}{\sqrt{3}}$

and $\overline{\overline{y}}^2 = \dfrac{I_x}{m} = \dfrac{\frac{1}{3}\rho b h^3}{\rho b h} = \dfrac{h^2}{3} \quad\Rightarrow\quad \overline{\overline{y}} = \dfrac{h}{\sqrt{3}}.$

12.5   Triple Integrals

1. $\iiint_B xyz^2\, dV = \int_0^1 \int_{-1}^2 \int_0^3 xyz^2\, dz\, dx\, dy = \int_0^1 \int_{-1}^2 xy \left[\frac{1}{3} z^3\right]_{z=0}^{z=3} dx\, dy = \int_0^1 \int_{-1}^2 9xy\, dx\, dy$

$= \int_0^1 \left[\frac{9}{2} x^2 y\right]_{x=-1}^{x=2} dy = \int_0^1 \frac{27}{2} y\, dy = \frac{27}{4} y^2 \Big]_0^1 = \frac{27}{4}$

3. $\int_0^2 \int_0^{z^2} \int_0^{y-z} (2x - y)\, dx\, dy\, dz = \int_0^2 \int_0^{z^2} \left[x^2 - xy\right]_{x=0}^{x=y-z} dy\, dz = \int_0^2 \int_0^{z^2} \left[(y-z)^2 - (y-z)y\right] dy\, dz$

$= \int_0^2 \int_0^{z^2} (z^2 - yz)\, dy\, dz = \int_0^2 \left[yz^2 - \frac{1}{2} y^2 z\right]_{y=0}^{y=z^2} dz = \int_0^2 \left(z^4 - \frac{1}{2} z^5\right) dz$

$= \left[\frac{1}{5} z^5 - \frac{1}{12} z^6\right]_0^2 = \frac{32}{5} - \frac{64}{12} = \frac{16}{15}$

5. $\int_0^{\pi/2} \int_0^y \int_0^x \cos(x+y+z)\, dz\, dx\, dy = \int_0^{\pi/2} \int_0^y \left[\sin(x+y+z)\right]_{z=0}^{z=x} dx\, dy$

$= \int_0^{\pi/2} \int_0^y \left[\sin(2x+y) - \sin(x+y)\right] dx\, dy$

$= \int_0^{\pi/2} \left[-\frac{1}{2}\cos(2x+y) + \cos(x+y)\right]_{x=0}^{x=y} dy$

$= \int_0^{\pi/2} \left[-\frac{1}{2}\cos 3y + \cos 2y + \frac{1}{2}\cos y - \cos y\right] dy$

$= \left[-\frac{1}{6}\sin 3y + \frac{1}{2}\sin 2y - \frac{1}{2}\sin y\right]_0^{\pi/2} = \frac{1}{6} - \frac{1}{2} = -\frac{1}{3}$

7. $\iiint_E y\, dV = \int_0^3 \int_0^x \int_{x-y}^{x+y} y\, dz\, dy\, dx = \int_0^3 \int_0^x [yz]_{z=x-y}^{z=x+y} dy\, dx = \int_0^3 \int_0^x 2y^2\, dy\, dx$

$= \int_0^3 \left[\frac{2}{3} y^3\right]_{y=0}^{y=x} dx = \int_0^3 \frac{2}{3} x^3\, dx = \frac{1}{6} x^4 \Big]_0^3 = \frac{81}{6} = \frac{27}{2}$

9. $\iiint_E \dfrac{z}{x^2 + z^2}\, dV = \int_1^4 \int_y^4 \int_0^z \dfrac{z}{x^2 + z^2}\, dx\, dz\, dy = \int_1^4 \int_y^4 \left[z \cdot \frac{1}{z} \tan^{-1}\frac{x}{z}\right]_{x=0}^{x=z} dz\, dy$

$= \int_1^4 \int_y^4 \left[\tan^{-1}(1) - \tan^{-1}(0)\right] dz\, dy = \int_1^4 \int_y^4 \left(\frac{\pi}{4} - 0\right) dz\, dy = \frac{\pi}{4} \int_1^4 [z]_{z=y}^{z=4}\, dy$

$= \frac{\pi}{4} \int_1^4 (4 - y)\, dy = \frac{\pi}{4} \left[4y - \frac{1}{2} y^2\right]_1^4 = \frac{\pi}{4} \left(16 - 8 - 4 + \frac{1}{2}\right) = \frac{9\pi}{8}$

11. Here $E = \{(x, y, z) \mid 0 \le x \le 1, 0 \le y \le \sqrt{x}, 0 \le z \le 1 + x + y\}$, so

$$\iiint_E 6xy \, dV = \int_0^1 \int_0^{\sqrt{x}} \int_0^{1+x+y} 6xy \, dz \, dy \, dx = \int_0^1 \int_0^{\sqrt{x}} \left[6xyz\right]_{z=0}^{z=1+x+y} dy \, dx$$

$$= \int_0^1 \int_0^{\sqrt{x}} 6xy(1 + x + y) \, dy \, dx = \int_0^1 \left[3xy^2 + 3x^2y^2 + 2xy^3\right]_{y=0}^{y=\sqrt{x}} dx$$

$$= \int_0^1 (3x^2 + 3x^3 + 2x^{5/2}) \, dx = \left[x^3 + \tfrac{3}{4}x^4 + \tfrac{4}{7}x^{7/2}\right]_0^1 = \tfrac{65}{28}$$

13.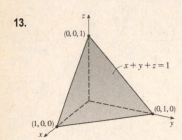

Here $T = \{(x, y, z) \mid 0 \le x \le 1, 0 \le y \le 1 - x, 0 \le z \le 1 - x - y\}$, so

$$\iiint_T x^2 \, dV = \int_0^1 \int_0^{1-x} \int_0^{1-x-y} x^2 \, dz \, dy \, dx = \int_0^1 \int_0^{1-x} x^2(1 - x - y) \, dy \, dx$$

$$= \int_0^1 \int_0^{1-x} (x^2 - x^3 - x^2y) \, dy \, dx = \int_0^1 \left[x^2 y - x^3 y - \tfrac{1}{2}x^2 y^2\right]_{y=0}^{y=1-x} dx$$

$$= \int_0^1 \left[x^2(1 - x) - x^3(1 - x) - \tfrac{1}{2}x^2(1 - x)^2\right] dx$$

$$= \int_0^1 \left(\tfrac{1}{2}x^4 - x^3 + \tfrac{1}{2}x^2\right) dx = \left[\tfrac{1}{10}x^5 - \tfrac{1}{4}x^4 + \tfrac{1}{6}x^3\right]_0^1$$

$$= \tfrac{1}{10} - \tfrac{1}{4} + \tfrac{1}{6} = \tfrac{1}{60}$$

15.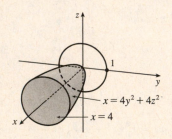

The projection of E on the yz-plane is the disk $y^2 + z^2 \le 1$. Using polar coordinates $y = r\cos\theta$ and $z = r\sin\theta$, we get

$$\iiint_E x \, dV = \iint_D \left[\int_{4y^2 + 4z^2}^4 x \, dx\right] dA = \tfrac{1}{2} \iint_D \left[4^2 - (4y^2 + 4z^2)^2\right] dA$$

$$= 8 \int_0^{2\pi} \int_0^1 (1 - r^4) \, r \, dr \, d\theta = 8 \int_0^{2\pi} d\theta \int_0^1 (r - r^5) \, dr$$

$$= 8(2\pi)\left[\tfrac{1}{2}r^2 - \tfrac{1}{6}r^6\right]_0^1 = \tfrac{16\pi}{3}$$

17. The plane $2x + y + z = 4$ intersects the xy-plane when

$2x + y + 0 = 4 \Rightarrow y = 4 - 2x$, so

$E = \{(x, y, z) \mid 0 \le x \le 2, 0 \le y \le 4 - 2x, 0 \le z \le 4 - 2x - y\}$ and

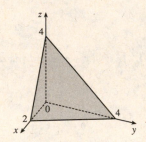

$$V = \int_0^2 \int_0^{4-2x} \int_0^{4-2x-y} dz \, dy \, dx = \int_0^2 \int_0^{4-2x} (4 - 2x - y) \, dy \, dx$$

$$= \int_0^2 \left[4y - 2xy - \tfrac{1}{2}y^2\right]_{y=0}^{y=4-2x} dx$$

$$= \int_0^2 \left[4(4 - 2x) - 2x(4 - 2x) - \tfrac{1}{2}(4 - 2x)^2\right] dx$$

$$= \int_0^2 (2x^2 - 8x + 8) \, dx = \left[\tfrac{2}{3}x^3 - 4x^2 + 8x\right]_0^2 = \tfrac{16}{3}$$

19. The plane $y + z = 1$ intersects the xy-plane in the line $y = 1$, so

$E = \{(x, y, z) \mid -1 \le x \le 1, x^2 \le y \le 1, 0 \le z \le 1 - y\}$ and

$$V = \iiint_E dV = \int_{-1}^1 \int_{x^2}^1 \int_0^{1-y} dz \, dy \, dx = \int_{-1}^1 \int_{x^2}^1 (1 - y) \, dy \, dx$$

$$= \int_{-1}^1 \left[y - \tfrac{1}{2}y^2\right]_{y=x^2}^{y=1} dx = \int_{-1}^1 \left(\tfrac{1}{2} - x^2 + \tfrac{1}{2}x^4\right) dx$$

$$= \left[\tfrac{1}{2}x - \tfrac{1}{3}x^3 + \tfrac{1}{10}x^5\right]_{-1}^1 = \tfrac{1}{2} - \tfrac{1}{3} + \tfrac{1}{10} + \tfrac{1}{2} - \tfrac{1}{3} + \tfrac{1}{10} = \tfrac{8}{15}$$

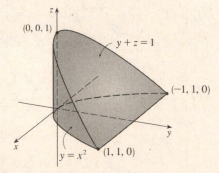

21. (a) The wedge can be described as the region

$$D = \left\{(x, y, z) \mid y^2 + z^2 \leq 1, 0 \leq x \leq 1, 0 \leq y \leq x\right\}$$
$$= \left\{(x, y, z) \mid 0 \leq x \leq 1, 0 \leq y \leq x, 0 \leq z \leq \sqrt{1-y^2}\right\}$$

So the integral expressing the volume of the wedge is

$$\iiint_D dV = \int_0^1 \int_0^x \int_0^{\sqrt{1-y^2}} dz\, dy\, dx.$$

(b) A CAS gives $\int_0^1 \int_0^x \int_0^{\sqrt{1-y^2}} dz\, dy\, dx = \frac{\pi}{4} - \frac{1}{3}$.

(Or use Formulas 30 and 87 from the Table of Integrals.)

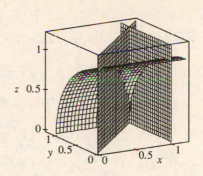

23. Here $f(x, y, z) = \cos(xyz)$ and $\Delta V = \frac{1}{2} \cdot \frac{1}{2} \cdot \frac{1}{2} = \frac{1}{8}$, so the Midpoint Rule gives

$$\iiint_B f(x, y, z)\, dV \approx \sum_{i=1}^{l} \sum_{j=1}^{m} \sum_{k=1}^{n} f(\bar{x}_i, \bar{y}_j, \bar{z}_k)\, \Delta V$$

$$= \frac{1}{8}\left[f\left(\tfrac{1}{4}, \tfrac{1}{4}, \tfrac{1}{4}\right) + f\left(\tfrac{1}{4}, \tfrac{1}{4}, \tfrac{3}{4}\right) + f\left(\tfrac{1}{4}, \tfrac{3}{4}, \tfrac{1}{4}\right) + f\left(\tfrac{1}{4}, \tfrac{3}{4}, \tfrac{3}{4}\right)\right.$$
$$\left. + f\left(\tfrac{3}{4}, \tfrac{1}{4}, \tfrac{1}{4}\right) + f\left(\tfrac{3}{4}, \tfrac{1}{4}, \tfrac{3}{4}\right) + f\left(\tfrac{3}{4}, \tfrac{3}{4}, \tfrac{1}{4}\right) + f\left(\tfrac{3}{4}, \tfrac{3}{4}, \tfrac{3}{4}\right)\right]$$

$$= \frac{1}{8}\left[\cos\tfrac{1}{64} + \cos\tfrac{3}{64} + \cos\tfrac{3}{64} + \cos\tfrac{9}{64} + \cos\tfrac{3}{64} + \cos\tfrac{9}{64} + \cos\tfrac{9}{64} + \cos\tfrac{27}{64}\right] \approx 0.985$$

25. $E = \left\{(x, y, z) \mid 0 \leq x \leq 1, 0 \leq z \leq 1-x, 0 \leq y \leq 2-2z\right\}$,

the solid bounded by the three coordinate planes and the planes

$z = 1 - x$, $y = 2 - 2z$.

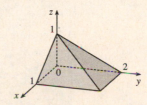

27.

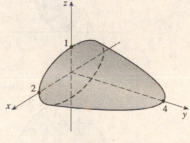

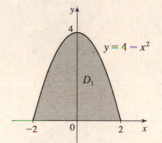

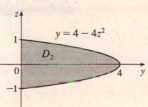

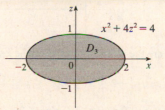

If D_1, D_2, D_3 are the projections of E on the xy-, yz-, and xz-planes, then

$$D_1 = \left\{(x, y) \mid -2 \leq x \leq 2, 0 \leq y \leq 4-x^2\right\} = \left\{(x, y) \mid 0 \leq y \leq 4, -\sqrt{4-y} \leq x \leq \sqrt{4-y}\right\}$$

$$D_2 = \left\{(y, z) \mid 0 \leq y \leq 4, -\tfrac{1}{2}\sqrt{4-y} \leq z \leq \tfrac{1}{2}\sqrt{4-y}\right\} = \left\{(y, z) \mid -1 \leq z \leq 1, 0 \leq y \leq 4-4z^2\right\}$$

$$D_3 = \left\{(x, z) \mid x^2 + 4z^2 \leq 4\right\}$$

[continued]

Therefore

$$E = \left\{ (x,y,z) \mid -2 \le x \le 2,\, 0 \le y \le 4 - x^2,\, -\tfrac{1}{2}\sqrt{4 - x^2 - y} \le z \le \tfrac{1}{2}\sqrt{4 - x^2 - y} \right\}$$

$$= \left\{ (x,y,z) \mid 0 \le y \le 4,\, -\sqrt{4 - y} \le x \le \sqrt{4 - y},\, -\tfrac{1}{2}\sqrt{4 - x^2 - y} \le z \le \tfrac{1}{2}\sqrt{4 - x^2 - y} \right\}$$

$$= \left\{ (x,y,z) \mid -1 \le z \le 1,\, 0 \le y \le 4 - 4z^2,\, -\sqrt{4 - y - 4z^2} \le x \le \sqrt{4 - y - 4z^2} \right\}$$

$$= \left\{ (x,y,z) \mid 0 \le y \le 4,\, -\tfrac{1}{2}\sqrt{4 - y} \le z \le \tfrac{1}{2}\sqrt{4 - y},\, -\sqrt{4 - y - 4z^2} \le x \le \sqrt{4 - y - 4z^2} \right\}$$

$$= \left\{ (x,y,z) \mid -2 \le x \le 2,\, -\tfrac{1}{2}\sqrt{4 - x^2} \le z \le \tfrac{1}{2}\sqrt{4 - x^2},\, 0 \le y \le 4 - x^2 - 4z^2 \right\}$$

$$= \left\{ (x,y,z) \mid -1 \le z \le 1,\, -\sqrt{4 - 4z^2} \le x \le \sqrt{4 - 4z^2},\, 0 \le y \le 4 - x^2 - 4z^2 \right\}$$

Then

$$\iiint_E f(x,y,z)\,dV = \int_{-2}^{2} \int_0^{4-x^2} \int_{-\sqrt{4-x^2-y}/2}^{\sqrt{4-x^2-y}/2} f(x,y,z)\,dz\,dy\,dx = \int_0^4 \int_{-\sqrt{4-y}}^{\sqrt{4-y}} \int_{-\sqrt{4-x^2-y}/2}^{\sqrt{4-x^2-y}/2} f(x,y,z)\,dz\,dx\,dy$$

$$= \int_{-1}^{1} \int_0^{4-4z^2} \int_{-\sqrt{4-y-4z^2}}^{\sqrt{4-y-4z^2}} f(x,y,z)\,dx\,dy\,dz = \int_0^4 \int_{-\sqrt{4-y}/2}^{\sqrt{4-y}/2} \int_{-\sqrt{4-y-4z^2}}^{\sqrt{4-y-4z^2}} f(x,y,z)\,dx\,dz\,dy$$

$$= \int_{-2}^{2} \int_{-\sqrt{4-x^2}/2}^{\sqrt{4-x^2}/2} \int_0^{4-x^2-4z^2} f(x,y,z)\,dy\,dz\,dx = \int_{-1}^{1} \int_{-\sqrt{4-4z^2}}^{\sqrt{4-4z^2}} \int_0^{4-x^2-4z^2} f(x,y,z)\,dy\,dx\,dz$$

29.

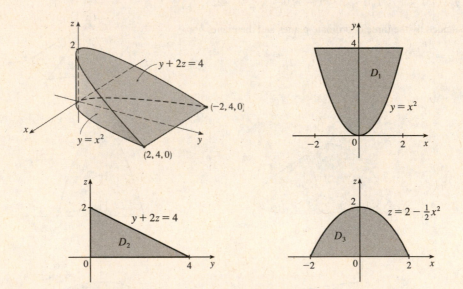

If D_1, D_2, and D_3 are the projections of E on the xy-, yz-, and xz-planes, then

$$D_1 = \left\{ (x,y) \mid -2 \le x \le 2,\, x^2 \le y \le 4 \right\} = \left\{ (x,y) \mid 0 \le y \le 4,\, -\sqrt{y} \le x \le \sqrt{y} \right\},$$

$$D_2 = \left\{ (y,z) \mid 0 \le y \le 4,\, 0 \le z \le 2 - \tfrac{1}{2}y \right\} = \left\{ (y,z) \mid 0 \le z \le 2,\, 0 \le y \le 4 - 2z \right\},\text{ and}$$

$$D_3 = \left\{ (x,z) \mid -2 \le x \le 2,\, 0 \le z \le 2 - \tfrac{1}{2}x^2 \right\} = \left\{ (x,z) \mid 0 \le z \le 2,\, -\sqrt{4 - 2z} \le x \le \sqrt{4 - 2z} \right\}$$

[continued]

Therefore
$$E = \left\{(x,y,z) \mid -2 \le x \le 2, x^2 \le y \le 4, 0 \le z \le 2 - \tfrac{1}{2}y\right\}$$
$$= \left\{(x,y,z) \mid 0 \le y \le 4, -\sqrt{y} \le x \le \sqrt{y}, 0 \le z \le 2 - \tfrac{1}{2}y\right\}$$
$$= \left\{(x,y,z) \mid 0 \le y \le 4, 0 \le z \le 2 - \tfrac{1}{2}y, -\sqrt{y} \le x \le \sqrt{y}\right\}$$
$$= \left\{(x,y,z) \mid 0 \le z \le 2, 0 \le y \le 4 - 2z, -\sqrt{y} \le x \le \sqrt{y}\right\}$$
$$= \left\{(x,y,z) \mid -2 \le x \le 2, 0 \le z \le 2 - \tfrac{1}{2}x^2, x^2 \le y \le 4 - 2z\right\}$$
$$= \left\{(x,y,z) \mid 0 \le z \le 2, -\sqrt{4-2z} \le x \le \sqrt{4-2z}, x^2 \le y \le 4 - 2z\right\}$$

Then
$$\iiint_E f(x,y,z)\,dV = \int_{-2}^{2}\int_{x^2}^{4}\int_{0}^{2-y/2} f(x,y,z)\,dz\,dy\,dx = \int_{0}^{4}\int_{-\sqrt{y}}^{\sqrt{y}}\int_{0}^{2-y/2} f(x,y,z)\,dz\,dx\,dy$$
$$= \int_{0}^{4}\int_{0}^{2-y/2}\int_{-\sqrt{y}}^{\sqrt{y}} f(x,y,z)\,dx\,dz\,dy = \int_{0}^{2}\int_{0}^{4-2z}\int_{-\sqrt{y}}^{\sqrt{y}} f(x,y,z)\,dx\,dy\,dz$$
$$= \int_{-2}^{2}\int_{0}^{2-x^2/2}\int_{x^2}^{4-2z} f(x,y,z)\,dy\,dz\,dx = \int_{0}^{2}\int_{-\sqrt{4-2z}}^{\sqrt{4-2z}}\int_{x^2}^{4-2z} f(x,y,z)\,dy\,dx\,dz$$

31.

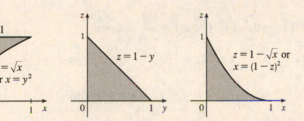

The diagrams show the projections of E on the xy-, yz-, and xz-planes.

Therefore

$$\int_{0}^{1}\int_{\sqrt{x}}^{1}\int_{0}^{1-y} f(x,y,z)\,dz\,dy\,dx = \int_{0}^{1}\int_{0}^{y^2}\int_{0}^{1-y} f(x,y,z)\,dz\,dx\,dy = \int_{0}^{1}\int_{0}^{1-z}\int_{0}^{y^2} f(x,y,z)\,dx\,dy\,dz$$
$$= \int_{0}^{1}\int_{0}^{1-y}\int_{0}^{y^2} f(x,y,z)\,dx\,dz\,dy = \int_{0}^{1}\int_{0}^{1-\sqrt{x}}\int_{\sqrt{x}}^{1-z} f(x,y,z)\,dy\,dz\,dx$$
$$= \int_{0}^{1}\int_{0}^{(1-z)^2}\int_{\sqrt{x}}^{1-z} f(x,y,z)\,dy\,dx\,dz$$

33.

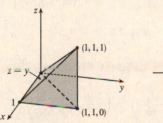
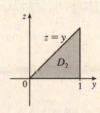
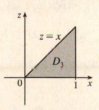

$$\int_{0}^{1}\int_{y}^{1}\int_{0}^{y} f(x,y,z)\,dz\,dx\,dy = \iiint_E f(x,y,z)\,dV \text{ where } E = \{(x,y,z) \mid 0 \le z \le y, y \le x \le 1, 0 \le y \le 1\}.$$

If D_1, D_2, and D_3 are the projections of E on the xy-, yz- and xz-planes then

$$D_1 = \{(x,y) \mid 0 \le y \le 1, y \le x \le 1\} = \{(x,y) \mid 0 \le x \le 1, 0 \le y \le x\},$$
$$D_2 = \{(y,z) \mid 0 \le y \le 1, 0 \le z \le y\} = \{(y,z) \mid 0 \le z \le 1, z \le y \le 1\}, \text{ and}$$
$$D_3 = \{(x,z) \mid 0 \le x \le 1, 0 \le z \le x\} = \{(x,z) \mid 0 \le z \le 1, z \le x \le 1\}.$$

[continued]

Thus we also have

$$E = \{(x,y,z) \mid 0 \le x \le 1, 0 \le y \le x, 0 \le z \le y\} = \{(x,y,z) \mid 0 \le y \le 1, 0 \le z \le y, y \le x \le 1\}$$

$$= \{(x,y,z) \mid 0 \le z \le 1, z \le y \le 1, y \le x \le 1\} = \{(x,y,z) \mid 0 \le x \le 1, 0 \le z \le x, z \le y \le x\}$$

$$= \{(x,y,z) \mid 0 \le z \le 1, z \le x \le 1, z \le y \le x\}.$$

Then

$$\int_0^1 \int_y^1 \int_0^y f(x,y,z)\,dz\,dx\,dy = \int_0^1 \int_0^x \int_0^y f(x,y,z)\,dz\,dy\,dx = \int_0^1 \int_0^y \int_y^1 f(x,y,z)\,dx\,dz\,dy$$

$$= \int_0^1 \int_z^1 \int_y^1 f(x,y,z)\,dx\,dy\,dz = \int_0^1 \int_0^x \int_z^x f(x,y,z)\,dy\,dz\,dx$$

$$= \int_0^1 \int_z^1 \int_z^x f(x,y,z)\,dy\,dx\,dz.$$

35. The region C is the solid bounded by a circular cylinder of radius 2 with axis the z-axis for $-2 \le z \le 2$. We can write

$\iiint_C (4 + 5x^2yz^2)\,dV = \iiint_C 4\,dV + \iiint_C 5x^2yz^2\,dV$, but $f(x,y,z) = 5x^2yz^2$ is an odd function with

respect to y. Since C is symmetrical about the xz-plane, we have $\iiint_C 5x^2yz^2\,dV = 0$. Thus

$\iiint_C (4 + 5x^2yz^2)\,dV = \iiint_C 4\,dV = 4 \cdot V(E) = 4 \cdot \pi(2)^2(4) = 64\pi$.

37. $m = \iiint_E \rho(x,y,z)\,dV = \int_0^1 \int_0^{\sqrt{x}} \int_0^{1+x+y} 2\,dz\,dy\,dx = \int_0^1 \int_0^{\sqrt{x}} 2(1+x+y)\,dy\,dx$

$= \int_0^1 \left[2y + 2xy + y^2\right]_{y=0}^{y=\sqrt{x}} dx = \int_0^1 \left(2\sqrt{x} + 2x^{3/2} + x\right) dx = \left[\frac{4}{3}x^{3/2} + \frac{4}{5}x^{5/2} + \frac{1}{2}x^2\right]_0^1 = \frac{79}{30}$

$M_{yz} = \iiint_E x\rho(x,y,z)\,dV = \int_0^1 \int_0^{\sqrt{x}} \int_0^{1+x+y} 2x\,dz\,dy\,dx = \int_0^1 \int_0^{\sqrt{x}} 2x(1+x+y)\,dy\,dx$

$= \int_0^1 \left[2xy + 2x^2y + xy^2\right]_{y=0}^{y=\sqrt{x}} dx = \int_0^1 (2x^{3/2} + 2x^{5/2} + x^2)\,dx = \left[\frac{4}{5}x^{5/2} + \frac{4}{7}x^{7/2} + \frac{1}{3}x^3\right]_0^1 = \frac{179}{105}$

$M_{xz} = \iiint_E y\rho(x,y,z)\,dV = \int_0^1 \int_0^{\sqrt{x}} \int_0^{1+x+y} 2y\,dz\,dy\,dx = \int_0^1 \int_0^{\sqrt{x}} 2y(1+x+y)\,dy\,dx$

$= \int_0^1 \left[y^2 + xy^2 + \frac{2}{3}y^3\right]_{y=0}^{y=\sqrt{x}} dx = \int_0^1 \left(x + x^2 + \frac{2}{3}x^{3/2}\right) dx = \left[\frac{1}{2}x^2 + \frac{1}{3}x^3 + \frac{4}{15}x^{5/2}\right]_0^1 = \frac{11}{10}$

$M_{xy} = \iiint_E z\rho(x,y,z)\,dV = \int_0^1 \int_0^{\sqrt{x}} \int_0^{1+x+y} 2z\,dz\,dy\,dx = \int_0^1 \int_0^{\sqrt{x}} \left[z^2\right]_{z=0}^{z=1+x+y} dy\,dx = \int_0^1 \int_0^{\sqrt{x}} (1+x+y)^2\,dy\,dx$

$= \int_0^1 \int_0^{\sqrt{x}} (1 + 2x + 2y + 2xy + x^2 + y^2)\,dy\,dx = \int_0^1 \left[y + 2xy + y^2 + xy^2 + x^2y + \frac{1}{3}y^3\right]_{y=0}^{y=\sqrt{x}} dx$

$= \int_0^1 \left(\sqrt{x} + \frac{7}{3}x^{3/2} + x + x^2 + x^{5/2}\right) dx = \left[\frac{2}{3}x^{3/2} + \frac{14}{15}x^{5/2} + \frac{1}{2}x^2 + \frac{1}{3}x^3 + \frac{2}{7}x^{7/2}\right]_0^1 = \frac{571}{210}$

Thus the mass is $\frac{79}{30}$ and the center of mass is $(\overline{x}, \overline{y}, \overline{z}) = \left(\dfrac{M_{yz}}{m}, \dfrac{M_{xz}}{m}, \dfrac{M_{xy}}{m}\right) = \left(\dfrac{358}{553}, \dfrac{33}{79}, \dfrac{571}{553}\right)$.

39. $m = \int_0^a \int_0^a \int_0^a (x^2 + y^2 + z^2)\,dx\,dy\,dz = \int_0^a \int_0^a \left[\frac{1}{3}x^3 + xy^2 + xz^2\right]_{x=0}^{x=a} dy\,dz = \int_0^a \int_0^a \left(\frac{1}{3}a^3 + ay^2 + az^2\right) dy\,dz$

$= \int_0^a \left[\frac{1}{3}a^3y + \frac{1}{3}ay^3 + ayz^2\right]_{y=0}^{y=a} dz = \int_0^a \left(\frac{2}{3}a^4 + a^2z^2\right) dz = \left[\frac{2}{3}a^4z + \frac{1}{3}a^2z^3\right]_0^a = \frac{2}{3}a^5 + \frac{1}{3}a^5 = a^5$

$M_{yz} = \int_0^a \int_0^a \int_0^a \left[x^3 + x(y^2 + z^2)\right] dx\,dy\,dz = \int_0^a \int_0^a \left[\frac{1}{4}a^4 + \frac{1}{2}a^2(y^2 + z^2)\right] dy\,dz$

$= \int_0^a \left(\frac{1}{4}a^5 + \frac{1}{6}a^5 + \frac{1}{2}a^3z^2\right) dz = \frac{1}{4}a^6 + \frac{1}{3}a^6 = \frac{7}{12}a^6 = M_{xz} = M_{xy}$ by symmetry of E and $\rho(x,y,z)$

Hence $(\overline{x}, \overline{y}, \overline{z}) = \left(\frac{7}{12}a, \frac{7}{12}a, \frac{7}{12}a\right)$.

41. (a) $m = \int_{-1}^{1} \int_{x^2}^{1} \int_{0}^{1-y} \sqrt{x^2 + y^2}\, dz\, dy\, dx$

(b) $(\overline{x}, \overline{y}, \overline{z})$ where $\overline{x} = \frac{1}{m} \int_{-1}^{1} \int_{x^2}^{1} \int_{0}^{1-y} x\sqrt{x^2 + y^2}\, dz\, dy\, dx,\ \overline{y} = \frac{1}{m} \int_{-1}^{1} \int_{x^2}^{1} \int_{0}^{1-y} y\sqrt{x^2 + y^2}\, dz\, dy\, dx,$ and

$\overline{z} = \frac{1}{m} \int_{-1}^{1} \int_{x^2}^{1} \int_{0}^{1-y} z\sqrt{x^2 + y^2}\, dz\, dy\, dx.$

(c) $I_z = \int_{-1}^{1} \int_{x^2}^{1} \int_{0}^{1-y} (x^2 + y^2)\sqrt{x^2 + y^2}\, dz\, dy\, dx = \int_{-1}^{1} \int_{x^2}^{1} \int_{0}^{1-y} (x^2 + y^2)^{3/2}\, dz\, dy\, dx$

43. (a) $m = \int_{0}^{1} \int_{0}^{\sqrt{1-x^2}} \int_{0}^{y} (1 + x + y + z)\, dz\, dy\, dx = \frac{3\pi}{32} + \frac{11}{24}$

(b) $(\overline{x}, \overline{y}, \overline{z}) = \left(m^{-1} \int_{0}^{1} \int_{0}^{\sqrt{1-x^2}} \int_{0}^{y} x(1 + x + y + z)\, dz\, dy\, dx, \right.$

$\qquad\qquad m^{-1} \int_{0}^{1} \int_{0}^{\sqrt{1-x^2}} \int_{0}^{y} y(1 + x + y + z)\, dz\, dy\, dx,$

$\qquad\qquad\qquad \left. m^{-1} \int_{0}^{1} \int_{0}^{\sqrt{1-x^2}} \int_{0}^{y} z(1 + x + y + z)\, dz\, dy\, dx \right)$

$\qquad\qquad = \left(\dfrac{28}{9\pi + 44}, \dfrac{30\pi + 128}{45\pi + 220}, \dfrac{45\pi + 208}{135\pi + 660} \right)$

(c) $I_z = \int_{0}^{1} \int_{0}^{\sqrt{1-x^2}} \int_{0}^{y} (x^2 + y^2)(1 + x + y + z)\, dz\, dy\, dx = \dfrac{68 + 15\pi}{240}$

45. $I_x = \int_{0}^{L} \int_{0}^{L} \int_{0}^{L} k(y^2 + z^2)\, dz\, dy\, dx = k \int_{0}^{L} \int_{0}^{L} \left(Ly^2 + \frac{1}{3}L^3 \right) dy\, dx = k \int_{0}^{L} \frac{2}{3}L^4\, dx = \frac{2}{3}kL^5.$

By symmetry, $I_x = I_y = I_z = \frac{2}{3}kL^5.$

47. $I_z = \iiint_E (x^2 + y^2)\, \rho(x, y, z)\, dV = \iint\limits_{x^2+y^2 \le a^2} \left[\int_{0}^{h} k(x^2 + y^2)\, dz \right] dA = \iint\limits_{x^2+y^2 \le a^2} k(x^2 + y^2)h\, dA$

$\qquad = kh \int_{0}^{2\pi} \int_{0}^{a} (r^2)\, r\, dr\, d\theta = kh \int_{0}^{2\pi} d\theta \int_{0}^{a} r^3\, dr = kh(2\pi)\left[\frac{1}{4}r^4 \right]_{0}^{a} = 2\pi kh \cdot \frac{1}{4}a^4 = \frac{1}{2}\pi kha^4$

49. $V(E) = L^3 \ \Rightarrow\ f_{\text{ave}} = \dfrac{1}{L^3} \int_{0}^{L} \int_{0}^{L} \int_{0}^{L} xyz\, dx\, dy\, dz = \dfrac{1}{L^3} \int_{0}^{L} x\, dx \int_{0}^{L} y\, dy \int_{0}^{L} z\, dz$

$\qquad = \dfrac{1}{L^3} \left[\dfrac{x^2}{2} \right]_{0}^{L} \left[\dfrac{y^2}{2} \right]_{0}^{L} \left[\dfrac{z^2}{2} \right]_{0}^{L} = \dfrac{1}{L^3} \dfrac{L^2}{2} \dfrac{L^2}{2} \dfrac{L^2}{2} = \dfrac{L^3}{8}$

51. (a) The triple integral will attain its maximum when the integrand $1 - x^2 - 2y^2 - 3z^2$ is positive in the region E and negative everywhere else. For if E contains some region F where the integrand is negative, the integral could be increased by excluding F from E, and if E fails to contain some part G of the region where the integrand is positive, the integral could be increased by including G in E. So we require that $x^2 + 2y^2 + 3z^2 \le 1$. This describes the region bounded by the ellipsoid $x^2 + 2y^2 + 3z^2 = 1$.

(b) The maximum value of $\iiint_E (1 - x^2 - 2y^2 - 3z^2)\, dV$ occurs when E is the solid region bounded by the ellipsoid $x^2 + 2y^2 + 3z^2 = 1$. The projection of E on the xy-plane is the planar region bounded by the ellipse $x^2 + 2y^2 = 1$, so

$$E = \left\{ (x, y, z) \mid -1 \le x \le 1, -\sqrt{\tfrac{1}{2}(1-x^2)} \le y \le \sqrt{\tfrac{1}{2}(1-x^2)}, -\sqrt{\tfrac{1}{3}(1-x^2-2y^2)} \le z \le \sqrt{\tfrac{1}{3}(1-x^2-2y^2)} \right\}$$

and

$$\iiint_E (1 - x^2 - 2y^2 - 3z^2)\, dV = \int_{-1}^{1} \int_{-\sqrt{\tfrac{1}{2}(1-x^2)}}^{\sqrt{\tfrac{1}{2}(1-x^2)}} \int_{-\sqrt{\tfrac{1}{3}(1-x^2-2y^2)}}^{\sqrt{\tfrac{1}{3}(1-x^2-2y^2)}} (1 - x^2 - 2y^2 - 3z^2)\, dz\, dy\, dx = \frac{4\sqrt{6}}{45}\, \pi$$

using a CAS.

12.6   Triple Integrals in Cylindrical Coordinates

1. (a)

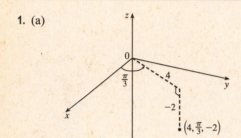

From Equations 1, $x = r \cos \theta = 4 \cos \dfrac{\pi}{3} = 4 \cdot \dfrac{1}{2} = 2$,

$y = r \sin \theta = 4 \sin \dfrac{\pi}{3} = 4 \cdot \dfrac{\sqrt{3}}{2} = 2\sqrt{3}$, $z = -2$, so the point is

$(2, 2\sqrt{3}, -2)$ in rectangular coordinates.

(b)

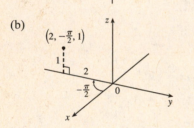

$x = 2 \cos\left(-\dfrac{\pi}{2}\right) = 0$, $y = 2 \sin\left(-\dfrac{\pi}{2}\right) = -2$,

and $z = 1$, so the point is $(0, -2, 1)$ in rectangular coordinates.

3. (a) From Equations 2 we have $r^2 = (-1)^2 + 1^2 = 2$ so $r = \sqrt{2}$; $\tan \theta = \dfrac{1}{-1} = -1$ and the point $(-1, 1)$ is in the second

quadrant of the xy-plane, so $\theta = \dfrac{3\pi}{4} + 2n\pi$; $z = 1$. Thus, one set of cylindrical coordinates is $\left(\sqrt{2}, \dfrac{3\pi}{4}, 1\right)$.

(b) $r^2 = (-2)^2 + (2\sqrt{3})^2 = 16$ so $r = 4$; $\tan \theta = \dfrac{2\sqrt{3}}{-2} = -\sqrt{3}$ and the point $(-2, 2\sqrt{3})$ is in the second quadrant of the

xy-plane, so $\theta = \dfrac{2\pi}{3} + 2n\pi$; $z = 3$. Thus, one set of cylindrical coordinates is $\left(4, \dfrac{2\pi}{3}, 3\right)$.

5. Since $\theta = \dfrac{\pi}{4}$ but r and z may vary, the surface is a vertical half-plane including the z-axis and intersecting the xy-plane in the

half-line $y = x$, $x \ge 0$.

7. $z = 4 - r^2 = 4 - (x^2 + y^2)$ or $4 - x^2 - y^2$, so the surface is a circular paraboloid with vertex $(0, 0, 4)$, axis the z-axis, and

opening downward.

9. (a) Substituting $x^2 + y^2 = r^2$ and $x = r \cos \theta$, the equation $x^2 - x + y^2 + z^2 = 1$ becomes $r^2 - r \cos \theta + z^2 = 1$ or

$z^2 = 1 + r \cos \theta - r^2$.

(b) Substituting $x = r \cos \theta$ and $y = r \sin \theta$, the equation $z = x^2 - y^2$ becomes

$z = (r \cos \theta)^2 - (r \sin \theta)^2 = r^2 (\cos^2 \theta - \sin^2 \theta)$ or $z = r^2 \cos 2\theta$.

11.

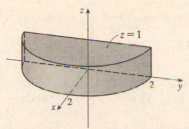

$0 \le r \le 2$ and $0 \le z \le 1$ describe a solid circular cylinder with radius 2, axis the z-axis, and height 1, but $-\pi/2 \le \theta \le \pi/2$ restricts the solid to the first and fourth quadrants of the xy-plane, so we have a half-cylinder.

13. We can position the cylindrical shell vertically so that its axis coincides with the z-axis and its base lies in the xy-plane. If we use centimeters as the unit of measurement, then cylindrical coordinates conveniently describe the shell as $6 \le r \le 7$, $0 \le \theta \le 2\pi$, $0 \le z \le 20$.

15.

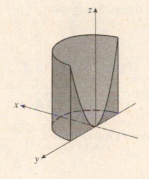

The region of integration is given in cylindrical coordinates by

$E = \{(r, \theta, z) \mid -\pi/2 \le \theta \le \pi/2, 0 \le r \le 2, 0 \le z \le r^2\}$. This represents the solid region above quadrants I and IV of the xy-plane enclosed by the circular cylinder $r = 2$, bounded above by the circular paraboloid $z = r^2$ ($z = x^2 + y^2$), and bounded below by the xy-plane ($z = 0$).

$$\int_{-\pi/2}^{\pi/2} \int_0^2 \int_0^{r^2} r \, dz \, dr \, d\theta = \int_{-\pi/2}^{\pi/2} \int_0^2 [rz]_{z=0}^{z=r^2} \, dr \, d\theta = \int_{-\pi/2}^{\pi/2} \int_0^2 r^3 \, dr \, d\theta$$

$$= \int_{-\pi/2}^{\pi/2} d\theta \int_0^2 r^3 \, dr = [\theta]_{-\pi/2}^{\pi/2} \left[\tfrac{1}{4}r^4\right]_0^2$$

$$= \pi(4 - 0) = 4\pi$$

17. In cylindrical coordinates, E is given by $\{(r, \theta, z) \mid 0 \le \theta \le 2\pi, 0 \le r \le 4, -5 \le z \le 4\}$. So

$$\iiint_E \sqrt{x^2 + y^2} \, dV = \int_0^{2\pi} \int_0^4 \int_{-5}^4 \sqrt{r^2} \, r \, dz \, dr \, d\theta = \int_0^{2\pi} d\theta \int_0^4 r^2 \, dr \int_{-5}^4 dz$$

$$= [\theta]_0^{2\pi} \left[\tfrac{1}{3}r^3\right]_0^4 [z]_{-5}^4 = (2\pi)\left(\tfrac{64}{3}\right)(9) = 384\pi$$

19. The paraboloid $z = 4 - x^2 - y^2 = 4 - r^2$ intersects the xy-plane in the circle $x^2 + y^2 = 4$ or $r^2 = 4 \quad \Rightarrow \quad r = 2$, so in cylindrical coordinates, E is given by $\{(r, \theta, z) \mid 0 \le \theta \le \pi/2, 0 \le r \le 2, 0 \le z \le 4 - r^2\}$. Thus

$$\iiint_E (x + y + z) \, dV = \int_0^{\pi/2} \int_0^2 \int_0^{4-r^2} (r\cos\theta + r\sin\theta + z) \, r \, dz \, dr \, d\theta$$

$$= \int_0^{\pi/2} \int_0^2 \left[r^2(\cos\theta + \sin\theta)z + \tfrac{1}{2}rz^2\right]_{z=0}^{z=4-r^2} \, dr \, d\theta$$

$$= \int_0^{\pi/2} \int_0^2 \left[(4r^2 - r^4)(\cos\theta + \sin\theta) + \tfrac{1}{2}r(4 - r^2)^2\right] \, dr \, d\theta$$

$$= \int_0^{\pi/2} \left[\left(\tfrac{4}{3}r^3 - \tfrac{1}{5}r^5\right)(\cos\theta + \sin\theta) - \tfrac{1}{12}(4 - r^2)^3\right]_{r=0}^{r=2} \, d\theta$$

$$= \int_0^{\pi/2} \left[\tfrac{64}{15}(\cos\theta + \sin\theta) + \tfrac{16}{3}\right] \, d\theta = \left[\tfrac{64}{15}(\sin\theta - \cos\theta) + \tfrac{16}{3}\theta\right]_0^{\pi/2}$$

$$= \tfrac{64}{15}(1 - 0) + \tfrac{16}{3} \cdot \tfrac{\pi}{2} - \tfrac{64}{15}(0 - 1) - 0 = \tfrac{8}{3}\pi + \tfrac{128}{15}$$

21. In cylindrical coordinates, E is bounded by the cylinder $r = 1$, the plane $z = 0$, and the cone $z = 2r$. So $E = \{(r, \theta, z) \mid 0 \le \theta \le 2\pi, 0 \le r \le 1, 0 \le z \le 2r\}$ and

$$\iiint_E x^2 \, dV = \int_0^{2\pi} \int_0^1 \int_0^{2r} r^2 \cos^2\theta \, r \, dz \, dr \, d\theta = \int_0^{2\pi} \int_0^1 [r^3 \cos^2\theta \, z]_{z=0}^{z=2r} \, dr \, d\theta = \int_0^{2\pi} \int_0^1 2r^4 \cos^2\theta \, dr \, d\theta$$

$$= \int_0^{2\pi} \left[\tfrac{2}{5}r^5 \cos^2\theta\right]_{r=0}^{r=1} \, d\theta = \tfrac{2}{5} \int_0^{2\pi} \cos^2\theta \, d\theta = \tfrac{2}{5} \int_0^{2\pi} \tfrac{1}{2}(1 + \cos 2\theta) \, d\theta = \tfrac{1}{5}\left[\theta + \tfrac{1}{2}\sin 2\theta\right]_0^{2\pi} = \tfrac{2\pi}{5}$$

23. In cylindrical coordinates, E is bounded below by the cone $z = r$ and above by the sphere $r^2 + z^2 = 2$ or $z = \sqrt{2 - r^2}$. The cone and the sphere intersect when $2r^2 = 2 \Rightarrow r = 1$, so $E = \left\{ (r, \theta, z) \mid 0 \le \theta \le 2\pi, 0 \le r \le 1, r \le z \le \sqrt{2 - r^2} \right\}$ and the volume is

$$\iiint_E dV = \int_0^{2\pi} \int_0^1 \int_r^{\sqrt{2 - r^2}} r \, dz \, dr \, d\theta = \int_0^{2\pi} \int_0^1 [rz]_{z=r}^{z=\sqrt{2 - r^2}} \, dr \, d\theta = \int_0^{2\pi} \int_0^1 \left(r\sqrt{2 - r^2} - r^2 \right) dr \, d\theta$$

$$= \int_0^{2\pi} d\theta \int_0^1 \left(r\sqrt{2 - r^2} - r^2 \right) dr = 2\pi \left[-\tfrac{1}{3}(2 - r^2)^{3/2} - \tfrac{1}{3}r^3 \right]_0^1$$

$$= 2\pi \left(-\tfrac{1}{3} \right) \left(1 + 1 - 2^{3/2} \right) = -\tfrac{2}{3}\pi \left(2 - 2\sqrt{2} \right) = \tfrac{4}{3}\pi \left(\sqrt{2} - 1 \right)$$

25. (a) The paraboloids intersect when $x^2 + y^2 = 36 - 3x^2 - 3y^2 \Rightarrow x^2 + y^2 = 9$, so the region of integration is $D = \left\{ (x, y) \mid x^2 + y^2 \le 9 \right\}$. Then, in cylindrical coordinates,

$E = \left\{ (r, \theta, z) \mid r^2 \le z \le 36 - 3r^2, 0 \le r \le 3, 0 \le \theta \le 2\pi \right\}$ and

$$V = \int_0^{2\pi} \int_0^3 \int_{r^2}^{36 - 3r^2} r \, dz \, dr \, d\theta = \int_0^{2\pi} \int_0^3 \left(36r - 4r^3 \right) dr \, d\theta = \int_0^{2\pi} \left[18r^2 - r^4 \right]_{r=0}^{r=3} d\theta = \int_0^{2\pi} 81 \, d\theta = 162\pi.$$

(b) For constant density K, $m = KV = 162\pi K$ from part (a). Since the region is homogeneous and symmetric, $M_{yz} = M_{xz} = 0$ and

$$M_{xy} = \int_0^{2\pi} \int_0^3 \int_{r^2}^{36-3r^2} (zK) r \, dz \, dr \, d\theta = K \int_0^{2\pi} \int_0^3 r \left[\tfrac{1}{2}z^2 \right]_{z=r^2}^{z=36-3r^2} dr \, d\theta$$

$$= \tfrac{K}{2} \int_0^{2\pi} \int_0^3 r \left((36 - 3r^2)^2 - r^4 \right) dr \, d\theta = \tfrac{K}{2} \int_0^{2\pi} d\theta \int_0^3 \left(8r^5 - 216r^3 + 1296r \right) dr$$

$$= \tfrac{K}{2} (2\pi) \left[\tfrac{8}{6}r^6 - \tfrac{216}{4}r^4 + \tfrac{1296}{2}r^2 \right]_0^3 = \pi K (2430) = 2430\pi K$$

Thus $(\overline{x}, \overline{y}, \overline{z}) = \left(\dfrac{M_{yz}}{m}, \dfrac{M_{xz}}{m}, \dfrac{M_{xy}}{m} \right) = \left(0, 0, \dfrac{2430\pi K}{162\pi K} \right) = (0, 0, 15)$.

27. The paraboloid $z = 4x^2 + 4y^2$ intersects the plane $z = a$ when $a = 4x^2 + 4y^2$ or $x^2 + y^2 = \tfrac{1}{4}a$. So, in cylindrical coordinates, $E = \left\{ (r, \theta, z) \mid 0 \le r \le \tfrac{1}{2}\sqrt{a}, 0 \le \theta \le 2\pi, 4r^2 \le z \le a \right\}$. Thus

$$m = \int_0^{2\pi} \int_0^{\sqrt{a}/2} \int_{4r^2}^a Kr \, dz \, dr \, d\theta = K \int_0^{2\pi} \int_0^{\sqrt{a}/2} \left(ar - 4r^3 \right) dr \, d\theta$$

$$= K \int_0^{2\pi} \left[\tfrac{1}{2}ar^2 - r^4 \right]_{r=0}^{r=\sqrt{a}/2} d\theta = K \int_0^{2\pi} \tfrac{1}{16}a^2 \, d\theta = \tfrac{1}{8}a^2 \pi K$$

Since the region is homogeneous and symmetric, $M_{yz} = M_{xz} = 0$ and

$$M_{xy} = \int_0^{2\pi} \int_0^{\sqrt{a}/2} \int_{4r^2}^a Krz \, dz \, dr \, d\theta = K \int_0^{2\pi} \int_0^{\sqrt{a}/2} \left(\tfrac{1}{2}a^2 r - 8r^5 \right) dr \, d\theta$$

$$= K \int_0^{2\pi} \left[\tfrac{1}{4}a^2 r^2 - \tfrac{4}{3}r^6 \right]_{r=0}^{r=\sqrt{a}/2} d\theta = K \int_0^{2\pi} \tfrac{1}{24}a^3 \, d\theta = \tfrac{1}{12}a^3 \pi K$$

Hence $(\overline{x}, \overline{y}, \overline{z}) = \left(0, 0, \tfrac{2}{3}a \right)$.

29. The region of integration is the region above the cone $z = \sqrt{x^2 + y^2}$, or $z = r$, and below the plane $z = 2$. Also, we have
$-2 \le y \le 2$ with $-\sqrt{4-y^2} \le x \le \sqrt{4-y^2}$ which describes a circle of radius 2 in the xy-plane centered at $(0,0)$. Thus,

$$\int_{-2}^{2} \int_{-\sqrt{4-y^2}}^{\sqrt{4-y^2}} \int_{\sqrt{x^2+y^2}}^{2} xz\, dz\, dx\, dy = \int_{0}^{2\pi} \int_{0}^{2} \int_{r}^{2} (r\cos\theta)\, z\, r\, dz\, dr\, d\theta = \int_{0}^{2\pi} \int_{0}^{2} \int_{r}^{2} r^2 (\cos\theta)\, z\, dz\, dr\, d\theta$$

$$= \int_{0}^{2\pi} \int_{0}^{2} r^2 (\cos\theta) \left[\tfrac{1}{2} z^2\right]_{z=r}^{z=2} dr\, d\theta = \tfrac{1}{2} \int_{0}^{2\pi} \int_{0}^{2} r^2 (\cos\theta)\left(4 - r^2\right) dr\, d\theta$$

$$= \tfrac{1}{2} \int_{0}^{2\pi} \cos\theta\, d\theta \int_{0}^{2} \left(4r^2 - r^4\right) dr = \tfrac{1}{2} \left[\sin\theta\right]_0^{2\pi} \left[\tfrac{4}{3}r^3 - \tfrac{1}{5}r^5\right]_0^2 = 0$$

31. (a) The mountain comprises a solid conical region C. The work done in lifting a small volume of material ΔV with density
$g(P)$ to a height $h(P)$ above sea level is $h(P)g(P)\,\Delta V$. Summing over the whole mountain we get

$W = \iiint_C h(P)g(P)\, dV$.

(b) Here C is a solid right circular cone with radius $R = 62{,}000$ ft, height $H = 12{,}400$ ft,

and density $g(P) = 200$ lb/ft^3 at all points P in C. We use cylindrical coordinates:

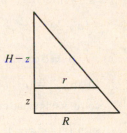

$$W = \int_{0}^{2\pi} \int_{0}^{H} \int_{0}^{R(1-z/H)} z \cdot 200r\, dr\, dz\, d\theta = 2\pi \int_{0}^{H} 200z \left[\tfrac{1}{2}r^2\right]_{r=0}^{r=R(1-z/H)} dz$$

$$= 400\pi \int_{0}^{H} z\, \frac{R^2}{2}\left(1 - \frac{z}{H}\right)^2 dz = 200\pi R^2 \int_{0}^{H} \left(z - \frac{2z^2}{H} + \frac{z^3}{H^2}\right) dz$$

$$= 200\pi R^2 \left[\frac{z^2}{2} - \frac{2z^3}{3H} + \frac{z^4}{4H^2}\right]_0^H = 200\pi R^2 \left(\frac{H^2}{2} - \frac{2H^2}{3} + \frac{H^2}{4}\right)$$

$\dfrac{r}{R} = \dfrac{H-z}{H} = 1 - \dfrac{z}{H}$

$$= \tfrac{50}{3}\pi R^2 H^2 = \tfrac{50}{3}\pi (62{,}000)^2 (12{,}400)^2 \approx 3.1 \times 10^{19} \text{ ft-lb}$$

12.7 Triple Integrals in Spherical Coordinates

1. (a)

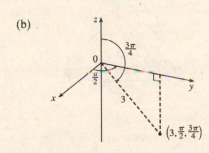

From Equations 1, $x = \rho\sin\phi\cos\theta = 6\sin\frac{\pi}{6}\cos\frac{\pi}{3} = 6 \cdot \frac{1}{2} \cdot \frac{1}{2} = \frac{3}{2}$,

$y = \rho\sin\phi\sin\theta = 6\sin\frac{\pi}{6}\sin\frac{\pi}{3} = 6 \cdot \frac{1}{2} \cdot \frac{\sqrt{3}}{2} = \frac{3\sqrt{3}}{2}$, and

$z = \rho\cos\phi = 6\cos\frac{\pi}{6} = 6 \cdot \frac{\sqrt{3}}{2} = 3\sqrt{3}$, so the point is $\left(\frac{3}{2}, \frac{3\sqrt{3}}{2}, 3\sqrt{3}\right)$ in

rectangular coordinates.

(b)

$x = 3\sin\frac{3\pi}{4}\cos\frac{\pi}{2} = 3 \cdot \frac{\sqrt{2}}{2} \cdot 0 = 0$,

$y = 3\sin\frac{3\pi}{4}\sin\frac{\pi}{2} = 3 \cdot \frac{\sqrt{2}}{2} \cdot 1 = \frac{3\sqrt{2}}{2}$, and

$z = 3\cos\frac{3\pi}{4} = 3\left(-\frac{\sqrt{2}}{2}\right) = -\frac{3\sqrt{2}}{2}$, so the point is $\left(0, \frac{3\sqrt{2}}{2}, -\frac{3\sqrt{2}}{2}\right)$ in

rectangular coordinates.

3. (a) From Equations 1 and 2, $\rho = \sqrt{x^2 + y^2 + z^2} = \sqrt{0^2 + (-2)^2 + 0^2} = 2$, $\cos\phi = \dfrac{z}{\rho} = \dfrac{0}{2} = 0 \;\Rightarrow\; \phi = \dfrac{\pi}{2}$, and

$\cos\theta = \dfrac{x}{\rho\sin\phi} = \dfrac{0}{2\sin(\pi/2)} = 0 \;\Rightarrow\; \theta = \dfrac{3\pi}{2}$ [since $y < 0$]. Thus spherical coordinates are $\left(2, \dfrac{3\pi}{2}, \dfrac{\pi}{2}\right)$.

(b) $\rho = \sqrt{1+1+2} = 2$, $\cos\phi = \dfrac{z}{\rho} = \dfrac{-\sqrt{2}}{2} \;\Rightarrow\; \phi = \dfrac{3\pi}{4}$, and

$\cos\theta = \dfrac{x}{\rho\sin\phi} = \dfrac{-1}{2\sin(3\pi/4)} = \dfrac{-1}{2\left(\sqrt{2}/2\right)} = -\dfrac{1}{\sqrt{2}} \;\Rightarrow\; \theta = \dfrac{3\pi}{4}$ [since $y > 0$]. Thus spherical coordinates

are $\left(2, \dfrac{3\pi}{4}, \dfrac{3\pi}{4}\right)$.

5. Since $\phi = \dfrac{\pi}{3}$, the surface is the top half of the right circular cone with vertex at the origin and axis the positive z-axis.

7. $\rho = \sin\theta\sin\phi \;\Rightarrow\; \rho^2 = \rho\sin\theta\sin\phi \;\Leftrightarrow\; x^2 + y^2 + z^2 = y \;\Leftrightarrow\; x^2 + y^2 - y + \tfrac{1}{4} + z^2 = \tfrac{1}{4} \;\Leftrightarrow\;$

$x^2 + (y - \tfrac{1}{2})^2 + z^2 = \tfrac{1}{4}$. Therefore, the surface is a sphere of radius $\tfrac{1}{2}$ centered at $\left(0, \tfrac{1}{2}, 0\right)$.

9. (a) $x = \rho\sin\phi\cos\theta$, $y = \rho\sin\phi\sin\theta$, and $z = \rho\cos\phi$, so the equation $z^2 = x^2 + y^2$ becomes

$(\rho\cos\phi)^2 = (\rho\sin\phi\cos\theta)^2 + (\rho\sin\phi\sin\theta)^2$ or $\rho^2\cos^2\phi = \rho^2\sin^2\phi$. If $\rho \neq 0$, this becomes $\cos^2\phi = \sin^2\phi$. ($\rho = 0$

corresponds to the origin which is included in the surface.) There are many equivalent equations in spherical coordinates,

such as $\tan^2\phi = 1$, $2\cos^2\phi = 1$, $\cos 2\phi = 0$, or even $\phi = \dfrac{\pi}{4}$, $\phi = \dfrac{3\pi}{4}$.

(b) $x^2 + z^2 = 9 \;\Leftrightarrow\; (\rho\sin\phi\cos\theta)^2 + (\rho\cos\phi)^2 = 9 \;\Leftrightarrow\; \rho^2\sin^2\phi\cos^2\theta + \rho^2\cos^2\phi = 9$ or

$\rho^2\left(\sin^2\phi\cos^2\theta + \cos^2\phi\right) = 9$.

11. $2 \leq \rho \leq 4$ represents the solid region between and including the spheres of

radii 2 and 4, centered at the origin. $0 \leq \phi \leq \dfrac{\pi}{3}$ restricts the solid to that

portion on or above the cone $\phi = \dfrac{\pi}{3}$, and $0 \leq \theta \leq \pi$ further restricts the

solid to that portion on or to the right of the xz-plane.

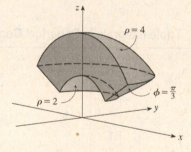

13. $\rho \leq 1$ represents the solid sphere of radius 1 centered at the origin.

$\dfrac{3\pi}{4} \leq \phi \leq \pi$ restricts the solid to that portion on or below the cone $\phi = \dfrac{3\pi}{4}$.

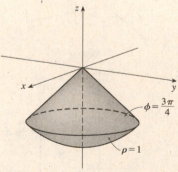

15. $z \geq \sqrt{x^2 + y^2}$ because the solid lies above the cone. Squaring both sides of this inequality gives $z^2 \geq x^2 + y^2 \;\Rightarrow\;$

$2z^2 \geq x^2 + y^2 + z^2 = \rho^2 \;\Rightarrow\; z^2 = \rho^2\cos^2\phi \geq \tfrac{1}{2}\rho^2 \;\Rightarrow\; \cos^2\phi \geq \tfrac{1}{2}$. The cone opens upward so that the inequality is

$\cos\phi \geq \frac{1}{\sqrt{2}}$, or equivalently $0 \leq \phi \leq \frac{\pi}{4}$. In spherical coordinates the sphere $z = x^2 + y^2 + z^2$ is $\rho\cos\phi = \rho^2 \Rightarrow$
$\rho = \cos\phi.$ $0 \leq \rho \leq \cos\phi$ because the solid lies below the sphere. The solid can therefore be described as the region in spherical coordinates satisfying $0 \leq \rho \leq \cos\phi, 0 \leq \phi \leq \frac{\pi}{4}$.

17.

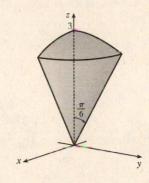

The region of integration is given in spherical coordinates by

$E = \{(\rho, \theta, \phi) \mid 0 \leq \rho \leq 3, 0 \leq \theta \leq \pi/2, 0 \leq \phi \leq \pi/6\}$. This represents the solid

region in the first octant bounded above by the sphere $\rho = 3$ and below by the cone

$\phi = \pi/6.$

$\int_0^{\pi/6} \int_0^{\pi/2} \int_0^3 \rho^2 \sin\phi \, d\rho \, d\theta \, d\phi = \int_0^{\pi/6} \sin\phi \, d\phi \int_0^{\pi/2} d\theta \int_0^3 \rho^2 \, d\rho$

$= \left[-\cos\phi\right]_0^{\pi/6} \left[\theta\right]_0^{\pi/2} \left[\frac{1}{3}\rho^3\right]_0^3$

$= \left(1 - \frac{\sqrt{3}}{2}\right)\left(\frac{\pi}{2}\right)(9) = \frac{9\pi}{4}\left(2 - \sqrt{3}\right)$

19. The solid E is most conveniently described if we use cylindrical coordinates:

$E = \left\{(r, \theta, z) \mid 0 \leq \theta \leq \frac{\pi}{2}, 0 \leq r \leq 3, 0 \leq z \leq 2\right\}.$ Then

$\iiint_E f(x, y, z) \, dV = \int_0^{\pi/2} \int_0^3 \int_0^2 f(r\cos\theta, r\sin\theta, z) \, r \, dz \, dr \, d\theta.$

21. In spherical coordinates, B is represented by $\{(\rho, \theta, \phi) \mid 0 \leq \rho \leq 5, 0 \leq \theta \leq 2\pi, 0 \leq \phi \leq \pi\}$. Thus

$\iiint_B (x^2 + y^2 + z^2)^2 \, dV = \int_0^{\pi} \int_0^{2\pi} \int_0^5 (\rho^2)^2 \rho^2 \sin\phi \, d\rho \, d\theta \, d\phi = \int_0^{\pi} \sin\phi \, d\phi \int_0^{2\pi} d\theta \int_0^5 \rho^6 \, d\rho$

$= \left[-\cos\phi\right]_0^{\pi} \left[\theta\right]_0^{2\pi} \left[\frac{1}{7}\rho^7\right]_0^5 = (2)(2\pi)\left(\frac{78,125}{7}\right)$

$= \frac{312,500}{7}\pi \approx 140,249.7$

23. In spherical coordinates, E is represented by $\{(\rho, \theta, \phi) \mid 2 \leq \rho \leq 3, 0 \leq \theta \leq 2\pi, 0 \leq \phi \leq \pi\}$ and

$x^2 + y^2 = \rho^2 \sin^2\phi \cos^2\theta + \rho^2 \sin^2\phi \sin^2\theta = \rho^2 \sin^2\phi \left(\cos^2\theta + \sin^2\theta\right) = \rho^2 \sin^2\phi.$ Thus

$\iiint_E (x^2 + y^2) \, dV = \int_0^{\pi} \int_0^{2\pi} \int_2^3 (\rho^2 \sin^2\phi) \rho^2 \sin\phi \, d\rho \, d\theta \, d\phi = \int_0^{\pi} \sin^3\phi \, d\phi \int_0^{2\pi} d\theta \int_2^3 \rho^4 \, d\rho$

$= \int_0^{\pi} (1 - \cos^2\phi) \sin\phi \, d\phi \left[\theta\right]_0^{2\pi} \left[\frac{1}{5}\rho^5\right]_2^3 = \left[-\cos\phi + \frac{1}{3}\cos^3\phi\right]_0^{\pi} (2\pi) \cdot \frac{1}{5}(243 - 32)$

$= \left(1 - \frac{1}{3} + 1 - \frac{1}{3}\right)(2\pi)\left(\frac{211}{5}\right) = \frac{1688\pi}{15}$

25. In spherical coordinates, E is represented by $\left\{(\rho, \theta, \phi) \mid 0 \leq \rho \leq 1, 0 \leq \theta \leq \frac{\pi}{2}, 0 \leq \phi \leq \frac{\pi}{2}\right\}$. Thus

$\iiint_E x e^{x^2 + y^2 + z^2} \, dV = \int_0^{\pi/2} \int_0^{\pi/2} \int_0^1 (\rho \sin\phi \cos\theta) e^{\rho^2} \rho^2 \sin\phi \, d\rho \, d\theta \, d\phi = \int_0^{\pi/2} \sin^2\phi \, d\phi \int_0^{\pi/2} \cos\theta \, d\theta \int_0^1 \rho^3 e^{\rho^2} \, d\rho$

$= \int_0^{\pi/2} \frac{1}{2}(1 - \cos 2\phi) \, d\phi \int_0^{\pi/2} \cos\theta \, d\theta \left(\left[\frac{1}{2}\rho^2 e^{\rho^2}\right]_0^1 - \int_0^1 \rho e^{\rho^2} \, d\rho\right)$

$\left[\text{integrate by parts with } u = \rho^2, dv = \rho e^{\rho^2} \, d\rho\right]$

$= \left[\frac{1}{2}\phi - \frac{1}{4}\sin 2\phi\right]_0^{\pi/2} \left[\sin\theta\right]_0^{\pi/2} \left[\frac{1}{2}\rho^2 e^{\rho^2} - \frac{1}{2}e^{\rho^2}\right]_0^1 = \left(\frac{\pi}{4} - 0\right)(1 - 0)\left(0 + \frac{1}{2}\right) = \frac{\pi}{8}$

27. (a) Since $\rho = 4\cos\phi$ implies $\rho^2 = 4\rho\cos\phi$, the equation is that of a sphere of radius 2 with center at $(0, 0, 2)$. Thus

$$V = \int_0^{2\pi}\int_0^{\pi/3}\int_0^{4\cos\phi}\rho^2\sin\phi\,d\rho\,d\phi\,d\theta = \int_0^{2\pi}\int_0^{\pi/3}\left[\tfrac{1}{3}\rho^3\right]_{\rho=0}^{\rho=4\cos\phi}\sin\phi\,d\phi\,d\theta = \int_0^{2\pi}\int_0^{\pi/3}\left(\tfrac{64}{3}\cos^3\phi\right)\sin\phi\,d\phi\,d\theta$$

$$= \int_0^{2\pi}\left[-\tfrac{16}{3}\cos^4\phi\right]_{\phi=0}^{\phi=\pi/3}d\theta = \int_0^{2\pi}-\tfrac{16}{3}\left(\tfrac{1}{16}-1\right)d\theta = 5\theta\Big]_0^{2\pi} = 10\pi$$

(b) By the symmetry of the problem $M_{yz} = M_{xz} = 0$. Then

$$M_{xy} = \int_0^{2\pi}\int_0^{\pi/3}\int_0^{4\cos\phi}\rho^3\cos\phi\sin\phi\,d\rho\,d\phi\,d\theta = \int_0^{2\pi}\int_0^{\pi/3}\cos\phi\sin\phi\left(64\cos^4\phi\right)d\phi\,d\theta$$

$$= \int_0^{2\pi}64\left[-\tfrac{1}{6}\cos^6\phi\right]_{\phi=0}^{\phi=\pi/3}d\theta = \int_0^{2\pi}\tfrac{21}{2}\,d\theta = 21\pi$$

Hence $(\overline{x}, \overline{y}, \overline{z}) = (0, 0, 2.1)$.

29. (a) By the symmetry of the region, $M_{yz} = 0$ and $M_{xz} = 0$. Assuming constant density K,

$$m = \iiint_E K\,dV = K\iiint_E dV = \tfrac{\pi}{8}K \text{ (from Example 4). Then}$$

$$M_{xy} = \iiint_E zK\,dV = K\int_0^{2\pi}\int_0^{\pi/4}\int_0^{\cos\phi}(\rho\cos\phi)\rho^2\sin\phi\,d\rho\,d\phi\,d\theta = K\int_0^{2\pi}\int_0^{\pi/4}\sin\phi\cos\phi\left[\tfrac{1}{4}\rho^4\right]_{\rho=0}^{\rho=\cos\phi}d\phi\,d\theta$$

$$= \tfrac{1}{4}K\int_0^{2\pi}\int_0^{\pi/4}\sin\phi\cos\phi\left(\cos^4\phi\right)d\phi\,d\theta = \tfrac{1}{4}K\int_0^{2\pi}d\theta\int_0^{\pi/4}\cos^5\phi\sin\phi\,d\phi$$

$$= \tfrac{1}{4}K\left[\theta\right]_0^{2\pi}\left[-\tfrac{1}{6}\cos^6\phi\right]_0^{\pi/4} = \tfrac{1}{4}K(2\pi)\left(-\tfrac{1}{6}\right)\left[\left(\tfrac{\sqrt{2}}{2}\right)^6-1\right] = -\tfrac{\pi}{12}K\left(-\tfrac{7}{8}\right) = \tfrac{7\pi}{96}K$$

Thus the centroid is $(\overline{x}, \overline{y}, \overline{z}) = \left(\dfrac{M_{yz}}{m}, \dfrac{M_{xz}}{m}, \dfrac{M_{xy}}{m}\right) = \left(0, 0, \dfrac{7\pi K/96}{\pi K/8}\right) = \left(0, 0, \tfrac{7}{12}\right)$.

(b) As in Exercise 23, $x^2 + y^2 = \rho^2\sin^2\phi$ and

$$I_z = \iiint_E (x^2+y^2)K\,dV = K\int_0^{2\pi}\int_0^{\pi/4}\int_0^{\cos\phi}(\rho^2\sin^2\phi)\rho^2\sin\phi\,d\rho\,d\phi\,d\theta = K\int_0^{2\pi}\int_0^{\pi/4}\sin^3\phi\left[\tfrac{1}{5}\rho^5\right]_{\rho=0}^{\rho=\cos\phi}d\phi\,d\theta$$

$$= \tfrac{1}{5}K\int_0^{2\pi}\int_0^{\pi/4}\sin^3\phi\cos^5\phi\,d\phi\,d\theta = \tfrac{1}{5}K\int_0^{2\pi}d\theta\int_0^{\pi/4}\cos^5\phi\left(1-\cos^2\phi\right)\sin\phi\,d\phi$$

$$= \tfrac{1}{5}K\left[\theta\right]_0^{2\pi}\left[-\tfrac{1}{6}\cos^6\phi + \tfrac{1}{8}\cos^8\phi\right]_0^{\pi/4}$$

$$= \tfrac{1}{5}K(2\pi)\left[-\tfrac{1}{6}\left(\tfrac{\sqrt{2}}{2}\right)^6 + \tfrac{1}{8}\left(\tfrac{\sqrt{2}}{2}\right)^8 + \tfrac{1}{6} - \tfrac{1}{8}\right] = \tfrac{2\pi}{5}K\left(\tfrac{11}{384}\right) = \tfrac{11\pi}{960}K$$

31. (a) The density function is $\rho(x, y, z) = K$, a constant, and by the symmetry of the problem $M_{xz} = M_{yz} = 0$. Then

$$M_{xy} = \int_0^{2\pi}\int_0^{\pi/2}\int_0^a K\rho^3\sin\phi\,\cos\phi\,d\rho\,d\phi\,d\theta = \tfrac{1}{2}\pi Ka^4\int_0^{\pi/2}\sin\phi\cdot\cos\phi\,d\phi = \tfrac{1}{8}\pi Ka^4.$$ But the mass is K(volume of

the hemisphere) $= \tfrac{2}{3}\pi Ka^3$, so the centroid is $\left(0, 0, \tfrac{3}{8}a\right)$.

(b) Place the center of the base at $(0, 0, 0)$; the density function is $\rho(x, y, z) = K$. By symmetry, the moments of inertia about any two such diameters will be equal, so we just need to find I_x:

$$I_x = \int_0^{2\pi}\int_0^{\pi/2}\int_0^a (K\rho^2\sin\phi)\rho^2\left(\sin^2\phi\sin^2\theta + \cos^2\phi\right)d\rho\,d\phi\,d\theta$$

$$= K\int_0^{2\pi}\int_0^{\pi/2}(\sin^3\phi\sin^2\theta + \sin\phi\cos^2\phi)\left(\tfrac{1}{5}a^5\right)d\phi\,d\theta$$

$$= \tfrac{1}{5}Ka^5\int_0^{2\pi}\left[\sin^2\theta\left(-\cos\phi + \tfrac{1}{3}\cos^3\phi\right) + \left(-\tfrac{1}{3}\cos^3\phi\right)\right]_{\phi=0}^{\phi=\pi/2}d\theta = \tfrac{1}{5}Ka^5\int_0^{2\pi}\left[\tfrac{2}{3}\sin^2\theta + \tfrac{1}{3}\right]d\theta$$

$$= \tfrac{1}{5}Ka^5\left[\tfrac{2}{3}\left(\tfrac{1}{2}\theta - \tfrac{1}{4}\sin 2\theta\right) + \tfrac{1}{3}\theta\right]_0^{2\pi} = \tfrac{1}{5}Ka^5\left[\tfrac{2}{3}(\pi - 0) + \tfrac{1}{3}(2\pi - 0)\right] = \tfrac{4}{15}Ka^5\pi$$

33. In spherical coordinates $z = \sqrt{x^2 + y^2}$ becomes $\cos\phi = \sin\phi$ or $\phi = \frac{\pi}{4}$. Then

$$V = \int_0^{2\pi}\int_0^{\pi/4}\int_0^1 \rho^2 \sin\phi\, d\rho\, d\phi\, d\theta = \int_0^{2\pi} d\theta \int_0^{\pi/4} \sin\phi\, d\phi \int_0^1 \rho^2\, d\rho = 2\pi\left(-\frac{\sqrt{2}}{2} + 1\right)\left(\frac{1}{3}\right) = \frac{1}{3}\pi(2 - \sqrt{2}),$$

$$M_{xy} = \int_0^{2\pi}\int_0^{\pi/4}\int_0^1 \rho^3 \sin\phi\cos\phi\, d\rho\, d\phi\, d\theta = 2\pi\left[-\frac{1}{4}\cos 2\phi\right]_0^{\pi/4}\left(\frac{1}{4}\right) = \frac{\pi}{8} \text{ and by symmetry } M_{yz} = M_{xz} = 0.$$

Hence $(\overline{x}, \overline{y}, \overline{z}) = \left(0, 0, \dfrac{3}{8(2 - \sqrt{2})}\right)$.

35. In cylindrical coordinates the paraboloid is given by $z = r^2$ and the plane by $z = 2r\sin\theta$ and they intersect in the circle

$r = 2\sin\theta$. Then $\iiint_E z\, dV = \int_0^\pi \int_0^{2\sin\theta} \int_{r^2}^{2r\sin\theta} rz\, dz\, dr\, d\theta = \frac{5\pi}{6}$ [using a CAS].

37. The region E of integration is the region above the cone $z = \sqrt{x^2 + y^2}$ and below the sphere $x^2 + y^2 + z^2 = 2$ in the first

octant. Because E is in the first octant we have $0 \le \theta \le \frac{\pi}{2}$. The cone has equation $\phi = \frac{\pi}{4}$ (as in Example 4), so $0 \le \phi \le \frac{\pi}{4}$,

and $0 \le \rho \le \sqrt{2}$. So the integral becomes

$$\int_0^{\pi/4}\int_0^{\pi/2}\int_0^{\sqrt{2}} (\rho\sin\phi\cos\theta)(\rho\sin\phi\sin\theta)\rho^2 \sin\phi\, d\rho\, d\theta\, d\phi$$

$$= \int_0^{\pi/4} \sin^3\phi\, d\phi \int_0^{\pi/2} \sin\theta\cos\theta\, d\theta \int_0^{\sqrt{2}} \rho^4\, d\rho = \left(\int_0^{\pi/4}(1 - \cos^2\phi)\sin\phi\, d\phi\right)\left[\frac{1}{2}\sin^2\theta\right]_0^{\pi/2}\left[\frac{1}{5}\rho^5\right]_0^{\sqrt{2}}$$

$$= \left[\frac{1}{3}\cos^3\phi - \cos\phi\right]_0^{\pi/4} \cdot \frac{1}{2} \cdot \frac{1}{5}\left(\sqrt{2}\right)^5 = \left[\frac{\sqrt{2}}{12} - \frac{\sqrt{2}}{2} - \left(\frac{1}{3} - 1\right)\right] \cdot \frac{2\sqrt{2}}{5} = \frac{4\sqrt{2}-5}{15}$$

39. The region of integration is the solid sphere $x^2 + y^2 + (z - 2)^2 \le 4$ or equivalently

$\rho^2 \sin^2\phi + (\rho\cos\phi - 2)^2 = \rho^2 - 4\rho\cos\phi + 4 \le 4 \quad \Rightarrow \quad \rho \le 4\cos\phi$, so $0 \le \theta \le 2\pi, 0 \le \phi \le \frac{\pi}{2}$, and

$0 \le \rho \le 4\cos\phi$. Also $(x^2 + y^2 + z^2)^{3/2} = (\rho^2)^{3/2} = \rho^3$, so the integral becomes

$$\int_0^{\pi/2}\int_0^{2\pi}\int_0^{4\cos\phi} (\rho^3)\rho^2\sin\phi\, d\rho\, d\theta\, d\phi = \int_0^{\pi/2}\int_0^{2\pi}\sin\phi\left[\frac{1}{6}\rho^6\right]_{\rho=0}^{\rho=4\cos\phi} d\theta\, d\phi = \frac{1}{6}\int_0^{\pi/2}\int_0^{2\pi}\sin\phi\left(4096\cos^6\phi\right)d\theta\, d\phi$$

$$= \frac{1}{6}(4096)\int_0^{\pi/2}\cos^6\phi\sin\phi\, d\phi \int_0^{2\pi} d\theta = \frac{2048}{3}\left[-\frac{1}{7}\cos^7\phi\right]_0^{\pi/2}\left[\theta\right]_0^{2\pi}$$

$$= \frac{2048}{3}\left(\frac{1}{7}\right)(2\pi) = \frac{4096\pi}{21}$$

41. In cylindrical coordinates, the equation of the cylinder is $r = 3, 0 \le z \le 10$.

The hemisphere is the upper part of the sphere radius 3, center $(0, 0, 10)$, equation

$r^2 + (z - 10)^2 = 3^2, z \ge 10$. In Maple, we can use the `coords=cylindrical` option

in a regular `plot3d` command. In Mathematica, we can use `ParametricPlot3D`.

43. If E is the solid enclosed by the surface $\rho = 1 + \frac{1}{5}\sin 6\theta\sin 5\phi$, it can be described in spherical coordinates as

$E = \left\{(\rho, \theta, \phi) \mid 0 \le \rho \le 1 + \frac{1}{5}\sin 6\theta\sin 5\phi, 0 \le \theta \le 2\pi, 0 \le \phi \le \pi\right\}$. Its volume is given by

$$V(E) = \iiint_E dV = \int_0^\pi\int_0^{2\pi}\int_0^{1 + (\sin 6\theta\sin 5\phi)/5} \rho^2\sin\phi\, d\rho\, d\theta\, d\phi = \frac{136\pi}{99} \text{ [using a CAS]}.$$

45. (a) From the diagram, $z = r \cot \phi_0$ to $z = \sqrt{a^2 - r^2}$, $r = 0$

to $r = a \sin \phi_0$ (or use $a^2 - r^2 = r^2 \cot^2 \phi_0$). Thus

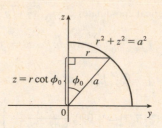

$$V = \int_0^{2\pi} \int_0^{a \sin \phi_0} \int_{r \cot \phi_0}^{\sqrt{a^2 - r^2}} r\, dz\, dr\, d\theta$$

$$= 2\pi \int_0^{a \sin \phi_0} \left(r \sqrt{a^2 - r^2} - r^2 \cot \phi_0 \right) dr$$

$$= \frac{2\pi}{3} \left[-(a^2 - r^2)^{3/2} - r^3 \cot \phi_0 \right]_0^{a \sin \phi_0}$$

$$= \frac{2\pi}{3} \left[-\left(a^2 - a^2 \sin^2 \phi_0\right)^{3/2} - a^3 \sin^3 \phi_0 \cot \phi_0 + a^3 \right]$$

$$= \frac{2}{3}\pi a^3 \left[1 - \left(\cos^3 \phi_0 + \sin^2 \phi_0 \cos \phi_0 \right) \right] = \frac{2}{3}\pi a^3 (1 - \cos \phi_0)$$

(b) The wedge in question is the shaded area rotated from $\theta = \theta_1$ to $\theta = \theta_2$.

Letting

V_{ij} = volume of the region bounded by the sphere of radius ρ_i

and the cone with angle ϕ_j ($\theta = \theta_1$ to θ_2)

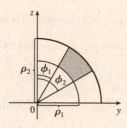

and letting V be the volume of the wedge, we have

$$V = (V_{22} - V_{21}) - (V_{12} - V_{11})$$

$$= \tfrac{1}{3}(\theta_2 - \theta_1)\left[\rho_2^3(1 - \cos \phi_2) - \rho_2^3(1 - \cos \phi_1) - \rho_1^3(1 - \cos \phi_2) + \rho_1^3(1 - \cos \phi_1) \right]$$

$$= \tfrac{1}{3}(\theta_2 - \theta_1)\left[(\rho_2^3 - \rho_1^3)(1 - \cos \phi_2) - (\rho_2^3 - \rho_1^3)(1 - \cos \phi_1) \right] = \tfrac{1}{3}(\theta_2 - \theta_1)\left[(\rho_2^3 - \rho_1^3)(\cos \phi_1 - \cos \phi_2) \right]$$

Or: Show that $V = \displaystyle\int_{\theta_1}^{\theta_2} \int_{\rho_1 \sin \phi_1}^{\rho_2 \sin \phi_2} \int_{r \cot \phi_2}^{r \cot \phi_1} r\, dz\, dr\, d\theta$.

(c) By the Mean Value Theorem with $f(\rho) = \rho^3$ there exists some $\tilde{\rho}$ with $\rho_1 \leq \tilde{\rho} \leq \rho_2$ such that

$f(\rho_2) - f(\rho_1) = f'(\tilde{\rho})(\rho_2 - \rho_1)$ or $\rho_1^3 - \rho_2^3 = 3\tilde{\rho}^2 \Delta \rho$. Similarly there exists ϕ with $\phi_1 \leq \tilde{\phi} \leq \phi_2$

such that $\cos \phi_2 - \cos \phi_1 = \left(-\sin \tilde{\phi} \right) \Delta \phi$. Substituting into the result from (b) gives

$$\Delta V = (\tilde{\rho}^2 \, \Delta \rho)(\theta_2 - \theta_1)(\sin \tilde{\phi}) \, \Delta \phi = \tilde{\rho}^2 \sin \tilde{\phi} \, \Delta \rho \, \Delta \phi \, \Delta \theta.$$

12.8 Change of Variables in Multiple Integrals

1. $x = 5u - v$, $y = u + 3v$.

The Jacobian is $\dfrac{\partial(x, y)}{\partial(u, v)} = \begin{vmatrix} \partial x/\partial u & \partial x/\partial v \\ \partial y/\partial u & \partial y/\partial v \end{vmatrix} = \begin{vmatrix} 5 & -1 \\ 1 & 3 \end{vmatrix} = 5(3) - (-1)(1) = 16.$

3. $x = e^{-r} \sin \theta$, $y = e^r \cos \theta$.

$\dfrac{\partial(x, y)}{\partial(r, \theta)} = \begin{vmatrix} \partial x/\partial r & \partial x/\partial \theta \\ \partial y/\partial r & \partial y/\partial \theta \end{vmatrix} = \begin{vmatrix} -e^{-r} \sin \theta & e^{-r} \cos \theta \\ e^r \cos \theta & -e^r \sin \theta \end{vmatrix} = e^{-r}e^r \sin^2 \theta - e^{-r}e^r \cos^2 \theta = \sin^2 \theta - \cos^2 \theta$ or $-\cos 2\theta$

5. $x = u/v,\ y = v/w,\ z = w/u$.

$$\frac{\partial(x,y,z)}{\partial(u,v,w)} = \begin{vmatrix} \partial x/\partial u & \partial x/\partial v & \partial x/\partial w \\ \partial y/\partial u & \partial y/\partial v & \partial y/\partial w \\ \partial z/\partial u & \partial z/\partial v & \partial z/\partial w \end{vmatrix} = \begin{vmatrix} 1/v & -u/v^2 & 0 \\ 0 & 1/w & -v/w^2 \\ -w/u^2 & 0 & 1/u \end{vmatrix}$$

$$= \frac{1}{v}\begin{vmatrix} 1/w & -v/w^2 \\ 0 & 1/u \end{vmatrix} - \left(-\frac{u}{v^2}\right)\begin{vmatrix} 0 & -v/w^2 \\ -w/u^2 & 1/u \end{vmatrix} + 0\begin{vmatrix} 0 & 1/w \\ -w/u^2 & 0 \end{vmatrix}$$

$$= \frac{1}{v}\left(\frac{1}{uw} - 0\right) + \frac{u}{v^2}\left(0 - \frac{v}{u^2 w}\right) + 0 = \frac{1}{uvw} - \frac{1}{uvw} = 0$$

7. The transformation maps the boundary of S to the boundary of the image R, so we first look at side S_1 in the uv-plane. S_1 is described by $v = 0,\ 0 \le u \le 3$, so $x = 2u + 3v = 2u$ and $y = u - v = u$. Eliminating u, we have $x = 2y,\ 0 \le x \le 6$. S_2 is the line segment $u = 3,\ 0 \le v \le 2$, so $x = 6 + 3v$ and $y = 3 - v$. Then $v = 3 - y \ \Rightarrow \ x = 6 + 3(3 - y) = 15 - 3y$, $6 \le x \le 12$. S_3 is the line segment $v = 2,\ 0 \le u \le 3$, so $x = 2u + 6$ and $y = u - 2$, giving $u = y + 2 \ \Rightarrow \ x = 2y + 10$, $6 \le x \le 12$. Finally, S_4 is the segment $u = 0,\ 0 \le v \le 2$, so $x = 3v$ and $y = -v \ \Rightarrow \ x = -3y,\ 0 \le x \le 6$.

The image of set S is the region R shown in the xy-plane, a parallelogram bounded by these four segments.

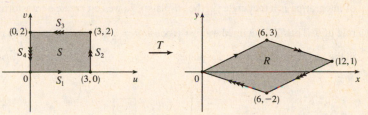

9. S_1 is the line segment $u = v,\ 0 \le u \le 1$, so $y = v = u$ and $x = u^2 = y^2$. Since $0 \le u \le 1$, the image is the portion of the parabola $x = y^2,\ 0 \le y \le 1$. S_2 is the segment $v = 1,\ 0 \le u \le 1$, thus $y = v = 1$ and $x = u^2$, so $0 \le x \le 1$. The image is the line segment $y = 1,\ 0 \le x \le 1$. S_3 is the segment $u = 0,\ 0 \le v \le 1$, so $x = u^2 = 0$ and $y = v \ \Rightarrow \ 0 \le y \le 1$. The image is the segment $x = 0,\ 0 \le y \le 1$. Thus, the image of S is the region R in the first quadrant bounded by the parabola $x = y^2$, the y-axis, and the line $y = 1$.

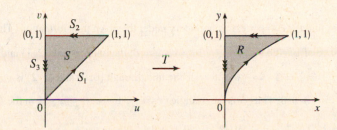

11. R is a parallelogram enclosed by the parallel lines $y = 2x - 1,\ y = 2x + 1$ and the parallel lines $y = 1 - x,\ y = 3 - x$. The first pair of equations can be written as $y - 2x = -1,\ y - 2x = 1$. If we let $u = y - 2x$ then these lines are mapped to the vertical lines $u = -1,\ u = 1$ in the uv-plane. Similarly, the second pair of equations can be written as $x + y = 1,\ x + y = 3$, and setting $v = x + y$ maps these lines to the horizontal lines $v = 1,\ v = 3$ in the uv-plane. Boundary curves are mapped to boundary curves under a transformation, so here the equations $u = y - 2x,\ v = x + y$ define a transformation T^{-1} that maps R in the xy-plane to the square S enclosed by the lines $u = -1,\ u = 1,\ v = 1,\ v = 3$ in the uv-plane. To find the transformation T that maps S to R we solve $u = y - 2x$, $v = x + y$ for x, y: Subtracting the first equation from the second gives $v - u = 3x$ $\Rightarrow$ $x = \frac{1}{3}(v - u)$ and adding twice the second equation to the first gives $u + 2v = 3y$ $\Rightarrow$ $y = \frac{1}{3}(u + 2v)$. Thus one possible transformation T (there are many) is given by $x = \frac{1}{3}(v - u)$, $y = \frac{1}{3}(u + 2v)$.

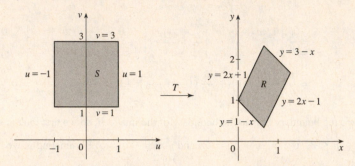

13. R is a portion of an annular region (see the figure) that is easily described in polar coordinates as
$R = \{(r, \theta) \mid 1 \le r \le \sqrt{2}, 0 \le \theta \le \pi/2\}$. If we converted a double integral over R to polar coordinates the resulting region of integration is a rectangle (in the $r\theta$-plane), so we can create a transformation T here by letting u play the role of r and v the role of θ. Thus T is defined by $x = u \cos v$, $y = u \sin v$ and T maps the rectangle $S = \{(u, v) \mid 1 \le u \le \sqrt{2}, 0 \le v \le \pi/2\}$ in the uv-plane to R in the xy-plane.

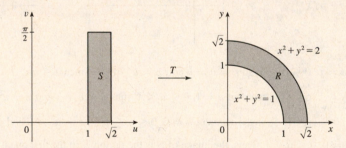

15. $\dfrac{\partial(x, y)}{\partial(u, v)} = \begin{vmatrix} 2 & 1 \\ 1 & 2 \end{vmatrix} = 3$ and $x - 3y = (2u + v) - 3(u + 2v) = -u - 5v$. To find the region S in the uv-plane that

corresponds to R we first find the corresponding boundary under the given transformation. The line through $(0, 0)$ and $(2, 1)$ is $y = \frac{1}{2}x$ which is the image of $u + 2v = \frac{1}{2}(2u + v)$ $\Rightarrow$ $v = 0$; the line through $(2, 1)$ and $(1, 2)$ is $x + y = 3$ which is the image of $(2u + v) + (u + 2v) = 3$ $\Rightarrow$ $u + v = 1$; the line through $(0, 0)$ and $(1, 2)$ is $y = 2x$ which is the image of $u + 2v = 2(2u + v)$ $\Rightarrow$ $u = 0$. Thus S is the triangle $0 \le v \le 1 - u$, $0 \le u \le 1$ in the uv-plane and

$$\iint_R (x - 3y)\, dA = \int_0^1 \int_0^{1-u} (-u - 5v)\, |3|\, dv\, du = -3 \int_0^1 \left[uv + \frac{5}{2}v^2 \right]_{v=0}^{v=1-u} du$$

$$= -3 \int_0^1 \left(u - u^2 + \frac{5}{2}(1 - u)^2 \right) du = -3 \left[\frac{1}{2}u^2 - \frac{1}{3}u^3 - \frac{5}{6}(1 - u)^3 \right]_0^1 = -3 \left(\frac{1}{2} - \frac{1}{3} + \frac{5}{6} \right) = -3$$

17. $\dfrac{\partial(x, y)}{\partial(u, v)} = \begin{vmatrix} 2 & 0 \\ 0 & 3 \end{vmatrix} = 6$, $x^2 = 4u^2$ and the planar ellipse $9x^2 + 4y^2 \le 36$ is the image of the disk $u^2 + v^2 \le 1$. Thus

$$\iint_R x^2\, dA = \iint_{u^2 + v^2 \le 1} (4u^2)(6)\, du\, dv = \int_0^{2\pi} \int_0^1 (24r^2 \cos^2 \theta)\, r\, dr\, d\theta = 24 \int_0^{2\pi} \cos^2 \theta\, d\theta \int_0^1 r^3\, dr$$

$$= 24 \left[\frac{1}{2}x + \frac{1}{4} \sin 2x \right]_0^{2\pi} \left[\frac{1}{4}r^4 \right]_0^1 = 24(\pi)\left(\frac{1}{4} \right) = 6\pi$$

19. $\dfrac{\partial(x,y)}{\partial(u,v)} = \begin{vmatrix} 1/v & -u/v^2 \\ 0 & 1 \end{vmatrix} = \dfrac{1}{v}$, $xy = u$, $y = x$ is the image of the parabola $v^2 = u$, $y = 3x$ is the image of the parabola

$v^2 = 3u$, and the hyperbolas $xy = 1$, $xy = 3$ are the images of the lines $u = 1$ and $u = 3$ respectively. Thus

$$\iint_R xy\, dA = \int_1^3 \int_{\sqrt{u}}^{\sqrt{3u}} u\left(\frac{1}{v}\right) dv\, du = \int_1^3 u\left(\ln\sqrt{3u} - \ln\sqrt{u}\,\right) du = \int_1^3 u \ln\sqrt{3}\, du = 4\ln\sqrt{3} = 2\ln 3.$$

21. (a) $\dfrac{\partial(x,y,z)}{\partial(u,v,w)} = \begin{vmatrix} a & 0 & 0 \\ 0 & b & 0 \\ 0 & 0 & c \end{vmatrix} = abc$ and since $u = \dfrac{x}{a}$, $v = \dfrac{y}{b}$, $w = \dfrac{z}{c}$ the solid enclosed by the ellipsoid is the image of the

ball $u^2 + v^2 + w^2 \le 1$. So

$$\iiint_E dV = \iiint_{u^2+v^2+w^2 \le 1} abc\, du\, dv\, dw = (abc)(\text{volume of the ball}) = \tfrac{4}{3}\pi abc$$

(b) If we approximate the surface of the earth by the ellipsoid $\dfrac{x^2}{6378^2} + \dfrac{y^2}{6378^2} + \dfrac{z^2}{6356^2} = 1$, then we can estimate

the volume of the earth by finding the volume of the solid E enclosed by the ellipsoid. From part (a), this is

$\iiint_E dV = \tfrac{4}{3}\pi(6378)(6378)(6356) \approx 1.083 \times 10^{12}$ km^3.

(c) The moment of intertia about the z-axis is $I_z = \iiint_E (x^2 + y^2)\,\rho(x,y,z)\, dV$, where E is the solid enclosed by

$\dfrac{x^2}{a^2} + \dfrac{y^2}{b^2} + \dfrac{z^2}{c^2} = 1$. As in part (a), we use the transformation $x = au$, $y = bv$, $z = cw$, so $\left|\dfrac{\partial(x,y,z)}{\partial(u,v,w)}\right| = abc$ and

$$I_z = \iiint_E (x^2 + y^2)\, k\, dV = \iiint_{u^2+v^2+w^2 \le 1} k(a^2u^2 + b^2v^2)(abc)\, du\, dv\, dw$$

$$= abck \int_0^\pi \int_0^{2\pi} \int_0^1 (a^2\rho^2 \sin^2\phi \cos^2\theta + b^2\rho^2 \sin^2\phi \sin^2\theta)\, \rho^2 \sin\phi\, d\rho\, d\theta\, d\phi$$

$$= abck \left[a^2 \int_0^\pi \int_0^{2\pi} \int_0^1 (\rho^2 \sin^2\phi \cos^2\theta)\, \rho^2 \sin\phi\, d\rho\, d\theta\, d\phi + b^2 \int_0^\pi \int_0^{2\pi} \int_0^1 (\rho^2 \sin^2\phi \sin^2\theta)\, \rho^2 \sin\phi\, d\rho\, d\theta\, d\phi \right]$$

$$= a^3 bck \int_0^\pi \sin^3\phi\, d\phi \int_0^{2\pi} \cos^2\theta\, d\theta \int_0^1 \rho^4\, d\rho + ab^3 ck \int_0^\pi \sin^3\phi\, d\phi \int_0^{2\pi} \sin^2\theta\, d\theta \int_0^1 \rho^4\, d\rho$$

$$= a^3 bck \left[\tfrac{1}{3}\cos^3\phi - \cos\phi\right]_0^\pi \left[\tfrac{1}{2}\theta + \tfrac{1}{4}\sin 2\theta\right]_0^{2\pi} \left[\tfrac{1}{5}\rho^5\right]_0^1 + ab^3 ck \left[\tfrac{1}{3}\cos^3\phi - \cos\phi\right]_0^\pi \left[\tfrac{1}{2}\theta - \tfrac{1}{4}\sin 2\theta\right]_0^{2\pi} \left[\tfrac{1}{5}\rho^5\right]_0^1$$

$$= a^3 bck \left(\tfrac{4}{3}\right)(\pi)\left(\tfrac{1}{5}\right) + ab^3 ck \left(\tfrac{4}{3}\right)(\pi)\left(\tfrac{1}{5}\right) = \tfrac{4}{15}\pi(a^2 + b^2)abck$$

23. Letting $u = x - 2y$ and $v = 3x - y$, we have $x = \tfrac{1}{5}(2v - u)$ and $y = \tfrac{1}{5}(v - 3u)$. Then $\dfrac{\partial(x,y)}{\partial(u,v)} = \begin{vmatrix} -1/5 & 2/5 \\ -3/5 & 1/5 \end{vmatrix} = \dfrac{1}{5}$

and R is the image of the rectangle enclosed by the lines $u = 0$, $u = 4$, $v = 1$, and $v = 8$. Thus

$$\iint_R \frac{x - 2y}{3x - y}\, dA = \int_0^4 \int_1^8 \frac{u}{v}\left|\frac{1}{5}\right| dv\, du = \frac{1}{5}\int_0^4 u\, du \int_1^8 \frac{1}{v}\, dv = \tfrac{1}{5}\left[\tfrac{1}{2}u^2\right]_0^4 \left[\ln|v|\right]_1^8 = \tfrac{8}{5}\ln 8.$$

25. Letting $u = y - x$, $v = y + x$, we have $y = \tfrac{1}{2}(u + v)$, $x = \tfrac{1}{2}(v - u)$. Then $\dfrac{\partial(x,y)}{\partial(u,v)} = \begin{vmatrix} -1/2 & 1/2 \\ 1/2 & 1/2 \end{vmatrix} = -\dfrac{1}{2}$ and R is the

image of the trapezoidal region with vertices $(-1, 1)$, $(-2, 2)$, $(2, 2)$, and $(1, 1)$. Thus

$$\iint_R \cos\left(\frac{y - x}{y + x}\right) dA = \int_1^2 \int_{-v}^{v} \cos\frac{u}{v}\left|-\frac{1}{2}\right| du\, dv = \frac{1}{2}\int_1^2 \left[v\sin\frac{u}{v}\right]_{u=-v}^{u=v} dv = \frac{1}{2}\int_1^2 2v\sin(1)\, dv = \tfrac{3}{2}\sin 1$$

27. Let $u = x + y$ and $v = -x + y$. Then $u + v = 2y$ $\Rightarrow$ $y = \frac{1}{2}(u + v)$ and $u - v = 2x$ $\Rightarrow$ $x = \frac{1}{2}(u - v)$.

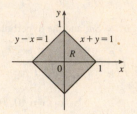

$$\frac{\partial(x, y)}{\partial(u, v)} = \begin{vmatrix} 1/2 & -1/2 \\ 1/2 & 1/2 \end{vmatrix} = \frac{1}{2}. \text{ Now } |u| = |x + y| \le |x| + |y| \le 1 \quad \Rightarrow \quad -1 \le u \le 1,$$

and $|v| = |-x + y| \le |x| + |y| \le 1$ $\Rightarrow$ $-1 \le v \le 1$. R is the image of the square

region with vertices $(1, 1)$, $(1, -1)$, $(-1, -1)$, and $(-1, 1)$.

So $\iint_R e^{x+y} \, dA = \frac{1}{2} \int_{-1}^{1} \int_{-1}^{1} e^u \, du \, dv = \frac{1}{2} \left[e^u \right]_{-1}^{1} \left[v \right]_{-1}^{1} = e - e^{-1}$.

12 Review

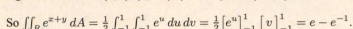

CONCEPT CHECK

1. (a) A double Riemann sum of f is $\sum_{i=1}^{m} \sum_{j=1}^{n} f(x_{ij}^*, y_{ij}^*) \, \Delta A$, where ΔA is the area of each subrectangle and (x_{ij}^*, y_{ij}^*) is a

   sample point in each subrectangle. If $f(x, y) \ge 0$, this sum represents an approximation to the volume of the solid that lies

   above the rectangle R and below the graph of f.

   (b) $\iint_R f(x, y) \, dA = \lim\limits_{m,n \to \infty} \sum_{i=1}^{m} \sum_{j=1}^{n} f(x_{ij}^*, y_{ij}^*) \, \Delta A$

   (c) If $f(x, y) \ge 0$, $\iint_R f(x, y) \, dA$ represents the volume of the solid that lies above the rectangle R and below the surface

   $z = f(x, y)$. If f takes on both positive and negative values, $\iint_R f(x, y) \, dA$ is the difference of the volume above R but

   below the surface $z = f(x, y)$ and the volume below R but above the surface $z = f(x, y)$.

   (d) We usually evaluate $\iint_R f(x, y) \, dA$ as an iterated integral according to Fubini's Theorem (see Theorem 12.1.10).

   (e) The Midpoint Rule for double integrals says that we approximate the double integral $\iint_R f(x, y) \, dA$ by the double

   Riemann sum $\sum_{i=1}^{m} \sum_{j=1}^{n} f(\overline{x}_i, \overline{y}_j) \, \Delta A$ where the sample points $(\overline{x}_i, \overline{y}_j)$ are the centers of the subrectangles.

2. (a) See (1) and (2) and the accompanying discussion in Section 12.2.

   (b) See (3) and the accompanying discussion in Section 12.2.

   (c) See (5) and the preceding discussion in Section 12.2.

   (d) See (6)–(11) in Section 12.2.

3. We may want to change from rectangular to polar coordinates in a double integral if the region R of integration is more easily

   described in polar coordinates. To accomplish this, we use $\iint_R f(x, y) \, dA = \int_{\alpha}^{\beta} \int_{a}^{b} f(r \cos\theta, r \sin\theta) \, r \, dr \, d\theta$ where R is

   given by $0 \le a \le r \le b$, $\alpha \le \theta \le \beta$.

4. (a) $m = \iint_D \rho(x, y) \, dA$

   (b) $M_x = \iint_D y\rho(x, y) \, dA$, $M_y = \iint_D x\rho(x, y) \, dA$

(c) The center of mass is $(\overline{x}, \overline{y})$ where $\overline{x} = \dfrac{M_y}{m}$ and $\overline{y} = \dfrac{M_x}{m}$.

(d) $I_x = \iint_D y^2 \rho(x, y) \, dA$, $I_y = \iint_D x^2 \rho(x, y) \, dA$, $I_0 = \iint_D (x^2 + y^2) \rho(x, y) \, dA$

5. (a) $\displaystyle\iiint_B f(x, y, z) \, dV = \lim_{\max \Delta x_i, \Delta y_j, \Delta z_k \to 0} \sum_{i=1}^{l} \sum_{j=1}^{m} \sum_{k=1}^{n} f(x_{ijk}^*, y_{ijk}^*, z_{ijk}^*) \, \Delta V_{ijk}$

(b) We usually evaluate $\iiint_B f(x, y, z) \, dV$ as an iterated integral according to Fubini's Theorem for Triple Integrals (see Theorem 12.5.4).

(c) See the paragraph following Example 12.5.1.

(d) See (5) and (6) and the accompanying discussion in Section 12.5.

(e) See (10) and the accompanying discussion in Section 12.5.

(f) See (11) and the preceding discussion in Section 12.5.

6. (a) $m = \iiint_E \rho(x, y, z) \, dV$

(b) $M_{yz} = \iiint_E x\rho(x, y, z) \, dV$, $M_{xz} = \iiint_E y\rho(x, y, z) \, dV$, $M_{xy} = \iiint_E z\rho(x, y, z) \, dV$.

(c) The center of mass is $(\overline{x}, \overline{y}, \overline{z})$ where $\overline{x} = \dfrac{M_{yz}}{m}$, $\overline{y} = \dfrac{M_{xz}}{m}$, and $\overline{z} = \dfrac{M_{xy}}{m}$.

(d) $I_x = \iiint_E (y^2 + z^2)\rho(x, y, z) \, dV$, $I_y = \iiint_E (x^2 + z^2)\rho(x, y, z) \, dV$, $I_z = \iiint_E (x^2 + y^2)\rho(x, y, z) \, dV$.

7. (a) See (1) and the discussion accompanying Figure 4 in Section 12.6.

(b) See (1) and Figures 2–4, and the accompanying discussion, in Section 12.7.

8. (a) See Formula 12.6.4 and the accompanying discussion.

(b) See Formula 12.7.3 and the accompanying discussion.

(c) We may want to change from rectangular to cylindrical or spherical coordinates in a triple integral if the region E of integration is more easily described in cylindrical or spherical coordinates or if the triple integral is easier to evaluate using cylindrical or spherical coordinates.

9. (a) $\dfrac{\partial(x, y)}{\partial(u, v)} = \begin{vmatrix} \partial x / \partial u & \partial x / \partial v \\ \partial y / \partial u & \partial y / \partial v \end{vmatrix} = \dfrac{\partial x}{\partial u} \dfrac{\partial y}{\partial v} - \dfrac{\partial x}{\partial v} \dfrac{\partial y}{\partial u}$

(b) See (9) and the accompanying discussion in Section 12.8.

(c) See (13) and the accompanying discussion in Section 12.8.

TRUE-FALSE QUIZ

1. This is true by Fubini's Theorem.

3. True by Equation 12.1.11.

5. True. By Equation 12.1.11 we can write $\int_0^1 \int_0^1 f(x)\,f(y)\,dy\,dx = \int_0^1 f(x)\,dx \int_0^1 f(y)\,dy$. But $\int_0^1 f(y)\,dy = \int_0^1 f(x)\,dx$ so

this becomes $\int_0^1 f(x)\,dx \int_0^1 f(x)\,dx = \left[\int_0^1 f(x)\,dx\right]^2$.

7. True: $\iint_D \sqrt{4 - x^2 - y^2}\,dA$ = the volume under the surface $x^2 + y^2 + z^2 = 4$ and above the xy-plane

$\qquad\qquad = \frac{1}{2}$ (the volume of the sphere $x^2 + y^2 + z^2 = 4$) $= \frac{1}{2} \cdot \frac{4}{3}\pi(2)^3 = \frac{16}{3}\pi$

9. The volume enclosed by the cone $z = \sqrt{x^2 + y^2}$ and the plane $z = 2$ is, in cylindrical coordinates,

$V = \int_0^{2\pi} \int_0^2 \int_r^2 r\,dz\,dr\,d\theta \neq \int_0^{2\pi} \int_0^2 \int_r^2 dz\,dr\,d\theta$, so the assertion is false.

EXERCISES

1. As shown in the contour map, we divide R into 9 equally sized subsquares, each with area $\Delta A = 1$. Then we approximate

$\iint_R f(x, y)\,dA$ by a Riemann sum with $m = n = 3$ and the sample points the upper right corners of each square, so

$$\iint_R f(x, y)\,dA \approx \sum_{i=1}^3 \sum_{j=1}^3 f(x_i, y_j)\,\Delta A$$

$$= \Delta A\,[f(1, 1) + f(1, 2) + f(1, 3) + f(2, 1) + f(2, 2) + f(2, 3) + f(3, 1) + f(3, 2) + f(3, 3)]$$

Using the contour lines to estimate the function values, we have

$$\iint_R f(x, y)\,dA \approx 1[2.7 + 4.7 + 8.0 + 4.7 + 6.7 + 10.0 + 6.7 + 8.6 + 11.9] \approx 64.0$$

3. $\int_1^2 \int_0^2 (y + 2xe^y)\,dx\,dy = \int_1^2 \left[xy + x^2 e^y\right]_{x=0}^{x=2} dy = \int_1^2 (2y + 4e^y)\,dy = \left[y^2 + 4e^y\right]_1^2$

$\qquad\qquad = 4 + 4e^2 - 1 - 4e = 4e^2 - 4e + 3$

5. $\int_0^1 \int_0^x \cos(x^2)\,dy\,dx = \int_0^1 \left[\cos(x^2)y\right]_{y=0}^{y=x} dx = \int_0^1 x\cos(x^2)\,dx = \frac{1}{2}\sin(x^2)\big]_0^1 = \frac{1}{2}\sin 1$

7. $\int_0^\pi \int_0^1 \int_0^{\sqrt{1-y^2}} y\sin x\,dz\,dy\,dx = \int_0^\pi \int_0^1 \left[(y\sin x)z\right]_{z=0}^{z=\sqrt{1-y^2}} dy\,dx = \int_0^\pi \int_0^1 y\sqrt{1-y^2}\sin x\,dy\,dx$

$\qquad\qquad = \int_0^\pi \left[-\frac{1}{3}(1-y^2)^{3/2}\sin x\right]_{y=0}^{y=1} dx = \int_0^\pi \frac{1}{3}\sin x\,dx = -\frac{1}{3}\cos x\big]_0^\pi = \frac{2}{3}$

9. The region R is more easily described by polar coordinates: $R = \{(r, \theta) \mid 2 \leq r \leq 4, 0 \leq \theta \leq \pi\}$. Thus

$\iint_R f(x, y)\,dA = \int_0^\pi \int_2^4 f(r\cos\theta, r\sin\theta)\,r\,dr\,d\theta$.

11.

$r = \sin 2\theta$

The region whose area is given by $\int_0^{\pi/2} \int_0^{\sin 2\theta} r\,dr\,d\theta$ is

$\{(r, \theta) \mid 0 \leq \theta \leq \frac{\pi}{2}, 0 \leq r \leq \sin 2\theta\}$, which is the region contained in the

loop in the first quadrant of the four-leaved rose $r = \sin 2\theta$.

13.

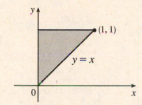

$$\int_0^1 \int_x^1 \cos(y^2)\, dy\, dx = \int_0^1 \int_0^y \cos(y^2)\, dx\, dy$$

$$= \int_0^1 \cos(y^2)\left[x\right]_{x=0}^{x=y} dy = \int_0^1 y \cos(y^2)\, dy$$

$$= \left[\tfrac{1}{2}\sin(y^2)\right]_0^1 = \tfrac{1}{2}\sin 1$$

15. $\iint_R y e^{xy}\, dA = \int_0^3 \int_0^2 y e^{xy}\, dx\, dy = \int_0^3 \left[e^{xy}\right]_{x=0}^{x=2} dy = \int_0^3 (e^{2y}-1)\, dy = \left[\tfrac{1}{2}e^{2y}-y\right]_0^3 = \tfrac{1}{2}e^6 - 3 - \tfrac{1}{2} = \tfrac{1}{2}e^6 - \tfrac{7}{2}$

17.

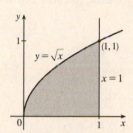

$$\iint_D \frac{y}{1+x^2}\, dA = \int_0^1 \int_0^{\sqrt{x}} \frac{y}{1+x^2}\, dy\, dx = \int_0^1 \frac{1}{1+x^2}\left[\tfrac{1}{2}y^2\right]_{y=0}^{y=\sqrt{x}} dx$$

$$= \tfrac{1}{2}\int_0^1 \frac{x}{1+x^2}\, dx = \left[\tfrac{1}{4}\ln(1+x^2)\right]_0^1 = \tfrac{1}{4}\ln 2$$

19.

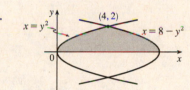

$$\iint_D y\, dA = \int_0^2 \int_{y^2}^{8-y^2} y\, dx\, dy$$

$$= \int_0^2 y[x]_{x=y^2}^{x=8-y^2}\, dy = \int_0^2 y(8 - y^2 - y^2)\, dy$$

$$= \int_0^2 (8y - 2y^3)\, dy = \left[4y^2 - \tfrac{1}{2}y^4\right]_0^2 = 8$$

21.

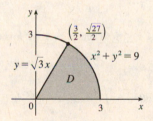

$$\iint_D \left(x^2+y^2\right)^{3/2} dA = \int_0^{\pi/3} \int_0^3 (r^2)^{3/2} r\, dr\, d\theta$$

$$= \int_0^{\pi/3} d\theta \int_0^3 r^4\, dr = \left[\theta\right]_0^{\pi/3}\left[\tfrac{1}{5}r^5\right]_0^3$$

$$= \frac{\pi}{3}\frac{3^5}{5} = \frac{81\pi}{5}$$

23. $\iiint_E xy\, dV = \int_0^3 \int_0^x \int_0^{x+y} xy\, dz\, dy\, dx = \int_0^3 \int_0^x xy\left[z\right]_{z=0}^{z=x+y} dy\, dx = \int_0^3 \int_0^x xy(x+y)\, dy\, dx$

$$= \int_0^3 \int_0^x (x^2 y + xy^2)\, dy\, dx = \int_0^3 \left[\tfrac{1}{2}x^2 y^2 + \tfrac{1}{3}xy^3\right]_{y=0}^{y=x} dx = \int_0^3 \left(\tfrac{1}{2}x^4 + \tfrac{1}{3}x^4\right) dx$$

$$= \tfrac{5}{6}\int_0^3 x^4\, dx = \left[\tfrac{1}{6}x^5\right]_0^3 = \tfrac{81}{2} = 40.5$$

25. $\iiint_E y^2 z^2\, dV = \int_{-1}^1 \int_{-\sqrt{1-y^2}}^{\sqrt{1-y^2}} \int_0^{1-y^2-z^2} y^2 z^2\, dx\, dz\, dy = \int_{-1}^1 \int_{-\sqrt{1-y^2}}^{\sqrt{1-y^2}} y^2 z^2 (1 - y^2 - z^2)\, dz\, dy$

$$= \int_0^{2\pi} \int_0^1 (r^2\cos^2\theta)(r^2\sin^2\theta)(1-r^2)\, r\, dr\, d\theta = \int_0^{2\pi}\int_0^1 \tfrac{1}{4}\sin^2 2\theta (r^5 - r^7)\, dr\, d\theta$$

$$= \int_0^{2\pi} \tfrac{1}{8}(1-\cos 4\theta)\left[\tfrac{1}{6}r^6 - \tfrac{1}{8}r^8\right]_{r=0}^{r=1} d\theta = \tfrac{1}{192}\left[\theta - \tfrac{1}{4}\sin 4\theta\right]_0^{2\pi} = \tfrac{2\pi}{192} = \tfrac{\pi}{96}$$

27. $\iiint_E yz\, dV = \int_{-2}^2 \int_0^{\sqrt{4-x^2}} \int_0^y yz\, dz\, dy\, dx = \int_{-2}^2 \int_0^{\sqrt{4-x^2}} \tfrac{1}{2}y^3\, dy\, dx = \int_0^\pi \int_0^2 \tfrac{1}{2}r^3(\sin^3\theta)\, r\, dr\, d\theta$

$$= \tfrac{16}{5}\int_0^\pi \sin^3\theta\, d\theta = \tfrac{16}{5}\left[-\cos\theta + \tfrac{1}{3}\cos^3\theta\right]_0^\pi = \tfrac{64}{15}$$

29. $V = \int_0^2 \int_1^4 (x^2 + 4y^2)\, dy\, dx = \int_0^2 \left[x^2 y + \tfrac{4}{3}y^3\right]_{y=1}^{y=4} dx = \int_0^2 (3x^2 + 84)\, dx = 176$

31.

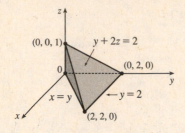

$$V = \int_0^2 \int_0^y \int_0^{(2-y)/2} dz\, dx\, dy = \int_0^2 \int_0^y \left(1 - \tfrac{1}{2}y\right) dx\, dy$$

$$= \int_0^2 \left(y - \tfrac{1}{2}y^2\right) dy = \tfrac{2}{3}$$

33. Using the wedge above the plane $z = 0$ and below the plane $z = mx$ and noting that we have the same volume for $m < 0$ as for $m > 0$ (so use $m > 0$), we have

$$V = 2\int_0^{a/3} \int_0^{\sqrt{a^2 - 9y^2}} mx\, dx\, dy = 2\int_0^{a/3} \tfrac{1}{2}m(a^2 - 9y^2)\, dy = m\left[a^2 y - 3y^3\right]_0^{a/3} = m\left(\tfrac{1}{3}a^3 - \tfrac{1}{9}a^3\right) = \tfrac{2}{9}ma^3.$$

35. (a) $m = \int_0^1 \int_0^{1-y^2} y\, dx\, dy = \int_0^1 (y - y^3)\, dy = \tfrac{1}{2} - \tfrac{1}{4} = \tfrac{1}{4}$

(b) $M_y = \int_0^1 \int_0^{1-y^2} xy\, dx\, dy = \int_0^1 \tfrac{1}{2}y(1 - y^2)^2\, dy = -\tfrac{1}{12}(1 - y^2)^3\Big]_0^1 = \tfrac{1}{12}$,

$M_x = \int_0^1 \int_0^{1-y^2} y^2\, dx\, dy = \int_0^1 (y^2 - y^4)\, dy = \tfrac{2}{15}$. Hence $(\overline{x}, \overline{y}) = \left(\tfrac{1}{3}, \tfrac{8}{15}\right)$.

(c) $I_x = \int_0^1 \int_0^{1-y^2} y^3\, dx\, dy = \int_0^1 (y^3 - y^5)\, dy = \tfrac{1}{12}$,

$I_y = \int_0^1 \int_0^{1-y^2} yx^2\, dx\, dy = \int_0^1 \tfrac{1}{3}y(1 - y^2)^3\, dy = -\tfrac{1}{24}(1 - y^2)^4\Big]_0^1 = \tfrac{1}{24}$,

$I_0 = I_x + I_y = \tfrac{1}{8}, \overline{\overline{y}}^2 = \frac{1/12}{1/4} = \tfrac{1}{3} \;\Rightarrow\; \overline{\overline{y}} = \tfrac{1}{\sqrt{3}}$, and $\overline{\overline{x}}^2 = \frac{1/24}{1/4} = \tfrac{1}{6} \;\Rightarrow\; \overline{\overline{x}} = \tfrac{1}{\sqrt{6}}$.

37. (a) The equation of the cone with the suggested orientation is $(h - z) = \tfrac{h}{a}\sqrt{x^2 + y^2}$, $0 \le z \le h$. Then $V = \tfrac{1}{3}\pi a^2 h$ is the volume of one frustum of a cone; by symmetry $M_{yz} = M_{xz} = 0$; and

$$M_{xy} = \iint\limits_{x^2+y^2 \le a^2} \int_0^{h - (h/a)\sqrt{x^2+y^2}} z\, dz\, dA = \int_0^{2\pi} \int_0^a \int_0^{(h/a)(a-r)} rz\, dz\, dr\, d\theta = \pi \int_0^a r\frac{h^2}{a^2}(a - r)^2\, dr$$

$$= \frac{\pi h^2}{a^2} \int_0^a (a^2 r - 2ar^2 + r^3)\, dr = \frac{\pi h^2}{a^2}\left(\frac{a^4}{2} - \frac{2a^4}{3} + \frac{a^4}{4}\right) = \frac{\pi h^2 a^2}{12}$$

Hence the centroid is $(\overline{x}, \overline{y}, \overline{z}) = \left(0, 0, \tfrac{1}{4}h\right)$.

(b) $I_z = \int_0^{2\pi} \int_0^a \int_0^{(h/a)(a-r)} r^3\, dz\, dr\, d\theta = 2\pi \int_0^a \frac{h}{a}(ar^3 - r^4)\, dr = \frac{2\pi h}{a}\left(\frac{a^5}{4} - \frac{a^5}{5}\right) = \frac{\pi a^4 h}{10}$

39. $x = r\cos\theta = 2\sqrt{3}\cos\frac{\pi}{3} = 2\sqrt{3}\cdot\tfrac{1}{2} = \sqrt{3}$, $y = r\sin\theta = 2\sqrt{3}\sin\frac{\pi}{3} = 2\sqrt{3}\cdot\frac{\sqrt{3}}{2} = 3$, $z = 2$, so in rectangular coordinates the point is $\left(\sqrt{3}, 3, 2\right)$. $\rho = \sqrt{r^2 + z^2} = \sqrt{12 + 4} = 4$, $\theta = \frac{\pi}{3}$, and $\cos\phi = \frac{z}{\rho} = \tfrac{1}{2}$, so $\phi = \frac{\pi}{3}$ and spherical coordinates are $\left(4, \frac{\pi}{3}, \frac{\pi}{3}\right)$.

41. $x = \rho\sin\phi\cos\theta = 8\sin\frac{\pi}{6}\cos\frac{\pi}{4} = 8\cdot\tfrac{1}{2}\cdot\frac{\sqrt{2}}{2} = 2\sqrt{2}$, $y = \rho\sin\phi\sin\theta = 8\sin\frac{\pi}{6}\sin\frac{\pi}{4} = 2\sqrt{2}$, and

$z = \rho\cos\phi = 8\cos\frac{\pi}{6} = 8\cdot\frac{\sqrt{3}}{2} = 4\sqrt{3}$. Thus rectangular coordinates for the point are $\left(2\sqrt{2}, 2\sqrt{2}, 4\sqrt{3}\right)$.

$r^2 = x^2 + y^2 = 8 + 8 = 16 \;\Rightarrow\; r = 4$, $\theta = \frac{\pi}{4}$, and $z = 4\sqrt{3}$, so cylindrical coordinates are $\left(4, \frac{\pi}{4}, 4\sqrt{3}\right)$.

43. $x^2 + y^2 + z^2 = 4$. In cylindrical coordinates, this becomes $r^2 + z^2 = 4$. In spherical coordinates, it becomes $\rho^2 = 4$ or $\rho = 2$.

45.

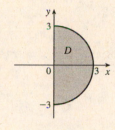

$$\int_0^3 \int_{-\sqrt{9-x^2}}^{\sqrt{9-x^2}} (x^3 + xy^2)\, dy\, dx = \int_0^3 \int_{-\sqrt{9-x^2}}^{\sqrt{9-x^2}} x(x^2 + y^2)\, dy\, dx$$

$$= \int_{-\pi/2}^{\pi/2} \int_0^3 (r\cos\theta)(r^2)\, r\, dr\, d\theta$$

$$= \int_{-\pi/2}^{\pi/2} \cos\theta\, d\theta \int_0^3 r^4\, dr$$

$$= \big[\sin\theta\big]_{-\pi/2}^{\pi/2} \big[\tfrac{1}{5}r^5\big]_0^3 = 2 \cdot \tfrac{1}{5}(243) = \tfrac{486}{5} = 97.2$$

47.

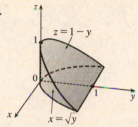

$$\int_{-1}^1 \int_{x^2}^1 \int_0^{1-y} f(x,y,z)\, dz\, dy\, dx = \int_0^1 \int_0^{1-z} \int_{-\sqrt{y}}^{\sqrt{y}} f(x,y,z)\, dx\, dy\, dz$$

49. Since $u = x - y$ and $v = x + y$, $x = \tfrac{1}{2}(u+v)$ and $y = \tfrac{1}{2}(v-u)$.

Thus $\dfrac{\partial(x,y)}{\partial(u,v)} = \begin{vmatrix} 1/2 & 1/2 \\ -1/2 & 1/2 \end{vmatrix} = \dfrac{1}{2}$ and $\displaystyle\iint_R \dfrac{x-y}{x+y}\, dA = \int_2^4 \int_{-2}^0 \dfrac{u}{v}\left(\dfrac{1}{2}\right) du\, dv = -\int_2^4 \dfrac{dv}{v} = -\ln 2$.

51. Let $u = y - x$ and $v = y + x$ so $x = y - u = (v - x) - u \;\Rightarrow\; x = \tfrac{1}{2}(v - u)$ and $y = v - \tfrac{1}{2}(v - u) = \tfrac{1}{2}(v + u)$.

$\left|\dfrac{\partial(x,y)}{\partial(u,v)}\right| = \left|\dfrac{\partial x}{\partial u}\dfrac{\partial y}{\partial v} - \dfrac{\partial x}{\partial v}\dfrac{\partial y}{\partial u}\right| = \left|-\tfrac{1}{2}\left(\tfrac{1}{2}\right) - \tfrac{1}{2}\left(\tfrac{1}{2}\right)\right| = \left|-\tfrac{1}{2}\right| = \tfrac{1}{2}$. R is the image under this transformation of the square

with vertices $(u,v) = (0,0)$, $(-2, 0)$, $(0, 2)$, and $(-2, 2)$. So

$$\iint_R xy\, dA = \int_0^2 \int_{-2}^0 \dfrac{v^2 - u^2}{4}\left(\dfrac{1}{2}\right) du\, dv = \tfrac{1}{8}\int_0^2 \big[v^2 u - \tfrac{1}{3}u^3\big]_{u=-2}^{u=0}\, dv = \tfrac{1}{8}\int_0^2 \left(2v^2 - \tfrac{8}{3}\right) dv = \tfrac{1}{8}\big[\tfrac{2}{3}v^3 - \tfrac{8}{3}v\big]_0^2 = 0$$

This result could have been anticipated by symmetry, since the integrand is an odd function of y and R is symmetric about the x-axis.

13 □ VECTOR CALCULUS

13.1 Vector Fields

1. $\mathbf{F}(x, y) = 0.3\,\mathbf{i} - 0.4\,\mathbf{j}$

All vectors in this field are identical, with length 0.5 and

parallel to $\langle 3, -4 \rangle$.

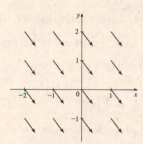

3. $\mathbf{F}(x, y) = -\frac{1}{2}\,\mathbf{i} + (y - x)\,\mathbf{j}$

The length of the vector $-\frac{1}{2}\,\mathbf{i} + (y - x)\,\mathbf{j}$ is

$\sqrt{\frac{1}{4} + (y - x)^2}$. Vectors along the line $y = x$ are

horizontal with length $\frac{1}{2}$.

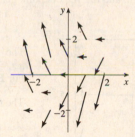

5. $\mathbf{F}(x, y) = \dfrac{y\,\mathbf{i} + x\,\mathbf{j}}{\sqrt{x^2 + y^2}}$

The length of the vector $\dfrac{y\,\mathbf{i} + x\,\mathbf{j}}{\sqrt{x^2 + y^2}}$ is 1.

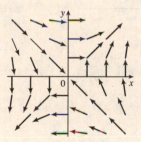

7. $\mathbf{F}(x, y, z) = \mathbf{k}$

All vectors in this field are parallel to the z-axis and have

length 1.

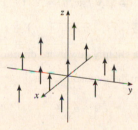

9. $\mathbf{F}(x, y, z) = x\,\mathbf{k}$

At each point (x, y, z), $\mathbf{F}(x, y, z)$ is a vector of length $|x|$.

For $x > 0$, all point in the direction of the positive z-axis,

while for $x < 0$, all are in the direction of the negative

z-axis. In each plane $x = k$, all the vectors are identical.

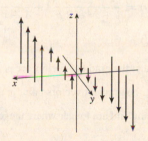

11. $\mathbf{F}(x, y) = \langle x, -y \rangle$ corresponds to graph IV. In the first quadrant all the vectors have positive x-components and negative y-components, in the second quadrant all vectors have negative x- and y-components, in the third quadrant all vectors have negative x-components and positive y-components, and in the fourth quadrant all vectors have positive x- and y-components. In addition, the vectors get shorter as we approach the origin.

13. $\mathbf{F}(x, y) = \langle y, y + 2 \rangle$ corresponds to graph I. As in Exercise 12, all vectors in quadrants I and II have positive x-components while all vectors in quadrants III and IV have negative x-components. Vectors along the line $y = -2$ are horizontal, and the vectors are independent of x (vectors along horizontal lines are identical).

15. $\mathbf{F}(x, y, z) = \mathbf{i} + 2\mathbf{j} + 3\mathbf{k}$ corresponds to graph IV, since all vectors have identical length and direction.

17. $\mathbf{F}(x, y, z) = x\mathbf{i} + y\mathbf{j} + 3\mathbf{k}$ corresponds to graph III; the projection of each vector onto the xy-plane is $x\mathbf{i} + y\mathbf{j}$, which points away from the origin, and the vectors point generally upward because their z-components are all 3.

19.

The vector field seems to have very short vectors near the line $y = 2x$.

For $\mathbf{F}(x, y) = \langle 0, 0 \rangle$ we must have $y^2 - 2xy = 0$ and $3xy - 6x^2 = 0$.

The first equation holds if $y = 0$ or $y = 2x$, and the second holds if

$x = 0$ or $y = 2x$. So both equations hold [and thus $\mathbf{F}(x, y) = \mathbf{0}$] along

the line $y = 2x$.

21. $f(x, y) = xe^{xy} \quad \Rightarrow$

$$\nabla f(x, y) = f_x(x, y)\mathbf{i} + f_y(x, y)\mathbf{j} = (xe^{xy} \cdot y + e^{xy})\mathbf{i} + (xe^{xy} \cdot x)\mathbf{j} = (xy + 1)e^{xy}\mathbf{i} + x^2 e^{xy}\mathbf{j}$$

23. $\nabla f(x, y, z) = f_x(x, y, z)\mathbf{i} + f_y(x, y, z)\mathbf{j} + f_z(x, y, z)\mathbf{k} = \dfrac{x}{\sqrt{x^2 + y^2 + z^2}}\mathbf{i} + \dfrac{y}{\sqrt{x^2 + y^2 + z^2}}\mathbf{j} + \dfrac{z}{\sqrt{x^2 + y^2 + z^2}}\mathbf{k}$

25. $f(x, y) = x^2 - y \quad \Rightarrow \quad \nabla f(x, y) = 2x\mathbf{i} - \mathbf{j}$.

The length of $\nabla f(x, y)$ is $\sqrt{4x^2 + 1}$. When $x \neq 0$, the vectors point away from the y-axis in a slightly downward direction with length that increases as the distance from the y-axis increases.

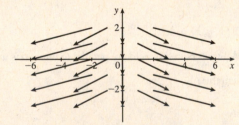

27. We graph $\nabla f(x, y) = \dfrac{2x}{1 + x^2 + 2y^2}\mathbf{i} + \dfrac{4y}{1 + x^2 + 2y^2}\mathbf{j}$ along with

a contour map of f.

The graph shows that the gradient vectors are perpendicular to the level curves. Also, the gradient vectors point in the direction in which f is increasing and are longer where the level curves are closer together.

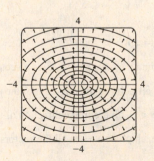

29. At $t = 3$ the particle is at $(2, 1)$ so its velocity is $\mathbf{V}(2, 1) = \langle 4, 3 \rangle$. After 0.01 units of time, the particle's change in

location should be approximately $0.01\,\mathbf{V}(2, 1) = 0.01\,\langle 4, 3 \rangle = \langle 0.04, 0.03 \rangle$, so the particle should be approximately at the

point $(2.04, 1.03)$.

31. (a) We sketch the vector field $\mathbf{F}(x, y) = x\,\mathbf{i} - y\,\mathbf{j}$ along with

several approximate flow lines. The flow lines appear to

be hyperbolas with shape similar to the graph of

$y = \pm 1/x$, so we might guess that the flow lines have

equations $y = C/x$.

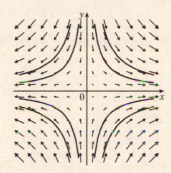

(b) If $x = x(t)$ and $y = y(t)$ are parametric equations of a flow line, then the velocity vector of the flow line at the

point (x, y) is $x'(t)\,\mathbf{i} + y'(t)\,\mathbf{j}$. Since the velocity vectors coincide with the vectors in the vector field, we have

$x'(t)\,\mathbf{i} + y'(t)\,\mathbf{j} = x\,\mathbf{i} - y\,\mathbf{j} \;\Rightarrow\; dx/dt = x, dy/dt = -y$. To solve these differential equations, we know

$dx/dt = x \;\Rightarrow\; dx/x = dt \;\Rightarrow\; \ln|x| = t + C \;\Rightarrow\; x = \pm e^{t+C} = Ae^t$ for some constant A, and

$dy/dt = -y \;\Rightarrow\; dy/y = -dt \;\Rightarrow\; \ln|y| = -t + K \;\Rightarrow\; y = \pm e^{-t+K} = Be^{-t}$ for some constant B. Therefore

$xy = Ae^t Be^{-t} = AB = \text{constant}$. If the flow line passes through $(1, 1)$ then $(1)\,(1) = \text{constant} = 1 \;\Rightarrow\; xy = 1 \;\Rightarrow\;$

$y = 1/x, x > 0$.

13.2 Line Integrals

1. $x = t^3$ and $y = t$, $0 \le t \le 2$, so by Formula 3

$$\int_C y^3\, ds = \int_0^2 t^3 \sqrt{\left(\frac{dx}{dt}\right)^2 + \left(\frac{dy}{dt}\right)^2}\, dt = \int_0^2 t^3 \sqrt{(3t^2)^2 + (1)^2}\, dt = \int_0^2 t^3 \sqrt{9t^4 + 1}\, dt$$

$$= \tfrac{1}{36} \cdot \tfrac{2}{3} \left(9t^4 + 1\right)^{3/2}\Big]_0^2 = \tfrac{1}{54}(145^{3/2} - 1) \text{ or } \tfrac{1}{54}\left(145\sqrt{145} - 1\right)$$

3. Parametric equations for C are $x = 4\cos t$, $\ y = 4\sin t$, $-\frac{\pi}{2} \le t \le \frac{\pi}{2}$. Then

$$\int_C xy^4\, ds = \int_{-\pi/2}^{\pi/2}(4\cos t)(4\sin t)^4 \sqrt{(-4\sin t)^2 + (4\cos t)^2}\, dt = \int_{-\pi/2}^{\pi/2} 4^5 \cos t \sin^4 t \sqrt{16(\sin^2 t + \cos^2 t)}\, dt$$

$$= 4^5 \int_{-\pi/2}^{\pi/2}(\sin^4 t \cos t)(4)\, dt = (4)^6 \left[\tfrac{1}{5}\sin^5 t\right]_{-\pi/2}^{\pi/2} = \tfrac{2 \cdot 4^6}{5} = 1638.4$$

5. If we choose x as the parameter, parametric equations for C are $x = x$, $\ y = \sqrt{x}$ for $1 \le x \le 4$ and

$$\int_C \left(x^2 y^3 - \sqrt{x}\right) dy = \int_1^4 \left[x^2 \cdot (\sqrt{x})^3 - \sqrt{x}\right] \frac{1}{2\sqrt{x}}\, dx = \tfrac{1}{2}\int_1^4 \left(x^3 - 1\right) dx$$

$$= \tfrac{1}{2}\left[\tfrac{1}{4}x^4 - x\right]_1^4 = \tfrac{1}{2}\left(64 - 4 - \tfrac{1}{4} + 1\right) = \tfrac{243}{8}$$

7.

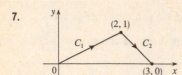

$C = C_1 + C_2$

On C_1: $x = x$, $y = \frac{1}{2}x$ $\Rightarrow$ $dy = \frac{1}{2}\,dx$, $0 \le x \le 2$.

On C_2: $x = x$, $y = 3 - x$ $\Rightarrow$ $dy = -dx$, $2 \le x \le 3$.

Then

$$\int_C (x + 2y)\,dx + x^2\,dy = \int_{C_1} (x + 2y)\,dx + x^2\,dy + \int_{C_2} (x + 2y)\,dx + x^2\,dy$$

$$= \int_0^2 \left[x + 2\left(\tfrac{1}{2}x\right) + x^2\left(\tfrac{1}{2}\right)\right]dx + \int_2^3 \left[x + 2(3 - x) + x^2(-1)\right]dx$$

$$= \int_0^2 \left(2x + \tfrac{1}{2}x^2\right)dx + \int_2^3 \left(6 - x - x^2\right)dx$$

$$= \left[x^2 + \tfrac{1}{6}x^3\right]_0^2 + \left[6x - \tfrac{1}{2}x^2 - \tfrac{1}{3}x^3\right]_2^3 = \tfrac{16}{3} - 0 + \tfrac{9}{2} - \tfrac{22}{3} = \tfrac{5}{2}$$

9. $x = 2\sin t$, $y = t$, $z = -2\cos t$, $0 \le t \le \pi$. Then by Formula 9,

$$\int_C xyz\,ds = \int_0^\pi (2\sin t)(t)(-2\cos t)\sqrt{\left(\tfrac{dx}{dt}\right)^2 + \left(\tfrac{dy}{dt}\right)^2 + \left(\tfrac{dz}{dt}\right)^2}\,dt$$

$$= \int_0^\pi -4t\sin t\,\cos t\,\sqrt{(2\cos t)^2 + (1)^2 + (2\sin t)^2}\,dt = \int_0^\pi -2t\sin 2t\,\sqrt{4(\cos^2 t + \sin^2 t) + 1}\,dt$$

$$= -2\sqrt{5}\int_0^\pi t\sin 2t\,dt = -2\sqrt{5}\left[-\tfrac{1}{2}t\cos 2t + \tfrac{1}{4}\sin 2t\right]_0^\pi \qquad \begin{bmatrix}\text{integrate by parts with} \\ u = t,\,dv = \sin 2t\,dt\end{bmatrix}$$

$$= -2\sqrt{5}\left(-\tfrac{\pi}{2} - 0\right) = \sqrt{5}\,\pi$$

11. Parametric equations for C are $x = t$, $y = 2t$, $z = 3t$, $0 \le t \le 1$. Then

$$\int_C xe^{yz}\,ds = \int_0^1 te^{(2t)(3t)}\sqrt{1^2 + 2^2 + 3^2}\,dt = \sqrt{14}\int_0^1 te^{6t^2}\,dt = \sqrt{14}\left[\tfrac{1}{12}e^{6t^2}\right]_0^1 = \tfrac{\sqrt{14}}{12}(e^6 - 1).$$

13. $\int_C xye^{yz}\,dy = \int_0^1 (t)(t^2)e^{(t^2)(t^3)}\cdot 2t\,dt = \int_0^1 2t^4 e^{t^5}\,dt = \tfrac{2}{5}e^{t^5}\Big]_0^1 = \tfrac{2}{5}(e^1 - e^0) = \tfrac{2}{5}(e - 1)$

15. Parametric equations for C are $x = 1 + 3t$, $y = t$, $z = 2t$, $0 \le t \le 1$. Then

$$\int_C z^2\,dx + x^2\,dy + y^2\,dz = \int_0^1 (2t)^2 \cdot 3\,dt + (1 + 3t)^2\,dt + t^2 \cdot 2\,dt = \int_0^1 \left(23t^2 + 6t + 1\right)dt$$

$$= \left[\tfrac{23}{3}t^3 + 3t^2 + t\right]_0^1 = \tfrac{23}{3} + 3 + 1 = \tfrac{35}{3}$$

17. (a) Along the line $x = -3$, the vectors of $\mathbf{F}$ have positive y-components, so since the path goes upward, the integrand $\mathbf{F} \cdot \mathbf{T}$ is always positive. Therefore $\int_{C_1} \mathbf{F} \cdot d\mathbf{r} = \int_{C_1} \mathbf{F} \cdot \mathbf{T}\,ds$ is positive.

(b) All of the (nonzero) field vectors along the circle with radius 3 are pointed in the clockwise direction, that is, opposite the direction to the path. So $\mathbf{F} \cdot \mathbf{T}$ is negative, and therefore $\int_{C_2} \mathbf{F} \cdot d\mathbf{r} = \int_{C_2} \mathbf{F} \cdot \mathbf{T}\,ds$ is negative.

19. $\mathbf{r}(t) = 11t^4\,\mathbf{i} + t^3\,\mathbf{j}$, so $\mathbf{F}(\mathbf{r}(t)) = (11t^4)(t^3)\,\mathbf{i} + 3(t^3)^2\,\mathbf{j} = 11t^7\,\mathbf{i} + 3t^6\,\mathbf{j}$ and $\mathbf{r}'(t) = 44t^3\,\mathbf{i} + 3t^2\,\mathbf{j}$. Then

$$\int_C \mathbf{F} \cdot d\mathbf{r} = \int_0^1 \mathbf{F}(\mathbf{r}(t)) \cdot \mathbf{r}'(t)\,dt = \int_0^1 (11t^7 \cdot 44t^3 + 3t^6 \cdot 3t^2)\,dt = \int_0^1 (484t^{10} + 9t^8)\,dt = \left[44t^{11} + t^9\right]_0^1 = 45.$$

21. $\int_C \mathbf{F} \cdot d\mathbf{r} = \int_0^1 \langle \sin t^3, \cos(-t^2), t^4 \rangle \cdot \langle 3t^2, -2t, 1 \rangle\,dt$

$$= \int_0^1 (3t^2 \sin t^3 - 2t\cos t^2 + t^4)\,dt = \left[-\cos t^3 - \sin t^2 + \tfrac{1}{5}t^5\right]_0^1 = \tfrac{6}{5} - \cos 1 - \sin 1$$

23. $\mathbf{F}(\mathbf{r}(t)) = (e^t)\left(e^{-t^2}\right)\mathbf{i} + \sin\left(e^{-t^2}\right)\mathbf{j} = e^{t-t^2}\,\mathbf{i} + \sin\left(e^{-t^2}\right)\mathbf{j},\ \mathbf{r}'(t) = e^t\,\mathbf{i} - 2te^{-t^2}\,\mathbf{j}.$ Then

$$\int_C \mathbf{F}\cdot d\mathbf{r} = \int_1^2 \mathbf{F}(\mathbf{r}(t))\cdot \mathbf{r}'(t)\,dt = \int_1^2 \left[e^{t-t^2}e^t + \sin\left(e^{-t^2}\right)\cdot\left(-2te^{-t^2}\right)\right]dt$$

$$= \int_1^2 \left[e^{2t-t^2} - 2te^{-t^2}\sin\left(e^{-t^2}\right)\right]dt \approx 1.9633$$

25. We graph $\mathbf{F}(x, y) = (x - y)\mathbf{i} + xy\,\mathbf{j}$ and the curve C. We see that most of the vectors starting on C point in roughly the same direction as C, so for these portions of C the tangential component $\mathbf{F}\cdot\mathbf{T}$ is positive. Although some vectors in the third quadrant which start on C point in roughly the opposite direction, and hence give negative tangential components, it seems reasonable that the effect of these portions of C is outweighed by the positive tangential components. Thus, we would expect $\int_C \mathbf{F}\cdot d\mathbf{r} = \int_C \mathbf{F}\cdot\mathbf{T}\,ds$ to be positive.

To verify, we evaluate $\int_C \mathbf{F}\cdot d\mathbf{r}$. The curve C can be represented by $\mathbf{r}(t) = 2\cos t\,\mathbf{i} + 2\sin t\,\mathbf{j},\ 0 \le t \le \frac{3\pi}{2}$, so $\mathbf{F}(\mathbf{r}(t)) = (2\cos t - 2\sin t)\mathbf{i} + 4\cos t\sin t\,\mathbf{j}$ and $\mathbf{r}'(t) = -2\sin t\,\mathbf{i} + 2\cos t\,\mathbf{j}.$ Then

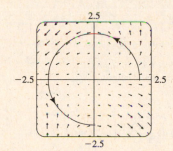

$$\int_C \mathbf{F}\cdot d\mathbf{r} = \int_0^{3\pi/2} \mathbf{F}(\mathbf{r}(t))\cdot \mathbf{r}'(t)\,dt$$

$$= \int_0^{3\pi/2}[-2\sin t(2\cos t - 2\sin t) + 2\cos t(4\cos t\sin t)]\,dt$$

$$= 4\int_0^{3\pi/2}(\sin^2 t - \sin t\cos t + 2\sin t\cos^2 t)\,dt$$

$$= 3\pi + \tfrac{2}{3}\qquad [\text{using a CAS}]$$

27. (a) $\int_C \mathbf{F}\cdot d\mathbf{r} = \int_0^1 \left\langle e^{t^2-1}, t^5\right\rangle\cdot\left\langle 2t, 3t^2\right\rangle dt = \int_0^1\left(2te^{t^2-1} + 3t^7\right)dt = \left[e^{t^2-1} + \tfrac{3}{8}t^8\right]_0^1 = \tfrac{11}{8} - 1/e$

(b) $\mathbf{r}(0) = \mathbf{0},\ \mathbf{F}(\mathbf{r}(0)) = \left\langle e^{-1}, 0\right\rangle;$

$\mathbf{r}\!\left(\tfrac{1}{\sqrt{2}}\right) = \left\langle\tfrac{1}{2}, \tfrac{1}{2\sqrt{2}}\right\rangle,\ \mathbf{F}\!\left(\mathbf{r}\!\left(\tfrac{1}{\sqrt{2}}\right)\right) = \left\langle e^{-1/2}, \tfrac{1}{4\sqrt{2}}\right\rangle;$

$\mathbf{r}(1) = \langle 1, 1\rangle,\ \mathbf{F}(\mathbf{r}(1)) = \langle 1, 1\rangle.$

In order to generate the graph with Maple, we use the `line` command in the `plottools` package to define each of the vectors. For example,

```
v1:=line([0,0],[exp(-1),0]):
```

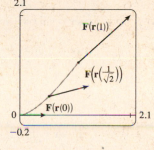

generates the vector from the vector field at the point $(0, 0)$ (but without an arrowhead) and gives it the name `v1`. To show everything on the same screen, we use the `display` command. In Mathematica, we use `ListPlot` (with the `PlotJoined -> True` option) to generate the vectors, and then `Show` to show everything on the same screen.

29. The part of the astroid that lies in the quadrant is parametrized by $x = \cos^3 t,\ y = \sin^3 t,\ 0 \le t \le \frac{\pi}{2}$.

Now $\dfrac{dx}{dt} = 3\cos^2 t\,(-\sin t)$ and $\dfrac{dy}{dt} = 3\sin^2 t\cos t$, so

$$\sqrt{\left(\frac{dx}{dt}\right)^2 + \left(\frac{dy}{dt}\right)^2} = \sqrt{9\cos^4 t\sin^2 t + 9\sin^4 t\cos^2 t} = 3\cos t\sin t\sqrt{\cos^2 t + \sin^2 t} = 3\cos t\sin t.$$

Therefore $\int_C x^3 y^5\,ds = \int_0^{\pi/2}\cos^9 t\sin^{15} t\,(3\cos t\sin t)\,dt = \dfrac{945}{16{,}777{,}216}\pi.$

31. We use the parametrization $x = 2\cos t$, $y = 2\sin t$, $-\frac{\pi}{2} \leq t \leq \frac{\pi}{2}$. Then

$$ds = \sqrt{\left(\frac{dx}{dt}\right)^2 + \left(\frac{dy}{dt}\right)^2}\, dt = \sqrt{(-2\sin t)^2 + (2\cos t)^2}\, dt = 2\, dt, \text{ so } m = \int_C k\, ds = 2k \int_{-\pi/2}^{\pi/2} dt = 2k(\pi),$$

$$\overline{x} = \frac{1}{2\pi k} \int_C xk\, ds = \frac{1}{2\pi} \int_{-\pi/2}^{\pi/2} (2\cos t)2\, dt = \frac{1}{2\pi}\left[4\sin t\right]_{-\pi/2}^{\pi/2} = \frac{4}{\pi}, \overline{y} = \frac{1}{2\pi k} \int_C yk\, ds = \frac{1}{2\pi} \int_{-\pi/2}^{\pi/2} (2\sin t)2\, dt = 0.$$

Hence $(\overline{x}, \overline{y}) = \left(\frac{4}{\pi}, 0\right)$.

33. (a) $\overline{x} = \frac{1}{m} \int_C x\rho(x, y, z)\, ds$, $\overline{y} = \frac{1}{m} \int_C y\rho(x, y, z)\, ds$, $\overline{z} = \frac{1}{m} \int_C z\rho(x, y, z)\, ds$ where $m = \int_C \rho(x, y, z)\, ds$.

(b) $m = \int_C k\, ds = k \int_0^{2\pi} \sqrt{4\sin^2 t + 4\cos^2 t + 9}\, dt = k\sqrt{13} \int_0^{2\pi} dt = 2\pi k\sqrt{13}$,

$$\overline{x} = \frac{1}{2\pi k\sqrt{13}} \int_0^{2\pi} 2k\sqrt{13}\sin t\, dt = 0, \overline{y} = \frac{1}{2\pi k\sqrt{13}} \int_0^{2\pi} 2k\sqrt{13}\cos t\, dt = 0,$$

$$\overline{z} = \frac{1}{2\pi k\sqrt{13}} \int_0^{2\pi} \left(k\sqrt{13}\right)(3t)\, dt = \frac{3}{2\pi}\left(2\pi^2\right) = 3\pi. \text{ Hence } (\overline{x}, \overline{y}, \overline{z}) = (0, 0, 3\pi).$$

35. From Example 3, $\rho(x, y) = k(1 - y)$, $x = \cos t$, $y = \sin t$, and $ds = dt$, $0 \leq t \leq \pi \Rightarrow$

$$I_x = \int_C y^2 \rho(x, y)\, ds = \int_0^\pi \sin^2 t\, [k(1 - \sin t)]\, dt = k \int_0^\pi (\sin^2 t - \sin^3 t)\, dt$$

$$= \frac{1}{2}k \int_0^\pi (1 - \cos 2t)\, dt - k \int_0^\pi (1 - \cos^2 t)\sin t\, dt \quad \begin{bmatrix} \text{Let } u = \cos t,\, du = -\sin t\, dt \\ \text{in the second integral} \end{bmatrix}$$

$$= k\left[\frac{\pi}{2} + \int_1^{-1}(1 - u^2)\, du\right] = k\left(\frac{\pi}{2} - \frac{4}{3}\right)$$

$$I_y = \int_C x^2 \rho(x, y)\, ds = k \int_0^\pi \cos^2 t\,(1 - \sin t)\, dt = \frac{k}{2} \int_0^\pi (1 + \cos 2t)\, dt - k \int_0^\pi \cos^2 t \sin t\, dt$$

$$= k\left(\frac{\pi}{2} - \frac{2}{3}\right), \text{ using the same substitution as above.}$$

37. $W = \int_C \mathbf{F} \cdot d\mathbf{r} = \int_0^{2\pi} \langle t - \sin t, 3 - \cos t \rangle \cdot \langle 1 - \cos t, \sin t \rangle\, dt$

$$= \int_0^{2\pi} (t - t\cos t - \sin t + \sin t\cos t + 3\sin t - \sin t\cos t)\, dt$$

$$= \int_0^{2\pi} (t - t\cos t + 2\sin t)\, dt = \left[\frac{1}{2}t^2 - (t\sin t + \cos t) - 2\cos t\right]_0^{2\pi} \quad \begin{bmatrix} \text{integrate by parts} \\ \text{in the second term} \end{bmatrix}$$

$$= 2\pi^2$$

39. $\mathbf{r}(t) = \langle 2t, t, 1 - t \rangle$, $0 \leq t \leq 1$.

$$W = \int_C \mathbf{F} \cdot d\mathbf{r} = \int_0^1 \langle 2t - t^2, t - (1 - t)^2, 1 - t - (2t)^2 \rangle \cdot \langle 2, 1, -1 \rangle\, dt$$

$$= \int_0^1 (4t - 2t^2 + t - 1 + 2t - t^2 - 1 + t + 4t^2)\, dt = \int_0^1 (t^2 + 8t - 2)\, dt = \left[\frac{1}{3}t^3 + 4t^2 - 2t\right]_0^1 = \frac{7}{3}$$

41. (a) $\mathbf{r}(t) = at^2\,\mathbf{i} + bt^3\,\mathbf{j} \Rightarrow \mathbf{v}(t) = \mathbf{r}'(t) = 2at\,\mathbf{i} + 3bt^2\,\mathbf{j} \Rightarrow \mathbf{a}(t) = \mathbf{v}'(t) = 2a\,\mathbf{i} + 6bt\,\mathbf{j}$, and force is mass times acceleration: $\mathbf{F}(t) = m\,\mathbf{a}(t) = 2ma\,\mathbf{i} + 6mbt\,\mathbf{j}$.

(b) $W = \int_C \mathbf{F} \cdot d\mathbf{r} = \int_0^1 (2ma\,\mathbf{i} + 6mbt\,\mathbf{j}) \cdot (2at\,\mathbf{i} + 3bt^2\,\mathbf{j})\, dt = \int_0^1 (4ma^2 t + 18mb^2 t^3)\, dt$

$$= \left[2ma^2 t^2 + \frac{9}{2}mb^2 t^4\right]_0^1 = 2ma^2 + \frac{9}{2}mb^2$$

43. Let $\mathbf{F} = 185\,\mathbf{k}$. To parametrize the staircase, let $x = 20\cos t$, $y = 20\sin t$, $z = \frac{90}{6\pi}t = \frac{15}{\pi}t$, $0 \leq t \leq 6\pi \Rightarrow$

$$W = \int_C \mathbf{F} \cdot d\mathbf{r} = \int_0^{6\pi} \langle 0, 0, 185 \rangle \cdot \langle -20\sin t, 20\cos t, \frac{15}{\pi} \rangle\, dt = (185)\frac{15}{\pi} \int_0^{6\pi} dt = (185)(90) \approx 1.67 \times 10^4 \text{ ft-lb}$$

45. Let $\mathbf{r}(t) = \langle x(t), y(t), z(t)\rangle$ and $\mathbf{v} = \langle v_1, v_2, v_3\rangle$. Then

$$\int_C \mathbf{v} \cdot d\mathbf{r} = \int_a^b \langle v_1, v_2, v_3\rangle \cdot \langle x'(t), y'(t), z'(t)\rangle \, dt = \int_a^b [v_1\, x'(t) + v_2\, y'(t) + v_3\, z'(t)]\, dt$$

$$= \left[v_1\, x(t) + v_2\, y(t) + v_3\, z(t)\right]_a^b = [v_1\, x(b) + v_2\, y(b) + v_3\, z(b)] - [v_1\, x(a) + v_2\, y(a) + v_3\, z(a)]$$

$$= v_1\,[x(b) - x(a)] + v_2\,[y(b) - y(a)] + v_3\,[z(b) - z(a)]$$

$$= \langle v_1, v_2, v_3\rangle \cdot \langle x(b) - x(a), y(b) - y(a), z(b) - z(a)\rangle$$

$$= \langle v_1, v_2, v_3\rangle \cdot [\langle x(b), y(b), z(b)\rangle - \langle x(a), y(a), z(a)\rangle] = \mathbf{v} \cdot [\mathbf{r}(b) - \mathbf{r}(a)]$$

47. (a) $\mathbf{r}(t) = \langle \cos t, \sin t\rangle$, $0 \le t \le 2\pi$, and let $\mathbf{F} = \langle a, b\rangle$. Then

$$W = \int_C \mathbf{F} \cdot d\mathbf{r} = \int_0^{2\pi} \langle a, b\rangle \cdot \langle -\sin t, \cos t\rangle \, dt = \int_0^{2\pi} (-a\sin t + b\cos t)\, dt = \left[a\cos t + b\sin t\right]_0^{2\pi}$$

$$= a + 0 - a + 0 = 0$$

(b) Yes. $\mathbf{F}(x, y) = k\,\mathbf{x} = \langle kx, ky\rangle$ and

$$W = \int_C \mathbf{F} \cdot d\mathbf{r} = \int_0^{2\pi} \langle k\cos t, k\sin t\rangle \cdot \langle -\sin t, \cos t\rangle \, dt = \int_0^{2\pi} (-k\sin t\,\cos t + k\sin t\,\cos t)\, dt = \int_0^{2\pi} 0\, dt = 0.$$

13.3 The Fundamental Theorem for Line Integrals

1. C appears to be a smooth curve, and since ∇f is continuous, we know f is differentiable. Then Theorem 2 says that the value

of $\int_C \nabla f \cdot d\mathbf{r}$ is simply the difference of the values of f at the terminal and initial points of C. From the graph, this is

$50 - 10 = 40$.

3. $\partial(2x - 3y)/\partial y = -3 = \partial(-3x + 4y - 8)/\partial x$ and the domain of $\mathbf{F}$ is $\mathbb{R}^2$ which is open and simply-connected, so by

Theorem 6 $\mathbf{F}$ is conservative. Thus, there exists a function f such that $\nabla f = \mathbf{F}$, that is, $f_x(x, y) = 2x - 3y$ and

$f_y(x, y) = -3x + 4y - 8$. But $f_x(x, y) = 2x - 3y$ implies $f(x, y) = x^2 - 3xy + g(y)$ and differentiating both sides of this

equation with respect to y gives $f_y(x, y) = -3x + g'(y)$. Thus $-3x + 4y - 8 = -3x + g'(y)$ so $g'(y) = 4y - 8$ and

$g(y) = 2y^2 - 8y + K$ where K is a constant. Hence $f(x, y) = x^2 - 3xy + 2y^2 - 8y + K$ is a potential function for $\mathbf{F}$.

5. $\partial(e^x \cos y)/\partial y = -e^x \sin y$, $\partial(e^x \sin y)/\partial x = e^x \sin y$. Since these are not equal, $\mathbf{F}$ is not conservative.

7. $\partial(ye^x + \sin y)/\partial y = e^x + \cos y = \partial(e^x + x\cos y)/\partial x$ and the domain of $\mathbf{F}$ is $\mathbb{R}^2$. Hence $\mathbf{F}$ is conservative so there

exists a function f such that $\nabla f = \mathbf{F}$. Then $f_x(x, y) = ye^x + \sin y$ implies $f(x, y) = ye^x + x\sin y + g(y)$ and

$f_y(x, y) = e^x + x\cos y + g'(y)$. But $f_y(x, y) = e^x + x\cos y$ so $g(y) = K$ and $f(x, y) = ye^x + x\sin y + K$ is a potential

function for $\mathbf{F}$.

9. $\partial(\ln y + 2xy^3)/\partial y = 1/y + 6xy^2 = \partial(3x^2y^2 + x/y)/\partial x$ and the domain of $\mathbf{F}$ is $\{(x, y) \mid y > 0\}$ which is open and simply

connected. Hence $\mathbf{F}$ is conservative so there exists a function f such that $\nabla f = \mathbf{F}$. Then $f_x(x, y) = \ln y + 2xy^3$ implies

$f(x, y) = x\ln y + x^2y^3 + g(y)$ and $f_y(x, y) = x/y + 3x^2y^2 + g'(y)$. But $f_y(x, y) = 3x^2y^2 + x/y$ so $g'(y) = 0 \Rightarrow$

$g(y) = K$ and $f(x, y) = x\ln y + x^2y^3 + K$ is a potential function for $\mathbf{F}$.

11. (a) $f_x(x, y) = xy^2$ implies $f(x, y) = \frac{1}{2}x^2y^2 + g(y)$ and $f_y(x, y) = x^2y + g'(y)$. But $f_y(x, y) = x^2y$ so $g'(y) = 0 \Rightarrow$

$g(y) = K$, a constant. We can take $K = 0$, so $f(x, y) = \frac{1}{2}x^2y^2$.

(b) The initial point of C is $\mathbf{r}(0) = (0, 1)$ and the terminal point is $\mathbf{r}(1) = (2, 1)$, so

$\int_C \mathbf{F} \cdot d\mathbf{r} = f(2, 1) - f(0, 1) = 2 - 0 = 2$.

13. (a) $f_x(x, y, z) = yz$ implies $f(x, y, z) = xyz + g(y, z)$ and so $f_y(x, y, z) = xz + g_y(y, z)$. But $f_y(x, y, z) = xz$ so

$g_y(y, z) = 0 \Rightarrow g(y, z) = h(z)$. Thus $f(x, y, z) = xyz + h(z)$ and $f_z(x, y, z) = xy + h'(z)$. But

$f_z(x, y, z) = xy + 2z$, so $h'(z) = 2z \Rightarrow h(z) = z^2 + K$. Hence $f(x, y, z) = xyz + z^2$ (taking $K = 0$).

(b) $\int_C \mathbf{F} \cdot d\mathbf{r} = f(4, 6, 3) - f(1, 0, -2) = 81 - 4 = 77$.

15. (a) $f_x(x, y, z) = yze^{xz}$ implies $f(x, y, z) = ye^{xz} + g(y, z)$ and so $f_y(x, y, z) = e^{xz} + g_y(y, z)$. But $f_y(x, y, z) = e^{xz}$ so

$g_y(y, z) = 0 \Rightarrow g(y, z) = h(z)$. Thus $f(x, y, z) = ye^{xz} + h(z)$ and $f_z(x, y, z) = xye^{xz} + h'(z)$. But

$f_z(x, y, z) = xye^{xz}$, so $h'(z) = 0 \Rightarrow h(z) = K$. Hence $f(x, y, z) = ye^{xz}$ (taking $K = 0$).

(b) $\mathbf{r}(0) = \langle 1, -1, 0 \rangle$, $\mathbf{r}(2) = \langle 5, 3, 0 \rangle$ so $\int_C \mathbf{F} \cdot d\mathbf{r} = f(5, 3, 0) - f(1, -1, 0) = 3e^0 + e^0 = 4$.

17. The functions $2xe^{-y}$ and $2y - x^2e^{-y}$ have continuous first-order derivatives on $\mathbb{R}^2$ and

$\dfrac{\partial}{\partial y}\left(2xe^{-y}\right) = -2xe^{-y} = \dfrac{\partial}{\partial x}\left(2y - x^2e^{-y}\right)$, so $\mathbf{F}(x, y) = 2xe^{-y}\,\mathbf{i} + \left(2y - x^2e^{-y}\right)\mathbf{j}$ is a conservative vector field by

Theorem 6 and hence the line integral is independent of path. Thus a potential function f exists, and $f_x(x, y) = 2xe^{-y}$

implies $f(x, y) = x^2e^{-y} + g(y)$ and $f_y(x, y) = -x^2e^{-y} + g'(y)$. But $f_y(x, y) = 2y - x^2e^{-y}$ so

$g'(y) = 2y \Rightarrow g(y) = y^2 + K$. We can take $K = 0$, so $f(x, y) = x^2e^{-y} + y^2$. Then

$\int_C 2xe^{-y}\,dx + (2y - x^2e^{-y})\,dy = f(2, 1) - f(1, 0) = 4e^{-1} + 1 - 1 = 4/e$.

19. $\mathbf{F}(x, y) = 2y^{3/2}\,\mathbf{i} + 3x\sqrt{y}\,\mathbf{j}$, $W = \int_C \mathbf{F} \cdot d\mathbf{r}$. Since $\partial(2y^{3/2})/\partial y = 3\sqrt{y} = \partial(3x\sqrt{y})/\partial x$, there exists a function f

such that $\nabla f = \mathbf{F}$. In fact, $f_x(x, y) = 2y^{3/2} \Rightarrow f(x, y) = 2xy^{3/2} + g(y) \Rightarrow f_y(x, y) = 3xy^{1/2} + g'(y)$. But

$f_y(x, y) = 3x\sqrt{y}$ so $g'(y) = 0$ or $g(y) = K$. We can take $K = 0 \Rightarrow f(x, y) = 2xy^{3/2}$. Thus

$W = \int_C \mathbf{F} \cdot d\mathbf{r} = f(2, 4) - f(1, 1) = 2(2)(8) - 2(1) = 30$.

21. We know that if the vector field (call it $\mathbf{F}$) is conservative, then around any closed path C, $\int_C \mathbf{F} \cdot d\mathbf{r} = 0$. But take C to be a

circle centered at the origin, oriented counterclockwise. All of the field vectors that start on C are roughly in the direction of

motion along C, so the integral around C will be positive. Therefore the field is not conservative.

23.

From the graph, it appears that $\mathbf{F}$ is conservative, since around all closed

paths, the number and size of the field vectors pointing in directions similar

to that of the path seem to be roughly the same as the number and size of the

vectors pointing in the opposite direction. To check, we calculate

$\dfrac{\partial}{\partial y}(\sin y) = \cos y = \dfrac{\partial}{\partial x}(1 + x\cos y)$. Thus $\mathbf{F}$ is conservative, by

Theorem 6.

25. Since $\mathbf{F}$ is conservative, there exists a function f such that $\mathbf{F} = \nabla f$, that is, $P = f_x$, $Q = f_y$, and $R = f_z$. Since P,

Q, and R have continuous first order partial derivatives, Clairaut's Theorem says that $\partial P / \partial y = f_{xy} = f_{yx} = \partial Q / \partial x$,

$\partial P / \partial z = f_{xz} = f_{zx} = \partial R / \partial x$, and $\partial Q / \partial z = f_{yz} = f_{zy} = \partial R / \partial y$.

27. $D = \{(x, y) \mid 0 < y < 3\}$ consists of those points between, but not

on, the horizontal lines $y = 0$ and $y = 3$.

(a) Since D does not include any of its boundary points, it is open. More

formally, at any point in D there is a disk centered at that point that

lies entirely in D.

(b) Any two points chosen in D can always be joined by a path that lies

entirely in D, so D is connected. (D consists of just one "piece.")

(c) D is connected and it has no holes, so it's simply-connected. (Every simple closed curve in D encloses only points that are

in D.)

29. $D = \{(x, y) \mid 1 \le x^2 + y^2 \le 4, \ y \ge 0\}$ is the semiannular region

in the upper half-plane between circles centered at the origin of radii

1 and 2 (including all boundary points).

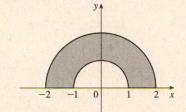

(a) D includes boundary points, so it is not open. [Note that at any

boundary point, $(1, 0)$ for instance, any disk centered there cannot lie

entirely in D.]

(b) The region consists of one piece, so it's connected.

(c) D is connected and has no holes, so it's simply-connected.

31. (a) $P = -\dfrac{y}{x^2 + y^2}$, $\dfrac{\partial P}{\partial y} = \dfrac{y^2 - x^2}{(x^2 + y^2)^2}$ and $Q = \dfrac{x}{x^2 + y^2}$, $\dfrac{\partial Q}{\partial x} = \dfrac{y^2 - x^2}{(x^2 + y^2)^2}$. Thus $\dfrac{\partial P}{\partial y} = \dfrac{\partial Q}{\partial x}$.

(b) C_1: $x = \cos t$, $y = \sin t$, $0 \le t \le \pi$, C_2: $x = \cos t$, $y = \sin t$, $t = 2\pi$ to $t = \pi$. Then

$$\int_{C_1} \mathbf{F} \cdot d\mathbf{r} = \int_0^\pi \frac{(-\sin t)(-\sin t) + (\cos t)(\cos t)}{\cos^2 t + \sin^2 t} \, dt = \int_0^\pi dt = \pi \text{ and } \int_{C_2} \mathbf{F} \cdot d\mathbf{r} = \int_{2\pi}^\pi dt = -\pi$$

Since these aren't equal, the line integral of $\mathbf{F}$ isn't independent of path. (Or notice that $\int_{C_3} \mathbf{F} \cdot d\mathbf{r} = \int_0^{2\pi} dt = 2\pi$ where

C_3 is the circle $x^2 + y^2 = 1$, and apply the contrapositive of Theorem 3.) This doesn't contradict Theorem 6, since the

domain of $\mathbf{F}$, which is $\mathbb{R}^2$ except the origin, isn't simply-connected.

13.4 Green's Theorem

1. (a) Parametric equations for C are $x = 2\cos t$, $y = 2\sin t$, $0 \le t \le 2\pi$. Then

$$\oint_C (x - y)\, dx + (x + y)\, dy = \int_0^{2\pi} [(2\cos t - 2\sin t)(-2\sin t) + (2\cos t + 2\sin t)(2\cos t)]\, dt$$

$$= \int_0^{2\pi} (4\sin^2 t + 4\cos^2 t)\, dt = \int_0^{2\pi} 4\, dt = 4t \Big]_0^{2\pi} = 8\pi$$

(b) Note that C as given in part (a) is a positively oriented, smooth, simple closed curve. Then by Green's Theorem,

$$\oint_C (x - y)\, dx + (x + y)\, dy = \iint_D \left[\frac{\partial}{\partial x}(x + y) - \frac{\partial}{\partial y}(x - y) \right] dA = \iint_D [1 - (-1)]\, dA = 2 \iint_D dA$$

$$= 2A(D) = 2\pi(2)^2 = 8\pi$$

3. (a)

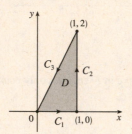

C_1: $x = t \;\Rightarrow\; dx = dt$, $y = 0 \;\Rightarrow\; dy = 0\, dt$, $0 \le t \le 1$.

C_2: $x = 1 \;\Rightarrow\; dx = 0\, dt$, $y = t \;\Rightarrow\; dy = dt$, $0 \le t \le 2$.

C_3: $x = 1 - t \;\Rightarrow\; dx = -dt$, $y = 2 - 2t \;\Rightarrow\; dy = -2\, dt$, $0 \le t \le 1$.

Thus

$$\oint_C xy\, dx + x^2 y^3\, dy = \oint_{C_1 + C_2 + C_3} xy\, dx + x^2 y^3\, dy$$

$$= \int_0^1 0\, dt + \int_0^2 t^3\, dt + \int_0^1 \left[-(1 - t)(2 - 2t) - 2(1 - t)^2 (2 - 2t)^3 \right] dt$$

$$= 0 + \left[\tfrac{1}{4} t^4 \right]_0^2 + \left[\tfrac{2}{3}(1 - t)^3 + \tfrac{8}{3}(1 - t)^6 \right]_0^1 = 4 - \tfrac{10}{3} = \tfrac{2}{3}$$

(b) $\oint_C xy\, dx + x^2 y^3\, dy = \iint_D \left[\frac{\partial}{\partial x}(x^2 y^3) - \frac{\partial}{\partial y}(xy) \right] dA = \int_0^1 \int_0^{2x} (2xy^3 - x)\, dy\, dx$

$$= \int_0^1 \left[\tfrac{1}{2} xy^4 - xy \right]_{y=0}^{y=2x} dx = \int_0^1 (8x^5 - 2x^2)\, dx = \tfrac{4}{3} - \tfrac{2}{3} = \tfrac{2}{3}$$

5.

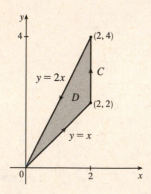

The region D enclosed by C is given by $\{(x, y) \mid 0 \le x \le 2, x \le y \le 2x\}$, so

$$\int_C xy^2\, dx + 2x^2 y\, dy = \iint_D \left[\frac{\partial}{\partial x}(2x^2 y) - \frac{\partial}{\partial y}(xy^2) \right] dA$$

$$= \int_0^2 \int_x^{2x} (4xy - 2xy)\, dy\, dx$$

$$= \int_0^2 \left[xy^2 \right]_{y=x}^{y=2x} dx$$

$$= \int_0^2 3x^3\, dx = \tfrac{3}{4} x^4 \Big]_0^2 = 12$$

7. $\int_C \left(y + e^{\sqrt{x}} \right) dx + (2x + \cos y^2)\, dy = \iint_D \left[\frac{\partial}{\partial x}(2x + \cos y^2) - \frac{\partial}{\partial y}\left(y + e^{\sqrt{x}} \right) \right] dA$

$$= \int_0^1 \int_{y^2}^{\sqrt{y}} (2 - 1)\, dx\, dy = \int_0^1 (y^{1/2} - y^2)\, dy = \tfrac{1}{3}$$

9. $\int_C y^3\, dx - x^3\, dy = \iint_D \left[\frac{\partial}{\partial x}(-x^3) - \frac{\partial}{\partial y}(y^3) \right] dA = \iint_D (-3x^2 - 3y^2)\, dA = \int_0^{2\pi} \int_0^2 (-3r^2)\, r\, dr\, d\theta$

$$= -3 \int_0^{2\pi} d\theta \int_0^2 r^3\, dr = -3(2\pi)(4) = -24\pi$$

11. $\mathbf{F}(x, y) = \langle y \cos x - xy \sin x, xy + x \cos x \rangle$ and the region D enclosed by C is given by

$\{(x, y) \mid 0 \le x \le 2, 0 \le y \le 4 - 2x\}$. C is traversed clockwise, so $-C$ gives the positive orientation.

$$\int_C \mathbf{F} \cdot d\mathbf{r} = -\int_{-C}(y \cos x - xy \sin x)\, dx + (xy + x \cos x)\, dy$$

$$= -\iint_D \left[\frac{\partial}{\partial x}(xy + x \cos x) - \frac{\partial}{\partial y}(y \cos x - xy \sin x) \right] dA$$

$$= -\iint_D (y - x \sin x + \cos x - \cos x + x \sin x)\, dA = -\int_0^2 \int_0^{4-2x} y\, dy\, dx$$

$$= -\int_0^2 \left[\tfrac{1}{2} y^2 \right]_{y=0}^{y=4-2x} dx = -\int_0^2 \tfrac{1}{2}(4 - 2x)^2\, dx = -\int_0^2 (8 - 8x + 2x^2)\, dx = -\left[8x - 4x^2 + \tfrac{2}{3}x^3 \right]_0^2$$

$$= -\left(16 - 16 + \tfrac{16}{3} - 0\right) = -\tfrac{16}{3}$$

13. $\mathbf{F}(x, y) = \langle y - \cos y, x \sin y \rangle$ and the region D enclosed by C is the disk with radius 2 centered at $(3, -4)$.

C is traversed clockwise, so $-C$ gives the positive orientation.

$$\int_C \mathbf{F} \cdot d\mathbf{r} = -\int_{-C}(y - \cos y)\, dx + (x \sin y)\, dy = -\iint_D \left[\frac{\partial}{\partial x}(x \sin y) - \frac{\partial}{\partial y}(y - \cos y) \right] dA$$

$$= -\iint_D (\sin y - 1 - \sin y)\, dA = \iint_D dA = \text{area of } D = \pi(2)^2 = 4\pi$$

15. Here $C = C_1 + C_2$ where

C_1 can be parametrized as $x = t, \ y = 1, \ -1 \le t \le 1$, and

C_2 is given by $x = -t, \ y = 2 - t^2, \ -1 \le t \le 1$.

Then the line integral is

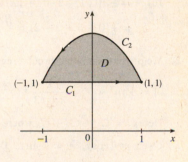

$$\oint_{C_1+C_2} y^2 e^x\, dx + x^2 e^y\, dy = \int_{-1}^1 [1 \cdot e^t + t^2 e \cdot 0]\, dt$$

$$+ \int_{-1}^1 [(2 - t^2)^2 e^{-t}(-1) + (-t)^2 e^{2-t^2}(-2t)]\, dt$$

$$= \int_{-1}^1 [e^t - (2 - t^2)^2 e^{-t} - 2t^3 e^{2-t^2}]\, dt = -8e + 48e^{-1}$$

according to a CAS. The double integral is

$$\iint_D \left(\frac{\partial Q}{\partial x} - \frac{\partial P}{\partial y} \right) dA = \int_{-1}^1 \int_1^{2-x^2} (2xe^y - 2ye^x)\, dy\, dx = -8e + 48e^{-1}, \text{ verifying Green's Theorem in this case.}$$

17. By Green's Theorem, $W = \int_C \mathbf{F} \cdot d\mathbf{r} = \int_C x(x + y)\, dx + xy^2\, dy = \iint_D (y^2 - x)\, dA$ where C is the path described in the

question and D is the triangle bounded by C. So

$$W = \int_0^1 \int_0^{1-x} (y^2 - x)\, dy\, dx = \int_0^1 \left[\tfrac{1}{3}y^3 - xy \right]_{y=0}^{y=1-x} dx = \int_0^1 \left(\tfrac{1}{3}(1 - x)^3 - x(1 - x) \right) dx$$

$$= \left[-\tfrac{1}{12}(1 - x)^4 - \tfrac{1}{2}x^2 + \tfrac{1}{3}x^3 \right]_0^1 = \left(-\tfrac{1}{2} + \tfrac{1}{3} \right) - \left(-\tfrac{1}{12} \right) = -\tfrac{1}{12}$$

19. Let C_1 be the arch of the cycloid from $(0, 0)$ to $(2\pi, 0)$, which corresponds to $0 \le t \le 2\pi$, and let C_2 be the segment from

$(2\pi, 0)$ to $(0, 0)$, so C_2 is given by $x = 2\pi - t, y = 0, 0 \le t \le 2\pi$. Then $C = C_1 \cup C_2$ is traversed clockwise, so $-C$ is

oriented positively. Thus $-C$ encloses the area under one arch of the cycloid and from (5) we have

$$A = -\oint_{-C} y\, dx = \int_{C_1} y\, dx + \int_{C_2} y\, dx = \int_0^{2\pi} (1 - \cos t)(1 - \cos t)\, dt + \int_0^{2\pi} 0\, (-dt)$$

$$= \int_0^{2\pi} (1 - 2\cos t + \cos^2 t)\, dt + 0 = \left[t - 2\sin t + \tfrac{1}{2}t + \tfrac{1}{4}\sin 2t \right]_0^{2\pi} = 3\pi$$

21. (a) Using Equation 13.2.8, we write parametric equations of the line segment as $x = (1-t)x_1 + tx_2$, $y = (1-t)y_1 + ty_2$,

$0 \le t \le 1$. Then $dx = (x_2 - x_1)\,dt$ and $dy = (y_2 - y_1)\,dt$, so

$$\int_C x\,dy - y\,dx = \int_0^1 [(1-t)x_1 + tx_2](y_2 - y_1)\,dt + [(1-t)y_1 + ty_2](x_2 - x_1)\,dt$$

$$= \int_0^1 (x_1(y_2 - y_1) - y_1(x_2 - x_1) + t[(y_2 - y_1)(x_2 - x_1) - (x_2 - x_1)(y_2 - y_1)])\,dt$$

$$= \int_0^1 (x_1 y_2 - x_2 y_1)\,dt = x_1 y_2 - x_2 y_1$$

(b) We apply Green's Theorem to the path $C = C_1 \cup C_2 \cup \cdots \cup C_n$, where C_i is the line segment that joins (x_i, y_i) to

(x_{i+1}, y_{i+1}) for $i = 1, 2, \ldots, n-1$, and C_n is the line segment that joins (x_n, y_n) to (x_1, y_1). From (5),

$\frac{1}{2}\int_C x\,dy - y\,dx = \iint_D dA$, where D is the polygon bounded by C. Therefore

$$\text{area of polygon} = A(D) = \iint_D dA = \frac{1}{2}\int_C x\,dy - y\,dx$$

$$= \frac{1}{2}\left(\int_{C_1} x\,dy - y\,dx + \int_{C_2} x\,dy - y\,dx + \cdots + \int_{C_{n-1}} x\,dy - y\,dx + \int_{C_n} x\,dy - y\,dx\right)$$

To evaluate these integrals we use the formula from (a) to get

$$A(D) = \frac{1}{2}[(x_1 y_2 - x_2 y_1) + (x_2 y_3 - x_3 y_2) + \cdots + (x_{n-1} y_n - x_n y_{n-1}) + (x_n y_1 - x_1 y_n)].$$

(c) $A = \frac{1}{2}[(0 \cdot 1 - 2 \cdot 0) + (2 \cdot 3 - 1 \cdot 1) + (1 \cdot 2 - 0 \cdot 3) + (0 \cdot 1 - (-1) \cdot 2) + (-1 \cdot 0 - 0 \cdot 1)]$

$= \frac{1}{2}(0 + 5 + 2 + 2) = \frac{9}{2}$

23. We orient the quarter-circular region as shown in the figure.

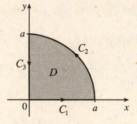

$A = \frac{1}{4}\pi a^2$ so $\overline{x} = \dfrac{1}{\pi a^2/2}\oint_C x^2\,dy$ and $\overline{y} = -\dfrac{1}{\pi a^2/2}\oint_C y^2\,dx$.

Here $C = C_1 + C_2 + C_3$ where C_1: $x = t$, $y = 0$, $0 \le t \le a$;

C_2: $x = a\cos t$, $y = a\sin t$, $0 \le t \le \frac{\pi}{2}$; and

C_3: $x = 0$, $y = a - t$, $0 \le t \le a$. Then

$$\oint_C x^2\,dy = \int_{C_1} x^2\,dy + \int_{C_2} x^2\,dy + \int_{C_3} x^2\,dy = \int_0^a 0\,dt + \int_0^{\pi/2} (a\cos t)^2(a\cos t)\,dt + \int_0^a 0\,dt$$

$$= \int_0^{\pi/2} a^3 \cos^3 t\,dt = a^3 \int_0^{\pi/2}(1 - \sin^2 t)\cos t\,dt = a^3\left[\sin t - \frac{1}{3}\sin^3 t\right]_0^{\pi/2} = \frac{2}{3}a^3$$

so $\overline{x} = \dfrac{1}{\pi a^2/2}\oint_C x^2\,dy = \dfrac{4a}{3\pi}$.

$$\oint_C y^2\,dx = \int_{C_1} y^2\,dx + \int_{C_2} y^2\,dx + \int_{C_3} y^2\,dx = \int_0^a 0\,dt + \int_0^{\pi/2}(a\sin t)^2(-a\sin t)\,dt + \int_0^a 0\,dt$$

$$= \int_0^{\pi/2}(-a^3 \sin^3 t)\,dt = -a^3\int_0^{\pi/2}(1 - \cos^2 t)\sin t\,dt = -a^3\left[\frac{1}{3}\cos^3 t - \cos t\right]_0^{\pi/2} = -\frac{2}{3}a^3,$$

so $\overline{y} = -\dfrac{1}{\pi a^2/2}\oint_C y^2\,dx = \dfrac{4a}{3\pi}$. Thus $(\overline{x}, \overline{y}) = \left(\dfrac{4a}{3\pi}, \dfrac{4a}{3\pi}\right)$.

25. By Green's Theorem, $-\frac{1}{3}\rho\oint_C y^3\,dx = -\frac{1}{3}\rho\iint_D(-3y^2)\,dA = \iint_D y^2\rho\,dA = I_x$ and

$\frac{1}{3}\rho\oint_C x^3\,dy = \frac{1}{3}\rho\iint_D(3x^2)\,dA = \iint_D x^2\rho\,dA = I_y$.

27. As in Example 5, let C' be a counterclockwise-oriented circle with center the origin and radius a, where a is chosen to be

small enough so that C' lies inside C, and D the region bounded by C and C'. Here

$$P = \frac{2xy}{(x^2 + y^2)^2} \quad \Rightarrow \quad \frac{\partial P}{\partial y} = \frac{2x(x^2 + y^2)^2 - 2xy \cdot 2(x^2 + y^2) \cdot 2y}{(x^2 + y^2)^4} = \frac{2x^3 - 6xy^2}{(x^2 + y^2)^3} \text{ and}$$

$Q = \dfrac{y^2 - x^2}{(x^2 + y^2)^2} \;\Rightarrow\; \dfrac{\partial Q}{\partial x} = \dfrac{-2x(x^2 + y^2)^2 - (y^2 - x^2) \cdot 2(x^2 + y^2) \cdot 2x}{(x^2 + y^2)^4} = \dfrac{2x^3 - 6xy^2}{(x^2 + y^2)^3}$. Thus, as in the example,

$$\int_C P\,dx + Q\,dy + \int_{-C'} P\,dx + Q\,dy = \iint_D \left(\frac{\partial Q}{\partial x} - \frac{\partial P}{\partial y} \right) dA = \iint_D 0\,dA = 0$$

and $\int_C \mathbf{F} \cdot d\mathbf{r} = \int_{C'} \mathbf{F} \cdot d\mathbf{r}$. We parametrize C' as $\mathbf{r}(t) = a\cos t\,\mathbf{i} + a\sin t\,\mathbf{j}, 0 \le t \le 2\pi$. Then

$$\int_C \mathbf{F} \cdot d\mathbf{r} = \int_{C'} \mathbf{F} \cdot d\mathbf{r} = \int_0^{2\pi} \frac{2\,(a\cos t)\,(a\sin t)\,\mathbf{i} + (a^2 \sin^2 t - a^2 \cos^2 t)\,\mathbf{j}}{\left(a^2 \cos^2 t + a^2 \sin^2 t\right)^2} \cdot \left(-a\sin t\,\mathbf{i} + a\cos t\,\mathbf{j}\right) dt$$

$$= \frac{1}{a} \int_0^{2\pi} \left(-\cos t \sin^2 t - \cos^3 t\right) dt = \frac{1}{a} \int_0^{2\pi} \left(-\cos t \sin^2 t - \cos t\,(1 - \sin^2 t)\right) dt$$

$$= -\frac{1}{a} \int_0^{2\pi} \cos t\,dt = -\frac{1}{a} \sin t\,\Big]_0^{2\pi} = 0$$

29. Since C is a simple closed path which doesn't pass through or enclose the origin, there exists an open region that doesn't contain the origin but does contain D. Thus $P = -y/(x^2 + y^2)$ and $Q = x/(x^2 + y^2)$ have continuous partial derivatives on this open region containing D and we can apply Green's Theorem. But by Exercise 13.3.31(a), $\partial P/\partial y = \partial Q/\partial x$, so $\oint_C \mathbf{F} \cdot d\mathbf{r} = \iint_D 0\,dA = 0$.

31. Using the first part of (5), we have that $\iint_R dx\,dy = A(R) = \int_{\partial R} x\,dy$. But $x = g(u, v)$, and $dy = \dfrac{\partial h}{\partial u}\,du + \dfrac{\partial h}{\partial v}\,dv$, and we orient ∂S by taking the positive direction to be that which corresponds, under the mapping, to the positive direction along ∂R, so

$$\int_{\partial R} x\,dy = \int_{\partial S} g(u, v) \left(\frac{\partial h}{\partial u}\,du + \frac{\partial h}{\partial v}\,dv \right) = \int_{\partial S} g(u, v)\,\frac{\partial h}{\partial u}\,du + g(u, v)\,\frac{\partial h}{\partial v}\,dv$$

$$= \pm \iint_S \left[\frac{\partial}{\partial u}\left(g(u, v)\,\frac{\partial h}{\partial v} \right) - \frac{\partial}{\partial v}\left(g(u, v)\,\frac{\partial h}{\partial u} \right) \right] dA \quad \text{[using Green's Theorem in the } uv\text{-plane]}$$

$$= \pm \iint_S \left(\frac{\partial g}{\partial u}\,\frac{\partial h}{\partial v} + g(u, v)\,\frac{\partial^2 h}{\partial u\,\partial v} - \frac{\partial g}{\partial v}\,\frac{\partial h}{\partial u} - g(u, v)\,\frac{\partial^2 h}{\partial v\,\partial u} \right) dA \quad \text{[using the Chain Rule]}$$

$$= \pm \iint_S \left(\frac{\partial x}{\partial u}\,\frac{\partial y}{\partial v} - \frac{\partial x}{\partial v}\,\frac{\partial y}{\partial u} \right) dA \quad \text{[by the equality of mixed partials]} \quad = \pm \iint_S \frac{\partial(x, y)}{\partial(u, v)}\,du\,dv$$

The sign is chosen to be positive if the orientation that we gave to ∂S corresponds to the usual positive orientation, and it is negative otherwise. In either case, since $A(R)$ is positive, the sign chosen must be the same as the sign of $\dfrac{\partial(x, y)}{\partial(u, v)}$.

Therefore $A(R) = \iint_R dx\,dy = \iint_S \left| \dfrac{\partial(x, y)}{\partial(u, v)} \right| du\,dv$.

13.5   Curl and Divergence

1. (a) $\operatorname{curl} \mathbf{F} = \nabla \times \mathbf{F} = \begin{vmatrix} \mathbf{i} & \mathbf{j} & \mathbf{k} \\ \partial/\partial x & \partial/\partial y & \partial/\partial z \\ x + yz & y + xz & z + xy \end{vmatrix}$

$$= \left[\frac{\partial}{\partial y}(z + xy) - \frac{\partial}{\partial z}(y + xz) \right]\mathbf{i} - \left[\frac{\partial}{\partial x}(z + xy) - \frac{\partial}{\partial z}(x + yz) \right]\mathbf{j} + \left[\frac{\partial}{\partial x}(y + xz) - \frac{\partial}{\partial y}(x + yz) \right]\mathbf{k}$$

$$= (x - x)\,\mathbf{i} - (y - y)\,\mathbf{j} + (z - z)\,\mathbf{k} = \mathbf{0}$$

(b) $\operatorname{div} \mathbf{F} = \nabla \cdot \mathbf{F} = \dfrac{\partial}{\partial x}(x + yz) + \dfrac{\partial}{\partial y}(y + xz) + \dfrac{\partial}{\partial z}(z + xy) = 1 + 1 + 1 = 3$

3. (a) $\operatorname{curl} \mathbf{F} = \nabla \times \mathbf{F} = \begin{vmatrix} \mathbf{i} & \mathbf{j} & \mathbf{k} \\ \partial/\partial x & \partial/\partial y & \partial/\partial z \\ xye^z & 0 & yze^x \end{vmatrix} = (ze^x - 0)\,\mathbf{i} - (yze^x - xye^z)\,\mathbf{j} + (0 - xe^z)\,\mathbf{k}$

$\qquad = ze^x\,\mathbf{i} + (xye^z - yze^x)\,\mathbf{j} - xe^z\,\mathbf{k}$

(b) $\operatorname{div} \mathbf{F} = \nabla \cdot \mathbf{F} = \dfrac{\partial}{\partial x}(xye^z) + \dfrac{\partial}{\partial y}(0) + \dfrac{\partial}{\partial z}(yze^x) = ye^z + 0 + ye^x = y(e^z + e^x)$

5. (a) $\operatorname{curl} \mathbf{F} = \nabla \times \mathbf{F} = \begin{vmatrix} \mathbf{i} & \mathbf{j} & \mathbf{k} \\ \partial/\partial x & \partial/\partial y & \partial/\partial z \\ \dfrac{x}{\sqrt{x^2 + y^2 + z^2}} & \dfrac{y}{\sqrt{x^2 + y^2 + z^2}} & \dfrac{z}{\sqrt{x^2 + y^2 + z^2}} \end{vmatrix}$

$\qquad = \dfrac{1}{(x^2 + y^2 + z^2)^{3/2}}\,[(-yz + yz)\,\mathbf{i} - (-xz + xz)\,\mathbf{j} + (-xy + xy)\,\mathbf{k}] = \mathbf{0}$

(b) $\operatorname{div} \mathbf{F} = \nabla \cdot \mathbf{F} = \dfrac{\partial}{\partial x}\left(\dfrac{x}{\sqrt{x^2 + y^2 + z^2}}\right) + \dfrac{\partial}{\partial y}\left(\dfrac{y}{\sqrt{x^2 + y^2 + z^2}}\right) + \dfrac{\partial}{\partial z}\left(\dfrac{z}{\sqrt{x^2 + y^2 + z^2}}\right)$

$\qquad = \dfrac{x^2 + y^2 + z^2 - x^2}{(x^2 + y^2 + z^2)^{3/2}} + \dfrac{x^2 + y^2 + z^2 - y^2}{(x^2 + y^2 + z^2)^{3/2}} + \dfrac{x^2 + y^2 + z^2 - z^2}{(x^2 + y^2 + z^2)^{3/2}} = \dfrac{2x^2 + 2y^2 + 2z^2}{(x^2 + y^2 + z^2)^{3/2}} = \dfrac{2}{\sqrt{x^2 + y^2 + z^2}}$

7. (a) $\operatorname{curl} \mathbf{F} = \nabla \times \mathbf{F} = \begin{vmatrix} \mathbf{i} & \mathbf{j} & \mathbf{k} \\ \partial/\partial x & \partial/\partial y & \partial/\partial z \\ e^x \sin y & e^y \sin z & e^z \sin x \end{vmatrix} = (0 - e^y \cos z)\,\mathbf{i} - (e^z \cos x - 0)\,\mathbf{j} + (0 - e^x \cos y)\,\mathbf{k}$

$\qquad = \langle -e^y \cos z,\, -e^z \cos x,\, -e^x \cos y \rangle$

(b) $\operatorname{div} \mathbf{F} = \nabla \cdot \mathbf{F} = \dfrac{\partial}{\partial x}(e^x \sin y) + \dfrac{\partial}{\partial y}(e^y \sin z) + \dfrac{\partial}{\partial z}(e^z \sin x) = e^x \sin y + e^y \sin z + e^z \sin x$

9. If the vector field is $\mathbf{F} = P\,\mathbf{i} + Q\,\mathbf{j} + R\,\mathbf{k}$, then we know $R = 0$. In addition, the y-component of each vector of $\mathbf{F}$ is 0, so

$Q = 0$, hence $\dfrac{\partial Q}{\partial x} = \dfrac{\partial Q}{\partial y} = \dfrac{\partial Q}{\partial z} = \dfrac{\partial R}{\partial x} = \dfrac{\partial R}{\partial y} = \dfrac{\partial R}{\partial z} = 0$. P increases as y increases, so $\dfrac{\partial P}{\partial y} > 0$, but P doesn't change in

the x- or z-directions, so $\dfrac{\partial P}{\partial x} = \dfrac{\partial P}{\partial z} = 0$.

(a) $\operatorname{div} \mathbf{F} = \dfrac{\partial P}{\partial x} + \dfrac{\partial Q}{\partial y} + \dfrac{\partial R}{\partial z} = 0 + 0 + 0 = 0$

(b) $\operatorname{curl} \mathbf{F} = \left(\dfrac{\partial R}{\partial y} - \dfrac{\partial Q}{\partial z}\right)\mathbf{i} + \left(\dfrac{\partial P}{\partial z} - \dfrac{\partial R}{\partial x}\right)\mathbf{j} + \left(\dfrac{\partial Q}{\partial x} - \dfrac{\partial P}{\partial y}\right)\mathbf{k} = (0 - 0)\,\mathbf{i} + (0 - 0)\,\mathbf{j} + \left(0 - \dfrac{\partial P}{\partial y}\right)\mathbf{k} = -\dfrac{\partial P}{\partial y}\mathbf{k}$

Since $\dfrac{\partial P}{\partial y} > 0$, $-\dfrac{\partial P}{\partial y}\mathbf{k}$ is a vector pointing in the negative z-direction.

11. $\operatorname{curl} \mathbf{F} = \nabla \times \mathbf{F} = \begin{vmatrix} \mathbf{i} & \mathbf{j} & \mathbf{k} \\ \partial/\partial x & \partial/\partial y & \partial/\partial z \\ y^2 z^3 & 2xyz^3 & 3xy^2 z^2 \end{vmatrix} = (6xyz^2 - 6xyz^2)\,\mathbf{i} - (3y^2 z^2 - 3y^2 z^2)\,\mathbf{j} + (2yz^3 - 2yz^3)\,\mathbf{k} = \mathbf{0}$

and $\mathbf{F}$ is defined on all of $\mathbb{R}^3$ with component functions which have continuous partial derivatives, so by Theorem 4,

$\mathbf{F}$ is conservative. Thus, there exists a function f such that $\mathbf{F} = \nabla f$. Then $f_x(x, y, z) = y^2 z^3$ implies

$f(x, y, z) = xy^2z^3 + g(y, z)$ and $f_y(x, y, z) = 2xyz^3 + g_y(y, z)$. But $f_y(x, y, z) = 2xyz^3$, so $g(y, z) = h(z)$ and

$f(x, y, z) = xy^2z^3 + h(z)$. Thus $f_z(x, y, z) = 3xy^2z^2 + h'(z)$ but $f_z(x, y, z) = 3xy^2z^2$ so $h(z) = K$, a constant.

Hence a potential function for **F** is $f(x, y, z) = xy^2z^3 + K$.

13. $\text{curl } \mathbf{F} = \nabla \times \mathbf{F} = \begin{vmatrix} \mathbf{i} & \mathbf{j} & \mathbf{k} \\ \partial/\partial x & \partial/\partial y & \partial/\partial z \\ 3xy^2z^2 & 2x^2yz^3 & 3x^2y^2z^2 \end{vmatrix}$

$= (6x^2yz^2 - 6x^2yz^2)\,\mathbf{i} - (6xy^2z^2 - 6xy^2z)\,\mathbf{j} + (4xyz^3 - 6xyz^2)\,\mathbf{k}$

$= 6xy^2z(1 - z)\,\mathbf{j} + 2xyz^2(2z - 3)\,\mathbf{k} \neq \mathbf{0}$

so **F** is not conservative.

15. $\text{curl } \mathbf{F} = \nabla \times \mathbf{F} = \begin{vmatrix} \mathbf{i} & \mathbf{j} & \mathbf{k} \\ \partial/\partial x & \partial/\partial y & \partial/\partial z \\ e^{yz} & xze^{yz} & xye^{yz} \end{vmatrix}$

$= [xyze^{yz} + xe^{yz} - (xyze^{yz} + xe^{yz})]\,\mathbf{i} - (ye^{yz} - ye^{yz})\,\mathbf{j} + (ze^{yz} - ze^{yz})\,\mathbf{k} = \mathbf{0}$

F is defined on all of $\mathbb{R}^3$, and the partial derivatives of the component functions are continuous, so **F** is conservative. Thus there exists a function f such that $\nabla f = \mathbf{F}$. Then $f_x(x, y, z) = e^{yz}$ implies $f(x, y, z) = xe^{yz} + g(y, z)$ $\Rightarrow$

$f_y(x, y, z) = xze^{yz} + g_y(y, z)$. But $f_y(x, y, z) = xze^{yz}$, so $g(y, z) = h(z)$ and $f(x, y, z) = xe^{yz} + h(z)$.

Thus $f_z(x, y, z) = xye^{yz} + h'(z)$ but $f_z(x, y, z) = xye^{yz}$ so $h(z) = K$ and a potential function for **F** is

$f(x, y, z) = xe^{yz} + K$.

17. **No.** Assume there is such a **G**. Then $\text{div}(\text{curl } \mathbf{G}) = \dfrac{\partial}{\partial x}(x\sin y) + \dfrac{\partial}{\partial y}(\cos y) + \dfrac{\partial}{\partial z}(z - xy) = \sin y - \sin y + 1 \neq 0$,

which contradicts Theorem 11.

19. $\text{curl } \mathbf{F} = \begin{vmatrix} \mathbf{i} & \mathbf{j} & \mathbf{k} \\ \partial/\partial x & \partial/\partial y & \partial/\partial z \\ f(x) & g(y) & h(z) \end{vmatrix} = (0 - 0)\,\mathbf{i} + (0 - 0)\,\mathbf{j} + (0 - 0)\,\mathbf{k} = \mathbf{0}$. Hence $\mathbf{F} = f(x)\,\mathbf{i} + g(y)\,\mathbf{j} + h(z)\,\mathbf{k}$

is irrotational.

For Exercises 23–29, let $\mathbf{F}(x, y, z) = P_1\,\mathbf{i} + Q_1\,\mathbf{j} + R_1\,\mathbf{k}$ and $\mathbf{G}(x, y, z) = P_2\,\mathbf{i} + Q_2\,\mathbf{j} + R_2\,\mathbf{k}$.

21. $\text{div}(\mathbf{F} + \mathbf{G}) = \text{div}\langle P_1 + P_2, Q_1 + Q_2, R_1 + R_2\rangle = \dfrac{\partial(P_1 + P_2)}{\partial x} + \dfrac{\partial(Q_1 + Q_2)}{\partial y} + \dfrac{\partial(R_1 + R_2)}{\partial z}$

$= \dfrac{\partial P_1}{\partial x} + \dfrac{\partial P_2}{\partial x} + \dfrac{\partial Q_1}{\partial y} + \dfrac{\partial Q_2}{\partial y} + \dfrac{\partial R_1}{\partial z} + \dfrac{\partial R_2}{\partial z} = \left(\dfrac{\partial P_1}{\partial x} + \dfrac{\partial Q_1}{\partial y} + \dfrac{\partial R_1}{\partial z}\right) + \left(\dfrac{\partial P_2}{\partial x} + \dfrac{\partial Q_2}{\partial y} + \dfrac{\partial R_2}{\partial z}\right)$

$= \text{div}\langle P_1, Q_1, R_1\rangle + \text{div}\langle P_2, Q_2, R_2\rangle = \text{div } \mathbf{F} + \text{div } \mathbf{G}$

23. $\text{div}(f\mathbf{F}) = \text{div}(f\langle P_1, Q_1, R_1\rangle) = \text{div}\langle fP_1, fQ_1, fR_1\rangle = \dfrac{\partial(fP_1)}{\partial x} + \dfrac{\partial(fQ_1)}{\partial y} + \dfrac{\partial(fR_1)}{\partial z}$

$= \left(f\dfrac{\partial P_1}{\partial x} + P_1\dfrac{\partial f}{\partial x}\right) + \left(f\dfrac{\partial Q_1}{\partial y} + Q_1\dfrac{\partial f}{\partial y}\right) + \left(f\dfrac{\partial R_1}{\partial z} + R_1\dfrac{\partial f}{\partial z}\right)$

$= f\left(\dfrac{\partial P_1}{\partial x} + \dfrac{\partial Q_1}{\partial y} + \dfrac{\partial R_1}{\partial z}\right) + \langle P_1, Q_1, R_1\rangle \cdot \left\langle \dfrac{\partial f}{\partial x}, \dfrac{\partial f}{\partial y}, \dfrac{\partial f}{\partial z}\right\rangle = f\,\text{div } \mathbf{F} + \mathbf{F} \cdot \nabla f$

25. $\text{div}(\mathbf{F} \times \mathbf{G}) = \nabla \cdot (\mathbf{F} \times \mathbf{G}) = \begin{vmatrix} \partial/\partial x & \partial/\partial y & \partial/\partial z \\ P_1 & Q_1 & R_1 \\ P_2 & Q_2 & R_2 \end{vmatrix} = \dfrac{\partial}{\partial x} \begin{vmatrix} Q_1 & R_1 \\ Q_2 & R_2 \end{vmatrix} - \dfrac{\partial}{\partial y} \begin{vmatrix} P_1 & R_1 \\ P_2 & R_2 \end{vmatrix} + \dfrac{\partial}{\partial z} \begin{vmatrix} P_1 & Q_1 \\ P_2 & Q_2 \end{vmatrix}$

$= \left[Q_1 \dfrac{\partial R_2}{\partial x} + R_2 \dfrac{\partial Q_1}{\partial x} - Q_2 \dfrac{\partial R_1}{\partial x} - R_1 \dfrac{\partial Q_2}{\partial x} \right] - \left[P_1 \dfrac{\partial R_2}{\partial y} + R_2 \dfrac{\partial P_1}{\partial y} - P_2 \dfrac{\partial R_1}{\partial y} - R_1 \dfrac{\partial P_2}{\partial y} \right]$

$+ \left[P_1 \dfrac{\partial Q_2}{\partial z} + Q_2 \dfrac{\partial P_1}{\partial z} - P_2 \dfrac{\partial Q_1}{\partial z} - Q_1 \dfrac{\partial P_2}{\partial z} \right]$

$= \left[P_2 \left(\dfrac{\partial R_1}{\partial y} - \dfrac{\partial Q_1}{\partial z} \right) + Q_2 \left(\dfrac{\partial P_1}{\partial z} - \dfrac{\partial R_1}{\partial x} \right) + R_2 \left(\dfrac{\partial Q_1}{\partial x} - \dfrac{\partial P_1}{\partial y} \right) \right]$

$- \left[P_1 \left(\dfrac{\partial R_2}{\partial y} - \dfrac{\partial Q_2}{\partial z} \right) + Q_1 \left(\dfrac{\partial P_2}{\partial z} - \dfrac{\partial R_2}{\partial x} \right) + R_1 \left(\dfrac{\partial Q_2}{\partial x} - \dfrac{\partial P_2}{\partial y} \right) \right]$

$= \mathbf{G} \cdot \text{curl } \mathbf{F} - \mathbf{F} \cdot \text{curl } \mathbf{G}$

27. $\text{curl}(\text{curl } \mathbf{F}) = \nabla \times (\nabla \times \mathbf{F}) = \begin{vmatrix} \mathbf{i} & \mathbf{j} & \mathbf{k} \\ \partial/\partial x & \partial/\partial y & \partial/\partial z \\ \partial R_1/\partial y - \partial Q_1/\partial z & \partial P_1/\partial z - \partial R_1/\partial x & \partial Q_1/\partial x - \partial P_1/\partial y \end{vmatrix}$

$= \left(\dfrac{\partial^2 Q_1}{\partial y \partial x} - \dfrac{\partial^2 P_1}{\partial y^2} - \dfrac{\partial^2 P_1}{\partial z^2} + \dfrac{\partial^2 R_1}{\partial z \partial x} \right) \mathbf{i} + \left(\dfrac{\partial^2 R_1}{\partial z \partial y} - \dfrac{\partial^2 Q_1}{\partial z^2} - \dfrac{\partial^2 Q_1}{\partial x^2} + \dfrac{\partial^2 P_1}{\partial x \partial y} \right) \mathbf{j}$

$+ \left(\dfrac{\partial^2 P_1}{\partial x \partial z} - \dfrac{\partial^2 R_1}{\partial x^2} - \dfrac{\partial^2 R_1}{\partial y^2} + \dfrac{\partial^2 Q_1}{\partial y \partial z} \right) \mathbf{k}$

Now let's consider $\text{grad}(\text{div } \mathbf{F}) - \nabla^2 \mathbf{F}$ and compare with the preceding. (Note that $\nabla^2 \mathbf{F}$ is defined on page 799.)

$\text{grad}(\text{div } \mathbf{F}) - \nabla^2 \mathbf{F} = \left[\left(\dfrac{\partial^2 P_1}{\partial x^2} + \dfrac{\partial^2 Q_1}{\partial x \partial y} + \dfrac{\partial^2 R_1}{\partial x \partial z} \right) \mathbf{i} + \left(\dfrac{\partial^2 P_1}{\partial y \partial x} + \dfrac{\partial^2 Q_1}{\partial y^2} + \dfrac{\partial^2 R_1}{\partial y \partial z} \right) \mathbf{j} + \left(\dfrac{\partial^2 P_1}{\partial z \partial x} + \dfrac{\partial^2 Q_1}{\partial z \partial y} + \dfrac{\partial^2 R_1}{\partial z^2} \right) \mathbf{k} \right]$

$- \left[\left(\dfrac{\partial^2 P_1}{\partial x^2} + \dfrac{\partial^2 P_1}{\partial y^2} + \dfrac{\partial^2 P_1}{\partial z^2} \right) \mathbf{i} + \left(\dfrac{\partial^2 Q_1}{\partial x^2} + \dfrac{\partial^2 Q_1}{\partial y^2} + \dfrac{\partial^2 Q_1}{\partial z^2} \right) \mathbf{j} \right.$

$\left. + \left(\dfrac{\partial^2 R_1}{\partial x^2} + \dfrac{\partial^2 R_1}{\partial y^2} + \dfrac{\partial^2 R_1}{\partial z^2} \right) \mathbf{k} \right]$

$= \left(\dfrac{\partial^2 Q_1}{\partial x \partial y} + \dfrac{\partial^2 R_1}{\partial x \partial z} - \dfrac{\partial^2 P_1}{\partial y^2} - \dfrac{\partial^2 P_1}{\partial z^2} \right) \mathbf{i} + \left(\dfrac{\partial^2 P_1}{\partial y \partial x} + \dfrac{\partial^2 R_1}{\partial y \partial z} - \dfrac{\partial^2 Q_1}{\partial x^2} - \dfrac{\partial^2 Q_1}{\partial z^2} \right) \mathbf{j}$

$+ \left(\dfrac{\partial^2 P_1}{\partial z \partial x} + \dfrac{\partial^2 Q_1}{\partial z \partial y} - \dfrac{\partial^2 R_1}{\partial x^2} - \dfrac{\partial^2 R_2}{\partial y^2} \right) \mathbf{k}$

Then applying Clairaut's Theorem to reverse the order of differentiation in the second partial derivatives as needed and comparing, we have $\text{curl curl } \mathbf{F} = \text{grad div } \mathbf{F} - \nabla^2 \mathbf{F}$ as desired.

29. (a) $\nabla r = \nabla \sqrt{x^2 + y^2 + z^2} = \dfrac{x}{\sqrt{x^2 + y^2 + z^2}}\mathbf{i} + \dfrac{y}{\sqrt{x^2 + y^2 + z^2}}\mathbf{j} + \dfrac{z}{\sqrt{x^2 + y^2 + z^2}}\mathbf{k} = \dfrac{x\mathbf{i} + y\mathbf{j} + z\mathbf{k}}{\sqrt{x^2 + y^2 + z^2}} = \dfrac{\mathbf{r}}{r}$

(b) $\nabla \times \mathbf{r} = \begin{vmatrix} \mathbf{i} & \mathbf{j} & \mathbf{k} \\ \dfrac{\partial}{\partial x} & \dfrac{\partial}{\partial y} & \dfrac{\partial}{\partial z} \\ x & y & z \end{vmatrix} = \left[\dfrac{\partial}{\partial y}(z) - \dfrac{\partial}{\partial z}(y)\right]\mathbf{i} + \left[\dfrac{\partial}{\partial z}(x) - \dfrac{\partial}{\partial x}(z)\right]\mathbf{j} + \left[\dfrac{\partial}{\partial x}(y) - \dfrac{\partial}{\partial y}(x)\right]\mathbf{k} = \mathbf{0}$

(c) $\nabla\left(\dfrac{1}{r}\right) = \nabla\left(\dfrac{1}{\sqrt{x^2 + y^2 + z^2}}\right)$

$= \dfrac{-\dfrac{1}{2\sqrt{x^2 + y^2 + z^2}}(2x)}{x^2 + y^2 + z^2}\mathbf{i} - \dfrac{\dfrac{1}{2\sqrt{x^2 + y^2 + z^2}}(2y)}{x^2 + y^2 + z^2}\mathbf{j} - \dfrac{\dfrac{1}{2\sqrt{x^2 + y^2 + z^2}}(2z)}{x^2 + y^2 + z^2}\mathbf{k}$

$= -\dfrac{x\mathbf{i} + y\mathbf{j} + z\mathbf{k}}{(x^2 + y^2 + z^2)^{3/2}} = -\dfrac{\mathbf{r}}{r^3}$

(d) $\nabla \ln r = \nabla \ln(x^2 + y^2 + z^2)^{1/2} = \tfrac{1}{2}\nabla \ln(x^2 + y^2 + z^2)$

$= \dfrac{x}{x^2 + y^2 + z^2}\mathbf{i} + \dfrac{y}{x^2 + y^2 + z^2}\mathbf{j} + \dfrac{z}{x^2 + y^2 + z^2}\mathbf{k} = \dfrac{x\mathbf{i} + y\mathbf{j} + z\mathbf{k}}{x^2 + y^2 + z^2} = \dfrac{\mathbf{r}}{r^2}$

31. By (13), $\oint_C f(\nabla g) \cdot \mathbf{n}\, ds = \iint_D \operatorname{div}(f\nabla g)\, dA = \iint_D [f \operatorname{div}(\nabla g) + \nabla g \cdot \nabla f]\, dA$ by Exercise 23. But $\operatorname{div}(\nabla g) = \nabla^2 g$.

Hence $\iint_D f\nabla^2 g\, dA = \oint_C f(\nabla g) \cdot \mathbf{n}\, ds - \iint_D \nabla g \cdot \nabla f\, dA$.

33. Let $f(x, y) = 1$. Then $\nabla f = \mathbf{0}$ and Green's first identity says $\iint_D \nabla^2 g\, dA = \oint_C (\nabla g) \cdot \mathbf{n}\, ds - \iint_D \mathbf{0} \cdot \nabla g\, dA \;\Rightarrow$

$\iint_D \nabla^2 g\, dA = \oint_C \nabla g \cdot \mathbf{n}\, ds$. But g is harmonic on D, so $\nabla^2 g = 0 \;\Rightarrow\; \oint_C \nabla g \cdot \mathbf{n}\, ds = 0$ and

$\oint_C D_{\mathbf{n}}g\, ds = \oint_C (\nabla g \cdot \mathbf{n})\, ds = 0$.

35. (a) We know that $\omega = v/d$, and from the diagram $\sin\theta = d/r \;\Rightarrow\; v = d\omega = (\sin\theta)r\omega = |\mathbf{w} \times \mathbf{r}|$. But $\mathbf{v}$ is perpendicular

to both $\mathbf{w}$ and $\mathbf{r}$, so that $\mathbf{v} = \mathbf{w} \times \mathbf{r}$.

(b) From (a), $\mathbf{v} = \mathbf{w} \times \mathbf{r} = \begin{vmatrix} \mathbf{i} & \mathbf{j} & \mathbf{k} \\ 0 & 0 & \omega \\ x & y & z \end{vmatrix} = (0 \cdot z - \omega y)\mathbf{i} + (\omega x - 0 \cdot z)\mathbf{j} + (0 \cdot y - x \cdot 0)\mathbf{k} = -\omega y\mathbf{i} + \omega x\mathbf{j}$

(c) $\operatorname{curl} \mathbf{v} = \nabla \times \mathbf{v} = \begin{vmatrix} \mathbf{i} & \mathbf{j} & \mathbf{k} \\ \partial/\partial x & \partial/\partial y & \partial/\partial z \\ -\omega y & \omega x & 0 \end{vmatrix}$

$= \left[\dfrac{\partial}{\partial y}(0) - \dfrac{\partial}{\partial z}(\omega x)\right]\mathbf{i} + \left[\dfrac{\partial}{\partial z}(-\omega y) - \dfrac{\partial}{\partial x}(0)\right]\mathbf{j} + \left[\dfrac{\partial}{\partial x}(\omega x) - \dfrac{\partial}{\partial y}(-\omega y)\right]\mathbf{k}$

$= [\omega - (-\omega)]\mathbf{k} = 2\omega\mathbf{k} = 2\mathbf{w}$

13.6 Parametric Surfaces and Their Areas

1. $\mathbf{r}(u,v) = (u+v)\,\mathbf{i} + (3-v)\,\mathbf{j} + (1+4u+5v)\,\mathbf{k} = \langle 0, 3, 1 \rangle + u\,\langle 1, 0, 4 \rangle + v\,\langle 1, -1, 5 \rangle$. From Example 3, we recognize

this as a vector equation of a plane through the point $(0, 3, 1)$ and containing vectors $\mathbf{a} = \langle 1, 0, 4 \rangle$ and $\mathbf{b} = \langle 1, -1, 5 \rangle$. If we

wish to find a more conventional equation for the plane, a normal vector to the plane is $\mathbf{a} \times \mathbf{b} = \begin{vmatrix} \mathbf{i} & \mathbf{j} & \mathbf{k} \\ 1 & 0 & 4 \\ 1 & -1 & 5 \end{vmatrix} = 4\mathbf{i} - \mathbf{j} - \mathbf{k}$

and an equation of the plane is $4(x-0) - (y-3) - (z-1) = 0$ or $4x - y - z = -4$.

3. $\mathbf{r}(s,t) = \langle s, t, t^2 - s^2 \rangle$, so the corresponding parametric equations for the surface are $x = s$, $y = t$, $z = t^2 - s^2$. For any

point (x, y, z) on the surface, we have $z = y^2 - x^2$. With no restrictions on the parameters, the surface is $z = y^2 - x^2$, which

we recognize as a hyperbolic paraboloid.

5. $\mathbf{r}(u,v) = \langle u^2, v^2, u+v \rangle$, $\ -1 \le u \le 1$, $\ -1 \le v \le 1$.

The surface has parametric equations $x = u^2$, $y = v^2$, $z = u+v$, $-1 \le u \le 1$, $-1 \le v \le 1$.

In Maple, the surface can be graphed by entering

`plot3d([u^2,v^2,u+v],u=-1..1,v=-1..1);.`

In Mathematica we use the `ParametricPlot3D` command.

If we keep u constant at u_0, $x = u_0^2$, a constant, so the

corresponding grid curves must be the curves parallel to the

yz-plane. If v is constant, we have $y = v_0^2$, a constant, so these

grid curves are the curves parallel to the xz-plane.

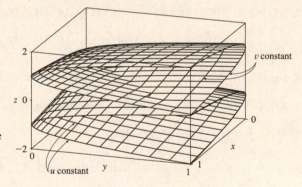

7. $\mathbf{r}(u,v) = \langle u\cos v, u\sin v, u^5 \rangle$.

The surface has parametric equations $x = u\cos v$, $y = u\sin v$,

$z = u^5$, $-1 \le u \le 1$, $0 \le v \le 2\pi$. Note that if $u = u_0$ is constant

then $z = u_0^5$ is constant and $x = u_0\cos v$, $y = u_0\sin v$ describe a

circle in x, y of radius $|u_0|$, so the corresponding grid curves are

circles parallel to the xy-plane. If $v = v_0$, a constant, the parametric

equations become $x = u\cos v_0$, $y = u\sin v_0$, $z = u^5$. Then

$y = (\tan v_0)x$, so these are the grid curves we see that lie in vertical

planes $y = kx$ through the z-axis.

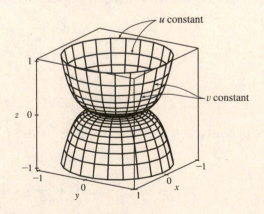

9. $x = \sin v$, $y = \cos u \sin 4v$, $z = \sin 2u \sin 4v$, $0 \le u \le 2\pi$, $-\frac{\pi}{2} \le v \le \frac{\pi}{2}$.

Note that if $v = v_0$ is constant, then $x = \sin v_0$ is constant, so the

corresponding grid curves must be parallel to the yz-plane. These

are the vertically oriented grid curves we see, each shaped like a

"figure-eight." When $u = u_0$ is held constant, the parametric

equations become $x = \sin v$, $y = \cos u_0 \sin 4v$,

$z = \sin 2u_0 \sin 4v$. Since z is a constant multiple of y, the

corresponding grid curves are the curves contained in planes

$z = ky$ that pass through the x-axis.

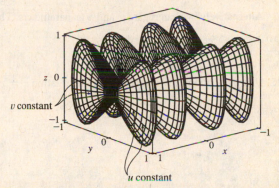

11. $\mathbf{r}(u, v) = u \cos v \, \mathbf{i} + u \sin v \, \mathbf{j} + v \, \mathbf{k}$. The parametric equations for the surface are $x = u \cos v$, $y = u \sin v$, $z = v$. We look at

the grid curves first; if we fix v, then x and y parametrize a straight line in the plane $z = v$ which intersects the z-axis. If u is

held constant, the projection onto the xy-plane is circular; with $z = v$, each grid curve is a helix. The surface is a spiraling

ramp, graph IV.

13. $\mathbf{r}(u, v) = \sin v \, \mathbf{i} + \cos u \sin 2v \, \mathbf{j} + \sin u \sin 2v \, \mathbf{k}$. Parametric equations for the surface are $x = \sin v$, $y = \cos u \sin 2v$,

$z = \sin u \sin 2v$. If $v = v_0$ is fixed, then $x = \sin v_0$ is constant, and $y = (\sin 2v_0) \cos u$ and $z = (\sin 2v_0) \sin u$ describe a

circle of radius $|\sin 2v_0|$, so each corresponding grid curve is a circle contained in the vertical plane $x = \sin v_0$ parallel to the

yz-plane. The only possible surface is graph II. The grid curves we see running lengthwise along the surface correspond to

holding u constant, in which case $y = (\cos u_0) \sin 2v$, $z = (\sin u_0) \sin 2v$ $\Rightarrow$ $z = (\tan u_0)y$, so each grid curve lies in a

plane $z = ky$ that includes the x-axis.

15. From Example 3, parametric equations for the plane through the point $(0, 0, 0)$ that contains the vectors $\mathbf{a} = \langle 1, -1, 0 \rangle$ and

$\mathbf{b} = \langle 0, 1, -1 \rangle$ are $x = 0 + u(1) + v(0) = u$, $y = 0 + u(-1) + v(1) = v - u$, $z = 0 + u(0) + v(-1) = -v$.

17. Solving the equation for x gives $x^2 = 1 + y^2 + \frac{1}{4}z^2$ $\Rightarrow$ $x = \sqrt{1 + y^2 + \frac{1}{4}z^2}$. (We choose the positive root since we want

the part of the hyperboloid that corresponds to $x \ge 0$.) If we let y and z be the parameters, parametric equations are $y = y$,

$z = z$, $x = \sqrt{1 + y^2 + \frac{1}{4}z^2}$.

19. Since the cone intersects the sphere in the circle $x^2 + y^2 = 2$, $z = \sqrt{2}$ and we want the portion of the sphere above this, we

can parametrize the surface as $x = x$, $y = y$, $z = \sqrt{4 - x^2 - y^2}$ where $x^2 + y^2 \le 2$.

Alternate solution: Using spherical coordinates, $x = 2 \sin \phi \cos \theta$, $y = 2 \sin \phi \sin \theta$, $z = 2 \cos \phi$ where $0 \le \phi \le \frac{\pi}{4}$ and

$0 \le \theta \le 2\pi$.

21. Parametric equations are $x = x$, $y = 4 \cos \theta$, $z = 4 \sin \theta$, $0 \le x \le 5$, $0 \le \theta \le 2\pi$.

23. The surface appears to be a portion of a circular cylinder of radius 3 with axis the x-axis. An equation of the cylinder is

$y^2 + z^2 = 9$, and we can impose the restrictions $0 \le x \le 5$, $y \le 0$ to obtain the portion shown. To graph the surface on a

CAS, we can use parametric equations $x = u$, $y = 3\cos v$, $z = 3\sin v$ with the parameter domain $0 \le u \le 5$, $\frac{\pi}{2} \le v \le \frac{3\pi}{2}$.
Alternatively, we can regard x and z as parameters. Then parametric equations are $x = x$, $z = z$, $y = -\sqrt{9 - z^2}$, where $0 \le x \le 5$ and $-3 \le z \le 3$.

25. Using Equations 3, we have the parametrization $x = x$, $\ y = e^{-x}\cos\theta$,
$z = e^{-x}\sin\theta$, $\ 0 \le x \le 3$, $\ 0 \le \theta \le 2\pi$.

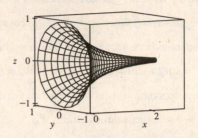

27. (a) Replacing $\cos u$ by $\sin u$ and $\sin u$ by $\cos u$ gives parametric equations

$x = (2 + \sin v)\sin u$, $y = (2 + \sin v)\cos u$, $z = u + \cos v$. From the graph, it
appears that the direction of the spiral is reversed. We can verify this observation by
noting that the projection of the spiral grid curves onto the xy-plane, given by

$x = (2 + \sin v)\sin u$, $y = (2 + \sin v)\cos u$, $z = 0$, draws a circle in the clockwise
direction for each value of v. The original equations, on the other hand, give circular
projections drawn in the counterclockwise direction. The equation for z is identical in
both surfaces, so as z increases, these grid curves spiral up in opposite directions for
the two surfaces.

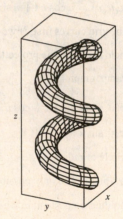

(b) Replacing $\cos u$ by $\cos 2u$ and $\sin u$ by $\sin 2u$ gives parametric equations

$x = (2 + \sin v)\cos 2u$, $y = (2 + \sin v)\sin 2u$, $z = u + \cos v$. From the graph, it
appears that the number of coils in the surface doubles within the same parametric
domain. We can verify this observation by noting that the projection of the spiral grid
curves onto the xy-plane, given by $x = (2 + \sin v)\cos 2u$, $y = (2 + \sin v)\sin 2u$,
$z = 0$ (where v is constant), complete circular revolutions for $0 \le u \le \pi$ while the
original surface requires $0 \le u \le 2\pi$ for a complete revolution. Thus, the new
surface winds around twice as fast as the original surface, and since the equation for z
is identical in both surfaces, we observe twice as many circular coils in the same
z-interval.

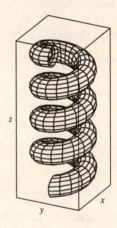

29. $\mathbf{r}(u, v) = (u + v)\,\mathbf{i} + 3u^2\,\mathbf{j} + (u - v)\,\mathbf{k}$.

$\mathbf{r}_u = \mathbf{i} + 6u\,\mathbf{j} + \mathbf{k}$ and $\mathbf{r}_v = \mathbf{i} - \mathbf{k}$, so $\mathbf{r}_u \times \mathbf{r}_v = -6u\,\mathbf{i} + 2\,\mathbf{j} - 6u\,\mathbf{k}$. Since the point $(2, 3, 0)$ corresponds to $u = 1$, $v = 1$, a
normal vector to the surface at $(2, 3, 0)$ is $-6\,\mathbf{i} + 2\,\mathbf{j} - 6\,\mathbf{k}$, and an equation of the tangent plane is $-6x + 2y - 6z = -6$ or
$3x - y + 3z = 3$.

31. $\mathbf{r}(u,v) = u\cos v\,\mathbf{i} + u\sin v\,\mathbf{j} + v\,\mathbf{k} \Rightarrow \mathbf{r}\left(1,\frac{\pi}{3}\right) = \left(\frac{1}{2},\frac{\sqrt{3}}{2},\frac{\pi}{3}\right).$

$\mathbf{r}_u = \cos v\,\mathbf{i} + \sin v\,\mathbf{j}$ and $\mathbf{r}_v = -u\sin v\,\mathbf{i} + u\cos v\,\mathbf{j} + \mathbf{k}$, so a normal vector to the surface at the point $\left(\frac{1}{2},\frac{\sqrt{3}}{2},\frac{\pi}{3}\right)$ is

$\mathbf{r}_u\left(1,\frac{\pi}{3}\right) \times \mathbf{r}_v\left(1,\frac{\pi}{3}\right) = \left(\frac{1}{2}\mathbf{i} + \frac{\sqrt{3}}{2}\mathbf{j}\right) \times \left(-\frac{\sqrt{3}}{2}\mathbf{i} + \frac{1}{2}\mathbf{j} + \mathbf{k}\right) = \frac{\sqrt{3}}{2}\mathbf{i} - \frac{1}{2}\mathbf{j} + \mathbf{k}$. Thus an equation of the tangent plane at

$\left(\frac{1}{2},\frac{\sqrt{3}}{2},\frac{\pi}{3}\right)$ is $\frac{\sqrt{3}}{2}\left(x - \frac{1}{2}\right) - \frac{1}{2}\left(y - \frac{\sqrt{3}}{2}\right) + 1\left(z - \frac{\pi}{3}\right) = 0$ or $\frac{\sqrt{3}}{2}x - \frac{1}{2}y + z = \frac{\pi}{3}$.

33. The surface S is given by $z = f(x,y) = 6 - 3x - 2y$ which intersects the xy-plane in the line $3x + 2y = 6$, so D is the

triangular region given by $\{(x,y) \mid 0 \le x \le 2, 0 \le y \le 3 - \frac{3}{2}x\}$. By Formula 9, the surface area of S is

$$A(S) = \iint_D \sqrt{1 + \left(\frac{\partial z}{\partial x}\right)^2 + \left(\frac{\partial z}{\partial y}\right)^2}\, dA$$

$$= \iint_D \sqrt{1 + (-3)^2 + (-2)^2}\, dA = \sqrt{14} \iint_D dA = \sqrt{14}\,A(D) = \sqrt{14}\left(\frac{1}{2}\cdot 2\cdot 3\right) = 3\sqrt{14}.$$

35. Here we can write $z = f(x,y) = \frac{1}{3} - \frac{1}{3}x - \frac{2}{3}y$ and D is the disk $x^2 + y^2 \le 3$, so by Formula 9 the area of the surface is

$$A(S) = \iint_D \sqrt{1 + \left(\frac{\partial z}{\partial x}\right)^2 + \left(\frac{\partial z}{\partial y}\right)^2}\, dA = \iint_D \sqrt{1 + \left(-\frac{1}{3}\right)^2 + \left(-\frac{2}{3}\right)^2}\, dA = \frac{\sqrt{14}}{3} \iint_D dA$$

$$= \frac{\sqrt{14}}{3}\,A(D) = \frac{\sqrt{14}}{3}\cdot \pi\left(\sqrt{3}\right)^2 = \sqrt{14}\,\pi$$

37. $z = f(x,y) = \frac{2}{3}(x^{3/2} + y^{3/2})$ and $D = \{(x,y) \mid 0 \le x \le 1, 0 \le y \le 1\}$. Then $f_x = x^{1/2}$, $f_y = y^{1/2}$ and

$$A(S) = \iint_D \sqrt{1 + \left(\sqrt{x}\right)^2 + \left(\sqrt{y}\right)^2}\, dA = \int_0^1\int_0^1 \sqrt{1 + x + y}\, dy\, dx$$

$$= \int_0^1 \left[\frac{2}{3}(x + y + 1)^{3/2}\right]_{y=0}^{y=1}\, dx = \frac{2}{3}\int_0^1 \left[(x+2)^{3/2} - (x+1)^{3/2}\right]\, dx$$

$$= \frac{2}{3}\left[\frac{2}{5}(x+2)^{5/2} - \frac{2}{5}(x+1)^{5/2}\right]_0^1 = \frac{4}{15}(3^{5/2} - 2^{5/2} - 2^{5/2} + 1) = \frac{4}{15}(3^{5/2} - 2^{7/2} + 1)$$

39. $z = f(x,y) = xy$ with $x^2 + y^2 \le 1$, so $f_x = y$, $f_y = x$ $\Rightarrow$

$$A(S) = \iint_D \sqrt{1 + y^2 + x^2}\, dA = \int_0^{2\pi}\int_0^1 \sqrt{r^2 + 1}\,r\, dr\, d\theta = \int_0^{2\pi}\left[\frac{1}{3}(r^2 + 1)^{3/2}\right]_{r=0}^{r=1}\, d\theta$$

$$= \int_0^{2\pi} \frac{1}{3}\left(2\sqrt{2} - 1\right)\, d\theta = \frac{2\pi}{3}\left(2\sqrt{2} - 1\right)$$

41. A parametric representation of the surface is $x = x$, $y = 4x + z^2$, $z = z$ with $0 \le x \le 1$, $0 \le z \le 1$.

Hence $\mathbf{r}_x \times \mathbf{r}_z = (\mathbf{i} + 4\mathbf{j}) \times (2z\,\mathbf{j} + \mathbf{k}) = 4\mathbf{i} - \mathbf{j} + 2z\,\mathbf{k}$.

Note: In general, if $y = f(x,z)$ then $\mathbf{r}_x \times \mathbf{r}_z = \frac{\partial f}{\partial x}\mathbf{i} - \mathbf{j} + \frac{\partial f}{\partial z}\mathbf{k}$ and $A(S) = \iint_D \sqrt{1 + \left(\frac{\partial f}{\partial x}\right)^2 + \left(\frac{\partial f}{\partial z}\right)^2}\, dA$. Then

$$A(S) = \int_0^1\int_0^1 \sqrt{17 + 4z^2}\, dx\, dz = \int_0^1 \sqrt{17 + 4z^2}\, dz$$

$$= \frac{1}{2}\left(z\sqrt{17 + 4z^2} + \frac{17}{2}\ln\left|2z + \sqrt{4z^2 + 17}\right|\right)\Big]_0^1 = \frac{\sqrt{21}}{2} + \frac{17}{4}\left[\ln\left(2 + \sqrt{21}\right) - \ln\sqrt{17}\right]$$

43. $\mathbf{r}_u = \langle v, 1, 1\rangle$, $\mathbf{r}_v = \langle u, 1, -1\rangle$ and $\mathbf{r}_u \times \mathbf{r}_v = \langle -2, u + v, v - u\rangle$. Then

$$A(S) = \iint_{u^2 + v^2 \le 1} \sqrt{4 + 2u^2 + 2v^2}\, dA = \int_0^{2\pi}\int_0^1 r\sqrt{4 + 2r^2}\, dr\, d\theta = \int_0^{2\pi} d\theta \int_0^1 r\sqrt{4 + 2r^2}\, dr$$

$$= 2\pi\left[\frac{1}{6}(4 + 2r^2)^{3/2}\right]_0^1 = \frac{\pi}{3}\left(6\sqrt{6} - 8\right) = \pi\left(2\sqrt{6} - \frac{8}{3}\right)$$

45. From Equation 9 we have $A(S) = \iint_D \sqrt{1 + (f_x)^2 + (f_y)^2} \, dA$. But if $|f_x| \le 1$ and $|f_y| \le 1$ then $0 \le (f_x)^2 \le 1$,

$0 \le (f_y)^2 \le 1 \;\;\Rightarrow\;\; 1 \le 1 + (f_x)^2 + (f_y)^2 \le 3 \;\;\Rightarrow\;\; 1 \le \sqrt{1 + (f_x)^2 + (f_y)^2} \le \sqrt{3}$. By Property 12.2.11,

$\iint_D 1 \, dA \le \iint_D \sqrt{1 + (f_x)^2 + (f_y)^2} \, dA \le \iint_D \sqrt{3} \, dA \;\;\Rightarrow\;\; A(D) \le A(S) \le \sqrt{3} \, A(D) \;\;\Rightarrow\;\;$
$\pi R^2 \le A(S) \le \sqrt{3} \pi R^2$.

47. $z = f(x, y) = e^{-x^2 - y^2}$ with $x^2 + y^2 \le 4$.

$A(S) = \iint_D \sqrt{1 + \left(-2xe^{-x^2 - y^2}\right)^2 + \left(-2ye^{-x^2 - y^2}\right)^2} \, dA = \iint_D \sqrt{1 + 4(x^2 + y^2)e^{-2(x^2 + y^2)}} \, dA$

$= \int_0^{2\pi} \int_0^2 \sqrt{1 + 4r^2 e^{-2r^2}} \, r \, dr \, d\theta = \int_0^{2\pi} d\theta \int_0^2 r \sqrt{1 + 4r^2 e^{-2r^2}} \, dr = 2\pi \int_0^2 r \sqrt{1 + 4r^2 e^{-2r^2}} \, dr \approx 13.9783$

49. (a) $A(S) = \iint_D \sqrt{1 + \left(\dfrac{\partial z}{\partial x}\right)^2 + \left(\dfrac{\partial z}{\partial y}\right)^2} \, dA = \int_0^6 \int_0^4 \sqrt{1 + \dfrac{4x^2 + 4y^2}{(1 + x^2 + y^2)^4}} \, dy \, dx$.

Using the Midpoint Rule with $f(x, y) = \sqrt{1 + \dfrac{4x^2 + 4y^2}{(1 + x^2 + y^2)^4}}$, $m = 3, n = 2$ we have

$A(S) \approx \sum_{i=1}^{3} \sum_{j=1}^{2} f(\overline{x}_i, \overline{y}_j) \, \Delta A = 4 \left[f(1, 1) + f(1, 3) + f(3, 1) + f(3, 3) + f(5, 1) + f(5, 3) \right] \approx 24.2055$

(b) Using a CAS we have $A(S) = \int_0^6 \int_0^4 \sqrt{1 + \dfrac{4x^2 + 4y^2}{(1 + x^2 + y^2)^4}} \, dy \, dx \approx 24.2476$. This agrees with the estimate in part (a)

to the first decimal place.

51. $z = 1 + 2x + 3y + 4y^2$, so

$A(S) = \iint_D \sqrt{1 + \left(\dfrac{\partial z}{\partial x}\right)^2 + \left(\dfrac{\partial z}{\partial y}\right)^2} \, dA = \int_1^4 \int_0^1 \sqrt{1 + 4 + (3 + 8y)^2} \, dy \, dx = \int_1^4 \int_0^1 \sqrt{14 + 48y + 64y^2} \, dy \, dx$.

Using a CAS, we have

$\int_1^4 \int_0^1 \sqrt{14 + 48y + 64y^2} \, dy \, dx = \frac{45}{8} \sqrt{14} + \frac{15}{16} \ln\left(11 \sqrt{5} + 3 \sqrt{14} \sqrt{5}\right) - \frac{15}{16} \ln\left(3 \sqrt{5} + \sqrt{14} \sqrt{5}\right)$

or $\frac{45}{8} \sqrt{14} + \frac{15}{16} \ln \dfrac{11 \sqrt{5} + 3\sqrt{70}}{3 \sqrt{5} + \sqrt{70}}$.

53. (a) $x = a \sin u \cos v, \; y = b \sin u \sin v, \; z = c \cos u \;\;\Rightarrow\;\;$

$\dfrac{x^2}{a^2} + \dfrac{y^2}{b^2} + \dfrac{z^2}{c^2} = (\sin u \cos v)^2 + (\sin u \sin v)^2 + (\cos u)^2$

$= \sin^2 u + \cos^2 u = 1$

and since the ranges of u and v are sufficient to generate the entire graph,

the parametric equations represent an ellipsoid.

(b)

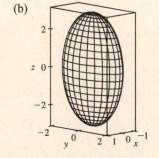

(c) From the parametric equations (with $a = 1$, $b = 2$, and $c = 3$),

we calculate $\mathbf{r}_u = \cos u \cos v \, \mathbf{i} + 2 \cos u \sin v \, \mathbf{j} - 3 \sin u \, \mathbf{k}$ and

$\mathbf{r}_v = -\sin u \sin v \, \mathbf{i} + 2 \sin u \cos v \, \mathbf{j}$. So $\mathbf{r}_u \times \mathbf{r}_v = 6 \sin^2 u \cos v \, \mathbf{i} + 3 \sin^2 u \sin v \, \mathbf{j} + 2 \sin u \cos u \, \mathbf{k}$, and the surface

area is given by $A(S) = \int_0^{2\pi} \int_0^{\pi} |\mathbf{r}_u \times \mathbf{r}_v| \, du \, dv = \int_0^{2\pi} \int_0^{\pi} \sqrt{36 \sin^4 u \cos^2 v + 9 \sin^4 u \sin^2 v + 4 \cos^2 u \sin^2 u} \, du \, dv$.

55. To find the region D: $z = x^2 + y^2$ implies $z + z^2 = 4z$ or $z^2 - 3z = 0$. Thus $z = 0$ or $z = 3$ are the planes where the

surfaces intersect. But $x^2 + y^2 + z^2 = 4z$ implies $x^2 + y^2 + (z-2)^2 = 4$, so $z = 3$ intersects the upper hemisphere.

Thus $(z-2)^2 = 4 - x^2 - y^2$ or $z = 2 + \sqrt{4 - x^2 - y^2}$. Therefore D is the region inside the circle $x^2 + y^2 + (3-2)^2 = 4$,

that is, $D = \{(x,y) \mid x^2 + y^2 \leq 3\}$.

$$A(S) = \iint_D \sqrt{1 + [(-x)(4 - x^2 - y^2)^{-1/2}]^2 + [(-y)(4 - x^2 - y^2)^{-1/2}]^2} \, dA$$

$$= \int_0^{2\pi} \int_0^{\sqrt{3}} \sqrt{1 + \frac{r^2}{4 - r^2}} \, r \, dr \, d\theta = \int_0^{2\pi} \int_0^{\sqrt{3}} \frac{2r \, dr}{\sqrt{4 - r^2}} \, d\theta = \int_0^{2\pi} \left[-2(4 - r^2)^{1/2} \right]_{r=0}^{r=\sqrt{3}} d\theta$$

$$= \int_0^{2\pi} (-2 + 4) \, d\theta = 2\theta \Big]_0^{2\pi} = 4\pi$$

57. If we revolve the curve $y = f(x)$, $a \leq x \leq b$ about the x-axis, where $f(x) \geq 0$, then from Equations 3 we know we can

parametrize the surface using $x = x$, $y = f(x) \cos\theta$, and $z = f(x) \sin\theta$, where $a \leq x \leq b$ and $0 \leq \theta \leq 2\pi$. Thus we can

say the surface is represented by $\mathbf{r}(x, \theta) = x\,\mathbf{i} + f(x) \cos\theta\,\mathbf{j} + f(x) \sin\theta\,\mathbf{k}$, with $a \leq x \leq b$ and $0 \leq \theta \leq 2\pi$. Then by (6),

the surface area is given by $A(S) = \iint_D |\mathbf{r}_x \times \mathbf{r}_\theta| \, dA$ where D is the rectangular parameter region $[a, b] \times [0, 2\pi]$. Here,

$\mathbf{r}_x(x, \theta) = \mathbf{i} + f'(x) \cos\theta\,\mathbf{j} + f'(x) \sin\theta\,\mathbf{k}$ and $\mathbf{r}_\theta(x) = -f(x) \sin\theta\,\mathbf{j} + f(x) \cos\theta\,\mathbf{k}$. So

$$\mathbf{r}_x \times \mathbf{r}_\theta = \begin{vmatrix} \mathbf{i} & \mathbf{j} & \mathbf{k} \\ 1 & f'(x)\cos\theta & f'(x)\sin\theta \\ 0 & -f(x)\sin\theta & f(x)\cos\theta \end{vmatrix} = \left[f(x)f'(x)\cos^2\theta + f(x)f'(x)\sin^2\theta \right] \mathbf{i} - f(x)\cos\theta\,\mathbf{j} - f(x)\sin\theta\,\mathbf{k}$$

$$= f(x)f'(x)\mathbf{i} - f(x)\cos\theta\,\mathbf{j} - f(x)\sin\theta\,\mathbf{k} \text{ and}$$

$$|\mathbf{r}_x \times \mathbf{r}_\theta| = \sqrt{[f(x)f'(x)]^2 + [f(x)]^2 \cos^2\theta + [f(x)]^2 \sin^2\theta}$$

$$= \sqrt{[f(x)]^2 ([f'(x)]^2 + 1)} = f(x)\sqrt{1 + [f'(x)]^2} \text{ [since } f(x) \geq 0].$$

Thus
$$A(S) = \iint_D |\mathbf{r}_x \times \mathbf{r}_\theta| \, dA = \int_a^b \int_0^{2\pi} f(x)\sqrt{1 + [f'(x)]^2} \, d\theta \, dx$$

$$= \int_a^b f(x)\sqrt{1 + [f'(x)]^2} \, [\theta]_0^{2\pi} \, dx = 2\pi \int_a^b f(x)\sqrt{1 + [f'(x)]^2} \, dx$$

59. $y = \sqrt{x} \implies 1 + \left(\dfrac{dy}{dx}\right)^2 = 1 + \left(\dfrac{1}{2\sqrt{x}}\right)^2 = 1 + \dfrac{1}{4x}$. So

$$S = \int_4^9 2\pi y \sqrt{1 + \left(\frac{dy}{dx}\right)^2} \, dx = \int_4^9 2\pi \sqrt{x} \sqrt{1 + \frac{1}{4x}} \, dx = 2\pi \int_4^9 \left(x + \tfrac{1}{4}\right) dx$$

$$= 2\pi \left[\tfrac{2}{3}\left(x + \tfrac{1}{4}\right)^{3/2} \right]_4^9 = \tfrac{4\pi}{3} \left[\tfrac{1}{8}(4x + 1)^{3/2} \right]_4^9 = \tfrac{\pi}{6}\left(37\sqrt{37} - 17\sqrt{17}\right)$$

13.7 Surface Integrals

1. The faces of the box in the planes $x = 0$ and $x = 2$ have surface area 24 and centers $(0, 2, 3)$, $(2, 2, 3)$. The faces in $y = 0$ and $y = 4$ have surface area 12 and centers $(1, 0, 3)$, $(1, 4, 3)$, and the faces in $z = 0$ and $z = 6$ have area 8 and centers $(1, 2, 0)$, $(1, 2, 6)$. For each face we take the point P_{ij}^* to be the center of the face and $f(x, y, z) = e^{-0.1(x+y+z)}$, so by Definition 1,

$$\iint_S f(x, y, z)\, dS \approx [f(0, 2, 3)](24) + [f(2, 2, 3)](24) + [f(1, 0, 3)](12)$$
$$+ [f(1, 4, 3)](12) + [f(1, 2, 0)](8) + [f(1, 2, 6)](8)$$
$$= 24(e^{-0.5} + e^{-0.7}) + 12(e^{-0.4} + e^{-0.8}) + 8(e^{-0.3} + e^{-0.9}) \approx 49.09$$

3. We can use the xz- and yz-planes to divide H into four patches of equal size, each with surface area equal to $\frac{1}{8}$ the surface area of a sphere with radius $\sqrt{50}$, so $\Delta S = \frac{1}{8}(4)\pi\left(\sqrt{50}\right)^2 = 25\pi$. Then $(\pm 3, \pm 4, 5)$ are sample points in the four patches, and using a Riemann sum as in Definition 1, we have

$$\iint_H f(x, y, z)\, dS \approx f(3, 4, 5)\, \Delta S + f(3, -4, 5)\, \Delta S + f(-3, 4, 5)\, \Delta S + f(-3, -4, 5)\, \Delta S$$
$$= (7 + 8 + 9 + 12)(25\pi) = 900\pi \approx 2827$$

5. $\mathbf{r}(u, v) = (u + v)\,\mathbf{i} + (u - v)\,\mathbf{j} + (1 + 2u + v)\,\mathbf{k}$, $0 \leq u \leq 2$, $0 \leq v \leq 1$ and

$\mathbf{r}_u \times \mathbf{r}_v = (\mathbf{i} + \mathbf{j} + 2\,\mathbf{k}) \times (\mathbf{i} - \mathbf{j} + \mathbf{k}) = 3\,\mathbf{i} + \mathbf{j} - 2\,\mathbf{k} \quad \Rightarrow \quad |\mathbf{r}_u \times \mathbf{r}_v| = \sqrt{3^2 + 1^2 + (-2)^2} = \sqrt{14}$. Then by Formula 2,

$$\iint_S (x + y + z)\, dS = \iint_D (u + v + u - v + 1 + 2u + v)\, |\mathbf{r}_u \times \mathbf{r}_v|\, dA = \int_0^1 \int_0^2 (4u + v + 1) \cdot \sqrt{14}\, du\, dv$$
$$= \sqrt{14} \int_0^1 \left[2u^2 + uv + u\right]_{u=0}^{u=2} dv = \sqrt{14} \int_0^1 (2v + 10)\, dv = \sqrt{14} \left[v^2 + 10v\right]_0^1 = 11\sqrt{14}$$

7. $\mathbf{r}(u, v) = \langle u \cos v, u \sin v, v\rangle$, $0 \leq u \leq 1$, $0 \leq v \leq \pi$ and

$\mathbf{r}_u \times \mathbf{r}_v = \langle \cos v, \sin v, 0\rangle \times \langle -u \sin v, u \cos v, 1\rangle = \langle \sin v, -\cos v, u\rangle \quad \Rightarrow$

$|\mathbf{r}_u \times \mathbf{r}_v| = \sqrt{\sin^2 v + \cos^2 v + u^2} = \sqrt{u^2 + 1}$. Then

$$\iint_S y\, dS = \iint_D (u \sin v)\, |\mathbf{r}_u \times \mathbf{r}_v|\, dA = \int_0^1 \int_0^\pi (u \sin v) \cdot \sqrt{u^2 + 1}\, dv\, du = \int_0^1 u\sqrt{u^2 + 1}\, du \int_0^\pi \sin v\, dv$$
$$= \left[\frac{1}{3}(u^2 + 1)^{3/2}\right]_0^1 [-\cos v]_0^\pi = \frac{1}{3}(2^{3/2} - 1) \cdot 2 = \frac{2}{3}(2\sqrt{2} - 1)$$

9. $z = 1 + 2x + 3y$ so $\dfrac{\partial z}{\partial x} = 2$ and $\dfrac{\partial z}{\partial y} = 3$. Then by Formula 4,

$$\iint_S x^2 yz\, dS = \iint_D x^2 yz \sqrt{\left(\frac{\partial z}{\partial x}\right)^2 + \left(\frac{\partial z}{\partial y}\right)^2 + 1}\, dA = \int_0^3 \int_0^2 x^2 y(1 + 2x + 3y)\sqrt{4 + 9 + 1}\, dy\, dx$$
$$= \sqrt{14} \int_0^3 \int_0^2 (x^2 y + 2x^3 y + 3x^2 y^2)\, dy\, dx = \sqrt{14} \int_0^3 \left[\tfrac{1}{2}x^2 y^2 + x^3 y^2 + x^2 y^3\right]_{y=0}^{y=2} dx$$
$$= \sqrt{14} \int_0^3 (10x^2 + 4x^3)\, dx = \sqrt{14} \left[\tfrac{10}{3}x^3 + x^4\right]_0^3 = 171\sqrt{14}$$

11. An equation of the plane through the points $(1, 0, 0)$, $(0, -2, 0)$, and $(0, 0, 4)$ is $4x - 2y + z = 4$, so S is the region in the plane $z = 4 - 4x + 2y$ over $D = \{(x, y) \mid 0 \leq x \leq 1, 2x - 2 \leq y \leq 0\}$. Thus by Formula 4,

$$\iint_S x\, dS = \iint_D x\sqrt{(-4)^2 + (2)^2 + 1}\, dA = \sqrt{21} \int_0^1 \int_{2x-2}^0 x\, dy\, dx = \sqrt{21} \int_0^1 [xy]_{y=2x-2}^{y=0} dx$$
$$= \sqrt{21} \int_0^1 (-2x^2 + 2x)\, dx = \sqrt{21} \left[-\tfrac{2}{3}x^3 + x^2\right]_0^1 = \sqrt{21} \left(-\tfrac{2}{3} + 1\right) = \frac{\sqrt{21}}{3}$$

13. S is the portion of the cone $z^2 = x^2 + y^2$ for $1 \le z \le 3$, or equivalently, S is the part of the surface $z = \sqrt{x^2 + y^2}$ over the region $D = \left\{ (x, y) \mid 1 \le x^2 + y^2 \le 9 \right\}$. Thus

$$\iint_S x^2 z^2 \, dS = \iint_D x^2 (x^2 + y^2) \sqrt{\left(\frac{x}{\sqrt{x^2+y^2}} \right)^2 + \left(\frac{y}{\sqrt{x^2+y^2}} \right)^2 + 1} \, dA$$

$$= \iint_D x^2 (x^2 + y^2) \sqrt{\frac{x^2+y^2}{x^2+y^2} + 1} \, dA = \iint_D \sqrt{2} \, x^2 (x^2 + y^2) \, dA = \sqrt{2} \int_0^{2\pi} \int_1^3 (r \cos \theta)^2 (r^2) \, r \, dr \, d\theta$$

$$= \sqrt{2} \int_0^{2\pi} \cos^2 \theta \, d\theta \int_1^3 r^5 \, dr = \sqrt{2} \left[\tfrac{1}{2}\theta + \tfrac{1}{4}\sin 2\theta \right]_0^{2\pi} \left[\tfrac{1}{6}r^6 \right]_1^3 = \sqrt{2} \, (\pi) \cdot \tfrac{1}{6}(3^6 - 1) = \frac{364\sqrt{2}}{3}\pi$$

15. Using x and z as parameters, we have $\mathbf{r}(x, z) = x\,\mathbf{i} + (x^2 + z^2)\,\mathbf{j} + z\,\mathbf{k}$, $x^2 + z^2 \le 4$. Then

$$\mathbf{r}_x \times \mathbf{r}_z = (\mathbf{i} + 2x\,\mathbf{j}) \times (2z\,\mathbf{j} + \mathbf{k}) = 2x\,\mathbf{i} - \mathbf{j} + 2z\,\mathbf{k} \text{ and } |\mathbf{r}_x \times \mathbf{r}_z| = \sqrt{4x^2 + 1 + 4z^2} = \sqrt{1 + 4(x^2 + z^2)}. \text{ Thus}$$

$$\iint_S y \, dS = \iint_{x^2+z^2 \le 4} (x^2 + z^2)\sqrt{1 + 4(x^2 + z^2)} \, dA = \int_0^{2\pi} \int_0^2 r^2 \sqrt{1 + 4r^2} \, r \, dr \, d\theta = \int_0^{2\pi} d\theta \int_0^2 r^2 \sqrt{1 + 4r^2} \, r \, dr$$

$$= 2\pi \int_0^2 r^2 \sqrt{1 + 4r^2} \, r \, dr \qquad [\text{let } u = 1 + 4r^2 \ \Rightarrow \ r^2 = \tfrac{1}{4}(u - 1) \text{ and } \tfrac{1}{8}du = r \, dr]$$

$$= 2\pi \int_1^{17} \tfrac{1}{4}(u - 1)\sqrt{u} \cdot \tfrac{1}{8} \, du = \tfrac{1}{16}\pi \int_1^{17} (u^{3/2} - u^{1/2}) \, du$$

$$= \tfrac{1}{16}\pi \left[\tfrac{2}{5}u^{5/2} - \tfrac{2}{3}u^{3/2} \right]_1^{17} = \tfrac{1}{16}\pi \left[\tfrac{2}{5}(17)^{5/2} - \tfrac{2}{3}(17)^{3/2} - \tfrac{2}{5} + \tfrac{2}{3} \right] = \frac{\pi}{60} \left(391\sqrt{17} + 1 \right)$$

17. Using spherical coordinates and Example 13.6.9 we have $\mathbf{r}(\phi, \theta) = 2\sin\phi\cos\theta\,\mathbf{i} + 2\sin\phi\sin\theta\,\mathbf{j} + 2\cos\phi\,\mathbf{k}$ and

$|\mathbf{r}_\phi \times \mathbf{r}_\theta| = 4\sin\phi$. Then $\iint_S (x^2 z + y^2 z) \, dS = \int_0^{2\pi} \int_0^{\pi/2} (4\sin^2\phi)(2\cos\phi)(4\sin\phi) \, d\phi \, d\theta = 16\pi \sin^4\phi \Big]_0^{\pi/2} = 16\pi$.

19. S is given by $\mathbf{r}(u, v) = u\,\mathbf{i} + \cos v\,\mathbf{j} + \sin v\,\mathbf{k}$, $0 \le u \le 3$, $0 \le v \le \pi/2$. Then

$$\mathbf{r}_u \times \mathbf{r}_v = \mathbf{i} \times (-\sin v\,\mathbf{j} + \cos v\,\mathbf{k}) = -\cos v\,\mathbf{j} - \sin v\,\mathbf{k} \text{ and } |\mathbf{r}_u \times \mathbf{r}_v| = \sqrt{\cos^2 v + \sin^2 v} = 1, \text{ so}$$

$$\iint_S (z + x^2 y) \, dS = \int_0^{\pi/2} \int_0^3 (\sin v + u^2 \cos v)(1) \, du \, dv = \int_0^{\pi/2} (3\sin v + 9\cos v) \, dv$$

$$= [-3\cos v + 9\sin v]_0^{\pi/2} = 0 + 9 + 3 - 0 = 12$$

21. From Exercise 5, $\mathbf{r}(u, v) = (u + v)\,\mathbf{i} + (u - v)\,\mathbf{j} + (1 + 2u + v)\,\mathbf{k}$, $0 \le u \le 2$, $0 \le v \le 1$, and $\mathbf{r}_u \times \mathbf{r}_v = 3\,\mathbf{i} + \mathbf{j} - 2\,\mathbf{k}$. Then

$$\mathbf{F}(\mathbf{r}(u, v)) = (1 + 2u + v)e^{(u+v)(u-v)}\,\mathbf{i} - 3(1 + 2u + v)e^{(u+v)(u-v)}\,\mathbf{j} + (u + v)(u - v)\,\mathbf{k}$$

$$= (1 + 2u + v)e^{u^2 - v^2}\,\mathbf{i} - 3(1 + 2u + v)e^{u^2 - v^2}\,\mathbf{j} + (u^2 - v^2)\,\mathbf{k}$$

Because the z-component of $\mathbf{r}_u \times \mathbf{r}_v$ is negative we use $-(\mathbf{r}_u \times \mathbf{r}_v)$ in Formula 9 for the upward orientation:

$$\iint_S \mathbf{F} \cdot d\mathbf{S} = \iint_D \mathbf{F} \cdot (-(\mathbf{r}_u \times \mathbf{r}_v)) \, dA = \int_0^1 \int_0^2 \left[-3(1 + 2u + v)e^{u^2 - v^2} + 3(1 + 2u + v)e^{u^2 - v^2} + 2(u^2 - v^2) \right] du \, dv$$

$$= \int_0^1 \int_0^2 2(u^2 - v^2) \, du \, dv = 2\int_0^1 \left[\tfrac{1}{3}u^3 - uv^2 \right]_{u=0}^{u=2} dv = 2\int_0^1 \left(\tfrac{8}{3} - 2v^2 \right) dv$$

$$= 2\left[\tfrac{8}{3}v - \tfrac{2}{3}v^3 \right]_0^1 = 2\left(\tfrac{8}{3} - \tfrac{2}{3} \right) = 4$$

23. $\mathbf{F}(x, y, z) = xy\,\mathbf{i} + yz\,\mathbf{j} + zx\,\mathbf{k}$, $z = g(x, y) = 4 - x^2 - y^2$, and D is the square $[0, 1] \times [0, 1]$, so by Equation 8

$$\iint_S \mathbf{F} \cdot d\mathbf{S} = \iint_D [-xy(-2x) - yz(-2y) + zx] \, dA = \int_0^1 \int_0^1 [2x^2 y + 2y^2(4 - x^2 - y^2) + x(4 - x^2 - y^2)] \, dy \, dx$$

$$= \int_0^1 \left(\tfrac{1}{3}x^2 + \tfrac{11}{3}x - x^3 + \tfrac{34}{15} \right) dx = \frac{713}{180}$$

25. $\mathbf{F}(x,y,z) = x\,\mathbf{i} - z\,\mathbf{j} + y\,\mathbf{k}$, $z = g(x,y) = \sqrt{4 - x^2 - y^2}$ and D is the quarter disk

$\{(x,y) \mid 0 \le x \le 2, 0 \le y \le \sqrt{4 - x^2}\,\}$. S has downward orientation, so by Formula 10,

$$\iint_S \mathbf{F} \cdot d\mathbf{S} = -\iint_D \left[-x \cdot \tfrac{1}{2}(4 - x^2 - y^2)^{-1/2}(-2x) - (-z) \cdot \tfrac{1}{2}(4 - x^2 - y^2)^{-1/2}(-2y) + y \right] dA$$

$$= -\iint_D \left(\frac{x^2}{\sqrt{4 - x^2 - y^2}} - \sqrt{4 - x^2 - y^2} \cdot \frac{y}{\sqrt{4 - x^2 - y^2}} + y \right) dA$$

$$= -\iint_D x^2(4 - (x^2 + y^2))^{-1/2}\, dA = -\int_0^{\pi/2} \int_0^2 (r \cos\theta)^2 (4 - r^2)^{-1/2}\, r\, dr\, d\theta$$

$$= -\int_0^{\pi/2} \cos^2\theta\, d\theta \int_0^2 r^3 (4 - r^2)^{-1/2}\, dr \qquad [\text{let } u = 4 - r^2 \;\Rightarrow\; r^2 = 4 - u \text{ and } -\tfrac{1}{2}\, du = r\, dr]$$

$$= -\int_0^{\pi/2} \left(\tfrac{1}{2} + \tfrac{1}{2} \cos 2\theta \right) d\theta \int_4^0 -\tfrac{1}{2}(4 - u)(u)^{-1/2}\, du$$

$$= -\left[\tfrac{1}{2}\theta + \tfrac{1}{4} \sin 2\theta \right]_0^{\pi/2} \left(-\tfrac{1}{2} \right) \left[8\sqrt{u} - \tfrac{2}{3} u^{3/2} \right]_4^0 = -\tfrac{\pi}{4} \left(-\tfrac{1}{2} \right) \left(-16 + \tfrac{16}{3} \right) = -\tfrac{4}{3}\pi$$

27. Let S_1 be the paraboloid $y = x^2 + z^2$, $0 \le y \le 1$ and S_2 the disk $x^2 + z^2 \le 1$, $y = 1$. Since S is a closed

surface, we use the outward orientation.

On S_1: $\mathbf{F}(\mathbf{r}(x,z)) = (x^2 + z^2)\,\mathbf{j} - z\,\mathbf{k}$ and $\mathbf{r}_x \times \mathbf{r}_z = 2x\,\mathbf{i} - \mathbf{j} + 2z\,\mathbf{k}$ (since the $\mathbf{j}$-component must be negative on S_1). Then

$$\iint_{S_1} \mathbf{F} \cdot d\mathbf{S} = \iint_{x^2 + z^2 \le 1} [-(x^2 + z^2) - 2z^2]\, dA = -\int_0^{2\pi} \int_0^1 (r^2 + 2r^2 \sin^2\theta)\, r\, dr\, d\theta$$

$$= -\int_0^{2\pi} \int_0^1 r^3 (1 + 2\sin^2\theta)\, dr\, d\theta = -\int_0^{2\pi} (1 + 1 - \cos 2\theta)\, d\theta \int_0^1 r^3\, dr$$

$$= -\left[2\theta - \tfrac{1}{2} \sin 2\theta \right]_0^{2\pi} \left[\tfrac{1}{4} r^4 \right]_0^1 = -4\pi \cdot \tfrac{1}{4} = -\pi$$

On S_2: $\mathbf{F}(\mathbf{r}(x,z)) = \mathbf{j} - z\,\mathbf{k}$ and $\mathbf{r}_z \times \mathbf{r}_x = \mathbf{j}$. Then $\iint_{S_2} \mathbf{F} \cdot d\mathbf{S} = \iint_{x^2 + z^2 \le 1} (1)\, dA = \pi$.

Hence $\iint_S \mathbf{F} \cdot d\mathbf{S} = -\pi + \pi = 0$.

29. Here S consists of the six faces of the cube as labeled in the figure. On S_1:

$\mathbf{F} = \mathbf{i} + 2y\,\mathbf{j} + 3z\,\mathbf{k}$, $\mathbf{r}_y \times \mathbf{r}_z = \mathbf{i}$ and $\iint_{S_1} \mathbf{F} \cdot d\mathbf{S} = \int_{-1}^1 \int_{-1}^1 dy\, dz = 4$;

S_2: $\mathbf{F} = x\,\mathbf{i} + 2\,\mathbf{j} + 3z\,\mathbf{k}$, $\mathbf{r}_z \times \mathbf{r}_x = \mathbf{j}$ and $\iint_{S_2} \mathbf{F} \cdot d\mathbf{S} = \int_{-1}^1 \int_{-1}^1 2\, dx\, dz = 8$;

S_3: $\mathbf{F} = x\,\mathbf{i} + 2y\,\mathbf{j} + 3\,\mathbf{k}$, $\mathbf{r}_x \times \mathbf{r}_y = \mathbf{k}$ and $\iint_{S_3} \mathbf{F} \cdot d\mathbf{S} = \int_{-1}^1 \int_{-1}^1 3\, dx\, dy = 12$;

S_4: $\mathbf{F} = -\mathbf{i} + 2y\,\mathbf{j} + 3z\,\mathbf{k}$, $\mathbf{r}_z \times \mathbf{r}_y = -\mathbf{i}$ and $\iint_{S_4} \mathbf{F} \cdot d\mathbf{S} = 4$;

S_5: $\mathbf{F} = x\,\mathbf{i} - 2\,\mathbf{j} + 3z\,\mathbf{k}$, $\mathbf{r}_x \times \mathbf{r}_z = -\mathbf{j}$ and $\iint_{S_5} \mathbf{F} \cdot d\mathbf{S} = 8$;

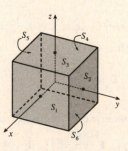

S_6: $\mathbf{F} = x\,\mathbf{i} + 2y\,\mathbf{j} - 3\,\mathbf{k}$, $\mathbf{r}_y \times \mathbf{r}_x = -\mathbf{k}$ and $\iint_{S_6} \mathbf{F} \cdot d\mathbf{S} = \int_{-1}^1 \int_{-1}^1 3\, dx\, dy = 12$.

Hence $\iint_S \mathbf{F} \cdot d\mathbf{S} = \sum_{i=1}^6 \iint_{S_i} \mathbf{F} \cdot d\mathbf{S} = 48$.

31. Here S consists of four surfaces: S_1, the top surface (a portion of the circular cylinder $y^2 + z^2 = 1$); S_2, the bottom surface

(a portion of the xy-plane); S_3, the front half-disk in the plane $x = 2$, and S_4, the back half-disk in the plane $x = 0$.

On S_1: The surface is $z = \sqrt{1 - y^2}$ for $0 \leq x \leq 2, -1 \leq y \leq 1$ with upward orientation, so

$$\iint_{S_1} \mathbf{F} \cdot d\mathbf{S} = \int_0^2 \int_{-1}^1 \left[-x^2(0) - y^2 \left(-\frac{y}{\sqrt{1 - y^2}} \right) + z^2 \right] dy\, dx = \int_0^2 \int_{-1}^1 \left(\frac{y^3}{\sqrt{1 - y^2}} + 1 - y^2 \right) dy\, dx$$

$$= \int_0^2 \left[-\sqrt{1 - y^2} + \tfrac{1}{3}(1 - y^2)^{3/2} + y - \tfrac{1}{3}y^3 \right]_{y=-1}^{y=1} dx = \int_0^2 \tfrac{4}{3}\, dx = \tfrac{8}{3}$$

On S_2: The surface is $z = 0$ with downward orientation, so

$$\iint_{S_2} \mathbf{F} \cdot d\mathbf{S} = \int_0^2 \int_{-1}^1 (-z^2)\, dy\, dx = \int_0^2 \int_{-1}^1 (0)\, dy\, dx = 0$$

On S_3: The surface is $x = 2$ for $-1 \leq y \leq 1, 0 \leq z \leq \sqrt{1 - y^2}$, oriented in the positive x-direction. Regarding y and z as parameters, we have $\mathbf{r}_y \times \mathbf{r}_z = \mathbf{i}$ and

$$\iint_{S_3} \mathbf{F} \cdot d\mathbf{S} = \int_{-1}^1 \int_0^{\sqrt{1 - y^2}} x^2\, dz\, dy = \int_{-1}^1 \int_0^{\sqrt{1 - y^2}} 4\, dz\, dy = 4A(S_3) = 2\pi$$

On S_4: The surface is $x = 0$ for $-1 \leq y \leq 1, 0 \leq z \leq \sqrt{1 - y^2}$, oriented in the negative x-direction. Regarding y and z as parameters, we use $-(\mathbf{r}_y \times \mathbf{r}_z) = -\mathbf{i}$ and

$$\iint_{S_4} \mathbf{F} \cdot d\mathbf{S} = \int_{-1}^1 \int_0^{\sqrt{1 - y^2}} x^2\, dz\, dy = \int_{-1}^1 \int_0^{\sqrt{1 - y^2}} (0)\, dz\, dy = 0$$

Thus $\iint_S \mathbf{F} \cdot d\mathbf{S} = \tfrac{8}{3} + 0 + 2\pi + 0 = 2\pi + \tfrac{8}{3}$.

33. We use Formula 4 with $z = 3 - 2x^2 - y^2 \;\Rightarrow\; \partial z/\partial x = -4x, \partial z/\partial y = -2y$. The boundaries of the region

$3 - 2x^2 - y^2 \geq 0$ are $-\sqrt{\tfrac{3}{2}} \leq x \leq \sqrt{\tfrac{3}{2}}$ and $-\sqrt{3 - 2x^2} \leq y \leq \sqrt{3 - 2x^2}$, so we use a CAS (with precision reduced to

seven or fewer digits; otherwise the calculation may take a long time) to calculate

$$\iint_S x^2 y^2 z^2\, dS = \int_{-\sqrt{3/2}}^{\sqrt{3/2}} \int_{-\sqrt{3 - 2x^2}}^{\sqrt{3 - 2x^2}} x^2 y^2 (3 - 2x^2 - y^2)^2 \sqrt{16x^2 + 4y^2 + 1}\, dy\, dx \approx 3.4895$$

35. If S is given by $y = h(x, z)$, then S is also the level surface $f(x, y, z) = y - h(x, z) = 0$.

$$\mathbf{n} = \frac{\nabla f(x, y, z)}{|\nabla f(x, y, z)|} = \frac{-h_x \mathbf{i} + \mathbf{j} - h_z \mathbf{k}}{\sqrt{h_x^2 + 1 + h_z^2}},$$ and $-\mathbf{n}$ is the unit normal that points to the left. Now we proceed as in the

derivation of (10), using Formula 4 to evaluate

$$\iint_S \mathbf{F} \cdot d\mathbf{S} = \iint_S \mathbf{F} \cdot \mathbf{n}\, dS = \iint_D (P\mathbf{i} + Q\mathbf{j} + R\mathbf{k}) \frac{\dfrac{\partial h}{\partial x}\mathbf{i} - \mathbf{j} + \dfrac{\partial h}{\partial z}\mathbf{k}}{\sqrt{\left(\dfrac{\partial h}{\partial x}\right)^2 + 1 + \left(\dfrac{\partial h}{\partial z}\right)^2}} \sqrt{\left(\dfrac{\partial h}{\partial x}\right)^2 + 1 + \left(\dfrac{\partial h}{\partial z}\right)^2}\, dA$$

where D is the projection of S onto the xz-plane. Therefore $\displaystyle \iint_S \mathbf{F} \cdot d\mathbf{S} = \iint_D \left(P\frac{\partial h}{\partial x} - Q + R\frac{\partial h}{\partial z} \right) dA$.

37. $m = \iint_S K\, dS = K \cdot 4\pi\left(\tfrac{1}{2}a^2\right) = 2\pi a^2 K$; by symmetry $M_{xz} = M_{yz} = 0$, and

$$M_{xy} = \iint_S zK\, dS = K \int_0^{2\pi} \int_0^{\pi/2} (a\cos\phi)(a^2\sin\phi)\, d\phi\, d\theta = 2\pi Ka^3 \left[-\tfrac{1}{4}\cos 2\phi \right]_0^{\pi/2} = \pi Ka^3.$$

Hence $(\overline{x}, \overline{y}, \overline{z}) = (0, 0, \tfrac{1}{2}a)$.

39. (a) $I_z = \iint_S (x^2 + y^2)\rho(x, y, z)\, dS$

(b) $I_z = \iint_S (x^2 + y^2)\left(10 - \sqrt{x^2 + y^2}\right) dS = \iint\limits_{1 \le x^2 + y^2 \le 16} (x^2 + y^2)\left(10 - \sqrt{x^2 + y^2}\right)\sqrt{2}\, dA$

$\quad = \int_0^{2\pi} \int_1^4 \sqrt{2}\,(10r^3 - r^4)\, dr\, d\theta = 2\sqrt{2}\,\pi\left(\frac{4329}{10}\right) = \frac{4329}{5}\sqrt{2}\,\pi$

41. The rate of flow through the cylinder is the flux $\iint_S \rho\mathbf{v} \cdot \mathbf{n}\, dS = \iint_S \rho\mathbf{v} \cdot d\mathbf{S}$. We use the parametric representation

$\mathbf{r}(u, v) = 2\cos u\,\mathbf{i} + 2\sin u\,\mathbf{j} + v\,\mathbf{k}$ for S, where $0 \le u \le 2\pi$, $0 \le v \le 1$, so $\mathbf{r}_u = -2\sin u\,\mathbf{i} + 2\cos u\,\mathbf{j}$, $\mathbf{r}_v = \mathbf{k}$, and the

outward orientation is given by $\mathbf{r}_u \times \mathbf{r}_v = 2\cos u\,\mathbf{i} + 2\sin u\,\mathbf{j}$. Then

$$\iint_S \rho\mathbf{v} \cdot d\mathbf{S} = \rho \int_0^{2\pi} \int_0^1 \left(v\,\mathbf{i} + 4\sin^2 u\,\mathbf{j} + 4\cos^2 u\,\mathbf{k}\right) \cdot (2\cos u\,\mathbf{i} + 2\sin u\,\mathbf{j})\, dv\, du$$

$$= \rho \int_0^{2\pi} \int_0^1 \left(2v\cos u + 8\sin^3 u\right) dv\, du = \rho \int_0^{2\pi} \left(\cos u + 8\sin^3 u\right) du$$

$$= \rho\left[\sin u + 8(-\tfrac{1}{3})(2 + \sin^2 u)\cos u\right]_0^{2\pi} = 0 \text{ kg/s}$$

43. S consists of the hemisphere S_1 given by $z = \sqrt{a^2 - x^2 - y^2}$ and the disk S_2 given by $0 \le x^2 + y^2 \le a^2$, $z = 0$.

On S_1: $\mathbf{E} = a\sin\phi\,\cos\theta\,\mathbf{i} + a\sin\phi\,\sin\theta\,\mathbf{j} + 2a\cos\phi\,\mathbf{k}$,

$\mathbf{T}_\phi \times \mathbf{T}_\theta = a^2\sin^2\phi\,\cos\theta\,\mathbf{i} + a^2\sin^2\phi\,\sin\theta\,\mathbf{j} + a^2\sin\phi\,\cos\phi\,\mathbf{k}$. Thus

$$\iint_{S_1} \mathbf{E} \cdot d\mathbf{S} = \int_0^{2\pi} \int_0^{\pi/2} (a^3\sin^3\phi + 2a^3\sin\phi\,\cos^2\phi)\, d\phi\, d\theta$$

$$= \int_0^{2\pi} \int_0^{\pi/2} (a^3\sin\phi + a^3\sin\phi\,\cos^2\phi)\, d\phi\, d\theta = (2\pi)a^3\left(1 + \tfrac{1}{3}\right) = \tfrac{8}{3}\pi a^3$$

On S_2: $\mathbf{E} = x\,\mathbf{i} + y\,\mathbf{j}$, and $\mathbf{r}_y \times \mathbf{r}_x = -\mathbf{k}$ so $\iint_{S_2} \mathbf{E} \cdot d\mathbf{S} = 0$. Hence the total charge is $q = \varepsilon_0 \iint_S \mathbf{E} \cdot d\mathbf{S} = \tfrac{8}{3}\pi a^3 \varepsilon_0$.

45. $K\nabla u = 6.5(4y\,\mathbf{j} + 4z\,\mathbf{k})$. S is given by $\mathbf{r}(x, \theta) = x\,\mathbf{i} + \sqrt{6}\,\cos\theta\,\mathbf{j} + \sqrt{6}\,\sin\theta\,\mathbf{k}$ and since we want the inward heat flow, we

use $\mathbf{r}_x \times \mathbf{r}_\theta = -\sqrt{6}\,\cos\theta\,\mathbf{j} - \sqrt{6}\,\sin\theta\,\mathbf{k}$. Then the rate of heat flow inward is given by

$\iint_S (-K\,\nabla u) \cdot d\mathbf{S} = \int_0^{2\pi} \int_0^4 -(6.5)(-24)\, dx\, d\theta = (2\pi)(156)(4) = 1248\pi$.

47. Let S be a sphere of radius a centered at the origin. Then $|\mathbf{r}| = a$ and $\mathbf{F}(\mathbf{r}) = c\mathbf{r}/|\mathbf{r}|^3 = (c/a^3)\,(x\,\mathbf{i} + y\,\mathbf{j} + z\,\mathbf{k})$. A

parametric representation for S is $\mathbf{r}(\phi, \theta) = a\sin\phi\,\cos\theta\,\mathbf{i} + a\sin\phi\,\sin\theta\,\mathbf{j} + a\cos\phi\,\mathbf{k}$, $0 \le \phi \le \pi$, $0 \le \theta \le 2\pi$. Then

$\mathbf{r}_\phi = a\cos\phi\,\cos\theta\,\mathbf{i} + a\cos\phi\,\sin\theta\,\mathbf{j} - a\sin\phi\,\mathbf{k}$, $\mathbf{r}_\theta = -a\sin\phi\,\sin\theta\,\mathbf{i} + a\sin\phi\,\cos\theta\,\mathbf{j}$, and the outward orientation is given

by $\mathbf{r}_\phi \times \mathbf{r}_\theta = a^2\sin^2\phi\,\cos\theta\,\mathbf{i} + a^2\sin^2\phi\,\sin\theta\,\mathbf{j} + a^2\sin\phi\,\cos\phi\,\mathbf{k}$. The flux of $\mathbf{F}$ across S is

$$\iint_S \mathbf{F} \cdot d\mathbf{S} = \int_0^\pi \int_0^{2\pi} \frac{c}{a^3}\,(a\sin\phi\,\cos\theta\,\mathbf{i} + a\sin\phi\,\sin\theta\,\mathbf{j} + a\cos\phi\,\mathbf{k})$$

$$\cdot (a^2\sin^2\phi\,\cos\theta\,\mathbf{i} + a^2\sin^2\phi\,\sin\theta\,\mathbf{j} + a^2\sin\phi\,\cos\phi\,\mathbf{k})\, d\theta\, d\phi$$

$$= \frac{c}{a^3} \int_0^\pi \int_0^{2\pi} a^3\,(\sin^3\phi + \sin\phi\,\cos^2\phi)\, d\theta\, d\phi = c\int_0^\pi \int_0^{2\pi} \sin\phi\, d\theta\, d\phi = 4\pi c$$

Thus the flux does not depend on the radius a.

13.8   Stokes' Theorem

1. The paraboloid $z = x^2 + y^2$ intersects the cylinder $x^2 + y^2 = 4$ in the circle $x^2 + y^2 = 4$, $z = 4$. This boundary curve C should be oriented in the counterclockwise direction when viewed from above, so a vector equation of C is

$\mathbf{r}(t) = 2\cos t\,\mathbf{i} + 2\sin t\,\mathbf{j} + 4\,\mathbf{k}$, $0 \le t \le 2\pi$. Then $\mathbf{r}'(t) = -2\sin t\,\mathbf{i} + 2\cos t\,\mathbf{j}$,

$\mathbf{F}(\mathbf{r}(t)) = (4\cos^2 t)(16)\,\mathbf{i} + (4\sin^2 t)(16)\,\mathbf{j} + (2\cos t)(2\sin t)(4)\,\mathbf{k} = 64\cos^2 t\,\mathbf{i} + 64\sin^2 t\,\mathbf{j} + 16\sin t\,\cos t\,\mathbf{k}$,

and by Stokes' Theorem,

$$\iint_S \text{curl}\,\mathbf{F} \cdot d\mathbf{S} = \int_C \mathbf{F} \cdot d\mathbf{r} = \int_0^{2\pi} \mathbf{F}(\mathbf{r}(t)) \cdot \mathbf{r}'(t)\,dt = \int_0^{2\pi} (-128\cos^2 t\,\sin t + 128\sin^2 t\,\cos t + 0)\,dt$$

$$= 128\left[\tfrac{1}{3}\cos^3 t + \tfrac{1}{3}\sin^3 t\right]_0^{2\pi} = 0$$

3. C is the square in the plane $z = -1$. Rather than evaluating a line integral around C we can use Equation 3:

$\iint_{S_1} \text{curl}\,\mathbf{F} \cdot d\mathbf{S} = \oint_C \mathbf{F} \cdot d\mathbf{r} = \iint_{S_2} \text{curl}\,\mathbf{F} \cdot d\mathbf{S}$ where S_1 is the original cube without the bottom and S_2 is the bottom face

of the cube. $\text{curl}\,\mathbf{F} = x^2 z\,\mathbf{i} + (xy - 2xyz)\,\mathbf{j} + (y - xz)\,\mathbf{k}$. For S_2, we choose $\mathbf{n} = \mathbf{k}$ so that C has the same orientation for

both surfaces. Then $\text{curl}\,\mathbf{F} \cdot \mathbf{n} = y - xz = x + y$ on S_2, where $z = -1$. Thus $\iint_{S_2} \text{curl}\,\mathbf{F} \cdot d\mathbf{S} = \int_{-1}^1 \int_{-1}^1 (x + y)\,dx\,dy = 0$

so $\iint_{S_1} \text{curl}\,\mathbf{F} \cdot d\mathbf{S} = 0$.

5. $\text{curl}\,\mathbf{F} = -2z\,\mathbf{i} - 2x\,\mathbf{j} - 2y\,\mathbf{k}$ and we take the surface S to be the planar region enclosed by C, so S is the portion of the plane

$x + y + z = 1$ over $D = \{(x, y) \mid 0 \le x \le 1, 0 \le y \le 1 - x\}$. Since C is oriented counterclockwise, we orient S upward.

Using Equation 13.7.10, we have $z = g(x, y) = 1 - x - y$, $P = -2z$, $Q = -2x$, $R = -2y$, and

$$\int_C \mathbf{F} \cdot d\mathbf{r} = \iint_S \text{curl}\,\mathbf{F} \cdot d\mathbf{S} = \iint_D [-(-2z)(-1) - (-2x)(-1) + (-2y)]\,dA$$

$$= \int_0^1 \int_0^{1-x} (-2)\,dy\,dx = -2\int_0^1 (1 - x)\,dx = -1$$

7. $\text{curl}\,\mathbf{F} = (xe^{xy} - 2x)\,\mathbf{i} - (ye^{xy} - y)\,\mathbf{j} + (2z - z)\,\mathbf{k}$ and we take S to be the disk $x^2 + y^2 \le 16$, $z = 5$. Since C is oriented

counterclockwise (from above), we orient S upward. Then $\mathbf{n} = \mathbf{k}$ and $\text{curl}\,\mathbf{F} \cdot \mathbf{n} = 2z - z$ on S, where $z = 5$. Thus

$$\oint_C \mathbf{F} \cdot d\mathbf{r} = \iint_S \text{curl}\,\mathbf{F} \cdot \mathbf{n}\,dS = \iint_S (2z - z)\,dS = \iint_S (10 - 5)\,dS = 5(\text{area of } S) = 5(\pi \cdot 4^2) = 80\pi$$

9. (a) The curve of intersection is an ellipse in the plane $x + y + z = 1$ with unit normal $\mathbf{n} = \frac{1}{\sqrt{3}}(\mathbf{i} + \mathbf{j} + \mathbf{k})$,

$\text{curl}\,\mathbf{F} = x^2\,\mathbf{j} + y^2\,\mathbf{k}$, and $\text{curl}\,\mathbf{F} \cdot \mathbf{n} = \frac{1}{\sqrt{3}}(x^2 + y^2)$. Then

$$\oint_C \mathbf{F} \cdot d\mathbf{r} = \iint_S \frac{1}{\sqrt{3}}(x^2 + y^2)\,dS = \iint_{x^2 + y^2 \le 9} (x^2 + y^2)\,dx\,dy = \int_0^{2\pi} \int_0^3 r^3\,dr\,d\theta = 2\pi\left(\tfrac{81}{4}\right) = \tfrac{81\pi}{2}$$

(b)

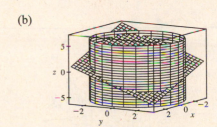

(c) One possible parametrization is $x = 3\cos t$, $y = 3\sin t$,

$z = 1 - 3\cos t - 3\sin t$, $0 \le t \le 2\pi$.

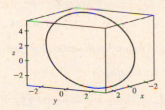

11. The boundary curve C is the circle $x^2 + y^2 = 16$, $z = 4$ oriented in the clockwise direction as viewed from above (since S is oriented downward). We can parametrize C by $\mathbf{r}(t) = 4\cos t\,\mathbf{i} - 4\sin t\,\mathbf{j} + 4\,\mathbf{k}$, $0 \le t \le 2\pi$, and then $\mathbf{r}'(t) = -4\sin t\,\mathbf{i} - 4\cos t\,\mathbf{j}$. Thus $\mathbf{F}(\mathbf{r}(t)) = 4\sin t\,\mathbf{i} + 4\cos t\,\mathbf{j} - 2\,\mathbf{k}$, $\mathbf{F}(\mathbf{r}(t))\cdot\mathbf{r}'(t) = -16\sin^2 t - 16\cos^2 t = -16$, and

$$\oint_C \mathbf{F}\cdot d\mathbf{r} = \int_0^{2\pi}\mathbf{F}(\mathbf{r}(t))\cdot\mathbf{r}'(t)\,dt = \int_0^{2\pi}(-16)\,dt = -16\,(2\pi) = -32\pi$$

Now curl $\mathbf{F} = 2\,\mathbf{k}$, and the projection D of S on the xy-plane is the disk $x^2 + y^2 \le 16$, so by Equation 13.7.10 with $z = g(x,y) = \sqrt{x^2+y^2}$ [and multiplying by -1 for the downward orientation] we have

$$\iint_S \text{curl}\,\mathbf{F}\cdot d\mathbf{S} = -\iint_D (-0 - 0 + 2)\,dA = -2 \cdot A(D) = -2 \cdot \pi(4^2) = -32\pi$$

13. The boundary curve C is the circle $x^2 + z^2 = 1$, $y = 0$ oriented in the counterclockwise direction as viewed from the positive y-axis. Then C can be described by $\mathbf{r}(t) = \cos t\,\mathbf{i} - \sin t\,\mathbf{k}$, $0 \le t \le 2\pi$, and $\mathbf{r}'(t) = -\sin t\,\mathbf{i} - \cos t\,\mathbf{k}$. Thus $\mathbf{F}(\mathbf{r}(t)) = -\sin t\,\mathbf{j} + \cos t\,\mathbf{k}$, $\mathbf{F}(\mathbf{r}(t))\cdot\mathbf{r}'(t) = -\cos^2 t$, and $\oint_C \mathbf{F}\cdot d\mathbf{r} = \int_0^{2\pi}(-\cos^2 t)\,dt = -\frac{1}{2}t - \frac{1}{4}\sin 2t\Big]_0^{2\pi} = -\pi$.

Now curl $\mathbf{F} = -\mathbf{i} - \mathbf{j} - \mathbf{k}$, and S can be parametrized (see Example 13.6.9) by $\mathbf{r}(\phi,\theta) = \sin\phi\cos\theta\,\mathbf{i} + \sin\phi\sin\theta\,\mathbf{j} + \cos\phi\,\mathbf{k}$, $0 \le \theta \le \pi$, $0 \le \phi \le \pi$. Then $\mathbf{r}_\phi \times \mathbf{r}_\theta = \sin^2\phi\cos\theta\,\mathbf{i} + \sin^2\phi\sin\theta\,\mathbf{j} + \sin\phi\cos\phi\,\mathbf{k}$ and

$$\iint_S \text{curl}\,\mathbf{F}\cdot d\mathbf{S} = \iint_{x^2+z^2\le 1}\text{curl}\,\mathbf{F}\cdot(\mathbf{r}_\phi\times\mathbf{r}_\theta)\,dA = \int_0^\pi\int_0^\pi(-\sin^2\phi\cos\theta - \sin^2\phi\sin\theta - \sin\phi\cos\phi)\,d\theta\,d\phi$$

$$= \int_0^\pi(-2\sin^2\phi - \pi\sin\phi\cos\phi)\,d\phi = \left[\tfrac{1}{2}\sin 2\phi - \phi - \tfrac{\pi}{2}\sin^2\phi\right]_0^\pi = -\pi$$

15. It is easier to use Stokes' Theorem than to compute the work directly. Let S be the planar region enclosed by the path of the particle, so S is the portion of the plane $z = \frac{1}{2}y$ for $0 \le x \le 1$, $0 \le y \le 2$, with upward orientation. curl $\mathbf{F} = 8y\,\mathbf{i} + 2z\,\mathbf{j} + 2y\,\mathbf{k}$ and

$$\oint_C \mathbf{F}\cdot d\mathbf{r} = \iint_S \text{curl}\,\mathbf{F}\cdot d\mathbf{S} = \iint_D \left[-8y\,(0) - 2z\left(\tfrac{1}{2}\right) + 2y\right]\,dA = \int_0^1\int_0^2\left(2y - \tfrac{1}{2}y\right)\,dy\,dx$$

$$= \int_0^1\int_0^2 \tfrac{3}{2}y\,dy\,dx = \int_0^1\left[\tfrac{3}{4}y^2\right]_{y=0}^{y=2}dx = \int_0^1 3\,dx = 3$$

17. Assume S is centered at the origin with radius a and let H_1 and H_2 be the upper and lower hemispheres, respectively, of S. Then $\iint_S\text{curl}\,\mathbf{F}\cdot d\mathbf{S} = \iint_{H_1}\text{curl}\,\mathbf{F}\cdot d\mathbf{S} + \iint_{H_2}\text{curl}\,\mathbf{F}\cdot d\mathbf{S} = \oint_{C_1}\mathbf{F}\cdot d\mathbf{r} + \oint_{C_2}\mathbf{F}\cdot d\mathbf{r}$ by Stokes' Theorem. But C_1 is the circle $x^2 + y^2 = a^2$ oriented in the counterclockwise direction while C_2 is the same circle oriented in the clockwise direction. Hence $\oint_{C_2}\mathbf{F}\cdot d\mathbf{r} = -\oint_{C_1}\mathbf{F}\cdot d\mathbf{r}$ so $\iint_S\text{curl}\,\mathbf{F}\cdot d\mathbf{S} = 0$ as desired.

13.9 The Divergence Theorem

1. div $\mathbf{F} = 3 + x + 2x = 3 + 3x$, so

$\iiint_E \text{div}\,\mathbf{F}\,dV = \int_0^1\int_0^1\int_0^1(3x + 3)\,dx\,dy\,dz = \frac{9}{2}$ (notice the triple integral is three times the volume of the cube plus three times $\overline{x}$).

To compute $\iint_S \mathbf{F}\cdot d\mathbf{S}$, on

S_1: $\mathbf{n} = \mathbf{i}$, $\mathbf{F} = 3\,\mathbf{i} + y\,\mathbf{j} + 2z\,\mathbf{k}$, and $\iint_{S_1}\mathbf{F}\cdot d\mathbf{S} = \iint_{S_1} 3\,dS = 3$;

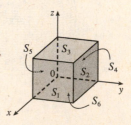

S_2: $\mathbf{F} = 3x\,\mathbf{i} + x\,\mathbf{j} + 2xz\,\mathbf{k}$, $\mathbf{n} = \mathbf{j}$ and $\iint_{S_2} \mathbf{F} \cdot d\mathbf{S} = \iint_{S_2} x\,dS = \frac{1}{2}$;

S_3: $\mathbf{F} = 3x\,\mathbf{i} + xy\,\mathbf{j} + 2x\,\mathbf{k}$, $\mathbf{n} = \mathbf{k}$ and $\iint_{S_3} \mathbf{F} \cdot d\mathbf{S} = \iint_{S_3} 2x\,dS = 1$;

S_4: $\mathbf{F} = \mathbf{0}$, $\iint_{S_4} \mathbf{F} \cdot d\mathbf{S} = 0$; S_5: $\mathbf{F} = 3x\,\mathbf{i} + 2x\,\mathbf{k}$, $\mathbf{n} = -\mathbf{j}$ and $\iint_{S_5} \mathbf{F} \cdot d\mathbf{S} = \iint_{S_5} 0\,dS = 0$;

S_6: $\mathbf{F} = 3x\,\mathbf{i} + xy\,\mathbf{j}$, $\mathbf{n} = -\mathbf{k}$ and $\iint_{S_6} \mathbf{F} \cdot d\mathbf{S} = \iint_{S_6} 0\,dS = 0$. Thus $\iint_S \mathbf{F} \cdot d\mathbf{S} = \frac{9}{2}$.

3. $\operatorname{div} \mathbf{F} = 0 + 1 + 0 = 1$, so $\iiint_E \operatorname{div}\mathbf{F}\,dV = \iiint_E 1\,dV = V(E) = \frac{4}{3}\pi \cdot 4^3 = \frac{256}{3}\pi$. S is a sphere of radius 4 centered at

the origin which can be parametrized by $\mathbf{r}(\phi, \theta) = \langle 4\sin\phi\cos\theta, 4\sin\phi\sin\theta, 4\cos\phi \rangle$, $0 \le \phi \le \pi$, $0 \le \theta \le 2\pi$ (similar to

Example 13.6.9). Then

$$\mathbf{r}_\phi \times \mathbf{r}_\theta = \langle 4\cos\phi\cos\theta, 4\cos\phi\sin\theta, -4\sin\phi \rangle \times \langle -4\sin\phi\sin\theta, 4\sin\phi\cos\theta, 0 \rangle$$

$$= \langle 16\sin^2\phi\cos\theta, 16\sin^2\phi\sin\theta, 16\cos\phi\sin\phi \rangle$$

and $\mathbf{F}(\mathbf{r}(\phi,\theta)) = \langle 4\cos\phi, 4\sin\phi\sin\theta, 4\sin\phi\cos\theta \rangle$. Thus

$\mathbf{F} \cdot (\mathbf{r}_\phi \times \mathbf{r}_\theta) = 64\cos\phi\sin^2\phi\cos\theta + 64\sin^3\phi\sin^2\theta + 64\cos\phi\sin^2\phi\cos\theta = 128\cos\phi\sin^2\phi\cos\theta + 64\sin^3\phi\sin^2\theta$

and

$$\iint_S \mathbf{F} \cdot d\mathbf{S} = \iint_D \mathbf{F} \cdot (\mathbf{r}_\phi \times \mathbf{r}_\theta)\,dA = \int_0^{2\pi}\int_0^\pi (128\cos\phi\sin^2\phi\cos\theta + 64\sin^3\phi\sin^2\theta)\,d\phi\,d\theta$$

$$= \int_0^{2\pi} \left[\frac{128}{3}\sin^3\phi\cos\theta + 64\left(-\frac{1}{3}(2+\sin^2\phi)\cos\phi\right)\sin^2\theta\right]_{\phi=0}^{\phi=\pi}\,d\theta$$

$$= \int_0^{2\pi} \frac{256}{3}\sin^2\theta\,d\theta = \frac{256}{3}\left[\frac{1}{2}\theta - \frac{1}{4}\sin 2\theta\right]_0^{2\pi} = \frac{256}{3}\pi$$

5. $\operatorname{div}\mathbf{F} = \frac{\partial}{\partial x}(xye^z) + \frac{\partial}{\partial y}(xy^2z^3) + \frac{\partial}{\partial z}(-ye^z) = ye^z + 2xyz^3 - ye^z = 2xyz^3$, so by the Divergence Theorem,

$$\iint_S \mathbf{F} \cdot d\mathbf{S} = \iiint_E \operatorname{div}\mathbf{F}\,dV = \int_0^3\int_0^2\int_0^1 2xyz^3\,dz\,dy\,dx = 2\int_0^3 x\,dx \int_0^2 y\,dy \int_0^1 z^3\,dz$$

$$= 2\left[\frac{1}{2}x^2\right]_0^3 \left[\frac{1}{2}y^2\right]_0^2 \left[\frac{1}{4}z^4\right]_0^1 = 2\left(\frac{9}{2}\right)(2)\left(\frac{1}{4}\right) = \frac{9}{2}$$

7. $\operatorname{div}\mathbf{F} = 3y^2 + 0 + 3z^2$, so using cylindrical coordinates with $y = r\cos\theta$, $z = r\sin\theta$, $x = x$ we have

$$\iint_S \mathbf{F} \cdot d\mathbf{S} = \iiint_E (3y^2 + 3z^2)\,dV = \int_0^{2\pi}\int_0^1\int_{-1}^2 (3r^2\cos^2\theta + 3r^2\sin^2\theta)\,r\,dx\,dr\,d\theta$$

$$= 3\int_0^{2\pi} d\theta \int_0^1 r^3\,dr \int_{-1}^2 dx = 3(2\pi)\left(\frac{1}{4}\right)(3) = \frac{9\pi}{2}$$

9. $\operatorname{div}\mathbf{F} = 2x\sin y - x\sin y - x\sin y = 0$, so by the Divergence Theorem, $\iint_S \mathbf{F} \cdot d\mathbf{S} = \iiint_E 0\,dV = 0$.

11. $\operatorname{div}\mathbf{F} = y^2 + 0 + x^2 = x^2 + y^2$ so

$$\iint_S \mathbf{F} \cdot d\mathbf{S} = \iiint_E (x^2 + y^2)\,dV = \int_0^{2\pi}\int_0^2\int_{r^2}^4 r^2 \cdot r\,dz\,dr\,d\theta = \int_0^{2\pi}\int_0^2 r^3(4 - r^2)\,dr\,d\theta$$

$$= \int_0^{2\pi} d\theta \int_0^2 (4r^3 - r^5)\,dr = 2\pi\left[r^4 - \frac{1}{6}r^6\right]_0^2 = \frac{32}{3}\pi$$

13. $\mathbf{F}(x, y, z) = x\sqrt{x^2 + y^2 + z^2}\,\mathbf{i} + y\sqrt{x^2 + y^2 + z^2}\,\mathbf{j} + z\sqrt{x^2 + y^2 + z^2}\,\mathbf{k}$, so

$\operatorname{div}\mathbf{F} = x \cdot \frac{1}{2}(x^2 + y^2 + z^2)^{-1/2}(2x) + (x^2 + y^2 + z^2)^{1/2} + y \cdot \frac{1}{2}(x^2 + y^2 + z^2)^{-1/2}(2y) + (x^2 + y^2 + z^2)^{1/2}$

$$+ z \cdot \frac{1}{2}(x^2 + y^2 + z^2)^{-1/2}(2z) + (x^2 + y^2 + z^2)^{1/2}$$

$$= (x^2 + y^2 + z^2)^{-1/2}\left[x^2 + (x^2 + y^2 + z^2) + y^2 + (x^2 + y^2 + z^2) + z^2 + (x^2 + y^2 + z^2)\right]$$

$$= \frac{4(x^2 + y^2 + z^2)}{\sqrt{x^2 + y^2 + z^2}} = 4\sqrt{x^2 + y^2 + z^2}.$$

[continued]

Then
$$\iint_S \mathbf{F} \cdot d\mathbf{S} = \iiint_E 4\sqrt{x^2 + y^2 + z^2}\, dV = \int_0^{\pi/2} \int_0^{2\pi} \int_0^1 4\sqrt{\rho^2} \cdot \rho^2 \sin\phi\, d\rho\, d\theta\, d\phi$$

$$= \int_0^{\pi/2} \sin\phi\, d\phi \int_0^{2\pi} d\theta \int_0^1 4\rho^3\, d\rho = \left[-\cos\phi\right]_0^{\pi/2} \left[\theta\right]_0^{2\pi} \left[\rho^4\right]_0^1 = (1)(2\pi)(1) = 2\pi$$

15. $\iint_S \mathbf{F} \cdot d\mathbf{S} = \iiint_E \sqrt{3 - x^2}\, dV = \int_{-1}^1 \int_{-1}^1 \int_0^{2 - x^4 - y^4} \sqrt{3 - x^2}\, dz\, dy\, dx = \frac{341}{60}\sqrt{2} + \frac{81}{20}\sin^{-1}\left(\frac{\sqrt{3}}{3}\right)$

17. For S_1 we have $\mathbf{n} = -\mathbf{k}$, so $\mathbf{F} \cdot \mathbf{n} = \mathbf{F} \cdot (-\mathbf{k}) = -x^2 z - y^2 = -y^2$ (since $z = 0$ on S_1). So if D is the unit disk, we get

$\iint_{S_1} \mathbf{F} \cdot d\mathbf{S} = \iint_{S_1} \mathbf{F} \cdot \mathbf{n}\, dS = \iint_D (-y^2)\, dA = -\int_0^{2\pi} \int_0^1 r^2 (\sin^2\theta)\, r\, dr\, d\theta = -\frac{1}{4}\pi$. Now since S_2 is closed, we can use

the Divergence Theorem. Since $\operatorname{div}\mathbf{F} = \frac{\partial}{\partial x}(z^2 x) + \frac{\partial}{\partial y}\left(\frac{1}{3}y^3 + \tan z\right) + \frac{\partial}{\partial z}(x^2 z + y^2) = z^2 + y^2 + x^2$, we use spherical

coordinates to get $\iint_{S_2} \mathbf{F} \cdot d\mathbf{S} = \iiint_E \operatorname{div}\mathbf{F}\, dV = \int_0^{2\pi} \int_0^{\pi/2} \int_0^1 \rho^2 \cdot \rho^2 \sin\phi\, d\rho\, d\phi\, d\theta = \frac{2}{5}\pi$. Finally

$\iint_S \mathbf{F} \cdot d\mathbf{S} = \iint_{S_2} \mathbf{F} \cdot d\mathbf{S} - \iint_{S_1} \mathbf{F} \cdot d\mathbf{S} = \frac{2}{5}\pi - \left(-\frac{1}{4}\pi\right) = \frac{13}{20}\pi$.

19. The vectors that end near P_1 are longer than the vectors that start near P_1, so the net flow is inward near P_1 and $\operatorname{div}\mathbf{F}(P_1)$ is

negative. The vectors that end near P_2 are shorter than the vectors that start near P_2, so the net flow is outward near P_2 and

$\operatorname{div}\mathbf{F}(P_2)$ is positive.

21.

From the graph it appears that for points above the x-axis, vectors starting near a

particular point are longer than vectors ending there, so divergence is positive.

The opposite is true at points below the x-axis, where divergence is negative.

$\mathbf{F}(x, y) = \langle xy, x + y^2 \rangle \quad \Rightarrow \quad \operatorname{div}\mathbf{F} = \frac{\partial}{\partial x}(xy) + \frac{\partial}{\partial y}(x + y^2) = y + 2y = 3y$.

Thus $\operatorname{div}\mathbf{F} > 0$ for $y > 0$, and $\operatorname{div}\mathbf{F} < 0$ for $y < 0$.

23. Since $\dfrac{\mathbf{x}}{|\mathbf{x}|^3} = \dfrac{x\,\mathbf{i} + y\,\mathbf{j} + z\,\mathbf{k}}{(x^2 + y^2 + z^2)^{3/2}}$ and $\dfrac{\partial}{\partial x}\left(\dfrac{x}{(x^2 + y^2 + z^2)^{3/2}}\right) = \dfrac{(x^2 + y^2 + z^2) - 3x^2}{(x^2 + y^2 + z^2)^{5/2}}$ with similar expressions

for $\dfrac{\partial}{\partial y}\left(\dfrac{y}{(x^2 + y^2 + z^2)^{3/2}}\right)$ and $\dfrac{\partial}{\partial z}\left(\dfrac{z}{(x^2 + y^2 + z^2)^{3/2}}\right)$, we have

$\operatorname{div}\left(\dfrac{\mathbf{x}}{|\mathbf{x}|^3}\right) = \dfrac{3(x^2 + y^2 + z^2) - 3(x^2 + y^2 + z^2)}{(x^2 + y^2 + z^2)^{5/2}} = 0$, except at $(0, 0, 0)$ where it is undefined.

25. $\iint_S \mathbf{a} \cdot \mathbf{n}\, dS = \iiint_E \operatorname{div}\mathbf{a}\, dV = 0$ since $\operatorname{div}\mathbf{a} = 0$.

27. $\iint_S \operatorname{curl}\mathbf{F} \cdot d\mathbf{S} = \iiint_E \operatorname{div}(\operatorname{curl}\mathbf{F})\, dV = 0$ by Theorem 13.5.11.

29. $\iint_S (f\nabla g) \cdot \mathbf{n}\, dS = \iiint_E \operatorname{div}(f\nabla g)\, dV = \iiint_E (f\nabla^2 g + \nabla g \cdot \nabla f)\, dV$ by Exercise 13.5.23.

13 Review

CONCEPT CHECK

1. See Definitions 1 and 2 in Section 13.1. A vector field can represent, for example, the wind velocity at any location in space, the speed and direction of the ocean current at any location, or the force vectors of Earth's gravitational field at a location in space.

2. (a) A conservative vector field $\mathbf{F}$ is a vector field which is the gradient of some scalar function f.

   (b) The function f in part (a) is called a potential function for $\mathbf{F}$, that is, $\mathbf{F} = \nabla f$.

3. (a) See Definition 13.2.2.

   (b) We normally evaluate the line integral using Formula 13.2.3.

   (c) The mass is $m = \int_C \rho(x,y)\, ds$, and the center of mass is $(\overline{x}, \overline{y})$ where $\overline{x} = \frac{1}{m} \int_C x\rho(x,y)\, ds, \overline{y} = \frac{1}{m} \int_C y\rho(x,y)\, ds$.

   (d) See (5) and (6) in Section 13.2 for plane curves; we have similar definitions when C is a space curve [see the equation preceding (10) in Section 13.2].

   (e) For plane curves, see Equations 13.2.7. We have similar results for space curves [see the equation preceding (10) in Section 13.2].

4. (a) See Definition 13.2.13.

   (b) If $\mathbf{F}$ is a force field, $\int_C \mathbf{F} \cdot d\mathbf{r}$ represents the work done by $\mathbf{F}$ in moving a particle along the curve C.

   (c) $\int_C \mathbf{F} \cdot d\mathbf{r} = \int_C P\, dx + Q\, dy + R\, dz$

5. See Theorem 13.3.2.

6. (a) $\int_C \mathbf{F} \cdot d\mathbf{r}$ is independent of path if the line integral has the same value for any two curves that have the same initial and terminal points.

   (b) See Theorem 13.3.4.

7. See the statement of Green's Theorem on page 788.

8. See Equations 13.4.5.

9. (a) $\operatorname{curl} \mathbf{F} = \left(\dfrac{\partial R}{\partial y} - \dfrac{\partial Q}{\partial z} \right) \mathbf{i} + \left(\dfrac{\partial P}{\partial z} - \dfrac{\partial R}{\partial x} \right) \mathbf{j} + \left(\dfrac{\partial Q}{\partial x} - \dfrac{\partial P}{\partial y} \right) \mathbf{k} = \nabla \times \mathbf{F}$

   (b) $\operatorname{div} \mathbf{F} = \dfrac{\partial P}{\partial x} + \dfrac{\partial Q}{\partial y} + \dfrac{\partial R}{\partial z} = \nabla \cdot \mathbf{F}$

   (c) For curl $\mathbf{F}$, see Figure 6 and the accompanying discussion on page 828. For div $\mathbf{F}$, see Equation 8 and the following discussion on page 834.

10. See Theorem 13.3.6; see Theorem 13.5.4.

11. (a) See (1) and (2) and the accompanying discussion in Section 13.6; See Figure 4 and the accompanying discussion on page 804.

(b) See Definition 13.6.6.

(c) See Equation 13.6.9.

12. (a) See (1) in Section 13.7.

(b) We normally evaluate the surface integral using Formula 13.7.2.

(c) See Formula 13.7.4.

(d) The mass is $m = \iint_S \rho(x, y, z) \, dS$ and the center of mass is $(\overline{x}, \overline{y}, \overline{z})$ where $\overline{x} = \frac{1}{m} \iint_S x \rho(x, y, z) \, dS$, $\overline{y} = \frac{1}{m} \iint_S y \rho(x, y, z) \, dS, \overline{z} = \frac{1}{m} \iint_S z \rho(x, y, z) \, dS$.

13. (a) See Figures 6 and 7 and the accompanying discussion in Section 13.7. A Möbius strip is a nonorientable surface; see Figures 4 and 5 and the accompanying discussion on page 817.

(b) See Definition 13.7.8.

(c) See Formula 13.7.9.

(d) See Formula 13.7.10.

14. See the statement of Stokes' Theorem on page 824.

15. See the statement of the Divergence Theorem on page 830.

16. In each theorem, we have an integral of a "derivative" over a region on the left side, while the right side involves the values of the original function only on the boundary of the region.

TRUE-FALSE QUIZ

1. False; div $\mathbf{F}$ is a scalar field.

3. True, by Theorem 13.5.3 and the fact that div $\mathbf{0} = 0$.

5. False. See Exercise 13.3.31. (But the assertion is true if D is simply-connected; see Theorem 13.3.6.)

7. False. For example, $\text{div}(y \, \mathbf{i}) = 0 = \text{div}(x \, \mathbf{j})$ but $y \, \mathbf{i} \neq x \, \mathbf{j}$.

9. True. See Exercise 13.5.22.

11. True. Apply the Divergence Theorem and use the fact that div $\mathbf{F} = 0$.

EXERCISES

1. (a) Vectors starting on C point in roughly the direction opposite to C, so the tangential component $\mathbf{F} \cdot \mathbf{T}$ is negative.

Thus $\int_C \mathbf{F} \cdot d\mathbf{r} = \int_C \mathbf{F} \cdot \mathbf{T}\, ds$ is negative.

(b) The vectors that end near P are shorter than the vectors that start near P, so the net flow is outward near P and div $\mathbf{F}(P)$ is positive.

3. $\int_C yz \cos x\, ds = \int_0^\pi (3\cos t)(3\sin t)\cos t\, \sqrt{(1)^2 + (-3\sin t)^2 + (3\cos t)^2}\, dt = \int_0^\pi (9\cos^2 t \sin t)\sqrt{10}\, dt$

$$= 9\sqrt{10}\left(-\tfrac{1}{3}\cos^3 t\right)\Big]_0^\pi = -3\sqrt{10}\,(-2) = 6\sqrt{10}$$

5. $\int_C y^3\, dx + x^2\, dy = \int_{-1}^1 \left[y^3(-2y) + (1-y^2)^2\right] dy = \int_{-1}^1 (-y^4 - 2y^2 + 1)\, dy$

$$= \left[-\tfrac{1}{5}y^5 - \tfrac{2}{3}y^3 + y\right]_{-1}^1 = -\tfrac{1}{5} - \tfrac{2}{3} + 1 - \tfrac{1}{5} - \tfrac{2}{3} + 1 = \tfrac{4}{15}$$

7. $C: x = 1 + 2t \;\Rightarrow\; dx = 2\, dt,\; y = 4t \;\Rightarrow\; dy = 4\, dt,\; z = -1 + 3t \;\Rightarrow\; dz = 3\, dt,\; 0 \le t \le 1.$

$$\int_C xy\, dx + y^2\, dy + yz\, dz = \int_0^1 [(1+2t)(4t)(2) + (4t)^2(4) + (4t)(-1+3t)(3)]\, dt$$

$$= \int_0^1 (116t^2 - 4t)\, dt = \left[\tfrac{116}{3}t^3 - 2t^2\right]_0^1 = \tfrac{116}{3} - 2 = \tfrac{110}{3}$$

9. $\mathbf{F}(\mathbf{r}(t)) = e^{-t}\,\mathbf{i} + t^2(-t)\,\mathbf{j} + (t^2 + t^3)\,\mathbf{k},\; \mathbf{r}'(t) = 2t\,\mathbf{i} + 3t^2\,\mathbf{j} - \mathbf{k}$ and

$\int_C \mathbf{F} \cdot d\mathbf{r} = \int_0^1 (2te^{-t} - 3t^5 - (t^2 + t^3))\, dt = \left[-2te^{-t} - 2e^{-t} - \tfrac{1}{2}t^6 - \tfrac{1}{3}t^3 - \tfrac{1}{4}t^4\right]_0^1 = \tfrac{11}{12} - \tfrac{4}{e}.$

11. $\frac{\partial}{\partial y}[(1+xy)e^{xy}] = 2xe^{xy} + x^2 ye^{xy} = \frac{\partial}{\partial x}\left[e^y + x^2 e^{xy}\right]$ and the domain of $\mathbf{F}$ is $\mathbb{R}^2$, so $\mathbf{F}$ is conservative. Thus there

exists a function f such that $\mathbf{F} = \nabla f$. Then $f_y(x,y) = e^y + x^2 e^{xy}$ implies $f(x,y) = e^y + xe^{xy} + g(x)$ and then

$f_x(x,y) = xye^{xy} + e^{xy} + g'(x) = (1+xy)e^{xy} + g'(x)$. But $f_x(x,y) = (1+xy)e^{xy}$, so $g'(x) = 0 \;\Rightarrow\; g(x) = K.$

Thus $f(x,y) = e^y + xe^{xy} + K$ is a potential function for $\mathbf{F}$.

13. Since $\frac{\partial}{\partial y}\left(4x^3 y^2 - 2xy^3\right) = 8x^3 y - 6xy^2 = \frac{\partial}{\partial x}\left(2x^4 y - 3x^2 y^2 + 4y^3\right)$ and the domain of $\mathbf{F}$ is $\mathbb{R}^2$, $\mathbf{F}$ is conservative.

Furthermore $f(x,y) = x^4 y^2 - x^2 y^3 + y^4$ is a potential function for $\mathbf{F}$. $t = 0$ corresponds to the point $(0,1)$ and $t = 1$

corresponds to $(1,1)$, so $\int_C \mathbf{F} \cdot d\mathbf{r} = f(1,1) - f(0,1) = 1 - 1 = 0.$

15. $C_1: \mathbf{r}(t) = t\,\mathbf{i} + t^2\,\mathbf{j},\; -1 \le t \le 1;$

$C_2: \mathbf{r}(t) = -t\,\mathbf{i} + \mathbf{j},\; -1 \le t \le 1.$

Then

$$\int_C xy^2\, dx - x^2 y\, dy = \int_{-1}^1 (t^5 - 2t^5)\, dt + \int_{-1}^1 t\, dt$$

$$= \left[-\tfrac{1}{6}t^6\right]_{-1}^1 + \left[\tfrac{1}{2}t^2\right]_{-1}^1 = 0$$

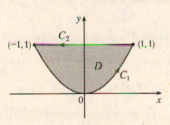

Using Green's Theorem, we have

$$\int_C xy^2\, dx - x^2 y\, dy = \iint_D \left[\frac{\partial}{\partial x}(-x^2 y) - \frac{\partial}{\partial y}(xy^2)\right] dA = \iint_D (-2xy - 2xy)\, dA = \int_{-1}^1 \int_{x^2}^1 -4xy\, dy\, dx$$

$$= \int_{-1}^1 \left[-2xy^2\right]_{y=x^2}^{y=1} dx = \int_{-1}^1 (2x^5 - 2x)\, dx = \left[\tfrac{1}{3}x^6 - x^2\right]_{-1}^1 = 0$$

17. $\int_C x^2 y \, dx - xy^2 \, dy = \iint\limits_{x^2+y^2 \le 4} \left[\frac{\partial}{\partial x} \left(-xy^2 \right) - \frac{\partial}{\partial y} \left(x^2 y \right) \right] dA = \iint\limits_{x^2+y^2 \le 4} \left(-y^2 - x^2 \right) dA = -\int_0^{2\pi} \int_0^2 r^3 \, dr \, d\theta = -8\pi$

19. If we assume there is such a vector field $\mathbf{G}$, then $\text{div}(\text{curl } \mathbf{G}) = 2 + 3z - 2xz$. But $\text{div}(\text{curl } \mathbf{F}) = 0$ for all vector fields $\mathbf{F}$. Thus such a $\mathbf{G}$ cannot exist.

21. For any piecewise-smooth simple closed plane curve C bounding a region D, we can apply Green's Theorem to

$\mathbf{F}(x, y) = f(x)\,\mathbf{i} + g(y)\,\mathbf{j}$ to get $\int_C f(x)\,dx + g(y)\,dy = \iint_D \left[\frac{\partial}{\partial x} g(y) - \frac{\partial}{\partial y} f(x) \right] dA = \iint_D 0 \, dA = 0.$

23. $\nabla^2 f = 0$ means that $\dfrac{\partial^2 f}{\partial x^2} + \dfrac{\partial^2 f}{\partial y^2} = 0$. Now if $\mathbf{F} = f_y\,\mathbf{i} - f_x\,\mathbf{j}$ and C is any closed path in D, then applying Green's

Theorem, we get

$$\int_C \mathbf{F} \cdot d\mathbf{r} = \int_C f_y \, dx - f_x \, dy = \iint_D \left[\frac{\partial}{\partial x} \left(-f_x \right) - \frac{\partial}{\partial y} \left(f_y \right) \right] dA$$

$$= -\iint_D (f_{xx} + f_{yy}) \, dA = -\iint_D 0 \, dA = 0$$

Therefore the line integral is independent of path, by Theorem 13.3.3.

25. $z = f(x, y) = x^2 + 2y$ with $0 \le x \le 1$, $0 \le y \le 2x$. Thus

$$A(S) = \iint_D \sqrt{1 + 4x^2 + 4} \, dA = \int_0^1 \int_0^{2x} \sqrt{5 + 4x^2} \, dy \, dx = \int_0^1 2x \sqrt{5 + 4x^2} \, dx = \tfrac{1}{6} (5 + 4x^2)^{3/2} \Big]_0^1 = \tfrac{1}{6} \left(27 - 5\sqrt{5} \right).$$

27. $z = f(x, y) = x^2 + y^2$ with $0 \le x^2 + y^2 \le 4$ so $\mathbf{r}_x \times \mathbf{r}_y = -2x\,\mathbf{i} - 2y\,\mathbf{j} + \mathbf{k}$ (using upward orientation). Then

$$\iint_S z \, dS = \iint\limits_{x^2+y^2 \le 4} (x^2 + y^2)\sqrt{4x^2 + 4y^2 + 1} \, dA$$

$$= \int_0^{2\pi} \int_0^2 r^3 \sqrt{1 + 4r^2} \, dr \, d\theta = \tfrac{1}{60}\pi \left(391\sqrt{17} + 1 \right)$$

(Substitute $u = 1 + 4r^2$ and use tables.)

29. Since the sphere bounds a simple solid region, the Divergence Theorem applies and

$$\iint_S \mathbf{F} \cdot d\mathbf{S} = \iiint_E \text{div }\mathbf{F} \, dV = \iiint_E (z - 2) \, dV = \iiint_E z \, dV - 2\iiint_E dV$$

$$= 0 \begin{bmatrix} \text{odd function in } z \\ \text{and } E \text{ is symmetric} \end{bmatrix} - 2 \cdot V(E) = -2 \cdot \tfrac{4}{3}\pi (2)^3 = -\tfrac{64}{3}\pi$$

Alternate solution: $\mathbf{F}(\mathbf{r}(\phi, \theta)) = 4\sin\phi \cos\theta \cos\phi\,\mathbf{i} - 4\sin\phi \sin\theta\,\mathbf{j} + 6\sin\phi \cos\theta\,\mathbf{k}$,

$\mathbf{r}_\phi \times \mathbf{r}_\theta = 4\sin^2\phi \cos\theta\,\mathbf{i} + 4\sin^2\phi \sin\theta\,\mathbf{j} + 4\sin\phi \cos\phi\,\mathbf{k}$, and

$\mathbf{F} \cdot (\mathbf{r}_\phi \times \mathbf{r}_\theta) = 16\sin^3\phi \cos^2\theta \cos\phi - 16\sin^3\phi \sin^2\theta + 24\sin^2\phi \cos\phi \cos\theta$. Then

$$\iint_S \mathbf{F} \cdot d\mathbf{S} = \int_0^{2\pi} \int_0^\pi (16\sin^3\phi \cos\phi \cos^2\theta - 16\sin^3\phi \sin^2\theta + 24\sin^2\phi \cos\phi \cos\theta) \, d\phi \, d\theta$$

$$= \int_0^{2\pi} \tfrac{4}{3}(-16\sin^2\theta) \, d\theta = -\tfrac{64}{3}\pi$$

31. Since $\text{curl }\mathbf{F} = \mathbf{0}$, $\iint_S (\text{curl }\mathbf{F}) \cdot d\mathbf{S} = 0$. We parametrize C: $\mathbf{r}(t) = \cos t\,\mathbf{i} + \sin t\,\mathbf{j}$, $0 \le t \le 2\pi$ and

$\oint_C \mathbf{F} \cdot d\mathbf{r} = \int_0^{2\pi} (-\cos^2 t \sin t + \sin^2 t \cos t) \, dt = \tfrac{1}{3}\cos^3 t + \tfrac{1}{3}\sin^3 t \Big]_0^{2\pi} = 0.$

33. The surface is given by $x + y + z = 1$ or $z = 1 - x - y$, $0 \le x \le 1$, $0 \le y \le 1 - x$ and $\mathbf{r}_x \times \mathbf{r}_y = \mathbf{i} + \mathbf{j} + \mathbf{k}$. Then

$$\oint_C \mathbf{F} \cdot d\mathbf{r} = \iint_S \operatorname{curl} \mathbf{F} \cdot d\mathbf{S} = \iint_D (-y\,\mathbf{i} - z\,\mathbf{j} - x\,\mathbf{k}) \cdot (\mathbf{i} + \mathbf{j} + \mathbf{k})\, dA = \iint_D (-1)\, dA = -(\text{area of } D) = -\tfrac{1}{2}.$$

35. $\iiint_E \operatorname{div} \mathbf{F}\, dV = \iiint\limits_{x^2 + y^2 + z^2 \le 1} 3\, dV = 3(\text{volume of sphere}) = 4\pi$. Then

$$\mathbf{F}(\mathbf{r}(\phi, \theta)) \cdot (\mathbf{r}_\phi \times \mathbf{r}_\theta) = \sin^3 \phi \cos^2 \theta + \sin^3 \phi \sin^2 \theta + \sin \phi \cos^2 \phi = \sin \phi \quad \text{and}$$

$$\iint_S \mathbf{F} \cdot d\mathbf{S} = \int_0^{2\pi} \int_0^{\pi} \sin \phi\, d\phi\, d\theta = (2\pi)(2) = 4\pi.$$

37. By the Divergence Theorem, $\iint_S \mathbf{F} \cdot \mathbf{n}\, dS = \iiint_E \operatorname{div} \mathbf{F}\, dV = 3(\text{volume of } E) = 3(8 - 1) = 21$.

39. Let $\mathbf{F} = \mathbf{a} \times \mathbf{r} = \langle a_1, a_2, a_3 \rangle \times \langle x, y, z \rangle = \langle a_2 z - a_3 y, a_3 x - a_1 z, a_1 y - a_2 x \rangle$. Then $\operatorname{curl} \mathbf{F} = \langle 2a_1, 2a_2, 2a_3 \rangle = 2\mathbf{a}$,

and $\iint_S 2\mathbf{a} \cdot d\mathbf{S} = \iint_S \operatorname{curl} \mathbf{F} \cdot d\mathbf{S} = \int_C \mathbf{F} \cdot d\mathbf{r} = \int_C (\mathbf{a} \times \mathbf{r}) \cdot d\mathbf{r}$ by Stokes' Theorem.

□ APPENDIXES

A Trigonometry

1. $210° = 210\left(\frac{\pi}{180}\right) = \frac{7\pi}{6}$ rad

3. $9° = 9\left(\frac{\pi}{180}\right) = \frac{\pi}{20}$ rad

5. $900° = 900\left(\frac{\pi}{180}\right) = 5\pi$ rad

7. 4π rad $= 4\pi\left(\frac{180}{\pi}\right) = 720°$

9. $\frac{5\pi}{12}$ rad $= \frac{5\pi}{12}\left(\frac{180}{\pi}\right) = 75°$

11. $-\frac{3\pi}{8}$ rad $= -\frac{3\pi}{8}\left(\frac{180}{\pi}\right) = -67.5°$

13. Using Formula 3, $a = r\theta = 36 \cdot \frac{\pi}{12} = 3\pi$ cm.

15. Using Formula 3, $\theta = a/r = \frac{1}{1.5} = \frac{2}{3}$ rad $= \frac{2}{3}\left(\frac{180}{\pi}\right) = \left(\frac{120}{\pi}\right)° \approx 38.2°$.

17.

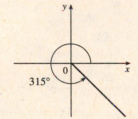

19.

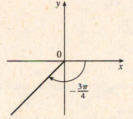

21.

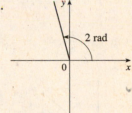

23.

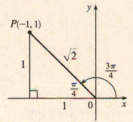

From the diagram we see that a point on the terminal side is $P(-1, 1)$. Therefore, taking $x = -1$, $y = 1$, $r = \sqrt{2}$ in the definitions of the trigonometric ratios, we have $\sin\frac{3\pi}{4} = \frac{1}{\sqrt{2}}$, $\cos\frac{3\pi}{4} = -\frac{1}{\sqrt{2}}$, $\tan\frac{3\pi}{4} = -1$, $\csc\frac{3\pi}{4} = \sqrt{2}$, $\sec\frac{3\pi}{4} = -\sqrt{2}$, and $\cot\frac{3\pi}{4} = -1$.

25.

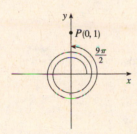

From the diagram we see that a point on the terminal line is $P(0, 1)$. Therefore taking $x = 0$, $y = 1$, $r = 1$ in the definitions of the trigonometric ratios, we have $\sin\frac{9\pi}{2} = 1$, $\cos\frac{9\pi}{2} = 0$, $\tan\frac{9\pi}{2} = y/x$ is undefined since $x = 0$, $\csc\frac{9\pi}{2} = 1$, $\sec\frac{9\pi}{2} = r/x$ is undefined since $x = 0$, and $\cot\frac{9\pi}{2} = 0$.

27.

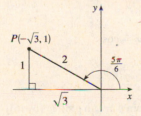

Using Figure 8 we see that a point on the terminal line is $P\left(-\sqrt{3}, 1\right)$. Therefore taking $x = -\sqrt{3}$, $y = 1$, $r = 2$ in the definitions of the trigonometric ratios, we have $\sin\frac{5\pi}{6} = \frac{1}{2}$, $\cos\frac{5\pi}{6} = -\frac{\sqrt{3}}{2}$, $\tan\frac{5\pi}{6} = -\frac{1}{\sqrt{3}}$, $\csc\frac{5\pi}{6} = 2$, $\sec\frac{5\pi}{6} = -\frac{2}{\sqrt{3}}$, and $\cot\frac{5\pi}{6} = -\sqrt{3}$.

29. $\sin\theta = y/r = \frac{3}{5}$ $\Rightarrow$ $y = 3, r = 5$, and $x = \sqrt{r^2 - y^2} = 4$ (since $0 < \theta < \frac{\pi}{2}$). Therefore taking $x = 4, y = 3, r = 5$ in the definitions of the trigonometric ratios, we have $\cos\theta = \frac{4}{5}$, $\tan\theta = \frac{3}{4}$, $\csc\theta = \frac{5}{3}$, $\sec\theta = \frac{5}{4}$, and $\cot\theta = \frac{4}{3}$.

31. $\frac{\pi}{2} < \phi < \pi$ $\Rightarrow$ ϕ is in the second quadrant, where x is negative and y is positive. Therefore $\sec\phi = r/x = -1.5 = -\frac{3}{2}$ $\Rightarrow$ $r = 3, x = -2$, and $y = \sqrt{r^2 - x^2} = \sqrt{5}$. Taking $x = -2, y = \sqrt{5}$, and $r = 3$ in the definitions of the trigonometric ratios, we have $\sin\phi = \frac{\sqrt{5}}{3}$, $\cos\phi = -\frac{2}{3}$, $\tan\phi = -\frac{\sqrt{5}}{2}$, $\csc\phi = \frac{3}{\sqrt{5}}$, and $\cot\theta = -\frac{2}{\sqrt{5}}$.

33. $\pi < \beta < 2\pi$ means that β is in the third or fourth quadrant where y is negative. Also since $\cot\beta = x/y = 3$ which is positive, x must also be negative. Therefore $\cot\beta = x/y = \frac{3}{1}$ $\Rightarrow$ $x = -3, y = -1$, and $r = \sqrt{x^2 + y^2} = \sqrt{10}$. Taking $x = -3, y = -1$ and $r = \sqrt{10}$ in the definitions of the trigonometric ratios, we have $\sin\beta = -\frac{1}{\sqrt{10}}$, $\cos\beta = -\frac{3}{\sqrt{10}}$, $\tan\beta = \frac{1}{3}$, $\csc\beta = -\sqrt{10}$, and $\sec\beta = -\frac{\sqrt{10}}{3}$.

35. $\sin 35° = \dfrac{x}{10}$ $\Rightarrow$ $x = 10\sin 35° \approx 5.73576$ cm

37. $\tan\dfrac{2\pi}{5} = \dfrac{x}{8}$ $\Rightarrow$ $x = 8\tan\dfrac{2\pi}{5} \approx 24.62147$ cm

39.

(a) From the diagram we see that $\sin\theta = \dfrac{y}{r} = \dfrac{a}{c}$, and $\sin(-\theta) = \dfrac{-a}{c} = -\dfrac{a}{c} = -\sin\theta$.

(b) Again from the diagram we see that $\cos\theta = \dfrac{x}{r} = \dfrac{b}{c} = \cos(-\theta)$.

41. (a) Using (12a) and (13a), we have

$$\tfrac{1}{2}\left[\sin(x+y) + \sin(x-y)\right] = \tfrac{1}{2}\left[\sin x\cos y + \cos x\sin y + \sin x\cos y - \cos x\sin y\right]$$
$$= \tfrac{1}{2}(2\sin x\cos y) = \sin x\cos y$$

(b) This time, using (12b) and (13b), we have

$$\tfrac{1}{2}\left[\cos(x+y) + \cos(x-y)\right] = \tfrac{1}{2}\left[\cos x\cos y - \sin x\sin y + \cos x\cos y + \sin x\sin y\right]$$
$$= \tfrac{1}{2}(2\cos x\cos y) = \cos x\cos y$$

(c) Again using (12b) and (13b), we have

$$\tfrac{1}{2}\left[\cos(x-y) - \cos(x+y)\right] = \tfrac{1}{2}\left[\cos x\cos y + \sin x\sin y - \cos x\cos y + \sin x\sin y\right]$$
$$= \tfrac{1}{2}(2\sin x\sin y) = \sin x\sin y$$

43. Using (12a), we have $\sin\left(\frac{\pi}{2} + x\right) = \sin\frac{\pi}{2}\cos x + \cos\frac{\pi}{2}\sin x = 1\cdot\cos x + 0\cdot\sin x = \cos x$.

45. Using (6), we have $\sin\theta\cot\theta = \sin\theta\cdot\dfrac{\cos\theta}{\sin\theta} = \cos\theta$.

47. $\sec y - \cos y = \dfrac{1}{\cos y} - \cos y$ [by (6)] $= \dfrac{1 - \cos^2 y}{\cos y} = \dfrac{\sin^2 y}{\cos y}$ [by (7)] $= \dfrac{\sin y}{\cos y}\sin y = \tan y\sin y$ [by (6)]

49. $\cot^2 \theta + \sec^2 \theta = \dfrac{\cos^2 \theta}{\sin^2 \theta} + \dfrac{1}{\cos^2 \theta}$ [by (6)] $= \dfrac{\cos^2 \theta \cos^2 \theta + \sin^2 \theta}{\sin^2 \theta \cos^2 \theta}$

$= \dfrac{\left(1 - \sin^2 \theta\right)\left(1 - \sin^2 \theta\right) + \sin^2 \theta}{\sin^2 \theta \cos^2 \theta}$ [by (7)] $= \dfrac{1 - \sin^2 \theta + \sin^4 \theta}{\sin^2 \theta \cos^2 \theta}$

$= \dfrac{\cos^2 \theta + \sin^4 \theta}{\sin^2 \theta \cos^2 \theta}$ [by (7)] $= \dfrac{1}{\sin^2 \theta} + \dfrac{\sin^2 \theta}{\cos^2 \theta} = \csc^2 \theta + \tan^2 \theta$ [by (6)]

51. Using (14a), we have $\tan 2\theta = \tan(\theta + \theta) = \dfrac{\tan \theta + \tan \theta}{1 - \tan \theta \tan \theta} = \dfrac{2 \tan \theta}{1 - \tan^2 \theta}$.

53. Using (15a) and (16a),

$\sin x \sin 2x + \cos x \cos 2x = \sin x \left(2 \sin x \cos x\right) + \cos x \left(2 \cos^2 x - 1\right) = 2 \sin^2 x \cos x + 2 \cos^3 x - \cos x$

$= 2 \left(1 - \cos^2 x\right) \cos x + 2 \cos^3 x - \cos x$ [by (7)]

$= 2 \cos x - 2 \cos^3 x + 2 \cos^3 x - \cos x = \cos x$

Or: $\sin x \sin 2x + \cos x \cos 2x = \cos(2x - x)$ [by 13(b)] $= \cos x$

55. $\dfrac{\sin \phi}{1 - \cos \phi} = \dfrac{\sin \phi}{1 - \cos \phi} \cdot \dfrac{1 + \cos \phi}{1 + \cos \phi} = \dfrac{\sin \phi \left(1 + \cos \phi\right)}{1 - \cos^2 \phi} = \dfrac{\sin \phi \left(1 + \cos \phi\right)}{\sin^2 \phi}$ [by (7)]

$= \dfrac{1 + \cos \phi}{\sin \phi} = \dfrac{1}{\sin \phi} + \dfrac{\cos \phi}{\sin \phi} = \csc \phi + \cot \phi$ [by (6)]

57. Using (12a),

$\sin 3\theta + \sin \theta = \sin(2\theta + \theta) + \sin \theta = \sin 2\theta \cos \theta + \cos 2\theta \sin \theta + \sin \theta$

$= \sin 2\theta \cos \theta + \left(2 \cos^2 \theta - 1\right) \sin \theta + \sin \theta$ [by (16a)]

$= \sin 2\theta \cos \theta + 2 \cos^2 \theta \sin \theta - \sin \theta + \sin \theta = \sin 2\theta \cos \theta + \sin 2\theta \cos \theta$ [by (15a)]

$= 2 \sin 2\theta \cos \theta$

59. Since $\sin x = \frac{1}{3}$ we can label the opposite side as having length 1,

the hypotenuse as having length 3, and use the Pythagorean Theorem

to get that the adjacent side has length $\sqrt{8}$. Then, from the diagram,

$\cos x = \frac{\sqrt{8}}{3}$. Similarly we have that $\sin y = \frac{3}{5}$. Now use (12a):

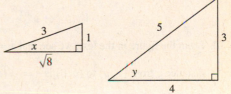

$\sin(x + y) = \sin x \cos y + \cos x \sin y = \frac{1}{3} \cdot \frac{4}{5} + \frac{\sqrt{8}}{3} \cdot \frac{3}{5} = \frac{4}{15} + \frac{3\sqrt{8}}{15} = \frac{4 + 6\sqrt{2}}{15}$.

61. Using (13b) and the values for $\cos x$ and $\sin y$ obtained in Exercise 59, we have

$\cos(x - y) = \cos x \cos y + \sin x \sin y = \frac{\sqrt{8}}{3} \cdot \frac{4}{5} + \frac{1}{3} \cdot \frac{3}{5} = \frac{8\sqrt{2} + 3}{15}$

63. Using (15a) and the values for $\sin y$ and $\cos y$ obtained in Exercise 59, we have

$\sin 2y = 2 \sin y \cos y = 2 \cdot \frac{3}{5} \cdot \frac{4}{5} = \frac{24}{25}$

65. $2 \cos x - 1 = 0 \iff \cos x = \frac{1}{2} \implies x = \frac{\pi}{3}, \frac{5\pi}{3}$ for $x \in [0, 2\pi]$.

67. $2 \sin^2 x = 1 \iff \sin^2 x = \frac{1}{2} \iff \sin x = \pm \frac{1}{\sqrt{2}} \implies x = \frac{\pi}{4}, \frac{3\pi}{4}, \frac{5\pi}{4}, \frac{7\pi}{4}$.

69. Using (15a), we have $\sin 2x = \cos x \iff 2 \sin x \cos x - \cos x = 0 \iff \cos x (2 \sin x - 1) = 0 \iff \cos x = 0$ or

$2 \sin x - 1 = 0 \implies x = \frac{\pi}{2}, \frac{3\pi}{2}$ or $\sin x = \frac{1}{2} \implies x = \frac{\pi}{6}$ or $\frac{5\pi}{6}$. Therefore, the solutions are $x = \frac{\pi}{6}, \frac{\pi}{2}, \frac{5\pi}{6}, \frac{3\pi}{2}$.

71. $\sin x = \tan x \iff \sin x - \tan x = 0 \iff \sin x - \dfrac{\sin x}{\cos x} = 0 \iff \sin x \left(1 - \dfrac{1}{\cos x}\right) = 0 \iff \sin x = 0$ or

$1 - \dfrac{1}{\cos x} = 0 \implies x = 0, \pi, 2\pi$ or $1 = \dfrac{1}{\cos x} \implies \cos x = 1 \implies x = 0, 2\pi$. Therefore the solutions are

$x = 0, \pi, 2\pi$.

73. We know that $\sin x = \frac{1}{2}$ when $x = \frac{\pi}{6}$ or $\frac{5\pi}{6}$, and from Figure 13(a), we see that $\sin x \leq \frac{1}{2} \Rightarrow 0 \leq x \leq \frac{\pi}{6}$ or

$\frac{5\pi}{6} \leq x \leq 2\pi$ for $x \in [0, 2\pi]$.

75. $\tan x = -1$ when $x = \frac{3\pi}{4}, \frac{7\pi}{4}$, and $\tan x = 1$ when $x = \frac{\pi}{4}$ or $\frac{5\pi}{4}$. From Figure 14 we see that $-1 < \tan x < 1 \Rightarrow$

$0 \leq x < \frac{\pi}{4}, \frac{3\pi}{4} < x < \frac{5\pi}{4}$, and $\frac{7\pi}{4} < x \leq 2\pi$.

77. $y = \cos\left(x - \frac{\pi}{3}\right)$. We start with the graph of $y = \cos x$
and shift it $\frac{\pi}{3}$ units to the right.

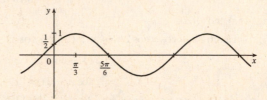

79. $y = \frac{1}{3}\tan\left(x - \frac{\pi}{2}\right)$. We start with the graph of
$y = \tan x$, shift it $\frac{\pi}{2}$ units to the right and compress it
to $\frac{1}{3}$ of its original vertical size.

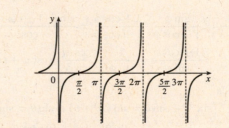

81. $y = |\sin x|$. We start with the graph of $y = \sin x$ and
reflect the parts below the x-axis about the x-axis.

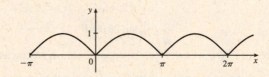

83. From the figure in the text, we see that $x = b\cos\theta$, $y = b\sin\theta$, and from the distance formula we have that the

distance c from (x, y) to $(a, 0)$ is $c = \sqrt{(x - a)^2 + (y - 0)^2} \Rightarrow$

$$c^2 = (b\cos\theta - a)^2 + (b\sin\theta)^2 = b^2\cos^2\theta - 2ab\cos\theta + a^2 + b^2\sin^2\theta$$
$$= a^2 + b^2(\cos^2\theta + \sin^2\theta) - 2ab\cos\theta = a^2 + b^2 - 2ab\cos\theta \quad [\text{by (7)}]$$

85. Using the Law of Cosines, we have $c^2 = 1^2 + 1^2 - 2(1)(1)\cos(\alpha - \beta) = 2[1 - \cos(\alpha - \beta)]$. Now, using the distance

formula, $c^2 = |AB|^2 = (\cos\alpha - \cos\beta)^2 + (\sin\alpha - \sin\beta)^2$. Equating these two expressions for c^2, we get

$2[1 - \cos(\alpha - \beta)] = \cos^2\alpha + \sin^2\alpha + \cos^2\beta + \sin^2\beta - 2\cos\alpha\cos\beta - 2\sin\alpha\sin\beta \Rightarrow$

$1 - \cos(\alpha - \beta) = 1 - \cos\alpha\cos\beta - \sin\alpha\sin\beta \Rightarrow \cos(\alpha - \beta) = \cos\alpha\cos\beta + \sin\alpha\sin\beta$.

87. In Exercise 86 we used the subtraction formula for cosine to prove the addition formula for cosine. Using that formula with

$x = \frac{\pi}{2} - \alpha$, $y = \beta$, we get $\cos\left[\left(\frac{\pi}{2} - \alpha\right) + \beta\right] = \cos\left(\frac{\pi}{2} - \alpha\right)\cos\beta - \sin\left(\frac{\pi}{2} - \alpha\right)\sin\beta \Rightarrow$

$\cos\left[\frac{\pi}{2} - (\alpha - \beta)\right] = \cos\left(\frac{\pi}{2} - \alpha\right)\cos\beta - \sin\left(\frac{\pi}{2} - \alpha\right)\sin\beta$. Now we use the identities given in the problem,

$\cos\left(\frac{\pi}{2} - \theta\right) = \sin\theta$ and $\sin\left(\frac{\pi}{2} - \theta\right) = \cos\theta$, to get $\sin(\alpha - \beta) = \sin\alpha\cos\beta - \cos\alpha\sin\beta$.

89. Using $A = \frac{1}{2}ab\sin\theta$, the area of the triangle is $\frac{1}{2}(10)(3)\sin 107° \approx 14.34457$ cm².

B   Sigma Notation

1. $\displaystyle\sum_{i=1}^{5}\sqrt{i}=\sqrt{1}+\sqrt{2}+\sqrt{3}+\sqrt{4}+\sqrt{5}$

3. $\displaystyle\sum_{i=4}^{6}3^{i}=3^{4}+3^{5}+3^{6}$

5. $\displaystyle\sum_{k=0}^{4}\frac{2k-1}{2k+1}=-1+\frac{1}{3}+\frac{3}{5}+\frac{5}{7}+\frac{7}{9}$

7. $\displaystyle\sum_{i=1}^{n}i^{10}=1^{10}+2^{10}+3^{10}+\cdots+n^{10}$

9. $\displaystyle\sum_{j=0}^{n-1}(-1)^{j}=1-1+1-1+\cdots+(-1)^{n-1}$

11. $1+2+3+4+\cdots+10=\displaystyle\sum_{i=1}^{10}i$

13. $\dfrac{1}{2}+\dfrac{2}{3}+\dfrac{3}{4}+\dfrac{4}{5}+\cdots+\dfrac{19}{20}=\displaystyle\sum_{i=1}^{19}\dfrac{i}{i+1}$

15. $2+4+6+8+\cdots+2n=\displaystyle\sum_{i=1}^{n}2i$

17. $1+2+4+8+16+32=\displaystyle\sum_{i=0}^{5}2^{i}$

19. $x+x^{2}+x^{3}+\cdots+x^{n}=\displaystyle\sum_{i=1}^{n}x^{i}$

21. $\displaystyle\sum_{i=4}^{8}(3i-2)=[3(4)-2]+[3(5)-2]+[3(6)-2]+[3(7)-2]+[3(8)-2]=10+13+16+19+22=80$

23. $\displaystyle\sum_{j=1}^{6}3^{j+1}=3^{2}+3^{3}+3^{4}+3^{5}+3^{6}+3^{7}=9+27+81+243+729+2187=3276$

(For a more general method, see Exercise 47.)

25. $\displaystyle\sum_{n=1}^{20}(-1)^{n}=-1+1-1+1-1+1-1+1-1+1-1+1-1+1-1+1-1+1-1+1=0$

27. $\displaystyle\sum_{i=0}^{4}\left(2^{i}+i^{2}\right)=(1+0)+(2+1)+(4+4)+(8+9)+(16+16)=61$

29. $\displaystyle\sum_{i=1}^{n}2i=2\sum_{i=1}^{n}i=2\cdot\frac{n(n+1)}{2}$ [by Theorem 3(c)] $=n(n+1)$

31. $\displaystyle\sum_{i=1}^{n}\left(i^{2}+3i+4\right)=\sum_{i=1}^{n}i^{2}+3\sum_{i=1}^{n}i+\sum_{i=1}^{n}4=\frac{n(n+1)(2n+1)}{6}+\frac{3n(n+1)}{2}+4n$

$$=\tfrac{1}{6}\left[(2n^{3}+3n^{2}+n)+(9n^{2}+9n)+24n\right]=\tfrac{1}{6}\left(2n^{3}+12n^{2}+34n\right)=\tfrac{1}{3}n\left(n^{2}+6n+17\right)$$

33. $\displaystyle\sum_{i=1}^{n}(i+1)(i+2)=\sum_{i=1}^{n}\left(i^{2}+3i+2\right)=\sum_{i=1}^{n}i^{2}+3\sum_{i=1}^{n}i+\sum_{i=1}^{n}2$

$$=\frac{n(n+1)(2n+1)}{6}+\frac{3n(n+1)}{2}+2n=\frac{n(n+1)}{6}\left[(2n+1)+9\right]+2n$$

$$=\frac{n(n+1)}{3}(n+5)+2n=\frac{n}{3}\left[(n+1)(n+5)+6\right]=\frac{n}{3}\left(n^{2}+6n+11\right)$$

35. $\displaystyle\sum_{i=1}^{n}\left(i^{3}-i-2\right)=\sum_{i=1}^{n}i^{3}-\sum_{i=1}^{n}i-\sum_{i=1}^{n}2=\left[\frac{n(n+1)}{2}\right]^{2}-\frac{n(n+1)}{2}-2n$

$$=\tfrac{1}{4}n(n+1)\left[n(n+1)-2\right]-2n=\tfrac{1}{4}n(n+1)(n+2)(n-1)-2n$$

$$=\tfrac{1}{4}n\left[(n+1)(n-1)(n+2)-8\right]=\tfrac{1}{4}n\left[(n^{2}-1)(n+2)-8\right]=\tfrac{1}{4}n\left(n^{3}+2n^{2}-n-10\right)$$

37. By Theorem 2(a) and Example 3, $\sum_{i=1}^{n} c = c \sum_{i=1}^{n} 1 = cn$.

39. $\sum_{i=1}^{n} \left[(i+1)^4 - i^4 \right] = \left(2^4 - 1^4 \right) + \left(3^4 - 2^4 \right) + \left(4^4 - 3^4 \right) + \cdots + \left[(n+1)^4 - n^4 \right]$

$$= (n+1)^4 - 1^4 = n^4 + 4n^3 + 6n^2 + 4n$$

On the other hand,

$$\sum_{i=1}^{n} \left[(i+1)^4 - i^4 \right] = \sum_{i=1}^{n} \left(4i^3 + 6i^2 + 4i + 1 \right) = 4 \sum_{i=1}^{n} i^3 + 6 \sum_{i=1}^{n} i^2 + 4 \sum_{i=1}^{n} i + \sum_{i=1}^{n} 1$$

$$= 4S + n(n+1)(2n+1) + 2n(n+1) + n \quad \left[\text{where } S = \sum_{i=1}^{n} i^3 \right]$$

$$= 4S + 2n^3 + 3n^2 + n + 2n^2 + 2n + n = 4S + 2n^3 + 5n^2 + 4n$$

Thus, $n^4 + 4n^3 + 6n^2 + 4n = 4S + 2n^3 + 5n^2 + 4n$, from which it follows that

$$4S = n^4 + 2n^3 + n^2 = n^2 \left(n^2 + 2n + 1 \right) = n^2 (n+1)^2 \text{ and } S = \left[\frac{n(n+1)}{2} \right]^2.$$

41. (a) $\sum_{i=1}^{n} \left[i^4 - (i-1)^4 \right] = \left(1^4 - 0^4 \right) + \left(2^4 - 1^4 \right) + \left(3^4 - 2^4 \right) + \cdots + \left[n^4 - (n-1)^4 \right] = n^4 - 0 = n^4$

(b) $\sum_{i=1}^{100} \left(5^i - 5^{i-1} \right) = \left(5^1 - 5^0 \right) + \left(5^2 - 5^1 \right) + \left(5^3 - 5^2 \right) + \cdots + \left(5^{100} - 5^{99} \right) = 5^{100} - 5^0 = 5^{100} - 1$

(c) $\sum_{i=3}^{99} \left(\frac{1}{i} - \frac{1}{i+1} \right) = \left(\frac{1}{3} - \frac{1}{4} \right) + \left(\frac{1}{4} - \frac{1}{5} \right) + \left(\frac{1}{5} - \frac{1}{6} \right) + \cdots + \left(\frac{1}{99} - \frac{1}{100} \right) = \frac{1}{3} - \frac{1}{100} = \frac{97}{300}$

(d) $\sum_{i=1}^{n} (a_i - a_{i-1}) = (a_1 - a_0) + (a_2 - a_1) + (a_3 - a_2) + \cdots + (a_n - a_{n-1}) = a_n - a_0$

43. $\lim_{n \to \infty} \sum_{i=1}^{n} \frac{1}{n} \left(\frac{i}{n} \right)^2 = \lim_{n \to \infty} \frac{1}{n^3} \sum_{i=1}^{n} i^2 = \lim_{n \to \infty} \frac{1}{n^3} \frac{n(n+1)(2n+1)}{6} = \lim_{n \to \infty} \frac{1}{6} \left(1 + \frac{1}{n} \right) \left(2 + \frac{1}{n} \right)$

$$= \frac{1}{6}(1)(2) = \frac{1}{3}$$

45. $\lim_{n \to \infty} \sum_{i=1}^{n} \frac{2}{n} \left[\left(\frac{2i}{n} \right)^3 + 5 \left(\frac{2i}{n} \right) \right] = \lim_{n \to \infty} \sum_{i=1}^{n} \left[\frac{16}{n^4} i^3 + \frac{20}{n^2} i \right] = \lim_{n \to \infty} \left[\frac{16}{n^4} \sum_{i=1}^{n} i^3 + \frac{20}{n^2} \sum_{i=1}^{n} i \right]$

$$= \lim_{n \to \infty} \left[\frac{16}{n^4} \frac{n^2(n+1)^2}{4} + \frac{20}{n^2} \frac{n(n+1)}{2} \right] = \lim_{n \to \infty} \left[\frac{4(n+1)^2}{n^2} + \frac{10n(n+1)}{n^2} \right]$$

$$= \lim_{n \to \infty} \left[4 \left(1 + \frac{1}{n} \right)^2 + 10 \left(1 + \frac{1}{n} \right) \right] = 4 \cdot 1 + 10 \cdot 1 = 14$$

47. Let $S = \sum_{i=1}^{n} ar^{i-1} = a + ar + ar^2 + \cdots + ar^{n-1}$. Multiplying both sides by r gives us

$rS = ar + ar^2 + \cdots + ar^{n-1} + ar^n$. Subtracting the first equation from the second, we find

$(r-1)S = ar^n - a = a(r^n - 1)$, so $S = \dfrac{a(r^n - 1)}{r - 1}$ (since $r \neq 1$).

49. $\sum_{i=1}^{n} \left(2i + 2^i \right) = 2 \sum_{i=1}^{n} i + \sum_{i=1}^{n} 2 \cdot 2^{i-1} = 2 \frac{n(n+1)}{2} + \frac{2(2^n - 1)}{2 - 1} = 2^{n+1} + n^2 + n - 2$.

For the first sum we have used Theorem 3(c), and for the second, Exercise 47 with $a = r = 2$.